COMMON CORE EDITION

Glencoe

ALGEBRA 1

Mc
Graw
Hill **Education**

Bothell, WA • Chicago, IL • Columbus, OH • New York, NY

connectED.mcgraw-hill.com   **Your Digital Math Portal**

Animation	Vocabulary	eGlossary	Personal Tutor	Virtual Manipulatives	Graphing Calculator	Audio	Foldables	Self-Check Practice	Worksheets

In this Common Core State Standards edition of *Glencoe Algebra 1*, students are challenged to develop 21st century skills such as critical thinking and creative problem solving while engaging with exciting careers within Science, Technology, Engineering, and Mathematics (STEM) related fields.

Photo Credits:

Cover Helicoid-punctured-warped-large. By Paul Nylander (http://bugman123.com)

Common Core State Standards © Copyright 2010. National Governors Association Center for Best Practices and Council of Chief State School Officers. All rights reserved.

TI-Nspire is a trademark of Texas Instruments.

TI-Nspire images used by permission.

Understanding by Design® is a registered trademark of the Association for Supervision and Curriculum Development ("ASCD").

The Math Forum®, Ask Dr. Math®, Dr. Math®, and T2T® are registered trademarks of Drexel University.

connectED.mcgraw-hill.com

Send all inquiries to:
McGraw-Hill Education
STEM Learning Solutions Center
8787 Orion Place
Columbus, OH 43240-4027

ISBN: 978-0-07-663923-6
MHID: 0-07-663923-1

Printed in the United States of America.

QVS 17 16 15 14

McGraw-Hill is committed to providing instructional materials in Science, Technology, Engineering, and Mathematics (STEM) that give all students a solid foundation, one that prepares them for college and careers in the 21st century.

Contents in Brief

Authors

Our lead authors ensure that the Macmillan/McGraw-Hill and Glencoe/McGraw-Hill mathematics programs are truly vertically aligned by beginning with the end in mind—success in Algebra 1 and beyond. By "backmapping" the content from the high school programs, all of our mathematics programs are well articulated in their scope and sequence.

Lead Authors

John A. Carter, Ph.D.
Principal
Adlai E. Stevenson High School
Lincolnshire, Illinois

Areas of Expertise: Using technology and manipulatives to visualize concepts; mathematics achievement of English-language learners

Gilbert J. Cuevas, Ph.D.
Professor of Mathematics Education
Texas State University–San Marcos
San Marcos, Texas

Areas of Expertise: Applying concepts and skills in mathematically rich contexts; mathematical representations; use of technology in the development of geometric thinking

Roger Day, Ph.D., NBCT
Mathematics Department Chairperson
Pontiac Township High School
Pontiac, Illinois

Areas of Expertise: Understanding and applying probability and statistics; mathematics teacher education

Carol Malloy, Ph.D.
Associate Professor Emerita
University of North Carolina at Chapel Hill
Chapel Hill, North Carolina

Areas of Expertise: Representations and critical thinking; student success in Algebra 1

Program Authors

Berchie Holliday, Ed.D.
National Mathematics Consultant
Silver Spring, Maryland

Areas of Expertise: Using mathematics to model and
understand real-world data; the effect of graphics on
mathematical understanding

Beatrice Moore Luchin
Mathematics Consultant
Houston, Texas

Areas of Expertise: Mathematical literacy; working with English
language learners

Contributing Authors

Dinah Zike FOLDABLES
Educational Consultant
Dinah-Might Activities, Inc.
San Antonio, Texas

Jay McTighe
Educational Author and Consultant
Columbia, Maryland

Consultants and Reviewers

These professionals were instrumental in providing valuable input and suggestions for improving the effectiveness of the mathematics instruction.

Lead Consultant

Viken Hovsepian
Professor of Mathematics
Rio Hondo College
Whittier, California

Consultants

Mathematical Content

Grant A. Fraser, Ph.D.
Professor of Mathematics
California State University, Los Angeles
Los Angeles, California

Arthur K. Wayman, Ph.D.
Professor of Mathematics Emeritus
California State University, Long Beach
Long Beach, California

Gifted and Talented

Shelbi K. Cole
Research Assistant
University of Connecticut
Storrs, Connecticut

College Readiness

Robert Lee Kimball, Jr.
Department Head, Math and Physics
Wake Technical Community College
Raleigh, North Carolina

Differentiation for English-Language Learners

Susana Davidenko
State University of New York
Cortland, New York

Alfredo Gómez
Mathematics/ESL Teacher
George W. Fowler High School
Syracuse, New York

Graphing Calculator

Ruth M. Casey
T³ National Instructor
Frankfort, Kentucky

Jerry Cummins
Former President
National Council of Supervisors of Mathematics
Western Springs, Illinois

Mathematical Fluency

Robert M. Capraro
Associate Professor
Texas A&M University
College Station, Texas

Pre-AP

Dixie Ross
Lead Teacher for Advanced Placement Mathematics
Pflugerville High School
Pflugerville, Texas

Reading and Writing

ReLeah Cossett Lent
Author and Educational Consultant
Morganton, Georgia

Lynn T. Havens
Director of Project CRISS
Kalispell, Montana

Reviewers

Sherri Abel
Mathematics Teacher
Eastside High School
Taylors, South Carolina

Kelli Ball, NBCT
Mathematics Teacher
Owasso 7th Grade Center
Owasso, Oklahoma

Cynthia A. Burke
Mathematics Teacher
Sherrard Junior High School
Wheeling, West Virginia

Patrick M. Cain, Sr.
Assistant Principal
Stanhope Elmore High School
Millbrook, Alabama

Robert D. Cherry
Mathematics Instructor
Wheaton Warrenville South
High School
Wheaton, Illinois

Tammy Cisco
8th Grade Mathematics/
  Algebra Teacher
Celina Middle School
Celina, Ohio

Amber L. Contrano
High School Teacher
Naperville Central High School
Naperville, Illinois

Catherine Creteau
Mathematics Department
Delaware Valley Regional
High School
Frenchtown, New Jersey

Glenna L. Crockett
Mathematics Department Chair
Fairland High School
Fairland, Oklahoma

Jami L. Cullen
Mathematics Teacher/Leader
Hilltonia Middle School
Columbus, Ohio

Franco DiPasqua
Director of K-12 Mathematics
West Seneca Central Schools
West Seneca, New York

Kendrick Fearson
Mathematics Department Chair
Amos P. Godby High School
Tallahassee, Florida

Lisa K. Gleason
Mathematics Teacher
Gaylord High School
Gaylord, Michigan

Debra Harley
Director of Math & Science
East Meadow School District
Westbury, New York

Tracie A. Harwood
Mathematics Teacher
Braden River High School
Bradenton, Florida

Bonnie C. Hill
Mathematics Department Chair
Triad High School
Troy, Illinois

Clayton Hutsler
Teacher
Goodwyn Junior High School
Montgomery, Alabama

Gureet Kaur
7th Grade Mathematics Teacher
Quail Hollow Middle School
Charlotte, North Carolina

Rima Seals Kelley, NBCT
Mathematics Teacher/
  Department Chair
Deerlake Middle School
Tallahassee, Florida

Holly W. Loftis
8th Grade Mathematics Teacher
Greer Middle School
Lyman, South Carolina

Katherine Lohrman
Teacher, Math Specialist, New
Teacher Mentor
John Marshall High School
Rochester, New York

Carol Y. Lumpkin
Mathematics Educator
Crayton Middle School
Columbia, South Carolina

Ron Mezzadri
Supervisor of Mathematics
  K–12
Fair Lawn Public Schools
Fair Lawn, New Jersey

Bonnye C. Newton
SOL Resource Specialist
Amherst County Public Schools
Amherst, Virginia

Kevin Olsen
Mathematics Teacher
River Ridge High School
New Port Richey, Florida

Kara Painter
Mathematics Teacher
Downers Grove South
High School
Downers Grove, Illinois

Sheila L. Ruddle, NBCT
Mathematics Teacher,
Grades 7 and 8
Pendleton County
Middle/High School
Franklin, West Virginia

Angela H. Slate
Mathematics Teacher/
  Grade 7, Pre-Algebra, Algebra
LeRoy Martin Middle School
Raleigh, North Carolina

Cathy Stellern
Mathematics Teacher
West High School
Knoxville, Tennessee

Dr. Maria J. Vlahos
Mathematics Division Head for
  Grades 6–12
Barrington High School
Barrington, Illinois

Susan S. Wesson
Mathematics Consultant/
  Teacher (Retired)
Pilot Butte Middle School
Bend, Oregon

Mary Beth Zinn
High School
Mathematics Teacher
Chippewa Valley High Schools
Clinton Township, Michigan

Online Guide

connectED.mcgraw-hill.com

The eStudentEdition allows you to access your math curriculum anytime, anywhere.

The icons found throughout your textbook provide you with the opportunity to connect the print textbook with online interactive learning.

Investigate

 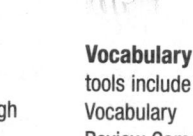 

Animations illustrate key concepts through step-by-step tutorials and videos.

Vocabulary tools include fun Vocabulary Review Games.

Multilingual eGlossary presents key vocabulary in 13 languages.

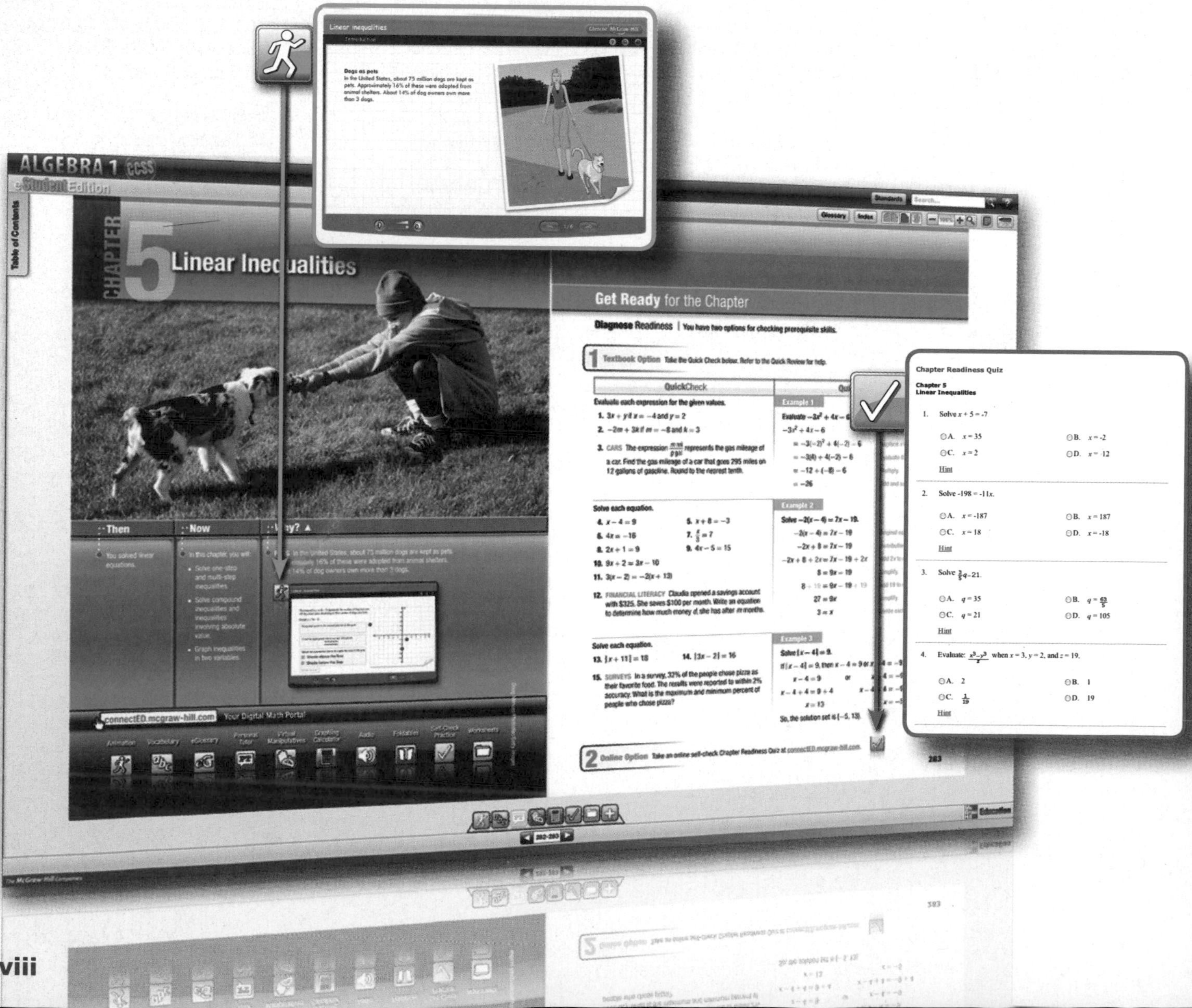

Learn

Personal Tutor
presents an experienced educator explaining step-by-step solutions to problems.

Virtual Manipulatives
are outstanding tools for enhancing understanding.

Graphing Calculator
provides other calculator keystrokes for each Graphing Technology Lab.

Audio
is provided to enhance accessibility.

Foldables
provide a unique way to enhance study skills.

Practice

Self-Check Practice
allows you to check your understanding and send results to your teacher.

Worksheets
provide additional practice.

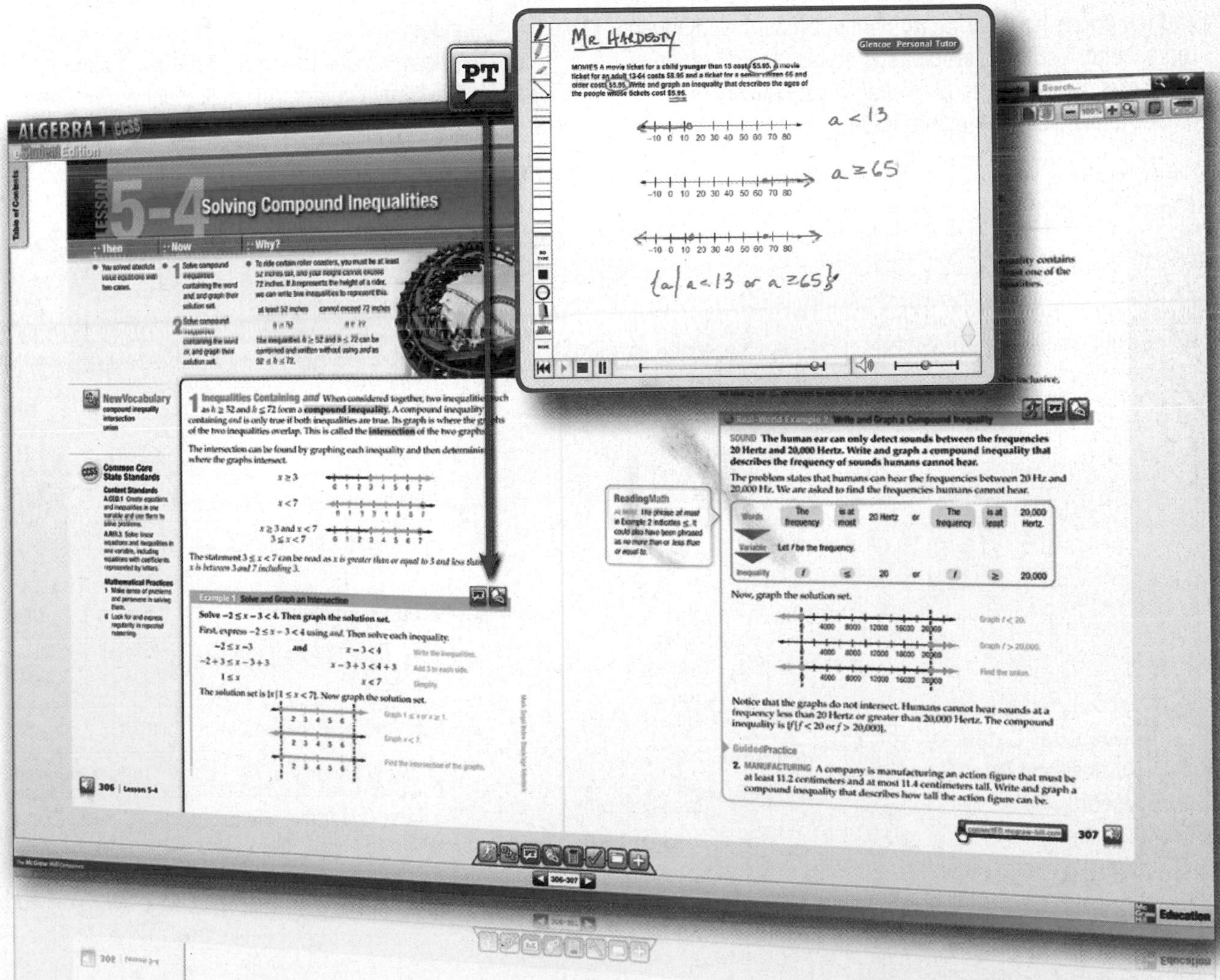

*W*ith American students fully prepared for the future, our communities will be best positioned to compete successfully in the global economy. — Common Core State Standards Initiative

What is the goal of the Common Core State Standards?

The mission of the *Common Core State Standards* is to provide a consistent, clear understanding of what you are expected to learn, so teachers and parents know what they need to do to help you. The standards are designed to be robust and relevant to the real world, reflecting the knowledge and skills needed for success in college and careers.

Who wrote the standards?
The National Governors Association Center for Best Practices and the Council of Chief State School Officers worked with representatives from participating states, a wide range of educators, content experts, researchers, national organizations, and community groups.

What are the major points of the standards?
The standards seek to develop both mathematical understanding and procedural skill. The *Standards for Mathematical Practice* describe varieties of expertise that you will develop in yourself. The *Standards for Mathematical Content* define what you should understand and be able to do at each level in your study of mathematics.

How do I decode the standards?
This diagram provides clarity for decoding the standard identifiers.

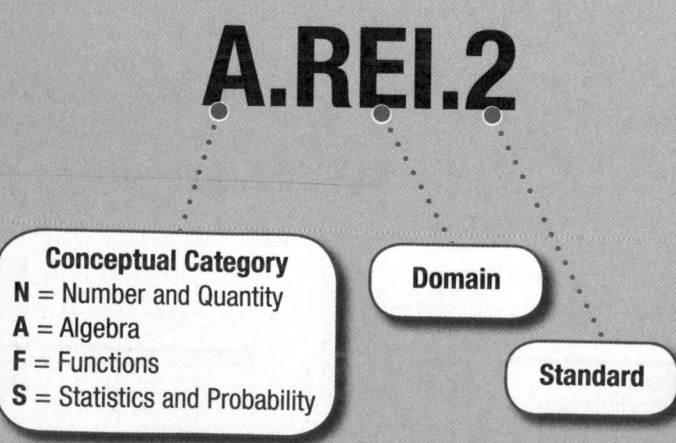

A.REI.2

Conceptual Category
N = Number and Quantity
A = Algebra
F = Functions
S = Statistics and Probability

Domain

Standard

Domain Names	Abbreviations
The **Real Number** System	RN
Quantities	Q
Seeing Structure in **Expressions**	SSE
Arithmetic with **Polynomials** and **Rational Expressions**	APR
Creating Equations	CED
Reasoning with **Equations** and **Inequalities**	REI
Interpreting Functions	IF
Building Functions	BF
Linear, Quadratic, and **Exponential Models**	LE
Interpreting Categorical and **Quantitative Data**	ID

CCSS **Mathematical Content**

connectED.mcgraw-hill.com   **Your Digital Math Portal**

 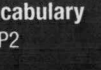 **Vocabulary** p. P2

 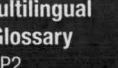 **Multilingual eGlossary** p. P2

 Personal Tutor p. P18

 Foldables p. P2

CHAPTER 1

Expressions, Equations, and Functions

Jupiter

CHAPTER 2 Linear Equations

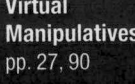
Virtual Manipulatives
pp. 27, 90

Graphing Calculator
pp. 55, 118

Foldables
pp. 4, 74

Self-Check Practice
pp. 3, 100

CHAPTER 3
Linear Functions

CCSS Mathematical Content

connectED.mcgraw-hill.com Your Digital Math Portal

Animation
pp. 150, 248

Vocabulary
pp. 152, 272

Multilingual eGlossary
pp. 152, 214

Personal Tutor
pp. 158, 248

Flame/Alamy

CHAPTER 4
Equations of Linear Functions

Virtual Manipulatives
pp. 158, 248

Graphing Calculator
pp. 169, 215

Foldables
pp. 152, 214

Self-Check Practice
pp. 207, 213

t Knudson/PhotoLink/Getty Images

CHAPTER 5
Linear Inequalities

connectED.mcgraw-hill.com   **Your Digital Math Portal**

 Animation
pp. 291, 332

 **Vocabulary**
pp. 284, 378

 Multilingual eGlossary
pp. 284, 334

 Personal Tutor
pp. 293, 342

CHAPTER 6

Systems of Linear Equations and Inequalities

Virtual Manipulatives
pp. 291, 336

Graphing Calculator
pp. 323, 342

Foldables
pp. 284, 334

Self-Check Practice
pp. 300, 383

CHAPTER 7

Exponents and Exponential Functions

Michael Dunning/Pho

 Animation pp. 398, 481

 Vocabulary pp. 390, 530

 **Multilingual eGlossary** pp. 390, 462

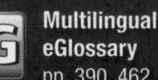

 **Personal Tutor** pp. 406, 473

CHAPTER 8

Quadratic Expressions and Equations

CCSS **Mathematical Content**

 **Virtual Manipulatives** pp. 424, 480

 Foldables pp. 390, 462

 Self-Check Practice pp. 389, 518

McGraw-Hill Education

CHAPTER 9

Quadratic Functions and Equations

Stephen Chernin/Getty Ima

🖐 connectED.mcgraw-hill.com   **Your Digital Math Portal**

 Animation pp. 574, 655

 Vocabulary pp. 542, 663

 Multilingual eGlossary pp. 542, 618

 Personal Tutor pp. 572, 627

CHAPTER 10
Radical Functions and Geometry

CCSS Mathematical Content

Getty Images

 Virtual Manipulatives pp. 548, 649

 **Graphing Calculator** pp. 572, 627

 **Foldables** pp. 542, 618

 **Self-Check Practice** pp. 611, 617

CHAPTER 11

Rational Functions and Equations

Doug Pensinger/

connectED.mcgraw-hill.com   **Your Digital Math Portal**

 Animation pp. 672, 793

 Vocabulary pp. 674, 746

 **Multilingual eGlossary** pp. 674, 746

 Personal Tutor pp. 697, 748

CHAPTER 12 Statistics and Probability

CCSS **Mathematical Content**

Student Handbook

0 Preparing for Algebra

∴ Now

○ **Chapter 0** contains lessons on topics from previous courses. You can use this chapter in various ways.

- Begin the school year by taking the Pretest. If you need additional review, complete the lessons in this chapter. To verify that you have successfully reviewed the topics, take the Posttest.

- As you work through the text, you may find that there are topics you need to review. When this happens, complete the individual lessons that you need.

- Use this chapter for reference. When you have questions about any of these topics, flip back to this chapter to review definitions or key concepts.

🖐 **connectED.mcgraw-hill.com** **Your Digital Math Portal**

Animation	Vocabulary	eGlossary	Personal Tutor	Virtual Manipulatives	Graphing Calculator	Audio	Foldables	Self-Check Practice	Worksheets

Get Started on the Chapter

You will review several concepts, skills, and vocabulary terms as you study Chapter 0. To get ready, identify important terms and organize your resources.

FOLDABLES StudyOrganizer

Throughout this text, you will be invited to use Foldables to organize your notes.

Why should you use them?

- They help you organize, display, and arrange information.

- They make great study guides, specifically designed for you.

- You can use them as your math journal for recording main ideas, problem-solving strategies, examples, or questions you may have.

- They give you a chance to improve your math vocabulary.

How should you use them?

- Write general information — titles, vocabulary terms, concepts, questions, and main ideas – on the front tabs of your Foldable.

- Write specific information — ideas, your thoughts, answers to questions, steps, notes, and definitions — under the tabs.

- Use the tabs for:
  - math concepts in parts, like types of triangles,
  - steps to follow, or
  - parts of a problem, like *compare* and *contrast* (2 parts) or *what*, *where*, *when*, *why*, and *how* (5 parts).

- You may want to store your Foldables in a plastic zipper bag that you have three-hole punched to fit in your notebook.

When should you use them?

- Set up your Foldable as you begin a chapter, or when you start learning a new concept.

- Write in your Foldable every day.

- Use your Foldable to review for homework, quizzes, and tests.

ReviewVocabulary

English		Español
integer	p. P7	entero
absolute value	p. P11	valor absolute
opposites	p. P11	opuestos
reciprocal	p. P18	recíproco
perimeter	p. P23	perímetro
circle	p. P24	círculo
diameter	p. P24	diámetro
center	p. P24	centro
circumference	p. P24	circunferencia
radius	p. P24	radio
area	p. P26	area
volume	p. P29	volumen
surface area	p. P31	area de superficie
probability	p. P33	probabilidad
sample space	p. P33	espacio muestral
complements	p. P33	complementos
tree diagram	p. P34	diagrama de árbol
odds	p. P35	probabilidades
mean	p. P37	media
median	p. P37	mediana
mode	p. P37	moda
range	p. P38	rango
quartile	p. P38	cuartil
interquartile range	p. P38	amplitud intercuartílica
outliers	p. P39	valores atípicos
bar graph	p. P41	gráfica de barras
histogram	p. P41	histograma
line graph	p. P42	gráfica lineal
circle graph	p. P42	gráfica circular
box-and-whisker plot	p. P43	diagrama de caja y patillas

Determine whether you need an estimate or an exact answer. Then solve.

1. **SHOPPING** Addison paid $1.29 for gum and $0.89 for a package of notebook paper. She gave the cashier a $5 bill. If the tax was $0.14, how much change should Addison receive?

2. **DISTANCE** Luis rode his bike 1.2 miles to his friend's house, then 0.7 mile to the video store, then 1.9 miles to the library. If he rode the same route back home, about how far did he travel in all?

Find each sum or difference.

3. $20 + (-7)$

4. $-15 + 6$

5. $-9 - 22$

6. $18.4 - (-3.2)$

7. $23.1 + (-9.81)$

8. $-5.6 + (-30.7)$

Find each product or quotient.

9. $11(-8)$

10. $-15(-2)$

11. $63 \div (-9)$

12. $-22 \div 11$

Replace each ● with <, >, or = to make a true sentence.

13. $\frac{7}{20} ● \frac{2}{5}$

14. $0.15 ● \frac{1}{8}$

15. Order 0.5, $-\frac{1}{7}$, -0.2, and $\frac{1}{3}$ from least to greatest.

Find each sum or difference. Write in simplest form.

16. $\frac{5}{6} + \frac{2}{3}$

17. $\frac{11}{12} - \frac{3}{4}$

18. $\frac{1}{2} + \frac{4}{9}$

19. $-\frac{3}{5} + \left(-\frac{1}{5}\right)$

Find each product or quotient.

20. $2.4(-0.7)$

21. $-40.5 \div (-8.1)$

Name the reciprocal of each number.

22. $\frac{4}{11}$

23. $-\frac{3}{7}$

Find each product or quotient. Write in simplest form.

24. $\frac{2}{21} \div \frac{1}{3}$

25. $\frac{1}{5} \cdot \frac{3}{20}$

26. $\frac{6}{25} \div \left(-\frac{3}{5}\right)$

27. $\frac{1}{9} \cdot \frac{3}{4}$

28. $-\frac{2}{21} \div \left(-\frac{2}{15}\right)$

29. $2\frac{1}{2} \cdot \frac{2}{15}$

Express each percent as a fraction in simplest form.

30. 20%

31. 7.5%

Use the percent proportion to find each number.

32. 18 is what percent of 72?

33. 35 is what percent of 200?

34. 24 is 60% of what number?

35. **TEST SCORES** James answered 14 items correctly on a 16-item quiz. What percent did he answer correctly?

36. **BASKETBALL** Emily made 75% of the baskets that she attempted. If she made 9 baskets, how many attempts did she make?

Find the perimeter and area of each figure.

37.
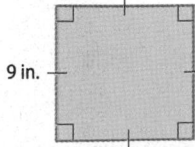
9 in.

38.
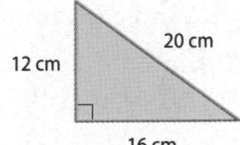
20 cm
12 cm
16 cm

39. A parallelogram has side lengths of 7 inches and 11 inches. Find the perimeter.

40. **GARDENS** Find the perimeter of the garden.

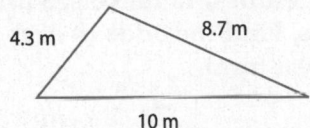
4.3 m
8.7 m
10 m

Find the circumference and area of each circle. Round to the nearest tenth.

41.
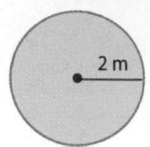
2 m

42.
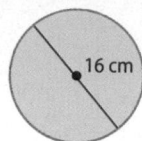
16 cm

43. BIRDS The floor of a birdcage is a circle with a circumference of about 47.1 inches. What is the diameter of the birdcage floor? Round to the nearest inch.

Find the volume and surface area of each rectangular prism given the measurements below.

44. $\ell = 3$ cm, $w = 1$ cm, $h = 3$ cm

45. $\ell = 6$ ft, $w = 2$ ft, $h = 5$ ft

46. Find the volume and surface area of the rectangular prism.

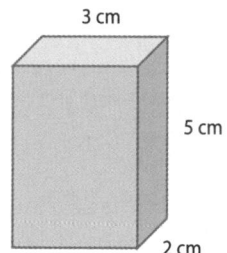
3 cm
5 cm
2 cm

One pencil is randomly selected from a case containing 3 red, 4 green, 2 black, and 6 blue pencils. Find each probability.

47. P(green)   **48.** P(red or blue)

49. Use a tree diagram to find the sample space for the event *a die is rolled, and a coin is tossed*. State the number of possible outcomes.

One coin is randomly selected from a jar containing 20 pennies, 15 nickels, 3 dimes, and 12 quarters. Find the odds of each outcome. Write in simplest form.

50. a penny   **51.** a penny or nickel

52. A coin is tossed 50 times. The results are shown in the table. Find the experimental probability of heads. Write as a fraction in simplest form.

Lands Face-Up	Number of Times
heads	22
tails	28

Find the mean, median, and mode for each set of data.

53. {10, 11, 18, 24, 30}

54. {4, 8, 9, 9, 10, 14, 16}

55. Find the range, median, lower quartile, and upper quartile for {16, 19, 21, 24, 25, 31, 35}.

56. SCHOOL Devonte's scores on his first four Spanish tests are 92, 85, 90, and 92. What test score must Devonte earn on the fifth test so that the mean will be exactly 90?

57. MUSIC The table shows the results of a survey in which students were asked to choose which of four instruments they would like to learn. Make a bar graph of the data.

Favorite Instrument	
Instrument	Number of Students
drums	8
guitar	12
piano	5
trumpet	7

58. Make a double box-and-whisker plot of the data.
A: 42, 50, 38, 59, 50, 44, 46, 62, 47, 35, 55, 56
B: 47, 49, 48, 49, 40, 54, 56, 42, 57, 45, 45, 46

59. EXPENSES The table shows how Dylan spent his money at the fair. What type of graph is the best way to display these data? Explain your reasoning and make a graph of the data.

Money Spent at the Fair	
How Spent	Amount ($)
rides	6
food	10
games	4

Plan for Problem Solving

··Objective

● Use the four-step problem-solving plan.

NewVocabulary
four-step problem-solving plan
defining a variable

Common Core State Standards

Mathematical Practices
1 Make sense of problems and persevere in solving them.

Using the **four-step problem-solving plan** can help you solve any word problem.

> **KeyConcept** Four-Step Problem Solving Plan
>
> **Step 1** Understand the Problem.   **Step 3** Solve the Problem.
> **Step 2** Plan the Solution.   **Step 4** Check the Solution.

Each step of the plan is important.

Step 1 **Understand the Problem**

To solve a verbal problem, first read the problem carefully and explore what the problem is about.
- Identify what information is given.
- Identify what you need to find.

Step 2 **Plan the Solution**

One strategy you can use is to write an equation. Choose a variable to represent one of the unspecified numbers in the problem. This is called **defining a variable**. Then use the variable to write expressions for the other unspecified numbers in the problem.

Step 3 **Solve the Problem**

Use the strategy you chose in Step 2 to solve the problem.

Step 4 **Check the Solution**

Check your answer in the context of the original problem.
- Does your answer make sense?
- Does it fit the information in the problem?

> **Example 1** Use the Four-Step Plan
>
> **FLOORS** Ling's hallway is 10 feet long and 4 feet wide. He paid $200 to tile his hallway floor. How much did Ling pay per square foot for the tile?
>
> **Understand** We are given the measurements of the hallway and the total cost of the tile. We are asked to find the cost of each square foot of tile.
>
> **Plan** Write an equation. Let f represent the cost of each square foot of tile. The area of the hallway is 10×4 or 40 ft^2.
>
> 40   times   the cost per square foot   equals   200.
> 40   · f =   200
>
> **Solve** $40 \cdot f = 200$. Find f mentally by asking, "What number times 40 is 200?"
> $f = 5$
> The tile cost $5 per square foot.
>
> **Check** If the tile costs $5 per square foot, then 40 square feet of tile costs $5 \cdot 40$ or $200. The answer makes sense.

When an exact value is needed, you can use estimation to check your answer.

Example 2 Use the Four-Step Plan

TRAVEL Emily's family drove 254.6 miles. Their car used 19 gallons of gasoline. Describe the car's gas mileage.

Understand We are given the total miles driven and how much gasoline was used. We are asked to find the gas mileage of the car.

Plan Write an equation. Let G represent the car's gas mileage.

gas mileage = number of miles ÷ number of gallons used

$G = 254.6 \div 19$

Solve $G = 254.6 \div 19$

$= 13.4$ mi/gal

The car's gas mileage is 13.4 miles per gallon.

Check Use estimation to check your solution.

260 mi ÷ 20 gal = 13 mi/gal

Since the solution 13.4 is close to the estimate, the answer is reasonable.

Exercises

Determine whether you need an estimate or an exact answer. Then use the four step problem-solving plan to solve.

1. **DRIVING** While on vacation, the Jacobson family drove 312.8 miles the first day, 177.2 miles the second day, and 209 miles the third day. About how many miles did they travel in all?

2. **PETS** Ms. Hernandez boarded her dog at a kennel for 4 days. It cost $18.90 per day, and she had a coupon for $5 off. What was the final cost for boarding her dog?

3. **MEASUREMENT** William is using a 1.75-liter container to fill a 14-liter container of water. About how many times will he need to fill the smaller container?

4. **SEWING** Fabric costs $5.15 per yard. The drama department needs 18 yards of the fabric for their new play. About how much should they expect to pay?

5. **FINANCIAL LITERACY** The table shows donations to help purchase a new tree for the school. How much money did the students donate in all?

Number of Students	Amount of Each Donation
20	$2.50
15	$3.25

6. **SHOPPING** Is $12 enough to buy a half gallon of milk for $2.30, a bag of apples for $3.99, and four cups of yogurt that cost $0.79 each? Explain.

Real Numbers

 NewVocabulary

positive number
negative number
natural number
whole number
integer
rational number
square root
principal square root
perfect square
irrational number
real number
graph
coordinate

A number line can be used to show the sets of natural numbers, whole numbers, integers, and rational numbers. Values greater than 0, or **positive numbers**, are listed to the right of 0, and values less than 0, or **negative numbers**, are listed to the left of 0.

natural numbers: 1, 2, 3, …

whole numbers: 0, 1, 2, 3, …

integers: … , −3, −2, −1, 0, 1, 2, 3, …

rational numbers: numbers that can be expressed in the form $\frac{a}{b}$, where a and b are integers and $b \neq 0$

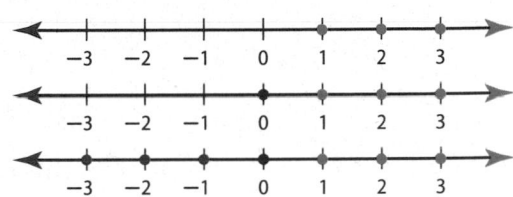

A **square root** is one of two equal factors of a number. For example, one square root of 64, written as $\sqrt{64}$, is 8 since 8 · 8 or 8^2 is 64. The nonnegative square root of a number is the **principal square root**. Another square root of 64 is −8 since $(−8) \cdot (−8)$ or $(−8)^2$ is also 64. A number like 64, with a square root that is a rational number, is called a **perfect square**. The square roots of a perfect square are rational numbers.

A number such as $\sqrt{3}$ is the square root of a number that is not a perfect square. It cannot be expressed as a terminating or repeating decimal; $\sqrt{3} \approx 1.73205\ldots$. Numbers that cannot be expressed as terminating or repeating decimals, or in the form $\frac{a}{b}$, where a and b are integers and $b \neq 0$, are called **irrational numbers**. Irrational numbers and rational numbers together form the set of **real numbers**.

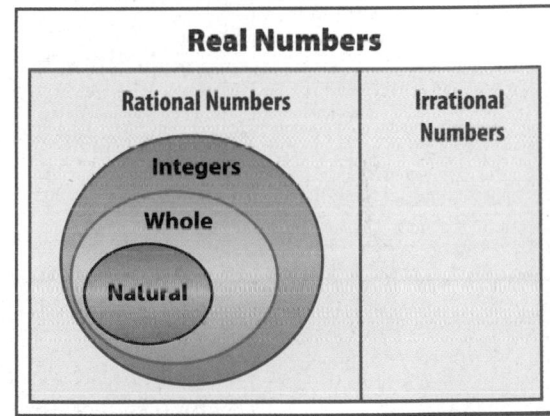

Example 1 Classify Real Numbers

Name the set or sets of numbers to which each real number belongs.

a. $\frac{5}{22}$

Because 5 and 22 are integers and $5 \div 22 = 0.2272727\ldots$ or $0.2\overline{27}$, which is a repeating decimal, this number is a rational number.

b. $\sqrt{81}$

Because $\sqrt{81} = 9$, this number is a natural number, a whole number, an integer, and a rational number.

c. $\sqrt{56}$

Because $\sqrt{56} = 7.48331477\ldots$, which is not a repeating or terminating decimal, this number is irrational.

To **graph** a set of numbers means to draw, or plot, the points named by those numbers on a number line. The number that corresponds to a point on a number line is called the **coordinate** of that point. The rational numbers and the irrational numbers complete the number line.

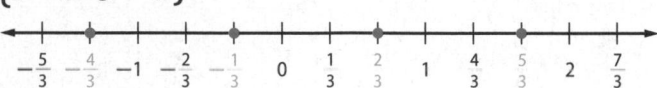

Example 2 Graph and Order Real Numbers

Graph each set of numbers on a number line. Then order the numbers from least to greatest.

a. $\left\{ \dfrac{5}{3}, -\dfrac{4}{3}, \dfrac{2}{3}, -\dfrac{1}{3} \right\}$

The number line shows marks at $-\dfrac{5}{3}$, $-\dfrac{4}{3}$, -1, $-\dfrac{2}{3}$, $-\dfrac{1}{3}$, 0, $\dfrac{1}{3}$, $\dfrac{2}{3}$, 1, $\dfrac{4}{3}$, $\dfrac{5}{3}$, 2, $\dfrac{7}{3}$.

From least to greatest, the order is $-\dfrac{4}{3}$, $-\dfrac{1}{2}$, $\dfrac{2}{3}$, and $\dfrac{5}{3}$.

b. $\left\{ 6\dfrac{4}{5}, \sqrt{49}, 6.\overline{3}, \sqrt{57} \right\}$

Express each number as a decimal. Then order the decimals.

$6\dfrac{4}{5} = 6.8 \qquad \sqrt{49} = 7 \qquad 6.\overline{3} = 6.33333333\ldots \qquad \sqrt{57} = 7.5468344\ldots$

The number line shows $6.\overline{3}$, $6\dfrac{4}{5}$, $\sqrt{49}$, $\sqrt{57}$ plotted between 6.0 and 8.0.

6.0 6.2 6.4 6.6 6.8 7.0 7.2 7.4 7.6 7.8 8.0

From least to greatest, the order is $6.\overline{3}$, $6\dfrac{4}{5}$, $\sqrt{49}$, and $\sqrt{57}$.

c. $\left\{ \sqrt{20}, 4.7, \dfrac{12}{3}, 4\dfrac{1}{3} \right\}$

$\sqrt{20} = 4.47213595\ldots \qquad 4.7 = 4.7 \qquad \dfrac{12}{3} = 4.0 \qquad 4\dfrac{1}{3} = 4.33333333\ldots$

The number line shows $\dfrac{3}{12}$, $4\dfrac{1}{3}$, $\sqrt{20}$, 4.7 plotted between 3.8 and 4.8.

3.8 3.9 4.0 4.1 4.2 4.3 4.4 4.5 4.6 4.7 4.8

From least to greatest, the order is $\dfrac{12}{3}$, $4\dfrac{1}{3}$, $\sqrt{20}$, and 4.7.

Any repeating decimal can be written as a fraction.

Example 3 Write Repeating Decimals as Fractions

Write $0.\overline{7}$ as a fraction in simplest form.

Step 1
$N = 0.777\ldots$	Let N represent the repeating decimal.
$10N = 10(0.777\ldots)$	Since only one digit repeats, multiply each side by 10.
$10N = 7.777\ldots$	Simplify.

Step 2 Subtract N from $10N$ to eliminate the part of the number that repeats.

$$10N = 7.777\ldots$$
$$-(N = 0.777\ldots)$$

$9N = 7$	Subtract.
$\dfrac{9N}{9} = \dfrac{7}{9}$	Divide each side by 9.
$N = \dfrac{7}{9}$	Simplify.

Perfect squares can be used to simplify square roots of rational numbers.

KeyConcept Perfect Square

Words   Rational numbers with square roots that are rational numbers.

Examples   25 is a perfect square since $\sqrt{25} = 5$.

144 is a perfect square since $\sqrt{144} = 12$.

Example 4 Simplify Roots

Simplify each square root.

a. $\sqrt{\dfrac{4}{121}}$

$\sqrt{\dfrac{4}{121}} = \sqrt{\left(\dfrac{2}{11}\right)^2}$ $2^2 = 4$ and $11^2 = 121$

$= \dfrac{2}{11}$ Simplify.

b. $-\sqrt{\dfrac{49}{256}}$

$-\sqrt{\dfrac{49}{256}} = -\sqrt{\left(\dfrac{7}{16}\right)^2}$ $7^2 = 49$ and $16^2 = 256$

$= -\dfrac{7}{16}$

You can estimate roots that are not perfect squares.

Example 5 Estimate Roots

Estimate each square root to the nearest whole number.

a. $\sqrt{15}$

Find the two perfect squares closest to 15. List some perfect squares.

1, 4, 9, 16, 25, 36, …

15 is between 9 and 16.

$9 < 15 < 16$ Write an inequality.

$\sqrt{9} < \sqrt{15} < \sqrt{16}$ Take the square root of each number.

$3 < \sqrt{15} < 4$ Simplify.

```
        3              4
  ──┼──────────────┼──
   √9          √15 √16
```

Since 15 is closer to 16 than 9, the best whole-number estimate for $\sqrt{15}$ is 4.

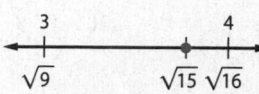

b. $\sqrt{130}$

Find the two perfect squares closest to 130. List some perfect squares.

81, 100, 121, 144

130 is between 121 and 144.

$121 < 130 < 144$ Write an inequality.

$\sqrt{121} < \sqrt{130} < \sqrt{144}$ Take the square root of each number.

$11 < \sqrt{130} < 12$ Simplify.

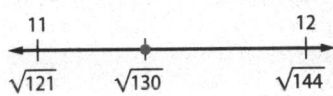

StudyTip

Draw a Diagram
Graphing points on a number line can help you analyze your estimate for accuracy.

Since 130 is closer to 121 than to 144, the best whole number estimate for $\sqrt{130}$ is 11.

CHECK $\sqrt{130} \approx 11.4018$ Use a calculator.

Rounded to the nearest whole number, $\sqrt{130}$ is 11. So the estimate is valid.

Exercises

Name the set or sets of numbers to which each real number belongs.

1. $-\sqrt{64}$ **2.** $\dfrac{8}{3}$ **3.** $\sqrt{28}$ **4.** $\dfrac{56}{7}$

5. $-\sqrt{22}$ **6.** $\dfrac{36}{6}$ **7.** $-\dfrac{5}{12}$ **8.** $\dfrac{18}{3}$

9. $\sqrt{10.24}$ **10.** $\dfrac{-54}{19}$ **11.** $\sqrt{\dfrac{82}{20}}$ **12.** $-\dfrac{72}{8}$

Graph each set of numbers on a number line. Then order the numbers from least to greatest.

13. $\left\{ \dfrac{7}{5}, -\dfrac{3}{5}, \dfrac{3}{4}, -\dfrac{6}{5} \right\}$ **14.** $\left\{ \dfrac{1}{2}, -\dfrac{7}{9}, \dfrac{1}{9}, -\dfrac{4}{9} \right\}$ **15.** $\left\{ 2\dfrac{1}{4}, \sqrt{7}, 2.\overline{3}, \sqrt{8} \right\}$

16. $\left\{ \dfrac{4}{5}, \sqrt{2}, 0.\overline{1}, \sqrt{3} \right\}$ **17.** $\left\{ -3.5, -\dfrac{15}{5}, -\sqrt{10}, -3\dfrac{3}{4} \right\}$ **18.** $\left\{ \sqrt{64}, 8.8, \dfrac{26}{3}, 8\dfrac{2}{7} \right\}$

Write each repeating decimal as a fraction in simplest form.

19. $0.\overline{5}$ **20.** $0.\overline{4}$

21. $0.\overline{13}$ **22.** $0.\overline{21}$

Simplify each square root.

23. $-\sqrt{25}$ **24.** $\sqrt{361}$ **25.** $\pm\sqrt{36}$

26. $\sqrt{0.64}$ **27.** $\pm\sqrt{1.44}$ **28.** $-\sqrt{6.25}$

29. $\sqrt{\dfrac{16}{49}}$ **30.** $\sqrt{\dfrac{169}{196}}$ **31.** $\sqrt{\dfrac{25}{324}}$

Estimate each root to the nearest whole number.

32. $\sqrt{112}$ **33.** $\sqrt{252}$ **34.** $\sqrt{415}$ **35.** $\sqrt{670}$

0-3 Operations with Integers

Objective

- Add, subtract, multiply, and divide integers.

NewVocabulary
absolute value
opposites
additive inverses

An integer is any number from the set $\{\ldots, -3, -2, -1, 0, 1, 2, 3, \ldots\}$. You can use a number line to add integers.

Example 1 Add Integers with the Same Sign

Use a number line to find $-3 + (-4)$.

Step 1 Draw an arrow from 0 to -3.

Step 2 Draw a second arrow 4 units to the left to represent adding -4.

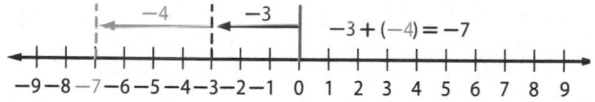

The second arrow ends at -7. So, $-3 + (-4) = -7$.

You can also use absolute value to add integers. The **absolute value** of a number is its distance from 0 on the number line.

Same Signs (+ + or − −)		Different Signs (+ − or − +)	
$3 + 5 = 8$	3 and 5 are positive. Their sum is positive.	$3 + (-5) = -2$	-5 has the greater absolute value. Their sum is negative.
$-3 + (-5) = -8$	-3 and -5 are negative. Their sum is negative.	$-3 + 5 = 2$	5 has the greater absolute value. Their sum is positive.

Example 2 Add Integers Using Absolute Value

Find $-11 + (-7)$.

$-11 + (-7) = -(|-11| + |-7|)$ Add the absolute values. Both numbers are negative, so the sum is negative.

$= -(11 + 7)$ Absolute values of nonzero numbers are always positive.

$= -18$ Simplify.

Every positive integer can be paired with a negative integer. These pairs are called **opposites**. A number and its opposite are **additive inverses**. Additive inverses can be used when you subtract integers.

Example 3 Subtract Positive Integers

Find $18 - 23$.

$18 - 23 = 18 + (-23)$ To subtract 23, add its inverse.

$= -(|-23| - |18|)$ Subtract the absolute values. Because $|-23|$ is greater than $|18|$, the result is negative.

$= -(23 - 18)$ Absolute values of nonzero numbers are always positive.

$= -5$ Simplify.

Same Signs (+ + or − −)		Different Signs (+ − or − +)	
$3(5) = 15$	3 and 5 are positive. Their product is positive.	$3(-5) = -15$	3 and −5 have different signs. Their product is negative.
$-3(-5) = 15$	−3 and −5 are negative. Their product is positive.	$-3(5) = -15$	−3 and 5 have different signs. Their product is negative.

Example 4 Multiply and Divide Integers

Find each product or quotient.

a. $4(-5)$

$4(-5) = -20$ different signs ⟶ negative product

b. $-51 \div (-3)$

$-51 \div (-3) = 17$ same sign ⟶ positive quotient

c. $-12(-14)$

$-12(-14) = 168$ same sign ⟶ positive product

d. $-63 \div 7$

$-63 \div 7 = -9$ different signs ⟶ negative quotient

Exercises

Find each sum or difference.

1. $-8 + 13$ **2.** $11 + (-19)$ **3.** $-19 - 8$

4. $-77 + (-46)$ **5.** $12 - 34$ **6.** $41 + (-56)$

7. $50 - 82$ **8.** $-47 - 13$ **9.** $-80 + 102$

Find each product or quotient.

10. $5(18)$ **11.** $60 \div 12$ **12.** $-12(15)$

13. $-64 \div (-8)$ **14.** $8(-22)$ **15.** $54 \div (-6)$

16. $30(14)$ **17.** $-23(5)$ **18.** $-200 \div 2$

19. WEATHER The outside temperature was −4°F in the morning and 13°F in the afternoon. By how much did the temperature increase?

20. DOLPHINS A dolphin swimming 24 feet below the ocean's surface dives 18 feet straight down. How many feet below the ocean's surface is the dolphin now?

21. MOVIES A movie theater gave out 50 coupons for $3 off each movie. What is the total amount of discounts provided by the theater?

22. WAGES Emilio earns $11 per hour. He works 14 hours a week. His employer withholds $32 from each paycheck for taxes. If he is paid weekly, what is the amount of his paycheck?

23. FINANCIAL LITERACY Talia is working on a monthly budget. Her monthly income is $500. She has allocated $200 for savings, $100 for vehicle expenses, and $75 for clothing. How much is available to spend on entertainment?

Adding and Subtracting Rational Numbers

∷ Objective

● Compare and order; add and subtract rational numbers.

You can use different methods to compare rational numbers. One way is to compare two fractions with common denominators. Another way is to compare decimals.

Example 1 Compare Rational Numbers

Replace ● with <, >, or = to make $\frac{2}{3}$ ● $\frac{5}{6}$ a true sentence.

Method 1 Write the fractions with the same denominator.

The least common denominator of $\frac{2}{3}$ and $\frac{5}{6}$ is 6.

$$\frac{2}{3} = \frac{4}{6}$$

$$\frac{5}{6} = \frac{5}{6}$$

Since $\frac{4}{6} < \frac{5}{6}, \frac{2}{3} < \frac{5}{6}$.

Method 2 Write as decimals.

Write $\frac{2}{3}$ and $\frac{5}{6}$ as decimals. You may want to use a calculator.

2 [÷] 3 [ENTER] .6666666667

So, $\frac{2}{3} = 0.\overline{6}$.

5 [÷] 6 [ENTER] .8333333333

So, $\frac{5}{6} = 0.8\overline{3}$.

Since $0.\overline{6} < 0.8\overline{3}, \frac{2}{3} < \frac{5}{6}$.

You can order rational numbers by writing all of the fractions as decimals.

Example 2 Order Rational Numbers

Order $5\frac{2}{9}$, $5\frac{3}{8}$, 4.9, and $-5\frac{3}{5}$ from least to greatest.

$$5\frac{2}{9} = 5.\overline{2} \qquad\qquad 5\frac{3}{8} = 5.375$$

$$4.9 = 4.9 \qquad\qquad -5\frac{3}{5} = -5.6$$

$-5.6 < 4.9 < 5.\overline{2} < 5.375$. So, from least to greatest, the numbers are $-5\frac{3}{5}$, 4.9, $5\frac{2}{9}$, and $5\frac{3}{8}$.

To add or subtract fractions with the same denominator, add or subtract the numerators and write the sum or difference over the denominator.

Example 3 Add and Subtract Like Fractions

Find each sum or difference. Write in simplest form.

a. $\frac{3}{5} + \frac{1}{5}$

$$\frac{3}{5} + \frac{1}{5} = \frac{3+1}{5}$$ The denominators are the same. Add the numerators.

$$= \frac{4}{5}$$ Simplify.

> **StudyTip**
>
> **Mental Math** If the denominators of the fractions are the same, you can use mental math to determine the sum or difference.

b. $\frac{7}{16} - \frac{1}{16}$

$$\frac{7}{16} - \frac{1}{16} = \frac{7-1}{16}$$ The denominators are the same. Subtract the numerators.

$$= \frac{6}{16}$$ Simplify.

$$= \frac{3}{8}$$ Rename the fraction.

c. $\frac{4}{9} - \frac{7}{9}$

$$\frac{4}{9} - \frac{7}{9} = \frac{4-7}{9}$$ The denominators are the same. Subtract the numerators.

$$= -\frac{3}{9}$$ Simplify.

$$= -\frac{1}{3}$$ Rename the fraction.

To add or subtract fractions with unlike denominators, first find the least common denominator (LCD). Rename each fraction with the LCD, and then add or subtract. Simplify if possible.

Example 4 Add and Subtract Unlike Fractions

Find each sum or difference. Write in simplest form.

a. $\frac{1}{2} + \frac{2}{3}$

$$\frac{1}{2} + \frac{2}{3} = \frac{3}{6} + \frac{4}{6}$$ The LCD for 2 and 3 is 6. Rename $\frac{1}{2}$ as $\frac{3}{6}$ and $\frac{2}{3}$ as $\frac{4}{6}$.

$$= \frac{3+4}{6}$$ Add the numerators.

$$= \frac{7}{6} \text{ or } 1\frac{1}{6}$$ Simplify.

b. $\frac{3}{8} - \frac{1}{3}$

$$\frac{3}{8} - \frac{1}{3} = \frac{9}{24} - \frac{8}{24}$$ The LCD for 8 and 3 is 24. Rename $\frac{3}{8}$ as $\frac{9}{24}$ and $\frac{1}{3}$ as $\frac{8}{24}$.

$$= \frac{9-8}{24}$$ Subtract the numerators.

$$= \frac{1}{24}$$ Simplify.

c. $\frac{2}{5} - \frac{3}{4}$

$$\frac{2}{5} - \frac{3}{4} = \frac{8}{20} - \frac{15}{20}$$ The LCD for 5 and 4 is 20. Rename $\frac{2}{5}$ as $\frac{8}{20}$ and $\frac{3}{4}$ as $\frac{15}{20}$.

$$= \frac{8-15}{20}$$ Subtract the numerators.

$$= -\frac{7}{20}$$ Simplify.

You can use a number line to add rational numbers.

Example 5 Add Decimals

Use a number line to find 2.5 + (−3.5).

Step 1 Draw an arrow from 0 to 2.5.

Step 2 Draw a second arrow 3.5 units to the left.

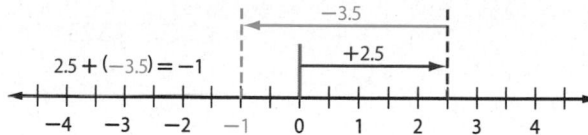

$2.5 + (-3.5) = -1$

The second arrow ends at −1.

So, $2.5 + (-3.5) = -1$.

You can also use absolute value to add rational numbers.

Same Signs (+ + or − −)		Different Signs (+ − or − +)	
3.1 + 2.5 = 5.6	3.1 and 2.5 are positive, so the sum is positive.	3.1 + (−2.5) = 0.6	3.1 has the greater absolute value, so the sum is positive.
−3.1 + (−2.5) = −5.6	−3.1 and −2.5 are negative, so the sum is negative.	−3.1 + 2.5 = −0.6	−3.1 has the greater absolute value, so the sum is negative.

Example 6 Use Absolute Value to Add Rational Numbers

Find each sum.

a. $-13.12 + (-8.6)$

$-13.12 + (-8.6) = -(|-13.12| + |-8.6|)$ Both numbers are negative, so the sum is negative.

$= -(13.12 + 8.6)$ Absolute values of nonzero numbers are always positive.

$= -21.72$ Simplify.

b. $\dfrac{7}{16} + \left(-\dfrac{3}{8}\right)$

$\dfrac{7}{16} + \left(-\dfrac{3}{8}\right) = \dfrac{7}{16} + \left(-\dfrac{6}{16}\right)$ The LCD is 16. Replace $-\dfrac{3}{8}$ with $-\dfrac{6}{16}$.

$= \left(\left|\dfrac{7}{16}\right| - \left|-\dfrac{6}{16}\right|\right)$ Subtract the absolute values. Because $\left|\dfrac{7}{16}\right|$ is greater than $\left|-\dfrac{6}{16}\right|$, the result is positive.

$= \dfrac{7}{16} - \dfrac{6}{16}$ Absolute values of nonzero numbers are always positive.

$= \dfrac{1}{16}$ Simplify.

To subtract a negative rational number, add its inverse.

Example 7 Subtract Decimals

Find $-32.25 - (-42.5)$.

$-32.25 - (-42.5) = -32.25 + 42.5$ To subtract -42.5, add its inverse.

$\qquad\qquad\qquad = |42.5| - |-32.25|$ Subtract the absolute values. Because $|42.5|$ is greater than $|-32.25|$, the result is positive.

$\qquad\qquad\qquad = 42.5 - 32.25$ Absolute values of nonzero numbers are always positive.

$\qquad\qquad\qquad = 10.25$ Simplify.

Exercises

Replace each ● with <, >, or = to make a true sentence.

1. $-\dfrac{5}{8} \bullet \dfrac{3}{8}$ **2.** $\dfrac{4}{5} \bullet 0.71$ **3.** $\dfrac{5}{6} \bullet 0.875$

4. $1.2 \bullet 1\dfrac{2}{9}$ **5.** $\dfrac{8}{15} \bullet 0.5\overline{3}$ **6.** $-\dfrac{7}{11} \bullet -\dfrac{2}{3}$

Order each set of rational numbers from least to greatest.

7. $3.8, 3.06, 3\dfrac{1}{6}, 3\dfrac{3}{4}$ **8.** $2\dfrac{1}{4}, 1\dfrac{7}{8}, 1.75, 2.4$

9. $0.11, -\dfrac{1}{9}, -0.5, \dfrac{1}{10}$ **10.** $-4\dfrac{3}{5}, -3\dfrac{2}{5}, -4.65, -4.09$

Find each sum or difference. Write in simplest form.

11. $\dfrac{2}{5} + \dfrac{1}{5}$ **12.** $\dfrac{3}{9} + \dfrac{4}{9}$ **13.** $\dfrac{5}{16} - \dfrac{4}{16}$

14. $\dfrac{6}{7} - \dfrac{3}{7}$ **15.** $\dfrac{2}{3} + \dfrac{1}{3}$ **16.** $\dfrac{5}{8} + \dfrac{7}{8}$

17. $\dfrac{4}{3} + \dfrac{4}{3}$ **18.** $\dfrac{7}{15} - \dfrac{2}{15}$ **19.** $\dfrac{1}{3} - \dfrac{2}{9}$

20. $\dfrac{1}{2} + \dfrac{1}{4}$ **21.** $\dfrac{1}{2} - \dfrac{1}{3}$ **22.** $\dfrac{3}{7} + \dfrac{5}{14}$

23. $\dfrac{7}{10} - \dfrac{2}{15}$ **24.** $\dfrac{3}{8} + \dfrac{1}{6}$ **25.** $\dfrac{13}{20} - \dfrac{2}{5}$

Find each sum or difference. Write in simplest form if necessary.

26. $-1.6 + (-3.8)$ **27.** $-32.4 + (-4.5)$ **28.** $-38.9 + 24.2$

29. $-9.16 - 10.17$ **30.** $26.37 + (-61.1)$ **31.** $72.5 - (-81.3)$

32. $43.2 + (-27.9)$ **33.** $79.3 - (-14)$ **34.** $1.34 - (-0.458)$

35. $-\dfrac{1}{6} - \dfrac{2}{3}$ **36.** $\dfrac{1}{2} - \dfrac{4}{5}$ **37.** $-\dfrac{2}{5} + \dfrac{17}{20}$

38. $-\dfrac{4}{5} + \left(-\dfrac{1}{3}\right)$ **39.** $-\dfrac{1}{12} - \left(-\dfrac{3}{4}\right)$ **40.** $-\dfrac{7}{8} - \left(-\dfrac{3}{16}\right)$

41. GEOGRAPHY About $\dfrac{7}{10}$ of the surface of Earth is covered by water. The rest of the surface is covered by land. How much of Earth's surface is covered by land?

Multiplying and Dividing Rational Numbers

Objective

● Multiply and divide rational numbers.

NewVocabulary
multiplicative inverses
reciprocals

The product or quotient of two rational numbers having the *same sign* is positive. The product or quotient of two rational numbers having *different signs* is negative.

Example 1 Multiply and Divide Decimals

Find each product or quotient.

a. 7.2(−0.2)

different signs ⟶ negative product

7.2(−0.2) = −1.44

b. −23.94 ÷ (−10.5)

same sign ⟶ positive quotient

−23.94 ÷ (−10.5) = 2.28

To multiply fractions, multiply the numerators and multiply the denominators. If the numerators and denominators have common factors, you can simplify before you multiply by canceling.

Example 2 Multiply Fractions

Find each product.

a. $\frac{2}{5} \cdot \frac{1}{3}$

$\frac{2}{5} \cdot \frac{1}{3} = \frac{2 \cdot 1}{5 \cdot 3}$ Multiply the numerators. Multiply the denominators.

$= \frac{2}{15}$ Simplify.

b. $\frac{3}{5} \cdot 1\frac{1}{2}$

$\frac{3}{5} \cdot 1\frac{1}{2} = \frac{3}{5} \cdot \frac{3}{2}$ Write $1\frac{1}{2}$ as an improper fraction.

$= \frac{3 \cdot 3}{5 \cdot 2}$ Multiply the numerators. Multiply the denominators.

$= \frac{9}{10}$ Simplify.

c. $\frac{1}{4} \cdot \frac{2}{9}$

$\frac{1}{4} \cdot \frac{2}{9} = \frac{1}{\overset{4}{\underset{2}{\cancel{4}}}} \cdot \frac{\overset{1}{\cancel{2}}}{9}$ Divide by the GCF, 2.

$= \frac{1 \cdot 1}{2 \cdot 9}$ or $\frac{1}{18}$ Multiply the numerators. Multiply the denominators and simplify.

Example 3 Multiply Fractions with Different Signs

Find $\left(-\frac{3}{4}\right)\left(\frac{3}{8}\right)$.

$\left(-\frac{3}{4}\right)\left(\frac{3}{8}\right) = -\left(\frac{3}{4} \cdot \frac{3}{8}\right)$ different signs ⟶ negative product

$= -\left(\frac{3 \cdot 3}{4 \cdot 8}\right)$ or $\frac{9}{32}$ Multiply the numerators. Multiply the denominators and simplify.

Two numbers whose product is 1 are called **multiplicative inverses** or **reciprocals**.

Example 4 Find the Reciprocal

Name the reciprocal of each number.

a. $\dfrac{3}{8}$

$\dfrac{3}{8} \cdot \dfrac{8}{3} = 1$ The product is 1.

The reciprocal of $\dfrac{3}{8}$ is $\dfrac{8}{3}$.

b. $2\dfrac{4}{5}$

$2\dfrac{4}{5} = \dfrac{14}{5}$ Write $2\dfrac{4}{5}$ as $\dfrac{14}{5}$.

$\dfrac{14}{5} \cdot \dfrac{5}{14} = 1$ The product is 1.

The reciprocal of $2\dfrac{4}{5}$ is $\dfrac{5}{14}$.

To divide one fraction by another fraction, multiply the dividend by the reciprocal of the divisor.

Example 5 Divide Fractions

Find each quotient.

a. $\dfrac{1}{3} \div \dfrac{1}{2}$

$\dfrac{1}{3} \div \dfrac{1}{2} = \dfrac{1}{3} \cdot \dfrac{2}{1}$ Multiply $\dfrac{1}{3}$ by $\dfrac{2}{1}$, the reciprocal of $\dfrac{1}{2}$.

$= \dfrac{2}{3}$ Simplify.

b. $\dfrac{3}{8} \div \dfrac{2}{3}$

$\dfrac{3}{8} \div \dfrac{2}{3} = \dfrac{3}{8} \cdot \dfrac{3}{2}$ Multiply $\dfrac{3}{8}$ by $\dfrac{3}{2}$, the reciprocal of $\dfrac{2}{3}$.

$= \dfrac{9}{16}$ Simplify.

c. $\dfrac{3}{4} \div 2\dfrac{1}{2}$

$\dfrac{3}{4} \div 2\dfrac{1}{2} = \dfrac{3}{4} \div \dfrac{5}{2}$ Write $2\dfrac{1}{2}$ as an improper fraction

$= \dfrac{3}{4} \cdot \dfrac{2}{5}$ Multiply $\dfrac{3}{4}$ by $\dfrac{2}{5}$, the reciprocal of $2\dfrac{1}{2}$.

$= \dfrac{6}{20}$ or $\dfrac{3}{10}$ Simplify.

d. $-\dfrac{1}{5} \div \left(-\dfrac{3}{10}\right)$

$-\dfrac{1}{5} \div \left(-\dfrac{3}{10}\right) = -\dfrac{1}{5} \cdot \left(-\dfrac{10}{3}\right)$ Multiply $-\dfrac{1}{5}$ by $-\dfrac{10}{3}$, the reciprocal of $-\dfrac{3}{10}$.

$= \dfrac{10}{15}$ or $\dfrac{2}{3}$ Same sign ⟶ positive quotient; simplify.

StudyTip

Use Estimation You can justify your answer by using estimation. $\dfrac{3}{8}$ is close to $\dfrac{1}{2}$ and $\dfrac{2}{3}$ is close to 1. So, the quotient is close to $\dfrac{1}{2}$ divided by 1 or $\dfrac{1}{2}$.

Find each product or quotient. Round to the nearest hundredth if necessary.

1. $6.5(0.13)$

2. $-5.8(2.3)$

3. $42.3 \div (-6)$

4. $-14.1(-2.9)$

5. $-78 \div (-1.3)$

6. $108 \div (-0.9)$

7. $0.75(-6.4)$

8. $-23.94 \div 10.5$

9. $-32.4 \div 21.3$

Find each product. Simplify before multiplying if possible.

10. $\frac{3}{4} \cdot \frac{1}{5}$

11. $\frac{2}{5} \cdot \frac{3}{7}$

12. $-\frac{1}{3} \cdot \frac{2}{5}$

13. $-\frac{2}{3} \cdot \left(-\frac{1}{11}\right)$

14. $2\frac{1}{2} \cdot \left(-\frac{1}{4}\right)$

15. $3\frac{1}{2} \cdot 1\frac{1}{2}$

16. $\frac{2}{9} \cdot \frac{1}{2}$

17. $\frac{3}{2} \cdot \left(-\frac{1}{3}\right)$

18. $\frac{1}{3} \cdot \frac{6}{5}$

19. $-\frac{9}{4} \cdot \frac{1}{18}$

20. $\frac{11}{3} \cdot \frac{9}{44}$

21. $\left(-\frac{30}{11}\right) \cdot \left(-\frac{1}{3}\right)$

22. $-\frac{3}{5} \cdot \frac{5}{6}$

23. $\left(-\frac{1}{3}\right)\left(-7\frac{1}{2}\right)$

24. $\frac{2}{7} \cdot 4\frac{2}{3}$

Name the reciprocal of each number.

25. $\frac{6}{7}$

26. $\frac{1}{22}$

27. $-\frac{14}{23}$

28. $2\frac{3}{4}$

29. $-5\frac{1}{3}$

30. $3\frac{3}{4}$

Find each quotient.

31. $\frac{2}{3} \div \frac{1}{3}$

32. $\frac{16}{9} \div \frac{4}{9}$

33. $\frac{3}{2} \div \frac{1}{2}$

34. $\frac{3}{7} \div \left(-\frac{1}{5}\right)$

35. $-\frac{9}{10} \div 3$

36. $\frac{1}{2} \div \frac{3}{5}$

37. $2\frac{1}{4} \div \frac{1}{2}$

38. $-1\frac{1}{3} \div \frac{2}{3}$

39. $\frac{11}{12} \div 1\frac{2}{3}$

40. $4 \div \left(-\frac{2}{7}\right)$

41. $-\frac{1}{3} \div \left(-1\frac{1}{5}\right)$

42. $\frac{3}{25} \div \frac{2}{15}$

43. PIZZA A large pizza at Pizza Shack has 12 slices. If Bobby ate $\frac{1}{4}$ of the pizza, how many slices of pizza did he eat?

44. MUSIC Samantha practices the flute for $4\frac{1}{2}$ hours each week. How many hours does she practice in a month?

45. BAND How many band uniforms can be made with $131\frac{3}{4}$ yards of fabric if each uniform requires $3\frac{7}{8}$ yards?

46. CARPENTRY How many boards, each 2 feet 8 inches long, can be cut from a board 16 feet long if there is no waste?

47. SEWING How many 9-inch ribbons can be cut from $1\frac{1}{2}$ yards of ribbon?

The Percent Proportion

A **percent** is a ratio that compares a number to 100. To write a percent as a fraction, express the ratio as a fraction with a denominator of 100. Fractions should be expressed in simplest form.

Example 1 Percents as Fractions

Express each percent as a fraction or mixed number.

a. 79%

$$79\% = \frac{79}{100}$$ Definition of percent

b. 107%

$$107\% = \frac{107}{100}$$ Definition of percent

$$= 1\frac{7}{100}$$ Simplify.

c. 0.5%

$$0.5\% = \frac{0.5}{100}$$ Definition of percent

$$= \frac{5}{1000}$$ Multiply the numerator and denominator by 10 to eliminate the decimal.

$$= \frac{1}{200}$$ Simplify.

In the **percent proportion**, the ratio of a part of something to the whole (base) is equal to the percent written as a fraction.

part $\longrightarrow$ $\dfrac{a}{b} = \dfrac{p}{100}$ $\longleftarrow$ percent
whole $\longrightarrow$

$$\begin{array}{ccc} \text{percent} & \text{whole} & \text{part} \\ \downarrow & \downarrow & \downarrow \end{array}$$

Example: 25%   of   40   is   10.

You can use the percent proportion to find the part.

Example 2 Find the Part

40% of 30 is what number?

$\dfrac{a}{b} = \dfrac{p}{100}$ The percent is 40, and the base is 30. Let a represent the part.

$\dfrac{a}{30} = \dfrac{40}{100}$ Replace b with 30 and p with 40.

$100a = 30(40)$ Find the cross products.

$100a = 1200$ Simplify.

$\dfrac{100a}{100} = \dfrac{1200}{100}$ Divide each side by 100.

$a = 12$ Simplify.

The part is 12. So, 40% of 30 is 12.

You can also use the percent proportion to find the percent of the base.

SURVEYS Kelsey took a survey of students in her lunch period. 42 out of the 70 students Kelsey surveyed said their family had a pet. What percent of the students had pets?

$$\frac{a}{b} = \frac{p}{100}$$ The part is 42, and the base is 70. Let p represent the percent.

$$\frac{42}{70} = \frac{p}{100}$$ Replace a with 42 and b with 70.

$$4200 = 70p$$ Find the cross products.

$$\frac{4200}{70} = \frac{70p}{70}$$ Divide each side by 70.

$$60 = p$$ Simplify.

The percent is 60, so $\frac{60}{100}$ or 60% of the students had pets.

StudyTip

Percent Proportion In percent problems, the whole, or base usually follows the word *of*.

67.5 is 75% of what number?

$$\frac{a}{b} = \frac{p}{100}$$ The percent is 75, and the part is 67.5. Let b represent the base.

$$\frac{67.5}{b} = \frac{75}{100}$$ Replace a with 67.5 and p with 75.

$$6750 = 75b$$ Find the cross products.

$$\frac{6750}{75} = \frac{75b}{75}$$ Divide each side by 75.

$$90 = b$$ Simplify.

The base is 90, so 67.5 is 75% of 90.

Exercises

Express each percent as a fraction or mixed number in simplest form.

1. 5% **2.** 60% **3.** 11%

4. 120% **5.** 78% **6.** 2.5%

7. 0.6% **8.** 0.4% **9.** 1400%

Use the percent proportion to find each number.

10. 25 is what percent of 125? **11.** 16 is what percent of 40?

12. 14 is 20% of what number? **13.** 50% of what number is 80?

14. What number is 25% of 18? **15.** Find 10% of 95.

16. What percent of 48 is 30? **17.** What number is 150% of 32?

18. 5% of what number is 3.5? **19.** 1 is what percent of 400?

20. Find 0.5% of 250. **21.** 49 is 200% of what number?

22. 15 is what percent of 12? **23.** 36 is what percent of 24?

24. **BASKETBALL** Madeline usually makes 85% of her shots in basketball. If she attempts 20, how many will she likely make?

25. **TEST SCORES** Brian answered 36 items correctly on a 40-item test. What percent did he answer correctly?

26. **CARD GAMES** Juanita told her dad that she won 80% of the card games she played yesterday. If she won 4 games, how many games did she play?

27. **SOLUTIONS** A glucose solution is prepared by dissolving 6 milliliters of glucose in 120 milliliters of pure solution. What is the percent of glucose in the resulting solution?

28. **DRIVER'S ED** Kara needs to get a 75% on her driving education test in order to get her license. If there are 35 questions on the test, how many does she need to answer correctly?

29. **HEALTH** The U.S. Food and Drug Administration requires food manufacturers to label their products with a nutritional label. The label shows the information from a package of macaroni and cheese.

   a. The label states that a serving contains 3 grams of saturated fat, which is 15% of the daily value recommended for a 2000-Calorie diet. How many grams of saturated fat are recommended for a 2000-Calorie diet?

   b. The 470 milligrams of sodium (salt) in the macaroni and cheese is 20% of the recommended daily value. What is the recommended daily value of sodium?

   c. For a healthy diet, the National Research Council recommends that no more than 30 percent of the total Calories come from fat. What percent of the Calories in a serving of this macaroni and cheese come from fat?

Nutrition Facts		
Serving Size 1 cup (228g)		
Servings per container 2		
Amount per serving		
Calories 250 Calories from Fat 110		
		%Daily value*
Total Fat 12g		18%
Saturated Fat 3g		15%
Cholesterol 30mg		10%
Sodium 470mg		20%
Total Carbohydrate 31g		10%
Dietary Fiber 0g		0%
Sugars 5g		
Protein 5g		
Vitamin A 4% • Vitamin C 2%		
Calcium 20% • Iron 4%		

30. **TEST SCORES** The table shows the number of points each student in Will's study group earned on a recent math test. There were 88 points possible on the test. Express all answers to the nearest tenth of a percent.

Name	Will	Penny	Cheng	Minowa	Rob
Score	72	68	81	87	75

   a. Find Will's percent correct on the test.

   b. Find Cheng's percent correct on the test.

   c. Find Rob's percent correct on the test.

   d. What was the highest percentage? The lowest?

31. **PET STORE** In a pet store, 15% of the animals are hamsters. If the store has 40 animals, how many of them are hamsters?

Perimeter

NewVocabulary
perimeter
circle
diameter
circumference
center
radius

Perimeter is the distance around a figure. Perimeter is measured in linear units.

Rectangle

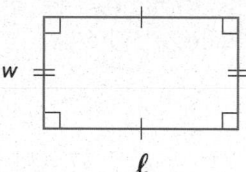

$P = 2(\ell + w)$ or
$P = 2\ell + 2w$

Parallelogram

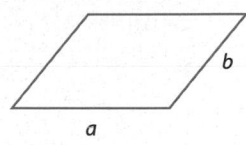

$P = 2(a + b)$ or
$P = 2a + 2b$

Square

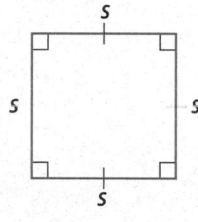

$P = 4s$

Triangle

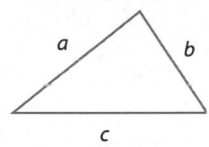

$P = a + b + c$

Example 1 Perimeters of Rectangles and Squares

Find the perimeter of each figure.

a. a rectangle with a length of 5 inches and a width of 1 inch

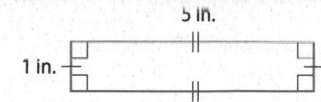

$P = 2(\ell + w)$ Perimeter formula

$= 2(5 + 1)$ $\ell = 5, w = 1$

$= 2(6)$ Add.

$= 12$ The perimeter is 12 inches.

b. a square with a side length of 7 centimeters

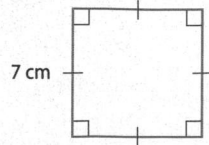

$P = 4s$ Perimeter formula

$= 4(7)$ Replace s with 7.

$= 28$ The perimeter is 28 centimeters.

Example 2 Perimeters of Parallelograms and Triangles

Find the perimeter of each figure.

a.

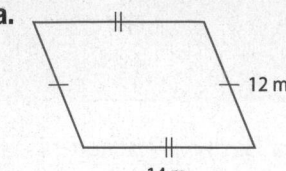

14 m

$P = 2(a + b)$ Perimeter formula

   $= 2(14 + 12)$ $a = 14, b = 12$

   $= 2(26)$ Add.

   $= 52$ Multiply.

The perimeter of the parallelogram is 52 meters.

b.

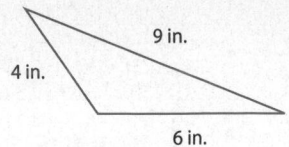

$P = a + b + c$ Perimeter formula

   $= 4 + 6 + 9$ $a = 4, b = 6, c = 9$

   $= 19$ Add.

The perimeter of the triangle is 19 inches.

A **circle** is the set of all points in a plane that are the same distance from a given point.

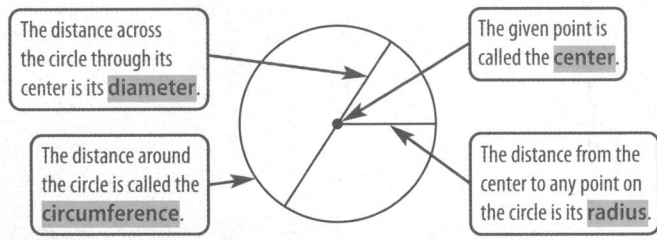

The distance across the circle through its center is its **diameter**.

The given point is called the **center**.

The distance around the circle is called the **circumference**.

The distance from the center to any point on the circle is its **radius**.

The formula for the circumference of a circle is $C = \pi d$ or $C = 2\pi r$.

Example 3 Circumference

Find each circumference to the nearest tenth.

a. The radius is 4 feet.

$C = 2\pi r$ Circumference formula

   $= 2\pi(4)$ Replace r with 4.

   $= 8\pi$ Simplify.

The exact circumference is 8π feet.

8 $\boxed{\pi}$ $\boxed{\text{ENTER}}$ **25.13274123**

The circumference is about 25.1 feet.

b. The diameter is 15 centimeters.

$C = \pi d$ Circumference formula

   $= \pi(15)$ Replace d with 15.

   $= 15\pi$ Simplify.

   ≈ 47.1 Use a calculator to evaluate 15π.

The circumference is about 47.1 centimeters.

c.

3 m

$C = 2\pi r$ Circumference formula

   $= 2\pi(3)$ Replace r with 3.

   $= 6\pi$ Simplify.

   ≈ 18.8 Use a calculator to evaluate 6π.

The circumference is about 18.8 meters.

Find the perimeter of each figure.

1.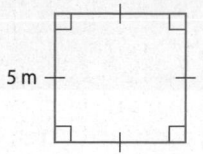
5 m

2.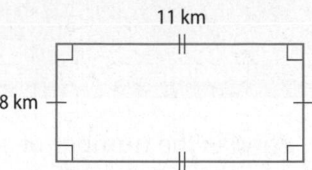
11 km
8 km

3.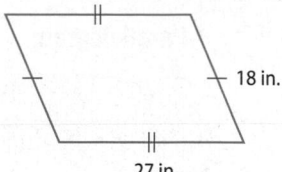
18 in.
27 in.

4.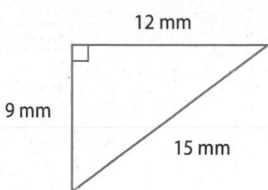
12 mm
9 mm
15 mm

5. a square with side length 8 inches

6. a rectangle with length 9 centimeters and width 3 centimeters

7. a triangle with sides 4 feet, 13 feet, and 12 feet

8. a parallelogram with side lengths $6\frac{1}{4}$ inches and 5 inches

9. a quarter-circle with a radius of 7 inches

Find the circumference of each circle. Round to the nearest tenth.

10.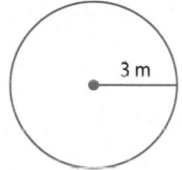
3 m

11.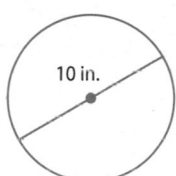
10 in.

12.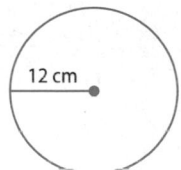
12 cm

13. GARDENS A square garden has a side length of 5.8 meters. What is the perimeter of the garden?

14. ROOMS A rectangular room is $12\frac{1}{2}$ feet wide and 14 feet long. What is the perimeter of the room?

15. CYCLING The tire for a 10-speed bicycle has a diameter of 27 inches. Find the distance traveled in 10 rotations of the tire. Round to the nearest tenth.

16. GEOGRAPHY Earth's circumference is approximately 25,000 miles. If you could dig a tunnel to the center of the Earth, how long would the tunnel be? Round to the nearest tenth mile.

Find the perimeter of each figure. Round to the nearest tenth.

17.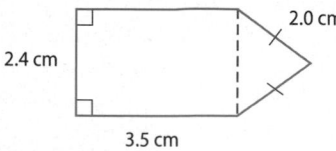
2.0 cm
2.4 cm
3.5 cm

18.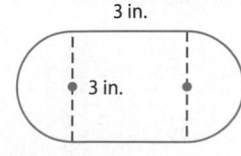
3 in.
3 in.

19.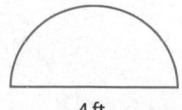
4 ft

20.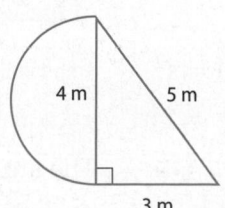
4 m
5 m
3 m

Area

Objective

● Find the area of two-dimensional figures.

NewVocabulary
area

Area is the number of square units needed to cover a surface. Area is measured in square units.

Rectangle

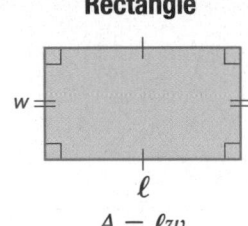

$A = \ell w$

Parallelogram

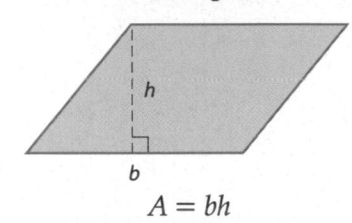

$A = bh$

Square

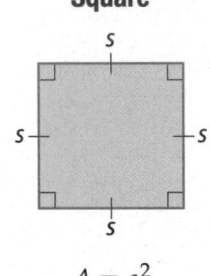

$A = s^2$

Triangle

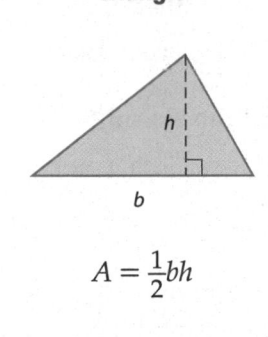

$A = \frac{1}{2}bh$

Example 1 Areas of Rectangles and Squares

Find the area of each figure.

a. a rectangle with a length of 7 yards and a width of 1 yard

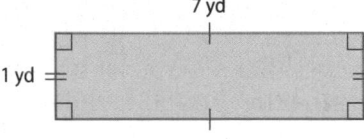

$A = \ell w$ Area formula

$ = 7(1)$ $\ell = 7, w = 1$

$ = 7$ The area of the rectangle is 7 square yards.

b. a square with a side length of 2 meters

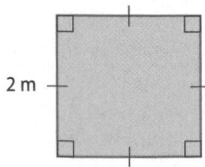

$A = s^2$ Area formula

$ = 2^2$ $s = 2$

$ = 4$ The area is 4 square meters.

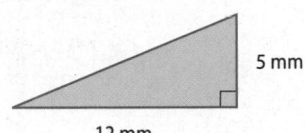

Example 2 Areas of Parallelograms and Triangles

Find the area of each figure.

a. a parallelogram with a base of 11 feet and a height of 9 feet

b. a triangle with a base of 12 millimeters and a height of 5 millimeters

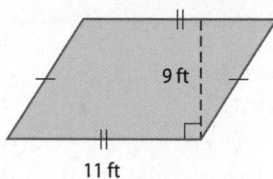

9 ft

11 ft

5 mm

12 mm

$A = bh$ Area formula

$= 11(9)$ $b = 11, h = 9$

$= 99$ Multiply.

The area is 99 square feet.

$A = \frac{1}{2}bh$ Area formula

$= \frac{1}{2}(12)(5)$ $b = 12, h = 5$

$= 30$ Multiply.

The area is 30 square millimeters.

The formula for the area of a circle is $A = \pi r^2$.

Example 3 Areas of Circles

Find the area of each circle to the nearest tenth.

a. a radius of 3 centimeters

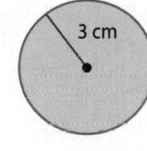

3 cm

$A = \pi r^2$ Area formula

$= \pi(3)^2$ Replace r with 3.

$= 9\pi$ Simplify.

≈ 28.3 Use a calculator to evaluate 9π.

The area is about 28.3 square centimeters.

b. a diameter of 21 meters

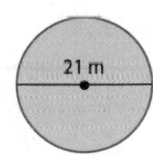

21 m

$A = \pi r^2$ Area formula

$= \pi(10.5)^2$ Replace r with 10.5.

$= 110.25\pi$ Simplify.

≈ 346.4 Use a calculator to evaluate 110.25π.

The area is about 346.4 square meters.

StudyTip

Mental Math You can use mental math to check your solutions. Square the radius and then multiply by 3.

Example 4 Estimate Area

Estimate the area of the polygon if each square represents 1 square mile.

One way to estimate the area is to count each square as one unit and each partial square as a half unit, no matter how large or small.

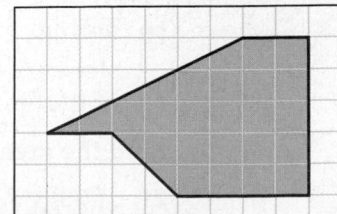

$A \approx$ squares + partial squares

$\approx 21(1) + 8(0.5)$ 21 whole squares and 8 partial squares

$\approx 21 + 4$ or 25

The area of the polygon is about 25 square miles.

Find the area of each figure.

1.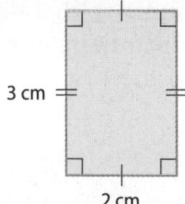
3 cm
2 cm

2.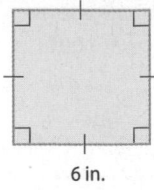
6 in.

3.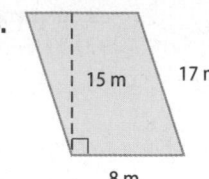
15 m   17 m
8 m

Find the area of each figure. Round to the nearest tenth if necessary.

4. a triangle with a base 12 millimeters and height 11 millimeters

5. a square with side length 9 feet

6. a rectangle with length 8 centimeters and width 2 centimeters

7. a triangle with a base 6 feet and height 3 feet

8. a quarter-circle with a diameter of 4 meters

9. a semi-circle with a radius of 3 inches

Find the area of each circle. Round to the nearest tenth.

10.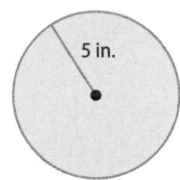
5 in.

11.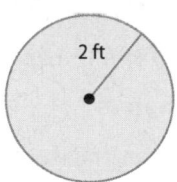
2 ft

12.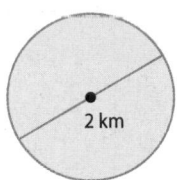
2 km

13. The radius is 4 centimeters.

14. The radius is 7.2 millimeters.

15. The diameter is 16 inches.

16. The diameter is 25 feet.

17. **CAMPING** The square floor of a tent has an area of 49 square feet. What is the side length of the tent?

Estimate the area of each polygon in square units.

18.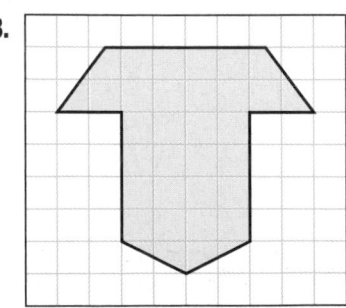

19.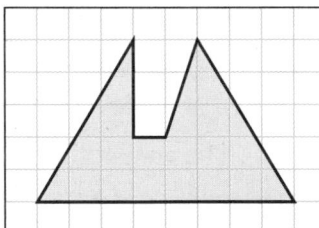

20. **HISTORY** Stonehenge is an ancient monument in Wiltshire, England. The giant stones of Stonehenge are arranged in a circle 30 meters in diameter. Find the area of the circle. Round to the nearest tenth square meter.

Find the area of each figure. Round to the nearest tenth.

21.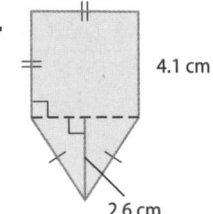
4.1 cm
2.6 cm

22.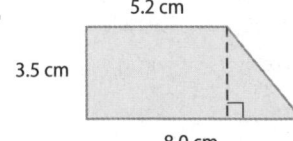
5.2 cm
3.5 cm
8.0 cm

23.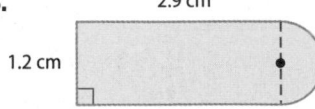
2.9 cm
1.2 cm

Volume

·· Objective

● Find the volumes of rectangular prisms and cylinders.

NewVocabulary
volume

Volume is the measure of space occupied by a solid. Volume is measured in cubic units.

To find the volume of a rectangular prism, multiply the length times the width times the height. The formula for the volume of a rectangular prism is shown below.

$$V = \ell \cdot w \cdot h$$

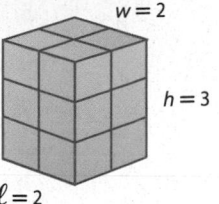

$w = 2$

$h = 3$

$\ell = 2$

The prism at the right has a volume of $2 \cdot 2 \cdot 3$ or 12 cubic units.

Example 1 Volumes of Rectangular Prisms

Find the volume of each rectangular prism.

a. The length is 8 centimeters, the width is 1 centimeter, and the height is 5 centimeters.

$V = \ell \cdot w \cdot h$ Volume formula
$= 8 \cdot 1 \cdot 5$ Replace ℓ with 8, w with 1, and h with 5.
$= 40$ Simplify.

The volume is 40 cubic centimeters.

b.

4 ft

2 ft

3 ft

The prism has a length of 4 feet, width of 2 feet, and height of 3 feet.

$V = \ell \cdot w \cdot h$ Volume formula
$= 4 \cdot 2 \cdot 3$ Replace ℓ with 4, w with 2, and h with 3.
$= 24$ Simplify.

The volume is 24 cubic feet.

The volume of a solid is the product of the area of the base and the height of the solid. For a cylinder, the area of the base is πr^2. So the volume is $V = \pi r^2 h$.

Example 2 Volume of a Cylinder

Find the volume of the cylinder.

$V = \pi r^2 h$ Volume of a cylinder

$= \pi(3^2)6$ $r = 3, h = 6$

$= 54\pi$ Simplify.

≈ 169.6 Use a calculator.

3 in.

6 in.

The volume is about 169.6 cubic inches.

Find the volume of each rectangular prism given the length, width, and height.

1. $\ell = 5$ cm, $w = 3$ cm, $h = 2$ cm

2. $\ell = 10$ m, $w = 10$ m, $h = 1$ m

3. $\ell = 6$ yd, $w = 2$ yd, $h = 4$ yd

4. $\ell = 2$ in., $w = 5$ in., $h = 12$ in.

5. $\ell = 13$ ft, $w = 9$ ft, $h = 12$ ft

6. $\ell = 7.8$ mm, $w = 0.6$ mm, $h = 8$ mm

Find the volume of each rectangular prism.

7.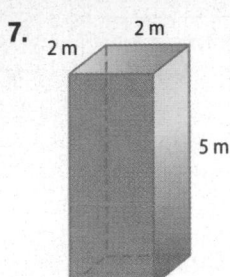

8.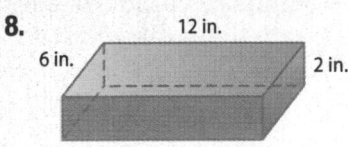

9. GEOMETRY A cube measures 3 meters on a side. What is its volume?

10. AQUARIUMS An aquarium is 8 feet long, 5 feet wide, and 5.5 feet deep. What is the volume of the aquarium?

11. COOKING What is the volume of a microwave oven that is 18 inches wide by 10 inches long with a depth of $11\frac{1}{2}$ inches?

12. BOXES A cardboard box is 32 inches long, 22 inches wide, and 16 inches tall. What is the volume of the box?

13. SWIMMING POOLS A children's rectangular pool holds 480 cubic feet of water. What is the depth of the pool if its length is 30 feet and its width is 16 feet?

14. BAKING A rectangular cake pan has a volume of 234 cubic inches. If the length of the pan is 9 inches and the width is 13 inches, what is the height of the pan?

15. GEOMETRY The volume of the rectangular prism at the right is 440 cubic centimeters. What is the width?

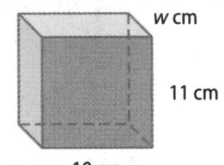

Find the volume of each cylinder. Round to the nearest tenth.

16.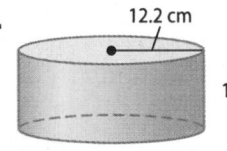

17.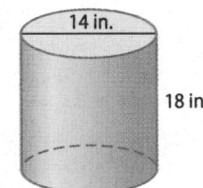

18.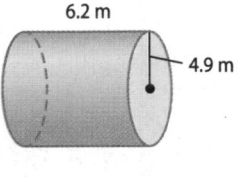

19. FIREWOOD Firewood is usually sold by a measure known as a *cord*. A full cord may be a stack $8 \times 4 \times 4$ feet or a stack $8 \times 8 \times 2$ feet.

  a. What is the volume of a full cord of firewood?

  b. A "short cord" of wood is $8 \times 4 \times$ the length of the logs. What is the volume of a short cord of $2\frac{1}{2}$-foot logs?

  c. If you have an area that is 12 feet long and 2 feet wide in which to store your firewood, how high will the stack be if it is a full cord of wood?

Surface Area

Objective

- Find the surface areas of rectangular prisms and cylinders.

NewVocabulary
surface area

Surface area is the sum of the areas of all the surfaces, or faces, of a solid. Surface area is measured in square units.

> **KeyConcept** Surface Area

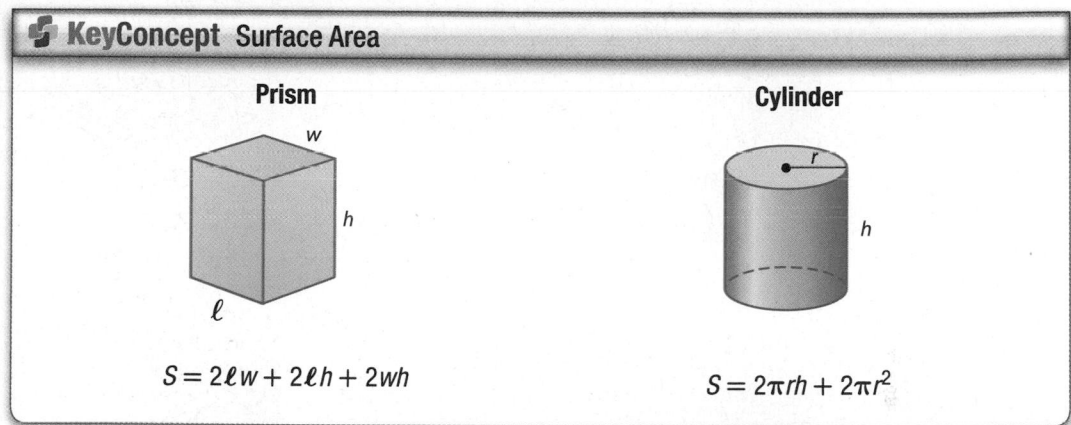

Prism

$$S = 2\ell w + 2\ell h + 2wh$$

Cylinder

$$S = 2\pi rh + 2\pi r^2$$

Example 1 Find Surface Areas

Find the surface area of each solid. Round to the nearest tenth if necessary.

a.

The prism has a length of 3 meters, width of 1 meter, and height of 5 meters.

$S = 2\ell w + 2\ell h + 2wh$	Surface area formula
$= 2(3)(1) + 2(3)(5) + 2(1)(5)$	$\ell = 3, w = 1, h = 5$
$= 6 + 30 + 10$	Multiply.
$= 46$	Add.

The surface area is 46 square meters.

b.

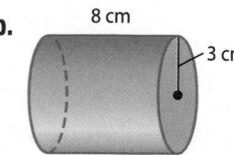

The height is 8 centimeters and the radius of the base is 3 centimeters. The surface area is the sum of the area of each base, $2\pi r^2$, and the area of the side, given by the circumference of the base times the height or $2\pi rh$.

$S = 2\pi rh + 2\pi r^2$	Formula for surface area of a cylinder.
$= 2\pi(3)(8) + 2\pi(3^2)$	$r = 3, h = 8$
$= 48\pi + 18\pi$	Simplify.
$\approx 207.3 \text{ cm}^2$	Use a calculator.

Find the surface area of each rectangular prism given the measurements below.

1. $\ell = 6$ in., $w = 1$ in., $h = 4$ in

2. $\ell = 8$ m, $w = 2$ m, $h = 2$ m

3. $\ell = 10$ mm, $w = 4$ mm, $h = 5$ mm

4. $\ell = 6.2$ cm, $w = 1$ cm, $h = 3$ cm

5. $\ell = 7$ ft, $w = 2$ ft, $h = \frac{1}{2}$ ft

6. $\ell = 7.8$ m, $w = 3.4$ m, $h = 9$ m

Find the surface area of each solid.

7.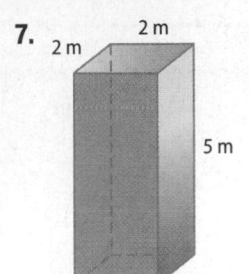

8.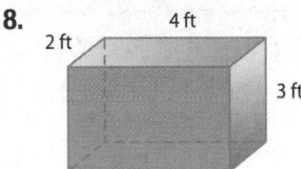

9.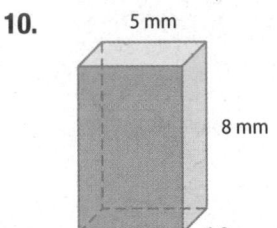

10.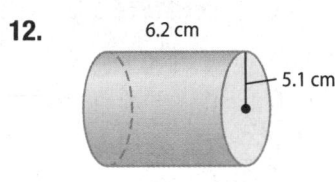

11.

12.

13. GEOMETRY What is the surface area of a cube with a side length of 2 meters?

14. GIFTS A gift box is a rectangular prism 14 inches long, 5 inches wide, and 4 inches high. If the box is to be covered in fabric, how much fabric is needed if there is no overlap?

15. BOXES A new refrigerator is shipped in a box 34 inches deep, 66 inches high, and $33\frac{1}{4}$ inches wide. What is the surface area of the box in square feet? Round to the nearest square foot. (*Hint:* 1 ft² = 144 in²)

16. PAINTING A cabinet is 6 feet high, 3 feet wide, and 2 feet long. The entire outside surface of the cabinet is being painted except for the bottom. What is the surface area of the cabinet that is being painted?

17. SOUP A soup can is 4 inches tall and has a diameter of $3\frac{1}{4}$ inches. How much paper is needed for the label on the can? Round your answer to the nearest tenth.

18. CRAFTS For a craft project, Sarah is covering all the sides of a box with stickers. The length of the box is 8 inches, the width is 6 inches, and the height is 4 inches. If each sticker has a length of 2 inches and a width of 4 inches, how many stickers does she need to cover the box?

Simple Probability and Odds

∷ Objective

- Find the probability and odds of simple events.

NewVocabulary

probability
sample space
equally likely
complements
tree diagram
odds

The **probability** of an event is the ratio of the number of favorable outcomes for the event to the total number of possible outcomes. When you roll a die, there are six possible outcomes: 1, 2, 3, 4, 5, or 6. This list of all possible outcomes is called the **sample space**.

When there are n outcomes and the probability of each one is $\frac{1}{n}$, we say that the outcomes are **equally likely**.

For example, when you roll a die, the 6 possible outcomes are equally likely because each outcome has a probability of $\frac{1}{6}$. The probability of an event is always between 0 and 1, inclusive. The closer a probability is to 1, the more likely it is to occur.

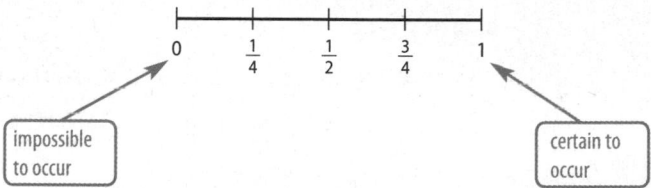

impossible to occur   certain to occur

Example 1   Find Probabilities

A die is rolled. Find each probability.

a. rolling a 1 or 5

There are six possible outcomes. There are two favorable outcomes, 1 and 5.

$$\text{probability} = \frac{\text{number of favorable outcomes}}{\text{total number of possible outcomes}} = \frac{2}{6}$$

So, $P(1 \text{ or } 5) = \frac{2}{6}$ or $\frac{1}{3}$.

b. rolling an even number

Three of the six outcomes are even numbers. So, there are three favorable outcomes.

Sample space: 1, 2, 3, 4, 5, 6   3 even numbers → $\frac{3}{6}$

6 total possible outcomes

So, $P(\text{even number}) = \frac{3}{6}$ or $\frac{1}{2}$.

The events for rolling a 1 and for *not* rolling a 1 are called **complements**.

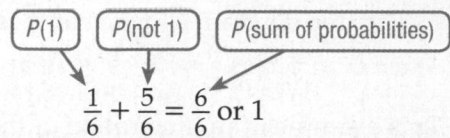

$P(1)$ $P(\text{not } 1)$ $P(\text{sum of probabilities})$

$$\frac{1}{6} + \frac{5}{6} = \frac{6}{6} \text{ or } 1$$

The sum of the probabilities for any two complementary events is always 1.

Example 2 Find Probabilities

A bowl contains 5 red chips, 7 blue chips, 6 yellow chips, and 10 green chips. One chip is randomly drawn. Find each probability.

a. blue

There are 7 blue chips and 28 total chips.

$P(\text{blue chip}) = \dfrac{7}{28}$ ← number of favorable outcomes
← number of possible outcomes

$= \dfrac{1}{4}$

The probability can be stated as $\dfrac{1}{4}$, 0.25, or 25%.

b. red or yellow

There are 5 + 6 or 11 chips that are red or yellow.

$P(\text{red or yellow}) = \dfrac{11}{28}$ ← number of favorable outcomes
← number of possible outcomes

≈ 0.39

The probability can be stated as $\dfrac{11}{28}$, about 0.39, or about 39%.

c. not green

There are 5 + 7 + 6 or 18 chips that are not green.

$P(\text{not green}) = \dfrac{18}{28}$ ← number of favorable outcomes
← number of possible outcomes

$= \dfrac{9}{14}$ or about 0.64

The probability can be stated as $\dfrac{9}{14}$, about 0.64, or about 64%.

> **StudyTip**
>
> **Alternate Method** A chip drawn will either be green or not green. So, another method for finding $P(\text{not green})$ is to find $P(\text{green})$ and subtract that probability from 1.

One method used for counting the number of possible outcomes is to draw a **tree diagram**. The last column of a tree diagram shows all of the possible outcomes.

Example 3 Use a Tree Diagram to Count Outcomes

School baseball caps come in blue, yellow, or white. The caps have either the school mascot or the school's initials. Use a tree diagram to determine the number of different caps possible.

Color	Design	Outcomes
blue	mascot	blue, mascot
	initials	blue, initials
yellow	mascot	yellow, mascot
	initials	yellow, initials
white	mascot	white, mascot
	initials	white, initials

The tree diagram shows that there are 6 different caps possible.

> **StudyTip**
>
> **Counting Outcomes** When counting possible outcomes, make a column in your tree diagram for each part of the event.

This example is an illustration of the **Fundamental Counting Principle**, which relates the number of outcomes to the number of choices.

> ## KeyConcept Fundamental Counting Principle
>
> **Words**   If event *M* can occur in *m* ways and is followed by event *N* that can occur in *n* ways, then the event *M* followed by *N* can occur in *m* • *n* ways.
>
> **Example**   If there are 4 possible sizes for fish tanks and 3 possible shapes, then there are 4 • 3 or 12 possible fish tanks.

Example 4 Use the Fundamental Counting Principle

a. An ice cream shop offers one, two, or three scoops of ice cream from among 12 different flavors. The ice cream can be served in a wafer cone, a sugar cone, or in a cup. Use the Fundamental Counting Principle to determine the number of choices possible.

There are 3 ways the ice cream is served, 3 different servings, and there are 12 different flavors of ice cream.

Use the Fundamental Counting Principle to find the number of possible choices.

number of scoops	number of flavors	number of serving options	number of choices of ordering ice cream
3 •	12 •	3 =	108

So, there are 108 different ways to order ice cream.

b. Jimmy needs to make a 3-digit password for his log-on name on a Web site. The password can include any digit from 0-9, but the digits may not repeat. How many possible 3-digit passwords are there?

If the first digit is a 4, then the next digit cannot be a 4.

We can use the Fundamental Counting Principle to find the number of possible passwords.

1st digit	2nd digit	3rd digit	number of passwords
10 •	9 •	8 =	720

So, there are 720 possible 3-digit passwords.

StudyTip

Odds The sum of the number of successes and the number of failures equals the size of the sample space, or the number of possible outcomes.

The **odds** of an event occurring is the ratio that compares the number of ways an event can occur (successes) to the number of ways it cannot occur (failures).

Example 5 Find the Odds

Find the odds of rolling a number less than 3.

There are six possible outcomes; 2 are successes and 4 are failures.

So, the odds of rolling a number less than 3 are $\frac{1}{2}$ or 1:2.

One coin is randomly selected from a jar containing 70 nickels, 100 dimes, 80 quarters, and 50 one-dollar coins. Find each probability.

1. P(quarter)

2. P(dime)

3. P(quarter or nickel)

4. P(value greater than $0.10)

5. P(value less than $1)

6. P(value at most $1)

One of the polygons below is chosen at random. Find each probability.

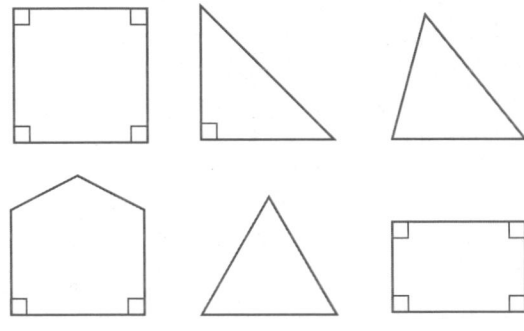

7. P(triangle)

8. P(pentagon)

9. P(not a quadrilateral)

10. P(more than 2 right angles)

Use a tree diagram to find the sample space for each event. State the number of possible outcomes.

11. The spinner at the right is spun and two coins are tossed.

12. At a restaurant, you choose two sides to have with breakfast. You can choose white or whole wheat toast. You can choose sausage links, sausage patties, or bacon.

13. How many different 3-character codes are there using A, B, or C for the first character, 8 or 9 for the second character, and 0 or 1 for the third character?

A bag is full of different colored marbles. The probability of randomly selecting a red marble from the bag is $\frac{1}{8}$. The probability of selecting a blue marble is $\frac{13}{24}$. Find each probability.

14. P(not red)

15. P(not blue)

Find the odds of each outcome if a computer randomly picks a letter in the name THE UNITED STATES OF AMERICA.

16. the letter A

17. the letter T

18. a vowel

19. a consonant

Margaret wants to order a sub at the local deli.

20. Find the number of possible orders of a sub with one topping and one dressing option.

21. Find the number of possible ham subs with mayonnaise, any combination of toppings or no toppings at all.

22. Find the number of possible orders of a sub with any combination of dressing and/or toppings.

Subs
ham, salami, roast beef, turkey, bologna, pepperoni

Dressing	Toppings
mayonnaise, mustard, vinegar, oil	lettuce, onions, peppers, olives

Measures of Center, Variation, and Position

Objective

- Find measures of central tendency, variation, and position.

 NewVocabulary

variable
data
measurement or quantitative data
categorical or qualitative data
univariate data
measures of center or central tendency
mean
median
mode
measures of spread or variation
range
quartile
measures of position
lower quartile
upper quartile
five-number summary
interquartile range
outlier

A **variable** is a characteristic of a group of people or objects that can assume different values called **data**. Data that have units and can be measured are called **measurement** or **quantitative data**. Data that can be organized into different categories are called **categorical** or **qualitative data**. Some examples of both types of data are listed below.

Measurement Data	Categorical Data
Times: 15 s, 20 s, 45 s, 19 s	Favorite color: blue, red, purple, green
Ages: 10 yr, 15 yr, 14 yr, 16 yr	Hair color: black, blonde, brown
Distance: 5 mi, 30 mi, 18 mi	Phone Numbers: 555-1234, 555-5678

Measurement data in one variable, called **univariate data**, are often summarized using a single number to represent what is average or typical. Measures of what is average are also called **measures of center** or **central tendency**. The most common measures of center are mean, median, and mode.

KeyConcept Measures of Center

- The **mean** is the sum of the values in a data set divided by the total number of values in the set.

- The **median** is the middle value or the mean of the two middle values in a set of data when the data are arranged in numerical order.

- The **mode** is the value or values that appear most often in a set of data. A set of data can have no mode, one mode, or more than one mode.

Example 1 Measures of Center

BASEBALL The table shows the number of hits Marcus made for his team. Find the mean, median, and mode.

Team Played	Hits
Badgers	3
Hornets	6
Bulldogs	5
Vikings	2
Rangers	3
Panthers	7

Mean: To find the mean, find the sum of all the hits and divide by the number of games in which he made these hits.

$$\text{mean} = \frac{3 + 6 + 5 + 2 + 3 + 7}{6} = \frac{26}{6} \text{ or about 4 hits}$$

Median: To find the median, order the numbers from least to greatest and find the middle value or values.

2, 3, 3, 5, 6, 7

$\frac{3 + 5}{2}$ or 4 hits Since there is an even number of values, find the mean of the middle two.

Mode: From the arrangement of the data values, we can see that the value that occurs most often in the set is 3, so the mode of the data set is 3 hits.

Marcus's mean and median number of hits for these games was 4, and his mode was 3 hits.

Two very different data sets can have the same mean, so statisticians also use **measures of spread** or **variation** to describe how widely the data values vary. One such measure is the **range**, which is the difference between the greatest and least values in a set of data.

PT

Example 2 Range

WALKING The times in minutes it took Olivia to walk to school each day this week are 18, 15, 15, 12, and 14. Find the range.

range = greatest value − least value   Definition of range

     = 18 − 12 or 6   The greatest value is 18, and the least value is 12.

The range of the times is 6 minutes.

Statisticians often talk about the position of a value relative to other values in a set. **Quartiles** are common **measures of position** that divide a data set arranged in ascending order into four groups, each containing about one fourth or 25% of the data. The median marks the second quartile Q_2 and separates the data into upper and lower halves. The first or **lower quartile** Q_1 is the median of the lower half, while the third or **upper quartile** Q_3 is the median of the upper half.

StudyTip

Calculating Quartiles
When the number of values in a set of data is odd, the median is not included in either half of the data when calculating Q_1 or Q_3.

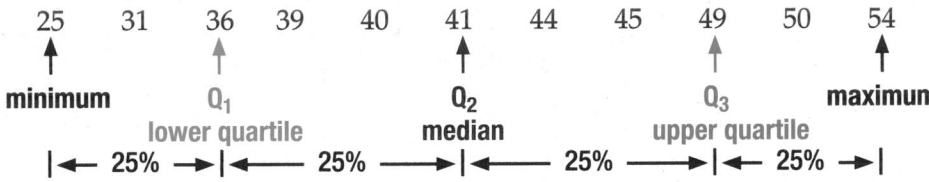

The three quartiles, along with the minimum and maximum values, are called a **five-number summary** of a data set.

PT

Example 3 Five-Number Summary

FUNDRAISER The number of boxes of donuts Aang sold for a fundraiser each day for the last 11 days were 22, 16, 35, 26, 14, 17, 28, 29, 21, 17, and 20. Find the minimum, lower quartile, median, upper quartile, and maximum of the data set. Then interpret this five-number summary.

Order the data from least to greatest. Use the list to determine the quartiles.

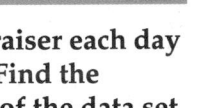

14, 16, 17, 17, 20, 21, 22, 26, 28, 29, 35

    Min. Q_1 Q_2 Q_3 Max.

The minimum is 14, the lower quartile is 17, the median is 21, the upper quartile is 28, and the maximum is 35. Over the last 11 days, Aang sold a minimum of 14 boxes and a maximum of 35 boxes. He sold fewer than 17 boxes 25% of the time, fewer than 21 boxes 50% of the time, and fewer than 28 boxes 75% of the time.

The difference between the upper and lower quartiles is called the **interquartile range**. The interquartile range, or IQR, contains about 50% of the values.

14, 16, 17, 17, 20, 21, 22, 26, 28, 29, 35

             Q_1 Q_3

     |← IQR = Q_1 − Q_3 or 11 →|

Before deciding on which measure of center best describes a data set, check for outliers. An **outlier** is an extremely high or extremely low value when compared with the rest of the values in the set. To check for outliers, look for data values that are beyond the upper or lower quartiles by more than 1.5 times the interquartile range.

Example 4 Effect of Outliers

TEST SCORES Students taking a make-up test received the following scores: 88, 79, 94, 90, 45, 71, 82, and 88.

a. Identify any outliers in the data.

First determine the median and upper and lower quartiles of the data.

$$45, \quad 71, \quad 79, \quad 82, \quad 88, \quad 88, \quad 90, \quad 94$$

$$Q_1 = \frac{71 + 79}{2} \text{ or } 75 \qquad Q_2 = \frac{82 + 88}{2} \text{ or } 85 \qquad Q_3 = \frac{88 + 90}{2} \text{ or } 89$$

Find the interquartile range.

$$\text{IQR} = Q_3 - Q_1 = 89 - 75 \text{ or } 14$$

Use the interquartile range to find the values beyond which any outliers would lie.

$Q_1 - 1.5(\text{IQR})$ and $Q_3 + 1.5(\text{IQR})$ Values beyond which outliers lie

$75 - 1.5(14)$ $89 + 1.5(14)$ $Q_1 = 75, Q_3 = 89,$ and IQR $= 14$

$\quad 54$ $\quad 110$ Simplify.

There are no scores greater than 110, but there is one score less than 54. The score of 45 can be considered an outlier for this data set.

b. Find the mean and median of the data set with and without the outlier. Describe what happens.

Data Set	Mean	Median
with outlier	$\frac{88 + 79 + 94 + 90 + 45 + 71 + 82 + 88}{8}$ or about 79.6	85
without outlier	$\frac{88 + 79 + 94 + 90 + 71 + 82 + 88}{7}$ or about 84.6	88

Removal of the outlier causes the mean and median to increase, but notice that the mean is affected more by the removal of the outlier than the median.

StudyTip

Interquartile Range
When the interquartile range is a small value, the data in the set are close together. A large interquartile range means that the data are spread out.

Exercises

Find the mean, median, mode, and range for each data set.

1. number of students helping at the cookie booth each hour: 3, 5, 8, 1, 4, 11, 3

2. weight in pounds of boxes loaded onto a semi truck: 201, 201, 200, 199, 199

3. car speeds in miles per hour observed by a highway patrol officer:
   60, 53, 53, 52, 53, 55, 55, 57

4. number of songs downloaded by students last week in Ms. Turner's class:
   3, 7, 21, 23, 63, 27, 29, 95, 23

5. ratings of an online video: 2, 5, 3.5, 4, 4.5, 1, 1, 4, 2, 1.5, 2.5, 2, 3, 3.5

6. **SCHOOL SUPPLIES** The table shows the cost of school supplies. Find the mean, median, mode and range of the costs.

Cost of School Supplies	
Supply	**Cost**
pencils	$0.50
pens	$2.00
paper	$2.00
pocket folder	$1.25
calculator	$5.25
notebook	$3.00
erasers	$2.50
markers	$3.50

7. **BOWLING** Sue's average for 9 games of bowling is 108. What is the lowest score she can receive for the tenth game to have an mean of 110?

8. **LAUNDRY** Two brands of laundry detergents were tested to determine how many times a shirt could be washed before it faded. The results for 6 shirts in number of washes follow.

   Brand A: 16, 15, 13, 14, 16, 16

   Brand B: 11, 16, 18, 12, 15, 18

   a. Find the mean and range for each brand.

   b. Which brand performed more consistently? Explain.

Find the minimum, lower quartile, median, upper quartile, and maximum values for each data set.

9. prices in dollars of smartphones: 311, 309, 312, 314, 399, 312

10. attendance at an event for the last nine years: 68, 99, 73, 65, 67, 62, 80, 81, 83

11. books a student checks out of the library: 17, 9, 10, 17, 18, 5, 2

12. ounces of soda dispensed into 36-ounce cups:
    36.1, 35.8, 35.2, 36.5, 36.0, 36.2, 35.7, 35.8, 35.9, 36.4, 35.6

13. ages of riders on a roller coaster:
    45, 17, 16, 22, 25, 19, 20, 21, 32, 37, 19, 21, 24, 20, 18, 22, 23, 19

14. **NUTRITION** The table shows the number of servings of fruit and vegetables that Cole eats one week. Find the minimum, median, lower quartile, upper quartile, and maximum number of servings. Then interpret this five-number summary.

Fruits and Vegetables	
Day	**Number of Servings**
Monday	5
Tuesday	7
Wednesday	5
Thursday	4
Friday	3
Saturday	3
Sunday	8

Find the mean and median of the data set, and then identify any outliers. If the set has an outlier, find the mean and median without the outlier, and state which measure is affected more by the removal of this value.

15. distance traveled in miles to visit relatives during winter break:
    210, 45, 10, 108, 452, 225, 35, 95, 140, 25, 65, 250

16. time spent on social networking Web sites in minutes per day:
    25, 35, 45, 30, 65, 50, 25, 100, 45, 35, 5, 105, 110, 190, 40, 30, 80

17. batting averages for the last 10 seasons: 0.267, 0.305, 0.304, 0.201, 0.284, 0.302, 0.311, 0.289, 0.300, 0.292

18. **CHALLENGE** The cost of 8 different pairs of pants at a department store are $39.99, $31.99, $19.99, $14.99, $19.99, $23.99, $36.99, and $26.99.

    a. Find the mean, median, mode, and range of the pants prices.

    b. Suppose each pair of pants needs to be hemmed at an additional cost of $8 per pair. Including these alteration costs, what are the mean, median, mode, and range of the pant prices?

    c. Suppose the original price of each pair of pants is discounted by 25%. Find the mean, median, mode, and range of the discounted pant prices.

    d. Make a conjecture as to the effect on the mean, median, mode, and range of a data set if the same value n is added to each value in the data set. What is the effect on these same measures if each item in a data set is multiplied by the same value n?

Representing Data

A **frequency table** uses tally marks to record and display frequencies of events. A **bar graph** compares categories of data with bars representing the frequencies.

Example 1 Make a Bar Graph

Make a bar graph to display the data.

Sport	Tally	Frequency
basketball	||||| ||||| |||||	15
football	||||| ||||| ||||| ||||| |||||	25
soccer	||||| ||||| ||||| |||	18
baseball	||||| ||||| ||||| |||||		21

Step 1 Draw a horizontal axis and a vertical axis. Label the axes as shown. Add a title.

Step 2 Draw a bar to represent each sport. The vertical scale is the number of students who chose each sport. The horizontal scale identifies the sport.

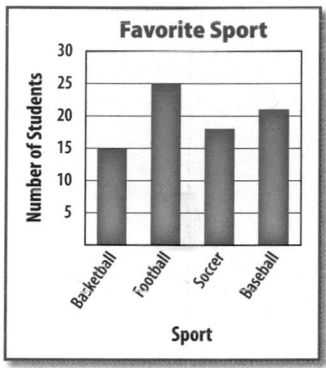

The **cumulative frequency** for each event is the sum of its frequency and the frequencies of all preceding events. A **histogram** is a type of bar graph used to display numerical data that have been organized into equal intervals.

Example 2 Make a Histogram and a Cumulative Frequency Histogram

Make histograms of the frequency and the cumulative frequency.

Age at Inauguration	40–44	45–49	50–54	55–59	60–64	65–69
U.S. Presidents	2	7	13	12	7	3

Find the cumulative frequency for each interval.

Age	< 45	< 50	< 55	< 60	< 65	< 70
Presidents	2	2 + 7 = 9	9 + 13 = 22	22 + 12 = 34	34 + 7 = 41	41 + 3 = 44

Make each histogram like a bar graph but with no space between the bars.

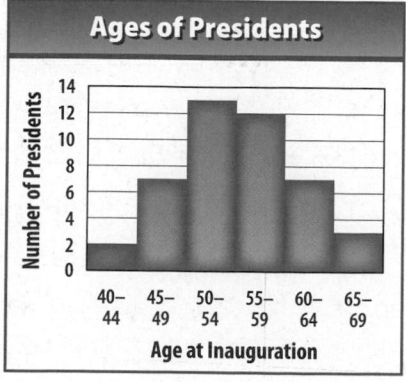

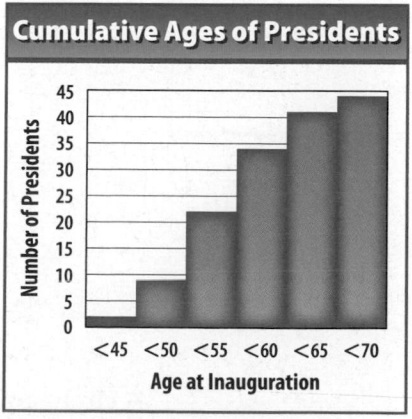

Another way to represent data is by using a line graph. A **line graph** usually shows how data change over a period of time.

Example 3 Make a Line Graph

Sales at the Marshall High School Store are shown in the table. Make a line graph of the data.

School Store Sales Amounts					
September	$670	December	$168	March	$412
October	$229	January	$290	April	$309
November	$300	February	$388	May	$198

Step 1 Draw a horizontal axis and a vertical axis and label them as shown. Include a title.

Step 2 Plot the points.

Step 3 Draw a line connecting each pair of consecutive points.

Data can also be organized and displayed by using a stem-and-leaf plot. In a **stem-and-leaf plot**, the digits of the least place value usually form the *leaves*, and the rest of the digits form the *stems*.

Real-World Example 4 Make a Stem-and-Leaf Plot

ANIMALS The speeds (mph) of 20 of the fastest land animals are listed at the right. Use the data to make a stem-and-leaf plot.

42	40	40	35	50
32	50	36	50	40
45	70	43	45	32
40	35	61	48	35

Source: *The World Almanac*

The least place value is ones. So, 32 miles per hour would have a stem of 3 and a leaf of 2.

Stem	Leaf
3	2 2 5 5 5 6
4	0 0 0 0 2 3 5 5 8
5	0 0 0
6	1
7	0

Key: 3|2 = 32

Real-World Link

The fastest animal on land is the cheetah. Cheetahs can run at speeds up to 60 miles per hour.

Source: Infoplease

A **circle graph** is a graph that shows the relationship between parts of the data and the whole. The circle represents all of the data.

Example 5 Make a Circle Graph

The table shows how Lily spent 8 hours of one day at summer camp. Make a circle graph of the data.

First, find the ratio that compares the number of hours for each activity to 8. Then multiply each ratio by 360° to find the number of degrees for each section of the graph.

Canoeing: $\frac{3}{8} \cdot 360° = 135°$

Crafts: $\frac{1}{8} \cdot 360° = 45°$

Eating: $\frac{2}{8} \cdot 360° = 90°$

Hiking: $\frac{2}{8} \cdot 360° = 90°$

Summer Camp	
Activity	**Hours**
canoeing	3
crafts	1
eating	2
hiking	2

Summer Camp

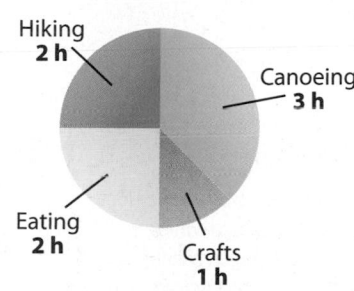

> **WatchOut!**
> Circle Graphs The sum of the measures of each section of a circle graph should be 360°.

A **box-and-whisker plot** is a graphical representation of the five-number summary of a data set. The box in a box-and-whisker plot represents the interquartile range.

Example 6 Make a Box-and-Whisker Plot

Draw a box-and-whisker plot for these data. Describe how the outlier affects the quartile points.

14, 30, 16, 20, 18, 16, 20, 18, 22, 13, 8

Step 1 Order the data from least to greatest. Then determine the maximum, minimum and the quartiles.

8, 13, **14**, 16, 16, **18**, 18, 20, **20**, 22, 30

↑ min. ↑ Q_1 ↑ Q_2 ↑ Q_3 ↑ max.

Determine the interquartile range.

$IQR = Q_3 - Q_1$

$= 20 - 14 \text{ or } 6$

Check to see if there are any outliers.

$14 - 1.5(6) = 5$ $20 + 1.5(6) = 29$

Numbers less than 5 or greater than 29 are outliers.

The only outlier is 30.

Step 2 Draw a number line that includes the minimum and maximum values in the data. Place dots above the number line to represent the three quartile points, any outliers, the minimum value that is not an outlier, and the maximum value that is not an outlier.

Step 3 Draw the box and the whiskers. The vertical rules go through the quartiles. The outliers are not connected to the whiskers.

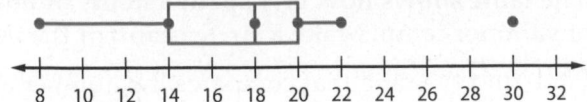

Step 4 Omit 30 from the data. Repeat Step 1 to determine Q_1, Q_2, and Q_3.

8, 13, **14**, 16, **16**, **18**, 18, **20**, 20, 22

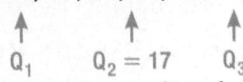

Removing the outlier does not affect Q_1 or Q_2 and thus does not affect the interquartile range. The value of Q_2 changes from 18 to 17.

Example 7 Compare Data

CLIMATE Lucas is going to go to college in either Dallas or Nashville. He wants to live in a place that does not get too cold. So he decides to compare the average monthly low temperatures of each city.

a. **Draw a double box-and-whisker plot for the data.**

Determine the quartiles and outliers for each city.

Dallas

36, 39, 41, 47, 49, 56, 58, 65, 69, 73, 76, 77

$Q_1 = 44 \qquad Q_2 = 57 \qquad Q_3 = 71$

Nashville

28, 31, 32, 39, 40, 47, 49, 57, 61, 65, 68, 70

$Q_1 = 35.5 \qquad Q_2 = 48 \qquad Q_3 = 63$

Average Monthly Low Temperatures (°F)		
Month	Dallas	Nashville
Jan.	36	28
Feb.	41	31
Mar.	49	39
Apr.	56	47
May	65	57
June	73	65
July	77	70
Aug.	76	68
Sept.	69	61
Oct.	58	49
Nov.	47	40
Dec.	39	32

Source: weather.com

There are no outliers. Draw the plots using the same number line.

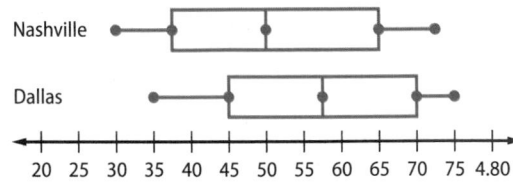

b. **Use the double box-and-whisker plot to compare the data.**

The interquartile range of temperatures for both cities is about the same. However, all quartiles of the Dallas temperatures are shifted to the right of those of Nashville, meaning Dallas has higher average low temperatures.

c. **One night in August, a weather reporter stated the low for Nashville as being "only 65." Is it appropriate for the weather reporter to use the word only in the statement? Is 65 an unusually low temperature for Nashville in August? Explain your answer.**

No, 65 is not an unusually low temperature for August in Nashville. It is lower than the average, but not by much.

When displaying data, some graphs are better choices than others.

Example 8 Select a Display

Which type of graph is the best way to display each set of data? Explain.

a. the results of the women's Olympic High Jump event from 1972 to 2008

Since the data would show change over time, a line graph would give the reader a clear picture of changes in height.

b. the percent of students in class who have 0, 1, 2, 3, or more than 3 pets

Since the data would show how parts are related to the whole, a circle graph would give the reader a clear picture of how different segments of the class relate to the whole class.

Exercises

1. SURVEYS Alana surveyed several students to find the number of hours of sleep they typically get each night. The results are shown in the table. Make a bar graph of the data.

Hours of Sleep					
Alana	8	Kwam	7.5	Tomas	7.75
Nick	8.25	Kate	7.25	Sharla	8.5

2. PLAYS The frequency table at the right shows the ages of people attending a high school play.

a. Make a histogram to display the data.
b. Make a cumulative frequency histogram showing the number of people attending who were less than 20, 40, 60, or 80 years old.

Age	Tally	Frequency
0–19	IIII IIII IIII IIII IIII IIII IIII IIII IIII II	47
20–39	IIII IIII IIII IIII IIII IIII IIII IIII III	43
40–59	IIII IIII IIII IIII IIII IIII I	31
60–79	IIII III	8

3. LAWN CARE Marcus started a lawn care service. The chart shows how much money he made over summer break. Make a line graph of the data.

Lawn Care Profits ($)								
Week	1	2	3	4	5	6	7	8
Profit	25	40	45	50	75	85	95	95

Use each set of data to make a stem-and-leaf plot and a box-and-whisker plot. Describe how the outliers affect the quartile points.

4. {65, 63, 69, 71, 73, 59, 60, 70, 72, 66, 71, 58}

5. {31, 30, 28, 26, 22, 34, 26, 31, 47, 32, 18, 33, 26, 23, 18}

6. FINANCIAL LITERACY The table shows how Ping spent his allowance of $40. Make a circle graph of the data.

Allowance	
How Spent	Amount ($)
savings	15
downloaded music	8
snacks	5
T-shirt	12

7. JOGGING The table shows the number of miles Hannah jogged each day for 10 days. Make a line graph of the data.

Day	1	2	3	4	5	6	7	8	9	10
Miles Jogged	2	2	3	3.5	4	4.5	2.5	3	4	5

8. BASKETBALL Two basketball teams are analyzing the number of points they scored in each game this season.

Lions: 48, 52, 55, 49, 53, 55, 51, 50, 46, 53, 47, 55, 50, 51, 60, 52, 57, 56, 58, 55
Eagles: 35, 39, 37, 40, 44, 42, 53, 42, 40, 44, 48, 46, 43, 47, 45, 41, 45, 43, 47, 48

a. Make a double box-and-whisker plot to display the data.

b. How does the number of points scored by the Lions compare to the number of points scored by the Eagles?

c. In the first game of the post season, a sports announcer reported the Lions scored a whopping 60 points. Is it appropriate for the announcer to use the word whopping in the statement? Is 60 an unusually high number of points for the Lions to score? Explain your answer.

9. TESTS Mr. O'Neil teaches two algebra classes. The test scores for the two classes are shown.

Third Period											
77	98	85	79	76	86	84	91	67	88	93	87
99	78	81	80	82	84	83	85	84	95	90	88

Sixth Period											
91	93	88	75	80	78	81	90	82	95	76	88
89	79	93	88	85	94	83	88	91	72	88	70

a. Make a double box-and-whisker plot to display the data.

b. Write a brief description of each data distribution.

c. How do the scores from the third period class compare to the scores from the sixth period class?

Which type of graph is the best way to display each set of data? Explain.

10. an organization's dollar contributions to 4 different charities

11. the prices of a college football ticket from 1990 to the present

12. the percent of glass, plastic, paper, steel, and aluminum in a recycling center

13. DISCUS The winning distances for the girls' discus throw at an annual track meet are shown below.

Year	1999	2000	2001	2002	2003	2004	2005	2006	2007	2008	2009	2010
Distance (m)	119	124	126	129	130	130	133	135	136	137	138	140

a. Make a stem-and-leaf plot to display the winning distances.

b. Make a histogram to display the winning distances.

c. What does the stem-and-leaf plot show you that the histogram does not?

d. If this trend continues, what would you expect the winning distance to be in 2030? Is your answer reasonable? Explain.

14. DRINKS Tate is buying drinks for a party. He is comparing 2-liter bottles to 12-packs of 12-ounce cans. The prices of 2-liter bottles are $0.99, $1.99, $1.87, $1.79, $1.29, $1.43, and $1.15. The prices of 12-packs are $2.50, $4.25, $3.34, $2.65, $3.19, $3.89, and 2.99.

a. Make a double box-and-whisker plot to display the data.

b. Notice that instead of comparing price per item it would be more beneficial to compare price per ounce. What is the price per ounce of each item if a 2-liter is approximately 67 ounces and a 12-pack is 144 ounces? Round to the nearest cent.

c. Make a new double box-and-whisker plot from the data obtained in part **b**.

d. Which is the better deal, the 12-packs of cans or the 2-liter bottles? Explain.

Determine whether you need an estimate or an exact answer. Then use the four-step problem-solving plan to solve.

1. **DISTANCE** Fabio rode his scooter 2.3 miles to his friend's house, then 0.7 mile to the grocery store, then 2.1 miles to the library. If he rode the same route back home, about how far did he travel in all?

2. **SHOPPING** The regular price of a T-shirt is $9.99. It is on sale for 15% off. Sales tax is 6%. If you give the cashier a $10 bill, how much change will you receive?

Find each sum or difference.

3. $-31 + (-4)$

4. $48 - 55$

5. $-71 - (-10)$

6. $31 - 42.9$

7. $-11.5 + 8.1$

8. $-0.38 - (-1.06)$

Find each product or quotient.

9. $-21(-5)$

10. $-81 \div (-3)$

11. $-120 \div 8$

12. $-39 \div -3$

Replace each ● with <, >, or = to make a true sentence.

13. $-0.62 ● -\frac{6}{7}$

14. $\frac{12}{44} ● \frac{8}{11}$

15. Order $4\frac{4}{5}$, 4.85, $2\frac{5}{8}$, and 2.6 from least to greatest.

Find each sum or difference. Write in simplest form.

16. $\frac{1}{7} + \frac{5}{7}$

17. $\frac{7}{8} - \frac{1}{8}$

18. $\frac{1}{6} + \left(-\frac{1}{2}\right)$

19. $-\frac{1}{12} - \left(-\frac{3}{4}\right)$

Find each product or quotient.

20. $-1.2(9.3)$

21. $-20.93 \div (-2.3)$

22. $10.5 \div (-1.2)$

23. $(-3.4)(-2.8)$

Name the reciprocal of each number.

24. 6

25. $1\frac{2}{5}$

26. $-2\frac{3}{7}$

27. $-\frac{1}{2}$

28. $\frac{4}{3}$

29. $5\frac{1}{3}$

Find each product or quotient. Write in simplest form.

30. $\frac{2}{5} \cdot \frac{5}{9}$

31. $\frac{4}{5} \div \frac{1}{5}$

32. $-\frac{7}{8} \cdot 2$

33. $\frac{1}{3} \div 2\frac{1}{4}$

34. $-6 \cdot \left(-\frac{3}{4}\right)$

35. $\frac{7}{18} \div \left(-\frac{14}{15}\right)$

36. **PICNIC** Joseph is mixing $5\frac{1}{2}$ gallons of orange drink for his class picnic. Every $\frac{1}{2}$ gallon requires 1 packet of orange drink mix. How many packets of orange drink mix does Joseph need?

Express each percent as a fraction in simplest form.

37. 6%

38. 140%

Use the percent proportion to find each number.

39. 50% of what number is 31?

40. What number is 110% of 51?

41. Find 8% of 95.

42. **SOLUTIONS** A solution is prepared by dissolving 24 milliliters of saline in 150 milliliters of pure solution. What is the percent of saline in the pure solution?

43. **SHOPPING** Marta got 60% off a pair of shoes. If the shoes cost $9.75 (before sales tax), what was the original price of the shoes?

Find the perimeter and area of each figure.

44.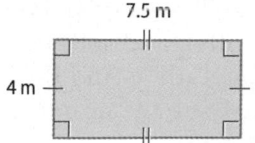
7.5 m
4 m

45.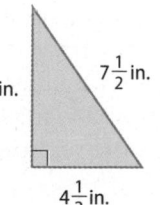
6 in.
$7\frac{1}{2}$ in.
$4\frac{1}{2}$ in.

46. A parallelogram has a base of 20 millimeters and a height of 6 millimeters. Find the area.

47. **GARDENS** Find the perimeter of the garden.

6.0 m
3.5 m
4.0 m

Find the circumference and area of each circle. Round to the nearest tenth.

48.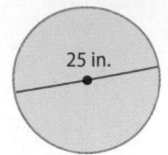
25 in.

49.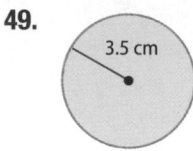
3.5 cm

50. PARKS A park has a circular area for a fountain that has a circumference of about 16 feet. What is the radius of the circular area? Round to the nearest tenth.

Find the volume and surface area of each rectangular prism given the measurements below.

51. $\ell = 1.5$ m, $w = 3$ m, $h = 2$ m

52. $\ell = 4$ in., $w = 1$ in., $h = \frac{1}{2}$ in.

53. Find the volume and surface area of the rectangular prism.

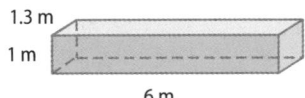
1.3 m
1 m
6 m

One marble is randomly selected from a jar containing 3 red, 4 green, 2 black, and 6 blue marbles. Find each probability.

54. P(red or blue)   **55.** P(green or red)

56. P(not black)   **57.** P(not blue)

58. A movie theater is offering snack specials. You can choose a small, medium, large, or jumbo popcorn with or without butter, and soda or bottled water. Use a tree diagram to find the sample space for the event. State the number of possible outcomes.

One coin is randomly selected from a jar containing 20 pennies, 15 nickels, 3 dimes, and 12 quarters. Find the odds of each outcome. Write in simplest form.

59. a dime

60. a value less than $0.25

61. a value greater than $0.10

62. a value less than $0.05

63. SCHOOL In a science class, each student must choose a lab project from a list of 15, write a paper on one of 6 topics, and give a presentation about one of 8 subjects. How many ways can students choose to do their assignments?

64. GAMES Marcos has been dealt seven different cards. How many different ways can he play his cards if he is required to play one card at a time?

Find the mean, median, and mode for each set of data.

65. {99, 88, 88, 92, 100}

66. {30, 22, 38, 41, 33, 41, 30, 24}

67. Find the range, median, lower quartile, and upper quartile for {77, 75, 72, 70, 79, 77, 70, 76}.

68. TESTS Kevin's scores on the first four science tests are 88, 92, 82, and 94. What score must he earn on the fifth test so that the mean will be 90?

69. FOOD The table shows the results of a survey in which students were asked to choose their favorite food. Make a bar graph of the data.

Favorite Foods	
Food	Number of Students
pizza	15
chicken nuggets	10
cheesy potatoes	8
ice cream	5

70. Make a double box-and-whisker plot of the data.
A: 26, 18, 26, 29, 18, 20, 35, 32, 31, 24, 26, 22
B: 16, 20, 16, 19, 21, 30, 25, 22, 21, 19, 16, 17

71. BUDGET The table shows how Kat spends her allowance. Which graph is the best way to display these data? Explain your reasoning and make a graph of the data.

Category	Amount ($)
Savings	25
Clothes	10
Entertainment	15

CHAPTER 1

Expressions, Equations, and Functions

Then

○ You have learned how to perform operations on whole numbers.

Now

○ In this chapter, you will:

- Write algebraic expressions.

- Use the order of operations.

- Solve equations.

- Represent and interpret relations and functions.

- Use function notation.

- Interpret the graphs of functions.

Why? ▲

○ **SCUBA DIVING** A scuba diving store rents air tanks and wet suits. An algebraic expression can be written to represent the total cost to rent this equipment. This expression can be evaluated to determine the total cost for a group of people to rent the equipment.

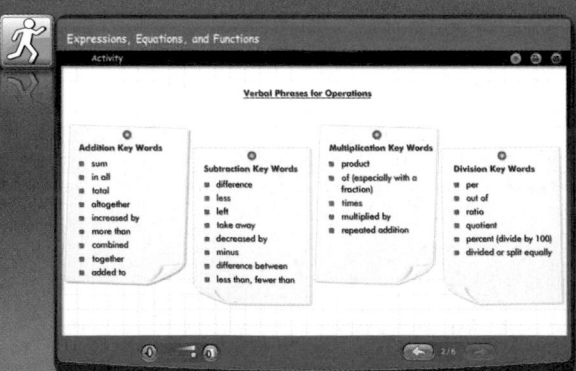

connectED.mcgraw-hill.com   **Your Digital Math Portal**

Animation   Vocabulary   eGlossary   Personal Tutor   Virtual Manipulatives   Graphing Calculator   Audio   Foldables   Self-Check Practice   Worksheets

Jupiterimages/Comstock Images/Alamy

Get Ready for the Chapter

Diagnose Readiness | You have two options for checking prerequisite skills.

1 Textbook Option Take the Quick Check below. Refer to the Quick Review for help.

QuickCheck	QuickReview

QuickCheck

Write each fraction in simplest form. If the fraction is already in simplest form, write *simplest form.*

1. $\frac{24}{36}$ 2. $\frac{34}{85}$ 3. $\frac{36}{12}$

4. $\frac{27}{45}$ 5. $\frac{11}{18}$ 6. $\frac{5}{65}$

7. $\frac{19}{1}$ 8. $\frac{16}{44}$ 9. $\frac{64}{88}$

10. **ICE CREAM** Fifty-four out of 180 customers said that cookie dough ice cream was their favorite flavor. What fraction of customers was this?

QuickReview

Example 1

Write $\frac{24}{40}$ in simplest form.

Find the greatest common factor (GCF) of 24 and 40.

factors of 24: 1, 2, 3, 4, 6, 8, 12, 24
factors of 40: 1, 2, 4, 5, 8, 10, 20, 40

The GCF of 24 and 40 is 8.

$\frac{24 \div 8}{40 \div 8} = \frac{3}{5}$ Divide the numerator and denominator by their GCF, 8.

Find the perimeter of each figure.

11.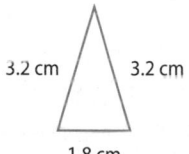

3.2 cm   3.2 cm

1.8 cm

12.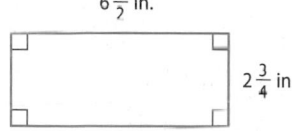

$6\frac{1}{2}$ in.

$2\frac{3}{4}$ in.

13. **FENCING** Jolon needs to fence a rectangular garden. The dimensions of the garden are 6 meters by 4 meters. How much fencing does Jolon need to purchase?

Example 2

Find the perimeter.

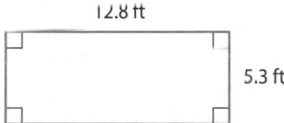

12.8 ft

5.3 ft

$P = 2\ell + 2w$

$= 2(12.8) + 2(5.3)$ $\ell = 12.8$ and $w = 5.3$

$= 25.6 + 10.6$ or 36.2 Simplify.

The perimeter is 36.2 feet.

Evaluate.

14. $6 \cdot \frac{2}{3}$ 15. $4.2 \cdot 8.1$ 16. $\frac{3}{8} \div \frac{1}{4}$

17. $5.13 \div 2.7$ 18. $3\frac{1}{5} \cdot \frac{3}{4}$ 19. $2.8 \cdot 0.2$

20. **CONSTRUCTION** A board measuring 7.2 feet must be cut into three equal pieces. Find the length of each piece.

Example 3

Find $2\frac{1}{4} \div 1\frac{1}{2}$.

$2\frac{1}{4} \div 1\frac{1}{2} = \frac{9}{4} \div \frac{3}{2}$ Write mixed numbers as improper fractions.

$= \frac{9}{4}\left(\frac{2}{3}\right)$ Multiply by the reciprocal.

$= \frac{18}{12}$ or $1\frac{1}{2}$ Simplify.

---

2 Online Option Take an online self-check Chapter Readiness Quiz at <u>connectED.mcgraw-hill.com</u>.

Get Started on the Chapter

You will learn several new concepts, skills, and vocabulary terms as you study Chapter 1. To get ready, identify important terms and organize your resources. You may wish to refer to Chapter 0 to review prerequisite skills.

FOLDABLES StudyOrganizer

Expressions, Equations, and Functions Make this Foldable to help you organize your Chapter 1 notes about expressions, equations, and functions. Begin with five sheets of plain paper.

1 **Fold** the sheets of paper in half along the width. Then cut along the crease.

2 **Staple** the ten half-sheets together to form a booklet.

3 **Cut** nine centimeters from the bottom of the top sheet, eight centimeters from the second sheet, and so on.

4 **Label** each of the tabs with a lesson number. The ninth tab is for Properties and the last tab is for Vocabulary.

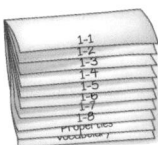

NewVocabulary

English		Español
algebraic expression	p. 5	expression algebraica
variable	p. 5	variable
term	p. 5	término
power	p. 5	potencia
coefficient	p. 28	coeficiente
equation	p. 33	ecuación
solution	p. 33	solución
identity	p. 35	identidad
relation	p. 40	relacíon
domain	p. 40	domino
range	p. 40	rango
independent variable	p. 42	variable independiente
dependent variable	p. 42	variable dependiente
function	p. 47	función
intercept	p. 56	intersección
line symmetry	p. 57	simetría
end behavior	p. 57	comportamiento final

ReviewVocabulary

additive inverse inverso aditivo a number and its opposite

multiplicative inverse inverso multiplicativo two numbers with a product of 1

perimeter perímetro the distance around a geometric figure

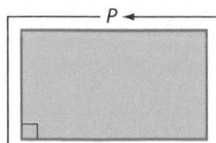

Variables and Expressions

| ∷Then | ∷Now | ∷Why? |

∷Then	∷Now	∷Why?
● You performed operations on integers.	**1** Write verbal expressions for algebraic expressions. **2** Write algebraic expressions for verbal expressions.	● Cassie and her friends are at a baseball game. The stadium is running a promotion where hot dogs are $0.10 each. Suppose d represents the number of hot dogs Cassie and her friends eat. Then $0.10d$ represents the cost of the hot dogs they eat.

 NewVocabulary
algebraic expression
variable
term
factor
product
power
exponent
base

 Common Core State Standards

Content Standards
A.SSE.1a Interpret parts of an expression, such as terms, factors, and coefficients.

A.SSE.2 Use the structure of an expression to identify ways to rewrite it.

Mathematical Practices
4 Model with mathematics.

1 Write Verbal Expressions An **algebraic expression** consists of sums and/or products of numbers and variables. In the algebraic expression $0.10d$, the letter d is called a variable. In algebra, **variables** are symbols used to represent unspecified numbers or values. Any letter may be used as a variable.

$$0.10d \qquad 2x + 4 \qquad 3 + \frac{z}{6} \qquad p \cdot q \qquad 4cd \div 3mn$$

A **term** of an expression may be a number, a variable, or a product or quotient of numbers and variables. For example, $0.10d$, $2x$ and 4 are each terms.

| The term that contains x or other letters is sometimes referred to as the *variable term*. | → $2x + 4$ ← | A term that does not have a variable is a *constant term*. |

In a multiplication expression, the quantities being multiplied are **factors**, and the result is the **product**. A raised dot or set of parentheses are often used to indicate a product. Here are several ways to represent the product of x and y.

$$xy \qquad x \cdot y \qquad x(y) \qquad (x)y \qquad (x)(y)$$

An expression like x^n is called a **power**. The word *power* can also refer to the exponent. The **exponent** indicates the number of times the base is used as a factor. In an expression of the form x^n, the **base** is x. The expression x^n is read "x to the nth power." When no exponent is shown, it is understood to be 1. For example, $a = a^1$.

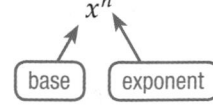
x^n base   exponent

Example 1 Write Verbal Expressions

Write a verbal expression for each algebraic expression.

a. $3x^4$

three times x to the fourth power

b. $5z^2 + 16$

5 times z to the second power plus sixteen

▸ **Guided Practice**

1A. $16u^2 - 3$ **1B.** $\frac{1}{2}a + \frac{6b}{7}$

StudyTip

CCSS Modeling When writing an expression to model a situation, begin by identifying the important quantities and their relationships.

2 **Write Algebraic Expressions** Another important skill is translating verbal expressions into algebraic expressions.

KeyConcept Translating Verbal to Algebraic Expressions

Operation	Verbal Phrases
Addition	more than, sum, plus, increased by, added to
Subtraction	less than, subtracted from, difference, decreased by, minus
Multiplication	product of, multiplied by, times, of
Division	quotient of, divided by

Example 2 Write Algebraic Expressions

Write an algebraic expression for each verbal expression.

a. a number t more than 6

The words *more than* suggest addition.
Thus, the algebraic expression is $6 + t$ or $t + 6$.

b. 10 less than the product of 7 and f

Less than implies subtraction, and *product* suggests multiplication.
So the expression is written as $7f - 10$.

c. two thirds of the volume v

The word *of* with a fraction implies that you should multiply.
The expression could be written as $\frac{2}{3}v$ or $\frac{2v}{3}$.

GuidedPractice

2A. the product of p and 6   **2B.** one third of the area a

Variables can represent quantities that are known and quantities that are unknown. They are also used in formulas, expressions, and equations.

Real-World Example 3 Write an Expression

SPORTS MARKETING **Mr. Martinez orders 250 key chains printed with his athletic team's logo and 500 pencils printed with their Web address. Write an algebraic expression that represents the cost of the order.**

Let k be the cost of each key chain and p be the cost of each pencil. Then the cost of the key chains is $250k$ and the cost of the pencils is $500p$. The cost of the order is represented by $250k + 500p$.

GuidedPractice

3. COFFEE SHOP Katie bakes 40 pastries and makes coffee for 200 people. Write an algebraic expression to represent this situation.

Masterfile

Example 1 **Write a verbal expression for each algebraic expression.**

1. $2m$ **2.** $\frac{2}{3}r^4$ **3.** $a^2 - 18b$

Example 2 **Write an algebraic expression for each verbal expression.**

4. the sum of a number and 14 **5.** 6 less a number t

6. 7 more than 11 times a number **7.** 1 minus the quotient of r and 7

8. two fifths of the square of a number j **9.** n cubed increased by 5

Example 3 **10. GROCERIES** Mr. Bailey purchased some groceries that cost d dollars. He paid with a $50 bill. Write an expression for the amount of change he will receive.

Practice and Problem Solving Extra Practice is on page R1.

Example 1 **Write a verbal expression for each algebraic expression.**

11. $4q$ **12.** $\frac{1}{8}y$ **13.** $15 + r$ **14.** $w - 24$

15. $3x^2$ **16.** $\frac{r^4}{9}$ **(17)** $2a + 6$ **18.** $r^4 \cdot t^3$

Example 2 **Write an algebraic expression for each verbal expression.**

19. x more than 7 **20.** a number less 35

21. 5 times a number **22.** one third of a number

23. f divided by 10 **24.** the quotient of 45 and r

25. three times a number plus 16 **26.** 18 decreased by 3 times d

27. k squared minus 11 **28.** 20 divided by t to the fifth power

Example 3 **29. GEOMETRY** The volume of a cylinder is π times the radius r squared multiplied by the height h. Write an expression for the volume.

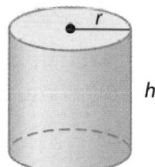

30. FINANCIAL LITERACY Jocelyn makes x dollars per hour working at the grocery store and n dollars per hour babysitting. Write an expression that describes her earnings if she babysat for 25 hours and worked at the grocery store for 15 hours.

Write a verbal expression for each algebraic expression.

31. $25 + 6x^2$ **32.** $6f^2 + 5f$ **33.** $\frac{3a^5}{2}$

34. CCSS SENSE-MAKING A certain smartphone family plan costs $55 per month plus additional usage costs. If x is the number of cell phone minutes used above the plan amount and y is the number of megabytes of data used above the plan amount, interpret the following expressions.

a. $0.25x$

b. $2y$

c. $0.25x + 2y + 55$

35 **DREAMS** It is believed that about $\frac{3}{4}$ of our dreams involve people that we know.

    a. Write an expression to describe the number of dreams that feature people you know if you have d dreams.

    b. Use the expression you wrote to predict the number of dreams that include people you know out of 28 dreams.

36. **SPORTS** In football, a touchdown is awarded 6 points and the team can then try for a point after a touchdown.

    a. Write an expression that describes the number of points scored on touchdowns T and points after touchdowns p by one team in a game.

    b. If a team wins a football game 27-0, write an equation to represent the possible number of touchdowns and points after touchdowns by the winning team.

    c. If a team wins a football game 21-7, how many possible number of touchdowns and points after touchdowns were scored during the game by both teams?

37. ⟳ **MULTIPLE REPRESENTATIONS** In this problem, you will explore the multiplication of powers with like bases.

    a. Tabular Copy and complete the table.

10^2	$\times$	10^1	$=$	$10 \times 10 \times 10$	$=$	10^3
10^2	$\times$	10^2	$-$	$10 \times 10 \times 10 \times 10$	$=$	10^4
10^2	$\times$	10^3	$=$	$10 \times 10 \times 10 \times 10 \times 10$	$=$	?
10^2	$\times$	10^4	$=$	?	$=$	?

    b. Algebraic Write an equation for the pattern in the table.

    c. Verbal Make a conjecture about the exponent of the product of two powers with like bases.

H.O.T. Problems   Use Higher-Order Thinking Skills

38. **REASONING** Explain the differences between an algebraic expression and a verbal expression.

39. **OPEN ENDED** Define a variable to represent a real-life quantity, such as time in minutes or distance in feet. Then use the variable to write an algebraic expression to represent one of your daily activities. Describe in words what your expression represents, and explain your reasoning.

40. **CCSS CRITIQUE** Consuelo and James are writing an algebraic expression for *three times the sum of n squared and 3*. Is either of them correct? Explain your reasoning.

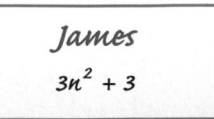

Consuelo	James
$3(n^2 + 3)$	$3n^2 + 3$

41. **CHALLENGE** For the cube, x represents a positive whole number. Find the value of x such that the volume of the cube and 6 times the area of one of its faces have the same value.

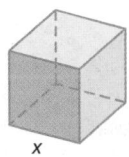

x

42. **WRITING IN MATH** Describe how to write an algebraic expression from a real-world situation. Include a definition of algebraic expression in your own words.

43. Which expression best represents the volume of the cube?

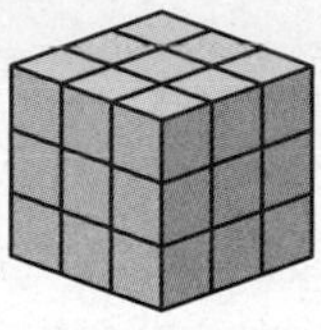

   A the product of three and five

   B three to the fifth power

   C three squared

   D three cubed

44. Which expression best represents the perimeter of the rectangle?

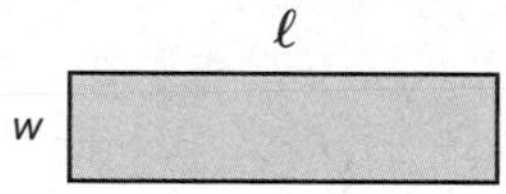

   F $2\ell w$

   G $\ell + w$

   H $2\ell + 2w$

   J $4(\ell + w)$

45. SHORT RESPONSE The yards of fabric needed to make curtains is 3 times the length of a window in inches, divided by 36. Write an expression that represents the yards of fabric needed in terms of the length of the window ℓ.

46. GEOMETRY Find the area of the rectangle.

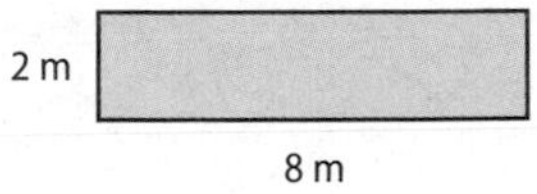

   A 14 square meters

   B 16 square meters

   C 50 square meters

   D 60 square meters

   2 m, 8 m

47. AMUSEMENT PARKS A roller coaster enthusiast club took a poll to see what each member's favorite ride was. Make a bar graph of the results. (Lesson 0-13)

Our Favorite Rides						
Ride	Big Plunge	Twisting Time	The Shiner	Raging Bull	The Bat	Teaser	The Adventure
Number of Votes	5	22	16	9	25	6	12

48. SPORTS The results for an annual 5K race are shown at the right. Make a box-and-whisker plot for the data. Write a sentence describing what the length of the box-and-whisker plot tells about the times for the race. (Lesson 0–13)

Find the mean, median, and mode for each set of data. (Lesson 0–12)

49. {7, 6, 5, 7, 4, 8, 2, 2, 7, 8}

50. {−1, 0, 5, 2, −2, 0, −1, 2, −1, 0}

51. {17, 24, 16, 3, 12, 11, 24, 15}

Annual 5K Race Results			
Joe	14:48	Carissa	19:58
Jessica	19:27	Jordan	14:58
Lupe	15:06	Taylor	20:47
Dante	20:39	Mi-Ling	15:48
Tia	15:54	Winona	21:35
Amber	20:49	Angel	16:10
Amanda	16:30	Catalina	20:21

52. SPORTS Lisa has a rectangular trampoline that is 6 feet long and 12 feet wide. What is the area of her trampoline in square feet? (Lesson 0–8)

Find each product or quotient. (Lesson 0–5)

53. $\dfrac{3}{5} \cdot \dfrac{7}{11}$

54. $\dfrac{4}{3} \div \dfrac{7}{6}$

55. $\dfrac{5}{6} \cdot \dfrac{8}{3}$

Evaluate each expression.

56. $\dfrac{3}{5} + \dfrac{4}{9}$

57. $5.67 - 4.21$

58. $\dfrac{5}{6} - \dfrac{8}{3}$

59. $10.34 + 14.27$

60. $\dfrac{11}{12} + \dfrac{5}{36}$

61. $37.02 - 15.86$

Order of Operations

:·Then	:·Now	:·Why?
● You expressed algebraic expressions verbally.	**1** Evaluate numerical expressions by using the order of operations. **2** Evaluate algebraic expressions by using the order of operations.	● The admission prices for SeaWorld Adventure Park in Orlando, Florida, are shown in the table. If four adults and three children go to the park, the expression below represents the cost of admission for the group. $4(78.95) + 3(68.95)$

Ticket	Price ($)
Adult	78.95
Child	68.95

NewVocabulary
evaluate
order of operations

Common Core State Standards

Content Standards
A.SSE.1b Interpret complicated expressions by viewing one or more of their parts as a single entity.

A.SSE.2 Use the structure of an expression to identify ways to rewrite it.

Mathematical Practices
7 Look for and make use of structure.

1 **Evaluate Numerical Expressions** To find the cost of admission, the expression $4(78.95) + 3(68.95)$ must be evaluated. To **evaluate** an expression means to find its value.

Example 1 Evaluate Expressions

Evaluate 3^5.

$$3^5 = 3 \cdot 3 \cdot 3 \cdot 3 \cdot 3 \qquad \text{Use 3 as a factor 5 times.}$$
$$= 243 \qquad\qquad\quad \text{Multiply.}$$

▶ **Guided**Practice

1A. 2^4 **1B.** 4^5 **1C.** 7^3

The numerical expression that represents the cost of admission contains more than one operation. The rule that lets you know which operation to perform first is called the **order of operations**.

KeyConcept Order of Operations

Step 1 Evaluate expressions inside grouping symbols.

Step 2 Evaluate all powers.

Step 3 Multiply and/or divide from left to right.

Step 4 Add and/or subtract from left to right.

Example 2 Order of Operations

Evaluate $16 - 8 \div 2^2 + 14$.

$$16 - 8 \div 2^2 + 14 = 16 - 8 \div 4 + 14 \qquad \text{Evaluate powers.}$$
$$= 16 - 2 + 14 \qquad\qquad \text{Divide 8 by 4.}$$
$$= 14 + 14 \qquad\qquad\quad \text{Subtract 2 from 16.}$$
$$= 28 \qquad\qquad\qquad\quad \text{Add 14 and 14.}$$

▶ **Guided**Practice

2A. $3 + 42 \cdot 2 - 5$ **2B.** $20 - 7 + 8^2 - 7 \cdot 11$

When one or more grouping symbols are used, evaluate within the innermost grouping symbols first.

Example 3 Expressions with Grouping Symbols

Evaluate each expression.

a. $4 \div 2 + 5(10 - 6)$

$$
\begin{aligned}
4 \div 2 + 5(10 - 6) &= 4 \div 2 + 5(4) && \text{Evaluate inside parentheses.}\\
&= 2 + 5(4) && \text{Divide 4 by 2.}\\
&= 2 + 20 && \text{Multiply 5 by 4.}\\
&= 22 && \text{Add 2 to 20.}
\end{aligned}
$$

b. $6\left[32 - (2 + 3)^2\right]$

$$
\begin{aligned}
6\left[32 - (2 + 3)^2\right] &= 6\left[32 - (5)^2\right] && \text{Evaluate innermost expression first.}\\
&= 6[32 - 25] && \text{Evaluate power.}\\
&= 6[7] && \text{Subtract 25 from 32.}\\
&= 42 && \text{Multiply.}
\end{aligned}
$$

c. $\dfrac{2^3 - 5}{15 + 9}$

$$
\begin{aligned}
\frac{2^3 - 5}{15 + 9} &= \frac{8 - 5}{15 + 9} && \text{Evaluate the power in the numerator.}\\[2mm]
&= \frac{3}{15 + 9} && \text{Subtract 5 from 8 in the numerator.}\\[2mm]
&= \frac{3}{24} \text{ or } \frac{1}{8} && \text{Add 15 and 9 in denominator, and simplify.}
\end{aligned}
$$

▶ **Guided**Practice

3A. $5 \cdot 4(10 - 8) + 20$ **3B.** $15 - \left[10 + (3 - 2)^2\right] + 6$ **3C.** $\dfrac{(4 + 5)^2}{3(7 - 4)}$

2 Evaluate Algebraic Expressions To evaluate an algebraic expression, replace the variables with their values. Then find the value of the numerical expression using the order of operations.

Example 4 Evaluate an Algebraic Expression

Evaluate $3x^2 + \left(2y + z^3\right)$ **if** $x = 4, y = 5, z = 3.$

$$
\begin{aligned}
3x^2 + \left(2y + z^3\right) \\
= 3(4)^2 + \left(2 \cdot 5 + 3^3\right) && \text{Replace } x \text{ with 4, } y \text{ with 5, and } z \text{ with 3.}\\
= 3(4)^2 + (2 \cdot 5 + 27) && \text{Evaluate } 3^3.\\
= 3(4)^2 + (10 + 27) && \text{Multiply 2 by 5.}\\
= 3(4)^2 + (37) && \text{Add 10 to 27.}\\
= 3(16) + 37 && \text{Evaluate } 4^2.\\
= 48 + 37 && \text{Multiply 3 by 16.}\\
= 85 && \text{Add 48 to 37.}
\end{aligned}
$$

▶ **Guided**Practice

Evaluate each expression.

4A. $a^2(3b + 5) \div c$ if $a = 2, b = 6, c = 4$ **4B.** $5d + (6f - g)$ if $d = 4, f = 3, g = 12$

Real-World Example 5 Write and Evaluate an Expression

ENVIRONMENTAL STUDIES Science on a Sphere (SOS)® demonstrates the effects of atmospheric storms, climate changes, and ocean temperature on the environment. The volume of a sphere is four thirds of π multiplied by the radius r to the third power.

a. Write an expression that represents the volume of a sphere.

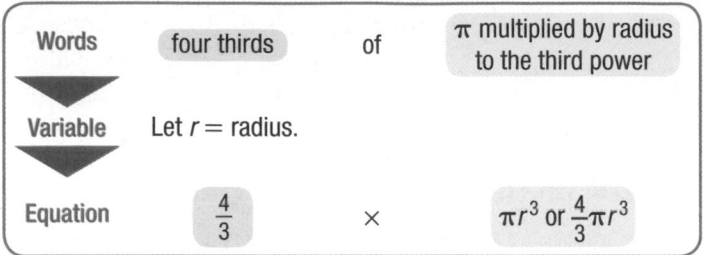

Words	four thirds	of	π multiplied by radius to the third power

Variable — Let r = radius.

Equation — $\dfrac{4}{3}$ × πr^3 or $\dfrac{4}{3}\pi r^3$

b. Find the volume of the 3-foot radius sphere used for SOS.

$$V = \frac{4}{3}\pi r^3 \qquad \text{Volume of a sphere}$$
$$= \frac{4}{3}\pi (3)^3 \qquad \text{Replace } r \text{ with 3.}$$
$$= \left(\frac{4}{3}\right)\pi (27) \qquad \text{Evaluate } 3^3 = 27.$$
$$= 36\pi \qquad \text{Multiply } \frac{4}{3} \text{ by 27.}$$

The volume of the sphere is 36π cubic feet.

GuidedPractice

5. **FOREST FIRES** According to the California Department of Forestry, an average of 539.2 fires each year are started by burning debris, while campfires are responsible for an average of 129.1 each year.

   A. Write an algebraic expression that represents the number of fires, on average, in d years of debris burning and c years of campfires.

   B. How many fires would there be in 5 years?

Check Your Understanding = Step-by-Step Solutions begin on page R13.

Examples 1–3 Evaluate each expression.

1. 9^2

2. 4^4

3. 3^5

4. $30 - 14 \div 2$

5. $5 \cdot 5 - 1 \cdot 3$

6. $(2 + 5)4$

7. $[8(2) - 4^2] + 7(4)$

8. $\dfrac{11 - 8}{1 + 7 \cdot 2}$

9. $\dfrac{(4 \cdot 3)^2}{9 + 3}$

Example 4 Evaluate each expression if $a = 4$, $b = 6$, and $c = 8$.

10. $8b - a$

11. $2a + (b^2 \div 3)$

12. $\dfrac{b(9 - c)}{a^2}$

Example 5 13. **BOOKS** Akira bought one new book for $20 and three used books for $4.95 each. Write and evaluate an expression to find how much money the books cost.

14. **CCSS REASONING** Koto purchased food for herself and her friends. She bought 4 cheeseburgers for $2.25 each, 3 French fries for $1.25 each, and 4 drinks for $4.00. Write and evaluate an expression to find how much the food cost.

Examples 1–3 Evaluate each expression.

15. 7^2 **16.** 14^3 **17.** 2^6

18. $35 - 3 \cdot 8$ **19.** $18 \div 9 + 2 \cdot 6$ **20.** $10 + 8^3 \div 16$

21. $24 \div 6 + 2^3 \cdot 4$ **22.** $(11 \cdot 7) - 9 \cdot 8$ **23.** $29 - 3(9 - 4)$

24. $(12 - 6) \cdot 5^2$ **25.** $3^5 - (1 + 10^2)$ **26.** $108 \div [3(9 + 3^2)]$

27. $[(6^3 - 9) \div 23]4$ **28.** $\dfrac{8 + 3^3}{12 - 7}$ **29.** $\dfrac{(1 + 6)9}{5^2 - 4}$

Example 4   Evaluate each expression if $g = 2$, $r = 3$, and $t = 11$.

30. $g + 6t$ **31.** $7 - gr$ **32.** $r^2 + (g^3 - 8)^5$

33 $(2t + 3g) \div 4$ **34.** $t^2 + 8rt + r^2$ **35.** $3g(g + r)^2 - 1$

Example 5   **36. GEOMETRY** Write an algebraic expression to represent the area of the triangle. Then evaluate it to find the area when $h = 12$ inches.

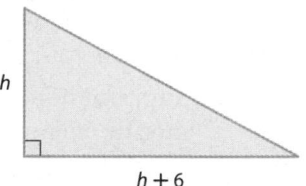

h

$h + 6$

37. AMUSEMENT PARKS In 2004, there were 3344 amusement parks and arcades. This decreased by 148 by 2009. Write and evaluate an expression to find the number of amusement parks and arcades in 2009.

38. CCSS STRUCTURE Kamilah sells tickets at Duke University's athletic ticket office. If p represents a preferred season ticket, b represents a blue zone ticket, and g represents a general admission ticket, interpret and then evaluate the following expressions.

a. $45b$ **b.** $15p + 35g$ **c.** $6p + 11b + 22g$

Duke University Football Ticket Prices	
Preferred Season Ticket	$100
Blue Zone	$80
General Admission	$70

Source: Duke University

Evaluate each expression.

39. 4^2 **40.** 12^3 **41.** 3^6

42. 11^5 **43.** $(3 - 4^2)^2 + 8$ **44.** $23 - 2(17 + 3^3)$

45. $3[4 - 8 + 4^2(2 + 5)]$ **46.** $\dfrac{2 \cdot 8^2 - 2^2 \cdot 8}{2 \cdot 8}$

47. $25 + \left[(16 - 3 \cdot 5) + \dfrac{12 + 3}{5}\right]$ **48.** $7^3 - \dfrac{2}{3}(13 \cdot 6 + 9)4$

Evaluate each expression if $a = 8$, $b = 4$, and $c = 16$.

49. $a^2bc - b^2$ **50.** $\dfrac{c^2}{b^2} + \dfrac{b^2}{a^2}$ **51.** $\dfrac{2b + 3c^2}{4a^2 - 2b}$

52. $\dfrac{3ab + c^2}{a}$ **53.** $\left(\dfrac{a}{b}\right)^2 - \dfrac{c}{a - b}$ **54.** $\dfrac{2a - b^2}{ab} + \dfrac{c - a}{b^2}$

55. SALES One day, 28 small and 12 large merchant spaces were rented. Another day, 30 small and 15 large spaces were rented. Write and evaluate an expression to show the total rent collected.

THE FLEA MARKET

MERCHANT SPACE RENTALS

Small space $7.00/day
Large space $9.75/day

Open Daily from 9:00–6:00

56. SHOPPING Evelina is shopping for back-to-school clothes. She bought 3 skirts, 2 pairs of jeans, and 4 sweaters. Write and evaluate an expression to find how much she spent, without including sales tax.

Clothing	
skirt	$25.99
jeans	$39.99
sweater	$22.99

57 PYRAMIDS The pyramid at the Louvre has a square base with a side of 35.42 meters and a height of 21.64 meters. The Great Pyramid in Egypt has a square base with a side of 230 meters and a height of 146.5 meters. The expression for the volume of a pyramid is $\frac{1}{3}Bh$, where B is the area of the base and h is the height.

  a. Draw both pyramids and label the dimensions.

  b. Write a verbal expression for the difference in volume of the two pyramids.

  c. Write an algebraic expression for the difference in volume of the two pyramids. Find the difference in volume.

58. FINANCIAL LITERACY A sales representative receives an annual salary s, an average commission each month c, and a bonus b for each sales goal that she reaches.

  a. Write an algebraic expression to represent her total earnings in one year if she receives four equal bonuses.

  b. Suppose her annual salary is $52,000 and her average commission is $1225 per month. If each of the four bonuses equals $1150, what does she earn annually?

H.O.T. Problems   Use Higher-Order Thinking Skills

59. ERROR ANALYSIS Tara and Curtis are simplifying $[4(10) - 3^2] + 6(4)$. Is either of them correct? Explain your reasoning.

Tara
$[4(10) - 3^2] + 6(4)$
$= [4(10) - 9] + 6(4)$
$= 4(1) + 6(4)$
$= 4 + 6(4)$
$= 4 + 24$
$= 28$

Curtis
$[4(10) - 3^2] + 6(4)$
$= [4(10) - 9] + 6(4)$
$= (40 - 9) + 6(4)$
$= 31 + 6(4)$
$= 31 + 24$
$= 55$

60. REASONING Explain how to evaluate $a[(b - c) \div d] - f$ if you were given values for $a, b, c, d,$ and f. How would you evaluate the expression differently if the expression was $a \cdot b - c \div d - f$?

61. CCSS PERSEVERANCE Write an expression using the whole numbers 1 to 5 using all five digits and addition and/or subtraction to create a numeric expression with a value of 3.

62. OPEN ENDED Write an expression that uses exponents, at least three different operations, and two sets of parentheses. Explain the steps you would take to evaluate the expression.

63. WRITING IN MATH Choose a geometric formula and explain how the order of operations applies when using the formula.

64. WRITING IN MATH Equivalent expressions have the same value. Are the expressions $(30 + 17) \times 10$ and $10 \times 30 + 10 \times 17$ equivalent? Explain why or why not.

65. Let m represent the number of miles. Which algebraic expression represents the number of feet in m miles?

   A $5280m$

   B $\dfrac{5280}{m}$

   C $m + 5280$

   D $5280 - m$

66. SHORT RESPONSE

Simplify: $\left[10 + 15(2^3)\right] \div \left[7(2^2) - 2\right]$

Step 1 $[10 + 15(8)] \div [7(4) - 2]$

Step 2 $[10 + 120] \div [28 - 2]$

Step 3 $130 \div 26$

Step 4 $\dfrac{1}{5}$

Which is the first *incorrect* step? Explain the error.

67. EXTENDED RESPONSE Consider the rectangle below.

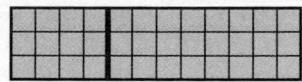

Part A Which expression models the area of the rectangle?

   F $4 + 3 \times 8$ **H** $3 \times 4 + 8$

   G $3 \times (4 + 8)$ **J** $3^2 + 8^2$

Part B Draw one or more rectangles to model each other expression.

68. GEOMETRY What is the perimeter of the triangle if $a = 9$ and $b = 10$?

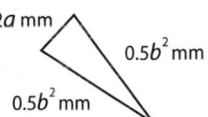

   A 164 mm   **C** 28 mm

   B 118 mm   **D** 4 mm

Write a verbal expression for each algebraic expression. (Lesson 1-1)

69. $14 - 9c$

70. $k^3 + 13$

71. $\dfrac{4 - v}{w}$

72. MONEY Destiny earns \$8 per hour babysitting and \$15 for each lawn she mows. Write an expression to show the amount of money she earns babysitting h hours and mowing m lawns. (Lesson 1-1)

Find the area of each figure. (Lesson 0-8)

73.

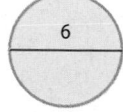

74.

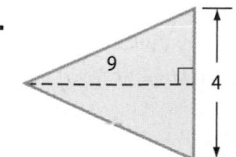

75.

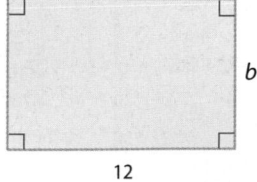

76. SCHOOL Aaron correctly answered 27 out of 30 questions on his last biology test. What percent of the questions did he answer correctly? (Lesson 0-6)

Find the value of each expression.

77. $5.65 - 3.08$

78. $6 \div \dfrac{4}{5}$

79. $4.85(2.72)$

80. $1\dfrac{1}{12} + 3\dfrac{2}{3}$

81. $\dfrac{4}{9} \cdot \dfrac{3}{2}$

82. $7\dfrac{3}{4} - 4\dfrac{7}{10}$

1-3 Properties of Numbers

:: Then	:: Now	:: Why?
● You used the order of operations to simplify expressions.	● **1** Recognize the properties of equality and identity. ● **2** Recognize the Commutative and Associative Properties.	● Natalie lives 32 miles away from the mall. The distance from her house to the mall is the same as the distance from the mall to her house. This is an example of the Reflexive Property.

 NewVocabulary
equivalent expressions
additive identity
multiplicative identity
multiplicative inverse
rcciprocal

 Common Core State Standards

Content Standards
A.SSE.1b Interpret complicated expressions by viewing one or more of their parts as a single entity.

A.SSE.2 Use the structure of an expression to identify ways to rewrite it.

Mathematical Practices
2 Reason abstractly and quantitatively.
3 Construct viable arguments and critique the reasoning of others.

1 Properties of Equality and Identity The expressions $4k + 8k$ and $12k$ are called **equivalent expressions** because they represent the same number. The properties below allow you to write an equivalent expression for a given expression.

KeyConcept Properties of Equality

Property	Words	Symbols	Examples
Reflexive Property	Any quantity is equal to itself.	For any number a, $a = a$.	$5 = 5$ $4 + 7 = 4 + 7$
Symmetric Property	If one quantity equals a second quantity, then the second quantity equals the first.	For any numbers a and b, if $a = b$, then $b = a$.	If $8 = 2 + 6$, then $2 + 6 = 8$.
Transitive Property	If one quantity equals a second quantity and the second quantity equals a third quantity, then the first quantity equals the third quantity.	For any numbers a, b, and c, if $a = b$ and $b = c$, then $a = c$.	If $6 + 9 = 3 + 12$ and $3 + 12 = 15$, then $6 + 9 = 15$.
Substitution Property	A quantity may be substituted for its equal in any expression.	If $a = b$, then a may be replaced by b in any expression.	If $n = 11$, then $4n = 4 \cdot 11$

The sum of any number and 0 is equal to the number. Thus, 0 is called the **additive identity**.

KeyConcept Addition Properties

Property	Words	Symbols	Examples
Additive Identity	For any number a, the sum of a and 0 is a.	$a + 0 = 0 + a = a$	$2 + 0 = 2$ $0 + 2 = 2$
Additive Inverse	A number and its opposite are additive inverses of each other.	$a + (-a) = 0$	$3 + (-3) = 0$ $4 - 4 = 0$

There are also special properties associated with multiplication. Consider the following equations.

$$4 \cdot n = 4$$

The solution of the equation is 1. Since the product of any number and 1 is equal to the number, 1 is called the **multiplicative identity**.

$$6 \cdot m = 0$$

The solution of the equation is 0. The product of any number and 0 is equal to 0. This is called the **Multiplicative Property of Zero**.

Two numbers whose product is 1 are called **multiplicative inverses** or **reciprocals**. Zero has no reciprocal because any number times 0 is 0.

KeyConcept Multiplication Properties

Property	Words	Symbols	Examples
Multiplicative Identity	For any number a, the product of a and 1 is a.	$a \cdot 1 = a$ $1 \cdot a = a$	$14 \cdot 1 = 14$ $1 \cdot 14 = 14$
Multiplicative Property of Zero	For any number a, the product of a and 0 is 0.	$a \cdot 0 = 0$ $0 \cdot a = 0$	$9 \cdot 0 = 0$ $0 \cdot 9 = 0$
Multiplicative Inverse	For every number $\frac{a}{b}$, where $a, b \neq 0$, there is exactly one number $\frac{b}{a}$ such that the product of $\frac{a}{b}$ and $\frac{b}{a}$ is 1.	$\frac{a}{b} \cdot \frac{b}{a} = 1$ $\frac{b}{a} \cdot \frac{a}{b} = 1$	$\frac{4}{5} \cdot \frac{5}{4} = \frac{20}{20}$ or 1 $\frac{5}{4} \cdot \frac{4}{5} = \frac{20}{20}$ or 1

Example 1 Evaluate Using Properties

Evaluate $7(4 - 3) - 1 + 5 \cdot \frac{1}{5}$. Name the property used in each step.

$7(4 - 3) - 1 + 5 \cdot \frac{1}{5} = 7(1) - 1 + 5 \cdot \frac{1}{5}$ Substitution: $4 - 3 = 1$

$= 7 - 1 + 5 \cdot \frac{1}{5}$ Multiplicative Identity: $7 \cdot 1 = 7$

$= 7 - 1 + 1$ Multiplicative Inverse: $5 \cdot \frac{1}{5} = 1$

$= 6 + 1$ Substitution: $7 - 1 = 6$

$= 7$ Substitution: $6 + 1 = 7$

▶ **Guided**Practice

Name the property used in each step.

1A. $2 \cdot 3 + (4 \cdot 2 - 8)$
$= 2 \cdot 3 + (8 - 8)$?
$= 2 \cdot 3 + (0)$?
$= 6 + 0$?
$= 6$?

1B. $7 \cdot \frac{1}{7} + 6(15 \div 3 - 5)$
$= 7 \cdot \frac{1}{7} + 6(5 - 5)$?
$= 7 \cdot \frac{1}{7} + 6(0)$?
$= 1 + 6(0)$?
$= 1 + 0$?
$= 1$?

2 Use Commutative and Associative Properties Nikki walks 2 blocks to her friend Sierra's house. They walk another 4 blocks to school. At the end of the day, Nikki and Sierra walk back to Sierra's house, and then Nikki walks home.

The distance from Nikki's house to school	equals	the distance from the school to Nikki's house.
2 + 4	=	4 + 2

This is an example of the **Commutative Property** for addition.

KeyConcept Commutative Property

Words	The order in which you add or multiply numbers does not change their sum or product.
Symbols	For any numbers a and b, $a + b = b + a$ and $a \cdot b = b \cdot a$.
Examples	$4 + 8 = 8 + 4$ $7 \cdot 11 = 11 \cdot 7$

An easy way to find the sum or product of numbers is to group, or associate, the numbers using the **Associative Property**.

KeyConcept Associative Property

Words	The way you group three or more numbers when adding or multiplying does not change their sum or product.
Symbols	For any numbers a, b, and c, $(a + b) + c = a + (b + c)$ and $(ab)c = a(bc)$.
Examples	$(3 + 5) + 7 = 3 + (5 + 7)$ $(2 \cdot 6) \cdot 9 = 2 \cdot (6 \cdot 9)$

Real-WorldLink

A child's birthday party may cost about $200 depending on the number of children invited.

Source: Family Corner

Real-World Example 2 Apply Properties of Numbers

PARTY PLANNING Eric makes a list of items that he needs to buy for a party and their costs. Find the total cost of these items.

Party Supplies	
Item	**Cost ($)**
balloons	6.75
decorations	14.00
food	23.25
beverages	20.50

Balloons		Decorations		Food		Beverages
6.75	+	14.00	+	23.25	+	20.50

$= 6.75 + 23.25 + 14.00 + 20.50$	Commutative (+)
$= (6.75 + 23.25) + (14.00 + 20.50)$	Associative (+)
$= 30.00 + 34.50$	Substitution
$= 64.50$	Substitution

The total cost is $64.50.

▶ **GuidedPractice**

2. FURNITURE Rafael is buying furnishings for his first apartment. He buys a couch for $300, lamps for $30.50, a rug for $25.50, and a table for $50. Find the total cost of these items.

Example 3 Use Multiplication Properties

Evaluate $5 \cdot 7 \cdot 4 \cdot 2$ using the properties of numbers. Name the property used in each step.

$$5 \cdot 7 \cdot 4 \cdot 2 = 5 \cdot 2 \cdot 7 \cdot 4 \qquad \text{Commutative } (\times)$$
$$= (5 \cdot 2) \cdot (7 \cdot 4) \qquad \text{Associative } (\times)$$
$$= 10 \cdot 28 \qquad \text{Substitution}$$
$$= 280 \qquad \text{Substitution}$$

▶ **Guided**Practice

Evaluate each expression using the properties of numbers. Name the property used in each step.

3A. $2.9 \cdot 4 \cdot 10$ **3B.** $\frac{5}{3} \cdot 25 \cdot 3 \cdot 2$

Check Your Understanding

○ = Step-by-Step Solutions begin on page R13. ✓

Example 1   Evaluate each expression. Name the property used in each step.

1. $(1 \div 5)5 \cdot 14$ **2.** $6 + 4(19 - 15)$ **3.** $5(14 - 5) + 6(3 + 7)$

4. FINANCIAL LITERACY Carolyn has 9 quarters, 4 dimes, 7 nickels, and 2 pennies, which can be represented as $9(25) + 4(10) + 7(5) + 2$. Evaluate the expression to find how much money she has. Name the property used in each step.

Examples 2–3 Evaluate each expression using the properties of numbers. Name the property used in each step.

5. $23 + 42 + 37$ **6.** $2.75 + 3.5 + 4.25 + 1.5$

7. $3 \cdot 7 \cdot 10 \cdot 2$ **8.** $\frac{1}{4} \cdot 24 \cdot \frac{2}{3}$

Practice and Problem Solving

Extra Practice is on page R1.

Example 1   Evaluate each expression. Name the property used in each step.

9 $3(22 - 3 \cdot 7)$ **10.** $7 + (9 - 3^2)$

11. $\frac{3}{4}[4 \div (7 - 4)]$ **12.** $[3 \div (2 \cdot 1)]\frac{2}{3}$

13. $2(3 \cdot 2 - 5) + 3 \cdot \frac{1}{3}$ **14.** $6 \cdot \frac{1}{6} + 5(12 \div 4 - 3)$

Example 2   **15. GEOMETRY** The expression $2 \cdot \frac{22}{7} \cdot 14^2 + 2 \cdot \frac{22}{7} \cdot 14 \cdot 7$ represents the approximate surface area of the cylinder at the right. Evaluate this expression to find the approximate surface area. Name the property used in each step.

7 in.

14 in.

16. **CCSS REASONING** A traveler checks into a hotel on Friday and checks out the following Tuesday morning. Use the table to find the total cost of the room including tax.

Hotel Rates Per Day		
Day	**Room Charge**	**Sales Tax**
Monday–Friday	$72	$5.40
Saturday–Sunday	$63	$5.10

 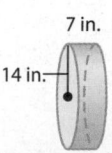

Examples 2–3 Evaluate each expression using properties of numbers. Name the property used in each step.

17. $25 + 14 + 15 + 36$

18. $11 + 7 + 5 + 13$

19. $3\frac{2}{3} + 4 + 5\frac{1}{3}$

20. $4\frac{4}{9} + 7\frac{2}{9}$

21. $4.3 + 2.4 + 3.6 + 9.7$

22. $3.25 + 2.2 + 5.4 + 10.75$

23. $12 \cdot 2 \cdot 6 \cdot 5$

24. $2 \cdot 8 \cdot 10 \cdot 2$

25. $0.2 \cdot 4.6 \cdot 5$

26. $3.5 \cdot 3 \cdot 6$

27. $1\frac{5}{6} \cdot 24 \cdot 3\frac{1}{11}$

28. $2\frac{3}{4} \cdot 1\frac{1}{8} \cdot 32$

29. SCUBA DIVING The sign shows the equipment rented or sold by a scuba diving store.

    a. Write two expressions to represent the total sales to rent 2 wet suits, 3 air tanks, 2 dive flags, and selling 5 underwater cameras.

    b. What are the total sales?

30. COOKIES Bobby baked 2 dozen chocolate chip cookies, 3 dozen sugar cookies, and a dozen oatmeal raisin cookies. How many total cookies did he bake?

THE DEEP
SCUBA SUPPLIES

SPECIALS	
Underwater Camera	$18.99

RENTALS	
Air Tanks	$ 7.50
Wet Suit	$10.95
Dive Flag	$ 5.00

Evaluate each expression if $a = -1$, $b = 4$, and $c = 6$.

31 $4a + 9b - 2c$

32. $-10c + 3a + a$

33. $a - b + 5a - 2b$

34. $8a + 5b - 11a - 7b$

35. $3c^2 + 2c + 2c^2$

36. $3a - 4a^2 + 2a$

37. FOOTBALL A football team is on the 35-yard line. The quarterback is sacked at the line of scrimmage. The team gains 0 yards, so they are still at the 35-yard line. Which identity or property does this represent? Explain.

Find the value of x. Then name the property used.

38. $8 = 8 + x$

39. $3.2 + x = 3.2$

40. $10x = 10$

41. $\frac{1}{2} \cdot x = \frac{1}{2} \cdot 7$

42. $x + 0 = 5$

43. $1 \cdot x = 3$

44. $5 \cdot \frac{1}{5} = x$

45. $2 + 8 = 8 + x$

46. $x + \frac{3}{4} = 3 + \frac{3}{4}$

47. $\frac{1}{3} \cdot x = 1$

48. GEOMETRY Write an expression to represent the perimeter of the triangle. Then find the perimeter if $x = 2$ and $y = 7$.

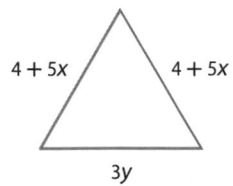

$4 + 5x$ $4 + 5x$

$3y$

49. SPORTS Tickets to a baseball game cost $25 each plus a $4.50 handling charge per ticket. If Sharon has a coupon for $10 off and orders 4 tickets, how much will she be charged?

50. CCSS PRECISION The table shows prices on children's clothing.

    a. Interpret the expression $5(8.99) + 2(2.99) + 7(5.99)$.

    b. Write and evaluate three different expressions that represent 8 pairs of shorts and 8 tops.

    c. If you buy 8 shorts and 8 tops, you receive a discount of 15%. Find the greatest and least amount of money you can spend on the 16 items at the sale.

Shorts	Shirts	Tank Tops
$7.99	$8.99	$6.99
$5.99	$4.99	$2.99

51. GEOMETRY A regular octagon measures $(3x + 5)$ units on each side. What is the perimeter if $x = 2$?

52. 🔄 **MULTIPLE REPRESENTATIONS** You can use *algebra tiles* to model and explore algebraic expressions. The rectangular tile has an area of x, with dimensions 1 by x. The small square tile has an area of 1, with dimensions 1 by 1.

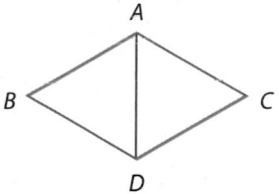

a. Concrete Make a rectangle with algebra tiles to model the expression $4(x + 2)$ as shown above. What are the dimensions of this rectangle? What is its area?

b. Analytical What are the areas of the green region and of the yellow region?

c. Verbal Complete this statement: $4(x + 2) =$ _?_. Write a convincing argument to justify your statement.

53 GEOMETRY A **proof** is a logical argument in which each statement you make is supported by a statement that is accepted as true. It is given that $\overline{AB} \cong \overline{CD}$, $\overline{AB} \cong \overline{BD}$, and $\overline{AB} \cong \overline{AC}$. Pedro wants to prove $\triangle ADB \cong \triangle ADC$. To do this, he must show that $\overline{AD} \cong \overline{AD}$, $\overline{AB} \cong \overline{DC}$ and $\overline{BD} \cong \overline{AC}$.

a. Copy the figure and label $\overline{AB} \cong \overline{CD}$, $\overline{AB} \cong \overline{BD}$, and $\overline{AB} \cong \overline{AC}$.

b. Explain how he can use the Reflexive and Transitive Properties to prove $\triangle ADB \cong \triangle ADC$.

c. If AC is x centimeters, write an equation for the perimeter of $ACDB$.

H.O.T. Problems   Use Higher-Order Thinking Skills

54. OPEN ENDED Write two equations showing the Transitive Property of Equality. Justify your reasoning.

55. (CCSS) **ARGUMENTS** Explain why 0 has no multiplicative inverse.

56. REASONING The sum of any two whole numbers is always a whole number. So, the set of whole numbers {0, 1, 2, 3, 4, … } is said to be closed under addition. This is an example of the **Closure Property**. State whether each statement is *true* or *false*. If false, justify your reasoning.

a. The set of whole numbers is closed under subtraction.

b. The set of whole numbers is closed under multiplication.

c. The set of whole numbers is closed under division.

57. CHALLENGE Does the Commutative Property *sometimes*, *always* or *never* hold for subtraction? Explain your reasoning.

58. REASONING Explain whether 1 can be an additive identity. Give an example to justify your answer.

59. WHICH ONE DOESN'T BELONG? Identify the equation that does not belong with the other three. Explain your reasoning.

| $x + 12 = 12 + x$ | $7h = h \cdot 7$ | $1 + a = a + 1$ | $(2j)k = 2(jk)$ |

60. WRITING IN MATH Determine whether the Commutative Property applies to division. Justify your answer.

61. A deck is shaped like a rectangle with a width of 12 feet and a length of 15 feet. What is the area of the deck?

A 3 ft²

B 27 ft²

C 108 ft²

D 180 ft²

62. GEOMETRY A box in the shape of a rectangular prism has a volume of 56 cubic inches. If the length of each side is multiplied by 2, what will be the approximate volume of the box?

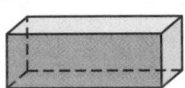

F 112 in³   **H** 336 in³

G 224 in³   **J** 448 in³

63. $27 \div 3 + (12 - 4) =$

A $\frac{-11}{5}$ **C** 17

B $\frac{27}{11}$ **D** 25

64. GRIDDED RESPONSE Ms. Beal had 1 bran muffin, 16 ounces of orange juice, 3 ounces of sunflower seeds, 2 slices of turkey, and half a cup of spinach. Find the total number of grams of protein she consumed.

Protein Content	
Food	Protein (g)
bran muffin (1)	3
orange juice (8 oz)	2
sunflower seeds (1 oz)	2
turkey (1 slice)	12
spinach (1 c)	5

Evaluate each expression. (Lesson 1-2)

65. $3 \cdot 5 + 1 - 2$

66. $14 \div 2 \cdot 6 - 5^2$

67. $\dfrac{3 \cdot 9^2 - 3^2 \cdot 9}{3 \cdot 9}$

68. GEOMETRY Write an expression for the perimeter of the figure. (Lesson 1-1)

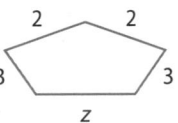

Find the perimeter and area of each figure. (Lessons 0-7 and 0-8)

69. a rectangle with length 5 feet and width 8 feet

70. a square with length 4.5 inches

71. SURVEY Andrew took a survey of his friends to find out their favorite type of music. Of the 34 friends surveyed, 22 said they liked rock music the best. What percent like rock music the best? (Lesson 0-6)

Name the reciprocal of each number. (Lesson 0-5)

72. $\frac{6}{17}$

73. $\frac{2}{23}$

74. $3\frac{4}{5}$

Find each product. Express in simplest form.

75. $\frac{12}{15} \cdot \frac{3}{14}$

76. $\frac{5}{7} \cdot \left(-\frac{4}{5}\right)$

77. $\frac{10}{11} \cdot \frac{21}{35}$

78. $\frac{63}{65} \cdot \frac{120}{126}$

79. $-\frac{4}{3} \cdot \left(-\frac{9}{2}\right)$

80. $\frac{1}{3} \cdot \frac{2}{5}$

1-3

Algebra Lab
Accuracy

All measurements taken in the real world are approximations. The greater the care with which a measurement is taken, the more accurate it will be. **Accuracy** refers to how close a measured value comes to the actual or desired value. For example, a fraction is more accurate than a rounded decimal.

CCSS Common Core State Standards
Content Standards
N.Q.3 Choose a level of accuracy appropriate to limitations on measurement when reporting quantities.
Mathematical Practices
6 Attend to precision.

Activity 1   When Is Close Good Enough?

Measure the length of your desktop. Record your results in centimeters, in meters, and in millimeters.

Analyze the Results

1. Did you round to the nearest whole measure? If so, when?
2. Did you round to the nearest half, tenth, or smaller? If so, when?
3. Which unit of measure was the most appropriate for this task?
4. Which unit of measure was the most accurate?

Deciding where to round a measurement depends on how the measurement will be used. But calculations should not be carried out to greater accuracy than that of the original data.

Activity 2   Decide Where to Round

a. **Elan has $13 that he wants to divide among his 6 nephews. When he types 13 ÷ 6 into his calculator, the number that appears is 2.166666667. Where should Elan round?**

Since Elan is rounding money, the smallest increment is a penny, so round to the hundredths place. This will give him 2.17, and $2.17 × 6 = $13.02. Elan will be two pennies short, so round to $2.16. Since $2.16 × 6 = $12.96, Elan can give each of his nephews $2.16.

b. **Dante's mother brings him a dozen cookies, but before she leaves she eats one and tells Dante he has to share with his two sisters. Dante types 11 ÷ 3 into his calculator and gets 3.666666667. Where should Dante round?**

After each sibling receives 3 cookies, there are two cookies left. In this case, it is more accurate to convert the decimal portion to a fraction and give each sibling $\frac{2}{3}$ of a cookie.

c. **Eva measures the dimensions of a box as 8.7, 9.52, and 3.16 inches. She multiplies these three numbers to find the measure of the volume. The result shown on her calculator is 261.72384. Where should Eva round?**

Eva should round to the tenths place, 261.7, because she was only accurate to the tenths place with one of her measures.

Exercises

5. Jessica wants to divide $23 six ways. Her calculator shows 3.833333333. Where should she round?

6. Ms. Harris wants to share 2 pizzas among 6 people. Her calculator shows 0.3333333333. Where should she round?

7. The measurements of an aquarium are 12.9, 7.67, and 4.11 inches. The measure of the volume is given by the product 406.65573. Where should the number be rounded?

Accuracy *Continued*

For most real-world measurements, a decision must be made on the level of accuracy needed or desired.

Activity 3   Find an Appropriate Level of Accuracy

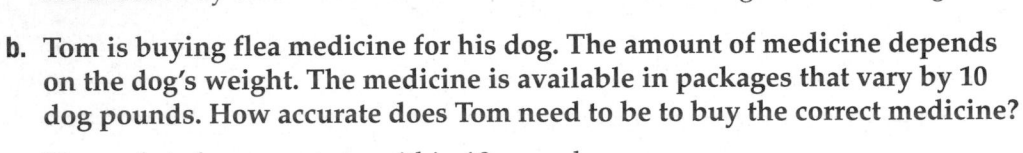

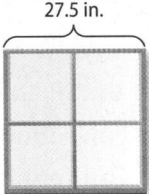

27.5 in.

a. Jon needs to buy a shade for the window opening shown, but the shades are only available in whole inch increments. What size shade should he buy?

He should buy the 27-inch shade because it will be enough to cover the glass.

b. Tom is buying flea medicine for his dog. The amount of medicine depends on the dog's weight. The medicine is available in packages that vary by 10 dog pounds. How accurate does Tom need to be to buy the correct medicine?

He needs to be accurate to within 10 pounds.

c. Tyrone is building a jet engine. How accurate do you think he needs to be with his measurements?

He needs to be very accurate, perhaps to the thousandth of an inch.

Exercises

8. Matt's table is missing a leg. He wants to cut a piece of wood to replace the leg. How accurate do you think he needs to be with his measurements?

For each situation, determine where the rounding should occur and give the rounded answer.

9. Sam wants to divide $111 seven ways. His calculator shows 15.85714286.

10. Kiri wants to share 3 pies among 11 people. Her calculator shows 0.2727272727.

11. Evan's calculator gives him the volume of his soccer ball as 137.2582774. Evan measured the radius of the ball to be 3.2 inches.

For each situation, determine the level of accuracy needed. Explain.

12. You are estimating the length of your school's basketball court. Which unit of measure should you use: 1 foot, 1 inch, or $\frac{1}{16}$ inch?

13. You are estimating the height of a small child. Which unit of measure should you use: 1 foot, 1 inch, or $\frac{1}{16}$ inch?

14. **TRAVEL** Curt is measuring the driving distance from one city to another. How accurate do you think he needs to be with his measurement?

15. **MEDICINE** A nurse is administering medicine to a patient based on his weight. How accurate do you think she needs to be with her measurements?

The Distributive Property

::·Then

- You explored Associative and Commutative Properties.

::·Now

1. Use the Distributive Property to evaluate expressions.

2. Use the Distributive Property to simplify expressions.

::·Why?

- John burns approximately 420 Calories per hour by inline skating. The chart below shows the time he spent inline skating in one week.

Day	Mon	Tue	Wed	Thu	Fri	Sat	Sun
Time (h)	1	$\frac{1}{2}$	0	1	0	2	$2\frac{1}{2}$

To determine the total number of Calories that he burned inline skating that week, you can use the Distributive Property.

 NewVocabulary
like terms
simplest form
coefficient

 Common Core State Standards

Content Standards
A.SSE.1a Interpret parts of an expression, such as terms, factors, and coefficients.

A.SSE.2 Use the structure of an expression to identify ways to rewrite it.

Mathematical Practices
1 Make sense of problems and persevere in solving them.

8 Look for and express regularity in repeated reasoning.

1 Evaluate Expressions There are two methods you could use to calculate the number of Calories John burned inline skating. You could find the total time spent inline skating and then multiply by the Calories burned per hour. Or you could find the number of Calories burned each day and then add to find the total.

Method 1 Rate Times Total Time

$$420\left(1 + \frac{1}{2} + 1 + 2 + 2\frac{1}{2}\right)$$
$$= 420(7)$$
$$= 2940$$

Method 2 Sum of Daily Calories Burned

$$420(1) + 420\left(\frac{1}{2}\right) + 420(1) + 420(2) + 420\left(2\frac{1}{2}\right)$$
$$= 420 + 210 + 420 + 840 + 1050$$
$$= 2940$$

Either method gives the same total of 2940 Calories burned. This is an example of the **Distributive Property**.

KeyConcept Distributive Property

Symbol	For any numbers a, b, and c, $a(b + c) = ab + ac$ and $(b + c)a = ba + ca$ and $a(b - c) = ab - ac$ and $(b - c)a = ba - ca$.
Examples	$3(2 + 5) = 3 \cdot 2 + 3 \cdot 5$ $4(9 - 7) = 4 \cdot 9 - 4 \cdot 7$ $3(7) = 6 + 15$ $4(2) = 36 - 28$ $21 = 21$ $8 = 8$

The Symmetric Property of Equality allows the Distributive Property to be written as follows.

If $a(b + c) = ab + ac$, then $ab + ac = a(b + c)$.

StudyTip

CCSS Sense-Making and Perseverance The four-step problem solving plan is a tool for making sense of any problem. When making and executing your plan, continually ask yourself, "Does this make sense?" Monitor and evaluate your progress and change course if necessary.

Real-World Example 1 Distribute Over Addition

SPORTS A group of 7 adults and 6 children are going to a University of South Florida Bulls baseball game. Use the Distributive Property to write and evaluate an expression for the total ticket cost.

USF Bulls Baseball Tickets	
Ticket	Cost ($)
Adult Single Game	5
Children Single Game (12 and under)	3
Groups of 10 or more Single Game	2
Senior Single Game (65 and over)	3

Source: USF

Understand You need to find the cost of each ticket and then find the total cost.

Plan $7 + 6$ or 13 people are going to the game, so the tickets are $2 each.

Solve Write an expression that shows the product of the cost of each ticket and the sum of adult tickets and children's tickets.

$$2(7 + 6) = 2(7) + 2(6) \quad \text{Distributive Property}$$
$$= 14 + 12 \quad \text{Multiply.}$$
$$= 26 \quad \text{Add.}$$

The total cost is $26.

Check The total number of tickets needed is 13 and they cost $2 each. Multiply 13 by 2 to get 26. Therefore, the total cost of tickets is $26.

GuidedPractice

1. **SPORTS** A group of 3 adults, an 11-year old, and 2 children under 10 years old are going to a baseball game. Write and evaluate an expression to determine the cost of tickets for the group.

You can use the Distributive Property to make mental math easier.

Example 2 Mental Math

Use the Distributive Property to rewrite $7 \cdot 49$. Then evaluate.

$$7 \cdot 49 = 7(50 - 1) \quad \text{Think: } 49 = 50 - 1$$
$$= 7(50) - 7(1) \quad \text{Distributive Property}$$
$$= 350 - 7 \quad \text{Multiply.}$$
$$= 343 \quad \text{Subtract.}$$

GuidedPractice

Use the Distributive Property to rewrite each expression. Then evaluate.

2A. $304(15)$

2B. $44 \cdot 2\frac{1}{2}$

2C. $210(5)$

2D. $52(17)$

2 **Simplify Expressions** You can use algebra tiles to investigate how the Distributive Property relates to algebraic expressions.

The rectangle at the right has 3 x-tiles and 6 1-tiles. The area of the rectangle is $x + 1 + 1 + x + 1 + 1 + x + 1 + 1$ or $3x + 6$. Therefore, $3(x + 2) = 3x + 6$.

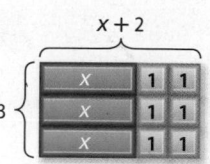

Example 3 Algebraic Expressions

Rewrite each expression using the Distributive Property. Then simplify.

a. $7(3w - 5)$

$7(3w - 5) = 7 \cdot 3w - 7 \cdot 5$ Distributive Property

         $= 21w - 35$ Multiply.

b. $(6v^2 + v - 3)4$

$(6v^2 + v - 3)4 = 6v^2(4) + v(4) - 3(4)$ Distributive Property

         $= 24v^2 + 4v - 12$ Multiply.

▶ **Guided**Practice

3A. $(8 + 4n)2$ **3B.** $-6(r + 3g - t)$

3C. $(2 - 5q)(-3)$ **3D.** $-4(-8 - 3m)$

Like terms are terms that contain the same variables, with corresponding variables having the same power.

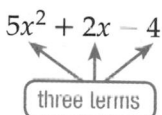   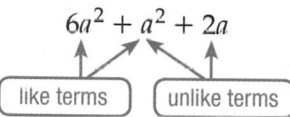

The Distributive Property and the properties of equality can be used to show that $4k + 8k = 12k$. In this expression, $4k$ and $8k$ are like terms.

$4k + 8k = (4 + 8)k$ Distributive Property

       $= 12k$ Substitution

An expression is in **simplest form** when it contains no like terms or parentheses.

Example 4 Combine Like Terms

a. Simplify $17u + 25u$.

$17u + 25u = (17 + 25)u$ Distributive Property

         $= 42u$ Substitution

b. Simplify $6t^2 + 3t - t$.

$6t^2 + 3t - t = 6t^2 + (3 - 1)t$ Distributive Property

         $= 6t^2 + 2t$ Substitution

▶ **Guided**Practice

Simplify each expression. If not possible, write *simplified*.

4A. $6n - 4n$ **4B.** $b^2 + 13b + 13$

4C. $4y^3 + 2y - 8y + 5$ **4D.** $7a + 4 - 6a^2 - 2a$

Example 5 Write and Simplify Expressions

Use the expression *twice the difference of 3x and y increased by five times the sum of x and 2y.*

a. **Write an algebraic expression for the verbal expression.**

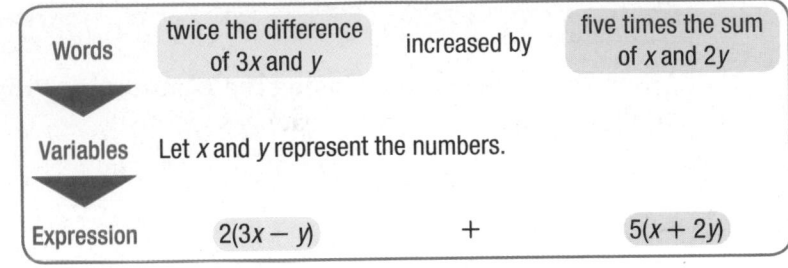

| Words | twice the difference of 3x and y | increased by | five times the sum of x and 2y |

Variables Let *x* and *y* represent the numbers.

| Expression | $2(3x - y)$ | $+$ | $5(x + 2y)$ |

b. **Simplify the expression, and indicate the properties used.**

$$2(3x - y) + 5(x + 2y) = 2(3x) - 2(y) + 5(x) + 5(2y) \qquad \text{Distributive Property}$$

$$= 6x - 2y + 5x + 10y \qquad \text{Multiply.}$$

$$= 6x + 5x - 2y + 10y \qquad \text{Commutative (+)}$$

$$= (6 + 5)x + (-2 + 10)y \qquad \text{Distributive Property}$$

$$= 11x + 8y \qquad \text{Substitution}$$

▶ **Guided**Practice

5. Use the expression *5 times the difference of q squared and r plus 8 times the sum of 3q and 2r.*

A. Write an algebraic expression for the verbal expression.

B. Simplify the expression, and indicate the properties used.

The **coefficient** of a term is the numerical factor. For example, in $6ab$, the coefficient is 6, and in $\frac{x^2}{3}$, the coefficient is $\frac{1}{3}$. In the term y, the coefficient is 1 since $1 \cdot y = y$ by the Multiplicative Identity Property.

ConceptSummary Properties of Numbers

The following properties are true for any numbers *a*, *b*, and *c*.

Properties	Addition	Multiplication
Commutative	$a + b = b + a$	$ab = ba$
Associative	$(a + b) + c = a + (b + c)$	$(ab)c = a(bc)$
Identity	0 is the identity. $a + 0 = 0 + a = a$	1 is the identity. $a \cdot 1 = 1 \cdot a = a$
Zero	—	$a \cdot 0 = 0 \cdot a = 0$
Distributive	$a(b + c) = ab + ac$ and $(b + c)a = ba + ca$	
Substitution	If $a = b$, then a may be substituted for b.	

Example 1 1. **PILOT** A pilot at an air show charges $25 per passenger for rides. If 12 adults and 15 children ride in one day, write and evaluate an expression to describe the situation.

Example 2 Use the Distributive Property to rewrite each expression. Then evaluate.

2. $14(51)$
3. $6\frac{1}{9}(9)$

Example 3 Use the Distributive Property to rewrite each expression. Then simplify.

4. $2(4 + t)$
5. $(g - 9)5$

Example 4 Simplify each expression. If not possible, write *simplified*.

6. $15m + m$
7. $3x^3 + 5y^3 + 14$
8. $(5m + 2m)10$

Example 5 Write an algebraic expression for each verbal expression. Then simplify, indicating the properties used.

9. 4 times the sum of 2 times x and six

10. one half of 4 times y plus the quantity of y and 3

Practice and Problem Solving

Extra Practice is on page R1.

Example 1 11. **TIME MANAGEMENT** Margo uses dots to track her activities on a calendar. Red dots represent homework, yellow dots represent work, and green dots represent track practice. In a typical week, she uses 5 red dots, 3 yellow dots, and 4 green dots. How many activities does Margo do in 4 weeks?

12. **CCSS REASONING** The Red Cross is holding blood drives in two locations. In one day, Center 1 collected 715 pints and Center 2 collected 1035 pints. Write and evaluate an expression to estimate the total number of pints of blood donated over a 3-day period.

Example 2 Use the Distributive Property to rewrite each expression. Then evaluate.

13. $(4 + 5)6$
14. $7(13 + 12)$
15. $6(6 - 1)$

16. $(3 + 8)15$
17. $14(8 - 5)$
18. $(9 - 4)19$

19. $4(7 - 2)$
20. $7(2 + 1)$
21. $7 \cdot 497$

22. $6(525)$
23. $36 \cdot 3\frac{1}{4}$
24. $\left(4\frac{2}{7}\right)21$

Example 3 Use the Distributive Property to rewrite each expression. Then simplify.

25. $2(x + 4)$
26. $(5 + n)3$

27. $(4 - 3m)8$
28. $-3(2x - 6)$

Example 4 Simplify each expression. If not possible, write *simplified*.

29. $13r + 5r$
30. $3x^3 - 2x^2$
31. $7m + 7 - 5m$

32. $5z^2 + 3z + 8z^2$
33. $(2 - 4n)17$
34. $11(4d + 6)$

35. $7m + 2m + 5p + 4m$
36. $3x + 7(3x + 4)$
37. $4(fg + 3g) + 5g$

Example 5 Write an algebraic expression for each verbal expression. Then simplify, indicating the properties used.

38. the product of 5 and m squared, increased by the sum of the square of m and 5

39. 7 times the sum of a squared and b minus 4 times the sum of a squared and b

40. GEOMETRY Find the perimeter of an isosceles triangle with side lengths of $5 + x$, $5 + x$, and xy. Write in simplest form.

41. GEOMETRY A regular hexagon measures $3x + 5$ units on each side. What is the perimeter in simplest form?

Simplify each expression.

42. $6x + 4y + 5x$

43. $3m + 5g + 6g + 11m$

44. $4a + 5a^2 + 2a^2 + a^2$

45. $5k + 3k^3 + 7k + 9k^3$

46. $6d + 4(3d + 5)$

47. $2(6x + 4) + 7x$

48. FOOD Kenji is picking up take-out food for his study group.

a. Interpret the expression $4(2.49) + 3(1.29) + 3(0.99) + 5(1.49)$.

b. How much would it cost if Kenji bought four of each item on the menu?

Menu	
Item	Cost ($)
sandwich	2.49
cup of soup	1.29
side salad	0.99
drink	1.49

Use the Distributive Property to rewrite each expression. Then simplify.

49. $\left(\frac{1}{3} - 2b\right)27$

50. $4(8p + 4q - 7r)$

51. $6(2c - cd^2 + d)$

Simplify each expression. If not possible, write *simplified***.**

52. $6x^2 + 14x - 9x$

53. $4y^3 + 3y^3 + y^4$

54. $a + \frac{a}{5} + \frac{2}{5}a$

55. MULTIPLE REPRESENTATIONS The area of the model is $2(x - 4)$ or $2x - 8$. The expression $2(x - 4)$ is in *factored form*.

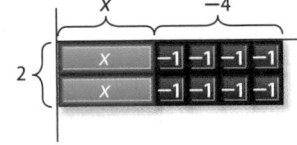

a. **Geometric** Use algebra tiles to form a rectangle with area $2x + 6$. Use the result to write $2x + 6$ in factored form.

b. **Tabular** Use algebra tiles to form rectangles to represent each area in the table. Record the factored form of each expression.

c. **Verbal** Explain how you could find the factored form of an expression.

Area	Factored Form
$2x + 6$	
$3x + 3$	
$3x - 12$	
$5x + 10$	

H.O.T. Problems   Use Higher-Order Thinking Skills

56. CCSS PERSEVERANCE Use the Distributive Property to simplify $6x^2[(3x - 4) + (4x + 2)]$.

57. REASONING Should the Distributive Property be a property of multiplication, addition, or both? Explain your answer.

58. WRITING IN MATH Why is it helpful to represent verbal expressions algebraically?

59. WRITING IN MATH Use the data about skating on page 25 to explain how the Distributive Property can be used to calculate quickly. Also, compare the two methods of finding the total Calories burned.

60. Which illustrates the Symmetric Property of Equality?

    A If $a = b$, then $b = a$.

    B If $a = b$, and $b = c$, then $a = c$.

    C If $a = b$, then $b = c$.

    D If $a = a$, then $a + 0 = a$.

61. Anna is three years younger than her sister Emily. Which expression represents Anna's age if we express Emily's age as y years?

    F $y + 3$ **H** $3y$

    G $y - 3$ **J** $\dfrac{3}{y}$

62. Which property is used below?
If $4xy^2 = 8y^2$ and $8y^2 = 72$, then $4xy^2 = 72$.

    A Reflexive Property

    B Substitution Property

    C Symmetric Property

    D Transitive Property

63. **SHORT RESPONSE** A drawer contains the socks in the chart. What is the probability that a randomly chosen sock is blue?

Color	Number
white	16
blue	12
black	8

Evaluate each expression. Name the property used in each step. (Lesson 1-3)

64. $14 + 23 + 8 + 15$ **65.** $0.24 \cdot 8 \cdot 7.05$ **66.** $1\frac{1}{4} \cdot 9 \cdot \frac{5}{6}$

67. **SPORTS** Braden runs 6 times a week for 30 minutes and lifts weights 3 times a week for 20 minutes. Write and evaluate an expression for the number of hours Braden works out in 4 weeks. (Lesson 1-2)

SPORTS Refer to the table showing Blanca's cross-country times for the first 8 meets of the season. Round answers to the nearest second. (Lesson 0-12)

68. Find the mean of the data.

69. Find the median of the data.

70. Find the mode of the data.

71. **SURFACE AREA** What is the surface area of the cube? (Lesson 0-10)

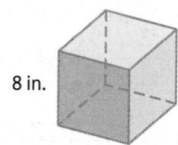

8 in.

Cross Country	
Meet	**Time**
1	22:31
2	22:21
3	21:48
4	22:01
5	21:48
6	20:56
7	20:34
8	20:15

Evaluate each expression.

72. $12(7 + 2)$ **73.** $11(5) - 8(5)$ **74.** $(13 - 9) \cdot 4$

75. $3(6) + 7(6)$ **76.** $(1 + 19) \cdot 8$ **77.** $16(5 + 7)$

Write a verbal expression for each algebraic expression. (Lesson 1-1)

1. $21 - x^3$

2. $3m^5 + 9$

Write an algebraic expression for each verbal expression. (Lesson 1-1)

3. five more than s squared

4. four times y to the fourth power

5. CAR RENTAL The XYZ Car Rental Agency charges a flat rate of \$29 per day plus \$0.32 per mile driven. Write an algebraic expression for the rental cost of a car for x days that is driven y miles. (Lesson 1-1)

Evaluate each expression. (Lesson 1-2)

6. $24 \div 3 - 2 \cdot 3$

7. $5 + 2^2$

8. $4(3 + 9)$

9. $36 - 2(1 + 3)^2$

10. $\dfrac{40 - 2^3}{4 + 3(2^2)}$

11. AMUSEMENT PARK The costs of tickets to a local amusement park are shown. Write and evaluate an expression to find the total cost for 5 adults and 8 children. (Lesson 1-2)

ADMIT ONE | Adult **$45** Children **$25** | ADMIT ONE

12. MULTIPLE CHOICE Write an algebraic expression to represent the perimeter of the rectangle shown below. Then evaluate it to find the perimeter when $w = 8$ cm. (Lesson 1-2)

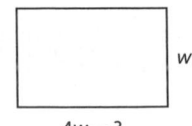

w

$4w - 3$

A 37 cm   **C** 74 cm

B 232 cm   **D** 45 cm

Evaluate each expression. Name the property used in each step. (Lesson 1-3)

13. $(8 - 2^3) + 21$

14. $3(1 \div 3) \cdot 9$

15. $[5 \div (3 \cdot 1)]\dfrac{3}{5}$

16. $18 + 35 + 32 + 15$

17. $0.25 \cdot 7 \cdot 4$

Use the Distributive Property to rewrite each expression. Then evaluate. (Lesson 1-4)

18. $3(5 + 2)$

19. $(9 - 6)12$

20. $8(7 - 4)$

Use the Distributive Property to rewrite each expression. Then simplify. (Lesson 1-4)

21. $4(x + 3)$

22. $(6 - 2y)7$

23. $-5(3m - 2)$

24. DVD SALES A video store chain has three locations. Use the information in the table below to write and evaluate an expression to estimate the total number of DVDs sold over a 4-day period. (Lesson 1-4)

Location	Daily Sales Numbers
Location 1	145
Location 2	211
Location 3	184

25. MULTIPLE CHOICE Rewrite the expression $(8 - 3p)(-2)$ using the Distributive Property. (Lesson 1-4)

F $16 - 6p$

G $-10p$

H $-16 + 6p$

J $10p$

Equations

··Then	··Now	··Why?
• You simplified expressions.	**1** Solve equations with one variable. **2** Solve equations with two variables.	• Mark's baseball team scored 3 runs in the first inning. At the top of the third inning, their score was 4. The open sentence below represents the change in their score. $$3 + r = 4$$ The solution is 1. The team got 1 run in the second inning.

 NewVocabulary
open sentence
equation
solving
solution
replacement set
set
element
solution set
identity

 Common Core State Standards

Content Standards
A.CED.1 Create equations and inequalities in one variable and use them to solve problems.

A.REI.3 Solve linear equations and inequalities in one variable, including equations with coefficients represented by letters.

Mathematical Practices
3 Construct viable arguments and critique the reasoning of others.

1 Solve Equations A mathematical statement that contains algebraic expressions and symbols is an **open sentence**. A sentence that contains an equals sign, =, is an **equation**.

$$\boxed{\text{expression}} \longrightarrow 3x + 7 \qquad 3x + 7 = 13 \longleftarrow \boxed{\text{equation}}$$

Finding a value for a variable that makes a sentence true is called **solving** the open sentence. This replacement value is a **solution**.

A set of numbers from which replacements for a variable may be chosen is called a **replacement set**. A **set** is a collection of objects or numbers that is often shown using braces. Each object or number in the set is called an **element**, or member. A **solution set** is the set of elements from the replacement set that make an open sentence true.

Example 1 Use a Replacement Set

Find the solution set of the equation $2q + 5 = 13$ if the replacement set is {2, 3, 4, 5, 6}.

Use a table to solve. Replace q in $2q + 5 = 13$ with each value in the replacement set.

Since the equation is true when $q = 4$, the solution of $2q + 5 = 13$ is $q = 4$.

The solution set is {4}.

q	$2q + 5 = 13$	True or False?
2	$2(2) + 5 = 13$	false
3	$2(3) + 5 = 13$	false
4	$2(4) + 5 = 13$	true
5	$2(5) + 5 = 13$	false
6	$2(6) + 5 = 13$	false

▶ **GuidedPractice**

Find the solution set for each equation if the replacement set is {0, 1, 2, 3}.

1A. $8m - 7 = 17$

1B. $28 = 4(1 + 3d)$

Paul Rees/ImageState

You can often solve an equation by applying the order of operations.

Standardized Test Example 2 Apply the Order of Operations

Solve $6 + (5^2 - 5) \div 2 = p$.

A 3   B 6   C 13   D 16

Read the Test Item

You need to apply the order of operations to the expression in order to solve for p.

Solve the Test Item

$6 + (5^2 - 5) \div 2 = p$	Original equation
$6 + (25 - 5) \div 2 = p$	Evaluate powers.
$6 + 20 \div 2 = p$	Subtract 5 from 25.
$6 + 10 = p$	Divide 20 by 2.
$16 = p$	Add.

The correct answer is D.

Test-TakingTip

Rewrite the Equation
If you are allowed to write in your testing booklet, it can be helpful to rewrite the equation with simplified terms.

▶ **Guided**Practice

2. Solve $t = 9^2 \div (5 - 2)$.

   F 3   G 6   H 14.2   J 27

Some equations have a unique solution. Other equations do not have a solution.

Example 3 Solutions of Equations

Solve each equation.

a. $7 - (4^2 - 10) + n = 10$

Simplify the equation first and then look for a solution.

$7 - (4^2 - 10) + n = 10$	Original equation
$7 - (16 - 10) + n = 10$	Evaluate powers.
$7 - 6 + n = 10$	Subtract 10 from 16.
$1 + n = 10$	Subtract 6 from 7.

The only value for n that makes the equation true is 9. Therefore, this equation has a unique solution of 9.

b. $n(3 + 2) + 6 = 5n + (10 - 3)$

$n(3 + 2) + 6 = 5n + (10 - 3)$	Original equation
$n(5) + 6 = 5n + (10 - 3)$	Add $3 + 2$.
$n(5) + 6 = 5n + 7$	Subtract 3 from 10.
$5n + 6 = 5n + 7$	Commutative ($\times$)

No matter what real value is substituted for n, the left side of the equation will always be one less than the right side. So, the equation will never be true. Therefore, there is no solution of this equation.

StudyTip

Guess and Check When the solution to an equation is not easy to see, substitute values for x and test the equation. Continue to test values until you get a true statement. For example, if $3x + 16 = 73$, test values for x.

$3(10) + 16 = 48$ too low

$3(20) + 16 = 76$ too high

$3(19) + 16 = 73$ ✓

▶ **Guided**Practice

3A. $(18 + 4) + m = (5 - 3)m$ **3B.** $8 \cdot 4 \cdot k + 9 \cdot 5 = (36 - 4)k - (2 \cdot 5)$

An equation that is true for every value of the variable is called an **identity**.

Example 4 Identities

Solve $(2 \cdot 5 - 8)(3h + 6) = [(2h + h) + 6]2$.

$(2 \cdot 5 - 8)(3h + 6) = [(2h + h) + 6]2$	Original Equation
$(10 - 8)(3h + 6) = [(2h + h) + 6]2$	Multiply $2 \cdot 5$.
$2(3h + 6) = [(2h + h) + 6]2$	Subtract 8 from 10.
$6h + 12 = [(2h + h) + 6]2$	Distributive Property
$6h + 12 = [3h + 6]2$	Add $2h + h$.
$6h + 12 = 6h + 12$	Distributive Property

No matter what value is substituted for h, the left side of the equation will always be equal to the right side. So, the equation will always be true. Therefore, the solution of this equation could be any real number.

GuidedPractice

Solve each equation.

4A. $12(10 - 7) + 9g = g(2^2 + 5) + 36$ **4B.** $2d + (2^3 - 5) = 10(5 - 2) + d(12 \div 6)$

4C. $3(b + 1) - 5 = 3b - 2$ **4D.** $5 - \frac{1}{2}(c - 6) = 4$

2 **Solve Equations with Two Variables** Some equations contain two variables. It is often useful to make a table of values and use substitution to find the corresponding values of the second variable.

Example 5 Equations Involving Two Variables

MOVIE RENTALS Mr. Hernandez pays $10 each month for movies delivered by mail. He can also rent movies in the store for $1.50 per title. Write and solve an equation to find the total amount Mr. Hernandez spends this month if he rents 3 movies from the store.

The cost of the movie plan is a flat rate. The variable is the number of movies he rents from the store. The total cost is the price of the plan plus $1.50 times the number of movies from the store. Let C be the total cost and m be the number of movies.

$C = 1.50m + 10$	Original equation
$= 1.50(3) + 10$	Substitute 3 for m.
$= 4.50 + 10$	Multiply.
$= 14.50$	

Mr. Hernandez spends $14.50 on movie rentals in one month.

GuidedPractice

5. TRAVEL Amelia drives an average of 65 miles per hour. Write and solve an equation to find the time it will take her to drive 36 miles.

Example 1 Find the solution set of each equation if the replacement set is {11, 12, 13, 14, 15}.

1. $n + 10 = 23$

2. $7 = \dfrac{c}{2}$

3. $29 = 3x - 7$

4. $(k - 8)12 = 84$

Example 2 **5. MULTIPLE CHOICE** Solve $\dfrac{d + 5}{10} = 2$.

     A 10   **B** 15   **C** 20   **D** 25

Examples 3–4 Solve each equation.

6. $x = 4(6) + 3$

7. $14 - 82 = w$

8. $5 + 22a = 2 + 10 \div 2$

9. $(2 \cdot 5) + \dfrac{c^3}{3} = c^3 \div (1^5 + 2) + 10$

Example 5 **10. RECYCLING** San Francisco has a recycling facility that accepts unused paint. Volunteers blend and mix the paint and give it away in 5-gallon buckets. Write and solve an equation to find the number of buckets of paint given away from the 30,000 gallons that are donated.

Practice and Problem Solving   Extra Practice is on page R1.

Example 1 Find the solution set of each equation if the replacement sets are y: {1, 3, 5, 7, 9} and z: {10, 12, 14, 16, 18}.

11. $z + 10 = 22$

12. $52 = 4z$

13. $\dfrac{15}{y} = 3$

14. $17 = 24 - y$

15. $2z - 5 = 27$

16. $4(y + 1) = 40$

17. $22 = \dfrac{60}{y} + 2$

18. $111 = z^2 + 11$

Examples 2–4 Solve each equation.

19. $a = 32 - 9(2)$

20. $w = 56 \div (2^2 + 3)$

21. $\dfrac{27 + 5}{16} = g$

22. $\dfrac{12 \cdot 5}{15 - 3} = y$

23. $r = \dfrac{9(6)}{(8 + 1)3}$

24. $a = \dfrac{4(14 - 1)}{3(6) - 5} + 7$

25. $(4 - 2^2 + 5)w = 25$

26. $7 + x - (3 + 32 \div 8) = 3$

27. $3^2 - 2 \cdot 3 + u = (3^3 - 3 \cdot 8)(2) + u$

28. $(3 \cdot 6 \div 2)v + 10 = 3^2 v + 9$

29. $6k + (3 \cdot 10 - 8) = (2 \cdot 3)k + 22$

30. $(3 \cdot 5)t + (21 - 12) = 15t + 3^2$

31 $(2^4 - 3 \cdot 5)q + 13 = (2 \cdot 9 - 4^2)q + \left(\dfrac{3 \cdot 4}{12} - 1\right)$

32. $\dfrac{3 \cdot 22}{18 + 4}r - \left(\dfrac{4^2}{9 + 7} - 1\right) = r + \left(\dfrac{8 \cdot 9}{3} \div 3\right)$

33. SCHOOL A conference room can seat a maximum of 85 people. The principal and two counselors need to meet with the school's juniors to discuss college admissions. If each student must bring a parent with them, how many students can attend each meeting? Assume that each student has a unique set of parents.

34. CCSS MODELING The perimeter of a regular octagon is 128 inches. Find the length of each side.

Example 5

35 **SPORTS** A 200-pound athlete who trains for four hours per day requires 2836 Calories for basic energy requirements. During training, the same athlete requires an additional 3091 Calories for extra energy requirements. Write an equation to find C, the total daily Calorie requirement for this athlete. Then solve the equation.

36. ENERGY An electric generator can power 3550 watts of electricity. Write and solve an equation to find how many 75-watt light bulbs a generator could power.

Make a table of values for each equation if the replacement set is $\{-2, -1, 0, 1, 2\}$.

37. $y = 3x - 2$

38. $3.25x + 0.75 = y$

Solve each equation using the given replacement set.

39. $t - 13 = 7; \{10, 13, 17, 20\}$

40. $14(x + 5) = 126; \{3, 4, 5, 6, 7\}$

41. $22 = \frac{n}{3}; \{62, 64, 66, 68, 70\}$

42. $35 = \frac{g - 8}{2}; \{78, 79, 80, 81\}$

Solve each equation.

43. $\frac{3(9) - 2}{1 + 4} = d$

44. $j = 15 \div 3 \cdot 5 - 4^2$

45. $c + (3^2 - 3) = 21$

46. $(3^3 - 3 \cdot 9) + (7 - 2^2)b = 24b$

47. CCSS SENSE-MAKING Blood flow rate can be expressed as $F = \frac{p_1 - p_2}{r}$, where F is the flow rate, p_1 and p_2 are the initial and final pressure exerted against the blood vessel's walls, respectively, and r is the resistance created by the size of the vessel.

a. Write and solve an equation to determine the resistance of the blood vessel for an initial pressure of 100 millimeters of mercury, a final pressure of 0 millimeters of mercury, and a flow rate of 5 liters per minute.

b. Use the equation to complete the table below.

Initial Pressure p_1 (mm Hg)	Final Pressure p_2 (mm Hg)	Resistance r (mm Hg/L/min)	Blood Flow Rate F (L/min)
100	0		5
100	0	30	
	5	40	4
90		10	6

Determine whether the given number is a solution of the equation.

48. $x + 6 = 15; 9$

49. $12 + y = 26; 14$

50. $2t - 10 = 4; 3$

51. $3r + 7 = -5; 2$

52. $6 + 4m = 18; 3$

53. $-5 + 2p = -11; -3$

54. $\frac{q}{2} = 20; 10$

55. $\frac{w - 4}{5} = -3; -11$

56. $\frac{g}{3} - 4 = 12; 48$

Make a table of values for each equation if the replacement set is $\{-2, -1, 0, 1, 2\}$.

57. $y = 3x + 5$

58. $-2x - 3 = y$

59. $y = \frac{1}{2}x + 2$

60. $4.2x - 1.6 = y$

61. GEOMETRY The length of a rectangle is 2 inches greater than the width. The length of the base of an isosceles triangle is 12 inches, and the lengths of the other two sides are 1 inch greater than the width of the rectangle.

a. Draw a picture of each figure and label the dimensions.

b. Write two expressions to find the perimeters of the rectangle and triangle.

c. Find the width of the rectangle if the perimeters of the figures are equal.

62. CONSTRUCTION The construction of a building requires 10 tons of steel per story.

    a. Define a variable and write an equation for the number of tons of steel required if the building has 15 stories.

    b. How many tons of steel are needed?

63 ⟳ **MULTIPLE REPRESENTATIONS** In this problem, you will further explore writing equations.

    a. Concrete Use centimeter cubes to build a tower similar to the one shown at the right.

    b. Tabular Copy and complete the table shown below. Record the number of layers in the tower and the number of cubes used in the table.

Layers	1	2	3	4	5	6	7
Cubes	?	?	?	?	?	?	?

    c. Analytical As the number of layers in the tower increases, how does the number of cubes in the tower change?

    d. Algebraic Write a rule that gives the number of cubes in terms of the number of layers in the tower.

H.O.T. Problems   Use Higher-Order Thinking Skills

64. REASONING Compare and contrast an expression and an equation.

65. OPEN ENDED Write an equation that is an identity.

66. REASONING Explain why an open sentence always has at least one variable.

67. CCSS CRITIQUE Tom and Li-Cheng are solving the equation $x = 4(3 - 2) + 6 \div 8$. Is either of them correct? Explain your reasoning.

Tom
$x = 4(3 - 2) + 6 \div 8$
$= 4(1) + 6 \div 8$
$= 4 + 6 \div 8$
$= 4 + \dfrac{6}{8}$
$= 4\dfrac{3}{4}$

Li-Cheng
$x = 4(3 - 2) + 6 \div 8$
$= 4(1) + 6 \div 8$
$= 4 + 6 \div 8$
$= 10 \div 8$
$= \dfrac{5}{4}$

68. CHALLENGE Find all of the solutions of $x^2 + 5 = 30$.

69. OPEN ENDED Write an equation that involves two or more operations with a solution of -7.

70. WRITING IN MATH Explain how you can determine that an equation has no real numbers as a solution. How can you determine that an equation has all real numbers as solutions?

71. Which of the following is *not* an equation?

A $y = 6x - 4$

B $\dfrac{a + 4}{2} = \dfrac{1}{4}$

C $(4 \cdot 3b) + (8 \div 2c)$

D $55 = 6 + d^2$

72. SHORT RESPONSE The expected attendance for the Drama Club production is 65% of the student body. If the student body consists of 300 students, how many students are expected to attend?

73. GEOMETRY A speedboat and a sailboat take off from the same port. The diagram shows their travel. What is the distance between the boats?

F 12 mi

G 15 mi

H 18 mi

J 24 mi

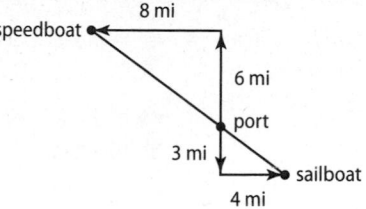

74. Michelle can read 1.5 pages per minute. How many pages can she read in two hours?

A 90 pages   **C** 150 pages

B 120 pages   **D** 180 pages

Spiral Review

75. ZOO A zoo has about 500 children and 750 adults visit each day. Write an expression to represent about how many visitors the zoo will have over a month. (Lesson 1-4)

Find the value of p in each equation. Then name the property that is used. (Lesson 1-3)

76. $7.3 + p = 7.3$

77. $12p = 1$

78. $1p = 4$

79. MOVING BOXES The figure shows the dimensions of the boxes Steve uses to pack. How many cubic inches can each box hold? (Lesson 0-9)

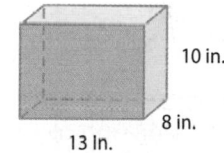

Express each percent as a fraction. (Lesson 0-6)

80. 35%

81. 15%

82. 28%

For each problem, determine whether you need an estimate or an exact answer. Then solve. (Lessons 0-6 and 0-1)

83. TRAVEL The distance from Raleigh, North Carolina, to Philadelphia, Pennsylvania, is approximately 428 miles. The average gas mileage of José's car is 45 miles per gallon. About how many gallons of gas will be needed to make the trip?

84. PART-TIME JOB An employer pays $8.50 per hour. If 20% of pay is withheld for taxes, what are the take-home earnings from 28 hours of work?

Skills Review

Find each sum or difference.

85. $1.14 + 5.6$

86. $4.28 - 2.4$

87. $8 - 6.35$

88. $\dfrac{4}{5} + \dfrac{1}{6}$

89. $\dfrac{2}{7} + \dfrac{3}{4}$

90. $\dfrac{6}{8} - \dfrac{1}{2}$

Relations

- You solved equations with one or two variables.

1 Represent relations.

2 Interpret graphs of relations.

- The deeper in the ocean you are, the greater pressure is on your body. This is because there is more water over you. The force of gravity pulls the water weight down, creating a greater pressure.

The equation that relates the total pressure of the water to the depth is $P = rgh$, where

$P =$ the pressure,
$r =$ the density of water,
$g =$ the acceleration due to gravity, and
$h =$ the height of water above you.

 NewVocabulary
coordinate system
coordinate plane
x- and y-axes
origin
ordered pair
x- and y-coordinates
relation
mapping
domain
range
independent variable
dependent variable

 Common Core State Standards

Content Standards
A.REI.10 Understand that the graph of an equation in two variables is the set of all its solutions plotted in the coordinate plane, often forming a curve (which could be a line).

F.IF.1 Understand that a function from one set (called the domain) to another set (called the range) assigns to each element of the domain exactly one element of the range. If f is a function and x is an element of its domain, then $f(x)$ denotes the output of f corresponding to the input x. The graph of f is the graph of the equation $y = f(x)$.

Mathematical Practices
1 Make sense of problems and persevere in solving them.

1 Represent a Relation This relationship between the depth and the pressure exerted can be represented by a line on a coordinate grid.

A **coordinate system** is formed by the intersection of two number lines, the *horizontal axis* and the *vertical axis*.

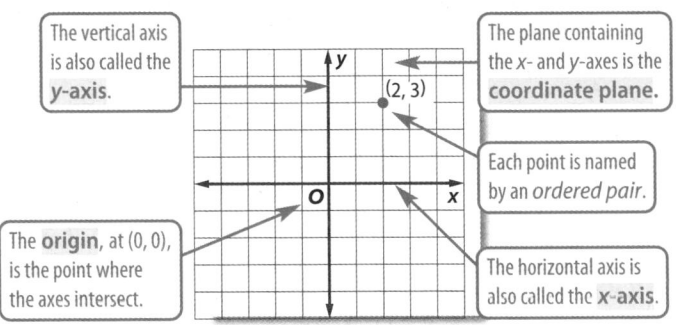

The vertical axis is also called the *y*-axis.

The plane containing the *x*- and *y*-axes is the **coordinate plane**.

(2, 3)

Each point is named by an *ordered pair*.

The **origin**, at (0, 0), is the point where the axes intersect.

The horizontal axis is also called the **x-axis**.

A point is represented on a graph using ordered pairs.

- An **ordered pair** is a set of numbers, or *coordinates*, written in the form (x, y).

- The x-value, called the **x-coordinate**, represents the horizontal placement of the point.

- The y-value, or **y-coordinate**, represents the vertical placement of the point.

A set of ordered pairs is called a **relation**. A relation can be represented in several different ways: as an equation, in a graph, with a table, or with a mapping.

A **mapping** illustrates how each element of the *domain* is paired with an element in the *range*. The set of the first numbers of the ordered pairs is the **domain**. The set of second numbers of the ordered pairs is the **range** of the relation. This mapping represents the ordered pairs $(-2, 4), (-1, 4), (0, 6) (1, 8),$ and $(2, 8)$.

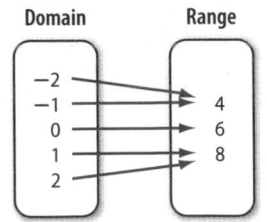

Domain   Range

−2
−1   4
0   6
1   8
2

Study the different representations of the same relation below.

Ordered Pairs	Table	Graph	Mapping

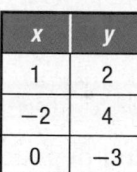

(1, 2)
(−2, 4)
(0, −3)

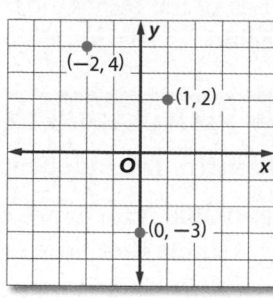

x	y
1	2
−2	4
0	−3

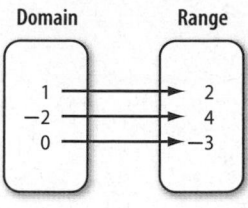

The x-values of a relation are members of the domain and the y-values of a relation are members of the range. In the relation above, the domain is {−2, 1, 0} and the range is {−3, 2, 4}.

Example 1 Representations of a Relation

a. Express {(2, 5), (−2, 3), (5, −2), (−1, −2)} as a table, a graph, and a mapping.

Table
Place the x-coordinates into the first column of the table. Place the corresponding y-coordinates in the second column of the table.

x	y
2	5
−2	3
5	−2
−1	−2

Graph
Graph each ordered pair on a coordinate plane.

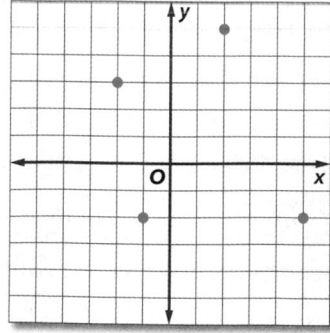

Mapping
List the x-values in the domain and the y-values in the range. Draw arrows from the x-values in the domain to the corresponding y-values in the range.

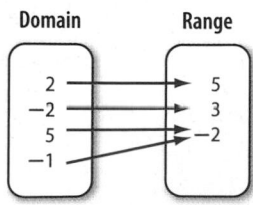

b. Determine the domain and the range of the relation.

The domain of the relation is {2, −2, 5, −1}. The range of the relation is {5, 3, −2}.

▶ **Guided**Practice

1A. Express {(4, −3), (3, 2), (−4, 1), (0, −3)} as a table, graph, and mapping.

1B. Determine the domain and range.

In a relation, the value of the variable that determines the output is called the **independent variable**. The variable with a value that is dependent on the value of the independent variable is called the **dependent variable**. The domain contains values of the independent variable. The range contains the values of the dependent variable.

Real-World Example 2 Independent and Dependent Variables

Identify the independent and dependent variables for each relation.

a. DANCE **The dance committee is selling tickets to the Fall Ball. The more tickets that they sell, the greater the amount of money they can spend for decorations.**

The number of tickets sold is the independent variable because it is unaffected by the money spent on decorations. The money spent on decorations is the dependent variable because it depends on the number of tickets sold.

b. MOVIES **Generally, the average price of going to the movies has steadily increased over time.**

Time is the independent variable because it is unaffected by the cost of attending the movies. The price of going to the movies is the dependent variable because it is affected by time.

GuidedPractice

Identify the independent and dependent variables for each relation.

2A. The air pressure inside a tire increases with the temperature.

2B. As the amount of rain decreases, so does the water level of the river.

Real-World Link

In 1948, a movie ticket cost $0.36. In 2008, the average ticket price in the United States was $7.18.

Source: National Association of Theatre Owners

2 Graphs of a Relation
A relation can be graphed without a scale on either axis. These graphs can be interpreted by analyzing their shape.

Example 3 Analyze Graphs

The graph represents the distance Francesca has ridden on her bike. Describe what happens in the graph.

As time increases, the distance increases until the graph becomes a horizontal line.

So, time is increasing but the distance remains constant. At this section Francesca stopped. Then she continued to ride her bike.

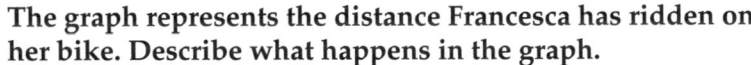

Bike Ride

GuidedPractice

Describe what is happening in each graph.

3A. **Driving to School**

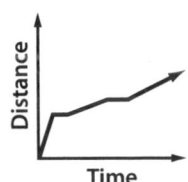

3B. **Change in Income**

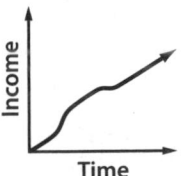

Erik Dreyer/Taxi/Getty Images

Example 1 Express each relation as a table, a graph, and a mapping. Then determine the domain and range.

1. {(4, 3), (−2, 2), (5, −6)}

2. {(5, −7), (−1, 4), (0, −5), (−2, 3)}

Example 2 Identify the independent and dependent variables for each relation.

3. Increasing the temperature of a compound inside a sealed container increases the pressure inside a sealed container.

4. Mike's cell phone is part of a family plan. If he uses more minutes than his share, then there are fewer minutes available for the rest of his family.

5. Julian is buying concert tickets for himself and his friends. The more concert tickets he buys the greater the cost.

6. A store is having a sale over Labor Day weekend. The more purchases, the greater the profits.

Example 3 **CCSS MODELING** Describe what is happening in each graph.

7. The graph represents the distance the track team runs during a practice.

8. The graph represents revenues generated through an online store.

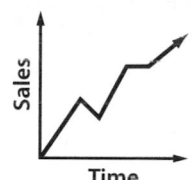

Practice and Problem Solving Extra Practice is on page R1.

Example 1 Express each relation as a table, a graph, and a mapping. Then determine the domain and range.

9. {(0, 0), (−3, 2), (6, 4), (−1, 1)}

10. {(5, 2), (5, 6), (3, −2), (0, −2)}

11. {(6, 1), (4, −3), (3, 2), (−1, −3)}

12. {(−1, 3), (3, −6), (−1, −8), (−3, −7)}

13. {(6, 7), (3, −2), (8, 8), (−6, 2), (2, −6)}

14. {(4, −3), (1, 3), (7, −2), (2, −2), (1, 5)}

Example 2 Identify the independent and dependent variables for each relation.

15 The Spanish classes are having a fiesta lunch. Each student that attends is to bring a Spanish side dish or dessert. The more students that attend, the more food there will be.

16. The faster you drive your car, the longer it will take to come to a complete stop.

Example 3 **CCSS MODELING** Describe what is happening in each graph.

17. The graph represents the height of a bungee jumper.

18. The graph represents the sales of lawn mowers.

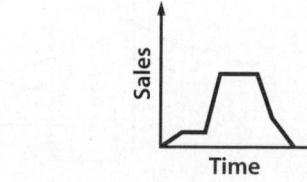

CCSS MODELING Describe what is happening in each graph.

19 The graph represents the value of a rare baseball card.

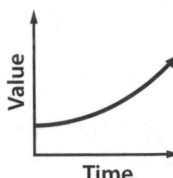

20. The graph represents the distance covered on an extended car ride.

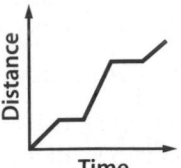

For Exercises 21–23, use the graph at the right.

21. Name the ordered pair at point *A* and explain what it represents.

22. Name the ordered pair at point *B* and explain what it represents.

23. Identify the independent and dependent variables for the relation.

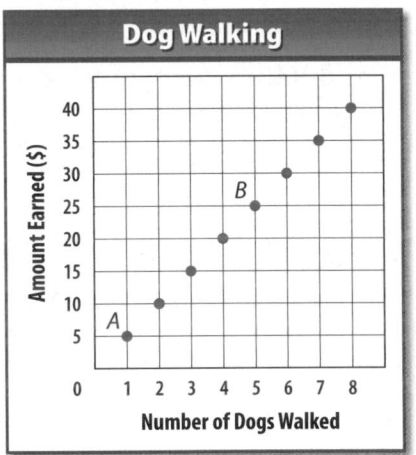

For Exercises 24–26, use the graph at the right.

24. Name the ordered pair at point *C* and explain what it represents.

25. Name the ordered pair at point *D* and explain what it represents.

26. Identify the independent and dependent variables.

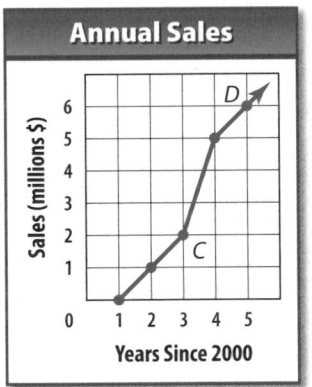

Express each relation as a set of ordered pairs. Describe the domain and range.

27.

Buying Aquarium Fish	
Number of Fish	Total Cost
1	$2.50
2	$4.50
5	$10.50
8	$16.50

28.

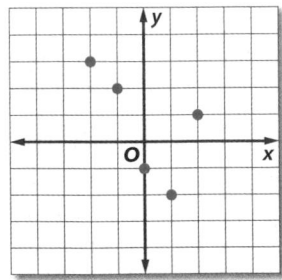

Express the relation in each table, mapping, or graph as a set of ordered pairs.

29.

x	y
4	−1
8	9
−2	−6
7	−3

30.

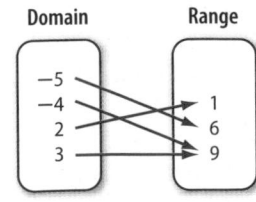

31.

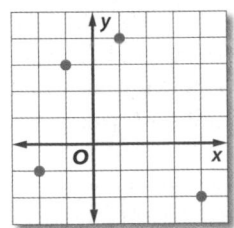

32. SPORTS In a triathlon, athletes swim 2.4 miles, bicycle 112 miles, and run 26.2 miles. Their total time includes transition time from one activity to the next. Which graph best represents a participant in a triathlon? Explain.

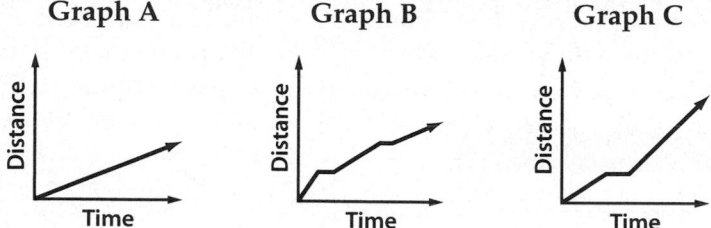

Graph A   Graph B   Graph C

Draw a graph to represent each situation.

33. ANTIQUES A grandfather clock that is over 100 years old has increased in value from when it was first purchased.

34. CAR A car depreciates in value. The value decreases quickly in the first few years.

35. REAL ESTATE A house typically increases in value over time.

36. EXERCISE An athlete alternates between running and walking during a workout.

37 PHYSIOLOGY A typical adult has about 2 pounds of water for every 3 pounds of body weight. This can be represented by the equation $w = 2\left(\dfrac{b}{3}\right)$, where w is the weight of water in pounds and b is the body weight in pounds.

   a. Make a table to show the relation between body and water weight for people weighing 100, 105, 110, 115, 120, 125, and 130 pounds. Round to the nearest tenth if necessary.

   b. What are the independent and dependent variables?

   c. State the domain and range, and then graph the relation.

   d. Reverse the independent and dependent variables. Graph this relation. Explain what the graph indicates in this circumstance.

H.O.T. Problems   Use Higher-Order Thinking Skills

38. OPEN ENDED Describe a real-life situation that can be represented using a relation and discuss how one of the quantities in the relation depends on the other. Then represent the relation in three different ways.

39. CHALLENGE Describe a real-world situation where it is reasonable to have a negative number included in the domain or range.

40. CCSS PRECISION Compare and contrast dependent and independent variables.

41. CHALLENGE The table presents a relation. Graph the ordered pairs. Then reverse the y-coordinate and the x-coordinate in each ordered pair. Graph these ordered pairs on the same coordinate plane. Graph the line $y = x$. Describe the relationship between the two sets of ordered pairs.

x	y
0	1
1	3
2	5
3	7

42. WRITING IN MATH Use the data about the pressure of water on page 40 to explain the difference between dependent and independent variables.

43. A school's cafeteria employees surveyed 250 students asking what beverage they drank with lunch. They used the data to create the table below.

Beverage	Number of Students
milk	38
chocolate milk	112
juice	75
water	25

What percent of the students surveyed preferred drinking juice with lunch?

A 25%

B 30%

C 35%

D 40%

44. Which of the following is equivalent to $6(3 - g) + 2(11 - g)$?

F $2(20 - g)$

G $8(14 - g)$

H $8(5 - g)$

J $40 - g$

45. SHORT RESPONSE Grant and Hector want to build a clubhouse at the midpoint between their houses. If Grant's house is at point G and Hector's house is at point H, what will be the coordinates of the clubhouse?

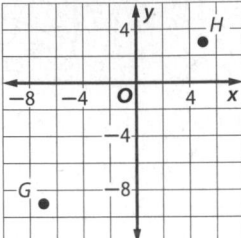

46. If $3b = 2b$, which of the following is true?

A $b = 0$

B $b = \frac{2}{3}$

C $b = 1$

D $b = \frac{3}{2}$

Solve each equation. (Lesson 1-5)

47. $6(a + 5) = 42$

48. $92 = k + 11$

49. $17 = \frac{45}{w} + 2$

50. HOT-AIR BALLOON A hot-air balloon owner charges $150 for a one-hour ride. If he gave 6 rides on Saturday and 5 rides on Sunday, write and evaluate an expression to describe his total income for the weekend. (Lesson 1-4)

51. LOLLIPOPS A bag of lollipops contains 19 cherry, 13 grape, 8 sour apple, 15 strawberry, and 9 orange flavored lollipops. What is the probability of drawing a sour apple flavored lollipop? (Lesson 0-11)

Find the perimeter of each figure. (Lesson 0-7)

52.

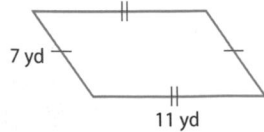

53.

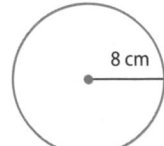

54.

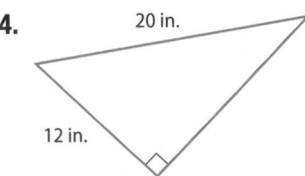

Evaluate each expression.

55. 8^2

56. $(-6)^2$

57. $(2.5)^2$

58. $(-1.8)^2$

59. $(3 + 4)^2$

60. $(1 - 4)^2$

Functions

- You solved equations with elements from a replacement set.

1 Determine whether a relation is a function.

2 Find function values.

- The distance a car travels from when the brakes are applied to the car's complete stop is the stopping distance. This includes time for the driver to react. The faster a car is traveling, the longer the stopping distance. The stopping distance is a function of the speed of the car.

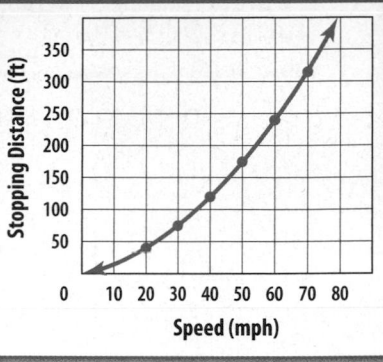

Stopping Distance of a Passenger Car

NewVocabulary
function
discrete function
continuous function
vertical line test
function notation
nonlinear function

Common Core State Standards

Content Standards

F.IF.1 Understand that a function from one set (called the domain) to another set (called the range) assigns to each element of the domain exactly one element of the range. If f is a function and x is an element of its domain, then $f(x)$ denotes the output of f corresponding to the input x. The graph of f is the graph of the equation $y = f(x)$.

F.IF.2 Use function notation, evaluate functions for inputs in their domains, and interpret statements that use function notation in terms of a context.

Mathematical Practices
3 Construct viable arguments and critique the reasoning of others.

1 **Identify Functions** A **function** is a relationship between input and output. In a function, there is exactly one output for each input.

KeyConcept Function

Words A function is a relation in which each element of the domain is paired with *exactly* one element of the range.

Examples

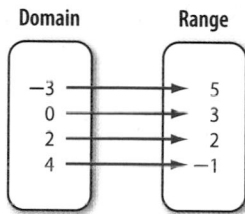

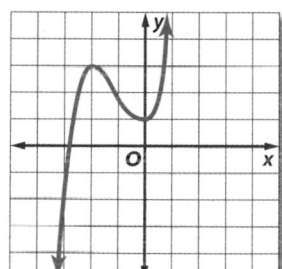

Example 1 Identify Functions

Determine whether each relation is a function. Explain.

a.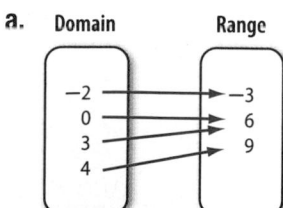

For each member of the domain, there is only one member of the range. So this mapping represents a function. It does not matter if more than one element of the domain is paired with one element of the range.

b.

Domain	1	3	5	1
Range	4	2	4	−4

The element 1 in the domain is paired with both 4 and −4 in the range. So, when x equals 1 there is more than one possible value for y. This relation is not a function.

▶ **Guided**Practice

1. {(2, 1), (3, −2), (3, 1), (2, −2)}

A graph that consists of points that are not connected is a **discrete function**. A function graphed with a line or smooth curve is a **continuous function**.

 **Real-World Example 2** Draw Graphs

ICE SCULPTING At an ice sculpting competition, each sculpture's height was measured to make sure that it was within the regulated height range of 0 to 6 feet. The measurements were as follows: Team 1, 4 feet; Team 2, 4.5 feet; Team 3, 3.2 feet; Team 4, 5.1 feet; Team 5, 4.8 feet.

a. Make a table of values showing the relation between the ice sculpting team and the height of their sculpture.

Team Number	1	2	3	4	5
Height (ft)	4	4.5	3.2	5.1	4.8

Real-World Link

The Icehotel, located in the Arctic Circle in Sweden, is a hotel made out of ice. The ice insulates the igloo-like hotel so the temperature is at least −8°C.

Source: Icehotel

b. Determine the domain and range of the function.

The domain of the function is {1, 2, 3, 4, 5} because this set represents values of the independent variable. It is unaffected by the heights.

The range of the function is {4, 4.5, 3.2, 5.1, 4.8} because this set represents values of the dependent variable. This value depends on the team number.

c. Write the data as a set of ordered pairs. Then graph the data.

Use the table. The team number is the independent variable and the height of the sculpture is the dependent variable. Therefore, the ordered pairs are (1, 4), (2, 4.5), (3, 3.2), (4, 5.1), and (5, 4.8).

Because the team numbers and their corresponding heights cannot be between the points given, the points should not be connected.

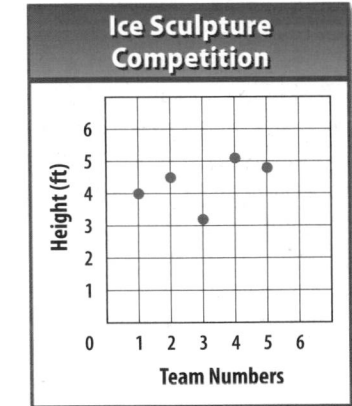

Ice Sculpture Competition

d. State whether the function is *discrete* or *continuous*. Explain your reasoning.

Because the points are not connected, the function is discrete.

▶ **Guided**Practice

2. A bird feeder will hold up to 3 quarts of seed. The feeder weighs 2.3 pounds when empty and 13.4 pounds when full.

A. Make a table that shows the bird feeder with 0, 1, 2, and 3 quarts of seed in it weighing 2.3, 6, 9.7, 13.4 pounds respectively.

B. Determine the domain and range of the function.

C. Write the data as a set of ordered pairs. Then graph the data.

D. State whether the function is *discrete* or *continuous*. Explain your reasoning.

You can use the **vertical line test** to see if a graph represents a function. If a vertical line intersects the graph more than once, then the graph is not a function. Otherwise, the relation is a function.

Function	Not a Function	Function

Recall that an equation is a representation of a relation. Equations can also represent functions. Every solution of the equation is represented by a point on a graph. The graph of an equation is the set of all its solutions, which often forms a curve or a line.

Example 3 Equations as Functions

Determine whether $-3x + y = 8$ is a function.

First make a table of values. Then graph the equation.

x	−1	0	1	2
y	5	4.5	11	14

Connect the points with a smooth graph to represent all of the solutions of the equation. The graph is a line. To use the vertical line test, place a pencil at the left of the graph to represent a vertical line. Slowly move the pencil across the graph.

For any value of x, the vertical line passes through no more than one point on the graph. So, the graph and the equation represent a function.

GuidedPractice **Determine whether each relation is a function.**

3A. $4x = 8$ **3B.** $4x = y + 8$

A function can be represented in different ways.

ConceptSummary Representations of a Function

Table	Mapping	Equation	Graph
<table>	Domain → Range	$f(x) = \frac{1}{2}x^2 - 1$	

x	y
−2	1
0	−1
2	1

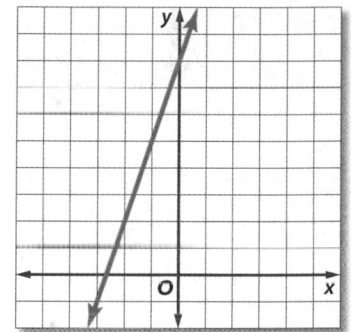

2 Find Function Values Equations that are functions can be written in a form called **function notation**. For example, consider $y = 3x - 8$.

Equation	Function Notation
$y = 3x - 8$	$f(x) = 3x - 8$

In a function, x represents the elements of the domain, and $f(x)$ represents the elements of the range. The graph of $f(x)$ is the graph of the equation $y = f(x)$. Suppose you want to find the value in the range that corresponds to the element 5 in the domain. This is written $f(5)$ and is read f of 5. The value $f(5)$ is found by substituting 5 for x in the equation.

Example 4 Function Values

For $f(x) = -4x + 7$, find each value.

a. $f(2)$

$$f(2) = -4(2) + 7 \qquad x = 2$$
$$= -8 + 7 \qquad \text{Multiply.}$$
$$= -1 \qquad \text{Add.}$$

b. $f(-3) + 1$

$$f(-3) + 1 = [-4(-3) + 7] + 1 \qquad x = -3$$
$$= 19 + 1 \qquad \text{Simplify.}$$
$$= 20 \qquad \text{Add.}$$

GuidedPractice

For $f(x) = 2x - 3$, find each value.

4A. $f(1)$ **4B.** $6 - f(5)$

4C. $f(-2)$ **4D.** $f(-1) + f(2)$

A function with a graph that is not a straight line is a **nonlinear function**.

Example 5 Nonlinear Function Values

If $h(t) = -16t^2 + 68t + 2$, find each value.

a. $h(4)$

$$h(4) = -16(4)^2 + 68(4) + 2 \qquad \text{Replace } t \text{ with 4.}$$
$$= -256 + 272 + 2 \qquad \text{Multiply.}$$
$$= 18 \qquad \text{Add.}$$

b. $2[h(g)]$

$$2[h(g)] = 2[-16(g)^2 + 68(g) + 2] \qquad \text{Replace } t \text{ with } g.$$
$$= 2(-16g^2 + 68g + 2) \qquad \text{Simplify.}$$
$$= -32g^2 + 136g + 4 \qquad \text{Distributive Property}$$

GuidedPractice

If $f(t) = 2t^3$, find each value.

5A. $f(4)$ **5B.** $3[f(t)] + 2$

5C. $f(-5)$ **5D.** $f(-3) - f(1)$

Examples 1, 3 Determine whether each relation is a function. Explain.

1.

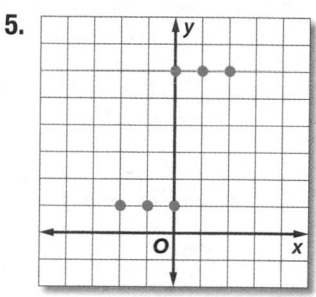

2.

Domain	Range
2	6
5	7
6	9
6	10

3. {(2, 2), (−1, 5), (5, 2), (2, −4)}

4. $y = \frac{1}{2}x - 6$

5.

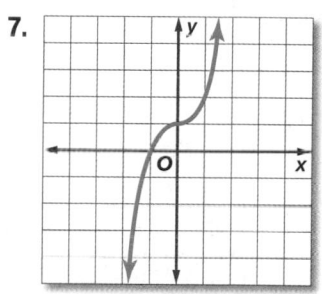

6.

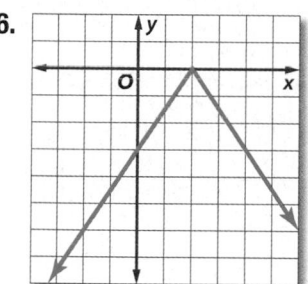

7.

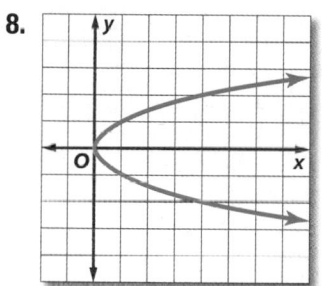

8.

Example 2

9. SCHOOL ENROLLMENT The table shows the total enrollment in U.S. public schools.

School Year	2004–05	2005–06	2006–07	2007–08
Enrollment (in thousands)	48,560	48,710	48,948	49,091

Source: *The World Almanac*

a. Write a set of ordered pairs representing the data in the table if x is the number of school years since 2004–2005.

b. Draw a graph showing the relationship between the year and enrollment.

c. Describe the domain and range of the data.

10. CCSS REASONING The cost of sending cell phone pictures is given by $y = 0.25x$, where x is the number of pictures that you send and y is the cost in dollars.

a. Write the equation in function notation. Interpret the function in terms of the context.

b. Find $f(5)$ and $f(12)$. What do these values represent?

c. Determine the domain and range of this function.

Examples 4–5 If $f(x) = 6x + 7$ and $g(x) = x^2 - 4$, find each value.

11. $f(-3)$

12. $f(m)$

13. $f(r - 2)$

14. $g(5)$

15. $g(a) + 9$

16. $g(-4t)$

17. $f(q + 1)$

18. $f(2) + g(2)$

19. $g(-b)$

Example 1 **Determine whether each relation is a function. Explain.**

20.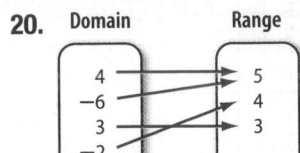

21.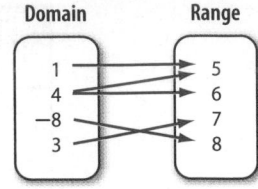

22.

Domain	Range
4	6
−5	3
6	−3
−5	5

23.

Domain	Range
−4	2
3	−5
4	2
9	−7
−3	−5

24.

25.

Example 2 26. **CCSS SENSE-MAKING** The table shows the median home prices in the United States, from 2007 to 2009.

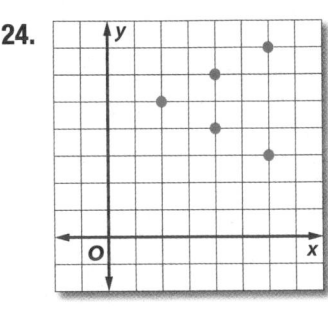

Year	Median Home Price (S)
2007	234,300
2008	213,200
2009	212,200

    a. Write a set of ordered pairs representing the data in the table.

    b. Draw a graph showing the relationship between the year and price.

    c. What is the domain and range for this data?

Example 3 **Determine whether each relation is a function.**

    27. $\{(5, -7), (6, -7), (-8, -1), (0, -1)\}$ 28. $\{(4, 5), (3, -2), (-2, 5), (4, 7)\}$

    29. $y = -8$ 30. $x = 15$

    31. $y = 3x - 2$ 32. $y = 3x + 2y$

Examples 4–5 **If $f(x) = -2x - 3$ and $g(x) = x^2 + 5x$, find each value.**

    33. $f(-1)$ 34. $f(6)$ 35. $g(2)$

    36. $g(-3)$ 37. $g(-2) + 2$ 38. $f(0) - 7$

    39. $f(4y)$ 40. $g(-6m)$ 41. $f(c - 5)$

    42. $f(r + 2)$ 43. $5[f(d)]$ 44. $3[g(n)]$

    45 **EDUCATION** The average national math test scores $f(t)$ for 17-year-olds can be represented as a function of the national science scores t by $f(t) = 0.8t + 72$.

    a. Graph this function. Interpret the function in terms of the context.

    b. What is the science score that corresponds to a math score of 308?

    c. What is the domain and range of this function?

Determine whether each relation is a function.

46.
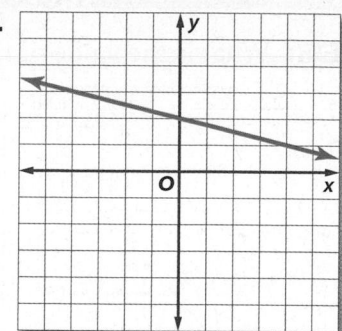

47
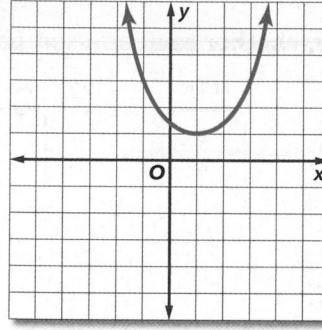

48. **BABYSITTING** Christina earns $7.50 an hour babysitting.

   a. Write an algebraic expression to represent the money Christina will earn if she works h hours.

   b. Choose five values for the number of hours Christina can babysit. Create a table with h and the amount of money she will make during that time.

   c. Use the values in your table to create a graph.

   d. Does it make sense to connect the points in your graph with a line? Why or why not?

H.O.T. Problems   Use Higher-Order Thinking Skills

49. **OPEN ENDED** Write a set of three ordered pairs that represent a function. Choose another display that represents this function.

50. **REASONING** The set of ordered pairs {(0, 1), (3, 2), (3, −5), (5, 4)} represents a relation between x and y. Graph the set of ordered pairs. Determine whether the relation is a function. Explain.

51. **CHALLENGE** Consider $f(x) = -4.3x - 2$. Write $f(g + 3.5)$ and simplify by combining like terms.

52. **WRITE A QUESTION** A classmate graphed a set of ordered pairs and used the vertical line test to determine whether it was a function. Write a question to help her decide if the same strategy can be applied to a mapping.

53. **CCSS PERSEVERANCE** If $f(3b - 1) = 9b - 1$, find one possible expression for $f(x)$.

54. **ERROR ANALYSIS** Corazon thinks $f(x)$ and $g(x)$ are representations of the same function. Maggie disagrees. Who is correct? Explain your reasoning.

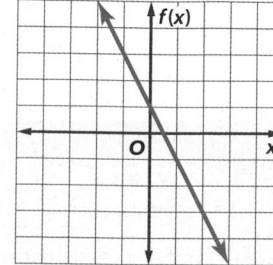

x	$g(x)$
−1	1
0	−1
1	−3
2	−5
3	−7

55. **WRITING IN MATH** How can you determine whether a relation represents a function?

56. Which point on the number line represents a number whose square is less than itself?

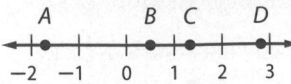

A A **C** C

B B **D** D

57. Determine which of the following relations is a function.

F $\{(-3, 2), (4, 1), (-3, 5)\}$

G $\{(2, -1), (4, -1), (2, 6)\}$

H $\{(-3, -4), (-3, 6), (8, -2)\}$

J $\{(5, -1), (3, -2), (-2, -2)\}$

58. GEOMETRY What is the value of x?

A 3 in.

B 4 in.

C 5 in.

D 6 in.

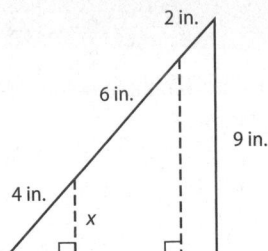

59. SHORT RESPONSE Camille made 16 out of 19 of her serves during her first volleyball game. She made 13 out of 16 of her serves during her second game. During which game did she make a greater percent of her serves?

Solve each equation. (Lesson 1-5)

60. $x = \dfrac{27 + 3}{10}$

61. $m = \dfrac{3^2 + 4}{7 - 5}$

62. $z = 32 + 4(-3)$

63. SCHOOL SUPPLIES The table shows the prices of some items Tom needs. If he needs 4 glue sticks, 10 pencils, and 4 notebooks, write and evaluate an expression to determine Tom's cost. (Lesson 1-4)

School Supplies Prices	
glue stick	$1.99
pencil	$0.25
notebook	$1.85

Write a verbal expression for each algebraic expression. (Lesson 1-1)

64. $4y + 2$

65. $\dfrac{2}{3}x$

66. $a^2 b + 5$

Find the volume of each rectangular prism. (Lesson 0-9)

67.

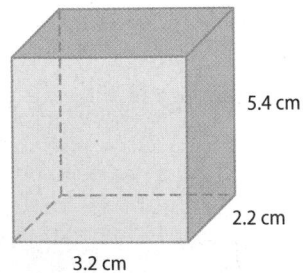

5.4 cm

2.2 cm

3.2 cm

68.

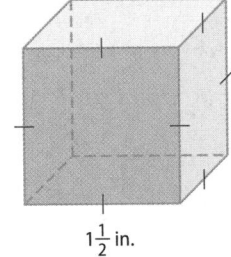

$1\frac{1}{2}$ in.

69.

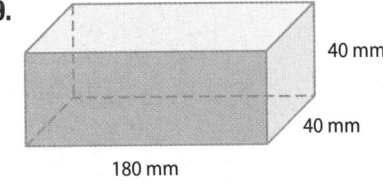

40 mm

40 mm

180 mm

Evaluate each expression.

70. If $x = 3$, then $6x - 5 = \underline{\ ?\ }$.

71. If $n = -1$, then $2n + 1 = \underline{\ ?\ }$.

72. If $p = 4$, then $3p + 4 = \underline{\ ?\ }$.

73. If $q = 7$, then $7q - 9 = \underline{\ ?\ }$

74. If $k = -11$, then $4k + 6 = \underline{\ ?\ }$

75. If $y = 10$, then $8y - 15 = \underline{\ ?\ }$

Graphing Technology Lab
Representing Functions

You can use TI-Nspire Technology to explore the different ways to represent a function.

CCSS **Common Core State Standards**
Content Standards
A.CED.2 Create equations in two or more variables to represent relationships between quantities; graph equations on coordinate axes with labels and scales.
Mathematical Practices
5 Use appropriate tools strategically.

Activity

Graph $f(x) = 2x + 3$ **on the TI-Nspire graphing calculator.**

Step 1 Add a new **Graphs** page.

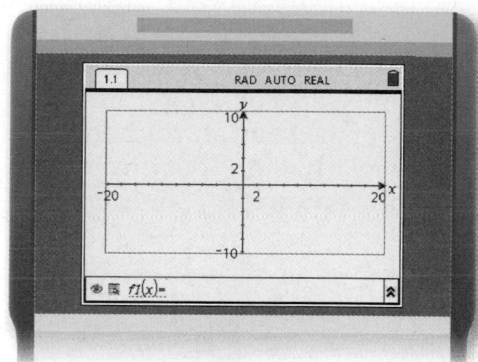

Step 2 Enter $2x + 3$ in the entry line.

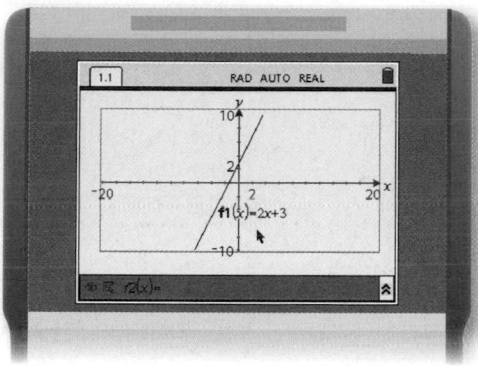

Represent the function as a table.

Step 3 Select the **Show Table** option from the **View** menu to add a table of values on the same display.

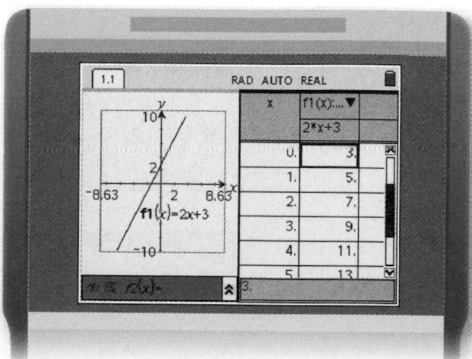

Step 4 Press **ctrl** and **tab** to toggle from the table to the graph. On the graph side, select the line and move it. Notice how the values in the table change.

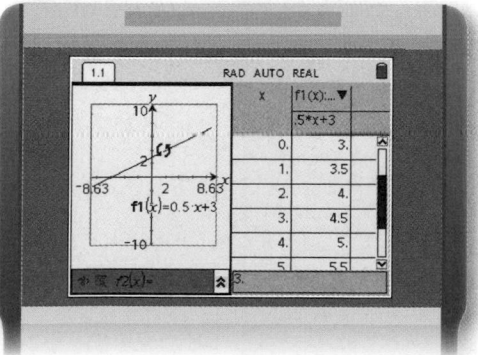

Analyze the Results

CCSS TOOLS Graph each function. Make a table of five ordered pairs that also represents the function.

1. $g(x) = -x - 3$

2. $h(x) = \frac{1}{3}x + 3$

3. $f(x) = -\frac{1}{2}x - 5$

4. $f(x) = 3x - \frac{1}{2}$

5. $g(x) = -2x + 5$

6. $h(x) = \frac{1}{5}x + 4$

| :: Then | :: Now | :: Why? |

- You identified functions and found function values.

1 Interpret intercepts, and symmetry of graphs of functions.

2 Interpret positive, negative, increasing, and decreasing behavior, extrema, and end behavior of graphs of functions.

- Sales of video games, including hardware, software, and accessories, have increased at times and decreased at other times over the years. Annual retail video game sales in the U.S. from 2000 to 2009 can be modeled by the graph of a nonlinear function.

 NewVocabulary
intercept
x-intercept
y-intercept
line symmetry
positive
negative
increasing
decreasing
extrema
relative maximum
relative minimum
end behavior

 Common Core State Standards

Content Standards
F.IF.4 For a function that models a relationship between two quantities, interpret key features of graphs and tables in terms of the quantities, and sketch graphs showing key features given a verbal description of the relationship.

Mathematical Practices
1 Make sense of problems and persevere in solving them.

1 Interpret Intercepts and Symmetry To interpret the graph of a function, estimate and interpret key features. The **intercepts** of a graph are points where the graph intersects an axis. The *y*-coordinate of the point at which the graph intersects the *y*-axis is called a **y-intercept**. Similarly, the *x*-coordinate of the point at which a graph intersects the *x*-axis is called an **x-intercept**.

Real-World Example 1 Interpret Intercepts

PHYSICS The graph shows the height *y* of an object as a function of time *x*. Identify the function as *linear* or *nonlinear*. Then estimate and interpret the intercepts.

Linear or Nonlinear: Since the graph is a curve and not a line, the graph is nonlinear.

y-Intercept: The graph intersects the *y*-axis at about (0, 15), so the *y*-intercept of the graph is about 15. This means that the object started at an initial height of about 15 meters above the ground.

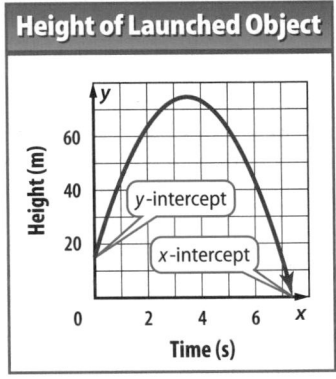

Height of Launched Object

x-Intercept(s): The graph intersects the *x*-axis at about (7.4, 0), so the *x*-intercept is about 7.4. This means that the object struck the ground after about 7.4 seconds.

▶ **Guided**Practice

1. The graph shows the temperature *y* of a medical sample thawed at a controlled rate. Identify the function as *linear* or *nonlinear*. Then estimate and interpret the intercepts.

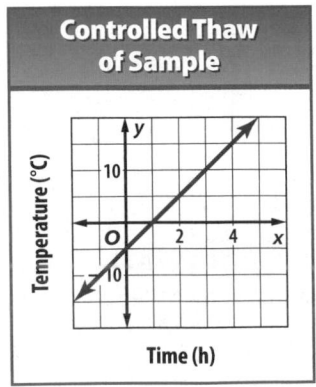

Controlled Thaw of Sample

The graphs of some functions exhibit another key feature: symmetry. A graph possesses **line symmetry** in the y-axis or some other vertical line if each half of the graph on either side of the line matches exactly.

Real-World Example 2 Interpret Symmetry

PHYSICS An object is launched. The graph shows the height y of the object as a function of time x. Describe and interpret any symmetry.

The right half of the graph is the mirror image of the left half in approximately the line $x = 3.5$ between approximately $x = 0$ and $x = 7$.

In the context of the situation, the symmetry of the graph tells you that the time it took the object to go up is equal to the time it took to come down.

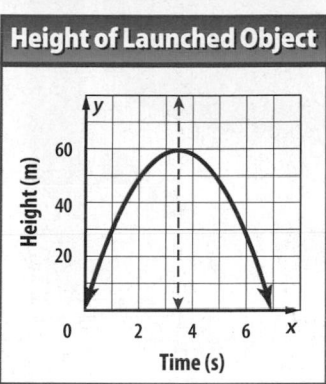

Height of Launched Object

> **StudyTip**
>
> Symmetry The graphs of most real-world functions do not exhibit symmetry over the entire domain. However, many have symmetry over smaller portions of the domain that are worth analyzing.

GuidedPractice

2. Describe and interpret any symmetry exhibited by the graph in Guided Practice 1.

2 **Interpret Extrema and End Behavior** Interpreting a graph also involves estimating and interpreting where the function is increasing, decreasing, positive, or negative, and where the function has any extreme values, either high or low.

KeyConcepts Positive, Negative, Increasing, Decreasing, Extrema, and End Behavior

A function is **positive** where its graph lies *above* the x-axis, and **negative** where its graph lies *below* the x-axis.

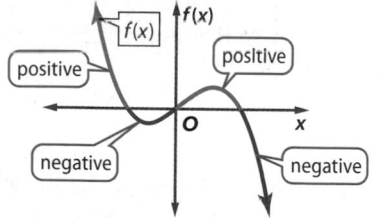

A function is **increasing** where the graph goes *up* and **decreasing** where the graph goes *down* when viewed from left to right.

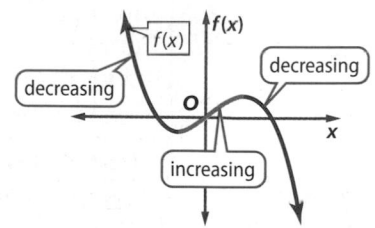

The points shown are the locations of relatively high or low function values called **extrema**. Point A is a **relative minimum**, since no other nearby points have a lesser y-coordinate. Point B is a **relative maximum**, since no other nearby points have a greater y-coordinate.

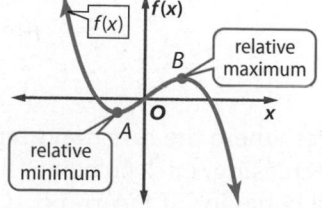

End behavior describes the values of a function at the positive and negative extremes in its domain.

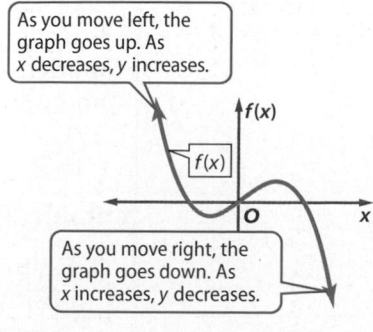

As you move left, the graph goes up. As x decreases, y increases.

As you move right, the graph goes down. As x increases, y decreases.

> **StudyTip**
>
> End Behavior The end behavior of some graphs can be described as approaching a specific y-value. In this case, a portion of the graph looks like a horizontal line.

VIDEO GAMES U.S. retail sales of video games from 2000 to 2009 can be modeled by the function graphed at the right. Estimate and interpret where the function is positive, negative, increasing, and decreasing, the *x*-coordinates of any relative extrema, and the end behavior of the graph.

Positive: between about $x = -0.6$ and $x = 10.4$

Negative: for about $x < -0.6$ and $x > 10.4$

This means that there were positive sales between about 2000 and 2010, but the model predicts negative sales after about 2010, indicating the unlikely collapse of the industry.

Real-World Link

The first successful commercially sold portable video game system was released in 1989 and sold for $120.

Source: *PCWorld*

Increasing: for about $x < 1.5$ and between about $x = 3$ and $x = 8$

Decreasing: between about $x = 2$ and $x = 3$ and for about $x > 8$

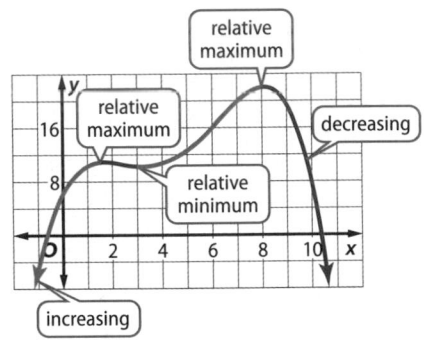

This means that sales increased from about 2000 to 2002, decreased from 2002 to 2003, increased from 2003 to 2008, and have been decreasing since 2008.

StudyTip

Constant A function is *constant* where the graph does not go up or down as the graph is viewed from left to right.

Relative Maximums: at about $x = 1.5$ and $x = 8$

Relative Minimum: at about $x = 3$

The extrema of the graph indicate that the industry experienced two relative peaks in sales during this period: one around 2002 of approximately $10.5 billion and another around 2008 of approximately $22 billion. A relative low of $10 billion in sales came in about 2003.

End Behavior:
As *x* increases or decreases, the value of *y* decreases.

The end behavior of the graph indicates negative sales several years prior to 2000 and several years after 2009, which is unlikely. This graph appears to only model sales well between 2000 and 2009 and can only be used to predict sales in 2010.

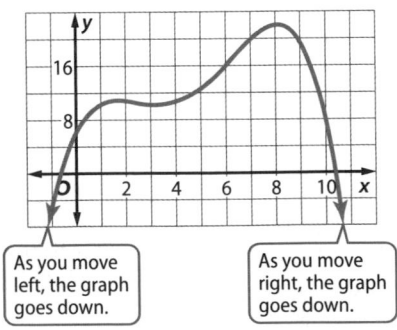

▶ **Guided**Practice

3. Estimate and interpret where the function graphed in Guided Practice 1 is positive, negative, increasing, or decreasing, the *x*-coordinate of any relative extrema, and the end behavior of the graph.

Examples 1–3 CCSS **SENSE-MAKING** Identify the function graphed as *linear* or *nonlinear*. Then estimate and interpret the intercepts of the graph, any symmetry, where the function is positive, negative, increasing, and decreasing, the *x*-coordinate of any relative extrema, and the end behavior of the graph.

1.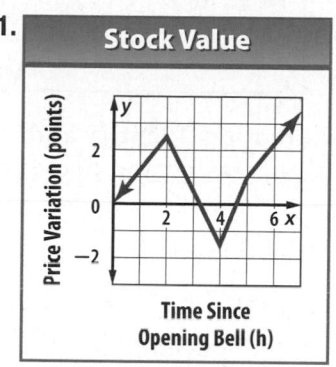

2.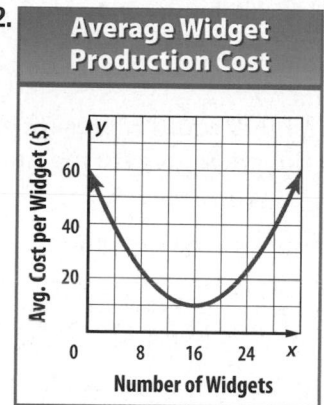

3.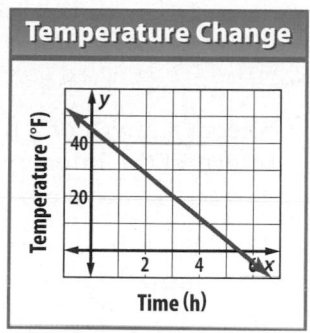

Practice and Problem Solving Extra Practice is on page R1.

Examples 1–3 CCSS **SENSE-MAKING** Identify the function graphed as *linear* or *nonlinear*. Then estimate and interpret the intercepts of the graph, any symmetry, where the function is positive, negative, increasing, and decreasing, the *x*-coordinate of any relative extrema, and the end behavior of the graph.

4.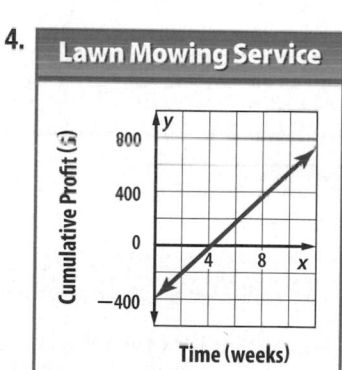

5.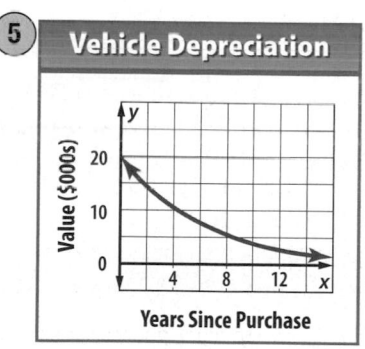

6.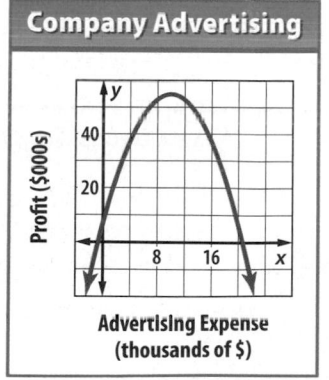

7.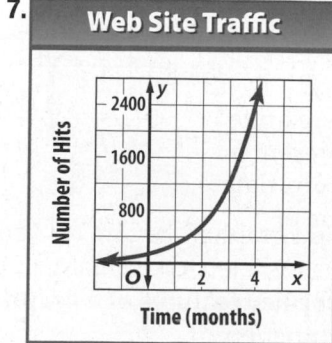

8.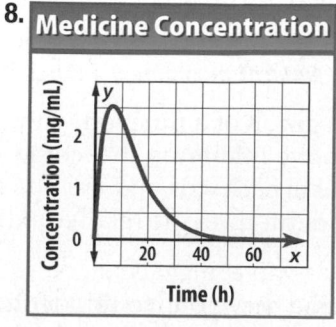

9.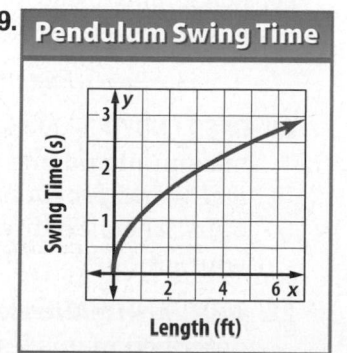

10. **FERRIS WHEEL** At the beginning of a Ferris wheel ride, a passenger cart is located at the same height as the center of the wheel. The position y in feet of this cart relative to the center t seconds after the ride starts is given by the function graphed at the right. Identify and interpret the key features of the graph. (*Hint:* Look for a pattern in the graph to help you describe its end behavior.)

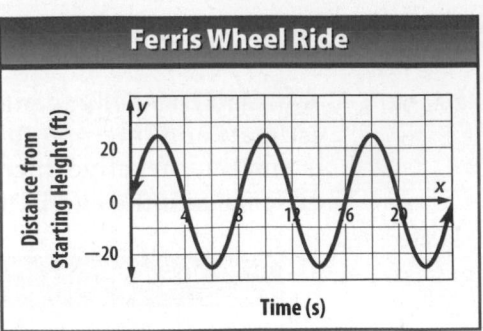

Ferris Wheel Ride

Sketch a graph of a function that could represent each situation. Identify and interpret the intercepts of the graph, where the graph is increasing and decreasing, and any relative extrema.

11. the height of a corn plant from the time the seed is planted until it reaches maturity 120 days later

12. the height of a football from the time it is punted until it reaches the ground 2.8 seconds later

13. the balance due on a car loan from the date the car was purchased until it was sold 4 years later

Sketch graphs of functions with the following characteristics.

14. The graph is linear with an x-intercept at -2. The graph is positive for $x < -2$, and negative for $x > -2$.

15. A nonlinear graph has x-intercepts at -2 and 2 and a y-intercept at -4. The graph has a relative minimum of -4 at $x = 0$. The graph is decreasing for $x < 0$ and increasing for $x > 0$.

16. A nonlinear graph has a y-intercept at 2, but no x-intercepts. The graph is positive and increasing for all values of x.

17. A nonlinear graph has x-intercepts at -8 and -2 and a y-intercept at 3. The graph has relative minimums at $x = -6$ and $x = 6$ and a relative maximum at $x = 2$. The graph is positive for $x < -8$ and $x > -2$ and negative between $x = -8$ and $x = -2$. As x decreases, y increases and as x increases, y increases.

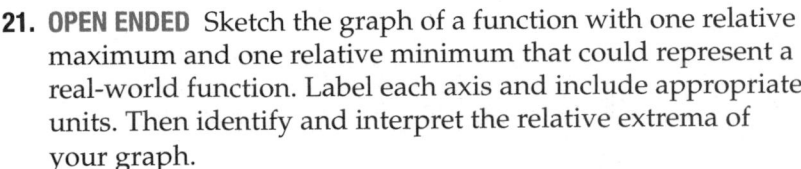

H.O.T. Problems Use Higher-Order Thinking Skills

18. **CCSS CRITIQUE** Katara thinks that all linear functions have exactly one x-intercept. Desmond thinks that a linear function can have at most one x-intercept. Is either of them correct? Explain your reasoning.

19. **CHALLENGE** Describe the end behavior of the graph shown.

20. **REASONING** Determine whether the following statement is *true* or *false*. Explain.

Functions have at most one y-intercept.

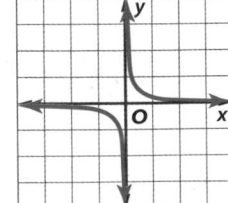

21. **OPEN ENDED** Sketch the graph of a function with one relative maximum and one relative minimum that could represent a real-world function. Label each axis and include appropriate units. Then identify and interpret the relative extrema of your graph.

22. **WRITING IN MATH** Describe how you would identify the key features of a graph described in this lesson using a table of values for a function.

23. Which sentence best describes the end behavior of the function shown?

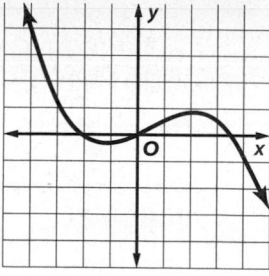

A As x increases, y increases, and as x decreases, y increases.

B As x increases, y increases, and as x decreases, y decreases.

C As x increases, y decreases, and as x decreases, y increases.

D As x increases, y decreases, and as x decreases, y decreases.

24. Which illustrates the Transitive Property of Equality?

F If $c = 1$, then $c \cdot \frac{1}{c} = 1$.

G If $c = d$ and $d = f$, then $c = f$.

H If $c = d$, then $d = c$.

J If $c = d$ and $d = c$, then $c = 1$.

25. Simplify the expression $5d(7 - 3) - 16d + 3 \cdot 2d$.

A $10d$ C $21d$

B $14d$ D $25d$

26. What is the probability of selecting a red card or an ace from a standard deck of cards?

F $\frac{1}{26}$ G $\frac{1}{2}$ H $\frac{7}{13}$ J $\frac{15}{26}$

Spiral Review

Determine whether each relation is a function. (Lesson 1-7)

27.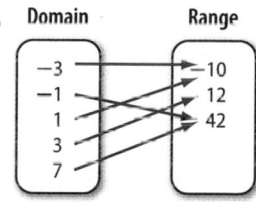

28. $\{(0, 2), (3, 5), (0, -1), (-2, 4)\}$

29.

x	y
17	6
18	6
19	5
20	4

30. GEOMETRY Express the relation in the graph at the right as a set of ordered pairs. Describe the domain and range. (Lesson 1-6)

Equilateral Triangles

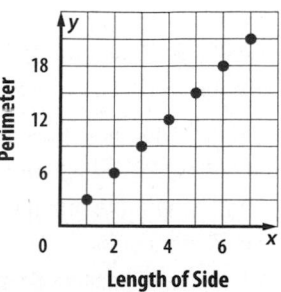

Use the Distributive Property to rewrite each expression. (Lesson 1-4)

31. $\frac{1}{2}d(2d + 6)$ **32.** $-h(6h - 1)$ **33.** $3z - 6x$

34. CLOTHING Robert has 30 socks in his sock drawer. 16 of the socks are white, 6 are black, 2 are red, and 6 are yellow. What is the probability that he randomly pulls out a black sock? (Lesson 0-11)

Skills Review

Evaluate each expression.

35. $(-7)^2$ **36.** 3.2^2 **37.** $(-4.2)^2$ **38.** $\left(\frac{1}{4}\right)^2$

Study Guide and Review

Study Guide

KeyConcepts

Order of Operations (Lesson 1-2)

- Evalute expressions inside grouping symbols.
- Evaluate all powers.
- Multiply and/or divide in order from left to right.
- Add or subtract in order from left to right.

Properties of Equality (Lessons 1-3 and 1-4)

- For any numbers a, b, and c:

  Reflexive: $a = a$

  Symmetric: If $a = b$, then $b = a$.

  Transitive: If $a = b$ and $b = c$, then $a = c$.

  Substitution: If $a = b$, then a may be replaced by b in any expression.

  Distributive: $a(b + c) = ab + ac$ and $a(b - c) = ab - ac$

  Commutative: $a + b = b + a$ and $ab = ba$

  Associative: $(a + b) + c = a + (b + c)$ and $(ab)c = a(bc)$

Solving Equations (Lesson 1-5)

- Apply order of operations and the properties of real numbers to solve equations.

Relations, Functions, and Interpreting Graphs of Functions (Lessons 1-6 through 1-8)

- Relations and functions can be represented by ordered pairs, a table, a mapping, or a graph.
- Use the vertical line test to determine if a relation is a function.
- End behavior describes the long-term behavior of a function on either end of its graph.
- Points where the graph of a function crosses an axis are called intercepts.
- A function is positive on a portion of its domain where its graph lies above the x-axis, and negative on a portion where its graph lies below the x-axis.

FOLDABLES StudyOrganizer

Be sure the Key Concepts are noted in your Foldable.

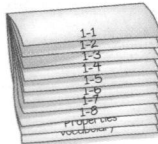

KeyVocabulary

algebraic expression (p. 5)	ordered pair (p. 40)
base (p. 5)	order of operations (p. 10)
coefficient (p. 28)	origin (p. 40)
coordinate system (p. 40)	power (p. 5)
dependent variable (p. 42)	range (p. 40)
domain (p. 40)	reciprocal (p. 17)
end behavior (p. 57)	relation (p. 40)
equation (p. 33)	relative maximum (p. 57)
exponent (p. 5)	relative minimum (p. 57)
function (p. 47)	replacement set (p. 33)
independent variable (p. 42)	simplest form (p. 27)
intercept (p. 56)	solution (p. 33)
like terms (p. 27)	term (p. 5)
line symmetry (p. 57)	variables (p. 5)
mapping (p. 40)	vertical line test (p. 49)

VocabularyCheck

State whether each sentence is *true* or *false*. If *false*, replace the underlined term to make a true sentence.

1. A <u>coordinate system</u> is formed by two intersecting number lines.

2. An <u>exponent</u> indicates the number of times the base is to be used as a factor.

3. An expression is <u>in simplest form</u> when it contains like terms and parentheses.

4. In an expression involving multiplication, the quantities being multiplied are called <u>factors</u>.

5. In a <u>function</u>, there is exactly one output for each input.

6. <u>Order of operations</u> tells us to perform multiplication before subtraction.

7. Since the product of any number and 1 is equal to the number, 1 is called the <u>multiplicative inverse</u>.

Lesson-by-Lesson Review

1-1 Variables and Expressions

Write a verbal expression for each algebraic expression.

8. $h - 7$ **9.** $3x^2$ **10.** $5 + 6m^3$

Write an algebraic expression for each verbal expression.

11. a number increased by 9

12. two thirds of a number d to the third power

13. 5 less than four times a number

Evaluate each expression.

14. 2^5 **15.** 6^3 **16.** 4^4

17. BOWLING Fantastic Pins Bowling Alley charges $2.50 for shoe rental plus $3.25 for each game. Write an expression representing the cost to rent shoes and bowl g games.

Example 1

Write a verbal expression for $4x + 9$.

nine more than four times a number x

Example 2

Write an algebraic expression for *the difference of twelve and two times a number cubed.*

Variable   Let x represent the number.

Expression $12 - 2x^3$

Example 3

Evaluate 3^4.

The base is 3 and the exponent is 4.

$3^4 = 3 \cdot 3 \cdot 3 \cdot 3$ Use 3 as a factor 4 times.

$\quad = 81$ Multiply.

1-2 Order of Operations

Evaluate each expression.

18. $24 - 4 \cdot 5$ **19.** $15 + 3^2 - 6$

20. $7 + 2(9 - 3)$ **21.** $8 \cdot 4 - 6 \cdot 5$

22. $\left[(2^5 - 5) \div 9\right]11$ **23.** $\dfrac{11 + 4^2}{5^2 - 4^2}$

Evaluate each expression if $a = 4$, $b = 3$, and $c = 9$.

24. $c + 3a$

25. $5b^2 \div c$

26. $\left(a^2 + 2bc\right) \div 7$

27. ICE CREAM The cost of a one-scoop sundae is $2.75, and the cost of a two-scoop sundae is $4.25. Write and evaluate an expression to find the total cost of 3 one-scoop sundaes and 2 two-scoop sundaes.

Example 4

Evaluate the expression $3(9 - 5)^2 \div 8$.

$3(9 - 5)^2 \div 8 = 3(4)^2 \div 8$ Work inside parentheses.

$\qquad = 3(16) \div 8$ Evaluate 4^2.

$\qquad = 48 \div 8$ Multiply.

$\qquad = 6$ Divide.

Example 5

Evaluate the expression $(5m - 2n) \div p^2$ if $m = 8$, $n = 4$, $p = 2$.

$(5m - 2n) \div p^2$

$\qquad = (5 \cdot 8 - 2 \cdot 4) \div 2^2$ Replace m with 8, n with 4, and p with 2.

$\qquad = (40 - 8) \div 2^2$ Multiply.

$\qquad = 32 \div 2^2$ Subtract.

$\qquad = 32 \div 4$ Evaluate 2^2.

$\qquad = 8$ Divide.

1-3 Properties of Numbers

Evaluate each expression using properties of numbers. Name the property used in each step.

28. $18 \cdot 3(1 \div 3)$

29. $[5 \div (8 - 6)]\dfrac{2}{5}$

30. $(16 - 4^2) + 9$

31. $2 \cdot \dfrac{1}{2} + 4(4 \cdot 2 - 7)$

32. $18 + 41 + 32 + 9$

33. $7\dfrac{2}{5} + 5 + 2\dfrac{3}{5}$

34. $8 \cdot 0.5 \cdot 5$

35. $5.3 + 2.8 + 3.7 + 6.2$

36. SCHOOL SUPPLIES Monica needs to purchase a binder, a textbook, a calculator, and a workbook for her algebra class. The binder costs \$9.25, the textbook \$32.50, the calculator \$18.75, and the workbook \$15.00. Find the total cost for Monica's algebra supplies.

Example 6

Evaluate $6(4 \cdot 2 - 7) + 5 \cdot \dfrac{1}{5}$. Name the property used in each step.

$6(4 \cdot 2 - 7) + 5 \cdot \dfrac{1}{5}$

$= 6(8 - 7) + 5 \cdot \dfrac{1}{5}$	Substitution
$= 6(1) + 5 \cdot \dfrac{1}{5}$	Substitution
$= 6 + 5 \cdot \dfrac{1}{5}$	Multiplicative Identity
$= 6 + 1$	Multiplicative Inverse
$= 7$	Substitution

1-4 The Distributive Property

Use the Distributive Property to rewrite each expression. Then evaluate.

37. $(2 + 3)6$

38. $5(18 + 12)$

39. $8(6 - 2)$

40. $(11 - 4)3$

41. $-2(5 - 3)$

42. $(8 - 3)4$

Rewrite each expression using the Distributive Property. Then simplify.

43. $3(x + 2)$

44. $(m + 8)4$

45. $6(d - 3)$

46. $-4(5 - 2t)$

47. $(9y - 6)(-3)$

48. $-6(4z + 3)$

49. TUTORING Write and evaluate an expression for the number of tutoring lessons Mrs. Green gives in 4 weeks.

Tutoring Schedule	
Day	**Students**
Monday	3
Tuesday	5
Wednesday	4

Example 7

Use the Distributive Property to rewrite the expression $5(3 + 8)$. Then evaluate.

$5(3 + 8) = 5(3) + 5(8)$	Distributive Property
$= 15 + 40$	Multiply.
$= 55$	Simplify.

Example 8

Rewrite the expression $6(x + 4)$ using the Distributive Property. Then simplify.

$6(x + 4) = 6 \cdot x + 6 \cdot 4$	Distributive Property
$= 6x + 24$	Simplify.

Example 9

Rewrite the expression $(3x - 2)(-5)$ using the Distributive Property. Then simplify.

$(3x - 2)(-5)$

$= (3x)(-5) - (2)(-5)$	Distributive Property
$= -15x + 10$	Simplify.

1-5 Equations

Find the solution set of each equation if the replacement sets are x: {1, 3, 5, 7, 9} and y: {6, 8, 10, 12, 14}.

50. $y - 9 = 3$　　　**51.** $14 + x = 21$

52. $4y = 32$　　　**53.** $3x - 11 = 16$

54. $\dfrac{42}{y} = 7$　　　**55.** $2(x - 1) = 8$

Solve each equation.

56. $a = 24 - 7(3)$

57. $z = 63 \div \left(3^2 - 2\right)$

58. AGE Shandra's age is four more than three times Sherita's age. Write an equation for Shandra's age. Solve if Sherita is 3 years old

Example 10

Solve the equation $5w - 19 = 11$ if the replacement set is w: {2, 4, 6, 8, 10}.

Replace w in $5w - 19 = 11$ with each value in the replacement set.

w	$5w - 19 = 11$	True or False?
2	$5(2) - 19 = 11$	false
4	$5(4) - 19 = 11$	false
6	$5(6) - 19 = 11$	true
8	$5(8) - 19 = 11$	false
10	$5(10) - 19 = 11$	false

Since the equation is true when $w = 6$, the solution of $5w - 19 = 11$ is $w = 6$.

1-6 Relations

Express each relation as a table, a graph, and a mapping. Then determine the domain and range.

59. {(1, 3), (2, 4), (3, 5), (4, 6)}

60. {(−1, 1), (0, −2), (3, 1), (4, −1)}

61. {(−2, 4), (−1, 3), (0, 2), (−1, 2)}

Express the relation shown in each table, mapping, or graph as a set of ordered pairs.

62.

x	y
5	3
3	−1
1	2
−1	0

63.

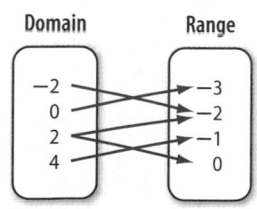

64. GARDENING On average, 7 plants grow for every 10 seeds of a certain type planted. Make a table to show the relation between seeds planted and plants growing for 50, 100, 150, and 200 seeds. Then state the domain and range and graph the relation.

Example 11

Express the relation {(−3, 4), (1, −2), (0, 1), (3, −1)} as a table, a graph, and a mapping.

Table

Place the x-coordinates into the first column. Place the corresponding y-coordinates in the second column.

x	y
−3	4
1	−2
0	1
3	−1

Graph

Graph each ordered pair on a coordinate plane.

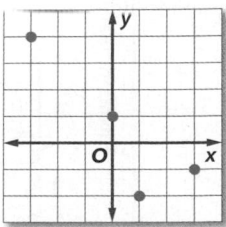

Mapping

List the x-values in the domain and the y-values in the range. Draw arrows from the x-values in set X to the corresponding y-values in set Y.

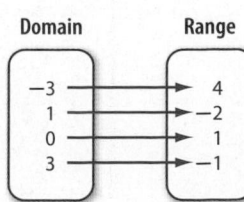

1-7 Functions

Determine whether each relation is a function.

65.

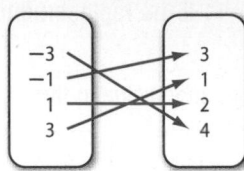

66.

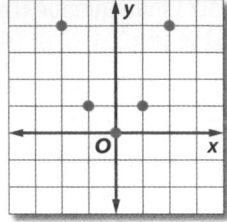

67. {(8, 4), (6, 3), (4, 2), (2, 1), (6, 0)}

If $f(x) = 2x + 4$ and $g(x) = x^2 - 3$, find each value.

68. $f(-3)$ **69.** $g(2)$ **70.** $f(0)$

71. $g(-4)$ **72.** $f(m + 2)$ **73.** $g(3p)$

74. GRADES A teacher claims that the relationship between number of hours studied for a test and test score can be described by $g(x) = 45 + 9x$, where x represents the number of hours studied. Graph this function.

Example 12

Determine whether $2x - y = 1$ represents a function.

First make a table of values. Then graph the equation.

x	y
-1	-3
0	-1
1	1
2	3
3	5

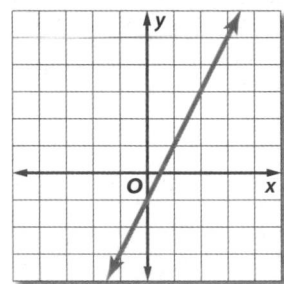

Using the vertical line test, it can be shown that $2x - y = 1$ does represent a function.

1-8 Interpreting Graphs of Functions

75. Identify the function graphed as *linear* or *nonlinear*. Then estimate and interpret the intercepts of the graph, any symmetry, where the function is positive, negative, increasing, and decreasing, the x-coordinate of any relative extrema, and the end behavior of the graph.

U.S. Patents Granted

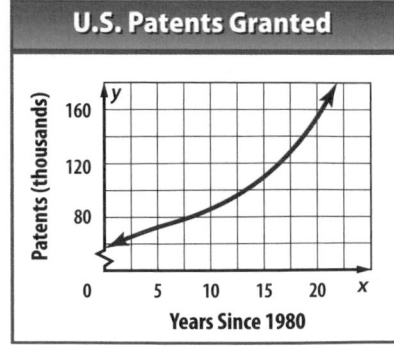

Example 13

POPULATION The population of Haiti from 1994 to 2010 can be modeled by the function graphed below. Estimate and interpret where the function is increasing, and decreasing, the x-coordinates of any relative extrema, and the end behavior of the graph.

Population of Haiti

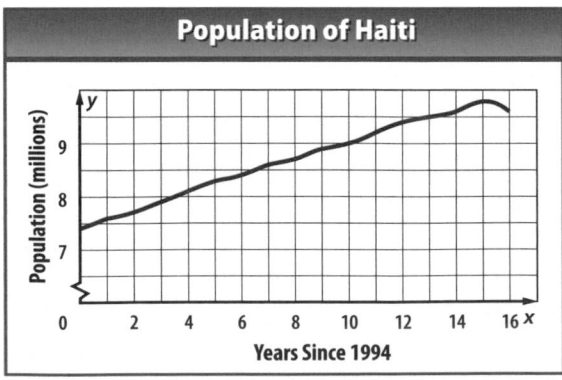

The population increased from 1994 to 2009 and decreased from 2009 to 2010. The relative maximum of the graph indicates that the population peaked in 2009.

As x increases or decreases, y decreases. The end behavior indicates a decline in population from 2009 to 2010.

Practice Test

Write an algebraic expression for each verbal expression.

1. six more than a number

2. twelve less than the product of three and a number

3. four divided by the difference between a number and seven

Evaluate each expression.

4. $32 \div 4 + 2^3 - 3$

5. $\dfrac{(2 \cdot 4)^2}{7 + 3^2}$

6. **MULTIPLE CHOICE** Find the value of the expression $a^2 + 2ab + b^2$ if $a = 6$ and $b = 4$.

   A 68

   B 92

   C 100

   D 121

Evaluate each expression. Name the property used in each step.

7. $13 + (16 - 4^2)$

8. $\dfrac{2}{9}[9 \div (7 - 5)]$

9. $37 + 29 + 13 + 21$

Rewrite each expression using the Distributive Property. Then simplify.

10. $4(x + 3)$

11. $(5p - 2)(-3)$

12. **MOVIE TICKETS** A company operates three movie theaters. The chart shows the typical number of tickets sold each week at the three locations. Write and evaluate an expression for the total typical number of tickets sold by all three locations in four weeks.

Location	Tickets Sold
A	438
B	374
C	512

Find the solution of each equation if the replacement sets are x: {1, 3, 5, 7, 9} and y: {2, 4, 6, 8, 10}.

13. $3x - 9 = 12$

14. $y^2 - 5y - 11 = 13$

15. **CELL PHONES** The ABC Cell Phone Company offers a plan that includes a flat fee of $29 per month plus a $0.12 charge per minute. Write an equation to find C, the total monthly cost for m minutes. Then solve the equation for $m = 50$.

Express the relation shown in each table, mapping, or graph as a set of ordered pairs.

16.

x	y
-2	4
1	2
3	0
4	-2

17.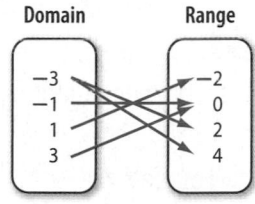

Domain   Range

18. **MULTIPLE CHOICE** Determine the domain and range for the relation {(2, 5), (−1, 3), (0, −1), (3, 3), (−4, −2)}.

   F D: {2, −1, 0, 3, −4}, R: {5, 3, −1, 3, −2}

   G D: {5, 3, −1, 3, −2}, R: {2, −1, 0, 3, 4}

   H D: {0, 1, 2, 3, 4}, R: {−4, −3, −2, −1, 0}

   J D: {2, −1, 0, 3, −4}, R: {2, −1, 0, 3, 4}

19. Determine whether the relation {(2, 3), (−1, 3), (0, 4), (3, 2), (−2, 3)} is a function.

If $f(x) = 5 - 2x$ and $g(x) = x^2 + 7x$, find each value.

20. $g(3)$

21. $f(-6y)$

22. Identify the function graphed as *linear* or *nonlinear*. Then estimate and interpret the intercepts of the graph, any symmetry, where the function is positive, negative, increasing, and decreasing, the x-coordinate of any relative extrema, and the end behavior of the graph.

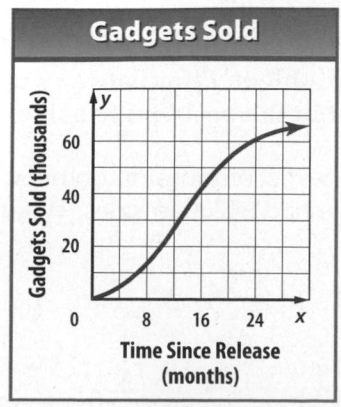

Preparing for Standardized Tests

Eliminate Unreasonable Answers

You can eliminate unreasonable answers to help you find the correct one when solving multiple choice test items. Doing so will save you time by narrowing down the list of possible correct answers.

Strategies for Eliminating Unreasonable Answers

Step 1

Read the problem statement carefully to determine exactly what you are being asked to find.

Ask yourself:

- What am I being asked to solve?

- What format (i.e., fraction, number, decimal, percent, type of graph) will the correct answer be?

- What units (if any) will the correct answer have?

Step 2

Carefully look over each possible answer choice and evaluate for reasonableness.

- Identify any answer choices that are clearly incorrect and eliminate them.

- Eliminate any answer choices that are not in the proper format.

- Eliminate any answer choices that do not have the correct units.

Step 3

Solve the problem and choose the correct answer from those remaining.
Check your answer.

Standardized Test Example

Read each problem. Eliminate any unreasonable answers. Then use the information in the problem to solve.

Jason earns 8.5% commission on his weekly sales at an electronics retail store. Last week he had $4200 in sales. What was his commission for the week?

A $332   C $425

B $357   D $441

Using mental math, you know that 10% of $4200 is $420. Since 8.5% is less than 10%, you know that Jason earned less than $420 in commission for his weekly sales. So, answer choices C and D can be eliminated because they are greater than $420. The correct answer is either A or B.

$4200 × 0.085 = $357

So, the correct answer is B.

Exercises

Read each problem. Eliminate any unreasonable answers. Then use the information in the problem to solve.

1. Coach Roberts expects 35% of the student body to turn out for a pep rally. If there are 560 students, how many does Coach Roberts expect to attend the pep rally?

   A 184

   B 196

   C 214

   D 390

2. Jorge and Sally leave school at the same time. Jorge walks 300 yards north and then 400 yards east. Sally rides her bike 600 yards south and then 800 yards west. What is the distance between the two students?

   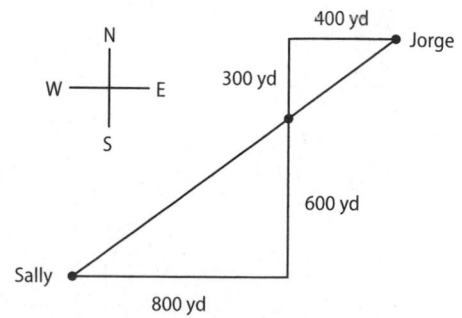

   F 500 yd

   G 750 yd

   H 1,200 yd

   J 1,500 yd

3. What is the range of the relation below?

   $\{(1, 2), (3, 4), (5, 6), (7, 8)\}$

   A all real numbers

   B all even numbers

   C {2, 4, 6, 8}

   D {1, 3, 5, 7}

4. The expression $3n + 1$ gives the total number of squares needed to make each figure of the pattern where n is the figure number. How many squares will be needed to make Figure 9?

   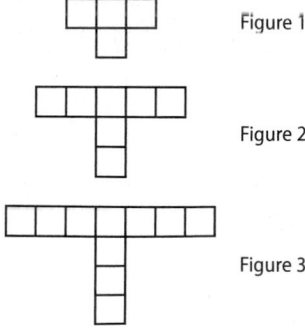

   Figure 1

   Figure 2

   Figure 3

   F 28 squares

   G 32.5 squares

   H 56 squares

   J 88.5 squares

5. The expression $3x - (2x + 4x - 6)$ is equivalent to

   A $-3x - 6$ **C** $3x + 6$

   B $-3x + 6$ **D** $3x - 6$

Multiple Choice

Read each question. Then fill in the correct answer on the answer document provided by your teacher or on a sheet of paper.

1. Evaluate the expression 2^6.

  A 12

  B 32

  C 64

  D 128

2. Which sentence best describes the end behavior of the function shown?

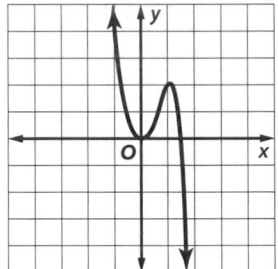

  F As x increases, y increases, and as x decreases, y increases.

  G As x increases, y increases, and as x decreases, y decreases.

  H As x increases, y decreases, and as x decreases, y increases.

  J As x increases, y decreases, and as x decreases, y decreases.

3. Let y represent the number of yards. Which algebraic expression represents the number of feet in y?

  A $y - 3$

  B $y + 3$

  C $3y$

  D $\dfrac{3}{y}$

4. What is the domain of the following relation?

$$\{(1, 3), (-6, 4), (8, 5)\}$$

  F $\{3, 4, 5\}$

  G $\{-6, 1, 8\}$

  H $\{-6, 1, 3, 4, 5, 8\}$

  J $\{1, 3, 4, 5, 8\}$

5. The table shows the number of some of the items sold at the concession stand at the first day of a soccer tournament. Estimate how many items were sold from the concession stand throughout the four days of the tournament.

Concession Sales Day 1 Results	
Item	Number Sold
Popcorn	78
Hot Dogs	80
Chip	48
Sodas	51
Bottled Water	92

  A 1350 items   **C** 1450 items

  B 1400 items   **D** 1500 items

6. There are 24 more cars than twice the number of trucks for sale at a dealership. If there are 100 cars for sale, how many trucks are there for sale at the dealership?

  F 28   **H** 34

  G 32   **J** 38

7. Refer to the relation in the table below. Which of the following values would result in the relation *not* being a function?

x	−6	−2	0	?	3	5
y	−1	8	3	−3	4	0

  A −1

  B 3

  C 7

  D 8

> **Test-TakingTip**
>
> Question 7 A function is a relation in which each element of the domain is paired with *exactly* one element of the range.

Record your answers on the answer sheet provided by your teacher or on a sheet of paper.

8. The edge of each box below is 1 unit long.

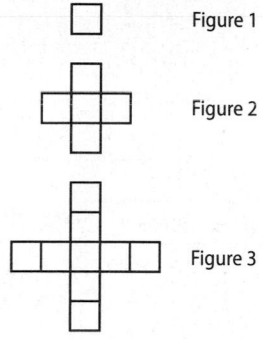

Figure 1

Figure 2

Figure 3

a. Make a table showing the perimeters of the first 3 figures in the pattern.

b. Look for a pattern in the perimeters of the shapes. Write an algebraic expression for the perimeter of Figure n.

c. What would be the perimeter of Figure 10 in the pattern?

9. The table shows the costs of certain items at a corner hardware store.

Item	Cost
box of nails	$3.80
box of screws	$5.25
claw hammer	$12.95
electric drill	$42.50

a. Write two expressions to represent the total cost of 3 boxes of nails, 2 boxes of screws, 2 hammers, and 1 electric drill.

b. What is the total cost of the items purchased?

10. GRIDDED RESPONSE Evaluate the expression below.

$$\frac{5^3 \cdot 4^2 - 5^2 \cdot 4^3}{5 \cdot 4}$$

11. Use the equation $y = 2(4 + x)$ to answer each question.

a. Complete the table for each value of x.

b. Plot the points from the table on a coordinate grid. What do you notice about the points?

c. Make a conjecture about the relationship between the change in x and the change in y.

x	y
1	
2	
3	
4	
5	
6	

Extended Response

Record your answers on a sheet of paper. Show your work.

12. The volume of a sphere is four-thirds the product of π and the radius cubed.

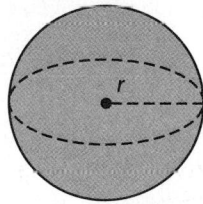

a. Write an expression for the volume of a sphere with radius r.

b. Find the volume of a sphere with a radius of 6 centimeters. Describe how you found your answer.

Need ExtraHelp?

If you missed Question...	1	2	3	4	5	6	7	8	9	10	11	12
Go to Lesson...	1-2	1-8	1-1	1-6	1-4	1-5	1-7	1-5	1-3	1-2	1-4	1-1

CHAPTER 2 Linear Equations

55
CAFE

UpperCut Images/Getty Images

Then

You learned to simplify algebraic expressions.

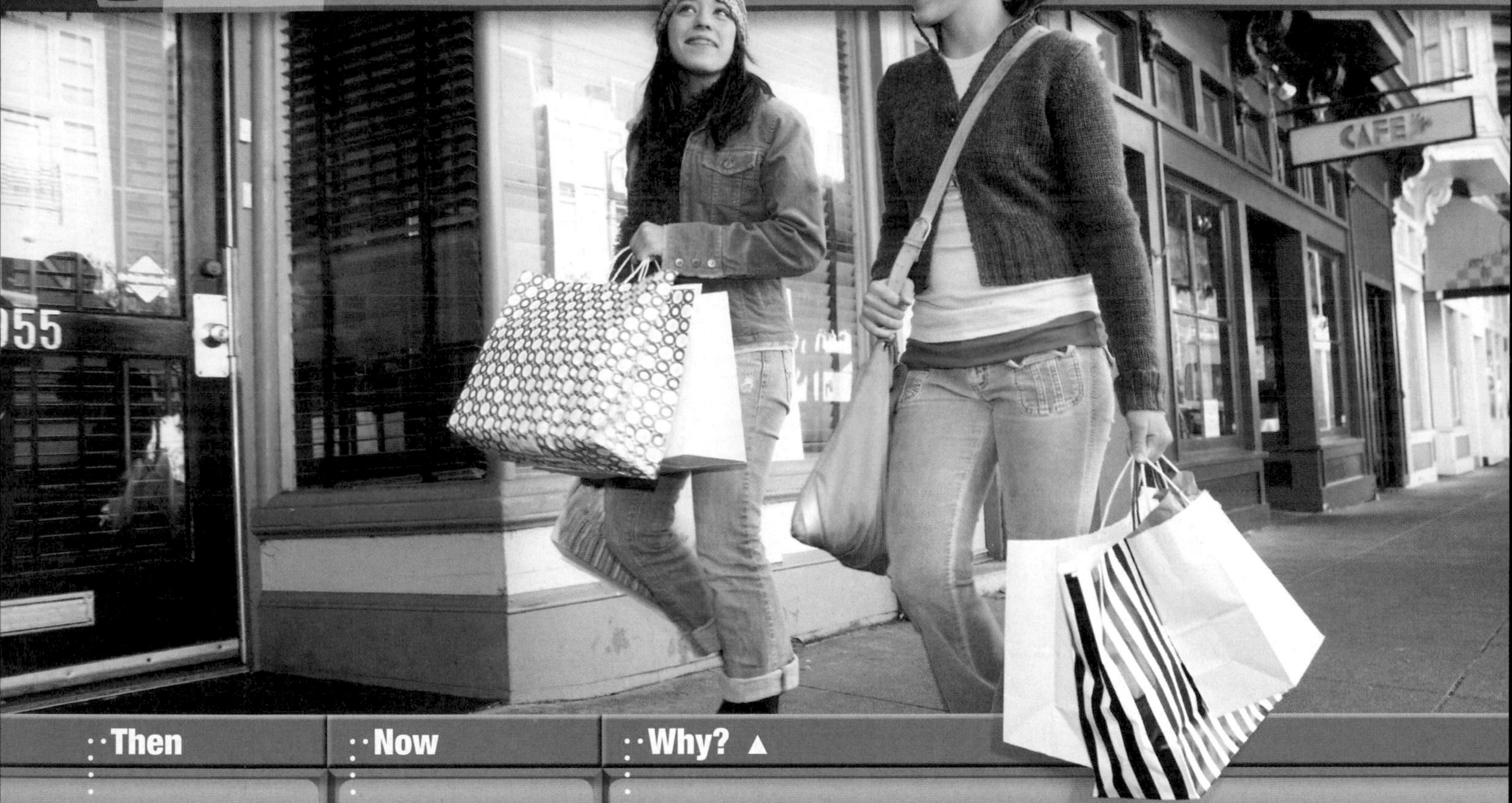

Now

In this chapter, you will:

- Create equations that describe relationships.
- Solve linear equations in one variable.
- Solve proportions.
- Use formulas to solve real-world problems.

Why? ▲

SHOPPING In recent years, the percent of change in sales per year at shopping malls in the U.S. averaged 5%. A store manager can use this data to set a sales goal for the upcoming year.

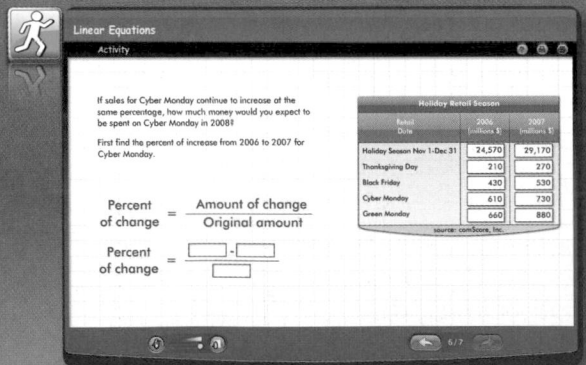

Linear Equations
Activity
If sales for Cyber Monday continue to increase at the same percentage, how much money would you expect to be spent on Cyber Monday in 2008?
First find the percent of increase from 2006 to 2007 for Cyber Monday.
Percent of change = Amount of change / Original amount
Percent of change
Holiday Retail Season
Retail Date
2006 (millions $)
2007 (millions $)
Holiday Season Nov 1-Dec 31 24,570 29,170
Thanksgiving Day 210 270
Black Friday 430 530
Cyber Monday 610 730
Green Monday 660 880
source: comScore, Inc.
6/7

connectED.mcgraw-hill.com **Your Digital Math Portal**

Animation | Vocabulary | eGlossary | Personal Tutor | Virtual Manipulatives | Graphing Calculator | Audio | Foldables | Self-Check Practice | Worksheets

Get Ready for the Chapter

Diagnose Readiness | You have two options for checking prerequisite skills.

1 **Textbook Option** Take the Quick Check below. Refer to the Quick Review for help.

QuickCheck	QuickReview

Write an algebraic expression for each verbal expression.

1. four less than three times a number n

2. a number d cubed less seven

3. the difference between two times b and eleven

Example 1

Write an algebraic expression for the phrase *the product of eight and w increased by nine.*

the product of eight and w increased by nine

$$8 \quad \cdot \quad w \quad\quad + \quad\quad 9$$

The expression is $8w + 9$.

Evaluate each expression.

4. $(9 - 4)^2 + 3$

5. $\dfrac{3 \cdot 8 - 12 \div 2}{3^2}$

6. $5(8 - 2) \div 3$

7. $\dfrac{1}{3}(21) + \dfrac{1}{8}(32)$

8. $72 \div 9 + 3 \cdot 2^3$

9. $\dfrac{11 - 3}{2} + 7$

10. $2\big[(5 - 3)^2 + 8\big] + (3 - 1) \div 2$

11. **BAKERY** Sue buys 1 carrot cake for $14, 6 large chocolate chip cookies for $1.50 each, and a dozen doughnuts for $0.45 each. How much money did Sue spend at the bakery?

Example 2

Evaluate $9 - \left[\dfrac{8 + 2^2}{2} - 2(5 \times 2 - 8)\right]$.

$9 - \left[\dfrac{8 + 2^2}{2} - 2(5 \times 2 - 8)\right]$ Original expression

$= 9 - \left[\dfrac{8 + 2^2}{2} - 2(2)\right]$ Evaluate inside the parentheses.

$= 9 - \left(\dfrac{8 + 2^2}{2} - 4\right)$ Multiply.

$= 9 - \left(\dfrac{8 + 4}{2} - 4\right)$ Evaluate the power.

$= 9 - (6 - 4)$ Add and then divide.

$= 7$ Simplify.

Find each percent.

12. What percent of 400 is 260?

13. Twelve is what percent of 60?

14. What percent of 25 is 75?

15. **ICE CREAM** What percent of the people surveyed prefer strawberry ice cream?

Favorite Flavor	Number of Responses
vanilla	82
chocolate	76
strawberry	42

Example 3

32 is what percent of 40?

$\dfrac{a}{b} = \dfrac{p}{100}$ Use the percent proportion.

$\dfrac{32}{40} = \dfrac{p}{100}$ Replace a with 32 and b with 40.

$32(100) = 40p$ Find the cross products.

$3200 = 40p$ Multiply.

$80 = p$ Divide each side by 40.

32 is 80% of 40.

---

2 **Online Option** Take an online self-check Chapter Readiness Quiz at <u>connectED.mcgraw-hill.com</u>.

Get Started on the Chapter

You will learn several new concepts, skills, and vocabulary terms as you study Chapter 2. To get ready, identify important terms and organize your resources. You may wish to refer to Chapter 0 to review prerequisite skills.

FOLDABLES StudyOrganizer

Linear Equations Make this Foldable to help you organize your Chapter 2 notes about linear equations. Begin with 5 sheets of grid paper.

1 **Fold** each sheet in half along the width.

2 **Unfold** each sheet and tape to form one long piece.

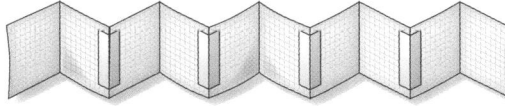

3 **Label** each page with the lesson number as shown. Refold to form a booklet.

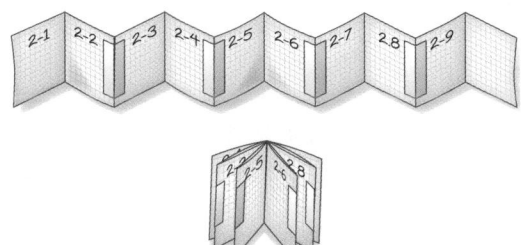

NewVocabulary

English		Español
formula	p. 76	fórmula
solve an equation	p. 83	resolver una ecuación
equivalent equations	p. 83	ecuaciones equivalentes
multi-step equation	p. 91	ecuación de varios pasos
identity	p. 98	identidad
ratio	p. 111	razón
proportion	p. 111	proporción
rate	p. 113	tasa
unit rate	p. 113	tasa unitaria
scale model	p. 114	modelo de escala
percent of change	p. 119	porcentaje de cambio
literal equation	p. 127	ecuación literal
dimensional analysis	p. 128	análisis dimensional
weighted average	p. 132	promedio ponderado

ReviewVocabulary

algebraic expression expresion algebraica an expression consisting of one or more numbers and variables along with one or more arithmetic operations

coordinate system sistema de coordenedas the grid formed by the intersection of two number lines, the horizontal axis and the vertical axis

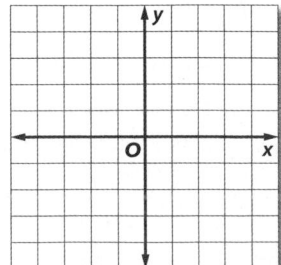

function función a relation in which each element of the domain is paired with exactly one element of the range

2-1 Writing Equations

:: Then	:: Now	:: Why?
• You evaluated and simplified algebraic expressions.	**1** Translate sentences into equations. **2** Translate equations into sentences.	• The Daytona 500 is widely considered to be the most important event of the NASCAR circuit. The distance around the track is 2.5 miles, and the race is a total of 500 miles. We can write an equation to determine how many laps it takes to finish the race.

NewVocabulary
formula

Common Core State Standards

Content Standards
A.CED.1 Create equations and inequalities in one variable and use them to solve problems.

Mathematical Practices
2 Reason abstractly and quantitatively.

1 **Write Verbal Expressions** To write an equation, identify the unknown for which you are looking and assign a variable to it. Then, write the sentence as an equation. Look for key words such as *is*, *is as much as*, *is the same as*, or *is identical to* that indicate where you should place the equals sign.

Consider the Daytona 500 example above.

Words	The length of each lap times the number of laps is the length of the race.
Variable	Let ℓ represent the number of laps in the race.
Equation	$2.5 \quad \times \quad \ell \quad = \quad 500$

Example 1 Translate Sentences into Equations

Translate each sentence into an equation.

a. Seven times a number squared is five times the difference of k and m.

Seven	times	n squared	is	five	times	the difference of k and m.
7	·	n^2	=	5	·	$(k - m)$

The equation is $7n^2 = 5(k - m)$.

b. Fifteen times a number subtracted from 80 is 25.

You can rewrite the verbal sentence so it is easier to translate. *Fifteen times a number subtracted from 80 is 25* is the same as *80 minus 15 times a number is 25.* Let n represent the number.

80	minus	15	times	a number	is	25.
80	−	15	·	n	=	25

The equation is $80 - 15n = 25$.

▶ **Guided**Practice

1A. Two plus the quotient of a number and 8 is the same as 16.

1B. Twenty-seven times k is h squared decreased by 9.

Translating sentences to algebraic expressions and equations is a valuable skill in solving real-world problems.

Real-World Example 2 Use the Four-Step Problem-Solving Plan

AIR TRAVEL Refer to the information at the left. In how many days will 261,000 flights have occurred in the United States?

Understand The information given in the problem is that there are approximately 87,000 flights per day in the United States. We are asked to find how many days it will take for 261,000 flights to have occurred.

Plan Write an equation. Let d represent the number of days needed.

87,000	times	the number of days	equals	261,000.
87,000	•	d	=	261,000

Solve $87,000\ d = 261,000$ Find d by asking, "What number times 87,000 is 261,000?"

$$d = 3$$

Check Check your answer by substituting 3 for d in the equation.

$$87,000(3) \stackrel{?}{=} 261,000 \qquad \text{Substitute 3 for } d.$$

$$261,000 = 261,000 \checkmark \qquad \text{Multiply.}$$

The answer makes sense and works for the original problem.

Guided Practice

2. **GOVERNMENT** There are 50 members in the North Carolina Senate. This is 70 fewer than the number in the North Carolina House of Representatives. How many members are in the North Carolina House of Representatives?

A rule for the relationship between certain quantities is called a **formula**. These equations use variables to represent numbers and form general rules.

Example 3 Write a Formula

GEOMETRY Translate the sentence into a formula.

The area of a triangle equals the product of $\frac{1}{2}$ the length of the base and the height.

Words	The	area of a triangle	equals	the product of $\frac{1}{2}$ the length of the base and the height.

Variables   Let A = area, b = base, and h = height.

Equation	A	=	$\frac{1}{2}bh$

The formula for the area of a triangle is $A = \frac{1}{2}bh$.

Guided Practice

3. **GEOMETRY** Translate the sentence into a formula.
   In a right triangle, the square of the measure of the hypotenuse c is equal to the sum of the squares of the measures of the legs, a and b.

2 Write Sentences from Equations
If you are given an equation, you can write a sentence or create your own word problem.

Example 4 Translate Equations into Sentences

Translate each equation into a sentence.

a. $6z - 15 = 45$

| $6z$ | $-$ | 15 | $=$ | 45 |
| Six times z | minus | fifteen | equals | forty-five. |

b. $y^2 + 3x = w$

| y^2 | $+$ | $3x$ | $=$ | w |
| The sum of y squared | and | three times x | is | w. |

▶ **Guided Practice**

4A. $15 = 25u^2 + 2$

4B. $\frac{3}{2}r - t^3 = 132$

When given a set of information, you can create a problem that relates a story.

Example 5 Write a Problem

Write a problem based on the given information.

t = the time that Maxine drove in each turn; $t + 4$ = the time that Tia drove in each turn; $2t + (t + 4) = 28$

Sample problem:

Maxine and Tia went on a trip, and they took turns driving. During her turn, Tia drove 4 hours more than Maxine. Maxine took 2 turns, and Tia took 1 turn. Together they drove for 28 hours. How many hours did Maxine drive?

▶ **Guided Practice**

5. p = Beth's salary; $0.1p$ = bonus; $p + 0.1p = 525$

Check Your Understanding

 = Step-by-Step Solutions begin on page R13.

Example 1

Translate each sentence into an equation.

1. Three times r less than 15 equals 6.

2. The sum of q and four times t is equal to 29.

3 A number n squared plus 12 is the same as the quotient of p and 4.

4. Half of j minus 5 is the sum of k and 13.

5. The sum of 8 and three times k equals the difference of 5 times k and 3.

6. Three fourths of w plus 5 is one half of w increased by nine.

7. The quotient of 25 and t plus 6 is the same as twice t plus 1.

8. Thirty-two divided by y is equal to the product of three and y minus four.

Example 2

9. FINANCIAL LITERACY Samuel has $1900 in the bank. He wishes to increase his account to a total of $2500 by depositing $30 per week from his paycheck. Write and solve an equation to find how many weeks he needs to reach his goal.

10. CCSS MODELING Miguel is earning extra money by painting houses. He charges a $200 fee plus $12 per can of paint needed to complete the job. Write and use an equation to find how many cans of paint he needs for a $260 job.

Translate each sentence into a formula.

Example 3

11. The perimeter of a regular pentagon is 5 times the length of each side.

12. The area of a circle is the product of π and the radius r squared.

13. Four times π times the radius squared is the surface area of a sphere.

14. One third the product of the length of the side squared and the height is the volume of a pyramid with a square base.

Example 4

Translate each equation into a sentence.

15. $7m - q = 23$

16. $6 + 9k + 5j = 54$

17. $3(g + 8) = 4h - 10$

18. $6d^2 - 7f = 8d + f^2$

Example 5

Write a problem based on the given information.

19. $g =$ gymnasts on a team; $3g = 45$

20. $c =$ cost of a notebook; $0.25c =$ markup; $c + 0.25c = 3.75$

Practice and Problem Solving

Extra Practice is on page R2.

Example 1

Translate each sentence into an equation.

21. The difference of f and five times g is the same as 25 minus f.

22. Three times b less than 100 is equal to the product of 6 and b.

23. Four times the sum of 14 and c is a squared.

Example 2

24. MUSIC A piano has 52 white keys. Write and use an equation to find the number of octaves on a piano keyboard.

25. GARDENING A flat of plants contains 12 plants. Yoshi wants a garden that has three rows with 10 plants per row. Write and solve an equation for the number of flats Yoshi should buy.

Example 3

Translate each sentence into a formula.

26. The perimeter of a rectangle is equal to 2 times the length plus twice the width.

27 Celsius temperature C is five ninths times the difference of the Fahrenheit temperature F and 32.

28. The density of an object is the quotient of its mass and its volume.

29. Simple interest is computed by finding the product of the principal amount p, the interest rate r, and the time t.

Example 4

Translate each equation into a sentence.

30. $j + 16 = 35$

31. $4m = 52$

32. $7(p + 23) = 102$

33. $r^2 - 15 = t + 19$

34. $\frac{2}{5}v + \frac{3}{4} = \frac{2}{3}x^2$

35. $\frac{1}{3} - \frac{4}{5}z = \frac{4}{3}y^3$

Example 5 **Write a problem based on the given information.**

36. q = quarts of strawberries; $2.50q = 10$

37. p = the principal amount; $0.12p$ = the interest charged; $p + 0.12p = 224$

38. m = number of movies rented; $10 + 1.50m = 14.50$

39. p = the number of players in the game; $5p + 7$ = number of cards in a deck

For Exercises 40–43, match each sentence with an equation.

  A. $g^2 = 2(g - 10)$ **C.** $g^3 = 24g + 4$

  B. $\frac{1}{2}g + 32 = 15 + 6g$ **D.** $3g^2 = 30 + 9g$

40. One half of g plus thirty-two is as much as the sum of fifteen and six times g.

41. A number g to the third power is the same as the product of 24 and g plus 4.

42. The square of g is the same as two times the difference of g and 10.

43. The product of 3 and the square of g equals the sum of thirty and the product of nine and g.

44. FINANCIAL LITERACY Tim's bank contains quarters, dimes, and nickels. He has three more dimes than quarters and 6 fewer nickels than quarters. If he has 63 coins, write and solve an equation to find how many quarters Tim has.

45 SHOPPING Pilar bought 17 items for her camping trip, including tent stakes, packets of drink mix, and bottles of water. She bought 3 times as many packets of drink mix as tent stakes. She also bought 2 more bottles of water than tent stakes. Write and solve an equation to discover how many tent stakes she bought.

46. **MULTIPLE REPRESENTATIONS** In this problem, you will explore how to translate relations with powers.

x	2	3	4	5	6
y	5	10	17	26	37

  a. Verbal Write a sentence to describe the relationship between x and y in the table.

  b. Algebraic Write an equation that represents the data in the table.

  c. Graphical Graph each ordered pair and draw the function. Describe the graph as discrete or continuous.

H.O.T. Problems   Use Higher-Order Thinking Skills

47. OPEN ENDED Write a problem about your favorite television show that uses the equation $x + 8 = 30$.

48. CCSS REASONING The surface area of a three-dimensional object is the sum of the areas of the faces. If ℓ represents the length of the side of a cube, write a formula for the surface area of the cube.

49. CHALLENGE Given the perimeter P and width w of a rectangle, write a formula to find the length ℓ.

50. **WRITING IN MATH** How can you translate a verbal sentence into an algebraic equation?

51. Which equation *best* represents the relationship between the number of hours an electrician works h and the total charges c?

Cost of Electrician	
Emergency House Call	$30 one time fee
Rate	$55/hour

A $c = 30 + 55$

B $c = 30h + 55$

C $c = 30 + 55h$

D $c = 30h + 55h$

52. A car traveled at 55 miles per hour for 2.5 hours and then at 65 miles per hour for 3 hours. How far did the car travel in all?

F 300.5 mi   **H** 330 mi

G 305 mi   **J** 332.5 mi

53. SHORT RESPONSE Suppose each dimension of rectangle *ABCD* is doubled. What is the perimeter of the new *ABCD*?

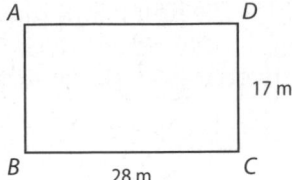

54. STATISTICS Stacy's first five science test scores were 95, 86, 83, 95, and 99. Which of the following is a true statement?

A The mode is the same as the median.

B The median is the same as the mean.

C The range is the same as the mode.

D The mode is the same as the mean.

55. POPULATION Identify the function graphed as *linear* or *nonlinear*. Then estimate and interpret the intercepts of the graph, any symmetry, where the function is positive, negative, increasing, and decreasing, the *x*-coordinate of any relative extrema, and the end behavior of the graph. (Lesson 1-8)

56. SHOPPING Cuties is having a sale on earrings that are regularly $29 for each pair. If you buy 2 pairs, you get 1 pair free. (Lesson 1-7)

  a. Make a table that shows the cost of buying 1 to 5 pairs of earrings.

  b. Write the data as a set of ordered pairs.

  c. Graph the data.

57. GEOMETRY Refer to the table below. (Lesson 1-6)

Polygon	triangle	quadrilateral	pentagon	hexagon	heptagon
Number of Sides	3	4	5	6	7
Interior Angle Sum	180	360	540	720	900

  a. Identify the independent and dependent variables.

  b. Identify the domain and range for this situation.

  c. State whether the function is *discrete* or *continuous*. Explain.

Evaluate each expression.

58. 9^2 **59.** 10^6 **60.** 3^5 **61.** 5^3

Algebra Lab
Solving Equations

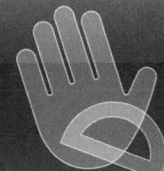

You can use **algebra tiles** to model solving equations. To **solve an equation** means to find the value of the variable that makes the equation true. An [x] tile represents the variable x. The [1] tile represents a positive 1. The [-1] tile represents a negative 1. And, the [-x] tile represents the variable negative x. The goal is to get the x-tile by itself on one side of the mat by using the rules stated below.

CCSS Common Core State Standards
Content Standards
A.REI.3 Solve linear equations and inequalities in one variable, including equations with coefficients represented by letters.
Mathematical Practices
8 Look for and express regularity in repeated reasoning.

Rules for Equation Models When Adding or Subtracting:

- You can remove or add the same number of identical algebra tiles to each side of the mat without changing the equation.

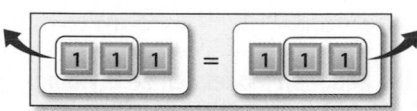

- One positive tile and one negative tile of the same unit are called a zero pair. Since $1 + (-1) = 0$, you can remove or add zero pairs to either side of the equation mat without changing the equation.

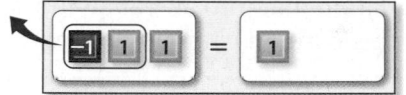

Activity 1   Addition Equation

Use an equation model to solve $x + 3 = -4$.

Step 1 Model the equation. Place 1 x-tile and 3 positive 1-tiles on one side of the mat. Place 4 negative 1-tiles on the other side of the mat.

Step 2 Isolate the x-term. Add 3 negative 1-tiles to each side. The resulting equation is $x = -7$.

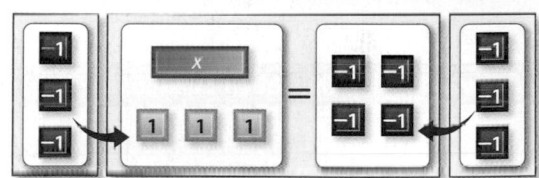

$$x + 3 = -4$$
$$x + 3 + (-3) = -4 + (-3)$$
$$x = -7$$

Activity 2   Subtraction Equation

Use an equation model to solve $x - 2 = 1$.

Step 1

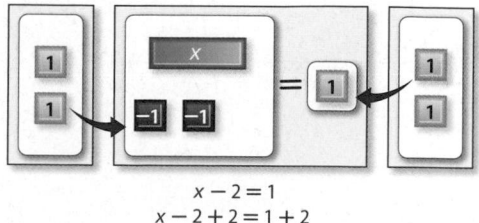

$$x - 2 = 1$$
$$x - 2 + 2 = 1 + 2$$

Place 1 x-tile and 2 negative 1-tiles on one side of the mat. Place 1 positive 1-tile on the other side of the mat. Then add 2 positive 1-tiles to each side.

Step 2

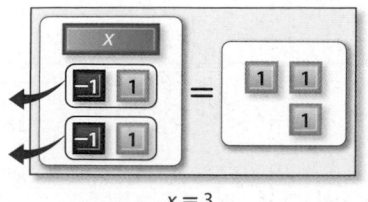

$$x = 3$$

Group the tiles to form zero pairs. Then remove all the zero pairs. The resulting equation is $x = 3$.

(continued on the next page)

Algebra Lab
Solving Equations *Continued*

Model and Analyze

Use algebra tiles to solve each equation.

1. $x + 4 = 9$ **2.** $x + (-3) = -4$ **3.** $x + 7 = -2$ **4.** $x + (-2) = 11$

5. **WRITING IN MATH** If $a = b$, what can you say about $a + c$ and $b + c$? about $a - c$ and $b - c$?

When solving multiplication equations, the goal is still to get the x-tile by itself on one side of the mat by using the rules for dividing.

Rules for the Equation Models When Dividing:

- You can group the tiles on each side of the equation mat into an equal number of groups without changing the equation.

- You can place an equal grouping on each side of the equation mat without changing the equation.

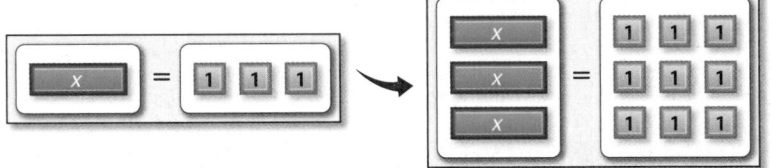

Activity 3   Multiplication Equations

Use an equation model to solve $3x = 12$.

Step 1 Model the equation. Place 3 x-tiles on one side of the mat. Place 12 positive 1-tiles on the other side of the mat.

Step 2 Isolate the x-term. Separate the tiles into 3 equal groups to match the 3 x-tiles. Each x-tile is paired with 4 positive 1-tiles. The resulting equation is $x = 4$.

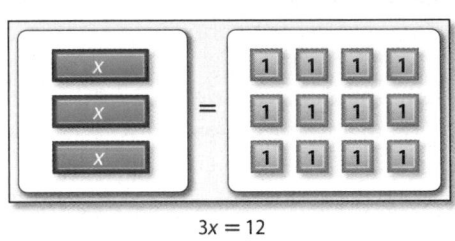

$$3x = 12$$
$$\frac{3x}{3} = \frac{12}{3}$$
$$x = 4$$

Model and Analyze

Use algebra tiles to solve each equation.

6. $5x = -15$ **7.** $-3x = -9$ **8.** $4x = 8$ **9.** $-6x = 18$

10. **MAKE A CONJECTURE** How would you use algebra tiles to solve $\frac{x}{4} = 5$? Discuss the steps you would take to solve this equation algebraically.

Solving One-Step Equations

∴ Then	∴ Now	∴ Why?
● You translated sentences into equations.	**1** Solve equations by using addition and subtraction. **2** Solve equations by using multiplication and division.	● A record for the most snow angels made at one time was set in Michigan when 3784 people participated. North Dakota had 8910 people register to break the record. To determine how many more people North Dakota had than Michigan, solve the equation $3784 + x = 8910$.

 NewVocabulary
solve an equation
equivalent equations

 Common Core State Standards

Content Standards
A.REI.1 Explain each step in solving a simple equation as following from the equality of numbers asserted at the previous step, starting from the assumption that the original equation has a solution. Construct a viable argument to justify a solution method.

A.REI.3 Solve linear equations and inequalities in one variable, including equations with coefficients represented by letters.

Mathematical Practices
6 Attend to precision.

1 Solve Equations Using Addition or Subtraction In an equation, the variable represents the number that satisfies the equation. To **solve an equation** means to find the value of the variable that makes the equation true.

The process of solving an equation requires assuming that the original equation has a solution and isolating the variable (with a coefficient of 1) on one side of the equation. Each step in this process results in equivalent equations. **Equivalent equations** have the same solution.

KeyConcept Addition Property of Equality

Words	If an equation is true and the same number is added to each side of the equation, the resulting equivalent equation is also true.
Symbols	For any real numbers a, b, and c, if $a = b$, then $a + c = b + c$.
Examples	$14 = 14$ $-3 = -3$ $14 + 3 = 14 + 3$ $+9 = +9$ $17 = 17$ $6 = 6$

Example 1 Solve by Adding

Solve $c - 22 = 54$.

Horizontal Method		**Vertical Method**
$c - 22 = 54$	Original equation	$c - 22 = 54$
$c - 22 + 22 = 54 + 22$	Add 22 to each side.	$+ 22 = + 22$
$c = 76$	Simplify.	$c = 76$

To check that 76 is the solution, substitute 76 for c in the original equation.

CHECK $c - 22 = 54$ Original equation
          $76 - 22 \stackrel{?}{=} 54$ Substitute 76 for c.
               $54 = 54\checkmark$ Subtract.

▶ **GuidedPractice**

1A. $113 = g - 25$ **1B.** $j - 87 = -3$

Similar to the Addition Property of Equality, the **Subtraction Property of Equality** can also be used to solve equations.

KeyConcept Subtraction Property of Equality

Words	If an equation is true and the same number is subtracted from each side of the equation, the resulting equivalent equation is also true.
Symbols	For any real numbers a, b, and c, if $a = b$, then $a - c = b - c$.

Examples

$$87 = 87$$
$$87 - 17 = 87 - 17$$
$$70 = 70$$

$$13 = 13$$
$$-28 = -28$$
$$\overline{-15 = -15}$$

Example 2 Solve by Subtracting

Solve $63 + m = 79$.

Horizontal Method		**Vertical Method**
$63 + m = 79$	Original equation	$63 + m = 79$
$63 - 63 + m = 79 - 63$	Subtract 63 from each side.	$\underline{-63 = -63}$
$m = 16$	Simplify.	$m = 16$

To check that 16 is the solution, replace m with 16 in the original equation.

CHECK $63 + m = 79$ Original equation
$63 + 16 \overset{?}{=} 79$ Substitution, $m = 16$
$79 = 79$ ✓ Simplify.

GuidedPractice

2A. $27 + k = 30$

2B. $-12 = p + 16$

2 Solve Equations Using Multiplication or Division In the equation $\frac{x}{3} = 9$, the variable x is divided by 3. To solve for x, undo the division by multiplying each side by 3. This is an example of the **Multiplication Property of Equality**.

KeyConcept Multiplication Property of Equality

Words	If an equation is true and each side is multiplied by the same nonzero number, the resulting equation is equivalent.
Symbols	For any real numbers a, b, and c, $c \neq 0$, if $a = b$, then $ac = bc$.
Example	If $x = 5$, then $3x = 15$.

Division Property of Equality

Words	If an equation is true and each side is divided by the same nonzero number, the resulting equation is equivalent.
Symbols	For any real numbers a, b, and c, $c \neq 0$, if $a = b$, then $\frac{a}{c} = \frac{b}{c}$.
Example	If $x = -20$, then $\frac{x}{5} = \frac{-20}{5}$ or -4.

The reciprocal of a number can be used to solve equations.

Example 3 Solve by Multiplying or Dividing

Solve each equation.

a. $\frac{2}{3}q = \frac{1}{2}$

$$\frac{2}{3}q = \frac{1}{2}$$ Original equation

$$\frac{3}{2}\left(\frac{2}{3}\right)q = \frac{3}{2}\left(\frac{1}{2}\right)$$ Multiply each side by $\frac{3}{2}$, the reciprocal of $\frac{2}{3}$.

$$q = \frac{3}{4}$$ Check the result.

b. $39 = -3r$

$$39 = -3r$$ Original equation

$$\frac{39}{-3} = \frac{-3r}{-3}$$ Divide each side by -3.

$$-13 = r$$ Check the result.

Review Vocabulary

reciprocal the multiplicative inverse of a number

▸ **Guided Practice**

3A. $\frac{3}{5}k = 6$

3B. $-\frac{1}{4} = \frac{2}{3}b$

We can also use reciprocals and properties of equality to solve real-world problems.

Real-World Example 4 Solve by Multiplying

SURVEYS Of a group of 13- to 15-year-old girls surveyed, 225, or about $\frac{9}{20}$ said they talk on the telephone while they watch television. About how many girls were surveyed?

Words	Nine twentieths times those surveyed	is	225.
▼			
Variable	Let g = the number of girls surveyed.		
▼			
Equation	$\frac{9}{20}g$	=	225

$$\frac{9}{20}g = 225$$ Original equation

$$\left(\frac{20}{9}\right)\frac{9}{20}g = \left(\frac{20}{9}\right)225$$ Multiply each side by $\frac{20}{9}$.

$$g = \frac{4500}{9}$$ $\left(\frac{20}{9}\right)\left(\frac{9}{20}\right) = 1$

$$g = 500$$ Simplify.

About 500 girls were surveyed.

Real-World Link

Almost half of 10- to 18-year-olds in the U.S. use a cell phone. Of those, 53% play games on their phones, more than 33% download games, 52% use the calendar/organizer, and nearly all teens with camera phones snap pictures.

Source: Lexdon Business Library

▸ **Guided Practice**

4. STAINED GLASS Allison is making a stained glass window. Her pattern requires that one fifth of the glass should be blue. She has 288 square inches of blue glass. If she intends to use all of her blue glass, how much glass will she need for the entire project?

Examples 1–3 Solve each equation. Check your solution.

1. $g + 5 = 33$

2. $104 = y - 67$

3. $\frac{2}{3} + w = 1\frac{1}{2}$

4. $-4 + t = -7$

5. $a + 26 = 35$

6. $-6 + c = 32$

7. $1.5 = y - (-5.6)$

8. $3 + g = \frac{1}{4}$

9. $x + 4 = \frac{3}{4}$

10. $\frac{t}{7} = -5$

11. $\frac{a}{36} = \frac{4}{9}$

12. $\frac{2}{3}n = 10$

13. $\frac{8}{9} = \frac{4}{5}k$

14. $12 = \frac{x}{-3}$

15. $-\frac{r}{4} = \frac{1}{7}$

Example 4

16. **FUNDRAISING** The television show "Idol Gives Back" raised money for relief organizations. During this show, viewers could call in and vote for their favorite performer. The parent company contributed $5 million for the 50 million votes cast. What did they pay for each vote?

17. **CCSS REASONING** Hana decides to buy her cat a bed from an online fund that gives $\frac{7}{8}$ of her purchase to shelters that care for animals. How much of Hana's money went to the animal shelter?

Online Price: $26.00
[1] [Add to Cart]
blue
yellow

Examples 1–3 Solve each equation. Check your solution.

18. $v - 9 = 14$

19. $44 = t - 72$

20. $-61 = d + (-18)$

21. $18 + z = 40$

22. $-4a = 48$

23. $12t = -132$

24. $18 - (-f) = 91$

25. $-16 - (-t) = -45$

26. $\frac{1}{3}v = -5$

27. $\frac{u}{8} = -4$

28. $\frac{a}{6} = -9$

29. $-\frac{k}{5} = \frac{7}{5}$

30. $\frac{3}{4} = w + \frac{2}{5}$

31. $-\frac{1}{2} + a = \frac{5}{8}$

32. $-\frac{t}{7} = \frac{1}{15}$

33. $-\frac{5}{7} = y - 2$

34. $v + 914 = -23$

35. $447 + x = -261$

36. $-\frac{1}{7}c = 21$

37. $-\frac{2}{3}h = -22$

38. $\frac{3}{5}q = -15$

39. $\frac{n}{8} = -\frac{1}{4}$

40. $\frac{c}{4} = -\frac{9}{8}$

41. $\frac{2}{3} + r = -\frac{4}{9}$

Example 4

42. **CATS** A domestic cat can run at speeds of 27.5 miles per hour when chasing prey. A cheetah can run 42.5 miles per hour faster when chasing prey. How fast can the cheetah go?

43. **CARS** The average time t it takes to manufacture a car in the United States is 24.9 hours. This is 8.1 hours longer than the average time it takes to manufacture a car in Japan. Write and solve an equation to find the average time in Japan.

Solve each equation. Check your solution.

44. $\dfrac{x}{9} = 10$

45. $\dfrac{b}{7} = -11$

46. $\dfrac{3}{4} = \dfrac{c}{24}$

47. $\dfrac{2}{3} = \dfrac{1}{8}y$

48. $\dfrac{2}{3}n = 14$

49. $\dfrac{3}{5}g = -6$

50. $4\dfrac{1}{5} = 3p$

51. $-5 = 3\dfrac{1}{2}x$

52. $6 = -\dfrac{1}{2}n$

53. $-\dfrac{2}{5} = -\dfrac{z}{45}$

54. $-\dfrac{8}{24} = \dfrac{5}{12}$

55. $-\dfrac{v}{5} = -45$

Write an equation for each sentence. Then solve the equation.

56. Six times a number is 132.

57. Two thirds equals negative eight times a number.

58. Five elevenths times a number is 55.

59. Four fifths is equal to ten sixteenths of a number.

60. Three and two thirds times a number equals two ninths.

61 Four and four fifths times a number is one and one fifth.

62. **CCSS PRECISION** Adelina is comparing prices for two brands of health and energy bars at the local grocery store. She wants to get the best price for each bar.

   a. Write an equation to find the price for each bar of the Feel Great brand.

   b. Write an equation to find the price of each bar for the Super Power brand.

   c. Which bar should Adelina buy? Explain.

63. **MEDIA** The world's largest passenger plane, the Airbus A380, was first used by Singapore Airlines in 2005. The following description appeared on a news Web site after the plane was introduced.

"That airline will see the A380 transporting some 555 passengers, 139 more than a similarly set-up 747."

How many passengers will a similarly set-up 747 transport?

64. **FUEL** In 2004, approximately 5 million cars and trucks were classified as flex-fuel, which means they could run on gasoline or ethanol. In 2009, that number increased to about 8 million. How many more cars and trucks were flex-fuel in 2009?

65. **CHEERLEADING** At a certain cheerleading competition, the maximum time per team, including the set up, is 3 minutes. The Ridgeview High School squad's performance time is 2 minutes and 34 seconds. How much time does the squad have left for their set up?

66. **COMIC BOOKS** An X-Men #1 comic book in mint condition recently sold for $45,000. An Action Comics #63 (Mile High), also in mint condition, sold for $15,000. How much more did the X-Men comic book sell for than the Action Comics book?

67. **MOVIES** A certain movie made $1.6 million in ticket sales. Its sequel made $0.8 million in ticket sales. How much more did the first movie make than the sequel?

68. **CAMERAS** An electronics store sells a certain digital camera for $126. This is $\dfrac{2}{3}$ of the price that a photography store charges. What is the cost of the camera at the photography store?

69 **BLOGS** In 2006, 57 million American adults read online blogs. However, 45 million fewer American adults say that they maintain their own blog. How many American adults maintain a blog?

70. SCIENCE CAREERS According to the Bureau of Labor and Statistics, approximately 140,000,000 people were employed in the United States in 2009.

   a. The number of people in production occupations times 20 is the number of working people. Write an equation to represent the number of people employed in production occupations in 2009. Then solve the equation.

   b. The number of people in repair occupations is 2,300,000 less than the number of people in production occupations. How many people are in repair occupations?

71. DANCES Student Council has a budget of $1000 for the homecoming dance. So far, they have spent $350 dollars for music.

   a. Write an equation to represent the amount of money left to spend. Then solve the equation.

   b. They then spent $225 on decorations. Write an equation to represent the amount of money left.

   c. If the Student Council spent their entire budget, write an equation to represent how many $6 tickets they must sell to make a profit.

H.O.T. Problems   Use Higher-Order Thinking Skills

72. WHICH ONE DOESN'T BELONG? Identify the equation that does not belong with the other three. Explain your reasoning.

$n + 14 = 27$	$12 + n = 25$	$n - 16 = 29$	$n - 4 = 9$

73. OPEN ENDED Write an equation involving addition and demonstrate two ways to solve it.

74. REASONING For which triangle is the height not $4\frac{1}{2}b$, where b is the length of the base?

Triangle	Base (cm)	Height (cm)
$\triangle ABC$	3.8	17.1
$\triangle MQP$	5.4	24.3
$\triangle RST$	6.3	28.5
$\triangle TRW$	1.6	7.2

75. CCSS STRUCTURE Determine whether each sentence is *sometimes*, *always*, or *never true*. Explain your reasoning.

   a. $x + x = x$ **b.** $x + 0 = x$

76. REASONING Determine the value for each statement below.

   a. If $x - 7 = 14$, what is the value of $x - 2$?

   b. If $t + 8 = -12$, what is the value of $t + 1$?

77. CHALLENGE Solve each equation for x. Assume that $a \neq 0$.

   a. $ax = 12$ **b.** $x + a = 15$ **c.** $-5 = x - a$ **d.** $\frac{1}{a}x = 10$

78. WRITING IN MATH Consider the Multiplication Property of Equality and the Division Property of Equality. Explain why they can be considered the same property. Which one do you think is easier to use?

79. Which of the following best represents the equation $w - 15 = 33$?

  A Jake added w ounces of water to his bottle, which originally contained 33 ounces of water. How much water did he add?

  B Jake added 15 ounces of water to his bottle, for a total of 33 ounces. How much water w was originally in the bottle?

  C Jake drank 15 ounces of water from his bottle and 33 ounces were left. How much water w was originally in the bottle?

  D Jake drank 15 ounces of water from his water bottle, which originally contained 33 ounces. How much water w was left?

80. SHORT RESPONSE Charlie's company pays him for every mile that he drives on his trip. When he drives 50 miles, he is paid $30. To the nearest tenth, how many miles did he drive if he was paid $275?

81. The table shows the results of a survey given to 500 international travelers. Based on the data, which statement is true?

Vacation Plans	
Destination	**Percent**
The Tropics	37
Europe	19
Asia	17
Other	17
No Vacation	10

  F Fifty have no vacation plans.

  G Fifteen are going to Asia.

  H One third are going to the tropics.

  J One hundred are going to Europe.

82. GEOMETRY The amount of water needed to fill a pool represents the pool's _____.

  A volume   **C** circumference

  B surface area   **D** perimeter

Translate each sentence into an equation. (Lesson 2-1)

83. The sum of twice r and three times k is identical to thirteen.

84. The quotient of t and forty is the same as twelve minus half of u.

85. The square of m minus the cube of p is sixteen.

86. TOYS Identify the function graphed as *linear* or *nonlinear*. Then estimate and interpret the intercepts of the graph, any symmetry, where the function is positive, negative, increasing, and decreasing, the x-coordinate of any relative extrema, and the end behavior of the graph. (Lesson 1-8)

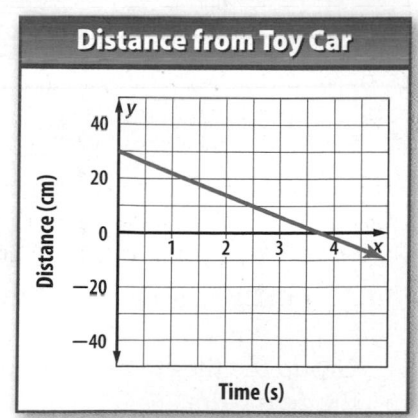

Distance from Toy Car

87. COMMUNICATION Sato communicates with friends for a project. He averages 5 hours using email, 8 hours on the phone, and 2 hours with them in person the first week. If this trend continues, write and evaluate an expression to predict how many hours he will spend communicating with friends over the next 12 weeks.

88. PETS The Poochie Pet supply store has the following items on sale. Write and evaluate an expression to find the total cost of purchasing 1 collar, 2 T-shirts, 3 kerchiefs, 1 leash, and 4 flying disks.

Item	Cost ($)
studded collar	4.50
kerchief	3.00
doggy T-shirt	6.25
leash	5.50
flying disk	3.25

2-3 Algebra Lab
Solving Multi-Step Equations

You can use algebra tiles to model solving multi-step equations.

CCSS Common Core State Standards
Content Standards
A.REI.3 Solve linear equations and inequalities in one variable, including equations with coefficients represented by letters.

Activity

Use an equation model to solve $4x + 3 = -5$.

Step 1 Model the equation.

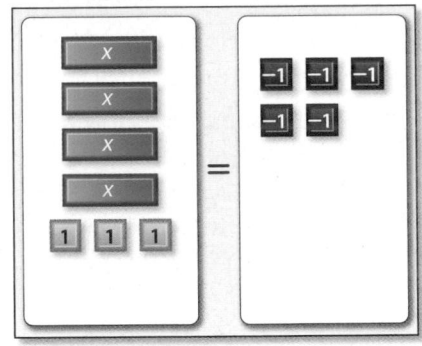

$$4x + 3 = -5$$

Place 4 x-tiles and 3 positive 1-tiles on one side of the mat. Place 5 negative 1-tiles on the other side.

Step 2 Isolate the x-term.

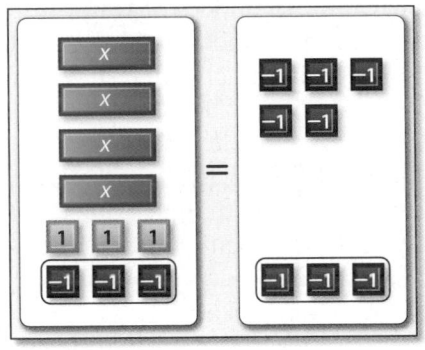

$$4x + 3 - 3 = -5 - 3$$

Since there are 3 positive 1-tiles with the x-tiles, add 3 negative 1-tiles to each side to form zero pairs.

Step 3 Remove zero pairs.

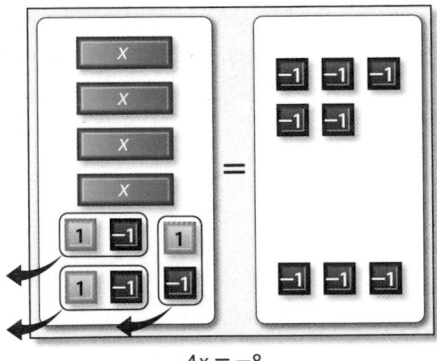

$$4x = -8$$

Group the tiles to form zero pairs and remove the zero pairs.

Step 4 Group the tiles.

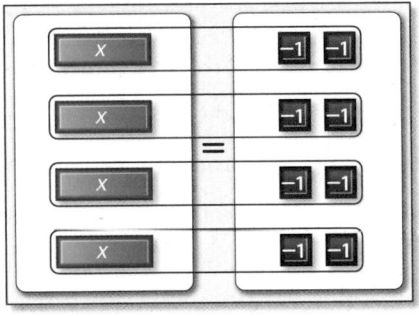

$$\frac{4x}{4} = \frac{-8}{4}$$
$$x = -2$$

Separate the remaining tiles into 4 equal groups to match the 4 x-tiles. Each x-tile is paired with 2 negative 1-tiles. The resulting equation is $x = -2$.

Model

Use algebra tiles to solve each equation.

1. $3x - 7 = -10$
2. $2x + 5 = 9$
3. $5x - 7 = 8$
4. $-7 = 3x + 8$
5. $5 + 4x = -11$
6. $3x + 1 = 7$
7. $11 = 2x - 5$
8. $7 + 6x = -11$

9. What would be your first step in solving $8x - 29 = 67$?

10. What steps would you use to solve $9x + 14 = -49$?

Solving Multi-Step Equations

FRANCE

Gand

Paris

Chablis

Tallard

:·Then	:·Now	:·Why?

● You solved one-step equations.

1 Solve equations involving more than one operation.

2 Solve equations involving consecutive integers.

● The Tour de France is the premier cycling event in the world. The map shows the 2007 Tour de France course. If the length of the shortest portion of the race can be represented by k, the expression $4k + 20$ is the length of the longest stage or 236 kilometers. This can be described by the equation $4k + 20 = 236$.

NewVocabulary
multi-step equation
consccutive integers
number theory

Common Core State Standards

Content Standards
A.REI.1 Explain each step in solving a simple equation as following from the equality of numbers asserted at the previous step, starting from the assumption that the original equation has a solution. Construct a viable argument to justify a solution method.

A.REI.3 Solve linear equations and inequalities in one variable, including equations with coefficients represented by letters.

Mathematical Practices
8 Look for and express regularity in repeated reasoning.

1 **Solve Multi-Step Equations** Since the above equation requires more than one step to solve, it is called a **multi-step equation**. To solve this equation, we must undo each operation by working backward.

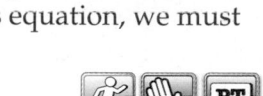

Example 1 Solve Multi-Step Equations

Solve each equation. Check your solution.

a. $11x - 4 = 29$

$11x - 4 = 29$	Original equation
$11x - 4 + 4 = 29 + 4$	Add 4 to each side.
$11x - 33$	Simplify.
$\dfrac{11x}{11} = \dfrac{33}{11}$	Divide each side by 11.
$x = 3$	Simplify.

b. $\dfrac{a + 7}{8} = 5$

$\dfrac{a + 7}{8} = 5$	Original equation
$8\left(\dfrac{a + 7}{8}\right) = 8(5)$	Multiply each side by 8.
$a + 7 = 40$	Simplify.
$\underline{-7 = -7}$	Subtract 7 from each side.
$a = 33$	Simplify.

You can check your solutions by substituting the results back into the original equations.

▶ **Guided**Practice

Solve each equation. Check your solution.

1A. $2a - 6 = 4$

1B. $\dfrac{n + 1}{-2} = 15$

Real-World Example 2 Write and Solve a Multi-Step Equation

SHOPPING Hiroshi is buying a pair of water skis that are on sale for $\frac{2}{3}$ of the original price. After he uses a $25 gift certificate, the total cost before taxes is $115. What was the original price of the skis? Write an equation for the problem. Then solve the equation.

| Words | Two thirds | of | the price | minus | 25 | is | 115. |

Variable Let $p =$ original price of the skis.

Equation $\frac{2}{3}$ · p − 25 = 115

$$\frac{2}{3}p - 25 = 115 \qquad \text{Original equation}$$

$$\frac{2}{3}p - 25 + 25 = 115 + 25 \qquad \text{Add 25 to each side.}$$

$$\frac{2}{3}p = 140 \qquad \text{Simplify.}$$

$$\frac{3}{2}\left(\frac{2}{3}p\right) = \frac{3}{2}(140) \qquad \text{Multiply each side by } \frac{3}{2}.$$

$$p = 210 \qquad \text{Simplify.}$$

The original price of the skis was $210.

▶ **Guided**Practice

2A. RETAIL A music store has sold $\frac{3}{5}$ of their hip-hop CDs, but 10 were returned. Now the store has 62 hip-hop CDs. How many were there originally?

2B. READING Len read $\frac{3}{4}$ of a graphic novel over the weekend. Monday, he read 22 more pages. If he has read 220 pages, how many pages does the book have?

2 **Solve Consecutive Integer Problems** **Consecutive integers** are integers in counting order, such as 4, 5, and 6 or n, $n + 1$, and $n + 2$. Counting by two will result in *consecutive even integers* if the starting integer n is even and *consecutive odd integers* if the starting integer n is odd.

ConceptSummary Consecutive Integers			
Type	**Words**	**Symbols**	**Example**
Consecutive Integers	Integers that come in counting order.	$n, n + 1, n + 2, \ldots$	$\ldots, -2, -1, 0, 1, 2, \ldots$
Consecutive Even Integers	Even integer followed by the next even integer.	$n, n + 2, n + 4, \ldots$	$\ldots, -2, 0, 2, 4, \ldots$
Consecutive Odd Integers	Odd integer followed by the next odd integer.	$n, n + 2, n + 4, \ldots$	$\ldots, -1, 1, -3, 5, \ldots$

Number theory is the study of numbers and the relationships between them.

Example 3 Solve a Consecutive Integer Problem

NUMBER THEORY Write an equation for the following problem. Then solve the equation and answer the problem.

Find three consecutive odd integers with a sum of −51.

Let n = the least odd integer.

Then $n + 2$ = the next greater odd integer, and $n + 4$ = the greatest of the three integers.

Words	The sum of three consecutive odd integers	is	−51.
Equation	$n + (n + 2) + (n + 4)$	=	−51.

$$n + (n + 2) + (n + 4) = -51 \qquad \text{Original equation}$$
$$3n + 6 = -51 \qquad \text{Simplify.}$$
$$\underline{-6 = -6} \qquad \text{Subtract 6 from each side.}$$
$$3n = -57 \qquad \text{Simplify.}$$
$$\frac{3n}{3} = \frac{-57}{3} \qquad \text{Divide each side by 3.}$$
$$n = -19 \qquad \text{Simplify.}$$

$n + 2 = -19 + 2$ or -17 $\qquad$ $n + 4 = -19 + 4$ or -15

The consecutive odd integers are −19, −17, and −15.

CHECK −19, −17, and −15 are consecutive odd integers.
$$-19 + (-17) + (-15) = -51 \checkmark$$

> **StudyTip**
>
> **CCSS** Regularity You can use the same expressions to represent either consecutive even integers or consecutive odd integers. It is the value of n (odd or even) that differs between the two expressions.

▶ **Guided**Practice

3. Write an equation for the following problem. Then solve the equation and answer the problem.

Find three consecutive integers with a sum of 21.

Check Your Understanding ◯ = Step-by-Step Solutions begin on page R13. ✓

Example 1   Solve each equation. Check your solution.

    ① $3m + 4 = -11$ **2.** $12 = -7f - 9$ **3.** $-3 = 2 + \frac{a}{11}$

    4. $\frac{3}{2}a - 8 = 11$ **5.** $8 = \frac{x-5}{7}$ **6.** $\frac{c+1}{-3} = -21$

Example 2   **7. NUMBER THEORY** Twelve decreased by twice a number equals −34. Write an equation for this situation and then find the number.

    8. BASEBALL Among the career home run leaders for Major League Baseball, Hank Aaron has 175 fewer than twice the number that Dave Winfield has. Hank Aaron hit 755 home runs. Write an equation for this situation. How many home runs did Dave Winfield hit in his career?

Example 3   Write an equation and solve each problem.

    9. Find three consecutive odd integers with a sum of 75.

    10. Find three consecutive integers with a sum of −36.

Example 1 Solve each equation. Check your solution.

11. $3t + 7 = -8$

12. $8 = 16 + 8n$

13. $-34 = 6m - 4$

14. $9x + 27 = -72$

15. $\dfrac{y}{5} - 6 = 8$

16. $\dfrac{f}{-7} - 8 = 2$

17. $1 + \dfrac{r}{9} = 4$

18. $\dfrac{k}{3} + 4 = -16$

19. $\dfrac{n-2}{7} = 2$

20. $14 = \dfrac{6+z}{-2}$

21. $-11 = \dfrac{a-5}{6}$

22. $\dfrac{22-w}{3} = -7$

Example 2 **23** **FINANCIAL LITERACY** The Cell+ Cellular Phone store offers the plans shown in the table. Raul chose the business plan and has budgeted $100 per month. Write an equation for this situation, and determine how many minutes per month he can use the phone and stay within budget.

Plan	Flat Monthly Fee	Anytime Minutes	Cost per Minute After Anytime Minutes
personal	$29.99	250	$0.20
business	$49.99	650	$0.15
executive	$59.99	1200	$0.10

Example 3 Write an equation and solve each problem.

24. Fourteen less than three fourths of a number is negative eight. Find the number.

25. Seventeen is thirteen subtracted from six times a number. What is the number?

26. Find three consecutive even integers with the sum of -84.

27. Find three consecutive odd integers with the sum of 141.

28. Find four consecutive integers with the sum of 54.

29. Find four consecutive integers with the sum of -142.

Solve each equation. Check your solution.

30. $-6m - 8 = 24$

31. $45 = 7 - 5n$

32. $\dfrac{2b}{3} + 6 = 24$

33. $\dfrac{5x}{9} - 11 = -51$

34. $65 = \dfrac{3}{4}c - 7$

35. $9 + \dfrac{2}{3}x = 81$

36. $-\dfrac{5}{2} = \dfrac{3}{4}z + \dfrac{1}{2}$

37. $\dfrac{5}{6}k + \dfrac{2}{3} = \dfrac{4}{3}$

38. $-\dfrac{1}{5} - \dfrac{4}{9}a = \dfrac{2}{15}$

39. $-\dfrac{y}{7} = \dfrac{3}{4} - \dfrac{b}{2}$

Write an equation and solve each problem.

40. **CCSS** **REASONING** The ages of three brothers are consecutive integers with the sum of 96. How old are the brothers?

 41. **VOLCANOES** Moving lava can build up and form beaches at the coast of an island. The growth of an island in a seaward direction may be modeled as $8y + 2$ centimeters, where y represents the number of years that the lava flows. An island has expanded 60 centimeters seaward. How long has the lava flowed?

Solve each equation. Check your solution.

42. $-5x - 4.8 = 6.7$

43 $3.7q + 26.2 = 111.67$

44. $0.6a + 9 = 14.4$

45. $\frac{c}{2} - 4.3 = 11.5$

46. $9 = \dfrac{-6p - (-3)}{-8}$

47. $3.6 - 2.4m = 12$

48. If $7m - 3 = 53$, what is the value of $11m + 2$?

49. If $13y + 25 = 64$, what is the value of $4y - 7$?

50. If $-5c + 6 = -69$, what is the value of $6c - 15$?

51. **AMUSEMENT PARKS** An amusement park offers a yearly membership of $275 that allows for free parking and admission to the park. Members can also use the water park for an additional $5 per day. Nonmembers pay $6 for parking, $15 for admission, and $9 for the water park.

   a. Write and solve an equation to find the number of visits it would take for the total cost to be the same for a member and a nonmember if they both use the water park at each visit.

   b. Make a table for the costs of members and nonmembers after 3, 6, 9, 12, and 15 visits to the park.

   c. Plot these points on a coordinate graph and describe what you see.

52. **SHOPPING** At The Family Farm, you can pick your own fruits and vegetables.

   a. The cost of a bag of potatoes is $1.50 less than $\frac{1}{2}$ of the price of apples. Write and solve an equation to find the cost of potatoes.

   b. The price of each zucchini is 3 times the price of winter squash minus $7. Write and solve an equation to find the cost of zucchini.

   c. Write an equation to represent the cost of a pumpkin using the cost of the blueberries.

The Family Farm	
Fruit	**Price ($)**
Apples	6.99/bag
Pumpkins	5.00 each
Blueberries	2.99/qt
Winter squash	2.99 each

H.O.T. Problems   Use Higher-Order Thinking Skills

53. **OPEN ENDED** Write a problem that can be modeled by the equation $2x + 40 = 60$. Then solve the equation and explain the solution in the context of the problem.

54. **CHALLENGE** Solve each equation for x. Assume that $a \neq 0$.

   a. $ax + 7 = 5$ **b.** $\frac{1}{a}x - 4 = 9$ **c.** $2 - ax = -8$

55. **REASONING** Determine whether each equation has a solution. Justify your answer.

   a. $\frac{a + 4}{5 + a} = 1$ **b.** $\frac{1 + b}{1 - b} = 1$ **c.** $\frac{c - 5}{5 - c} = 1$

56. **CCSS REGULARITY** Determine whether the following statement is *sometimes*, *always*, or *never* true. Explain your reasoning.

    The sum of three consecutive odd integers equals an even integer.

57. **WRITING IN MATH** Write a paragraph explaining the order of the steps that you would take to solve a multi-step equation.

58. Which is the best estimate for the number of minutes on the calling card advertised below?

$10 **Prepaid Calling Card**

Only 5.4¢ per Minute

A 10 min   **C** 50 min

B 20 min   **D** 200 min

59. GRIDDED RESPONSE The scale factor for two similar triangles is 2 : 3. The perimeter of the smaller triangle is 56 cm. What is the perimeter of the larger triangle in centimeters?

60. Mr. Morrison is draining his cylindrical pool. The pool has a radius of 10 feet and a standard height of 4.5 feet. If the pool water is pumped out at a constant rate of 5 gallons per minute, about how long will it take to drain the pool? ($1 \text{ ft}^3 = 7.5 \text{ gal}$)

F 37.8 min   **H** 25.4 h

G 7 h   **J** 35.3 h

61. STATISTICS Look at the golf scores for the five players in the table.

Player	1	2	3	4	5
Score	80	91	103	79	78

Which of these is the range of the golf scores?

A 10   **C** 35

B 25   **D** 40

Spiral Review

62. GAS MILEAGE A midsize car with a 4-cylinder engine travels 34 miles on a gallon of gas. This is 10 miles more than a luxury car with an 8-cylinder engine travels on a gallon of gas. How many miles does a luxury car travel on a gallon of gas? (Lesson 2-2)

63. DEER In a recent year, 1286 female deer were born in Clark County. That is 93 fewer than the number of male deer born. How many male deer were born that year? (Lesson 2-2)

Translate each equation into a verbal sentence. (Lesson 2-1)

64. $f - 15 = 6$

65. $3h + 7 = 20$

66. $k^2 + 18 = 54 - m$

67. $3p = 8p - r$

68. $\frac{3}{5}t + \frac{1}{3} = t$

69. $\frac{1}{2}v = \frac{2}{3}v + 4$

70. GEOGRAPHY The Pacific Ocean covers about 46% of Earth. If P represents the surface area of the Pacific Ocean and E represents the surface area of Earth, write an equation for this situation. (Lesson 2-1)

Find the value of n in each equation. Then name the property that is used. (Lesson 1-3)

71. $1.5 + n = 1.5$

72. $8n = 1$

73. $4 - n = 0$

74. $1 = 2n$

Skills Review

Evaluate each expression.

75. $5 + 3(4^2)$

76. $\frac{38 - 12}{2 \cdot 13}$

77. $[5(1 + 1)]^3$

78. $[8(2) - 4^2] + 7(4)$

Solving Equations with the Variable on Each Side

:: Then	:: Now	:: Why?
• You solved multi-step equations.	**1** Solve equations with the variable on each side. **2** Solve equations involving grouping symbols.	• The equation $y = 1.3x + 19$ represents the number of times Americans eat in their cars each year, where x is the number of years since 1985, and y is the number of times that they eat in their car. The equation $y = -1.3x + 93$ represents the number of times Americans eat in restaurants each year, where x is the number of years since 1985, and y is the number of times that they eat in a restaurant. The equation $1.3x + 19 = -1.3x + 93$ represents the year when the number of times Americans eat in their cars will equal the number of times Americans eat in restaurants.

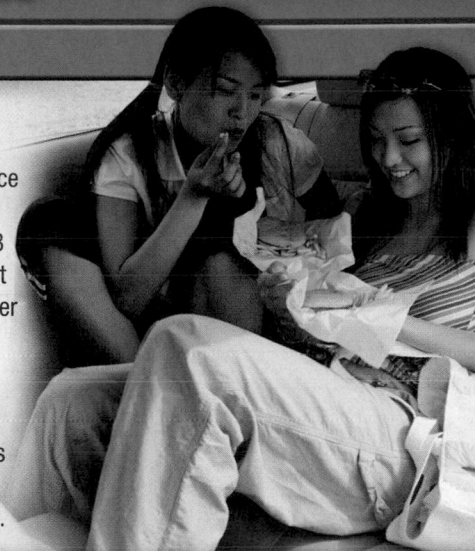

 NewVocabulary
identity

 **Common Core State Standards**

Content Standards
A.REI.1 Explain each step in solving a simple equation as following from the equality of numbers asserted at the previous step, starting from the assumption that the original equation has a solution. Construct a viable argument to justify a solution method.

A.REI.3 Solve linear equations and inequalities in one variable, including equations with coefficients represented by letters.

Mathematical Practices
1 Make sense of problems and persevere in solving them.
5 Use appropriate tools strategically.

1 Variables on Each Side To solve an equation that has variables on each side, use the Addition or Subtraction Property of Equality to write an equivalent equation with the variable terms on one side.

Example 1 Solve an Equation with Variables on Each Side

Solve $2 + 5k = 3k - 6$. Check your solution.

$2 + 5k = 3k - 6$	Original equation
$\underline{-3k = -3k}$	Subtract $3k$ from each side.
$2 + 2k = -6$	Simplify.
$\underline{-2 \quad\quad = -2}$	Subtract 2 from each side.
$2k = -8$	Simplify.
$\dfrac{2k}{2} = \dfrac{-8}{2}$	Divide each side by 2.
$k = -4$	Simplify.

CHECK

$2 + 5k = 3k - 6$	Original equation
$2 + 5(-4) \stackrel{?}{=} 3(-4) - 6$	Substitution, $k = -4$
$2 + -20 \stackrel{?}{=} -12 - 6$	Multiply.
$-18 = -18$ ✔	Simplify.

▶ **GuidedPractice**

Solve each equation. Check your solution.

1A. $3w + 2 = 7w$

1B. $5a + 2 = 6 - 7a$

1C. $\dfrac{x}{2} + 1 = \dfrac{1}{4}x - 6$

1D. $1.3c = 3.3c + 2.8$

Gen Nishino/Taxi/Getty Images

2 Grouping Symbols
If equations contain grouping symbols such as parentheses or brackets, use the Distributive Property first to remove the grouping symbols.

Example 2 Solve an Equation with Grouping Symbols

Solve $6(5m - 3) = \frac{1}{3}(24m + 12)$.

$6(5m - 3) = \frac{1}{3}(24m + 12)$	Original equation
$30m - 18 = 8m + 4$	Distributive Property
$30m - 18 - 8m = 8m + 4 - 8m$	Subtract $8m$ from each side.
$22m - 18 = 4$	Simplify.
$22m - 18 + 18 = 4 + 18$	Add 18 to each side.
$22m = 22$	Simplify.
$\frac{22m}{22} = \frac{22}{22}$	Divide each side by 22.
$m = 1$	Simplify.

> **StudyTip**
>
> Solving an Equation
> You may want to eliminate the terms with a variable from one side before eliminating a constant.

▶ **Guided Practice**

Solve each equation. Check your solution.

2A. $8s - 10 = 3(6 - 2s)$

2B. $7(n - 1) = -2(3 + n)$

Some equations may have no solution. That is, there is no value of the variable that will result in a true equation. Some equations are true for all values of the variables. These are called **identities**.

Example 3 Find Special Solutions

Solve each equation.

a. $5x + 5 = 3(5x - 4) - 10x$

$5x + 5 = 3(5x - 4) - 10x$	Original equation
$5x + 5 = 15x - 12 - 10x$	Distributive Property
$5x + 5 = 5x - 12$	Simplify.
$\underline{-5x \quad\quad = -5x}$	Subtract $5x$ from each side.
$5 \neq -12$	

Since $5 \neq -12$, this equation has no solution.

> **ReadingMath**
>
> No Solution The symbol that represents no solution is $\varnothing$.

b. $3(2b - 1) - 7 = 6b - 10$

$3(2b - 1) - 7 = 6b - 10$	Original equation
$6b - 3 - 7 = 6b - 10$	Distributive Property
$6b - 10 = 6b - 10$	Simplify.
$0 = 0$	Subtract $6b - 10$ from each side.

Since the expressions on each side of the equation are the same, this equation is an identity. It is true for all values of b.

▶ **Guided Practice**

3A. $7x + 5(x - 1) = -5 + 12x$

3B. $6(y - 5) = 2(10 + 3y)$

The steps for solving an equation can be summarized as follows.

ConceptSummary Steps for Solving Equations

Step 1 Simplify the expressions on each side. Use the Distributive Property as needed.

Step 2 Use the Addition and/or Subtraction Properties of Equality to get the variables on one side and the numbers without variables on the other side. Simplify.

Step 3 Use the Multiplication or Division Property of Equality to solve.

There are many situations in which you must simplify expressions with grouping symbols in order to solve an equation.

Standardized Test Example 4 Write an Equation

Find the value of x so that the figures have the same area.

10 cm

x cm

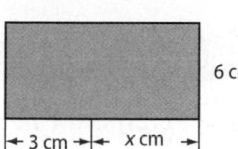

6 cm

← 3 cm → ← x cm →

A 3   **C** 6.5

B 4.5   **D** 7

Test-TakingTip

CCSS Tools There is often more than one way to solve a problem. In this example, you can write an algebraic equation and solve for x. Or you can substitute each answer choice into the formulas to find the correct answer.

Read the Test Item

The area of the first rectangle is $10x$, and the area of the second is $6(3 + x)$. The equation $10x = 6(3 + x)$ represents this situation.

Solve the Test Item

A $\quad 10x = 6(3 + x)$

$10(3) \stackrel{?}{=} 6(3 + 3)$

$30 \stackrel{?}{=} 6(6)$

$30 \neq 36$ ✗

B $\quad 10x = 6(3 + x)$

$10(4.5) \stackrel{?}{=} 6(3 + 4.5)$

$45 \stackrel{?}{=} 6(7.5)$

$45 = 45$ ✓

Since the value 4.5 results in a true statement, you do not need to check 6.5 and 7. The answer is B.

▸ **Guided**Practice

4. Find the value of x so that the figures have the same perimeter.

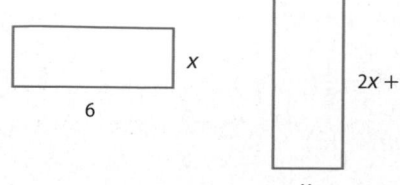

x

6

$2x + 2$

x

F 1.5   **G** 2   **H** 3.2   **J** 4

Examples 1–3 Solve each equation. Check your solution.

1. $13x + 2 = 4x + 38$

2. $\frac{2}{3} + \frac{1}{6}q = \frac{5}{6}q + \frac{1}{3}$

3. $6(n + 4) = -18$

4. $7 = -11 + 3(b + 5)$

5. $5 + 2(n + 1) = 2n$

6. $7 - 3r = r - 4(2 + r)$

7. $14v + 6 = 2(5 + 7v) - 4$

8. $5h - 7 = 5(h - 2) + 3$

Example 4

9. **MULTIPLE CHOICE** Find the value of x so that the figures have the same perimeter.

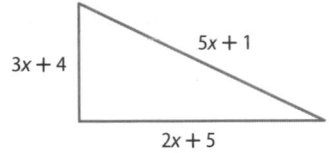

A 4   B 5   C 6   D 7

Practice and Problem Solving **Extra Practice is on page R2.**

Examples 1–3 Solve each equation. Check your solution.

10. $7c + 12 = -4c + 78$

11. $2m - 13 = -8m + 27$

12. $9x - 4 = 2x + 3$

13. $6 + 3t = 8t - 14$

14. $\frac{b - 4}{6} = \frac{b}{2}$

15. $\frac{5v - 4}{10} = \frac{4}{5}$

16. $8 = 4(r + 4)$

17. $6(n + 5) = 66$

18. $5(g + 8) - 7 = 103$

19. $12 - \frac{4}{5}(x + 15) = 4$

20. $3(3m - 2) = 2(3m + 3)$

21. $6(3a + 1) - 30 = 3(2a - 4)$

Example 4

22. **GEOMETRY** Find the value of x so the rectangles have the same area.

23. **NUMBER THEORY** Four times the lesser of two consecutive even integers is 12 less than twice the greater number. Find the integers.

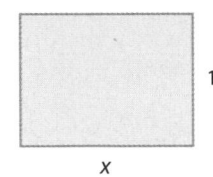
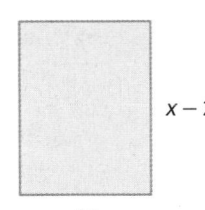

24. **CCSS SENSE-MAKING** Two times the least of three consecutive odd integers exceeds three times the greatest by 15. What are the integers?

Solve each equation. Check your solution.

25. $2x = 2(x - 3)$

26. $\frac{2}{5}h - 7 = \frac{12}{5}h - 2h + 3$

27. $-5(3 - q) + 4 = 5q - 11$

28. $2(4r + 6) = \frac{2}{3}(12r + 18)$

29. $\frac{3}{5}f + 24 = 4 - \frac{1}{5}f$

30. $\frac{1}{12} + \frac{3}{8}y = \frac{5}{12} + \frac{5}{8}y$

31. $\frac{2m}{5} = \frac{1}{3}(2m - 12)$

32. $\frac{1}{8}(3d - 2) = \frac{1}{4}(d + 5)$

33. $6.78j - 5.2 = 4.33j + 2.15$

34. $14.2t - 25.2 = 3.8t + 26.8$

35. $3.2k - 4.3 = 12.6k + 14.5$

36. $5[2p - 4(p + 5)] = 25$

37. NUMBER THEORY Three times the lesser of two consecutive even integers is 6 less than six times the greater number. Find the integers.

38. MONEY Chris has saved twice the number of quarters that Nora saved plus 6. The number of quarters Chris saved is also five times the difference of the number of quarters and 3 that Nora has saved. Write and solve an equation to find the number of quarters they each have saved.

39 DVD A company that replicates DVDs spends $1500 per day in building overhead plus $0.80 per DVD in supplies and labor. If the DVDs sell for $1.59 per disk, how many DVDs must the company sell each day before it makes a profit?

40. MOBILE PHONES The table shows the number of mobile phone subscribers for two states for a recent year. How long will it take for the numbers of subscribers to be the same?

State	Mobile Phone Subscribers (thousands)	New Subscribers Each Year (thousands)
Alabama	3765	325
Wisconsin	3842	292

41. MULTIPLE REPRESENTATIONS In this problem, you will explore $2x + 4 = -x - 2$.

a. Graphical Make a table of values with five points for $y = 2x + 4$ and $y = -x - 2$. Graph the points from the tables.

b. Algebraic Solve $2x + 4 = -x - 2$.

c. Verbal Explain how the solution you found in part **b** is related to the intersection point of the graphs in part **a**.

H.O.T. Problems Use Higher-Order Thinking Skills

42. REASONING Solve $5x + 2 = ax - 1$ for x. Assume that $a \neq 0$. Describe each step.

43. CHALLENGE Write an equation with the variable on each side of the equals sign, at least one fractional coefficient, and a solution of -6. Discuss the steps you used.

44. OPEN ENDED Create an equation with at least two grouping symbols for which there is no solution.

45. CCSS CRITIQUE Determine whether each solution is correct. If the solution is not correct, describe the error and give the correct solution.

a.
$$2(g + 5) = 22$$
$$2g + 5 = 22$$
$$2g + 5 - 5 = 22$$
$$2g = 17$$
$$g = 8.5$$

b.
$$5d = 2d - 18$$
$$5d - 2d = 2d - 18 - 2d$$
$$3d = -18$$
$$d = -6$$

c.
$$-6z + 13 = 7z$$
$$-6z + 13 - 6z = 7z - 6z$$
$$13 = z$$

46. CHALLENGE Find the value of k for which each equation is an identity.
a. $k(3x - 2) = 4 - 6x$
b. $15y - 10 + k = 2(ky - 1) - y$

47. WRITING IN MATH Compare and contrast solving equations with variables on both sides of the equation to solving one-step or multi-step equations with a variable on one side of the equation.

48. A hang glider, 25 meters above the ground, starts to descend at a constant rate of 2 meters per second. Which equation shows the height h after t seconds of descent?

  A $h = 25t + 2t$

  B $h = -25t + 2$

  C $h = 2t + 25$

  D $h = -2t + 25$

49. GEOMETRY Two rectangular walls each with a length of 12 feet and a width of 23 feet need to be painted. It costs $0.08 per square foot for paint. How much will it cost to paint the walls?

  F $22.08   **H** $34.50

  G $23.04   **J** $44.16

50. SHORT RESPONSE Maddie works at Game Exchange. They are having a sale as shown.

Item	Price	Special
video games	$20	Buy 2 get 1 Free
DVDs	$15	Buy 1 get 1 Free

Her employee discount is 15%. If sales tax is 7.25%, how much does she spend for a total of 4 video games?

51. Solve $\frac{4}{5}x + 7 = \frac{3}{15}x - 3$.

  A $-16\frac{2}{3}$ **C** -10

  B $-14\frac{4}{9}$ **D** $-6\frac{2}{3}$

Solve each equation. Check your solution. (Lesson 2-3)

52. $5n + 6 = -4$

53. $-1 = 7 + 3c$

54. $\frac{1}{2}z + 7 = 16 - \frac{3}{5}z$

55. $\frac{2}{5}x + 6 = \frac{2}{3}x + 10$

56. $\frac{a}{7} - 3 = -2$

57. $9 + \frac{y}{5} = 6$

58. WORLD RECORDS In 1998, Winchell's House of Donuts in Pasadena, California, made the world's largest donut. It weighed 5000 pounds and had a circumference of 298.3 feet. What was the donut's diameter to the nearest tenth? (*Hint:* $C = \pi d$) (Lesson 2-2)

59. ZOO At a zoo, the cost of admission is posted on the sign. Find the cost of admission for two adults and two children. (Lesson 1-4)

ZOO ADMISSION

Adults $9.75

Children $7.25

Find the value of n. Then name the property used in each step. (Lesson 1-3)

60. $25n = 25$

61. $n \cdot 1 = 2$

62. $12 \cdot n = 12 \cdot 6$

63. $n + 0 = \frac{2}{3}$

64. $4 \cdot \frac{1}{4} = n$

65. $(10 - 8)(7) = 2(n)$

Translate each sentence into an equation.

66. Twice a number t decreased by eight equals seventy.

67. Five times the sum of m and k is the same as seven times k.

68. Half of p is the same as p minus 3.

Evaluate each expression.

69. $-9 - (-14)$

70. $-10 + (20)$

71. $-15 - 9$

72. $5(14)$

73. $-55 \div (-5)$

74. $-25(-5)$

2-5 Solving Equations Involving Absolute Value

·· Then	·· Now	·· Why?

- You solved equations with the variable on each side.

- **1** Evaluate absolute value expressions.

- **2** Solve absolute value equations.

- In 2007, a telephone poll was conducted to determine the reading habits of people in the U.S. People in this survey were allowed to select more than one type of book.

  The survey had a margin of error of ±3%. This means that the results could be three points higher or lower. So, the percent of people who read religious material could be as high as 69% or as low as 63%.

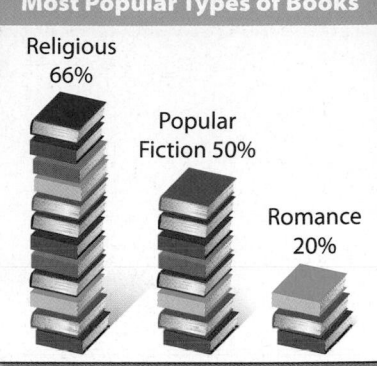

Most Popular Types of Books

Religious 66%

Popular Fiction 50%

Romance 20%

Source: CNN

Common Core State Standards

Content Standards
A.REI.1 Explain each step in solving a simple equation as following from the equality of numbers asserted at the previous step, starting from the assumption that the original equation has a solution. Construct a viable argument to justify a solution method.

A.REI.3 Solve linear equations and inequalities in one variable, including equations with coefficients represented by letters.

Mathematical Practices
3 Construct viable arguments and critique the reasoning of others.

7 Look for and make use of structure.

1 Absolute Value Expressions Expressions with absolute values define an upper and lower range in which a value must lie. Expressions involving absolute value can be evaluated using the given value for the variable.

Example 1 Expressions with Absolute Value

Evaluate $|m + 6| - 14$ if $m = 4$.

$$|m + 6| - 14 = |4 + 6| - 14 \qquad \text{Replace } m \text{ with 4.}$$
$$= |10| - 14 \qquad 4 + 6 = 10$$
$$= 10 - 14 \qquad |10| = 10$$
$$= -4 \qquad \text{Simplify.}$$

▶ **Guided Practice**

1. Evaluate $23 - |3 - 4r|$ if $r = 2$.

2 Absolute Value Equations The margin of error in the example at the top of the page is an example of absolute value. The distance between 66 and 69 on a number line is the same as the distance between 63 and 66.

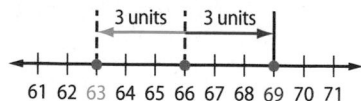

3 units | 3 units

61 62 63 64 65 66 67 68 69 70 71

There are three types of open sentences involving absolute value, $|x| = n$, $|x| < n$, and $|x| > n$. In this lesson, we will consider only the first type. Look at the equation $|x| = 4$. This means that the distance between 0 and x is 4.

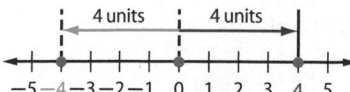

4 units | 4 units

−5 −4 −3 −2 −1 0 1 2 3 4 5

If $|x| = 4$, then $x = -4$ or $x = 4$. Thus, the solution set is $\{-4, 4\}$.

For each absolute value equation, we must consider both cases. To solve an absolute value equation, first isolate the absolute value on one side of the equals sign if it is not already by itself.

KeyConcept Absolute Value Equations

Words	When solving equations that involve absolute values, there are two cases to consider.
	Case 1 The expression inside the absolute value symbol is positive or zero.
	Case 2 The expression inside the absolute value symbol is negative.
Symbols	For any real numbers a and b, if $|a| = b$ and $b \geq 0$, then $a = b$ or $a = -b$.
Example	$|d| = 10$, so $d = 10$ or $d = -10$.

Example 2 Solve Absolute Value Equations

Solve each equation. Then graph the solution set.

a. $|f + 5| = 17$

$$|f + 5| = 17 \qquad \text{Original equation}$$

Case 1

$$f + 5 = 17$$
$$f + 5 - 5 = 17 - 5 \qquad \text{Subtract 5 from each side.}$$
$$f = 12 \qquad \text{Simplify.}$$

Case 2

$$f + 5 = -17$$
$$f + 5 - 5 = -17 - 5$$
$$f = -22$$

−25 −20 −15 −10 −5   0   5   10   15   20   25

b. $|b - 1| = -3$

$|b - 1| = -3$ means the distance between b and 1 is -3. Since distance cannot be negative, the solution is the empty set ∅.

−5 −4 −3 −2 −1   0   1   2   3   4   5

GuidedPractice

2A. $|y + 2| = 4$

2B. $|3n - 4| = -1$

Absolute value equations occur in real-world situations that describe a range within which a value must lie.

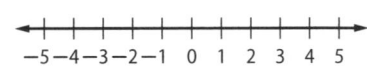

Real-World Example 3 Solve an Absolute Value Equation

SNAKES The temperature of an enclosure for a pet snake should be about 80°F, give or take 5°. Find the maximum and minimum temperatures.

You can use a number line to solve.

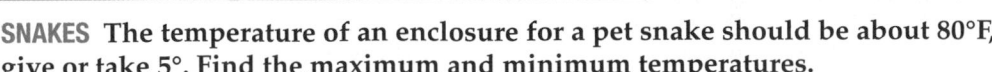

The distance from 80 to 75 is 5 units.
The distance from 80 to 85 is 5 units.

The solution set is {75, 85}. The maximum and minimum temperatures are 85° and 75°.

> **Guided**Practice

> **3. ICE CREAM** Ice cream should be stored at 5°F with an allowance for 5°. Write and solve an equation to find the maximum and minimum temperatures at which the ice cream should be stored.

When given two points on a graph, you can write an absolute value equation for the graph.

StudyTip

Find the Midpoint To find the point midway between two points, add the values together and divide by 2. For Example 4, $11 + 19 = 30$, $30 \div 2 = 15$. So 15 is the point halfway between 11 and 19.

Example 4 Write an Absolute Value Equation

Write an equation involving absolute value for the graph.

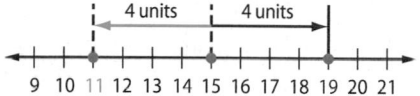

Find the point that is the same distance from 11 and from 19. This is the midpoint between 11 and 19, which is 15.

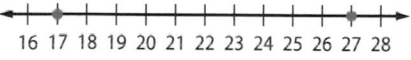

The distance from 15 to 11 is 4 units.
The distance from 15 to 19 is 4 units.

So an equation is $|x - 15| = 4$.

> **Guided**Practice

> **4.** Write an equation involving absolute value for the graph.
>
> (number line from 16 to 28)

Check Your Understanding

 = Step-by-Step Solutions begin on page R13.

Example 1 Evaluate each expression if $f = 3$, $g = -4$, and $h = 5$.

1. $|3 - h| + 13$ **2.** $16 - |g + 9|$ **3.** $|f + g| - h$

Example 2 Solve each equation. Then graph the solution set.

4. $|n + 7| = 5$ **5.** $|3z - 3| = 9$ **6.** $|4n - 1| = -6$

7. $|b + 4| = 2$ **8.** $|2t - 4| = 8$ **9.** $|5h + 2| = -8$

Example 3 **10. FINANCIAL LITERACY** For a company to invest in a product, they must believe they will receive a 12% return on investment (ROI) plus or minus 3%. Write an equation to find the least and the greatest ROI they believe they will receive.

Example 4 Write an equation involving absolute value for each graph.

(11) (number line from −4 to 6)   **12.** (number line from −10 to 6)

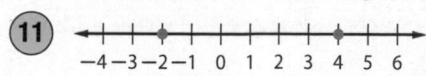

Example 1   Evaluate each expression if $a = -2$, $b = -3$, $c = 2$, $x = 2.1$, $y = 3$, and $z = -4.2$.

(13) $|2x + z| + 2y$

14. $4a - |3b + 2c|$

15. $-|5a + c| + |3y + 2z|$

16. $-a + |2x - a|$

17. $|y - 2z| - 3$

18. $3|3b - 8c| - 3$

19. $|2x - z| + 6b$

20. $-3|z| + 2(a + y)$

21. $-4|c - 3| + 2|z - a|$

Example 2   Solve each equation. Then graph the solution set.

22. $|n - 3| = 5$

23. $|f + 10| = 1$

24. $|v - 2| = -5$

25. $|4t - 8| = 20$

26. $|8w + 5| = 21$

27. $|6y - 7| = -1$

28. $\left|\frac{1}{2}x + 5\right| = -3$

29. $|-2y + 6| = 6$

30. $\left|\frac{3}{4}a - 3\right| = 9$

Example 3   **31. SURVEY** The circle graph at the right shows the results of a survey that asked, "How likely is it that you will be rich some day?" If the margin of error is ±4%, what is the range of the percent of teens who say it is very likely that they will be rich?

32. CHEERLEADING For competition, the cheerleading team is preparing a dance routine that must last 4 minutes, with a variation of ±5 seconds.

a. Find the least and greatest possible times for the routine in minutes and seconds.

b. Find the least and greatest possible times in seconds.

Example 4   Write an equation involving absolute value for each graph.

33.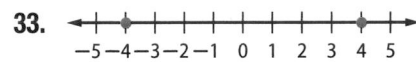

34.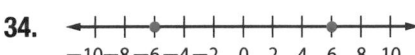

35.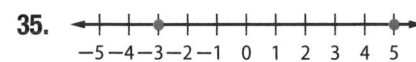

36.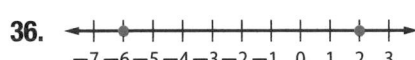

Solve each equation. Then graph the solution set.

37. $\left|-\frac{1}{2}b - 2\right| = 10$

38. $|-4d + 6| = 12$

39. $|5f - 3| = 12$

40. $2|h| - 3 = 8$

41. $4 - 3|q| = 10$

42. $\frac{4}{|p|} + 12 = 14$

43. CCSS SENSE-MAKING The 4×400 relay is a race where 4 runners take turns running 400 meters, or one lap around the track.

a. If a runner runs the first leg in 52 seconds plus or minus 2 seconds, write an equation to find the fastest and slowest times.

b. If the runners of the second and third legs run their laps in 53 seconds plus or minus 1 second, write an equation to find the fastest and slowest times.

c. Suppose the runner of the fourth leg is the fastest on the team. If he runs an average of 50.5 seconds plus or minus 1.5 seconds, what are the team's fastest and slowest times?

44. FASHION To allow for a model's height, a designer is willing to use models that require him to change hems either up or down 2 inches. The length of the skirts is 20 inches.

   a. Write an absolute value equation that represents the length of the skirts.

   b. What is the range of the lengths of the skirts?

   c. If a 20-inch skirt was fitted for a model that is 5 feet 9 inches tall, will the designer use a 6-foot-tall model?

45. CCSS PRECISION Speedometer accuracy can be affected by many details such as tire diameter and axle ratio. For example, there is variation of ±3 miles per hour when calibrated at 50 miles per hour.

   a. What is the range of actual speeds of the car if calibrated at 50 miles per hour?

   b. A speedometer calibrated at 45 miles per hour has an accepted variation of ±1 mile per hour. What can we conclude from this?

Write an equation involving absolute value for each graph.

46.

47.

48.

49.

50.

51.

52. MUSIC A CD will record an hour and a half of music plus or minus 3 minutes for time between tracks.

   a. Write an absolute value equation that represents the recording time.

   b. What is the range of time in minutes that the CD could run?

   c. Graph the possible times on a number line.

53 ACOUSTICS The Red Rocks Amphitheater located in the Red Rock Park near Denver, Colorado, is the only naturally occurring amphitheater. The acoustic qualities here are such that a maximum of 20,000 people, plus or minus 1000, can hear natural voices clearly.

   a. Write an equation involving an absolute value that represents the number of people that can hear natural voices at Red Rocks Amphitheater.

   b. Find the maximum and minimum number of people that can hear natural voices clearly in the amphitheater.

   c. What is the range of people in part **b**?

54. BOOK CLUB The members of a book club agree to read within ten pages of the last page of the chapter. The chapter ends on page 203.

    a. Write an absolute value equation that represents the pages where club members could stop reading.

    b. Write the range of the pages where the club members could stop reading.

55 SCHOOL Teams from Washington and McKinley High Schools are competing in an academic challenge. A correct response on a question earns 10 points and an incorrect response loses 10 points. A team earns 0 points on an unattempted question. There are 5 questions in the math section.

    a. What are the maximum and minimum scores a team can earn on the math section?

    b. Suppose the McKinley team has 160 points at the start of the math section. Write and solve an equation that represents the maximum and minimum scores the team could have at the end of the math section.

    c. What are all of the possible scores that a school can earn on the math section?

H.O.T. Problems   Use Higher-Order Thinking Skills

56. OPEN ENDED Describe a real-world situation that could be represented by the absolute value equation $|x - 4| = 10$.

CCSS STRUCTURE Determine whether the following statements are *sometimes*, *always*, or *never* true, if c is an integer. Explain your reasoning.

57. The value of $|x + 1|$ is greater than zero.

58. The solution of $|x + c| = 0$ is greater than 0.

59. The inequality $|x| + c < 0$ has no solution.

60. The value of $|x + c| + c$ is greater than zero.

61. REASONING Explain why an absolute value can never be negative.

62. CHALLENGE Use the sentence $x = 7 \pm 4.6$.

    a. Describe the values of x that make the sentence true.

    b. Translate the sentence into an equation involving absolute value.

63. ERROR ANALYSIS Alex and Wesley are solving $|x + 5| = -3$. Is either of them correct? Explain your reasoning.

Alex	Wesley
$\|x + 5\| = 3$ *or* $\|x + 5\| = -3$	$\|x + 5\| = -3$
$x + 5 = 3$ $x + 5 = -3$	The solution is ∅.
$\underline{\quad -5 \quad -5 \quad}$ $\underline{\quad -5 \quad -5 \quad}$	
$x = -2$ $x = -8$	

64. WRITING IN MATH Explain why there are either two, one, or no solutions for absolute value equations. Demonstrate an example of each possibility.

65. Which equation represents the second step of the solution process?

Step 1: $4(2x + 7) - 6 = 3x$

Step 2: _____

Step 3: $5x + 28 - 6 = 0$

Step 4: $5x = -22$

Step 5: $x = -4.4$

A $4(2x - 6) + 7 = 3x$
B $4(2x + 1) = 3x$
C $8x + 7 - 6 = 3x$
D $8x + 28 - 6 = 3x$

66. GEOMETRY The area of a circle is 25π square centimeters. What is the circumference?

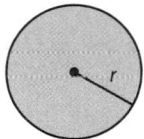

F 625π cm
G 50π cm
H 25π cm
J 10π cm

67. Tanya makes $5 an hour and 15% commission of the total dollar value on cosmetics she sells. Suppose Tanya's commission is increased to 17%. How much money will she make if she sells $300 worth of product and works 30 hours?

A $201   **C** $255
B $226   **D** $283

68. EXTENDED RESPONSE John's mother has agreed to take him driving every day for two weeks. On the first day, John drives for 20 minutes. Each day after that, John drives 5 minutes more than the day before.

a. Write an expression for the minutes John drives on the nth day. Explain.

b. For how many minutes will John drive on the last day? Show your work.

c. John's driver's education teacher requires that each student drive for 30 hours with an adult outside of class. Will John's sessions with his mother fulfill this requirement?

Spiral Review

Write and solve an equation for each sentence. (Lesson 2-4)

69. One half of a number increased by 16 is four less than two thirds of the number.

70. The sum of one half of a number and 6 equals one third of the number.

71. SHOE If ℓ represents the length of a man's foot in inches, the expression $2\ell - 12$ can be used to estimate his shoe size. What is the approximate length of a man's foot if he wears a size 8? (Lesson 2-3)

Skills Review

Write an equation for each problem. Then solve the equation.

72. Seven times a number equals -84. What is the number?

73. Two fifths of a number equals -24. Find the number.

74. Negative 117 is nine times a number. Find the number.

75. Twelve is one fifth of a number. What is the number?

Mid-Chapter Quiz
Lessons 2-1 through 2-5

Translate each sentence into an equation. (Lesson 2-1)

1. The sum of three times a and four is the same as five times a.

2. One fourth of m minus six is equal to two times the sum of m and 9.

3. The product of five and w is the same as w to the third power.

4. **MARBLES** Drew has 50 red, green, and blue marbles. He has six more red marbles than blue marbles and four fewer green marbles than blue marbles. Write and solve an equation to determine how many blue marbles Drew has. (Lesson 2-2)

Solve each equation. Check your solution. (Lesson 2-2)

5. $p + 8 = 13$

6. $-26 = b - 3$

7. $\frac{t}{6} = 3$

8. **MULTIPLE CHOICE** Solve the equation $\frac{3}{5}a = \frac{1}{4}$. (Lesson 2-2)

   A -3

   B $\frac{3}{20}$

   C $\frac{5}{12}$

   D 2

Solve each equation. Check your solution. (Lesson 2-3)

9. $2x + 5 = 13$

10. $-21 = 7 - 4y$

11. $\frac{m}{6} - 3 = 8$

12. $-4 = \frac{d + 3}{5}$

13. **FISH** The average length of a yellow-banded angelfish is 12 inches. This is 4.8 times as long as an average common goldfish. (Lesson 2-3)

    a. Write an equation you could use to find the length of the average common goldfish.

    b. What is the length of an average common goldfish?

Write an equation and solve each problem. (Lesson 2-3)

14. Three less than three fourths of a number is negative 9. Find the number.

15. Thirty is twelve added to six times a number. What is the number?

16. Find four consecutive integers with a sum of 106.

Solve each equation. Check your solution. (Lesson 2-4)

17. $8p + 3 = 5p + 9$

18. $\frac{3}{4}w + 6 = 9 - \frac{1}{4}w$

19. $\frac{z + 6}{3} = \frac{2z}{4}$

20. **PERIMETER** Find the value of x so that the triangles have the same perimeter. (Lesson 2-4)

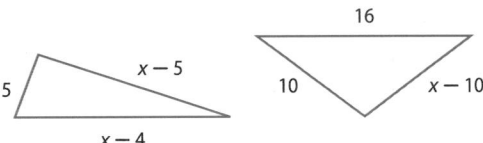

21. **PRODUCTION** ABC Sporting Goods Company produces baseball gloves. Their fixed monthly production cost is $8000 with a per glove cost of $5. XYZ Sporting Goods Company also produces baseball gloves. Their fixed monthly production cost is $10,000 with a per glove cost of $3. Find the value of x, the number of gloves produced monthly, so that the total monthly production cost is the same for both companies. (Lesson 2-4)

Evaluate each expression if $x = -4$, $y = 7$, and $z = -9$.
(Lesson 2-5)

22. $|3x - 2| + 2y$

23. $|-4y + 2z| - 7z$

24. **MULTIPLE CHOICE** Solve $|6m - 3| = 9$. (Lesson 2-5)

   F $\{2\}$　　　　　　　　H $\{-3, 6\}$

   G $\{-1, 2\}$　　　　　　J $\{-3, 3\}$

25. **COFFEE** Some say to brew an excellent cup of coffee, you must have a brewing temperature of 200° F, plus or minus 5 degrees. Write and solve an equation describing the maximum and minimum brewing temperatures for an excellent cup of coffee.

Ratios and Proportions

∷ **Then**
- You evaluated percents by using a proportion.

∷ **Now**
1 Compare ratios.

2 Solve proportions.

∷ **Why?**
- Ratios allow us to compare many items by using a common reference. The table below shows the number of restaurants a certain popular fast food chain has per 10,000 people in the United States as well as other countries. This allows us to compare the number of these restaurants using an equal reference.

Countries	United States	New Zealand	Canada	Australia	Japan	Singapore
Number of Restaurants per 10,000 People	0.433	0.369	0.352	0.349	0.282	0.273

NewVocabulary
ratio
proportion
means
extremes
rate
unit rate
scale
scale model

Common Core State Standards

Content Standards
A.REI.1 Explain each step in solving a simple equation as following from the equality of numbers asserted at the previous step, starting from the assumption that the original equation has a solution. Construct a viable argument to justify a solution method.

A.REI.3 Solve linear equations and inequalities in one variable, including equations with coefficients represented by letters.

Mathematical Practices
6 Attend to precision.

1 Ratios and Proportions The comparison between the number of restaurants and the number of people is a ratio. A **ratio** is a comparison of two numbers by division. The ratio of x to y can be expressed in the following ways.

$$x \text{ to } y \qquad x : y \qquad \frac{x}{y}$$

Suppose you wanted to determine the number of restaurants per 100,000 people in Australia. Notice that this ratio is equal to the original ratio.

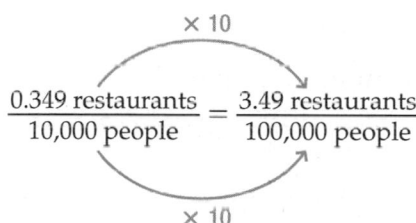

$$\frac{0.349 \text{ restaurants}}{10{,}000 \text{ people}} = \frac{3.49 \text{ restaurants}}{100{,}000 \text{ people}}$$

An equation stating that two ratios are equal is called a **proportion**. So, we can state that $\frac{0.349}{10{,}000} = \frac{3.49}{100{,}000}$ is a proportion.

Example 1 Determine Whether Ratios Are Equivalent

Determine whether $\frac{2}{3}$ and $\frac{16}{24}$ are equivalent ratios. Write *yes* or *no*. Justify your answer.

$$\frac{2}{3} = \frac{2}{3} \qquad \frac{16}{24} = \frac{2}{3}$$
(÷1)   (÷8)

When expressed in simplest form, the ratios are equivalent.

▶ **Guided**Practice

Determine whether each pair of ratios are equivalent ratios. Write *yes* or *no*. Justify your answer.

1A. $\frac{6}{10}, \frac{2}{5}$

1B. $\frac{1}{6}, \frac{5}{30}$

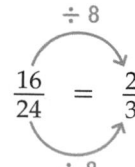

There are special names for the terms in a proportion.

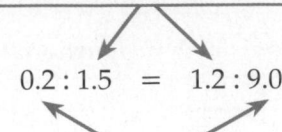

1.5 and 1.2 are called the **means**. They are the middle terms of the proportion.

$$0.2 : 1.5 = 1.2 : 9.0$$

0.2 and 9.0 are called the **extremes**. They are the first and last terms of the proportion.

KeyConcept Means-Extremes Property of Proportion

Words

In a proportion, the product of the extremes is equal to the product of the means.

Symbols

If $\dfrac{a}{b} = \dfrac{c}{d}$ and $b, d \neq 0$, then $ad = bc$.

Examples

Since $\dfrac{2}{4} = \dfrac{1}{2}$, $2(2) = 4(1)$ or $4 = 4$.

Another way to determine whether two ratios form a proportion is to use cross products. If the cross products are equal, then the ratios form a proportion.

This is the same as multiplying the means, and multiplying the extremes.

Example 2 Cross Products

Use cross products to determine whether each pair of ratios forms a proportion.

a. $\dfrac{2}{3.5}, \dfrac{8}{14}$

$\dfrac{2}{3.5} \overset{?}{=} \dfrac{8}{14}$ Original proportion

$2(14) \overset{?}{=} 3.5(8)$ Cross products

$28 = 28 \checkmark$ Simplify.

The cross products are equal, so the ratios form a proportion.

b. $\dfrac{0.3}{1.5}, \dfrac{0.5}{2.0}$

$\dfrac{0.3}{1.5} \overset{?}{=} \dfrac{0.5}{2.0}$ Original proportion

$0.3(2.0) \overset{?}{=} 1.5(0.5)$ Cross products

$0.6 \neq 0.75 \;✗$ Simplify.

The cross products are not equal, so the ratios do not form a proportion.

▶ **GuidedPractice**

2A. $\dfrac{0.2}{1.8}, \dfrac{1}{0.9}$

2B. $\dfrac{15}{36}, \dfrac{35}{42}$

2 Solve Proportions To solve proportions, use cross products.

Example 3 Solve a Proportion

Solve each proportion. If necessary, round to the nearest hundredth.

a. $\frac{x}{10} = \frac{3}{5}$

$\frac{x}{10} = \frac{3}{5}$ Original proportion

$x(5) = 10(3)$ Find the cross products.

$5x = 30$ Simplify.

$\frac{5x}{5} = \frac{30}{5}$ Divide each side by 5.

$x = 6$ Simplify.

b. $\frac{x-2}{14} = \frac{2}{7}$

$\frac{x-2}{14} = \frac{2}{7}$ Original proportion

$(x-2)7 = 14(2)$ Find the cross products.

$7x - 14 = 28$ Simplify.

$7x = 42$ Add 14 to each side.

$x = 6$ Divide each side by 7.

GuidedPractice

3A. $\frac{r}{8} = \frac{25}{40}$ **3B.** $\frac{x+4}{5} = \frac{3}{8}$

The ratio of two measurements having different units of measure is called a **rate**. For example, a price of \$9.99 per 10 songs is a rate. A rate that tells how many of one item is being compared to 1 of another item is called a **unit rate**.

Real-World Example 4 Rate of Growth

RETAIL In the past two years, a retailer has opened 232 stores. If the rate of growth remains constant, how many stores will the retailer open in the next 3 years?

Understand Let r represent the number of retail stores.

Plan Write a proportion for the problem.

$$\frac{232 \text{ retail stores}}{2 \text{ years}} = \frac{r \text{ retail stores}}{3 \text{ years}}$$

Solve $\frac{232}{2} = \frac{r}{3}$ Original proportion

$232(3) = 2r$ Find the cross products.

$696 = 2r$ Simplify.

$\frac{696}{2} = \frac{2r}{2}$ Divide each side by 2.

$348 = r$ Simplify.

The retailer will open 348 stores in 3 years.

Check If the clothing retailer continues to open 232 stores every 2 years, then in the next 3 years, it will open 348 stores.

4. **EXERCISE** It takes 7 minutes for Isabella to walk around the gym track twice. At this rate, how many times can she walk around the track in a half hour?

A rate called a **scale** is used to make a **scale model** of something too large or too small to be convenient at actual size.

● **Real-World Example 5** Scale and Scale Models

MOUNTAIN TRAIL The Ramsey Cascades Trail is about $1\frac{1}{8}$ inches long on a map with scale 3 inches = 10 miles. What is the actual length of the trail?

Let ℓ represent the actual length.

scale → $\dfrac{3}{10} = \dfrac{1\frac{1}{8}}{\ell}$ ← scale
actual → $\phantom{\dfrac{3}{10}}$ ← actual

$$3(\ell) = 1\frac{1}{8}(10) \qquad \text{Find the cross products.}$$

$$3\ell = \frac{45}{4} \qquad \text{Simplify.}$$

$$3\ell \div 3 = \frac{45}{4} \div 3 \qquad \text{Divide each side by 3.}$$

$$\ell = \frac{15}{4} \text{ or } 3\frac{3}{4} \qquad \text{Simplify.}$$

Real-World Link

The Great Smoky Mountains National Park in Tennessee is home to several waterfalls. The Ramsey Cascades is 100 feet tall. It is the tallest in the park.

Source: National Park Service

The actual length is about $3\frac{3}{4}$ miles.

▶ **Guided**Practice

5. **AIRPLANES** On a model airplane, the scale is 5 centimeters = 2 meters. If the model's wingspan is 28.5 centimeters, what is the actual wingspan?

Check Your Understanding ◯ = Step-by-Step Solutions begin on page R13.

Examples 1–2 **Determine whether each pair of ratios are equivalent ratios. Write *yes* or *no*.**

1. $\dfrac{3}{7}, \dfrac{9}{14}$

2. $\dfrac{7}{8}, \dfrac{42}{48}$

3 $\dfrac{2.8}{4.4}, \dfrac{1.4}{2.1}$

Example 3 **Solve each proportion. If necessary, round to the nearest hundredth.**

4. $\dfrac{n}{9} = \dfrac{6}{27}$

5. $\dfrac{4}{u} = \dfrac{28}{35}$

6. $\dfrac{3}{8} = \dfrac{b}{10}$

Example 4 7. **RACE** Jennie ran the first 6 miles of a marathon in 58 minutes. If she is able to maintain the same pace, how long will it take her to finish the 26.2 miles?

Example 5 8. ◯**CCSS** **PRECISION** On a map of North Carolina, Raleigh and Asheville are about 8 inches apart. If the scale is 1 inch = 12 miles, how far apart are the cities?

Examples 1–2 Determine whether each pair of ratios are equivalent ratios. Write *yes* or *no*.

9. $\dfrac{9}{11}, \dfrac{81}{99}$

10. $\dfrac{3}{7}, \dfrac{18}{42}$

11. $\dfrac{8.4}{9.2}, \dfrac{8.8}{9.6}$

12. $\dfrac{4}{3}, \dfrac{6}{8}$

13. $\dfrac{29.2}{10.4}, \dfrac{7.3}{2.6}$

14. $\dfrac{39.68}{60.14}, \dfrac{6.4}{9.7}$

Example 3 Solve each proportion. If necessary, round to the nearest hundredth.

15. $\dfrac{3}{8} = \dfrac{15}{a}$

16. $\dfrac{t}{2} = \dfrac{6}{12}$

17. $\dfrac{4}{9} = \dfrac{13}{q}$

18. $\dfrac{15}{35} = \dfrac{g}{7}$

19. $\dfrac{7}{10} = \dfrac{m}{14}$

20. $\dfrac{8}{13} = \dfrac{v}{21}$

21. $\dfrac{w}{2} = \dfrac{4.5}{6.8}$

22. $\dfrac{1}{0.19} = \dfrac{12}{n}$

23. $\dfrac{2}{0.21} = \dfrac{8}{n}$

24. $\dfrac{2.4}{3.6} = \dfrac{k}{1.8}$

25 $\dfrac{t}{0.3} = \dfrac{1.7}{0.9}$

26. $\dfrac{7}{1.066} = \dfrac{z}{9.65}$

27. $\dfrac{x-3}{5} = \dfrac{6}{10}$

28. $\dfrac{7}{x+9} = \dfrac{21}{36}$

29. $\dfrac{10}{15} = \dfrac{4}{x-5}$

Example 4 **30. CAR WASH** The B-Clean Car Wash washed 128 cars in 3 hours. At that rate, how many cars can they wash in 8 hours?

Example 5 **31. GEOGRAPHY** On a map of Florida, the distance between Jacksonville and Tallahassee is 2.6 centimeters. If 2 centimeters = 120 miles, what is the distance between the two cities?

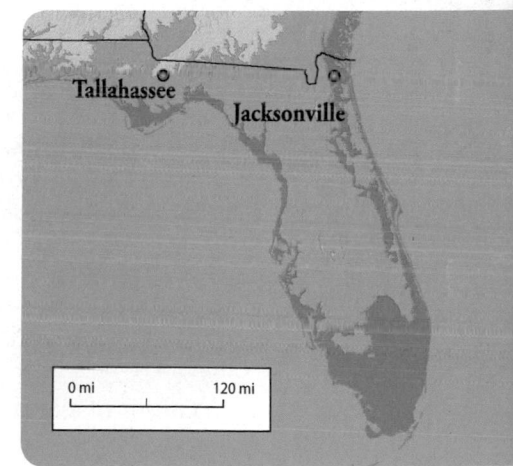

32. **CCSS PRECISION** An artist used interlocking building blocks to build a scale model of Kennedy Space Center, Florida. In the model, 1 inch equals 1.67 feet of an actual space shuttle. The model is 110.3 inches tall. How tall is the actual space shuttle? Round to the nearest tenth.

33. MENU On Monday, a restaurant made $545 from selling 110 hamburgers. If they sold 53 hamburgers on Tuesday, how much did they make?

Solve each proportion. If necessary, round to the nearest hundredth.

34. $\dfrac{6}{14} = \dfrac{7}{x-3}$

35. $\dfrac{7}{4} = \dfrac{f-4}{8}$

36. $\dfrac{3-y}{4} = \dfrac{1}{9}$

37. $\dfrac{4v+7}{15} = \dfrac{6v+2}{10}$

38. $\dfrac{9b-3}{9} = \dfrac{5b+5}{3}$

39. $\dfrac{2n-4}{5} = \dfrac{3n+3}{10}$

40. ATHLETES At Piedmont High School, 3 out of every 8 students are athletes. If there are 1280 students at the school, how many are not athletes?

41. BRACES Two out of five students in the ninth grade have braces. If there are 325 students in the ninth grade, how many have braces?

42. PAINT Joel used a half gallon of paint to cover 84 square feet of wall. He has 932 square feet of wall to paint. How many gallons of paint should he purchase?

43 **MOVIE THEATERS** Use the table at the right.

a. Write a ratio of the number of indoor theaters to the total number of theaters for each year.

b. Do any two of the ratios you wrote for part **a** form a proportion? If so, explain the real-world meaning of the proportion.

44. DIARIES In a survey, 36% of the students said that they kept an electronic diary. There were 900 students who kept an electronic diary. How many students were in the survey?

Year	Indoor	Drive-In	Total
2003	35,361	634	35,995
2004	36,012	640	36,652
2005	37,092	648	37,740
2006	37,776	649	38,425
2007	38,159	635	38,794
2008	38,201	633	38,834
2009	38,605	628	39,233

Source: North American Theater Owners

45. 🔄 **MULTIPLE REPRESENTATIONS** In this problem, you will explore how changing the lengths of the sides of a shape by a factor changes the perimeter of that shape.

a. **Geometric** Draw a square *ABCD*. Draw a square *MNPQ* with sides twice as long as *ABCD*. Draw a square *FGHJ* with sides half as long as *ABCD*.

b. **Tabular** Complete the table below using the appropriate measures.

ABCD		MNPQ		FGHJ	
Side length		Side length		Side length	
Perimeter		Perimeter		Perimeter	

c. **Verbal** Make a conjecture about the change in the perimeter of a square if the side length is increased or decreased by a factor.

H.O.T. Problems Use Higher-Order Thinking Skills

46. **CCSS** **STRUCTURE** In 2007, organic farms occupied 2.6 million acres in the United States and produced goods worth about $1.7 billion. Divide one of these numbers by the other and explain the meaning of the result.

47. REASONING Compare and contrast ratios and rates.

48. CHALLENGE If $\frac{a+1}{b-1} = \frac{5}{1}$ and $\frac{a-1}{b+1} = \frac{1}{1}$, find the value of $\frac{b}{a}$. (*Hint:* Choose values of *a* and *b* for which the proportions are true and evaluate $\frac{b}{a}$.)

49. WRITING IN MATH On a road trip, Marcus reads a highway sign and then looks at his gas gauge.

Marcus's gas tank holds 10 gallons and his car gets 32 miles per gallon at his current speed of 65 miles per hour. If he maintains this speed, will he make it to Atlanta without having to stop and get gas? Explain your reasoning.

50. WRITING IN MATH Describe how businesses can use ratios. Write about a real-world situation in which a business would use a ratio.

51. In the figure, $x : y = 2 : 3$ and $y : z = 3 : 5$. If $x = 10$, find the value of z.

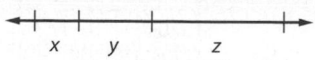

A 15

B 20

C 25

D 30

52. GRIDDED RESPONSE A race car driver records the finishing times for recent practice trials.

Trial	Time (seconds)
1	5.09
2	5.10
3	4.95
4	4.91
5	5.05

What is the mean time, in seconds, for the trials?

53. GEOMETRY If $\triangle LMN$ is similar to $\triangle LPO$, what is z?

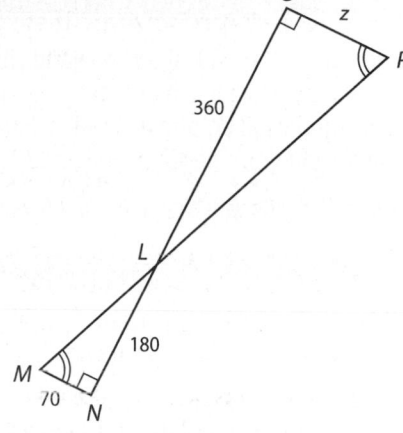

F 240

G 140

H 120

J 70

54. Which equation below illustrates the Commutative Property?

A $(3x + 4y) + 2z = 3x + (4y + 2z)$

B $7(x + y) = 7x + 7y$

C $xyz = yxz$

D $x + 0 = x$

Spiral Review

Solve each equation. (Lesson 2-5)

55. $|x + 5| = -8$

56. $|b + 9| = 2$

57. $|2p - 3| = 17$

58. $|5c - 8| = 12$

59. HEALTH When exercising, a person's pulse rate should not exceed a certain limit. This maximum rate is represented by the expression $0.8(220 - a)$, where a is age in years. Find the age of a person whose maximum pulse rate is 122 more than their age. (Lesson 2-4)

Solve each equation. Check your solution. (Lesson 2-3)

60. $15 = 4a - 5$

61. $7g - 14 = -63$

62. $9 + \dfrac{y}{5} = 6$

63. $\dfrac{t}{8} - 6 = -12$

64. GEOMETRY Find the area of $\triangle ABC$ if each small triangle has a base of 5.2 inches and a height of 4.5 inches. (Lesson 1-4)

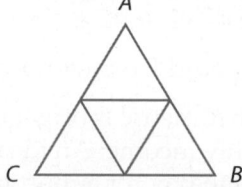

Evaluate each expression. (Lesson 1-2)

65. $3 + 16 \div 8 \cdot 5$

66. $4^2 \cdot 3 - 5(6 + 3)$

Skills Review

Solve each equation.

67. $4p = 22$

68. $5h = 33$

69. $1.25y = 4.375$

70. $9.8m = 30.87$

Spreadsheet Lab
Descriptive Modeling

When using numbers to model a real-world situation, it is often helpful to have a metric. A **metric** is a rule for assigning a number to some characteristic or attribute. For example, teachers use metrics to determine grades. Each teacher determines an appropriate metric for assessing a student's performance and assigning a grade.

You can use a spreadsheet to calculate different metrics.

CCSS Common Core State Standards
Content Standards
N.Q.2 Define appropriate quantities for the purpose of descriptive modeling.

Activity

Dorrie wants to buy a house. She has the following expenses: rent of $650, credit card monthly bills of $320, a car payment of $410, and a student loan payment of $115. Dorrie has a yearly salary of $46,500. Use a spreadsheet to find Dorrie's debt-to-income ratio.

Step 1 Enter Dorrie's debts in column B.

Step 2 Add her debts using a function in cell B6. Go to Insert and then Function. Then choose Sum. The sum of 1495 appears in B6.

Step 3 Now insert Dorrie's salary in column C. Remember to find her monthly salary by dividing the yearly salary by 12.

A mortgage company will use the debt-to-income ratio as a metric to determine if Dorrie qualifies for a loan. The **debt-to-income ratio** is calculated as *how much she owes per month* divided by *how much she earns each month*.

Step 4 Enter a formula to find the debt-to-income ratio in cell C6. In the formula bar, enter =B6/C2.

The ratio of about 0.39 appears. An ideal ratio would be 0.36 or less. A ratio higher than 0.36 would cause an increased interest rate or may require a higher down payment.

The spreadsheet shows a debt-to-income ratio of about 0.39. Dorrie should try to eliminate or reduce some debts or try to earn more money in order to lower her debt-to-income ratio.

Lab 2-6 B Spreadsheet.xls ⬜ ⧉ ☒

◇	A	B	C
1	Type of Debt	Expenses	Salary
2	Rent	650	3875
3	Credit Cards	320	
4	Car Payment	410	
5	Student Loan	115	
6		1495	0.385806
7			

Sheet 1 / Sheet 2 / Sheet 3

Exercises

1. How could Dorrie improve her debt-to-income ratio?

2. Another metric mortgage companies use is the ratio of monthly mortgage to total monthly income. An ideal ratio is 0.28. Using this metric, how much could Dorrie afford to pay for a mortgage each month?

3. How effective are each of these metrics as measures of whether Dorrie can afford to buy a house? Explain your reasoning.

4. **CCSS MODELING** Metrics are used to compare athletes. For example, ERAs are used to compare pitchers. Find a metric and evaluate its effectiveness for modeling. Compare it to other metrics, and then define your own metric.

Image Source/Punchstock

Percent of Change

:·Then
- You solved proportions.

:·Now
1. Find the percent of change.
2. Solve problems involving percent of change.

:·Why?
- Every year, millions of people volunteer their time to improve their community. The difference in the number of volunteers from one year to the next can be used to determine a percent to represent the increase or decrease in volunteers.

NewVocabulary
percent of change
percent of increase
percent of decrease

Common Core State Standards

Content Standards
N.Q.1 Use units as a way to understand problems and to guide the solution of multi-step problems; choose and interpret units consistently in formulas; choose and interpret the scale and the origin in graphs and data displays.

A.REI.3 Solve linear equations and inequalities in one variable, including equations with coefficients represented by letters.

Mathematical Practices
8 Look for and express regularity in repeated reasoning.

1 Percent of Change **Percent of change** is the ratio of the change in an amount to the original amount expressed as a percent. If the new number is greater than the original number, the percent of change is a **percent of increase.** If the new number is less than the original number, the percent of change is a **percent of decrease.**

Example 1 Percent of Change

Determine whether each percent of change is a percent of *increase* or a percent of *decrease.* Then find the percent of change.

a. original: 20
final: 23

Subtract the original amount from the final amount to find the amount of change: $23 - 20 = 3$.

Since the new amount is greater than the original, this is a percent of increase.

Use the original number, 20, as the base.

$$\text{change} \longrightarrow \frac{3}{20} = \frac{r}{100} \longleftarrow \text{original amount}$$

$$3(100) = r(20)$$

$$300 = 20r$$

$$\frac{300}{20} = \frac{20r}{20}$$

$$15 = r$$

The percent of increase is 15%.

b. original: 25
final: 17

Subtract the original amount from the final amount to find the amount of change: $17 - 25 = -8$.

Since the new amount is less than the original, this is a percent of decrease.

Use the original number, 25, as the base.

$$\text{change} \longrightarrow \frac{-8}{25} = \frac{r}{100} \longleftarrow \text{original amount}$$

$$-8(100) = r(25)$$

$$-800 = 25r$$

$$\frac{-800}{25} = \frac{25r}{25}$$

$$-32 = r$$

The percent of decrease is 32%.

▶ **Guided**Practice

1A. original: 66
new: 30

1B. original: 9.8
new: 12.1

1C. original: 24
new: 40

1D. original: 500
new: 131

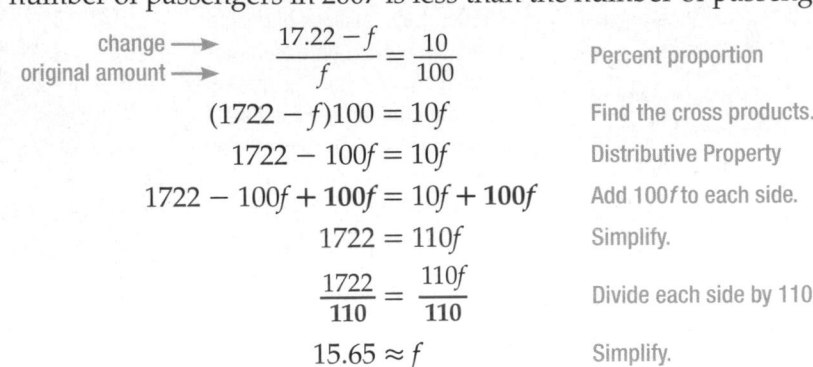

Real-World Example 2 Percent of Change

CRUISE The total number of passengers on cruise ships increased 10% from 2007 to 2009. If there were 17.22 million passengers in 2009, how many were there in 2007?

Let f = the number of passengers in 2009. Since 10% is a percent of increase, the number of passengers in 2007 is less than the number of passengers in 2009.

$$\begin{array}{ll} \dfrac{17.22 - f}{f} = \dfrac{10}{100} & \text{Percent proportion} \\ (1722 - f)100 = 10f & \text{Find the cross products.} \\ 1722 - 100f = 10f & \text{Distributive Property} \\ 1722 - 100f + \mathbf{100f} = 10f + \mathbf{100f} & \text{Add 100}f\text{ to each side.} \\ 1722 = 110f & \text{Simplify.} \\ \dfrac{1722}{110} = \dfrac{110f}{110} & \text{Divide each side by 110.} \\ 15.65 \approx f & \text{Simplify.} \end{array}$$

change ⟶ and original amount ⟶ label the fraction $\dfrac{17.22 - f}{f}$.

There were approximately 15.65 million passengers in 2007.

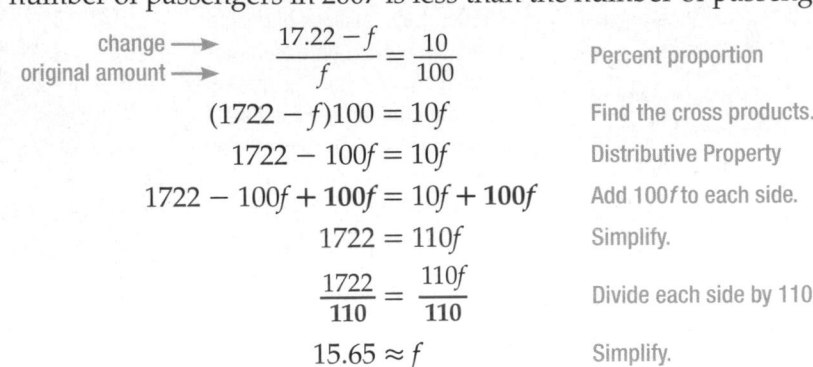

Real-WorldLink

In 2009, the total revenue earned by the North American cruise industry was more than $15.16 billion.

Source: Cruise Market Watch

▶ **Guided**Practice

2. **TUITION** A recent percent of increase in tuition at Northwestern University, in Evanston, Illinois, was 5.4%. If the new cost is $33,408 per year, find the original cost per year.

2 **Solve Problems** Two applications of percent of change are sales tax and discounts. Sales tax is an example of a percent of increase. Discount is an example of a percent of decrease.

Example 3 Sales Tax

SHOPPING Marta is purchasing wire and beads to make jewelry. Her merchandise is $28.62 before tax. If the tax is 7.25% of the total sales, what is the final cost?

Step 1 Find the tax.

The tax is 7.25% of the price of the merchandise.

$$7.25\% \text{ of } \$28.62 = 0.0725 \times 28.62 \qquad 7.25\% = 0.0725$$
$$= 2.07495 \qquad \text{Use a calculator.}$$

Step 2 Find the cost with tax.

Round $2.07495 to $2.07 since tax is always rounded to the nearest cent. Add this amount to the original price: $28.62 + $2.07 = $30.69.

The total cost of Marta's jewelry supplies is $30.69.

▶ **Guided**Practice

3. **SHOPPING** A new DVD costs $24.99. If the sales tax is 6.85%, what is the total cost?

To find a discounted amount, you will follow similar steps to those for sales tax.

Example 4 Discounts

DISCOUNT Since Tyrell has earned good grades in school, he qualifies for the Good Student Discount on his car insurance. His monthly payment without the discount is $85. If the discount is 20%, what will he pay each month?

Step 1 Find the discount.

The discount is 20% of the original payment.

$$20\% \text{ of } \$85 = 0.20 \times 85 \qquad 20\% = 0.20$$
$$= 17 \qquad \text{Use a calculator.}$$

Step 2 Find the cost after discount.

Subtract $17 from the original payment: $85 − $17 = $68.

With the Good Student Discount, Tyrell will pay $68 per month.

▶ **Guided**Practice

4. SALES A picture frame originally priced at $14.89 is on sale for 40% off. What is the discounted price?

StudyTip

CCSS Regularity When translating a problem from word sentences to math sentences, the word "is" translates to =, and the word "of" translates to ×.

Check Your Understanding

⬤ = **Step-by-Step Solutions begin on page R13.**

Example 1 State whether each percent of change is a percent of *increase* or a percent of *decrease*. Then find the percent of change. Round to the nearest whole percent.

1 original: 78
new: 125

2. original: 41
new: 24

3. original: 6 candles
new: 8 candles

4. original: 35 computers
new: 32 computers

Example 2 **5. GEOGRAPHY** The distance from Phoenix to Tucson is 120 miles. The distance from Phoenix to Flagstaff is about 21.7% longer. To the nearest mile, what is the distance from Phoenix to Flagstaff?

Example 3 Find the total price of each item.

6. dress: $22.50
sales tax: 7.5%

7. video game: $35.99
sales tax: 6.75%

8. PROM A limo costs $85 to rent for 3 hours plus a 7% sales tax. What is the total cost to rent a limo for 6 hours?

9. GAMES A computer game costs $49.95 plus a 6.25% sales tax. What is the total cost of the game?

Example 4 Find the discounted price of each item.

10. guitar: $95.00
discount: 15%

11. DVD: $22.95
discount: 25%

12. SKATEBOARD A skateboard costs $99.99. If you have a coupon for 20% off, how much will you save?

13. CCSS MODELING Tickets to the county fair are $8 for an adult and $5 for a child. If you have a 15% discount card, how much will 2 adult tickets and 2 child tickets cost?

Example 1 State whether each percent of change is a percent of *increase* or a percent of *decrease*. Then find the percent of change. Round to the nearest whole percent.

14. original: 35
new: 40

15 original: 16
new: 10

16. original: 27
new: 73

17. original: 92
new: 21

18. original: 21.2 grams
new: 10.8 grams

19. original: 11 feet
new: 25 feet

20. original: $68
new: $76

21. original: 21 hours
new: 40 hours

Example 2 **22. GASOLINE** The average cost of regular gasoline in North Carolina increased by 73% from 2006 to 2007. If the average cost of a gallon of gas in 2006 was $2.069, what was the average cost in 2007? Round to the nearest cent.

23. CARS Beng is shopping for a car. The cost of a new car is $15,500. This is 25% greater than the cost of a used car. What is the cost of the used car?

Example 3 Find the total price of each item.

24. messenger bag: $28.00
tax: 7.25%

25. software: $45.00
tax: 5.5%

26. vase: $5.50
tax: 6.25%

27. book: $25.95
tax: 5.25%

28. magazine: $3.50
tax: 5.75%

29. pillow: $9.99
tax: 6.75%

Example 4 Find the discounted price of each item.

30. computer: $1099.00
discount: 25%

31. CD player: $89.99
discount: 15%

32. athletic shoes: $59.99
discount: 40%

33. jeans: $24.50
discount: 33%

34. jacket: $125.00
discount: 25%

35. belt: $14.99
discount: 20%

Find the final price of each item.

36. sweater: $14.99
discount: 12%
tax: 6.25%

37. printer: $60.00
discount: 25%
tax: 6.75%

38. board game: $25.00
discount: 15%
tax: 7.5%

39. CONSUMER PRICE INDEX An *index* measures the percent change of a value from a base year. An index of 115 means that there was a 15% increase from the base year. In 2000, the consumer price index of dairy products was 160.7. In 2007, it was 194.0. Determine the percent of change.

40. FINANCIAL LITERACY The current price of each share of a technology company is $135. If this represents a 16.2% increase over the past year, what was the price per share a year ago?

41. **CCSS MODELING** A group of girls are shopping for dresses to wear to the spring dance. One finds a dress priced $75 with a 20% discount. A second girl finds a dress priced $85 with a 30% discount.

   a. Find the amount of discount for each dress.

   b. Which girl is getting the better price for the dress?

42. RECREATIONAL SPORTS In 1995, there were 73,567 youth softball teams. By 2007, there were 86,049. Determine the percent of increase.

43 **CCSS** TOOLS Which grocery item had the greatest percent increase in cost from 2000 to 2007?

Average Retail Prices of Selected Grocery Items		
Grocery Item	Cost in 2000 ($ per pound)	Cost in 2007 ($ per pound)
milk (gallon)	2.79	3.87
turkey (whole)	0.99	1.01
chicken (whole)	1.08	1.17
ground beef	1.63	2.23
apples	0.82	1.12
iceberg lettuce	0.85	0.95
peanut butter	1.96	1.88

Source: Statistical Abstract of the United States

44. **MULTIPLE REPRESENTATIONS** In this problem, you will explore patterns in percentages.

   a. **Tabular** Copy and complete the following table.

1% of	500	is 5.	100% of		is 20.		% of 80 is 20.
2% of		is 5.	50% of		is 20.		% of 40 is 20.
4% of		is 5.	25% of		is 20.		% of 20 is 20.
8% of		is 5.	12.5% of		is 20.		% of 10 is 20.

   b. **Verbal** Describe the patterns in the second and fifth columns.

   c. **Analytical** Use the patterns to write the fifth row of the table.

H.O.T. Problems Use Higher-Order Thinking Skills

45. **OPEN ENDED** Write a real-world problem to find the total price of an item including sales tax.

46. **REASONING** If you have 75% of a number n, what percent of decrease is it from the number n? If you have 40% of a number a, what percent of decrease do you have from the number a? What pattern do you notice? Is this always true?

47. **ERROR ANALYSIS** Maddie and Xavier are solving for the percent change if the original amount was $25 and the new amount is $28. Is either of them correct? Explain your reasoning.

   Maddie
   $$\frac{3}{28} = \frac{r}{100}$$
   $$3(100) = 28r$$
   $$300 = 28r$$
   $$10.7 = r$$

   Xavier
   $$\frac{3}{25} = \frac{r}{100}$$
   $$3(100) = 25r$$
   $$300 = 25r$$
   $$12 = r$$

48. **CHALLENGE** Determine whether the following statement is *sometimes*, *always*, or *never* true. *The percent of change is less than 100%.*

49. **WRITING IN MATH** When is percent of change used in the real world? Explain how to find a percent of change between two values.

50. GEOMETRY The rectangle has a perimeter of P centimeters. Which equation could be used to find the length ℓ of the rectangle?

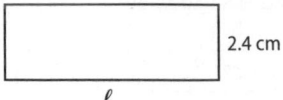

2.4 cm

ℓ

A $P = 2.4\ell$ **C** $P = 2.4 + 2\ell$

B $P = 4.8 + \ell$ **D** $P = 4.8 + 2\ell$

51. SHORT RESPONSE Henry is painting a room with four walls that are 12 feet by 14 feet. A gallon of paint costs $18 and covers 350 square feet. If he uses two coats of paint, how much will it cost him to paint the room?

52. The number of students at Franklin High School increased from 840 to 910 over a 5-year period. What was the percent of increase?

F 8.3%

G 14.0%

H 18.5%

J 92.3%

53. PROBABILITY Two dice are rolled. What is the probability that the sum is 10?

A $\dfrac{1}{3}$ **B** $\dfrac{1}{6}$ **C** $\dfrac{1}{12}$ **D** $\dfrac{1}{36}$

Spiral Review

54. TRAVEL The Chan's minivan requires 5 gallons of gasoline to travel 120 miles. How many gallons of gasoline will they need to travel 360 miles? (Lesson 2-6)

Evaluate each expression if $x = -2$, $y = 6$, and $z = 4$. (Lesson 2-5)

55. $|3 - x| + 7$

56. $12 - |z + 9|$

57. $|y + x| - z + 4$

Solve each equation. Round to the nearest hundredth. Check your solution. (Lesson 2-4)

58. $1.03p - 4 = -2.15p + 8.72$

59. $18 - 3.8t = 7.36 - 1.9t$

60. $5.4w + 8.2 = 9.8w - 2.8$

61. $2[d + 3(d - 1)] = 18$

Solve each equation. Check your solution. (Lesson 2-3)

62. $5n + 6 = -4$

63. $-11 = 7 + 3c$

64. $15 = 4a - 5$

65. $-14 + 7g = -63$

66. RIVERS The Congo River in Africa is 2900 miles long. That is 310 miles longer than the Niger River, which is also in Africa. (Lesson 2-2)

    a. Write an equation you could use to find the length of the Niger River.

    b. What is the length of the Niger River?

67. FOOD Cameron purchased x pounds of apples for $0.99 per pound and y pounds of oranges for $1.29 per pound. Write an algebraic expression that represents the cost of the purchase. (Lesson 1-1)

Skills Review

Translate each equation into a sentence.

68. $d - 14 = 5$

69. $2f + 6 = 19$

70. $y - 12 = y + 8$

71. $3a + 5 = 27 - 2a$

72. $-6c^2 - 4c = 25$

73. $d^4 + 64 = 3d^3 + 77$

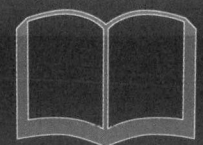

A **percentile** is a measure that is often used to report test data, such as standardized test scores. It tells us what percent of the total scores were below a given score.

- Percentiles measure rank from the bottom.

- There is no 0 percentile rank. The lowest score is at the 1st percentile.

- There is no 100th percentile rank. The highest score is at the 99th percentile.

Activity

A talent show was held for the twenty finalists in the Teen Idol contest. Each performer received a score from 0 through 30 with 30 being the highest. What is Victor's percentile rank?

Step 1 Write one score on each of 20 slips of paper.

Step 2 Arrange the slips vertically from greatest to least score.

Step 3 Find Victor's percentile rank.

Victor had a score of 28. There are 18 scores below his score. To find his percentile rank, use the following formula.

$$\frac{\text{number of scores below 28}}{\text{total number of scores}} \cdot 100 = \frac{18}{20} \cdot 100 \text{ or } 90$$

Victor scored at the 90th percentile in the contest.

Name	Score	Name	Score
Arnold	17	Ishi	27
Benito	9	James	20
Brooke	25	Kat	16
Carmen	21	Malik	10
Daniel	14	Natalie	26
Delia	29	Pearl	4
Fernando	15	Twyla	6
Heather	12	Victor	28
Horatio	5	Warren	22
Ingrid	11	Yolanda	18

Analyze the Results

1. Find the median, lower quartile, and upper quartile of the scores.

2. Which performer was at the 50th percentile? the 25th percentile? the 75th percentile?

3. Compare and contrast the values for the median, lower quartile, and upper quartile and the scores for the 25th, 50th, and 75th percentiles.

4. While Victor scored at the 90th percentile, what percent of the 30 possible points did he score?

5. **CCSS ARGUMENTS** Compare and contrast the percentile rank and the percent score.

6. Are there any outliers in the data that could alter the results of our computations?

7. **Deciles** are values that divide a set of data into ten equal-sized parts. The 1st decile contains data up to but not including the 10th percentile; the 2nd decile contains data from the 10th percentile up to but not including the 20th percentile, and so on.

   a. Which contestants' scores fall in the 6th decile?

   b. In which decile are Heather and Daniel?

Literal Equations and Dimensional Analysis

:·Then
- You solved equations with variables on each side.

:·Now
1. Solve equations for given variables.
2. Use formulas to solve real-world problems.

:·Why?
- Each year, more people use credit cards to make everyday purchases. If the entire balance is not paid by the due date, compound interest is applied. The formula for computing the balance of an account with compound interest added annually is $A = P(1 + r)^t$.
  - *A* represents the amount of money in the account including the interest,
  - *P* is the amount in the account before interest is added,
  - *r* is the interest rate written as a decimal,
  - *t* is the time in years.

 NewVocabulary
literal equation
dimensional analysis
unit analysis

 Common Core State Standards

Content Standards
A.CED.4 Rearrange formulas to highlight a quantity of interest, using the same reasoning as in solving equations.

A.REI.3 Solve linear equations and inequalities in one variable, including equations with coefficients represented by letters.

Mathematical Practices
6 Attend to precision.

1 Solve for a Specific Variable Some equations such as the one above contain more than one variable. At times, you will need to solve these equations for one of the variables.

Example 1 Solve for a Specific Variable

Solve $4m - 3n = 8$ for m.

$4m - 3n = 8$	Original equation
$4m - 3n + 3n = 8 + 3n$	Add $3n$ to each side.
$4m = 8 + 3n$	Simplify.
$\dfrac{4m}{4} = \dfrac{8 + 3n}{4}$	Divide each side by 4.
$m = \dfrac{8}{4} + \dfrac{3}{4}n$	Simplify.
$m = 2 + \dfrac{3}{4}n$	Simplify.

▸ **Guided**Practice

Solve each equation for the variable indicated.

1A. $15 = 3n + 6p$, for n

1B. $\dfrac{k - 2}{5} = 11j$, for k

1C. $28 = t(r + 4)$, for t

1D. $a(q - 8) = 23$, for q

Sometimes we need to solve equations for a variable that is on both sides of the equation. When this happens, you must get all terms with that variable onto one side of the equation. It is then helpful to use the Distributive Property to isolate the variable for which you are solving.

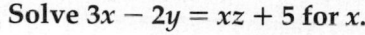

StudyTip

Solving for a Specific Variable When an equation has more than one variable, it can be helpful to highlight the variable for which you are solving on your paper.

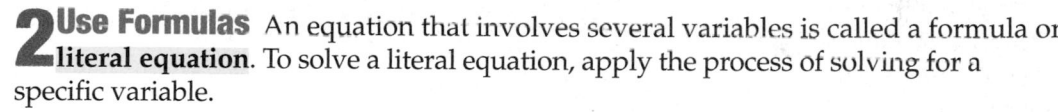

Example 2 Solve for a Specific Variable

Solve $3x - 2y = xz + 5$ for x.

$3x - 2y = xz + 5$	Original equation
$3x - 2y + 2y = xz + 5 + 2y$	Add $2y$ to each side.
$3x - xz = xz - xz + 5 + 2y$	Subtract xz from each side.
$3x - xz = 5 + 2y$	Simplify.
$x(3 - z) = 5 + 2y$	Distributive Property
$\dfrac{x(3 - z)}{3 - z} = \dfrac{5 + 2y}{3 - z}$	Divide each side by $3 - z$.
$x = \dfrac{5 + 2y}{3 - z}$	Simplify.

Since division by 0 is undefined, $3 - z \neq 0$ so $z \neq 3$.

GuidedPractice

Solve each equation for the variable indicated.

2A. $d + 5c = 3d - 1$, for d

2B. $6q - 18 = qr + t$, for q

2 Use Formulas An equation that involves several variables is called a formula or **literal equation**. To solve a literal equation, apply the process of solving for a specific variable.

Real-World Example 3 Use Literal Equations

YO-YOS Use the information about the largest yo-yo at the left. The formula for the circumference of a circle is $C = 2\pi r$, where C represents circumference and r represents radius.

a. Solve the formula for r.

$C = 2\pi r$	Formula for circumference
$\dfrac{C}{2\pi} = \dfrac{2\pi r}{2\pi}$	Divide each side by 2π.
$\dfrac{C}{2\pi} = r$	Simplify.

b. Find the radius of the yo-yo.

$\dfrac{C}{2\pi} = r$	Formula for radius
$\dfrac{32.7}{2\pi} = r$	$C = 32.7$
$5.2 \approx r$	Use a calculator.

The yo-yo has a radius of about 5.2 feet.

Real-WorldLink

The largest yo-yo in the world is 32.7 feet in circumference. It was launched by crane from a height of 189 feet.

Source: *Guinness Book of World Records*

GuidedPractice

3. GEOMETRY The formula for the volume of a rectangular prism is $V = \ell wh$, where ℓ is the length, w is the width, and h is the height.

A. Solve the formula for w.

B. Find the width of a rectangular prism that has a volume of 79.04 cubic centimeters, a length of 5.2 centimeters, and a height of 4 centimeters.

When using formulas, you may want to use dimensional analysis. **Dimensional analysis** or **unit analysis** is the process of carrying units throughout a computation.

Example 4 Use Dimensional Analysis

RUNNING A 10K run is 10 kilometers long. If 1 meter = 1.094 yards, use dimensional analysis to find the length of the race in miles. (*Hint*: 1 mi = 1760 yd)

Since the given conversion relates meters to yards, first convert 10 kilometers to meters. Then multiply by the conversion factor such that the unit meters are divided out. To convert from yards to miles, multiply by $\frac{1 \text{ mi}}{1760 \text{ yd}}$.

length of run	×	kilometers to meters	×	meters to yards	×	yards to miles
10 km	×	$\frac{1000 \text{ m}}{1 \text{ km}}$	×	$\frac{1.094 \text{ yd}}{1 \text{ m}}$	×	$\frac{1 \text{ mi}}{1760 \text{ yd}}$

Notice how the units cancel, leaving the unit to which you are converting.

$$10 \text{ km} \times \frac{1000 \text{ m}}{1 \text{ km}} \times \frac{1.094 \text{ yd}}{1 \text{ m}} \times \frac{1 \text{ mi}}{1760 \text{ yd}} = \frac{10{,}940 \text{ mi}}{1760}$$

$$\approx 6.2 \text{ mi}$$

A 10K race is approximately 6.2 miles.

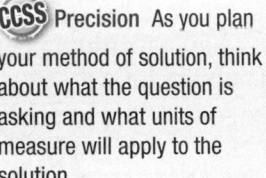

StudyTip

CCSS Precision As you plan your method of solution, think about what the question is asking and what units of measure will apply to the solution.

▶ **Guided**Practice

4. A car travels a distance of 100 feet in about 2.8 seconds. What is the velocity of the car in miles per hour? Round to the nearest whole number.

Check Your Understanding

◯ **= Step-by-Step Solutions begin on page R13.**

Examples 1–2 Solve each equation or formula for the variable indicated.

1 $5a + c = -8a$, for a

2. $7h + f = 2h + g$, for g

3. $\dfrac{k + m}{-7} = n$, for k

4. $q = p(r + s)$, for p

Example 3

5. PACKAGING A soap company wants to use a cylindrical container to hold their new liquid soap.

  a. Solve the formula for h.

  b. What is the height of a container if the volume is 56.52 cubic inches and the radius is 1.5 inches? Round to the nearest tenth.

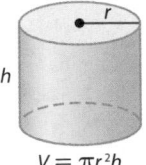

$V = \pi r^2 h$

Example 4

6. SHOPPING Scott found a rare video game on an online auction site priced at 35 Australian dollars. If the exchange rate is \$1 U.S. = \$1.24 Australian, find the cost of the game in United States dollars. Round to the nearest cent.

7. CCSS PRECISION A fisheye lens has a minimum focus range of 13.5 centimeters. If 1 centimeter is equal in length to about 0.39 inch, what is the minimum focus range of the lens in feet?

Examples 1–2 Solve each equation or formula for the variable indicated.

8. $u = vw + z$, for v

9 $x = b - cd$, for c

10. $fg - 9h = 10j$, for g

11. $10m - p = -n$, for m

12. $r = \frac{2}{3}t + v$, for t

13. $\frac{5}{9}v + w = z$, for v

14. $\frac{10ac - x}{11} = -3$, for a

15. $\frac{df + 10}{6} = g$, for f

Example 3

16. FITNESS The formula to compute a person's body mass index is $B = 703 \cdot \frac{w}{h^2}$. B represents the body mass index, w is the person's weight in pounds, and h represents the person's height in inches.

  a. Solve the formula for w.

  b. What is the weight to the nearest pound of a person who is 64 inches tall and has a body mass index of 21.45?

17. PHYSICS Acceleration is the measure of how fast a velocity is changing. The formula for acceleration is $a = \frac{v_f - v_i}{t}$. a represents the acceleration rate, v_f is the final velocity, v_i is the initial velocity, and t represents the time in seconds.

  a. Solve the formula for v_f.

  b. What is the final velocity of a runner who is accelerating at 2 feet per second squared for 3 seconds with an initial velocity of 4 feet per second?

Example 4

18. SWIMMING If each lap in a pool is 100 meters long, how many laps equal one mile? Round to the nearest tenth. (*Hint*: 1 foot ≈ 0.3048 meter)

19. CCSS PRECISION How many liters of gasoline are needed to fill a 13.2-gallon tank? There are about 1.06 quarts per 1 liter. Round to the nearest tenth.

Solve each equation or formula for the variable indicated.

20. $-14n + q = rt - 4n$, for n

21. $18t + 11v = w - 13t$, for t

22. $ax + z = aw - y$, for a

23. $10c - f = -13 + cd$, for c

Select an appropriate unit from the choices below and convert the rate to that unit.

ft/s	mph	mm/s	km/s

24. a car traveling at 36 ft/s

25. a snail moving at 3.6 m/h

26. a person walking at 3.4 mph

27. a satellite moving at 234,000 m/min

28. DANCING The formula $P = \frac{1.2W}{H^2}$ represents the amount of pressure exerted on the floor by a ballroom dancer's heel. In this formula, P is the pressure in pounds per square inch, W is the weight of a person wearing the shoe in pounds, and H is the width of the heel of the shoe in inches.

  a. Solve the formula for W.

  b. Find the weight of the dancer if the heel is 3 inches wide and the pressure exerted is 30 pounds per square inch.

Write an equation and solve for the variable indicated.

29. Seven less than a number t equals another number r plus 6. Solve for t.

30. Ten plus eight times a number a equals eleven times number d minus six. Solve for a.

31. Nine tenths of a number g is the same as seven plus two thirds of another number k. Solve for k.

32. Three fourths of a number p less two is five sixths of another number r plus five. Solve for r.

33 **GIFTS** Ashley has 214 square inches of paper to wrap a gift box. The surface area S of the box can be found by using the formula $S = 2w(\ell + h) + 2\ell h$, where w is the width of the box, ℓ is the length of the box, and h is the height. If the length of the box is 7 inches and the width is 6 inches, how tall can Ashley's box be?

34. **DRIVING** A car is driven x miles a year and averages m miles per gallon.

   a. Write a formula for g, the number of gallons used in a year.

   b. If the average price of gas is p dollars per gallon, write a formula for the total gas cost c in dollars for driving this car each year.

   c. Car A averages 15 miles per gallon on the highway, while Car B averages 35 miles per gallon on the highway. If you average 15,000 miles each year, how much money would you save on gas per week by using Car B instead of Car A if the cost of gas averages $3 per gallon? Explain.

H.O.T. Problems   Use Higher-Order Thinking Skills

35. **CHALLENGE** The circumference of an NCAA women's basketball is 29 inches, and the rubber coating is $\frac{3}{16}$ inch thick. Use the formula $v = \frac{4}{3}\pi r^3$, where v represents the volume and r is the radius of the inside of the ball, to determine the volume of the air inside the ball. Round to the nearest whole number.

36. **REASONING** Select an appropriate unit to describe the highway speed of a car and the speed of a crawling caterpillar. Can the same unit be used for both? Explain.

37. **ERROR ANALYSIS** Sandrea and Fernando are solving $4a - 5b = 7$ for b. Is either of them correct? Explain.

Sandrea	Fernando
$4a - 5b = 7$	$4a - 5b = 7$
$-5b = 7 - 4a$	$5b = 7 - 4a$
$\dfrac{-5b}{-5} = \dfrac{7 - 4a}{-5}$	$\dfrac{5b}{5} = \dfrac{7 - 4a}{5}$
$b = \dfrac{7 - 4a}{-5}$	$b = \dfrac{7 - 4a}{5}$

38. **OPEN ENDED** Write a formula for A, the area of a geometric figure such as a triangle or rectangle. Then solve the formula for a variable other than A.

39. **CCSS PERSEVERANCE** Solve each equation or formula for the variable indicated.

   a. $n = \dfrac{x + y - 1}{xy}$ for x

   b. $\dfrac{x + y}{x - y} = \dfrac{1}{2}$ for y

40. **WRITING IN MATH** Why is it helpful to be able to represent a literal equation in different ways?

41. Eula is investing $6000, part at 4.5% interest and the rest at 6% interest. If d represents the amount invested at 4.5%, which expression represents the amount of interest earned in one year by the amount paying 6%?

A $0.06d$

B $0.06(d - 6000)$

C $0.06(d + 6000)$

D $0.06(6000 - d)$

42. Todd drove from Boston to Cleveland, a distance of 616 miles. His breaks, gasoline, and food stops took 2 hours. If his trip took 16 hours altogether, what was Todd's average speed?

F 38.5 mph

G 40 mph

H 44 mph

J 47.5 mph

43. SHORT RESPONSE Brian has 3 more books than Erika. Jasmine has triple the number of books that Brian has. Altogether Brian, Erika, and Jasmine have 22 books. How many books does Jasmine have?

44. GEOMETRY Which of the following best describes a plane?

A a location having neither size nor shape

B a flat surface made up of points having no depth

C made up of points and has no thickness or width

D a boundless, three-dimensional set of all points

Find the final price of each item. (Lesson 2-7)

45. lamp: $120.00
discount: 20%
tax: 6%

46. dress: $70.00
discount: 30%
tax: 7%

47. camera: $58.00
discount: 25%
tax: 6.5%

48. jacket: $82.00
discount: 15%
tax: 6%

49. comforter: $67.00
discount: 20%
tax: 6.25%

50. lawnmower: $720.00
discount: 35%
tax: 7%

Solve each proportion. If necessary, round to the nearest hundredth. (Lesson 2-6)

51. $\frac{3}{4.5} = \frac{x}{2.5}$

52. $\frac{?}{0.36} = \frac{7}{p}$

53. $\frac{m}{9} = \frac{2.8}{4.9}$

54. JOBS Laurie mows lawns to earn extra money. She can mow at most 30 lawns in one week. She profits $15 on each lawn she mows. Identify a reasonable domain and range for this situation and draw a graph. (Lesson 1-6)

55. ENTERTAINMENT Each member of the pit orchestra is selling tickets for the school musical. The trombone section sold 50 floor tickets and 90 balcony tickets. Write and evaluate an expression to find how much money the trombone section collected. (Lesson 1-4)

School Musical

Tickets
Floor $7.50
Balcony $5.00

Solve each equation.

56. $8k + 9 = 7k + 6$

57. $3 - 4q = 10q + 10$

58. $\frac{3}{4}n + 16 = 2 - \frac{1}{8}n$

59. $\frac{1}{4} - \frac{2}{3}y = \frac{3}{4} - \frac{1}{3}y$

60. $4(2a - 1) = -10(a - 5)$

61. $2(w - 3) + 5 = 3(w - 1)$

Weighted Averages

:: Then	:: Now	:: Why?
● You translated sentences into equations.	**1** Solve mixture problems. **2** Solve uniform motion problems.	● Baseball players' performance is measured in large part by statistics. Slugging average (SLG) is a weighted average that measures the power of a hitter. The slugging average is calculated by using the following formula.

$$SLG = \frac{1B + (2 \times 2B) + (3 \times 3B) + (4 \times HR)}{\text{at bats}}$$

 NewVocabulary
weighted average
mixture problem
uniform motion problem
rate problem

 Common Core State Standards

Content Standards
A.REI.1 Explain each step in solving a simple equation as following from the equality of numbers asserted at the previous step, starting from the assumption that the original equation has a solution. Construct a viable argument to justify a solution method.

A.REI.3 Solve linear equations and inequalities in one variable, including equations with coefficients represented by letters.

Mathematical Practices
1 Make sense of problems and persevere in solving them.
4 Model with mathematics.

1 Weighted Averages The batter's slugging percentage is an example of a weighted average. The **weighted average** M of a set of data is found by multiplying each data value by its weight and then finding the mean of the new data set.

Mixture problems are problems in which two or more parts are combined into a whole. They are solved using weighted averages. In a mixture problem, the units are usually the number of gallons or pounds and the value is the cost, value, or concentration per unit.

Real-World Example 1 Mixture Problem

RETAIL A tea company sells blended tea for $25 per pound. To make blackberry tea, dried blackberries that cost $10.50 per pound are blended with black tea that costs $35 per pound. How many pounds of black tea should be added to 5 pounds of dried blackberries to make blackberry tea?

Step 1 Let w be the weight of the black tea. Make a table to organize the information.

	Number of Units (lb)	Price per Unit ($)	Total Price (price)(units)
Dried Blackberries	5	10.50	10.50(5)
Black Tea	w	35	$35w$
Blackberry Tea	$5 + w$	25	$25(5 + w)$

Write an equation using the information in the table.

Price of blackberries	plus	price of tea	equals	price of blackberry tea.
10.50(5)	+	$35w$	=	$25(5 + w)$

Step 2 Solve the equation.

$10.50(5) + 35w = 25(5 + w)$	Original equation
$52.5 + 35w = 125 + 25w$	Distributive Property
$52.5 + 35w - 25w = 125 + 25w - 25w$	Subtract $25w$ from each side.
$52.5 + 10w = 125$	Simplify.
$52.5 - 52.5 + 10w = 125 - 52.5$	Subtract 52.5 from each side.
$10w = 72.5$	Simplify.
$w = 7.25$	Divide each side by 10.

Rim Light/PhotoLink/Photodisc/Getty Images

To make the blackberry tea, 7.25 pounds of black tea will need to be added to the dried blackberries.

▶ **Guided**Practice

1. **COFFEE** How many pounds of Premium coffee beans should be mixed with 2 pounds of Supreme coffee to make the Blend coffee?

Premium   Supreme   Blend

Sometimes mixture problems are expressed in terms of percents.

⬤ Real-World Example 2   Percent Mixture Problem

FRUIT PUNCH Mrs. Matthews has 16 cups of punch that is 3% pineapple juice. She also has a punch that is 33% pineapple juice. How many cups of the 33% punch will she need to add to the 3% punch to obtain a punch that is 20% pineapple juice?

Step 1 Let x = the amount of 33% solution to be added. Make a table.

	Amount of Punch (cups)	Amount of Pineapple Juice
3% Punch	16	0.03(16)
33% Punch	x	0.33x
20% Punch	16 + x	0.20(16 + x)

Write an equation using the information in the table.

Amount of pineapple juice in 3% punch	plus	amount of pineapple juice in 33% punch	equals	amount of pineapple juice in 20% punch.
0.03(16)	+	0.33x	=	0.20(16 + x)

Step 2 Solve the equation.

$$0.03(16) + 0.33x = 0.20(16 + x)$$ Original equation

$$0.48 + 0.33x = 3.2 + 0.20x$$ Simplify.

$$0.48 + 0.33x - \mathbf{0.20x} = 3.2 + 0.20x - \mathbf{0.20x}$$ Subtract 0.20x from each side.

$$0.48 + 0.13x = 3.2$$ Simplify.

$$0.48 - \mathbf{0.48} + 0.13x = 3.2 - \mathbf{0.48}$$ Subtract 0.48 from each side.

$$0.13x = 2.72$$ Simplify.

$$\frac{0.13x}{\mathbf{0.13}} = \frac{2.72}{\mathbf{0.13}}$$ Divide each side by 0.13.

$$x \approx 20.9$$ Round to the nearest tenth.

Mrs. Matthews should add about 20.9 cups of the 33% punch to the 16 cups of the 3% punch.

▶ **Guided**Practice

2. **ANTIFREEZE** One type of antifreeze is 40% glycol, and another type of antifreeze is 60% glycol. How much of each kind should be used to make 100 gallons of antifreeze that is 48% glycol?

2 **Uniform Motion Problems** **Uniform motion problems** or **rate problems** are problems in which an object moves at a certain speed or rate. The formula $d = rt$ is used to solve these problems. In the formula, d represents distance, r represents rate, and t represents time.

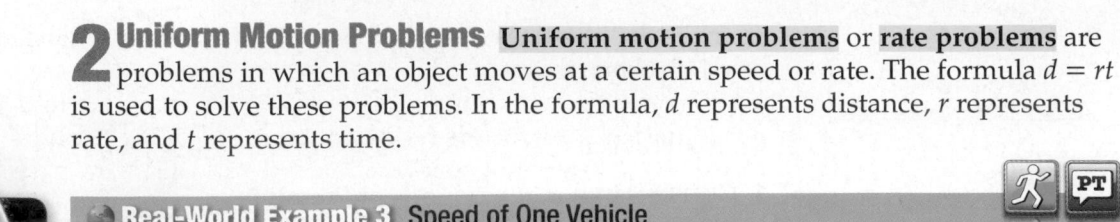

Real-World Example 3 Speed of One Vehicle

INLINE SKATING It took Travis and Tony 40 minutes to skate 5 miles. The return trip took them 30 minutes. What was their average speed for the trip?

Understand We know that the boys did not travel the same amount of time on each portion of their trip. So, we will need to find the weighted average of their speeds. We are asked to find their average speed for both portions of the trip.

Plan First find the rate of the going portion, and then the return portion of the trip. Because the rate is in miles per hour we convert 40 minutes to about 0.667 hours and 30 minutes to 0.5 hours.

Going

$$r = \frac{d}{t} \qquad \text{Formula for rate}$$

$$\approx \frac{5 \text{ miles}}{0.667 \text{ hour}} \text{ or about 7.5 miles per hour} \qquad \text{Substitution } d = 5 \text{ mi}, t = 0.667 \text{ h}$$

Return

$$r = \frac{d}{t} \qquad \text{Formula for rate}$$

$$= \frac{5 \text{ miles}}{0.5 \text{ hour}} \text{ or 10 miles per hour} \qquad \text{Substitution } d = 5 \text{ mi}, t = 0.5 \text{ h}$$

Because we are looking for a weighted average we cannot just average their speeds. We need to find the weighted average for the round trip.

Solve $M = \dfrac{(\textbf{rate of going})(\text{time of going}) + (\textbf{rate of return})(\text{time of return})}{\text{time of going} + \text{time of return}}$

$$\approx \frac{(7.5)(0.667) + (10)(0.5)}{0.667 + 0.5} \qquad \text{Substitution}$$

$$\approx \frac{10.0025}{1.167} \text{ or about 8.6} \qquad \text{Simplify.}$$

Their average speed was about 8.6 miles per hour.

Check Our solution of 8.6 miles per hour is between the going portion rate, 7.5 miles per hour, and the return rate, 10 miles per hour. So, we know that our answer is reasonable.

▶ **Guided**Practice

3. EXERCISE Austin jogged 2.5 miles in 16 minutes and then walked 1 mile in 10 minutes. What was his average speed?

The formula $d = rt$ can also be used to solve real-world problems involving two vehicles in motion.

Real-World Example 4 Speeds of Two Vehicles

FREIGHT TRAINS Two trains are 550 miles apart heading toward each other on parallel tracks. Train A is traveling east at 35 miles per hour, while Train B travels west at 45 miles per hour. When will the trains pass each other?

Step 1 Draw a diagram.

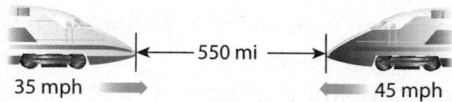

35 mph   ← 550 mi →   45 mph

Step 2 Let t = the number of hours until the trains pass each other. Make a table.

	r	t	d = rt
Train A	35	t	35t
Train B	45	t	45t

Step 3 Write and solve an equation.

Distance traveled by Train A	plus	distance traveled by Train B	equals	550 miles.
35t	+	45t	=	550

$$35t + 45t = 550 \qquad \text{Original equation}$$
$$80t = 550 \qquad \text{Simplify.}$$
$$\frac{80t}{80} = \frac{550}{80} \qquad \text{Divide each side by 80.}$$
$$t = 6.875 \qquad \text{Simplify.}$$

The trains will pass each other in about 6.875 hours.

GuidedPractice

4. CYCLING Two cyclists begin traveling in opposite directions on a circular bike trail that is 5 miles long. One cyclist travels 12 miles per hour, and the other travels 18 miles per hour. How long will it be before they meet?

Check Your Understanding

◯ = Step-by-Step Solutions begin on page R13.

Example 1

(1) FOOD Tasha ordered soup and salad for lunch. If Tasha ordered 10 ounces of soup for lunch and the total cost was $3.30, how many ounces of salad did Tasha order?

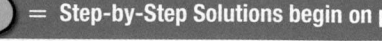
 15¢/ounce    20¢/ounce

Example 2

2. CHEMISTRY Margo has 40 milliliters of 25% solution. How many milliliters of 60% solution should she add to obtain the required 30% solution?

Example 3

3. TRAVEL A boat travels 16 miles due north in 2 hours and 24 miles due west in 2 hours. What is the average speed of the boat?

4. EXERCISE Felisa jogged 3 miles in 25 minutes and then jogged 3 more miles in 30 minutes. What was her average speed in miles per minute?

Example 4

5. CYCLING A cyclist begins traveling 18 miles per hour. At the same time and at the same starting point, an inline skater follows the cyclist's path and begins traveling 6 miles per hour. After how much time will they be 24 miles apart?

Example 1

6. CANDY A candy store wants to create a mix using two hard candies. One is priced at $5.45 per pound, and the other is priced at $7.33 per pound. How many pounds of the $7.33 candy should be mixed with 11 pounds of the $5.45 candy to sell the mixture for $6.14 per pound?

7 BUSINESS Party Supplies Inc. sells metallic balloons for $2 each and helium balloons for $3.50 per bunch. Yesterday, they sold 36 more metallic balloons than the number of bunches of helium balloons. The total sales for both types of balloons were $281. Let b represent the number of metallic balloons sold.

a. Copy and complete the table representing the problem.

	Number	Price	Total Price
Metallic Balloons	b		
Bunches of Helium Balloons	$b - 36$		

b. Write an equation to represent the problem.

c. How many metallic balloons were sold?

d. How many bunches of helium balloons were sold?

8. FINANCIAL LITERACY Lakeisha spent $4.57 on color and black-and-white copies for her project. She made 7 more black-and-white copies than color copies. How many color copies did she make?

Type of Copy	Cost per Page
color	$0.44
black-and-white	$0.07

Example 2

9. FISH Rosamaria is setting up a 20-gallon saltwater fish tank that needs to have a salt content of 3.5%. If Rosamaria has water that has 2.5% salt and water that has 3.7% salt, how many gallons of the water with 3.7% salt content should Rosamaria use?

10. CHEMISTRY Hector is performing a chemistry experiment that requires 160 milliliters of 40% sulfuric acid solution. He has a 25% sulfuric acid solution and a 50% sulfuric acid solution. How many milliliters of each solution should he mix to obtain the needed solution?

Example 3

11. TRAVEL A boat travels 36 miles in 1.5 hours and then 14 miles in 0.75 hour. What is the average speed of the boat?

12. CCSS MODELING A person walked 1.5 miles in 28 minutes and then jogged 1.2 more miles in 10 minutes. What was the average speed in miles per minute?

Example 4

13. AIRLINERS Two airliners are 1600 miles apart and heading toward each other at different altitudes. The first plane is traveling north at 620 miles per hour, while the second is traveling south at 780 miles per hour. When will the planes pass each other?

14. SAILING A ship is sailing due east at 20 miles per hour when it passes the lighthouse. At the same time a ship is sailing due west at 15 miles per hour when it passes a point. The point is 175 miles east of the lighthouse. When will these ships pass each other?

15. CHEMISTRY A lab technician has 40 gallons of a 15% iodine solution. How many gallons of a 40% iodine solution must he add to make a 20% iodine solution?

16. GRADES At Westbridge High School, a student's grade point average (GPA) is based on the student's grade and the class credit rating. Brittany's grades for this quarter are shown. Find Brittany's GPA if a grade of A equals 4 and a B equals 3.

Class	Credit Rating	Grade
Algebra 1	1	A
Science	1	A
English	1	B
Spanish	1	A
Music	$\frac{1}{2}$	B

17. SPORTS In a triathlon, Steve swam 0.5 mile in 15 minutes, biked 20 miles in 90 minutes, and ran 4 miles in 30 minutes. What was Steve's average speed for the triathlon in miles per hour?

18. MUSIC Amalia has 10 songs on her digital media player. If 3 songs are 5 minutes long, 3 are 4 minutes long, 2 are 2 minutes long, and 2 are 3.5 minutes long, what is the average length of the songs?

19 DISTANCE Garcia is driving to Florida for vacation. The trip is a total of 625 miles.

a. How far can he drive in 6 hours at 65 miles per hour?

b. If Garcia maintains a speed of 65 miles per hour, how long will it take him to drive to Florida?

20. TRAVEL Two buses leave Smithville at the same time, one traveling north and the other traveling south. The northbound bus travels at 50 miles per hour, and the southbound bus travels at 65 miles per hour. Let t represent the amount of time since their departure.

a. Copy and complete the table representing the situation.

	r	t	$d = rt$
Northbound bus	?	?	?
Southbound bus	?	?	?

b. Write an equation to find when the buses will be 345 miles apart.

c. Solve the equation. Explain how you found your answer.

21. TRAVEL A subway travels 60 miles per hour from Glendale to Midtown. Another subway, traveling at 45 miles per hour, takes 11 minutes longer for the same trip. How far apart are Glendale and Midtown?

H.O.T. Problems   Use Higher-Order Thinking Skills

22. OPEN ENDED Write a problem that depicts motion in opposite directions.

23. CCSS ARGUMENTS Describe the conditions so that adding a 50% solution to a 100% solution would produce a 75% solution.

24. CHALLENGE Find five consecutive odd integers from least to greatest in which the sum of the first and the fifth is one less than three times the fourth.

25. CHALLENGE Describe a situation involving mixtures that could be represented by $1.00x + 0.15(36) = 0.50(x + 36)$.

26. WRITING IN MATH Describe how a gallon of 25% solution is added to an unknown amount of 10% solution to get a 15% solution.

27. If $2x + y = 5$, what is the value of $4x$?

  A $10 - y$

  B $10 - 2y$

  C $\dfrac{5 - y}{2}$

  D $\dfrac{10 - y}{2}$

28. Which expression is equivalent to $7x^2 3x^{-4}$?

  F $21x^{-8}$

  G $21x^2$

  H $21x^{-6}$

  J $21x^{-2}$

29. GEOMETRY What is the base of the triangle if the area is 56 square meters?

  A 4 m

  B 8 m

  C 16 m

  D 28 m

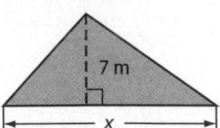

30. SHORT RESPONSE Brianne makes blankets for a baby store. She works on the blankets 30 hours per week. The store pays her $9.50 per hour plus 30% of the profit. If her hourly rate is increased by $0.75 and her commission is raised to 40%, how much will she earn for a week in which there was a $300 profit?

Solve each equation or formula for x. (Lesson 2-8)

31. $2bx - b = -5$

32. $3x - r = r(-3 + x)$

33. $A = 2\pi r^2 + 2\pi rx$

34. SKIING Yuji is registering for ski camp. The cost of the camp is $1254, but there is a sales tax of 7%. What is the total cost of the camp including tax? (Lesson 2-7)

Translate each equation into a sentence. (Lesson 2-1)

35. $\dfrac{n}{-6} = 2n + 1$

36. $18 - 5h = 13h$

37. $2x^2 + 3 = 21$

Refer to the graph.

38. Name the ordered pair at point A and explain what it represents. (Lesson 1-6)

39. Name the ordered pair at point B and explain what it represents. (Lesson 1-6)

40. Identify the independent and dependent variables for the function. (Lesson 1-6)

41. BASEBALL Tickets to a baseball game cost $18.95, $12.95, or $9.95. A hot dog and soda combo costs $5.50. The Madison family is having a reunion. They buy 10 tickets in each price category and plan to buy 30 combos. What is the total cost for the tickets and meals? (Lesson 1-4)

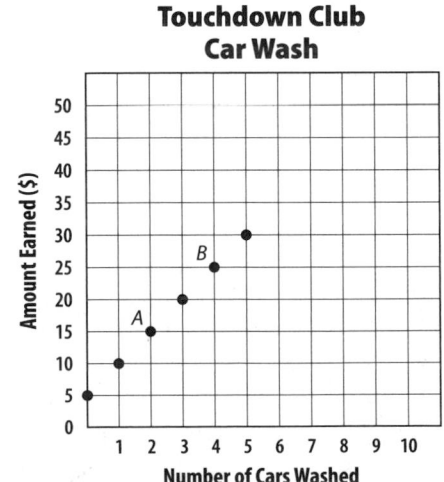

Touchdown Club Car Wash

Solve each equation.

42. $a - 8 = 15$

43. $9m - 11 = -29$

44. $18 - 2k = 24$

45. $5 - 8y = 61$

46. $7 = \dfrac{h}{2} + 3$

47. $\dfrac{n}{6} + 1 = 5$

Study Guide

KeyConcepts

Writing Equations (Lesson 2-1)

- Identify the unknown you are looking for and assign a variable to it. Then, write the sentence as an equation.

Solving Equations (Lessons 2-2 to 2-4)

- Addition and Subtraction Properties of Equality:
  If an equation is true and the same number is added to or subtracted from each side, the resulting equation is true.

- Multiplication and Division Properties of Equality:
  If an equation is true and each side is multiplied or divided by the same nonzero number, the resulting equation is true.

- Steps for Solving Equations:

  Step 1 Simplify the expression on each side. Use the Distributive Property as needed.

  Step 2 Use the Addition and/or Subtraction Properties of Equality to get the variables on one side and the numbers without variables on the other side.

  Step 3 Use the Multiplication or Division Property of Equality to solve.

Absolute Value Equations (Lesson 2-5)

- For any real numbers a and b, if $|a| = b$ and $b \geq 0$, then $a = b$ or $a = -b$.

Ratios and Proportions (Lesson 2-6)

- The Means-Extremes Property of Proportion states that in a proportion, the product of the extremes is equal to the product of the means.

Percent of Change (Lesson 2-7)

- percent of change $= \dfrac{\text{the change in an amount}}{\text{the original amount}}$ expressed as a percent

Weighted Averages (Lesson 2-9)

- the weighted average M of a set of data
  $= \dfrac{\text{sum of (units} \times \text{the value per unit)}}{\text{the total number of units}}$

FOLDABLES StudyOrganizer

Be sure the Key Concepts are noted in your Foldable.

KeyVocabulary

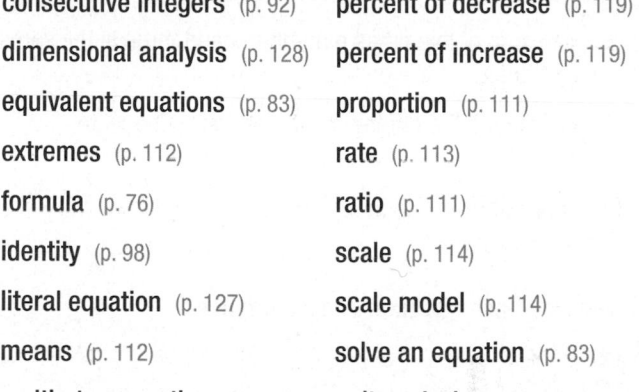

consecutive integers (p. 92)

dimensional analysis (p. 128)

equivalent equations (p. 83)

extremes (p. 112)

formula (p. 76)

identity (p. 98)

literal equation (p. 127)

means (p. 112)

multi-step equations (p. 91)

number theory (p. 92)

percent of change (p. 119)

percent of decrease (p. 119)

percent of increase (p. 119)

proportion (p. 111)

rate (p. 113)

ratio (p. 111)

scale (p. 114)

scale model (p. 114)

solve an equation (p. 83)

unit analysis (p. 128)

unit rate (p. 113)

weighted average (p. 132)

VocabularyCheck

State whether each sentence is *true* or *false*. If *false*, replace the underlined term to make a true sentence.

1. In order to write an equation to solve a problem, identify the unknown for which you are looking and assign a(n) <u>number</u> to it.

2. To <u>solve an equation</u> means to find the value of the variable that makes the equation true.

3. The numbers 10, 12, and 14 are an example of <u>consecutive even integers</u>.

4. The <u>absolute value</u> of any number is simply the distance the number is away from zero on a number line.

5. A(n) <u>equation</u> is a comparison of two numbers by division.

6. An equation stating that two ratios are equal is called a(n) <u>proportion</u>.

7. If the new number is less than the original number, the percent of change is a percent of <u>increase</u>.

8. The <u>weighted average</u> of a set of data is the sum of the product of the number of units and the value per unit divided by the sum of the number of units.

Lesson-by-Lesson Review

2-1 Writing Equations

Translate each sentence into an equation.

9. The sum of five times a number x and three is the same as fifteen.

10. Four times the difference of b and six is equal to b squared.

11. One half of m cubed is the same as four times m minus nine.

Translate each equation into a sentence.

12. $3p + 8 = 20$

13. $h^2 - 5h + 6 = 0$

14. $\frac{3}{4}w^2 + \frac{2}{3}w - \frac{1}{5} = 2$

15. **FENCING** Adrianne wants to create an outdoor rectangular kennel. The length will be three feet more than twice the width. Write and use an equation to find the length and the width of the kennel if Adrianne has 54 feet of fencing.

Example 1

Translate the following sentence into an equation.

Six times the sum of a number n and four is the same as the difference between two times n to the second power and ten.

$6(n + 4) = 2n^2 - 10$

Example 2

Translate $3d^2 - 9d + 8 = 4(d + 2)$ into a sentence.

Three times a number d squared minus nine times d increased by eight is equal to four times the sum of d and two.

2-2 Solving One-Step Equations

Solve each equation. Check your solution.

16. $x - 9 = 4$

17. $-6 + g = -11$

18. $\frac{5}{9} + w = \frac{7}{9}$

19. $3.8 = m + 1.7$

20. $\frac{a}{12} = 5$

21. $8y = 48$

22. $\frac{2}{5}b = -4$

23. $-\frac{t}{16} = -\frac{7}{8}$

24. **AGE** Max is four years younger than his sister Brenda. Max is 16 years old. Write and solve an equation to find Brenda's age.

Example 3

Solve $x - 13 = 9$. Check your solution.

$$x - 13 = 9 \qquad \text{Original equation}$$
$$x - 13 + 13 = 9 + 13 \qquad \text{Add 13 to each side.}$$
$$x = 22 \qquad -13 + 13 = 0 \text{ and } 9 + 13 = 22$$

To check that 22 is the solution, substitute 22 for x in the original equation.

CHECK $x - 13 = 9$ Original equation

$22 - 13 \stackrel{?}{=} 9$ Substitute 22 for x.

$9 = 9 \checkmark$ Subtract.

2-3 Solving Multi-Step Equations

Solve each equation. Check your solution.

25. $2d - 4 = 8$

26. $-9 = 3t + 6$

27. $14 = -8 - 2k$

28. $\frac{n}{4} - 7 = -2$

29. $\frac{r + 4}{3} = 7$

30. $-18 = \frac{9 - a}{2}$

31. $6g - 3.5 = 8.5$

32. $0.2c + 4 = 6$

33. $\frac{f}{3} - 9.2 = 3.5$

34. $4 = \frac{-3u - (-7)}{-8}$

35. CONSECUTIVE INTEGERS Find three consecutive odd integers with a sum of 63.

36. CONSECUTIVE INTEGERS Find three consecutive integers with a sum of −39.

2-4 Solving Equations with the Variable on Each Side

Solve each equation. Check your solution.

37. $8m + 7 = 5m + 16$

38. $2h - 14 = -5h$

39. $21 + 3j = 9 - 3j$

40. $\frac{x - 3}{4} = \frac{x}{2}$

41. $\frac{6r - 7}{10} = \frac{r}{4}$

42. $3(p + 4) = 33$

43. $-2(b - 3) - 4 = 18$

44. $4(3w - 2) = 8(2w + 3)$

Write an equation and solve each problem.

45. Find the sum of three consecutive odd integers if the sum of the first two integers is equal to twenty-four less than four times the third integer.

46. TRAVEL Mr. Jones drove 480 miles to a business meeting. His travel time to the meeting was 8 hours and from the meeting was 7.5 hours. Find his rate of travel for each leg of the trip.

2-5 Solving Equations Involving Absolute Value

Evaluate each expression if $m = -8$, $n = 4$, and $p = -12$.

47. $|3m - n|$

48. $|-2p + m| - 3n$

49. $-3|6n - 2p|$

50. $4|7m + 3p| + 4n$

Solve each equation. Then graph the solution set.

51. $|x - 6| = 11$

52. $|-4w + 2| = 14$

53. $\left|\frac{1}{3}d - 6\right| = 15$

54. $\left|\frac{2b}{3} + 8\right| = 20$

Example 7

Solve $|y - 9| = 16$. Then graph the solution set.

Case 1

$$y - 9 = 16 \qquad \text{Original equation}$$

$$y - 9 + 9 = 16 + 9 \qquad \text{Add 9 to each side.}$$

$$y = 25 \qquad \text{Simplify.}$$

Case 2

$$y - 9 = -16 \qquad \text{Original equation}$$

$$y - 9 + 9 = -16 + 9 \qquad \text{Add 9 to each side.}$$

$$y = -7 \qquad \text{Simplify.}$$

The solution set is $\{-7, 25\}$.

Graph the points on a number line.

2-6 Ratios and Proportions

Determine whether each pair of ratios are equivalent ratios. Write *yes* or *no*.

55. $\frac{27}{45}, \frac{3}{5}$

56. $\frac{18}{32}, \frac{3}{4}$

Solve each proportion. If necessary, round to the nearest hundredth.

57. $\frac{4}{9} = \frac{a}{45}$

58. $\frac{3}{8} = \frac{21}{t}$

59. $\frac{9}{12} = \frac{g}{16}$

60. CONSTRUCTION A new gym is being built at Greenfield Middle School. The length of the gym as shown on the builder's blueprints is 12 inches. Find the actual length of the new gym.

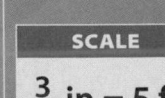

SCALE

$\frac{3}{4}$ in = 5 ft

Example 8

Determine whether $\frac{7}{9}$ and $\frac{42}{54}$ are equivalent ratios. Write *yes* or *no*. Justify your answer.

First, simplify each ratio. $\frac{7}{9}$ is already in simplest form.

$$\frac{42}{54} = \frac{42 \div 6}{54 \div 6} = \frac{7}{9}$$

When expressed in simplest form, the ratios are equivalent. The answer is yes.

Example 9

Solve $\frac{r}{8} = \frac{3}{4}$. If necessary, round to the nearest hundredth.

$$\frac{r}{8} = \frac{3}{4} \qquad \text{Original equation}$$

$$r(4) = 3(8) \qquad \text{Find the cross products.}$$

$$4r = 24 \qquad \text{Simplify.}$$

$$\frac{4r}{4} = \frac{24}{4} \qquad \text{Divide each side by 4.}$$

$$r = 6 \qquad \text{Simplify.}$$

State whether each percent of change is a percent of *increase* or a percent of *decrease*. Then find the percent of change. Round to the nearest whole percent.

61. original: 40, new: 50

62. original: 36, new: 24

63. original: $72, new: $60

Find the total price of each item.

64. boots: $64, tax: 7%

65. video game: $49, tax: 6.5%

66. hockey skates: $199, tax: 5.25%

Find the discounted price of each item.

67. digital media player: $69.00, discount: 20%

68. jacket: $129, discount: 15%

69. backpack: $45, discount: 25%

70. ATTENDANCE An amusement park recorded attendance of 825,000 one year. The next year, the attendance increased to 975,000. Determine the percent of increase in attendance.

Example 10

State whether the percent of change is a percent of *increase* or a percent of *decrease*. Then find the percent of change. Round to the nearest whole percent.

original: 80
final: 60

Subtract the original amount from the final amount to find the amount of change. $60 - 80 = -20$. Since the new amount is less than the original, this is a percent of decrease.

Use the original number, 80, as the base.

$$\underset{\text{original amount}}{\overset{\text{change}}{\longrightarrow}} \frac{20}{80} = \frac{r}{100} \qquad \text{Percent proportion}$$

$$20(100) = r(80) \qquad \text{Find cross products.}$$

$$2000 = 80r \qquad \text{Simplify.}$$

$$\frac{2000}{00} = \frac{80r}{00} \qquad \text{Divide each side by 80.}$$

$$25 = r \qquad \text{Simplify.}$$

The percent of decrease is 25%.

Solve each equation or formula for the variable indicated.

71. $3x + 2y = 9$, for y

72. $P = 2\ell + 2w$, for ℓ

73. $-5m + 9n = 15$, for m

74. $14w + 15x = y - 21w$, for w

75. $m = \frac{2}{5}y + n$, for y

76. $7d - 3c = f + 2d$, for d

77. GEOMETRY The formula for the area of a trapezoid is $A = \frac{1}{2}h(a + b)$, where h represents the height and a and b represent the lengths of the bases. Solve for h.

Example 11

Solve $6p - 8n = 12$ for p.

$$6p - 8n = 12 \qquad \text{Original equation}$$

$$6p - 8n + 8n = 12 + 8n \qquad \text{Add } 8n \text{ to each side.}$$

$$6p = 12 + 8n \qquad \text{Simplify.}$$

$$\frac{6p}{6} = \frac{12 + 8n}{6} \qquad \text{Divide each side by 6.}$$

$$\frac{6p}{6} = \frac{12}{6} + \frac{8}{6}n \qquad \text{Simplify.}$$

$$p = 2 + \frac{4}{3}n \qquad \text{Simplify.}$$

2-9 Weighted Averages

78. CANDY Michael is mixing two types of candy for a party. The chocolate pieces cost $0.40 per ounce, and the hard candy costs $0.20 per ounce. Michael purchases 20 ounces of the chocolate pieces, and the total cost of his candy was $11. How many ounces of hard candy did he purchase?

79. TRAVEL A car travels 100 miles east in 2 hours and 30 miles north in half an hour. What is the average speed of the car?

80. FINANCIAL LITERACY A candle supply store sells votive wax and low-shrink wax. How many pounds of low-shrink wax should be mixed with 8 pounds of votive wax to obtain a blend that sells for $0.98 a pound?

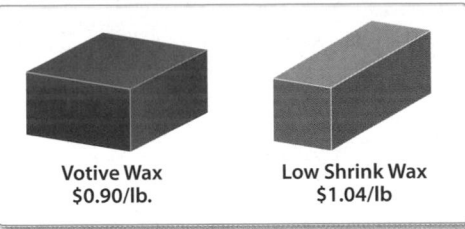

Votive Wax
$0.90/lb.

Low Shrink Wax
$1.04/lb

Example 12

METALS An alloy of metals is 25% copper. Another alloy is 50% copper. How much of each should be used to make 1000 grams of an alloy that is 45% copper?

Let x = the amount of the 25% copper alloy. Write and solve an equation.

$0.25x + 0.50(1000 - x) = 0.45(1000)$	Original Equation
$0.25x + 500 - 0.50x = 450$	Distributive Property
$-0.25x + 500 = 450$	Simplify.
$-0.25x + 500 - 500 = 450 - 500$	Subtract 500 from each side.
$-0.25x = -50$	Simplify.
$\dfrac{-0.25x}{-0.25} = \dfrac{-50}{-0.25}$	Divide each side by −0.25.
$x = 200$	Simplify.

200 grams of the 25% alloy and 800 grams of the 50% alloy should be used.

Translate each sentence into an equation.

1. The sum of six and four times d is the same as d minus nine.

2. Three times the difference of two times m and five is equal to eight times m to the second power increased by four.

Solve each equation. Check your solutions.

3. $x - 5 = -11$

4. $\frac{2}{3} = w + \frac{1}{4}$

5. $\frac{t}{6} = -3$

Solve each equation. Check your solution.

6. $2a - 5 = 13$

7. $\frac{p}{4} - 3 = 9$

8. **MULTIPLE CHOICE** At Mama Mia Pizza, the price of a large pizza is determined by $P = 9 + 1.5x$, where x represents the number of toppings added to a cheese pizza. Daniel spent $13.50 on a large pizza. How many toppings did he get?

   A 0

   B 1

   C 3

   D 5

Solve each equation. Check your solution.

9. $5y - 4 = 9y + 8$

10. $3(2k - 2) = -2(4k - 11)$

11. **GEOMETRY** Find the value of x so that the figures have the same perimeter.

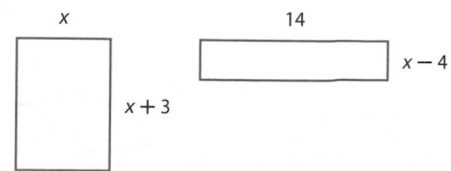

12. Evaluate the expression $|3t - 2u| + 5v$ if $t = 2$, $u = -5$, and $v = -3$.

Solve each equation. Then graph the solution set.

13. $|p - 4| = 6$

14. $|2b + 5| = 9$

Solve each proportion. If necessary, round to the nearest hundredth.

15. $\frac{a}{3} = \frac{16}{24}$

16. $\frac{9}{k + 3} = \frac{3}{5}$

17. **MULTIPLE CHOICE** Akiko uses 2 feet of thread for every three squares that she sews for her quilt. How many squares can she sew if she has 38 feet of thread?

   F 19

   G 57

   H 76

   J 228

18. State whether the percent of change is a percent of *increase* or a percent of *decrease*. Then find the percent of change. Round to the nearest whole percent.

   original: 54   new: 45

19. Find the total price of a sweatshirt that is priced at $48 and taxed at 6.5%.

20. **SHOPPING** Kirk wants to purchase a wide-screen TV. He sees an advertisement for a TV that was originally priced at $3200 and is 20% off. Find the discounted price of the TV.

21. Solve $5x - 3y = 9$ for y.

22. Solve $A = \frac{1}{2}bh$ for h.

23. **CHEMISTRY** Deon has 12 milliliters of a 5% solution. He also has a solution that has a concentration of 30%. How many milliliters of the 30% solution does Deon need to add to the 5% solution to obtain a 20% solution?

24. **BICYCLING** Shanee bikes 5 miles to the park in 30 minutes and 3 miles to the library in 45 minutes. What was her average speed?

25. **MAPS** On a map of North Carolina, the distance between Charlotte and Wilmington is 14.75 inches. If 2 inches equals 24 miles, what is the approximate distance between the two cities?

Preparing for Standardized Tests

Gridded Response Questions

In addition to multiple-choice, short-answer, and extended-response questions, you will likely encounter gridded-response questions on standardized tests. For gridded-response questions, you must print your answer on an answer sheet and mark in the correct circles on the grid to match your answer.

Strategies for Solving Gridded Response Questions

Step 1

Read the problem carefully.

- **Ask yourself:** "What information is given?" "What do I need to find?" "How do I solve this type of problem?"
- **Solve the Problem:** Use the information given in the problem to solve.
- **Check your answer:** If time permits, check your answer to make sure you have solved the problem correctly.

Step 2

Write your answer in the answer boxes.

- Print only one digit or symbol in each answer box.
- Do not write any digits or symbols outside the answer boxes.
- You may write your answer with the first digit in the left answer box, or with the last digit in the right answer box. You may leave blank any boxes you do not need on the right or the left side of your answer.

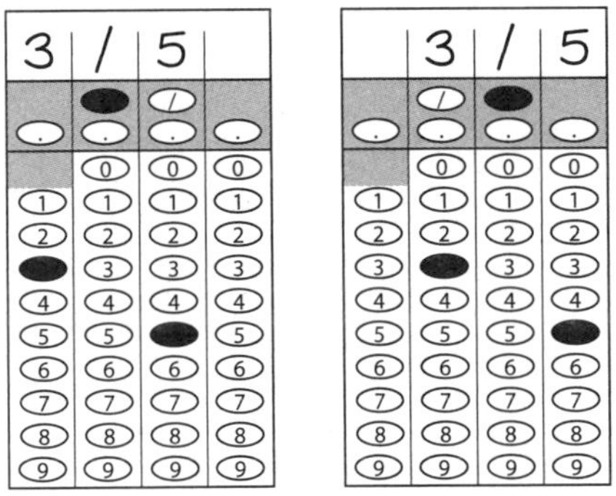

Step 3

Fill in the grid.

- Fill in only one bubble for every answer box that you have written in. Be sure not to fill in a bubble under a blank answer box.
- Fill in each bubble completely and clearly.

Standardized Test Example

Read the problem. Identify what you need to know. Then use the information in the problem to solve.

GRIDDED RESPONSE Ashley is 3 years older than her sister, Tina. Combined, the sum of their ages is 27 years. How old is Ashley?

Read the problem carefully. You are told that Ashley is 3 years older than her sister and that their ages combined equal 27 years. You need to find Ashley's age.

Solve the Problem

Words	Ashley's age plus Tina's age is equal to 27 years.
Variable	Let a represent Ashley's age. Then Tina's age is $a - 3$, since she is 3 years younger than Ashley.
Equation	a + $(a - 3)$ = 27

Solve the equation for a.

$a + (a - 3) = 27$ Original equation.

$2a - 3 = 27$ Add like terms.

$2a = 30$ Add 3 to each side.

$a = 15$ Divide each side by 2.

Since we let a represent Ashley's age, we know that she is 15 years old.

Fill in the Grid

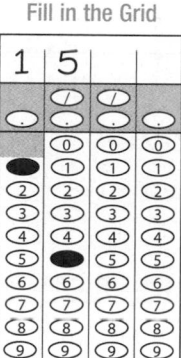

Exercises

Read each problem. Identify what you need to know. Then use the information in the problem to solve. Copy and complete an answer grid on your paper.

1. Orlando has $1350 in the bank. He wants to increase his balance to a total of $2550 by depositing $40 each week from his paycheck. How many weeks will he need to save in order to reach his goal?

2. Fourteen less than three times a number is equal to 40. Find the number.

3. The table shows the regular prices and sale prices of certain items at a department store this week. What is the percent of discount during the sale?

Item	Regular Price ($)	Sale Price ($)
pillows	25	20
sweaters	30	24
entertainment center	125	100

4. Maureen is driving from Raleigh, North Carolina, to Charlotte, North Carolina, to visit her brother at college. If she averages 65 miles per hour on the trip, then the equation $\frac{d}{2.65} = 65$ can be solved for the distance d. What is the distance to the nearest mile from Raleigh to Charlotte?

5. Find the value of x so that the figures below have the same area.

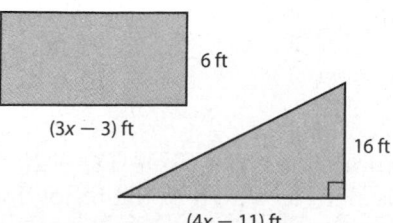

6. The sum of three consecutive whole numbers is 18. What is the greatest of the numbers?

Multiple Choice

Read each question. Then fill in the correct answer on the answer document provided by your teacher or on a sheet of paper.

1. Which point on the number line best represents the position of $\sqrt{8}$?

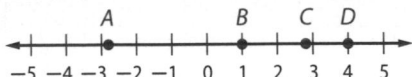

A -2.8

C 2.8

B 1

D 4

2. Find the value of x so that the figures have the same area.

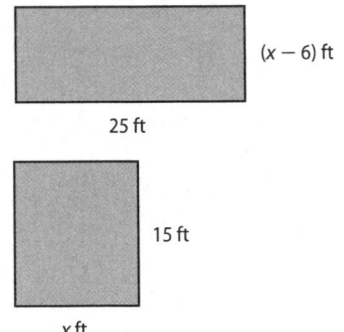

F 10

H 13

G 12

J 15

3. The elevation of Black Mountain is 27 feet more than 16 times the lowest point in the state. If the elevation of the lowest point in the state is 257 feet, what is the elevation of Black Mountain?

A $4{,}085$ feet

C $4{,}139$ feet

B $4{,}103$ feet

D $4{,}215$ feet

4. The expression $(3x^2 + 5x - 12) - 2(x^2 + 4x + 9)$ is equivalent to which of the following?

F $x^2 - 3x - 30$

G $x^2 + 13x + 6$

H $5x^2 + x - 18$

J $x^2 + 3x - 21$

5. The amount of soda, in fluid ounces, dispensed from a machine must satisfy the equation $|a - 0.4| = 20$. Which of the following graphs shows the acceptable minimum and maximum amounts that can be dispensed from the machine?

A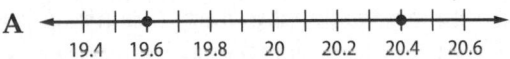

B

C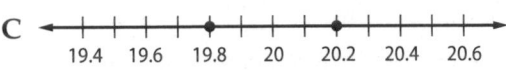

D

6. If a and b represent integers, $ab = ba$ is an example of which property?

F Associative Property

G Commutative Property

H Distributive Property

J Closure Property

7. The sum of one fifth of a number and three is equal to half of the number. What is the number?

A 5

C 15

B 10

D 20

8. Aaron charges $15 to mow the lawn and $10 per hour for other gardening work. Which expression represents his earnings?

F $10h$

G $15h$

H $15h + 10$

J $15 + 10h$

Test-Taking Tip

Question 2 Use the figures and the formula for area to set up an equation. The product of the length and width of each figure should be equal.

Short Response/Gridded Response

Record your answers on the answer sheet provided by your teacher or on a sheet of paper.

9. The formula for the lateral area of a cylinder is $A = 2\pi rh$, where r is the radius and h is the height. Solve the equation for h.

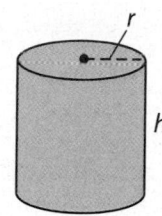

10. **GRIDDED RESPONSE** Solve the proportion $\frac{x}{18} = \frac{7}{21}$.

11. **GRIDDED RESPONSE** The table shows the cost of renting a moving van. If Miguel budgeted $75, how many miles could he drive the van and maintain his budget?

Moving Van Rentals	
Flat Fee	$50 for up to 300 miles
Variable Fee	$0.20 per mile over 300

12. Find the height of a soup can if the area of the label is 302 square centimeters and the radius of the can is 4 centimeters. Round to the nearest whole number.

13. **GRIDDED RESPONSE** Lara's car needed a particular part that costs $75. The mechanic charges $50 per hour to install the part. If the total cost was $350, how many hours did it take to install the part?

14. Lucinda is buying a set of patio furniture that is on sale for $\frac{4}{5}$ of the original price. After she uses a $50 gift certificate, the total cost before sales tax is $222. What was the original price of the patio furniture?

Extended Response

Record your answers on a sheet of paper. Show your work.

15. The city zoo offers a yearly membership that costs $120. A yearly membership includes free parking. Members can also purchase a ride pass for an additional $2 per day that allows them unlimited access to the rides in the park. Nonmembers pay $12 for admission to the park, $5 for parking, and $5 for a ride pass.

   a. Write an equation that could be solved for the number of visits it would take for the total cost to be the same for a member and a nonmember if they both purchase a ride pass each day. Solve the equation.

   b. What would the total cost be for members and nonmembers after this number of visits?

   c. Georgena is deciding whether or not to purchase a yearly membership. Explain how she could use the results above to help make her decision.

Need ExtraHelp?

If you missed Question...	1	2	3	4	5	6	7	8	9	10	11	12	13	14	15
Go to Lesson...	0-2	2-4	2-3	1-4	2-5	1-3	2-4	1-1	2-8	2-6	2-3	2-8	2-3	2-3	2-4

3 Linear Functions

··Then

○ You solved linear equations algebraically.

··Now

○ In this chapter you will:

- Identify linear equations, intercepts, and zeros.

- Graph and write linear equations.

- Use rate of change to solve problems.

··Why? ▲

○ **AMUSEMENT PARKS** The Magic Kingdom in Orlando, Florida, is one of the most popular amusement parks in the world. Yearly attendance figures increase steadily each year. Quantities like populations that change with respect to time can be described using rate of change. Often you can represent these situations with linear functions.

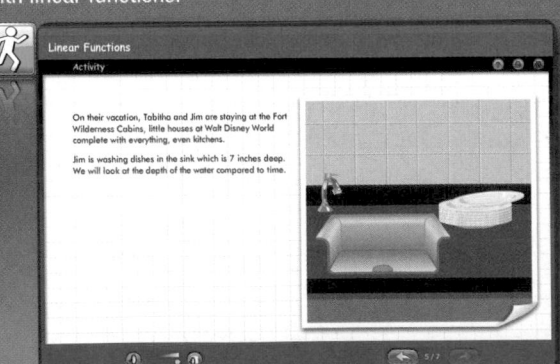

Linear Functions
Activity

On their vacation, Tabitha and Jim are staying at the Fort Wilderness Cabins, little houses at Walt Disney World complete with everything, even kitchens.

Jim is washing dishes in the sink which is 7 inches deep. We will look at the depth of the water compared to time.

5/7

connectED.mcgraw-hill.com   **Your Digital Math Portal**

Animation	Vocabulary	eGlossary	Personal Tutor	Virtual Manipulatives	Graphing Calculator	Audio	Foldables	Self-Check Practice	Worksheets

Get Ready for the Chapter

Diagnose Readiness | You have two options for checking prerequisite skills.

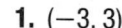 **Textbook Option** Take the Quick Check below. Refer to the Quick Review for help.

QuickCheck	QuickReview

QuickCheck

Graph each ordered pair on a coordinate grid.

1. $(-3, 3)$ **2.** $(-2, 1)$ **3.** $(3, 0)$

4. $(-5, 5)$ **5.** $(0, 6)$ **6.** $(2, -1)$

Write the ordered pair for each point.

7. A **8.** B

9. C **10.** D

11. F **12.** G

QuickReview

Example 1

Graph $(3, -2)$ on a coordinate grid.

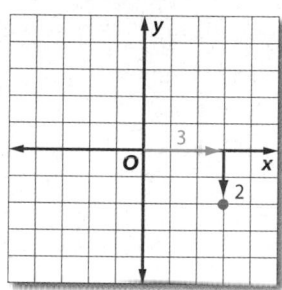

Solve each equation for y.

13. $3x + y = 1$ **14.** $8 - y = x$

15. $5x - 2y = 12$ **16.** $3x + 4y = 10$

17. $3 - \frac{1}{2}y = 5x$ **18.** $\frac{y + 1}{3} = x + 2$

Example 2

Solve $x - 2y = 8$ for y.

$$x - 2y = 8 \qquad \text{Original equation}$$
$$x - x - 2y - 8 - x \qquad \text{Subtract } x \text{ from each side.}$$
$$-2y = 8 - x \qquad \text{Simplify.}$$
$$\frac{-2y}{-2} = \frac{8 - x}{-2} \qquad \text{Divide each side by } -2.$$
$$y = \frac{1}{2}x - 4 \qquad \text{Simplify.}$$

Evaluate $\frac{a - b}{c - d}$ for each set of values.

19. $a = 7, b = 6, c = 9, d = 5$

20. $a = -3, b = 0, c = 3, d = -1$

21. $a = -5, b = -5, c = 5, d = 8$

22. $a = -6, b = 3, c = 8, d = 2$

23. MOVIES A movie made $297.2 million in 22 weeks. How much did the movie make on average each week?

Example 3

Evaluate $\frac{a - b}{c - d}$ for $a = 3, b = 5, c = -2,$ and $d = -6$.

$$\frac{a - b}{c - d} \qquad \text{Original expression}$$
$$= \frac{3 - 5}{-2 - (-6)} \qquad \text{Substitute 3 for } a, \text{ 5 for } b, -2 \text{ for } c,$$
$$\qquad \text{and } -6 \text{ for } d.$$
$$= \frac{-2}{4} \qquad \text{Simplify.}$$
$$= \frac{-2 \div 2}{4 \div 2} \qquad \text{Divide } -2 \text{ and 4 by their GCF, 2.}$$
$$= \frac{-1}{2} \text{ or } -\frac{1}{2} \qquad \text{Simplify. The signs are different so the quotient is negative.}$$

 Online Option Take an online self-check Chapter Readiness Quiz at <u>connectED.mcgraw-hill.com</u>.

Get Started on the Chapter

You will learn several new concepts, skills, and vocabulary terms as you study Chapter 3. To get ready, identify important terms and organize your resources. You may wish to refer to Chapter 0 to review prerequisite skills.

FOLDABLES StudyOrganizer

Linear Functions Make this Foldable to help you organize your Chapter 3 notes about graphing relations and functions. Begin with four sheets of grid paper.

1 **Fold** each sheet of grid paper in half from top to bottom.

2 **Cut** along fold. Staple the eight half-sheets together to form a booklet.

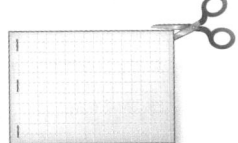

3 **Cut** tabs into margin. The top tab is 4 lines wide, the next tab is 8 lines wide, and so on. When you reach the bottom of a sheet, start the next tab at the top of the page.

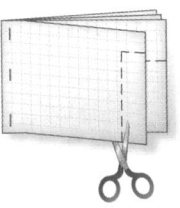

4 **Label** each of the tabs with a lesson number. Use the extra pages for vocabulary.

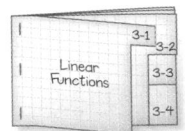

NewVocabulary

English		Español
linear equation	p. 155	ecuación lineal
standard form	p. 155	forma estándar
constant	p. 155	constante
x-intercept	p. 156	intersección x
y-intercept	p. 156	intersección y
linear function	p. 163	función lineal
parent function	p. 163	críe la función
family of graphs	p. 163	la familia de gráficas
root	p. 163	raíz
rate of change	p. 172	tasa de cambio
slope	p. 174	pendiente
direct variation	p. 182	variación directa
constant of variation	p. 182	constante de variación
arithmetic sequence	p. 189	sucesión aritmética
inductive reasoning	p. 196	razonamiento inductivo
deductive reasoning	p. 196	razonamiento deductivo

ReviewVocabulary

origin origen
the point where the two axes in a coordinate plane intersect with coordinates (0, 0)

x-axis eje x the horizontal number line on a coordinate plane

y-axis eje y the vertical number line on a coordinate plane

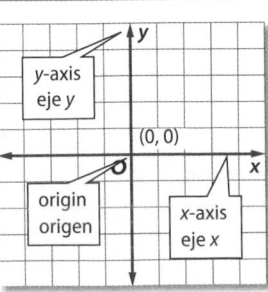

Algebra Lab
Analyzing Linear Graphs

Analyzing a graph can help you learn about the relationship between two quantities. A **linear function** is a function for which the graph is a line. There are four types of linear graphs. Let's analyze each type.

CCSS Common Core State Standards
Content Standards
F.IF.4 For a function that models a relationship between two quantities, interpret key features of graphs and tables in terms of the quantities, and sketch graphs showing key features given a verbal description of the relationship.

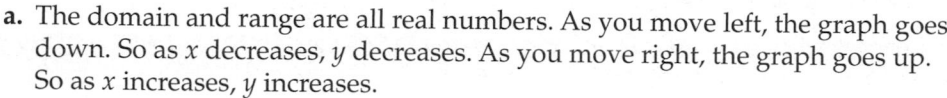

Activity 1 Line that Slants Up

Analyze the function graphed at the right.

a. **Describe the domain, range, and end behavior.**

b. **Describe the intercepts and any maximum or minimum points.**

c. **Identify where the function is positive, negative, increasing, and decreasing.**

d. **Describe any symmetry.**

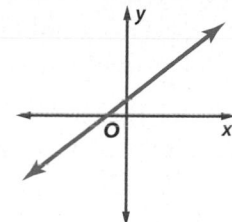

a. The domain and range are all real numbers. As you move left, the graph goes down. So as x decreases, y decreases. As you move right, the graph goes up. So as x increases, y increases.

b. There is one x-intercept and one y-intercept. There are no maximum or minimum points.

c. The function value is 0 at the x-intercept. The function values are negative to the left of the x-intercept and positive to the right. The function goes up from left to right, so it is increasing on the entire domain.

d. The graph has no symmetry.

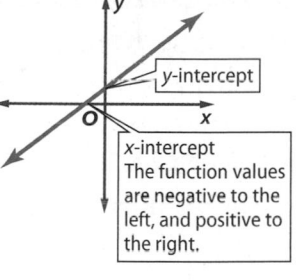

Lines that slant down from left to right have some different key features.

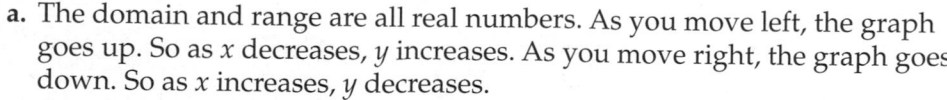

Activity 2 Line that Slants Down

Analyze the function graphed at the right.

a. **Describe the domain, range, and end behavior.**

b. **Describe the intercepts and any maximum or minimum points.**

c. **Identify where the function is positive, negative, increasing, and decreasing.**

d. **Describe any symmetry.**

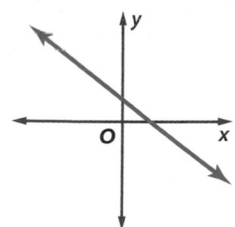

a. The domain and range are all real numbers. As you move left, the graph goes up. So as x decreases, y increases. As you move right, the graph goes down. So as x increases, y decreases.

b. There is one x-intercept and one y-intercept. There are no maximum or minimum points.

c. The function values are positive to the left of the x-intercept and negative to the right.
The function goes down from left to right, so it is decreasing on the entire domain.

d. The graph has no symmetry.

Algebra Lab
Analyzing Linear Graphs *Continued*

Horizontal lines represent special functions called **constant functions**.

Activity 3   Horizontal Line

Analyze the function graphed at the right.

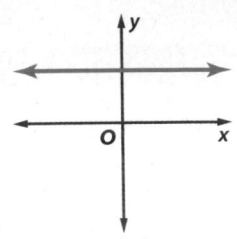

a. The domain is all real numbers, and the range is one value. As you move left or right, the graph stays constant. So as x decreases or increases, y is constant.

b. The graph does not intersect the x-axis, so there is no x-intercept. The graph has one y-intercept. There are no maximum or minimum points.

c. The function values are all positive. The function is constant on the entire domain.

d. The graph is symmetric about any vertical line.

Vertical lines represent linear relations that are *not* functions.

Activity 4   Vertical Line

Analyze the relation graphed at the right.

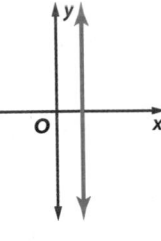

a. The domain is one value, and the range is all real numbers. This relation is not a function. Because you cannot move left or right on the graph, there is no end behavior.

b. There is one x-intercept and no y-intercept. There are no maximum or minimum points.

c. The y-values are positive above the x-axis and negative below. Because you cannot move left or right on the graph, the relation is neither increasing nor decreasing.

d. The graph is symmetric about itself.

Analyze the Results

1. Compare and contrast the key features of lines that slant up and lines that slant down.

2. How would the key features of a horizontal line below the x-axis differ from the features of a line above the x-axis?

3. Consider lines that pass through the origin.

   a. How do the key features of a line that slants up and passes through the origin compare to the key features of the line in Activity 1?

   b. Compare the key features of a line that slants down and passes through the origin to the key features of the line in Activity 2.

   c. Describe a horizontal line that passes through the origin and a vertical line that passes through the origin. Compare their key features to those of the lines in Activities 3 and 4.

4. **CCSS** **TOOLS** Place a pencil on a coordinate plane to represent a line. Move the pencil to represent different lines and evaluate each conjecture.

   a. *True* or *false*: A line can have more than one x-intercept.

   b. *True* or *false*: If the end behavior of a line is that as x increases, y increases, then the function values are increasing over the entire domain.

   c. *True* or *false*: Two different lines can have the same x- and y-intercepts.

Sketch a linear graph that fits each description.

5. as x increases, y decreases

6. one x-intercept and one y-intercept

7. has symmetry

8. is not a function

Graphing Linear Equations

- You represented relationships among quantities using equations.

1 Identify linear equations, intercepts, and zeros.

2 Graph linear equations.

- Recycling one ton of waste paper saves an average of 17 trees, 7000 gallons of water, 3 barrels of oil, and about 3.3 cubic yards of landfill space.

  The relationship between the amount of paper recycled and the number of trees saved can be expressed with the equation $y = 17x$, where y represents the number of trees and x represents the tons of paper recycled.

 NewVocabulary
linear equation
standard form
constant
x-intercept
y-intercept

 Common Core State Standards

Content Standards
F.IF.4 For a function that models a relationship between two quantities, interpret key features of graphs and tables in terms of the quantities, and sketch graphs showing key features given a verbal description of the relationship.

F.IF.7a Graph linear and quadratic functions and show intercepts, maxima, and minima.

Mathematical Practices
8 Look for and express regularity in repeated reasoning.

1 Linear Equations and Intercepts A **linear equation** is an equation that forms a line when it is graphed. Linear equations are often written in the form $Ax + By = C$. This is called the **standard form** of a linear equation. In this equation, C is called a **constant**, or a number. Ax and By are variable terms.

> **KeyConcept** Standard Form of a Linear Equation
>
> | Words | The standard form of a linear equation is $Ax + By = C$, where $A \geq 0$, A and B are not both zero, and A, B, and C are integers with a greatest common factor of 1. |
> |---|---|
> | Examples | In $3x + 2y = 5$, $A = 3$, $B = 2$, and $C = 5$. In $x = -7$, $A = 1$, $B = 0$, and $C = -7$. |

Example 1 Identify Linear Equations

Determine whether each equation is a linear equation. Write the equation in standard form.

a. $y = 4 - 3x$

Rewrite the equation so that it appears in standard form.

$y = 4 - 3x$	Original equation
$y + 3x = 4 - 3x + 3x$	Add $3x$ to each side.
$3x + y = 4$	Simplify.

The equation is now in standard form where $A = 3$, $B = 1$, and $C = 4$. This is a linear equation.

b. $6x - xy = 4$

Since the term xy has two variables, the equation cannot be written in the form $Ax + By = C$. Therefore, this is not a linear equation.

▸ **Guided**Practice

1A. $\frac{1}{3}y = -1$

1B. $y = x^2 - 4$

A linear equation can be represented on a coordinate graph. The *x*-coordinate of the point at which the graph of an equation crosses the *x*-axis is an **x-intercept**. The *y*-coordinate of the point at which the graph crosses the *y*-axis is called a **y-intercept**.

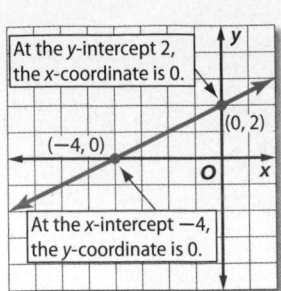

At the *y*-intercept 2, the *x*-coordinate is 0.

(0, 2)

(−4, 0)

At the *x*-intercept −4, the *y*-coordinate is 0.

The graph of a linear equation has at most one *x*-intercept and one *y*-intercept, unless it is the equation $x = 0$ or $y = 0$, in which case every number is a *y*-intercept or an *x*-intercept, respectively.

Standardized Test Example 2 Find Intercepts from a Graph

Find the *x*- and *y*-intercepts of the line graphed at the right.

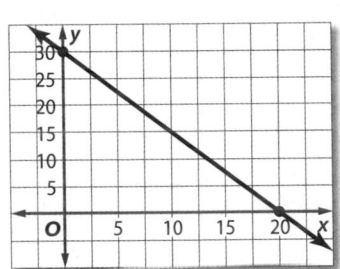

A *x*-intercept is 0; *y*-intercept is 30.

B *x*-intercept is 20; *y*-intercept is 30.

C *x*-intercept is 20; *y*-intercept is 0.

D *x*-intercept is 30; *y*-intercept is 20.

Read the Test Item

We need to determine the *x*- and *y*-intercepts of the line in the graph.

ReadingMath

Intercepts Usually, the individual coordinates are called the *x*-intercept and the *y*-intercept. The *x*-intercept 20 is located at (20, 0). The *y*-intercept 30 is located at (0, 30).

Solve the Test Item

Step 1 Find the *x*-intercept. Look for the point where the line crosses the *x*-axis.

The line crosses at (20, 0). The *x*-intercept is 20 because it is the *x*-coordinate of the point where the line crosses the *x*-axis.

Step 2 Find the *y*-intercept. Look for the point where the line crosses the *y*-axis.

The line crosses at (0, 30). The *y*-intercept is 30 because it is the *y*-coordinate of the point where the line crosses the *y*-axis.

Thus, the answer is B.

GuidedPractice

2. **HEALTH** Find the *x*- and *y*-intercepts of the graph.

   F *x*-intercept is 0; *y*-intercept is 150.

   G *x*-intercept is 150; *y*-intercept is 0.

   H *x*-intercept is 150; no *y*-intercept.

   J No *x*-intercept; *y*-intercept is 150.

Gym Membership

Total Cost ($)

Number of Months

Real-World Example 3 Find Intercepts from a Table

SWIMMING POOL A swimming pool is being drained at a rate of 720 gallons per hour. The table shows the function relating the volume of water in a pool and the time in hours that the pool has been draining.

Draining a Pool	
Time (h)	Volume (gal)
x	*y*
0	10,080
2	8640
6	5760
10	2880
12	1440
14	0

a. Find the *x*- and *y*-intercepts of the graph of the function.

x-intercept = 14   14 is the value of *x* when *y* = 0.
y-intercept = 10,080   10,080 is the value of *y* when *x* = 0.

b. Describe what the intercepts mean in this situation.

The *x*-intercept 14 means that after 14 hours, the water has a volume of 0 gallons, or the pool is completely drained.

The *y*-intercept 10,080 means that the pool contained 10,080 gallons of water at time 0, or before it started to drain. This is shown in the graph.

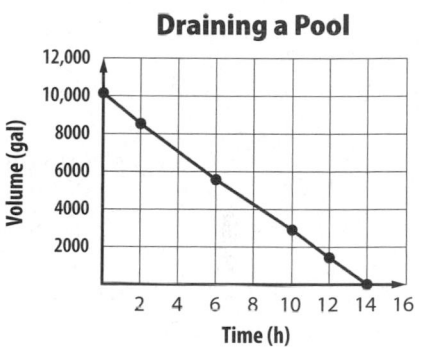

Draining a Pool

GuidedPractice

3. DRIVING The table shows the function relating the distance to an amusement park in miles and the time in hours the Torres family has driven. Find the *x*- and *y*-intercepts. Describe what the intercepts mean in this situation.

Time (h)	Distance (mi)
0	248
1	186
2	124
3	62
4	0

2 Graph Linear Equations By first finding the *x*- and *y*-intercepts, you have the ordered pairs of two points through which the graph of the linear equation passes. This information can be used to graph the line because only two points are needed to graph a line.

Example 4 Graph by Using Intercepts

Graph $2x + 4y = 16$ by using the *x*- and *y*-intercepts.

To find the *x*-intercept, let $y = 0$.

$2x + 4y = 16$ Original equation

$2x + 4(0) = 16$ Replace *y* with 0.

$2x = 16$ Simplify.

$x = 8$ Divide each side by 2.

The *x*-intercept is 8. This means that the graph intersects the *x*-axis at (8, 0).

To find the *y*-intercept, let $x = 0$.

$2x + 4y = 16$	Original equation
$2(0) + 4y = 16$	Replace *x* with 0.
$4y = 16$	Simplify.
$y = 4$	Divide each side by 4.

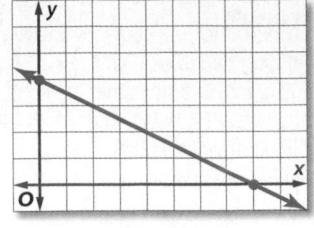

The *y*-intercept is 4. This means the graph intersects the *y*-axis at $(0, 4)$.

Plot these two points and then draw a line through them.

▶ **Guided Practice**

Graph each equation by using the *x*- and *y*-intercepts.

4A. $-x + 2y = 3$ **4B.** $y = -x - 5$

StudyTip

Equivalent Equations
Rewriting equations by solving for *y* may make it easier to find values for *y*.

$$-x + 2y = 3 \rightarrow y = \frac{x + 3}{2}$$

Note that the graph in Example 4 has both an *x*- and a *y*-intercept. Some lines have an *x*-intercept and no *y*-intercept or vice versa. The graph of $y = b$ is a horizontal line that only has a *y*-intercept (unless $b = 0$). The intercept occurs at $(0, b)$. The graph of $x = a$ is a vertical line that only has an *x*-intercept (unless $a = 0$). The intercept occurs at $(a, 0)$.

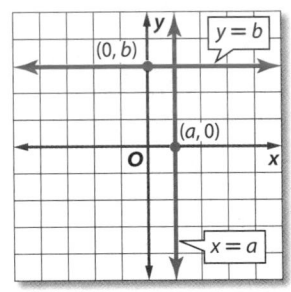

Every ordered pair that makes an equation true represents a point on the graph. So, the graph of an equation represents all of its solutions. Any ordered pair that does not make the equation true represents a point that is not on the line.

Example 5 Graph by Making a Table

Graph $y = \frac{1}{3}x + 2$.

The domain is all real numbers. Select values from the domain and make a table. When the *x*-coefficient is a fraction, select a number from the domain that is a multiple of the denominator. Create ordered pairs and graph them.

x	$\frac{1}{3}x + 2$	*y*	(*x*, *y*)
-3	$\frac{1}{3}(-3) + 2$	1	$(-3, 1)$
0	$\frac{1}{3}(0) + 2$	2	$(0, 2)$
3	$\frac{1}{3}(3) + 2$	3	$(3, 3)$
6	$\frac{1}{3}(6) + 2$	4	$(6, 4)$

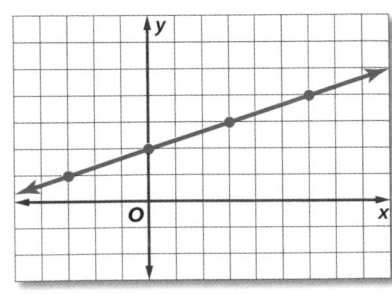

▶ **Guided Practice**

Graph each equation by making a table.

5A. $2x - y = 2$ **5B.** $x = 3$ **5C.** $y = -2$

Example 1   Determine whether each equation is a linear equation. Write *yes* or *no*.
If yes, write the equation in standard form.

    1. $x = y - 5$ **2.** $-2x - 3 = y$ **3.** $-4y + 6 = 2$ **4.** $\frac{2}{3}x - \frac{1}{3}y = 2$

Examples 2–3 Find the *x*- and *y*-intercepts of the graph of each linear function.
Describe what the intercepts mean.

5.

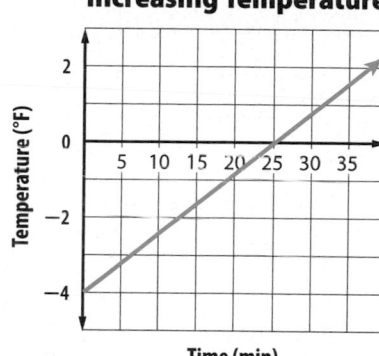

Increasing Temperature

Time (min)

6.

Position of Scuba Diver	
Time (s)	**Depth (m)**
x	**y**
0	−24
3	−18
6	−12
9	−6
12	0

Example 4   Graph each equation by using the *x*- and *y*-intercepts.

    7. $y = 4 + x$ **8.** $2x - 5y = 1$

Example 5   Graph each equation by making a table.

    9. $x + 2y = 4$ **10.** $-3 + 2y = -5$ **11.** $y = 3$

12. **CCSS** **REASONING** The equation $5x + 10y = 60$ represents the
number of children *x* and adults *y* who can attend the rodeo
for $60.

    a. Use the *x*- and *y*-intercepts to graph the equation.

    b. Describe what these values mean.

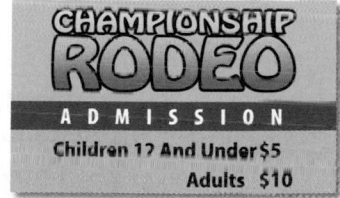

CHAMPIONSHIP RODEO

A D M I S S I O N

Children 12 And Under $5

Adults $10

Example 1   Determine whether each equation is a linear equation. Write *yes* or *no*. If yes, write
the equation in standard form.

    13 $5x + y^2 = 25$ **14.** $8 + y = 4x$ **15.** $9xy - 6x = 7$

    16. $4y^2 + 9 = -4$ **17.** $12x = 7y - 10y$ **18.** $y = 4x + x$

Example 2   Find the *x*- and *y*-intercepts of the graph of each linear function.

19.

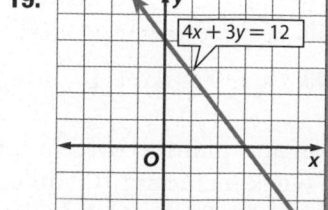

$4x + 3y = 12$

20.

x	y
−3	−1
−2	0
−1	1
0	2
1	3

Example 3 Find the *x*- and *y*-intercepts of each linear function. Describe what the intercepts mean.

21. **Descent of Eagle**

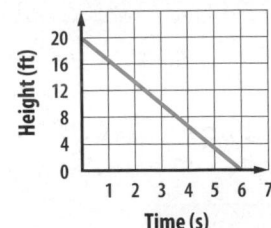

22.

Eva's Distance from Home	
Time (min)	Distance (mi)
x	*y*
0	4
2	3
4	2
6	1
8	0

Example 4 Graph each equation by using the *x*- and *y*-intercepts.

23. $y = 4 + 2x$ **24.** $5 - y = -3x$ **25.** $x = 5y + 5$

26. $x + y = 4$ **27.** $x - y = -3$ **28.** $y = 8 - 6x$

Example 5 Graph each equation by making a table.

29. $x = -2$ **30.** $y = -4$ **31.** $y = -8x$

32. $3x = y$ **33.** $y - 8 - -x$ **34.** $x = 10 - y$

35. **TV RATINGS** The number of people who watch a singing competition can be given by $p = 0.15v$, where *p* represents the number of people in millions who saw the show and *v* is the number of potential viewers in millions.

a. Make a table of values for the points (v, p).

b. Graph the equation.

c. Use the graph to estimate the number of people who saw the show if there are 14 million potential viewers.

d. Explain why it would not make sense for *v* to be a negative number.

Determine whether each equation is a linear equation. Write *yes* or *no*. If yes, write the equation in standard form.

36. $x + \frac{1}{y} = 7$ **37.** $\frac{x}{2} = 10 + \frac{2y}{3}$

38. $7n - 8m = 4 - 2m$ **39.** $3a + b - 2 = b$

40. $2r - 3rt + 5t = 1$ **41.** $\frac{3m}{4} = \frac{2n}{3} - 5$

42. **FINANCIAL LITERACY** James earns a monthly salary of $1200 and a commission of $125 for each car he sells.

a. Graph an equation that represents how much James earns in a month in which he sells *x* cars.

b. Use the graph to estimate the number of cars James needs to sell in order to earn $5000.

Graph each equation.

43. $2.5x - 4 = y$ **44.** $1.25x + 7.5 = y$ **45.** $y + \frac{1}{5}x = 3$

46. $\frac{2}{3}x + y = -7$ **47.** $2x - 3 = 4y + 6$ **48.** $3y - 7 = 4x + 1$

49. **CCSS REASONING** Mrs. Johnson is renting a car for vacation and plans to drive a total of 800 miles. A rental car company charges $153 for the week including 700 miles and $0.23 for each additional mile. If Mrs. Johnson has only $160 to spend on the rental car, can she afford to rent a car? Explain your reasoning.

50. AMUSEMENT PARKS An amusement park charges $50 for admission before 6 P.M. and $20 for admission after 6 P.M. On Saturday, the park took in a total of $20,000.

    a. Write an equation that represents the number of admissions that may have been sold. Let x represent the admissions sold before 6 P.M., and let y represent the admissions sold after 6 P.M.

    b. Graph the equation.

    c. Find the x- and y-intercepts of the graph. What does each intercept represent?

Find the x-intercept and y-intercept of the graph of each equation.

(51) $5x + 3y = 15$ **52.** $2x - 7y = 14$ **53.** $2x - 3y = 5$

54. $6x + 2y = 8$ **55.** $y = \frac{1}{4}x - 3$ **56.** $y = \frac{2}{3}x + 1$

57. ONLINE GAMES The percent of teens who play online games can be modeled by $p = \frac{15}{4}t + 66$. p is the percent of students, and t represents time in years since 2000.

    a. Graph the equation.

    b. Use the graph to estimate the percent of students playing the games in 2008.

58. MULTIPLE REPRESENTATIONS In this problem, you will explore x- and y-intercepts of graphs of linear equations.

    a. Graphical If possible, use a straightedge to draw a line on a coordinate plane with each of the following characteristics.

x- and y-intercept	x-intercept, no y-intercept	exactly 2 x-intercepts	no x-intercept, y-intercept	exactly 2 y-intercepts

    b. Analytical For which characteristics were you able to create a line and for which characteristics were you unable to create a line? Explain.

    c. Verbal What must be true of the x- and y-intercepts of a line?

H.O.T. Problems   Use Higher-Order Thinking Skills

59. CCSS REGULARITY Copy and complete each table. State whether any of the tables show a linear relationship. Explain.

Perimeter of a Square	
Side Length	Perimeter
1	
2	
3	
4	

Area of a Square	
Side Length	Area
1	
2	
3	
4	

Volume of a Cube	
Side Length	Volume
1	
2	
3	
4	

60. REASONING Compare and contrast the graphs of $y = 2x + 1$ with the domain $\{1, 2, 3, 4\}$ and $y = 2x + 1$ with the domain of all real numbers.

OPEN ENDED Give an example of a linear equation of the form $Ax + By = C$ for each condition. Then describe the graph of the equation.

61. $A = 0$ **62.** $B = 0$ **63.** $C = 0$

64. WRITING IN MATH Explain how to find the x-intercept and y-intercept of a graph and summarize how to graph a linear equation.

65. Sancho can ride 8 miles on his bicycle in 30 minutes. At this rate, about how long would it take him to ride 30 miles?

  A 8 hours

  B 6 hours 32 minutes

  C 2 hours

  D 1 hour 53 minutes

66. GEOMETRY Which is a true statement about the relation graphed?

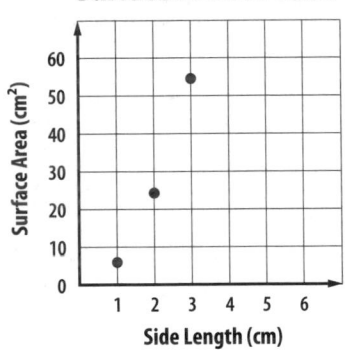

Surface Area of Cube

  F The relation is not a function.

  G Surface area is the independent quantity.

  H The surface area of a cube is a function of the side length.

  J As the side length of a cube increases, the surface area decreases.

67. SHORT RESPONSE Selena deposited $2000 into a savings account that pays 1.5% interest compounded annually. If she does not deposit any more money into her account, how much will she earn in interest at the end of one year?

68. A candle burns as shown in the graph.

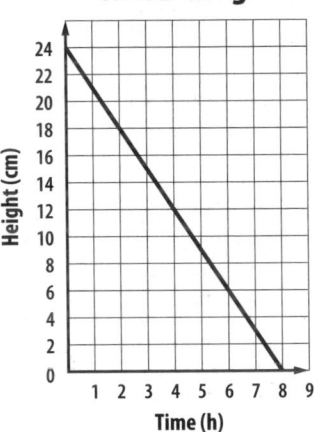

Candle Height

If the height of the candle is 8 centimeters, approximately how long has the candle been burning?

  A 0 hours   **C** 64 minutes

  B 24 minutes   **D** $5\frac{1}{2}$ hours

Spiral Review

69. FUNDRAISING The Madison High School Marching Band sold solid-color gift wrap for $4 per roll and print gift wrap for $6 per roll. The total number of rolls sold was 480, and the total amount of money collected was $2340. How many rolls of each kind of gift wrap were sold? (Lesson 2-9)

Solve each equation or formula for the variable specified. (Lesson 2-8)

70. $S = \frac{n}{2}(A + t)$, for A

71. $2g - m = 5 - gh$, for g

72. $\frac{y + a}{3} = c$, for y

73. $4z + b = 2z + c$, for z

Skills Review

Evaluate each expression if $x = 2$, $y = 5$, **and** $z = 7$.

74. $3x^2 - 4y$

75. $\frac{x - y^2}{2z}$

76. $\left(\frac{y}{z}\right)^2 + \frac{xy}{2}$

77. $z^2 - y^3 + 5x^2$

Solving Linear Equations by Graphing

∴ Then	∴ Now	∴ Why?
● You graphed linear equations by using tables and finding roots, zeros, and intercepts.	● **1** Solve linear equations by graphing. **2** Estimate solutions to a linear equation by graphing.	● The cost of braces can vary widely. The graph shows the balance of the cost of treatments as payments are made. This is modeled by the function $b = -85p + 5100$, where p represents the number of $85 payments made, and b is the remaining balance.

Orthodontic Payments

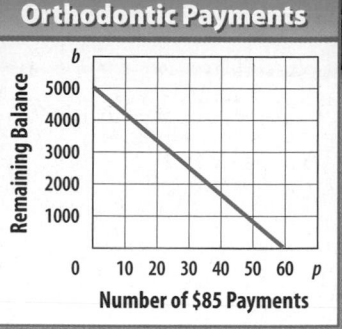

NewVocabulary

linear function
parent function
family of graphs
root
zeros

Common Core State Standards

Content Standards

A.REI.10 Understand that the graph of an equation in two variables is the set of all its solutions plotted in the coordinate plane, often forming a curve (which could be a line).

F.IF.7a Graph linear and quadratic functions and show intercepts, maxima, and minima.

Mathematical Practices

4 Model with mathematics.

1 **Solve by Graphing** A **linear function** is a function for which the graph is a line. The simplest linear function is $f(x) = x$ and is called the **parent function** of the family of linear functions. A **family of graphs** is a group of graphs with one or more similar characteristics.

⚛ KeyConcept Linear Function

Parent function:	$f(x) = x$
Type of graph:	line
Domain:	all real numbers
Range:	all real numbers

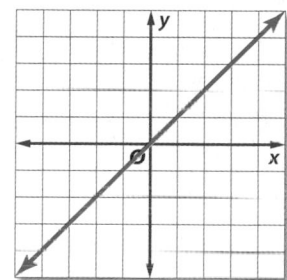

The solution or **root** of an equation is any value that makes the equation true. A linear equation has at most one root. You can find the root of an equation by graphing its related function. To write the related function for an equation, replace 0 with $f(x)$.

Linear Equation	Related Function
$2x - 8 = 0$	$f(x) = 2x - 8$ or $y = 2x - 8$

Values of x for which $f(x) = 0$ are called **zeros** of the function f. The zero of a function is located at the x-intercept of the function. The root of an equation is the value of the x-intercept. So:

- 4 is the x-intercept of $2x - 8 = 0$.
- 4 is the solution of $2x - 8 = 0$.
- 4 is the root of $2x - 8 = 0$.
- 4 is the zero of $f(x) = 2x - 8$.

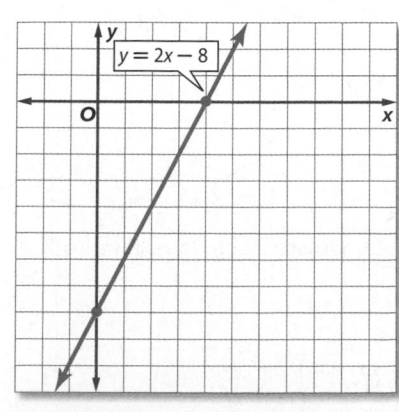

$y = 2x - 8$

Example 1 Solve an Equation with One Root

Solve each equation.

a. $0 = \frac{1}{3}x - 2$

Method 1 Solve algebraically.

$0 = \frac{1}{3}x - 2$	Original equation
$0 + 2 = \frac{1}{3}x - 2 + 2$	Add 2 to each side.
$3(2) = 3\left(\frac{1}{3}x\right)$	Multiply each side by 3.
$6 = x$	Solve.

The solution is 6.

b. $3x + 1 = -2$

Method 2 Solve by graphing.

Find the related function. Rewrite the equation with 0 on the right side.

$3x + 1 = -2$	Original equation
$3x + 1 + 2 = -2 + 2$	Add 2 to each side.
$3x + 3 = 0$	Simplify.

The related function is $f(x) = 3x + 3$. To graph the function, make a table.

x	$f(x) = 3x + 3$	$f(x)$	$(x, f(x))$
-2	$f(-2) = 3(-2) + 3$	-3	$(-2, -3)$
-1	$f(-1) = 3(-1) + 3$	0	$(-1, 0)$

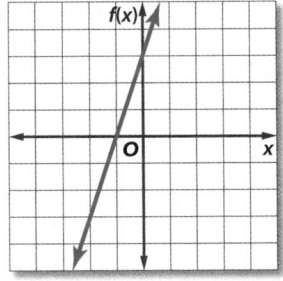

The graph intersects the x-axis at -1.
So, the solution is -1.

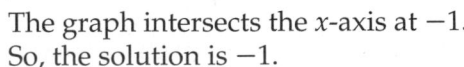

> **Study**Tip
>
> Zeros from tables
> The zero is located at the x-intercept, so the value of y will equal 0. When looking at a table, the zero is the x-value when $y = 0$.

GuidedPractice

1A. $0 = \frac{2}{5}x + 6$

1B. $-1.25x + 3 = 0$

For equations with the same variable on each side of the equation, use addition or subtraction to get the terms with variables on one side. Then solve.

Example 2 Solve an Equation with No Solution

Solve each equation.

a. $3x + 7 = 3x + 1$

Method 1 Solve algebraically.

$3x + 7 = 3x + 1$	Original equation
$3x + 7 - 1 = 3x + 1 - 1$	Subtract 1 from each side.
$3x + 6 = 3x$	Simplify.
$3x - 3x + 6 = 3x - 3x$	Subtract $3x$ from each side.
$6 = 0$	Simplify.

The related function is $f(x) = 6$. The root of a linear equation is the value of x when $f(x) = 0$. Since $f(x)$ is always equal to 6, this equation has no solution.

b. $2x - 4 = 2x - 6$

Method 2 | Solve by graphing.

$$2x - 4 = 2x - 6 \qquad \text{Original equation}$$
$$2x - 4 + 6 = 2x - 6 + 6 \qquad \text{Add 6 to each side.}$$
$$2x + 2 = 2x \qquad \text{Simplify.}$$
$$2x - 2x + 2 = 2x - 2x \qquad \text{Subtract } 2x \text{ from each side.}$$
$$2 = 0 \qquad \text{Simplify.}$$

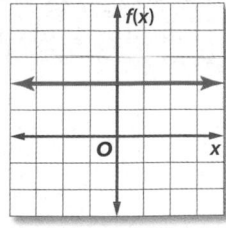

Graph the related function, which is $f(x) = 2$. The graph does not intersect the x-axis. Thus, there is no solution.

▶ **Guided**Practice

2A. $4x + 3 = 4x - 5$ **2B.** $2 - 3x = 6 - 3x$

---

2 **Estimate Solutions by Graphing** Graphing may provide only an estimate. In these cases, solve algebraically to find the exact solution.

● **Real-World Example 3** Estimate by Graphing

CARNIVAL RIDES Emily is going to a local carnival. The function $m = 20 - 0.75r$ represents the amount of money m she has left after r rides. Find the zero of this function. Describe what this value means in this context.

Make a table of values.

r	$m = 20 - 0.75r$	m	(r, m)
0	$m = 20 - 0.75(0)$	20	$(0, 20)$
5	$m = 20 - 0.75(5)$	16.25	$(5, 16.25)$

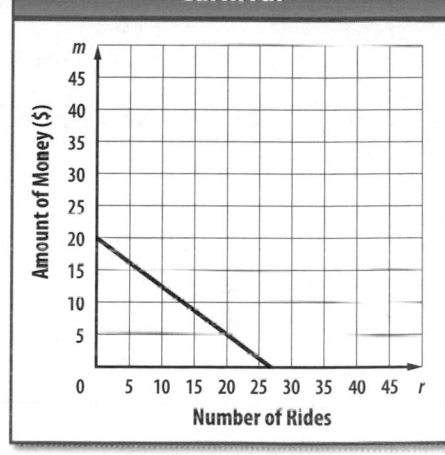

Carnival

The graph appears to intersect the r-axis at 27.

Next, solve algebraically to check.

$$m = 20 - 0.75r \qquad \text{Original equation}$$
$$0 = 20 - 0.75r \qquad \text{Replace } m \text{ with 0.}$$
$$0 + \mathbf{0.75r} = 20 - 0.75r + \mathbf{0.75r} \qquad \text{Add } 0.75r \text{ to each side.}$$
$$0.75r = 20 \qquad \text{Simplify.}$$
$$\frac{0.75r}{0.75} = \frac{20}{0.75} \qquad \text{Divide each side by 0.75.}$$
$$r \approx 26.67 \qquad \text{Simplify and round to the nearest hundredth.}$$

The zero of this function is about 26.67. Since Emily cannot ride part of a ride, she can ride 26 rides before she will run out of money.

▶ **Guided**Practice

3. FINANCIAL LITERACY Antoine's class is selling candy to raise money for a class trip. They paid $45 for the candy, and they are selling each candy bar for $1.50. The function $y = 1.50x - 45$ represents their profit y when they sell x candy bars. Find the zero and describe what it means in the context of this situation.

Real-World Career

Entertainment Manager
An entertainment manager supervises tech tests, calls show cues, schedules performances and performers, coaches employees and guest talent, and manages expenses. Entertainment managers need a college degree in a field such as communication or theater.

Comstock/PunchStock

Examples 1–2 Solve each equation by graphing. Verify your answer algebraically.

1. $-2x + 6 = 0$

2. $-x - 3 = 0$

3. $4x - 2 = 0$

4. $9x + 3 = 0$

5. $2x - 5 = 2x + 8$

6. $4x + 11 = 4x - 24$

7. $3x - 5 = 3x - 10$

8. $-6x + 3 = -6x + 5$

Example 3

9. NEWSPAPERS The function $w = 30 - \frac{3}{4}n$ represents the weight w in pounds of the papers in Tyrone's newspaper delivery bag after he delivers n newspapers. Find the zero and explain what it means in the context of this situation.

Practice and Problem Solving   Extra Practice is on page R3.

Solve each equation by graphing. Verify your answer algebraically.

10. $0 = x - 5$

11. $0 = x + 3$

12. $5 - 8x = 16 - 8x$

13. $3x - 10 = 21 + 3x$

14. $4x - 36 = 0$

15. $0 = 7x + 10$

16. $2x + 22 = 0$

17 $5x - 5 = 5x + 2$

18. $-7x + 35 = 20 - 7x$

19. $-4x - 28 = 3 - 4x$

20. $0 = 6x - 8$

21. $12x + 132 = 12x - 100$

Example 3

22. TEXTING Sean is sending texts to his friends. The function $y = 160 - x$ represents the number of characters y the message can hold after he has typed x characters. Find the zero and explain what it means in the context of this situation.

23. GIFT CARDS For her birthday Kwan receives a $50 gift card to download songs. The function $m = -0.50d + 50$ represents the amount of money m that remains on the card after a number of songs d are downloaded. Find the zero and explain what it means in the context of this situation.

Solve each equation by graphing. Verify your answer algebraically.

24. $-7 = 4x + 1$

25. $4 - 2x = 20$

26. $2 - 5x = -23$

27. $10 - 3x = 0$

28. $15 + 6x = 0$

29. $0 = 13x + 34$

30. $0 = 22x - 10$

31. $25x - 17 = 0$

32. $0 = \frac{1}{2} + \frac{2}{3}x$

33. $0 = \frac{3}{4} - \frac{2}{5}x$

34. $13x + 117 = 0$

35. $24x - 72 = 0$

36. SEA LEVEL Parts of New Orleans lie 0.5 meter below sea level. After d days of rain the equation $w = 0.3d - 0.5$ represents the water level w in meters. Find the zero, and explain what it means in the context of this situation.

 37. CCSS MODELING An artist completed an ice sculpture when the temperature was $-10°C$. The equation $t = 1.25h - 10$ shows the temperature h hours after the sculpture's completion. If the artist completed the sculpture at 8:00 A.M., at what time will it begin to melt?

Solve each equation by graphing. Verify your answer algebraically.

38. $7 - 3x = 8 - 4x$

39. $19 + 3x = 13 + x$

40. $16x + 6 = 14x + 10$

41. $15x - 30 = 5x - 50$

42. $\frac{1}{2}x - 5 = 3x - 10$

43. $3x - 11 = \frac{1}{3}x - 8$

44. HAIR PRODUCTS Chemical hair straightening makes curly hair straight and smooth. The percent of the process left to complete is modeled by $p = -12.5t + 100$, where t is the time in minutes that the solution is left on the hair, and p represents the percent of the process left to complete.

   a. Find the zero of this function.

   b. Make a graph of this situation.

   c. Explain what the zero represents in this context.

   d. State the possible domain and range of this function.

45 MUSIC DOWNLOADS In this problem, you will investigate the change between two quantities.

   a. Copy and complete the table.

Number of Songs Downloaded	Total Cost ($)	Total Cost / Number of Songs Downloaded
2	4	
4	8	
6	12	

   b. As the number of songs downloaded increases, how does the total cost change?

   c. Interpret the value of the total cost divided by the number of songs downloaded.

H.O.T. Problems   Use Higher-Order Thinking Skills

46. ERROR ANALYSIS Clarissa and Koko solve $3x + 5 = 2x + 4$ by graphing the related function. Is either of them correct? Explain your reasoning.

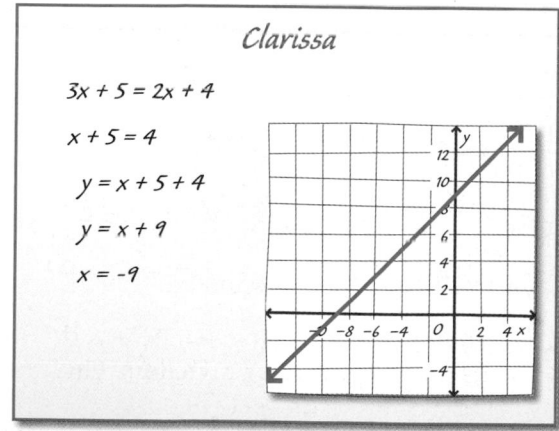

Clarissa

$3x + 5 = 2x + 4$

$x + 5 = 4$

$y = x + 5 + 4$

$y = x + 9$

$x = -9$

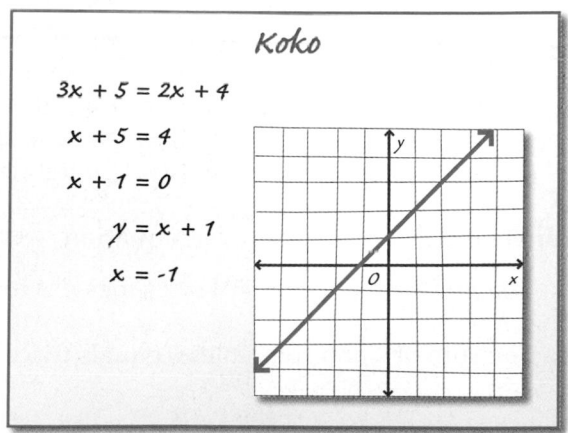

Koko

$3x + 5 = 2x + 4$

$x + 5 = 4$

$x + 1 = 0$

$y = x + 1$

$x = -1$

47. CHALLENGE Find the solution of $\frac{2}{3}(x + 3) = \frac{1}{2}(x + 5)$ by graphing. Verify your solution algebraically.

48. CCSS TOOLS Explain when it is better to solve an equation using algebraic methods and when it is better to solve by graphing.

49. OPEN ENDED Write a linear equation that has a root of $-\frac{3}{4}$. Write its related function.

50. WRITING IN MATH Summarize how to solve a linear equation algebraically and graphically.

51. What are the x- and y-intercepts of the graph of the function?

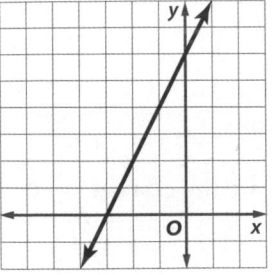

   A $-3, 6$ **C** $3, -6$

   B $6, -3$ **D** $-6, 3$

52. The table shows the cost C of renting a pontoon boat for h hours.

Hours	1	2	3
Cost ($)	7.25	14.5	21.75

Which equation best represents the data?

   F $C = 7.25h$ **H** $C = 21.75 - 7.25h$

   G $C = h + 7.25$ **J** $C = 7.25h + 21.75$

53. Which is the best estimate for the x-intercept of the graph of the linear function represented in the table?

x	y
0	5
1	3
2	1
3	−1
4	−3

   A between 0 and 1

   B between 2 and 3

   C between 1 and 2

   D between 3 and 4

54. EXTENDED RESPONSE Mr. Kauffmann has the following options for a backyard pool.

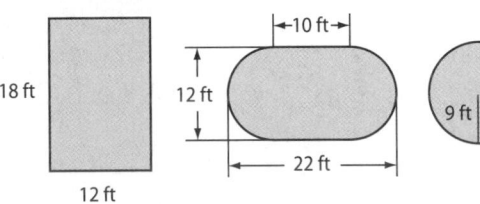

If each pool has the same depth, which pool would give the greatest area to swim? Explain your reasoning.

Find the x- and y-intercepts of the graph of each linear equation. (Lesson 3-1)

55. $y = 2x + 10$ **56.** $3y = 6x - 9$ **57.** $4x - 14y = 28$

58. FOOD If 2% milk contains 2% butterfat and whipping cream contains 9% butterfat, how much whipping cream and 2% milk should be mixed to obtain 35 gallons of milk with 4% butterfat? (Lesson 2-9)

Translate each sentence into an equation. (Lesson 2-1)

59. The product of 3 and m plus 2 times n is the same as the quotient of 4 and p.

60. The sum of x and five times y equals twice z minus 7.

Simplify.

61. $\dfrac{25}{10}$ **62.** $\dfrac{-4}{-12}$ **63.** $\dfrac{6}{-12}$ **64.** $\dfrac{-36}{8}$

Evaluate $\dfrac{a-b}{c-d}$ for the given values.

65. $a = 6, b = 2, c = 9, d = 3$ **66.** $a = -8, b = 4, c = 5, d = -3$ **67.** $a = 4, b = -7, c = -1, d = -2$

The power of a graphing calculator is the ability to graph different types of equations accurately and quickly. By entering one or more equations in the calculator you can view features of a graph, such as the x-intercept, y-intercept, the origin, intersections, and the coordinates of specific points.

Often linear equations are graphed in the **standard viewing window**, which is [−10, 10] by [−10, 10] with a scale of 1 on each axis. To quickly choose the standard viewing window on a TI-83/84 Plus, press $\boxed{\text{ZOOM}}$ 6.

 Common Core State Standards
Content Standards
N.Q.1 Use units as a way to understand problems and to guide the solution of multi-step problems; choose and interpret units consistently in formulas; choose and interpret the scale and the origin in graphs and data displays.
F.IF.7a Graph linear and quadratic functions and show intercepts, maxima, and minima.
Mathematical Practices
5 Use appropriate tools strategically.

Activity 1   Graph a Linear Equation

Graph $3x - y = 4$.

Step 1 Enter the equation in the Y= list.

- The Y= list shows the equation or equations that you will graph.

- Equations must be entered with the y isolated on one side of the equation. Solve the equation for y, then enter it into the calculator.

$$3x - y = 4 \qquad \text{Original equation}$$

$$3x - y - 3x = 4 - 3x \qquad \text{Subtract } 3x \text{ from each side.}$$

$$-y = -3x + 4 \qquad \text{Simplify.}$$

$$y = 3x - 4 \qquad \text{Multiply each side by } -1.$$

KEYSTROKES: $\boxed{\text{Y=}}$ 3 $\boxed{\text{X,T,}\theta\text{,}n}$ $\boxed{-}$ 4

The equals sign appears shaded for graphs that are selected to be displayed.

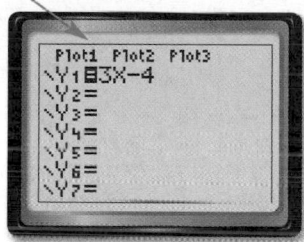

Step 2 Graph the equation in the standard viewing window.

- Graph the selected equation.

KEYSTROKES: $\boxed{\text{ZOOM}}$ 6

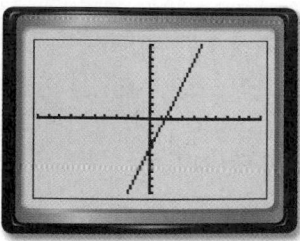

[−10, 10] scl: 1 by [−10, 10] scl: 1

Sometimes a complete graph is not displayed using the standard viewing window. A **complete graph** includes all of the important characteristics of the graph on the screen including the origin and the x- and y-intercepts. Note that the graph above is a complete graph because all of these points are visible.

When a complete graph is not displayed using the standard viewing window, you will need to change the viewing window to accommodate these important features. Use what you have learned about intercepts to help you choose an appropriate viewing window.

(continued on the next page)

Graphing Technology Lab
Graphing Linear Functions Continued

Activity 2   Graph a Complete Graph

Graph $y = 5x - 14$.

Step 1 Enter the equation in the Y= list and graph in the standard viewing window.

- Clear the previous equation from the Y= list. Then enter the new equation and graph.

KEYSTROKES: Y= CLEAR 5 X,T,θ,n − 14 ZOOM 6

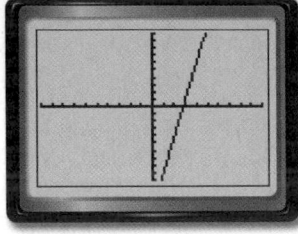

[−10, 10] scl: 1 by [−10, 10] scl: 1

Step 2 Modify the viewing window and graph again.

- The origin and the x-intercept are displayed in the standard viewing window. But notice that the y-intercept is outside of the viewing window.

Find the y-intercept.

$y = 5x - 14$ Original equation

$= 5(0) - 14$ Replace x with 0.

$= -14$ Simplify.

This window allows the complete graph, including the y-intercept, to be displayed.

Since the y-intercept is −14, choose a viewing window that includes a number less than −14. The window [−10, 10] by [−20, 5] with a scale of 1 on each axis is a good choice.

[−10, 10] scl: 1 by [−20, 5] scl: 1

KEYSTROKES: WINDOW −10 ENTER 10 ENTER 1 ENTER −20 ENTER 5 ENTER 1 GRAPH

Exercises

Use a graphing calculator to graph each equation in the standard viewing window. Sketch the result.

1. $y = x + 5$
2. $y = 5x + 6$
3. $y = 9 - 4x$
4. $3x + y = 5$
5. $x + y = -4$
6. $x - 3y = 6$

CCSS SENSE-MAKING Graph each equation in the standard viewing window. Determine whether the graph is complete. If the graph is not complete, adjust the viewing window and graph the equation again.

7. $y = 4x + 7$
8. $y = 9x - 5$
9. $y = 2x - 11$
10. $4x - y = 16$
11. $6x + 2y = 23$
12. $x + 4y = -36$

Consider the linear equation $y = 3x + b$.

13. Choose several different positive and negative values for b. Graph each equation in the standard viewing window.

14. For which values of b is the complete graph in the standard viewing window?

15. How is the value of b related to the y-intercept of the graph of $y = 3x + b$?

Algebra Lab
Rate of Change of a Linear Function

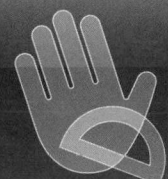

In mathematics, you can measure the steepness of a line using a ratio.

CCSS Common Core State Standards
Content Standards
F.IF.6 Calculate and interpret the average rate of change of a function (presented symbolically or as a table) over a specified interval. Estimate the rate of change from a graph.
F.LE.1a Prove that linear functions grow by equal differences over equal intervals, and that exponential functions grow by equal factors over equal intervals.

Set Up the Lab

- Stack three books on your desk.
- Lean a ruler on the books to create a ramp.
- Tape the ruler to the desk.
- Measure the rise and the run. Record your data in a table like the one at the right.
- Calculate and record the ratio $\frac{\text{rise}}{\text{run}}$.

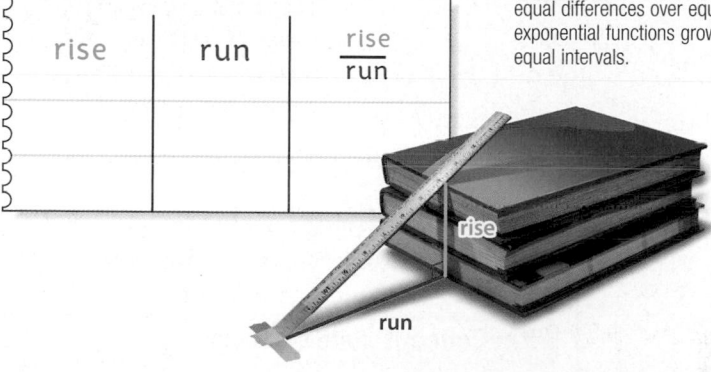

Activity

Step 1	Step 2
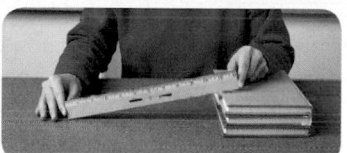	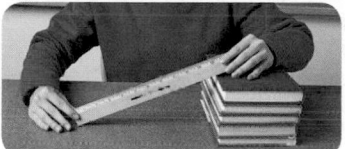
Move the books to make the ramp steeper. Measure and record the rise and the **run**. Calculate and record $\frac{\text{rise}}{\text{run}}$.	Add books to the stack to make the ramp even steeper. Measure, calculate, and record your data in the table.

Analyze the Results

1. Examine the ratios you recorded. How did they change as the ramp became steeper?

2. **MAKE A PREDICTION** Suppose you want to construct a skateboard ramp that is not as steep as the one shown at the right. List three different sets of $\frac{\text{rise}}{\text{run}}$ measurements that will result in a less steep ramp. Verify your predictions by calculating the ratio $\frac{\text{rise}}{\text{run}}$ for each ramp.

3. Copy the coordinate graph shown and draw a line through the origin with a $\frac{\text{rise}}{\text{run}}$ ratio greater than the original line. Then draw a line through the origin with a ratio less than that of the original line. Explain using the words *rise* and *run* why the lines you drew have a ratio greater or less than the original line.

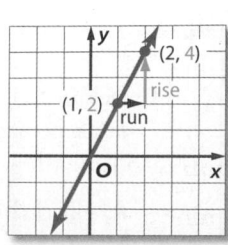

18 in.

24 in.

$m = \frac{18}{24} = \frac{3}{4}$

4. We have seen what happens on the graph as the $\frac{\text{rise}}{\text{run}}$ ratio gets closer to zero. What would you predict will happen when the ratio is zero? Explain your reasoning. Give an example to support your prediction.

Rate of Change and Slope

∷Then

- You graphed ordered pairs in the coordinate plane.

∷Now

1 Use rate of change to solve problems.

2 Find the slope of a line.

∷Why?

- The Daredevil Drop at Wet 'n Wild Emerald Pointe in Greensboro, North Carolina, is a thrilling ride that drops you 76 feet down a steep water chute. A *rate of change* of the ride might describe the distance a rider has fallen over a length of time.

 NewVocabulary
rate of change
slope

 Common Core State Standards

Content Standards

F.IF.6 Calculate and interpret the average rate of change of a function (presented symbolically or as a table) over a specified interval. Estimate the rate of change from a graph.

F.LE.1a Prove that linear functions grow by equal differences over equal intervals, and that exponential functions grow by equal factors over equal intervals.

Mathematical Practices
2 Reason abstractly and quantitatively.

1 Rate of Change **Rate of change** is a ratio that describes, on average, how much one quantity changes with respect to a change in another quantity.

KeyConcept Rate of Change

If x is the independent variable and y is the dependent variable, then

$$\text{rate of change} = \frac{\text{change in } y}{\text{change in } x}.$$

● Real-World Example 1 Find Rate of Change

ENTERTAINMENT Use the table to find the rate of change. Then explain its meaning.

Number of Computer Games	Total Cost ($)
x	y
2	78
4	156
6	234

$$\text{rate of change} = \frac{\text{change in } y}{\text{change in } x} \leftarrow \text{dollars} \atop \leftarrow \text{games}$$

$$= \frac{\text{change in cost}}{\text{change in number of games}}$$

$$= \frac{156 - 78}{4 - 2}$$

$$= \frac{78}{2} \text{ or } \frac{39}{1}$$

The rate of change is $\frac{39}{1}$ or 39. This means that the cost per game is $39.

▶ **Guided**Practice

1. **REMODELING** The table shows how the tiled surface area changes with the number of floor tiles.

   A. Find the rate of change.

   B. Explain the meaning of the rate of change.

Number of Floor Tiles	Area of Tiled Surface (in²)
x	y
3	48
6	96
9	144

So far, you have seen rates of change that are *constant*. Many real-world situations involve rates of change that are not constant.

Real-World Example 2 Compare Rates of Change

AMUSEMENT PARKS The graph shows the number of people who visited U.S. theme parks in recent years.

a. **Find the rates of change for 2000–2002 and 2002–2004.**

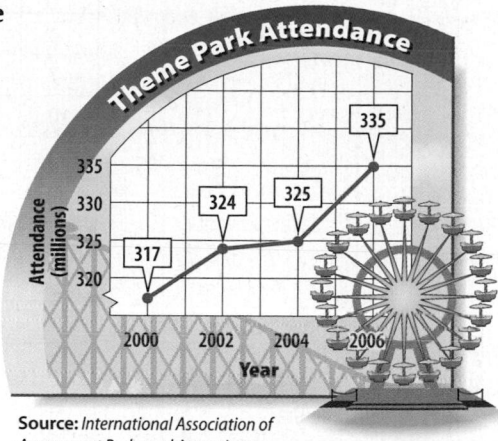

Theme Park Attendance

Source: *International Association of Amusement Parks and Attractions*

2000–2002:

$$\frac{\text{change in attendance}}{\text{change in time}} = \frac{324 - 317}{2002 - 2000} \quad \begin{matrix} \leftarrow \text{people} \\ \leftarrow \text{years} \end{matrix} \qquad \text{Substitute.}$$

$$= \frac{7}{2} \text{ or } 3.5 \qquad \text{Simplify.}$$

Over this 2-year period, attendance increased by 7 million, for a rate of change of 3.5 million per year.

2002–2004:

$$\frac{\text{change in attendance}}{\text{change in time}} = \frac{325 - 324}{2004 - 2002} \qquad \text{Substitute.}$$

$$= \frac{1}{2} \text{ or } 0.5 \qquad \text{Simplify.}$$

Over this 2-year period, attendance increased by 1 million, for a rate of change of 0.5 million per year.

b. **Explain the meaning of the rate of change in each case.**

For 2000–2002, on average, 3.5 million more people went to a theme park each year than the last.

For 2002–2004, on average, 0.5 million more people attended theme parks each year than the last.

c. **How are the different rates of change shown on the graph?**

There is a greater vertical change for 2000–2002 than for 2002–2004. Therefore, the section of the graph for 2000–2002 is steeper.

▶ **Guided**Practice

2. Refer to the graph above. Without calculating, find the 2-year period that has the least rate of change. Then calculate to verify your answer.

StudyTip

CCSS Reasoning A positive rate of change indicates an increase over time. A negative rate of change indicates that a quantity is decreasing.

A rate of change is constant for a function when the rate of change is the same between any pair of points on the graph of the function. Linear functions have a constant rate of change.

Example 3 Constant Rates of Change

Determine whether each function is linear. Explain.

a.

x	y
1	−6
4	−8
7	−10
10	−12
13	−14

b.

x	y
−3	10
−1	12
1	16
3	18
5	22

StudyTip

Linear or Nonlinear Function? Notice that the changes in *x* and *y* are not the same. For the rate of change to be linear, the change in *x*-values must be constant and the change in *y*-values must be constant.

x	y	rate of change
1	−6	$\dfrac{-8-(-6)}{4-1}$ or $-\dfrac{2}{3}$
4	−8	$\dfrac{-10-(-8)}{7-4}$ or $-\dfrac{2}{3}$
7	−10	$\dfrac{-12-(-10)}{10-7}$ or $-\dfrac{2}{3}$
10	−12	$\dfrac{-14-(-12)}{13-10}$ or $-\dfrac{2}{3}$
13	−14	

x	y	rate of change
−3	10	$\dfrac{12-10}{-1-(-3)}$ or 1
−1	12	$\dfrac{16-12}{1-(-1)}$ or 2
1	16	$\dfrac{18-16}{3-1}$ or 1
3	18	$\dfrac{22-18}{5-3}$ or 2
5	22	

The rate of change is constant. Thus, the function is linear.

This rate of change is not constant. Thus, the function is not linear.

GuidedPractice

3A.

x	y
−3	11
−2	15
−1	19
1	23
2	27

3B.

x	y
12	−4
9	1
6	6
3	11
0	16

2 **Find Slope** The **slope** of a nonvertical line is the ratio of the change in the *y*-coordinates (rise) to the change in the *x*-coordinates (run) as you move from one point to another.

It can be used to describe a rate of change. Slope describes how steep a line is. The greater the absolute value of the slope, the steeper the line.

The graph shows a line that passes through (−1, 3) and (2, −2).

$$\textbf{slope} = \frac{\textbf{rise}}{\textbf{run}}$$

$$= \frac{\textbf{change in }y\textbf{-coordinates}}{\textbf{change in }x\textbf{-coordinates}}$$

$$= \frac{-2-3}{2-(-1)} \text{ or } -\frac{5}{3}$$

So, the slope of the line is $-\dfrac{5}{3}$.

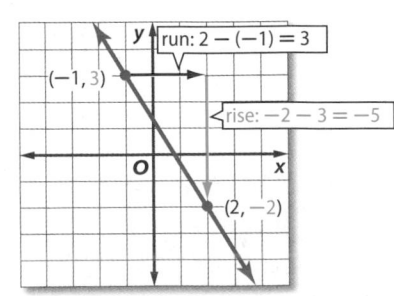

Because a linear function has a constant rate of change, any two points on a nonvertical line can be used to determine its slope.

KeyConcept Slope

Words The slope of a nonvertical line is the ratio of the rise to the run.

Symbols The slope *m* of a nonvertical line through any two points, (x_1, y_1) and (x_2, y_2), can be found as follows.

$$m = \frac{y_2 - y_1}{x_2 - x_1} \quad \begin{matrix} \leftarrow \text{ change in } y \\ \leftarrow \text{ change in } x \end{matrix}$$

Graph

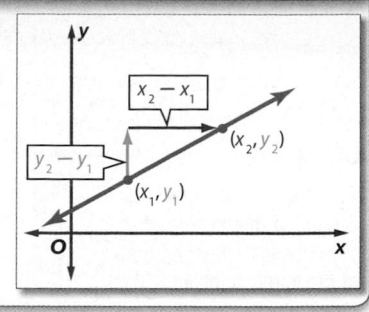

The slope of a line can be positive, negative, zero, or undefined. If the line is not horizontal or vertical, then the slope is either positive or negative.

Example 4 Positive, Negative and Zero Slope

Find the slope of a line that passes through each pair of points.

a. $(-2, 0)$ and $(1, 5)$

$$m = \frac{y_2 - y_1}{x_2 - x_1} \qquad \frac{\text{rise}}{\text{run}}$$

$$= \frac{5 - 0}{1 - (-2)} \qquad (-2, 0) = (x_1, y_1) \text{ and } (1, 5) = (x_2, y_2)$$

$$= \frac{5}{3} \qquad \text{Simplify.}$$

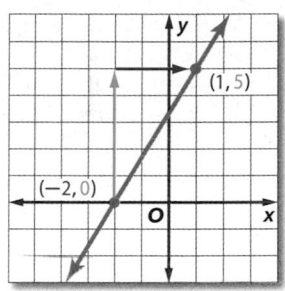

b. $(-3, 4)$ and $(2, -3)$

$$m = \frac{y_2 - y_1}{x_2 - x_1} \qquad \frac{\text{rise}}{\text{run}}$$

$$= \frac{-3 - 4}{2 - (-3)} \qquad (-3, 4) = (x_1, y_1) \text{ and } (2, -3) = (x_2, y_2)$$

$$= \frac{-7}{5} \text{ or } -\frac{7}{5} \qquad \text{Simplify.}$$

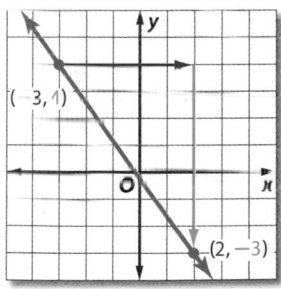

c. $(-3, -1)$ and $(2, -1)$

$$m = \frac{y_2 - y_1}{x_2 - x_1} \qquad \frac{\text{rise}}{\text{run}}$$

$$= \frac{-1 - (-1)}{2 - (-3)} \qquad \text{Substitute.}$$

$$= \frac{0}{5} \text{ or } 0 \qquad \text{Simplify.}$$

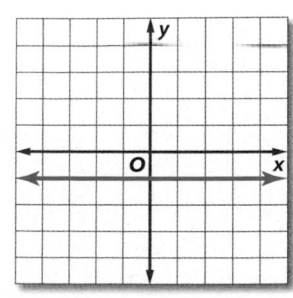

▶ **Guided**Practice

Find the slope of the line that passes through each pair of points.

4A. $(3, 6), (4, 8)$ **4B.** $(-4, -2), (0, -2)$ **4C.** $(-4, 2), (-2, 10)$

4D. $(6, 7), (-2, 7)$ **4E.** $(-2, 2), (-6, 4)$ **4F.** $(4, 3), (-1, 11)$

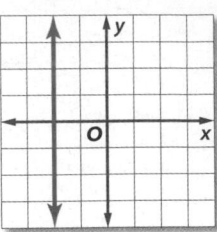

Example 5 Undefined Slope

Find the slope of the line that passes through $(-2, 4)$ and $(-2, -3)$.

$$m = \frac{y_2 - y_1}{x_2 - x_1} \qquad \frac{\text{rise}}{\text{run}}$$

$$= \frac{-3 - 4}{-2 - (-2)} \qquad \text{Substitute.}$$

$$= \frac{-7}{0} \text{ or undefined} \qquad \text{Simplify.}$$

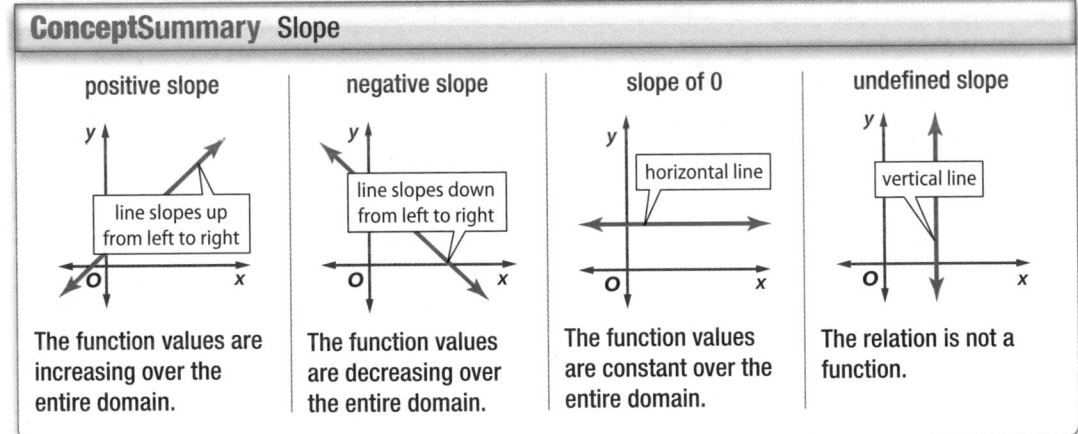

StudyTip

Zero and Undefined Slopes If the change in *y*-values is 0, then the graph of the line is horizontal. If the change in *x*-values is 0, then the slope is undefined. This graph is a vertical line.

▶ **Guided**Practice

Find the slope of the line that passes through each pair of points.

5A. $(6, 3), (6, 7)$ **5B.** $(-3, 2), (-3, -1)$

The graphs of lines with different slopes are summarized below.

ConceptSummary Slope

positive slope	negative slope	slope of 0	undefined slope
line slopes up from left to right	line slopes down from left to right	horizontal line	vertical line
The function values are increasing over the entire domain.	The function values are decreasing over the entire domain.	The function values are constant over the entire domain.	The relation is not a function.

Example 6 Find Coordinates Given the Slope

Find the value of *r* so that the line through $(1, 4)$ and $(-5, r)$ has a slope of $\frac{1}{3}$.

$$m = \frac{y_2 - y_1}{x_2 - x_1} \qquad \text{Slope Formula}$$

$$\frac{1}{3} = \frac{r - 4}{-5 - 1} \qquad \text{Let } (1, 4) = (x_1, y_1) \text{ and } (-5, r) = (x_2, y_2).$$

$$\frac{1}{3} = \frac{r - 4}{-6} \qquad \text{Subtract.}$$

$$3(r - 4) = 1(-6) \qquad \text{Find the cross products.}$$

$$3r - 12 = -6 \qquad \text{Distributive Property}$$

$$3r = 6 \qquad \text{Add 12 to each side and simplify.}$$

$$r = 2 \qquad \text{Divide each side by 3 and simplify.}$$

So, the line goes through $(-5, 2)$.

▶ **Guided**Practice

Find the value of *r* so the line that passes through each pair of points has the given slope.

6A. $(-2, 6), (r, -4); m = -5$ **6B.** $(r, -6), (5, -8); m = -8$

Example 1 Find the rate of change represented in each table or graph.

1.

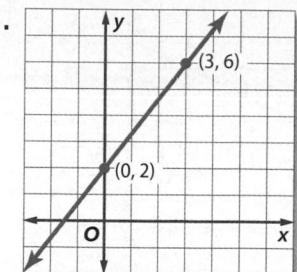

Points: (3, 6) and (0, 2)

2.

x	y
3	−6
5	2
7	10
9	18
11	26

Example 2 **3.** CCSS **SENSE-MAKING** Refer to the graph at the right.

  a. Find the rate of change of prices from 2006 to 2008. Explain the meaning of the rate of change.

  b. Without calculating, find a two-year period that had a greater rate of change than 2006–2008. Explain.

  c. Between which years would you guess the new stadium was built? Explain your reasoning.

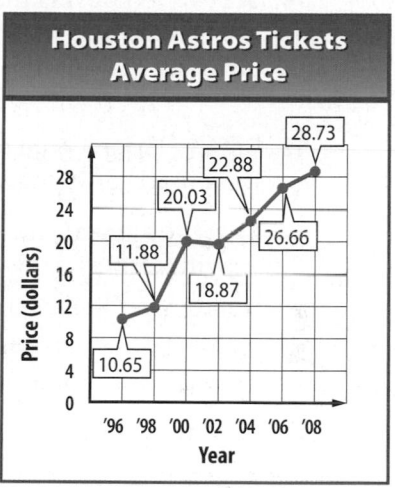

Houston Astros Tickets Average Price

Values on graph: 10.65, 11.88, 20.03, 18.87, 22.88, 26.66, 28.73

Source: *Team Marketing Report*

Example 3 Determine whether each function is linear. Write *yes* or *no*. Explain.

4.

x	−7	−4	−1	2	5
y	5	4	3	2	1

5.

x	8	12	16	20	24
y	7	5	3	0	−2

Examples 4–5 Find the slope of the line that passes through each pair of points.

6. (5, 3), (6, 9)

7. (−4, 3), (−2, 1)

8. (6, −2), (8, 3)

9. (1, 10), (−8, 3)

10. (−3, 7), (−3, 4)

11. (5, 2), (−6, 2)

Example 6 Find the value of r so the line that passes through each pair of points has the given slope.

12. (−4, r), (−8, 3), $m = -5$

13. (5, 2), (−7, r), $m = \frac{5}{6}$

Practice and Problem Solving Extra Practice is on page R3.

Example 1 Find the rate of change represented in each table or graph.

14.

x	y
5	2
10	3
15	4
20	5

15

x	y
1	15
2	9
3	3
4	−3

 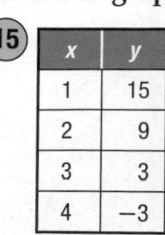

Example 1 Find the rate of change represented in each table or graph.

16.

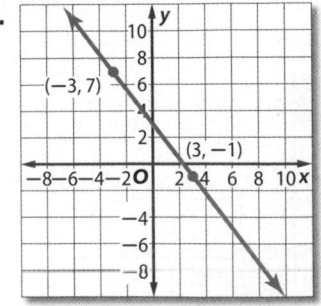

(−3, 7)
(3, −1)

17.

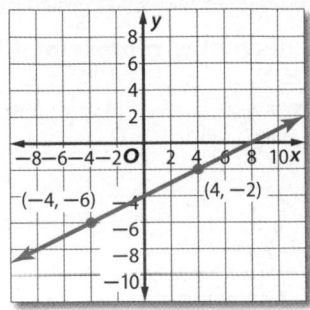

(−4, −6)
(4, −2)

Example 2 **18. SPORTS** What was the annual rate of change from 2004 to 2008 for women participating in collegiate lacrosse? Explain the meaning of the rate of change.

Year	Number of Women
2004	5545
2008	6830

19. RETAIL The average retail price in the spring of 2009 for a used car is shown in the table at the right.

Age (years)	Value ($)
2	17,378
3	16,157

  a. Write a linear function to model the price of the car with respect to age.

  b. Interpret the meaning of the slope of the line.

  c. Assuming a constant rate of change predict the average retail price for a 7-year-old car.

Example 3 Determine whether each function is linear. Write *yes* or *no*. Explain.

20.

x	4	2	0	−2	−4
y	−1	1	3	5	7

21.

x	−7	−5	−3	−1	0
y	11	14	17	20	23

22.

x	−0.2	0	0.2	0.4	0.6
y	0.7	0.4	0.1	0.3	0.6

23.

x	$\frac{1}{2}$	$\frac{3}{2}$	$\frac{5}{2}$	$\frac{7}{2}$	$\frac{9}{2}$
y	$\frac{1}{2}$	1	$\frac{3}{2}$	2	$\frac{5}{2}$

Examples 4–5 Find the slope of the line that passes through each pair of points.

24. $(4, 3), (−1, 6)$ **25** $(8, −2), (1, 1)$ **26.** $(2, 2), (−2, −2)$

27. $(6, −10), (6, 14)$ **28.** $(5, −4), (9, −4)$ **29.** $(11, 7), (−6, 2)$

30. $(−3, 5), (3, 6)$ **31.** $(−3, 2), (7, 2)$ **32.** $(8, 10), (−4, −6)$

33. $(−8, 6), (−8, 4)$ **34.** $(−12, 15), (18, −13)$ **35.** $(−8, −15), (−2, 5)$

Example 6 Find the value of *r* so the line that passes through each pair of points has the given slope.

36. $(12, 10), (−2, r), m = −4$ **37.** $(r, −5), (3, 13), m = 8$

38. $(3, 5), (−3, r), m = \frac{3}{4}$ **39.** $(−2, 8), (r, 4), m = −\frac{1}{2}$

CCSS TOOLS Use a ruler to estimate the slope of each object.

40.

41.

42. DRIVING When driving up a certain hill, you rise 15 feet for every 1000 feet you drive forward. What is the slope of the road?

Find the slope of the line that passes through each pair of points.

43.

x	y
4.5	−1
5.3	2

44.

x	y
0.75	1
0.75	−1

45.

x	y
$2\frac{1}{2}$	$-1\frac{1}{2}$
$-\frac{1}{2}$	$\frac{1}{2}$

46. ⟳ **MULTIPLE REPRESENTATIONS** In this problem, you will investigate why the slope of a line through any two points on that line is constant.

  a. Visual Sketch a line ℓ that contains points A, B, A' and B' on a coordinate plane.

  b. Geometric Add segments to form right triangles ABC and $A'B'C'$ with right angles at C and C'. Describe $\overline{AC}$ and $\overline{A'C'}$, and $\overline{BC}$ and $\overline{B'C'}$.

  c. Verbal How are triangles ABC and $A'B'C'$ related? What does that imply for the slope between any two distinct points on line ℓ?

47 **BASKETBALL** The table shown below shows the average points per game (PPG) Michael Redd has scored in each of his first 9 seasons with the NBA's Milwaukee Bucks.

Season	1	2	3	4	5	6	7	8	9
PPG	2.2	11.4	15.1	21.7	23.0	25.4	26.7	22.7	21.2

  a. Make a graph of the data. Connect each pair of adjacent points with a line.

  b. Use the graph to determine in which period Michael Redd's PPG increased the fastest. Explain your reasoning.

  c. Discuss the difference in the rate of change from season 1 through season 4, from season 4 through season 7, from season 7 through season 9.

H.O.T. Problems Use Higher-Order Thinking Skills

48. REASONING Why does the Slope Formula not work for vertical lines? Explain.

49. OPEN ENDED Use what you know about rate of change to describe the function represented by the table.

Time (wk)	Height of Plant (in.)
4	9.0
6	13.5
8	18.0

50. CHALLENGE Find the value of d so the line that passes through (a, b) and (c, d) has a slope of $\frac{1}{2}$.

51. WRITING IN MATH Explain how the rate of change and slope are related and how to find the slope of a line.

52. **CCSS** **ARGUMENTS** Kyle and Luna are finding the value of a so the line that passes through $(10, a)$ and $(−2, 8)$ has a slope of $\frac{1}{4}$. Is either of them correct? Explain.

Kyle

$$\frac{-2 - 10}{8 - a} = \frac{1}{4}$$

$$1(8 - a) = 4(-12)$$

$$8 - a = -48$$

$$a = 56$$

Luna

$$\frac{8 - a}{-2 - 10} = \frac{1}{4}$$

$$4(8 - a) = 1(-12)$$

$$32 - 4a = -12$$

$$a = 11$$

53. The cost of prints from an online photo processor is given by $C(p) = 29.99 + 0.13p$. $29.99 is the cost of the membership, and p is the number of 4-inch by 6-inch prints. What does the slope represent?

   A cost per print

   B cost of the membership

   C cost of the membership and 1 print

   D number of prints

54. Danita bought a computer for $1200 and its value depreciated linearly. After 2 years, the value was $250. What was the amount of yearly depreciation?

   F $950

   G $475

   H $250

   J $225

55. SHORT RESPONSE The graph represents how much the Wright Brothers National Monument charges visitors. How much does the park charge each visitor?

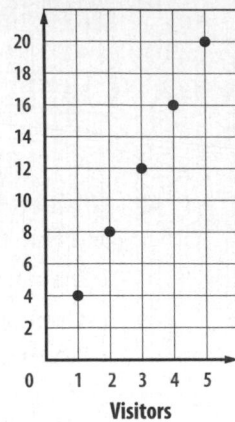

Wright Brothers National Monument

56. PROBABILITY At a gymnastics camp, 1 gymnast is chosen at random from each team. The Flipstars Gymnastics Team consists of 5 eleven-year-olds, 7 twelve-year-olds, 10 thirteen-year-olds, and 8 fourteen-year-olds. What is the probability that the age of the gymnast chosen is an odd number?

   A $\frac{1}{30}$ **B** $\frac{1}{15}$ **C** $\frac{1}{2}$ **D** $\frac{3}{5}$

Spiral Review

Solve each equation by graphing. (Lesson 3-2)

57. $3x + 18 = 0$ **58.** $8x - 32 = 0$ **59.** $0 = 12x - 48$

Find the x- and y-intercepts of the graph of each linear function. (Lesson 3-1)

60.

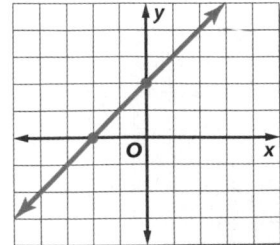

61.

x	y
−3	−4
−2	−2
−1	0
0	2
1	4

62. HOMECOMING Dance tickets are $9 for one person and $15 for two people. If a group of seven students wishes to go to the dance, write and solve an equation that would represent the least expensive price p of their tickets. (Lesson 1-3)

Skills Review

Find each quotient.

63. $8 \div \frac{2}{3}$ **64.** $\frac{3}{8} \div \frac{1}{4}$ **65.** $\frac{5}{8} \div 2$

66. $\frac{12 \cdot 6}{9}$ **67.** $\frac{2 \cdot 15}{6}$ **68.** $\frac{18 \cdot 5}{15}$

Determine whether each equation is a linear equation. Write *yes* or *no*. If yes, write the equation in standard form. (Lesson 3-1)

1. $y = -4x + 3$

2. $x^2 + 3y = 8$

3. $\frac{1}{4}x - \frac{3}{4}y = -1$

Graph each equation using the *x*- and *y*-intercepts. (Lesson 3-1)

4. $y = 3x - 6$

5. $2x + 5y = 10$

Graph each equation by making a table. (Lesson 3-1)

6. $y = -2x$

7. $x = 8 - y$

8. BOOK SALES The equation $5x + 12y = 240$ describes the total amount of money collected when selling *x* paperback books at $5 per book and *y* hardback books at $12 per book. Graph the equation using the *x*- and *y*-intercepts. (Lesson 3-1)

Find the root of each equation. (Lesson 3-2)

9. $x + 8 = 0$

10. $4x - 24 = 0$

11. $18 + 8x = 0$

12. $\frac{3}{5}x - \frac{1}{2} = 0$

Solve each equation by graphing. (Lesson 3-2)

13. $-5x + 35 = 0$

14. $14x - 84 = 0$

15. $118 + 11x = -3$

16. MULTIPLE CHOICE The function $y = -15 + 3x$ represents the outside temperature, in degrees Fahrenheit, in a small Alaskan town where *x* represents the number of hours after midnight. The function is accurate for *x* values representing midnight through 4:00 P.M. Find the zero of this function. (Lesson 3-2)

A 0

B 3

C 5

D −15

17. Find the rate of change represented in the table. (Lesson 3-3)

x	y
1	2
4	6
7	10
10	14

Find the slope of the line that passes through each pair of points. (Lesson 3-3)

18. (2, 6), (4, 12)

19. (1, 5), (3, 8)

20. (−3, 4), (2, −6)

21. $\left(\frac{1}{3}, \frac{3}{4}\right), \left(\frac{2}{3}, \frac{1}{4}\right)$

22. MULTIPLE CHOICE Find the value of *r* so the line that passes through the pair of points has the given slope. (Lesson 3-3)

$$(-4, 8), (r, 12), m = \frac{4}{3}$$

F −4

G −1

H 0

J 3

23. Find the slope of the line that passes through the pair of points. (Lesson 3-3)

x	y
2.6	−2
3.1	4

24. POPULATION GROWTH The graph shows the population growth in Heckertsville since 2003. (Lesson 3-3)

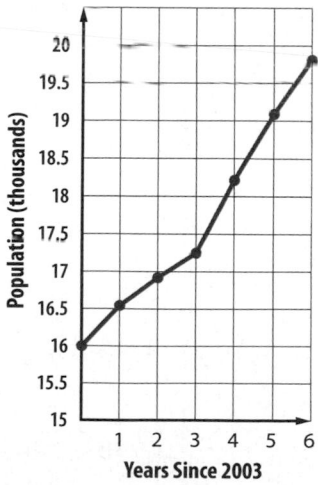

a. For which time period is the rate of change the greatest?

b. Explain the meaning of the slope from 2003 to 2009.

Direct Variation

Then	Now	Why?
● You found rates of change of linear functions.	**1** Write and graph direct variation equations. **2** Solve problems involving direct variation.	● Bianca is saving her money to buy a designer purse that costs $295. To help raise the money, she charges $8 per hour to babysit her neighbors' child. The slope of the line that represents the amount of money Bianca earns is 8, and the rate of change is constant.

 NewVocabulary
direct variation
constant of variation
constant of proportionality

 Common Core State Standards

Content Standards
A.REI.10 Understand that the graph of an equation in two variables is the set of all its solutions plotted in the coordinate plane, often forming a curve (which could be a line).

F.IF.7a Graph linear and quadratic functions and show intercepts, maxima, and minima.

Mathematical Practices
1 Make sense of problems and persevere in solving them.
6 Attend to precision.

1 **Direct Variation Equations** A **direct variation** is described by an equation of the form $y = kx$, where $k \neq 0$. The equation $y = kx$ illustrates a constant rate of change, and k is the **constant of variation**, also called the **constant of proportionality**.

Example 1 Slope and Constant of Variation

Name the constant of variation for each equation. Then find the slope of the line that passes through each pair of points.

a.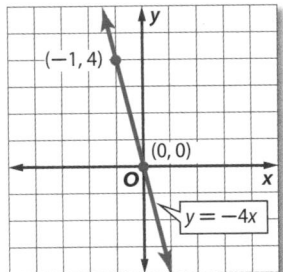

b.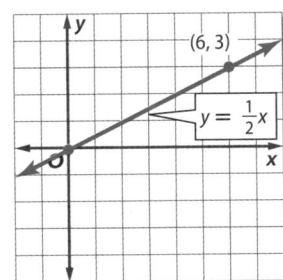

The constant of variation is -4.

$$m = \frac{y_2 - y_1}{x_2 - x_1} \quad \text{Slope Formula}$$

$$= \frac{4 - 0}{-1 - 0} \quad \begin{array}{l}(x_1, y_1) = (0, 0)\\(x_2, y_2) = (-1, 4)\end{array}$$

$$= -4 \quad \text{The slope is } -4.$$

The constant of variation is $\frac{1}{2}$.

$$m = \frac{y_2 - y_1}{x_2 - x_1} \quad \text{Slope Formula}$$

$$= \frac{3 - 0}{6 - 0} \quad \begin{array}{l}(x_1, y_1) = (0, 0)\\(x_2, y_2) = (6, 3)\end{array}$$

$$= \frac{1}{2} \quad \text{The slope is } \frac{1}{2}.$$

▶ **Guided Practice**

1A. Name the constant of variation for $y = \frac{1}{4}x$. Then find the slope of the line that passes through $(0, 0)$ and $(4, 1)$, two points on the line.

1B. Name the constant of variation for $y = -2x$. Then find the slope of the line that passes through $(0, 0)$ and $(1, -2)$, two points on the line.

The slope of the graph of $y = kx$ is k. Since $0 = k(0)$, the graph of $y = kx$ always passes through the origin. Therefore the x- and y-intercepts are zero.

Example 2 Graph a Direct Variation

Graph $y = -6x$.

Step 1 Write the slope as a ratio.

$$-6 = \frac{-6}{1} \qquad \frac{\text{rise}}{\text{run}}$$

Step 2 Graph $(0, 0)$.

Step 3 From the point $(0, 0)$, move down 6 units and right 1 unit. Draw a dot.

Step 4 Draw a line containing the points.

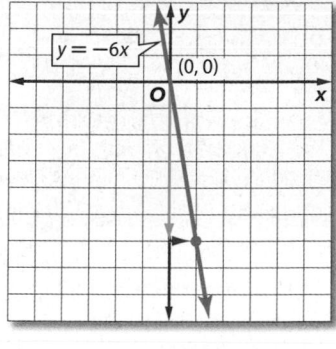

StudyTip

Constant of Variation
A line with a positive constant of variation will go up from left to right and a line with a negative constant of variation will go down from left to right.

Guided Practice

2A. $y = 6x$ **2B.** $y = \frac{2}{3}x$ **2C.** $y = -5x$ **2D.** $y = -\frac{3}{4}x$

The graphs of all direct variation equations share some common characteristics.

ConceptSummary Direct Variation Graphs

- Direct variation equations are of the form $y = kx$, where $k \neq 0$.
- The graph of $y = kx$ always passes through the origin.
- The slope is positive if $k > 0$.

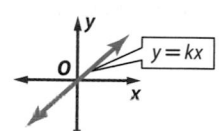

- The slope is negative if $k < 0$.

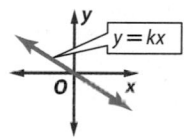

If the relationship between the values of y and x can be described by a direct variation equation, then we say that y varies directly as x.

Example 3 Write and Solve a Direct Variation Equation

Suppose y varies directly as x, and $y = 72$ when $x = 8$.

a. Write a direct variation equation that relates x and y.

$$
\begin{array}{ll}
y = kx & \text{Direct variation formula} \\
72 = k(8) & \text{Replace } y \text{ with 72 and } x \text{ with 8.} \\
9 = k & \text{Divide each side by 8.}
\end{array}
$$

Therefore, the direct variation equation is $y = 9x$.

b. Use the direct variation equation to find x when $y = 63$.

$$
\begin{array}{ll}
y = 9x & \text{Direct variation formula} \\
63 = 9x & \text{Replace } y \text{ with 63.} \\
7 = x & \text{Divide each side by 9.}
\end{array}
$$

Therefore, $x = 7$ when $y = 63$.

Guided Practice

3. Suppose y varies directly as x, and $y = 98$ when $x = 14$. Write a direct variation equation that relates x and y. Then find y when $x = -4$.

2 Direct Variation Problems
One of the most common applications of direct variation is the formula $d = rt$. Distance d varies directly as time t, and the rate r is the constant of variation.

TRAVEL The distance a jet travels varies directly as the number of hours it flies. A jet traveled 3420 miles in 6 hours.

a. Write a direct variation equation for the distance d flown in time t.

Words	Distance	equals	rate	times	time.
Variable	Let r = rate.				
Equation	3420	=	r	×	6

Solve for the rate.

$3420 = r(6)$ Original equation

$\dfrac{3420}{6} = \dfrac{r(6)}{6}$ Divide each side by 6.

$570 = r$ Simplify.

Therefore, the direct variation equation is $d = 570t$. The airliner flew at a rate of 570 miles per hour.

b. Graph the equation.

The graph of $d = 570t$ passes through the origin with slope 570.

$m = \dfrac{570}{1}$ $\dfrac{\text{rise}}{\text{run}}$

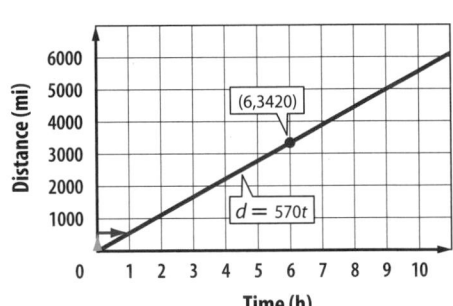

Distance Flown

c. Estimate how many hours it will take for an airliner to fly 6500 miles.

$d = 570t$ Original equation

$6500 = 570t$ Replace d with 6500.

$\dfrac{6500}{570} = \dfrac{570t}{570}$ Divide each side by 570.

$t \approx 11.4$ Simplify.

It would take the airliner approximately 11.4 hours to fly 6500 miles.

▸ **Guided**Practice

4. HOT-AIR BALLOONS A hot-air balloon's height varies directly as the balloon's ascent time in minutes.

  A. Write a direct variation for the distance d ascended in time t.

  B. Graph the equation.

  C. Estimate how many minutes it would take to ascend 2100 feet.

  D. About how many minutes would it take to ascend 3500 feet?

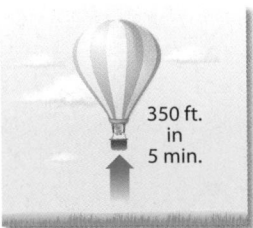

350 ft. in 5 min.

Real-World Link

In 2006, domestic airlines transported over 660 million passengers an average distance of 724 miles per flight.

Source: Bureau of Transportation Statistics

Problem-Solving Tip

CCSS Precision Notice that the question asks for an estimate, not an exact answer.

Example 1   Name the constant of variation for each equation. Then find the slope of the line that passes through each pair of points.

1.
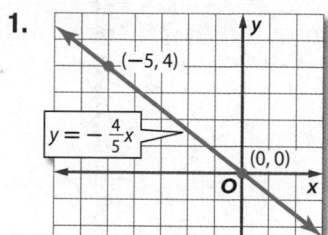
$(-5, 4)$

$y = -\dfrac{4}{5}x$

$(0, 0)$

2.
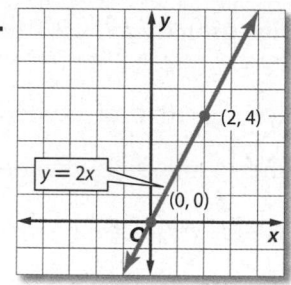
$(2, 4)$

$y = 2x$

$(0, 0)$

Example 2   Graph each equation.

3. $y = -x$ **4.** $y = \dfrac{3}{4}x$ **5.** $y = -8x$ **6.** $y = -\dfrac{8}{5}$

Example 3   Suppose y varies directly as x. Write a direct variation equation that relates x and y. Then solve.

7. If $y = 15$ when $x = 12$, find y when $x = 32$.

8. If $y = -11$ when $x = 6$, find x when $y = 44$.

Example 4   **9.** **CCSS REASONING** You find that the number of messages you receive on your message board varies directly as the number of messages you post. When you post 5 messages, you receive 12 messages in return.

a. Write a direct variation equation relating your posts to the messages received. Then graph the equation.

b. Find the number of messages you need to post to receive 96 messages.

Practice and Problem Solving   Extra Practice is on page R3.

Example 1   Name the constant of variation for each equation. Then find the slope of the line that passes through each pair of points.

10.
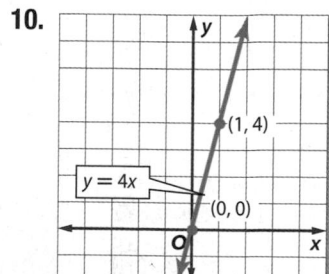
$(1, 4)$

$y = 4x$

$(0, 0)$

11
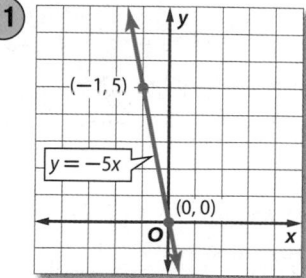
$(-1, 5)$

$y = -5x$

$(0, 0)$

12.
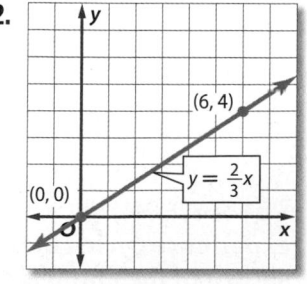
$(6, 4)$

$(0, 0)$ $y = \dfrac{2}{3}x$

13.
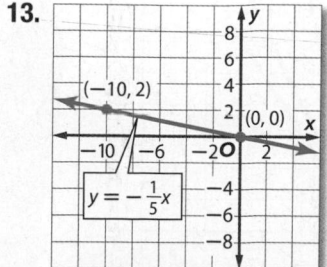
$(-10, 2)$

$(0, 0)$

$y = -\dfrac{1}{5}x$

14.
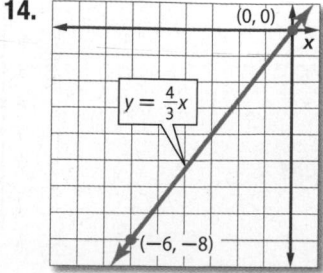
$(0, 0)$

$y = \dfrac{4}{3}x$

$(-6, -8)$

15.
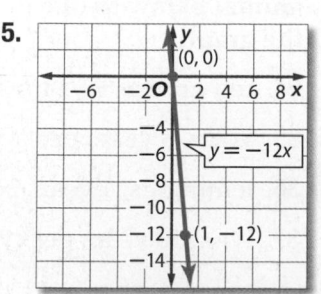
$(0, 0)$

$y = -12x$

$(1, -12)$

Example 2   **Graph each equation.**

16. $y = 10x$ **17.** $y = -7x$ **18.** $y = x$ **19.** $y = \frac{7}{6}x$

20. $y = \frac{1}{6}x$ **21.** $y = \frac{2}{9}x$ **22.** $y = \frac{6}{5}x$ **23.** $y = -\frac{5}{4}x$

Example 3   **Suppose y varies directly as x. Write a direct variation equation that relates x and y. Then solve.**

24. If $y = 6$ when $x = 10$, find x when $y = 18$.

25 If $y = 22$ when $x = 8$, find y when $x = -16$.

26. If $y = 4\frac{1}{4}$ when $x = \frac{3}{4}$, find y when $x = 4\frac{1}{2}$.

27. If $y = 12$ when $x = \frac{6}{7}$, find x when $y = 16$.

Example 4   **28. SPORTS** The distance a golf ball travels at an altitude of 7000 feet varies directly with the distance the ball travels at sea level, as shown.

Hitting a Golf Ball		
Altitude (ft)	0 (sea level)	7000
Distance (yd)	200	210

a. Write and graph an equation that relates the distance a golf ball travels at an altitude of 7000 feet y with the distance at sea level x.

b. What would be a person's average driving distance at 7000 feet if his average driving distance at sea level is 180 yards?

29. FINANCIAL LITERACY Depreciation is the decline in a car's value over the course of time. The table below shows the values of a car with an average depreciation.

Age of Car (years)	1	2	3	4	5
Value (dollars)	12,000	10,200	8400	6600	4800

a. Write an equation that relates the age x of the car to the value y that it lost after each year.

b. Find the age of the car if the value is $300.

Suppose y varies directly as x. Write a direct variation equation that relates x and y. Then solve.

30. If $y = 3.2$ when $x = 1.6$, find y when $x = 19$.

31. If $y = 15$ when $x = \frac{3}{4}$, find x when $y = 25$.

32. If $y = 4.5$ when $x = 2.5$, find y when $x = 12$.

33. If $y = -6$ when $x = 1.6$, find y when $x = 8$.

 SENSE-MAKING Certain endangered species experience cycles in their populations as shown in the graph at the right. Match each animal below to one of the colored lines in the graph.

34. red grouse, 8 years per cycle

35. voles, 3 years per cycle

36. lemmings, 4 years per cycle

37. lynx, 10 years per cycle

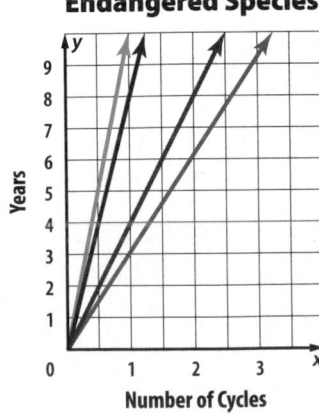

Population Cycles of Endangered Species

Comstock Images/Alamy

In Exercises 38–40, write and graph a direct variation equation that relates the variables.

38. PHYSICAL SCIENCE The weight W of an object is 9.8 m/s^2 times the mass of the object m.

39 MUSIC Music downloads are $0.99 per song. The total cost of d songs is T.

40. GEOMETRY The circumference of a circle C is approximately 3.14 times the diameter d.

41. MULTIPLE REPRESENTATIONS In this problem, you will investigate the family of direct variation functions.

    a. Graphical Graph $y = x$, $y = 3x$, and $y = 5x$ on the same coordinate plane.

    b. Algebraic Describe the relationship among the constant of variation, the slope of the line, and the rate of change of the graph.

    c. Verbal Make a conjecture about how you can determine without graphing which of two direct variation equations has the steeper graph.

42. TRAVEL A map of North Carolina is scaled so that 3 inches represents 93 miles. How far apart are Raleigh and Charlotte if they are 1.8 inches apart on the map?

43. INTERNET A company will design and maintain a Web site for your company for $9.95 per month. Write a direct variation equation to find the total cost C for having a Web page for n months.

44. BASEBALL Before their first game, high school student Todd McCormick warmed all 5200 seats in a new minor league stadium, by sitting in every seat. He started at 11:50 A.M. and finished around 3 P.M.

    a. Write a direct variation equation relating the number of seats to time. What is the meaning of the constant of variation in this situation?

    b. About how many seats had Todd sat in by 1:00 P.M.?

    c. How long would you expect it to take Todd to sit in all of the seats at a major league stadium with more than 40,000 seats?

H.O.T. Problems   Use Higher-Order Thinking Skills

45. WHICH ONE DOESN'T BELONG? Identify the equation that does not belong. Explain.

$$9 = rt \qquad 9a = 0 \qquad z = \frac{1}{9}x \qquad w = \frac{9}{t}$$

46. REASONING How are the constant of variation and the slope related in a direct variation equation? Explain your reasoning.

47. OPEN ENDED Model a real-world situation using a direct variation equation. Graph the equation and describe the rate of change.

48. CCSS STRUCTURE Suppose y varies directly as x. If the value of x is doubled, then the value of y is also *always*, *sometimes* or *never* doubled. Explain your reasoning.

49. ERROR ANALYSIS Eddy says the slope between any two points on the graph of a direct variation equation $y = kx$ is $\frac{1}{k}$. Adelle says the slope depends on the points chosen. Is either of them correct? Explain.

50. WRITING IN MATH How can you identify the graph of a direct variation equation?

51. Patricia pays $1.19 each to download songs to her digital media player. If n is the number of downloaded songs, which equation represents the cost C in dollars?

A $C = 1.19n$

B $n = 1.19C$

C $C = 1.19 \div n$

D $C = n + 1.19$

52. Suppose that y varies directly as x, and $y = 8$ when $x = 6$. What is the value of y when $x = 8$?

F 6

G 12

H $10\frac{2}{3}$

J 16

53. What is the relationship between the input (x) and output (y)?

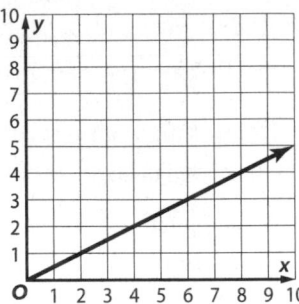

A The output is two more than the input.

B The output is two less than the input.

C The output is twice the input.

D The output is half the input.

54. **SHORT RESPONSE** A telephone company charges $40 per month plus $0.07 per minute. How much would a month of service cost a customer if the customer talked for 200 minutes?

55. **TELEVISION** The graph shows the average number of television channels American households receive. What was the annual rate of change from 2004 to 2008? Explain the meaning of the rate of change. (Lesson 3-3)

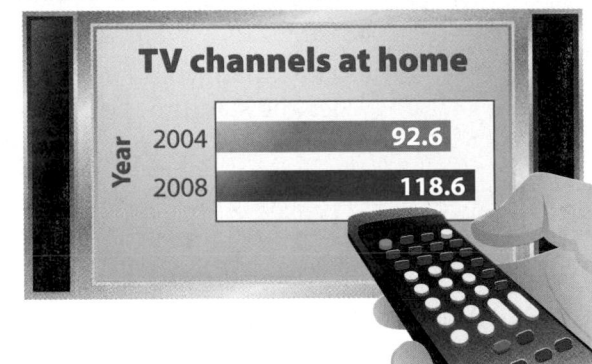

Solve each equation by graphing. (Lesson 3-2)

56. $0 = 18 - 9x$

57. $2x + 14 = 0$

58. $-4x + 16 = 0$

59. $-5x - 20 = 0$

60. $8x - 24 = 0$

61. $12x - 144 = 0$

Evaluate each expression if $a = 4$, $b = -2$, and $c = -4$. (Lesson 2-5)

62. $|2a + c| + 1$

63. $4a - |3b + 2|$

64. $-|a + 1| + |3c|$

65. $-a + |2 - a|$

66. $|c - 2b| - 3$

67. $-2|3b - 8|$

Find each difference.

68. $13 - (-1)$

69. $4 - 16$

70. $-3 - 3$

71. $-8 - (-2)$

72. $16 - (-10)$

73. $-8 - 4$

3-5 Arithmetic Sequences as Linear Functions

- You indentified linear functions.

1 Recognize arithmetic sequences.

2 Relate arithmetic sequences to linear functions.

- During a 2000-meter race, the coach of a women's crew team recorded the team's times at several intervals.
  - At 400 meters, the time was 1 minute 32 seconds.
  - At 800 meters, it was 3 minutes 4 seconds.
  - At 1200 meters, it was 4 minutes 36 seconds.
  - At 1600 meters, it was 6 minutes 8 seconds.

  They completed the race with a time of 7 minutes 40 seconds.

NewVocabulary
sequence
terms of the sequence
arithmetic sequence
common difference

Common Core State Standards

Content Standards
F.BF.2 Write arithmetic and geometric sequences both recursively and with an explicit formula, use them to model situations, and translate between the two forms.

F.LE.2 Construct linear and exponential functions, including arithmetic and geometric functions, given a graph, a description of a relationship, or two input-output pairs (include reading these from a table).

Mathematical Practices
8 Look for and express regularity in repeated reasoning.

1 **Recognize Arithmetic Sequences** You can relate the pattern of team times to linear functions. A **sequence** is a set of numbers, called the **terms of the sequence**, in a specific order. Look for a pattern in the information given for the women's crew team. Make a table to analyze the data.

Distance (m)	400	800	1200	1600	2000
Time (min : sec)	1:32	3:04	4:36	6:08	7:40

$$+ 1{:}32 \quad + 1{:}32 \quad + 1{:}32 \quad + 1{:}32$$

As the distance increases in regular intervals, the time increases by 1 minute 32 seconds. Since the difference between successive terms is constant, this is an **arithmetic sequence**. The difference between the terms is called the **common difference** d.

KeyConcept Arithmetic Sequence

Words	An arithmetic sequence is a numerical pattern that increases or decreases at a constant rate called the *common difference*.
Examples	3, 5, 7, 9, 11, ... 33, 29, 25, 21, 17, ...
	$+2 +2 +2 +2$ $-4 \ -4 \ -4 \ -4$
	$d = 2$ $d = -4$

The three dots used with sequences are called an *ellipsis*. The ellipsis indicates that there are more terms in the sequence that are not listed.

Aurora Open/Ty Milford/Getty Images

Example 1 Identify Arithmetic Sequences

Determine whether each sequence is an arithmetic sequence. Explain.

a. $-4, -2, 0, 2, \ldots$

$$-4 \quad -2 \quad 0 \quad 2$$
$$+2 \quad +2 \quad +2$$

The difference between terms in the sequence is constant. Therefore, this sequence is arithmetic.

b. $\frac{1}{2}, \frac{5}{8}, \frac{3}{4}, \frac{13}{16}, \ldots$

$$\frac{1}{2} \quad \frac{5}{8} \quad \frac{3}{4} \quad \frac{13}{16}$$
$$+\frac{1}{8} \quad +\frac{1}{8} \quad +\frac{1}{16}$$

This is not an arithmetic sequence. The difference between terms is not constant.

> **Guided**Practice

1A. $-26, -22, -18, -14, \ldots$

1B. $1, 4, 9, 25, \ldots$

You can use the common difference of an arithmetic sequence to find the next term.

Example 2 Find the Next Term

Find the next three terms of the arithmetic sequence $15, 9, 3, -3, \ldots$.

Step 1 Find the common difference by subtracting successive terms.

$$15 \quad 9 \quad 3 \quad -3$$
$$-6 \quad -6 \quad -6$$

The common difference is -6.

Step 2 Add -6 to the last term of the sequence to get the next term.

$$-3 \quad -9 \quad -15 \quad -21$$
$$-6 \quad -6 \quad -6$$

The next three terms in the sequence are $-9, -15,$ and -21.

> **Guided**Practice

2. Find the next four terms of the arithmetic sequence $9.5, 11.0, 12.5, 14.0, \ldots$.

Each term in an arithmetic sequence can be expressed in terms of the first term a_1 and the common difference d.

Term	Symbol	In Terms of a_1 and d	Numbers
first term	a_1	a_1	8
second term	a_2	$a_1 + d$	$8 + 1(3) = 11$
third term	a_3	$a_1 + 2d$	$8 + 2(3) = 14$
fourth term	a_4	$a_1 + 3d$	$8 + 3(3) = 17$
$\vdots$	$\vdots$	$\vdots$	$\vdots$
nth term	a_n	$a_1 + (n-1)d$	$8 + (n-1)(3)$

KeyConcept nth Term of an Arithmetic Sequence

The nth term of an arithmetic sequence with first term a_1 and common difference d is given by $a_n = a_1 + (n-1)d$, where n is a positive integer.

Example 3 Find the *n*th Term

a. Write an equation for the *n*th term of the arithmetic sequence −12, −8, −4, 0, … .

Step 1 Find the common difference.

$$-12 \quad -8 \quad -4 \quad 0$$
$$+4 \quad +4 \quad +4$$

The common difference is 4.

Step 2 Write an equation.

$$a_n = a_1 + (n - 1)d \qquad \text{Formula for the } n\text{th term}$$
$$= -12 + (n - 1)4 \qquad a_1 = -12 \text{ and } d = 4$$
$$= -12 + 4n - 4 \qquad \text{Distributive Property}$$
$$= 4n - 16 \qquad \text{Simplify.}$$

> **StudyTip**
>
> nth Terms Since *n* represents the number of the term, the inputs for *n* are the counting numbers.

b. Find the 9th term of the sequence.

Substitute 9 for *n* in the formula for the *n*th term.

$$a_n = 4n - 16 \qquad \text{Formula for the } n\text{th term}$$
$$a_9 = 4(9) - 16 \qquad n = 9$$
$$a_9 = 36 - 16 \qquad \text{Multiply.}$$
$$a_9 = 20 \qquad \text{Simplify.}$$

c. Graph the first five terms of the sequence.

n	4*n* − 16	a_n	(*n*, a_n)
1	4(1) − 16	−12	(1, −12)
2	4(2) − 16	−8	(2, −8)
3	4(3) − 16	−4	(3, −4)
4	4(4) − 16	0	(4, 0)
5	4(5) − 16	4	(5, 4)

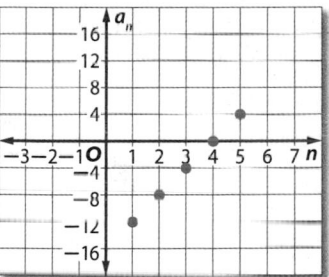

d. Which term of the sequence is 32?

In the formula for the *n*th term, substitute 32 for a_n.

$$a_n = 4n - 16 \qquad \text{Formula for the } n\text{th term}$$
$$32 = 4n - 16 \qquad a_n = 32$$
$$32 + 16 = 4n - 16 + 16 \qquad \text{Add 16 to each side.}$$
$$48 = 4n \qquad \text{Simplify.}$$
$$12 = n \qquad \text{Divide each side by 4.}$$

▶ **Guided**Practice

Consider the arithmetic sequence 3, −10, −23, −36, … .

3A. Write an equation for the *n*th term of the sequence.

3B. Find the 15th term in the sequence.

3C. Graph the first five terms of the sequence.

3D. Which term of the sequence is −114?

2 Arithmetic Sequences and Functions
As you can see from Example 3, the graph of the first five terms of the arithmetic sequence lie on a line. An arithmetic sequence is a linear function in which n is the independent variable, a_n is the dependent variable, and d is the slope. The formula can be rewritten as the function $f(n) = (n - 1)d + a_1$, where n is a counting number.

While the domain of most linear functions are all real numbers, in Example 3 the domain of the function is the set of counting numbers and the range of the function is the set of integers on the line.

Real-World Example 4 Arithmetic Sequences as Functions

INVITATIONS Marisol is mailing invitations to her quinceañera. The arithmetic sequence $0.42, $0.84, $1.26, $1.68, ... represents the cost of postage.

a. Write a function to represent this sequence.

The first term, a_1, is 0.42. Find the common difference.

$$0.42 \quad 0.84 \quad 1.26 \quad 1.68$$
$$+0.42 \quad +0.42 \quad +0.42$$

The common difference is 0.42.

$a_n = a_1 + (n - 1)d$	Formula for the nth term
$= 0.42 + (n - 1)0.42$	$a_1 = 0.42$ and $d = 0.42$
$= 0.42 + 0.42n - 0.42$	Distributive Property
$= 0.42n$	Simplify.

The function is $f(n) = 0.42n$.

b. Graph the function and determine the domain.

The rate of change of the function is 0.42. Make a table and plot points.

n	$f(n)$
1	0.42
2	0.84
3	1.26
4	1.68
5	2.10

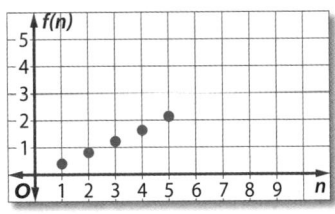

The domain of a function is the number of invitations Marisol mails. So, the domain is {1, 2, 3, 4, ...}.

GuidedPractice

4. TRACK The chart below shows the length of Martin's long jumps.

Jump	1	2	3	4
Length (ft)	8	9.5	11	12.5

A. Write a function to represent this arithmetic sequence.

B. Then graph the function.

Real-WorldLink

When a Latina turns 15, her family may host a quinceañera for her birthday. The quinceañera is a traditional Hispanic ceremony and reception that signifies the transition from childhood to adulthood.

Source: Quince Girl

Check Your Understanding   ◯ = Step-by-Step Solutions begin on page R13.

Example 1   Determine whether each sequence is an arithmetic sequence. Write *yes* or *no*. Explain.

1. 18, 16, 15, 13, …   **2.** 4, 9, 14, 19, …

Example 2   Find the next three terms of each arithmetic sequence.

3. 12, 9, 6, 3, …   **4.** −2, 2, 6, 10, …

Example 3   Write an equation for the *n*th term of each arithmetic sequence. Then graph the first five terms of the sequence.

5. 15, 13, 11, 9, …   **6.** −1, −0.5, 0, 0.5, …

Example 4   **7. SAVINGS** Kaia has $525 in a savings account. After one month she has $580 in the account. The next month the balance is $635. The balance after the third month is $690. Write a function to represent the arithmetic sequence. Then graph the function.

Practice and Problem Solving   Extra Practice is on page R3.

Example 1   Determine whether each sequence is an arithmetic sequence. Write *yes* or *no*. Explain.

8. −3, 1, 5, 9, …   **9.** $\frac{1}{2}, \frac{3}{4}, \frac{5}{8}, \frac{7}{16}, \dots$

10. −10, −7, −4, 1, …   **11.** −12.3, −9.7, −7.1, −4.5, …

Example 2   Find the next three terms of each arithmetic sequence.

12. 0.02, 1.08, 2.14, 3.2, …   **13.** 6, 12, 18, 24, …

14. 21, 19, 17, 15, …   **(15)** $-\frac{1}{2}, 0, \frac{1}{2}, 1, \dots$

16. $2\frac{1}{3}, 2\frac{2}{3}, 3, 3\frac{1}{3}, \dots$ **17.** $\frac{7}{12}, 1\frac{1}{3}, 2\frac{1}{12}, 2\frac{5}{6}, \dots$

Example 3   Write an equation for the *n*th term of the arithmetic sequence. Then graph the first five terms in the sequence.

18. −3, −8, −13, −18, …   **19.** −2, 3, 8, 13, …

20. −11, −15, −19, −23, …   **21.** −0.75, −0.5, −0.25, 0, …

Example 4   **22. AMUSEMENT PARKS** Shiloh and her friends spent the day at an amusement park. In the first hour, they rode two rides. After 2 hours, they had ridden 4 rides. They had ridden 6 rides after 3 hours.

   a. Write a function to represent the arithmetic sequence.

   b. Graph the function and determine the domain.

23. CCSS MODELING The table shows how Ryan is paid at his lumber yard job.

Number of 10-ft 2×4 Planks Cut	1	2	3	4	5	6	7
Amount Paid in Commission ($)	8	16	24	32	40	48	56

   a. Write a function to represent Ryan's commission.

   b. Graph the function and determine the domain.

24. The graph is a representation of an arithmetic sequence.

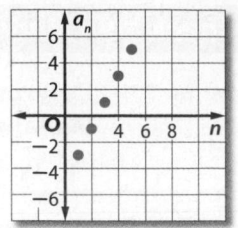

    a. List the first five terms.

    b. Write the formula for the nth term.

    c. Write the function.

25 **NEWSPAPERS** A local newspaper charges by the number of words for advertising. Write a function to represent the advertising costs.

DAILY NEWS ADVERTISING

10 words $7.50	20 words $10.00
15 words $8.75	25 words $11.25

26. The fourth term of an arithmetic sequence is 8. If the common difference is 2, what is the first term?

27. The common difference of an arithmetic sequence is -5. If a_{12} is 22, what is a_1?

28. The first four terms of an arithmetic sequence are 28, 20, 12, and 4. Which term of the sequence is -36?

29. **CARS** Jamal's odometer of his car reads 24,521. If Jamal drives 45 miles every day, what will the odometer reading be after 25 days?

30. **YEARBOOKS** The yearbook staff is unpacking a box of school yearbooks. The arithmetic sequence 281, 270, 259, 248 … represents the total number of ounces that the box weighs as each yearbook is taken out of the box.

    a. Write a function to represent this sequence.

    b. Determine the weight of each yearbook.

    c. If the box weighs at least 17 ounces empty and 292 ounces when it is full, how many yearbooks were in the box?

31. **SPORTS** To train for an upcoming marathon, Olivia plans to run 3 miles per day for the first week and then increase the daily distance by a half mile each of the following weeks.

    a. Write an equation to represent the nth term of the sequence.

    b. If the pattern continues, during which week will she run 10 miles per day?

    c. Is it reasonable to think that this pattern will continue indefinitely? Explain.

H.O.T. Problems   Use Higher-Order Thinking Skills

32. **OPEN ENDED** Create an arithmetic sequence with a common difference of -10.

33. **CCSS PERSEVERANCE** Find the value of x that makes $x + 8$, $4x + 6$, and $3x$ the first three terms of an arithmetic sequence.

34. **REASONING** Compare and contrast the domain and range of the linear functions described by $Ax + By = C$ and $a_n = a_1 + (n - 1)d$.

35. **CHALLENGE** Determine whether each sequence is an arithmetic sequence. Write *yes* or *no*. Explain. If yes, find the common difference and the next three terms.

    a. $2x + 1, 3x + 1, 4x + 1…$ **b.** $2x, 4x, 8x, …$

36. **WRITING IN MATH** How are graphs of arithmetic sequences and linear functions similar? different?

37. GRIDDED RESPONSE The population of Westerville is about 35,000. Each year the population increases by about 400. This can be represented by the following equation, where n represents the number of years from now and p represents the population.

$$p = 35,000 + 400n$$

In how many years will the Westerville population be about 38,200?

38. Which relation is a function?

A $\{(-5, 6), (4, -3), (2, -1), (4, 2)\}$

B $\{(3, -1), (3, -5), (3, 4), (3, 6)\}$

C $\{(-2, 3), (0, 3), (-2, -1), (-1, 2)\}$

D $\{(-5, 6), (4, -3), (2, -1), (0, 2)\}$

39. Find the formula for the nth term of the arithmetic sequence.

$$-7, -4, -1, 2, \ldots$$

F $a_n = 3n - 4$

G $a_n = -7n + 10$

H $a_n = 3n - 10$

J $a_n = -7n + 4$

40. STATISTICS A class received the following scores on the ACT. What is the difference between the median and the mode in the scores?

18, 26, 20, 30, 25, 21, 32, 19, 22, 29, 29, 27, 24

A 1   C 3

B 2   D 4

Spiral Review

Name the constant of variation for each direct variation. Then find the slope of the line that passes through each pair of points. (Lesson 3-4)

41.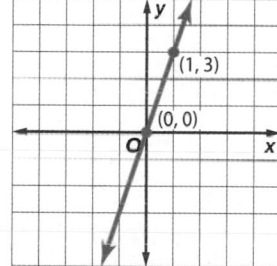

42.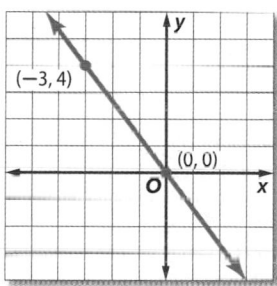

Find the slope of the line that passes through each pair of points. (Lesson 3-3)

43. $(5, 3), (-2, 6)$ **44.** $(9, 2), (-3, -1)$ **45.** $(2, 8), (-2, -4)$

Solve each equation. Check your solution. (Lesson 2-3)

46. $5x + 7 = -8$ **47.** $8 = 2 + 3n$ **48.** $12 = \dfrac{c - 6}{2}$

49. SPORTS The most popular sports for high school girls are basketball and softball. Write and use an equation to find how many more girls play on basketball teams than on softball teams. (Lesson 2-1)

Basketball
453,000 girls

Softball
369,000 girls

Skills Review

Graph each point on the same coordinate plane.

50. $A(2, 5)$ **51.** $B(-2, 1)$ **52.** $C(-3, -1)$

53. $D(0, 4)$ **54.** $F(5, -3)$ **55.** $G(-5, 0)$

Algebra Lab
Inductive and Deductive Reasoning

If Jolene is not feeling well, she may go to a doctor. The doctor will ask her questions about how she is feeling and possibly run other tests. Based on her symptoms, the doctor can diagnose Jolene's illness. This is an example of inductive reasoning. **Inductive reasoning** is used to derive a general rule after observing many events.

CCSS Common Core State Standards
Mathematical Practices
3 Construct viable arguments and critique the reasoning of others.

To use inductive reasoning:

Step 1 Observe many examples.

Step 2 Look for a pattern.

Step 3 Make a conjecture.

Step 4 Check the conjecture.

Step 5 Discover a likely conclusion.

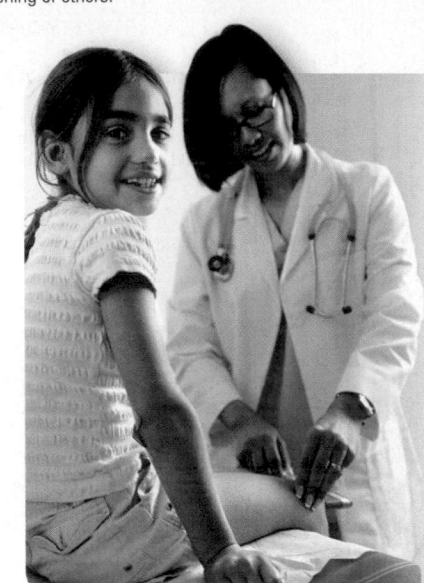

With **deductive reasoning**, you come to a conclusion by accepting facts. The results of the tests ordered by the doctor may support the original diagnosis or lead to a different conclusion. This is an example of deductive reasoning. There is no conjecturing involved. Consider the two statements below.

1) If the strep test is positive, then the patient has strep throat.

2) Jolene tested positive for strep.

If these two statements are accepted as facts, then the obvious conclusion is that Jolene has strep throat. This is an example of deductive reasoning.

Exercises

1. Explain the difference between *inductive* and *deductive* reasoning. Then give an example of each.

2. When a detective reaches a conclusion about the height of a suspect from the distance between footprints, what kind of reasoning is being used? Explain.

3. When you examine a finite number of terms in a sequence of numbers and decide that it is an arithmetic sequence, what kind of reasoning are you using? Explain.

4. Suppose you have found the common difference for an arithmetic sequence based on analyzing a finite number of terms, what kind of reasoning do you use to find the 100th term in the sequence?

5. **CCSS PERSEVERANCE**
   a. Copy and complete the table.

3^1	3^2	3^3	3^4	3^5	3^6	3^7	3^8	3^9
3	9	27						

   b. Write the sequence of numbers representing the numbers in the ones place.

   c. Find the number in the ones place for the value of 3^{100}. Explain your reasoning. State the type of reasoning that you used.

3-6 Proportional and Nonproportional Relationships

Then	**Now**	**Why?**

Then
- You recognized arithmetic sequences and related them to linear functions.

Now
1. Write an equation for a proportional relationship.
2. Write an equation for a nonproportional relationship.

Why?
Heather is planting flats of flowers. The table shows the number of flowers that she has planted and the amount of time that she has been working in the garden.

Number of flowers planted (*p*)	1	6	12	18
Number of minutes working (*f*)	5	30	60	90

The relationship between the flowers planted and the time that Heather worked in minutes can be graphed. Let *p* represent the number of flowers planted. Let *t* represent the number of minutes that Heather has worked.

When the ordered pairs are graphed, they form a linear pattern. This pattern can be described by an equation.

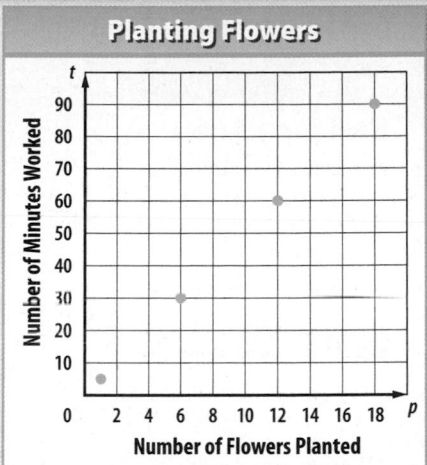

Planting Flowers

Number of Minutes Worked (vertical axis)
Number of Flowers Planted (horizontal axis)

Common Core State Standards

Content Standards
F.LE.1b Recognize situations in which one quantity changes at a constant rate per unit interval relative to another.

F.LE.2 Construct linear and exponential functions, including arithmetic and geometric sequences, given a graph, a description of a relationship, or two input-output pairs (include reading these from a table).

Mathematical Practices
1 Make sense of problems and persevere in solving them.

7 Look for and make use of structure.

1 Proportional Relationships If the relationship between the domain and range of a relation is linear, the relationship can be described by a linear equation. If the equation is of the form $y = kx$, then the relationship is proportional. In a proportional relationship, the graph will pass through (0, 0). So, direct variations are proportional relationships.

KeyConcept Proportional Relationship

Words A relationship is proportional if its equation is of the form $y = kx$, $k \neq 0$. The graph passes through (0, 0).

Example $y = 3x$

x	0	1	2	3	4
y	0	3	6	9	12

The ratio of the value of *x* to the value of *y* is constant when $x \neq 0$.

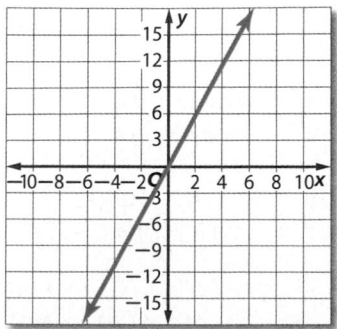

Proportional relationships are useful when analyzing real-world data. The pattern can be described using a table, a graph, and an equation.

Real-World Example 1 Proportional Relationships

BONUS PAY Marcos is a personal trainer at a gym. In addition to his salary, he receives a bonus for each client he sees.

Number of Clients	1	2	3	4	5
Bonus Pay ($)	45	90	135	180	225

a. Graph the data. What can you deduce from the pattern about the relationship between the number of clients and the bonus pay?

The graph demonstrates a linear relationship between the number of clients and the bonus pay.

The graph also passes through the point (0, 0) because when Marcos sees 0 clients, he does not receive any bonus money. Therefore, the relationship is proportional.

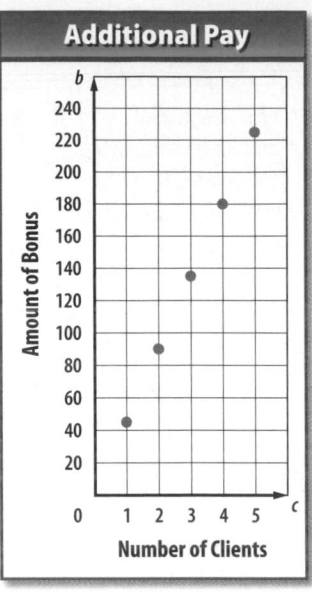

Additional Pay

b. Write an equation to describe this relationship.

Look for a pattern that can be described in an equation.

$+1 \quad +1 \quad +1 \quad +1$

Number of Clients	1	2	3	4	5
Bonus Pay ($)	45	90	135	180	225

$+45 \quad +45 \quad +45 \quad +45$

The difference between the values for the number of clients c is 1. The difference in the values for the bonus pay b is 45. This suggests that the k-value is $\frac{45}{1}$ or 45. So the equation is $b = 45c$. You can check this equation by substituting values for c into the equation.

CHECK If $c = 1$, then $b = 45(1)$ or 45. ✓
          If $c = 5$, then $b = 45(5)$ or 225. ✓

c. Use this equation to predict the amount of Marcos's bonus if he sees 8 clients.

$b = 45c$ Original equation
$ = 45(8)$ or 360 $c = 8$

Marcos will receive a bonus of $360 if he sees 8 clients.

GuidedPractice

1. CHARITY A professional soccer team is donating money to a local charity for each goal they score.

Number of Goals	1	2	3	4	5
Donation ($)	75	150	225	300	375

A. Graph the data. What can you deduce from the pattern about the relationship between the number of goals and the money donated?

B. Write an equation to describe this relationship.

C. Use this equation to predict how much money will be donated for 12 goals.

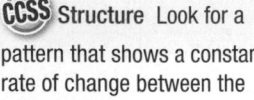

Real-World Link

Attendance at fitness clubs has steadily grown over the past fifteen years. Members' ages are expanding to a range of 15–34 on average.

Source: International Health, Raquet, and Sportsclub Association

StudyTip

CCSS Structure Look for a pattern that shows a constant rate of change between the terms.

2 **Nonproportional Relationships** Some linear equations can represent a nonproportional relationship. If the ratio of the value of x to the value of y is different for select ordered pairs that are on the line, the equation is nonproportional and the graph will not pass through $(0, 0)$.

Example 2 Nonproportional Relationships

Write an equation in function notation for the graph.

Understand You are asked to write an equation of the relation that is graphed in function notation.

Plan Find the difference between the x-values and the difference between the y-values.

Solve Select points from the graph and place them in a table.

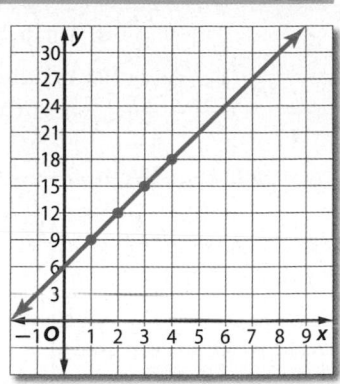

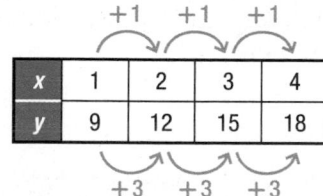

Notice that

$$\frac{1}{9} \neq \frac{2}{12} \neq \frac{3}{15} \neq \frac{4}{18}.$$

The difference between the x-values is 1, while the difference between the y-values is 3. This suggests that $y = 3x$ or $f(x) = 3x$.

If $x = 1$, then $y = 3(1)$ or 3. But the y-value for $x = 1$ is 9. Let's try some other values and see if we can detect a pattern.

x	1	2	3	4
3x	3	6	9	12
y	9	12	15	18

y is always 6 more than 3x.

This pattern shows that 6 should be added to one side of the equation. Thus, the equation is $y = 3x + 6$ or $f(x) = 3x + 6$.

Check Compare the ordered pairs from the table to the graph. The points correspond. ✓

> **StudyTip**
>
> **Graphs of Lines** A value added to or subtracted from one side of the equation $y = ax$ will cause a shift along the y-axis for the graph of the line.

▶ **Guided**Practice

2. Write an equation in function notation for the relation shown in the table.

A.

x	1	2	3	4
y	3	2	1	0

B. Write an equation in function notation for the graph.

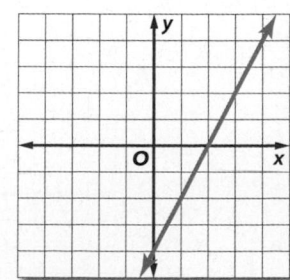

Example 1   **1. GEOMETRY** The table shows the perimeter of a square with sides of a given length.

Side Length (in.)	1	2	3	4	5
Perimeter (in.)	4	8	12	16	20

   a. Graph the data.

   b. Write an equation to describe the relationship.

   c. What conclusion can you make regarding the relationship between the side and the perimeter?

Example 2   Write an equation in function notation for each relation.

2. 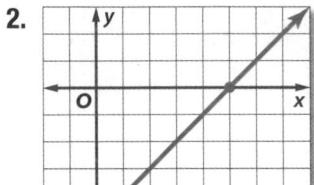   **3.**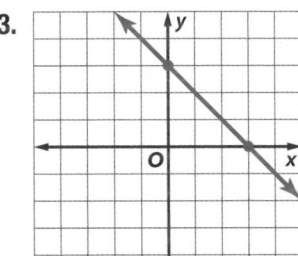

Example 1   **4.** **CCSS** **STRUCTURE** The table shows the pages of comic books read.

Books Read	1	2	3	4	5
Pages Read	35	70	105	140	175

   a. Graph the data.

   b. Write an equation to describe the relationship.

   c. Find the number of pages read if 8 comic books were read.

Example 2   Write an equation in function notation for each relation.

5 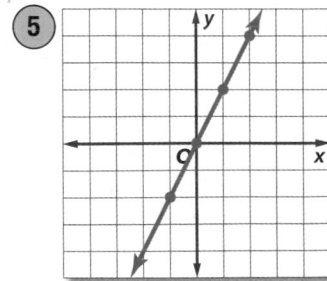   **6.**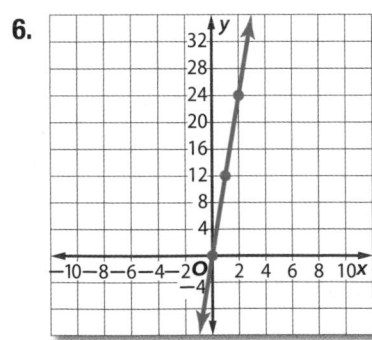

7. 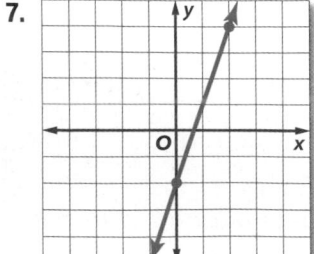   **8.**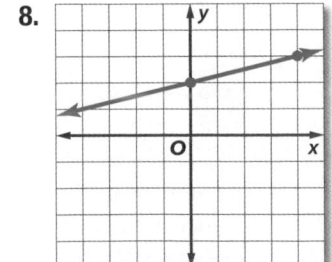

For each arithmetic sequence, determine the related function. Then determine if the function is *proportional* or *nonproportional*. Explain.

9. 0, 3, 6, …

10. −4, 0, 4, …

11. BOWLING Marielle is bowling with her friends. The table shows prices for renting a pair of shoes and bowling. Write an equation to represent the total price y if Marielle buys x games.

Games Bowled	Total Price ($)
2	7.00
4	11.50
6	16.00
8	20.50

12. SNOWFALL The total snowfall each hour of a winter snowstorm is shown in the table below.

Hour	1	2	3	4
Inches of Snowfall	1.65	3.30	4.95	6.60

 a. Write an equation to fit the data in the table.

 b. Describe the relationship between the hour and inches of snowfall.

13 FUNDRAISER The Cougar Pep Squad wants to sell T-shirts in the bookstore for the spring dance. The cost in dollars to order T-shirts in their school colors is represented by the equation $C = 2t + 3$.

 a. Make a table of values that represents this relationship.

 b. Rewrite the equation in function notation.

 c. Graph the function.

 d. Describe the relationship between the number of T-shirts and the cost.

H.O.T. Problems   Use Higher-Order Thinking Skills

14. CCSS CRITIQUE Quentin thinks that $f(x)$ and $g(x)$ are both proportional. Claudia thinks they are not proportional. Is either of them correct? Explain your reasoning.

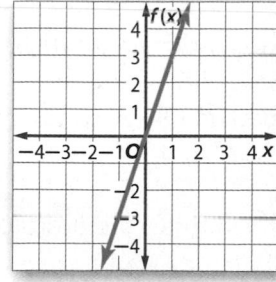

x	g(x)
−2	−7
−1	−4
0	−1
1	2
2	5

15. OPEN ENDED Create an arithmetic sequence in which the first term is 4. Explain the pattern that you used. Write an equation that represents your sequence.

16. CHALLENGE Describe how inductive reasoning can be used to write an equation from a pattern.

17. REASONING A **counterexample** is a specific case that shows that a statement is false. Provide a counterexample to the following statement. *The related function of an arithmetic sequence is always proportional.* Explain your reasoning.

18. WRITING IN MATH Compare and contrast proportional relationships with nonproportional relationships.

19. What is the slope of a line that contains the point $(1, -5)$ and has the same y-intercept as $2x - y = 9$?

A -9	**C** 2
B -7	**D** 4

20. SHORT RESPONSE $\triangle FGR$ is an isosceles triangle. What is the measure of $\angle G$?

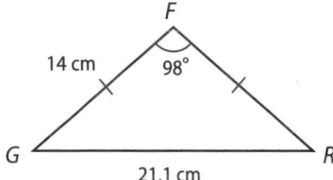

21. Luis deposits $25 each week into a savings account from his part-time job. If he has $350 in savings now, how much will he have in 12 weeks?

F $600	**H** $650
G $625	**J** $675

22. GEOMETRY Omar and Mackenzie want to build a pulley system by attaching one end of a rope to their 8-foot-tall tree house and anchoring the other end to the ground 28 feet away from the base of the tree house. How long, to the nearest foot, does the piece of rope need to be?

A 26 ft	**C** 28 ft
B 27 ft	**D** 29 ft

Spiral Review

Find the next three terms in each sequence. (Lesson 3-5)

23. $3, 13, 23, 33, \ldots$

24. $-2, -1.4, -0.8, -0.2, \ldots$

25. $\frac{3}{4}, \frac{7}{8}, 1, \frac{9}{8}, \ldots$

Suppose y varies directly as x. Write a direct variation equation that relates x and y. Then solve. (Lesson 3-4)

26. If $y = 45$ when $x = 9$, find y when $x = 7$.

27. If $y = -7$ when $x = -1$, find x when $y = -84$.

28. GENETICS About $\frac{2}{25}$ of the male population in the world cannot distinguish red from green. If there are 14 boys in the ninth grade who cannot distinguish red from green, about how many ninth-grade boys are there in all? Write and solve an equation to find the answer. (Lesson 2-2)

29. GEOMETRY The volume V of a cone equals one third times the product of π, the square of the radius r of the base, and the height h. (Lesson 2-1)

a. Write the formula for the volume of a cone.

b. Find the volume of a cone if r is 10 centimeters and h is 30 centimeters.

Skills Review

Solve each equation for y.

30. $3x = y + 7$

31. $2y = 6x - 10$

32. $9y + 2x = 12$

Graph each equation.

33. $y = x - 8$

34. $x - y = -4$

35. $2x + 4y = 8$

Study Guide

KeyConcepts

Graphing Linear Equations (Lesson 3-1)

- The standard form of a linear equation is $Ax + By = C$, where $A \geq 0$, A and B are not both zero, and A, B, and C are integers whose greatest common factor is 1.

Solving Linear Equations by Graphing (Lesson 3-2)

- Values of x for which $f(x) = 0$ are called zeros of the function f. A zero of a function is located at an x-intercept of the graph of the function.

Rate of Change and Slope (Lesson 3-3)

- If x is the independent variable and y is the dependent variable, then rate of change equals

$$\frac{\text{change in } y}{\text{change in } x}.$$

- The slope of a line is the ratio of the rise to the run.

$$m = \frac{y_2 - y_1}{x_2 - x_1}$$

Direct Variation (Lesson 3-4)

- A direct variation is described by an equation of the form $y = kx$, where $k \neq 0$.

Arithmetic Sequences (Lesson 3-5)

- The nth term a_n of an arithmetic sequence with first term a_1 and common difference d is given by $a_n = a_1 + (n - 1)d$, where n is a positive integer.

Proportional and Nonproportional Relationships (Lesson 3-6)

- In a proportional relationship, the graph will pass through (0, 0).
- In a nonproportional relationship, the graph will not pass through (0, 0)

FOLDABLES StudyOrganizer

Be sure the Key Concepts are noted in your Foldable.

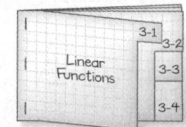

KeyVocabulary

arithmetic sequence (p. 189)	rate of change (p. 172)
common difference (p. 189)	root (p. 163)
constant (p. 155)	sequence (p. 189)
constant of variation (p. 182)	slope (p. 174)
deductive reasoning (p. 196)	standard form (p. 155)
direct variation (p. 182)	terms of the sequence (p. 189)
inductive reasoning (p. 196)	x-intercept (p. 156)
linear equation (p. 155)	y-intercept (p. 156)
linear function (p. 163)	zero of a function (p. 163)

VocabularyCheck

State whether each sentence is *true* or *false*. If *false*, replace the underlined word or number to make a true sentence.

1. The x-coordinate of the point at which the graph of an equation crosses the x-axis is an <u>x-intercept</u>.

2. A <u>linear equation</u> is an equation of a line.

3. The difference between successive terms of an arithmetic sequence is the <u>constant of variation</u>.

4. The <u>regular form</u> of a linear equation is $Ax + By = C$.

5. Values of x for which $f(x) = 0$ are called <u>zeros</u> of the function f.

6. Any two points on a nonvertical line can be used to determine the <u>slope</u>.

7. The slope of the line $y = 5$ is <u>5</u>.

8. The graph of any direct variation equation passes through <u>(0, 1)</u>.

9. A ratio that describes, on average, how much one quantity changes with respect to a change in another quantity is a <u>rate of change</u>.

10. In the linear equation $4x + 3y = 12$, the constant term is <u>12</u>.

Lesson-by-Lesson Review

3-1 Graphing Linear Equations

Find the *x*-intercept and *y*-intercept of the graph of each linear function.

11.

x	y
−8	0
−4	3
0	6
4	9
8	12

12.

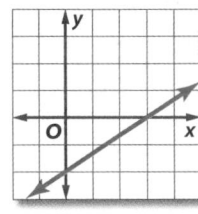

Graph each equation.

13. $y = -x + 2$

14. $x + 5y = 4$

15. $2x - 3y = 6$

16. $5x + 2y = 10$

17. **SOUND** The distance *d* in kilometers that sound waves travel through water is given by $d = 1.6t$, where *t* is the time in seconds.

   a. Make a table of values and graph the equation.

   b. Use the graph to estimate how far sound can travel through water in 7 seconds.

Example 1

Graph $3x - y = 4$ by using the *x*- and *y*-intercepts.

Find the *x*-intercept.

$3x - y = 4$

$3x - 0 = 4$ Let $y = 0$.

$3x = 4$

$x = \dfrac{4}{3}$

Find the *y*-intercept.

$3x - y = 4$

$3(0) - y = 4$ Let $x = 0$.

$-y = 4$

$y = -4$

x-intercept: $\dfrac{4}{3}$

y-intercept: -4

The graph intersects the *x*-axis at $\left(\dfrac{4}{3}, 0\right)$ and the *y*-axis at $(0, -4)$. Plot these points. Then draw the line through them.

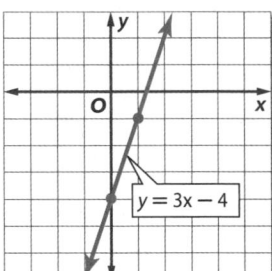

3-2 Solving Linear Equations by Graphing

Find the root of each equation.

18. $0 = 2x + 8$

19. $0 = 4x - 24$

20. $3x - 5 = 0$

21. $6x + 3 = 0$

Solve each equation by graphing.

22. $0 = 16 - 8x$

23. $0 = 21 + 3x$

24. $-4x - 28 = 0$

25. $25x - 225 = 0$

26. **FUNDRAISING** Sean's class is selling boxes of popcorn to raise money for a class trip. Sean's class paid $85 for the popcorn, and they are selling each box for $1. The function $y = x - 85$ represents their profit *y* for each box of popcorn sold *x*. Find the zero and describe what it means in this situation.

Example 2

Solve $3x + 1 = -2$ by graphing.

The first step is to find the related function.

$3x + 1 = -2$ Original equation

$3x + 1 + 2 = -2 + 2$ Add 2 to each side.

$3x + 3 = 0$ Simplify.

The related function is $y = 3x + 3$.

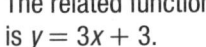

The graph intersects the *x*-axis at -1. So, the solution is -1.

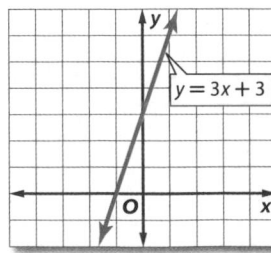

3-3 Rate of Change and Slope

Find the rate of change represented in each table or graph.

27.

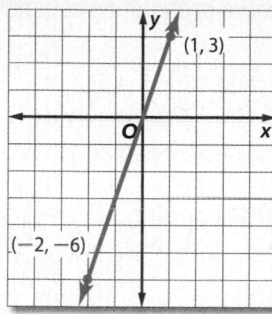

28.

x	y
−2	−3
0	−3
4	−3
12	−3

Find the slope of the line that passes through each pair of points.

29. $(0, 5)$, $(6, 2)$

30. $(−6, 4)$, $(−6, −2)$

31. PHOTOS The average cost of online photos decreased from $0.50 per print to $0.15 per print between 2002 and 2009. Find the average rate of change in the cost. Explain what it means.

Example 3

Find the slope of the line that passes through $(0, −4)$ and $(3, 2)$.

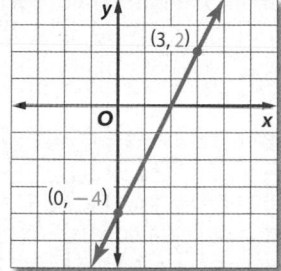

Let $(0, −4) = (x_1, y_1)$ and $(3, 2) = (x_2, y_2)$.

$m = \dfrac{y_2 - y_1}{x_2 - x_1}$ Slope formula

$= \dfrac{2 - (-4)}{3 - 0}$ $x_1 = 0, x_2 = 3, y_1 = -4, y_2 = 2$

$= \dfrac{6}{3}$ or 2   Simplify.

3-4 Direct Variation

Graph each equation.

32. $y = x$ **33.** $y = \dfrac{4}{3}x$ **34.** $y = −2x$

Suppose y varies directly as x. Write a direct variation equation that relates x and y. Then solve.

35. If $y = 15$ when $x = 2$, find y when $x = 8$.

36. If $y = −6$ when $x = 9$, find x when $y = −3$.

37. If $y = 4$ when $x = −4$, find y when $x = 7$.

38. JOBS Suppose you earn $127 for working 20 hours.

   a. Write a direct variation equation relating your earnings to the number of hours worked.

   b. How much would you earn for working 35 hours?

Example 4

Suppose y varies directly as x, and $y = −24$ when $x = 8$.

   a. Write a direct variation equation that relates x and y.

     $y = kx$ Direct variation equation

     $−24 = k(8)$ Substitute −24 for y and 8 for x.

     $\dfrac{-24}{8} = \dfrac{k(8)}{8}$ Divide each side by 8.

     $−3 = k$ Simplify.

   So, the direct variation equation is $y = −3x$.

   b. Use the direct variation equation to find x when $y = −18$.

     $y = −3x$ Direct variation equation

     $−18 = −3x$ Replace y with −18.

     $\dfrac{-18}{-3} = \dfrac{-3x}{-3}$ Divide each side by −3.

     $6 = x$ Simplify.

   Therefore, $x = 6$ when $y = −18$.

3-5 Arithmetic Sequences as Linear Functions

Find the next three terms of each arithmetic sequence.

39. 6, 11, 16, 21, ...   **40.** 1.4, 1.2, 1.0, ...

Write an equation for the *n*th term of each arithmetic sequence.

41. $a_1 = 6$, $d = 5$

42. 28, 25, 22, 19, ...

43. SCIENCE The table shows the distance traveled by sound in water. Write an equation for this sequence. Then find the time for sound to travel 72,300 feet.

Time (s)	1	2	3	4
Distance ft)	4820	9640	14,460	19,280

Example 5

Find the next three terms of the arithmetic sequence 10, 23, 36, 49, ...

Find the common difference.

10   23   36   49

+13 +13 +13

So, $d = 13$.

Add 13 to the last term of the sequence. Continue adding 13 until the next three terms are found.

49   62   75   88

+13 +13 +13

The next three terms are 62, 75, and 88.

3-6 Proportional and Nonproportional Relationships

44. Write an equation in function notation for this relation.

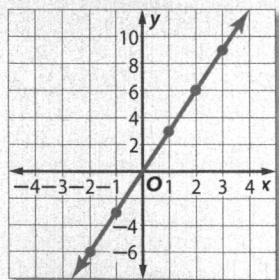

45. ANALYZE TABLES The table shows the cost of picking your own strawberries at a farm.

Number of Pounds	1	2	3	4
Total Cost ($)	1.25	2.50	3.75	5.00

  a. Graph the data.

  b. Write an equation in function notation to describe this relationship.

  c. How much would it cost to pick 6 pounds of strawberries?

Example 6

Write an equation in function notation for this relation.

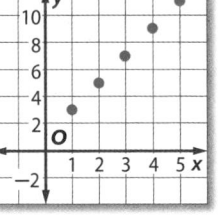

Make a table of ordered pairs for several points on the graph.

x	1	2	3	4	5
y	3	5	7	9	11

The difference in *y*-values is twice the difference of *x* values. This suggests that $y = 2x$. However, $3 \neq 2(1)$. Compare the values of *y* to the values of $2x$.

x	1	2	3	4	5
2x	2	4	6	8	10
y	3	5	7	9	11

The difference between *y* and $2x$ is always 1. So the equation is $y = 2x + 1$. Since this relation is also a function, it can be written as $f(x) = 2x + 1$.

1. **TEMPERATURE** The equation to convert Celsius temperature C to Kelvin temperature K is shown.

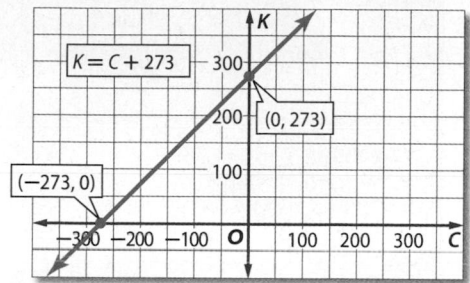

   a. State the independent and dependent variables. Explain.

   b. Determine the C- and K-intercepts and describe what the intercepts mean in this situation.

Graph each equation.

2. $y = x + 2$

3. $y = 4x$

4. $x + 2y = -1$

5. $-3x = 5 - y$

Solve each equation by graphing.

6. $4x + 2 = 0$

7. $0 = 6 - 3x$

8. $5x + 2 = 3$

9. $12x = 4x + 16$

Find the slope of the line that passes through each pair of points.

10. $(5, 8), (-3, 7)$

11. $(5, -2), (3, -2)$

12. $(-4, 7), (8, -1)$

13. $(6, -3), (6, 4)$

14. **MULTIPLE CHOICE** Which is the slope of the linear function shown in the graph?

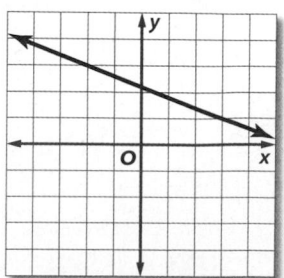

   A $-\dfrac{5}{2}$

   B $-\dfrac{2}{5}$

   C $\dfrac{2}{5}$

   D $\dfrac{5}{2}$

Suppose y varies directly as x. Write a direct variation equation that relates x and y. Then solve.

15. If $y = 6$ when $x = 9$, find x when $y = 12$.

16. When $y = -8$, $x = 8$. What is x when $y = -6$?

17. If $y = -5$ when $x = -2$, what is y when $x = 14$?

18. If $y = 2$ when $x = -12$, find y when $x = -4$.

19. **BIOLOGY** The number of pints of blood in a human body varies directly with the person's weight. A person who weighs 120 pounds has about 8.4 pints of blood in his or her body.

   a. Write and graph an equation relating weight and amount of blood in a person's body.

   b. Predict the weight of a person whose body holds 12 pints of blood.

Find the next three terms of each arithmetic sequence.

20. $0, -15, -30, -45, -60, \ldots$

21. $5, 8, 11, 14, \ldots$

Determine whether each sequence is an arithmetic sequence. If it is, state the common difference.

22. $-40, -32, -24, -16, \ldots$

23. $0.75, 1.5, 3, 6, 12, \ldots$

24. $5, 17, 29, 41, \ldots$

25. **MULTIPLE CHOICE** In each figure, only one side of each regular pentagon is shared with another pentagon. The length of each side is 1 centimeter. If the pattern continues, what is the perimeter of a figure that has 6 pentagons?

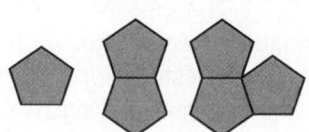

   F 30 cm

   G 25 cm

   H 20 cm

   J 15 cm

3 Preparing for Standardized Tests

Reading Math Problems

The first step to solving any math problem is to read the problem. When reading a math problem to get the information you need to solve, it is helpful to use special reading strategies.

Strategies for Reading Math Problems

Step 1

Read the problem quickly to gain a general understanding of it.

- **Ask yourself:** "What do I know?" "What do I need to find out?"

- **Think:** "Is there enough information to solve the problem? Is there extra information?"

- **Highlight:** If you are allowed to write in your test booklet, underline or highlight important information. Cross out any information you don't need.

Step 2

Reread the problem to identify relevant facts.

- **Analyze:** Determine how the facts are related.

- **Key Words:** Look for keywords to solve the problem.

- **Vocabulary:** Identify mathematical terms. Think about the concepts and how they are related.

- **Plan:** Make a plan to solve the problem.

- **Estimate:** Quickly estimate the answer.

Step 3

Identify any obvious wrong answers.

- **Eliminate:** Eliminate any choices that are very different from your estimate.

- **Units of Measure:** Identify choices that are possible answers based on the units of measure in the question. For example, if the question asks for area, only answers in square units will work.

Step 4

Look back after solving the problem.

Check: Make sure you have answered the question.

Ableimages/Digital Vision/Getty Images

Read the problem. Identify what you need to know. Then use the information in the problem to solve.

Jamal, Gina, Lisa, and Renaldo are renting a car for a road trip. The cost of renting the car is given by the function $C = 12.5 + 21d$, where C is the total cost for renting the car for d days. What does the slope of the function represent?

A number of people

C number of days

B cost per day

D miles per gallon

Read the problem carefully. The number of people going on the trip is not needed information. You need to know what the slope of the function represents.

Slope is a ratio. The word "per" in answers B and D imply that they are both ratios. Since choices A and C are not ratios, eliminate them.

The problem says that C represents the cost of renting the car. So the slope cannot represent the miles per gallon of the car. The slope must represent the cost per day.

The correct answer is B.

Exercises

Read each problem. Identify what you need to know. Then use the information in the problem to solve.

1. What does the x-intercept mean in the context of the situation given below?

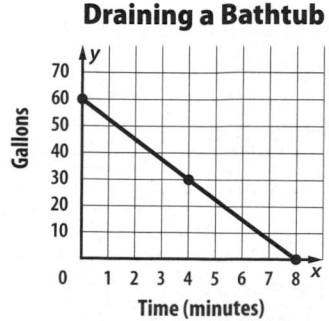

Draining a Bathtub

A amount of time needed to drain the bathtub

B number of gallons in the tub when the drain plug is pulled

C number of gallons in the tub after x minutes

D amount of water drained each minute

2. The amount of money raised by a charity carwash varies directly as the number of cars washed. When 11 cars are washed, $79.75 is raised. How many cars must be washed to raise $174.00?

F 10 cars

H 22 cars

G 16 cars

J 24 cars

3. The function $C = 25 + 0.45(x - 450)$ represents the cost of a monthly cell phone bill, when x minutes are used. Which statement best represents the formula for the cost of the bill?

A The cost consists of a flat fee of $0.45 and $25 for each minute used over 450.

B The cost consists of a flat fee of $450 and $0.45 for each minute used over 25.

C The cost consists of a flat fee of $25 and $0.45 for each minute used over 450.

D The cost consists of a flat fee of $25 and $0.45 for each minute used.

Multiple Choice

Read each question. Then fill in the correct answer on the answer document provided by your teacher or on a sheet of paper.

1. Horatio is purchasing a computer cable for $15.49. If the sales tax rate in his state is 5.25%, what is the total cost of the purchase?

 A $16.42   **C** $15.73

 B $16.30   **D** $15.62

2. What is the value of the expression below?
$$3^2 + 5^3 - 2^5$$

 F 14   **H** 102

 G 34   **J** 166

3. What is the slope of the linear function graphed below?

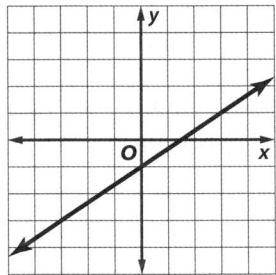

 A $-\dfrac{1}{3}$ **C** $\dfrac{2}{3}$

 B $\dfrac{1}{2}$ **D** $\dfrac{3}{2}$

4. Find the rate of change for the linear function represented in the table.

Hours Worked	1	2	3	4
Money Earned ($)	5.50	11.00	16.50	22.00

 F increase $6.50/h

 G increase $5.50/h

 H decrease $5.50/h

 J decrease $6.50/h

5. Suppose that y varies directly as x, and $y = 14$ when $x = 4$. What is the value of y when $x = 9$?

 A 25.5   **C** 29.5

 B 27.5   **D** 31.5

6. Write an equation for the nth term of the arithmetic sequence shown below.
$$-2, 1, 4, 7, 10, 13, \ldots$$

 F $a_n = 2n - 1$ **H** $a_n = 3n + 2$

 G $a_n = 2n + 4$ **J** $a_n = 3n - 5$

7. The table shows the labor charges of an electrician for jobs of different lengths.

Number of Hours (n)	Labor Charges (c)
1	$60
2	$85
3	$110
4	$135

Which function represents the situation?

 A $C(n) = 25n + 35$ **C** $C(n) = 35n + 25$

 B $C(n) = 25n + 30$ **D** $C(n) = 35n + 40$

8. Find the value of x so that the figures have the same area.

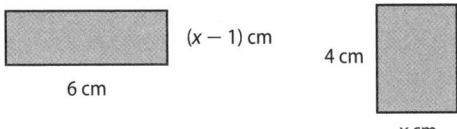

 F 3   **H** 5

 G 4   **J** 6

9. The table shows the total amount of rain during a storm. Write a formula to find out how much rain will fall after a given hour.

Hours (h)	1	2	3	4
Inches (n)	0.45	9.9	1.35	1.8

 A $h = 0.45n$ **C** $h = 0.9n$

 B $n = 0.45h$ **D** $h = 1.8n$

Test-Taking Tip

Question 3 You can *eliminate unreasonable answers* to multiple choice items. The line slopes up from left to right, so the slope is positive. Answer choice A can be eliminated.

Short Response/Gridded Response

Record your answers on the answer sheet provided by your teacher or on a sheet of paper.

10. The scale on a map is 1.5 inches = 6 miles. If two cities are 4 inches apart on the map, what is the actual distance between the cities?

11. Write a direct variation equation to represent the graph below.

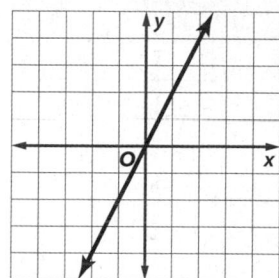

12. Justine bought a car for $18,500 and its value depreciated linearly. After 3 years, the value was $14,150. What is the amount of yearly depreciation?

13. GRIDDED RESPONSE Use the graph to determine the solution to the equation $-\frac{1}{3}x + 1 = 0$?

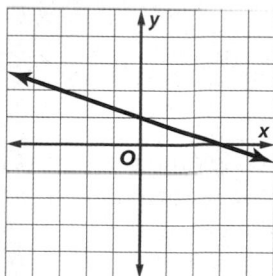

14. Write an expression that represents the total surface area (including the top and bottom) of a tower of n cubes each having a side length of s. (Do not include faces that cover each other.)

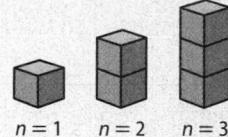

$n = 1 \qquad n = 2 \qquad n = 3$

15. GRIDDED RESPONSE There are 120 members in the North Carolina House of Representatives. This is 70 more than the number of members in the North Carolina Senate. How many members are in the North Carolina Senate?

Extended Response

Record your answers on a sheet of paper. Show your work.

16. A hot air balloon was at a height of 60 feet above the ground when it began to ascend. The balloon climbed at a rate of 15 feet per minute.

a. Make a table that shows the height of the hot air balloon after climbing for 1, 2, 3, and 4 minutes.

b. Let t represent the time in minutes since the balloon began climbing. Write an algebraic equation for a sequence that can be used to find the height, h, of the balloon after t minutes.

c. Use your equation from part b to find the height, in feet, of the hot air balloon after climbing for 8 minutes.

Need ExtraHelp?

If you missed Question...	1	2	3	4	5	6	7	8	9	10	11	12	13	14	15	16
Go to Lesson...	2-7	1-2	3-3	3-3	3-4	3-5	3-6	2-4	3-4	2-6	3-4	3-3	3-2	0-10	2-1	3-5

Equations of Linear Functions

Then

○ You graphed linear functions.

Now

○ In this chapter, you will:

- Write and graph linear equations in various forms.

- Use scatter plots and lines of fit, and write equations of best-fit lines using linear regression.

- Find inverse linear functions.

Why? ▲

○ **OIL** The amount of oil drilled at an oil field changes from year to year. From the yearly data, patterns emerge. Rate of change can be applied to these data to determine a linear model. This can be used to predict the amount of oil drilled in future years.

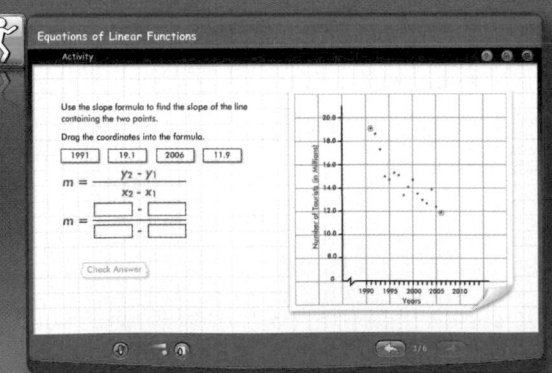

Kent Knudson/PhotoLink/Getty Images

connectED.mcgraw-hill.com **Your Digital Math Portal**

Animation	Vocabulary	eGlossary	Personal Tutor	Virtual Manipulatives	Graphing Calculator	Audio	Foldables	Self-Check Practice	Worksheets

Get Ready for the Chapter

Diagnose Readiness | You have two options for checking prerequisite skills.

1 **Textbook Option** Take the Quick Check below. Refer to the Quick Review for help.

QuickCheck	QuickReview

Evaluate $3a^2 - 2ab + c$ **for the values given.**

1. $a = 2, b = 1, c = 5$

2. $a = -3, b = -2, c = 3$

3. $a = -1, b = 0, c = 11$

4. $a = 5, b = -3, c = -9$

5. **CAR RENTAL** The cost of renting a car is given by $49x + 0.3y$. Let x represent the number of days rented, and let y represent the number of miles driven. Find the cost for a five-day rental over 125 miles.

Example 1

Evaluate $2(m - n)^2 + 3p$ for $m = 5$, $n = 2$, and $p = -3$.

$2(m - n)^2 + 3p$	Original expression
$= 2(5 - 2)^2 + 3(-3)$	Substitute.
$= 2(3)^2 + 3(-3)$	Subtract.
$= 2(9) + 3(-3)$	Evaluate power.
$= 18 + (-9)$	Multiply.
$= 9$	Add.

Solve each equation for the given variable.

6. $x + y = 5$ for y

7. $2x - 4y = 6$ for x

8. $y - 2 = x + 3$ for y

9. $4x - 3y = 12$ for x

10. **GEOMETRY** The formula for the perimeter of a rectangle is $P = 2w + 2\ell$, where w represents width and ℓ represents length. Solve for w.

Example 2

Solve $5x + 15y = 9$ for x.

$5x + 15y = 9$	Original equation
$5x + 15y - 15y = 9 - 15y$	Subtract 15y from each side.
$5x = 9 - 15y$	Simplify.
$\dfrac{5x}{5} = \dfrac{9 - 15y}{5}$	Divide each side by 5.
$x = \dfrac{9}{5} - 3y$	Simplify.

Write the ordered pair for each point.

11. A

12. B

13. C

14. D

15. E

16. F

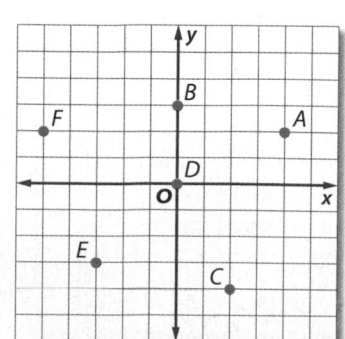

Example 3

Write the ordered pair for A.

Step 1 Begin at point A.

Step 2 Follow along a vertical line to the x-axis. The x-coordinate is -4.

Step 3 Follow along a horizontal line to the y-axis. The y-coordinate is 2.

The ordered pair for point A is $(-4, 2)$.

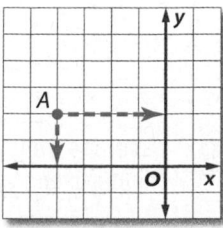

2 **Online Option** Take an online self-check Chapter Readiness Quiz at <u>connectED.mcgraw-hill.com</u>.

Get Started on the Chapter

You will learn several new concepts, skills, and vocabulary terms as you study Chapter 4. To get ready, identify important terms and organize your resources. You may wish to refer to Chapter 0 to review prerequisite skills.

FOLDABLES StudyOrganizer

Equations of Linear Functions Make this Foldable to help you organize your Chapter 4 notes about linear functions. Begin with one sheet of 11" by 17" paper.

1 **Fold** each end of the paper in about 2 inches.

2 **Fold** along the width and the length. Unfold. Cut along the fold line from the top to the center.

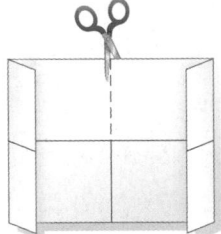

3 **Fold** the top flaps down. Then fold in half and turn to form a folder. Staple the flaps down to form pockets.

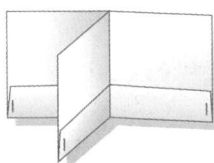

4 **Label** the front with the chapter title.

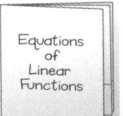

NewVocabulary

English		Español
slope-intercept form	p. 216	forma pendiente-intersección
linear extrapolation	p. 228	extrapolación lineal
point-slope form	p. 233	forma punto-pendiente
parallel lines	p. 239	rectas paralelas
perpendicular lines	p. 240	rectas perpendiculares
scatter plot	p. 247	gráfica de dispersión
line of fit	p. 248	recta de ajuste
linear interpolation	p. 249	interpolación lineal
best-fit line	p. 255	recta de ajuste óptimo
linear regression	p. 255	retroceso lineal
correlation coefficient	p. 255	coeficiente de correlación
median-fit line	p. 258	línea de mediana-ataque
inverse relation	p. 263	relación inversa
inverse function	p. 264	función inversa

ReviewVocabulary

coefficient coeficiente the numerical factor of a term

function función a relation in which each element of the domain is paired with exactly one element of the range

Domain → Range
-3 → 2
2 → 3
0 → 5
1 → 6
4

ratio razón a comparison of two numbers by division

Graphing Technology Lab
Investigating
Slope-Intercept Form

Set Up the Lab

- Cut a small hole in a top corner of a plastic sandwich bag. Hang the bag from the end of the force sensor.

- Connect the force sensor to your data collection device.

Activity Collect Data

Step 1 Use the sensor to collect the weight with 0 washers in the bag. Record the data pair in the calculator.

Step 2 Place one washer in the plastic bag. Wait for the bag to stop swinging, then measure and record the weight.

Step 3 Repeat the experiment, adding different numbers of washers to the bag. Each time, record the number of washers and the weight.

Analyze the Results

1. The domain contains values of the independent variable, number of washers. The range contains values of the dependent variable, weight. Use the graphing calculator to create a scatter plot using the ordered pairs (washers, weight).

2. Write a sentence that describes the points on the graph.

3. Describe the position of the point on the graph that represents the trial with no washers in the bag.

4. The rate of change can be found by using the formula for slope.

$$\frac{\text{rise}}{\text{run}} = \frac{\text{change in weight}}{\text{change in number of washers}}$$

Find the rate of change in the weight as more washers are added.

5. Explain how the rate of change is shown on the graph.

Make a Conjecture

The graph shows sample data from a washer experiment. Describe the graph for each situation.

6. a bag that hangs weighs 0.8 N when empty and increases in weight at the rate of the sample

7. a bag that has the same weight when empty as the sample and increases in weight at a faster rate

8. a bag that has the same weight when empty as the sample and increases in weight at a slower rate

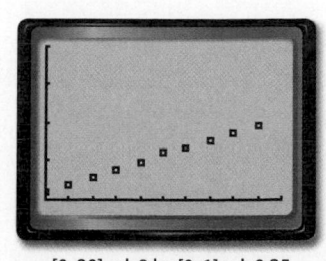

[0, 20] scl: 2 by [0, 1] scl: 0.25

Ed Imaging

Graphing Equations in Slope-Intercept Form

∷ Then	∷ Now	∷ Why?
● You found rates of change and slopes.	**1** Write and graph linear equations in slope-intercept from. **2** Model real-world data with equations in slope-intercept form.	● Jamil has 500 songs on his digital media player. He joins a music club that lets him download 30 songs per month for a monthly fee. The number of songs that Jamil could eventually have in his player if he does not delete any songs is represented by $y = 30x + 500$.

 NewVocabulary
slope-intercept form
constant function

 Common Core State Standards

Content Standards
F.IF.7a Graph linear and quadratic functions and show intercepts, maxima, and minima.

S.ID.7 Interpret the slope (rate of change) and the intercept (constant term) of a linear model in the context of the data.

Mathematical Practices
2 Reason abstractly and quantitatively.

8 Look for and express regularity in repeated reasoning.

1 **Slope-Intercept Form** An equation of the form $y = mx + b$, where m is the slope and b is the y-intercept, is in **slope-intercept form**. The variables m and b are called *parameters* of the equation. Changing either value changes the equation's graph.

◆ KeyConcept Slope-Intercept Form

Words	The slope-intercept form of a linear equation is $y = mx + b$, where m is the slope and b is the y-intercept.
Example	$y = mx + \mathbf{b}$ $y = 2x + 6$ slope⬈ ⬉y-intercept

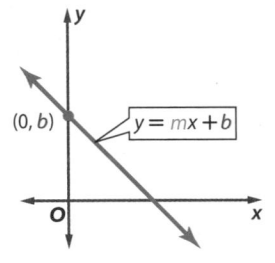

Example 1 Write and Graph an Equation

Write an equation in slope-intercept form for the line with a slope of $\frac{3}{4}$ and a y-intercept of -2. Then graph the equation.

$y = mx + b$ Slope-intercept form

$y = \frac{3}{4}x + (-2)$ Replace m with $\frac{3}{4}$ and b with -2.

$y = \frac{3}{4}x - 2$ Simplify.

Now graph the equation.

Step 1 Plot the y-intercept $(0, -2)$.

Step 2 The slope is $\frac{\text{rise}}{\text{run}} = \frac{3}{4}$. From $(0, -2)$, move up 3 units and right 4 units. Plot the point.

Step 3 Draw a line through the two points.

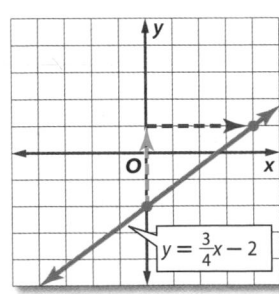

> **Guided**Practice

Write an equation of a line in slope intercept form with the given slope and y-intercept. Then graph the equation.

1A. slope: $-\frac{1}{2}$, y-intercept: 3   **1B.** slope: -3, y-intercept: -8

When an equation is not written in slope-intercept form, it may be easier to rewrite it before graphing.

Example 2 Graph Linear Equations

Graph $3x + 2y = 6$.

Rewrite the equation in slope-intercept form.

$3x + 2y = 6$	Original equation
$3x + 2y - 3x = 6 - 3x$	Subtract $3x$ from each side.
$2y = 6 - 3x$	Simplify.
$2y = -3x + 6$	$6 - 3x = 6 + (-3x)$ or $-3x + 6$
$\dfrac{2y}{2} = \dfrac{-3x + 6}{2}$	Divide each side by 2.
$y = -\dfrac{3}{2}x + 3$	Slope-intercept form

Now graph the equation. The slope is $-\dfrac{3}{2}$, and the y-intercept is 3.

Step 1 Plot the y-intercept $(0, 3)$.

Step 2 The slope is $\dfrac{\text{rise}}{\text{run}} = -\dfrac{3}{2}$. From $(0, 3)$, move down 3 units and right 2 units. Plot the point.

Step 3 Draw a line through the two points.

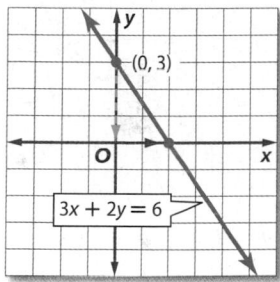

▶ **Guided**Practice

Graph each equation.

2A. $3x - 4y = 12$ **2B.** $-2x + 5y = 10$

StudyTip

Counting and Direction When counting rise and run, a negative sign may be associated with the value in the numerator or denominator. If with the numerator, begin by counting down for the rise. If with the denominator, count left when counting the run. The resulting line will be the same.

Except for the graph of $y = 0$, which lies on the x axis, horizontal lines have a slope of 0. They are graphs of **constant functions**, which can be written in slope-intercept form as $y = 0x + b$ or $y = b$, where b is any number. Constant functions do not cross the x-axis. Their domain is all real numbers, and their range is b.

Example 3 Graph Linear Equations

Graph $y = -3$.

Step 1 Plot the y-intercept $(0, -3)$.

Step 2 The slope is 0. Draw a line through the points with y-coordinate -3.

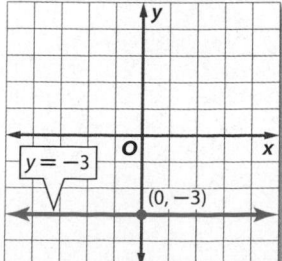

▶ **Guided**Practice

Graph each equation.

3A. $y = 5$ **3B.** $2y = 1$

Vertical lines have no slope. So, equations of vertical lines cannot be written in slope-intercept form.

There are times when you will need to write an equation when given a graph. To do this, locate the y-intercept and use the rise and run to find another point on the graph. Then write the equation in slope-intercept form.

Standardized Test Example 4 Write an Equation in Slope-Intercept Form

Which of the following is an equation in slope-intercept form for the line shown?

A $y = -3x + 1$

B $y = -3x + 3$

C $y = -\frac{1}{3}x + 1$

D $y = -\frac{1}{3}x + 3$

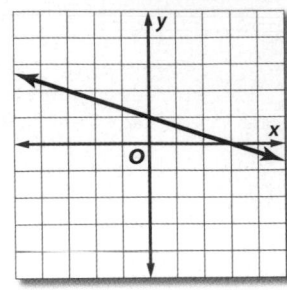

Test-TakingTip

Eliminating Choices
Analyze the graph to determine the slope and the y-intercept. Then you can save time by eliminating answer choices that do not match the graph.

Read the Test Item

You need to find the slope and y-intercept of the line to write the equation.

Solve the Test Item

Step 1 The line crosses the y-axis at (0, 1), so the y-intercept is 1. The answer is either A or C.

Step 2 To get from (0, 1) to (3, 0), go down 1 unit and 3 units to the right. The slope is $-\frac{1}{3}$.

Step 3 Write the equation.

$y = mx + b$

$y = -\frac{1}{3}x + 1$

CHECK The graph also passes through $(-3, 2)$. If the equation is correct, this should be a solution.

$y = -\frac{1}{3}x + 1$

$2 \stackrel{?}{=} -\frac{1}{3}(-3) + 1$

$2 \stackrel{?}{=} 1 + 1$

$2 = 2 \checkmark$ The answer is C.

▶ **GuidedPractice**

4. Which of the following is an equation in slope-intercept form for the line shown?

F $y = \frac{1}{4}x - 1$

G $y = \frac{1}{4}x + 4$

H $y = 4x - 1$

J $y = 4x + 4$

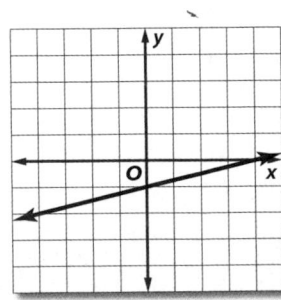

2 Modeling Real-World Data Real-world data can be modeled by a linear equation if there is a constant rate of change. The rate of change represents the slope. The y-intercept is the point where the value of the independent variable is 0.

Real-World Example 5 Write and Graph a Linear Equation

SPORTS Use the information at the left about high school sports.

a. Write a linear equation to find the number of girls in high school sports after 1997.

Words	Number of girls competing	equals	rate of change	times	number of years	plus	amount at start.
Variables	Let G = number of girls competing.		Let n = number of years since 1997.				
Equation	G	=	0.06	×	n	+	2.6

The equation is $G = 0.06n + 2.6$.

Real-World Link

In 1997, about 2.6 million girls competed in high school sports. The number of girls competing in high school sports has increased by an average of 0.06 million per year since 1997.

Source: National Federation of High School Associations

b. **Graph the equation.**

The y-intercept is where the data begins. So, the graph passes through $(0, 2.6)$.

The rate of change is the slope, so the slope is 0.06.

c. **Estimate the number of girls competing in 2017.**

The year 2017 is 20 years after 1997.

$G = 0.06n + 2.6$ Write the equation.

$ = 0.06(20) + 2.6$ Replace n with 20.

$ = 3.8$ Simplify.

There will be about 3.8 million girls competing in high school sports in 2017.

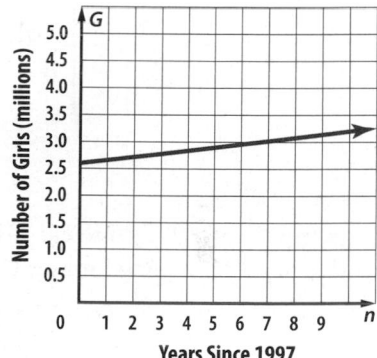

▶ **Guided**Practice

5. **FUNDRAISERS** The band boosters are selling sandwiches for $5 each. They bought $1160 in ingredients.

   A. Write an equation for the profit P made on n sandwiches.

   B. Graph the equation.

   C. Find the total profit if 1400 sandwiches are sold.

Check Your Understanding

 = Step-by-Step Solutions begin on page R13.

Example 1   Write an equation of a line in slope-intercept form with the given slope and y-intercept. Then graph the equation.

1. slope: 2, y-intercept: 4
2. slope: -5, y-intercept: 3
3. slope: $\frac{3}{4}$, y-intercept: -1
4. slope: $-\frac{5}{7}$, y-intercept: $-\frac{2}{3}$

Examples 2–3 Graph each equation.

5. $-4x + y = 2$
6. $2x + y = -6$
7. $-3x + 7y = 21$
8. $6x - 4y = 16$
9. $y = -1$
10. $15y = 3$

Example 4   Write an equation in slope-intercept form for each graph shown.

11.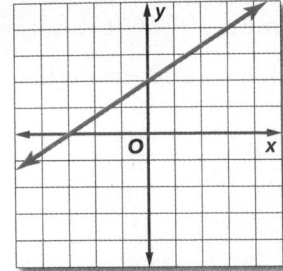

12.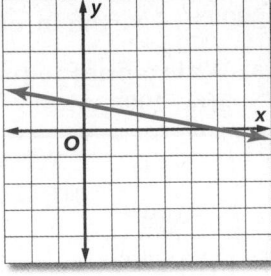

13.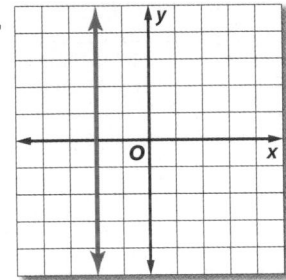

14.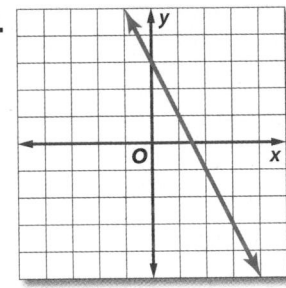

Example 5   15. **FINANCIAL LITERACY** Rondell is buying a new stereo system for his car using Jack's Stereo layaway plan.

Jack's Stereo Layaway Plan
$75 down and
$10 each week

   a. Write an equation for the total amount S that he has paid after w weeks.

   b. Graph the equation.

   c. Find out how much Rondell will have paid after 8 weeks.

16. **CCSS REASONING** Ana is driving from her home in Miami, Florida, to her grandmother's house in New York City. On the first day, she will travel 240 miles to Orlando, Florida, to pick up her cousin. Then they will travel 350 miles each day.

   a. Write an equation that models the total number of miles m Ana has traveled, if d represents the number of days after she picks up her cousin.

   b. Graph the equation.

   c. How long will the drive take if the total length of the trip is 1343 miles?

Practice and Problem Solving

Extra Practice is on page R4.

Example 1   Write an equation of a line in slope-intercept form with the given slope and y-intercept. Then graph the equation.

(17) slope: 5, y-intercept: 8

18. slope: 3, y-intercept: 10

19. slope: -4, y-intercept: 6

20. slope: -2, y-intercept: 8

21. slope: 3, y-intercept: -4

22. slope: 4, y-intercept: -6

Examples 2–3 Graph each equation.

23. $-3x + y = 6$

24. $-5x + y = 1$

25. $-2x + y = -4$

26. $y = 7x - 7$

27. $5x + 2y = 8$

28. $4x + 9y = 27$

29. $y = 7$

30. $y = -\dfrac{2}{3}$

31. $21 = 7y$

32. $3y - 6 = 2x$

Example 4 Write an equation in slope-intercept form for each graph shown.

33.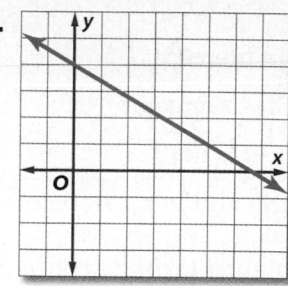

34.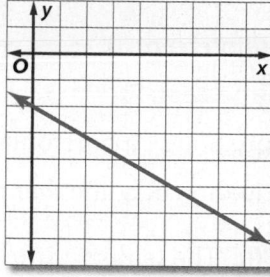

35.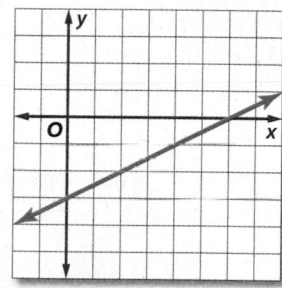

36.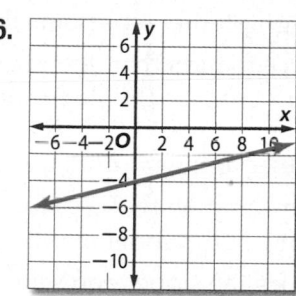

Example 5

37 **MANATEES** In 1991, 1267 manatees inhabited Florida's waters. The manatee population has increased at a rate of 123 manatees per year.

   a. Write an equation for the manatee population, P, t years since 1991.

   b. Graph this equation.

   c. In 2006, the manatee was removed from Florida's endangered species list. What was the manatee population in 2006?

Write an equation of a line in slope-intercept form with the given slope and y-intercept.

38. slope: $\frac{1}{2}$, y-intercept: -3 **39.** slope: $\frac{2}{3}$, y-intercept: -5

40. slope: $-\frac{5}{6}$, y-intercept: 5 **41.** slope: $-\frac{3}{7}$, y-intercept: 2

42. slope: 1, y-intercept: 4 **43.** slope: 0, y-intercept: 5

Graph each equation.

44. $y = \frac{3}{4}x - 2$ **45.** $y = \frac{5}{3}x + 4$ **46.** $3x + 8y = 32$

47. $5x - 6y = 36$ **48.** $-4x + \frac{1}{2}y = -1$ **49.** $3x - \frac{1}{4}y = 2$

50. TRAVEL A rental company charges $8 per hour for a mountain bike plus a $5 fee for a helmet.

   a. Write an equation in slope-intercept form for the total rental cost C for a helmet and a bicycle for t hours.

   b. Graph the equation.

   c. What would the cost be for 2 helmets and 2 bicycles for 8 hours?

51. CCSS REASONING For Illinois residents, the average tuition at Chicago State University is $157 per credit hour. Fees cost $218 per year.

   a. Write an equation in slope-intercept form for the tuition T for c credit hours.

   b. Find the cost for a student who is taking 32 credit hours.

Write an equation of a line in slope-intercept form with the given slope and y-intercept.

52. slope: -1, y-intercept: 0

53. slope: 0.5, y-intercept: 7.5

54. slope: 0, y-intercept: 7

55. slope: -1.5, y-intercept: -0.25

56. Write an equation of a horizontal line that crosses the y-axis at $(0, -5)$.

57. Write an equation of a line that passes through the origin and has a slope of 3.

58. TEMPERATURE The temperature dropped rapidly overnight. Starting at 80°F, the temperature dropped 3° per minute.

    a. Draw a graph that represents this drop from 0 to 8 minutes.

    b. Write an equation that describes this situation. Describe the meaning of each variable as well as the slope and y-intercept.

59. FITNESS Refer to the information at the right.

    a. Write an equation that represents the cost C of a membership for m months.

    b. What does the slope represent?

    c. What does the C-intercept represent?

    d. What is the cost of a two-year membership?

GET FIT GYM
startup fee $145
$45 monthly fee

60. MAGAZINES A teen magazine began with a circulation of 500,000 in its first year. Since then, the circulation has increased an average of 33,388 per year.

    a. Write an equation that represents the circulation c after t years.

    b. What does the slope represent?

    c. What does the c-intercept represent?

    d. If the magazine began in 1944, and this trend continues, in what year will the circulation reach 3,000,000?

61. SMART PHONES A telecommunications company sold 3305 smart phones in the first year of production. Suppose, on average, they expect to sell 25 phones per day.

    a. Write an equation for the number of smart phones P sold t years after the first year of production, assuming 365 days per year.

    b. If sales continue at this rate, how many years will it take for the company to sell 100,000 phones?

H.O.T. Problems   Use Higher-Order Thinking Skills

62. OPEN ENDED Draw a graph representing a real-world linear function and write an equation for the graph. Describe what the graph represents.

63. REASONING Determine whether the equation of a vertical line can be written in slope-intercept form. Explain your reasoning.

64. CHALLENGE Summarize the characteristics that the graphs $y = 2x + 3$, $y = 4x + 3$, $y = -x + 3$, and $y = -10x + 3$ have in common.

65. CCSS REGULARITY If given an equation in standard form, explain how to determine the rate of change.

66. WRITING IN MATH Explain how you would use a given y-intercept and the slope to predict a y-value for a given x-value without graphing.

67. A music store has x CDs in stock. If 350 are sold and $3y$ are added to stock, which expression represents the number of CDs in stock?

A $350 + 3y - x$ **C** $x + 350 + 3y$

B $x - 350 + 3y$ **D** $3y - 350 - x$

68. PROBABILITY The table shows the result of a survey of favorite activities. What is the probability that a student's favorite activity is sports or drama club?

Extracurricular Activity	Students
art club	24
band	134
choir	37
drama club	46
mock trial	19
school paper	26
sports	314

F $\dfrac{3}{8}$ **G** $\dfrac{4}{9}$ **H** $\dfrac{3}{5}$ **J** $\dfrac{2}{3}$

69. A recipe for fruit punch calls for 2 ounces of orange juice for every 8 ounces of lemonade. If Jennifer uses 64 ounces of lemonade, which proportion can she use to find x, the number of ounces of orange juice needed?

A $\dfrac{2}{x} = \dfrac{64}{6}$ **C** $\dfrac{2}{8} = \dfrac{x}{64}$

B $\dfrac{8}{x} = \dfrac{64}{2}$ **D** $\dfrac{6}{2} = \dfrac{x}{64}$

70. EXTENDED RESPONSE The table shows the results of a canned food drive. 1225 cans were collected, and the 12th-grade class collected 55 more cans than the 10th-grade class. How many cans each did the 10th- and 12th-grade classes collect? Show your work.

Grade	Cans
9	340
10	x
11	280
12	y

For each arithmetic sequence, determine the related function. Then determine if the function is *proportional* or *nonproportional*. (Lesson 3-6)

71. 3, 7, 11, ...   **72.** 8, 6, 4, ...   **73.** 0, 3, 6, ...   **74.** 1, 2, 3, ...

75. GAME SHOWS Contestants on a game show win money by answering 10 questions. (Lesson 3-5)

     a. Find the value of the 10th question.

     b. If all questions are answered correctly, how much are the winnings?

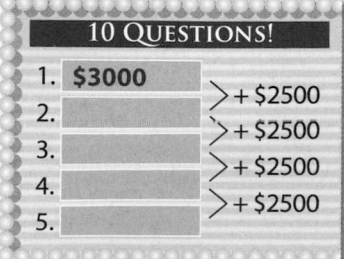

Suppose y varies directly as x. Write a direct variation equation that relates x and y. Then solve. (Lesson 3-4)

76. If $y = 10$ when $x = 5$, find y when $x = 6$.

77. If $y = -16$ when $x = 4$, find x when $y = 20$.

78. If $y = 6$ when $x = 18$, find y when $x = -12$.

79. If $y = 12$ when $x = 15$, find x when $y = -6$.

Find the slope of the line that passes through each pair of points.

80. $(2, 3)$, $(9, 7)$ **81.** $(-3, 6)$, $(2, 4)$ **82.** $(2, 6)$, $(-1, 3)$ **83.** $(-3, 3)$, $(1, 3)$

Graphing Technology Lab
The Family of Linear Graphs

A family of people is related by birth, marriage, or adoption. Often people in families share characteristics. The graphs in a family share at least one characteristic. Graphs in the linear family are all lines, with the simplest graph in the family being that of the parent function $y = x$. This parent function is also known as the **identity function**. Its graph contains all points with coordinates (a, a). Its domain and range are all real numbers.

Parent Graph
Identity Function

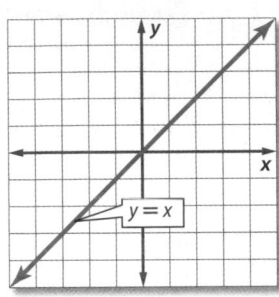

You can use a graphing calculator to investigate how changing the parameters m and b in $y = mx + b$ affects the graphs in the family of linear functions.

CCSS Common Core State Standards
Content Standards
F.BF.3 Identify the effect on the graph of replacing $f(x)$ by $f(x) + k$, $k\,f(x)$, $f(kx)$, and $f(x + k)$ for specific values of k (both positive and negative); find the value of k given the graphs. Experiment with cases and illustrate an explanation of the effects on the graph using technology.
S.ID.7 Interpret the slope (rate of change) and the intercept (constant term) of a linear model in the context of the data.
Mathematical Practices
7 Look for and make use of structure.

Activity 1 Changing b in $y = mx + b$

Graph $y = x$, $y = x + 4$, and $y = x - 2$ in the standard viewing window.

Enter the equations in the **Y=** list as **Y1**, **Y2**, and **Y3**. Then graph the equations.

KEYSTROKES: Y= X,T,θ,n ENTER X,T,θ,n +
4 ENTER X,T,θ,n − 2 ENTER
ZOOM 6

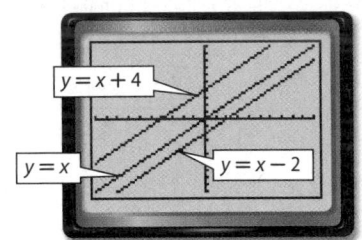

[−10, 10] scl: 1 by [−10, 10] scl: 1

1A. How do the slopes of the graphs compare?

1B. Compare the graph of $y = x + 4$ and the graph of $y = x$. How would you obtain the graph of $y = x + 4$ from the graph of $y = x$?

1C. How would you obtain the graph of $y = x - 2$ from the graph of $y = x$?

Changing the y-intercept, b, translates, or moves, a linear function up or down the y-axis. Changing m in $y = mx + b$ affects the graphs in a different way. First, investigate positive values of m.

Activity 2 Changing m in $y = mx + b$, Positive Values

Graph $y = x$, $y = 2x$, and $y = \frac{1}{3}x$ in the standard viewing window.

Enter the equations in the **Y=** list and graph.

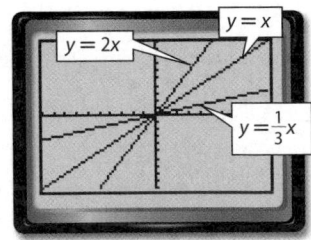

[−10, 10] scl: 1 by [−10, 10] scl: 1

2A. How do the y-intercepts of the graphs compare?

2B. Compare the graph of $y = 2x$ and the graph of $y = x$.

2C. Which is steeper, the graph of $y = \frac{1}{3}x$ or the graph of $y = x$?

Does changing m to a negative value affect the graph differently than changing it to a positive value?

Graph $y = x$, $y = -x$, $y = -3x$, and $y = -\frac{1}{2}x$ in the standard viewing window.

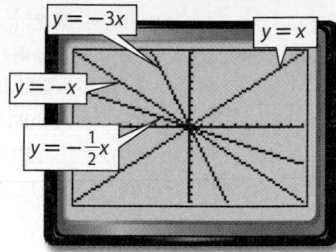

Enter the equations in the **Y=** list and graph.

3A. How are the graphs with negative values of m different than graphs with a positive m?

3B. Compare the graphs of $y = -x$, $y = -3x$, and $y = -\frac{1}{2}x$. Which is steepest?

[−10, 10] scl: 1 by [−10, 10] scl: 1

Analyze the Results

CCSS SENSE-MAKING AND PERSEVERANCE Graph each set of equations on the same screen. Describe the similarities or differences.

1. $y = 2x$
$y = 2x + 3$
$y = 2x - 7$

2. $y = x + 1$
$y = 2x + 1$
$y = \frac{1}{4}x + 1$

3. $y = x + 4$
$y = 2x + 4$
$y = \frac{3}{4}x + 4$

4. $y = 0.5x + 2$
$y = 0.5x - 5$
$y = 0.5x + 4$

5. $y = -2x - 2$
$y = -4.2x - 2$
$y = -\frac{1}{3}x - 2$

6. $y = 3x$
$y = 3x + 6$
$y = 3x - 7$

7. Families of graphs have common characteristics. What do the graphs of all equations of the form $y = mx + b$ have in common?

8. How does the value of b affect the graph of $y = mx + b$?

9. What is the result of changing the value of m on the graph of $y = mx + b$ if m is positive?

10. How can you determine which graph is steepest by examining the following equations?
$y = 3x$, $y = -4x - 7$, $y = \frac{1}{2}x + 4$

11. Explain how knowing about the effects of m and b can help you sketch the graph of an equation.

12. The equation $y = k$ can also be a parent graph. Graph $y = 5$, $y = 2$, and $y = -4$ on the same screen. Describe the similarities or differences among the graphs.

Extension

Nonlinear functions can also be defined in terms of a family of graphs. Graph each set of equations on the same screen. Describe the similarities or differences.

13. $y = x^2$
$y = -3x^2$
$y = (-3x)^2$

14. $y = x^2$
$y = x^2 + 3$
$y = (x - 2)^2$

15. $y = x^2$
$y = 2x^2 + 4$
$y = (3x)^2 - 5$

16. Describe the similarities and differences in the classes of functions $f(x) = x^2 + c$ and $f(x) = (x + c)^2$, where c is any real number.

Writing Equations in Slope-Intercept Form

∴ Then
- You graphed lines given the slope and the *y*-intercept.

∴ Now
- **1** Write an equation of a line in slope-intercept form given the slope and one point.
- **2** Write an equation of a line in slope-intercept form given two points.

∴ Why?
- In 2006, the attendance at the Columbus Zoo and Aquarium was about 1.6 million. In 2009, the zoo's attendance was about 2.2 million. You can find the average rate of change for these data. Then you can write an equation that would model the average attendance at the zoo for a given year.

 NewVocabulary
constraint
linear extrapolation

 **Common Core State Standards**

Content Standards
F.BF.1 Write a function that describes a relationship between two quantities.
a. Determine an explicit expression, a recursive process, or steps for calculation from a context.
b. Combine standard function types using arithmetic operations.
F.LE.2 Construct linear and exponential functions, including arithmetic and geometric sequences, given a graph, a description of a relationship, or two input-output pairs (include reading these from a table).

Mathematical Practices
3 Construct viable arguments and critique the reasoning of others.
6 Attend to precision.

1 Write an Equation Given the Slope and a Point The next example shows how to write an equation of a line if you are given a slope and a point other than the *y*-intercept.

Example 1 Write an Equation Given the Slope and a Point

Write an equation of the line that passes through (2, 1) with a slope of 3.

You are given the slope but not the *y*-intercept.

Step 1 Find the *y*-intercept.

$$y = mx + b \qquad \text{Slope-intercept form}$$
$$1 = 3(2) + b \qquad \text{Replace } m \text{ with 3, } y \text{ with 1, and } x \text{ with 2.}$$
$$1 = 6 + b \qquad \text{Simplify.}$$
$$1 - 6 = 6 + b - 6 \qquad \text{Subtract 6 from each side.}$$
$$-5 = b \qquad \text{Simplify.}$$

Step 2 Write the equation in slope-intercept form.

$$y = mx + b \qquad \text{Slope-intercept form}$$
$$y = 3x - 5 \qquad \text{Replace } m \text{ with 3 and } b \text{ with } -5.$$

Therefore, the equation of the line is $y = 3x - 5$.

▸ **Guided Practice**

Write an equation of a line that passes through the given point and has the given slope.

1A. (−2, 5), slope 3

1B. (4, −7), slope −1

2 Write an Equation Given Two Points If you are given two points through which a line passes, you can use them to find the slope first. Then follow the steps in Example 1 to write the equation.

Example 2 Write an Equation Given Two Points

Write an equation of the line that passes through each pair of points.

a. (3, 1) and (2, 4)

Step 1 Find the slope of the line containing the given points.

$$m = \frac{y_2 - y_1}{x_2 - x_1}$$
Slope Formula

$$= \frac{4 - 1}{2 - 3}$$
$(x_1, y_1) = (3, 1)$ and $(x_2, y_2) = (2, 4)$

$$= \frac{3}{-1} \text{ or } -3$$
Simplify.

Step 2 Use either point to find the y-intercept.

$y = mx + b$ Slope-intercept form

$4 = (-3)(2) + b$ Replace m with -3, x with 2, and y with 4.

$4 = -6 + b$ Simplify.

$4 - (-6) = -6 + b - (-6)$ Subtract -6 from each side.

$10 = b$ Simplify.

Step 3 Write the equation in slope-intercept form.

$y = mx + b$ Slope-intercept form

$y = -3x + 10$ Replace m with -3 and b with 10.

Therefore, the equation is $y = -3x + 10$.

b. (−4, −2) and (−5, −6)

Step 1 Find the slope of the line containing the given points.

$$m = \frac{y_2 - y_1}{x_2 - x_1}$$
Slope Formula

$$= \frac{-6 - (-2)}{-5 - (-4)}$$
$(x_1, y_1) = (-4, -2)$ and $(x_2, y_2) = (-5, -6)$

$$= \frac{-4}{-1} \text{ or } 4$$
Simplify.

Step 2 Use either point to find the y-intercept.

$y = mx + b$ Slope-intercept form

$-2 = 4(-4) + b$ Replace m with 4, x with -4, and y with -2.

$-2 = -16 + b$ Simplify.

$-2 - (-16) = -16 + b - (-16)$ Subtract -16 from each side.

$14 = b$ Simplify.

Step 3 Write the equation in slope-intercept form.

$y = mx + b$ Slope-intercept form

$y = 4x + 14$ Replace m with 4 and b with 14.

Therefore, the equation is $y = 4x + 14$.

▶ **GuidedPractice**

Write an equation of the line that passes through each pair of points.

2A. $(-1, 12), (4, -8)$ **2B.** $(5, -8), (-7, 0)$

In mathematics, a **constraint** is a condition that a solution must satisfy. Equations can be viewed as constraints in a problem situation. The solutions of the equation meet the constraints of the problem.

Real-WorldCareer

Ground Crew
Airline ground crew responsibilities include checking tickets, helping passengers with luggage, and making sure that baggage is loaded properly and secure. This job usually requires a high school diploma or GED.

Source: Airline Jobs

Real-World Example 3 Use Slope-Intercept Form

FLIGHTS The table shows the number of domestic flights in the U.S. from 2004 to 2008. Write an equation that could be used to predict the number of flights if it continues to decrease at the same rate.

Year	Flights (millions)
2004	9.97
2005	10.04
2006	9.71
2007	9.84
2008	9.37

Understand You know the number of flights for 2004–2008.

Plan Let x represent the number of years since 2000, and let y represent the number of flights. Write an equation of the line that passes through $(4, 9.97)$ and $(8, 9.37)$.

Solve Find the slope.

$$m = \frac{y_2 - y_1}{x_2 - x_1} \qquad \text{Slope formula}$$

$$= \frac{9.37 - 9.97}{8 - 4} \qquad \text{Let } (x_1, y_1) = (4, 9.97) \text{ and } (x_2, y_2) = (8, 9.37).$$

$$= -\frac{0.6}{4} \text{ or } -0.15 \qquad \text{Simplify.}$$

Use $(8, 9.37)$ to find the y-intercept of the line.

$$y = mx + b \qquad \text{Slope-intercept form}$$
$$9.37 = -0.15(8) + b \qquad \text{Replace } y \text{ with } 9.37, m \text{ with } -0.15, \text{ and } x \text{ with } 8.}$$
$$9.37 = -1.2 + b \qquad \text{Simplify.}$$
$$10.57 = b \qquad \text{Add 1.2 to each side.}$$

Write the equation using $m = -0.15$ and $b = 10.57$.

$$y = mx + b \qquad \text{Slope-intercept form}$$
$$y = -0.15x + 10.57 \qquad \text{Replace } m \text{ with } -0.15 \text{ and } b \text{ with } 10.57.}$$

Check Check your result by using the coordinates of the other point.

$$y = -0.15x + 10.57 \qquad \text{Original equation}$$
$$9.97 \stackrel{?}{=} -0.15(4) + 10.57 \qquad \text{Replace } y \text{ with } 9.97 \text{ and } x \text{ with } 4.}$$
$$9.97 = 9.97 \checkmark \qquad \text{Simplify.}$$

▶ **Guided**Practice

3. FINANCIAL LITERACY In addition to his weekly salary, Ethan is paid $16 per delivery. Last week, he made 5 deliveries, and his total pay was $215. Write a linear equation to find Ethan's total weekly pay T if he makes d deliveries.

You can use a linear equation to make predictions about values that are beyond the range of the data. This process is called **linear extrapolation**.

Problem-Solving Tip

CCSS Precision Deciding whether an answer is reasonable is useful when an exact answer is not neccessary.

Real-World Example 4 Predict from Slope-Intercept Form

FLIGHTS Estimate the number of domestic flights in 2020.

$$y = -0.15x + 10.57 \qquad \text{Original equation}$$
$$= -0.15(20) + 10.57 \text{ or } 7.57 \text{ million} \qquad \text{Replace } x \text{ with } 20.}$$

▶ **Guided**Practice

4. MONEY Use the equation in Guided Practice 3 to predict how much money Ethan will earn in a week if he makes 8 deliveries.

Stephan Goerlich/age fotostock

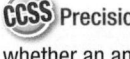

Example 1   Write an equation of the line that passes through the given point and has the given slope.

    1. $(3, -3)$, slope 3   **2.** $(2, 4)$, slope 2

    3. $(1, 5)$, slope -1 **4.** $(-4, 6)$, slope -2

Example 2   Write an equation of the line that passes through each pair of points.

    5. $(4, -3)$, $(2, 3)$ **6.** $(-7, -3)$, $(-3, 5)$

    7. $(-1, 3)$, $(0, 8)$ **8.** $(-2, 6)$, $(0, 0)$

Examples 3, 4   **9. WHITEWATER RAFTING** Ten people from a local youth group went to Black Hills Whitewater Rafting Tour Company for a one-day rafting trip. The group paid $425.

Guide's **FEE** *plus* **$35.00** per person for **1-day** trip

    a. Write an equation in slope-intercept form to find the total cost C for p people.

    b. How much would it cost for 15 people?

Practice and Problem Solving   Extra Practice is on page R4.

Example 1   Write an equation of the line that passes through the given point and has the given slope.

    10. $(3, 1)$, slope 2   **11** $(-1, 4)$, slope -1 **12.** $(1, 0)$, slope 1

    13. $(7, 1)$, slope 8   **14.** $(2, 5)$, slope -2 **15.** $(2, 6)$, slope 2

Example 2   Write an equation of the line that passes through each pair of points.

    16. $(9, -2)$, $(4, 3)$ **17.** $(-2, 5)$, $(5, -2)$ **18.** $(-5, 3)$, $(0, -7)$

    19. $(3, 5)$, $(2, -2)$ **20.** $(-1, -3)$, $(-2, 3)$ **21.** $(-2, -4)$, $(2, 3)$

Examples 3, 4   **22. CCSS MODELING** Greg is driving a remote control car at a constant speed. He starts the timer when the car is 5 feet away. After 2 seconds the car is 35 feet away.

    a. Write a linear equation to find the distance d of the car from Greg.

    b. Estimate the distance the car has traveled after 10 seconds.

    23. ZOOS Refer to the beginning of the lesson.

    a. Write a linear equation to find the attendance (in millions) y after x years. Let x be the number of years since 2000.

    b. Estimate the zoo's attendance in 2020.

    24. BOOKS In 1904, a dictionary cost 30¢. Since then the cost of a dictionary has risen an average of 6¢ per year.

    a. Write a linear equation to find the cost C of a dictionary y years after 1904.

    b. If this trend continues, what will the cost of a dictionary be in 2020?

Write an equation of the line that passes through the given point and has the given slope.

    25. $(4, 2)$, slope $\frac{1}{2}$ **26.** $(3, -2)$, slope $\frac{1}{3}$ **27.** $(6, 4)$, slope $-\frac{3}{4}$

    28. $(2, -3)$, slope $\frac{2}{3}$ **29.** $(2, -2)$, slope $\frac{2}{7}$ **30.** $(-4, -2)$, slope $-\frac{3}{5}$

31. DOGS In 2001, there were about 56.1 thousand golden retrievers registered in the United States. In 2002, the number was 62.5 thousand.

a. Write a linear equation to find the number of thousands of golden retrievers G that will be registered in year t, where $t = 0$ is the year 2000.

b. Graph the equation.

c. Estimate the number of golden retrievers that will be registered in 2017.

32. GYM MEMBERSHIPS A local recreation center offers a yearly membership for $265. The center offers aerobics classes for an additional $5 per class.

a. Write an equation that represents the total cost of the membership.

b. Carly spent $500 one year. How many aerobics classes did she take?

33. SUBSCRIPTION A magazine offers an online subscription that allows you to view up to 25 archived articles free. To view 30 archived articles, you pay $49.15. To view 33 archived articles, you pay $57.40.

a. What is the cost of each archived article for which you pay a fee?

b. What is the cost of the magazine subscription?

Write an equation of the line that passes through the given points.

34. $(5, -2), (7, 1)$ **35** $(5, -3), (2, 5)$ **36.** $\left(\frac{5}{4}, 1\right), \left(-\frac{1}{4}, \frac{3}{4}\right)$ **37.** $\left(\frac{5}{12}, -1\right), \left(-\frac{3}{4}, \frac{1}{6}\right)$

Determine whether the given point is on the line. Explain why or why not.

38. $(3, -1); y = \frac{1}{3}x + 5$ **39.** $(6, -2); y = \frac{1}{2}x - 5$

For Exercises 40–42, determine which equation best represents each situation. Explain the meaning of each variable.

| **A** $y = -\frac{1}{3}x + 72$ | **B** $y = 2x + 225$ | **C** $y = 8x + 4$ |

40. CONCERTS Tickets to a concert cost $8 each plus a processing fee of $4 per order.

41. FUNDRAISING The freshman class has $225. They sell raffle tickets at $2 each to raise money for a field trip.

42. POOLS The current water level of a swimming pool in Tucson, Arizona, is 6 feet. The rate of evaporation is $\frac{1}{3}$ inch per day.

43. CCSS SENSE-MAKING A manufacturer implemented a program to reduce waste. In 1998 they sent 946 tons of waste to landfills. Each year after that, they reduced their waste by an average 28.4 tons.

a. How many tons were sent to the landfill in 2010?

b. In what year will it become impossible for this trend to continue? Explain.

44. COMBINING FUNCTIONS The parents of a college student open an account for her with a deposit of $5000, and they set up automatic deposits of $100 to the account every week.

a. Write a function $d(t)$ to express the amount of money in the account t weeks after the initial deposit.

b. The student plans on spending $600 the first week and $250 in each of the following weeks for room and board and other expenses. Write a function $w(t)$ to express the amount of money taken out of the account each week.

c. Find $B(t) = d(t) - w(t)$. What does this new function represent?

d. Will the student run out of money? If so, when?

45 **CONCERT TICKETS** Jackson is ordering tickets for a concert online. There is a processing fee for each order, and the tickets are $52 each. Jackson ordered 5 tickets and the cost was $275.

    a. Determine the processing fee. Write a linear equation to represent the total cost C for t tickets.

    b. Make a table of values for at least three other numbers of tickets.

    c. Graph this equation. Predict the cost of 8 tickets.

46. **MUSIC** A music store is offering a Frequent Buyers Club membership. The membership costs $22 per year, and then a member can buy CDs at a reduced price. If a member buys 17 CDs in one year, the cost is $111.25.

    a. Determine the cost of each CD for a member.

    b. Write a linear equation to represent the total cost y of a one year membership, if x CDs are purchased.

    c. Graph this equation.

H.O.T. Problems   Use Higher-Order Thinking Skills

47. **ERROR ANALYSIS** Tess and Jacinta are writing an equation of the line through $(3, -2)$ and $(6, 4)$. Is either of them correct? Explain your reasoning.

Tess	Jacinta
$m = \dfrac{4-(-2)}{6-3} = \dfrac{6}{3}$ or 2	$m = \dfrac{4-(-2)}{6-3} = \dfrac{6}{3}$ or 2
$y = mx + b$	$y = mx + b$
$6 = 2(4) + b$	$-2 = 2(3) + b$
$6 = 8 + b$	$-2 = 6 + b$
$-2 = b$	$-8 = b$
$y = 2x - 2$	$y = 2x - 8$

48. **CHALLENGE** Consider three points, $(3, 7)$, $(-6, 1)$ and $(9, p)$, on the same line. Find the value of p and explain your steps.

49. **REASONING** Consider the standard form of a linear equation, $Ax + By = C$.

    a. Rewrite the equation in slope-intercept form.

    b. What is the slope?

    c. What is the y-intercept?

    d. Is this true for all real values of A, B, and C?

50. **OPEN ENDED** Create a real-world situation that fits the graph at the right. Define the two quantities and describe the functional relationship between them. Write an equation to represent this relationship and describe what the slope and y-intercept mean.

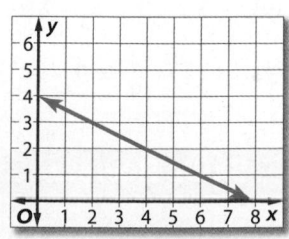

51. **WRITING IN MATH** Linear equations are useful in predicting future events. Describe some factors in real-world situations that might affect the reliability of the graph in making any predictions.

52. **CCSS ARGUMENTS** What information is needed to write the equation of a line? Explain.

53. Which equation *best* represents the graph?

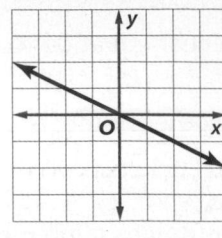

   A $y = 2x$

   B $y = -2x$

   C $y = \frac{1}{2}x$

   D $y = -\frac{1}{2}x$

54. Roberto receives an employee discount of 12%. If he buys a $355 item at the store, what is his discount to the nearest dollar?

   F $3   **H** $30

   G $4   **J** $43

55. GEOMETRY The midpoints of the sides of the large square are joined to form a smaller square. What is the area of the smaller square?

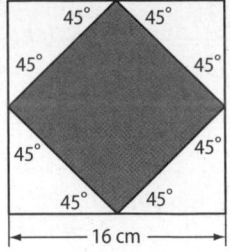

   A 64 cm^2

   B 128 cm^2

   C 248 cm^2

   D 256 cm^2

56. SHORT RESPONSE If $\frac{5(x + 4)}{2} + 7 = 37$, what is the value of $3x - 9$?

Graph each equation. (Lesson 4-1)

57. $y = 3x + 2$ **58.** $y = -4x + 2$ **59.** $3y = 2x + 6$

60. $y = \frac{1}{2}x + 6$ **61.** $3x + y = -1$ **62.** $2x + 3y = 6$

Write an equation in function notation for each relation. (Lesson 3-6)

63.

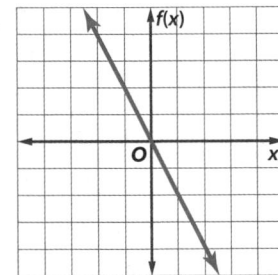

64.

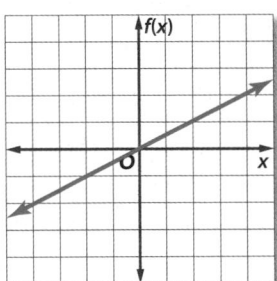

65. METEOROLOGY The distance d in miles that the sound of thunder travels in t seconds is given by the equation $d = 0.21t$. (Lesson 3-4)

   a. Graph the equation.

   b. Use the graph to estimate how long it will take you to hear thunder from a storm 3 miles away.

Solve each equation. Check your solution. (Lesson 2-3)

66. $-5t - 2.2 = -2.9$ **67.** $-5.5a - 43.9 = 77.1$ **68.** $4.2r + 7.14 = 12.6$

69. $-14 - \frac{n}{9} = 9$ **70.** $\frac{-8b - (-9)}{-10} = 17$ **71.** $9.5x + 11 - 7.5x = 14$

Find the value of r so the line through each pair of points has the given slope.

72. $(6, -2), (r, -6), m = 4$ **73.** $(8, 10), (r, 4), m = 6$ **74.** $(7, -10), (r, 4), m = -3$

75. $(6, 2), (9, r), m = -1$ **76.** $(9, r), (6, 3), m = -\frac{1}{3}$ **77.** $(5, r), (2, -3), m = \frac{4}{3}$

4-3 Writing Equations in Point-Slope Form

∵Then	∵Now	∵Why?
● You wrote linear equations given either one point and the slope or two points.	**1** Write equations of lines in point-slope form. **2** Write linear equations in different forms.	● Most humane societies have foster homes for newborn puppies, kittens, and injured or ill animals. During the spring and summer, a large shelter can place 3000 animals in homes each month. If a shelter had 200 animals in foster homes at the beginning of spring, the number of animals in foster homes at the end of the summer could be represented by $y = 3000x + 200$, where x is the number of months and y is the number of animals.

 NewVocabulary
point-slope form

 Common Core State Standards

Content Standards
F.IF.2 Use function notation, evaluate functions for inputs in their domains, and interpret statements that use function notation in terms of a context.

F.LE.2 Construct linear and exponential functions, including arithmetic and geometric sequences, given a graph, a description of a relationship, or two input-output pairs (include reading these from a table).

Mathematical Practices
2 Reason abstractly and quantitatively.

1 **Point-Slope Form** An equation of a line can be written in **point-slope form** when given the coordinates of one known point on a line and the slope of that line.

KeyConcept Point-Slope Form

Words	The linear equation $y - y_1 = m(x - x_1)$ is written in point-slope form, where (x_1, y_1) is a given point on a nonvertical line and m is the slope of the line.
Symbols	$y - y_1 = m(x - x_1)$

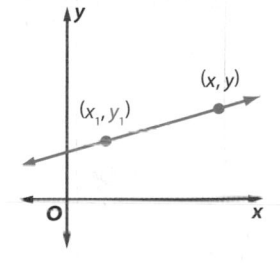

Example 1 Write and Graph an Equation in Point-Slope Form

Write an equation in point-slope form for the line that passes through $(3, -2)$ with a slope of $\frac{1}{4}$. Then graph the equation.

$y - y_1 = m(x - x_1)$ Point-slope form

$y - (-2) = \frac{1}{4}(x - 3)$ $(x_1, y_1) = (3, -2)$, $m = \frac{1}{4}$

$y + 2 = \frac{1}{4}(x - 3)$ Simplify.

Plot the point at $(3, -2)$ and use the slope to find another point on the line. Draw a line through the two points.

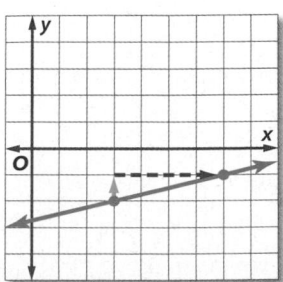

▶ **GuidedPractice**

1. Write an equation in point-slope form for the line that passes through $(-2, 1)$ with a slope of -6. Then graph the equation.

2 Forms of Linear Equations
If you are given the slope and the coordinates of one or two points, you can write the linear equation in the following ways.

ConceptSummary Writing Equations

Given the Slope and One Point

Step 1 Substitute the value of m and let the x and y coordinates be (x_1, y_1). Or, substitute the value of m, x, and y into the slope-intercept form and solve for b.

Step 2 Rewrite the equation in the needed form.

Given Two Points

Step 1 Find the slope.

Step 2 Choose one of the two points to use.

Step 3 Follow the steps for writing an equation given the slope and one point.

Example 2 Standard Form

Write $y - 1 = -\frac{2}{3}(x - 5)$ in standard form.

$y - 1 = -\frac{2}{3}(x - 5)$ Original equation

$3(y - 1) = 3\left(-\frac{2}{3}\right)(x - 5)$ Multiply each side by 3 to eliminate the fraction.

$3(y - 1) = -2(x - 5)$ Simplify.

$3y - 3 = -2x + 10$ Distributive Property

$3y = -2x + 13$ Add 3 to each side.

$2x + 3y = 13$ Add $2x$ to each side.

Guided Practice

2. Write $y - 1 = 7(x + 5)$ in standard form.

To find the y-intercept of an equation, rewrite the equation in slope-intercept form.

Example 3 Slope-Intercept Form

Write $y + 3 = \frac{3}{2}(x + 1)$ in slope-intercept form.

$y + 3 = \frac{3}{2}(x + 1)$ Original equation

$y + 3 = \frac{3}{2}x + \frac{3}{2}$ Distributive Property

$y = \frac{3}{2}x - \frac{3}{2}$ Subtract 3 from each side.

Guided Practice

3. Write $y + 6 = -3(x - 4)$ in slope-intercept form.

Being able to use a variety of forms of linear equations can be useful in other subjects as well.

Example 4 Point-Slope Form and Standard Form

GEOMETRY The figure shows square *RSTU*.

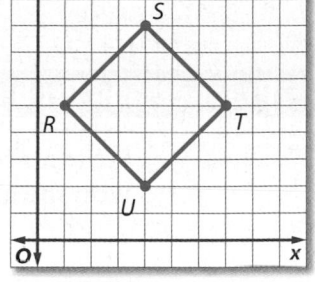

a. Write an equation in point-slope form for the line containing side $\overline{TU}$.

Step 1 Find the slope of $\overline{TU}$.

$$m = \frac{y_2 - y_1}{x_2 - x_1}$$ Slope Formula

$$= \frac{5 - 2}{7 - 4} \text{ or } 1$$ $(x_1, y_1) = (4, 2)$ and $(x_2, y_2) = (7, 5)$

Step 2 You can select either point for (x_1, y_1) in the point-slope form.

$$y - y_1 = m(x - x_1)$$ Point-slope form

$$y - 2 = 1(x - 4)$$ $(x_1, y_1) = (4, 2)$

$$y - 5 = 1(x - 7)$$ $(x_1, y_1) = (7, 5)$

b. Write an equation in standard form for the same line.

$y - 2 = 1(x - 4)$	Original equation	$y - 5 = 1(x - 7)$
$y - 2 = 1x - 4$	Distributive Property	$y - 5 = 1x - 7$
$y = 1x - 2$	Add to each side.	$y = 1x - 2$
$-1x + y = -2$	Subtract 1x from each side.	$-1x + y = -2$
$x - y = 2$	Multiply each side by −1.	$x - y = 2$

▶ **Guided**Practice

4A. Write an equation in point-slope form of the line containing side $\overline{ST}$.

4B. Write an equation in standard form of the line containing $\overline{ST}$.

Check Your Understanding

 = Step-by-Step Solutions begin on page R13.

Example 1   Write an equation in point-slope form for the line that passes through the given point with the slope provided. Then graph the equation.

① $(-2, 5)$, slope -6 **2.** $(-2, -8)$, slope $\frac{5}{6}$ **3.** $(4, 3)$, slope $-\frac{1}{2}$

Example 2   Write each equation in standard form.

4. $y + 2 = \frac{7}{8}(x - 3)$ **5.** $y + 7 = -5(x + 3)$ **6.** $y + 2 = \frac{5}{3}(x + 6)$

Example 3   Write each equation in slope-intercept form.

7. $y - 10 = 4(x + 6)$ **8.** $y - 7 = -\frac{3}{4}(x + 5)$ **9.** $y - 9 = x + 4$

Example 4   **10. GEOMETRY** Use right triangle *FGH*.

 a. Write an equation in point-slope form for the line containing $\overline{GH}$.

 b. Write the standard form of the line containing $\overline{GH}$.

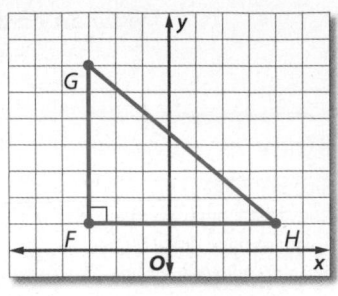

Practice and Problem Solving

Extra Practice is on page R4.

Example 1   Write an equation in point-slope form for the line that passes through each point with the given slope. Then graph the equation.

11. $(5, 3)$, $m = 7$ **12.** $(2, -1)$, $m = -3$

13. $(-6, -3)$, $m = -1$ **14.** $(-7, 6)$, $m = 0$

15. $(-2, 11)$, $m = \frac{4}{3}$ **16.** $(-6, -8)$, $m = -\frac{5}{8}$

17. $(-2, -9)$, $m = -\frac{7}{5}$ **18.** $(-6, 0)$, horizontal line

Example 2   Write each equation in standard form.

19. $y - 10 = 2(x - 8)$ **20.** $y - 6 = -3(x + 2)$

21. $y - 9 = -6(x + 9)$ **22.** $y + 4 = \frac{2}{3}(x + 7)$

23. $y + 7 = \frac{9}{10}(x + 3)$ **24.** $y + 7 = -\frac{3}{2}(x + 1)$

25. $2y + 3 = -\frac{1}{3}(x - 2)$ **26.** $4y - 5x = 3(4x - 2y + 1)$

Example 3   Write each equation in slope-intercept form.

27. $y - 6 = -2(x - 7)$ **28.** $y - 11 = 3(x + 4)$

29. $y + 5 = -6(x + 7)$ **30.** $y - 1 = \frac{4}{5}(x + 5)$

31. $y + 2 = \frac{1}{6}(x - 4)$ **32.** $y + 6 = -\frac{3}{4}(x + 8)$

33. $y + 3 = -\frac{1}{3}(2x + 6)$ **34.** $y + 4 = 3(3x + 3)$

Example 4   **35 MOVIE RENTALS** The number of copies of a movie rented at a video kiosk decreased at a constant rate of 5 copies per week. The 6th week after the movie was released, 4 copies were rented. How many copies were rented during the second week?

36. CCSS REASONING A company offers premium cable for $39.95 per month plus a one-time setup fee. The total cost for setup and 6 months of service is $264.70.

 a. Write an equation in point-slope form to find the total price *y* for any number of months *x*. (*Hint*: The point (6, 264.70) is a solution to the equation.)

 b. Write the equation in slope-intercept form.

 c. What is the setup fee?

Write an equation for the line described in standard form.

37. through $(-1, 7)$ and $(8, -2)$ **38.** through $(-4, 3)$ with *y*-intercept 0

39. with *x*-intercept 4 and *y*-intercept 5

Write an equation in point-slope form for each line.

40.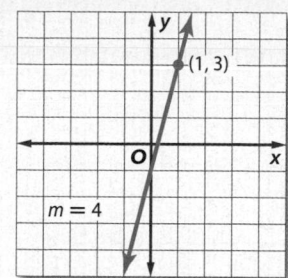

41.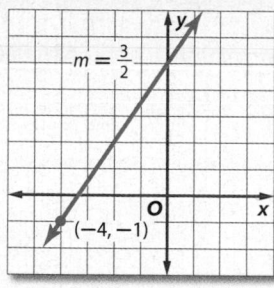

42.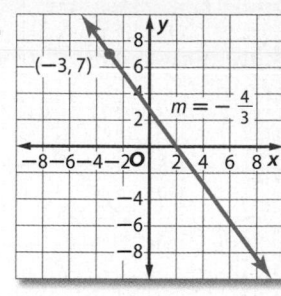

Write each equation in slope-intercept form.

43 $y + \dfrac{3}{5} = x - \dfrac{2}{5}$

44. $y - \dfrac{7}{2} = \dfrac{1}{2}(x - 4)$

45. $y + \dfrac{1}{3} = \dfrac{5}{6}\left(x + \dfrac{2}{5}\right)$

46. Write an equation in point-slope form, slope-intercept form, and standard form for a line that passes through $(-2, 8)$ with slope $\dfrac{8}{5}$.

47. Line ℓ passes through $(-9, 4)$ with slope $\dfrac{4}{7}$. Write an equation in point-slope form, slope-intercept form, and standard form for line ℓ.

48. WEATHER The barometric pressure is 598 millimeters of mercury (mmHg) at an altitude of 1.8 kilometers and 577 millimeters of mercury at 2.1 kilometers.

   a. Write a formula for the barometric pressure as a function of the altitude.

   b. What is the altitude if the pressure is 657 millimeters of mercury?

H.O.T. Problems Use Higher-Order Thinking Skills

49. WHICH ONE DOESN'T BELONG? Identify the equation that does not belong. Explain your reasoning.

   | $y - 5 = 3(x - 1)$ | $y + 1 - 3(x + 1)$ | $y + 4 = 3(x + 1)$ | $y - 8 = 3(x - 2)$ |

50. CCSS CRITIQUE Juana thinks that $f(x)$ and $g(x)$ have the same slope but different intercepts. Sabrina thinks that $f(x)$ and $g(x)$ describe the same line. Is either of them correct? Explain your reasoning.

   The graph of $g(x)$ is the line that passes through $(3, -7)$ and $(-6, 4)$.

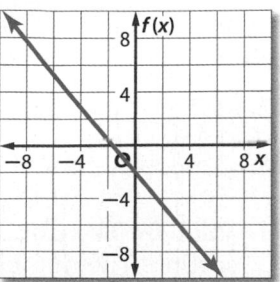

51. OPEN ENDED Describe a real-life scenario that has a constant rate of change and a value of y for a particular value of x. Represent this situation using an equation in point-slope form, an equation in standard form, and an equation in slope-intercept form.

52. REASONING Write an equation for the line that passes through $(-4, 8)$ and $(3, -7)$. What is the slope? Where does the line intersect the x-axis? the y-axis?

53. CHALLENGE Write an equation in point-slope form for the line that passes through the points (f, g) and (h, j).

54. WRITING IN MATH Why do we represent linear equations in more than one form?

55. Which statement is *most* strongly supported by the graph?

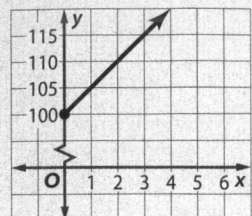

A You have $100 and spend $5 weekly.

B You have $100 and save $5 weekly.

C You need $100 for a new CD player and save $5 weekly.

D You need $100 for a new CD player and spend $5 weekly.

56. SHORT RESPONSE A store offers customers a $5 gift certificate for every $75 they spend. How much would a customer have to spend to earn $35 worth of gift certificates?

57. GEOMETRY Which triangle is similar to $\triangle ABC$?

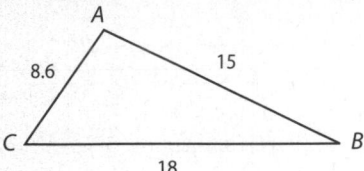

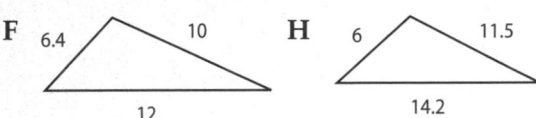

58. In a class of 25 students, 6 have blue eyes, 15 have brown hair, and 3 have blue eyes and brown hair. How many students have neither blue eyes nor brown hair?

A 4   C 10

B 7   D 22

Spiral Review

Write an equation of the line that passes through each pair of points. (Lesson 4-2)

59. $(4, 2), (-2, -4)$

60. $(3, -2), (6, 4)$

61. $(-1, 3), (2, -3)$

62. $(2, -2), (3, 2)$

63. $(7, -2), (-4, -2)$

64. $(0, 5), (-3, 5)$

Write an equation in slope-intercept form of the line with the given slope and y-intercept. (Lesson 4-1)

65. slope: -2, y-intercept: 6

66. slope: 3, y-intercept: -5

67. slope: $\frac{1}{2}$, y-intercept: 3

68. slope: $-\frac{3}{5}$, y-intercept: 12

69. slope: 0, y-intercept: 3

70. slope: -1, y-intercept: 0

71. THEATER The Coral Gables Actors' Playhouse has 7 rows of seats in the orchestra section. The number of seats in the rows forms an arithmetic sequence, as shown in the table. On opening night, 368 tickets were sold for the orchestra section. Was the section oversold? (Lesson 3-5)

Rows	Number of Seats
7	76
6	68
5	60

Skills Review

Solve each equation or formula for the variable specified.

72. $y = mx + b$, for m

73. $v = r + at$, for a

74. $km + 5x = 6y$, for m

75. $4b - 5 = -t$, for b

Parallel and Perpendicular Lines

Purestock/Getty Images

:: Then

- You wrote equations in point-slope form.

:: Now

- **1** Write an equation of the line that passes through a given point, parallel to a given line.

- **2** Write an equation of the line that passes through a given point, perpendicular to a given line.

:: Why?

- Notice the squares, rectangles and lines in the piece of art shown at the right. Some of the lines intersect forming right angles. Other lines do not intersect at all.

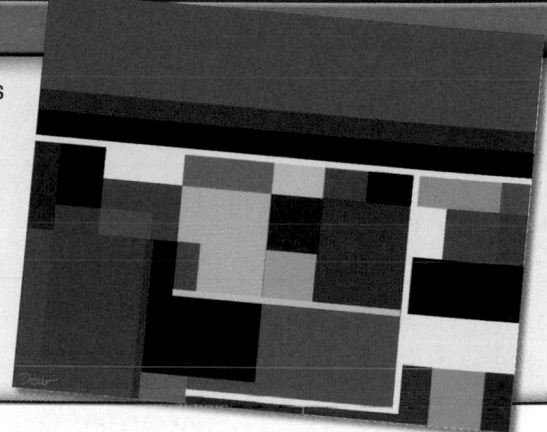

NewVocabulary
parallel lines
perpendicular lines

 Common Core State Standards

Content Standards

F.LE.2 Construct linear and exponential functions, including arithmetic and geometric sequences, given a graph, a description of a relationship, or two input-output pairs (include reading these from a table).

S.ID.7 Interpret the slope (rate of change) and the intercept (constant term) of a linear model in the context of the data.

Mathematical Practices
5 Use appropriate tools strategically.

1 Parallel Lines Lines in the same plane that do not intersect are called **parallel lines**. Nonvertical parallel lines have the same slope.

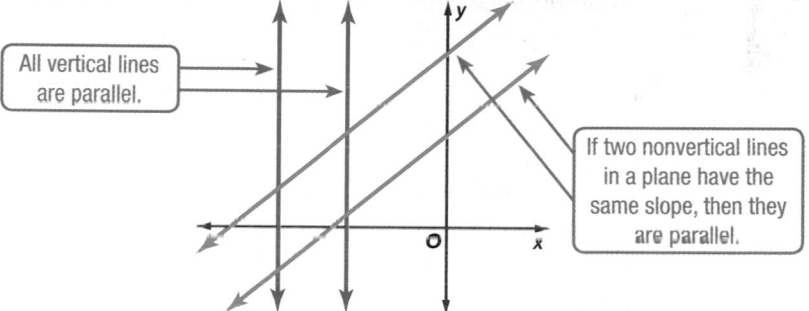

All vertical lines are parallel.

If two nonvertical lines in a plane have the same slope, then they are parallel.

You can write an equation of a line parallel to a given line if you know a point on the line and an equation of the given line. First find the slope of the given line. Then, substitute the point provided and the slope from the given line into the point-slope form.

Example 1 Parallel Line Through a Given Point

Write an equation in slope-intercept form for the line that passes through $(-3, 5)$ and is parallel to the graph of $y = 2x - 4$.

Step 1 The slope of the line with equation $y = 2x - 4$ is 2. The line parallel to $y = 2x - 4$ has the same slope, 2.

Step 2 Find the equation in slope-intercept form.

$y - y_1 = m(x - x_1)$	Point-slope form
$y - 5 = 2[x - (-3)]$	Replace m with 2 and (x_1, y_1) with $(-3, 5)$.
$y - 5 = 2(x + 3)$	Simplify.
$y - 5 = 2x + 6$	Distributive Property
$y - 5 + 5 = 2x + 6 + 5$	Add 5 to each side.
$y = 2x + 11$	Write the equation in slope-intercept form.

Guided Practice

1. Write an equation in point-slope form for the line that passes through $(4, -1)$ and is parallel to the graph of $y = \frac{1}{4}x + 7$.

2 Perpendicular Lines

Lines that intersect at right angles are called **perpendicular lines**. The slopes of nonvertical perpendicular lines are opposite reciprocals. That is, if the slope of a line is 4, the slope of the line perpendicular to it is $-\frac{1}{4}$.

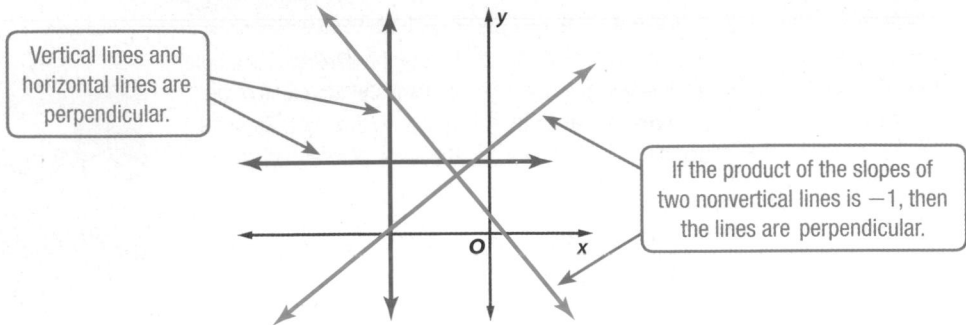

Vertical lines and horizontal lines are perpendicular.

If the product of the slopes of two nonvertical lines is -1, then the lines are perpendicular.

You can use slope to determine whether two lines are perpendicular.

Real-World Example 2 Slopes of Perpendicular Lines

DESIGN The outline of a company's new logo is shown on a coordinate plane.

a. Is ∠DFE a right angle in the logo?

If $\overline{BE}$ and $\overline{AD}$ are perpendicular, then ∠DFE is a right angle. Find the slopes of $\overline{BE}$ and $\overline{AD}$.

slope of $\overline{BE}$: $m = \frac{1-3}{7-2}$ or $-\frac{2}{5}$

slope of $\overline{AD}$: $m = \frac{6-1}{4-2}$ or $\frac{5}{2}$

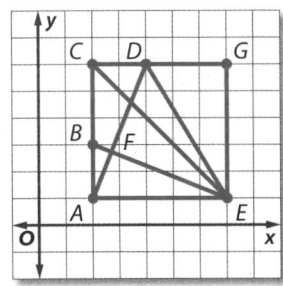

The line segments are perpendicular because $-\frac{2}{5} \times \frac{5}{2} = -1$. Therefore, ∠DFE is a right angle.

b. Is each pair of opposite sides parallel?

If a pair of opposite sides are parallel, then they have the same slope.

slope of $\overline{AC}$: $m = \frac{6-1}{2-2}$ or undefined

Since $\overline{AC}$ and $\overline{GE}$ are both parallel to the y-axis, they are vertical and are therefore parallel.

slope of $\overline{CG}$: $m = \frac{6-6}{7-2}$ or 0

Since $\overline{CG}$ and $\overline{AE}$ are both parallel to the x-axis, they are horizontal and are therefore parallel.

▶ **Guided**Practice

2. CONSTRUCTION On the plans for a treehouse, a beam represented by $\overline{QR}$ has endpoints $Q(-6, 2)$ and $R(-1, 8)$. A connecting beam represented by $\overline{ST}$ has endpoints $S(-3, 6)$ and $T(-8, 5)$. Are the beams perpendicular? Explain.

You can determine whether the graphs of two linear equations are parallel or perpendicular by comparing the slopes of the lines.

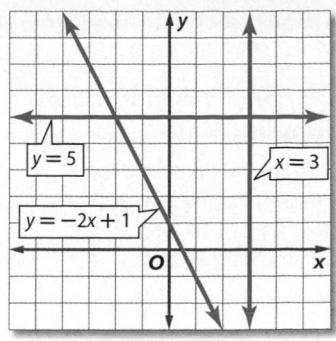

Example 3 Parallel or Perpendicular Lines

Determine whether the graphs of $y = 5$, $x = 3$, and $y = -2x + 1$ are *parallel* or *perpendicular*. Explain.

Graph each line on a coordinate plane.

From the graph, you can see that $y = 5$ is parallel to the *x*-axis and $x = 3$ is parallel to the *y*-axis. Therefore, they are perpendicular. None of the lines are parallel.

▶ **Guided**Practice

3. Determine whether the graphs of $6x - 2y = -2$, $y = 3x - 4$, and $y = 4$ are *parallel* or *perpendicular*. Explain.

You can write the equation of a line perpendicular to a given line if you know a point on the line and the equation of the given line.

Example 4 Perpendicular Line Through a Given Point

StudyTip

CCSS Tools Graph the given equation on a coordinate grid and plot the given point. Using a ruler, draw a line perpendicular to the given line that passes through the point.

Write an equation in slope-intercept form for the line that passes through (4, 6) and is perpendicular to the graph of $2x + 3y = 12$.

Step 1 Find the slope of the given line by solving the equation for *y*.

$$2x + 3y = 12 \qquad \text{Original equation}$$

$$2x - 2x + 3y = -2x + 12 \qquad \text{Subtract 2x from each side.}$$

$$3y = -2x + 12 \qquad \text{Simplify.}$$

$$\frac{3y}{3} = \frac{-2x + 12}{3} \qquad \text{Divide each side by 3.}$$

$$y = -\frac{2}{3}x + 4 \qquad \text{Simplify.}$$

The slope is $-\frac{2}{3}$.

Step 2 The slope of the perpendicular line is the opposite reciprocal of $-\frac{2}{3}$ or $\frac{3}{2}$. Find the equation of the perpendicular line.

$$y - y_1 = m(x - x_1) \qquad \text{Point-slope form}$$

$$y - 6 = \frac{3}{2}[x - (-4)] \qquad (x_1, y_1) = (-4, 6) \text{ and } m = \frac{3}{2}$$

$$y - 6 = \frac{3}{2}(x + 4) \qquad \text{Simplify.}$$

$$y - 6 = \frac{3}{2}x + 6 \qquad \text{Distributive Property}$$

$$y - 6 + 6 = \frac{3}{2}x + 6 + 6 \qquad \text{Add 6 to each side.}$$

$$y = \frac{3}{2}x + 12 \qquad \text{Simplify.}$$

▶ **Guided**Practice

4. Write an equation in slope-intercept form for the line that passes through (4, 7) and is perpendicular to the graph of $y = \frac{2}{3}x - 1$.

ConceptSummary Parallel and Perpendicular Lines

	Parallel Lines	Perpendicular Lines
Words	Two nonvertical lines are parallel if they have the same slope.	Two nonvertical lines are perpendicular if the product of their slopes is −1.
Symbols	$\overleftrightarrow{AB} \parallel \overleftrightarrow{CD}$	$\overleftrightarrow{EF} \perp \overleftrightarrow{GH}$
Models	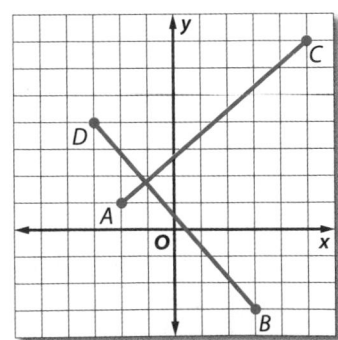	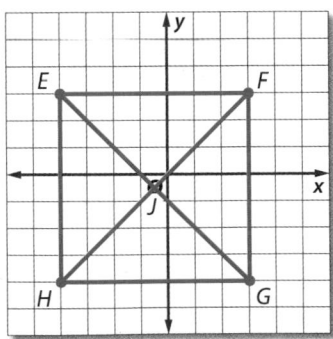

ReadingMath

Parallel and Perpendicular Lines The symbol for parallel is ∥. The symbol for perpendicular is ⊥.

Check Your Understanding

○ — Step-by-Step Solutions begin on page R13.

Example 1 Write an equation in slope-intercept form for the line that passes through the given point and is parallel to the graph of the given equation.

1. $(-1, 2), y = \frac{1}{2}x - 3$

2. $(0, 4), y = -4x + 5$

Example 2

3. GARDENS A garden is in the shape of a quadrilateral with vertices $A(-2, 1)$, $B(3, -3)$, $C(5, 7)$, and $D(-3, 4)$. Two paths represented by $\overline{AC}$ and $\overline{BD}$ cut across the garden. Are the paths perpendicular? Explain.

4. CCSS PRECISION A square is a quadrilateral that has opposite sides parallel, consecutive sides that are perpendicular, and diagonals that are perpendicular. Determine whether the quadrilateral is a square. Explain.

Example 3 Determine whether the graphs of the following equations are *parallel* or *perpendicular*. Explain.

5 $y = -2x, 2y = x, 4y = 2x + 4$

6. $y = \frac{1}{2}x, 3y = x, y = -\frac{1}{2}x$

Example 4 Write an equation in slope-intercept form for the line that passes through the given point and is perpendicular to the graph of the equation.

7. $(-2, 3), y = -\frac{1}{2}x - 4$

8. $(-1, 4), y = 3x + 5$

9. $(2, 3), 2x + 3y = 4$

10. $(3, 6), 3x - 4y = -2$

Example 1 Write an equation in slope-intercept form for the line that passes through the given point and is parallel to the graph of the given equation.

11. $(3, -2), y = x + 4$ **12.** $(4, -3), y = 3x - 5$ **13.** $(0, 2), y = -5x + 8$

14. $(-4, 2), y = -\frac{1}{2}x + 6$ **15.** $(-2, 3), y = -\frac{3}{4}x + 4$ **16.** $(9, 12), y = 13x - 4$

Example 2 **17. GEOMETRY** A trapezoid is a quadrilateral that has exactly one pair of parallel opposite sides. Is *ABCD* a trapezoid? Explain your reasoning.

18. GEOMETRY *CDEF* is a kite. Are the diagonals of the kite perpendicular? Explain your reasoning.

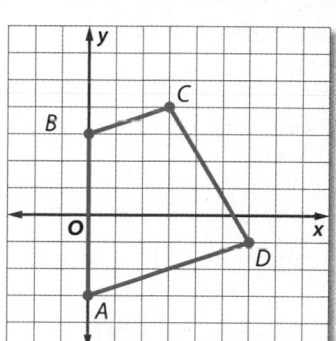

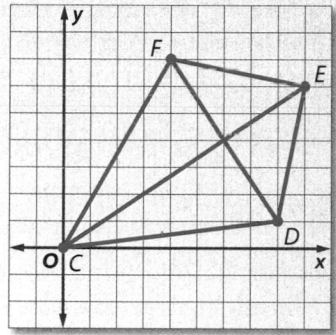

19. Determine whether the graphs of $y = -6x + 4$ and $y = \frac{1}{6}x$ are perpendicular. Explain.

20. MAPS On a map, Elmwood Drive passes through $R(4, -11)$ and $S(0, -9)$, and Taylor Road passes through $J(6, -2)$ and $K(4, -5)$. If they are straight lines, are the two streets perpendicular? Explain.

Example 3 **CCSS PERSEVERANCE** Determine whether the graphs of the following equations are *parallel* or *perpendicular*. Explain.

21. $2x - 8y = -24, 4x + y = -2, x - 4y = 4$

22. $3x - 9y - 9, 3y - x + 12, 2x - 6y - 12$

Example 4 Write an equation in slope-intercept form for the line that passes through the given point and is perpendicular to the graph of the equation.

23 $(-3, -2), y = -2x + 4$ **24.** $(-5, 2), y = \frac{1}{2}x - 3$ **25.** $(-4, 5), y = \frac{1}{3}x + 6$

26. $(2, 6), y = -\frac{1}{4}x + 3$ **27.** $(3, 8), y = 5x - 3$ **28.** $(4, -2), y = 3x + 5$

Write an equation in slope-intercept form for a line perpendicular to the graph of the equation that passes through the *x*-intercept of that line.

29. $y = -\frac{1}{2}x - 4$ **30.** $y = \frac{2}{3}x - 6$ **31.** $y = 5x + 3$

32. Write an equation in slope-intercept form for the line that is perpendicular to the graph of $3x + 2y = 8$ and passes through the *y*-intercept of that line.

Determine whether the graphs of each pair of equations are *parallel*, *perpendicular*, or *neither*.

33. $y = 4x + 3$ **34.** $y = -2x$ **35.** $3x + 5y = 10$
  $4x + y = 3$ $2x + y = 3$ $5x - 3y = -6$

36. $-3x + 4y = 8$ **37.** $2x + 5y = 15$ **38.** $2x + 7y = -35$
  $-4x + 3y = -6$ $3x + 5y = 15$ $4x + 14y = -42$

39. Write an equation of the line that is parallel to the graph of $y = 7x - 3$ and passes through the origin.

40. EXCAVATION Scientists excavating a dinosaur mapped the site on a coordinate plane. If one bone lies from $(-5, 8)$ to $(10, -1)$ and a second bone lies from $(-10, -3)$ to $(-5, -6)$, are the bones parallel? Explain.

41 **ARCHAEOLOGY** In the ruins of an ancient civilization, an archaeologist found pottery at $(2, 6)$ and hair accessories at $(4, -1)$. A pole is found with one end at $(7, 10)$ and the other end at $(14, 12)$. Is the pole perpendicular to the line through the pottery and the hair accessories? Explain.

42. GRAPHICS To create a design on a computer, Andeana must enter the coordinates for points on the design. One line segment she drew has endpoints of $(-2, 1)$ and $(4, 3)$. The other coordinates that Andeana entered are $(2, -7)$ and $(8, -3)$. Could these points be the vertices of a rectangle? Explain.

43. **MULTIPLE REPRESENTATIONS** In this problem, you will explore parallel and perpendicular lines.

   a. Graphical Graph the points $A(-3, 3)$, $B(3, 5)$, and $C(-4, 0)$ on a coordinate plane.

   b. Analytical Determine the coordinates of a fourth point D that would form a parallelogram. Explain your reasoning.

   c. Analytical What is the minimum number of points that could be moved to make the parallelogram a rectangle? Describe which points should be moved, and explain why.

H.O.T. Problems   Use Higher-Order Thinking Skills

44. CHALLENGE If the line through $(-2, 4)$ and $(5, d)$ is parallel to the graph of $y = 3x + 4$, what is the value of d?

45. REASONING Which key features of the graphs of two parallel lines are the same, and which are different? Which key features of the graphs of two perpendicular lines are the same, and which are different?

46. OPEN ENDED Graph a line that is parallel and a line that is perpendicular to $y = 2x - 1$.

Example 3   **47.** **CCSS** **CRITIQUE** Carmen and Chase are finding an equation of the line that is perpendicular to the graph of $y = \frac{1}{3}x + 2$ and passes through the point $(-3, 5)$. Is either of them correct? Explain your reasoning.

Carmen	Chase
$y - 5 = -3[x - (-3)]$	$y - 5 = 3[x - (-3)]$
$y - 5 = -3(x + 3)$	$y - 5 = 3(x + 3)$
$y = -3x - 9 + 5$	$y = 3x + 9 + 5$
$y = -3x - 4$	$y = 3x + 14$

48. WRITING IN MATH Illustrate how you can determine whether two lines are parallel or perpendicular. Write an equation for the graph that is parallel and an equation for the graph that is perpendicular to the line shown. Explain your reasoning.

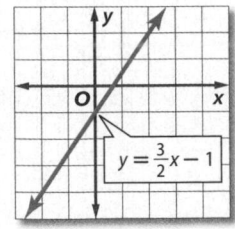

$y = \frac{3}{2}x - 1$

49. Which of the following is an algebraic translation of the following phrase?

5 less than the quotient of a number and 8

A $5 - \frac{n}{8}$ **C** $5 - \frac{8}{n}$

B $\frac{n}{8} - 5$ **D** $\frac{8}{n} - 5$

50. A line through which two points would be parallel to a line with a slope of $\frac{3}{4}$?

F $(0, 5)$ and $(-4, 2)$ **H** $(0, 0)$ and $(0, -2)$

G $(0, 2)$ and $(-4, 1)$ **J** $(0, -2)$ and $(-4, -2)$

51. Which equation best fits the data in the table?

x	y
1	5
2	7
3	9
4	11

A $y = x + 4$

B $y = 2x + 3$

C $y = 7$

D $y = 4x - 5$

52. SHORT RESPONSE Tyler is filling his 6000-gallon pool at a constant rate. After 4 hours, the pool contained 800 gallons. How many total hours will it take to completely fill the pool?

Write each equation in standard form. (Lesson 4-3)

53. $y - 13 = 4(x - 2)$

54. $y - 5 = -2(x + 2)$

55. $y + 3 = -5(x + 1)$

56. $y + 7 = \frac{1}{2}(x + 2)$

57. $y - 1 = \frac{5}{6}(x - 4)$

58. $y - 2 = -\frac{2}{5}(x - 8)$

59. CANOE RENTAL Latanya and her friends rented a canoe for 3 hours and paid a total of $45. (Lesson 4-2)

  a. Write a linear equation to find the total cost C of renting the canoe for h hours.

  b. How much would it cost to rent the canoe for 8 hours?

Canoe Rentals
Daily rates
plus **$10 per hour**

Write an equation of the line that passes through each point with the given slope. (Lesson 4-2)

60. $(5, -2), m = 3$

61. $(-5, 4), m = -5$

62. $(3, 0), m = -2$

63. $(3, 5), m = 2$

64. $(-3, -1), m = -3$

65. $(-2, 4), m = -5$

Simplify each expression. If not possible, write *simplified*. (Lesson 1-4)

66. $13m + m$

67. $14a^2 + 13b^2 + 27$

68. $3(x + 2x)$

69. FINANCIAL LITERACY At a Farmers' Market, merchants can rent a small table for $5.00 and a large table for $8.50. One time, 25 small and 10 large tables were rented. Another time, 35 small and 12 large were rented. (Lesson 1-2)

  a. Write an algebraic expression to show the total amount of money collected.

  b. Evaluate the expression.

Express each relation as a graph. Then determine the domain and range.

70. $\{(3, 8), (3, 7), (2, -9), (1, -9), (-5, -3)\}$

71. $\{(3, 4), (4, 3), (2, 2), (5, -4), (-4, 5)\}$

72. $\{(0, 2), (-5, 1), (0, 6), (-1, 9), (-4, -5)\}$

73. $\{(-7, 6), (-3, -4), (4, -5), (-2, 6), (-3, 2)\}$

Mid-Chapter Quiz
Lessons 4-1 through 4-4

Write an equation in slope-intercept form for each graph shown. (Lesson 4-1)

1.

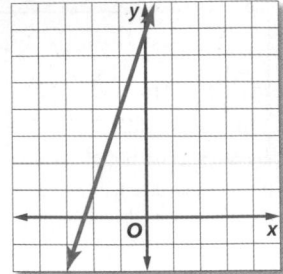

2.

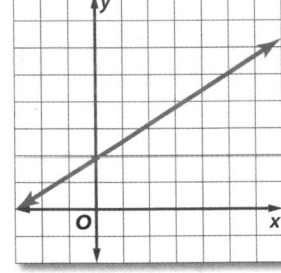

Graph each equation. (Lesson 4-1)

3. $y = 2x + 3$

4. $y = \frac{1}{3}x - 2$

5. BOATS Write an equation in slope-intercept form for the total rental cost C for a pontoon boat used for t hours. (Lesson 4-1)

Pontoon Boat Rentals
$60 per hour
plus
$20 cleaning fee

Write an equation of the line with the given conditions. (Lesson 4-2)

6. $(2, 5)$; slope 3

7. $(-3, -1)$, slope $\frac{1}{2}$

8. $(-3, 4), (1, 12)$

9. $(-1, 6), (2, 4)$

10. $(2, 1)$, slope 0

11. MULTIPLE CHOICE Write an equation of the line that passes through the point $(0, 0)$ and has slope -4. (Lesson 4-2)

A $y = x - 4$ **C** $y = -4x$

B $y = x + 4$ **D** $y = 4 - x$

Write an equation in point-slope form for the line that passes through each point with the given slope. (Lesson 4-3)

12. $(1, 4), m = 6$ **13.** $(-2, -1), m = -3$

14. Write an equation in point-slope form for the line that passes through the point $(8, 3)$, $m = -2$. (Lesson 4-3)

15. Write $y + 3 = \frac{1}{2}(x - 5)$ in standard form. (Lesson 4-3)

16. Write $y + 4 = -7(x - 3)$ in slope-intercept form. (Lesson 4-3)

Write each equation in standard form. (Lesson 4-3)

17. $y - 5 = -2(x - 3)$ **18.** $y + 4 = \frac{2}{3}(x - 3)$

Write each equation in slope-intercept form. (Lesson 4-3)

19. $y - 3 = 4(x + 3)$ **20.** $y + 1 = \frac{1}{2}(x - 8)$

21. MULTIPLE CHOICE Determine whether the graphs of the pair of equations are *parallel*, *perpendicular*, or *neither*. (Lesson 4-4)

$$y = -6x + 8$$
$$3x + \frac{1}{2}y = -3$$

F parallel

G perpendicular

H neither

J not enough information

Write an equation in slope-intercept form for the line that passes through the given point and is perpendicular to the graph of the equation. (Lesson 4-4)

22. $(3, -4)$; $y = -\frac{1}{3}x - 5$

23. $(0, -3)$; $y = -2x + 4$

24. $(-4, -5)$; $-4x + 5y = -6$

25. $(-1, -4)$; $-x - 2y = 0$

Scatter Plots and Lines of Fit

- You wrote linear equations given a point and the slope.

1. Investigate relationships between quantities by using points on scatter plots.

2. Use lines of fit to make and evaluate predictions.

- The graph shows the number of people from the United States who travel to other countries. The points do not all lie on the same line; however, you may be able to draw a line that is close to all of the points. That line would show a linear relationship between the year x and the number of travelers each year y. Generally, international travel has increased.

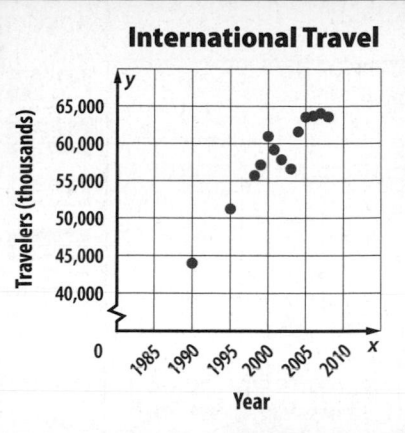

International Travel

NewVocabulary
bivariate data
scatter plot
line of fit
linear interpolation

Common Core State Standards

Content Standards

S.ID.6a Fit a function to the data; use functions fitted to data to solve problems in the context of the data. Use given functions or choose a function suggested by the context. Emphasize linear, quadratic, and exponential models.

S.ID.6c Fit a linear function for a scatter plot that suggests a linear association.

Mathematical Practices

1 Make sense of problems and persevere in solving them.

4 Model with mathematics.

1 Investigate Relationships Using Scatter Plots
Data with two variables are called **bivariate data**. A **scatter plot** shows the relationship between a set of data with two variables, graphed as ordered pairs on a coordinate plane. Scatter plots are used to investigate a relationship between two quantities.

ConceptSummary Scatter Plots

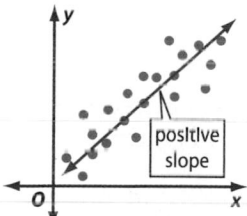

Positive Correlation

positive slope

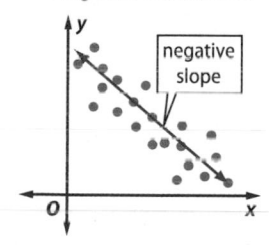

Negative Correlation

negative slope

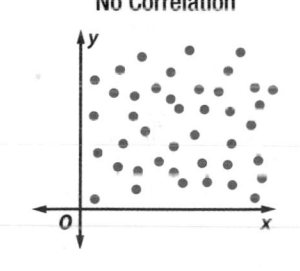

No Correlation

As x increases, y increases   As x decreases, y decreases x and y are not related

Real-World Example 1   Evaluate a Correlation

WAGES Determine whether the graph shows a *positive*, *negative*, or *no* correlation. If there is a positive or negative correlation, describe its meaning in the situation.

The graph shows a positive correlation. As the number of hours worked increases, the wages usually increase.

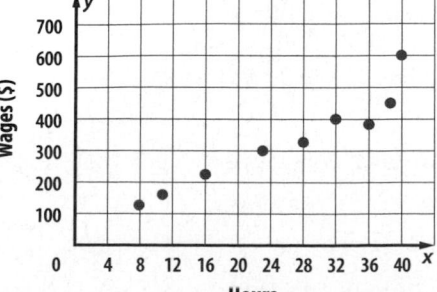

Wages

> **Guided**Practice

1. Refer to the graph on international travel. Determine whether the graph shows a *positive*, *negative*, or *no* correlation. If there is a positive or negative correlation, describe its meaning.

2 Use Lines of Fit
Scatter plots can show whether there is a trend in a set of data. When the data points all lie close to a line, a **line of fit** or *trend line* can model the trend.

KeyConcept Using a Linear Function to Model Data

Step 1 Make a scatter plot. Determine whether any relationship exists in the data.

Step 2 Draw a line that seems to pass close to most of the data points.

Step 3 Use two points on the line of fit to write an equation for the line.

Step 4 Use the line of fit to make predictions.

Real-World Example 2 Write a Line of Fit

ROLLER COASTERS The table shows the largest vertical drops of nine roller coasters in the United States and the number of years after 1988 that they were opened. Identify the independent and the dependent variables. Is there a relationship in the data? If so, predict the vertical drop in a roller coaster built 30 years after 1988.

Years Since 1988	1	3	5	8	12	12	12	13	15
Vertical Drop (ft)	151	155	225	230	306	300	255	255	400

Source: Ultimate Roller Coaster

Step 1 Make a scatter plot.

The independent variable is the year, and the dependent variable is the vertical drop. As the number of years increases, the vertical drop of roller coasters increases. There is a positive correlation between the two variables.

Vertical Drops of Roller Coasters

Step 2 Draw a line of fit.

No one line will pass through all of the data points. Draw a line that passes close to the points. A line of fit is shown.

Step 3 Write the slope-intercept form of an equation for the line of fit.

The line of fit passes close to (2, 150) and the data point (12, 300).

Find the slope.

$$m = \frac{y_2 - y_1}{x_2 - x_1} \qquad (x_1, y_1) = (2, 150),$$
$$\qquad\qquad\qquad (x_2, y_2) = (12, 300)$$

$$= \frac{300 - 150}{12 - 2}$$

$$= \frac{150}{10} \text{ or } 15$$

Use $m = 15$ and either the point-slope form or the slope-intercept form to write the equation of the line of fit.

$$y - y_1 = m(x - x_1)$$

$$y - 150 = 15(x - 2)$$

$$y - 150 = 15x - 30$$

$$y = 15x + 120$$

A slope of 15 means that the vertical drops increased an average of 15 feet per year. To predict the vertical drop of a roller coaster built 30 years after 1988, substitute 30 for x in the equation. The vertical drop is 15(30) + 120 or 570 feet.

2. MUSIC The table shows the dollar value in millions for the sales of CDs for the year. Make a scatter plot and determine what relationship exists, if any.

Year	2000	2001	2002	2003	2004	2005	2006	2007	2008
Sales	13,215	12,909	12,044	11,233	11,447	10,520	9373	7452	5471

ReadingMath

Interpolation and Extrapolation The Latin prefix *inter-* means between, and the Latin prefix *extra-* means beyond.

In Lesson 4-2, you learned that linear extrapolation is used to predict values *outside* the range of the data. You can also use a linear equation to predict values *inside* the range of the data. This is called **linear interpolation**.

🌐 Real-World Example 3 Use Interpolation or Extrapolation

TRAVEL Use the scatter plot to find the approximate number of United States travelers to international countries in 1996.

Step 1 Draw a line of fit. The line should be as close to as many points as possible.

Step 2 Write the slope-intercept form of the equation. The line of fit passes through (0, 44,623) and (18, 63,554).

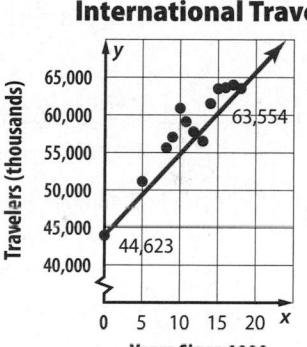

International Travel

Source: *Statistical Abstract of the United States*

Find the slope.

$$m = \frac{y_2 - y_1}{x_2 - x_1} \qquad \text{Slope Formula}$$

$$= \frac{63,554 - 44,623}{18 - 0} \qquad \begin{array}{l}(x_1, y_1) = (0, 44,623),\\ (x_2, y_2) = (18, 63,554)\end{array}$$

$$= \frac{18,931}{18} \qquad \text{Simplify.}$$

Use $m = \dfrac{18,931}{18}$ and either the point-slope form or the slope-intercept form to write the equation of the line of fit.

$$y - y_1 = m(x - x_1)$$

$$y - 44,623 = \frac{18,931}{18}(x - 0)$$

$$y - 44,623 = \frac{18,931}{18}x$$

$$y = \frac{18,931}{18}x + 44,623$$

Step 3 Evaluate the function for $x = 1996 - 1990$ or 6.

$$y = \frac{18,931}{18}x + 44,623 \qquad \text{Equation of best-fit line}$$

$$= \frac{18,931}{18}(6) + 44,623 \qquad x = 6$$

$$= 6310\frac{1}{3} + 44,623 \text{ or } 50,933\frac{1}{3} \qquad \text{Add.}$$

In 1996, there were approximately 50,933 thousand or 50,933,000 people who traveled from the United States to international countries.

▶ **Guided**Practice

3. MUSIC Use the equation for the line of fit for the data in Guided Practice 2 to estimate CD sales in 2015.

Example 1 Determine whether each graph shows a *positive*, *negative*, or *no* correlation. If there is a positive or negative correlation, describe its meaning in the situation.

1. Free Throws

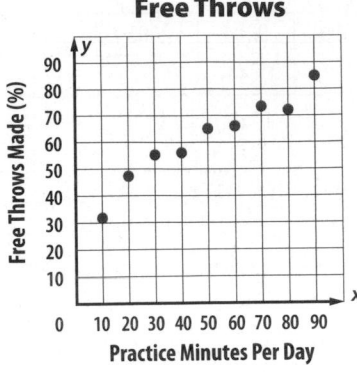

2. Lemonade Sales

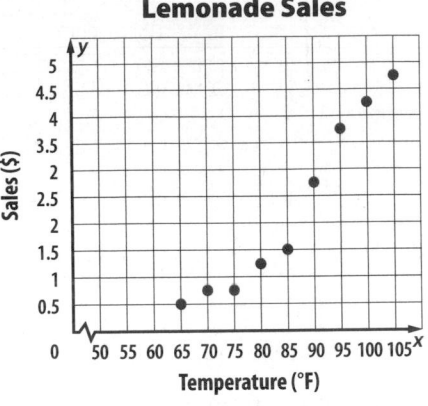

Example 2 **3. CCSS SENSE-MAKING** The table shows the median age of females when they were first married.

a. Make a scatter plot and determine what relationship exists, if any, in the data. Identify the independent and the dependent variables.

b. Draw a line of fit for the scatter plot.

c. Write an equation in slope-intercept form for the line of fit.

Example 3 **d.** Predict what the median age of females when they are first married will be in 2016.

e. Do you think the equation can give a reasonable estimate for the year 2056? Explain.

Year	Age
1996	24.8
1997	25.0
1998	25.0
1999	25.1
2000	25.1
2001	25.1
2002	25.3
2003	25.3
2004	25.3
2005	25.5
2006	25.9

Source: U.S. Bureau of Census

Practice and Problem Solving Extra Practice is on page R4.

Example 1 Determine whether each graph shows a *positive*, *negative*, or *no* correlation. If there is a positive or negative correlation, describe its meaning in the situation.

4. Game Tickets at the Fair

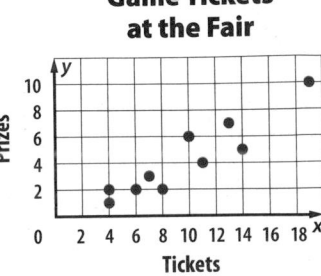

5 NBA 3-Point Percentage

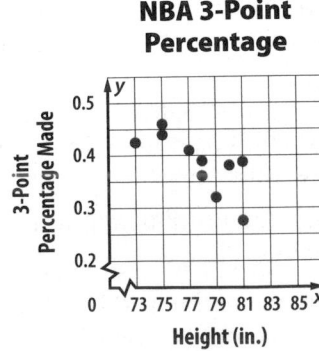

6.

Salaries

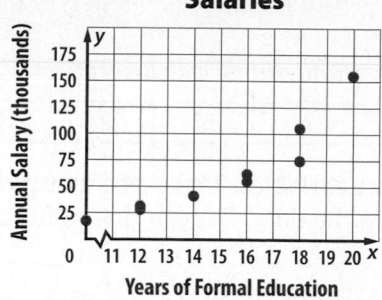

7

Gas Mileage of Various Vehicles

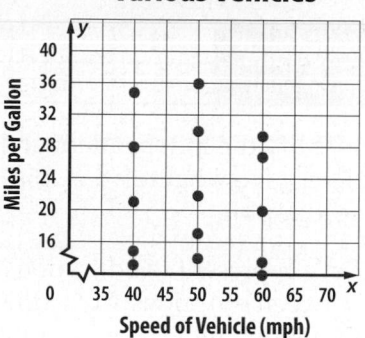

Examples 2–3 **8. MILK** Refer to the scatter plot of gallons of milk consumption per person for selected years.

Consumption of Milk in Gallons

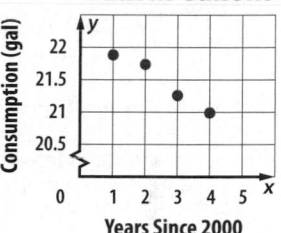

   a. Use the points (2, 21.75) and (4, 21) to write the slope-intercept form of an equation for the line of fit.

   b. Predict the milk consumption in 2020.

   c. Predict in what year milk consumption will be 10 gallons.

   d. Is it reasonable to use the equation to estimate the consumption of milk for any year? Explain.

9. FOOTBALL Use the scatter plot.

Buffalo Bills Average Game Attendance

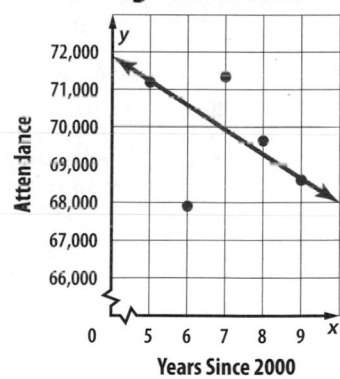

   a. Use the points (5, 71,205) and (9, 68,611) to write the slope-intercept form of an equation for the line of fit shown in the scatter plot.

   b. Predict the average attendance at a game in 2020.

   c. Can you use the equation to make a decision about the average attendance in any given year in the future? Explain.

10. **CCSS** **SENSE-MAKING** The Body Mass Index (BMI) is a measure of body fat using height and weight. The heights and weights of twelve men with normal BMI are given in the table at the right.

   a. Make a scatter plot comparing the height in inches to the weight in pounds.

   b. Draw a line of fit for the data.

   c. Write the slope-intercept form of an equation for the line of fit.

   d. Predict the normal weight for a man who is 84 inches tall.

   e. A man's weight is 188 pounds. Use the equation of the line of fit to predict the height of the man.

Height (in.)	Weight (lb)
62	115
63	124
65	120
67	134
67	140
68	138
68	144
68	152
69	147
72	155
73	168
73	166

11 **GEYSERS** The time to the next eruption of Old Faithful can be predicted by using the duration of the current eruption.

Duration (min)	1.5	2	2.5	3	3.5	4	4.5	5
Interval (min)	48	55	70	72	74	82	93	100

a. Identify the independent and the dependent variables. Make a scatter plot and determine what relationship, if any, exists in the data. Draw a line of fit for the scatter plot.

b. Let x represent the duration of the previous interval. Let y represent the time between eruptions. Write the slope-intercept form of the equation for the line of fit. Predict the interval after a 7.5-minute eruption.

c. Make a critical judgment about using the equation to predict the duration of the next eruption. Would the equation be a useful model?

12. COLLECT DATA Use a tape measure to measure both the foot size and the height in inches of ten individuals.

a. Record your data in a table.

b. Make a scatter plot and draw a line of fit for the data.

c. Write an equation for the line of fit.

d. Make a conjecture about the relationship between foot size and height.

H.O.T. Problems   Use Higher-Order Thinking Skills

13. OPEN ENDED Describe a real-life situation that can be modeled using a scatter plot. Decide whether there is a *positive*, *negative*, or *no* correlation. Explain what this correlation means.

14. WHICH ONE DOESN'T BELONG? Analyze the following situations and determine which one does not belong.

hours worked and amount of money earned	height of an athlete and favorite color
seedlings that grow an average of 2 centimeters each week	number of photos stored on a camera and capacity of camera

15. **CCSS ARGUMENTS** Determine which line of fit is better for the scatter plot. Explain your reasoning.

16. REASONING What can make a scatter plot and line of fit more useful for accurate predictions? Does an accurate line of fit always predict what will happen in the future? Explain.

17. WRITING IN MATH Make a scatter plot that shows the height of a person and age. Explain how you could use the scatter plot to predict the age of a person given his or her height. How can the information from a scatter plot be used to identify trends and make decisions?

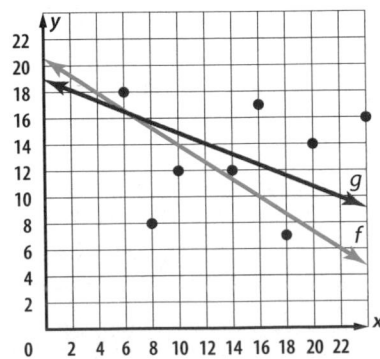

18. Which equation best describes the relationship between the values of x and y in the table?

A $y = x - 5$
B $y = 2x - 5$
C $y = 3x - 7$
D $y = 4x - 7$

x	y
−1	−7
0	−5
2	−1
4	3

19. STATISTICS Mr. Hernandez collected data on the heights and average stride lengths of a random sample of high school students. He then made a scatter plot. What kind of correlation did he most likely see?

F positive   **H** negative
G constant   **J** no

20. GEOMETRY Mrs. Aguilar's rectangular bedroom measures 13 feet by 11 feet. She wants to purchase carpet for the bedroom that costs $2.95 per square foot, including tax. How much will the carpet cost?

A $70.80
B $141.60
C $145.95
D $421.85

21. SHORT RESPONSE Nikia bought a one-month membership to a fitness center for $35. Each time she goes, she rents a locker for $0.25. If she spent $40.50 at the fitness center last month, how many days did she go?

Determine whether the graphs of each pair of equations are *parallel, perpendicular,* or *neither.* (Lesson 4-4)

22. $y = -2x + 11$
$y + 2x = 23$

23. $3y = 2x + 14$
$2x + 3y = 2$

24. $y = -5x$
$y = 5x - 18$

25. $y = 3x + 2$
$y = -\frac{1}{3}x - 2$

Write each equation in standard form. (Lesson 4-3)

26. $y - 13 = 4(x - 2)$

27. $y - 5 = -2(x + 2)$

28. $y + 3 = -5(x + 1)$

29. $y + 7 = \frac{1}{2}(x + 2)$

30. $y - 1 = \frac{5}{6}(x - 4)$

31. $y - 2 = -\frac{2}{5}(x - 8)$

Graph each equation. (Lesson 4-1)

32. $y = 2x + 3$

33. $4x + y = -1$

34. $3x + 4y = 7$

Find the slope of the line that passes through each pair of points. (Lesson 3-3)

35. $(3, 4), (10, 8)$

36. $(-4, 7), (3, 5)$

37. $(3, 7), (-2, 4)$

38. $(-3, 2), (-3, 4)$

39. $(-2, -6), (-1, 10)$

40. $(1, -5), (-3, -5)$

41. DRIVING Latisha drove 248 miles in 4 hours. At that rate, how long will it take her to drive an additional 93 miles? (Lesson 2-6)

Express each relation as a graph. Then determine the domain and range.

42. $\{(4, 5), (5, 4), (-2, -2), (4, -5), (-5, 4)\}$

43. $\{(7, 6), (3, 4), (4, 5), (-2, 6), (-3, 2)\}$

Algebra Lab
Correlation and Causation

CCSS **Common Core State Standards**
Content Standards
S.ID.9 Distinguish between correlation and causation.

You may be considering attending a college or technical school in the future. What factors cause tuition to rise—increased building costs, higher employee salaries, or the amount of bottled water consumed?

Let's see how bottled water and college tuition are related. The table shows the average college tuition and fees for public colleges and the per person U.S. consumption of bottled water per year for 2003 through 2007.

Year	2003	2004	2005	2006	2007
Water Consumed (gallons)	21.6	23.2	25.4	27.6	29.3
Tuition ($)	4645	5126	5492	5804	6191

Source: *Beverage Marketing Corporation* and *College Board*

Activity Correlation and Causation

Follow the steps to learn about correlation and causation.

Step 1 Graph the ordered pairs (gallons, tuition) to create a scatter plot. For example, one ordered pair is (21.6, 4645). Describe the graph.

Step 2 Is the correlation *positive* or *negative*? Explain.

Step 3 Do you think drinking more bottled water *causes* college tuition costs to rise? Explain.

Step 4 **Causation** occurs when a change in one variable produces a change in another variable. Correlation can be observed between many variables, but causation can only be determined from data collected from a controlled experiment. Describe an experiment that could illustrate causation.

Exercises

For each exercise, determine whether each situation illustrates *correlation* or *causation*. Explain your reasoning, including other factors that might be involved.

1. A survey showed that sleeping with the light on was positively correlated to nearsightedness.

2. A controlled experiment showed a positive correlation between the number of cigarettes smoked and the probability of developing lung cancer.

3. A random sample of students found that owning a cell phone had a negative correlation with riding the bus to school.

4. A controlled experiment showed a positive correlation between the number of hours using headphones when listening to music and the level of hearing loss.

5. DeQuan read in the newspaper that shark attacks are positively correlated with monthly ice cream sales.

Regression and Median-Fit Lines

:: Then	:: Now	:: Why?
● You used lines of fit and scatter plots to evaluate trends and make predictions.	**1** Write equations of best-fit lines using linear regression. **2** Write equations of median-fit lines.	● The table shows the total attendance, in millions of people, at the Minnesota State Fair from 2005 to 2009. You can use a graphing calculator to find the equation of a *best-fit line* and use it to make predictions about future attendance at the fair.

Year	Attendance (millions)
2005	1.633
2006	1.681
2007	1.682
2008	1.693
2009	1.790

NewVocabulary
best-fit line
linear regression
correlation coefficient
residual
median-fit line

Common Core State Standards

Content Standards

S.ID.6 Represent data on two quantitative variables on a scatter plot, and describe how the variables are related.

a. Fit a function to the data; use functions fitted to data to solve problems in the context of the data. Use given functions or choose a function suggested by the context. Emphasize linear, quadratic, and exponential models.

b. Informally assess the fit of a function by plotting and analyzing residuals.

c. Fit a linear function for a scatter plot that suggests a linear association.

S.ID.8 Compute (using technology) and interpret the correlation coefficient of a linear fit.

Mathematical Practices
5 Use appropriate tools strategically.

1 Best-Fit Lines You have learned how to find and write equations for lines of fit by hand. Many calculators use complex algorithms that find a more precise line of fit called the **best-fit line**. One algorithm is called **linear regression**.

Your calculator may also compute a number called the **correlation coefficient**. This number will tell you if your correlation is positive or negative and how closely the equation is modeling the data. The closer the correlation coefficient is to 1 or −1, the more closely the equation models the data.

Real-World Example 1 Best-Fit Line

MOVIES The table shows the amount of money made by movies in the United States. Use a graphing calculator to write an equation for the best-fit line for that data.

Year	2000	2001	2002	2003	2004	2005	2006	2007	2008	2009
Income ($ billion)	7.48	8.13	9.19	9.35	9.27	8.95	9.25	9.65	9.85	10.21

Before you begin, make sure that your Diagnostic setting is on. You can find this under the **CATALOG** menu. Press **D** and then scroll down and click **DiagnosticOn**. Then press ENTER.

Step 1 Enter the data by pressing STAT and selecting the **Edit** option. Let the year 2000 be represented by 0. Enter the years since 2000 into List 1 (**L1**). These will represent the *x*-values. Enter the income ($ billion) into List 2 (**L2**). These will represent the *y*-values.

Step 2 Perform the regression by pressing STAT and selecting the **CALC** option. Scroll down to **LinReg (ax+b)** and press ENTER twice.

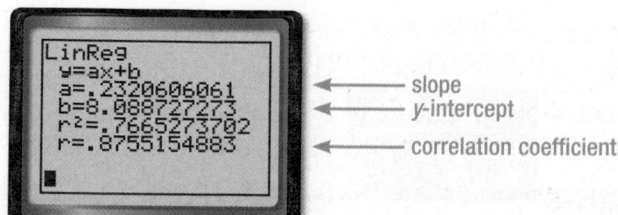

LinReg
y=ax+b
a=.2320606061 ← slope
b=8.088727273 ← *y*-intercept
r²=.7665273702
r=.8755154883 ← correlation coefficient

Step 3 Write the equation of the regression line by rounding the *a* and *b* values on the screen. The form that we chose for the regression was $ax + b$, so the equation is $y = 0.23x + 8.09$. The correlation coefficient is about 0.8755, which means that the equation models the data fairly well.

▶ **Guided**Practice

Write an equation of the best-fit line for the data in each table. Name the correlation coefficient. Round to the nearest ten-thousandth. Let *x* be the number of years since 2003.

1A. HOCKEY The table shows the number of goals of leading scorers for the Mustang Girls Hockey Team.

Year	2003	2004	2005	2006	2007	2008	2009	2010
Goals	30	23	41	35	31	43	33	45

1B. HOCKEY The table gives the number of goals scored by the team each season.

Year	2003	2004	2005	2006	2007	2008	2009	2010
Goals	63	44	55	63	81	85	93	84

We know that not all of the points will lie on the best-fit line. The difference between an observed *y*-value and its predicted *y*-value (found on the best-fit line) is called a **residual**. Residuals measure how much the data deviate from the regression line. When residuals are plotted on a scatter plot they can help to assess how well the best-fit line describes the data. If the best-fit line is a good fit, there is no pattern in the residual plot.

● **Real-World Example 2** Graph and Analyze a Residual Plot

HOCKEY **Graph and analyze the residual plot for the data for Guided Practice 1A. Determine if the best-fit line models the data well.**

After calculating the best-fit line in Guided Practice 1A, you can obtain the residual plot of the data. Turn on **Plot2** under the **STAT PLOT** menu and choose ⸬. Use **L1** for the **Xlist** and **RESID** for the **Ylist**. You can obtain **RESID** by pressing 2nd [STAT] and selecting **RESID** from the list of names. Graph the scatter plot of the residuals by pressing ZOOM and choosing **ZoomStat**.

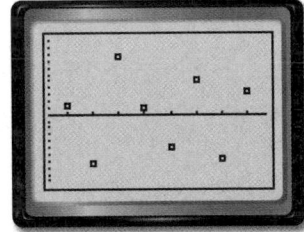

[0, 8] scl: 1 by [−10, 10] scl: 2

The residuals appear to be randomly scattered and centered about the line $y = 0$. Thus, the best-fit line seems to model the data well.

▶ **Guided**Practice

2. UNEMPLOYMENT Graph and analyze the residual plot for the following data comparing graduation rates and unemployment rates.

Graduation Rate	73	85	64	81	68	82
Unemployment Rate	6.9	4.1	3.2	5.5	4.3	5.1

A residual is positive when the observed value is above the line, negative when the observed value is below the line, and zero when it is on the line. One common measure of goodness of fit is the sum of squared vertical distances from the points to the line. The best-fit line, which is also called the *least-squares regression line*, minimizes the sum of the squares of those distances.

We can use points on the best-fit line to estimate values that are not in the data. Recall that when we estimate values that are between known values, this is called *linear interpolation*. When we estimate a number outside of the range of the data, it is called *linear extrapolation*.

Real-World Example 3 Use Interpolation and Extrapolation

PAINTBALL The table shows the points received by the top ten paintball teams at a tournament. Estimate how many points the 20th-ranked team received.

Rank	1	2	3	4	5	6	7	8	9	10
Score	100	89	96	99	97	98	78	70	64	80

Write an equation of the best-fit line for the data. Then extrapolate to find the missing value.

Step 1 Enter the data from the table into the lists. Let the ranks be the x-values and the scores be the y-values. Then graph the scatter plot.

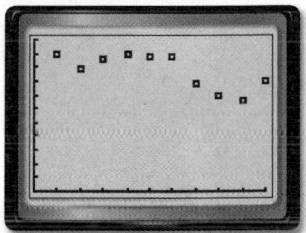

[0, 10] scl: 1 by [0, 110] scl: 10

Step 2 Perform the linear regression using the data in the lists. Find the equation of the best-fit line.

The equation is about $y = -3.32x + 105.3$.

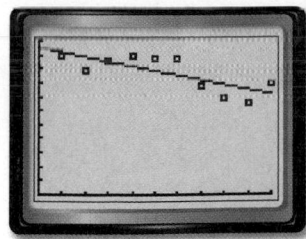

[0, 10] scl: 1 by [0, 110] scl: 10

Step 3 Graph the best-fit line. Press Y= VARS and choose **Statistics**. From the EQ menu, choose **RegEQ**. Then press GRAPH.

Step 4 Use the graph to predict the points that the 20th-ranked team received. Change the viewing window to include the x-value to be evaluated. Press 2nd [CALC] ENTER 20 ENTER to find that when $x = 20$, $y \approx 39$. It is estimated that the 20th ranked team received 39 points.

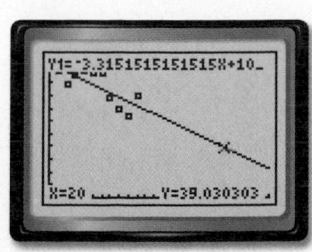

[0, 25] scl: 1 by [0, 110] scl: 1

ONLINE GAMES Use linear interpolation to estimate the percent of Americans that play online games for the following ages.

Age	15	20	30	40	50
Percent	81	54	37	29	25

Source: Pew Internet & American Life Survey

3A. 35 years   **3B.** 18 years

2 Median-Fit Lines A second type of fit line that can be found using a graphing calculator is a **median-fit line**. The equation of a median-fit line is calculated using the medians of the coordinates of the data points.

Example 4 Median-Fit Line

PAINTBALL Find and graph the equation of a median-fit line for the data in Example 3. Then predict the score of the 15th ranked team.

Step 1 Reenter the data if it is not in the lists. Clear the **Y=** list and graph the scatter plot.

[0, 10] scl: 1 by [0, 110] scl: 10

Step 2 To find the median-fit equation, press the STAT key and select the **CALC** option. Scroll down to the **Med-Med** option and press ENTER. The value of a is the slope, and the value of b is the y-intercept.

The equation for the median-fit line is about $y = -3.71x + 108.26$.

Step 3 Copy the equation to the **Y=** list and graph. Use the **value** option to find the value of y when $x = 15$.

The 15th place team scored about 53 points.

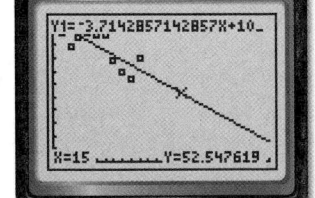

[0, 25] scl: 1 by [0, 110] scl: 1

Notice that the equations for the regression line and the median-fit line are very similar.

4. Use the data from Guided Practice 3 and a median-fit line to estimate the numbers of 18- and 35-year-olds who play online games. Compare these values with the answers from the regression line.

Real-WorldLink

Paintball is more popular with 12- to 17-year-olds than any other age group. In a recent year, 3,649,000 teens participated in paintball while 2,195,000 18- to 24-year-olds participated.

Source: *Statistical Abstract* of the *United States*

Alan Thornton/Stone/Getty Images

Examples 1, 2 **1. POTTERY** A local university is keeping track of the number of art students who use the pottery studio each day.

Day	1	2	3	4	5	6	7
Students	10	15	18	15	13	19	20

    a. Write an equation of the regression line and find the correlation coefficient.

    b. Graph the residual plot and determine if the regression line models the data well.

Example 3 **2. COMPUTERS** The table below shows the percent of Americans with a broadband connection at home in a recent year. Use linear extrapolation and a regression equation to estimate the percentage of 60-year-olds with broadband at home.

Age	25	30	35	40	45	50
Percent	40	42	36	35	36	32

Example 4 **3. VACATION** The Smiths want to rent a house on the lake that sleeps eight people. The cost of the house per night is based on how close it is to the water.

Distance from Lake (mi)	0.0 (houseboat)	0.3	0.5	1.0	1.25	1.5	2.0
Price/Night ($)	785	325	250	200	150	140	100

    a. Find and graph an equation for the median-fit line.

    b. What would you estimate is the cost of a rental 1.75 miles from the lake?

Practice and Problem Solving

Extra Practice is on page R4.

Example 1 **Write an equation of the regression line for the data in each table. Then find the correlation coefficient.**

    4. SKYSCRAPERS The table ranks the ten tallest buildings in the world.

Rank	1	2	3	4	5	6	7	8	9	10
Stories	101	88	110	88	88	80	69	102	78	70

    5 **MUSIC** The table gives the number of annual violin auditions held by a youth symphony each year since 2004. Let x be the number of years since 2004.

Year	2004	2005	2006	2007	2008	2009	2010
Auditions	22	19	25	37	32	35	42

Example 2 **6. RETAIL** The table gives the sales at a clothing chain since 2004. Let x be the number of years since 2004.

Year	2004	2005	2006	2007	2008	2009	2010
Sales (Millions of Dollars)	6.84	7.6	10.9	15.4	17.6	21.2	26.5

    a. Write an equation of the regression line.

    b. Graph and analyze the residual plot.

Examples 3, 4 (7) MARATHON The number of entrants in the Boston Marathon every five years since 1975 is shown. Let x be the number of years since 1975.

Year	1975	1980	1985	1990	1995	2000	2005	2010
Entrants	2395	5417	5594	9412	9416	17,813	20,453	26,735

  a. Find an equation for the median-fit line.

  b. According to the equation, how many entrants were there in 2003?

8. CAMPING A campground keeps a record of the number of campsites rented the week of July 4 for several years. Let x be the number of years since 2000.

Year	2002	2003	2004	2005	2006	2007	2008	2009	2010
Sites Rented	34	45	42	53	58	47	57	65	59

  a. Find an equation for the regression line.

  b. Predict the number of campsites that will be rented in 2012.

  c. Predict the number of campsites that will be rented in 2020.

9. ICE CREAM An ice cream company keeps a count of the tubs of chocolate ice cream delivered to each of their stores in a particular area.

  a. Find an equation for the median-fit line.

  b. Graph the points and the median-fit line.

  c. How many tubs would be delivered to a 1500-square-foot store? a 5000-square-foot store?

Store Size (ft²)	2100	2225	3135	3569	4587
Tubs (hundreds)	110	102	215	312	265

10. CCSS SENSE-MAKING The prices of the eight top-selling brands of jeans at Jeanie's Jeans are given in the table below.

Sales Rank	1	2	3	4	5	6	7	8
Price ($)	43	44	50	61	64	135	108	78

  a. Find the equation for the regression line.

  b. According to the equation, what would be the price of a pair of the 12th best-selling brand?

  c. Is this a reasonable prediction? Explain.

11. STATE FAIRS Refer to the beginning of the lesson.

  a. Graph a scatter plot of the data, where $x = 1$ represents 2005. Then find and graph the equation for the best-fit line.

  b. Graph and analyze the residual plot.

  c. Predict the total attendance in 2020.

12. FIREFIGHTERS The table shows statistics from the U.S. Fire Administration.

   a. Find an equation for the median-fit line.

   b. Graph the points and the median-fit line.

   c. Does the median-fit line give you an accurate picture of the number of firefighters? Explain.

Age	Number of Firefighters
18	40,919
25	245,516
35	330,516
45	296,665
55	167,087
65	54,559

13. ATHLETICS The table shows the number of participants in high school athletics.

Year Since 1970	1	10	20	30	35
Athletes	3,960,932	5,356,913	5,298,671	6,705,223	7,159,904

   a. Find an equation for the regression line.

   b. According to the equation, how many participated in 1988?

14. ART A count was kept on the number of paintings sold at an auction by the year in which they were painted. Let x be the number of years since 1950.

Year Painted	1950	1955	1960	1965	1970	1975
Paintings Solds	8	5	25	21	9	22

   a. Find the equation for the linear regression line.

   b. How many paintings were sold that were painted in 1961?

   c. Is the linear regression equation an accurate model of the data? Explain why or why not.

H.O.T. Problems   Use Higher-Order Thinking Skills

15. CCSS ARGUMENTS Below are the results of the World Superpipe Championships in 2008.

Men	Score	Rank	Women	Score
Shaun White	93.00	1	Torah Bright	96.67
Mason Aguirre	90.33	2	Kelly Clark	93.00
Janne Korpi	85.33	3	Soko Yamaoka	85.00
Luke Mitrani	85.00	4	Ellery Hollingsworth	79.33
Keir Dillion	81.33	5	Sophie Rodriguez	71.00

Find an equation of the regression line for each, and graph them on the same coordinate plane. Compare and contrast the men's and women's graphs.

16. REASONING For a class project, the scores that 10 randomly selected students earned on the first 8 tests of the school year are given. Explain how to find a line of best fit. Could it be used to predict the scores of other students? Explain your reasoning.

17. OPEN ENDED For 10 different people, measure their heights and the lengths of their heads from chin to top. Use these data to generate a linear regression equation and a median-fit equation. Make a prediction using both of the equations.

18. ✏ WRITING IN MATH How are lines of fit and linear regression similar? different?

19. **GEOMETRY** Sam is putting a border around a poster. x represents the poster's width, and y represents the poster's length. Which equation represents how much border Sam will use if he doubles the length and the width?

    A $4xy$ **C** $4(x + y)$

    B $(x + y)^4$ **D** $16(x + y)$

20. **SHORT RESPONSE** Tatiana wants to run 5 miles at an average pace of 9 minutes per mile. After 4 miles, her average pace is 9 minutes 10 seconds. In how many minutes must she complete the final mile to reach her goal?

21. What is the slope of the line that passes through $(1, 3)$ and $(-3, 1)$?

    F -2 **H** $\frac{1}{2}$

    G $-\frac{1}{2}$ **J** 2

22. What is an equation of the line that passes through $(0, 1)$ and has a slope of 3?

    A $y = 3x - 1$

    B $y = 3x - 2$

    C $y = 3x + 4$

    D $y = 3x + 1$

23. **USED CARS** Gianna wants to buy a specific make and model of a used car. She researched prices from dealers and private sellers and made the graph shown. (Lesson 4-5)

    a. Describe the relationship in the data.

    b. Use the line of fit to predict the price of a car that is 7 years old.

    c. Is it reasonable to use this line of fit to predict the price of a 10-year-old car? Explain.

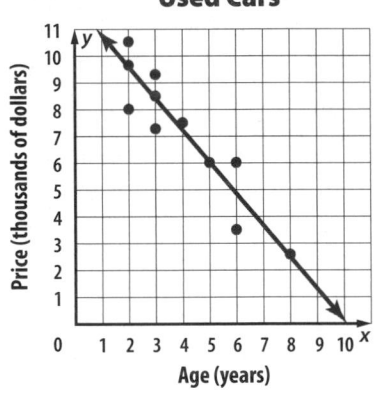

Used Cars

24. **GEOMETRY** A quadrilateral has sides with equations $y = -2x$, $2x + y = 6$, $y = \frac{1}{2}x + 6$, and $x - 2y = 9$. Is the figure a rectangle? Explain your reasoning. (Lesson 4-4)

Write each equation in standard form. (Lesson 4-3)

25. $y - 2 = 3(x - 1)$ 26. $y - 5 = 6(x + 1)$

27. $y + 2 = -2(x - 5)$ 28. $y + 3 = \frac{1}{2}(x + 4)$

29. $y - 1 = \frac{2}{3}(x + 9)$ 30. $y + 3 = -\frac{1}{4}(x + 2)$

Find the slope of the line that passes through each pair of points. (Lesson 3-3)

31. $(3, 4), (10, 8)$ 32. $(-4, 7), (3, 5)$ 33. $(3, 7), (-2, 4)$ 34. $(-3, 2), (-3, 4)$

If $f(x) = x^2 - x + 1$, find each value.

35. $f(-1)$ 36. $f(5) - 3$ 37. $f(a)$ 38. $f(b^2)$

Graph each equation.

39. $y = x + 2$ 40. $x + 5y = 4$ 41. $2x - 3y = 6$ 42. $5x + 2y = 6$

Inverse Linear Functions

- You represented relations as tables, graphs, and mappings.

1. Find the inverse of a relation.
2. Find the inverse of a linear function.

- Randall is writing a report on Santiago, Chile, and he wants to include a brief climate analysis. He found a table of temperatures recorded in degrees Celsius. He knows that a formula for converting degrees Fahrenheit to degrees Celsius is $C(x) = \frac{5}{9}(x - 32)$. He will need to find the *inverse* function to convert from degrees Celsius to degrees Fahrenheit.

Average Temp (°C)		
Month	Min	Max
Jan	12	29
March	9	27
May	5	18
July	3	15
Sept	6	29
Nov	9	26

 NewVocabulary
inverse relation
inverse function

 Common Core State Standards

Content Standards
A.CED.2 Create equations in two or more variables to represent relationships between quantities; graph equations on coordinate axes with labels and scales.

F.BF.4a Solve an equation of the form $f(x) = c$ for a simple function f that has an inverse and write an expression for the inverse.

Mathematical Practices
6 Attend to precision.

1 **Inverse Relations** An **inverse relation** is the set of ordered pairs obtained by exchanging the x-coordinates with the y-coordinates of each ordered pair in a relation. If (5, 3) is an ordered pair of a relation, then (3, 5) is an ordered pair of the inverse relation.

🔑 KeyConcept Inverse Relations

Words If one relation contains the element (a, b), then the inverse relation will contain the element (b, a).

Example A and B are inverse relations.

A		B
$(-3, -16)$	⟶	$(-16, -3)$
$(-1, 4)$	⟶	$(4, -1)$
$(2, 14)$	⟶	$(14, 2)$
$(5, 32)$	⟶	$(32, 5)$

Notice that the domain of a relation becomes the range of its inverse, and the range of the relation becomes the domain of its inverse.

PT

Example 1 Inverse Relations

Find the inverse of each relation.

a. $\{(4, -10), (7, -19), (-5, 17), (-3, 11)\}$

To find the inverse, exchange the coordinates of the ordered pairs.

$(4, -10) \rightarrow (-10, 4)$ $(-5, 17) \rightarrow (17, -5)$
$(7, -19) \rightarrow (-19, 7)$ $(-3, 11) \rightarrow (11, -3)$

The inverse is $\{(-10, 4), (-19, 7), (17, -5), (11, -3)\}$.

b.

x	-4	-1	5	9
y	-13	-8.5	0.5	6.5

Write the coordinates as ordered pairs. Then exchange the coordinates of each pair.

$(-4, -13) \rightarrow (-13, -4)$ $(5, 0.5) \rightarrow (0.5, 5)$
$(-1, -8.5) \rightarrow (-8.5, -1)$ $(9, 6.5) \rightarrow (6.5, 9)$

The inverse is $\{(-13, -4), (-8.5, -1), (0.5, 5), (6.5, 9)\}$.

> **Guided**Practice

1A. {(−6, 8), (−15, 11), (9, 3), (0, 6)}

1B.

x	−10	−4	−3	0
y	5	11	12	15

The graphs of relations can be used to find and graph inverse relations.

Example 2 Graph Inverse Relations

Graph the inverse of the relation.

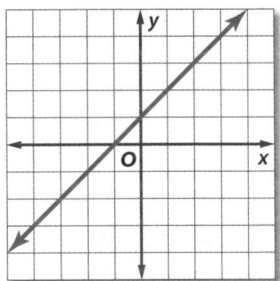

StudyTip

CCSS Precision Only two points are necessary to graph the inverse of a line, but several should be used to avoid possible error.

The graph of the relation passes through the points at (−4, −3), (−2, −1), (0, 1), (2, 3), and (3, 4). To find points through which the graph of the inverse passes, exchange the coordinates of the ordered pairs. The graph of the inverse passes through the points at (−3, −4), (−1, −2), (1, 0), (3, 2), and (4, 3). Graph these points and then draw the line that passes through them.

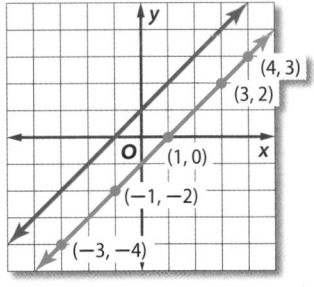

> **Guided**Practice

Graph the inverse of each relation.

2A.

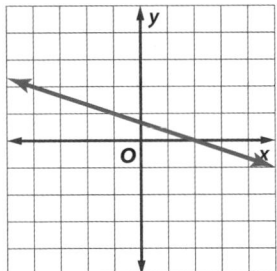

2B.

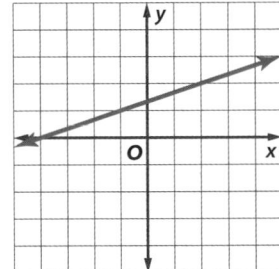

The graphs from Example 2 are graphed on the right with the line $y = x$. Notice that the graph of an inverse is the graph of the original relation reflected in the line $y = x$. For every point (x, y) on the graph of the original relation, the graph of the inverse will include the point (y, x).

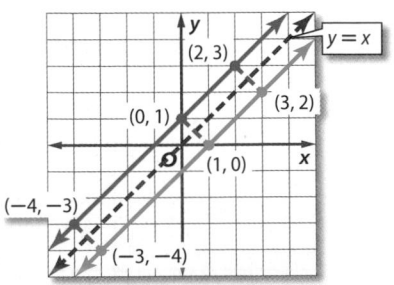

2 Inverse Functions A linear relation that is described by a function has an **inverse function** that can generate ordered pairs of the inverse relation. The inverse of the linear function $f(x)$ can be written as $f^{-1}(x)$ and is read *f of x inverse* or *the inverse of f of x.*

To find the inverse function $f^{-1}(x)$ of the linear function $f(x)$, complete the following steps.

Step 1 Replace $f(x)$ with y in the equation for $f(x)$.

Step 2 Interchange y and x in the equation.

Step 3 Solve the equation for y.

Step 4 Replace y with $f^{-1}(x)$ in the new equation.

Example 3 Find Inverse Linear Functions

Find the inverse of each function.

a. $f(x) = 4x - 8$

Step 1	$f(x) = 4x - 8$	Original equation
	$y = 4x - 8$	Replace $f(x)$ with y.
Step 2	$x = 4y - 8$	Interchange y and x.
Step 3	$x + 8 = 4y$	Add 8 to each side.
	$\dfrac{x + 8}{4} = y$	Divide each side by 4.
Step 4	$\dfrac{x + 8}{4} = f^{-1}(x)$	Replace y with $f^{-1}(x)$.

The inverse of $f(x) = 4x - 8$ is $f^{-1}(x) = \dfrac{x + 8}{4}$ or $f^{-1}(x) = \frac{1}{4}x + 2$.

CHECK Graph both functions and the line $y = x$ on the same coordinate plane. $f^{-1}(x)$ appears to be the reflection of $f(x)$ in the line $y = x$. ✓

> **Watch Out!**
> Notation The -1 in $f^{-1}(x)$ is *not* an exponent.

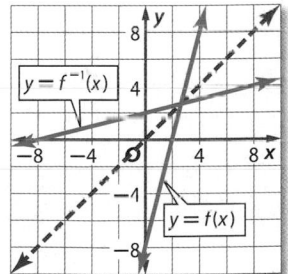

b. $f(x) = -\frac{1}{2}x + 11$

Step 1	$f(x) = -\frac{1}{2}x + 11$	Original equation
	$y = -\frac{1}{2}x + 11$	Replace $f(x)$ with y.
Step 2	$x = -\frac{1}{2}y + 11$	Interchange y and x.
Step 3	$x - 11 = -\frac{1}{2}y$	Subtract 11 from each side.
	$-2(x - 11) = y$	Multiply each side by -2.
	$-2x + 22 = y$	Distributive Property
Step 4	$-2x + 22 = f^{-1}(x)$	Replace y with $f^{-1}(x)$.

The inverse of $f(x) = -\frac{1}{2}x + 11$ is $f^{-1}(x) = -2x + 22$.

▸ **Guided Practice**

3A. $f(x) = 4x - 12$

3B. $f(x) = \frac{1}{3}x + 7$

Real-World Example 4 Use an Inverse Function

TEMPERATURE Refer to the beginning of the lesson. Randall wants to convert the temperatures from degrees Celsius to degrees Fahrenheit.

a. Find the inverse function $C^{-1}(x)$.

Step 1	$C(x) = \frac{5}{9}(x - 32)$	Original equation
	$y = \frac{5}{9}(x - 32)$	Replace $C(x)$ with y.
Step 2	$x = \frac{5}{9}(y - 32)$	Interchange y and x.
Step 3	$\frac{9}{5}x = y - 32$	Multiply each side by $\frac{9}{5}$.
	$\frac{9}{5}x + 32 = y$	Add 32 to each side.
Step 4	$\frac{9}{5}x + 32 = C^{-1}(x)$	Replace y with $C^{-1}(x)$.

The inverse function of $C(x)$ is $C^{-1}(x) = \frac{9}{5}x + 32$.

b. What do x and $C^{-1}(x)$ represent in the context of the inverse function?

x represents the temperature in degrees Celsius. $C^{-1}(x)$ represents the temperature in degrees Fahrenheit.

c. Find the average temperatures for July in degrees Fahrenheit.

The average minimum and maximum temperatures for July are 3° C and 15° C, respectively. To find the average minimum temperature, find $C^{-1}(3)$.

$C^{-1}(x) = \frac{9}{5}x + 32$	Original equation
$C^{-1}(3) = \frac{9}{5}(3) + 32$	Substitute 3 for x.
$= 37.4$	Simplify.

To find the average maximum temperature, find $C^{-1}(15)$.

$C^{-1}(x) = \frac{9}{5}x + 32$	Original equation
$C^{-1}(15) = \frac{9}{5}(15) + 32$	Substitute 15 for x.
$= 59$	Simplify.

The average minimum and maximum temperatures for July are 37.4° F and 59° F, respectively.

Real-World Link

The winter months in Chile occur during the summer months in the U.S. due to Chile's location in the southern hemisphere. The average daily high temperature of Santiago during its winter months is about 60° F.

Source: World Weather Information Service

▶ **Guided**Practice

4. RENTAL CAR Peggy rents a car for the day. The total cost $C(x)$ in dollars is given by $C(x) = 19.99 + 0.3x$, where x is the number of miles she drives.

   A. Find the inverse function $C^{-1}(x)$.

   B. What do x and $C^{-1}(x)$ represent in the context of the inverse function?

   C. How many miles did Peggy drive if her total cost was $34.99?

Example 1 Find the inverse of each relation.

1. {(4, −15), (−8, −18), (−2, −16.5), (3, −15.25)}

2.

x	−3	0	1	6
y	11.8	3.7	1	−12.5

Example 2 Graph the inverse of each relation.

3.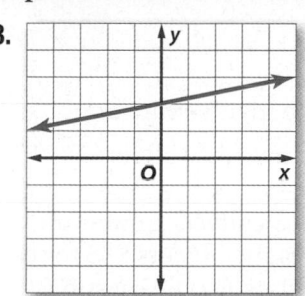

4.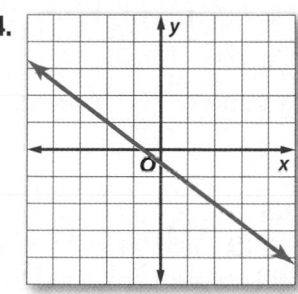

Example 3 Find the inverse of each function.

5. $f(x) = -2x + 7$

6. $f(x) = \frac{2}{3}x + 6$

Example 4

7. **CCSS REASONING** Dwayne and his brother purchase season tickets to the Cleveland Crusaders games. The ticket package requires a one-time purchase of a personal seat license costing $1200 for two seats. A ticket to each game costs $70. The cost $C(x)$ in dollars for Dwayne for the first season is $C(x) = 600 + 70x$, where x is the number of games Dwayne attends.

  a. Find the inverse function.

  b. What do x and $C^{-1}(x)$ represent in the context of the inverse function?

  c. How many games did Dwayne attend if his total cost for the season was $950?

Practice and Problem Solving

Extra Practice is on page R4.

Example 1 Find the inverse of each relation.

8. {(−5, 13), (6, 10.8), (3, 11.4), (−10, 14)}

⑨ {(−4, −49), (8, 35), (−1, −28), (4, 7)}

10.

x	y
−8	−36.4
−2	−15.4
1	−4.9
5	9.1
11	30.1

11.

x	y
−3	7.4
−1	4
1	0.6
3	−2.8
5	−6.2

Example 2 Graph the inverse of each relation.

12.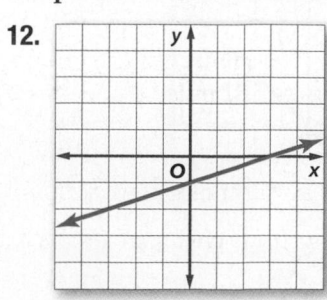

13.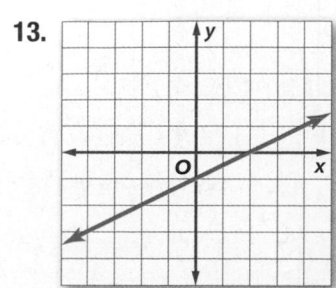

Example 3 Find the inverse of each function.

14. $f(x) = 25 + 4x$

15 $f(x) = 17 - \frac{1}{3}x$

16. $f(x) = 4(x + 17)$

17. $f(x) = 12 - 6x$

18. $f(x) = \frac{2}{5}x + 10$

19. $f(x) = -16 - \frac{4}{3}x$

Example 4 **20. DOWNLOADS** An online music subscription service allows members to download songs for $0.99 each after paying a monthly service charge of $3.99. The total monthly cost $C(x)$ of the service in dollars is $C(x) = 3.99 + 0.99x$, where x is the number of songs downloaded.

a. Find the inverse function.

b. What do x and $C^{-1}(x)$ represent in the context of the inverse function?

c. How many songs were downloaded if a member's monthly bill is $27.75?

21. LANDSCAPING At the start of the mowing season, Chuck collects a one-time maintenance fee of $10 from his customers. He charges the Fosters $35 for each cut. The total amount collected from the Fosters in dollars for the season is $C(x) = 10 + 35x$, where x is the number of times Chuck mows the Fosters' lawn.

a. Find the inverse function.

b. What do x and $C^{-1}(x)$ represent in the context of the inverse function?

c. How many times did Chuck mow the Fosters' lawn if he collected a total of $780 from them?

Write the inverse of each equation in $f^{-1}(x)$ notation.

22. $3y - 12x = -72$

23. $x + 5y = 15$

24. $-42 + 6y = x$

25. $3y + 24 = 2x$

26. $-7y + 2x = -28$

27. $3y - x = 3$

CCSS TOOLS Match each function with the graph of its inverse.

A.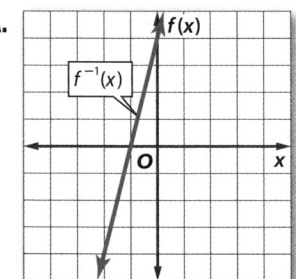

B.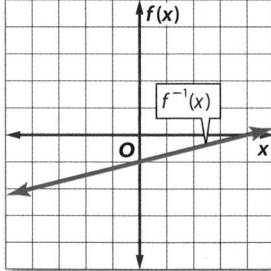

C.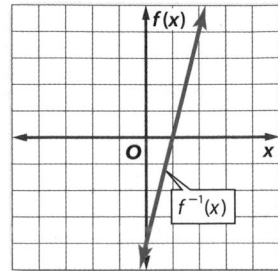

D.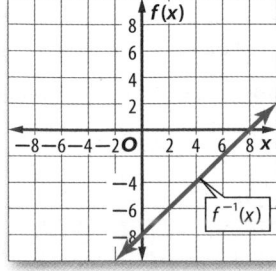

28. $f(x) = x + 4$

29. $f(x) = 4x + 4$

30. $f(x) = \frac{1}{4}x + 1$

31. $f(x) = \frac{1}{4}x - 1$

Write an equation for the inverse function $f^{-1}(x)$ that satisfies the given conditions.

32. slope of $f(x)$ is 7; graph of $f^{-1}(x)$ contains the point (13, 1)

33. graph of $f(x)$ contains the points $(-3, 6)$ and $(6, 12)$

34. graph of $f(x)$ contains the point (10, 16); graph of $f^{-1}(x)$ contains the point $(3, -16)$

35. slope of $f(x)$ is 4; $f^{-1}(5) = 2$

36. CELL PHONES Mary Ann pays a monthly fee for her cell phone package which includes 700 minutes. She gets billed an additional charge for every minute she uses the phone past the 700 minutes. During her first month, Mary Ann used 26 additional minutes and her bill was $37.79. During her second month, Mary Ann used 38 additional minutes and her bill was $41.39.

   a. Write a function that represents the total monthly cost $C(x)$ of Mary Ann's cell phone package, where x is the number of additional minutes used.

   b. Find the inverse function.

   c. What do x and $C^{-1}(x)$ represent in the context of the inverse function?

   d. How many additional minutes did Mary Ann use if her bill for her third month was $48.89?

37. **MULTIPLE REPRESENTATIONS** In this problem, you will explore the domain and range of inverse functions.

   a. Algebraic Write a function for the area $A(x)$ of the rectangle shown.

   b. Graphical Graph $A(x)$. Describe the domain and range of $A(x)$ in the context of the situation.

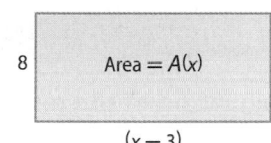

   c. Algebraic Write the inverse of $A(x)$. What do x and $A^{-1}(x)$ represent in the context of the situation?

   d. Graphical Graph $A^{-1}(x)$. Describe the domain and range of $A^{-1}(x)$ in the context of the situation.

   e. Logical Determine the relationship between the domains and ranges of $A(x)$ and $A^{-1}(x)$.

H.O.T. Problems   Use Higher-Order Thinking Skills

38. CHALLENGE If $f(x) = 5x + a$ and $f^{-1}(10) = -1$, find a.

39. CHALLENGE If $f(x) = \frac{1}{a}x + 7$ and $f^{-1}(x) = 2x - b$, find a and b.

CCSS ARGUMENTS Determine whether the following statements are *sometimes*, *always*, or *never* true. Explain your reasoning.

40. If $f(x)$ and $g(x)$ are inverse functions, then $f(a) = b$ and $g(b) = a$.

41. If $f(a) = b$ and $g(b) = a$, then $f(x)$ and $g(x)$ are inverse functions.

42. OPEN ENDED Give an example of a function and its inverse. Verify that the two functions are inverses by graphing the functions and the line $y = x$ on the same coordinate plane.

43. WRITING IN MATH Explain why it may be helpful to find the inverse of a function.

44. Which equation represents a line that is perpendicular to the graph and passes through the point at $(2, 0)$?

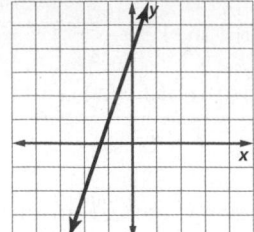

A $y = 3x - 6$

B $y = -3x + 6$

C $y = -\frac{1}{3}x + \frac{2}{3}$

D $y = \frac{1}{3}x - \frac{2}{3}$

45. A giant tortoise travels at a rate of 0.17 mile per hour. Which equation models the time t it would take the giant tortoise to travel 0.8 mile?

F $t = \frac{0.8}{0.17}$

H $t = \frac{0.17}{0.8}$

G $t = (0.17)(0.8)$

J $0.8 = \frac{0.17}{t}$

46. GEOMETRY If $\triangle JKL$ is similar to $\triangle JNM$ what is the value of a?

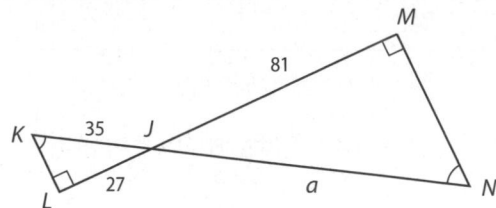

A 62.5

B 105

C 125

D 155.5

47. GRIDDED RESPONSE What is the difference in the value of $2.1(x + 3.2)$, when $x = 5$ and when $x = 3$?

Write an equation of the regression line for the data in each table. (Lesson 4-6)

48.

x	1	3	5	7	9
y	3	8	15	18	21

49.

x	3	5	7	9	11
y	7.2	23.5	41.2	56.4	73.1

50.

x	1	2	3	4	5
y	21	33	39	54	64

51.

x	2	4	6	8	10
y	1.4	2.4	2.9	3.3	4.2

52. TESTS Determine whether the graph at the right shows a *positive*, *negative*, or *no* correlation. If there is a correlation, describe its meaning. (Lesson 4-5)

Test Scores

Suppose y varies directly as x. (Lesson 3-4)

53. If $y = 2.5$ when $x = 0.5$, find y when $x = 20$.

54. If $y = -6.6$ when $x = 9.9$, find y when $x = 6.6$.

55. If $y = 2.6$ when $x = 0.25$, find y when $x = 1.125$.

56. If $y = 6$ when $x = 0.6$, find x when $y = 12$.

Solve each equation.

57. $104 = k - 67$

58. $-4 + x = -7$

59. $\frac{m}{7} = -11$

60. $\frac{2}{3}p = 14$

61. $-82 = 18 - n$

62. $\frac{9}{t} = -27$

4-7

Algebra Lab
Drawing Inverses

You can use patty paper to draw the graph of an inverse relation by reflecting the original graph in the line $y = x$.

CCSS **Common Core State Standards**
Content Standards
F.BF.4a Solve an equation of the form $f(x) = c$ for a simple function f that has an inverse and write an expression for the inverse.

Activity Draw an Inverse

Consider the graphs shown.

Step 1 Trace the graphs onto a square of patty paper, waxed paper, or tracing paper.

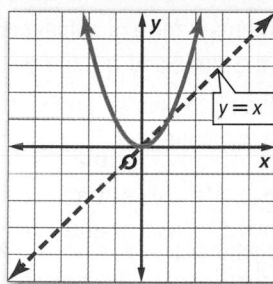

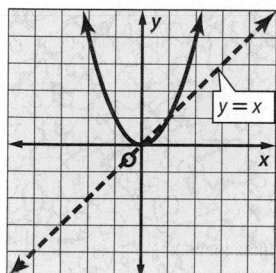

Step 2 Flip the patty paper over and lay it on the original graph so that the traced $y = x$ is on the original $y = x$.

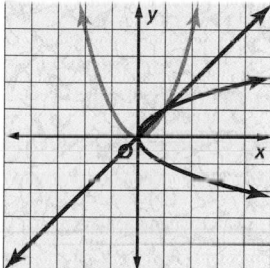

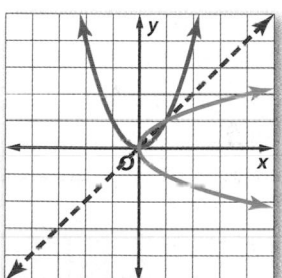

Notice that the result is the reflection of the graph in the line $y = x$ or the inverse of the graph.

Analyze The Results

1. Is the graph of the original relation a function? Explain.

2. Is the graph of the inverse relation a function? Explain.

3. What are the domain and range of the original relation? of the inverse relation?

4. If the domain of the original relation is restricted to $D = \{x \mid x \geq 0\}$, is the inverse relation a function? Explain.

5. If the graph of a relation is a function, what can you conclude about the graph of its inverse?

6. **CHALLENGE** The vertical line test can be used to determine whether a relation is a function. Write a rule that can be used to determine whether a function has an inverse that is also a function.

Study Guide

Slope-Intercept Form (Lessons 4-1 and 4-2)

- The slope-intercept form of a linear equation is $y = mx + b$, where m is the slope and b is the y-intercept.
- If you are given two points through which a line passes, use them to find the slope first.

Point-Slope Form (Lesson 4-3)

- The linear equation $y - y_1 = m(x - x_1)$ is written in point-slope form, where (x_1, y_1) is a given point on a nonvertical line and m is the slope of the line.

Parallel and Perpendicular Lines (Lesson 4-4)

- Nonvertical parallel lines have the same slope.
- Lines that intersect at right angles are called perpendicular lines. The slopes of perpendicular lines are opposite reciprocals.

Scatter Plots and Lines of Fit (Lesson 4-5)

- Data with two variables are called bivariate data.
- A scatter plot is a graph in which two sets of data are plotted as ordered pairs in a coordinate plane.

Regression and Median-Fit Lines (Lesson 4-6)

- A graphing calculator can be used to find regression lines and median-fit lines.

Inverse Linear Functions (Lesson 4-7)

- An inverse relation is the set of ordered pairs obtained by exchanging the x-coordinates with the y-coordinates of each ordered pair of a relation.
- A linear function $f(x)$ has an inverse function that can be written as $f^{-1}(x)$ and is read *f of x inverse* or *the inverse of f of x.*

Be sure the Key Concepts are noted in your Foldable.

Equations of Linear Functions

best-fit line (p. 255)	linear interpolation (p. 249)
bivariate data (p. 247)	linear regression (p. 255)
constant function (p. 217)	line of fit (p. 248)
constraint (p. 228)	median-fit line (p. 258)
correlation coefficient (p. 255)	parallel lines (p. 239)
identity function (p. 224)	perpendicular lines (p. 240)
inverse function (p. 264)	point-slope form (p. 233)
inverse relation (p. 263)	scatter plot (p. 247)
linear extrapolation (p. 228)	slope-intercept form (p. 216)

State whether each sentence is *true* or *false*. If *false*, replace the underlined term to make a true sentence.

1. The y-intercept is the y-coordinate of the point where the graph crosses the y-axis.

2. The process of using a linear equation to make predictions about values that are beyond the range of the data is called linear regression.

3. An inverse relation is the set of ordered pairs obtained by exchanging the x-coordinates with the y-coordinates of each ordered pair of a relation.

4. The correlation coefficient describes whether the correlation between the variables is positive or negative and how closely the regression equation is modeling the data.

5. Lines in the same plane that do not intersect are called parallel lines.

6. Lines that intersect at acute angles are called perpendicular lines.

7. A(n) constant function can generate ordered pairs for an inverse relation.

8. The range of a relation is the range of its inverse function.

9. An equation of the form $y = mx + b$ is in point-slope form.

Lesson-by-Lesson Review

4-1 Graphing Equations in Slope-Intercept Form

Write an equation of a line in slope-intercept form with the given slope and *y*-intercept. Then graph the equation.

10. slope: 3, *y*-intercept: 5

11. slope: −2, *y*-intercept: −9

12. slope: $\frac{2}{3}$, *y*-intercept: 3

13. slope: $-\frac{5}{8}$, *y*-intercept: −2

Graph each equation.

14. $y = 4x - 2$

15. $y = -3x + 5$

16. $y = \frac{1}{2}x + 1$

17. $3x + 4y = 8$

18. **SKI RENTAL** Write an equation in slope-intercept form for the total cost of skiing for *h* hours with one lift ticket.

> **Slippery Slope**
> **Ski Lodge**
> Lift Ticket $15/day
> Ski Rental $5/hour

Example 1

Write an equation of a line in slope-intercept form with slope −5 and *y*-intercept −3. Then graph the equation.

$$y = mx + b \qquad \text{Slope-intercept form}$$
$$y = -5x + (-3) \qquad m = -5 \text{ and } b = -3$$
$$y = -5x - 3 \qquad \text{Simplify.}$$

To graph the equation, plot the *y*-intercept (0, −3).

Then move up 5 units and left 1 unit. Plot the point. Draw a line through the two points.

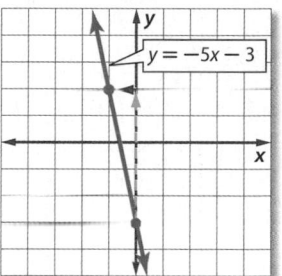

4-2 Writing Equations in Slope-Intercept Form

Write an equation of the line that passes through the given point and has the given slope.

19. (1, 2), slope 3

20. (2, −6), slope −4

21. (−3, −1), slope $\frac{2}{5}$

22. (5, −2), slope $-\frac{1}{3}$

Write an equation of the line that passes through the given points.

23. (2, −1), (5, 2)

24. (−4, 3), (1, 13)

25. (3, 5), (5, 6)

26. (2, 4), (7, 2)

27. **CAMP** In 2005, a camp had 450 campers. Five years later, the number of campers rose to 750. Write a linear equation that represents the number of campers that attend camp.

Example 2

Write an equation of the line that passes through (3, 2) with a slope of 5.

Step 1 Find the *y*-intercept.

$$y = mx + b \qquad \text{Slope-intercept form}$$
$$2 = 5(3) + b \qquad m = 5, y = 2, \text{ and } x = 3$$
$$2 = 15 + b \qquad \text{Simplify.}$$
$$-13 = b \qquad \text{Subtract 15 from each side.}$$

Step 2 Write the equation in slope-intercept form.

$$y = mx + b \qquad \text{Slope-intercept form}$$
$$y = 5x - 13 \qquad m = 5 \text{ and } b = -13$$

4-3 Writing Equations in Point-Slope Form

Write an equation in point-slope form for the line that passes through the given point with the slope provided.

28. (6, 3), slope 5

29. (−2, 1), slope −3

30. (−4, 2), slope 0

Write each equation in standard form.

31. $y - 3 = 5(x - 2)$

32. $y - 7 = -3(x + 1)$

33. $y + 4 = \frac{1}{2}(x - 3)$

34. $y - 9 = -\frac{4}{5}(x + 2)$

Write each equation in slope-intercept form.

35. $y - 2 = 3(x - 5)$

36. $y - 12 = -2(x - 3)$

37. $y + 3 = 5(x + 1)$

38. $y - 4 = \frac{1}{2}(x + 2)$

Example 3

Write an equation in point-slope form for the line that passes through (3, 4) with a slope of −2.

$$y - y_1 = m(x - x_1) \quad \text{Point-slope form}$$

$$y - 4 = -2(x - 3) \quad \text{Replace } m \text{ with } -2 \text{ and } (x_1, y_1) \text{ with (3, 4).}$$

Example 4

Write $y + 6 = -4(x - 3)$ in standard form.

$$y + 6 = -4(x - 3) \quad \text{Original equation}$$

$$y + 6 = -4x + 12 \quad \text{Distributive Property}$$

$$4x + y + 6 = 12 \quad \text{Add } 4x \text{ to each side.}$$

$$4x + y = 6 \quad \text{Subtract 6 from each side.}$$

4-4 Parallel and Perpendicular Lines

Write an equation in slope-intercept form for the line that passes through the given point and is parallel to the graph of each equation.

39. (2, 5), $y = x - 3$

40. (0, 3), $y = 3x + 5$

41. (−4, 1), $y = -2x - 6$

42. (−5, −2), $y = -\frac{1}{2}x + 4$

Write an equation in slope-intercept form for the line that passes through the given point and is perpendicular to the graph of the given equation.

43. (2, 4), $y = 3x + 1$

44. (1, 3), $y = -2x - 4$

45. (−5, 2), $y = \frac{1}{3}x + 4$

46. (3, 0), $y = -\frac{1}{2}x$

Example 5

Write an equation in slope-intercept form for the line that passes through (−2, 4) and is parallel to the graph of $y = 6x - 3$.

The slope of the line with equation $y = 6x - 3$ is 6. The line parallel to $y = 6x - 3$ has the same slope, 6.

$$y - y_1 = m(x - x_1) \quad \text{Point-slope form}$$

$$y - 4 = 6[x - (-2)] \quad \text{Substitute.}$$

$$y - 4 = 6(x + 2) \quad \text{Simplify.}$$

$$y - 4 = 6x + 12 \quad \text{Distributive Property}$$

$$y = 6x + 16 \quad \text{Add 4 to each side.}$$

47. Determine whether the graph shows a *positive*, *negative*, or *no* correlation. If there is a positive or negative correlation, describe its meaning.

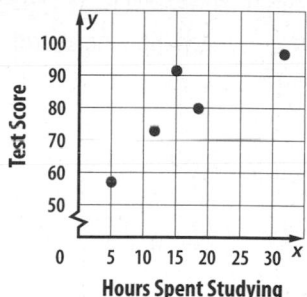

48. ATTENDANCE A scatter plot of data compares the number of years since a business has opened and its annual number of sales. It contains the ordered pairs (2, 650) and (5, 1280). Write an equation in slope-intercept form for the line of fit for this situation.

Example 6

The scatter plot displays the number of texts and the number of calls made daily. Write an equation for the line of fit.

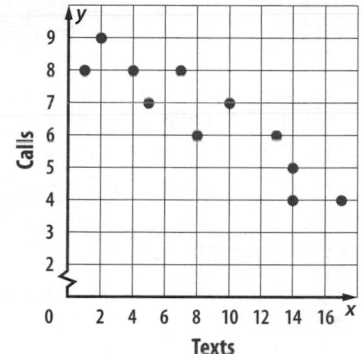

First, find the slope using (2, 9) and (17, 4).

$m = \dfrac{4 - 9}{17 - 2} = \dfrac{-5}{15}$ or $-\dfrac{1}{3}$ Substitute and simplify.

Then find the y intercept.

$9 = -\dfrac{1}{3}(2) + b$ Substitute.

$9\dfrac{2}{3} = b$ Add $\dfrac{2}{3}$ to each side.

Write the equation. $y = -\dfrac{1}{3}x + 9\dfrac{2}{3}$

---

49. SALE The table shows the number of purchases made at an outerwear store during a sale. Write an equation of the regression line. Then estimate the daily purchases on day 10 of the sale.

Days Since Sale Began	1	2	3	4	5	6	7
Daily Purchases	15	21	32	30	40	38	51

50. MOVIES The table shows ticket sales at a certain theater during the first week after a movie opened. Write an equation of the regression line. Then estimate the daily ticket sales on the 15th day.

Days Since Movie Opened	1	2	3	4	5	6	7
Daily Ticket Sales	85	92	89	78	65	68	55

Example 7

ATTENDANCE The table shows the annual attendance at an amusement park. Write an equation of the regression line for the data.

Years Since 2004	0	1	2	3	4	5	6
Attendance (thousands)	75	80	72	68	65	60	53

Step 1 Enter the data by pressing [STAT] and selecting the **Edit** option.

Step 2 Perform the regression by pressing [STAT] and selecting the **CALC** option. Scroll down to **LinReg (ax + b)** and press [ENTER].

Step 3 Write the equation of the regression line by rounding the a- and b-values on the screen.
$y = -4.04x + 79.68$

4-7 Inverse Linear Functions

Find the inverse of each relation.

51. $\{(7, 3.5), (6.2, 8), (-4, 2.7), (-12, 1.4)\}$

52. $\{(1, 9), (13, 26), (-3, 4), (-11, -2)\}$

53.

X	Y
-4	2.7
-1	3.8
0	4.1
3	7.2

54.

X	Y
-12	4
-8	0
-4	-4
0	-8

Find the inverse of each function.

55. $f(x) = \frac{5}{11}x + 10$

56. $f(x) = 3x + 8$

57. $f(x) = -4x - 12$

58. $f(x) = \frac{1}{4}x - 7$

59. $f(x) = -\frac{2}{3}x + \frac{1}{4}$

60. $f(x) = -3x + 3$

Example 8

Find the inverse of the relation.

$$\{(5, -3), (11, 2), (-6, 12), (4, -2)\}$$

To find the inverse, exchange the coordinates of the ordered pairs.

$(5, -3) \rightarrow (-3, 5)$ $(-6, 12) \rightarrow (12, -6)$

$(11, 2) \rightarrow (2, 11)$ $(4, -2) \rightarrow (-2, 4)$

The inverse is $\{(-3, 5), (2, 11), (12, -6), (-2, 4)\}$.

Example 9

Find the inverse of $f(x) = \frac{1}{4}x + 9$.

$f(x) = \frac{1}{4}x + 9$	Original equation
$y = \frac{1}{4}x + 9$	Replace $f(x)$ with y.
$x = \frac{1}{4}y + 9$	Interchange y and x.
$x - 9 = \frac{1}{4}y$	Subtract 9 from each side.
$4(x - 9) = y$	Multiply each side by 4.
$4x - 36 = y$	Distributive Property
$4x - 36 = f^{-1}(x)$	Replace y with $f^{-1}(x)$.

Practice Test

1. Graph $y = 2x - 3$.

2. **MULTIPLE CHOICE** A popular pizza parlor charges $12 for a large cheese pizza plus $1.50 for each additional topping. Write an equation in slope-intercept form for the total cost C of a pizza with t toppings.

   A $C = 12t + 1.50$

   B $C = 13.50t$

   C $C = 12 + 1.50t$

   D $C = 1.50t - 12$

Write an equation of a line in slope-intercept form that passes through the given point and has the given slope.

3. $(-4, 2)$; slope -3

4. $(3, -5)$; slope $\frac{2}{3}$

Write an equation of the line in slope-intercept form that passes through the given points.

5. $(1, 4), (3, 10)$

6. $(2, 5), (-2, 8)$

7. $(0, 4), (-3, 0)$

8. $(7, -1), (9, -4)$

9. **PAINTING** The data in the table show the size of a room in square feet and the time it takes to paint the room in minutes.

Room Size	100	150	200	400	500
Painting Time	160	220	270	500	680

   a. Use the points $(100, 160)$ and $(500, 680)$ to write an equation in slope-intercept form.

   b. Predict the amount of time required to paint a room measuring 750 square feet.

10. **SALARY** The table shows the relationship between years of experience and teacher salary.

Years Experience	1	5	10	15	20
Salary (thousands of dollars)	28	31	42	49	64

   a. Write an equation for the best-fit line.

   b. Find the correlation coefficient and explain what it tells us about the relationship between experience and salary.

Write an equation in slope-intercept form for the line that passes through the given point and is parallel to the graph of each equation.

11. $(2, -3), y = 4x - 9$

12. $(-5, 1), y = -3x + 2$

Write an equation in slope-intercept form for the line that passes through the given point and is perpendicular to the graph of the equation.

13. $(1, 4), y = -2x + 5$ 14. $(-3, 6), y = \frac{1}{4}x + 2$

15. **MULTIPLE CHOICE** The graph shows the relationship between outside temperature and daily ice cream cone sales. What type of correlation is shown?

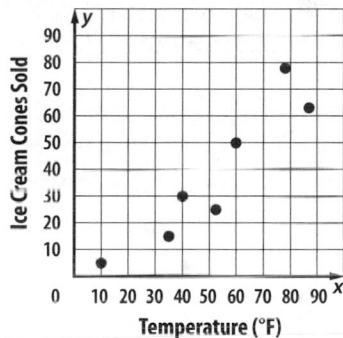

   F positive correlation

   G negative correlation

   H no correlation

   J not enough information

16. **ADOPTION** The table shows the number of children from Ethiopia adopted by U.S. citizens.

Years Since 2000	5	6	7	8	9
Number of Children	442	731	1254	1724	2277

   a. Write the slope-intercept form of the equation for the line of fit.

   b. Predict the number of children from Ethiopia who will be adopted in 2025.

Find the inverse of each function.

17. $f(x) = -5x - 30$

18. $f(x) = 4x + 10$

19. $f(x) = \frac{1}{6}x - 2$

20. $f(x) = \frac{3}{4}x + 12$

Short Answer Questions

Short answer questions require you to provide a solution to the problem, along with a method, explanation, and/or justification used to arrive at the solution.

Strategies for Solving Short Answer Questions

Step 1

Short answer questions are typically graded using a **rubric**, or a scoring guide. The following is an example of a short answer question scoring rubric.

Scoring Rubric	
Criteria	**Score**
Full Credit: The answer is correct and a full explanation is provided that shows each etep.	2
Partial Credit: • The answer is correct, but the explanation is incomplete. • The answer is incorrect, but the explanation is correct.	1
No Credit: Either an answer is not provided or the answer does not make sense.	0

Step 2

In solving short answer questions, remember to...

- explain your reasoning or state your approach to solving the problem.

- show all of your work or steps.

- check your answer if time permits.

Standardized Test Example

Read the problem. Identify what you need to know. Then use the information in the problem to solve. Show your work.

The table shows production costs for building different numbers of skateboards. Determine the missing value, x, that will result in a linear model.

Skateboards Built	Production Costs
14	$325
28	$500
x	$375
22	$425

Read the problem carefully. You are given several data points and asked to find the missing value that results in a linear model.

Example of a 2-point response:

Set up a coordinate grid and plot the three given points:
(14, 325), (28, 500), (22, 425).

Then draw a straight line through them and find the *x*-value that produces a *y*-value of 375.

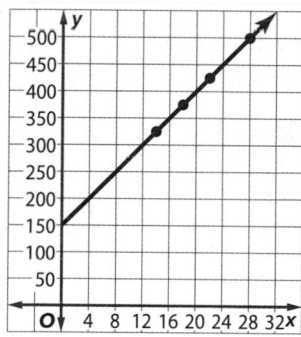

So, building 18 skateboards would result in production costs of $375. These data form a linear model.

The steps, calculations, and reasoning are clearly stated. The student also arrives at the correct answer. So, this response is worth the full 2 points.

Exercises

Read each problem. Identify what you need to know. Then use the information in the problem to solve. Show your work.

1. Given points $M(-1, 7)$, $N(3, -5)$, $O(6, 1)$, and $P(-3, -2)$, determine two segments that are perpendicular to each other.

2. Write the equation of a line that is parallel to $4x + 2y = 8$ and has a *y*-intercept of 5.

3. Three vertices of a quadrilateral are shown on the coordinate grid. Determine a fourth vertex that would result in a trapezoid.

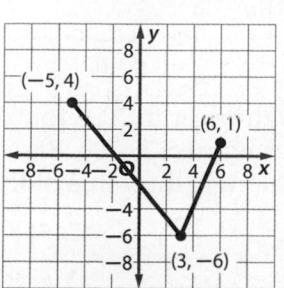

Multiple Choice

Read each question. Then fill in the correct answer on the answer document provided by your teacher or on a sheet of paper.

1. What is the rate of change represented in the graph?

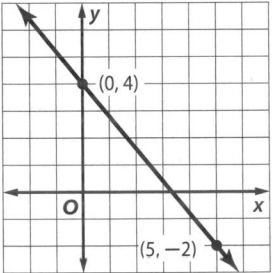

A $-\dfrac{2}{5}$

B $-\dfrac{5}{6}$

C $-\dfrac{6}{5}$

D $-\dfrac{5}{2}$

2. The table below shows the cost for renting a bicycle at a bike shop located in Venice Beach. What is a function that can represent this sequence?

Number of Hours	Cost ($)
1	10
2	14
3	18
4	22

F $f(n) = 4n + 10$

G $f(n) = 4n + 6$

H $f(n) = 10n + 4$

J $f(n) = 10n - 6$

3. Jaime bought a car in 2005 for $28,500. By 2008, the car was worth $23,700. Based on a linear model, what will the value of the car be in 2012?

A $17,300

B $17,550

C $18,100

D $18,475

4. If the graph of a line has a positive slope and a negative y-intercept, what happens to the x-intercept if the slope and the y-intercept are doubled?

F The x-intercept becomes four times as great.

G The x-intercept becomes twice as great.

H The x-intercept becomes one-fourth as great.

J The x-intercept remains the same.

5. Which absolute value equation has the graph below as its solution?

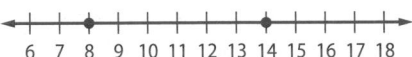

A $|x - 3| = 11$

B $|x - 4| = 12$

C $|x - 11| = 3$

D $|x - 12| = 4$

6. The table below shows the relationship between certain temperatures in degrees Fahrenheit and degrees Celsius. Which of the following linear equations correctly models this relationship?

Celsius (C)	Fahrenheit (F)
10°	50°
15°	59°
20°	68°
25°	77°
30°	86°

F $F = \dfrac{8}{5}C + 35$

G $F = \dfrac{4}{5}C + 42$

H $F = \dfrac{9}{5}C + 32$

J $F = \dfrac{12}{5}C + 26$

Test-TakingTip

Question 3 Find the average annual depreciation between 2005 and 2008. Then extend the pattern to find the car's value in 2012.

Record your answers on the answer sheet provided by your teacher or on a sheet of paper.

7. What is the equation of the line graphed below?

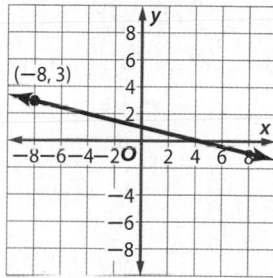

Express your answer in point slope form using the point $(-8, 3)$.

8. GRIDDED RESPONSE The linear equation below is a best fit model for the peak depth of the Mad River when x inches of rain fall. What would you expect the peak depth of the river to be after a storm that produces $1\frac{3}{4}$ inches of rain? Round your answer to the nearest tenth of a foot if necessary.

$$y = 2.5x + 14.8$$

9. Jacob formed an advertising company in 1992. Initially, the company only had 14 employees. In 2008, the company had grown to a total of 63 employees. Find the percent of change in the number of employees working at Jacob's company. Round to the nearest tenth of a percent if necessary.

10. The table shows the total amount of rain during a storm.

Hours	1	2	3	4
Inches	0.45	0.9	1.35	1.8

   a. Write an equation to fit the data in the table.

   b. Describe the relationship between the hour and the amount of rain received.

11. An electrician charges a $25 consultation fee plus $35 per hour for labor.

   a. Copy and complete the following table showing the charges for jobs that take 1, 2, 3, 4, or 5 hours.

Hours, h	Total Cost, C
1	
2	
3	
4	
5	

   b. Write an equation in slope-intercept form for the total cost of a job that takes h hours.

   c. If the electrician bills in quarter hours, how much would it cost for a job that takes 3 hours 15 minutes to complete?

Record your answer on a sheet of paper. Show your work.

12. Explain how you can determine whether two lines are parallel or perpendicular.

Need ExtraHelp?

If you missed Question...	1	2	3	4	5	6	7	8	9	10	11	12
Go to Lesson...	3-3	3-5	4-5	3-1	2-5	4-2	4-3	4-5	2-7	3-6	4-2	4-4

Linear Inequalities

Then

○ You solved linear equations.

Now

○ In this chapter, you will:

- Solve one-step and multi-step inequalities.

- Solve compound inequalities and inequalities involving absolute value.

- Graph inequalities in two variables.

Why? ▲

○ **PETS** In the United States, about 75 million dogs are kept as pets. Approximately 16% of these were adopted from animal shelters. About 14% of dog owners own more than 3 dogs.

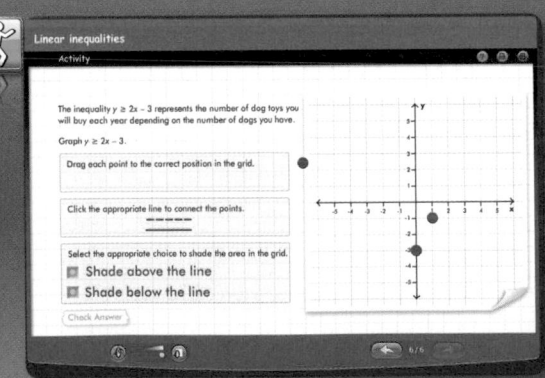

Linear inequalities
Activity

The inequality $y \geq 2x - 3$ represents the number of dog toys you will buy each year depending on the number of dogs you have.

Graph $y \geq 2x - 3$.

Drag each point to the correct position in the grid.

Click the appropriate line to connect the points.

Select the appropriate choice to shade the area in the grid.
☑ Shade above the line
☑ Shade below the line

Check Answer

 connectED.mcgraw-hill.com   **Your Digital Math Portal**

Animation	Vocabulary	eGlossary	Personal Tutor	Virtual Manipulatives	Graphing Calculator	Audio	Foldables	Self-Check Practice	Worksheets

Get Ready for the Chapter

Diagnose Readiness | You have two options for checking prerequisite skills.

1 **Textbook Option** Take the Quick Check below. Refer to the Quick Review for help.

| **Quick**Check | **Quick**Review |

QuickCheck

Evaluate each expression for the given values.

1. $3x + y$ if $x = -4$ and $y = 2$

2. $-2m + 3k$ if $m = -8$ and $k = 3$

3. **CARS** The expression $\frac{m \text{ mi}}{g \text{ gal}}$ represents the gas mileage of a car. Find the gas mileage of a car that goes 295 miles on 12 gallons of gasoline. Round to the nearest tenth.

QuickReview

Example 1

Evaluate $-3x^2 + 4x - 6$ if $x = -2$.

$-3x^2 + 4x - 6$	Original expression
$= -3(-2)^2 + 4(-2) - 6$	Replace x with -2.
$= -3(4) + 4(-2) - 6$	Evaluate the power.
$= -12 + (-8) - 6$	Multiply.
$= -26$	Add and subtract.

Solve each equation.

4. $x - 4 = 9$ 5. $x + 8 = -3$

6. $4x = -16$ 7. $\frac{x}{3} = 7$

8. $2x + 1 = 9$ 9. $4x - 5 = 15$

10. $9x + 2 = 3x - 10$

11. $3(x - 2) = -2(x + 13)$

12. **FINANCIAL LITERACY** Claudia opened a savings account with $325. She saves $100 per month. Write an equation to determine how much money d, she has after m months.

Example 2

Solve $-2(x - 4) = 7x - 19$.

$-2(x - 4) = 7x - 19$	Original equation
$-2x + 8 = 7x - 19$	Distributive Property
$-2x + 8 + 2x = 7x - 19 + 2x$	Add $2x$ to each side.
$8 = 9x - 19$	Simplify.
$8 + 19 = 9x - 19 + 19$	Add 19 to each side.
$27 = 9x$	Simplify.
$3 = x$	Divide each side by 3.

Solve each equation.

13. $|x + 11| = 18$ 14. $|3x - 2| = 16$

15. **SURVEYS** In a survey, 32% of the people chose pizza as their favorite food. The results were reported to within 2% accuracy. What is the maximum and minimum percent of people who chose pizza?

Example 3

Solve $|x - 4| = 9$.

If $|x - 4| = 9$, then $x - 4 = 9$ or $x - 4 = -9$.

$x - 4 = 9$	or	$x - 4 = -9$
$x - 4 + 4 = 9 + 4$		$x - 4 + 4 = -9 + 4$
$x = 13$		$x = -5$

So, the solution set is $\{-5, 13\}$.

2 **Online Option** Take an online self-check Chapter Readiness Quiz at <u>connectED.mcgraw-hill.com</u>.

Get Started on the Chapter

You will learn several new concepts, skills, and vocabulary terms as you study Chapter 5. To get ready, identify important terms and organize your resources. You may wish to refer to Chapter 0 to review prerequisite skills.

FOLDABLES StudyOrganizer

Linear Inequalities Make this Foldable to help you organize your Chapter 5 notes about linear inequalities. Begin with a sheet of 11" by 17" paper.

1 **Fold** each side so the edges meet in the center.

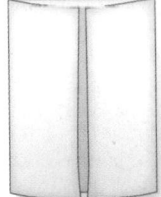

2 **Fold** in half.

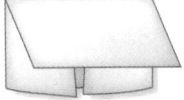

3 **Unfold** and cut from each end until you reach the vertical line.

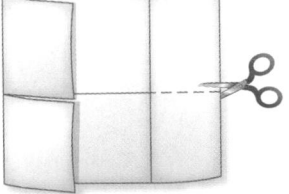

4 **Label** the front of each flap.

NewVocabulary

English		Español
inequality	p. 285	desigualdad
set-builder notation	p. 286	notación de construcción de conjuntos
compound inequality	p. 306	desigualdad compuesta
intersection	p. 306	intersección
union	p. 307	unión
boundary	p. 317	frontera
half-plane	p. 317	semiplano
closed half-plane	p. 317	semiplano cerrada
open half-plane	p. 317	semiplano abierto

ReviewVocabulary

equivalent equations ecuaciones equivalentes equations that have the same solution

linear equation ecuación lineal an equation in the form $Ax + By = C$, with a graph consisting of points on a straight line

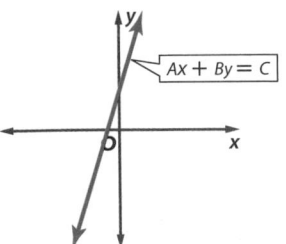

solution set conjunto solución the set of elements from the replacement set that makes an open sentence true

Solving Inequalities by Addition and Subtraction

:: Then	:: Now	:: Why?
● You solved equations by using addition and subtraction.	**1** Solve linear inequalities by using addition. **2** Solve linear inequalities by using subtraction.	● The data in the table show that the recommended daily allowance of Calories for girls 11–14 years old is less than that of girls between 15–18 years old.

Calories	
Girls 11–14 Years	**Girls 15–18**
1845	2110

Source: Vital Health Zone

$$1845 < 2110$$

If a 13-year-old girl and a 16-year-old girl each eat 150 more Calories in a day than is suggested, the 16-year-old will still eat more Calories.

$$1845 + 150 \underline{\ ?\ } 2110 + 150$$
$$1995 < 2260$$

 NewVocabulary
inequality
set-builder notation

 Common Core State Standards

Content Standards
A.CED.1 Create equations and inequalities in one variable and use them to solve problems.

A.REI.3 Solve linear equations and inequalities in one variable, including equations with coefficients represented by letters.

Mathematical Practices
2 Reason abstractly and quantitatively.
4 Model with mathematics.

1 **Solve Inequalities by Addition** An open sentence that contains $<$, $>$, $\leq$, or $\geq$ is an **inequality**. The example above illustrates the Addition Property of Inequalities.

KeyConcept Addition Property of Inequalities

Words	If the same number is added to each side of a true inequality, the resulting inequality is also true.
Symbols	For all numbers a, b, and c, the following are true. 1. If $a > b$, then $a + c > b + c$. 2. If $a < b$, then $a + c < b + c$.

This property is also true for $\geq$ and $\leq$.

Example 1 Solve by Adding

Solve $x - 12 \geq 8$. Check your solution.

$$x - 12 \geq 8 \qquad \text{Original inequality}$$
$$x - 12 + 12 \geq 8 + 12 \qquad \text{Add 12 to each side.}$$
$$x \geq 20 \qquad \text{Simplify.}$$

The solution is the set {all numbers greater than or equal to 20}.

CHECK To check, substitute three different values into the original inequality: 20, a number less than 20, and a number greater than 20.

▶ **Guided**Practice

Solve each inequality. Check your solution.

1A. $22 > m - 8$ **1B.** $d - 14 \geq -19$

A more concise way of writing a solution set is to use **set-builder notation**. In set-builder notation, the solution set in Example 1 is $\{x \mid x \geq 20\}$.

This solution set can be graphed on a number line. Be sure to check if the endpoint of the graph of an inequality should be a circle or a dot. If the endpoint is not included in the graph, use a circle, otherwise use a dot.

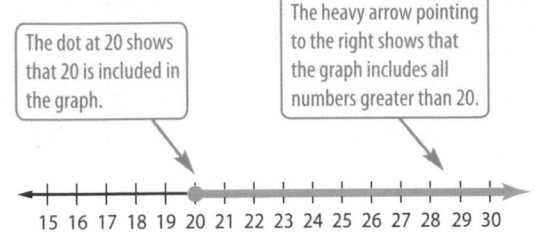

The dot at 20 shows that 20 is included in the graph.

The heavy arrow pointing to the right shows that the graph includes all numbers greater than 20.

15 16 17 18 19 20 21 22 23 24 25 26 27 28 29 30

2 Solve Inequalities by Subtraction Subtraction can also be used to solve inequalities.

KeyConcept Subtraction Property of Inequalities

Words If the same number is subtracted from each side of a true inequality, the resulting inequality is also true.

Symbols For all numbers a, b, and c, the following are true.

1. If $a > b$, then $a - c > b - c$.

2. If $a < b$, then $a - c < b - c$.

This property is also true for $\geq$ and $\leq$.

Standardized Test Example 2 Solve by Subtracting

Solve $m + 19 > 56$.

A $\{m \mid m < 41\}$ **B** $\{m \mid m < 37\}$ **C** $\{m \mid m > 37\}$ **D** $\{m \mid m > 41\}$

Read the Test Item

You need to find the solution set for the inequality.

Solve the Test Item

Step 1 Solve the inequality.

$$m + 19 > 56 \qquad \text{Original inequality}$$
$$m + 19 - 19 > 56 - 19 \qquad \text{Subtract 19 from each side.}$$
$$m > 37 \qquad \text{Simplify.}$$

Step 2 Write in set-builder notation: $\{m \mid m > 37\}$. The answer is C.

GuidedPractice

2. Solve $p + 8 \leq 18$.

F $\{p \mid p \geq 10\}$ **G** $\{p \mid p \leq 10\}$ **H** $\{p \mid p \leq 26\}$ **J** $\{p \mid p \geq 126\}$

Terms that are constants are not the only terms that can be subtracted. Terms with variables can also be subtracted from each side to solve inequalities.

Example 3 Variables on Each Side

Solve $3a + 6 \leq 4a$. Then graph the solution set on a number line.

$$3a + 6 \leq 4a \qquad \text{Original inequality}$$
$$3a - 3a + 6 \leq 4a - 3a \qquad \text{Subtract } 3a \text{ from each side.}$$
$$6 \leq a \qquad \text{Simplify.}$$

Since $6 \leq a$ is the same as $a \geq 6$, the solution set is $\{a \mid a \geq 6\}$.

GuidedPractice

Solve each inequality. Then graph the solution set on a number line.

3A. $9n - 1 < 10n$ **3B.** $5h \leq 12 + 4h$

Verbal problems containing phrases like *greater than* or *less than* can be solved by using inequalities. The chart shows some other phrases that indicate inequalities.

ConceptSummary Phrases for Inequalities

$<$	$>$	$\leq$	$\geq$
less than fewer than	greater than more than	at most, no more than, less than or equal to	at least, no less than, greater than or equal to

Real-World Example 4 Use an Inequality to Solve a Problem

PETS Felipe needs for the temperature of his leopard gecko's basking spot to be at least 82°F. Currently the basking spot is 62.5°F. How much warmer does the basking spot need to be?

Words	The current temperature needs to be at least 82°F.
Variable	Let t = the number of degrees that the temperature needs to rise.
Inequality	$62.5 + t$ $\geq$ 82

$$62.5 + t \geq 82 \qquad \text{Original inequality}$$
$$62.5 + t - 62.5 \geq 82 - 62.5 \qquad \text{Subtract 62.5 from each side.}$$
$$t \geq 19.5 \qquad \text{Simplify.}$$

Felipe needs to raise the temperature of the basking spot 19.5°F or more.

GuidedPractice

4. SHOPPING Sanjay has $65 to spend at the mall. He bought a T-shirt for $18 and a belt for $14. If Sanjay wants a pair of jeans, how much can he spend?

Examples 1–3 Solve each inequality. Then graph the solution set on a number line.

1. $x - 3 > 7$

2. $5 \geq 7 + y$

3. $g + 6 < 2$

4. $11 \leq p + 4$

5. $10 > n - 1$

6. $k + 24 > -5$

7. $8r + 6 < 9r$

8. $8n \geq 7n - 3$

Example 4 Define a variable, write an inequality, and solve each problem. Check your solution.

9. Twice a number increased by 4 is at least 10 more than the number.

10. Three more than a number is less than twice the number.

11. **AMUSEMENT** A thrill ride swings passengers back and forth, a little higher each time up to 137 feet. Suppose the height of the swing after 30 seconds is 45 feet. How much higher will the ride swing?

Practice and Problem Solving

Extra Practice is on page R5.

Examples 1–3 Solve each inequality. Then graph the solution set on a number line.

12. $m - 4 < 3$

13. $p - 6 \geq 3$

14. $r - 8 \leq 7$

15. $t - 3 > -8$

16. $b + 2 \geq 4$

17. $13 > 18 + r$

18. $5 + c \leq 1$

19. $-23 \geq q - 30$

20. $11 + m \geq 15$

21. $h - 26 < 4$

22. $8 \leq r - 14$

23. $-7 > 20 + c$

24. $2a \leq -4 + a$

25. $z + 4 \geq 2z$

26. $w - 5 \leq 2w$

27. $3y + 6 \leq 2y$

28. $6x + 5 \geq 7x$

29. $-9 + 2a < 3a$

Example 4 Define a variable, write an inequality, and solve each problem. Check your solution.

30. Twice a number is more than the sum of that number and 9.

31. The sum of twice a number and 5 is at most 3 less than the number.

32. The sum of three times a number and −4 is at least twice the number plus 8.

33. Six times a number decreased by 8 is less than five times the number plus 21.

CCSS MODELING Define a variable, write an inequality, and solve each problem. Then interpret your solution.

34. **FINANCIAL LITERACY** Keisha is babysitting at $8 per hour to earn money for a car. So far she has saved $1300. The car that Keisha wants to buy costs at least $5440. How much money does Keisha still need to earn to buy the car?

35. **TECHNOLOGY** A recent survey found that more than 21 million people between the ages of 12 and 17 use the Internet. Of those, about 16 million said they use the Internet at school. How many teens that are online do not use the Internet at school?

36. **MUSIC** A DJ added 20 more songs to his digital media player, making the total more than 61. How many songs were originally on the player?

37. TEMPERATURE The water temperature in a swimming pool increased 4°F this morning. The temperature is now less than 81°F. What was the water temperature this morning?

38. BASKETBALL A player's goal was to score at least 150 points this season. So far, she has scored 123 points. If there is one game left, how many points must she score to reach her goal?

39. SPAS Samantha received a $75 gift card for a local day spa for her birthday. She plans to get a haircut and a manicure. How much money will be left on her gift card after her visit?

Service	Cost ($)
haircut	at least 32
manicure	at least 26

40. VOLUNTEER Kono knows that he can only volunteer up to 25 hours per week. If he has volunteered for the times recorded at the right, how much more time can Kono volunteer this week?

Center	Time (h)
Shelter	3 h 15 min
Kitchen	2 h 20 min

Solve each inequality. Check your solution, and then graph it on a number line.

41. $c + (-1.4) \geq 2.3$

42. $9.1g + 4.5 < 10.1g$

43. $k + \frac{3}{4} > \frac{1}{3}$

44. $\frac{3}{2}p - \frac{2}{3} \leq \frac{4}{9} + \frac{1}{2}p$

45. MULTIPLE REPRESENTATIONS In this problem, you will explore multiplication and division in inequalities.

   a. Geometric Suppose a balance has 12 pounds on the left side and 18 pounds on the right side. Draw a picture to represent this situation.

   b. Numerical Write an inequality to represent the situation.

   c. Tabular Create a table showing the result of doubling, tripling, or quadrupling the weight on each side of the balance. Create a second table showing the result of reducing the weight on each side of the balance by a factor of $\frac{1}{2}$, $\frac{1}{3}$, or $\frac{1}{4}$. Include a column in each table for the inequality representing each situation.

   d. Verbal Describe the effect multiplying or dividing each side of an inequality by the same positive value has on the inequality.

CCSS REASONING If $m + 7 \geq 24$, then complete each inequality.

46. $m \geq \underline{\ ?\ }$

47. $m + \underline{\ ?\ } \geq 27$

48. $m - 5 \geq \underline{\ ?\ }$

49. $m - \underline{\ ?\ } \geq 14$

50. $m - 19 \geq \underline{\ ?\ }$

51. $m + \underline{\ ?\ } \geq 43$

H.O.T. Problems   Use Higher-Order Thinking Skills

52. REASONING Compare and contrast the graphs of $a < 4$ and $a \leq 4$.

53. CHALLENGE Suppose $b > d + \frac{1}{3}$, $c + 1 < a - 4$, and $d + \frac{5}{8} > a + 2$. Order $a, b, c,$ and d from least to greatest.

54. OPEN ENDED Write three linear inequalities that are equivalent to $y < -3$.

55. WRITING IN MATH Summarize the process of solving and graphing linear inequalities.

56. WRITING IN MATH Explain why $x - 2 > 5$ has the same solution set as $x > 7$.

57. Which equation represents the relationship shown?

x	y
1	1
2	9
3	17
4	25
5	33
6	41

A $y = 7x - 8$
B $y = 7x + 8$
C $y = 8x - 7$
D $y = 8x + 7$

58. What is the solution set of the inequality $7 + x < 5$?

F $\{x \mid x < 2\}$ **H** $\{x \mid x < -2\}$
G $\{x \mid x > 2\}$ **J** $\{x \mid x > -2\}$

59. Francisco has $3 more than $\frac{1}{4}$ the number of dollars that Kayla has. Which expression represents how much money Francisco has?

A $3\left(\frac{1}{4}k\right)$ **C** $3 - \frac{1}{4}k$
B $\frac{1}{4}k + 3$ **D** $\frac{1}{4} + 3k$

60. GRIDDED RESPONSE The mean score for 10 students on the chemistry final exam was 178. However, the teacher had made a mistake and recorded one student's score as ten points less than the actual score. What should the mean score be?

Find the inverse of each function. (Lesson 4-7)

61. $f(x) = 7x - 28$

62. $f(x) = \frac{2}{5}x + 12$

63. $f(x) = -\frac{1}{3}x - 8$

64. $f(x) = 12x + 16$

Write the slope-intercept form of an equation for the line that passes through the given point and is perpendicular to the graph of each equation. (Lesson 4-4)

65. $(-2, 0), y = x - 6$

66. $(-3, 1), y = -3x + 7$

67. $(1, -3), y = \frac{1}{2}x + 4$

68. $(-2, 7), 2x - 5y = 3$

69. TRAVEL On an island cruise in Hawaii, each passenger is given a lei. A crew member hands out 3 red, 3 blue, and 3 green leis in that order. If this pattern is repeated, what color lei will the 50th person receive? (Lesson 3-6)

Find the nth term of each arithmetic sequence described. (Lesson 3-5)

70. $a_1 = 52, d = 12, n = 102$

71. $-9, -7, -5, -3, \ldots$ for $n = 18$

72. $0.5, 1, 1.5, 2, \ldots$ for $n = 50$

73. JOBS Refer to the time card shown. Write a direct variation equation relating your pay to the hours worked and find your pay if you work 30 hours. (Lesson 3-4)

Weekly Time Card

Day	Hours
FRIDAY	2.0
SATURDAY	3.5
SUNDAY	2.0
TOTAL HOURS	7.5
PAY	$52.50

Solve each equation.

74. $8y = 56$

75. $4p = -120$

76. $-3a = -21$

77. $2c = \frac{1}{5}$

78. $\frac{r}{2} = 21$

79. $-\frac{3}{4}g = -12$

80. $\frac{2}{5}w = -4$

81. $-6x = \frac{2}{3}$

Algebra Lab
Solving Inequalities

You can use algebra tiles to solve inequalities.

CCSS **Common Core State Standards**
Content Standards
A.REI.3 Solve linear equations and inequalities in one variable, including equations with coefficients represented by letters.

Activity   Solve Inequalities

Solve $-2x \leq 4$.

Step 1 Use a self-adhesive note to cover the equals sign on the equation mat. Then write a $\leq$ symbol on the note. Model the inequality.

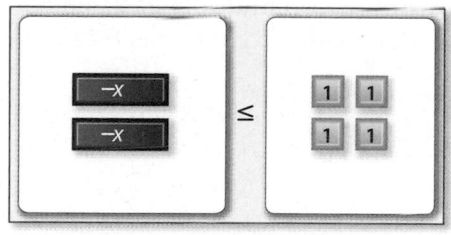

$$-2x \leq 4$$

Step 2 Since you do not want to solve for a negative x-tile, eliminate the negative x-tiles by adding 2 positive x-tiles to each side. Remove the zero pairs.

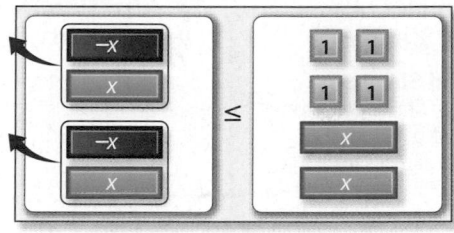

$$-2x + 2x \leq 4 + 2x$$

Step 3 Add 4 negative 1-tiles to each side to isolate the x-tiles. Remove the zero pairs.

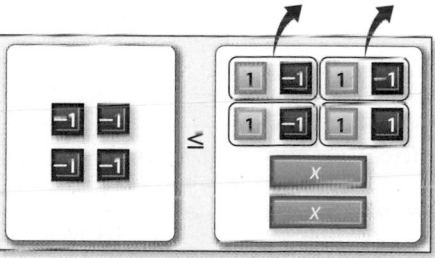

$$-4 \leq 2x$$

Step 4 Separate the tiles into 2 groups.

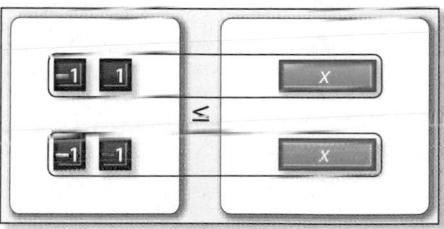

$$-2 \leq x \text{ or } x \geq -2$$

Model and Analyze

Use algebra tiles to solve each inequality.

1. $-3x < 9$ **2.** $-4x > -4$ **3.** $-5x \geq 15$ **4.** $-6x \leq -12$

5. In Exercises 1–4, is the coefficient of x in each inequality positive or negative?

6. Compare the inequality symbols and locations of the variable in Exercises 1–4 with those in their solutions. What do you find?

7. Model the solution for $3x \leq 12$. How is this different from solving $-3x \leq 12$?

8. Write a rule for solving inequalities involving multiplication and division. (*Hint:* Remember that dividing by a number is the same as multiplying by its reciprocal.)

Solving Inequalities by Multiplication and Division

·· Then

- You solved equations by using multiplication and division.

·· Now

1. Solve linear inequalities by using multiplication.

2. Solve linear inequalities by using division.

·· Why?

- Terrell received a gift card for $20 of music downloads. If each download costs $0.89, the number of downloads he can purchase can be represented by the inequality $0.89d \leq 20$.

Common Core State Standards

Content Standards
A.CED.1 Create equations and inequalities in one variable and use them to solve problems.

A.REI.3 Solve linear equations and inequalities in one variable, including equations with coefficients represented by letters.

Mathematical Practices
1 Make sense of problems and persevere in solving them.
6 Attend to precision.

1 Solve Inequalities by Multiplication If you multiply each side of an inequality by a positive number, then the inequality remains true.

$4 > 2$	Original inequality
$4(3) \overset{?}{\underline{}} 2(3)$	Multiply each side by 3.
$12 > 6$	Simplify.

Notice that the direction of the inequality remains the same.

If you multiply each side of an inequality by a negative number, the inequality symbol changes direction.

$7 < 9$	Original inequality
$7(-2) \overset{?}{\underline{}} 9(-2)$	Multiply each side by -2.
$-14 > -18$	Simplify.

These examples demonstrate the **Multiplication Property of Inequalities**.

↯ KeyConcept Multiplication Property of Inequalities

Words	Symbols	Examples
If both sides of an inequality that is true are multiplied by a positive number, the resulting inequality is also true.	For any real numbers a and b and any positive real number c, if $a > b$, then $ac > bc$. And, if $a < b$, then $ac < bc$.	$6 > 3.5$ $6(2) > 3.5(2)$ $12 > 7$ and $2.1 < 5$ $2.1(0.5) < 5(0.5)$ $1.05 < 2.5$
If both sides of an inequality that is true are multiplied by a negative number, the direction of the inequality sign is reversed to make the resulting inequality also true.	For any real numbers a and b and any negative real number c, if $a > b$, then $ac < bc$. And, if $a < b$, then $ac > bc$.	$7 > 4.5$ $7(-3) < 4.5(-3)$ $-21 < -13.5$ and $3.1 < 5.2$ $3.1(-4) > 5.2(-4)$ $-12.4 > -20.8$

This property also holds for inequalities involving $\leq$ and $\geq$.

Real-World Example 1 Write and Solve an Inequality

SURVEYS Of the students surveyed at Madison High School, fewer than eighty-four said they have never purchased an item online. This is about one eighth of those surveyed. How many students were surveyed?

Understand You know the number of students who have never purchased an item online and the portion this is of the number of students surveyed.

Plan Let n = the number of students surveyed. Write an open sentence that represents this situation.

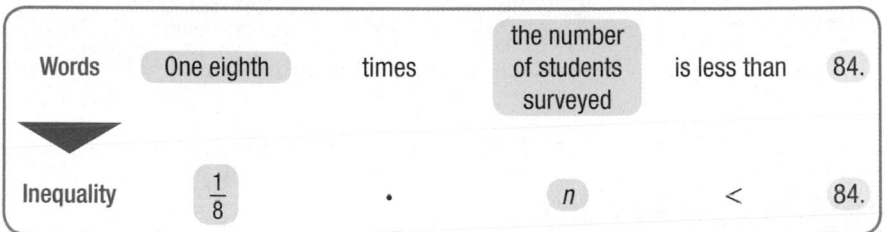

Words	One eighth	times	the number of students surveyed	is less than	84.
Inequality	$\frac{1}{8}$	$\cdot$	n	$<$	84.

Solve Solve for n.

$$\frac{1}{8}n < 84 \qquad \text{Original inequality}$$

$$(8)\frac{1}{8}n < (8)84 \qquad \text{Multiply each side by 8.}$$

$$n < 672 \qquad \text{Simplify.}$$

Check Check the endpoint with 672 and the direction of the inequality with a value less than 672.

$$\frac{1}{8}(672) \overset{?}{=} 84 \qquad \text{Check endpoint.} \qquad \frac{1}{8}(0) \overset{?}{<} 84 \qquad \text{Check direction.}$$

$$84 = 84 \checkmark \qquad\qquad\qquad\qquad 0 < 84 \checkmark$$

The solution set is $\{n \mid n < 672\}$, so fewer than 672 students were surveyed.

▶ **Guided**Practice

1. BIOLOGY Mount Kinabalue in Malaysia has the greatest concentration of wild orchids on Earth. It contains more than 750 species, or about one fourth of all orchid species in Malaysia. How many orchid species are there in Malaysia?

You can also use multiplicative inverses with the Multiplication Property of Inequalities to solve an inequality.

Example 2 Solve by Multiplying

Solve $-\frac{3}{7}r < 21$. **Graph the solution on a number line.**

$$-\frac{3}{7}r < 21 \qquad \text{Original inequality}$$

$$\left(-\frac{7}{3}\right)\left(-\frac{3}{7}r\right) > \left(-\frac{7}{3}\right)21 \qquad \text{Multiply each side by } -\frac{7}{3}. \text{ Reverse the inequality symbol.}$$

$$r > -49 \qquad \text{Simplify. Check by substituting values.}$$

The solution set is $\{r \mid r > -49\}$.

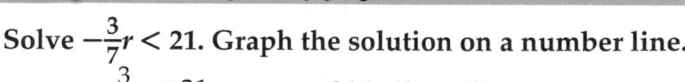
-51 -50 -49 -47 -45 -43 -41

▶ **Guided**Practice

Solve each inequality. Check your solution.

2A. $-\frac{n}{6} \leq 8$ **2B.** $-\frac{4}{3}p > -10$ **2C.** $\frac{1}{5}m \geq -3$ **2D.** $\frac{3}{8}t < 5$

2 Solve Inequalities by Division

If you divide each side of an inequality by a positive number, then the inequality remains true.

$-10 < -5$	Original inequality
$\dfrac{-10}{5} \; ? \; \dfrac{-5}{5}$	Divide each side by -5.
$-2 < -1$	Simplify.

Notice that the direction of the inequality remains the same. If you divide each side of an inequality by a negative number, the inequality symbol changes direction.

$15 < 18$	Original inequality
$\dfrac{15}{-3} \; ? \; \dfrac{18}{-3}$	Divide each side by -3.
$-5 > -6$	Simplify.

These examples demonstrate the **Division Property of Inequalities**.

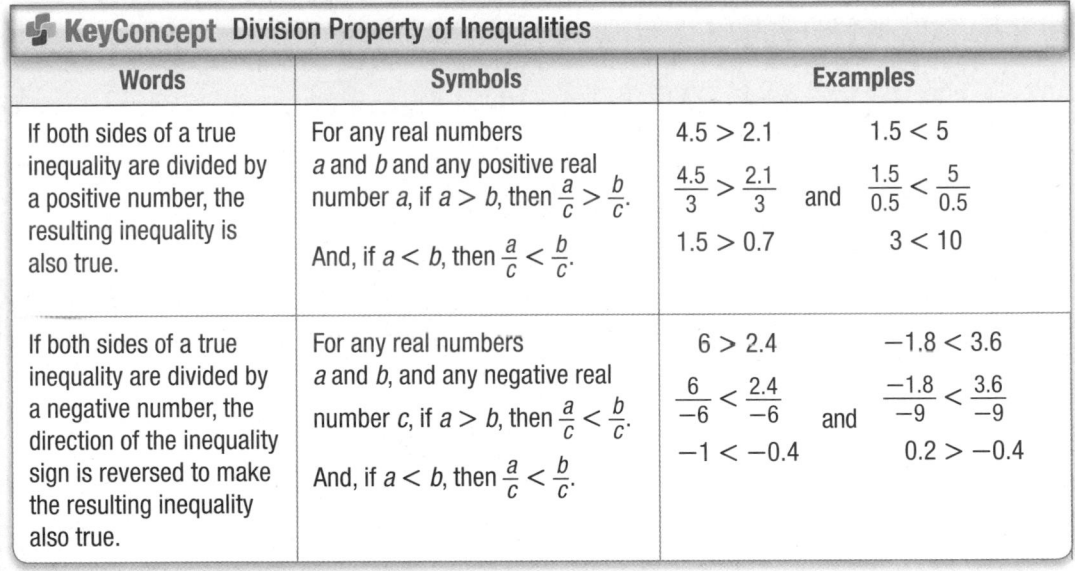

KeyConcept Division Property of Inequalities

Words	Symbols	Examples
If both sides of a true inequality are divided by a positive number, the resulting inequality is also true.	For any real numbers a and b and any positive real number a, if $a > b$, then $\frac{a}{c} > \frac{b}{c}$. And, if $a < b$, then $\frac{a}{c} < \frac{b}{c}$.	$4.5 > 2.1$ $\quad$ $1.5 < 5$ $\dfrac{4.5}{3} > \dfrac{2.1}{3}$ and $\dfrac{1.5}{0.5} < \dfrac{5}{0.5}$ $1.5 > 0.7$ $\quad$ $3 < 10$
If both sides of a true inequality are divided by a negative number, the direction of the inequality sign is reversed to make the resulting inequality also true.	For any real numbers a and b, and any negative real number c, if $a > b$, then $\frac{a}{c} < \frac{b}{c}$. And, if $a < b$, then $\frac{a}{c} < \frac{b}{c}$.	$6 > 2.4$ $\quad$ $-1.8 < 3.6$ $\dfrac{6}{-6} < \dfrac{2.4}{-6}$ and $\dfrac{-1.8}{-9} < \dfrac{3.6}{-9}$ $-1 < -0.4$ $\quad$ $0.2 > -0.4$

This property also holds true for inequalities involving $\leq$ and $\geq$.

Example 3 Divide to Solve an Inequality

Solve each inequality. Graph the solution on a number line.

a. $60t > 8$

$60t > 8$	Original inequality
$\dfrac{60t}{60} > \dfrac{8}{60}$	Divide each side by 60.
$t > \dfrac{2}{15}$	Simplify.

$\left\{ t \; \middle| \; t > \dfrac{2}{15} \right\}$

b. $-7d \leq 147$

$-7d \leq 147$	Original inequality
$\dfrac{-7d}{-7} \geq \dfrac{147}{-7}$	Divide each side by -7.
$d \geq -21$	Simplify.

$\{d \mid d \geq -21\}$

Guided Practice

3A. $8p \leq 58$

3C. $-12h > 15$

3B. $-42 \geq 6r$

3D. $-\dfrac{1}{2}n \leq 6$

Example 1

1. **FUNDRAISING** The Jefferson Band Boosters raised more than $5500 from sales of their $15 band DVD. Define a variable, and write an inequality to represent the number of DVDs they sold. Solve the inequality and interpret your solution.

Examples 2–3 Solve each inequality. Graph the solution on a number line.

2. $30 > \frac{1}{2}n$ 3. $-\frac{3}{4}r \le -6$ 4. $-\frac{c}{6} \ge 7$ 5. $\frac{h}{2} < -5$

6. $9t > 108$ 7. $-84 < 7v$ 8. $-28 \le -6x$ 9. $40 \ge -5z$

Practice and Problem Solving Extra Practice is on page R5.

Example 1 Define a variable, write an inequality, and solve each problem. Then interpret your solution.

10. **CELL PHONE PLAN** Mario purchases a prepaid phone plan for $50 at $0.13 per minute. How many minutes can Mario talk on this plan?

11. **FINANCIAL LITERACY** Rodrigo needs at least $560 to pay for his spring break expenses, and he is saving $25 from each of his weekly paychecks. How long will it be before he can pay for his trip?

Examples 2–3 Solve each inequality. Graph the solution on a number line.

12. $\frac{1}{4}m \le -17$ (13) $\frac{1}{2}a < 20$ 14. $-11 > -\frac{c}{11}$

15. $-2 \ge -\frac{d}{34}$ 16. $-10 \le \frac{x}{-2}$ 17. $-72 < \frac{f}{-6}$

18. $\frac{2}{3}h > 14$ 19. $-\frac{3}{4}j \ge 12$ 20. $-\frac{1}{6}n \le -18$

21. $6p \le 96$ 22. $4r < 64$ 23. $32 > -2y$

24. $-26 < 26t$ 25. $-6v > -72$ 26. $-33 \ge -3z$

27. $4b \le -3$ 28. $-2d < 5$ 29. $-7f > 5$

30. **CHEERLEADING** To remain on the cheerleading squad, Lakita must attend at least $\frac{3}{5}$ of the study table sessions offered. She attends 15 sessions. If Lakita met the requirements, what is the maximum number of study table sessions?

31. **BRACELETS** How many bracelets can Caitlin buy for herself and her friends if she wants to spend no more than $22?

32. **CCSS PRECISION** The National Honor Society at Pleasantville High School wants to raise at least $500 for a local charity. Each student earns $0.50 for every quarter of a mile walked in a walk-a-thon. How many miles will the students need to walk?

33. **MUSEUM** The American history classes are planning a trip to a local museum. Admission is $8 per person. Determine how many people can go for $260.

34. **GASOLINE** If gasoline costs $3.15 per gallon, how many gallons of gasoline, to the nearest tenth, can Jan buy for $24?

Match each inequality to the graph of its solution.

35. $-\frac{2}{3}h \leq 9$ **36.** $25j \geq 8$ **37.** $3.6p < -4.5$ **38.** $2.3 < -5t$

a.

b.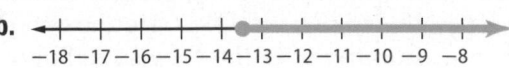

c.

d.

39. **CANDY** Fewer than 42 employees at a factory stated that they preferred fudge over fruit candy. This is about two thirds of the employees. How many employees are there?

40. **TRAVEL** A certain travel agency employs more than 275 people at all of its branches. Approximately three fifths of all the people are employed at the west branch. How many people work at the west branch?

41. **MULTIPLE REPRESENTATIONS** The equation for the volume of a pyramid is $\frac{1}{3}$ the area of the base times the height.

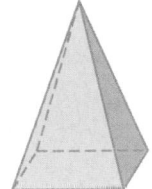

  a. Geometric Draw a pyramid with a square base b cm long and a height of h cm.

  b. Numerical Suppose the pyramid has a volume of 72 cm³. Write an equation to find the height.

  c. Tabular Create a table showing the value of h when $b = 1, 3, 6, 9,$ and 12.

  d. Numerical Write an inequality for the possible lengths of b such that $b < h$. Write an inequality for the possible lengths of h such that $b > h$.

H.O.T. Problems   Use Higher-Order Thinking Skills

42. **ERROR ANALYSIS** Taro and Jamie are solving $6d \geq -84$. Is either of them correct? Explain your reasoning.

> **Taro**
> $6d \geq -84$
> $\frac{6d}{6} \geq \frac{-84}{6}$
> $d \geq -14$

> **Jamie**
> $6d \geq -84$
> $\frac{6d}{6} \leq \frac{-84}{6}$
> $d \leq -14$

43. **CHALLENGE** Solve each inequality for x. Assume that $a > 0$.

  a. $-ax < 5$ **b.** $\frac{1}{a}x \geq 8$ **c.** $-6 \geq ax$

44. **CCSS STRUCTURE** Determine whether $x^2 > 1$ and $x > 1$ are equivalent. Explain.

45. **REASONING** Explain whether the statement *If $a > b$, then $\frac{1}{a} > \frac{1}{b}$ is sometimes, always,* or *never* true.

46. **OPEN ENDED** Create a real-world situation to represent the inequality $-\frac{5}{8} \geq x$.

47. **WRITING IN MATH** How are solving linear inequalities and linear equations similar? different?

48. Juan's international calling card costs 9¢ for each minute. Which inequality can be used to find how long he can talk to a friend if he does not want to spend more than $2.50 on the call?

A $0.09 \geq 2.50m$

B $0.09 \leq 2.50m$

C $0.09m \geq 2.50$

D $0.09m \leq 2.50$

49. SHORT RESPONSE Find the value of x.

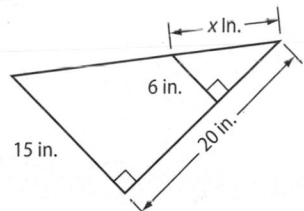

50. What is the greatest rate of decrease of this function?

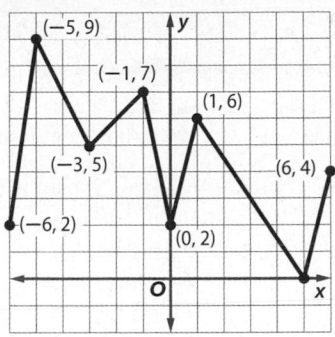

F -5 **H** -2

G -3 **J** 1

51. What is the value of x if $4x - 3 = -2x$?

A -2 **C** $\dfrac{1}{2}$

B $-\dfrac{1}{2}$ **D** 2

Spiral Review

Solve each inequality. Check your solution, and then graph it on a number line. (Lesson 5-1)

52. $a + 8 > 4$

53. $6y \geq 5y + 21$

54. $b + \dfrac{1}{2} < 1$

Find the inverse of each function. (Lesson 4-7)

55. $f(x) = -6x + 18$

56. $f(x) = \dfrac{3}{7}x + 9$

57. $f(r) = 4x - 5$

58. HOME DECOR Pam is having blinds installed at her home. The cost c of installation for any number of blinds b can be described by $c = 25 + 6.5b$. Graph the equation and determine how much it would cost if Pam has 8 blinds installed. (Lesson 3-1)

59. RESCUE A boater radioed for a helicopter to pick up a sick crew member. At that time, the boat and the helicopter were at the positions shown. How long will it take for the helicopter to reach the boat? (Lesson 2-9)

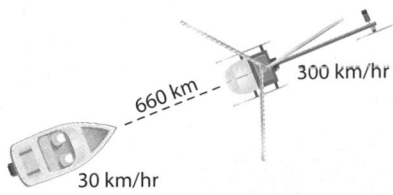

Solve each equation. (Lesson 2-5)

60. $|x + 3| = 10$

61. $|2x - 8| = 6$

62. $|3x + 1| = -2$

Skills Review

Solve each equation.

63. $4y + 11 = 19$

64. $2x - 7 = 9 + 4x$

65. $\dfrac{1}{4} + 2x = 4x - 8$

66. $\dfrac{1}{3}(6w - 3) = 3w + 12$

67. $\dfrac{7r + 5}{2} = 13$

68. $\dfrac{1}{2}a = \dfrac{a - 3}{4}$

Solving Multi-Step Inequalities

⋮ Then	**⋮ Now**	**⋮ Why?**
● You solved multi-step equations.	**1** Solve linear inequalities involving more than one operation. **2** Solve linear inequalities involving the Distributive Property.	● A salesperson may make a base monthly salary and earn a commission on each of her sales. To find the number of sales she needs to make to pay her monthly bills, you can use a multi-step inequality.

 Common Core State Standards

Content Standards
A.CED.1 Create equations and inequalities in one variable and use them to solve problems.

A.REI.3 Solve linear equations and inequalities in one variable, including equations with coefficients represented by letters.

Mathematical Practices
7 Look for and make use of structure.

1 Solve Multi-Step Inequalities Multi–step inequalities can be solved by undoing the operations in the same way you would solve a multi-step equation.

Real-World Example 1   Solve a Multi-Step Inequality

SALES Write and solve an inequality to find the sales Mrs. Jones needs if she earns a monthly salary of $2000 plus a 10% commission on her sales. Her goal is to make at least $4000 per month. What sales does she need to meet her goal?

base salary + (commission × sales) ≥ income needed

$$2000 + 0.10x \geq 4000 \qquad \text{Substitution}$$
$$0.10x \geq 2000 \qquad \text{Subtract 2000 from each side.}$$
$$x \geq 20,000 \qquad \text{Divide each side by 0.10.}$$

She must make at least $20,000 in sales to meet her monthly goal.

▶ **Guided**Practice

1. **FINANCIAL LITERACY** The Print Shop advertises a special to print 400 flyers for less than the competition. The price includes a $3.50 set-up fee. If the competition charges $35.50, what does the Print Shop charge for each flyer?

When multiplying or dividing by a negative number, the direction of the inequality symbol changes. This holds true for multi-step inequalities.

Example 2   Inequality Involving a Negative Coefficient

Solve $-11y - 13 > 42$. Graph the solution on a number line.

$$-11y - 13 > 42 \qquad \text{Original inequality}$$
$$-11y > 55 \qquad \text{Add 13 to each side and simplify.}$$
$$\frac{-11y}{-11} < \frac{55}{-11} \qquad \text{Divide each side by } -11, \text{ and reverse the inequality.}$$
$$y < -5 \qquad \text{Simplify.}$$

The solution set is $\{y \mid y < -5\}$.

⟵‖‖‖‖⊕‖‖‖‖⟶
−10 −8 −6 −4 −2 0 2

▶ **Guided**Practice   **Solve each inequality.**

2A. $23 \geq 10 - 2w$ **2B.** $43 > -4y + 11$

Examples 1
and 2

CCSS STRUCTURE Solve each inequality. Graph the solution on a number line.

12. $5b - 1 \geq -11$

13. $21 > 15 + 2a$

14. $-9 \geq \frac{2}{5}m + 7$

15. $\frac{w}{8} - 13 > -6$

16. $-a + 6 \leq 5$

17. $37 < 7 - 10w$

18. $8 - \frac{z}{3} \geq 11$

19. $-\frac{5}{4}p + 6 < 12$

20. $3b - 6 \geq 15 + 24b$

21. $15h + 30 < 10h - 45$

Example 3

Define a variable, write an inequality, and solve each problem. Check your solution.

22. Three fourths of a number decreased by nine is at least forty-two.

23. Two thirds of a number added to six is at least twenty-two.

24. Seven tenths of a number plus 14 is less than forty-nine.

25. Eight times a number minus twenty-seven is no more than the negative of that number plus eighteen.

26. Ten is no more than 4 times the sum of twice a number and three.

27. Three times the sum of a number and seven is greater than five times the number less thirteen.

28. The sum of nine times a number and fifteen is less than or equal to the sum of twenty-four and ten times the number.

Examples 4
and 5

CCSS STRUCTURE Solve each inequality. Graph the solution on a number line.

29. $-3(7n + 3) < 6n$

30. $21 \geq 3(a - 7) + 9$

31. $2y + 4 > 2(3 + y)$

32. $3(2 - b) < 10 - 3(b - 6)$

33. $7 + t \leq 2(t + 3) + 2$

34. $8a + 2(1 - 5a) \leq 20$

Define a variable, write an inequality, and solve each problem. Then interpret your solution.

35. **CARS** A car salesperson is paid a base salary of $35,000 a year plus 8% of sales. What are the sales needed to have an annual income greater than $65,000?

36. **ANIMALS** Keith's dog weighs 90 pounds. A healthy weight for his dog would be less than 75 pounds. If Keith's dog can lose an average of 1.25 pounds per week on a certain diet, after how long will the dog reach healthy weight?

37. Solve $6(m - 3) > 5(2m + 4)$. Show each step and justify your work.

38. Solve $8(a - 2) \leq 10(a + 2)$. Show each step and justify your work.

39. **MUSICAL** A high school drama club is performing a musical to benefit a local charity. Tickets are $5 each. They also received donations of $565. They want to raise at least $1500.

a. Write an inequality that describes this situation. Then solve the inequality.

b. Graph the solution.

40. **ICE CREAM** Benito has $6 to spend. A sundae costs $3.25 plus $0.65 per topping. Write and solve an inequality to find how many toppings he can order.

41. SCIENCE The normal body temperature of a camel is 97.7°F in the morning. If it has had no water by noon, its body temperature can be greater than 104°F.

   a. Write an inequality that represents a camel's body temperature at noon if the camel had no water.

   b. If C represents degrees Celsius, then $F = \frac{9}{5}C + 32$. Write and solve an inequality to find the camel's body temperature at noon in degrees Celsius.

42. NUMBER THEORY Find all sets of three consecutive positive even integers with a sum no greater than 36.

43. NUMBER THEORY Find all sets of four consecutive positive odd integers with a sum that is less than 42.

Solve each inequality. Check your solution.

44. $2(x - 4) \le 2 + 3(x - 6)$

45. $\dfrac{2x - 4}{6} \ge -5x + 2$

46. $5.6z + 1.5 < 2.5z - 4.7$

47. $0.7(2m - 5) \ge 21.7$

GRAPHING CALCULATOR Use a graphing calculator to solve each inequality.

48. $3x + 7 > 4x + 9$

49. $13x - 11 \le 7x + 37$

50. $2(x - 3) < 3(2x + 2)$

51. $\frac{1}{2}x - 9 < 2x$

52. $2x - \frac{2}{3} \ge x - 22$

53. $\frac{1}{3}(4x + 3) \ge \frac{2}{3}x + 2$

54. 🔄 **MULTIPLE REPRESENTATIONS** In this problem, you will solve compound inequalities. A number x is greater than 4, and the same number is less than 9.

   a. Numerical Write two separate inequalities for the statement.

   b. Graphical Graph the solution set for the first inequality in red. Graph the solution set for the second inequality in blue. Highlight where they overlap.

   c. Tabular Make a table using ten points from your number line, including points from each section. Use one column for each inequality and a third column titled "Both are True." Complete the table by writing true or false.

   d. Verbal Describe the relationship between the colored regions of the graph and the chart.

   e. Logical Make a prediction of what the graph of $4 < x < 9$ looks like.

H.O.T. Problems   Use Higher-Order Thinking Skills

55. **CCSS REASONING** Explain how you could solve $-3p + 7 \ge -2$ without multiplying or dividing each side by a negative number.

56. **CHALLENGE** If $ax + b < ax + c$ is true for all real values of x, what will be the solution of $ax + b > ax + c$? Explain how you know.

57. **CHALLENGE** Solve each inequality for x. Assume that $a > 0$.

   a. $ax + 4 \ge -ax - 5$

   b. $2 - ax < x$

   c. $-\frac{2}{a}x + 3 > -9$

58. **WHICH ONE DOESN'T BELONG?** Name the inequality that does not belong. Explain.

$4y + 9 > -3$	$3y - 4 > 5$	$-2y + 1 < -5$	$-5y + 2 < -13$

59. **WRITING IN MATH** Explain when the solution set of an inequality will be the empty set or the set of all real numbers. Show an example of each.

60. What is the solution set of the inequality $4t + 2 < 8t - (6t - 10)$?

A $\{t \mid t < -6.5\}$ **C** $\{t \mid t < 4\}$

B $\{t \mid t > -6.5\}$ **D** $\{t \mid t > 4\}$

61. GEOMETRY The section of Liberty Ave. between 5th St. and King Ave. is temporarily closed. Traffic is being detoured right on 5th St., left on King Ave. and then back on Liberty Ave. How long is the closed section of Liberty Ave.?

F 100 ft

G 120 ft

H 144 ft

J 180 ft

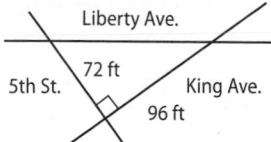

62. SHORT RESPONSE Rhiannon is paid \$52 for working 4 hours. At this rate, how many hours will it take her to earn \$845?

63. GEOMETRY Classify the triangle.

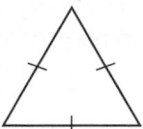

A right

B parallel

C obtuse

D equilateral

Solve each inequality. Check your solution. (Lesson 5-2)

64. $\dfrac{y}{2} \le -5$

65. $12b > -48$

66. $-\dfrac{2}{3}t \le -30$

Solve each inequality. Check your solution, and graph it on a number line. (Lesson 5-1)

67. $6 - h > -8$

68. $p - 9 < 2$

69. $3 \ge 4 - m$

Solve each equation by graphing. Verify your answer algebraically. (Lesson 3-2)

70. $2x - 7 = 4x + 9$

71. $5 + 3x = 7x - 11$

72. $2(x - 3) = 5x + 12$

73. THEME PARKS In a recent year, 70.9 million people visited the top 5 theme parks in North America. That represents an increase of about 1.14% in the number of visitors from the prior year. About how many people visited the top 5 theme parks in North America in the prior year? (Lesson 2-7)

If $f(x) = 4x - 3$ and $g(x) = 2x^2 + 5$, find each value. (Lesson 1-7)

74. $f(-2)$

75. $g(2) - 5$

76. $f(c + 3)$

77. COSMETOLOGY On average, a barber received a tip of \$4 for each of 12 haircuts. Write and evaluate an expression to determine the total amount that she earned. (Lesson 1-4)

Graph each set of numbers on a number line.

78. $\{-4, -2, 2, 4\}$

79. $\{-3, 0, 1, 5\}$

80. {integers less than 3}

81. {integers greater than or equal to -2}

82. {integers between -3 and 4}

83. {integers less than -1}

Solve each inequality. Then graph it on a number line.
(Lesson 5-1)

1. $x - 8 > 4$

2. $m + 2 \geq 6$

3. $p - 4 < -7$

4. $12 \leq t - 9$

5. CONCERTS Lupe's allowance for the month is $60. She wants to go to a concert for which a ticket costs $45.
(Lesson 5-1)

 a. Write and solve an inequality that shows how much money she can spend that month after buying a concert ticket.

 b. She spends $9.99 on music downloads and $2 on lunch in the cafeteria. Write and solve an inequality that shows how much she can spend after these purchases and the concert ticket.

Define a variable, write an inequality, and solve each problem. Check your solution. (Lesson 5-1)

6. The sum of a number and -2 is no more than 6.

7. A number decreased by 4 is more than -1.

8. Twice a number increased by 3 is less than the number decreased by 4.

9. MULTIPLE CHOICE Jane is saving money to buy a new cell phone that costs no more than $90. So far, she has saved $52. How much more money does Jane need to save? (Lesson 5-1)

 A $38

 B more than $38

 C no more than $38

 D at least $38

Solve each inequality. Check your solution. (Lesson 5-2)

10. $\frac{1}{3}y \geq 5$

11. $4 < \frac{c}{5}$

12. $-8x > 24$

13. $2m \leq -10$

14. $\frac{x}{2} < \frac{5}{8}$

15. $-9a \geq -45$

16. $\frac{w}{6} > -3$

17. $\frac{k}{7} < -2$

18. ANIMALS The world's heaviest flying bird is the great bustard. A male bustard can be up to 4 feet long and weigh up to 40 pounds. (Lesson 5-2)

 a. Write inequalities to describe the ranges of lengths and weights of male bustards.

 b. Male bustards are usually about four times as heavy as females. Write and solve an inequality that describes the range of weights of female bustards.

19. GARDENING Bill is building a fence around a square garden to keep deer out. He has 60 feet of fencing. Find the maximum length of a side of the garden. (Lesson 5-2)

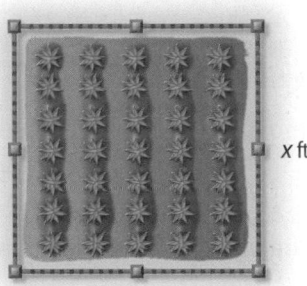

x ft

Solve each inequality. Check your solution. (Lesson 5-3)

20. $4a - 2 > 14$

21. $2x + 11 \leq 5x - 10$

22. $-p + 4 < -9$

23. $\frac{d}{4} + 1 \geq -3$

24. $-2(4b + 1) < -3b + 8$

Define a variable, write an inequality, and solve each problem. Check your solution. (Lesson 5-3)

25. Three times a number increased by 8 is no more than the number decreased by 4.

26. Two thirds of a number plus 5 is greater than 17.

27. MULTIPLE CHOICE Shoe rental costs $2, and each game bowled costs $3. How many games can Kyle bowl without spending more than $15? (Lesson 5-3)

 F 2

 G 3

 H 4

 J 5

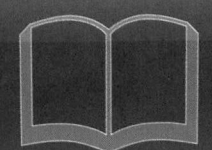
A compound statement is made up of two simple statements connected by the word *and* or *or*. Before you can determine whether a compound statement is true or false, you must understand what the words *and* and *or* mean.

A spider has eight legs, *and* a dog has five legs.

For a compound statement connected by the word *and* to be true, both simple statements must be true.

A spider has eight legs. ⟶ true

A dog has five legs. ⟶ false

Since one of the statements is false, the compound statement is false.

A compound statement connected by the word *or* may be *exclusive* or *inclusive*. For example, the statement "With your lunch, you may have milk *or* juice," Is exclusive. In everyday language, *or* means one or the other, but not both. However, in mathematics, *or* is inclusive. It means one or the other or both.

A spider has eight legs, *or* a dog has five legs.

For a compound statement connected by the word *or* to be true, at least one of the simple statements must be true. Since it is true that a spider has eight legs, the compound statement is true.

Exercises

Is each compound statement *true* or *false*? Explain.

1. Most top 20 movies in 2007 were rated PG-13, *or* most top 20 movies in 2005 were rated G.

2. In 2008 more top 20 movies were rated PG than were rated G, *and* more were rated PG than rated PG-13.

3. For the years shown most top 20 movies are rated PG-13, *and* the least top 20 movies are rated G.

4. No top 20 movies in 2008 were rated G, *or* most top 20 movies in 2008 were *not* rated PG.

5. $11 < 5$ or $9 < 7$
6. $-2 > 0$ and $3 < 7$
7. $5 > 0$ and $-3 < 0$
8. $-2 > -3$ or $0 = 0$
9. $8 \neq 8$ or $-2 > -5$
10. $5 > 10$ and $4 > -2$

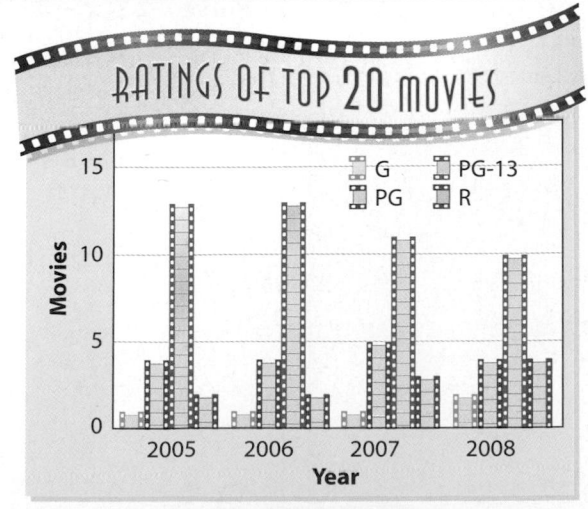

Source: National Association of Theater Owners

Solving Compound Inequalities

:·Then	:·Now	:·Why?

- You solved absolute value equations with two cases.

- **1** Solve compound inequalities containing the word *and*, and graph their solution set.

- **2** Solve compound inequalities containing the word *or*, and graph their solution set.

- Visitors 4 years of age and older have to purchase a ticket to visit a certain history museum. Visitors less than 18 years of age can purchase a youth ticket. If *v* represents the age of a visitor, we can write two inequalities to represent when a youth ticket can be purchased.

4 years of age and older   less than 18 years of age

$$v \geq 4 \qquad\qquad\qquad v < 18$$

The inequalities $v \geq 4$ and $v < 18$ can be combined and written without using *and* as $4 \leq v < 18$.

 NewVocabulary
compound inequality
intersection
union

 Common Core State Standards

Content Standards
A.CED.1 Create equations and inequalities in one variable and use them to solve problems.

A.REI.3 Solve linear equations and inequalities in one variable, including equations with coefficients represented by letters.

Mathematical Practices
1 Make sense of problems and persevere in solving them.
8 Look for and express regularity in repeated reasoning.

1 Inequalities Containing *and* When considered together, two inequalities such as $h \geq 52$ and $h \leq 72$ form a **compound inequality**. A compound inequality containing *and* is only true if both inequalities are true. Its graph is where the graphs of the two inequalities overlap. This is called the **intersection** of the two graphs.

The intersection can be found by graphing each inequality and then determining where the graphs intersect.

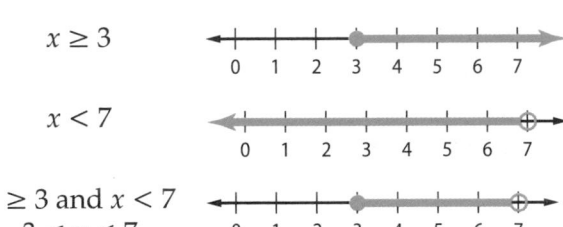

$x \geq 3$

$x < 7$

$x \geq 3$ and $x < 7$
$3 \leq x < 7$

The statement $3 \leq x < 7$ can be read as *x is greater than or equal to 3 and less than 7* or *x is between 3 and 7 including 3*.

Example 1 Solve and Graph an Intersection

Solve $-2 \leq x - 3 < 4$. Then graph the solution set.

First, express $-2 \leq x - 3 < 4$ using *and*. Then solve each inequality.

$-2 \leq x - 3$	**and**	$x - 3 < 4$	Write the inequalities.
$-2 + 3 \leq x - 3 + 3$		$x - 3 + 3 < 4 + 3$	Add 3 to each side.
$1 \leq x$		$x < 7$	Simplify.

The solution set is $\{x \mid 1 \leq x < 7\}$. Now graph the solution set.

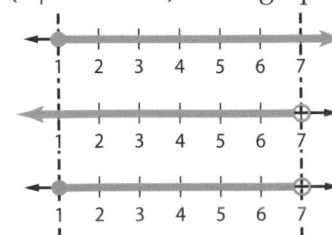

Graph $1 \leq x$ or $x \geq 1$.

Graph $x < 7$.

Find the intersection of the graphs.

Solve each compound inequality. Then graph the solution set.

1A. $y - 3 \geq -11$ and $y - 3 \leq -8$ **1B.** $6 \leq r + 7 < 10$

2 Inequalities Containing _or_ Another type of compound inequality contains the word _or_. A compound inequality containing _or_ is true if at least one of the inequalities is true. Its graph is the **union** of the graphs of two inequalities.

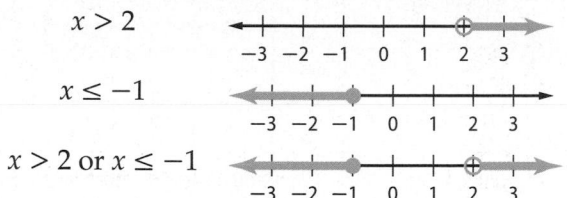

When solving problems involving inequalities, _within_ is meant to be inclusive, so use $\geq$ or $\leq$. _Between_ is meant to be exclusive, so use $<$ or $>$.

Real-World Example 2 Write and Graph a Compound Inequality

SOUND The human ear can only detect sounds between the frequencies 20 Hertz and 20,000 Hertz. Write and graph a compound inequality that describes the frequency of sounds humans cannot hear.

The problem states that humans can hear the frequencies between 20 Hz and 20,000 Hz. We are asked to find the frequencies humans cannot hear.

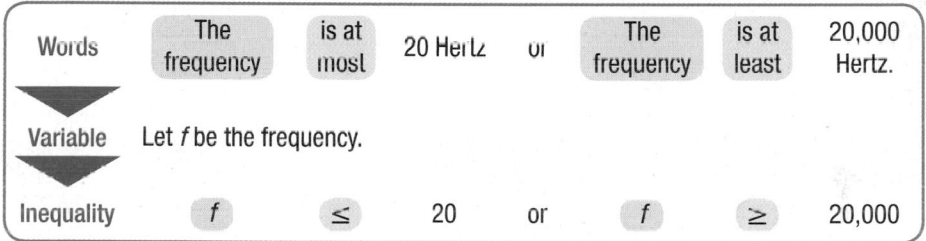

Now, graph the solution set.

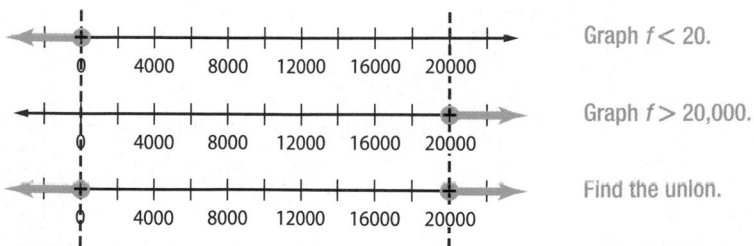

Graph $f < 20$.

Graph $f > 20,000$.

Find the union.

Notice that the graphs do not intersect. Humans cannot hear sounds at a frequency less than 20 Hertz or greater than 20,000 Hertz. The compound inequality is $\{f \mid f < 20 \text{ or } f > 20,000\}$.

> **GuidedPractice**

2. MANUFACTURING A company is manufacturing an action figure that must be at least 11.2 centimeters and at most 11.4 centimeters tall. Write and graph a compound inequality that describes how tall the action figure can be.

Example 3 Solve and Graph a Union

Solve $-2m + 7 \leq 13$ **or** $5m + 12 > 37$. **Then graph the solution set.**

$-2m + 7 \leq 13$		**or**	$5m + 12 > 37$
$-2m + 7 - 7 \leq 13 - 7$	Subtract.		$5m + 12 - 12 > 37 - 12$
$-2m \leq 6$	Simplify.		$5m > 25$
$\dfrac{-2m}{-2} \geq \dfrac{6}{-2}$	Divide.		$\dfrac{5m}{5} > \dfrac{25}{5}$
$m \geq -3$	Simplify.		$m > 5$

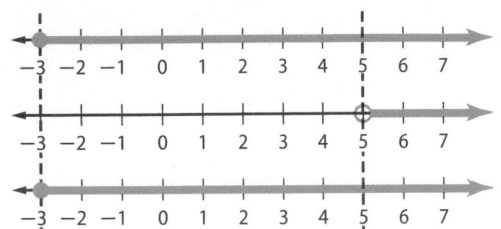

Graph $m \geq -3$.

Graph $m > 5$.

Find the union.

Notice that the graph of $m \geq -3$ contains every point in the graph of $m > 5$. So, the union is the graph of $m \geq -3$. The solution set is $\{m \mid m \geq -3\}$.

GuidedPractice

Solve each compound inequality. Then graph the solution set.

3A. $a + 1 < 4$ or $a - 1 \geq 3$ **3B.** $x \leq 9$ or $2 + 4x < 10$

Check Your Understanding

 = Step-by-Step Solutions begin on page R13.

Examples 1, 3 Solve each compound inequality. Then graph the solution set.

1. $4 \leq p - 8$ and $p - 14 \leq 2$ **2.** $r + 6 < -8$ or $r - 3 > -10$

3. $4a + 7 \geq 31$ or $a > 5$ **4.** $2 \leq g + 4 < 7$

Example 2 **5.** ⒸⒸⓈⓈ **SENSE-MAKING** The recommended air pressure for the tires of a mountain bike is at least 35 pounds per square inch (psi), but no more than 80 pounds per square inch. If a bike's tires have 24 pounds per square inch, what is the recommended range of air that should be put into the tires?

Practice and Problem Solving

Extra Practice is on page R5.

Examples 1, 3 Solve each compound inequality. Then graph the solution set.

6. $f - 6 < 5$ and $f - 4 \geq 2$ **7.** $n + 2 \leq -5$ and $n + 6 \geq -6$

8. $y - 1 \geq 7$ or $y + 3 < -1$ **9.** $t + 14 \geq 15$ or $t - 9 < -10$

10. $-5 < 3p + 7 \leq 22$ **11.** $-3 \leq 7c + 4 < 18$

12. $5h - 4 \geq 6$ and $7h + 11 < 32$ **13.** $22 \geq 4m - 2$ or $5 - 3m \leq -13$

14. $-4a + 13 \geq 29$ and $10 < 6a - 14$ **15.** $-y + 5 \geq 9$ or $3y + 4 < -5$

Example 2

16. **SPEED** The posted speed limit on an interstate highway is shown. Write an inequality that represents the sign. Graph the inequality.

17. **NUMBER THEORY** Find all sets of two consecutive positive odd integers with a sum that is at least 8 and less than 24.

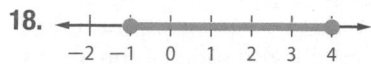

Write a compound inequality for each graph.

18. (number line from −2 to 4, closed dots at −1 and 4)

19. (number line from −4 to 2, open circle at −3, closed dot at 2)

20. (number line from −1 to 4, open circle at 0, closed dot at 3)

21. (number line from −6 to 0, open circles at −5 and −4)

22. (number line from 1 to 7, closed dots at 3 and 6)

23. (number line from −4 to 2, closed dot at −3, open circle at 1)

Solve each compound inequality. Then graph the solution set.

24. $3b + 2 < 5b - 6 \le 2b + 9$

25. $-2a + 3 \ge 6a - 1 > 3a - 10$

26. $10m - 7 < 17m$ or $-6m > 36$

27. $5n - 1 < -16$ or $-3n - 1 < 8$

28. **COUPON** Juanita has a coupon for 10% off any digital camera at a local electronics store. She is looking at digital cameras that range in price from $100 to $250.

 a. How much are the cameras after the coupon is used?

 b. If the tax amount is 6.5%, how much should Juanita expect to spend?

Define a variable, write an inequality, and solve each problem. Then check your solution.

29. Eight less than a number is no more than 14 and no less than 5.

30. The sum of 3 times a number and 4 is between −8 and 10.

31. The product of −5 and a number is greater than 35 or less than 10.

32. One half a number is greater than 0 and less than or equal to 1.

33. **SNAKES** Most snakes live where the temperature ranges from 75°F to 90°F, inclusive. Write an inequality for temperatures where snakes will *not* thrive.

34. **FUNDRAISING** Yumas is selling gift cards to raise money for a class trip. He can earn prizes depending on how many cards he sells. So far, he has sold 34 cards. How many more does he need to sell to earn a prize in category 4?

Cards	Prize
1–15	1
16–30	2
31–45	3
46–60	4
+61	5

35. **TURTLES** Atlantic sea turtle eggs that incubate below 23°C or above 33°C rarely hatch. Write the temperature requirements in two ways: as a pair of simple inequalities, and as a compound inequality.

36. **CCSS STRUCTURE** The *Triangle Inequality Theorem* states that the sum of the measures of any two sides of a triangle is greater than the measure of the third side.

 a. Write and solve three inequalities to express the relationships among the measures of the sides of the triangle shown at the right.

 b. What are four possible lengths for the third side of the triangle?

 c. Write a compound inequality for the possible values of *x*.

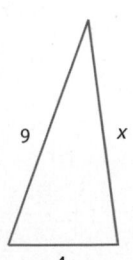

37 **HURRICANES** The Saffir-Simpson Hurricane Scale rates hurricanes on a scale from 1 to 5 based on their wind speed.

Category	Wind Speed (mph)	Example (year)
1	74–95	Gaston (2004)
2	96–110	Frances (2004)
3	111–130	Ivan (2004)
4	131–155	Charley (2004)
5	> 155	Andrew (1992)

a. Write a compound inequality for the wind speeds of a category 3 and a category 4 hurricane.

b. What is the intersection of the two graphs of the inequalities you found in part **a**?

38. **MULTIPLE REPRESENTATIONS** In this problem, you will investigate measurements. The **absolute error** of a measurement is equal to one half the unit of measure. The **relative error** of a measure is the ratio of the absolute error to the expected measure.

a. **Tabular** Copy and complete the table.

Measure	Absolute Error	Relative Error
14.3 cm	$\frac{1}{2}(0.1) = 0.05$ cm	$\dfrac{\text{absolute error}}{\text{expected measure}} = \dfrac{0.05 \text{ cm}}{14.3 \text{ cm}}$ ≈ 0.0035 or 0.4%
1.85 cm		
61.2 cm		
237 cm		

b. **Analytical** You measured a length of 12.8 centimeters. Compute the absolute error and then write the range of possible measures.

c. **Logical** To what precision would you have to measure a length in centimeters to have an absolute error of less than 0.05 centimeter?

d. **Analytical** To find the relative error of an area or volume calculation, add the relative errors of each linear measure. If the measures of the sides of a rectangular box are 6.5 centimeters, 7.2 centimeters, and 10.25 centimeters, what is the relative error of the volume of the box?

H.O.T. Problems Use Higher-Order Thinking Skills

39. **ERROR ANALYSIS** Chloe and Jonas are solving $3 < 2x - 5 < 7$. Is either of them correct? Explain your reasoning.

> **Chloe**
>
> $3 < 2x - 5 < 7$
>
> $3 < 2x < 12$
>
> $\dfrac{3}{2} < x < 6$

> **Jonas**
>
> $3 < 2x - 5 < 7$
>
> $8 < 2x < 7$
>
> $4 < x < \dfrac{7}{2}$

40. **CCSS PERSEVERANCE** Solve each inequality for x. Assume a is constant and $a > 0$.

a. $-3 < ax + 1 \leq 5$

b. $-\frac{1}{a}x + 6 < 1$ or $2 - ax > 8$

41. **OPEN ENDED** Create an example of a compound inequality containing *or* that has infinitely many solutions.

42. **CHALLENGE** Determine whether the following statement is *always, sometimes,* or *never* true. Explain. *The graph of a compound inequality that involves an* or *statement is bounded on the left and right by two values of x.*

43. **WRITING IN MATH** Give an example of a compound inequality you might encounter at an amusement park. Does the example represent an intersection or a union?

44. What is the solution set of the inequality $-7 < x + 2 < 4$?

A $\{x \mid -5 < x < 6\}$ **C** $\{x \mid -9 < x < 2\}$

B $\{x \mid -5 < x < 2\}$ **D** $\{x \mid -9 < x < 6\}$

45. GEOMETRY What is the surface area of the rectangular solid?

F 249.6 cm^2

G 278.4 cm^2

H 313.6 cm^2

J 371.2 cm^2

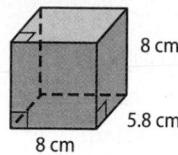

8 cm

5.8 cm

8 cm

46. GRIDDED RESPONSE What is the next term in the sequence?

$$\frac{13}{2}, \frac{18}{5}, \frac{23}{8}, \frac{28}{11}, \frac{33}{14}, \cdots$$

47. After paying a $15 membership fee, members of a video club can rent movies for $2. Nonmembers can rent movies for $4. What is the least number of movies which must be rented for it to be less expensive for members?

A 9   **C** 7

B 8   **D** 6

Spiral Review

48. BABYSITTING Marilyn earns $150 per month delivering newspapers plus $7 an hour babysitting. If she wants to earn at least $300 this month, how many hours will she have to babysit? (Lesson 5-3)

49. MAGAZINES Carlos has earned more than $260 selling magazine subscriptions. Each subscription was sold for $12. How many did Carlos sell? (Lesson 5-2)

50. PUNCH Raquel is mixing lemon-lime soda and a fruit juice blend that is 45% juice. If she uses 3 quarts of soda, how many quarts of fruit juice must be added to produce punch that is 30% juice? (Lesson 2-9)

Solve each proportion. If necessary, round to the nearest hundredth. (Lesson 2-6)

51. $\dfrac{14}{x} = \dfrac{20}{8}$

52. $\dfrac{0.47}{6} = \dfrac{1.41}{m}$

53. $\dfrac{16}{7} = \dfrac{9}{b}$

54. $\dfrac{2 + y}{5} = \dfrac{10}{3}$

55. $\dfrac{8}{9} = \dfrac{2r - 3}{4}$

56. $\dfrac{6 - 2y}{8} = \dfrac{2}{18}$

Determine whether each relation is a function. Explain. (Lesson 1-7)

57.

Domain	2	6	10	7
Range	5	0	5	0

58.

Domain	-5	2	-3	2
Range	-10	-7	-5	-3

59. $\{(-4, 11), (-2, 7), (1, 3), (-4, -1)\}$

60. $\{(2, 7), (5, -3), (7, 6), (10, 7)\}$

Evaluate each expression. (Lesson 1-2)

61. $5 + (4 - 2^2)$

62. $\dfrac{3}{8}[8 \div (7 - 4)]$

63. $2(4 \cdot 9 - 3) + 5 \cdot \dfrac{1}{5}$

Skills Review

Solve each equation.

64. $4p - 2 = -6$

65. $18 = 5p + 3$

66. $9 = 1 + \dfrac{m}{7}$

67. $1.5a - 8 = 11$

68. $20 = -4c - 8$

69. $\dfrac{b + 4}{-2} = -17$

70. $\dfrac{n - 3}{8} = 20$

71. $6y - 16 = 44$

72. $130 = 11k + 9$

Inequalities Involving Absolute Value

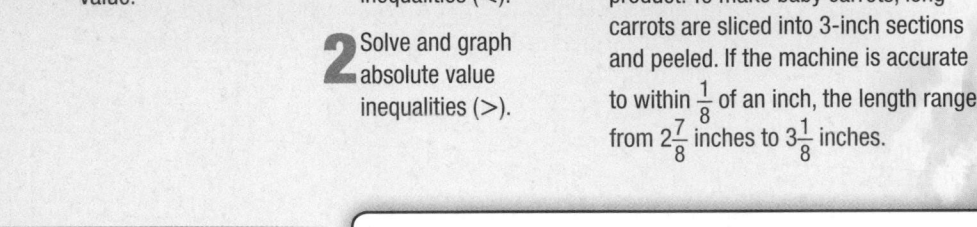

- You solved equations involving absolute value.

1. Solve and graph absolute value inequalities (<).

2. Solve and graph absolute value inequalities (>).

- Some companies use absolute value inequalities to control the quality of their product. To make baby carrots, long carrots are sliced into 3-inch sections and peeled. If the machine is accurate to within $\frac{1}{8}$ of an inch, the length ranges from $2\frac{7}{8}$ inches to $3\frac{1}{8}$ inches.

Common Core State Standards

Content Standards

A.CED.1 Create equations and inequalities in one variable and use them to solve problems.

A.REI.3 Solve linear equations and inequalities in one variable, including equations with coefficients represented by letters.

Mathematical Practices

3 Construct viable arguments and critique the reasoning of others.

7 Look for and make use of structure.

1 Absolute Value Inequalities (<)

The inequality $|x| < 3$ means that the distance between x and 0 is less than 3.

So, $x > -3$ and $x < 3$. The solution set is $\{x \mid -3 < x < 3\}$.

When solving absolute value inequalities, there are two cases to consider.

Case 1 The expression inside the absolute value symbols is nonnegative.

Case 2 The expression inside the absolute value symbols is negative.

The solution is the intersection of the solutions of these two cases.

Example 1 Solve Absolute Value Inequalities (<)

Solve each inequality. Then graph the solution set.

a. $|m + 2| < 11$

Rewrite $|m + 2| < 11$ for Case 1 *and* Case 2.

Case 1 $m + 2$ is nonnegative. **and** **Case 2** $m + 2$ is negative.

$$m + 2 < 11 \qquad\qquad -(m + 2) < 11$$
$$m + 2 - 2 < 11 - 2 \qquad\qquad m + 2 > -11$$
$$m < 9 \qquad\qquad m + 2 - 2 > -11 - 2$$
$$\qquad\qquad m > -13$$

So, $m < 9$ and $m > -13$. The solution set is $\{m \mid -13 < m < 9\}$.

b. $|y - 1| < -2$

$|y - 1|$ cannot be negative. So it is not possible for $|y - 1|$ to be less than -2. Therefore, there is no solution, and the solution set is the empty set, Ø.

▶ **Guided**Practice

1A. $|n - 8| \leq 2$ **1B.** $|2c - 5| < -3$

Stockbyte/Veer

Real-World Example 2 Apply Absolute Value Inequalities

INTERNET A recent survey showed that 65% of young adults watched online video clips. The margin of error was within 3 percentage points. Find the range of young adults who use video sharing sites.

The difference between the actual number of viewers and the number from the survey is less than or equal to 3. Let x be the actual number of viewers. Then $|x - 65| \leq 3$.

Solve each case of the inequality.

Case 1 $x - 65$ is nonnegative.   **and**   **Case 2** $x - 65$ is negative.

$$x - 65 \leq 3 \qquad\qquad\qquad -(x - 65) \leq 3$$
$$x - 65 + 65 \leq 3 + 65 \qquad\qquad x - 65 \geq -3$$
$$x \leq 68 \qquad\qquad\qquad\qquad x \geq 62$$

The range of young adults who use video sharing sites is $\{x \mid 62 \leq x \leq 68\}$.

Real-WorldLink

One in five Americans use the Internet to view videos. Young adults tend to watch funny videos, while other age groups tend to watch the news.

Source: Pew Internet and American Life Project

▸ **Guided**Practice

2. CHEMISTRY The melting point of ice is 0°C. During a chemistry experiment, Jill observed ice melting within 2°C of this measurement. Write the range of temperatures that Jill observed.

2 Absolute Value Inequalities (>) The inequality $|x| > 3$ means that the distance between x and 0 is greater than 3.

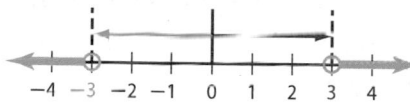

So, $x < -3$ or $x > 3$. The solution set is $\{x \mid x < -3 \text{ or } x > 3\}$.

As in the previous example, we must consider both cases.

Case 1 The expression inside the absolute value symbols is nonnegative.

Case 2 The expression inside the absolute value symbols is negative.

Example 3 Solve Absolute Value Inequalities (>)

Solve $|3n + 6| \geq 12$. Then graph the solution set.

Rewrite $|3n + 6| \geq 12$ for Case 1 *or* Case 2.

Case 1 $3n + 6$ is nonnegative.   **or**   **Case 2** $3n + 6$ is negative.

$$3n + 6 \geq 12 \qquad\qquad\qquad -(3n + 6) \geq 12$$
$$3n + 6 - 6 \geq 12 - 6 \qquad\qquad 3n + 6 \leq -12$$
$$3n \geq 6 \qquad\qquad\qquad\qquad 3n \leq -18$$
$$n \geq 2 \qquad\qquad\qquad\qquad n \leq -6$$

So, $n \geq 2$ or $n \leq -6$. The solution set is $\{n \mid n \geq 2 \text{ or } n \leq -6\}$.

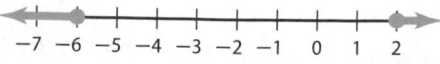

StudyTip

CCSS Structure For $|a| \geq b$, where a is any linear expression in one variable and b is a negative number, the solution set will always be the set of all real numbers. Since $|a|$ is always greater than or equal to zero, $|a|$ is always greater than or equal to b.

▸ **Guided**Practice

Solve each inequality. Then graph the solution set.

3A. $|2k + 1| > 7$ **3B.** $|r - 6| \geq -5$

Examples 1–3 Solve each inequality. Then graph the solution set.

1. $|a - 5| < 3$
2. $|u + 3| < 7$
3. $|t + 4| \leq -2$
4. $|c + 2| > -2$
5. $|n + 5| \geq 3$
6. $|p - 2| \geq 8$

Example 2
7. **FINANCIAL LITERACY** Jerome bought stock in his favorite fast-food restaurant chain at $70.85. However, it has fluctuated up to $0.75 in a day. Find the range of prices for which the stock could trade in a day.

Practice and Problem Solving   Extra Practice is on page R5.

Examples 1–3 Solve each inequality. Then graph the solution set.

8. $|x + 8| < 16$
9. $|r + 1| \leq 2$
10. $|2c - 1| \leq 7$
11. $|3h - 3| < 12$
12. $|m + 4| < -2$
13. $|w + 5| < -8$
14. $|r + 2| > 6$
15. $|k - 4| > 3$
16. $|2h - 3| \geq 9$
17. $|4p + 2| \geq 10$
18. $|5v + 3| > -9$
19. $|-2c - 3| > -4$

Example 2
20. **SCUBA DIVING** The pressure of a scuba tank should be within 500 pounds per square inch (psi) of 2500 psi. Write the range of optimum pressures.

Solve each inequality. Then graph the solution set.

21. $|4n + 3| \geq 18$
22. $|5t - 2| \leq 6$
23. $\left|\dfrac{3h + 1}{2}\right| < 8$
24. $\left|\dfrac{2p - 8}{4}\right| \geq 9$
25. $\left|\dfrac{7c + 3}{2}\right| \leq -5$
26. $\left|\dfrac{2g + 3}{2}\right| > -7$
27. $|-6r - 4| < 8$
28. $|-3p - 7| > 5$
29. $|-h + 1.5| < 3$

30. **MUSIC DOWNLOADS** Kareem is allowed to download $10 worth of music each month. This month he has spent within $3 of his allowance.

   a. What is the range of money he has spent on music downloads this month?

   b. Graph the range of the money that he spent.

31. **CHEMISTRY** Water can be present in our atmosphere as a solid, liquid, or gas. Water freezes at 32°F and vaporizes at 212°F.

   a. Write the range of temperatures in which water is not a liquid.

   b. Graph this range.

   c. Write the absolute value inequality that describes this situation.

CCSS REGULARITY Write an open sentence involving absolute value for each graph.

32.
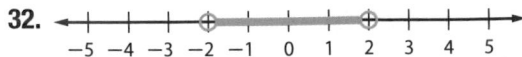

33.

34.

35.
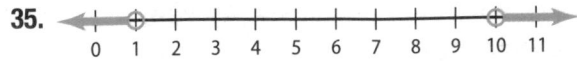

36. ANIMALS A sheep's normal body temperature is 39°C. However, a healthy sheep may have body temperatures 1°C above or below this temperature. What is the range of body temperatures for a sheep?

37 MINIATURE GOLF Ginger's score was within 5 strokes of her average score of 52. Determine the range of scores for Ginger's game.

Express each statement using an inequality involving absolute value. Do *not* solve.

38. The pH of a swimming pool must be within 0.3 of a pH of 7.5.

39. The temperature inside a refrigerator should be within 1.5 degrees of 38°F.

40. Ramona's bowling score was within 6 points of her average score of 98.

41. The cruise control of a car should keep the speed within 3 miles per hour of 55.

42. MULTIPLE REPRESENTATIONS In this problem, you will investigate the graphs of linear inequalities on a coordinate plane.

a. Tabular Copy and complete the table. Substitute the x and $f(x)$ values for each point into each inequality. Mark whether the resulting statement is *true* or *false*.

Point	$f(x) \geq x - 1$	true/false	$f(x) \leq x - 1$	true/false
$(-4, 2)$				
$(-2, 2)$				
$(0, 2)$				
$(2, 2)$				
$(4, 2)$				

b. Graphical Graph $f(x) = x - 1$.

c. Graphical Plot each point from the table that made $f(x) \geq x - 1$ a true statement on the graph in red. Plot each point that made $f(x) \leq x - 1$ a true statement in blue.

d. Logical Make a conjecture about what the graphs of $f(x) \geq x - 1$ and $f(x) \leq x - 1$ look like. Complete the table with other points to verify your conjecture.

e. Logical Use what you discovered to describe the graph of a linear inequality.

H.O.T. Problems   Use Higher-Order Thinking Skills

43. ERROR ANALYSIS Lucita sketched a graph of her solution to $|2a - 3| > 1$. Is she correct? Explain your reasoning.

44. REASONING The graph of an absolute value inequality is *sometimes, always,* or *never* the union of two graphs. Explain.

45. CCSS ARGUMENTS Demonstrate why the solution of $|t| > 0$ is not all real numbers. Explain your reasoning.

46. WRITING IN MATH How are symbols used to represent mathematical ideas? Use an example to justify your reasoning.

47. WRITING IN MATH Explain how to determine whether an absolute value inequality uses a compound inequality with *and* or a compound inequality with *or*. Then summarize how to solve absolute value inequalities.

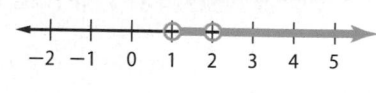

48. The formula for acceleration in a circle is $a = \frac{v^2}{r}$. Which of the following shows the equation solved for r?

A $r = v$

C $r = av^2$

B $r = \frac{v^2}{a}$

D $r = \frac{\sqrt{a}}{v}$

49. An engraver charges a $3 set-up fee and $0.25 per word. Which table shows the total price p for w words?

F

w	p
15	$3
20	$4.25
25	$5.50
30	$7.75

H

w	p
15	$3.75
20	$5
25	$6.25
30	$8.50

G

w	p
15	$6.75
20	$7
25	$7.25
30	$7.50

J

w	p
15	$6.75
20	$8
25	$9.25
30	$10.50

50. SHORT RESPONSE The table shows the items in stock at the school store the first day of class. What is the probability that an item chosen at random was a notebook?

Item	Number Purchased
pencil	57
pen	38
eraser	6
folder	25
notebook	18

51. Solve for n.

$$|2n - 3| = 5$$

A $\{-4, -1\}$

B $\{-1, 4\}$

C $\{1, 1\}$

D $\{4, 4\}$

Solve each compound inequality. Then graph the solution set. (Lesson 5-4)

52. $b + 3 < 11$ and $b + 2 > -3$

53. $6 \le 2t - 4 \le 8$

54. $2c - 3 \ge 5$ or $3c + 7 \le -5$

55. FINANCIAL LITERACY Jackson's bank charges him a monthly service fee of $6 for his checking account and $2 for each out-of-network ATM withdrawal. Jackson's account balance is $87. Write and solve an inequality to find how many out-of-network ATM withdrawals of $20 Jackson can make without overdrawing his account. (Lesson 5-3)

56. GEOMETRY One angle of a triangle measures 10° more than the second. The measure of the third angle is twice the sum of the measure of the first two angles. Find the measure of each angle. (Lesson 2-4)

Solve each equation. Then check your solution. (Lesson 2-2)

57. $c - 7 = 11$

58. $2w = 24$

59. $9 + p = -11$

60. $\frac{t}{5} = 20$

Graph each equation.

61. $y = 4x - 1$

62. $y - x = 3$

63. $2x - y = -4$

64. $3y + 2x = 6$

65. $4y = 4x - 16$

66. $2y - 2x = 8$

67. $-9 = -3x - y$

68. $-10 = 5y - 2x$

Graphing Inequalities in Two Variables

● You graphed linear equations.

1 Graph linear inequalities on the coordinate plane.

2 Solve inequalities by graphing.

● Hannah has budgeted $35 every three months for car maintenance. From this she must buy oil costing $3 and filters that cost $7 each. How much oil and how many filters can Hannah buy and stay within her budget?

 NewVocabulary
boundary
half-plane
closed half-plane
open half-plane

 Common Core State Standards

Content Standards
A.CED.3 Represent constraints by equations or inequalities, and by systems of equations and/or inequalities, and interpret solutions as viable or nonviable options in a modeling context.

A.REI.12 Graph the solutions to a linear inequality in two variables as a halfplane (excluding the boundary in the case of a strict inequality), and graph the solution set to a system of linear inequalities in two variables as the intersection of the corresponding half-planes.

Mathematical Practices
5 Use appropriate tools strategically.

1 Graph Linear Inequalities The graph of a linear inequality is the set of points that represent all of the possible solutions of that inequality. An equation defines a **boundary**, which divides the coordinate plane into two **half-planes**.

The boundary may or may not be included in the solution. When it is included, the solution is a **closed half-plane**. When not included, the solution is an **open half-plane**.

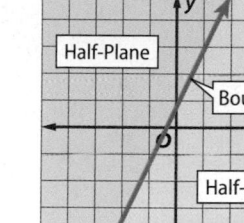

KeyConcept Graphing Linear Inequalities

Step 1 Graph the boundary. Use a solid line when the inequality contains ≤ or ≥. Use a dashed line when the inequality contains < or >.

Step 2 Use a test point to determine which half-plane should be shaded.

Step 3 Shade the half-plane that contains the solution.

Example 1 Graph an Inequality (< or >)

Graph $3x - y < 2$.

Step 1 First, solve for y in terms of x.
$$3x - y < 2$$
$$-y < -3x + 2$$
$$y > 3x - 2$$
Then, graph $y = 3x - 2$. Because the inequality involves >, graph the boundary with a dashed line.

Step 2 Select $(0, 0)$ as a test point.
$3x - y < 2$	Original inequality
$3(0) - 0 < 2$	$x = 0$ and $y = 0$
$0 < 2$	true

Step 3 So, the half-plane containing the origin is the solution. Shade this half-plane.

 **GuidedPractice** Graph each inequality.

1A. $y > \frac{1}{2}x + 3$

1B. $x - 1 > y$

Joos Mind/Photographer's Choice/Getty Images

Example 2 Graph an Inequality (≤ or ≥)

Graph $x + 5y \leq 10$.

Step 1 Solve for y in terms of x.

$x + 5y \leq 10$	Original inequality
$5y \leq -x + 10$	Subtract x from each side and simplify.
$y \leq -\dfrac{1}{5}x + 2$	Divide each side by 5.

Graph $y = -\dfrac{1}{5}x + 2$. Because the inequality symbol is ≤, graph the boundary with a solid line.

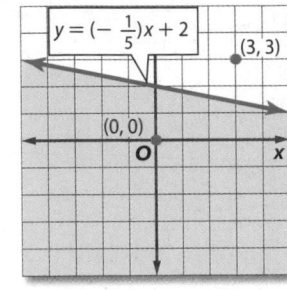

Step 2 Select a test point. Let's use (3, 3). Substitute the values into the original inequality.

$x + 5y \leq 10$	Original inequality
$3 + 5(3) \leq 10$	$x = 3$ and $y = 3$
$18 \nleq 10$	Simplify.

Step 3 Since this statement is false, shade the other half-plane.

▶ **Guided**Practice

Graph each inequality.

2A. $x - y \leq 3$ **2B.** $2x + 3y \geq 18$

2 **Solve Linear Inequalities** We can use a coordinate plane to solve inequalities with one variable.

Example 3 Solve Inequalities From Graphs

Use a graph to solve $3x + 5 < 14$.

Step 1 First graph the boundary, which is the related equation. Replace the inequality sign with an equals sign, and solve for x.

$3x + 5 < 14$	Original inequality
$3x + 5 = 14$	Change < to =.
$3x = 9$	Subtract 5 from each side and simplify.
$x = 3$	Divide each side by 3.

Graph $x = 3$ with a dashed line.

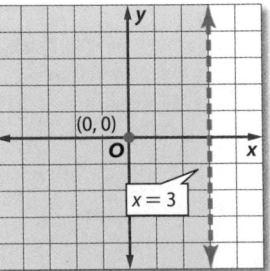

Step 2 Choose (0, 0) as a test point. These values in the original inequality give us $5 < 14$.

Step 3 Since this statement is true, shade the half-plane that contains the point (0, 0).

Notice that the x-intercept of the graph is at 3. Since the half-plane to the left of the x-intercept is shaded, the solution is $x < 3$.

▶ **Guided**Practice

Use a graph to solve each inequality.

3A. $4x - 3 \geq 17$ **3B.** $-2x + 6 > 12$

An inequality can be viewed as a constraint in a problem situation. Each solution of the inequality represents a combination that meets the constraint. In real-world problems, the domain and range are often restricted to nonnegative or whole numbers.

Real-World Example 4   Write and Solve an Inequality

CLASS PICNIC A yearbook company promises to give the junior class a picnic if they spend at least $28,000 on yearbooks and class rings. Each yearbook costs $35, and each class ring costs $140. How many yearbooks and class rings must the junior class buy to get their picnic?

Understand   You know the cost of each item and the minimum amount the class needs to spend.

Plan   Let x = the number of yearbooks and y = the number of class rings the class must buy. Write an inequality.

$35	times	the number of yearbooks	plus	$140	times	the number of rings	is at least	$28,000.
35	·	x	+	140	·	y	$\geq$	28,000

Solve   Solve for y in terms of x.

$35x + 140y - 35x \geq 28,000 - 35x$ Subtract $35x$ from each side.

$140y \geq -35x + 28,000$ Divide each side by 140.

$\dfrac{140y}{140} \geq \dfrac{-35x}{140} + \dfrac{28000}{140}$ Simplify.

$y \geq -0.25x + 200$ Simplify.

Because the yearbook company cannot sell a negative number of items, the domain and range must be nonnegative numbers. Graph the boundary with a solid line. If we test $(0, 0)$, the result is $0 \geq 28,000$, which is false. Shade the closed half-plane that does not include $(0, 0)$.

One solution is (500, 100), or 500 yearbooks and 100 class rings.

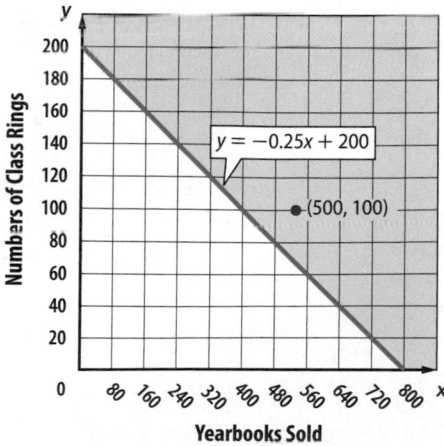

Check   If we test $(500, 100)$, the result is $100 \geq 75$, which is true. Because the company cannot sell a fraction of an item, only points with whole-number coordinates can be solutions.

Problem-SolvingTip

Use a Graph You can use a graph to visualize data, analyze trends, and make predictions.

> **GuidedPractice**

4. MARATHONS Neil wants to run a marathon at a pace of at least 6 miles per hour. Write and graph an inequality for the miles y he will run in x hours.

Manchan/Photodisc/Getty Images

Examples 1–2 Graph each inequality.

1. $y > x + 3$ **2.** $y \geq -8$ **3.** $x + y > 1$

4. $y \leq x - 6$ **5.** $y < 2x - 4$ **6.** $x - y \leq 4$

Example 3   Use a graph to solve each inequality.

7. $7x + 1 < 15$ **8.** $-3x - 2 \geq 11$

9. $3y - 5 \leq 34$ **10.** $4y - 21 > 1$

Example 4   **11. FINANCIAL LITERACY** The surf shop has a weekly overhead of $2300.

    a. Write an inequality to represent the number of skimboards and longboards the shop sells each week to make a profit.

    b. How many skimboards and longboards must the shop sell each week to make a profit?

Practice and Problem Solving   Extra Practice is on page R5.

Examples 1–2 Graph each inequality.

12. $y < x - 3$ **13.** $y > x + 12$ **14.** $y \geq 3x - 1$

15. $y \leq -4x + 12$ **16.** $6x + 3y > 12$ **17.** $2x + 2y < 18$

18. $5x + y > 10$ **19.** $2x + y < -3$ **20.** $-2x + y \geq -4$

21. $8x + y \leq 6$ **22.** $10x + 2y \leq 14$ **23.** $-24x + 8y \geq -48$

Example 3   Use a graph to solve each inequality.

24. $10x - 8 < 22$ **25.** $20x - 5 > 35$ **26.** $4y - 77 \geq 23$

27. $5y + 8 \leq 33$ **28.** $35x + 25 < 6$ **29.** $14x - 12 > -31$

Example 4   **30. CCSS MODELING** Sybrina is decorating her bedroom. She has $300 to spend on paint and bed linens. A gallon of paint costs $14, while a set of bed linens costs $60.

    a. Write an inequality for this situation.

    b. How many gallons of paint and bed linen sets can Sybrina buy and stay within her budget?

Use a graph to solve each inequality.

31. $3x + 2 < 0$ **32.** $4x - 1 > 3$ **33.** $-6x - 8 \geq -4$

34. $-5x + 1 < 3$ **35.** $-7x + 13 < 10$ **36.** $-4x - 4 \leq -6$

37 SOCCER The girls' soccer team wants to raise $2000 to buy new goals. How many of each item must they sell to buy the goals?

    a. Write an inequality that represents this situation.

    b. Graph this inequality.

    c. Make a table of values that shows at least five possible solutions.

    d. Plot the solutions from part **c**.

Graph each inequality. Determine which of the ordered pairs are part of the solution set for each inequality.

38. $y \geq 6$; $\{(0, 4), (-2, 7), (4, 8), (-4, -8), (1, 6)\}$

39 $x < -4$; $\{(2, 1), (-3, 0), (0, -3), (-5, -5), (-4, 2)\}$

40. $2x - 3y \leq 1$; $\{(2, 3), (3, 1), (0, 0), (0, -1), (5, 3)\}$

41. $5x + 7y \geq 10$; $\{(-2, -2), (1, -1), (1, 1), (2, 5), (6, 0)\}$

42. $-3x + 5y < 10$; $\{(3, -1), (1, 1), (0, 8), (-2, 0), (0, 2)\}$

43. $2x - 2y \geq 4$; $\{(0, 0), (0, 7), (7, 5), (5, 3), (2, -5)\}$

44. RECYCLING Mr. Jones would like to spend no more than $37.50 per week on recycling. A curbside recycling service will remove up to 50 pounds of plastic bottles and paper products per week. They charge $0.25 per pound of plastic and $0.75 per pound of paper products.

    a. Write an inequality that describes the number of pounds of each product that can be included in the curbside service.

    b. Write an inequality that describes Mr. Jones' weekly cost for the service if he stays within his budget.

    c. Graph an inequality for the weekly costs for the service.

45. ⚏ **MULTIPLE REPRESENTATIONS** Use inequalities A and B to investigate graphing compound inequalities on a coordinate plane.

    A. $7(y + 6) \leq 21x + 14$ **B.** $-3y \leq 3x - 12$

    a. Numerical Solve each inequality for y.

    b. Graphical Graph both inequalities on one graph. Shade the half-plane that makes A true in red. Shade the half-plane that makes B true in blue.

    c. Verbal What does the overlapping region represent?

H.O.T. Problems   Use Higher-Order Thinking Skills

46. ERROR ANALYSIS Reiko and Kristin are solving $4y \leq \frac{8}{3}x$ by graphing. Is either of them correct? Explain your reasoning.

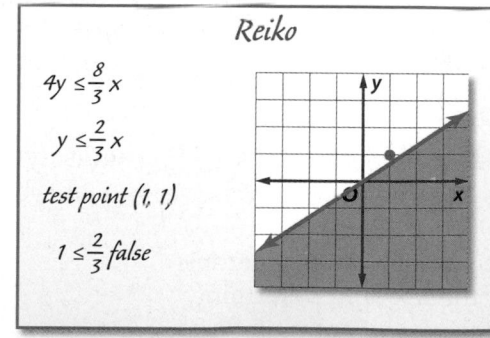
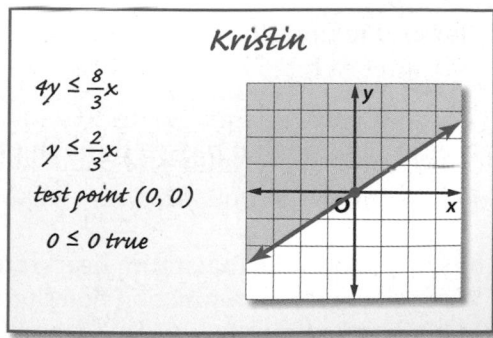

47. Ⓒ**CSS TOOLS** Write a linear inequality for which $(-1, 2)$, $(0, 1)$, and $(3, -4)$ are solutions but $(1, 1)$ is not.

48. REASONING Explain why a point on the boundary should not be used as a test point.

49. OPEN ENDED Write a two-variable inequality with a restricted domain and range to represent a real-world situation. Give the domain and range, and explain why they are restricted.

50. WRITING IN MATH Summarize the steps to graph an inequality in two variables.

51. What is the domain of this function?

A $\{x \mid 0 \leq x \leq 3\}$

B $\{x \mid 0 \leq x \leq 9\}$

C $\{y \mid 0 \leq y \leq 9\}$

D $\{y \mid 0 \leq y \leq 3\}$

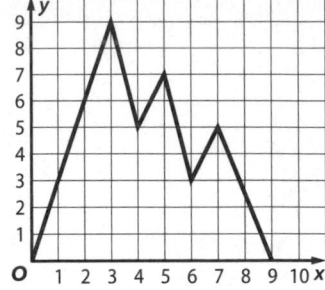

52. **EXTENDED RESPONSE** An arboretum will close for the winter when all of the trees have lost their leaves. The table shows the number of trees each day that still have leaves.

Day	5	10	15	20
Trees with Leaves	325	260	195	130

a. Write an equation that represents the number of trees with leaves y after d days.

b. Find the y-intercept. What does it mean in the context of this problem?

c. After how many days will the arboretum close? Explain how you got your answer.

53. Which inequality best represents the statement below?

A jar contains 832 gumballs. Ebony's guess was within 46 pieces.

F $|g - 832| \leq 46$

G $|g + 832| \leq 46$

H $|g - 832| \geq 46$

J $|g + 832| \geq 46$

54. **GEOMETRY** If the rectangular prism has a volume of 10,080 cm³, what is the value of x?

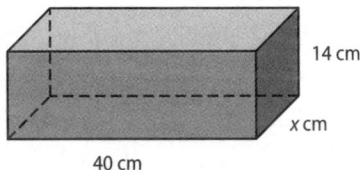

A 12

B 14

C 16

D 18

Solve each open sentence. (Lesson 5-5)

55. $|y - 2| > 4$

56. $|t - 6| \leq 5$

57. $|3 + d| < -4$

Solve each compound inequality. (Lesson 5-4)

58. $4c - 4 < 8c - 16 < 6c - 6$

59. $5 < \frac{1}{2}p + 3 < 8$

60. $0.5n \geq -7$ or $2.5n + 2 \leq 9$

Write an equation of the line that passes through each pair of points. (Lesson 4-2)

61. $(1, -3)$ and $(2, 5)$

62. $(-2, -4)$ and $(-7, 3)$

63. $(-6, -8)$ and $(-8, -5)$

64. **FITNESS** The table shows the maximum heart rate to maintain during aerobic activities. Write an equation in function notation for the relation. Determine what would be the maximum heart rate to maintain in aerobic training for an 80-year-old. (Lesson 3-5)

Age (yr)	20	30	40	50	60	70
Pulse rate (beats/min)	175	166	157	148	139	130

65. **WORK** The formula $s = \dfrac{w - 10r}{m}$ is used to find keyboarding speeds. In the formula, s represents the speed in words per minute, w the number of words typed, r the number of errors, and m the number of minutes typed. Solve for r.

Graphing Technology Lab
Graphing Inequalities

You can use a graphing calculator to investigate the graphs of inequalities.

 Common Core State Standards
Content Standards
A.REI.12 Graph the solutions to a linear inequality in two variables as a halfplane (excluding the boundary in the case of a strict inequality), and graph the solution set to a system of linear inequalities in two variables as the intersection of the corresponding half-planes.
Mathematical Practices
5 Use appropriate tools strategically.

Activity 1   Less Than

Graph $y \leq 2x + 5$.

Clear all functions from the **Y=** list.

KEYSTROKES: [Y=] [CLEAR]

Graph $y \leq 2x + 5$ in a standard viewing window.

KEYSTROKES: 2 [X,T,θ,n] [+] 5 ◄ ◄ ◄ ◄ ◄ ◄ [ENTER]
[ENTER] [ENTER] [ZOOM] 6

All ordered pairs for which y is *less than or equal* to $2x + 5$ lie *below or on* the line and are solutions.

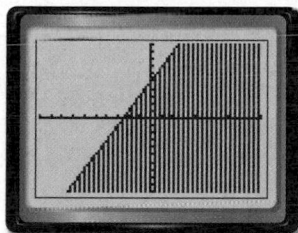

[−10, 10] scl: 1 by [−10, 10] scl: 1

Activity 2   Greater Than

Graph $y − 2x \geq 5$.

Clear the graph that is currently displayed.

KEYSTROKES: [Y=] [CLEAR]

Rewrite $y − 2x \geq 5$ as $y \geq 2x + 5$ and graph it.

KEYSTROKES: 2 [X,T,θ,n] [+] 5 ◄ ◄ ◄ ◄ ◄ ◄ [ENTER]
[ENTER] [ZOOM] 6

All ordered pairs for which y is *greater than or equal to* $2x + 5$ lie *above or on* the line and are solutions.

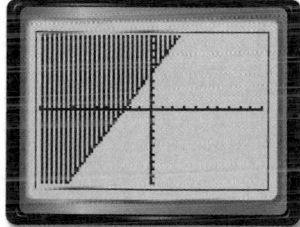

[−10, 10] scl: 1 by [−10, 10] scl: 1

Exercises

1. Compare and contrast the two graphs shown above.

2. Graph $y \geq −3x + 1$ in the standard viewing window. Using your graph, name four solutions of the inequality.

3. Suppose student water park tickets cost $16, and adult water park tickets cost $20. You would like to buy at least 10 tickets but spend no more than $200.

   a. Let x = number of student tickets and y = number of adult tickets. Write two inequalities, one representing the total number of tickets and the other representing the total cost of the tickets.

   b. Graph the inequalities. Use the viewing window [0, 20] scl: 1 by [0, 20] scl: 1.

   c. Name four possible combinations of student and adult tickets.

Study Guide and Review

Study Guide

KeyConcepts

Solving One-Step Inequalities (Lessons 5-1 and 5-2)

For all numbers a, b, and c, the following are true.

- If $a > b$ and c is positive, $ac > bc$.
- If $a > b$ and c is negative, $ac < bc$.

Multi-Step and Compound Inequalities (Lessons 5-3 and 5-4)

- Multi-step inequalities can be solved by undoing the operations in the same way you would solve a multi-step equation.
- A compound inequality containing *and* is only true if both inequalities are true.
- A compound inequality containing *or* is true if at least one of the inequalities is true.

Absolute Value Inequalities (Lesson 5-5)

- The absolute value of any number x is its distance from zero on a number line and is written as $|x|$. If $x \geq 0$, then $|x| = x$. If $x < 0$, then $|x| = -x$.
- If $|x| < n$ and $n > 0$, then $-n < x < n$.
- If $|x| > n$ and $n > 0$, then $x > n$ or $x < -n$.

Inequalities in Two Variables (Lesson 5-6)

To graph an inequality:

Step 1 Graph the boundary. Use a solid line when the inequality contains $\leq$ or $\geq$. Use a dashed line when the inequality contains $<$ or $>$.

Step 2 Use a test point to determine which half-plane should be shaded.

Step 3 Shade the half-plane.

FOLDABLES StudyOrganizer

Be sure the Key Concepts are noted in your Foldable.

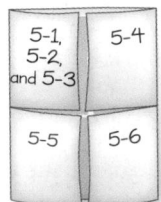

KeyVocabulary

boundary (p. 317)

closed half-plane (p. 317)

compound inequality (p. 306)

half-plane (p. 317)

inequality (p. 285)

intersection (p. 306)

open half-plane (p. 317)

set-builder notation (p. 286)

union (p. 307)

VocabularyCheck

State whether each sentence is *true* or *false*. If *false*, replace the underlined term to make a true sentence.

1. Set-builder notation is a <u>less</u> concise way of writing a solution set.

2. There are <u>two</u> types of compound inequalities.

3. The graph of a compound inequality containing *and* shows the <u>union</u> of the individual graphs.

4. A compound inequality containing *or* is true if one or both of the inequalities is true. Its graph is the <u>union</u> of the graphs of the two inequalities.

5. The graph of an inequality of the form $y < ax + b$ is a region on the coordinate plane called a <u>half-plane</u>.

6. A <u>point</u> defines the boundary of an open half-plane.

7. The <u>boundary</u> is the graph of the equation of the line that defines the edge of each half-plane.

8. The solution set to the inequality $y \geq x$ includes the <u>boundary</u>.

9. When solving an inequality, <u>multiplying</u> each side by a negative number reverses the inequality symbol.

10. The graph of a compound inequality that contains <u>*and*</u> is the intersection of the graphs of the two inequalities.

Lesson-by-Lesson Review

5-1 Solving Inequalities by Addition and Subtraction

Solve each inequality. Then graph it on a number line.

11. $w - 4 > 9$

12. $x + 8 \leq 3$

13. $6 + h < 1$

14. $-5 < a + 2$

15. $13 - p \geq 15$

16. $y + 1 \leq 8$

17. FIELD TRIP A bus can hold 44 people. If there are 35 students in Samantha's class, how many more people can ride on the bus?

Example 1

Solve $x - 9 < -4$. Then graph it on a number line.

$x - 9 < -4$ Original inequality

$x - 9 + 9 < -4 + 9$ Add 9 to each side.

$x < 5$ Simplify.

The solution set is $\{x \mid x < 5\}$.

$$\begin{array}{c} \leftarrow\!\!+\!\!+\!\!+\!\!+\!\!+\!\!+\!\!+\!\!+\!\!+\!\!\oplus\!\!\rightarrow \\ -5\,-4\,-3\,-2\,-1\ 0\ 1\ 2\ 3\ 4\ 5 \end{array}$$

5-2 Solving Inequalities by Multiplication and Division

Solve each inequality. Graph the solution on a number line.

18. $\frac{1}{3}x > 6$

19. $\frac{1}{5}g \geq -4$

20. $4p < 32$

21. $-55 \leq -5w$

22. $-2m > 100$

23. $\frac{2}{3}t < -48$

24. MOVIE RENTAL Jack has no more than $24 to spend on DVDs for a party. Each DVD rents for $4. Find the maximum number of DVDs Jack can rent for his party.

Example 2

Solve $-14h < 56$. Check your solution.

$-14h < 56$ Original inequality

$\dfrac{-14h}{-14} > \dfrac{56}{-14}$ Divide each side by -14.

$h > -4$ Simplify.

$\{h \mid h > -4\}$

CHECK To check, substitute three different values into the original inequality: -4, a number less than -4, and a number greater than -4.

5-3 Solving Multi-Step Inequalities

Solve each inequality. Graph the solution on a number line.

25. $3h - 7 < 14$

26. $4 + 5b > 34$

27. $18 \leq -2x + 8$

28. $\frac{t}{3} - 6 > -4$

29. Four times a number decreased by 6 is less than -2. Define a variable, write an inequality, and solve for the number.

30. TICKET SALES The drama club collected $160 from ticket sales for the spring play. They need to collect at least $400 to pay for new lighting for the stage. If tickets sell for $3 each, how many more tickets need to be sold?

Example 3

Solve $-6y - 13 > 29$. Check your solution.

$-6y - 13 > 29$ Original inequality

$-6y - 13 + 13 > 29 + 13$ Add 13 to each side.

$-6y > 42$ Simplify.

$\dfrac{-6y}{-6} < \dfrac{42}{-6}$ Divide each side by -6 and change $>$ to $<$.

$y < -7$ Simplify.

The solution set is $\{y \mid y < -7\}$.

CHECK $-6y - 13 > 29$ Original inequality

$-6(-10) - 13 \overset{?}{>} 29$ Substitute -10 for y.

$47 > 29 \checkmark$ Simplify.

5-4 Solving Compound Inequalities

Solve each compound inequality. Then graph the solution set.

31. $m - 3 < 6$ and $m + 2 > 4$

32. $-4 < 2t - 6 < 8$

33. $3x + 2 \leq 11$ or $5x - 8 > 22$

34. KITES A kite can be flown in wind speeds no less than 7 miles per hour and no more than 16 miles per hour. Write an inequality for the wind speeds at which the kite can fly.

Example 4

Solve $-3w + 4 > -8$ and $2w - 11 > -19$. Then graph the solution set.

$$-3w + 4 > -8 \quad\quad \text{and} \quad\quad 2w - 11 > -19$$
$$w < 4 \quad\quad\quad\quad\quad\quad\quad w > -4$$

To graph the solution set, graph $w < 4$ and graph $w > -4$. Then find the intersection.

5-5 Inequalities Involving Absolute Value

Solve each inequality. Then graph the solution set.

35. $|x - 4| < 9$

36. $|p + 2| > 7$

37. $|2c + 3| \leq 11$

38. $|f - 9| \geq 2$

39. $|3d - 1| \leq 8$

40. $\left|\dfrac{4b - 2}{3}\right| < 12$

41. $\left|\dfrac{2t + 6}{2}\right| > 10$

42. $|-4y - 3| < 13$

43. $|m + 19| \leq 1$

44. $|-k - 7| \geq 4$

Example 5

Solve $|x - 6| < 9$. Then graph the solution set.

Case 1 $x - 6$ is nonnegative.

$$x - 6 < 9$$
$$x < 15$$

Case 2 $x - 6$ is negative.

$$-(x - 6) < 9$$
$$x > -3$$

The solution set is $\{x \mid -3 < x < 15\}$.

5-6 Graphing Inequalities in Two Variables

Graph each inequality.

45. $y > x - 3$

46. $y < 2x + 1$

47. $3x - y \leq 4$

48. $y \geq -2x + 6$

49. $5x - 2y < 10$

50. $x + y \geq 1$

Graph each inequality. Determine which of the ordered pairs are part of the solution set for each inequality.

51. $y \leq 4$; $\{(3, 6), (1, 2), (-4, 8), (3, -2), (1, 7)\}$

52. $-2x + 3y \geq 12$; $\{(-2, 2), (-1, 1), (0, 4), (2, 2)\}$

53. BAKERY Ben has $24 to spend on cookies and cupcakes. Write and graph an inequality that represents what Ben can buy.

$2

$3

Example 6

Graph $2x - y > 3$.

Solve for y in terms of x.

$$2x - y > 3 \quad\quad \text{Original inequality}$$
$$-y > -2x + 3 \quad\quad \text{Subtract } 2x \text{ from each side.}$$
$$y < 2x - 3 \quad\quad \text{Multiply each side by } -1.$$

Graph the boundary using a dashed line. Choose $(0, 0)$ as a test point.

$$2(0) - 0 \overset{?}{>} 3$$
$$0 \not> 3$$

Since 0 is not greater than 3, shade the plane that does not contain $(0, 0)$.

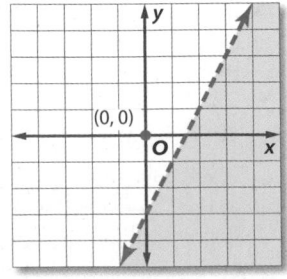

Solve each inequality. Then graph it on a number line.

1. $x - 9 < -4$

2. $6p \geq 5p - 3$

3. MULTIPLE CHOICE Drew currently has 31 comic books in his collection. His friend Connor has 58 comic books. How many more comic books does Drew need to add to his collection in order to have a larger collection than Connor?

    A no more than 21

    B 27

    C at least 28

    D more than 30

Solve each inequality. Graph the solution on a number line.

4. $\frac{1}{5}h > 3$

5. $7w \leq -42$

6. $-\frac{2}{3}t \geq 24$

7. $-9m < -36$

8. $3c - 7 < 11$

9. $\frac{g}{4} + 3 \leq -9$

10. $-2(x - 4) > 5x - 13$

11. ZOO The 8th grade science class is going to the zoo. The class can spend up to $300 on admission.

Zoo Admission	
Visitor	**Cost**
student	$8
adult	$10

    a. Write an inequality for this situation.

    b. If there are 32 students in the class and 1 adult will attend for every 8 students, can the entire class go to the zoo?

Solve each compound inequality. Then graph the solution set.

12. $y - 8 < -3$ or $y + 5 > 19$

13. $-11 \leq 2h - 5 \leq 13$

14. $3z - 2 > -5$ and $7z + 4 < -17$

Define a variable, write an inequality, and solve the problem. Check your solution.

15. The difference of a number and 4 is no more than 8.

16. Nine times a number decreased by four is at least twenty-three.

17. MULTIPLE CHOICE Write a compound inequality for the graph shown below.

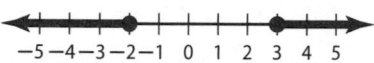

    F $-2 \leq x < 3$ **H** $x < -2$ or $x \geq 3$

    G $x \leq -2$ or $x \geq 3$ **J** $-2 < x \leq 3$

Solve each inequality. Then graph the solution set.

18. $|p - 5| < 3$

19. $|2f + 7| \geq 21$

20. $|-4m + 3| \leq 15$

21. $\left|\frac{x - 3}{4}\right| > 5$

22. RETAIL A sporting goods store is offering a $15 coupon on any pair of shoes.

    a. The most and least expensive pairs of shoes are $149.95 and $24.95. What is the range of costs for customers with coupons?

    b. When buying a pair of $109.95 shoes, you can use a coupon or a 15% discount. Which option is best?

Graph each inequality.

23. $y < 4x - 1$

24. $2x + 3y \geq 12$

25. Graph $y > -2x + 5$. Then determine which of the ordered pairs in $\{(-2, 0), (-1, 5), (2, 3), (7, 3)\}$ are in the solution set.

26. PRESCHOOL Mrs. Jones is buying new books and puzzles for her preschool classroom. Each book costs $6, and each puzzle costs $4. Write and graph an inequality to determine how many books and puzzles she can buy for $96.

Preparing for Standardized Tests

Write and Solve an Inequality

Many multiple-choice items will require writing and solving inequalities.
Follow the steps below to help you successfully solve these types of problems.

Strategies for Writing and Solving Inequalities

Step 1

Read the problem statement carefully.

Ask yourself:

- What am I being asked to solve?
- What information is given in the problem?
- What are the unknowns for which I need to solve?

Step 2

Translate the problem statement into an inequality.

- Assign variables to the unknown(s).
- Write the word sentence as a mathematical number sentence looking for words such as *greater than, less than, no more than, up to*, or *at least* to indicate the type of inequality as well as where to place the inequality sign.

Step 3

Solve the inequality.

- Solve for the unknowns in the inequality.
- Remember that multiplying or dividing each side by a negative number reverses the direction of the inequality.
- Check your answer to make sure it makes sense.

Standardized Test Example

Read the problem. Identify what you need to know. Then use the information in the problem to solve. Show your work.

Pedro has earned scores of 89, 74, 79, 85, and 88 on his tests this semester. He needs a test average of at least 85 in order to earn an A for the semester. There will be one more test given this semester.

A Write an inequality to model the situation.

B What score must he have on his final test to earn an A for the semester?

Rob Gage/Taxi/Getty Images

Read the problem carefully. You are given Pedro's first 5 test scores and told that he needs an average of *at least* 85 after his next test to earn an A for the semester.

a. Write the inequality.

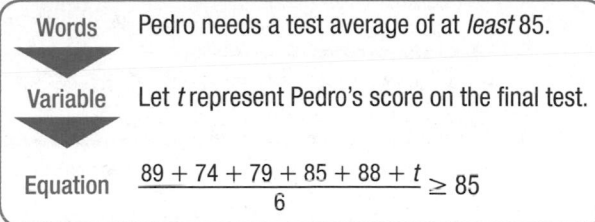

Words	Pedro needs a test average of at *least* 85.
Variable	Let *t* represent Pedro's score on the final test.
Equation	$\dfrac{89 + 74 + 79 + 85 + 88 + t}{6} \geq 85$

b. Solve the inequality for *t*.

$$\frac{89 + 74 + 79 + 85 + 88 + t}{6} \geq 85$$
$$89 + 74 + 79 + 85 + 88 + t \geq 85(6)$$
$$415 + t \geq 510$$
$$t \geq 95$$

So, Pedro's final test score must be greater than or equal to 95 in order for him to earn an A for the semester.

Exercises

Read each problem. Identify what you need to know. Then use the information in the problem to solve.

1. Craig has $20 to order a pizza. The pizza costs $12.50 plus $0.95 per topping. If there is also a $3 delivery fee, how many toppings can Craig order?

2. To join an archery club, Nina had to pay an initiation fee of $75, plus $40 per year in membership dues.

   a. Write an equation to model the total cost, *y*, of belonging to the club for *x* years.

   b. How many years will it take her to spend more than $400 to belong to the club?

3. The area of the triangle below is no more than 84 square millimeters. What is the height of the triangle?

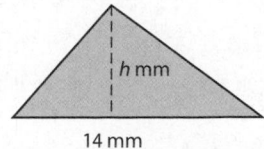

h mm

14 mm

4. Rosa earns $200 a month delivering newspapers, plus an average of $11 per hour babysitting. If her goal is to earn at least $295 this month, how many hours will she have to babysit?

5. To earn money for a new bike, Ethan is selling some of his baseball cards. He has saved $245. If the bike costs $1400, and he can sell 154 cards, for how much money will he need to sell each card to reach his goal?

6. In a certain lacrosse league, there can be no more than 22 players on each team, and no more than 10 teams per age group. There are 6 age groups.

   a. Write an inequality to represent this situation.

   b. What is the greatest number of players that can play lacrosse in this league?

7. Sarah has $120 to shop for herself and to buy some gifts for 6 of her friends. She has purchased a shirt for herself for $32. Assuming that she spends an equal amount on each friend, what is the maximum that she can spend per person?

Multiple Choice

Read each question. Then fill in the correct answer on the answer document provided by your teacher or on a sheet of paper.

1. Miguel received a $100 gift certificate for a graduation gift. He wants to buy a CD player that costs $38 and CDs that cost $12 each. Which of the following inequalities represents how many CDs Miguel can buy?

    A $n \leq 6$

    B $n \geq 5$

    C $n < 5$

    D $n \leq 5$

2. Craig is paid time-and-a-half for any additional hours over 40 that he works.

Time	Pay Rate
Up to 40 hours	$12.80/hr
Additional hours worked over 40	$19.20/hr

    If Craig's goal is to earn at least $600 next week, what is the minimum number of hours he needs to work?

    F   43 hours   H   45 hours

    G   44 hours   J   46 hours

3. Which equation has a slope of $-\frac{2}{3}$ and a y-intercept of 6?

    A $y = 6x + \frac{2}{3}$ C $y = -\frac{2}{3}x + 6$

    B $y = -\frac{2}{3}x - 6$ D $y = 6x - \frac{2}{3}$

4. The highest score that is on record on a video game is 10,219 points. The lowest score on record is 257 points. Which of the following inequalities best shows the range of scores recorded on the game?

    F $x \leq 10{,}219$

    G $x \geq 257$

    H $257 < x < 10{,}219$

    J $257 \leq x \leq 10{,}219$

5. The current temperature is 82°. If the temperature rises more than 4 degrees, there will be a new record high for the date. Which number line represents the temperatures that would set a new record high?

    A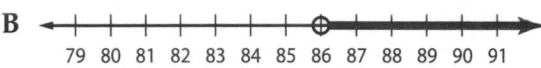

    B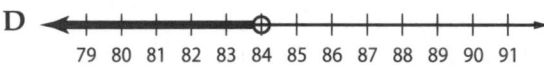

    C
    79 80 81 82 83 84 85 86 87 88 89 90 91

    D
    79 80 81 82 83 84 85 86 87 88 89 90 91

6. The girls' volleyball team is selling T-shirts and pennants to raise money for new uniforms. The team hopes to raise more than $250.

Item	Price
T-shirt	$10
Pennant	$4

    Which of the following combinations of items sold would meet this goal?

    F   16 T-shirts and 20 pennants

    G   20 T-shirts and 12 pennants

    H   18 T-shirts and 18 pennants

    J   15 T-shirts and 20 pennants

7. What type of line does not have a defined slope?

    A   horizontal   C   perpendicular

    B   parallel   D   vertical

8. Which expression below illustrates the Associative Property?

    F $abc = bac$

    G $2(x - 3) = 2x - 6$

    H $(p + 3) - t = p + (3 - t)$

    J $5 + (-5) = 0$

Test-TakingTip

Question 2 You can check your answer by finding Craig's earnings for the hours worked.

Record your answers on the answer sheet provided by your teacher or on a sheet of paper.

9. Solve $-4 < 3x + 8 \leq 23$.

10. **GRIDDED RESPONSE** Tien is saving money for a new television. She needs to save at least $720 to pay for her expenses. Each week Tien saves $50 toward her new television. How many weeks will it take so she can pay for the television?

11. Write an inequality that best represents the graph.

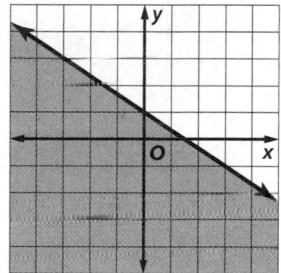

12. Solve $|x - 4| < 2$.

13. **GRIDDED RESPONSE** Daniel wants to ship a set of golf clubs and several boxes of golf balls in a box that can hold up to 20 pounds. If the set of clubs weighs 9 pounds and each box of golf balls weighs 12 ounces, how many boxes of golf balls can Daniel ship?

14. Graph the solution set for the inequality $3x - 6 \leq 4x - 4 \leq 3x + 1$.

15. Write an equation that represents the data in the table.

x	y
3	12.5
4	16
5	19.5
6	23
7	26.5

16. A sporting goods company near the beach rents bicycles for $10 plus $5 per hour. Write an equation in slope-intercept form that shows the total cost, y, of renting a bicycle for x hours. How much would it cost Emily to rent a bicycle for 6 hours?

Extended Response

Record your answers on a sheet of paper. Show your work.

17. Theresa is saving money for a vacation. She needs to save at least $640 to pay for her expenses. Each week, she puts $35 towards her vacation savings.

   a. Let w represent the number of weeks Theresa saves money. Write an inequality to model the situation.

   b. Solve the inequality from part a. What is the minimum number of weeks Theresa must save money in order to reach her goal?

   c. If Theresa were to save $45 each week instead, by how many weeks would the minimum savings time be decreased?

Need ExtraHelp?																
If you missed Question...	1	2	3	4	5	6	7	8	9	10	11	12	13	14	15	16	17
Go to Lesson...	5-3	5-3	4-1	5-4	5-1	5-6	3-3	1-3	5-4	5-2	5-6	5-5	5-3	5-4	3-5	4-2	5-2

6 Systems of Linear Equations and Inequalities

·· Then

○ You solved linear equations in one variable.

·· Now

○ In this chapter, you will:

- Solve systems of linear equations by graphing, substitution, and elimination.

- Solve systems of linear inequalities by graphing.

·· Why? ▲

○ **MUSIC** $1500 worth of tickets were sold for a marching band competition. Adult tickets were $12 each, and student tickets were $8 each. If you knew how many total tickets were sold, you could use a system of equations to determine how many adult tickets and how many student tickets were sold.

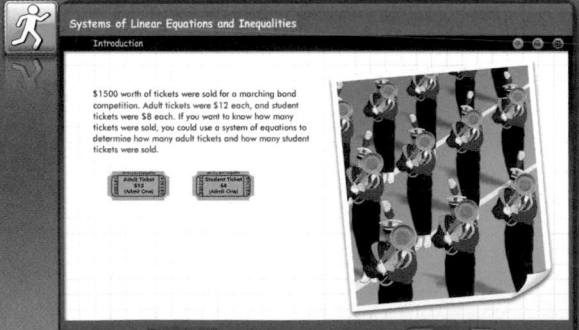

ﮩ **connectED.mcgraw-hill.com** | **Your Digital Math Portal**

Animation	Vocabulary	eGlossary	Personal Tutor	Virtual Manipulatives	Graphing Calculator	Audio	Foldables	Self-Check Practice	Worksheets

Get Ready for the Chapter

Diagnose Readiness | You have two options for checking prerequisite skills.

1 **Textbook Option** Take the Quick Check below. Refer to the Quick Review for help.

QuickCheck	QuickReview

Name the ordered pair for each point on the coordinate plane.

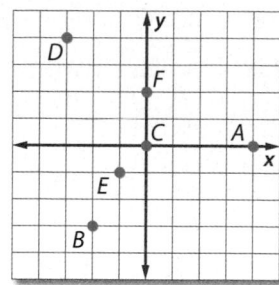

1. A **2.** D

3. B **4.** C

5. E **6.** F

Example 1

Name the ordered pair for Q on the coordinate plane.

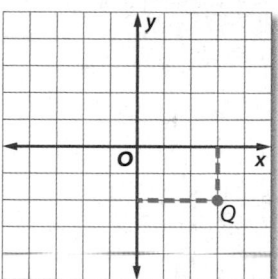

Follow a vertical line from the point to the x-axis. This gives the x-coordinate, 3.

Follow a horizontal line from the point to the y-axis. This gives the y-coordinate, -2.

The ordered pair is $(3, -2)$.

Solve each equation or formula for the variable specified.

7. $2x + 4y = 12$, for x

8. $x = 3y - 9$, for y

9. $m - 2n = 6$, for m

10. $y = mx + b$, for x

11. $P = 2\ell + 2w$, for ℓ

12. $5x - 10y = 40$, for y

13. **GEOMETRY** The formula for the area of a triangle is $A = \frac{1}{2}bh$, where A represents the area, b is the base, and h is the height of the triangle. Solve the equation for b.

Example 2

Solve $12x + 3y = 36$ for y.

$12x + 3y = 36$	Original equation
$12x + 3y - 12x = 36 - 12x$	Subtract 12x from each side.
$3y = 36 - 12x$	Simplify.
$\dfrac{3y}{3} = \dfrac{36 - 12x}{3}$	Divide each side by 3.
$y = 12 - 4x$	Simplify.

---

2 **Online Option** Take an online self-check Chapter Readiness Quiz at <u>connectED.mcgraw-hill.com</u>.

Get Started on the Chapter

You will learn several new concepts, skills, and vocabulary terms as you study Chapter 6. To get ready, identify important terms and organize your resources. You may wish to refer to Chapter 0 to review prerequisite skills.

FOLDABLES StudyOrganizer

Systems of Linear Equations and Inequalities Make this Foldable to help you organize your Chapter 6 notes about solving systems of equations and inequalities. Begin with a sheet of notebook paper.

1 Fold lengthwise to the holes.

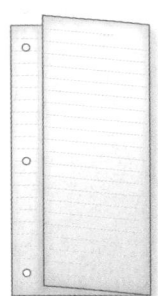

2 Cut 6 tabs.

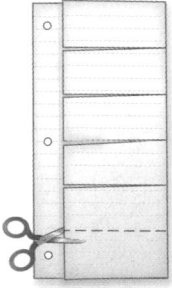

3 Label the tabs using the lesson titles.

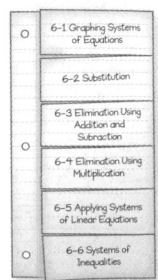

NewVocabulary

English		Español
system of equations	p. 335	sistema de ecuaciones
consistent	p. 335	consistente
independent	p. 335	independiente
dependent	p. 335	dependiente
inconsistent	p. 335	inconsistente
substitution	p. 344	sustitución
elimination	p. 350	eliminación
matrix	p. 370	matriz
element	p. 370	elemento
dimension	p. 370	dimensión
augmented matrix	p. 370	matriz ampliada
row reduction	p. 371	reducción de fila
identity matrix	p. 371	matriz
system of inequalities	p. 372	sistema de desigualdades

ReviewVocabulary

domain dominio the set of the first numbers of the ordered pairs in a relation

intersection intersección the graph of a compound inequality containing *and*; the solution is the set of elements common to both graphs

proportion proporción an equation stating that two ratios are equal

Proportion

$$\frac{24}{30} = \frac{4}{5}$$

$\div 6$

$\div 6$

Graphing Systems of Equations

:: Then	:: Now	:: Why?
● You graphed linear equations.	**1** Determine the number of solutions a system of linear equations has. **2** Solve systems of linear equations by graphing.	● The cost to begin production on a band's CD is $1500. Each CD costs $4 to produce and will sell for $10. The band wants to know how many CDs they will have to sell to earn a profit.

The cost to begin production on a band's CD is $1500. Each CD costs $4 to produce and will sell for $10. The band wants to know how many CDs they will have to sell to earn a profit.

Graphing a system can show when a company makes a profit. The cost of producing the CD can be modeled by the equation $y = 4x + 1500$, where y represents the cost of production and x is the number of CDs produced.

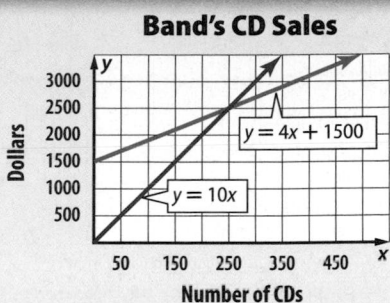

Band's CD Sales

$y = 4x + 1500$

$y = 10x$

NewVocabulary
system of equations
consistent
independent
dependent
inconsistent

Common Core State Standards

Content Standards

A.CED.3 Represent constraints by equations or inequalities, and by systems of equations and/or inequalities, and interpret solutions as viable or nonviable options in a modeling context.

A.REI.6 Solve systems of linear equations exactly and approximately (e.g., with graphs), focusing on pairs of linear equations in two variables.

Mathematical Practices

3 Construct viable arguments and critique the reasoning of others.

8 Look for and express regularity in repeated reasoning.

1 **Possible Number of Solutions** The income from the CDs sold can be modeled by the equation $y = 10x$, where y represents the total income of selling the CDs, and x is the number of CDs sold.

If we graph these equations, we can see at which point the band begins making a profit. The point where the two graphs intersect is where the band breaks even. This happens when the band sells 250 CDs. If the band sells more than 250 CDs, they will make a profit.

The two equations, $y = 4x + 1500$ and $y = 10x$, form a **system of equations**. The ordered pair that is a solution of both equations is the solution of the system. A system of two linear equations can have one solution, an infinite number of solutions, or no solution.

- If a system has at least one solution, it is said to be **consistent**. The graphs intersect at one point or are the same line.

- If a consistent system has exactly one solution, it is said to be **independent**. If it has an infinite number of solutions, it is **dependent**. This means that there are unlimited solutions that satisfy both equations.

- If a system has no solution, it is said to be **inconsistent**. The graphs are parallel.

ConceptSummary Possible Solutions			
Number of Solutions	exactly one	infinite	no solution
Terminology	consistent and independent	consistent and dependent	inconsistent
Graph			

PT

StudyTip

Number of Solutions
When both equations are of the form $y = mx + b$, the values of m and b can determine the number of solutions.

Compare m and b	Number of Solutions
different m values	one
same m value, but different b values	none
same m value, and same b value	infinite

Example 1 Number of Solutions

Use the graph at the right to determine whether each system is *consistent* or *inconsistent* and if it is *independent* or *dependent*.

a. $y = -2x + 3$
$y = x - 5$

Since the graphs of these two lines intersect at one point, there is exactly one solution. Therefore, the system is consistent and independent.

b. $y = -2x - 5$
$y = -2x + 3$

Since the graphs of these two lines are parallel, there is no solution of the system. Therefore, the system is inconsistent.

$y = 2x + 3$
$y = x - 5$
$y = -2x - 5$
$y = -2x + 3$

▶ **Guided**Practice

1A. $y = 2x + 3$
$y = -2x - 5$

1B. $y = x - 5$
$y = -2x - 5$

2 Solve by Graphing One method of solving a system of equations is to graph the equations carefully on the same coordinate grid and find their point of intersection. This point is the solution of the system.

Example 2 Solve by Graphing

Graph each system and determine the number of solutions that it has. If it has one solution, name it.

a. $y = -3x + 10$
$y = x - 2$

The graphs appear to intersect at the point (3, 1). You can check this by substituting 3 for x and 1 for y.

CHECK $y = -3x + 10$	Original equation
$1 \stackrel{?}{=} -3(3) + 10$	Substitution
$1 \stackrel{?}{=} -9 + 10$	Multiply.
$1 = 1$ ✔	
$y = x - 2$	Original equation
$1 \stackrel{?}{=} 3 - 2$	Substitution
$1 = 1$ ✔	Multiply.

The solution is (3, 1).

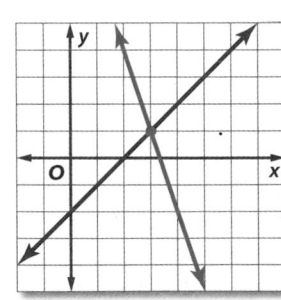

b. $2x - y = -1$
$4x - 2y = 6$

The lines have the same slope but different y-intercepts, so the lines are parallel. Since they do not intersect, there is no solution of this system. The system is inconsistent.

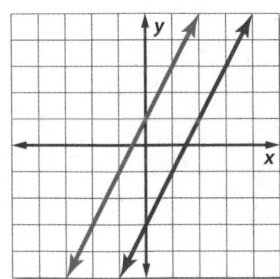

ReviewVocabulary

parallel lines never intersect and have the same slope

GuidedPractice

Graph each system and determine the number of solutions that it has. If it has one solution, name it.

2A. $x - y = 2$
$3y + 2x = 9$

2B. $y = -2x - 3$
$6x + 3y = -9$

We can use what we know about systems of equations to solve many real-world problems involving constraints that are modeled by two or more different functions.

Real-World Example 3 Write and Solve a System of Equations

SPORTS The number of girls participating in high school soccer and track and field has steadily increased over the past few years. Use the information in the table to predict the approximate year when the number of girls participating in these two sports will be the same.

High School Sport	Number of Girls Participating in 2008 (thousands)	Average rate of increase (thousands per year)
soccer	345	8
track and field	458	3

Source: National Federation of State High School Associations

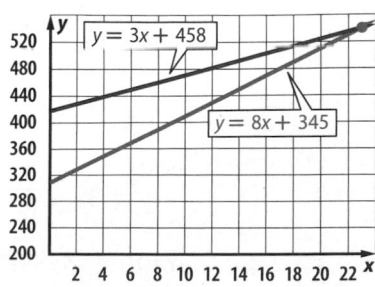

Real-World Link

In 2008, 3.1 million girls participated in high school sports. This was an all-time high for female participation.

Source: National Federation of State High School Associations

Words	Number of girls participating	equals	rate of increase	times	number of years after 2008	plus	number participating in 2008.

Variables Let y = number of girls competing. Let x = number of years after 2008.

Equations	Soccer:	y	$=$	8	$\bullet$	x	$+$	345
	Track and field:	y	$=$	3	$\bullet$	x	$+$	458

Graph $y = 8x + 345$ and $y = 3x + 458$. The graphs appear to intersect at approximately (22.5, 525).

CHECK Use substitution to check this answer.

$y = 8x + 345$ $\qquad$ $y = 3x + 458$

$525 \stackrel{?}{=} 8(22.5) + 345$ $\qquad$ $525 \stackrel{?}{=} 3(22.5) + 458$

$525 = 525$ ✓ $\qquad$ $525 \approx 525.5$ ✓

The solution means that approximately 22 years after 2008, or in 2030, the number of girls participating in high school soccer and track and field will be the same, about 525,000.

GuidedPractice

3. VIDEO GAMES Joe and Josh each want to buy a video game. Joe has \$14 and saves \$10 a week. Josh has \$26 and saves \$7 a week. In how many weeks will they have the same amount?

Check Your Understanding

● = Step-by-Step Solutions begin on page R13.

Example 1

Use the graph at the right to determine whether each system is *consistent* or *inconsistent* and if it is *independent* or *dependent*.

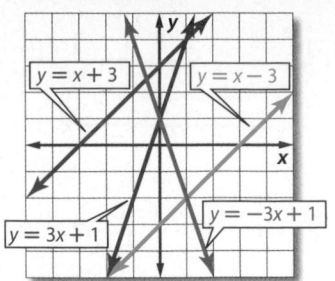

1. $y = -3x + 1$
$y = 3x + 1$

2. $y = 3x + 1$
$y = x - 3$

3. $y = x - 3$
$y = x + 3$

4. $y = x + 3$
$x - y = -3$

5. $x - y = -3$
$y = -3x + 1$

6. $y = -3x + 1$
$y = x - 3$

Example 2

Graph each system and determine the number of solutions that it has. If it has one solution, name it.

7. $y = x + 4$
$y = -x - 4$

8. $y = x + 3$
$y = 2x + 4$

Example 3

9. **CCSS MODELING** Alberto and Ashanti are reading a graphic novel.

    a. Write an equation to represent the pages each boy has read.

    b. Graph each equation.

    c. How long will it be before Alberto has read more pages than Ashanti? Check and interpret your solution.

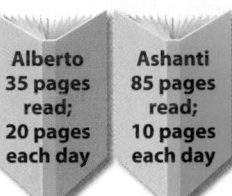

Alberto	Ashanti
35 pages read; 20 pages each day	85 pages read; 10 pages each day

Practice and Problem Solving

Extra Practice is on page R6.

Example 1

Use the graph at the right to determine whether each system is *consistent* or *inconsistent* and if it is *independent* or *dependent*.

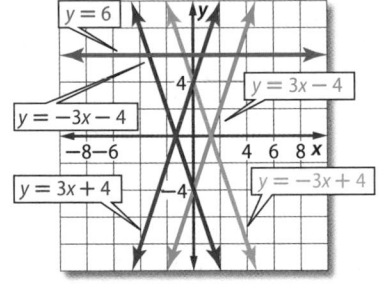

10. $y = 6$
$y = 3x + 4$

11. $y = 3x + 4$
$y = -3x + 4$

12. $y = -3x + 4$
$y = -3x - 4$

13 $y = -3x - 4$
$y = 3x - 4$

14. $3x - y = -4$
$y = 3x + 4$

15. $3x - y = 4$
$3x + y = 4$

Example 2

Graph each system and determine the number of solutions that it has. If it has one solution, name it.

16. $y = -3$
$y = x - 3$

17. $y = 4x + 2$
$y = -2x - 3$

18. $y = x - 6$
$y = x + 2$

19. $x + y = 4$
$3x + 3y = 12$

20. $x - y = -2$
$-x + y = 2$

21. $x + 2y = 3$
$x = 5$

22. $2x + 3y = 12$
$2x - y = 4$

23. $2x + y = -4$
$y + 2x = 3$

24. $2x + 2y = 6$
$5y + 5x = 15$

Example 3   **25. SCHOOL DANCE** Akira and Jen are competing to see who can sell the most tickets for the Winter Dance. On Monday, Akira sold 22 and then sold 30 per day after that. Jen sold 53 on Monday and then sold 20 per day after that.

    a. Write equations for the number of tickets each person has sold.

    b. Graph each equation.

    c. Solve the system of equations. Check and interpret your solution.

26. CCSS MODELING If x is the number of years since 2000 and y is the percent of people using travel services, the following equations represent the percent of people using travel agents and the percent of people using the Internet to plan travel.

    Travel agents: $y = -2x + 30$ Internet: $y = 6x + 41$

    a. Graph the system of equations.

    b. Estimate the year travel agents and the Internet were used equally.

Graph each system and determine the number of solutions that it has. If it has one solution, name it.

27 $y = \frac{1}{2}x$ **28.** $y = 6x + 6$ **29.** $y = 2x - 17$

    $y = x + 2$ $y = 3x + 6$ $y = x - 10$

30. $8x - 4y = 16$ **31.** $3x + 5y = 30$ **32.** $-3x + 4y = 24$

    $-5x - 5y = 5$ $3x + y = 18$ $4x - y = 7$

33. $2x - 8y = 6$ **34.** $4x - 6y = 12$ **35.** $2x + 3y = 10$

    $x - 4y = 3$ $-2x + 3y = -6$ $4x + 6y = 12$

36. $3x + 2y = 10$ **37.** $3y - x = -2$ **38.** $\frac{8}{5}y = \frac{2}{5}x + 1$

    $2x + 3y = 10$ $y - \frac{1}{3}x = 2$ $\frac{2}{5}y = \frac{1}{10}x + \frac{1}{4}$

39. $\frac{1}{3}x + \frac{1}{3}y = 1$ **40.** $\frac{3}{4}x + \frac{1}{2}y = \frac{1}{4}$ **41.** $\frac{5}{6}x + \frac{2}{3}y = \frac{1}{2}$

    $x + y = 1$ $\frac{2}{3}x + \frac{1}{6}y = \frac{1}{2}$ $\frac{2}{5}x + \frac{1}{5}y = \frac{3}{5}$

42. PHOTOGRAPHY Suppose x represents the number of cameras sold and y represents the number of years since 2000. Then the number of digital cameras sold each year since 2000, in millions, can be modeled by the equation $y = 12.5x + 10.9$. The number of film cameras sold each year since 2000, in millions, can be modeled by the equation $y = -9.1x + 78.8$.

    a. Graph each equation.

    b. In which year did digital camera sales surpass film camera sales?

    c. In what year did film cameras stop selling altogether?

    d. What are the domain and range of each of the functions in this situation?

Graph each system and determine the number of solutions that it has. If it has one solution, name it.

43. $2y = 1.2x - 10$ **44.** $x = 6 - \frac{3}{8}y$

    $4y = 2.4x$ $4 = \frac{2}{3}x + \frac{1}{4}y$

45 **WEB SITES** Personal publishing site *Lookatme* had 2.5 million visitors in 2009. Each year after that, the number of visitors rose by 13.1 million. Online auction site *Buyourstuff* had 59 million visitors in 2009, but each year after that the number of visitors fell by 2 million.

  a. Write an equation for each of the companies.

  b. Make a table of values for 5 years for each of the companies.

  c. Graph each equation.

  d. When will *Lookatme* and *Buyourstuff's* sites have the same number of visitors?

  e. Name the domain and range of these functions in this situation.

46. ⬡ **MULTIPLE REPRESENTATIONS** In this problem, you will explore different methods for finding the intersection of the graphs of two linear equations.

  a. **Algebraic** Use algebra to solve the equation $\frac{1}{2}x + 3 = -x + 12$.

  b. **Graphical** Use a graph to solve $y = \frac{1}{2}x + 3$ and $y = -x + 12$.

  c. **Analytical** How is the equation in part **a** related to the system in part **b**?

  d. **Verbal** Explain how to use the graph in part **b** to solve the equation in part **a**.

H.O.T. Problems   Use Higher-Order Thinking Skills

47. **ERROR ANALYSIS** Store A is offering a 10% discount on the purchase of all electronics in their store. Store B is offering $10 off all the electronics in their store. Francisca and Alan are deciding which offer will save them more money. Is either of them correct? Explain your reasoning.

Francisca	Alan
You can't determine which store has the better offer unless you know the price of the items you want to buy.	Store A has the better offer because 10% of the sale price is a greater discount than $10.

48. **CHALLENGE** Use graphing to find the solution of the system of equations $2x + 3y = 5$, $3x + 4y = 6$, and $4x + 5y = 7$.

49. **CCSS** **ARGUMENTS** Determine whether a system of two linear equations with $(0, 0)$ and $(2, 2)$ as solutions *sometimes*, *always*, or *never* has other solutions. Explain.

50. **WHICH ONE DOESN'T BELONG?** Which one of the following systems of equations doesn't belong with the other three? Explain your reasoning.

$4x - y = 5$	$-x + 4y = 8$	$4x + 2y = 14$	$3x - 2y = 1$
$-2x + y = -1$	$3x - 6y = 6$	$12x + 6y = 18$	$2x + 3y = 18$

51. **OPEN ENDED** Write three equations such that they form three systems of equations with $y = 5x - 3$. The three systems should be inconsistent, consistent and independent, and consistent and dependent, respectively.

52. **WRITING IN MATH** Describe the advantages and disadvantages to solving systems of equations by graphing.

53. SHORT RESPONSE Certain bacteria can reproduce every 20 minutes, doubling the population. If there are 450,000 bacteria in a population at 9:00 A.M., how many bacteria will be in the population at 2:00 P.M.?

54. GEOMETRY An 84-centimeter piece of wire is cut into equal segments and then attached at the ends to form the edges of a cube. What is the volume of the cube?

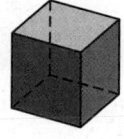

  A 294 cm^3 **C** 1158 cm^3

  B 343 cm^3 **D** 2744 cm^3

55. What is the solution of the inequality $-9 < 2x + 3 < 15$?

  F $-x \geq 0$ **H** $-6 < x < 6$

  G $x \leq 0$ **J** $-5 < x < 5$

56. What is the solution of the system of equations?

$$x + 2y = -1$$
$$2x + 4y = -2$$

  A $(-1, -1)$ **C** no solution

  B $(2, 1)$ **D** infinitely many solutions

Spiral Review

Graph each inequality. (Lesson 5-6)

57. $3x + 6y > 0$

58. $4x - 2y < 0$

59. $3y - x \leq 9$

60. $4y - 3x \geq 12$

61. $y < -4x - 8$

62. $3x - 1 > y$

63. LIBRARY To get a grant from the city's historical society, the number of history books must be within 25 of 1500. What is the range of the number of historical books that must be in the library? (Lesson 5-5)

64. SCHOOL Camilla's scores on three math tests are shown in the table. The fourth and final test of the grading period is tomorrow. She needs an average of at least 92 to receive an A for the grading period. (Lesson 5-3)

  a. If m represents her score on the fourth math test, write an inequality to represent this situation.

  b. If Camilla wants an A in math, what must she score on the test?

  c. Is your solution reasonable? Explain.

Test	Score
1	91
2	95
3	88

Write the slope-intercept form of an equation for the line that passes through the given point and is perpendicular to the graph of the equation. (Lesson 4-4)

65. $(-3, 1)$, $y = \frac{1}{3}x + 2$

66. $(6, -2)$, $y = \frac{3}{5}x - 4$

67. $(2, -2)$, $2x + y = 5$

68. $(-3, -3)$, $-3x + y = 6$

Skills Review

Find the solution of each equation using the given replacement set.

69. $f - 14 = 8$; $\{12, 15, 19, 22\}$

70. $15(n + 6) = 165$; $\{3, 4, 5, 6, 7\}$

71. $23 = \frac{d}{4}$; $\{91, 92, 93, 94, 95\}$

72. $36 = \frac{t - 9}{2}$; $\{78, 79, 80, 81\}$

Evaluate each expression if $a = 2$, $b = -3$, and $c = 11$.

73. $a + 6b$

74. $7 - ab$

75. $(2c + 3a) \div 4$

76. $b^2 + (a^3 - 8)5$

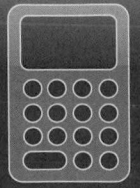

Graphing Technology Lab
Systems of Equations

You can use a graphing calculator to graph and solve a system of equations.

Activity 1 Solve a System of Equations

Solve the system of equations. State the decimal solution to the nearest hundredth.

$$5.23x + y = 7.48$$
$$6.42x - y = 2.11$$

CCSS Common Core State Standards
Content Standards
A.REI.6 Solve systems of linear equations exactly and approximately (e.g., with graphs), focusing on pairs of linear equations in two variables.
A.REI.11 Explain why the *x*-coordinates of the points where the graphs of the equations $y = f(x)$ and $y = g(x)$ intersect are the solutions of the equation $f(x) = g(x)$; find the solutions approximately, e.g., using technology to graph the functions, make tables of values, or find successive approximations. Include cases where $f(x)$ and/or $g(x)$ are linear, polynomial, rational, absolute value, exponential, and logarithmic functions.
Mathematical Practices
5 Use appropriate tools strategically.

Step 1 Solve each equation for *y* to enter them into the calculator.

$5.23x + y = 7.48$	First equation
$5.23x + y - 5.23x = 7.48 - 5.23x$	Subtract 5.23x from each side.
$y = 7.48 - 5.23x$	Simplify.
$6.42x - y = 2.11$	Second equation
$6.42x - y - 6.42x = 2.11 - 6.42x$	Subtract 6.42x from each side.
$-y = 2.11 - 6.42x$	Simplify.
$(-1)(-y) = (-1)(2.11 - 6.42x)$	Multiply each side by −1.
$y = -2.11 + 6.42x$	Simplify.

Step 2 Enter these equations in the **Y=** list and graph in the standard viewing window.

KEYSTROKES: [Y=] 7.48 [−] 5.23 [X,T,θ,n]
[ENTER] [(−)] 2.11 [+]
6.42 [X,T,θ,n] [ZOOM] 6

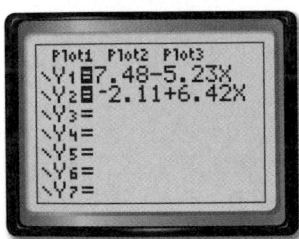

Step 3 Use the **CALC** menu to find the point of intersection.

KEYSTROKES: [2nd] [CALC] 5 [ENTER] [ENTER]
[ENTER]

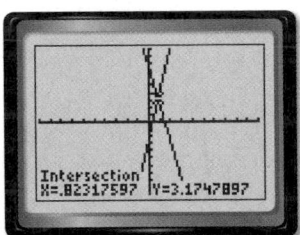

[−10, 10] scl: 1 by [−10, 10] scl: 1

The solution is approximately (0.82, 3.17).

When you solve a system of equations with $y = f(x)$ and $y = g(x)$, the solution is an ordered pair that satisfies both equations. The solution always occurs when $f(x) = g(x)$. Thus, the *x*-coordinate of the solution is the value of *x* where $f(x) = g(x)$.

One method you can use to solve an equation with one variable is by graphing and solving a system of equations based on the equation. To do this, write a system using both sides of the equation. Then use a graphing calculator to solve the system.

Activity 2 Use a System to Solve a Linear Equation

Use a system of equations to solve $5x + 6 = -4$.

Step 1 Write a system of equations. Set each side of the equation equal to y.

$y = 5x + 6$ First equation
$y = -4$ Second equation

Step 2 Enter these equations in the **Y=** list and graph.

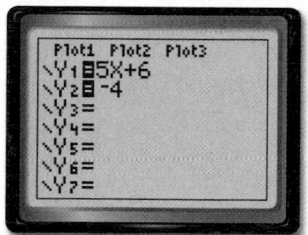

Step 3 Use the **CALC** menu to find the point of intersection.

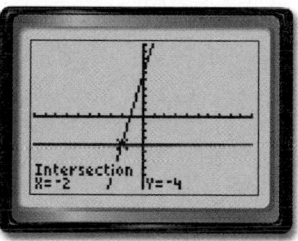

[−10, 10] scl: 1 by [−10, 10] scl: 1

The solution is −2.

Exercises

Use a graphing calculator to solve each system of equations. Write decimal solutions to the nearest hundredth.

1. $y = 2x - 3$
$y = -0.4x + 5$

2. $y = 6x + 1$
$y = -3.2x - 4$

3. $x + y = 9.35$
$5x - y = 8.75$

4. $2.32x - y = 6.12$
$4.5x + y = -6.05$

5. $5.2x - y = 4.1$
$1.5x + y = 6.7$

6. $1.8 = 5.4x - y$
$y = -3.8 - 6.2x$

7. $7x - 2y = 16$
$11x + 6y = 32.3$

8. $3x + 2y = 16$
$5x + y = 9$

9. $0.62x + 0.35y = 1.60$
$-1.38x + y = 8.24$

10. $75x - 100y = 400$
$33x - 10y = 70$

Use a graphing calculator to solve each equation. Write decimal solutions to the nearest hundredth.

11. $4x - 2 = -6$

12. $3 = 1 + \frac{x}{2}$

13. $\frac{x + 4}{-2} = -1$

14. $\frac{3}{2}x + \frac{1}{2} = 2x - 3$

15. $4x - 9 = 7 + 7x$

16. $-2 + 10x = 8x - 1$

17. WRITING IN MATH Explain why you can solve an equation like $r = ax + b$ by solving the system of equations $y = r$ and $y = ax + b$.

Then	Now	Why?
• You solved systems of equations by graphing.	**1** Solve systems of equations by using substitution. **2** Solve real-world problems involving systems of equations by using substitution.	• Two movies were released at the same time. Movie A earned $31 million in its opening week, but fell to $15 million the following week. Movie B opened earning $21 million and fell to $11 million the following week. If the earnings for each movie continue to decrease at the same rate, when will they earn the same amount?

NewVocabulary
substitution

Common Core State Standards

Content Standards
A.CED.3 Represent constraints by equations or inequalities, and by systems of equations and/or inequalities, and interpret solutions as viable or nonviable options in a modeling context.

A.REI.6 Solve systems of linear equations exactly and approximately (e.g., with graphs), focusing on pairs of linear equations in two variables.

Mathematical Practices
2 Reason abstractly and quantitatively.

1 Solve by Substitution You can use a system of equations to find when the movie earnings are the same. One method of finding an exact solution of a system of equations is called **substitution**.

KeyConcept Solving by Substitution

Step 1 When necessary, solve at least one equation for one variable.

Step 2 Substitute the resulting expression from Step 1 into the other equation to replace the variable. Then solve the equation.

Step 3 Substitute the value from Step 2 into either equation, and solve for the other variable. Write the solution as an ordered pair.

Example 1 Solve a System by Substitution

Use substitution to solve the system of equations.

$y = 2x + 1$ ← **Step 1** The first equation is already solved for y.
$3x + y = -9$

Step 2 Substitute $2x + 1$ for y in the second equation.

$3x + y = -9$	Second equation
$3x + 2x + 1 = -9$	Substitute $2x + 1$ for y.
$5x + 1 = -9$	Combine like terms.
$5x = -10$	Subtract 1 from each side.
$x = -2$	Divide each side by 5.

Step 3 Substitute -2 for x in either equation to find y.

$y = 2x + 1$	First equation
$= 2(-2) + 1$	Substitute -2 for x.
$= -3$	Simplify.

The solution is $(-2, -3)$.

CHECK You can check your solution by graphing.

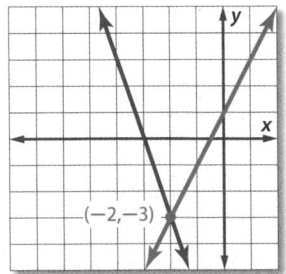

(−2,−3)

▶ **Guided**Practice

1A. $y = 4x - 6$
    $5x + 3y = -1$

1B. $2x + 5y = -1$
    $y = 3x + 10$

If a variable is not isolated in one of the equations in a system, solve an equation for a variable first. Then you can use substitution to solve the system.

Example 2 Solve and then Substitute

Use substitution to solve the system of equations.
$x + 2y = 6$
$3x - 4y = 28$

Step 1 Solve the first equation for x since the coefficient is 1.

$x + 2y = 6$	First equation
$x + 2y - 2y = 6 - 2y$	Subtract $2y$ from each side.
$x = 6 - 2y$	Simplify.

Step 2 Substitute $6 - 2y$ for x in the second equation to find the value of y.

$3x - 4y = 28$	Second equation
$3(6 - 2y) - 4y = 28$	Substitute $6 - 2y$ for x.
$18 - 6y - 4y = 28$	Distributive Property
$18 - 10y = 28$	Combine like terms.
$18 - 10y - 18 = 28 - 18$	Subtract 18 from each side.
$-10y = 10$	Simplify.
$y = -1$	Divide each side by -10.

Step 3 Find the value of x.

$x + 2y = 6$	First equation
$x + 2(-1) = 6$	Substitute -1 for y.
$x - 2 = 6$	Simplify.
$x = 8$	Add 2 to each side.

▶ **Guided**Practice

2A. $4x + 5y = 11$
　　　$y - 3x = -13$

2B. $x - 3y = -9$
　　　$5x - 2y = 7$

Generally, if you solve a system of equations and the result is a false statement such as $3 = -2$, there is no solution. If the result is an identity, such as $3 = 3$, then there are an infinite number of solutions.

Example 3 No Solution or Infinitely Many Solutions

Use substitution to solve the system of equations.
$y = 2x - 4$
$-6x + 3y = -12$

Substitute $2x - 4$ for y in the second equation.

$-6x + 3y = -12$	Second equation
$-6x + 3(2x - 4) = -12$	Substitute $2x - 4$ for y.
$-6x + 6x - 12 = -12$	Distributive Property
$-12 = -12$	Combine like terms.

This statement is an identity. Thus, there are an infinite number of solutions.

▶ **Guided**Practice Use substitution to solve each system of equations.

3A. $2x - y = 8$
$y = 2x - 3$

3B. $4x - 3y = 1$
$6y - 8x = -2$

2 Solve Real-World Problems You can use substitution to find the solution of a real-world problem involving constraints modeled by a system of equations.

● Real-World Example 4 Write and Solve a System of Equations

MUSIC A store sold a total of **125 car stereo systems and speakers in one week. The stereo systems sold for $104.95, and the speakers sold for $18.95. The sales from these two items totaled $6926.75. How many of each item were sold?**

Number of Units Sold	c	t	125
Sales ($)	104.95c	18.95t	6926.75

Let c = the number of car stereo systems sold, and let t = the number of speakers sold.

So, the two equations are $c + t = 125$ and $104.95c + 18.95t = 6926.75$.

Notice that $c + t = 125$ represents combinations of car stereo systems and speakers with a sum of 125. The equation $104.95c + 18.95t = 6926.75$ represents the combinations of car stereo systems and speakers with a sales of $6926.75. The solution of the system of equations represents the option that meets both of the constraints.

Step 1 Solve the first equation for c.

$c + t = 125$ First equation

$c + t - t = 125 - t$ Subtract t from each side.

$c = 125 - t$ Simplify.

Step 2 Substitute $125 - t$ for c in the second equation.

$104.95c + 18.95t = 6926.75$ Second equation

$104.95(125 - t) + 18.95t = 6926.75$ Substitute $125 - t$ for c.

$13{,}118.75 - 104.95t + 18.95t = 6926.75$ Distributive Property

$13{,}118.75 - 86t = 6926.75$ Combine like terms.

$-86t = -6192$ Subtract 13,118.75 from each side.

$t = 72$ Divide each side by -86.

Step 3 Substitute 72 for t in either equation to find the value of c.

$c + t = 125$ First equation

$c + 72 = 125$ Substitute 72 for t.

$c = 53$ Subtract 72 from each side.

The store sold 53 car stereo systems and 72 speakers.

▶ **Guided**Practice

4. BASEBALL As of 2009, the New York Yankees and the Cincinnati Reds together had won a total of 32 World Series. The Yankees had won 5.4 times as many as the Reds. How many World Series had each team won?

Real-WorldLink

Sound Engineering Technician Sound engineering technicians record, synchronize, mix, and reproduce music, voices, and sound effects in recording studios, sporting arenas, and theater, movie, or video productions. They need to have at least a 2-year associate's degree in electronics.

Examples 1–3 Use substitution to solve each system of equations.

1. $y = x + 5$
$3x + y = 25$

2. $x = y - 2$
$4x + y = 2$

3. $3x + y = 6$
$4x + 2y = 8$

4. $2x + 3y = 4$
$4x + 6y = 9$

5. $x - y = 1$
$3x = 3y + 3$

6. $2x - y = 6$
$-3y = -6x + 18$

Example 4

7. **GEOMETRY** The sum of the measures of angles X and Y is 180°. The measure of angle X is 24° greater than the measure of angle Y.

   a. Define the variables, and write equations for this situation.

   b. Find the measure of each angle.

Practice and Problem Solving

Extra Practice is on page R6.

Examples 1–3 Use substitution to solve each system of equations.

8. $y = 5x + 1$
$4x + y = 10$

9. $y = 4x + 5$
$2x + y = 17$

10. $y = 3x - 34$
$y = 2x - 5$

11. $y = 3x - 2$
$y = 2x - 5$

12. $2x + y = 3$
$4x + 4y = 8$

13. $3x + 4y = -3$
$x + 2y = -1$

14. $y = -3x + 4$
$-6x - 2y = -8$

15. $-1 = 2x - y$
$8x - 4y = -4$

16. $x - y - 1$
$-x + y = -1$

17. $y = -4x + 11$
$3x + y = 9$

18. $y = -3x + 1$
$2x + y = 1$

19. $3x + y = -5$
$6x + 2y - 10$

20. $5x - y - 5$
$-x + 3y = 13$

21. $2x + y - 4$
$-2x + y = -4$

22. $-5x + 4y = 20$
$10x - 8y = -40$

Example 4

23. **ECONOMICS** In 2000, the demand for nurses was 2,000,000, while the supply was only 1,890,000. The projected demand for nurses in 2020 is 2,810,414, while the supply is only projected to be 2,001,998.

   a. Define the variables, and write equations to represent these situations.

   b. Use substitution to determine during which year the supply of nurses was equal to the demand.

24. **CCSS REASONING** The table shows the approximate number of tourists in two areas of the world during a recent year and the average rates of change in tourism.

Destination	Number of Tourists	Average Rates of Change in Tourists (millions per year)
South America and the Caribbean	40.3 million	increase of 0.8
Middle East	17.0 million	increase of 1.8

   a. Define the variables, and write an equation for each region's tourism rate.

   b. If the trends continue, in how many years would you expect the number of tourists in the regions to be equal?

25 SPORTS The table shows the winning times for the Triathlon World Championship.

Year	Men's	Women's
2000	1:51:39	1:54:43
2009	1:44:51	1:59:14

a. The times are in hours, minutes, and seconds. Rewrite the times rounded to the nearest minute.

b. Let the year 2000 be 0. Assume that the rate of change remains the same for years after 2000. Write an equation to represent each of the men's and women's winning times y in any year x.

c. If the trend continues, when would you expect the men's and women's winning times to be the same? Explain your reasoning.

26. CONCERT TICKETS Booker is buying tickets online for a concert. He finds tickets for himself and his friends for $65 each plus a one-time fee of $10. Paula is looking for tickets to the same concert. She finds them at another Web site for $69 and a one-time fee of $13.60.

a. Define the variables, and write equations to represent this situation.

b. Create a table of values for 1 to 5 tickets for each person's purchase.

c. Graph each of these equations.

d. Use the graph to determine who received the better deal? Explain why.

27. ERROR ANALYSIS In the system $a + b = 7$ and $1.29a + 0.49b = 6.63$, a represents pounds of apples and b represents pounds of bananas. Guillermo and Cara are finding and interpreting the solution. Is either of them correct? Explain.

Guillermo
$1.29a + 0.49b = 6.63$
$1.29a + 0.49(a + 7) = 6.63$
$1.29 + 0.49a + 3.43 = 6.63$
$0.49a = 3.2$
$a = 1.9$
$a + b = 7$, so $b = 5$. The solution (2, 5) means that 2 pounds of apples and 5 pounds of bananas were bought.

Cara
$1.29a + 0.49b = 6.63$
$1.29(7 - b) + 0.49b = 6.63$
$9.03 - 1.29b + 0.49b = 6.63$
$-0.8b = -2.4$
$b = 3$
The solution $b = 3$ means that 3 pounds of apples and 3 pounds of bananas were bought.

28. CCSS PERSEVERANCE A local charity has 60 volunteers. The ratio of boys to girls is 7:5. Find the number of boy and the number of girl volunteers.

29. REASONING Compare and contrast the solution of a system found by graphing and the solution of the same system found by substitution.

30. OPEN ENDED Create a system of equations that has one solution. Illustrate how the system could represent a real-world situation and describe the significance of the solution in the context of the situation.

31. WRITING IN MATH Explain how to determine what to substitute when using the substitution method of solving systems of equations.

32. The debate team plans to make and sell trail mix. They can spend $34.

Item	Cost Per Pound
sunflower seeds	$4.00
raisins	$1.50

The pounds of raisins in the mix are to be 3 times the pounds of sunflower seeds. Which system can be used to find r, the pounds of raisins, and p, pounds of sunflower seeds, they should buy?

A $3p = r$
$4p + 1.5r = 34$

C $3r = p$
$4p + 1.5r = 34$

B $3p = r$
$4r + 1.5p = 34$

D $3r = p$
$4r + 1.5p = 34$

33. GRIDDED RESPONSE The perimeters of two similar polygons are 250 centimeters and 300 centimeters, respectively. What is the scale factor between the first and second polygons?

34. Based on the graph, which statement is true?

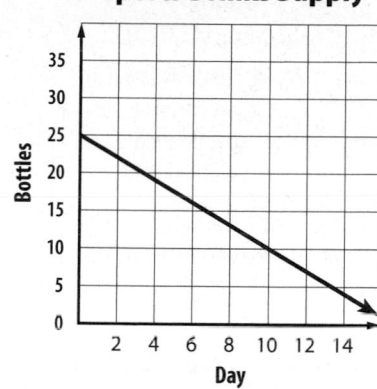

Sports Drinks Supply

F Mary started with 30 bottles.
G On day 10, Mary will have 10 bottles left.
H Mary will be out of sports drinks on day 14.
J Mary drank 5 bottles the first two days.

35. If p is an integer, which of the following is the solution set for $2|p| = 16$?

A $\{0, 8\}$

C $\{-8, 8\}$

B $\{-8, 0\}$

D $\{-8, 0, 8\}$

Graph each system and determine how many solutions it has. If it has one solution, name it. (Lesson 6-1)

36. $y = -5$
$3x + y = 1$

37. $x = 1$
$2x - y = 7$

38. $y = x + 5$
$y = x - 2$

39. $x + y = 1$
$3y + 3x = 3$

40. ENTERTAINMENT Coach Ross wants to take the soccer team out for pizza after their game. Her budget is at most $70. (Lesson 5-6)

a. Using the sign, write an inequality that represents this situation.

b. Are there any restrictions on the variables? Explain.

Welcome to Rini's Pizza

Large Pizza $12
Pitcher of Soft Drinks $2

Solve each inequality. Check your solution. (Lesson 5-3)

41. $6v + 1 \geq -11$

42. $24 > 18 + 2n$

43. $-11 \geq \frac{2}{5}q + 5$

44. $\frac{a}{8} - 10 > -3$

45. $-3t + 9 \leq 0$

46. $54 > -10 - 8n$

Rewrite each product using the Distributive Property. Then simplify.

47. $10b + 5(3 + 9b)$

48. $5(3t^2 + 4) - 8t$

49. $7h^2 + 4(3h + h^2)$

50. $-2(7a + 5b) + 5(2a - 7b)$

Elimination Using Addition and Subtraction

:·Then

- You solved systems of equations by using substitution.

:·Now

1. Solve systems of equations by using elimination with addition.

2. Solve systems of equations by using elimination with subtraction.

:·Why?

- In Chicago, Illinois, there are two more months a when the mean high temperature is below 70°F than there are months b when it is above 70°F. The system of equations, $a + b = 12$ and $a - b = 2$, represents this situation.

 NewVocabulary
elimination

 **Common Core State Standards**

Content Standards
A.CED.2 Create equations in two or more variables to represent relationships between quantities; graph equations on coordinate axes with labels and scales.

A.REI.6 Solve systems of linear equations exactly and approximately (e.g., with graphs), focusing on pairs of linear equations in two variables.

Mathematical Practices
7 Look for and make use of structure.

1 Elimination Using Addition If you add these equations, the variable b will be eliminated. Using addition or subtraction to solve a system is called **elimination**.

KeyConcept Solving by Elimination

Step 1 Write the system so like terms with the same or opposite coefficients are aligned.

Step 2 Add or subtract the equations, eliminating one variable. Then solve the equation.

Step 3 Substitute the value from Step 2 into one of the equations and solve for the other variable. Write the solution as an ordered pair.

Example 1 Elimination Using Addition

Use elimination to solve the system of equations.

$4x + 6y = 32$
$3x - 6y = 3$ ⟵ **Step 1** $6y$ and $-6y$ have opposite coefficients.

Step 2 Add the equations.

$$\begin{aligned} 4x + 6y &= 32 \\ (+)\ 3x - 6y &= \ \ 3 \\ \hline 7x \quad\ \ \, &= 35 \end{aligned}$$ The variable y is eliminated.

$\dfrac{7x}{7} = \dfrac{35}{7}$ Divide each side by 7.

$x = 5$ Simplify.

Step 3 Substitute 5 for x in either equation to find the value of y.

$4x + 6y = 32$ First equation

$4(5) + 6y = 32$ Replace x with 5.

$20 + 6y = 32$ Multiply.

$20 + 6y - 20 = 32 - 20$ Subtract 20 from each side.

$6y = 12$ Simplify.

$\dfrac{6y}{6} = \dfrac{12}{6}$ Divide each side by 6.

$y = 2$ Simplify.

The solution is $(5, 2)$.

1A. $-4x + 3y = -3$
$4x - 5y = 5$

1B. $4y + 3x = 22$
$3x - 4y = 14$

We can use elimination to find specific numbers that are described as being related to each other.

Example 2 Write and Solve a System of Equations

Negative three times one number plus five times another number is −11. Three times the first number plus seven times the other number is −1. Find the numbers.

Negative three times one number	plus	five times another number	is	−11.
$-3x$	$+$	$5y$	$=$	-11

Three times the first number	plus	seven times the other number	is	−1.
$3x$	$+$	$7y$	$=$	-1

Steps 1 and 2 Write the equations vertically and add.

$$-3x + 5y = -11$$
$$\underline{(+)\ 3x + 7y = \ \ -1}$$

$$12y = -12 \qquad \text{The variable } x \text{ is eliminated.}$$

$$\frac{12y}{12} = \frac{-12}{12} \qquad \text{Divide each side by 12.}$$

$$y = -1 \qquad \text{Simplify.}$$

Step 3 Substitute −1 for y in either equation to find the value of x.

$$3x + 7y = -1 \qquad \text{Second equation}$$
$$3x + 7(-1) = -1 \qquad \text{Replace } y \text{ with } -1.$$
$$3x + (-7) = -1 \qquad \text{Simplify.}$$
$$3x + (-7) + 7 = -1 + 7 \qquad \text{Add 7 to each side.}$$
$$3x = 6 \qquad \text{Simplify.}$$
$$\frac{3x}{3} = \frac{6}{3} \qquad \text{Divide each side by 3.}$$
$$x = 2 \qquad \text{Simplify.}$$

The numbers are 2 and −1.

CHECK
$$-3x + 5y = -11 \qquad \text{First equation}$$
$$-3(2) + 5(-1) \stackrel{?}{=} -11 \qquad \text{Substitute 2 for } x \text{ and } -1 \text{ for } y.$$
$$-11 = -11 \checkmark \qquad \text{Simplify.}$$
$$3x + 7y = -1 \qquad \text{Second equation}$$
$$3(2) + 7(-1) \stackrel{?}{=} -1 \qquad \text{Substitute 2 for } x \text{ and } -1 \text{ for } y.$$
$$-1 = -1 \checkmark \qquad \text{Simplify.}$$

> **Study**Tip
>
> **Coefficients** When the coefficients of a variable are the same, subtracting the equations will eliminate the variable. When the coefficients are opposites, adding the equations will eliminate the variable.

> **Problem-Solving**Tip
>
> **CCSS** Perseverance
>
> Checking your answers in both equations of a system helps ensure there are no calculation errors.

> **Guided**Practice

2. The sum of two numbers is −10. Negative three times the first number minus the second number equals 2. Find the numbers.

2 Elimination Using Subtraction
Sometimes we can eliminate a variable by subtracting one equation from another.

Standardized Test Example 3

Solve the system of equations.
$$2t + 5r = 6$$
$$9r + 2t = 22$$

A $(-7, 15)$ **B** $\left(7, \dfrac{8}{9}\right)$ **C** $(4, -7)$ **D** $\left(4, -\dfrac{2}{5}\right)$

Read the Test Item

Since both equations contain $2t$, use elimination by subtraction.

Solve the Test Item

Step 1 Subtract the equations.

$$\begin{aligned} 5r + 2t = 6 & \qquad \text{Write the system so like terms are aligned.} \\ \underline{(-)\ 9r + 2t = 22} & \\ -4r = -16 & \qquad \text{The variable } t \text{ is eliminated.} \\ r = 4 & \qquad \text{Simplify.} \end{aligned}$$

Step 2 Substitute 4 for r in either equation to find the value of t.

$$\begin{aligned} 5r + 2t &= 6 & \text{First equation} \\ 5(4) + 2t &= 6 & r = 4 \\ 20 + 2t &= 6 & \text{Simplify.} \\ 20 + 2t - 20 &= 6 - 20 & \text{Subtract 20 from each side.} \\ 2t &= -14 & \text{Simplify.} \\ t &= -7 & \text{Simplify.} \end{aligned}$$

The solution is $(4, -7)$. The correct answer is C.

GuidedPractice

3. Solve the system of equations.
$$8b + 3c = 11$$
$$8b + 7c = 7$$

F $(1.5, -1)$ **G** $(1.75, -1)$ **H** $(1.75, 1)$ **J** $(1.5, 1)$

Real-World Example 4 Write and Solve a System of Equations

JOBS Cheryl and Jackie work at an ice cream shop. Cheryl earns $8.50 per hour and Jackie earns $7.50 per hour. During a typical week, Cheryl and Jackie earn $299.50 together. One week, Jackie doubles her work hours, and the girls earn $412. How many hours does each girl work during a typical week?

Understand You know how much Cheryl and Jackie each earn per hour and how much they earned together.

Plan Let $c =$ Cheryl's hours and $j =$ Jackie's hours.

Cheryl's pay	plus	Jackie's pay	equals	$299.50.
$8.50c$	$+$	$7.50j$	$=$	299.50

Cheryl's pay	plus	Jackie's pay	equals	$412.
$8.50c$	$+$	$7.50(2)j$	$=$	412

Photodisc/Alamy

Solve Subtract the equations to eliminate one of the variables. Then solve for the other variable.

$$8.50c + 7.50j = 299.50 \qquad \text{Write the equations vertically.}$$
$$(-)\ 8.50c + 7.50(2)j = 412$$
___

$$8.50c + 7.50j = 299.50$$
$$(-)\ 8.50c + \quad 15j = 412 \qquad \text{Simplify.}$$
___
$$\qquad\qquad -7.50j = -112.50 \qquad \text{Subtract. The variable } c \text{ is eliminated.}$$

$$\frac{-7.50j}{-7.50} = \frac{-112.50}{-7.50} \qquad \text{Divide each side by } -7.50.$$

$$j = 15 \qquad \text{Simplify.}$$

Now substitute 15 for j in either equation to find the value of c.

$$8.50c + 7.50j = 299.50 \qquad \text{First equation}$$
$$8.50c + 7.50(\mathbf{15}) = 299.50 \qquad \text{Substitute 15 for } j.$$
$$8.50c + 112.50 = 299.50 \qquad \text{Simplify.}$$
$$8.50c = 187 \qquad \text{Subtract 112.50 from each side.}$$
$$c = 22 \qquad \text{Divide each side by 8.50.}$$

Check Substitute both values into the other equation to see if the equation holds true. If $c = 22$ and $j = 15$, then $8.50(22) + 15(15)$ or 412.

Cheryl works 22 hours, while Jackie works 15 hours during a typical week.

▶ **Guided**Practice

4. PARTIES Tamera and Adelina are throwing a birthday party for their friend. Tamera invited 5 fewer friends than Adelina. Together they invited 47 guests. How many guests did each girl invite?

Check Your Understanding

 = Step-by-Step Solutions begin on page R13.

Examples 1, 3 Use elimination to solve each system of equations.

1. $5m - p = 7$
$7m - p = 11$

2. $8x + 5y = 38$
$-8x + 2y = 4$

3. $7f + 3g = -6$
$7f - 2g = -31$

4. $6a - 3b = 27$
$2a - 3b = 11$

Example 2
5. CCSS REASONING The sum of two numbers is 24. Five times the first number minus the second number is 12. What are the two numbers?

Example 4
6. RECYCLING The recycling and reuse industry employs approximately 1,025,000 more workers than the waste management industry. Together they provide 1,275,000 jobs. How many jobs does each industry provide?

Examples 1, 3 Use elimination to solve each system of equations.

7. $-v + w = 7$
$v + w = 1$

8. $y + z = 4$
$y - z = 8$

9. $-4x + 5y = 17$
$4x + 6y = -6$

10. $5m - 2p = 24$
$3m + 2p = 24$

11. $a + 4b = -4$
$a + 10b = -16$

12. $6r - 6t = 6$
$3r - 6t = 15$

13. $6c - 9d = 111$
$5c - 9d = 103$

14. $11f + 14g = 13$
$11f + 10g = 25$

15. $9x + 6y = 78$
$3x - 6y = -30$

16. $3j + 4k = 23.5$
$8j - 4k = 4$

17. $-3x - 8y = -24$
$3x - 5y = 4.5$

18. $6x - 2y = 1$
$10x - 2y = 5$

Example 2   **19.** The sum of two numbers is 22, and their difference is 12. What are the numbers?

20. Find the two numbers with a sum of 41 and a difference of 9.

21 Three times a number minus another number is -3. The sum of the numbers is 11. Find the numbers.

22. A number minus twice another number is 4. Three times the first number plus two times the second number is 12. What are the numbers?

Example 4   **23.** **TOURS** The Blackwells and Joneses are going to Hershey's Really Big 3D Show in Pennsylvania. Find the adult price and the children's price of the show.

Family	Number of Adults	Number of Children	Total Cost
Blackwell	2	5	$31.65
Jones	2	3	$23.75

Use elimination to solve each system of equations.

24. $4(x + 2y) = 8$
$4x + 4y = 12$

25. $3x - 5y = 11$
$5(x + y) = 5$

26. $4x + 3y = 6$
$3x + 3y = 7$

27. $6x - 7y = -26$

$6x + 5y = 10$

28. $\frac{1}{2}x + \frac{2}{3}y = 2\frac{3}{4}$

$\frac{1}{4}x - \frac{2}{3}y = 6\frac{1}{4}$

29. $\frac{3}{5}x + \frac{1}{2}y = 8\frac{1}{3}$

$-\frac{3}{5}x + \frac{3}{4}y = 8\frac{1}{3}$

30. **CCSS SENSE-MAKING** The total height of an office building b and the granite statue that stands on top of it g is 326.6 feet. The difference in heights between the building and the statue is 295.4 feet.

a. How tall is the statue?

b. How tall is the building?

31. **BIKE RACING** Professional Mountain Bike Racing currently has 66 teams. The number of non-U.S. teams is 30 more than the number of U.S. teams.

a. Let x represent the number of non-U.S. teams and y represent the number of U.S. teams. Write a system of equations that represents the number of U.S. teams and non-U.S. teams.

b. Use elimination to find the solution of the system of equations.

c. Interpret the solution in the context of the situation.

d. Graph the system of equations to check your solution.

32. SHOPPING Let x represent the number of years since 2004 and y represent the number of catalogs.

Catalogs	Number in 2004	Growth Rate (number per year)
online	7440	1293
print	3805	−1364

Source: MediaPost Publications

a. Write a system of equations to represent this situation.

b. Use elimination to find the solution to the system of equations.

c. Analyze the solution in terms of the situation. Determine the reasonableness of the solution.

33 ⛊ **MULTIPLE REPRESENTATIONS** Collect 9 pennies and 9 paper clips. For this game, you use objects to score points. Each paper clip is worth 1 point and each penny is worth 3 points. Let p represent the number of pennies and c represent the number of paper clips.

$$9 \text{ points} = \text{🪙🪙} + \text{📎📎📎} = 3p + c = 3(2) + 3$$

a. Concrete Choose a combination of 9 objects and find your score.

b. Analytical Write and solve a system of equations to find the number of paper clips and pennies used for 15 points, if 9 total objects are used.

c. Tabular If 9 total objects are used, make a table showing the number of paper clips used and the total number of points when the number of pennies is 0, 1, 2, 3, 4, or 5.

d. Verbal Does the result in the table match the results in part **b**? Explain.

H.O.T. Problems Use Higher-Order Thinking Skills

34. REASONING Describe the solution of a system of equations if after you added two equations the result was $0 = 0$.

35. REASONING What is the solution of a system of equations if the sum of the equations is $0 = 2$?

36. OPEN ENDED Create a system of equations that can be solved by using addition to eliminate one variable. Formulate a general rule for creating such systems.

37. CCSS STRUCTURE The solution of a system of equations is $(-3, 2)$. One equation in the system is $x + 4y = 5$. Find a second equation for the system. Explain how you derived this equation.

38. CHALLENGE The sum of the digits of a two-digit number is 8. The result of subtracting the units digit from the tens digit is −4. Define the variables and write the system of equations that you would use to find the number. Then solve the system and find the number.

39. WRITING IN MATH Describe when it would be most beneficial to use elimination to solve a system of equations.

40. SHORT RESPONSE Martina is on a train traveling at a speed of 188 mph between two cities 1128 miles apart. If the train has been traveling for an hour, how many more hours is her train ride?

41. GEOMETRY Ms. Miller wants to tile her rectangular kitchen floor. She knows the dimensions of the floor. Which formula should she use to find the area?

A $A = \ell w$ **C** $P = 2\ell + 2w$

B $V = Bh$ **D** $c^2 = a^2 + b^2$

42. If the pattern continues, what is the 8th number in the sequence?

$$2, 3, \frac{9}{2}, \frac{27}{4}, \frac{81}{8}, \ldots$$

F $\frac{2187}{64}$ **G** $\frac{2245}{64}$ **H** $\frac{2281}{64}$ **J** $\frac{2445}{64}$

43. What is the solution of this system of equations?

$$x + 4y = 1$$
$$2x - 3y = -9$$

A $(2, -8)$ **C** no solution

B $(-3, 1)$ **D** infinitely many solutions

Spiral Review

Use substitution to solve each system of equations. If the system does not have exactly one solution, state whether it has no solution or infinitely many solutions. (Lesson 6-2)

44. $y = 6x$
$2x + 3y = 40$

45. $x = 3y$
$2x + 3y = 45$

46. $x = 5y + 6$
$x = 3y - 2$

47. $y = 3x + 2$
$y = 4x - 1$

48. $3c = 4d + 2$
$c = d - 1$

49. $z = v + 4$
$2z - v = 6$

50. FINANCIAL LITERACY Gregorio and Javier each want to buy a bicycle. Gregorio has already saved $35 and plans to save $10 per week. Javier has $26 and plans to save $13 per week. (Lesson 6-1)

  a. In how many weeks will Gregorio and Javier have saved the same amount of money?

  b. How much will each person have saved at that time?

51. GEOMETRY A *parallelogram* is a quadrilateral in which opposite sides are parallel. Determine whether *ABCD* is parallelogram. Explain your reasoning. (Lesson 4-4)

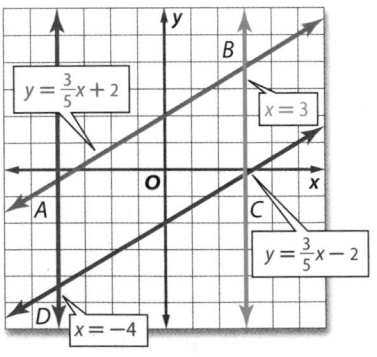

Solve each equation. Check your solution. (Lesson 2-2)

52. $6u = -48$

53. $75 = -15p$

54. $\frac{2}{3}a = 8$

55. $-\frac{3}{4}d = 15$

Skills Review

Simplify each expression. If not possible, write *simplified*.

56. $6q - 3 + 7q + 1$

57. $7w^2 - 9w + 4w^2$

58. $10(2 + r) + 3r$

59. $5y - 7(y + 5)$

Elimination Using Multiplication

:: Then
- You used elimination with addition and subtraction to solve systems of equations.

:: Now
1. Solve systems of equations by using elimination with multiplication.
2. Solve real-world problems involving systems of equations.

:: Why?
- The table shows the number of cars at Scott's Auto Repair Shop for each type of service.

  The manager has allotted 1110 minutes for body work and 570 minutes for engine work. The system $3r + 4m = 1110$ and $2r + 2m = 570$ can be used to find the average time for each service.

Item	Repairs	Maintenance
body	3	4
engine	2	2

Common Core State Standards

Content Standards

A.REI.5 Prove that, given a system of two equations in two variables, replacing one equation by the sum of that equation and a multiple of the other produces a system with the same solutions.

A.REI.6 Solve systems of linear equations exactly and approximately (e.g., with graphs), focusing on pairs of linear equations in two variables.

Mathematical Practices

1 Make sense of problems and persevere in solving them.

---

1 Elimination Using Multiplication
In the system above, neither variable can be eliminated by adding or subtracting. You can use multiplication to solve.

KeyConcept Solving by Elimination

Step 1 Multiply at least one equation by a constant to get two equations that contain opposite terms.

Step 2 Add the equations, eliminating one variable. Then solve the equation.

Step 3 Substitute the value from Step 2 into one of the equations and solve for the other variable. Write the solution as an ordered pair.

Example 1 Multiply One Equation to Eliminate a Variable

Use elimination to solve the system of equations.

$5x + 6y = -8$
$2x + 3y = -5$

Steps 1 and 2

$5x + 6y = -8$
$2x + 3y = -5$ → Multiply each term by −2.

$5x + 6y = -8$
$(+) -4x - 6y = 10$ Add.
$\overline{x = 2}$ y is eliminated.

Step 3

$2x + 3y = -5$ Second equation
$2(2) + 3y = -5$ Substitution, $x = 2$
$4 + 3y = -5$ Simplify.
$3y = -9$ Subtract 4 from each side and simplify.
$y = -3$ Divide each side by 3 and simplify.

The solution is $(2, -3)$.

GuidedPractice

1A. $6x - 2y = 10$
$3x - 7y = -19$

1B. $9r + q = 13$
$3r + 2q = -4$

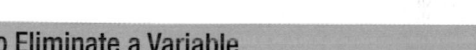
Exactostock/SuperStock

Sometimes you have to multiply each equation by a different number in order to solve the system.

Example 2 Multiply Both Equations to Eliminate a Variable

Use elimination to solve the system of equations.

$4x + 2y = 8$
$3x + 3y = 9$

Method 1 Eliminate x.

$$4x + 2y = 8 \quad \boxed{\text{Multiply by 3.}}$$
$$3x + 3y = 9 \quad \boxed{\text{Multiply by } -4.}$$

$$12x + 6y = 24$$
$$(+) -12x - 12y = -36 \qquad \text{Add equations.}$$
$$\overline{\qquad\quad -6y = -12} \qquad x \text{ is eliminated.}$$
$$\frac{-6y}{-6} = \frac{-12}{-6} \qquad \text{Divide each side by } -6.$$
$$y = 2 \qquad \text{Simplify.}$$

Now substitute 2 for y in either equation to find the value of x.

$$3x + 3y = 9 \qquad \text{Second equation}$$
$$3x + 3(2) = 9 \qquad \text{Substitute 2 for } y.$$
$$3x + 6 = 9 \qquad \text{Simplify.}$$
$$3x = 3 \qquad \text{Subtract 6 from each side and simplify.}$$
$$\frac{3x}{3} = \frac{3}{3} \qquad \text{Divide each side by 3.}$$
$$x = 1 \qquad \text{The solution is (1, 2).}$$

Method 2 Eliminate y.

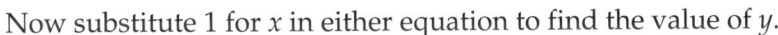

$$4x + 2y = 8 \quad \boxed{\text{Multiply by 3.}}$$
$$3x + 3y = 9 \quad \boxed{\text{Multiply by } -2.}$$

$$12x + 6y = 24$$
$$(+) -6x - 6y = -18 \qquad \text{Add equations.}$$
$$\overline{\qquad\quad 6x \quad\;\;\; = 6} \qquad y \text{ is eliminated.}$$
$$\frac{6x}{6} = \frac{6}{6} \qquad \text{Divide each side by 6.}$$
$$x = 1 \qquad \text{Simplify.}$$

Now substitute 1 for x in either equation to find the value of y.

$$3x + 3y = 9 \qquad \text{Second equation}$$
$$3(1) + 3y = 9 \qquad \text{Substitute 1 for } x.$$
$$3 + 3y = 9 \qquad \text{Simplify.}$$
$$3y = 6 \qquad \text{Subtract 3 from each side and simplify.}$$
$$\frac{3y}{3} = \frac{6}{3} \qquad \text{Divide each side by 3.}$$
$$y = 2 \qquad \text{Simplify.}$$

The solution is (1, 2), which matches the result obtained with Method 1.

CHECK Substitute 1 for x and 2 for y in the first equation.

$$4x + 2y = 8 \qquad \text{Original equation}$$
$$4(1) + 2(2) \stackrel{?}{=} 8 \qquad \text{Substitute (1, 2) for } (x, y).$$
$$4 + 4 \stackrel{?}{=} 8 \qquad \text{Multiply.}$$
$$8 = 8 \checkmark \qquad \text{Add.}$$

GuidedPractice

2A. $5x - 3y = 6$
$\quad\;\; 2x + 5y = -10$

2B. $6a + 2b = 2$
$\quad\;\; 4a + 3b = 8$

StudyTip

Choosing a Variable to Eliminate Unless the problem is asking for the value of a specific variable, you may use multiplication to eliminate either variable.

Math HistoryLink

Leonardo Pisano (1170–1250) Leonardo Pisano is better known by his nickname *Fibonacci*. His book introduced the Hindu-Arabic place-valued decimal system. Systems of linear equations are studied in this work.

2 Solve Real-World Problems Sometimes it is necessary to use multiplication before elimination in real-world problem solving too.

Real-World Example 3 Solve a System of Equations

FLIGHT A personal aircraft traveling with the wind flies 520 miles in 4 hours. On the return trip, the airplane takes 5 hours to travel the same distance. Find the speed of the airplane if the air is still.

You are asked to find the speed of the airplane in still air.

Let a = the rate of the airplane if the air is still.
Let w = the rate of the wind.

	r	t	d	$r \cdot t = d$
With the Wind	$a + w$	4	520	$(a + w)4 = 520$
Against the Wind	$a - w$	5	520	$(a - w)5 = 520$

So, our two equations are $4a + 4w = 520$ and $5a - 5w = 520$.

$4a + 4w = 520$ [Multiply by 5.]
$5a - 5w = 520$ [Multiply by 4.]

$$20a + 20w = 2600$$
$$(+) \ 20a - 20w = 2080$$
$$\overline{40a = 4680}$$ *w* is eliminated.

$$\frac{40a}{40} = \frac{4680}{40}$$ Divide each side by 40.

$$a = 117$$ Simplify.

The rate of the airplane in still air is 117 miles per hour.

▸ **Guided**Practice

3. CANOEING A canoeist travels 4 miles downstream in 1 hour. The return trip takes the canoeist 1.5 hours. Find the rate of the boat in still water.

Check Your Understanding

◯ **= Step-by-Step Solutions begin on page R13.**

Examples 1–2 Use elimination to solve each system of equations.

1. $2x - y = 4$
$7x + 3y = 27$

2. $2x + 7y = 1$
$x + 5y = 2$

③ $4x + 2y = -14$
$5x + 3y = -17$

4. $9a - 2b = -8$
$-7a + 3b = 12$

Example 3

5. CCSS SENSE-MAKING A kayaking group with a guide travels 16 miles downstream, stops for a meal, and then travels 16 miles upstream. The speed of the current remains constant throughout the trip. Find the speed of the kayak in still water.

Leave	10:00 A.M.
Stop for meal	12:00 noon
Return	1:00 P.M.
Finish	5:00 P.M.

6. PODCASTS Steve subscribed to 10 podcasts for a total of 340 minutes. He used his two favorite tags, Hobbies and Recreation and Soliloquies. Each of the Hobbies and Recreation episodes lasted about 32 minutes. Each Soliloquies episode lasted 42 minutes. To how many of each tag did Steve subscribe?

Examples 1–2 Use elimination to solve each system of equations.

7. $x + y = 2$
$-3x + 4y = 15$

8. $x - y = -8$
$7x + 5y = 16$

9. $x + 5y = 17$
$-4x + 3y = 24$

10. $6x + y = -39$
$3x + 2y = -15$

11. $2x + 5y = 11$
$4x + 3y = 1$

12. $3x - 3y = -6$
$-5x + 6y = 12$

13. $3x + 4y = 29$
$6x + 5y = 43$

14. $8x + 3y = 4$
$-7x + 5y = -34$

15. $8x + 3y = -7$
$7x + 2y = -3$

16. $4x + 7y = -80$
$3x + 5y = -58$

17. $12x - 3y = -3$
$6x + y = 1$

18. $-4x + 2y = 0$
$10x + 3y = 8$

Example 3

19. **NUMBER THEORY** Seven times a number plus three times another number equals negative one. The sum of the two numbers is negative three. What are the numbers?

20. **FOOTBALL** A field goal is 3 points and the extra point after a touchdown is 1 point. In a recent post-season, Adam Vinatieri of the Indianapolis Colts made a total of 21 field goals and extra point kicks for 49 points. Find the number of field goals and extra points that he made.

Use elimination to solve each system of equations.

21. $2.2x + 3y = 15.25$
$4.6x + 2.1y = 18.325$

22. $-0.4x + 0.25y = -2.175$
$2x + y = 7.5$

23. $\frac{1}{4}x + 4y = 2\frac{3}{4}$
$3x + \frac{1}{2}y = 9\frac{1}{4}$

24. $\frac{2}{5}x + 6y = 24\frac{1}{5}$
$3x + \frac{1}{2}y = 3\frac{1}{2}$

25. **CCSS MODELING** A staffing agency for in-home nurses and support staff places necessary personnel at locations on a daily basis. Each placed nurse works 240 minutes per day at a daily rate of $90. Each support staff employee works 360 minutes per day at a daily rate of $120.

   a. On a given day, 3000 total minutes are worked by the nurses and support staff that were placed. Write an equation that represents this relationship.

   b. On the same day, earnings for placed nurses and support staff totaled $1050. Write an equation that represents this relationship.

   c. Solve the system of equations, and interpret the solution in the context of the situation.

26. **GEOMETRY** The graphs of $x + 2y = 6$ and $2x + y = 9$ contain two of the sides of a triangle. A vertex of the triangle is at the intersection of the graphs.

   a. What are the coordinates of the vertex?

   b. Draw the graph of the two lines. Identify the vertex of the triangle.

   c. The line that forms the third side of the triangle is the line $x - y = -3$. Draw this line on the previous graph.

   d. Name the other two vertices of the triangle.

27 **ENTERTAINMENT** At an entertainment center, two groups of people bought batting tokens and miniature golf games, as shown in the table.

Group	Number of Batting Tokens	Number of Miniature Golf Games	Total Cost
A	16	3	$30
B	22	5	$43

    a. Define the variables, and write a system of linear equations from this situation.

    b. Solve the system of equations, and explain what the solution represents.

28. **TESTS** Mrs. Henderson discovered that she had accidentally reversed the digits of a test score and did not give a student 36 points. Mrs. Henderson told the student that the sum of the digits was 14 and agreed to give the student his correct score plus extra credit if he could determine his actual score. What was his correct score?

H.O.T. Problems   Use Higher-Order Thinking Skills

29. **REASONING** Explain how you could recognize a system of linear equations with infinitely many solutions.

30. **CCSS CRITIQUE** Jason and Daniela are solving a system of equations. Is either of them correct? Explain your reasoning.

Jason

$2r + 7t = 11$

$r - 9t = -7$

$2r + 7t = 11$

$(-)\ 2r - 18t = -14$

$25t = 25$

$t = 1$

$2r + 7t = 11$

$2r + 7(1) = 11$

$2r + 7 = 11$

$2r = 4$

$\dfrac{2r}{2} = \dfrac{4}{2}$

$r = 2$

The solution is (2, 1).

Daniela

$2r + 7t = 11$

$(-)\ r - 9t = -7$

$r = 18$

$2r + 7t = 11$

$2(18) + 7t = 11$

$36 + 7t = 11$

$7t = -25$

$\dfrac{7t}{7} = -\dfrac{25}{7}$

$t = -3.6$

The solution is (18, -3.6).

31. **OPEN ENDED** Write a system of equations that can be solved by multiplying one equation by -3 and then adding the two equations together.

32. **CHALLENGE** The solution of the system $4x + 5y = 2$ and $6x - 2y = b$ is $(3, a)$. Find the values of a and b. Discuss the steps that you used.

33. **WRITING IN MATH** Why is substitution sometimes more helpful than elimination, and vice versa?

34. What is the solution of this system of equations?

$2x - 3y = -9$
$-x + 3y = 6$

A $(3, 3)$ C $(-3, 1)$

B $(-3, 3)$ D $(1, -3)$

35. A buffet has one price for adults and another for children. The Taylor family has two adults and three children, and their bill was $40.50. The Wong family has three adults and one child. Their bill was $38. Which system of equations could be used to determine the price for an adult and for a child?

F $x + y = 40.50$ H $2x + 3y = 40.50$
  $x + y = 38$ $x + 3y = 38$

G $2x + 3y = 40.50$ J $2x + 2y = 40.50$
  $3x + y = 38$ $3x + y = 38$

36. SHORT RESPONSE A customer at the paint store has ordered 3 gallons of ivy green paint. Melissa mixes the paint in a ratio of 3 parts blue to one part yellow. How many quarts of blue paint does she use?

37. PROBABILITY The table shows the results of a number cube being rolled. What is the experimental probability of rolling a 3?

Outcome	Frequency
1	4
2	8
3	2
4	0
5	5
6	1

A $\dfrac{2}{3}$ B $\dfrac{1}{3}$ C 0.2   D 0.1

Spiral Review

Use elimination to solve each system of equations. (Lesson 6-3)

38. $f + g = -3$
    $f - g = 1$

39. $6g + h = -7$
    $6g + 3h = -9$

40. $5j + 3k = -9$
    $3j + 3k = -3$

41. $2x - 4z = 6$
    $x - 4z = -3$

42. $-5c - 3v = 9$
    $5c + 2v = -6$

43. $4b - 6n = -36$
    $3b - 6n = -36$

44. JOBS Brandy and Adriana work at an after-school child care center. Together they cared for 32 children this week. Brandy cared for 0.6 times as many children as Adriana. How many children did each girl care for? (Lesson 6-2)

Solve each inequality. Then graph the solution set. (Lesson 5-5)

45. $|m - 5| \le 8$ **46.** $|q + 11| < 5$ **47.** $|2w + 9| > 11$ **48.** $|2r + 1| \ge 9$

Skills Review

Translate each sentence into a formula.

49. The area A of a triangle equals one half times the base b times the height h.

50. The circumference C of a circle equals the product of 2, π, and the radius r.

51. The volume V of a rectangular box is the length ℓ times the width w multiplied by the height h.

52. The volume of a cylinder V is the same as the product of π and the radius r to the second power multiplied by the height h.

53. The area of a circle A equals the product of π and the radius r squared.

54. Acceleration A equals the increase in speed s divided by time t in seconds.

Use the graph to determine whether each system is *consistent* or *inconsistent* and if it is *independent* or *dependent*. (Lesson 6-1)

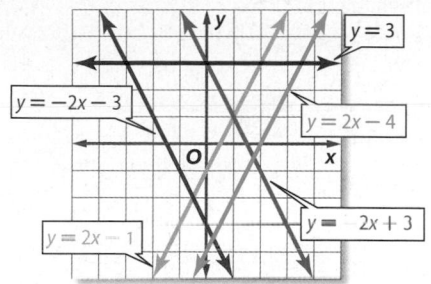

1. $y = 2x - 1$
   $y = -2x + 3$

2. $y = -2x + 3$
   $y = -2x - 3$

Graph each system and determine the number of solutions that it has. If it has one solution, name it. (Lesson 6-1)

3. $y = 2x - 3$
   $y = x + 4$

4. $x + y = 6$
   $x - y = 4$

5. $x + y = 8$
   $3x + 3y = 24$

6. $x - 4y = -6$
   $y = -1$

7. $3x + 2y = 12$
   $3x + 2y = 6$

8. $2x + y = -4$
   $5x + 3y = -6$

Use substitution to solve each system of equations.
(Lesson 6-2)

9. $y = x + 4$
   $2x + y = 16$

10. $y = -2x - 3$
    $x + y = 9$

11. $x + y = 6$
    $x - y = 8$

12. $y = -4x$
    $6x - y = 30$

13. **FOOD** The cost of two meals at a restaurant is shown in the table below. (Lesson 6-2)

Meal	Total Cost
3 tacos, 2 burritos	$7.40
4 tacos, 1 burrito	$6.45

a. Define variables to represent the cost of a taco and the cost of a burrito.

b. Write a system of equations to find the cost of a single taco and a single burrito.

c. Solve the systems of equations, and explain what the solution means.

d. How much would a customer pay for 2 tacos and 2 burritos?

14. **AMUSEMENT PARKS** The cost of two groups going to an amusement park is shown in the table. (Lesson 6-3)

Group	Total Cost
4 adults, 2 children	$184
4 adults, 3 children	$200

a. Define variables to represent the cost of an adult ticket and the cost of a child ticket.

b. Write a system of equations to find the cost of an adult ticket and a child ticket.

c. Solve the system of equations, and explain what the solution means.

d. How much will a group of 3 adults and 5 children be charged for admission?

15. **MULTIPLE CHOICE** Angelina spent $16 for 12 pieces of candy to take to a meeting. Each chocolate bar costs $2, and each lollipop costs $1. Determine how many of each she bought. (Lesson 6-3)

A 6 chocolate bars, 6 lollipops

B 4 chocolate bars, 8 lollipops

C 7 chocolate bars, 5 lollipops

D 3 chocolate bars, 9 lollipops

Use elimination to solve each system of equations.
(Lessons 6-3 and 6-4)

16. $x + y = 9$
    $x - y = -3$

17. $x + 3y = 11$
    $x + 7y = 19$

18. $9x - 24y = -6$
    $3x + 4y = 10$

19. $-5x + 2y = -11$
    $5x - 7y = 1$

20. **MULTIPLE CHOICE** The Blue Mountain High School Drama Club is selling tickets to their spring musical. Adult tickets are $4 and student tickets are $1. A total of 285 tickets are sold for $765. How many of each type of ticket are sold? (Lesson 6-4)

F 145 adult, 140 student

G 120 adult, 165 student

H 180 adult, 105 student

J 160 adult, 125 student

Applying Systems of Linear Equations

··Then

- You solved systems of equations by using substitution and elimination.

··Now

1. Determine the best method for solving systems of equations.

2. Apply systems of equations.

··Why?

- In speed skating, competitors race two at a time on a double track. Indoor speed skating rinks have two track sizes for race events: an official track and a short track.

Speed Skating Tracks	
official track	x
short track	y

The total length of the two tracks is 511 meters. The official track is 44 meters less than four times the short track. The total length is represented by $x + y = 511$. The length of the official track is represented by $x = 4y - 44$.

You can solve the system of equations to find the length of each track.

Common Core State Standards

Content Standards
A.REI.6 Solve systems of linear equations exactly and approximately (e.g., with graphs), focusing on pairs of linear equations in two variables.

Mathematical Practices
2 Reason abstractly and quantitatively.
4 Model with mathematics.

1 **Determine the Best Method** You have learned five methods for solving systems of linear equations. The table summarizes the methods and the types of systems for which each method works best.

ConceptSummary Solving Systems of Equations

Method	The Best Time to Use
Graphing	To estimate solutions, since graphing usually does not give an exact solution.
Substitution	If one of the variables in either equation has a coefficient of 1 or −1.
Elimination Using Addition	If one of the variables has opposite coefficients in the two equations.
Elimination Using Subtraction	If one of the variables has the same coefficient in the two equations.
Elimination Using Multiplication	If none of the coefficients are 1 or −1 and neither of the variables can be eliminated by simply adding or subtracting the equations.

Substitution and elimination are algebraic methods for solving systems of equations. An algebraic method is best for an exact solution. Graphing, with or without technology, is a good way to estimate a solution.

A system of equations can be solved using each method. To determine the best approach, analyze the coefficients of each term in each equation.

Example 1 Choose the Best Method

Determine the best method to solve the system of equations. Then solve the system.

$$4x - 4y = 8$$
$$-8x + y = 19$$

Understand To determine the best method to solve the system of equations, look closely at the coefficients of each term.

Plan Neither the coefficients of *x* nor *y* are the same or additive inverses, so you cannot add or subtract to eliminate a variable. Since the coefficient of *y* in the second equation is 1, you can use substitution.

Solve First, solve the second equation for *y*.

$$-8x + y = 19 \qquad \text{Second equation}$$
$$-8x + y + 8x = 19 + 8x \qquad \text{Add } 8x \text{ to each side.}$$
$$y = 19 + 8x \qquad \text{Simplify.}$$

Next, substitute $19 + 8x$ for *y* in the first equation.

$$4x - 4y = 8 \qquad \text{First equation}$$
$$4x - 4(19 + 8x) = 8 \qquad \text{Substitution}$$
$$4x - 76 - 32x = 8 \qquad \text{Distributive Property}$$
$$-28x - 76 = 8 \qquad \text{Simplify.}$$
$$-28x - 76 + 76 = 8 + 76 \qquad \text{Add 76 to each side.}$$
$$-28x = 84 \qquad \text{Simplify.}$$
$$\frac{-28x}{-28} = \frac{84}{-28} \qquad \text{Divide each side by } -28.$$
$$x = -3 \qquad \text{Simplify.}$$

Last, substitute -3 for *x* in the second equation.

$$-8x + y = 19 \qquad \text{Second equation}$$
$$-8(-3) + y = 19 \qquad x = -3$$
$$y = -5 \qquad \text{Simplify.}$$

The solution of the system of equations is $(-3, -5)$.

Check Use a graphing calculator to check your solution. If your algebraic solution is correct, then the graphs will intersect at $(-3, -5)$.

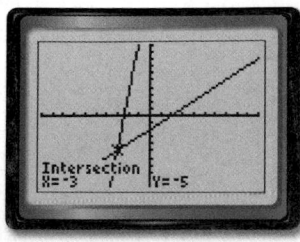

[−10, 10] scl: 1 [−10, 10] scl: 1

GuidedPractice

1A. $5x + 7y = 2$
$\quad -2x + 7y = 9$

1B. $3x - 4y = -10$
$\quad 5x + 8y = -2$

1C. $x - y = 9$
$\quad 7x + y = 7$

1D. $5x - y = 17$
$\quad 3x + 2y = 5$

2 Apply Systems of Linear Equations

When applying systems of linear equations to problems, it is important to analyze each solution in the context of the situation.

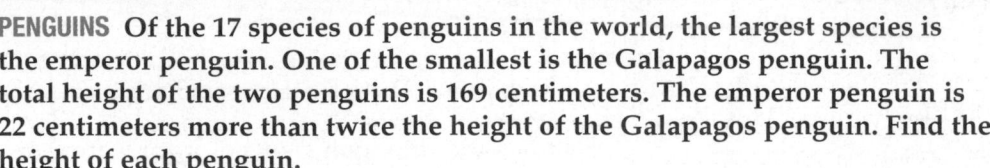

Real-World Example 2 Apply Systems of Linear Equations

PENGUINS Of the 17 species of penguins in the world, the largest species is the emperor penguin. One of the smallest is the Galapagos penguin. The total height of the two penguins is 169 centimeters. The emperor penguin is 22 centimeters more than twice the height of the Galapagos penguin. Find the height of each penguin.

The total height of the two species can be represented by $p + g = 169$, where p represents the height of the emperor penguin and g the height of the Galapagos penguin. Next write an equation to represent the height of the emperor penguin.

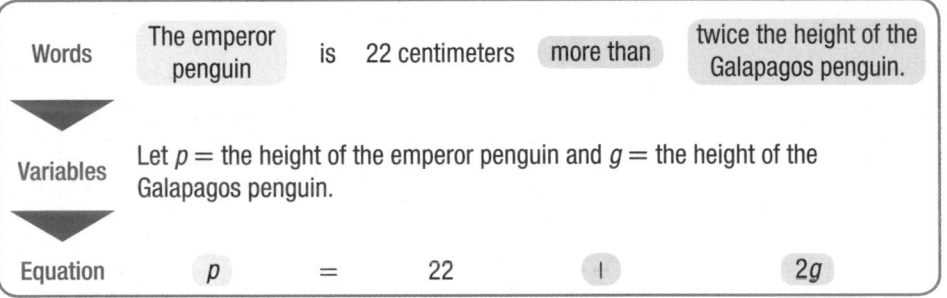

| Words | The emperor penguin | is | 22 centimeters | more than | twice the height of the Galapagos penguin. |

Variables Let $p =$ the height of the emperor penguin and $g =$ the height of the Galapagos penguin.

| Equation | p | $=$ | 22 | $+$ | $2g$ |

First rewrite the second equation.

$$p = 22 + 2g \qquad \text{Second equation}$$
$$p - 2g = 22 \qquad \text{Subtract } 2g \text{ from each side.}$$

You can use elimination by subtraction to solve this system of equations.

$$p + g = 169 \qquad \text{First equation}$$
$$(-)\, p - 2g = 22 \qquad \text{Subtract the second equation.}$$
$$\overline{ 3g = 147} \qquad \text{Eliminate } p.$$
$$\frac{3g}{3} = \frac{147}{3} \qquad \text{Divide each side by 3.}$$
$$g = 49 \qquad \text{Simplify.}$$

Next substitute 49 for g in one of the equations.

$$p = 22 + 2g \qquad \text{Second equation}$$
$$= 22 + 2(49) \qquad g = 49$$
$$= 120 \qquad \text{Simplify.}$$

The height of the emperor penguin is 120 centimeters, and the height of the Galapagos penguin is 49 centimeters.

Does the solution make sense in the context of the problem?

Check by verifying the given information. The penguins' heights added together would be $120 + 49$ or 169 centimeters and $22 + 2(49)$ is 120 centimeters.

▸ **Guided**Practice

2. **VOLUNTEERING** Jared has volunteered 50 hours and plans to volunteer 3 hours in each coming week. Clementine is a new volunteer who plans to volunteer 5 hours each week. Write and solve a system of equations to find how long it will be before they will have volunteered the same number of hours.

Joel Simon/Digital Vision/Getty Images

Check Your Understanding
= Step-by-Step Solutions begin on page R13.

Example 1 Determine the best method to solve each system of equations. Then solve the system.

1. $2x + 3y = -11$
$-8x - 5y = 9$

2. $3x + 4y = 11$
$2x + y = -1$

3. $3x - 4y = -5$
$-3x + 2y = 3$

4. $3x + 7y = 4$
$5x - 7y = -12$

Example 2 5. **SHOPPING** At a sale, Salazar bought 4 T-shirts and 3 pairs of jeans for $181. At the same store, Jenna bought 1 T-shirt and 2 pairs of jeans for $94. The T-shirts were all the same price, and the jeans were all the same price.

    a. Write a system of equations that can be used to represent this situation.

    b. Determine the best method to solve the system of equations.

    c. Solve the system.

Practice and Problem Solving
Extra Practice is on page R6.

Example 1 Determine the best method to solve each system of equations. Then solve the system.

6. $-3x + y = -3$
$4x + 2y - 14$

7. $2x + 6y = -8$
$x - 3y = 8$

8. $3x - 4y = -5$
$-3x - 6y = -5$

9. $5x + 8y = 1$
$-2x + 8y = -6$

10. $y + 4x = 3$
$y = -4x - 1$

(11) $-5x + 4y = 7$
$-5x - 3y = 14$

Example 2 12. **FINANCIAL LITERACY** For a Future Teachers of America fundraiser, Denzell sold food as shown in the table. He sold 11 more subs than pizzas and earned a total of $233. Write and solve a system of equations to represent this situation. Then describe what the solution means.

Item	Selling Price
pizza	$5.00
sub	$3.00

13. **DVDs** Manuela has a total of 40 DVDs of movies and television shows. The number of movies is 4 less than 3 times the number of television shows. Write and solve a system of equations to find the numbers of movies and television shows that she has on DVD.

14. **CAVES** The Caverns of Sonora have two different tours: the Crystal Palace tour and the Horseshoe Lake tour. The total length of both tours is 3.25 miles. The Crystal Palace tour is a half-mile less than twice the distance of the Horseshoe Lake tour. Determine the length of each tour.

15. **CCSS MODELING** The *break-even point* is the point at which income equals expenses. Ridgemont High School is paying $13,200 for the writing and research of their yearbook plus a printing fee of $25 per book. If they sell the books for $40 each, how many will they have to sell to break even? Explain.

16. **PAINTBALL** Clara and her friends are planning a trip to a paintball park. Find the cost of lunch and the cost of each paintball. What would be the cost for 400 paintballs and lunch?

Paintball in the Park

- $25 for 500 paintballs
- $16 for 200 paintballs

Lunch is included

17 RECYCLING Mara and Ling each recycled aluminum cans and newspaper, as shown in the table. Mara earned $3.77, and Ling earned $4.65.

Materials	Pounds Recycled	
	Mara	Ling
aluminum cans	9	9
newspaper	26	114

a. Define variables and write a system of linear equations from this situation.

b. What was the price per pound of aluminum? Determine the reasonableness of your solution.

18. BOOKS The library is having a book sale. Hardcover books sell for $4 each, and paperback books are $2 each. If Connie spends $26 for 8 books, how many hardcover books did she buy?

19. MUSIC An online music club offers individual songs for one price or entire albums for another. Kendrick pays $14.90 to download 5 individual songs and 1 album. Geoffrey pays $21.75 to download 3 individual songs and 2 albums.

a. How much does the music club charge to download a song?

b. How much does the music club charge to download an entire album?

20. CANOEING Malik canoed against the current for 2 hours and then with the current for 1 hour before resting. Julio traveled against the current for 2.5 hours and then with the current for 1.5 hours before resting. If they traveled a total of 9.5 miles against the current, 20.5 miles with the current, and the current is 3 miles per hour, how fast do Malik and Julio travel in still water?

H.O.T. Problems   Use Higher-Order Thinking Skills

21. OPEN ENDED Formulate a system of equations that represents a situation in your school. Describe the method that you would use to solve the system. Then solve the system and explain what the solution means.

22. **CCSS** REASONING In a system of equations, x represents the time spent riding a bike, and y represents the distance traveled. You determine the solution to be $(-1, 7)$. Use this problem to discuss the importance of analyzing solutions in the context of real-world problems.

23. CHALLENGE Solve the following system of equations by using three different methods. Show your work.

$$4x + y = 13$$
$$6x - y = 7$$

24. WRITE A QUESTION A classmate says that elimination is the best way to solve a system of equations. Write a question to challenge his conjecture.

25. WHICH ONE DOESN'T BELONG? Which system is different? Explain.

$x - y = 3$ $x + \frac{1}{2}y = 1$	$-x + y = 0$ $5x = 2y$	$y = x - 4$ $y = \frac{2}{x}$	$y = x + 1$ $y = 3x$

26. WRITING IN MATH How do you know what method to use when solving a system of equations?

27. If $5x + 3y = 12$ and $4x - 5y = 17$, what is y?

    A -1 **B** 3 **C** $(-1, 3)$ **D** $(3, -1)$

28. STATISTICS The scatter plot shows the number of hay bales used on the Bostwick farm during the last year.

Hay Bales Used

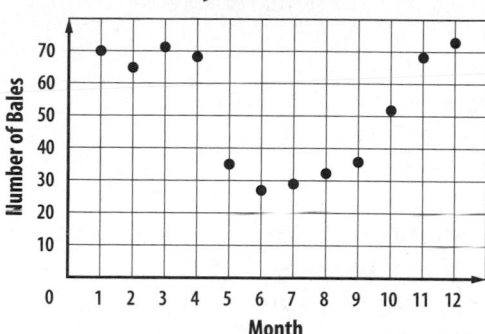

Which is an invalid conclusion?

  F The Bostwicks used less hay in the summer than they did in the winter.

  G The Bostwicks used about 629 bales of hay during the year.

  H On average, the Bostwicks used about 52 bales each month.

  J The Bostwicks used the most hay in February.

29. SHORT RESPONSE At noon, Cesar cast a shadow 0.15 foot long. Next to him a streetlight cast a shadow 0.25 foot long. If Cesar is 6 feet tall, how tall is the streetlight?

30. The graph shows the solution to which of the following systems of equations?

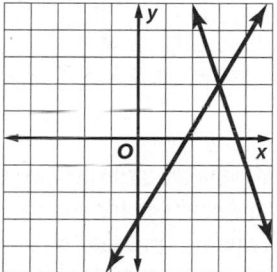

  A $y = -3x + 11$
     $3y = 5x - 9$

  B $y = 5x - 15$
     $2y = x + 7$

  C $y = -3x + 11$
     $2y = 4x - 5$

  D $y = 5x - 15$
     $3y = 2x + 18$

Spiral Review

Use elimination to solve each system of equations. (Lesson 6-4)

31. $x + y = 3$
    $3x - 4y = -12$

32. $-4x + 2y = 0$
    $2x - 3y = 16$

33. $4x + 2y = 10$
    $5x - 3y = 7$

34. TRAVELING A youth group is traveling in two vans to visit an aquarium. The number of people in each van and the cost of admission for that van are shown. What are the adult and student prices? (Lesson 6-3)

Van	Number of Adults	Number of Students	Total Cost
A	2	5	$77
B	2	7	$95

Graph each inequality. (Lesson 5-6)

35. $y < 4$ **36.** $x \geq 3$ **37.** $7x + 12y > 0$ **38.** $y - 3x \leq 4$

Skills Review

Find each sum or difference.

39. $(-3.81) + (-8.5)$ **40.** $12.625 + (-5.23)$ **41.** $21.65 + (-15.05)$

42. $(-4.27) + 1.77$ **43.** $(-78.94) - 14.25$ **44.** $(-97.623) - (-25.14)$

Algebra Lab
Using Matrices to Solve Systems of Equations

A **matrix** is a rectangular arrangement of numbers, called **elements**, in rows and columns enclosed in brackets. Usually named using an uppercase letter, a matrix can be described by its **dimensions** or by the number of rows and columns in the matrix. A matrix with m rows and n columns is an $m \times n$ matrix (read "m by n").

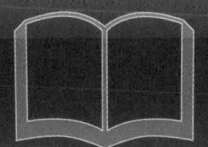
CCSS Common Core State Standards
Content Standards
A.REI.6 Solve systems of linear equations exactly and approximately (e.g., with graphs), focusing on pairs of linear equations in two variables.

$$A = \begin{bmatrix} 7 & -9 & 5 & 3 \\ -1 & 3 & -3 & 6 \\ 0 & -4 & 8 & 2 \end{bmatrix}$$

A is a 3 × 4 matrix.

3 rows

The element 2 is in Row 3, Column 4.

4 columns

You can use an augmented matrix to solve a system of equations. An **augmented matrix** consists of the coefficients and the constant terms of a system of equations. Make sure that the coefficients of the x-terms are listed in one column, the coefficients of the y-terms are in another column, and the constant terms are in a third column. The coefficients and constant terms are usually separated by a dashed line.

Linear System

$$x - 3y = 8$$
$$-9x + 2y = -4$$

Augmented Matrix

$$\begin{bmatrix} 1 & -3 & | & 8 \\ -9 & 2 & | & -4 \end{bmatrix}$$

Activity 1 Write an Augmented Matrix

Write an augmented matrix for each system of equations.

a. $-2x + 7y = 11$
$6x - 4y = 2$

Place the coefficients of the equations and the constant terms into a matrix.

$-2x + 7y = 11$
$6x - 4y = 2$ $\longrightarrow$ $\begin{bmatrix} -2 & 7 & | & 11 \\ 6 & -4 & | & 2 \end{bmatrix}$

b. $x - 2y = 5$
$y = -4$

$x - 2y = 5$
$y = -4$ $\longrightarrow$ $\begin{bmatrix} 1 & -2 & | & 5 \\ 0 & 1 & | & -4 \end{bmatrix}$

You can solve a system of equations by using an augmented matrix. By performing row operations, you can change the form of the matrix. The operations are the same as the ones used when working with equations.

KeyConcept Elementary Row Operations

The following operations can be performed on an augmented matrix.

• Interchange any two rows.

• Multiply all entries in a row by a nonzero constant.

• Replace one row with the sum of that row and a multiple of another row.

Row operations produce a matrix equivalent to the original system. **Row reduction** is the process of performing elementary row operations on an augmented matrix to solve a system.

The goal is to get the coefficients portion of the matrix to have the form $\begin{bmatrix} 1 & 0 \\ 0 & 1 \end{bmatrix}$, which is called the **identity matrix**. The first row will give you the solution for x, because the coefficient of y is 0. The second row will give you the solution for y, because the coefficient of x is 0.

Activity 2 Use Row Operations to Solve a System

Use an augmented matrix to solve the system of equations.
$-5x + 3y = 6$
$x - y = 4$

Step 1 Write the augmented matrix: $\begin{bmatrix} -5 & 3 & | & 6 \\ 1 & -1 & | & 4 \end{bmatrix}$.

Step 2 Notice that the first element in the second row is 1. Interchange the rows so 1 can be in the upper left-hand corner.

$\begin{bmatrix} -5 & 3 & | & 6 \\ 1 & -1 & | & 4 \end{bmatrix}$ Interchange R_1 and R_2. → $\begin{bmatrix} 1 & -1 & | & 4 \\ -5 & 3 & | & 6 \end{bmatrix}$

Step 3 To make the first element in the second row a 0, multiply the first row by 5 and add the result to row 2.

$\begin{bmatrix} 1 & -1 & | & 4 \\ -5 & 3 & | & 6 \end{bmatrix}$ $5R_1 + R_2$ → $\begin{bmatrix} 1 & -1 & | & 4 \\ 0 & -2 & | & 26 \end{bmatrix}$ $1(5) + (-5) = 0; -1(5) + 3 = -2;$ $4(5) + 6 = 26$

Step 4 To make the second element in the second row a 1, multiply the second row by $\frac{1}{2}$.

$\begin{bmatrix} 1 & -1 & | & 4 \\ 0 & -2 & | & 26 \end{bmatrix}$ $-\frac{1}{2}R_2$ → $\begin{bmatrix} 1 & -1 & | & 4 \\ 0 & 1 & | & -13 \end{bmatrix}$ $0\left(-\frac{1}{2}\right) = 0; -2\left(-\frac{1}{2}\right) = 1;$ $26\left(-\frac{1}{2}\right) = -13$

Step 5 To make the second element in the second row a 0, add the rows together.

$\begin{bmatrix} 1 & -1 & | & 4 \\ 0 & 1 & | & -13 \end{bmatrix}$ $R_2 + R_1$ → $\begin{bmatrix} 1 & 0 & | & -9 \\ 0 & 1 & | & -13 \end{bmatrix}$ $1 + 0 = 1; -1 + 1 = 0;$ $4 + (-13) = -9$

The solution is $(-9, -13)$.

Model and Analyze
Write an augmented matrix for each system of equations. Then solve the system.

1. $x + y = -3$
   $x - y = 1$

2. $x - y = -2$
   $2x + 2y = 12$

3. $3x - 4y = -27$
   $x + 2y = 11$

4. $x + 4y = -6$
   $2x - 5y = 1$

5. $x - 3y = -2$
   $4x + y = 31$

6. $x + 2y = 3$
   $-3x + 3y = 27$

Systems of Inequalities

:: Then

- You graphed and solved linear inequalities.

:: Now

1. Solve systems of linear inequalities by graphing.
2. Apply systems of linear inequalities.

:: Why?

- Jacui is beginning an exercise program that involves an intense cardiovascular workout. Her trainer recommends that for a person her age, her heart rate should stay within the following range as she exercises.

  - It should be higher than 102 beats per minute.
  - It should not exceed 174 beats per minute.

  The graph shows the maximum and minimum target heart rate for people ages 0 to 30 as they exercise. If the preferred range is in light green, how old do you think Jacui is?

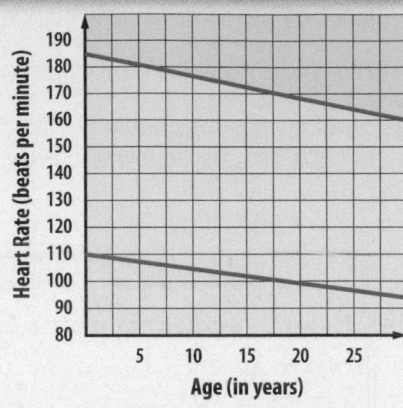

NewVocabulary
system of inequalities

Common Core State Standards

Content Standards
A.REI.12 Graph the solutions to a linear inequality in two variables as a halfplane (excluding the boundary in the case of a strict inequality), and graph the solution set to a system of linear inequalities in two variables as the intersection of the corresponding half-planes.

Mathematical Practices
1 Make sense of problems and persevere in solving them.
6 Attend to precision.

1 Systems of Inequalities

The graph above is a graph of two inequalities. A set of two or more inequalities with the same variables is called a **system of inequalities**.

The solution of a system of inequalities with two variables is the set of ordered pairs that satisfy all of the inequalities in the system. The solution set is represented by the overlap, or intersection, of the graphs of the inequalities.

Example 1 Solve by Graphing

Solve the system of inequalities by graphing.

$y > -2x + 1$
$y \leq x + 3$

The graph of $y = -2x + 1$ is dashed and is not included in the graph of the solution. The graph of $y = x + 3$ is solid and is included in the graph of the solution.

The solution of the system is the set of ordered pairs in the intersection of the graphs of $y > -2x + 1$ and $y \leq x + 3$. This region is shaded in green.

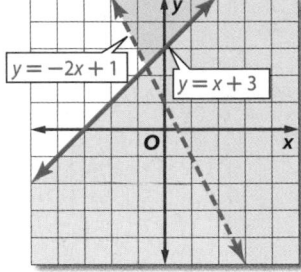

When graphing more than one region, it is helpful to use two different colored pencils or two different patterns for each region. This will make it easier to see where the regions intersect and find possible solutions.

Guided Practice

1A. $y \leq 3$
$x + y \geq 1$

1B. $2x + y \geq 2$
$2x + y < 4$

1C. $y \geq -4$
$3x + y \leq 2$

1D. $x + y > 2$
$-4x + 2y < 8$

Sometimes the regions never intersect. When this happens, there is no solution because there are no points in common.

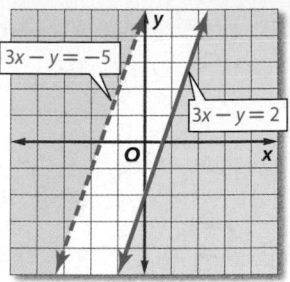

StudyTip

Parallel Boundaries
A system of equations represented by parallel lines does not have a solution. However, a system of inequalities with parallel boundaries can have a solution. For example:

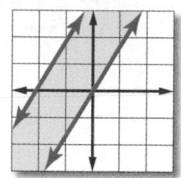

Example 2 No Solution

Solve the system of inequalities by graphing.

$3x - y \geq 2$
$3x - y < -5$

The graphs of $3x - y = 2$ and $3x - y = -5$ are parallel lines. The two regions do not intersect at any point, so the system has no solution.

GuidedPractice

2A. $y > 3$
　　$y < 1$

2B. $x + 6y \leq 2$
　　$y \geq -\frac{1}{6}x + 7$

---

2 **Apply Systems of Inequalities** When using a system of inequalities to describe constraints on the possible combinations in a real-world problem, sometimes only whole-number solutions will make sense.

Real-World Example 3 Whole-Number Solutions

ELECTIONS Monifa is running for student council. The election rules say that for the election to be valid, at least 80% of the 900 students must vote. Monifa knows that she needs more than 330 votes to win.

a. Define the variables, and write a system of inequalities to represent this situation. Then graph the system.

Let r = the number of votes required by the election rules; 80% of 900 students is 720 students. So $r \geq 720$.

Let v = the number of votes that Monifa needs to win. So $v > 330$.

The system of inequalities is $r \geq 720$ and $v > 330$.

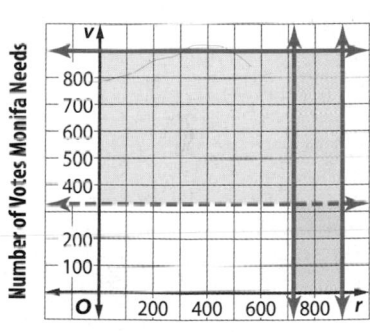

Number of Votes Required

b. Name one viable option.

Only whole-number solutions make sense in this problem. One possible solution is (800, 400); 800 students voted and Monifa received 400 votes.

Real-WorldLink

Student government might be a good activity for you if you like to bring about change, plan events, and work with others.

GuidedPractice

3. FUNDRAISING The Theater Club is selling shirts. They have only enough supplies to print 120 shirts. They will sell sweatshirts for $22 and T-shirts for $15, with a goal of at least $2000 in sales.

A. Define the variables, and write a system of inequalities to represent this situation.

B. Then graph the system.

C. Name one possible solution.

D. Is (45, 30) a solution? Explain.

Examples 1–2 Solve each system of inequalities by graphing.

1. $x \geq 4$
$y \leq x - 3$

2. $y > -2$
$y \leq x + 9$

3. $y < 3x + 8$
$y \geq 4x$

4. $3x - y \geq -1$
$2x + y \geq 5$

5. $y \leq 2x - 7$
$y \geq 2x + 7$

6. $y > -2x + 5$
$y \geq -2x + 10$

7. $2x + y \leq 5$
$2x + y \leq 7$

8. $5x - y < -2$
$5x - y > 6$

Example 3

9. AUTO RACING At a racecar driving school there are safety requirements.

   a. Define the variables, and write a system of inequalities to represent the height and weight requirements in this situation. Then graph the system.

   b. Name one possible solution.

   c. Is (50, 180) a solution? Explain.

UR FAST DRIVING SCHOOL
RULES TO QUALIFY
18 years of age or older
Good physical condition
Under 6 ft 7 in. tall
Under 295 lb

Practice and Problem Solving   Extra Practice is on page R6.

Examples 1–2 Solve each system of inequalities by graphing.

10. $y < 6$
$y > x + 3$

11 $y \geq 0$
$y \leq x - 5$

12. $y \leq x + 10$
$y > 6x + 2$

13. $y < 5x - 2$
$y > -6x + 2$

14. $2x - y \leq 6$
$x - y \geq -1$

15. $3x - y > -5$
$5x - y < 9$

16. $y \geq x + 10$
$y \leq x - 3$

17. $y < 5x - 5$
$y > 5x + 9$

18. $y \geq 3x - 5$
$3x - y > -4$

19. $4x + y > -1$
$y < -4x + 1$

20. $3x - y \geq -2$
$y < 3x + 4$

21. $y > 2x - 3$
$2x - y \geq 1$

22. $5x - y < -6$
$3x - y \geq 4$

23. $x - y \leq 8$
$y < 3x$

24. $4x + y < -2$
$y > -4x$

Example 3

25. ICE RINKS Ice resurfacers are used for rinks of at least 1000 square feet and up to 17,000 square feet. The price ranges from as little as $10,000 to as much as $150,000.

   a. Define the variables, and write a system of inequalities to represent this situation. Then graph the system.

   b. Name one possible solution.

   c. Is (15,000, 30,000) a solution? Explain.

26. CCSS MODELING Josefina works between 10 and 30 hours per week at a pizzeria. She earns $6.50 an hour, but can earn tips when she delivers pizzas.

   a. Write a system of inequalities to represent the dollars d she could earn for working h hours in a week.

   b. Graph this system.

   c. If Josefina received $17.50 in tips and earned a total of $180 for the week, how many hours did she work?

Solve each system of inequalities by graphing.

27. $x + y \geq 1$
$x + y \leq 2$

28. $3x - y < -2$
$3x - y < 1$

29. $2x - y \leq -11$
$3x - y \geq 12$

30. $y < 4x + 13$
$4x - y \geq 1$

31. $4x - y < -3$
$y \geq 4x - 6$

32. $y \leq 2x + 7$
$y < 2x - 3$

33. $y > -12x + 1$
$y \leq 9x + 2$

34. $2y \geq x$
$x - 3y > -6$

35. $x - 5y > -15$
$5y \geq x - 5$

36. CLASS PROJECT An economics class formed a company to sell school supplies. They would like to sell at least 20 notebooks and 50 pens per week, with a goal of earning at least $60 per week.

School Supplies

Notebooks........$2.50

Pens $1.25

a. Define the variables, and write a system of inequalities to represent this situation.

b. Graph the system.

c. Name one possible solution.

37. FINANCIAL LITERACY Opal makes $15 per hour working for a photographer. She also coaches a competitive soccer team for $10 per hour. Opal needs to earn at least $90 per week, but she does not want to work more than 20 hours per week.

a. Define the variables, and write a system of inequalities to represent this situation.

b. Graph this system.

c. Give two possible solutions to describe how Opal can meet her goals.

d. Is (2, 2) a solution? Explain.

H.O.T. Problems Use Higher-Order Thinking Skills

38. CHALLENGE Create a system of inequalities equivalent to $|x| \leq 4$.

39. REASONING State whether the following statement is *sometimes*, *always*, or *never* true. Explain your answer with an example or counterexample.

Systems of inequalities with parallel boundaries have no solutions.

40. REASONING Describe the graph of the solution of this system without graphing.
$6x - 3y \leq -5$
$6x - 3y \geq -5$

41. OPEN ENDED One inequality in a system is $3x - y > 4$. Write a second inequality so that the system will have no solution.

42. CCSS PRECISION Graph the system of inequalities. Estimate the area of the solution.
$y \geq 1$
$y \leq x + 4$
$y \leq -x + 4$

43. WRITING IN MATH Refer to the beginning of the lesson. Explain what each colored region of the graph represents. Explain how shading in various colors can help to clearly show the solution set of a system of inequalities.

44. EXTENDED RESPONSE To apply for a scholarship, you must have a minimum of 20 hours of community service and a grade-point average of at least 3.75. Another scholarship requires at least 40 hours of community service and a minimum grade-point average of 3.0.

   a. Write a system of inequalities to represent the credentials you must have to apply for both scholarships.

   b. Graph the system of inequalities.

   c. If you are eligible for both scholarships, give one possible solution.

45. GEOMETRY What is the measure of $\angle 1$?

25°   1   62°

   A 83°   **C** 90°

   B 87°   **D** 93°

46. GEOMETRY What is the volume of the triangular prism?

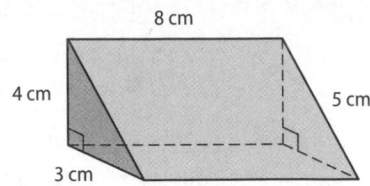

8 cm   4 cm   5 cm   3 cm

   F 120 cm³   **H** 48 cm³

   G 96 cm³   **J** 30 cm³

47. Ten pounds of fresh tomatoes make about 15 cups of cooked tomatoes. How many cups of cooked tomatoes does one pound of fresh tomatoes make?

   A $1\frac{1}{2}$ cups

   B 3 cups

   C 4 cups

   D 5 cups

Spiral Review

48. CHEMISTRY Orion Labs needs to make 500 gallons of 34% acid solution. The only solutions available are a 25% acid solution and a 50% acid solution. Write and solve a system of equations to find the number of gallons of each solution that should be mixed to make the 34% solution. (Lesson 6-5)

Use elimination to solve each system of equations. (Lesson 6-4)

49. $x + y = 7$
   $2x + y = 11$

50. $a - b = 9$
   $7a + b = 7$

51. $q + 4r = -8$
   $3q + 2r = 6$

52. ENTERTAINMENT A group of 11 adults and children bought tickets for the baseball game. If the total cost was $156, how many of each type of ticket did they buy? (Lesson 6-4)

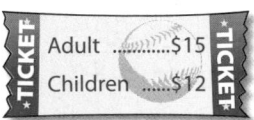

Adult $15
Children $12

Graph each inequality. (Lesson 5-6)

53. $4x - 2 \geq 2y$

54. $9x - 3y < 0$

55. $2y \leq -4x - 6$

Skills Review

Evaluate each expression.

56. 3^3

57. 2^4

58. $(-4)^3$

Graphing Technology Lab
Systems of Inequalities

You can use TI-Nspire technology to explore systems of inequalities. To prepare your calculator, add a new **Graphs** page from the Home screen.

CCSS **Common Core State Standards**
Content Standards
A.REI.12 Graph the solutions to a linear inequality in two variables as a half-plane (excluding the boundary in the case of a strict inequality), and graph the solution set to a system of linear inequalities in two variables as the intersection of the corresponding half-planes.

Activity   Graph Systems of Inequalities

Mr. Jackson owns a car washing and detailing business. It takes 20 minutes to wash a car and 60 minutes to detail a car. He works at most 8 hours per day and does at most 4 details per day. Write a system of linear inequalities to represent this situation.

First, write a linear inequality that represents the time it takes for car washing and car detailing. Let x represent the number of car washes, and let y represent the number of car details. Then $20x + 60y \leq 480$.

To graph this using a graphing calculator, solve for y.

$20x + 60y \leq 480$	Original inequality
$60y \leq -20x + 480$	Subtract 20x from each side and simplify.
$y \leq -\frac{1}{3}x + 8$	Divide each side by 60 and simplify.

Mr. Jackson does at most 4 details per day. This means that $y \leq 4$.

Step 1 Adjust the viewing window and then graph $y \leq 4$. Use the **Window Settings** option from the **Window/Zoom** menu to adjust the window to -4 to 30 for x and -2 to 10 for y. Keep the scales as **Auto**. Then enter **del** < 4 **enter**.

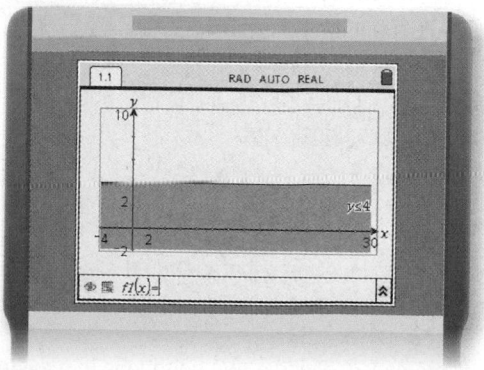

Step 2 Graph $y \leq -\frac{1}{3}x + 8$. Press **tab del** $\leq$ and then enter $-\frac{1}{3}x + 8$.

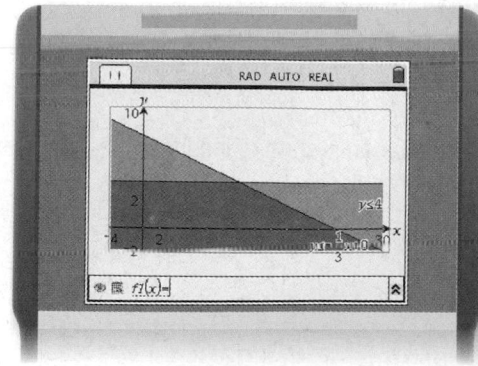

The darkest shaded region of the graph represents the solutions.

Analyze the Results

1. If Mr. Jackson charges $75 for each car he details and $25 for each car wash, what is the maximum amount of money he could earn in one day?

2. What is the greatest number of car washes that Mr. Jackson could do in a day? Explain your reasoning.

Study Guide

KeyConcepts

Systems of Equations (Lessons 6-1 through 6-5)

- A system with a graph of two intersecting lines has one solution and is *consistent and independent*.

- Graphing a system of equations can only provide approximate solutions. For exact solutions, you must use algebraic methods.

- In the substitution method, one equation is solved for a variable and the expression substituted into the second equation to find the value of another variable.

- In the elimination method, one variable is eliminated by adding or subtracting the equations.

- Sometimes multiplying one or both equations by a constant makes it easier to use the elimination method.

- The best method for solving a system of equations depends on the coefficients of the variables.

Systems of Inequalities (Lesson 6-6)

- A system of inequalities is a set of two or more inequalities with the same variables.

- The solution of a system of inequalities is the intersection of the graphs.

FOLDABLES StudyOrganizer

Be sure the Key Concepts are noted in your Foldable.

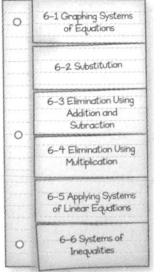

KeyVocabulary

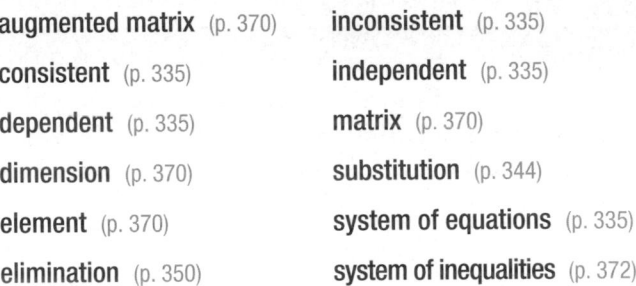

augmented matrix (p. 370)

consistent (p. 335)

dependent (p. 335)

dimension (p. 370)

element (p. 370)

elimination (p. 350)

inconsistent (p. 335)

independent (p. 335)

matrix (p. 370)

substitution (p. 344)

system of equations (p. 335)

system of inequalities (p. 372)

VocabularyCheck

State whether each sentence is *true* or *false*. If *false,* replace the underlined term to make a true sentence.

1. If a system has at least one solution, it is said to be <u>consistent</u>.

2. If a consistent system has exactly <u>two</u> solution(s), it is said to be independent.

3. If a consistent system has an infinite number of solutions, it is said to be <u>inconsistent</u>.

4. If a system has no solution, it is said to be <u>inconsistent</u>.

5. <u>Substitution</u> involves substituting an expression from one equation for a variable in the other.

6. In some cases, <u>dividing</u> two equations in a system together will eliminate one of the variables. This process is called elimination.

7. A set of two or more inequalities with the same variables is called a <u>system of equations</u>.

8. When the graphs of the inequalities in a system of inequalities <u>do not intersect</u>, there are no solutions to the system.

Lesson-by-Lesson Review

6-1 Graphing Systems of Equations

Graph each system and determine the number of solutions that it has. If it has one solution, name it.

9. $x - y = 1$
   $x + y = 5$

10. $y = 2x - 4$
    $4x + y = 2$

11. $2x - 3y = -6$
    $y = -3x + 2$

12. $-3x + y = -3$
    $y = x - 3$

13. $x + 2y = 6$
    $3x + 6y = 8$

14. $3x + y = 5$
    $6x = 10 - 2y$

15. **MAGIC NUMBERS** Sean is trying to find two numbers with a sum of 14 and a difference of 4. Define two variables, write a system of equations, and solve by graphing.

Example 1

Graph the system and determine the number of solutions it has. If it has one solution, name it.

$y = 2x + 2$
$y = -3x - 3$

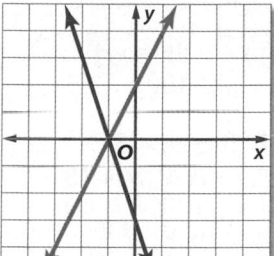

The lines appear to intersect at the point $(-1, 0)$. You can check this by substituting -1 for x and 0 for y.

CHECK
$y = 2x + 2$	Original equation
$0 \stackrel{?}{=} 2(-1) + 2$	Substitution
$0 \stackrel{?}{=} -2 + 2$	Multiply.
$0 = 0$ ✔	
$y = -3x - 3$	Original equation
$0 \stackrel{?}{=} -3(-1) - 3$	Substitution
$0 \stackrel{?}{=} 3 - 3$	Multiply.
$0 = 0$ ✔	

The solution is $(-1, 0)$.

6-2 Substitution

Use substitution to solve each system of equations.

16. $x + y = 3$
    $x = 2y$

17. $x + 3y = -28$
    $y = -5x$

18. $3x + 2y = 16$
    $x = 3y - 2$

19. $x - y = 8$
    $y = -3x$

20. $y = 5x - 3$
    $x + 2y = 27$

21. $x + 3y = 9$
    $x + y = 1$

22. **GEOMETRY** The perimeter of a rectangle is 48 inches. The length is 6 inches greater than the width. Define the variables, and write equations to represent this situation. Solve the system by using substitution.

Example 2

Use substitution to solve the system.

$3x - y = 18$
$y = x - 4$

$3x - y = 18$	First equation
$3x - (x - 4) = 18$	Substitute $x - 4$ for y.
$2x + 4 = 18$	Simplify.
$2x = 14$	Subtract 4 from each side.
$x = 7$	Divide each side by 2.

Use the value of x and either equation to find the value for y.

$y = x - 4$	Second equation
$= 7 - 4$ or 3	Substitute and simplify.

The solution is $(7, 3)$.

6-3 Elimination Using Addition and Subtraction

Use elimination to solve each system of equations.

23. $x + y = 13$
$x - y = 5$

24. $-3x + 4y = 21$
$3x + 3y = 14$

25. $x + 4y = -4$
$x + 10y = -16$

26. $2x + y = -5$
$x - y = 2$

27. $6x + y = 9$
$-6x + 3y = 15$

28. $x - 4y = 2$
$3x + 4y = 38$

29. $2x + 2y = 4$
$2x - 8y = -46$

30. $3x + 2y = 8$
$x + 2y = 2$

31. BASEBALL CARDS Cristiano bought 24 baseball cards for $50. One type cost $1 per card, and the other cost $3 per card. Define the variables, and write equations to find the number of each type of card he bought. Solve by using elimination.

Example 3

Use elimination to solve the system of equations.

$3x - 5y = 11$
$x + 5y = -3$

$\begin{array}{ll}
3x - 5y = 11 & \\
(+) \quad x + 5y = -3 & \\
\hline
4x \qquad = 8 & \text{The variable } y \text{ is eliminated.} \\
\qquad x = 2 & \text{Divide each side by 4.}
\end{array}$

Now, substitute 2 for x in either equation to find the value of y.

$\begin{array}{ll}
3x - 5y = 11 & \text{First equation} \\
3(2) - 5y = 11 & \text{Substitute.} \\
6 - 5y = 11 & \text{Multiply.} \\
-5y = 5 & \text{Subtract 6 from each side.} \\
y = -1 & \text{Divide each side by } -5.
\end{array}$

The solution is $(2, -1)$.

6-4 Elimination Using Multiplication

Use elimination to solve each system of equations.

32. $x + y = 4$
$-2x + 3y = 7$

33. $x - y = -2$
$2x + 4y = 38$

34. $3x + 4y = 1$
$5x + 2y = 11$

35. $-9x + 3y = -3$
$3x - 2y = -4$

36. $8x - 3y = -35$
$3x + 4y = 33$

37. $2x + 9y = 3$
$5x + 4y = 26$

38. $-7x + 3y = 12$
$2x - 8y = -32$

39. $8x - 5y = 18$
$6x + 6y = -6$

40. BAKE SALE On the first day, a total of 40 items were sold for $356. Define the variables, and write a system of equations to find the number of cakes and pies sold. Solve by using elimination.

MONARCH
MIDDLE SCHOOL

Bake Sale

Pies $10

Cakes $8

Example 4

Use elimination to solve the system of equations.

$3x + 6y = 6$
$2x + 3y = 5$

Notice that if you multiply the second equation by -2, the coefficients of the y-terms are additive inverses.

$\begin{array}{l}
3x + 6y = 6 \\
2x + 3y = 5
\end{array}$ **Multiply by -2.** $\quad \begin{array}{l}
3x + 6y = 6 \\
(+) -4x - 6y = -10 \\
\hline
-x \qquad = -4 \\
\qquad x = 4
\end{array}$

Now, substitute 4 for x in either equation to find the value of y.

$\begin{array}{ll}
2x + 3y = 5 & \text{Second equation} \\
2(4) + 3y = 5 & \text{Substitution} \\
8 + 3y = 5 & \text{Multiply.} \\
3y = -3 & \text{Subtract 8 from both sides.} \\
y = -1 & \text{Divide each side by 3.}
\end{array}$

The solution is $(4, -1)$.

Determine the best method to solve each system of equations. Then solve the system.

41. $y = x - 8$
  $y = -3x$

42. $y = -x$
  $y = 2x$

43. $x + 3y = 12$
  $x = -6y$

44. $x + y = 10$
  $x - y = 18$

45. $3x + 2y = -4$
  $5x + 2y = -8$

46. $6x + 5y = 9$
  $-2x + 4y = 14$

47. $3x + 4y = 26$
  $2x + 3y = 19$

48. $11x - 6y = 3$
  $5x - 8y = -25$

49. **COINS** Tionna has saved dimes and quarters in her piggy bank. Define the variables, and write a system of equations to determine the number of dimes and quarters. Then solve the system using the best method for the situation.

$4.00
25 coins

50. **FAIR** At a county fair, the cost for 4 slices of pizza and 2 orders of French fries is $21.00. The cost of 2 slices of pizza and 3 orders of French fries is $16.50. To find out how much a single slice of pizza and an order of French fries costs, define the variables and write a system of equations to represent the situation. Determine the best method to solve the system of equations. Then solve the system. (Lesson 6-5)

Example 5

Determine the best method to solve the system of equations. Then solve the system.

$3x + 5y = 4$
$4x + y = -6$

The coefficient of y is 1 in the second equation. So solving by substitution is a good method. Solve the second equation for y.

$4x + y = -6$	Second equation
$y = -6 - 4x$	Subtract 4x from each side.

Substitute $-6 - 4x$ for y in the first equation.

$3x + 5(-6 - 4x) = 4$	Substitute.
$3x - 30 - 20x = 4$	Distributive Property
$-17x - 30 = 4$	Simplify.
$-17x = 34$	Add 30 to each side.
$x = -2$	Divide by -17.

Last, substitute -2 for x in either equation to find y.

$4x + y = -6$	Second equation
$4(-2) + y = -6$	Substitute.
$-8 + y = -6$	Multiply.
$y = 2$	Add 8 to each side.

The solution is $(-2, 2)$.

6-6 Systems of Inequalities

Solve each system of inequalities by graphing.

51. $x > 3$
$y < x + 2$

52. $y \le 5$
$y > x - 4$

53. $y < 3x - 1$
$y \ge -2x + 4$

54. $y \le -x - 3$
$y \ge 3x - 2$

55. **JOBS** Kishi makes $7 an hour working at the grocery store and $10 an hour delivering newspapers. She cannot work more than 20 hours per week. Graph two inequalities that Kishi can use to determine how many hours she needs to work at each job if she wants to earn at least $90 per week.

Example 6

Solve the system of inequalities by graphing.

$y < 3x + 1$
$y \ge -2x + 3$

The solution set of the system is the set of ordered pairs in the intersection of the two graphs. This portion is shaded in the graph below.

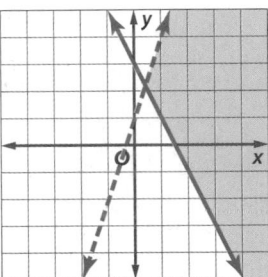

Graph each system and determine the number of solutions that it has. If it has one solution, name it.

1. $y = 2x$
$y = 6 - x$

2. $y = x - 3$
$y = -2x + 9$

3. $x - y = 4$
$x + y = 10$

4. $2x + 3y = 4$
$2x + 3y = -1$

Use substitution to solve each system of equations.

5. $y = x + 8$
$2x + y = -10$

6. $x = -4y - 3$
$3x - 2y = 5$

7. GARDENING Corey has 42 feet of fencing around his garden. The garden is rectangular in shape, and its length is equal to twice the width minus 3 feet. Define the variables, and write a system of equations to find the length and width of the garden. Solve the system by using substitution.

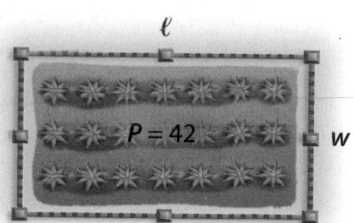

8. MULTIPLE CHOICE Use elimination to solve the system.

$$6x - 4y = 6$$
$$-6x + 3y = 0$$

A $(5, 6)$

B $(-3, -6)$

C $(1, 0)$

D $(4, -8)$

9. SHOPPING Shelly has $175 to shop for jeans and sweaters. Each pair of jeans costs $25, each sweater costs $20, and she buys 8 items. Determine the number of pairs of jeans and sweaters Shelly bought.

Use elimination to solve each system of equations.

10. $x + y = 13$
$x - y = 5$

11. $3x + 7y = 2$
$3x - 4y = 13$

12. $x + y = 8$
$x - 3y = -4$

13. $2x + 6y = 18$
$3x + 2y = 13$

14. MAGAZINES Julie subscribes to a sports magazine and a fashion magazine. She received 24 issues this year. The number of fashion issues is 6 less than twice the number of sports issues. Define the variables, and write a system of equations to find the number of issues of each magazine.

Determine the best method to solve each system of equations. Then solve the system.

15. $y = 3x$
$x + 2y = 21$

16. $x + y = 12$
$y = x - 4$

17. $x + y = 15$
$x - y = 9$

18. $3x + 5y = 7$
$2x - 3y = 11$

19. OFFICE SUPPLIES At a sale, Ricardo bought 24 reams of paper and 4 inkjet cartridges for $320. Britney bought 2 reams of paper and 1 inkjet cartridge for $50. The reams of paper were all the same price and the inkjet cartridges were all the same price. Write a system of equations to represent this situation. Determine the best method to solve the system of equations. Then solve the system.

Solve each system of inequalities by graphing.

20. $x > 2$
$y < 4$

21. $x + y \le 5$
$y \ge x + 2$

22. $3x - y > 9$
$y > -2x$

23. $y \ge 2x + 3$
$-4x - 3y > 12$

Preparing for Standardized Tests

Guess and Check

It is very important to pace yourself and keep track of how much time you have when taking a standardized test. If time is running short, or if you are unsure how to solve a problem, the guess and check strategy may help you determine the correct answer quickly.

Strategies for Guessing and Checking

Step 1

Carefully look over each possible answer choice, and evaluate for reasonableness. Eliminate unreasonable answers.

Ask yourself:

- Are there any answer choices that are clearly incorrect?

- Are there any answer choices that are not in the proper format?

- Are there any answer choices that do not have the proper units for the correct answer?

Step 2

For the remaining answer choices, use the guess and check method.

- **Equations:** If you are solving an equation, substitute the answer choice for the variable and see if this results in a true number sentence.

- **Inequalities:** Likewise, you can substitute the answer choice for the variable and see if it satisfies the inequality.

- **System of Equations:** Find the answer choice that satisfies both equations of the system.

Step 3

Choose an answer choice and see if it satisfies the constraints of the problem statement. Identify the correct answer.

- If the answer choice you are testing does not satisfy the problem, move on to the next reasonable guess and check it.

- When you find the correct answer choice, stop. You do not have to check the other answer choices.

Read the problem. Identify what you need to know. Then use the information in the problem to solve.

Solve $\begin{cases} 4x - 8y = 20 \\ -3x + 5y = -14 \end{cases}$.

A $(5, 0)$ **C** $(3, -1)$

B $(4, -2)$ **D** $(-6, -5)$

The solution of a system of equations is an ordered pair, (x, y). Since all four answer choices are of this form, they are all possible correct answers and must be checked. Begin with the first answer choice and substitute it in each equation. Continue until you find the ordered pair that satisfies both equations of the system.

	First Equation	Second Equation
Guess: (5, 0)	$4x - 8y = 20$ $4(5) - 8(0) = 20$ ✓	$-3x + 5y = -14$ $-3(5) + 5(0) \neq -14$ ✗

	First Equation	Second Equation
Guess: (4, -2)	$4x - 8y = 20$ $4(4) - 8(-2) \neq 20$ ✗	$-3x + 5y = -14$ $-3(4) + 5(-2) \neq -14$ ✗

	First Equation	Second Equation
Guess: (3, -1)	$4x - 8y = 20$ $4(3) - 8(-1) = 20$ ✓	$-3x + 5y = -14$ $-3(3) + 5(-1) = -14$ ✓

The ordered pair $(3, -1)$ satisfies both equations of the system. So, the correct answer is C.

Exercises

Read each problem. Eliminate any unreasonable answers. Then use the information in the problem to solve.

1. Gina bought 5 hot dogs and 3 soft drinks at the ball game for $11.50. Renaldo bought 4 hot dogs and 2 soft drinks for $8.50. How much does a single hot dog and a single drink cost?

   A hot dogs: $1.25   **C** hot dogs: $1.50
        soft drinks: $1.50   soft drinks: $1.25

   B hot dogs: $1.25   **D** hot dogs: $1.50
        soft drinks: $1.75   soft drinks: $1.75

2. The bookstore hopes to sell at least 30 binders and calculators each week. The store also hopes to have sales revenue of at least $200 in binders and calculators. How many binders and calculators could be sold to meet both of these sales goals?

Store Prices	
Item	**Price**
binders	$3.65
calculators	$14.80

   F 25 binders,   **H** 22 binders,
        5 calculators   9 calculators

   G 12 binders,   **J** 28 binders,
        15 calculators   6 calculators

Multiple Choice

Read each question. Then fill in the correct answer on the answer document provided by your teacher or on a sheet of paper.

1. Which of the following terms *best* describes the system of equations shown in the graph?

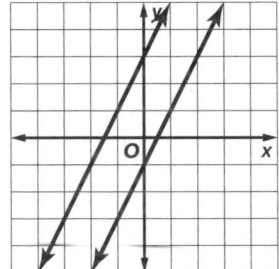

  A consistent

  B consistent and dependent

  C consistent and independent

  D inconsistent

2. Use substitution to solve the system of equations below.

$$\begin{cases} y = 4x - 7 \\ 3x - 2y = -1 \end{cases}$$

  F $(3, 5)$ **H** $(5, -2)$

  G $(4, -1)$ **J** $(-6, 2)$

3. Which ordered pair is the solution of the system of linear equations shown below?

$$\begin{cases} 3x - 8y = -50 \\ 3x - 5y = -38 \end{cases}$$

  A $\left(\dfrac{5}{8}, \dfrac{3}{2}\right)$ **C** $\left(-\dfrac{2}{7}, \dfrac{4}{9}\right)$

  B $(4, -9)$ **D** $(-6, 4)$

4. A home goods store received $881 from the sale of 4 table saws and 9 electric drills. If the receipts from the saws exceeded the receipts from the drills by $71, what is the price of an electric drill?

  F $45   **H** $108

  G $59   **J** $119

5. A region is defined by this system.

$$y > -\frac{1}{2}x - 1$$

$$y > -x + 3$$

In which quadrant(s) of the coordinate plane is the region located?

  A I and IV only   **C** I, II, and IV only

  B III only   **D** II and III only

6. Which of the following terms *best* describes the system of equations shown in the graph?

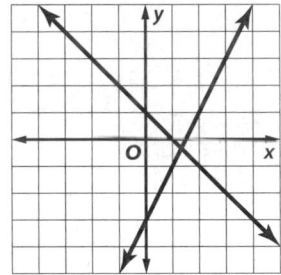

  F consistent

  G consistent and independent

  H consistent and dependent

  J inconsistent

7. Use elimination to solve the system of equations below.

$$3x + 2y = -2$$

$$2x - 2y = -18$$

  A $(1, 3)$ **C** $(-2, -3)$

  B $(7, -4)$ **D** $(-4, 5)$

8. What is the solution of the following system of equations?

$$\begin{cases} y = 6x - 1 \\ y = 6x + 1 \end{cases}$$

  F $(2, 11)$ **H** $(7, 5)$

  G $(-3, -14)$ **J** no solution

> **Test-TakingTip**
>
> **Question 8** You can subtract the second equation from the first equation to eliminate the *x*-variable. Then solve for *y*.

Short Response/Gridded Response

Record your answers on the answer sheet provided by your teacher or on a sheet of paper.

9. **GRIDDED RESPONSE** Angie and her sister have $15 to spend on pizza. A medium pizza costs $11.50 plus $0.75 per topping. What is the maximum number of toppings Angie and her sister can get on their pizza?

10. Write an inequality for the graph below.

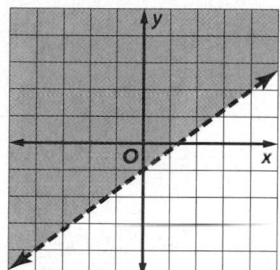

11. **GRIDDED RESPONSE** Christy is taking a road trip. After she drives 12 more miles, she will have driven at least half of the 108 mile trip. What is the least number of miles she has driven so far?

12 Write an equation in slope-intercept form with a slope of $-\frac{2}{3}$ and a y-intercept of 6.

13. A rental company charges $9.50 per hour for a scooter plus a $15 fee. Write an equation in slope-intercept form for the total rental cost C of renting a scooter for h hours.

14. **GRIDDED RESPONSE** A computer supplies store is having a storewide sale this weekend. An inkjet printer that normally sells for $179.00 is on sale for $143.20. What is the percent discount of the sale price?

15. In 1980, the population of Kentucky was about 3.66 million people. By 2000, this number had grown to about 4.04 million people. What was the annual rate of change in population from 1980 to 2000?

16. Joseph's cell phone service charges him $0.15 per text. Write an equation that represents the cost C of his cell phone service for t texts sent each month.

17. A store is offering a $15 mail-in-rebate on all printers. If Mark is looking at printers that range from $45 to $89, how much can he expect to pay?

Extended Response

Record your answers on a sheet of paper. Show your work.

18. The table shows how many canned goods were collected during the first day of a charity food drive.

Food Drive Day 1 Results	
Class	Number Collected
10th graders	78
11th graders	80
12th graders	92

a. Estimate how many canned goods will be collected during the 5-day food drive. Explain your answer.

b. Is this estimate a reasonable expectation? Explain.

Need ExtraHelp?

If you missed Question...	1	2	3	4	5	6	7	8	9	10	11	12	13	14	15	16	17	18
Go to Lesson...	6-1	6-2	6-3	6-3	6-6	6-1	6-3	6-3	5-3	5-6	5-3	4-2	4-2	2-7	2-7	3-4	5-4	1-4

7

Exponents and Exponential Functions

··Then

○ You evaluated expressions involving exponents.

··Now

○ In this chapter, you will:

- Simplify and perform operations on expressions involving exponents.

- Extend the properties of integer exponents to rational exponents.

- Use scientific notation.

- Graph and use exponential functions.

··Why? ▲

○ **SPACE** The Very Large Array is an arrangement of 27 radio antennas in a Y pattern. The data the antennas collect is used by astronomers around the world to study the planets and stars. Astrophysicists use and apply properties of exponents to model the distance and orbit of celestial bodies.

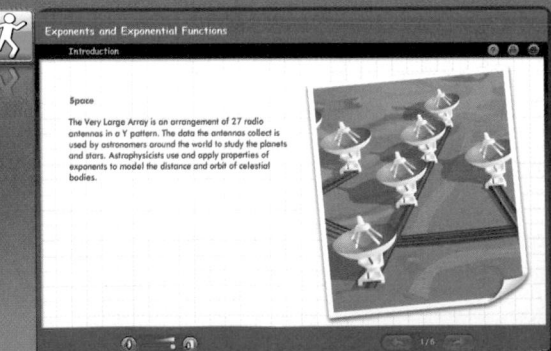

connectED.mcgraw-hill.com **Your Digital Math Portal**

Animation	Vocabulary	eGlossary	Personal Tutor	Virtual Manipulatives	Graphing Calculator	Audio	Foldables	Self-Check Practice	Worksheets

Diagnose Readiness | You have two options for checking prerequisite skills.

1 **Textbook Option** Take the Quick Check below. Refer to the Quick Review for help.

QuickCheck	QuickReview

Write each expression using exponents.

1. $4 \cdot 4 \cdot 4 \cdot 4 \cdot 4$

2. $y \cdot y \cdot y$

3. $6 \cdot 6$

4. $2 \cdot 2 \cdot 2 \cdot 2 \cdot 2 \cdot 2 \cdot 2 \cdot 2 \cdot 2$

5. $b \cdot b \cdot b \cdot b \cdot b \cdot b$

6. $m \cdot m \cdot m \cdot p \cdot p \cdot p \cdot p \cdot p \cdot p$

7. $\frac{1}{3} \cdot \frac{1}{3} \cdot \frac{1}{3} \cdot \frac{1}{3} \cdot \frac{1}{3} \cdot \frac{1}{3} \cdot \frac{1}{3} \cdot \frac{1}{3}$

8. $\frac{x}{y} \cdot \frac{x}{y} \cdot \frac{x}{y} \cdot \frac{x}{y} \cdot \frac{w}{z} \cdot \frac{w}{z}$

Example 1

Write $5 \cdot 5 \cdot 5 \cdot 5 + x \cdot x \cdot x$ using exponents.

4 factors of 5 is 5^4.

3 factors of x is x^3.

So, $5 \cdot 5 \cdot 5 \cdot 5 + x \cdot x \cdot x = 5^4 + x^3$.

Find the area or volume of each figure.

9.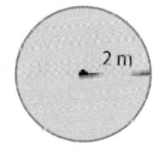

10.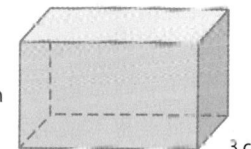

11. **PHOTOGRAPHY** A photo is 4 inches by 6 inches. What is the area of the photo?

Example 2

Find the volume of the figure.

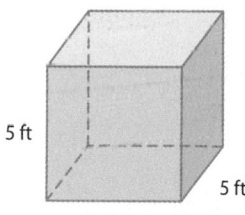

$V = \ell w h$ Volume of a rectangular prism

$= 5 \cdot 5 \cdot 5$ or 125 $\ell = 5$, $w = 5$, and $h = 5$

The volume is 125 cubic feet.

Evaluate each expression.

12. 2^3

13. $(-5)^2$

14. 3^3

15. $(-4)^3$

16. $\left(\frac{2}{3}\right)^2$

17. $\left(\frac{1}{2}\right)^4$

18. **SCHOOL** The probability of guessing correctly on 5 true-false questions is $\left(\frac{1}{2}\right)^5$. Express this probability as a fraction without exponents.

Example 3

Evaluate $\left(\frac{5}{7}\right)^2$.

$\left(\frac{5}{7}\right)^2 = \frac{5^2}{7^2}$ Power of a Quotient

$= \frac{25}{49}$ Simplify.

2 **Online Option** Take an online self-check Chapter Readiness Quiz at <u>connectED.mcgraw-hill.com</u>.

Get Started on the Chapter

You will learn several new concepts, skills, and vocabulary terms as you study Chapter 7. To get ready, identify important terms and organize your resources. You may wish to refer to Chapter 0 to review prerequisite skills.

FOLDABLES StudyOrganizer

Exponents and Exponential Functions Make this Foldable to help you organize your Chapter 7 notes about exponents and exponential functions. Begin with nine sheets of notebook paper.

1 **Arrange** the paper into a stack.

2 **Staple** along the left side. Starting with the second sheet of paper, cut along the right side to form tabs.

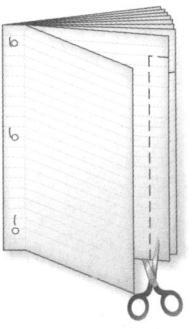

3 **Label** the cover sheet "Exponents and Exponential Functions" and label each tab with a lesson number.

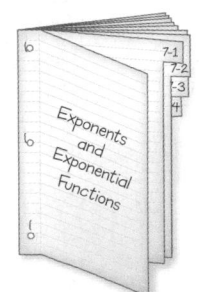

NewVocabulary

English		Español
monomial	p. 391	monomio
constant	p. 391	constante
zero exponent	p. 399	cero exponente
negative exponent	p. 400	exponente negativo
order of magnitude	p. 401	orden de magnitud
rational exponent	p. 406	exponent racional
cube root	p. 407	raíz cúbica
nth root	p. 407	raíz enésima
exponential equation	p. 409	ecuación exponencial
scientific notation	p. 414	notación científica
exponential function	p. 424	función exponencial
exponential growth	p. 424	crecimiento exponencial
exponential decay	p. 424	desintegración exponencial
compound interest	p. 433	interés es compuesta
geometric sequence	p. 438	secuencia geométrica
common ratio	p. 438	proporción común
recursive formula	p. 445	fórmula recursiva

ReviewVocabulary

base base In an expression of the form x^n, the base is x.

Distributive Property Propiedad distributiva
For any numbers a, b, and c, $a(b + c) = ab + ac$ and $a(b - c) = ab - ac$.

exponent exponente
In an expression of the form x^n, the exponent is n. It indicates the number of times x is used as a factor.

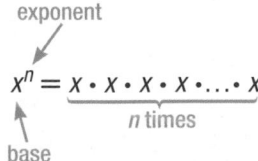

$$x^n = \underbrace{x \cdot x \cdot x \cdot x \cdot \ldots \cdot x}_{n \text{ times}}$$

Multiplication Properties of Exponents

:: Then

- You evaluated expressions with exponents.

:: Now

1. Multiply monomials using the properties of exponents.

2. Simplify expressions using the multiplication properties of exponents.

:: Why?

- Many formulas contain *monomials*. For example, the formula for the horsepower of a car is $H = w\left(\dfrac{v}{234}\right)^3$. *H* represents the horsepower produced by the engine, *w* equals the weight of the car with passengers, and *v* is the velocity of the car at the end of a quarter of a mile. As the velocity increases, the horsepower increases.

NewVocabulary
monomial
constant

Common Core State Standards

Content Standards
A.SSE.2 Use the structure of an expression to identify ways to rewrite it.

F.IF.8b Use the properties of exponents to interpret expressions for exponential functions.

Mathematical Practices
8 Look for and express regularity in repeated reasoning.

---

1 **Multiply Monomials** A **monomial** is a number, a variable, or the product of a number and one or more variables with nonnegative integer exponents. It has only one term. In the formula to calculate the horsepower of a car, the term $w\left(\dfrac{v}{234}\right)^3$ is a monomial.

An expression that involves division by a variable, like $\dfrac{ab}{c}$, is not a monomial.

A **constant** is a monomial that is a real number. The monomial $3x$ is an example of a *linear expression* since the exponent of x is 1. The monomial $2x^2$ is a *nonlinear expression* since the exponent is a positive number other than 1.

Example 1 Identify Monomials

Determine whether each expression is a monomial. Write *yes* or *no*. Explain your reasoning.

a. 10   Yes; this is a constant, so it is a monomial.

b. $f + 24$ No; this expression has addition, so it has more than one term.

c. h^2 Yes; this expression is a product of variables.

d. j Yes; single variables are monomials.

▸ **Guided**Practice

1A. $-x + 5$ **1B.** $23abcd^2$

1C. $\dfrac{xyz^2}{2}$ **1D.** $\dfrac{mp}{n}$

Recall that an expression of the form x^n is called a *power* and represents the result of multiplying x by itself n times. x is the *base*, and n is the *exponent*. The word *power* is also used sometimes to refer to the exponent.

$$\underset{\text{base}}{\underbrace{3}}\overset{\text{exponent}}{}{}^{4} = \overset{\text{4 factors}}{\overbrace{3 \cdot 3 \cdot 3 \cdot 3}} = 81$$

By applying the definition of a power, you can find the product of powers. Look for a pattern in the exponents.

$$\overset{\text{2 factors}}{\overbrace{2^2}} \cdot \overset{\text{4 factors}}{\overbrace{2^4}} = \underset{2 + 4 = 6 \text{ factors}}{\underbrace{2 \cdot 2 \cdot 2 \cdot 2 \cdot 2 \cdot 2}} \qquad \overset{\text{3 factors}}{\overbrace{4^3}} \cdot \overset{\text{2 factors}}{\overbrace{4^2}} = \underset{3 + 2 = 5 \text{ factors}}{\underbrace{4 \cdot 4 \cdot 4 \cdot 4 \cdot 4}}$$

These examples demonstrate the property for the product of powers.

KeyConcept Product of Powers

Words   To multiply two powers that have the same base, add their exponents.

Symbols   For any real number a and any integers m and p, $a^m \cdot a^p = a^{m+p}$.

Examples $b^3 \cdot b^5 = b^{3+5}$ or b^8 $\qquad$ $g^4 \cdot g^6 = g^{4+6}$ or g^{10}

Example 2 Product of Powers

Simplify each expression.

a. $\left(6n^3\right)\left(2n^7\right)$

$\left(6n^3\right)\left(2n^7\right) = (6 \cdot 2)\left(n^3 \cdot n^7\right)$ $\qquad$ Group the coefficients and the variables.

$ = (6 \cdot 2)\left(n^{3+7}\right)$ $\qquad$ Product of Powers

$ = 12n^{10}$ $\qquad$ Simplify.

b. $\left(3pt^3\right)\left(p^3t^4\right)$

$\left(3pt^3\right)\left(p^3t^4\right) = (3 \cdot 1)\left(p \cdot p^3\right)\left(t^3 \cdot t^4\right)$ $\qquad$ Group the coefficients and the variables.

$ = (3 \cdot 1)\left(p^{1+3}\right)\left(t^{3+4}\right)$ $\qquad$ Product of Powers

$ = 3p^4t^7$ $\qquad$ Simplify.

> **StudyTip**
>
> **Coefficients and Powers of 1** A variable with no exponent or coefficient shown can be assumed to have an exponent and coefficient of 1. For example, $x = 1x^1$.

▶ **GuidedPractice**

2A. $\left(3y^4\right)\left(7y^5\right)$ $\qquad\qquad$ **2B.** $\left(-4rx^2t^3\right)\left(-6r^5x^2t\right)$

We can use the Product of Powers Property to find the power of a power. In the following examples, look for a pattern in the exponents.

$$\left(3^2\right)^4 = \overset{\text{4 factors}}{\overbrace{\left(3^2\right)\left(3^2\right)\left(3^2\right)\left(3^2\right)}} \qquad \left(r^4\right)^3 = \overset{\text{3 factors}}{\overbrace{\left(r^4\right)\left(r^4\right)\left(r^4\right)}}$$
$$= 3^{2+2+2+2} \qquad\qquad\qquad = r^{4+4+4}$$
$$= 3^8 \qquad\qquad\qquad\qquad = r^{12}$$

These examples demonstrate the property for the power of a power.

KeyConcept Power of a Power

Words   To find the power of a power, multiply the exponents.

Symbols   For any real number a and any integers m and p, $\left(a^m\right)^p = a^{m \cdot p}$.

Examples $\left(b^3\right)^5 = b^{3 \cdot 5}$ or b^{15} $\qquad$ $\left(g^6\right)^7 = g^{6 \cdot 7}$ or g^{42}

Standardized Test Example 3 Power of a Power

Simplify $\left[(2^3)^2\right]^4$.

A 2^{24}　　　**B** 2^{12}　　　**C** 2^{10}　　　**D** 2^9

Read the Test Item

You need to apply the power of a power rule.

Solve the Test Item

$$\left[(2^3)^2\right]^4 = (2^{3\cdot 2})^4 \qquad \text{Power of a Power}$$

$$= (2^6)^4 \qquad \text{Simplify.}$$

$$= 2^{6\cdot 4} \text{ or } 2^{24} \qquad \text{Power of a Power}$$

The correct choice is A.

GuidedPractice

3. Simplify $\left[(2^2)^2\right]^4$.

　F 2^8　　　**G** 2^{10}　　　**H** 2^{16}　　　**J** 2^{24}

We can use the Product of Powers Property and the Power of a Power Property to find the power of a product. Look for a pattern in the exponents below.

$$
\begin{array}{ll}
\overbrace{}^{3\text{ factors}} & \overbrace{}^{3\text{ factors}}\\
(tw)^3 = (tw)(tw)(tw) & (2yz^2)^3 = (2yz^2)(2yz^2)(2yz^2)\\
\quad = (t\cdot t\cdot t)(w\cdot w\cdot w) & \quad = (2\cdot 2\cdot 2)(y\cdot y\cdot y)(z^2\cdot z^2\cdot z^2)\\
\quad = t^3 w^3 & \quad = 2^3 y^3 z^6 \text{ or } 8y^3 z^6
\end{array}
$$

These examples demonstrate the property for the power of a product.

KeyConcept Power of a Product

Words	To find the power of a product, find the power of each factor and multiply.
Symbols	For any real numbers a and b and any integer m, $(ab)^m = a^m b^m$.
Example	$(-2xy^3)^5 = (-2)^5 x^5 y^{15}$ or $-32x^5 y^{15}$

Example 4 Power of a Product

GEOMETRY Express the area of the circle as a monomial.

$$\text{Area} = \pi r^2 \qquad \text{Formula for the area of a circle}$$

$$= \pi (2xy^2)^2 \qquad \text{Replace } r \text{ with } 2xy^2.$$

$$= \pi (2^2 x^2 y^4) \qquad \text{Power of a Product}$$

$$= 4x^2 y^4 \pi \qquad \text{Simplify.}$$

The area of the circle is $4x^2 y^4 \pi$ square units.

[circle diagram with radius labeled $2xy^2$]

GuidedPractice

4A. Express the area of a square with sides of length $3xy^2$ as a monomial.

4B. Express the area of a triangle with height $4a$ and base $5ab^2$ as a monomial.

2 **Simplify Expressions** We can combine and use these properties to simplify expressions involving monomials.

> **KeyConcept** Simplify Expressions
>
> To simplify a monomial expression, write an equivalent expression in which:
>
> • each variable base appears exactly once,
>
> • there are no powers of powers, and
>
> • all fractions are in simplest form.

StudyTip

Simplify When simplifying expressions with multiple grouping symbols, begin at the innermost expression and work outward.

Example 5 Simplify Expressions

Simplify $(3xy^4)^2[(-2y)^2]^3$.

$$(3xy^4)^2[(-2y)^2]^3 = (3xy^4)^2(-2y)^6 \qquad \text{Power of a Power}$$
$$= (3)^2 x^2 (y^4)^2 (-2)^6 y^6 \qquad \text{Power of a Product}$$
$$= 9x^2 y^8 (64) y^6 \qquad \text{Power of a Power}$$
$$= 9(64) x^2 \cdot y^8 \cdot y^6 \qquad \text{Commutative}$$
$$= 576 x^2 y^{14} \qquad \text{Product of Powers}$$

▶ **Guided**Practice

5. Simplify $\left(\frac{1}{2} a^2 b^2\right)^3 [(-4b)^2]^2$.

Check Your Understanding  ◯ = **Step-by-Step Solutions begin on page R13.**

Example 1 Determine whether each expression is a monomial. Write *yes* or *no*. Explain your reasoning.

1. 15	**2.** $2 - 3a$	**3.** $\frac{5c}{d}$
4. $-15g^2$	**5.** $\frac{r}{2}$	**6.** $7b + 9$

Examples 2–3 Simplify each expression.

7. $k(k^3)$	**8.** $m^4(m^2)$	**9** $2q^2(9q^4)$
10. $(5u^4v)(7u^4v^3)$	**11.** $[(3^2)^2]^2$	**12.** $(xy^4)^6$
13. $(4a^4b^9c)^2$	**14.** $(-2f^2g^3h^2)^3$	**15.** $(-3p^5t^6)^4$

Example 4 **16. GEOMETRY** The formula for the surface area of a cube is $SA = 6s^2$, where SA is the surface area and s is the length of any side.

   a. Express the surface area of the cube as a monomial.

   b. What is the surface area of the cube if $a = 3$ and $b = 4$?

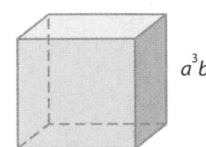

Example 5 Simplify each expression.

   17. $(5x^2y)^2(2xy^3z)^3(4xyz)$ **18.** $(-3d^2f^3g)^2[(-3d^2f)^3]^2$

   19. $(-2g^3h)(-3gj^4)^2(-ghj)^2$ **20.** $(-7ab^4c)^3[(2a^2c)^2]^3$

Example 1   Determine whether each expression is a monomial. Write *yes* or *no*. Explain your reasoning.

21. 122   **22.** $3a^4$ **23.** $2c + 2$

24. $\dfrac{-2g}{4h}$ **25.** $\dfrac{5k}{10}$ **26.** $6m + 3n$

Examples 2–3 Simplify each expression.

㉗ $(q^2)(2q^4)$ **28.** $(-2u^2)(6u^6)$ **29.** $(9w^2x^8)(w^6x^4)$

30. $(y^6z^9)(6y^4z^2)$ **31.** $(b^8c^6d^5)(7b^6c^2d)$ **32.** $(14fg^2h^2)(-3f^4g^2h^2)$

33. $(j^5k^7)^4$ **34.** $(n^3p)^4$ **35.** $[(2^2)^2]^2$

36. $[(3^2)^2]^4$ **37.** $[(4r^2t)^3]^2$ **38.** $[(-2xy^2)^3]^2$

Example 4   **GEOMETRY** Express the area of each triangle as a monomial.

39.

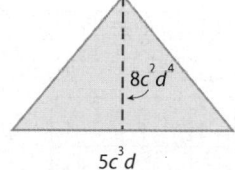

$8c^2d^4$

$5c^3d$

40.

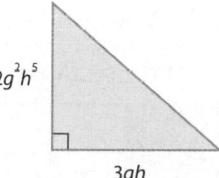

$2g^2h^5$

$3gh$

Example 5   Simplify each expression.

41. $(2a^3)^4(a^3)^3$ **42.** $(c^3)^2(-3c^5)^2$

43. $(2gh^4)^3[(-2g^4h)^3]^2$ **44.** $(5k^2m)^3[(4km^4)^2]^2$

45. $(p^5r^2)^4(-7p^3r^4)^2(6pr^3)$ **46.** $(5x^2y)^2(2xy^3z)^3(4xyz)$

47. $(5a^2b^3c^4)(6a^3b^4c^2)$ **48.** $(10xy^5z^3)(3x^4y^6z^3)$

49. $(0.5x^3)^2$ **50.** $(0.4h^5)^3$

51. $\left(-\dfrac{3}{4}c\right)^3$ **52.** $\left(\dfrac{4}{5}a^2\right)^2$

53. $(8y^3)(-3x^2y^2)\left(\dfrac{3}{8}xy^4\right)$ **54.** $\left(\dfrac{4}{7}m\right)^?(49m)(17p)\left(\dfrac{1}{34}p^5\right)$

55. $(-3r^3w^4)^3(2rw)^2(-3r^2)^3(4rw^2)^3(2r^2w^3)^4$

56. $(3ub^2c)^?(-2u^2b^4)^2(a^4c^2)^3(a^2b^4c^5)^2(2a^3b^2c^4)^3$

57. **FINANCIAL LITERACY** Cleavon has money in an account that earns 3% simple interest. The formula for computing simple interest is $I = Prt$, where I is the interest earned, P represents the principal that he put into the account, r is the interest rate (in decimal form), and t represents time in years.

   a. Cleavon makes a deposit of $2c and leaves it for 2 years. Write a monomial that represents the interest earned.

   b. If c represents a birthday gift of $250, how much will Cleavon have in this account after 2 years?

   CCSS **TOOLS** Express the volume of each solid as a monomial.

58.

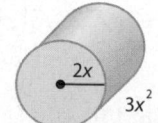

$2x$

$3x^2$

59.

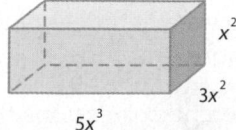

x^2

$3x^2$

$5x^3$

60.

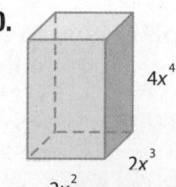

$4x^4$

$2x^3$

$2x^2$

61 **PACKAGING** For a commercial art class, Aiko must design a new container for individually wrapped pieces of candy. The shape that she chose is a cylinder. The formula for the volume of a cylinder is $V = \pi r^2 h$.

a. The radius that Aiko would like to use is $2p^3$, and the height is $4p^3$. Write a monomial that represents the volume of her container.

b. Make a table for five possible measures for the radius and height of a cylinder having the same volume.

c. What is the volume of Aiko's container if the height is doubled?

62. **ENERGY** Albert Einstein's formula $E = mc^2$ shows that if mass is accelerated enough, it could be converted into usable energy. Energy E is measured in joules, mass m in kilograms, and the speed c of light is about 300 million meters per second.

a. Complete the calculations to convert 3 kilograms of gasoline completely into energy.

b. What happens to the energy if the amount of gasoline is doubled?

63. **MULTIPLE REPRESENTATIONS** In this problem, you will explore exponents.

a. Tabular Copy and use a calculator to complete the table.

Power	3^4	3^3	3^2	3^1	3^0	3^{-1}	3^{-2}	3^{-3}	3^{-4}
Value						$\frac{1}{3}$	$\frac{1}{9}$	$\frac{1}{27}$	$\frac{1}{81}$

b. Analytical What do you think the values of 5^0 and 5^{-1} are? Verify your conjecture using a calculator.

c. Analytical Complete: For any nonzero number a and any integer n, $a^{-n} =$ _____.

d. Verbal Describe the value of a nonzero number raised to the zero power.

H.O.T. Problems   Use Higher-Order Thinking Skills

64. **CCSS PERSEVERANCE** For any nonzero real numbers a and b and any integers m and t, simplify the expression $\left(-\dfrac{a^m}{b^t}\right)^{2t}$ and describe each step.

65. **REASONING** Copy the table below.

Equation	Related Expression	Power of x	Linear or Nonlinear
$y = x$			
$y = x^2$			
$y = x^3$			

a. For each equation, write the related expression and record the power of x.

b. Graph each equation using a graphing calculator.

c. Classify each graph as *linear* or *nonlinear*.

d. Explain how to determine whether an equation, or its related expression, is linear or nonlinear without graphing.

66. **OPEN ENDED** Write three different expressions that can be simplified to x^6.

67. **WRITING IN MATH** Write two formulas that have monomial expressions in them. Explain how each is used in a real-world situation.

68. Which of the following is not a monomial?

  A $-6xy$ **C** $-\dfrac{1}{2b^3}$

  B $\dfrac{1}{2}a^2$ **D** $5gh^4$

69. GEOMETRY The accompanying diagram shows the transformation of $\triangle XYZ$ to $\triangle X'Y'Z'$.

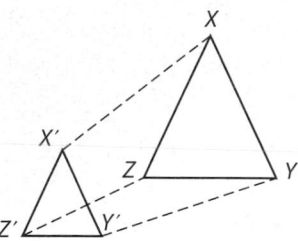

  This transformation is an example of a

  F dilation

  G line reflection

  H rotation

  J translation

70. CARS In 2002, the average price of a new domestic car was \$19,126. In 2008, the average price was \$28,715. Based on a linear model, what is the predicted average price for 2014?

  A \$45,495   **C** \$35,906

  B \$38,304   **D** \$26,317

71. SHORT RESPONSE If a line has a positive slope and a negative y-intercept, what happens to the x-intercept if the slope and the y-intercept are both doubled?

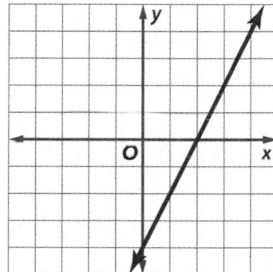

Solve each system of inequalities by graphing. (Lesson 6-6)

72. $y < 4x$

    $2x + 3y \geq -21$

73. $y \geq 2$

    $2y + 2x \leq 4$

74. $y > -2x - 1$

    $2y \leq 3x + 2$

75. $3x + 2y < 10$

    $2x + 12y < -6$

76. SPORTS In the 2006 Winter Olympic Games, the total number of gold and silver medals won by the U.S. was 18. The total points scored for gold and silver medals was 45. Write and solve a system of equations to find how many gold and silver medals were won by the U.S. (Lesson 6-5)

77. DRIVING Tires should be kept within 2 pounds per square inch (psi) of the manufacturer's recommended tire pressure. If the recommendation for a tire is 30 psi, what is the range of acceptable pressures? (Lesson 5-5)

78. BABYSITTING Alexis charges \$10 plus \$4 per hour to babysit. Alexis needs at least \$40 more to buy a television for which she is saving. Write an inequality for this situation. Will she be able to get her television if she babysits for 5 hours? (Lesson 5-6)

Find each quotient.

79. $-64 \div (-8)$ **80.** $-78 \div 1.3$ **81.** $42.3 \div (-6)$

82. $-23.94 \div 10.5$ **83.** $-32.5 \div (-2.5)$ **84.** $-98.44 \div 4.6$

Division Properties of Exponents

:: Then

- You multiplied monomials using the properties of exponents.

:: Now

1. Divide monomials using the properties of exponents.
2. Simplify expressions containing negative and zero exponents.

:: Why?

- The tallest redwood tree is 112 meters or about 10^2 meters tall. The average height of a redwood tree is 15 meters. The closest power of ten to 15 is 10^1, so an average redwood is about 10^1 meters tall. The ratio of the tallest tree's height to the average tree's height is $\frac{10^2}{10^1}$ or 10^1. This means the tallest redwood tree is approximately 10 times as tall as the average redwood tree.

 NewVocabulary
zero exponent
negative exponent
order of magnitude

 Common Core State Standards

Content Standards
A.SSE.2 Use the structure of an expression to identify ways to rewrite it.

F.IF.8b Use the properties of exponents to interpret expressions for exponential functions.

Mathematical Practices
2 Reason abstractly and quantitatively.

1 Divide Monomials We can use the principles for reducing fractions to find quotients of monomials like $\frac{10^2}{10^1}$. In the following examples, look for a pattern in the exponents.

$$\frac{2^7}{2^4} = \frac{\overset{1}{\cancel{2}} \cdot \overset{1}{\cancel{2}} \cdot \overset{1}{\cancel{2}} \cdot \overset{1}{\cancel{2}} \cdot 2 \cdot 2 \cdot 2}{\underset{1}{\cancel{2}} \cdot \underset{1}{\cancel{2}} \cdot \underset{1}{\cancel{2}} \cdot \underset{1}{\cancel{2}}} = 2 \cdot 2 \cdot 2 \text{ or } 2^3$$

7 factors / 4 factors

$$\frac{t^4}{t^3} = \frac{\overset{1}{\cancel{t}} \cdot \overset{1}{\cancel{t}} \cdot \overset{1}{\cancel{t}} \cdot t}{\underset{1}{\cancel{t}} \cdot \underset{1}{\cancel{t}} \cdot \underset{1}{\cancel{t}}} = t$$

4 factors / 3 factors

These examples demonstrate the Quotient of Powers Rule.

KeyConcept Quotient of Powers

Words	To divide two powers with the same base, subtract the exponents.
Symbols	For any nonzero number a, and any integers m and p, $\frac{a^m}{a^p} = a^{m-p}$.
Examples	$\frac{c^{11}}{c^8} = c^{11-8}$ or c^3 $\qquad$ $\frac{r^5}{r^2} = r^{5-2} = r^3$

Example 1 Quotient of Powers

Simplify $\dfrac{g^3 h^5}{g h^2}$. Assume that no denominator equals zero.

$\dfrac{g^3 h^5}{g h^2} = \left(\dfrac{g^3}{g}\right)\left(\dfrac{h^5}{h^2}\right)$ $\qquad$ Group powers with the same base.

$\qquad = \left(g^{3-1}\right)\left(h^{5-2}\right)$ $\qquad$ Quotient of Powers

$\qquad = g^2 h^3$ $\qquad$ Simplify.

▶ **Guided**Practice

Simplify each expression. Assume that no denominator equals zero.

1A. $\dfrac{x^3 y^4}{x^2 y}$ $\qquad\qquad\qquad\qquad$ **1B.** $\dfrac{k^7 m^{10} p}{k^5 m^3 p}$

We can use the Product of Powers Rule to find the powers of quotients for monomials. In the following example, look for a pattern in the exponents.

$$\left(\frac{3}{4}\right)^3 = \overbrace{\left(\frac{3}{4}\right)\left(\frac{3}{4}\right)\left(\frac{3}{4}\right)}^{3 \text{ factors}} = \underbrace{\frac{3 \cdot 3 \cdot 3}{4 \cdot 4 \cdot 4}}_{3 \text{ factors}} = \frac{3^3}{4^3}$$

$$\left(\frac{c}{d}\right)^2 = \overbrace{\left(\frac{c}{d}\right)\left(\frac{c}{d}\right)}^{2 \text{ factors}} = \underbrace{\frac{c \cdot c}{d \cdot d}}_{2 \text{ factors}} = \frac{c^2}{d^2}$$

StudyTip

Power Rules with Variables
The power rules apply to variables as well as numbers. For example,
$\left(\frac{3a}{4b}\right)^3 = \frac{(3a)^3}{(4b)^3}$ or $\frac{27a^3}{64b^3}$.

KeyConcept Power of a Quotient

Words To find the power of a quotient, find the power of the numerator and the power of the denominator.

Symbols For any real numbers a and $b \neq 0$, and any integer m, $\left(\frac{a}{b}\right)^m = \frac{a^m}{b^m}$.

Examples $\left(\frac{3}{5}\right)^4 = \frac{3^4}{5^4}$ $\left(\frac{r}{t}\right)^5 = \frac{r^5}{t^5}$

Real-World Career

Astronomer An astronomer studies the universe and analyzes space travel and satellite communications. To be a technician or research assistant, a bachelor's degree is required.

Example 2 Power of a Quotient

Simplify $\left(\frac{3p^3}{7}\right)^2$.

$$\left(\frac{3p^3}{7}\right)^2 = \frac{(3p^3)^2}{7^2} \qquad \text{Power of a Quotient}$$

$$= \frac{3^2(p^3)^2}{7^2} \qquad \text{Power of a Product}$$

$$= \frac{9p^6}{49} \qquad \text{Power of a Power}$$

GuidedPractice

Simplify each expression.

2A. $\left(\frac{3x^4}{4}\right)^3$ **2B.** $\left(\frac{5x^5y}{6}\right)^2$ **2C.** $\left(\frac{2y^2}{3z^3}\right)^2$ **2D.** $\left(\frac{4x^3}{5y^4}\right)^3$

A calculator can be used to explore expressions with 0 as the exponent. There are two methods to explain why a calculator gives a value of 1 for 3^0.

Method 1

$$\frac{3^5}{3^5} = 3^{5-5} \qquad \text{Quotient of Powers}$$

$$= 3^0 \qquad \text{Simplify.}$$

Method 2

$$\frac{3^5}{3^5} = \frac{\cancel{3} \cdot \cancel{3} \cdot \cancel{3} \cdot \cancel{3} \cdot \cancel{3}}{\cancel{3} \cdot \cancel{3} \cdot \cancel{3} \cdot \cancel{3} \cdot \cancel{3}} \qquad \text{Definition of powers}$$

$$= 1 \qquad \text{Simplify.}$$

Since $\frac{3^5}{3^5}$ can only have one value, we can conclude that $3^0 = 1$. A **zero exponent** is any nonzero number raised to the zero power.

KeyConcept Zero Exponent Property

Words	Any nonzero number raised to the zero power is equal to 1.
Symbols	For any nonzero number a, $a^0 = 1$.
Examples	$15^0 = 1$ $\qquad \left(\dfrac{b}{c}\right)^0 = 1$ $\qquad \left(\dfrac{2}{7}\right)^0 = 1$

Example 3 Zero Exponent

Simplify each expression. Assume that no denominator equals zero.

a. $\left(-\dfrac{4n^2q^5r^2}{9n^3q^2r}\right)^0$

$\left(-\dfrac{4n^2q^5r^2}{9n^3q^2r}\right)^0 = 1 \qquad a^0 = 1$

b. $\dfrac{x^5y^0}{x^3}$

$\dfrac{x^5y^0}{x^3} = \dfrac{x^5(1)}{x^3} \qquad a^0 = 1$

$\qquad\quad = x^2 \qquad$ Quotient of Powers

StudyTip

Zero Exponent Be careful of parentheses. The expression $(5x)^0$ is 1 but $5x^0 = 5$.

GuidedPractice

3A. $\dfrac{b^4c^2d^0}{b^2c}$

3B. $\left(\dfrac{2f^4g^7h^3}{15f^3g^9h^6}\right)^0$

2 Negative Exponents Any nonzero real number raised to a negative power is a **negative exponent**. To investigate the meaning of a negative exponent, we can simplify expressions like $\dfrac{c^2}{c^5}$ using two methods.

Method 1

$\dfrac{c^2}{c^5} = c^{2-5}$ $\qquad$ Quotient of Powers

$\quad\ = c^{-3}$ $\qquad$ Simplify.

Method 2

$\dfrac{c^2}{c^5} = \dfrac{\cancel{c} \cdot \cancel{c}}{\cancel{c} \cdot \cancel{c} \cdot c \cdot c \cdot c}$ $\qquad$ Definition of powers

$\quad\ = \dfrac{1}{c^3}$ $\qquad$ Simplify.

Since $\dfrac{c^2}{c^5}$ can only have one value, we can conclude that $c^{-3} = \dfrac{1}{c^3}$.

KeyConcept Negative Exponent Property

Words	For any nonzero number a and any integer n, a^{-n} is the reciprocal of a^n. Also, the reciprocal of a^{-n} is a^n.
Symbols	For any nonzero number a and any integer n, $a^{-n} = \dfrac{1}{a^n}$.
Examples	$2^{-4} = \dfrac{1}{2^4} = \dfrac{1}{16}$ $\qquad \dfrac{1}{j^{-4}} = j^4$

An expression is considered simplified when it contains only positive exponents, each base appears exactly once, there are no powers of powers, and all fractions are in simplest form.

Example 4 Negative Exponents

Simplify each expression. Assume that no denominator equals zero.

a. $\dfrac{n^{-5}p^4}{r^{-2}}$

$\dfrac{n^{-5}p^4}{r^{-2}} = \left(\dfrac{n^{-5}}{1}\right)\left(\dfrac{p^4}{1}\right)\left(\dfrac{1}{r^{-2}}\right)$ Write as a product of fractions.

$= \left(\dfrac{1}{n^5}\right)\left(\dfrac{p^4}{1}\right)\left(\dfrac{r^2}{1}\right)$ $a^{-n} = \dfrac{1}{a^n}$ and $\dfrac{1}{a^{-n}} = a^n$

$= \dfrac{p^4 r^2}{n^5}$ Multiply.

b. $\dfrac{5r^{-3}t^4}{-20r^2t^7u^{-5}}$

$\dfrac{5r^{-3}t^4}{-20r^2t^7u^{-5}} = \left(\dfrac{5}{-20}\right)\left(\dfrac{r^{-3}}{r^2}\right)\left(\dfrac{t^4}{t^7}\right)\left(\dfrac{1}{u^{-5}}\right)$ Group powers with the same base.

$= \left(-\dfrac{1}{4}\right)(r^{-3-2})(t^{4-7})(u^5)$ Quotient of Powers and Negative Exponents Property

$= -\dfrac{1}{4}r^{-5}t^{-3}u^5$ Simplify.

$= -\dfrac{1}{4}\left(\dfrac{1}{r^5}\right)\left(\dfrac{1}{t^3}\right)(u^5)$ Negative Exponent Property

$= -\dfrac{u^5}{4r^5t^3}$ Multiply.

c. $\dfrac{2a^2b^3c^{-5}}{10a^{-3}b^{-1}c^{-4}}$

$\dfrac{2a^2b^3c^{-5}}{10a^{-3}b^{-1}c^{-4}} = \left(\dfrac{2}{10}\right)\left(\dfrac{a^2}{a^{-3}}\right)\left(\dfrac{b^3}{b^{-1}}\right)\left(\dfrac{c^{-5}}{c^{-4}}\right)$ Group powers with the same base.

$= \left(\dfrac{1}{5}\right)(a^{2-(-3)})(b^{3-(-1)})(c^{-5-(-4)})$ Quotient of Powers and Negative Exponents Property

$= \dfrac{1}{5}a^5b^4c^{-1}$ Simplify.

$= \dfrac{1}{5}(a^5)(b^4)\left(\dfrac{1}{c}\right)$ Negative Exponent Property

$= \dfrac{a^5b^4}{5c}$ Multiply.

GuidedPractice

Simplify each expression. Assume that no denominator equals zero.

4A. $\dfrac{v^{-3}wx^2}{wy^{-6}}$ **4B.** $\dfrac{32a^{-8}b^3c^{-4}}{4a^3b^5c^{-2}}$ **4C.** $\dfrac{5j^{-3}k^2m^{-6}}{25k^{-4}m^{-2}}$

Order of magnitude is used to compare measures and to estimate and perform rough calculations. The **order of magnitude** of a quantity is the number rounded to the nearest power of 10. For example, the power of 10 closest to 95,000,000,000 is 10^{11}, or 100,000,000,000. So the order of magnitude of 95,000,000,000 is 10^{11}.

Real-World Example 5 Apply Properties of Exponents

HEIGHT Suppose the average height of a man is about 1.7 meters, and the average height of an ant is 0.0008 meter. How many orders of magnitude as tall as an ant is a man?

Understand We must find the order of magnitude of the heights of the man and ant. Then find the ratio of the orders of magnitude of the man's height to that of the ant's height.

Plan Round each height to the nearest power of ten. Then find the ratio of the height of the man to the height of the ant.

Solve The average height of a man is close to 1 meter. So, the order of magnitude is 10^0 meter. The average height of an ant is about 0.001 meter. So, the order of magnitude is 10^{-3} meters.

The ratio of the height of a man to the height of an ant is about $\frac{10^0}{10^{-3}}$.

$$\frac{10^0}{10^{-3}} = 10^{0 - (-3)} \qquad \text{Quotient of Powers}$$

$$= 10^3 \qquad 0 - (-3) = 0 + 3 \text{ or } 3$$

$$= 1000 \qquad \text{Simplify.}$$

So, a man is approximately 1000 times as tall as an ant, or a man is 3 orders of magnitude as tall as an ant.

Check The ratio of the man's height to the ant's height is $\frac{1.7}{0.0008} = 2125$. The order of magnitude of 2125 is 10^3. ✓

GuidedPractice

5. **ASTRONOMY** The order of magnitude of the mass of Earth is about 10^{27}. The order of magnitude of the Milky Way galaxy is about 10^{44}. How many orders of magnitude as big is the Milky Way galaxy as Earth?

Real-World Link

There are over 14,000 species of ants living all over the world. Some ants can carry objects that are 50 times their own weight.

Source: Maine Animal Coalition

Check Your Understanding

○ = Step-by-Step Solutions begin on page R13.

Examples 1–4 Simplify each expression. Assume that no denominator equals zero.

1. $\frac{t^5 u^4}{t^2 u}$

2. $\frac{a^6 b^4 c^{10}}{a^3 b^2 c}$

 $\frac{m^6 r^5 p^3}{m^5 r^2 p^3}$

4. $\frac{b^4 c^6 f^8}{b^4 c^3 f^5}$

5. $\frac{g^8 h^2 m}{h g^7}$

6. $\frac{r^4 t^7 v^2}{t^7 v^2}$

7. $\frac{x^3 y^2 z^6}{z^5 x^2 y}$

8. $\frac{n^4 q^4 w^6}{q^2 n^3 w}$

9. $\left(\frac{2a^3 b^5}{3}\right)^2$

10. $\frac{r^3 v^{-2}}{t^{-7}}$

11. $\left(\frac{2c^3 d^5}{5g^2}\right)^5$

12. $\left(-\frac{3xy^4 z^2}{x^3 yz^4}\right)^0$

13. $\left(\frac{3f^4 gh^4}{32f^3 g^4 h}\right)^0$

14. $\frac{4r^2 v^0 t^5}{2rt^3}$

15. $\frac{f^{-3} g^2}{h^{-4}}$

16. $\frac{-8x^2 y^8 z^{-5}}{12x^4 y^{-7} z^7}$

17. $\frac{2a^2 b^{-7} c^{10}}{6a^{-3} b^2 c^{-3}}$

Example 5 18. **FINANCIAL LITERACY** The gross domestic product (GDP) for the United States in 2008 was $14.204 trillion, and the GDP per person was $47,580. Use order of magnitude to approximate the population of the United States in 2008.

Examples 1–4 Simplify each expression. Assume that no denominator equals zero.

19. $\dfrac{m^4 p^2}{m^2 p}$

20. $\dfrac{p^{12} t^3 r}{p^2 t r}$

21. $\dfrac{3m^{-3} r^4 p^2}{12 t^4}$

22. $\dfrac{c^4 d^4 f^3}{c^2 d^4 f^3}$

23. $\left(\dfrac{3xy^4}{5z^2}\right)^2$

24. $\left(\dfrac{3t^6 u^2 v^5}{9tuv^{21}}\right)^0$

25. $\left(\dfrac{p^2 t^7}{10}\right)^3$

26. $\dfrac{x^{-4} y^9}{z^{-2}}$

27. $\dfrac{a^7 b^8 c^8}{a^5 b c^7}$

28. $\left(\dfrac{3np^3}{7q^2}\right)^2$

(29) $\left(\dfrac{2r^3 t^6}{5u^9}\right)^4$

30. $\left(\dfrac{3m^5 r^3}{4p^8}\right)^4$

31. $\left(-\dfrac{5f^9 g^4 h^2}{fg^2 h^3}\right)^0$

32. $\dfrac{p^{12} t^7 r^2}{p^2 t^7 r}$

33. $\dfrac{p^4 t^{-3}}{r^{-2}}$

34. $-\dfrac{5c^2 d^5}{8cd^5 f^0}$

35. $\dfrac{-2f^3 g^2 h^0}{8f^2 g^2}$

36. $\dfrac{12m^{-4} p^2}{-15m^3 p^{-9}}$

37. $\dfrac{k^4 m^3 p^2}{k^2 m^2}$

38. $\dfrac{14f^{-3} g^2 h^{-7}}{21k^3}$

39. $\dfrac{39t^4 uv^{-2}}{13t^{-3} u^7}$

40. $\left(\dfrac{a^{-2} b^4 c^5}{a^{-4} b^{-4} c^3}\right)^2$

41. $\dfrac{r^3 t^{-1} x^{-5}}{tx^5}$

42. $\dfrac{g^0 h^7 j^{-2}}{g^{-5} h^0 j^{-2}}$

Example 5

43. **INTERNET** In a recent year, there were approximately 3.95 million Internet hosts. Suppose there were 208 million Internet users. Determine the order of magnitude for the Internet hosts and Internet users. Using the orders of magnitude, how many Internet users were there compared to Internet hosts?

44. **PROBABILITY** The probability of rolling a die and getting an even number is $\frac{1}{2}$. If you roll the die twice, the probability of getting an even number both times is $\left(\frac{1}{2}\right)\left(\frac{1}{2}\right)$ or $\left(\frac{1}{2}\right)^2$.

   a. What does $\left(\frac{1}{2}\right)^4$ represent?

   b. Write an expression to represent the probability of rolling a die d times and getting an even number every time. Write the expression as a power of 2.

Simplify each expression. Assume that no denominator equals zero.

45. $\dfrac{-4w^{12}}{12w^3}$

46. $\dfrac{13r^7}{39r^4}$

47. $\dfrac{(4k^3 m^2)^3}{(5k^2 m^{-3})^{-2}}$

48. $\dfrac{3wy^{-2}}{(w^{-1} y)^3}$

49. $\dfrac{20qr^{-2} t^{-5}}{4q^0 r^4 t^{-2}}$

50. $\dfrac{-12c^3 d^0 f^{-2}}{6c^5 d^{-3} f^4}$

51. $\dfrac{(2g^3 h^{-2})^2}{(g^2 h^0)^{-3}}$

52. $\dfrac{(5pr^{-2})^{-2}}{(3p^{-1} r)^3}$

53. $\left(\dfrac{-3x^{-6} y^{-1} z^{-2}}{6x^{-2} yz^{-5}}\right)^{-2}$

54. $\left(\dfrac{2a^{-2} b^4 c^2}{-4a^{-2} b^{-5} c^{-7}}\right)^{-1}$

55. $\dfrac{(16x^2 y^{-1})^0}{(4x^0 y^{-4} z)^{-2}}$

56. $\left(\dfrac{4^0 c^2 d^3 f}{2c^{-4} d^{-5}}\right)^{-3}$

57. **CCSS SENSE-MAKING** The processing speed of an older desktop computer is about 10^8 instructions per second. A new computer can process about 10^{10} instructions per second. The newer computer is how many times as fast as the older one?

58. ASTRONOMY The brightness of a star is measured in magnitudes. The lower the magnitude, the brighter the star. A magnitude 9 star is 2.51 times as bright as a magnitude 10 star. A magnitude 8 star is $2.51 \cdot 2.51$ or 2.51^2 times as bright as a magnitude 10 star.

   a. How many times as bright is a magnitude 3 star as a magnitude 10 star?

   b. Write an expression to compare a magnitude m star to a magnitude 10 star.

   c. A full moon is considered to be magnitude -13, approximately. Does your expression make sense for this magnitude? Explain.

59 PROBABILITY The probability of rolling a die and getting a 3 is $\frac{1}{6}$. If you roll the die twice, the probability of getting a 3 both times is $\frac{1}{6} \cdot \frac{1}{6}$ or $\left(\frac{1}{6}\right)^2$.

   a. Write an expression to represent the probability of rolling a die d times and getting a 3 each time.

   b. Write the expression as a power of 6.

60. ✦ MULTIPLE REPRESENTATIONS To find the area of a circle, use $A = \pi r^2$. The formula for the area of a square is $A = s^2$.

   a. Algebraic Find the ratio of the area of the circle to the area of the square.

   b. Algebraic If the radius of the circle and the length of each side of the square are doubled, find the ratio of the area of the circle to the square.

   c. Tabular Copy and complete the table.

Radius	Area of Circle	Area of Square	Ratio
r			
$2r$			
$3r$			
$4r$			
$5r$			
$6r$			

   d. Analytical What conclusion can be drawn from this?

H.O.T. Problems Use Higher-Order Thinking Skills

61. REASONING Is $x^y \cdot x^z = x^{yz}$ *sometimes*, *always*, or *never* true? Explain.

62. OPEN ENDED Name two monomials with a quotient of $24a^2b^3$.

63. CHALLENGE Use the Quotient of Powers Property to explain why $x^{-n} = \frac{1}{x^n}$.

64. CCSS REGULARITY Write a convincing argument to show why $3^0 = 1$.

65. WRITING IN MATH Explain how to use the Quotient of Powers property and the Power of a Quotient property.

66. What is the perimeter of the figure in meters?

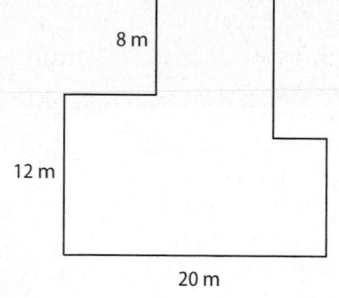

8 m

12 m

20 m

   A 40 meters
   B 80 meters
   C 160 meters
   D 400 meters

67. In researching her science project, Leigh learned that light travels at a constant rate and that it takes 500 seconds for light to travel the 93 million miles from the Sun to Earth. Mars is 142 million miles from the Sun. About how many seconds will it take for light to travel from the Sun to Mars?

   F 235 seconds
   G 327 seconds
   H 642 seconds
   J 763 seconds

68. EXTENDED RESPONSE Jessie and Jonas are playing a game using the spinners below. Each spinner is equally likely to stop on any of the four numbers. In the game, a player spins both spinners and calculates the product of the two numbers on which the spinners have stopped.

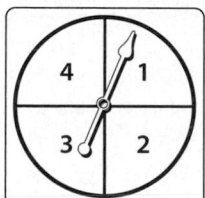 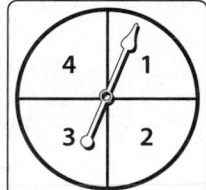

   a. What product has the greatest probability of occurring?

   b. What is the probability of that product occurring?

69. Simplify $(4^{-2} \cdot 5^0 \cdot 64)^3$.

   A $\dfrac{1}{64}$ C 320

   B 64   D 1024

70. GEOMETRY A rectangular prism has a width of $7x^3$ units, a length of $4x^2$ units, and a height of $3x$ units. What is the volume of the prism? (Lesson 7-1)

Solve each system of inequalities by graphing. (Lesson 6-6)

71. $y \geq 1$
$x < -1$

72. $y \geq -3$
$y - x < 1$

73. $y < 3x + 2$
$y \geq -2x + 4$

74. $y - 2x < 2$
$y - 2x > 4$

Solve each inequality. Check your solution. (Lesson 5-3)

75. $5(2h - 6) > 4h$

76. $22 \geq 4(b - 8) + 10$

77. $5(u - 8) \leq 3(u + 10)$

78. $8 + t \leq 3(t + 4) + 2$

79. $9n + 3(1 - 6n) \leq 21$

80. $-6(b + 5) > 3(b - 5)$

81. GRADES In a high school science class, a test is worth three times as much as a quiz. What is the student's average grade? (Lesson 2-9)

Science Grades	
Tests	Quizzes
85	82
92	75
	95

Evaluate each expression.

82. 9^2

83. 11^2

84. 10^6

85. 10^4

86. 3^5

87. 5^3

88. 12^3

89. 4^6

Rational Exponents

- You used the laws of exponents to find products and quotients of monomials.

1 Evaluate and rewrite expressions involving rational exponents.

2 Solve equations involving expressions with rational exponents.

- It's important to protect your skin with sunscreen to prevent damage. The sun protection factor (SPF) of a sunscreen indicates how well it protects you. Sunscreen with an SPF of f absorbs about p percent of the UV-B rays, where $p = 50f^{0.2}$.

 NewVocabulary
rational exponent
cube root
*n*th root
exponential equation

 Common Core State Standards

Content Standards

N.RN.1 Explain how the definition of the meaning of rational exponents follows from extending the properties of integer exponents to those values, allowing for a notation for radicals in terms of rational exponents.

N.RN.2 Rewrite expressions involving radicals and rational exponents using the properties of exponents.

Mathematical Practices
5 Use appropriate tools strategically.

1 **Rational Exponents** You know that an exponent represents the number of times that the base is used as a factor. But how do you evaluate an expression with an exponent that is not an integer like the one above? Let's investigate **rational exponents** by assuming that they behave like integer exponents.

$$\left(b^{\frac{1}{2}}\right)^2 = b^{\frac{1}{2}} \cdot b^{\frac{1}{2}}$$ Write as a multiplication expression.

$$= b^{\frac{1}{2} + \frac{1}{2}}$$ Product of Powers

$$= b^1 \text{ or } b$$ Simplify.

Thus, $b^{\frac{1}{2}}$ is a number with a square equal to b. So $b^{\frac{1}{2}} = \sqrt{b}$.

KeyConcept $b^{\frac{1}{2}}$

Words   For any nonnegative real number b, $b^{\frac{1}{2}} = \sqrt{b}$.

Examples $16^{\frac{1}{2}} = \sqrt{16}$ or 4 $38^{\frac{1}{2}} = \sqrt{38}$

Example 1 Radical and Exponential Forms

Write each expression in radical form, or write each radical in exponential form.

a. $25^{\frac{1}{2}}$

$25^{\frac{1}{2}} = \sqrt{25}$ Definition of $b^{\frac{1}{2}}$

$= 5$ Simplify.

b. $\sqrt{18}$

$\sqrt{18} = 18^{\frac{1}{2}}$ Definition of $b^{\frac{1}{2}}$

c. $5x^{\frac{1}{2}}$

$5x^{\frac{1}{2}} = 5\sqrt{x}$ Definition of $b^{\frac{1}{2}}$

d. $\sqrt{8p}$

$\sqrt{8p} = (8p)^{\frac{1}{2}}$ Definition of $b^{\frac{1}{2}}$

▶ **Guided**Practice

1A. $a^{\frac{1}{2}}$ **1B.** $\sqrt{22}$ **1C.** $(7w)^{\frac{1}{2}}$ **1D.** $2\sqrt{x}$

You know that to find the square root of a number a you find a number with a square of a. In the same way, you can find other roots of numbers. If $a^3 = b$, then a is the **cube root** of b, and if $a^n = b$ for a positive integer n, then a is an **nth root** of b.

KeyConcept nth Root

Words	For any real numbers a and b and any positive integer n, if $a^n = b$, then a is an nth root of b.
Symbols	If $a^n = b$, then $\sqrt[n]{b} = a$.
Example	Because $2^4 = 16$, 2 is a fourth root of 16; $\sqrt[4]{16} = 2$.

Since $3^2 = 9$ and $(-3)^2 = 9$, both 3 and -3 are square roots of 9. Similarly, since $2^4 = 16$ and $(-2)^4 = 16$, both 2 and -2 are fourth roots of 16. The positive roots are called *principal roots*. Radical symbols indicate principal roots, so $\sqrt[4]{16} = 2$.

Example 2 nth roots

Simplify.

a. $\sqrt[3]{27}$

$$\sqrt[3]{27} = \sqrt[3]{3 \cdot 3 \cdot 3}$$
$$= 3$$

b. $\sqrt[5]{32}$

$$\sqrt[5]{32} = \sqrt[5]{2 \cdot 2 \cdot 2 \cdot 2 \cdot 2}$$
$$= 2$$

GuidedPractice

2A. $\sqrt[3]{64}$

2B. $\sqrt[4]{10,000}$

Like square roots, nth roots can be represented by rational exponents.

$$\left(b^{\frac{1}{n}}\right)^n = \underbrace{b^{\frac{1}{n}} \cdot b^{\frac{1}{n}} \cdot \ldots \cdot b^{\frac{1}{n}}}_{n \text{ factors}}$$ Write as a multiplication expression.

$$= b^{\frac{1}{n} + \frac{1}{n} + \ldots + \frac{1}{n}}$$ Product of Powers

$$= b^1 \text{ or } b$$ Simplify.

Thus, $b^{\frac{1}{n}}$ is a number with an nth power equal to b. So $b^{\frac{1}{n}} = \sqrt[n]{b}$.

KeyConcept $b^{\frac{1}{n}}$

Words	For any positive real number b and any integer $n > 1$, $b^{\frac{1}{n}} = \sqrt[n]{b}$.
Example	$8^{\frac{1}{3}} = \sqrt[3]{8} = \sqrt[3]{2 \cdot 2 \cdot 2}$ or 2

Example 3 Evaluate $b^{\frac{1}{n}}$ Expressions

Simplify.

a. $125^{\frac{1}{3}}$

$$125^{\frac{1}{3}} = \sqrt[3]{125} \qquad b^{\frac{1}{n}} = \sqrt[n]{b}$$

$$= \sqrt[3]{5 \cdot 5 \cdot 5} \qquad 125 = 5^3$$

$$= 5 \qquad\qquad \text{Simplify.}$$

b. $1296^{\frac{1}{4}}$

$$1296^{\frac{1}{4}} = \sqrt[4]{1296} \qquad b^{\frac{1}{n}} = \sqrt[n]{b}$$

$$= \sqrt[4]{6 \cdot 6 \cdot 6 \cdot 6} \qquad 1296 = 6^4$$

$$= 6 \qquad\qquad \text{Simplify.}$$

▸ **Guided**Practice

3A. $27^{\frac{1}{3}}$

3B. $256^{\frac{1}{4}}$

The Power of a Power property allows us to extend the definition of $b^{\frac{1}{n}}$ to $b^{\frac{m}{n}}$.

$$b^{\frac{m}{n}} = \left(b^{\frac{1}{n}}\right)^m \qquad\qquad \text{Power of a Power}$$

$$= \left(\sqrt[n]{b}\right)^m \text{ or } \sqrt[n]{b^m} \qquad b^{\frac{1}{n}} = \sqrt[n]{b}$$

KeyConcept $b^{\frac{m}{n}}$

Words   For any positive real number b and any integers m and $n > 1$,
$$b^{\frac{m}{n}} = \left(\sqrt[n]{b}\right)^m \text{ or } \sqrt[n]{b^m}.$$

Example $8^{\frac{2}{3}} = \left(\sqrt[3]{8}\right)^2 = 2^2 \text{ or } 4$

Example 4 Evaluate $b^{\frac{m}{n}}$ Expressions

Simplify.

a. $64^{\frac{2}{3}}$

$$64^{\frac{2}{3}} = \left(\sqrt[3]{64}\right)^2 \qquad b^{\frac{m}{n}} = \left(\sqrt[n]{b}\right)^m$$

$$= \left(\sqrt[3]{4 \cdot 4 \cdot 4}\right)^2 \qquad 64 = 4^3$$

$$= 4^2 \text{ or } 16 \qquad \text{Simplify.}$$

b. $36^{\frac{3}{2}}$

$$36^{\frac{3}{2}} = \left(\sqrt[2]{36}\right)^3 \qquad b^{\frac{m}{n}} = \left(\sqrt[n]{b}\right)^m$$

$$= 6^3 \qquad \sqrt{36} = 6$$

$$= 216 \qquad \text{Simplify.}$$

▸ **Guided**Practice

4A. $27^{\frac{2}{3}}$

4B. $256^{\frac{5}{4}}$

2 Solve Exponential Equations

In an **exponential equation**, variables occur as exponents. The Power Property of Equality and the other properties of exponents can be used to solve exponential equations.

KeyConcept Power Property of Equality

Words	For any real number $b > 0$ and $b \neq 1$, $b^x = b^y$ if and only if $x = y$.
Examples	If $5^x = 5^3$, then $x = 3$. If $n = \frac{1}{2}$, then $4^n = 4^{\frac{1}{2}}$.

Example 5 Solve Exponential Equations

Solve each equation.

a. $6^x = 216$

$6^x = 216$	Original equation
$6^x = 6^3$	Rewrite 216 as 6^3.
$x = 3$	Property of Equality

CHECK
$$6^x = 216$$
$$6^3 \stackrel{?}{=} 216$$
$$216 = 216 \checkmark$$

b. $25^{x-1} = 5$

$25^{x-1} = 5$	Original equation
$(5^2)^{x-1} = 5$	Rewrite 25 as 5^2.
$5^{2x-2} = 5^1$	Power of a Power, Distributive Property
$2x - 2 = 1$	Power Property of Equality
$2x = 3$	Add 2 to each side.
$x = \frac{3}{2}$	Divide each side by 2.

CHECK
$$25^{x-1} = 5$$
$$25^{\frac{3}{2}-1} \stackrel{?}{=} 5$$
$$25^{\frac{1}{2}} = 5 \checkmark$$

> **Guided**Practice

5A. $5^x = 125$

5B. $12^{2x+3} = 144$

Real-World Example 6 Solve Exponential Equations

SUNSCREEN Refer to the beginning of the lesson. Find the SPF that absorbs 100% of UV-B rays.

$p = 50f^{0.2}$	Original equation
$100 = 50f^{0.2}$	$p = 100$
$2 = f^{0.2}$	Divide each side by 50.
$2 = f^{\frac{1}{5}}$	$0.2 = \frac{1}{5}$
$(2^5)^{\frac{1}{5}} = f^{\frac{1}{5}}$	$2 = 2^1 = (2^5)^{\frac{1}{5}}$
$2^5 = f$	Power Property of Equality
$32 = f$	Simplify.

> **Guided**Practice

6. CHEMISTRY The radius r of the nucleus of an atom of mass number A is $r = 1.2A^{\frac{1}{3}}$ femtometers. Find A if $r = 3.6$ femtometers.

Example 1   Write each expression in radical form, or write each radical in exponential form.

1. $12^{\frac{1}{2}}$ **2.** $3x^{\frac{1}{2}}$ **3.** $\sqrt{33}$ **4.** $\sqrt{8n}$

Examples 2–4 Simplify.

5. $\sqrt[3]{512}$ **6.** $\sqrt[5]{243}$ **7.** $343^{\frac{1}{3}}$ **8.** $\left(\frac{1}{16}\right)^{\frac{1}{4}}$

9. $343^{\frac{2}{3}}$ **10.** $81^{\frac{3}{4}}$ **(11)** $216^{\frac{4}{3}}$ **12.** $\left(\frac{1}{49}\right)^{\frac{3}{2}}$

Example 5   Solve each equation.

13. $8^x = 4096$ **14.** $3^{3x+1} = 81$ **15.** $4^{x-3} = 32$

Example 6   **16.** **CCSS** **TOOLS** A weir is used to measure water flow in a channel. For a rectangular broad crested weir, the flow Q in cubic feet per second is related to the weir length L in feet and height H of the water by $Q = 1.6LH^{\frac{3}{2}}$. Find the water height for a weir that is 3 feet long and has flow of 38.4 cubic feet per second.

Practice and Problem Solving

Extra Practice is on page R7.

Example 1   Write each expression in radical form, or write each radical in exponential form.

17. $15^{\frac{1}{2}}$ **18.** $24^{\frac{1}{2}}$ **19.** $4k^{\frac{1}{2}}$ **20.** $(12y)^{\frac{1}{2}}$

21. $\sqrt{26}$ **22.** $\sqrt{44}$ **23.** $2\sqrt{ab}$ **24.** $\sqrt{3xyz}$

Examples 2–4 Simplify.

25. $\sqrt[3]{8}$ **26.** $\sqrt[5]{1024}$ **27.** $\sqrt[3]{216}$ **28.** $\sqrt[4]{10{,}000}$

29. $\sqrt[3]{0.001}$ **30.** $\sqrt[4]{\frac{16}{81}}$ **31.** $1331^{\frac{1}{3}}$ **32.** $64^{\frac{1}{6}}$

33. $3375^{\frac{1}{3}}$ **34.** $512^{\frac{1}{9}}$ **35.** $\left(\frac{1}{81}\right)^{\frac{1}{4}}$ **36.** $\left(\frac{3125}{32}\right)^{\frac{1}{5}}$

37. $8^{\frac{2}{3}}$ **38.** $625^{\frac{3}{4}}$ **39.** $729^{\frac{5}{6}}$ **40.** $256^{\frac{3}{8}}$

41. $125^{\frac{4}{3}}$ **42.** $49^{\frac{5}{2}}$ **43.** $\left(\frac{9}{100}\right)^{\frac{3}{2}}$ **44.** $\left(\frac{8}{125}\right)^{\frac{4}{3}}$

Example 5

Solve each equation.

45. $3^x = 243$ **46.** $12^x = 144$ **47.** $16^x = 4$

48. $27^x = 3$ **49.** $9^x = 27$ **50.** $32^x = 4$

51. $2^{x-1} = 128$ **52.** $4^{2x+1} = 1024$ **53.** $6^{x-4} = 1296$

54. $9^{2x+3} = 2187$ **55.** $4^{3x} = 512$ **56.** $128^{3x} = 8$

Example 6

57. CONSERVATION Water collected in a rain barrel can be used to water plants and reduce city water use. Water flowing from an open rain barrel has velocity $v = 8h^{\frac{1}{2}}$, where v is in feet per second and h is the height of the water in feet. Find the height of the water if it is flowing at 16 feet per second.

58. ELECTRICITY The radius r in millimeters of a platinum wire L centimeters long with resistance 0.1 ohm is $r = 0.059L^{\frac{1}{2}}$. How long is a wire with radius 0.236 millimeter?

Write each expression in radical form, or write each radical in exponential form.

59. $17^{\frac{1}{3}}$ **60.** $q^{\frac{1}{4}}$ **61.** $7b^{\frac{1}{3}}$ **62.** $m^{\frac{2}{3}}$

63. $\sqrt[3]{29}$ **64.** $\sqrt[5]{h}$ **65.** $2\sqrt[3]{a}$ **66.** $\sqrt[3]{xy^2}$

Simplify.

67. $\sqrt[3]{0.027}$ **68.** $\sqrt[4]{\dfrac{n^4}{16}}$ **69.** $a^{\frac{1}{3}} \cdot a^{\frac{2}{3}}$ **70.** $c^{\frac{1}{2}} \cdot c^{\frac{3}{2}}$

71. $(8^2)^{\frac{2}{3}}$ **72.** $\left(y^{\frac{3}{4}}\right)^{\frac{1}{2}}$ **73.** $9^{-\frac{1}{2}}$ **74.** $16^{-\frac{3}{2}}$

75. $(3^2)^{-\frac{3}{2}}$ **76.** $\left(81^{\frac{1}{4}}\right)^{-2}$ **77.** $k^{-\frac{1}{2}}$ **78.** $\left(d^{\frac{4}{3}}\right)^{0}$

Solve each equation.

79. $2^{5x} = 8^{2x-4}$ **80.** $81^{2x-3} = 9^{x+3}$ **81.** $2^{4x} = 32^{x+1}$

82. $16^x = \dfrac{1}{2}$ **83.** $25^x = \dfrac{1}{125}$ **84.** $6^{8-x} = \dfrac{1}{216}$

85. CCSS MODELING The frequency f in hertz of the nth key on a piano is $f = 440\left(2^{\frac{1}{12}}\right)^{n-49}$.

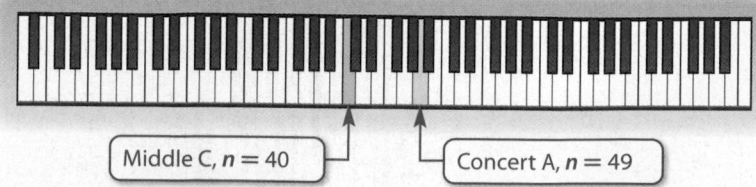

Middle C, $n = 40$ Concert A, $n = 49$

a. What is the frequency of Concert A?

b. Which note has a frequency of 220 Hz?

86. RANDOM WALKS Suppose you go on a walk where you choose the direction of each step at random. The path of a molecule in a liquid or a gas, the path of a foraging animal, and a fluctuating stock price are all modeled as random walks. The number of possible random walks w of n steps where you choose one of d directions at each step is $w = d^n$.

   a. How many steps have been taken in a 2-direction random walk if there are 4096 possible walks?

   b. How many steps have been taken in a 4-direction random walk if there are 65,536 possible walks?

   c. If a walk of 7 steps has 2187 possible walks, how many directions could be taken at each step?

87 SOCCER The radius r of a ball that holds V cubic units of air is modeled by $r = 0.62V^{\frac{1}{3}}$. What are the possible volumes of each size soccer ball?

Soccer Ball Dimensions	
Size	Diameter (in.)
3	7.3–7.6
4	8.0–8.3
5	8.6–9.0

88. ⬧ **MULTIPLE REPRESENTATIONS** In this problem, you will explore the graph of an exponential function.

   a. TABULAR Copy and complete the table below.

x	-2	$-\frac{3}{2}$	-1	$-\frac{1}{2}$	0	$\frac{1}{2}$	1	$\frac{3}{2}$	2
$f(x) = 4^x$									

   b. GRAPHICAL Graph $f(x)$ by plotting the points and connecting them with a smooth curve.

   c. VERBAL Describe the shape of the graph of $f(x)$. What are its key features? Is it linear?

H.O.T. Problems   *Use Higher-Order Thinking Skills*

89. OPEN ENDED Write two different expressions with rational exponents equal to $\sqrt{2}$.

90. **CCSS ARGUMENTS** Determine whether each statement is *always*, *sometimes*, or *never* true. Assume that x is a nonnegative real number. Explain your reasoning.

   a. $x^2 = x^{\frac{1}{2}}$

   b. $x^{-2} = x^{\frac{1}{2}}$

   c. $x^{\frac{1}{3}} = x^{\frac{1}{2}}$

   d. $\sqrt{x} = x^{\frac{1}{2}}$

   e. $\left(x^{\frac{1}{2}}\right)^2 = x$

   f. $x^{\frac{1}{2}} \cdot x^2 = x$

91. CHALLENGE For what values of x is $x = x^{\frac{1}{3}}$?

92. ERROR ANALYSIS Anna and Jamal are solving $128^x = 4$. Is either of them correct? Explain your reasoning.

Anna	Jamal
$128^x = 4$	$128^x = 4$
$(2^7)^x = 2^2$	$(2^7)^x = 4$
$2^{7x} = 2^2$	$2^{7x} = 4^1$
$7x = 2$	$7x = 1$
$x = \frac{2}{7}$	$x = \frac{1}{7}$

93. WRITING IN MATH Explain why 2 is the principal fourth root of 16.

94. What is the value of $16^{\frac{3}{4}} + 9^{\frac{3}{2}}$?

   A 5   **C** 25

   B 11   **D** 35

95. At a movie theater, the costs for various numbers of popcorn and hot dogs are shown.

Hot Dogs	Boxes of Popcorn	Total Cost
1	1	$8.50
2	4	$21.60

Which pair of equations can be used to find p, the cost of a box of popcorn, and h, the cost of a hot dog?

  F $p + h = 8.5$ **H** $p + h = 8.5$
       $p + 2h = 10.8$ $2p + 4h = 21.6$

  G $p + h = 8.5$ **J** $p + h = 8.5$
       $2h + 4p = 21.6$ $2p + 2h = 21.6$

96. SHORT RESPONSE Find the dimensions of the rectangle if its perimeter is 52 inches.

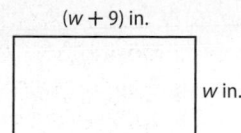

$(w + 9)$ in.

w in.

97. If $3^4 = 9^x$, then $x =$

   A 1

   B 2

   C 4

   D 5

Simplify each expression. Assume that no denominator equals zero. (Lesson 7-2)

98. $\dfrac{a^3b^5}{ab^0}$

99. $\dfrac{c^8d^{11}}{c^4d^5}$

100. $\dfrac{4r^3y^3z^6}{xyz^5}$

101. $\dfrac{a^5b^3c}{a^5bc}$

102. $\left(\dfrac{3m^4}{4p^2}\right)^2$

103. $\left(\dfrac{3df^2}{9d^2f}\right)^0$

104. GARDENING Felipe is planting a flower garden that is shaped like a trapezoid as shown at the right. Use the formula $A = \frac{1}{2}h(b_1 + b_2)$ to find the area of the garden. (Lesson 7-1)

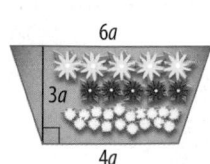

$6a$

$3a$

$4a$

Write each equation in slope-intercept form. (Lesson 4-2)

105. $y - 2 = 3(x - 1)$

106. $y - 5 = 6(x + 1)$

107. $y + 2 = -2(x + 5)$

108. $y + 3 = \frac{1}{2}(x + 4)$

109. $y - 1 = \frac{2}{3}(x + 9)$

110. $y + 3 = -\frac{1}{4}(x + 2)$

Find each power.

111. 10^3

112. 10^5

113. 10^{-1}

114. 10^{-4}

Scientific Notation

::Then	::Now	::Why?
● You used the laws of exponents to find products and quotients of monomials.	**1** Express numbers in scientific notation. **2** Find products and quotients of numbers expressed in scientific notation.	● Space tourism is a multibillion dollar industry. For a price of $20 million, a civilian can travel on a rocket or shuttle and visit the International Space Station (ISS) for a week.

 NewVocabulary
scientific notation

 Common Core State Standards

Content Standards
A.SSE.2 Use the structure of an expression to identify ways to rewrite it.

Mathematical Practices
3 Construct viable arguments and critique the reasoning of others.
6 Attend to precision.

1 **Scientific Notation** Very large and very small numbers such as $20 million can be cumbersome to use in calculations. For this reason, numbers are often expressed in scientific notation. A number written in **scientific notation** is of the form $a \times 10^n$, where $1 \le a < 10$ and n is an integer.

KeyConcept Standard Form to Scientific Notation

Step 1 Move the decimal point until it is to the right of the first nonzero digit. The result is a real number a.

Step 2 Note the number of places n and the direction that you moved the decimal point.

Step 3 If the decimal point is moved left, write the number as $a \times 10^n$. If the decimal point is moved right, write the number as $a \times 10^{-n}$.

Step 4 Remove the unnecessary zeros.

Example 1 Standard Form to Scientific Notation

Express each number in scientific notation.

a. 201,000,000

    Step 1 201,000,000 ⟶ 2.01000000 $a = 2.01000000$

    Step 2 The decimal point moved 8 places to the left, so $n = 8$.

    Step 3 $201{,}000{,}000 = 2.01000000 \times 10^8$

    Step 4 2.01×10^8

b. 0.000051

    Step 1 0.000051 ⟶ 00005.1 $a = 00005.1$

    Step 2 The decimal point moved 5 places to the right, so $n = 5$.

    Step 3 $0.000051 = 00005.1 \times 10^{-5}$

    Step 4 5.1×10^{-5}

▶ **Guided**Practice

1A. 68,700,000,000   **1B.** 0.0000725

You can also rewrite numbers in scientific notation in standard form.

KeyConcept Scientific Notation to Standard Form

Step 1 In $a \times 10^n$, note whether $n > 0$ or $n < 0$.

Step 2 If $n > 0$, move the decimal point n places right.
If $n < 0$, move the decimal point $-n$ places left.

Step 3 Insert zeros, decimal point, and commas as needed for place value.

Example 2 Scientific Notation to Standard Form

Express each number in standard form.

a. 6.32×10^9

  Step 1 The exponent is 9, so $n = 9$.

  Step 2 Since $n > 0$, move the decimal point 9 places to the right.
  $6.32 \times 10^9 \longrightarrow 6320000000$

  Step 3 $6.32 \times 10^9 = 6,320,000,000$ Rewrite; insert commas.

b. 4×10^{-7}

  Step 1 The exponent is -7, so $n = -7$.

  Step 2 Since $n < 0$, move the decimal point 7 places to the left.
  $4 \times 10^{-7} \longrightarrow 0000004$

  Step 3 $4 \times 10^{-7} = 0.0000004$ Rewrite; insert a 0 before the decimal point.

▶ **Guided**Practice

2A. 3.201×10^6 **2B.** 9.03×10^{-5}

2 Product and Quotients in Scientific Notation You can use scientific notation to simplify multiplying and dividing very large and very small numbers.

Example 3 Multiply with Scientific Notation

Evaluate $(3.5 \times 10^{-3})(7 \times 10^5)$. Express the result in both scientific notation and standard form.

$(3.5 \times 10^{-3})(7 \times 10^5)$ Original expression
$= (3.5 \times 7)(10^{-3} \times 10^5)$ Commutative and Associative Properties
$= 24.5 \times 10^2$ Product of Powers
$= (2.45 \times 10^1) \times 10^2$ $24.5 = 2.45 \times 10$
$= 2.45 \times 10^3$ or 2450 Product of Powers

▶ **Guided**Practice

Evaluate each product. Express the results in both scientific notation and standard form.

3A. $(6.5 \times 10^{12})(8.7 \times 10^{-15})$ **3B.** $(7.8 \times 10^{-4})^2$

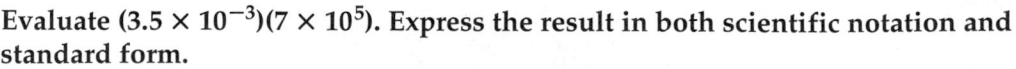

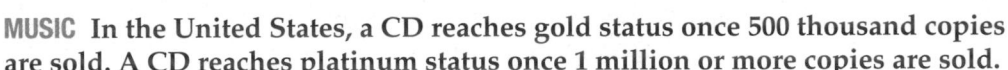

Example 4 Divide with Scientific Notation

Evaluate $\dfrac{3.066 \times 10^8}{7.3 \times 10^3}$. Express the result in both scientific notation and standard form.

$$\dfrac{3.066 \times 10^8}{7.3 \times 10^3} = \left(\dfrac{3.066}{7.3}\right)\left(\dfrac{10^8}{10^3}\right) \qquad \text{Product rule for fractions}$$

$$= 0.42 \times 10^5 \qquad \text{Quotient of Powers}$$

$$= 4.2 \times 10^{-1} \times 10^5 \qquad 0.42 = 4.2 \times 10^{-1}$$

$$= 4.2 \times 10^4 \qquad \text{Product of Powers}$$

$$= 42{,}000 \qquad \text{Standard form}$$

StudyTip

Quotient of Powers
Recall that the Quotient of Powers Property is only valid for powers that have the same base. Since 10^8 and 10^3 have the same base, the property applies.

GuidedPractice

Evaluate each quotient. Express the results in both scientific notation and standard form.

4A. $\dfrac{2.3958 \times 10^3}{1.98 \times 10^8}$

4B. $\dfrac{1.305 \times 10^3}{1.45 \times 10^{-4}}$

Real-World Example 5 Use Scientific Notation

MUSIC In the United States, a CD reaches gold status once 500 thousand copies are sold. A CD reaches platinum status once 1 million or more copies are sold.

a. Express the number of copies of CDs that need to be sold to reach each status in standard notation.

gold status: 500 thousand = 500,000; platinum status: 1 million = 1,000,000

b. Write each number in scientific notation.

gold status: $500{,}000 = 5 \times 10^5$; platinum status: $1{,}000{,}000 = 1 \times 10^6$

c. How many copies of a CD have sold if it has gone platinum 13 times? Write your answer in scientific notation and standard form.

A CD reaches platinum status once it sells 1 million records. Since the CD has gone platinum 13 times, we need to multiply by 13.

$$(13)(1 \times 10^6) \qquad \text{Original expression}$$

$$= (13 \times 1)(10^6) \qquad \text{Associative Property}$$

$$= 13 \times 10^6 \qquad 13 \times 1 = 13$$

$$= (1.3 \times 10^1) \times 10^6 \qquad 13 = 1.3 \times 10$$

$$= 1.3 \times 10^7 \qquad \text{Product of Powers}$$

$$= 13{,}000{,}000 \qquad \text{Standard form}$$

Real-World Link

The platinum award was created in 1976. In 2004, the criteria for the award was extended to digital sales. The top-selling artist of all time is the Beatles with 170 million units sold.

Source: Recording Industry Association of America

GuidedPractice

5. SATELLITE RADIO Suppose a satellite radio company earned $125.4 million in one year.

 A. Write this number in standard form.

 B. Write this number in scientific notation.

 C. If the following year the company earned 2.5 times the amount earned the previous year, determine the amount earned. Write your answer in scientific notation and standard form.

akg-images

Example 1   Express each number in scientific notation.

1. 185,000,000   2. 1,902,500,000
3. 0.000564   4. 0.00000804

MONEY Express each number in scientific notation.

5. Teens spend $13 billion annually on clothing.

6. Teens have an influence on their families' spending habit. They control about $1.5 billion of discretionary income.

Example 2   Express each number in standard form.

7. 1.98×10^7 8. 4.052×10^6
9. 3.405×10^{-8} 10. 6.8×10^{-5}

Example 3   Evaluate each product. Express the results in both scientific notation and standard form.

11. $(1.2 \times 10^3)(1.45 \times 10^{12})$ 12. $(7.08 \times 10^{14})(5 \times 10^{-9})$
13. $(5.18 \times 10^2)(9.1 \times 10^{-5})$ 14. $(2.18 \times 10^{-2})^2$

Example 4   Evaluate each quotient. Express the results in both scientific notation and standard form.

15. $\dfrac{1.035 \times 10^8}{2.3 \times 10^4}$ 16. $\dfrac{2.542 \times 10^5}{4.1 \times 10^{-10}}$

17. $\dfrac{1.445 \times 10^{-7}}{1.7 \times 10^5}$ 18. $\dfrac{2.05 \times 10^{-8}}{4 \times 10^{-2}}$

Example 5   19. **CCSS PRECISION** Salvador bought an air purifier to help him deal with his allergies. The filter in the purifier will stop particles as small as one hundredth of a micron. A micron is one millionth of a millimeter.

a. Write one hundredth and one micron in standard form.

b. Write one hundredth and one micron in scientific notation.

c. What is the smallest size particle in meters that the filter will stop? Write the result in both standard form and scientific notation.

Practice and Problem Solving   **Extra Practice is on page R7.**

Example 1   Express each number in scientific notation.

20. 1,220,000   ㉑ 58,600,000   22. 1,405,000,000,000
23. 0.0000013   24. 0.000056   25. 0.000000000709

EMAIL Express each number in scientific notation.

26. Approximately 100 million emails sent to the President are put into the National Archives.

27. By 2015, the email security market will generate $6.5 billion.

Example 2   Express each number in standard form.

28. 1×10^{12} 29. 9.4×10^7 30. 8.1×10^{-3}
31. 5×10^{-4} 32. 8.73×10^{11} 33. 6.22×10^{-6}

Example 2   **INTERNET** Express each number in standard form.

     34. About 2.1×10^7 people aged 12 to 17 use the Internet.

     35. Approximately 1.1×10^7 teens go online daily.

Evaluate each product or quotient. Express the results in both scientific notation and standard form.

     36. $(3.807 \times 10^3)(5 \times 10^2)$

     37. $\dfrac{9.6 \times 10^3}{1.2 \times 10^{-4}}$

     38. $\dfrac{2.88 \times 10^3}{1.2 \times 10^{-5}}$

     39 $(6.5 \times 10^7)(7.2 \times 10^{-2})$

     40. $(9.5 \times 10^{-18})(9 \times 10^9)$

     41. $\dfrac{8.8 \times 10^3}{4 \times 10^{-4}}$

     42. $\dfrac{9.15 \times 10^{-3}}{6.1 \times 10}$

     43. $(1.4 \times 10^6)^2$

     44. $(2.58 \times 10^2)(3.6 \times 10^6)$

     45. $\dfrac{5.6498 \times 10^{10}}{8.2 \times 10^4}$

     46. $\dfrac{1.363 \times 10^{16}}{2.9 \times 10^6}$

     47. $(5 \times 10^3)(1.8 \times 10^{-7})$

     48. $(2.3 \times 10^{-3})^2$

     49. $\dfrac{6.25 \times 10^{-4}}{1.25 \times 10^2}$

     50. $\dfrac{3.75 \times 10^{-9}}{1.5 \times 10^{-4}}$

     51. $(7.2 \times 10^7)^2$

     52. $\dfrac{8.6 \times 10^4}{2 \times 10^{-6}}$

     53. $(6.3 \times 10^{-5})^2$

Example 5   **54.** **ASTRONOMY** The distance between Earth and the Sun varies throughout the year. Earth is closest to the Sun in January when the distance is 91.4 million miles. In July, the distance is greatest at 94.4 million miles.

     a. Write 91.4 million in both standard form and in scientific notation.

     b. Write 94.4 million in both standard form and in scientific notation.

     c. What is the percent increase in distance from January to July? Round to the nearest tenth of a percent.

Evaluate each product or quotient. Express the results in both scientific notation and standard form.

     55. $(4.65 \times 10^{-2})(5.91 \times 10^6)$

     56. $\dfrac{2.548 \times 10^5}{2.8 \times 10^{-2}}$

     57. $\dfrac{2.135 \times 10^5}{3.5 \times 10^{12}}$

     58. $(3.16 \times 10^{-2})^2$

     59. $(2.01 \times 10^{-4})(8.9 \times 10^{-3})$

     60. $\dfrac{5.184 \times 10^{-5}}{7.2 \times 10^3}$

     61. $(9.04 \times 10^6)(5.2 \times 10^{-4})$

     62. $\dfrac{1.032 \times 10^{-4}}{8.6 \times 10^{-5}}$

LIGHT The speed of light is approximately 3×10^8 **meters per second.**

     63. Write an expression to represent the speed of light in kilometers per second.

     64. Write an expression to represent the speed of light in kilometers per hour.

     65. Make a table to show how many kilometers light travels in a day, a week, a 30-day month, and a 365-day year. Express your results in scientific notation.

     66. **CCSS MODELING** A recent cell phone study showed that company A's phone processes up to 7.95×10^5 bits of data every second. Company B's phone processes up to 1.41×10^6 bits of data every second. Evaluate and interpret $\dfrac{1.41 \times 10^6}{7.95 \times 10^5}$.

67 **EARTH** The population of Earth is about 6.623×10^9. The land surface of Earth is 1.483×10^8 square kilometers. What is the population density for the land surface area of Earth?

68. RIVERS A drainage basin separated from adjacent basins by a ridge, hill, or mountain is known as a watershed. The watershed of the Amazon River is 2,300,000 square miles. The watershed of the Mississippi River is 1,200,000 square miles.

   a. Write each of these numbers in scientific notation.

   b. How many times as large is the Amazon River watershed as the Mississippi River watershed?

 69. AGRICULTURE In a recent year, farmers planted approximately 92.9 million acres of corn. They also planted 64.1 million acres of soybeans and 11.1 million acres of cotton.

   a. Write each of these numbers in scientific notation and in standard form.

   b. How many times as much corn was planted as soybeans? Write your results in standard form and in scientific notation. Round your answer to four decimal places.

   c. How many times as much corn was planted as cotton? Write your results in standard form and in scientific notation. Round your answer to four decimal places.

70. REASONING Which is greater, 100^{10} or 10^{100}? Explain your reasoning.

71. ERROR ANALYSIS Syreeta and Pete are solving a division problem with scientific notation. Is either of them correct? Explain your reasoning.

Syreeta	Pete
$\dfrac{3.65 \times 10^{-12}}{5 \times 10^{5}} = 0.73 \times 10^{-17}$	$\dfrac{3.65 \times 10^{-12}}{5 \times 10^{5}} = 0.73 \times 10^{-17}$
$= 7.3 \times 10^{-16}$	$= 7.3 \times 10^{-18}$

72. CHALLENGE Order these numbers from least to greatest without converting them to standard form.

$$5.46 \times 10^{-3}, \; 6.54 \times 10^3, \; 4.56 \times 10^{-4}, \; -5.64 \times 10^4, \; -4.65 \times 10^5$$

73. CCSS ARGUMENTS Determine whether the statement is *always*, *sometimes*, or *never* true. Give examples or a counterexample to verify your reasoning.

When multiplying two numbers written in scientific notation, the resulting number can have no more than two digits to the left of the decimal point.

74. OPEN ENDED Write two numbers in scientific notation with a product of 1.3×10^{-3}. Then name two numbers in scientific notation with a quotient of 1.3×10^{-3}.

75. WRITING IN MATH Write the steps that you would use to divide two numbers written in scientific notation. Then describe how you would write the results in standard form. Demonstrate by finding $\frac{a}{b}$ for $a = 2 \times 10^3$ and $b = 4 \times 10^5$.

76. Which number represents 0.05604×10^8 written in standard form?

  A 0.0000000005604   **C** 5,604,000

  B 560,400   **D** 50,604,000

77. Toni left school and rode her bike home. The graph below shows the relationship between her distance from the school and time.

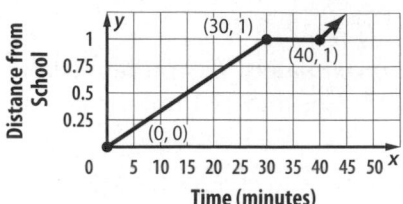

Which explanation could account for the section of the graph from $x = 30$ to $x = 40$?

  F Toni rode her bike down a hill.

  G Toni ran all the way home.

  H Toni stopped at a friend's house on her way home.

  J Toni returned to school to get her mathematics book.

78. SHORT RESPONSE In his first four years of coaching football, Coach Delgato's team won 5 games the first year, 10 games the second year, 8 games the third year, and 7 games the fourth year. How many games does the team need to win during the fifth year to have an average of 8 wins per year?

79. The table shows the relationship between Calories and grams of fat contained in an order of fried chicken from various restaurants.

Calories	305	410	320	500	510	440
Fat (g)	28	34	28	41	42	38

Assuming that the data can best be described by a linear model, about how many grams of fat would you expect to be in a 275-Calorie order of fried chicken?

  A 22

  B 25

  C 28

  D 30

Spiral Review

80. HEALTH A ponderal index p is a measure of a person's body based on height h in centimeters and mass m in kilograms. One such formula is $p = 100m^{\frac{1}{3}}h^{-1}$. If a person who is 182 centimeters tall has a ponderal index of about 2.2, how much does the person weigh in kilograms? (Lesson 7-3)

Simplify. Assume that no denominator is equal to zero. (Lesson 7-2)

81. $\dfrac{8^9}{8^6}$

82. $\dfrac{6^5}{6^3}$

83. $\dfrac{r^8 t^{12}}{r^2 t^7}$

84. $\left(\dfrac{3a^4b^4}{8c^2}\right)^4$

85. $\left(\dfrac{5d^3g^2}{3h^4}\right)^2$

86. $\left(\dfrac{4n^2p^4}{8p^3}\right)^3$

87. CHEMISTRY Lemon juice is 10^2 times as acidic as tomato juice. Tomato juice is 10^3 times as acidic as egg whites. How many times as acidic is lemon juice as egg whites? (Lesson 7-2)

Skills Review

Evaluate $a(b^x)$ for each of the given values.

88. $a = 1, b = 2, x = 4$

89. $a = 4, b = 1, x = 7$

90. $a = 5, b = 3, x = 0$

91. $a = 0, b = 6, x = 8$

92. $a = -2, b = 3, x = 1$

93. $a = -3, b = 5, x = 2$

Simplify each expression. (Lesson 7-1)

1. $(x^3)(4x^5)$

2. $(m^2p^5)^3$

3. $[(2xy^3)^2]^3$

4. $(6ab^3c^4)(-3a^2b^3c)$

5. **MULTIPLE CHOICE** Express the volume of the solid as a monomial. (Lesson 7-1)

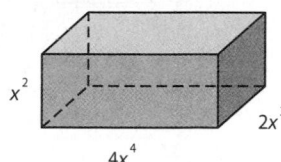

A $6x^9$

B $8x^9$

C $8x^{24}$

D $7x^{24}$

Simplify each expression. Assume that no denominator equals 0. (Lesson 7-2)

6. $\left(\dfrac{2a^4b^3}{c^6}\right)^3$

7. $\dfrac{2xy^0}{6x}$

8. $\dfrac{m^7n^4p}{m^3n^3p}$

9. $\dfrac{p^4t^{-2}}{r^{-5}}$

10. **ASTRONOMY** Physicists estimate that the number of stars in the universe has an order of magnitude of 10^{21}. The number of stars in the Milky Way galaxy is around 100 billion. Using orders of magnitude, how many times as many stars are there in the universe as the Milky Way? (Lesson 7-2)

Write each expression in radical form, or write each radical in exponential form. (Lesson 7-3)

11. $42^{\frac{1}{2}}$

12. $11x^{\frac{1}{2}}$

13. $(11g)^{\frac{1}{2}}$

14. $\sqrt{55}$

15. $\sqrt{5k}$

16. $4\sqrt{p}$

Simplify. (Lesson 7-3)

17. $\sqrt[3]{729}$

18. $\sqrt[4]{625}$

19. $1331^{\frac{1}{3}}$

20. $\left(\dfrac{16}{81}\right)^{\frac{1}{4}}$

21. $8^{\frac{2}{3}}$

22. $625^{\frac{3}{4}}$

23. $216^{\frac{5}{3}}$

24. $\left(\dfrac{1}{4}\right)^{\frac{3}{2}}$

Solve each equation. (Lesson 7-3)

25. $4^x = 4096$

26. $5^{2x+1} = 125$

27. $4^{x-3} = 128$

Express each number in scientific notation. (Lesson 7-4)

28. 0.00000054

29. 0.0042

30. 234,000

31. 418,000,000

Express each number in standard form. (Lesson 7-4)

32. 4.1×10^{-3}

33. 2.74×10^6

34. 3×10^9

35. 9.1×10^{-5}

Evaluate each product or quotient. Express the results in scientific notation. (Lesson 7-4)

36. $(2.13 \times 10^2)(3 \times 10^5)$

37. $(7.5 \times 10^6)(2.5 \times 10^{-2})$

38. $\dfrac{7.5 \times 10^8}{2.5 \times 10^4}$

39. $\dfrac{6.6 \times 10^5}{2 \times 10^{-3}}$

40. **MAMMALS** A blue whale has been caught that was 4.2×10^5 pounds. The smallest mammal is a bumblebee bat, which is about 0.0044 pound. (Lesson 7-4)

   a. Write the whale's weight in standard form.

   b. Write the bat's weight in scientific notation.

   c. How many orders of magnitude as big as a blue whale is a bumblebee bat?

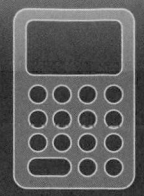

Graphing Technology Lab
Family of Exponential Functions

An **exponential function** is a function of the form $y = ab^x$, where $a \neq 0$, $b > 0$, and $b \neq 1$. You have studied the effects of changing parameters in linear functions. You can use a graphing calculator to analyze how changing the parameters a and b affects the graphs in the family of exponential functions.

CCSS Common Core State Standards
Content Standards
F.IF.7e Graph exponential and logarithmic functions, showing intercepts and end behavior, and trigonometric functions, showing period, midline, and amplitude.
F.BF.3 Identify the effect on the graph of replacing $f(x)$ by $f(x) + k$, $kf(x)$, $f(kx)$, and $f(x + k)$ for specific values of k (both positive and negative); find the value of k given the graphs. Experiment with cases and illustrate an explanation of the effects on the graph using technology.

Activity 1 b in $y = b^x$, $b > 1$

Graph the set of equations on the same screen.
Describe any similarities and differences among the graphs.

$y = 2^x$, $y = 3^x$, $y = 6^x$

Enter the equations in the $\boxed{Y=}$ list and graph.

There are many similarities in the graphs. The domain for each function is all real numbers, and the range is all positive real numbers. The functions are increasing over the entire domain. The graphs do not display any line symmetry.

Use the $\boxed{\text{ZOOM}}$ feature to investigate the key features of the graphs.

Zooming in twice on a point near the origin allows closer inspection of the graphs. The y-intercept is 1 for all three graphs.

Tracing along the graphs reveals that there are no x-intercepts, no maxima and no minima.

The graphs are different in that the graphs for the equations in which b is greater are steeper.

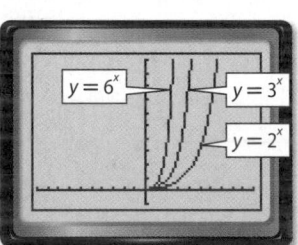

[−10, 10] scl: 1 by [−10, 100] scl: 10

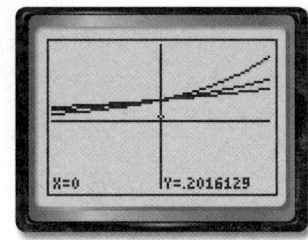

[−0.625, 0.625] scl: 1 by
[−3.25..., 3.63...] scl: 10

The effect of b on the graph is different when $0 < b < 1$.

Activity 2 b in $y = b^x$, $0 < b < 1$

Graph the set of equations on the same screen.
Describe any similarities and differences among the graphs.

$y = \left(\frac{1}{2}\right)^x$, $y = \left(\frac{1}{3}\right)^x$, $y = \left(\frac{1}{6}\right)^x$

The domain for each function is all real numbers, and the range is all positive real numbers. The function values are all positive and the functions are decreasing over the entire domain. The graphs display no line symmetry. There are no x-intercepts, and the y-intercept is 1 for all three graphs. There are no maxima or minima.

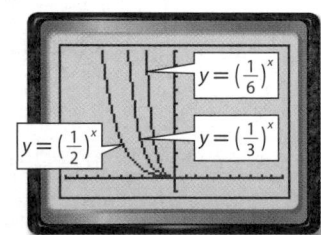

[−10, 10] scl: 1 by [−10, 100] scl: 10

However, the graphs in which b is lesser are steeper.

Activity 3 a in $y = ab^x$, $a > 0$

Graph each set of equations on the same screen. Describe any similarities and differences among the graphs.

$y = 2^x$, $y = 3(2^x)$, $y = \frac{1}{6}(2^x)$

The domain for each function is all real numbers, and the range is all positive real numbers. The functions are increasing over the entire domain. The graphs do not display any line symmetry.

Use the ZOOM feature to investigate the key features of the graphs.

Zooming in twice on a point near the origin allows closer inspection of the graphs.

Tracing along the graphs reveals that there are no x-intercepts, no maxima and no minima.

However, the graphs in which a is greater are steeper. The y-intercept is 1 in the graph of $y = 2^x$, 3 in $y = 3(2^x)$, and $\frac{1}{6}$ in $y = \frac{1}{6}(2^x)$.

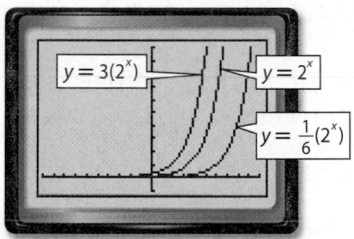

[−10, 10] scl: 1 by [−10, 100] scl: 10

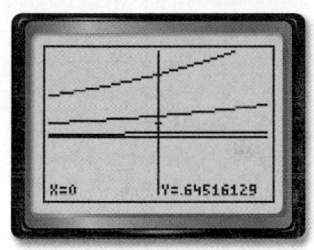

[−0.625, 0.625] scl: 1 by
[−2.79..., 4.08...] scl: 10

Activity 4 a in $y = ab^x$, $a < 0$

Graph each set of equations on the same screen. Describe any similarities and differences among the graphs.

$y = -2^x$, $y = -3(2^x)$, $y = -\frac{1}{6}(2^x)$

The domain for each function is all real numbers, and the range is all negative real numbers. The functions are decreasing over the entire domain. The graphs do not display any line symmetry.

There are no x-intercepts, no maxima and no minima.

However, the graphs in which the absolute value of a is greater are steeper. The y-intercept is -1 in the graph of $y = -2^x$, -3 in $y = -3(2^x)$, and $-\frac{1}{6}$ in $y = -\frac{1}{6}(2^x)$.

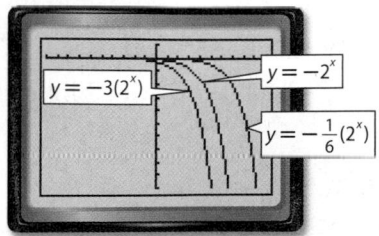

[−10, 10] scl: 1 by [−100, 10] scl: 10

Model and Analyze

1. How does b affect the graph of $y = ab^x$ when $b > 1$ and when $0 < b < 1$? Give examples.

2. How does a affect the graph of $y = ab^x$ when $a > 0$ and when $a < 0$? Give examples.

3. **CCSS REGULARITY** Make a conjecture about the relationship of the graphs of $y = 3^x$ and $y = \left(\frac{1}{3}\right)^x$. Verify your conjecture by graphing both functions.

Exponential Functions

:: Then	:: Now	:: Why?
• You evaluated numerical expressions involving exponents.	• 1 Graph exponential functions. 2 Identify data that display exponential behavior.	• Tarantulas can appear scary with their large hairy bodies and legs, but they are harmless to humans. The graph shows a tarantula spider population that increases over time. Notice that the graph is not linear.

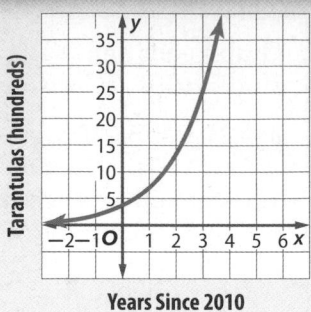

The graph represents the function $y = 3(2)^x$. This is an example of an *exponential* function.

Years Since 2010

1 Graph Exponential Functions An **exponential function** is a function of the form $y = ab^x$, where $a \neq 0$, $b > 0$, and $b \neq 1$. Notice that the base is a constant and the exponent is a variable. Exponential functions are nonlinear.

KeyConcept Exponential Function

Words	An exponential function is a function that can be described by an equation of the form $y = ab^x$, where $a \neq 0$, $b > 0$, and $b \neq 1$.
Examples	$y = 2(3)^x$ $\qquad\qquad$ $y = 4^x$ $\qquad\qquad$ $y = \left(\frac{1}{2}\right)^x$

Example 1 Graph with $a > 0$ and $b > 1$

Graph $y = 3^x$. Find the y-intercept, and state the domain and range.

The graph crosses the y-axis at 1, so the y-intercept is 1. The domain is all real numbers, and the range is all positive real numbers.

Notice that the graph approaches the x-axis but there is no x-intercept. The graph is increasing on the entire domain.

x	3^x	y
−2	3^{-2}	$\frac{1}{9}$
−1	3^{-1}	$\frac{1}{3}$
0	3^0	1
$\frac{1}{2}$	$3^{\frac{1}{2}}$	≈1.73
1	3^1	3
2	3^2	9

> **Guided**Practice

1. Graph $y = 7^x$. Find the y-intercept, and state the domain and range.

Functions of the form $y = ab^x$, where $a > 0$ and $b > 1$, are called **exponential growth functions** and all have the same shape as the graph in Example 1. Functions of the form $y = ab^x$, where $a > 0$ and $0 < b < 1$ are called **exponential decay functions** and also have the same general shape.

Example 2 Graph with *a* > 0 and 0 < *b* < 1

Graph $y = \left(\frac{1}{3}\right)^x$. **Find the** *y*-**intercept, and state the domain and range.**

The *y*-intercept is 1. The domain is all real numbers, and the range is all positive real numbers. Notice that as *x* increases, the *y*-values decrease less rapidly.

x	$\left(\frac{1}{3}\right)^x$	y
−2	$\left(\frac{1}{3}\right)^{-2}$	9
0	$\left(\frac{1}{3}\right)^{0}$	1
2	$\left(\frac{1}{3}\right)^{2}$	$\frac{1}{9}$

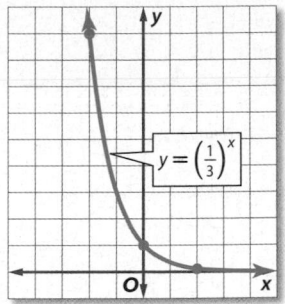

▶ **Guided**Practice

2. Graph $y = \left(\frac{1}{2}\right)^x - 1$. Find the *y*-intercept, and state the domain and range.

The key features of the graphs of exponential functions can be summarized as follows.

KeyConcept Graphs of Exponential Functions

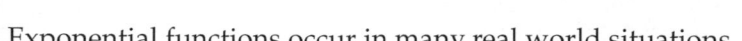

Exponential Growth Functions	**Exponential Decay Functions**
Equation: $f(x) = ab^x$, $a > 0$, $b > 1$	**Equation:** $f(x) = ab^x$, $a > 0$, $0 < b < 1$
Domain, Range: all reals; all positive reals	**Domain, Range:** all reals; all negative reals
Intercepts: one *y*-intercept, no *x*-intercepts	**Intercepts:** one *y*-intercept, no *x*-intercepts
End behavior: as *x* increases, $f(x)$ increases; as *x* decreases, $f(x)$ approaches 0	**End behavior:** as *x* increases, $f(x)$ approaches 0; as *x* decreases, $f(x)$ increases

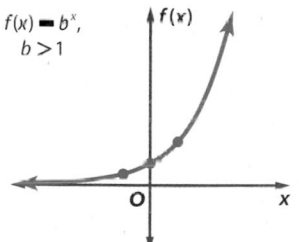

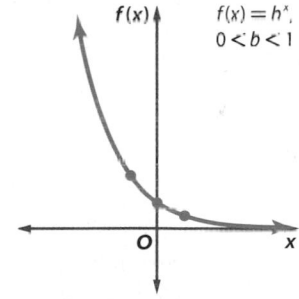

Exponential functions occur in many real world situations.

🌐 **Real-World Example 3** Use Exponential Functions to Solve Problems

SODA The function $C = 179(1.029)^t$ models the amount of soda consumed in the world, where *C* is the amount consumed in billions of liters and *t* is the number of years since 2000.

a. Graph the function. What values of *C* and *t* are meaningful in the context of the problem?

Since *t* represents time, $t > 0$. At $t = 0$, the consumption is 179 billion liters. Therefore, in the context of this problem, $C > 179$ is meaningful.

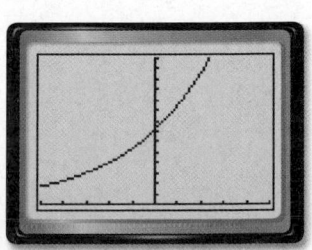

[−50, 50] scl: 10 by [0, 350] scl: 25

b. How much soda was consumed in 2005?

$$C = 179(1.029)^t \qquad \text{Original equation}$$
$$= 179(1.029)^5 \qquad t = 5$$
$$\approx 206.5 \qquad \text{Use a calculator.}$$

The world soda consumption in 2005 was approximately 206.5 billion liters.

▶ **Guided**Practice

3. BIOLOGY A certain bacteria population doubles every 20 minutes. Beginning with 10 cells in a culture, the population can be represented by the function $B = 10(2)^t$, where B is the number of bacteria cells and t is the time in 20 minute increments. How many will there be after 2 hours?

2 **Identify Exponential Behavior** Recall from Lesson 3-3 that linear functions have a constant rate of change. Exponential functions do not have constant rates of change, but they do have constant ratios.

Problem-Solving Tip

Make an Organized List
Making an organized list of x-values and corresponding y-values is helpful in graphing the function. It can also help you identify patterns in the data.

Example 4 **Identify Exponential Behavior**

Determine whether the set of data shown below displays exponential behavior. Write *yes* or *no*. Explain why or why not.

x	0	5	10	15	20	25
y	64	32	16	8	4	2

Method 1 Look for a pattern.

The domain values are at regular intervals of 5. Look for a common factor among the range values.

$$64 \quad 32 \quad 16 \quad 8 \quad 4 \quad 2$$
$$\times\frac{1}{2} \quad \times\frac{1}{2} \quad \times\frac{1}{2} \quad \times\frac{1}{2} \quad \times\frac{1}{2}$$

The range values differ by the common factor of $\frac{1}{2}$.

Since the domain values are at regular intervals and the range values differ by a positive common factor, the data are probably exponential. Its equation may involve $\left(\frac{1}{2}\right)^x$.

Method 2 Graph the data.

Plot the points and connect them with a smooth curve. The graph shows a rapidly decreasing value of y as x increases. This is a characteristic of exponential behavior in which the base is between 0 and 1.

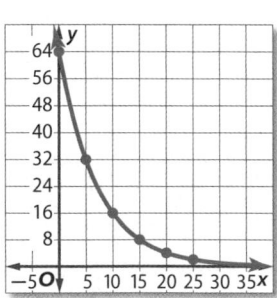

▶ **Guided**Practice

4. Determine whether the set of data shown below displays exponential behavior. Write *yes* or *no*. Explain why or why not.

x	0	3	6	9	12	15
y	12	16	20	24	28	32

Examples 1–2 Graph each function. Find the *y*-intercept and state the domain and range.

1. $y = 2^x$

2. $y = -5^x$

3. $y = -\left(\dfrac{1}{5}\right)^x$

4. $y = 3\left(\dfrac{1}{4}\right)^x$

5. $f(x) = 6^x + 3$

6. $f(x) = 2 - 2^x$

Example 3

7. BIOLOGY The function $f(t) = 100(1.05)^t$ models the growth of a fruit fly population, where $f(t)$ is the number of flies and t is time in days.

   a. What values for the domain and range are reasonable in the context of this situation? Explain.

   b. After two weeks, approximately how many flies are in this population?

Example 4 Determine whether the set of data shown below displays exponential behavior. Write *yes* or *no*. Explain why or why not.

8.

x	1	2	3	4	5	6
y	−4	−2	0	2	4	6

9.

x	2	4	6	8	10	12
y	1	4	16	64	256	1024

Practice and Problem Solving   Extra Practice is on page R7.

Examples 1–2 Graph each function. Find the *y*-intercept and state the domain and range.

10. $y = 2 \cdot 8^x$

11. $y = 2 \cdot \left(\dfrac{1}{6}\right)^x$

12. $y = \left(\dfrac{1}{12}\right)^x$

13. $y = -3 \cdot 9^x$

14. $y = -4 \cdot 10^x$

15. $y = 3 \cdot 11^x$

16. $y - 4^x + 3$

17. $y = \dfrac{1}{2}(2^x - 8)$

18. $y = 5(3^x) + 1$

19. $y = -2(3^x) + 5$

Example 3

20. CCSS MODELING A population of bacteria in a culture increases according to the model $p = 300(2.7)^{0.02t}$, where t is the number of hours and $t = 0$ corresponds to 9:00 A.M.

   a. Use this model to estimate the number of bacteria at 11 A.M.

   b. Graph the function and name the *p*-intercept. Describe what the *p*-intercept represents, and describe a reasonable domain and range for this situation.

Example 4 Determine whether the set of data shown below displays exponential behavior. Write *yes* or *no*. Explain why or why not.

21.

x	−4	0	4	8	12
y	2	−4	8	−16	32

22.

x	−6	−3	0	3
y	5	10	15	20

23.

x	−8	−6	−4	−2
y	0.25	0.5	1	2

24.

x	20	30	40	50	60
y	1	0.4	0.16	0.064	0.0256

25 PHOTOGRAPHY Jameka is enlarging a photograph to make a poster for school. She will enlarge the picture repeatedly at 150%. The function $P = 1.5^x$ models the new size of the picture being enlarged, where x is the number of enlargements. How many times as big is the picture after 4 enlargements?

26. FINANCIAL LITERACY Daniel deposited $500 into a savings account and after 8 years, his investment is worth $807.07. The equation $A = d(1.005)^{12t}$ models the value of Daniel's investment A after t years with an initial deposit d.

  a. What would the value of Daniel's investment be if he had deposited $1000?

  b. What would the value of Daniel's investment be if he had deposited $250?

  c. Interpret $d(1.005)^{12t}$ to explain how the amount of the original deposit affects the value of Daniel's investment.

Identify each function as *linear*, *exponential*, or *neither*.

27. 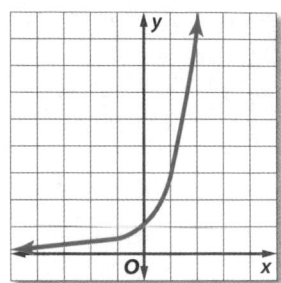   **28.** 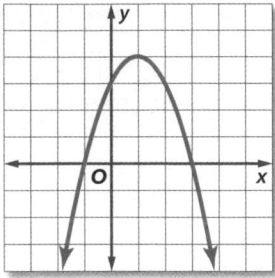   **29.**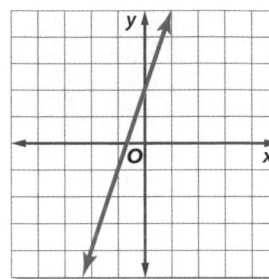

30. $y = 4^x$ **31.** $y = 2x(x - 1)$ **32.** $5x + y = 8$

33. GRADUATION The number of graduates at a high school has increased by a factor of 1.055 every year since 2001. In 2001, 110 students graduated. The function $N = 110(1.055)^t$ models the number of students N expected to graduate t years after 2001. How many students will graduate in 2012?

Describe the graph of each equation as a transformation of the graph of $y = 2^x$.

34. $y = 2^x + 6$ **35.** $y = 3(2)^x$ **36.** $y = -\frac{1}{4}(2)^x$

37. $y = -3 + 2^x$ **38.** $y = \left(\frac{1}{2}\right)^x$ **39.** $y = -5(2)^x$

40. DEER The deer population at a national park doubles every year. In 2000, there were 25 deer in the park. The function $N = 25(2)^t$ models the number of deer N in the park t years after 2000. What will the deer population be in 2015?

H.O.T. Problems   Use Higher-Order Thinking Skills

41. CCSS PERSEVERANCE Write an exponential function for which the graph passes through the points at $(0, 3)$ and $(1, 6)$.

42. REASONING Determine whether the graph of $y = ab^x$, where $a \neq 0$, $b > 0$, and $b \neq 1$, *sometimes*, *always*, or *never* has an x-intercept. Explain your reasoning.

43. OPEN ENDED Find an exponential function that represents a real-world situation, and graph the function. Analyze the graph, and explain why the situation is modeled by an exponential function rather than a linear function.

44. REASONING Use tables and graphs to compare and contrast an exponential function $f(x) = ab^x + c$, where $a \neq 0$, $b > 0$, and $b \neq 1$, and a linear function $g(x) = ax + c$. Include intercepts, intervals where the functions are increasing, decreasing, positive, or negative, relative maxima and minima, symmetry, and end behavior.

45. WRITING IN MATH Explain how to determine whether a set of data displays exponential behavior.

46. SHORT RESPONSE What are the x-intercepts of the function graphed below?

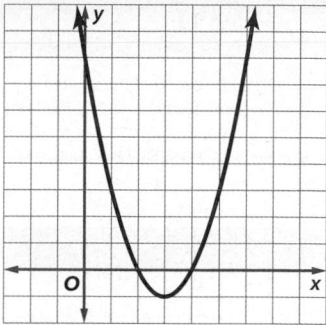

47. Hinto invested $300 into a savings account. The equation $A = 300(1.005)^{12t}$ models the amount in Hinto's account A after t years. How much will be in Hinto's account after 7 years?

    A $25,326   **C** $385.01

    B $456.11   **D** $301.52

48. GEOMETRY Ayana placed a circular piece of paper on a square picture as shown below. If the picture extends 4 inches beyond the circle on each side, what is the perimeter of the square picture?

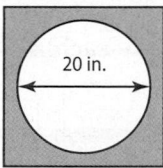

20 in.

    F 64 in.   **H** 94 in.

    G 80 in.   **J** 112 in.

49. The points with coordinates $(0, -3)$ and $(2, 7)$ are on line ℓ. Line p contains $(3, -1)$ and is perpendicular to line ℓ. What is the x-coordinate of the point where ℓ and p intersect?

    A $\dfrac{1}{2}$ **B** $-\dfrac{2}{5}$

    C $-\dfrac{1}{2}$ **D** -3

Spiral Review

Evaluate each product. Express the results in both scientific notation and standard form. (Lesson 7-4)

50. $(1.9 \times 10^2)(4.7 \times 10^6)$

51. $(4.5 \times 10^{-3})(5.6 \times 10^4)$

52. $(3.8 \times 10^{-4})(6.4 \times 10^{-8})$

Simplify. (Lesson 7-3)

53. $\sqrt[3]{343}$

54. $\sqrt[6]{729}$

55. $\left(\dfrac{1}{32}\right)^{\frac{1}{5}}$

66. $729^{\frac{5}{6}}$

57. $216^{\frac{5}{3}}$

58. $\left(\dfrac{1}{81}\right)^{\frac{3}{2}}$

59. DEMOLITION DERBY When a car hits an object, the damage is measured by the collision impact. For a certain car the collision impact I is given by $I = 2v^2$, where v represents the speed in kilometers per minute. What is the collision impact if the speed of the car is 4 kilometers per minute? (Lesson 7-1)

Use elimination to solve each system of equations. (Lesson 6-3)

60. $x + y = -3$
    $x - y = 1$

61. $3a + b = 5$
    $2a + b = 10$

62. $3x - 5y = 16$
    $-3x + 2y = -10$

Skills Review

Find the next three terms of each arithmetic sequence.

63. $1, 3, 5, 7, \ldots$

64. $-6, -4, -2, 0, \ldots$

65. $6.5, 9, 11.5, 14, \ldots$

66. $10, 3, -4, -11, \ldots$

67. $\dfrac{1}{2}, \dfrac{5}{4}, 2, \dfrac{11}{4}, \ldots$

68. $1, \dfrac{3}{4}, \dfrac{1}{2}, \dfrac{1}{4}, \ldots$

Graphing Technology Lab
Solving Exponential Equations and Inequalities

You can use TI-Nspire Technology to solve exponential equations and inequalities by graphing and by using tables.

CCSS Common Core State Standards
Content Standards
A.REI.11 Explain why the *x*-coordinates of the points where the graphs of the equations $y = f(x)$ and $y = g(x)$ intersect are the solutions of the equation $f(x) = g(x)$; find the solutions approximately, e.g., using technology to graph the functions, make tables of values, or find successive approximations. Include cases where $f(x)$ and/or $g(x)$ are linear, polynomial, rational, absolute value, exponential, and logarithmic functions.
Mathematical Practices
5 Use appropriate tools strategically.

Activity 1 Graph an Exponential Equation

Graph $y = 3^x + 4$ using a graphing calculator.

Step 1 Add a new **Graphs** page.

Step 2 Enter $3^x + 4$ as f1(**x**).

Step 3 Use the **Window Settings** option from the **Window/Zoom** menu to adjust the window so that x is from -10 to 10 and y is from -100 to 100. Keep the scales as **Auto**.

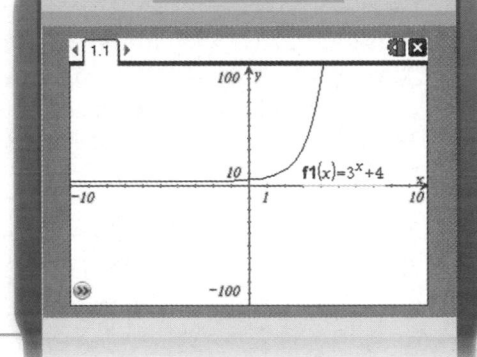

To solve an equation by graphing, graph both sides of the equation and locate the point(s) of intersection.

Activity 2 Solve an Exponential Equation by Graphing

Solve $2^{x-2} = \dfrac{3}{4}$.

Step 1 Add a new **Graphs** page.

Step 2 Enter 2^{x-2} as f1(**x**) and $\dfrac{3}{4}$ as f2(**x**).

Step 3 Use the **Intersection Point(s)** tool from the **Points & Lines** menu to find the intersection of the two graphs. Select the graph of **f1(x) enter** and then the graph of **f2(x) enter**.

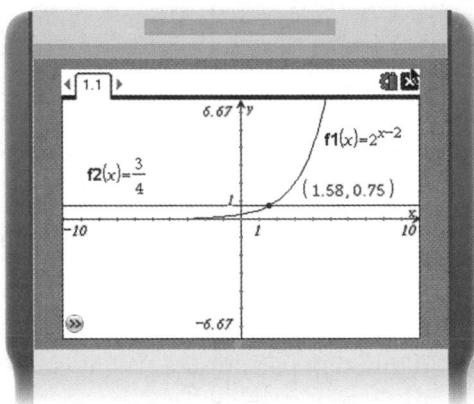

The graphs intersect at about (1.58, 0.75). Therefore, the solution of $2^{x-2} = \dfrac{3}{4}$ is 1.58.

Exercises

CCSS TOOLS Use a graphing calculator to solve each equation.

1. $\left(\dfrac{1}{3}\right)^{x-1} = \dfrac{3}{4}$

2. $2^{2x-1} = 2x$

3. $\left(\dfrac{1}{2}\right)^{2x} = 2^{2x}$

4. $5^{\frac{1}{3}x+2} = -x$

5. $\left(\dfrac{1}{8}\right)^{2x} = -2x + 1$

6. $2^{\frac{1}{4}x-1} = 3^{x+1}$

7. $2^{3x-1} = 4^x$

8. $4^{2x-3} = 5^{-x+1}$

9. $3^{2x-4} = 2^x + 1$

Activity 3　Solve an Exponential Equation by Using a Table

Solve $2\left(\dfrac{1}{2}\right)^{x+2} = \dfrac{1}{4}$ using a table.

Step 1　Add a new **Lists & Spreadsheet** page.

Step 2　Label column A as x. Enter values from -4 to 4 in cells A1 to A9.

Step 3　In column B in the formula row, enter the left side of the rational equation. In column C of the formula row, enter $= \dfrac{1}{4}$. Specify **Variable Reference** when prompted.

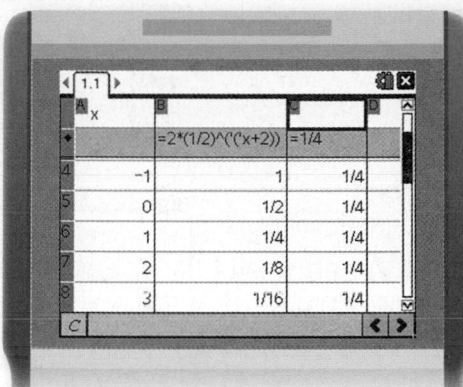

Scroll until you see where the values in Columns B and C are equal.

This occurs at $x = 1$. Therefore, the solution of $2\left(\dfrac{1}{2}\right)^{x+2} = \dfrac{1}{4}$ is 1.

You can also use a graphing calculator to solve exponential inequalities.

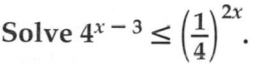

Activity 4　Solve an Exponential Inequality

Solve $4^{x-3} \le \left(\dfrac{1}{4}\right)^{2x}$.

Step 1　Add a new **Graphs** page.

Step 2　Enter the left side of the inequality into f1(**x**). Press **del**, select ≥, and enter 4^{x-3}. Enter the right side of the inequality into f2(**x**). Press **tab del** ≤, and enter $\left(\dfrac{1}{4}\right)^{2x}$.

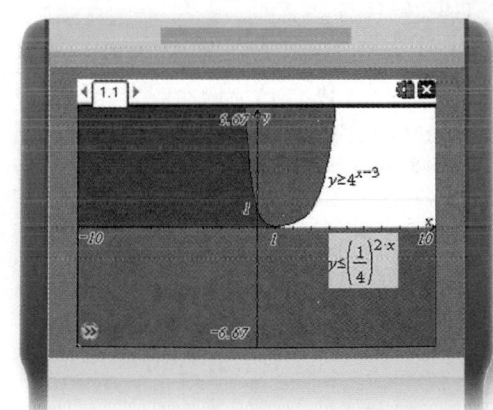

The x-values of the points in the region where the shading overlap is the solution set of the original inequality. Therefore the solution of $4^{x-3} \le \left(\dfrac{1}{4}\right)^{2x}$ is $x \le 1$.

Exercises

CCSS **TOOLS**　Use a graphing calculator to solve each equation or inequality.

10. $\left(\dfrac{1}{3}\right)^{3x} = 3^x$

11. $\left(\dfrac{1}{6}\right)^{2x} = 4^x$

12. $3^{1-x} \le 4^x$

13. $4^{3x} \le 2x + 1$

14. $\left(\dfrac{1}{4}\right)^{x} > 2^{x+4}$

15. $\left(\dfrac{1}{3}\right)^{x-1} \ge 2^x$

7-6 Growth and Decay

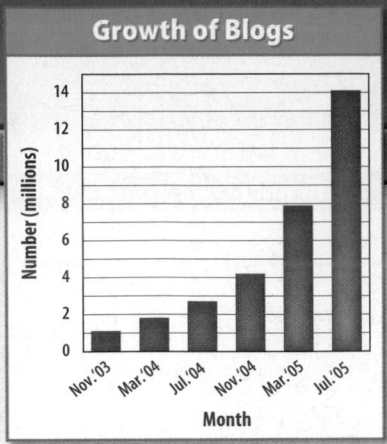
·:· Then

● You analyzed exponential functions.

·:· Now

1 Solve problems involving exponential growth.

2 Solve problems involving exponential decay.

·:· Why?

● The number of Weblogs or blogs increased at a monthly rate of about 13.7% over 21 months. The average number of blogs per month can be modeled by $y = 1.1(1 + 0.137)^t$ or $y = 1.1(1.137)^t$, where y represents the total number of blogs in millions and t is the number of months since November 2003.

NewVocabulary
compound interest

Common Core State Standards

Content Standards
F.IF.8b Use the properties of exponents to interpret expressions for exponential functions.

F.LE.2 Construct linear and exponential functions, including arithmetic and geometric sequences, given a graph, a description of a relationship, or two input-output pairs (include reading these from a table).

Mathematical Practices
4 Model with mathematics.

1 **Exponential Growth** The equation for the number of blogs is in the form $y = a(1 + r)^t$. This is the general equation for exponential growth.

KeyConcept Equation for Exponential Growth

a is the initial amount. t is time.

$$y = a(1 + r)^t$$

y is the final amount.

r is the rate of change expressed as a decimal, $r > 0$.

Real-World Example 1 Exponential Growth

CONTEST The prize for a radio station contest begins with a $100 gift card. Once a day, a name is announced. The person has 15 minutes to call or the prize increases by 2.5% for the next day.

a. Write an equation to represent the amount of the gift card in dollars after t days with no winners.

$y = a(1 + r)^t$ Equation for exponential growth

$y = 100(1 + 0.025)^t$ $a = 100$ and $r = 2.5\%$ or 0.025

$y = 100(1.025)^t$ Simplify.

In the equation $y = 100(1.025)^t$, y is the amount of the gift card and t is the number of days since the contest began.

b. How much will the gift card be worth if no one wins after 10 days?

$y = 100(1.025)^t$ Equation for amount of gift card

$= 100(1.025)^{10}$ $t = 10$

≈ 128.01 Use a calculator.

In 10 days, the gift card will be worth $128.01.

▶ **Guided**Practice

1. **TUITION** A college's tuition has risen 5% each year since 2000. If the tuition in 2000 was $10,850, write an equation for the amount of the tuition t years after 2000. Predict the cost of tuition for this college in 2015.

Compound interest is interest earned or paid on both the initial investment and previously earned interest. It is an application of exponential growth.

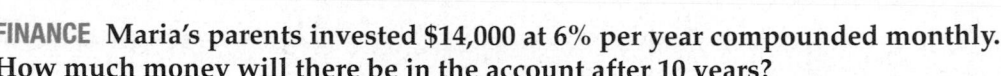

KeyConcept Equation for Compound Interest

A is the current amount.

n is the number of times the interest is compounded each year, and *t* is time in years.

$$A = P\left(1 + \frac{r}{n}\right)^{nt}$$

P is the principal or initial amount.

r is the annual interest rate expressed as a decimal, $r > 0$.

Real-World Example 2 Compound Interest

FINANCE Maria's parents invested $14,000 at 6% per year compounded monthly. How much money will there be in the account after 10 years?

$$A = P\left(1 + \frac{r}{n}\right)^{nt}$$ Compound interest equation

$$= 14{,}000\left(1 + \frac{0.06}{12}\right)^{12(10)}$$ $P = 14{,}000$, $r = 6\%$ or 0.06, $n = 12$, and $t = 10$

$$= 14{,}000(1.005)^{120}$$ Simplify.

$$\approx 25{,}471.55$$ Use a calculator.

There will be about $25,471.55 in 10 years.

GuidedPractice

2. **FINANCE** Determine the amount of an investment if $300 is invested at an interest rate of 3.5% compounded monthly for 22 years.

Real-World Career

Financial Advisor Financial advisors help people plan their financial futures. A good financial advisor has mathematical, problem-solving, and communication skills. A bachelor's degree is strongly preferred but not required.

---

2 Exponential Decay In exponential decay, the original amount decreases by the same percent over a period of time. A variation of the growth equation can be used as the general equation for exponential decay.

StudyTip

Growth and Decay
Since *r* is added to 1, the value inside the parentheses will be greater than 1 for exponential growth functions. For exponential decay functions, this value will be less than 1 since *r* is subtracted from 1.

KeyConcept Equation for Exponential Decay

a is the initial amount.   *t* is time.

$$y = a(1 - r)^t$$

y is the final amount.

r is the rate of decay expressed as a decimal, $0 < r < 1$.

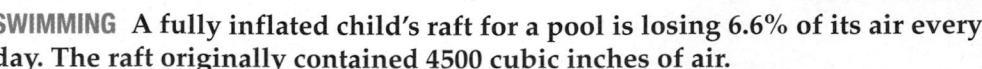

Real-World Example 3 Exponential Decay

SWIMMING A fully inflated child's raft for a pool is losing 6.6% of its air every day. The raft originally contained 4500 cubic inches of air.

a. **Write an equation to represent the loss of air.**

$$y = a(1 - r)^t$$ Equation for exponential decay

$$= 4500(1 - 0.066)^t$$ $a = 4500$ and $r = 6.6\%$ or 0.066

$$= 4500(0.934)^t$$ Simplify.

$y = 4500(0.934)^t$, where y is the air in the raft in cubic inches after t days.

b. Estimate the amount of air in the raft after 7 days.

$y = 4500(0.934)^t$ Equation for air loss

$= 4500(0.934)^7$ $t = 7$

≈ 2790 Use a calculator.

The amount of air in the raft after 7 days will be about 2790 cubic inches.

▶ **Guided**Practice

3. POPULATION The population of Campbell County, Kentucky, has been decreasing at an average rate of about 0.3% per year. In 2000, its population was 88,647. Write an equation to represent the population since 2000. If the trend continues, predict the population in 2010.

Check Your Understanding

 = Step-by-Step Solutions begin on page R13.

Example 1
1. SALARY Ms. Acosta received a job as a teacher with a starting salary of $34,000. According to her contract, she will receive a 1.5% increase in her salary every year. How much will Ms. Acosta earn in 7 years?

Example 2
2. MONEY Paul invested $400 into an account with a 5.5% interest rate compounded monthly. How much will Paul's investment be worth in 8 years?

Example 3
3. ENROLLMENT In 2000, 2200 students attended Polaris High School. The enrollment has been declining 2% annually.

     a. Write an equation for the enrollment of Polaris High School t years after 2000.

     b. If this trend continues, how many students will be enrolled in 2015?

Practice and Problem Solving

Extra Practice is on page R7.

Example 1
4. MEMBERSHIPS The Work-Out Gym sold 550 memberships in 2001. Since then the number of memberships sold has increased 3% annually.

     a. Write an equation for the number of memberships sold at Work-Out Gym t years after 2001.

     b. If this trend continues, predict how many memberships the gym will sell in 2020.

5. COMPUTERS The number of people who own computers has increased 23.2% annually since 1990. If half a million people owned a computer in 1990, predict how many people will own a computer in 2015.

6. COINS Camilo purchased a rare coin from a dealer for $300. The value of the coin increases 5% each year. Determine the value of the coin in 5 years.

Example 2
(7) INVESTMENTS Theo invested $6600 at an interest rate of 4.5% compounded monthly. Determine the value of his investment in 4 years.

8. COMPOUND INTEREST Paige invested $1200 at an interest rate of 5.75% compounded quarterly. Determine the value of her investment in 7 years.

9. CCSS PRECISION Brooke is saving money for a trip to the Bahamas that costs $295.99. She puts $150 into a savings account that pays 7.25% interest compounded quarterly. Will she have enough money in the account after 4 years? Explain.

Example 3
10. INVESTMENTS Jin's investment of $4500 has been losing its value at a rate of 2.5% each year. What will his investment be worth in 5 years?

11 **POPULATION** In the years from 2010 to 2015, the population of the District of Columbia is expected decrease about 0.9% annually. In 2010, the population was about 530,000. What is the population of the District of Columbia expected to be in 2015?

12. CARS Leonardo purchases a car for $18,995. The car depreciates at a rate of 18% annually. After 6 years, Manuel offers to buy the car for $4500. Should Leonardo sell the car? Explain.

13. HOUSING The median house price in the United States increased an average of 1.4% each year between 2005 and 2007. Assume that this pattern continues.

   a. Write an equation for the median house price for t years after 2007.

   b. Predict the median house price in 2018.

Median House Price	
2005	$240,900
2006	$246,500
2007	$247,900

Source: *Real Estate Journal*

14. ELEMENTS A radioactive element's half-life is the time it takes for one half of the element's quantity to decay. The half-life of Plutonium-241 is 14.4 years. The number of grams A of Plutonium-241 left after t years can be modeled by $A = p(0.5)^{\frac{t}{14.4}}$, where p is the original amount of the element.

   a. How much of a 0.2-gram sample remains after 72 years?

   b. How much of a 5.4-gram sample remains after 1095 days?

15. COMBINING FUNCTIONS A swimming pool holds a maximum of 20,500 gallons of water. It evaporates at a rate of 0.5% per hour. The pool currently contains 19,000 gallons of water.

   a. Write an exponential function $w(t)$ to express the amount of water remaining in the pool after time t where t is the number of hours after the pool has reached 19,000 gallons.

   b. At this same time, a hose is turned on to refill the pool at a net rate of 300 gallons per hour. Write a function $p(t)$ where t is time in hours the hose is running to express the amount of water that is pumped into the pool.

   c. Find $C(t) = p(t) + w(t)$. What does this new function represent?

   d. Use the graph of $C(t)$ to determine how long the hose must run to fill the pool to its maximum capacity.

H.O.T. Problems   Use Higher-Order Thinking Skills

16. REASONING Determine the growth rate (as a percent) of a population that quadruples every year. Explain.

17. CCSS PRECISION Santos invested $1200 into an account with an interest rate of 8% compounded monthly. Use a calculator to approximate how long it will take for Santos's investment to reach $2500.

18. REASONING The amount of water in a container doubles every minute. After 8 minutes, the container is full. After how many minutes was the container half full? Explain.

19. **WRITING IN MATH** What should you consider when using exponential models to make decisions?

20. WRITING IN MATH Compare and contrast the exponential growth formula and the exponential decay formula.

21. GEOMETRY The parallelogram has an area of 35 square inches. Find the height h of the parallelogram.

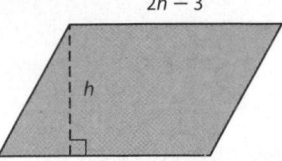

$2h - 3$

h

A 3.5 inches   **C** 5 inches

B 4 inches   **D** 7 inches

22. Which is greater than $64^{\frac{1}{3}}$?

F 2^2 **H** $64^{\frac{1}{2}}$

G $64^{\frac{1}{6}}$ **J** 64^{-3}

23. Thi purchased a car for \$22,900. The car depreciated at an annual rate of 16%. Which of the following equations models the value of Thi's car after 5 years?

A $A = 22{,}900(1.16)^5$

B $A = 22{,}900(0.16)^5$

C $A = 16(22{,}900)^5$

D $A = 22{,}900(0.84)^5$

24. GRIDDED RESPONSE A deck measures 12 feet by 18 feet. If a painter charges \$2.65 per square foot, including tax, how much will it cost in dollars to have the deck painted?

Spiral Review

Graph each function. Find the y-intercept and state the domain and range. (Lesson 7-5)

25. $y = 3^x$ **26.** $y = \left(\frac{1}{2}\right)^x$ **27.** $y = 6^x$

Evaluate each product. Express the results in both scientific notation and standard form. (Lesson 7-4)

28. $(4.2 \times 10^3)(3.1 \times 10^{10})$ **29.** $(6.02 \times 10^{23})(5 \times 10^{-14})$ **30.** $(7 \times 10^5)^2$

31. $(1.1 \times 10^{-2})^2$ **32.** $(9.1 \times 10^{-2})(4.2 \times 10^{-7})$ **33.** $(3.14 \times 10^2)(6.1 \times 10^{-3})$

34. EVENT PLANNING A hall does not charge a rental fee as long as at least \$4000 is spent on food. For the prom, the hall charges \$28.95 per person for a buffet. How many people must attend the prom to avoid a rental fee for the hall? (Lesson 5-2)

Determine whether the graphs of each pair of equations are *parallel*, *perpendicular*, or *neither*. (Lesson 4-4)

35. $y = -2x + 11$
    $y + 2x = 23$

36. $3y = 2x + 14$
    $-3x - 2y = 2$

37. $y = -5x$
    $y = 5x - 18$

38. AGES The table shows equivalent ages for horses and humans. Write an equation that relates human age to horse age and find the equivalent horse age for a human who is 16 years old. (Lesson 3-4)

Horse age (x)	0	1	2	3	4	5
Human age (y)	0	3	6	9	12	15

Find the total price of each item. (Lesson 2-7)

39. umbrella: \$14.00
tax: 5.5%

40. sandals: \$29.99
tax: 5.75%

41. backpack: \$35.00
tax: 7%

Skills Review

Graph each set of ordered pairs.

42. $(3, 0), (0, 1), (-4, -6)$ **43.** $(0, -2), (-1, -6), (3, 4)$ **44.** $(2, 2), (-2, -3), (-3, -6)$

7-6 Algebra Lab
Transforming Exponential Expressions

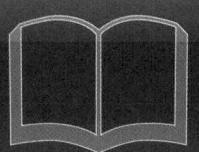

You can use the properties of rational exponents to transform exponential functions into other forms in order to solve real-world problems.

CCSS Common Core State Standards
Content Standards
A.SSE.3c Use the properties of exponents to transform expressions for exponential functions.
F.IF.8b Use the properties of exponents to interpret expressions for exponential functions.

Activity Write Equivalent Exponential Expressions

Monique is trying to decide between two savings account plans. Plan A offers a monthly compounding interest rate of 0.25%, while Plan B offers 2.5% interest compounded annually. Which is the better plan? Explain.

In order to compare the plans, we must compare rates with the same compounding frequency. One way to do this is to compare the approximate monthly interest rates of each plan, also called the *effective* monthly interest rate. While you can use the compound interest formula to find this rate, you can also use the properties of exponents.

Write a function to represent the amount A Monique would earn after t years with Plan B. For convenience, let the initial amount of Monique's investment be $1.

$\quad y = a(1 + r)^t$ Equation for exponential growth

$A(t) = 1(1 + 0.025)^t$ $y = A(t)$, $a = 1$, $r = 2.5\%$ or 0.025

$\quad\quad = 1.025^t$ Simplify.

Now write a function equivalent to $A(t)$ that represents 12 compoundings per year, with a power of $12t$, instead of 1 per year, with a power of $1t$.

$A(t) = 1.025^{1t}$ Original function

$\quad\quad = 1.025^{\left(\frac{1}{12} \cdot 12\right)t}$ $1 = \frac{1}{12} \cdot 12$

$\quad\quad = \left(1.025^{\frac{1}{12}}\right)^{12t}$ Power of a Power

$\quad\quad \approx 1.0021^{12t}$ $(1.025)^{\frac{1}{12}} = \sqrt[12]{1.025}$ or about 1.0021

From this equivalent function, we can determine that the effective monthly interest by Plan B is about 0.0021 or about 0.21% per month. This rate is less than the monthly interest rate of 0.25% per month offered by Plan A, so Plan A is the better plan.

Model and Analyze

1. Use the compound interest formula $A = P\left(1 + \frac{r}{n}\right)^{nt}$ to determine the effective monthly interest rate for Plan B. How does this rate compare to the rate calculated using the method in the Activity above?

2. Write a function to represent the amount A Monique would earn after t months by Plan A. Then use the properties of exponents to write a function equivalent to $A(t)$ that represents the amount earned after t years.

3. From the expression you wrote in Exercise 2, identify the effective annual interest rate by Plan A. Use this rate to explain why Plan A is the better plan.

4. Suppose Plan A offered a quarterly compounded interest rate of 1.5%. Use the properties of exponents to explain which is the better plan.

Geometric Sequences as Exponential Functions

:⋅Then	:⋅Now	:⋅Why?
● You related arithmetic sequences to linear functions.	**1** Identify and generate geometric sequences. **2** Relate geometric sequences to exponential functions.	● You send a chain email to a friend who forwards the email to five more people. Each of these five people forwards the email to five more people. The number of new email generated forms a geometric sequence.

Karen Moskowitz/Taxi/Getty Images

 NewVocabulary
geometric sequence
common ratio

 Common Core State Standards

Content Standards

F.BF.2 Write arithmetic and geometric sequences both recursively and with an explicit formula, use them to model situations, and translate between the two forms.

F.LE.1 Distinguish between situations that can be modeled with linear functions and with exponential functions.

a. Prove that linear functions grow by equal differences over equal intervals, and that exponential functions grow by equal factors over equal intervals.

b. Recognize situations in which one quantity changes at a constant rate per unit interval relative to another.

c. Recognize situations in which a quantity grows or decays by a constant percent rate per unit interval relative to another.

Mathematical Practices
7 Look for and make use of structure.

1 **Recognize Geometric Sequences** The first person generates 5 emails. If each of these people sends the email to 5 more people, 25 emails are generated. If each of the 25 people sends 5 emails, 125 emails are generated. The sequence of emails generated, 1, 5, 25, 125, … is an example of a **geometric sequence**.

In a geometric sequence, the first term is nonzero and each term after the first is found by multiplying the previous term by a nonzero constant r called the **common ratio**. The common ratio can be found by dividing any term by its previous term.

Example 1 Identify Geometric Sequences

Determine whether each sequence is *arithmetic*, *geometric*, or *neither*. Explain.

a. 256, 128, 64, 32, …

Find the ratios of consecutive terms.

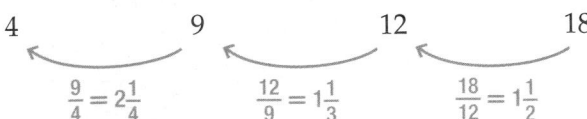

$$\frac{128}{256} = \frac{1}{2} \qquad \frac{64}{128} = \frac{1}{2} \qquad \frac{32}{64} = \frac{1}{2}$$

Since the ratios are constant, the sequence is geometric. The common ratio is $\frac{1}{2}$.

b. 4, 9, 12, 18, …

Find the ratios of consecutive terms.

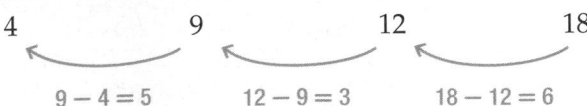

$$\frac{9}{4} = 2\frac{1}{4} \qquad \frac{12}{9} = 1\frac{1}{3} \qquad \frac{18}{12} = 1\frac{1}{2}$$

The ratios are not constant, so the sequence is not geometric.

Find the differences of consecutive terms.

4   9   12   18

$$9 - 4 = 5 \qquad 12 - 9 = 3 \qquad 18 - 12 = 6$$

There is no common difference, so the sequence is not arithmetic. Thus, the sequence is neither geometric nor arithmetic.

▶ **Guided**Practice

1A. 1, 3, 9, 27, …   **1B.** −20, −15, −10, −5, …   **1C.** 2, 8, 14, 22, …

Once the common ratio is known, more terms of a sequence can be generated. The formula can be rewritten as $a_n = a_1 r^{n-1}$, where n is a counting number and r is the common ratio.

Example 2 Find Terms of Geometric Sequences

Find the next three terms in each geometric sequence.

a. 1, −4, 16, −64, …

Step 1 Find the common ratio.

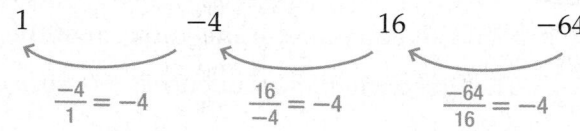

$$\frac{-4}{1} = -4 \qquad \frac{16}{-4} = -4 \qquad \frac{-64}{16} = -4$$

Step 2 Multiply each term by the common ratio to find the next three terms.

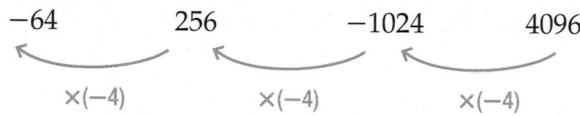

$$\times(-4) \qquad \times(-4) \qquad \times(-4)$$

The next three terms are 256, −1024, and 4096.

b. 9, 3, 1, $\frac{1}{3}$ …

Step 1 Find the common ratio.

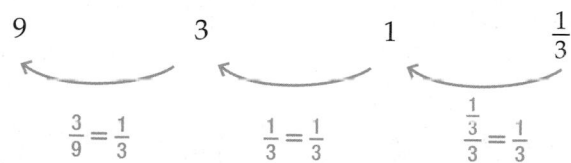

$$\frac{3}{9} = \frac{1}{3} \qquad \frac{1}{3} = \frac{1}{3} \qquad \frac{\frac{1}{3}}{3} = \frac{1}{3}$$

The value of r is $\frac{1}{3}$.

Step 2 Multiply each term by the common ratio to find the next three terms.

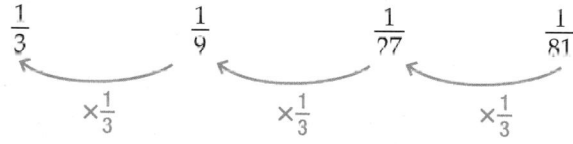

$$\times\frac{1}{3} \qquad \times\frac{1}{3} \qquad \times\frac{1}{3}$$

The next three terms are $\frac{1}{9}, \frac{1}{27}$, and $\frac{1}{81}$.

> **Guided**Practice

2A. −3, 15, −75, 375, …   **2B.** 24, 36, 54, 81, …

2 Geometric Sequences and Functions Finding the nth term of a geometric sequence would be tedious if we used the above method. The table below shows a rule for finding the nth term of a geometric sequence.

Position, n	1	2	3	4	…	n
Term, a_n	a_1	$a_1 r$	$a_1 r^2$	$a_1 r^3$	…	$a_1 r^{n-1}$

Notice that the common ratio between the terms is r. The table shows that to get the nth term, you multiply the first term by the common ratio r raised to the power $n - 1$. A geometric sequence can be defined by an exponential function in which n is the independent variable, a_n is the dependent variable, and r is the base. The domain is the counting numbers.

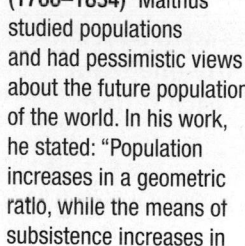

Math History Link

Thomas Robert Malthus (1766–1834) Malthus studied populations and had pessimistic views about the future population of the world. In his work, he stated: "Population increases in a geometric ratio, while the means of subsistence increases in an arithmetic ratio."

Mary Evans Picture Library

KeyConcept nth term of a Geometric Sequence

The nth term a_n of a geometric sequence with first term a_1 and common ratio r is given by the following formula, where n is any positive integer and $a_1, r \neq 0$.

$$a_n = a_1 r^{n-1}$$

Example 3 Find the nth Term of a Geometric Sequence

a. **Write an equation for the nth term of the sequence $-6, 12, -24, 48, \ldots$.**

The first term of the sequence is -6. So, $a_1 = -6$. Now find the common ratio.

$$-6 \qquad 12 \qquad -24 \qquad 48$$

$$\frac{12}{-6} = -2 \qquad \frac{-24}{12} = -2 \qquad \frac{48}{-24} = -2$$

The common ratio is -2.

$a_n = a_1 r^{n-1}$ Formula for nth term

$a_n = -6(-2)^{n-1}$ $a_1 = -6$ and $r = 2$

b. **Find the ninth term of this sequence.**

$a_n = a_1 r^{n-1}$ Formula for nth term

$a_9 = -6(-2)^{9-1}$ For the nth term, $n = 9$.

$ = -6(-2)^8$ Simplify.

$ = -6(256)$ $(-2)^8 = 256$

$ = -1536$

WatchOut!

Negative Common Ratio If the common ratio is negative, as in Example 3, make sure to enclose the common ratio in parentheses. $(-2)^8 \neq -2^8$

▶ **Guided**Practice

3. Write an equation for the nth term of the geometric sequence $96, 48, 24, 12, \ldots$. Then find the tenth term of the sequence.

Real-World Example 4 Graph a Geometric Sequence

BASKETBALL The NCAA women's basketball tournament begins with 64 teams. In each round, one half of the teams are left to compete, until only one team remains. Draw a graph to represent how many teams are left in each round.

Compared to the previous rounds, one half of the teams remain. So, $r = \frac{1}{2}$. Therefore, the geometric sequence that models this situation is $64, 32, 16, 8, 4, 2, 1$. So in round two, 32 teams compete, in round three 16 teams compete and so forth. Use this information to draw a graph.

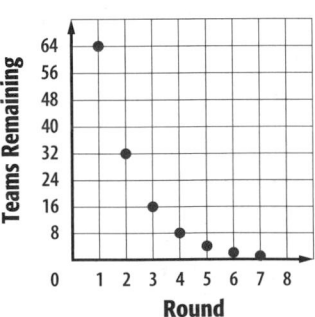

Real-WorldLink

The first NCAA Division I women's basketball tournament was held in 1982. The University of Tennessee has won the most national titles with 8 titles as of 2012.

Source: NCAA Sports

▶ **Guided**Practice

4. **TENNIS** A tennis ball is dropped from a height of 12 feet. Each time the ball bounces back to 80% of the height from which it fell. Draw a graph to represent the height of the ball after each bounce.

Check Your Understanding

= Step-by-Step Solutions begin on page R13.

Example 1 Determine whether each sequence is *arithmetic, geometric,* or *neither.* Explain.

1. 200, 40, 8, …   **2.** 2, 4, 16, …   **3.** −6, −3, 0, 3, …   **4.** 1, −1, 1, −1, …

Example 2 Find the next three terms in each geometric sequence.

5. 10, 20, 40, 80, …   **6.** 100, 50, 25, …   **7.** 4, −1, $\frac{1}{4}$, …   **8.** −7, 21, −63, …

Example 3 Write an equation for the *n*th term of each geometric sequence, and find the indicated term.

9. the fifth term of −6, −24, −96, …

10. the seventh term of −1, 5, −25, …

11. the tenth term of 72, 48, 32, …

12. the ninth term of 112, 84, 63, …

Example 4 **13. EXPERIMENT** In a physics class experiment, Diana drops a ball from a height of 16 feet. Each bounce has 70% the height of the previous bounce. Draw a graph to represent the height of the ball after each bounce.

Practice and Problem Solving

Extra Practice is on R7.

Example 1 Determine whether each sequence is *arithmetic, geometric,* or *neither.* Explain.

14. 4, 1, 2, …   **15.** 10, 20, 30, 40, …   **16.** 4, 20, 100, …

17. 212, 106, 53, …   **18.** −10, −8, −6, −4, …   **19.** 5,   10, 20, 40, …

Example 2 Find the next three terms in each geometric sequence.

20. 2, −10, 50, …   **21** 36, 12, 4, …   **22.** 4, 12, 36, …

23. 400, 100, 25, …   **24.** −6, −42, −294, …   **25.** 1024, −128, 16, …

Example 3 **26.** The first term of a geometric series is 1 and the common ratio is 9. What is the 8th term of the sequence?

27. The first term of a geometric series is 2 and the common ratio is 4. What is the 14th term of the sequence?

28. What is the 15th term of the geometric sequence −9, 27, −81, …?

29. What is the 10th term of the geometric sequence 6, −24, 96, …?

Example 4 **30. PENDULUM** The first swing of a pendulum is shown. On each swing after that, the arc length is 60% of the length of the previous swing. Draw a graph that represents the arc length after each swing.

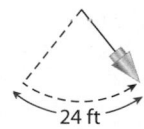

24 ft

31. Find the eighth term of a geometric sequence for which $a_3 = 81$ and $r = 3$.

32. CCSS REASONING At an online mapping site, Mr. Mosley notices that when he clicks a spot on the map, the map zooms in on that spot. The magnification increases by 20% each time.

   a. Write a formula for the *n*th term of the geometric sequence that represents the magnification of each zoom level. (*Hint:* The common ratio is not just 0.2.)

   b. What is the fourth term of this sequence? What does it represent?

33 **ALLOWANCE** Danielle's parents have offered her two different options to earn her allowance for a 9-week period over the summer. She can either get paid $30 each week or $1 the first week, $2 for the second week, $4 for the third week, and so on.

    a. Does the second option form a geometric sequence? Explain.

    b. Which option should Danielle choose? Explain.

34. **SIERPINSKI'S TRIANGLE** Consider the inscribed equilateral triangles at the right. The perimeter of each triangle is one half of the perimeter of the next larger triangle. What is the perimeter of the smallest triangle?

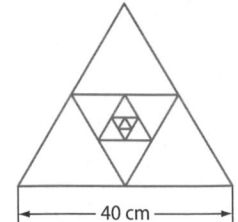

40 cm

35. If the second term of a geometric sequence is 3 and the third term is 1, find the first and fourth terms of the sequence.

36. If the third term of a geometric sequence is −12 and the fourth term is 24, find the first and fifth terms of the sequence.

37. **EARTHQUAKES** The Richter scale is used to measure the force of an earthquake. The table shows the increase in magnitude for the values on the Richter scale.

    a. Copy and complete the table. Remember that the rate of change is the change in y divided by the change in x.

Richter Number (x)	Increase in Magnitude (y)	Rate of Change (slope)
1	1	—
2	10	9
3	100	
4	1000	
5	10,000	

    b. Plot the ordered pairs (Richter number, increase in magnitude).

    c. Describe the graph that you made of the Richter scale data. Is the rate of change between any two points the same?

    d. Write an exponential equation that represents the Richter scale.

H.O.T. Problems   Use Higher-Order Thinking Skills

38. **CHALLENGE** Write a sequence that is both geometric and arithmetic. Explain your answer.

39. **CCSS CRITIQUE** Haro and Matthew are finding the ninth term of the geometric sequence −5, 10, −20, … . Is either of them correct? Explain your reasoning.

Haro	Matthew
$r = \dfrac{10}{-5}$ or -2	$r = \dfrac{10}{-5}$ or -2
$a_9 = -5\,(-2)^{9-1}$	$a_9 = -5 \cdot (-2)^{9-1}$
$= -5(512)$	$= -5 \cdot -256$
$= -2560$	$= 1280$

40. **REASONING** Write a sequence of numbers that form a pattern but are neither arithmetic nor geometric. Explain the pattern.

41. **WRITING IN MATH** How are graphs of geometric sequences and exponential functions similar? different?

42. **WRITING IN MATH** Summarize how to find a specific term of a geometric sequence.

43. Find the eleventh term of the sequence 3, −6, 12, −24, … .

   A 6144   **C** 33

   B 3072   **D** −6144

44. What is the total amount of the investment shown in the table below if interest is compounded monthly?

Principal	$500
Length of Investment	4 years
Annual Interest Rate	5.25%

   F $613.56   **H** $616.56

   G $616.00   **J** $718.75

45. SHORT RESPONSE Gloria has $6.50 in quarters and dimes. If she has 35 coins in total, how many of each coin does she have?

46. What are the domain and range of the function $y = 4(3^x) - 2$?

   A D = {all real numbers}, R = $\{y \mid y > -2\}$

   B D = {all real numbers}, R = $\{y \mid y > 0\}$

   C D = {all integers}, R = $\{y \mid y > -2\}$

   D D = {all integers}, R = $\{y \mid y > 0\}$

Spiral Review

Find the next three terms in each geometric sequence. (Lesson 7-6)

47. 2, 6, 18, 54, …

48. −5, −10, −20, −40, …

49. 1, $-\frac{1}{2}$, $\frac{1}{4}$, $-\frac{1}{8}$, …

50. −3, 1.5, −0.75, 0.375, …

51. 1, 0.6, 0.36, 0.216, …

52. 4, 6, 9, 13.5, …

Graph each function. Find the y-intercept and state the domain and range. (Lesson 7-5)

53. $y = \left(\frac{1}{4}\right)^x - 5$

54. $y = 2(4)^x$

55. $y = \frac{1}{2}(3^x)$

56. LANDSCAPING A blue spruce grows an average of 6 inches per year. A hemlock grows an average of 4 inches per year. If a blue spruce is 4 feet tall and a hemlock is 6 feet tall, write a system of equations to represent their growth. Find and interpret the solution in the context of the situation. (Lesson 6-2)

57. MONEY City Bank requires a minimum balance of $1500 to maintain free checking services. If Mr. Hayashi is going to write checks for the amounts listed in the table, how much money should he start with in order to have free checking? (Lesson 5-1)

Check	Amount
750	$1300
751	$947

Write an equation in slope-intercept form of the line with the given slope and y-intercept. (Lesson 4-2)

58. slope: 4, y-intercept: 2

59. slope: −3, y-intercept: $-\frac{2}{3}$

60. slope: $-\frac{1}{4}$, y-intercept: −5

61. slope: $\frac{1}{2}$, y-intercept: −9

62. slope: $-\frac{2}{5}$, y-intercept: $\frac{3}{4}$

63. slope: −6, y-intercept: −7

Skills Review

Simplify each expression. If not possible, write *simplified*.

64. $3u + 10u$

65. $5a - 2 + 6a$

66. $6m^2 - 8m$

67. $4w^2 + w + 15w^2$

68. $13(5 + 4a)$

69. $(4t - 6)16$

7-7 Algebra Lab
Average Rate of Change of Exponential Functions

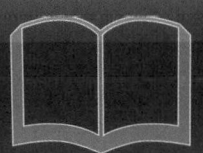

You know that the rate of change of a linear function is the same for any two points on the graph. The rate of change of an exponential function is not constant.

CCSS **Common Core State Standards**
Content Standards
F.IF.6 Calculate and interpret the average rate of change of a function (presented symbolically or as a table) over a specified interval. Estimate the rate of change from a graph.

Activity Evaluating Investment Plans

John has $2000 to invest in one of two plans. Plan 1 offers to increase his principal by $75 each year, while Plan 2 offers to pay 3.6% interest compounded monthly. The dollar value of each investment after t years is given by $A_1 = 2000 + 75t$ and $A_2 = 2000(1.003)^{12t}$, respectively. Use the function values, the average rate of change, and the graphs of the equations to interpret and compare the plans.

Step 1 Copy and complete the table below by finding the missing values for A_1 and A_2.

t	0	1	2	3	4	5
A_1						
A_2						

Step 2 Find the average rate of change for each plan from $t = 0$ to 1, $t = 3$ to 4, and $t = 0$ to 5.

Plan 1: $\dfrac{2075 - 2000}{1 - 0}$ or 75 $\dfrac{2300 - 2225}{4 - 3}$ or 75 $\dfrac{2375 - 2000}{5 - 0}$ or 75

Plan 2: $\dfrac{2073.2 - 2000}{1 - 0}$ or 73.2 $\dfrac{2309.27 - 2227.74}{4 - 3}$ or about 82 $\dfrac{2393.79 - 2000}{5 - 0}$ or about 79

Step 3 Graph the ordered pairs for each function. Connect each set of points with a smooth curve.

Step 4 Use the graph and the rates of change to compare the plans. Both graphs have a rate of change for the first year of about $75 per year. From year 3 to 4, Plan 1 continues to increase at $75 per year, but Plan 2 grows at a rate of more than $81 per year. The average rate of change over the first five years for Plan 1 is $75 per year and for Plan 2 is over $78 per year. This indicates that as the number of years increases, the investment in Plan 2 grows at an increasingly faster pace. This is supported by the widening gap between their graphs.

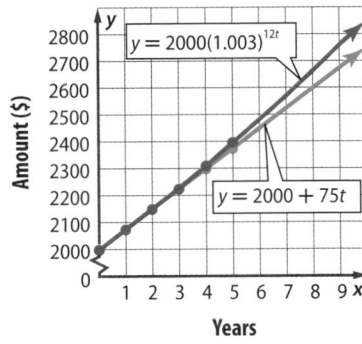

Exercises

The value of a company's piece of equipment decreases over time due to depreciation. The function $y = 16{,}000(0.985)^{2t}$ represents the value after t years.

1. What is the average rate of change over the first five years?

2. What is the average rate of change of the value from year 5 to year 10?

3. What conclusion about the value can we make based on these average rates of change?

4. **CCSS REGULARITY** Copy and complete the table for $y = x^4$.

x	−3	−2	−1	0	1	2	3
y							

Compare and interpret the average rate of change for $x = -3$ to 0 and for $x = 0$ to 3.

Recursive Formulas

- You wrote explicit formulas to represent arithmetic and geometric sequences.

1 Use a recursive formula to list terms in a sequence.

2 Write recursive formulas for arithmetic and geometric sequences.

- Clients of a shuttle service get picked up from their homes and driven to premium outlet stores for shopping. The total cost of the service depends on the total number of customers. The costs for the first six customers are shown.

Number of Customers	Cost ($)
1	25
2	35
3	45
4	55
5	65
6	75

NewVocabulary
recursive formula

Common Core State Standards

Content Standards
F.IF.3 Recognize that sequences are functions, sometimes defined recursively, whose domain is a subset of the integers.

F.BF.2 Write arithmetic and geometric sequences both recursively and with an explicit formula, use them to model situations, and translate between the two forms.

Mathematical Practices
3 Construct viable arguments and critique the reasoning of others.

1 **Using Recursive Formulas** An explicit formula allows you to find any term a_n of a sequence by using a formula written in terms of n. For example, $a_n = 2n$ can be used to find the fifth term of the sequence 2, 4, 6, 8, …: $a_5 = 2(5)$ or 10.

A **recursive formula** allows you to find the nth term of a sequence by performing operations to one or more of the preceding terms. Since each term in the sequence above is 2 greater than the term that preceded it, we can add 2 to the fourth term to find that the fifth term is $8 + 2$ or 10. We can then write a recursive formula for a_n.

$$
\begin{aligned}
a_1 &= && = 2 \\
a_2 &= a_1 + 2 \text{ or } 2 + 2 && = 4 \\
a_3 &= a_2 + 2 \text{ or } 4 + 2 && = 6 \\
a_4 &= a_3 + 2 \text{ or } 6 + 2 && = 8 \\
&\vdots && \vdots \\
a_n &= a_{n-1} + 2
\end{aligned}
$$

A recursive formula for the sequence above is $a_1 = 2$, $a_n = a_{n-1} + 2$, for $n \geq 2$ where n is an integer. The term denoted a_{n-1} represents the term immediately before a_n. Notice that the first term a_1 is given, along with the domain for n.

Example 1 Use a Recursive Formula

Find the first five terms of the sequence in which $a_1 = 7$ and $a_n = 3a_{n-1} - 12$, if $n \geq 2$.

Use $a_1 = 7$ and the recursive formula to find the next four terms.

$$
\begin{aligned}
a_2 &= 3a_{2-1} - 12 & n &= 2 \\
&= 3a_1 - 12 & &\text{Simplify.} \\
&= 3(7) - 12 \text{ or } 9 & a_1 &= 7
\end{aligned}
$$

$$
\begin{aligned}
a_4 &= 3a_{4-1} - 12 & n &= 4 \\
&= 3a_3 - 12 & &\text{Simplify.} \\
&= 3(15) - 12 \text{ or } 33 & a_3 &= 15
\end{aligned}
$$

$$
\begin{aligned}
a_3 &= 3a_{3-1} - 12 & n &= 3 \\
&= 3a_2 - 12 & &\text{Simplify.} \\
&= 3(9) - 12 \text{ or } 15 & a_2 &= 9
\end{aligned}
$$

$$
\begin{aligned}
a_5 &= 3a_{5-1} - 12 & n &= 5 \\
&= 3a_4 - 12 & &\text{Simplify.} \\
&= 3(33) - 12 \text{ or } 87 & a_4 &= 33
\end{aligned}
$$

The first five terms are 7, 9, 15, 33, and 87.

GuidedPractice

1. Find the first five terms of the sequence in which $a_1 = -2$ and $a_n = (-3)a_{n-1} + 4$, if $n \geq 2$.

2 Writing Recursive Formulas To write a recursive formula for an arithmetic or geometric sequence, complete the following steps.

KeyConcept Writing Recursive Formulas

Step 1 Determine if the sequence is arithmetic or geometric by finding a common difference or a common ratio.

Step 2 Write a recursive formula.

| Arithmetic Sequences | $a_n = a_{n-1} + d$, where d is the common difference |
| Geometric Sequences | $a_n = r \cdot a_{n-1}$, where r is the common ratio |

Step 3 State the first term and domain for n.

Example 2 Write Recursive Formulas

Write a recursive formula for each sequence.

a. 17, 13, 9, 5, ...

Step 1 First subtract each term from the term that follows it.

$$13 - 17 = -4 \qquad 9 - 13 = -4 \qquad 5 - 9 = -4$$

There is a common difference of -4. The sequence is arithmetic.

Step 2 Use the formula for an arithmetic sequence.

$a_n = a_{n-1} + d$ Recursive formula for arithmetic sequence

$a_n = a_{n-1} + (-4)$ $d = -4$

Step 3 The first term a_1 is 17, and $n \geq 2$.

A recursive formula for the sequence is $a_1 = 17$, $a_n = a_{n-1} - 4$, $n \geq 2$.

b. 6, 24, 96, 384, ...

Step 1 First subtract each term from the term that follows it.

$$24 - 6 = 18 \qquad 96 - 24 = 72 \qquad 384 - 96 = 288$$

There is no common difference. Check for a common ratio by dividing each term by the term that precedes it.

$$\frac{24}{6} = 4 \qquad\qquad \frac{96}{24} = 4 \qquad\qquad \frac{384}{96} = 4$$

There is a common ratio of 4. The sequence is geometric.

Step 2 Use the formula for a geometric sequence.

$a_n = r \cdot a_{n-1}$ Recursive formula for geometric sequence

$a_n = 4a_{n-1}$ $r = 4$

Step 3 The first term a_1 is 6, and $n \geq 2$.

A recursive formula for the sequence is $a_1 = 6$, $a_n = 4a_{n-1}$, $n \geq 2$.

GuidedPractice

2A. 4, 10, 25, 62.5, ...   **2B.** 9, 36, 63, 90, ...

A sequence can be represented by both an explicit formula and a recursive formula.

Example 3 Write Recursive and Explicit Formulas

COST Refer to the beginning of the lesson. Let n be the number of customers.

a. Write a recursive formula for the sequence.

> **Steps 1 and 2** First subtract each term from the term that follows it.
>
> $35 - 25 = 10$ $45 - 35 = 10$ $55 - 45 = 10$
>
> There is a common difference of 10. The sequence is arithmetic.

> **Step 3** Use the formula for an arithmetic sequence.
>
> $a_n = a_{n-1} + d$ Recursive formula for arithmetic sequence
> $a_n = a_{n-1} + 10$ $d = 10$

> **Step 4** The first term a_1 is 25, and $n \geq 2$.

A recursive formula for the sequence is $a_1 = 25, a_n = a_{n-1} + 10, n \geq 2$.

b. Write an explicit formula for the sequence.

> **Step 1** The common difference is 10.

> **Step 2** Use the formula for the nth term of an arithmetic sequence.
>
> $a_n = a_1 + (n-1)d$ Formula for the nth term
> $\quad = 25 + (n-1)10$ $a_1 = 25$ and $d = 10$
> $\quad = 25 + 10n - 10$ Distributive Property
> $\quad = 10n + 15$ Simplify.

An explicit formula for the sequence is $a_n = 10n + 15$.

> **GuidedPractice**
>
> **3. SAVINGS** The money that Ronald has in his savings account earns interest each year. He does not make any withdrawals or additional deposits. The account balance at the beginning of each year is $10,000, $10,300, $10,609, $10,927.27, and so on. Write a recursive formula and an explicit formula for the sequence.

If several successive terms of a sequence are needed, a recursive formula may be useful, whereas if just the nth term of a sequence is needed, an explicit formula may be useful. Thus, it is sometimes beneficial to translate between the two forms.

Example 4 Translate between Recursive and Explicit Formulas

a. Write a recursive formula for $a_n = 6n + 3$.

$a_n = 6n + 3$ is an explicit formula for an arithmetic sequence with $d = 6$ and $a_1 = 6(1) + 3$ or 9. Therefore, a recursive formula for a_n is $a_1 = 9, a_n = a_{n-1} + 6, n \geq 2$.

b. Write an explicit formula for $a_1 = 120, a_n = 0.8a_{n-1}, n \geq 2$.

$a_n = 0.8a_{n-1}$ is a recursive formula for a geometric sequence with $a_1 = 120$ and $r = 0.8$. Therefore, an explicit formula for a_n is $a_n = 120(0.8)^{n-1}$.

> **GuidedPractice**
>
> **4A.** Write a recursive formula for $a_n = 4(3)^{n-1}$.
>
> **4B.** Write an explicit formula for $a_1 = -16, a_n = a_{n-1} - 7, n \geq 2$.

Real-World Career

Transportation The number of jobs in the transportation industry is expected to grow by an estimated 1.1 million between 2004 and 2014. The specific fields dictate the educational requirements, which include a high school diploma and some form of specialized training.

Source: United States Department of Labor

StudyTip

Geometric Sequence Recall that the formula for the nth term of a geometric sequence is $a_n = a_1 r^{n-1}$.

John A. Rizzo/Photodisc/Getty Images

Example 1 Find the first five terms of each sequence.

1. $a_1 = 16, a_n = a_{n-1} - 3, n \geq 2$

2. $a_1 = -5, a_n = 4a_{n-1} + 10, n \geq 2$

Example 2 Write a recursive formula for each sequence.

3. 1, 6, 11, 16, …

4. 4, 12, 36, 108, …

Example 3 **5. BALL** A ball is dropped from an initial height of 10 feet. The maximum heights the ball reaches on the first three bounces are shown.

a. Write a recursive formula for the sequence.

b. Write an explicit formula for the sequence.

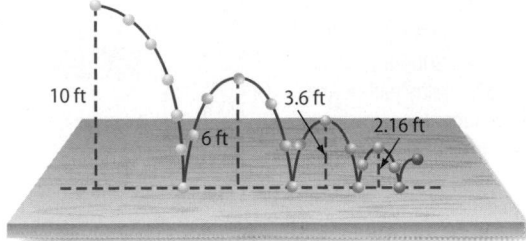

Example 4 For each recursive formula, write an explicit formula. For each explicit formula, write a recursive formula.

6. $a_1 = 4, a_n = a_{n-1} + 16, n \geq 2$

7 $a_n = 5n + 8$

8. $a_n = 15(2)^{n-1}$

9. $a_1 = 22, a_n = 4a_{n-1}, n \geq 2$

Practice and Problem Solving Extra Practice is on page R7.

Example 1 Find the first five terms of each sequence.

10. $a_1 = 23, a_n = a_{n-1} + 7, n \geq 2$

11. $a_1 = 48, a_n = -0.5a_{n-1} + 8, n \geq 2$

12. $a_1 = 8, a_n = 2.5a_{n-1}, n \geq 2$

13. $a_1 = 12, a_n = 3a_{n-1} - 21, n \geq 2$

14. $a_1 = 13, a_n = -2a_{n-1} - 3, n \geq 2$

15. $a_1 = \frac{1}{2}, a_n = a_{n-1} + \frac{3}{2}, n \geq 2$

Example 2 Write a recursive formula for each sequence.

16. 12, −1, −14, −27, …

17. 27, 41, 55, 69, …

18. 2, 11, 20, 29, …

19. 100, 80, 64, 51.2, …

20. 40, −60, 90, −135, …

21. 81, 27, 9, 3, …

Example 3 **22. CCSS MODELING** A landscaper is building a brick patio. Part of the patio includes a pattern constructed from triangles. The first four rows of the pattern are shown.

a. Write a recursive formula for the sequence.

b. Write an explicit formula for the sequence.

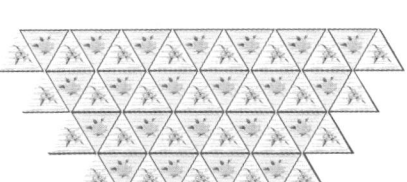

15 bricks
13 bricks
11 bricks
9 bricks

Example 4 For each recursive formula, write an explicit formula. For each explicit formula, write a recursive formula.

23. $a_n = 3(4)^{n-1}$

24. $a_1 = -2, a_n = a_{n-1} - 12, n \geq 2$

25. $a_1 = 38, a_n = \frac{1}{2}a_{n-1}, n \geq 2$

26. $a_n = -7n + 52$

27 TEXTING Barbara received a chain text that she forwarded to five of her friends. Each of her friends forwarded the text to five more friends, and so on.

 a. Find the first five terms of the sequence representing the number of people who receive the text in the nth round.

 b. Write a recursive formula for the sequence.

 c. If Barbara represents a_1, find a_8.

28. GEOMETRY Consider the pattern below. The number of blue boxes increases according to a specific pattern.

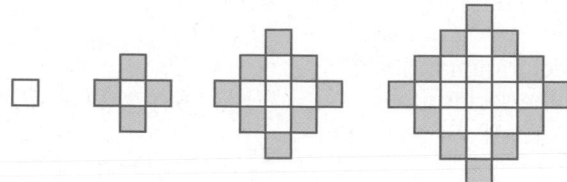

 a. Write a recursive formula for the sequence of the number of blue boxes in each figure.

 b. If the first box represents a_1, find the number of blue boxes in a_8.

29. TREE The growth of a certain type of tree slows as the tree continues to age. The heights of the tree over the past four years are shown.

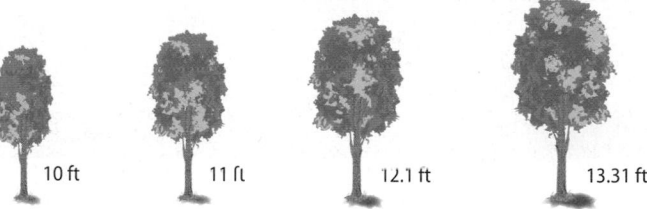

10 ft   11 ft   12.1 ft   13.31 ft

 a. Write a recursive formula for the height of the tree.

 b. If the pattern continues, how tall will the tree be in two more years? Round your answer to the nearest tenth of a foot.

30. MULTIPLE REPRESENTATIONS The Fibonacci sequence is neither arithmetic nor geometric and can be defined by a recursive formula. The first terms are $1, 1, 2, 3, 5, 8, \ldots$

 a. **Logical** Determine the relationship between the terms of the sequence. What are the next five terms in the sequence?

 b. **Algebraic** Write a formula for the nth term if $a_1 = 1$, $a_2 = 1$, and $n \geq 3$.

 c. **Algebraic** Find the 15th term.

 d. **Analytical** Explain why the Fibonacci sequence is not an arithmetic sequence.

H.O.T. Problems   Use Higher-Order Thinking Skills

31. ERROR ANALYSIS Patrick and Lynda are working on a math problem that involves the sequence $2, -2, 2, -2, 2, \ldots$. Patrick thinks that the sequence can be written as a recursive formula. Lynda believes that the sequence can be written as an explicit formula. Is either of them correct? Explain.

32. CHALLENGE Find a_1 for the sequence in which $a_4 = 1104$ and $a_n = 4a_{n-1} + 16$.

33. CCSS ARGUMENTS Determine whether the following statement is *true* or *false*. Justify your reasoning.

There is only one recursive formula for every sequence.

34. CHALLENGE Find a recursive formula for $4, 9, 19, 39, 79, \ldots$

35. WRITING IN MATH Explain the difference between an explicit formula and a recursive formula.

36. Find a recursive formula for the sequence 12, 24, 36, 48, … .

 A $a_1 = 12, a_n = 2a_{n-1}, n \geq 2.$

 B $a_1 = 12, a_n = 4a_{n-1} - 24, n \geq 2.$

 C $a_1 = 12, a_n = a_{n-1} + 12, n \geq 2.$

 D $a_1 = 12, a_n = 12a_{n-1} + 12, n \geq 2.$

37. GEOMETRY The area of a rectangle is $36m^4n^6$ square feet. The length of the rectangle is $6m^3n^3$ feet. What is the width of the rectangle?

 F $216m^7n^9$ ft

 G $6mn^3$ ft

 H $42m^7n^3$ ft

 J $30mn^3$ ft

38. Find an inequality for the graph shown.

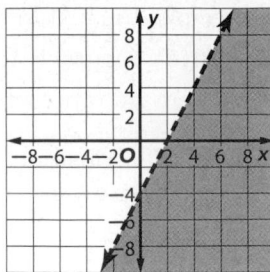

 A $y > 2x - 4$ **C** $y < 2x - 4$

 B $y \geq 2x - 4$ **D** $y \leq 2x - 4$

39. Write an equation of the line that passes through $(-2, -20)$ and $(4, 58)$.

 F $y = 13x + 6$ **H** $y = 19x + 18$

 G $y = 19x - 18$ **J** $y = 13x - 6$

Find the next three terms in each geometric sequence. (Lesson 7-7)

40. 675, 225, 75, …

41. 16, −24, 36, …

42. 6, 18, 54, …

43. 512, −256, 128, …

44. 125, 25, 5, …

45. 12, 60, 300, …

46. INVESTMENT Nicholas invested $2000 with a 5.75% interest rate compounded monthly. How much money will Nicholas have after 5 years? (Lesson 7-6)

47. TOURS The Snider family and the Rollins family are traveling together on a trip to visit a candy factory. The number of people in each family and the total cost are shown in the table below. Find the adult and children's admission prices. (Lesson 6-3)

Family	Number of Adults	Number of Children	Total Cost
Snider	2	3	$58
Rollins	2	1	$38

Write each equation in standard form. (Lesson 4-3)

48. $y + 6 = -3(x + 2)$

49. $y - 12 = 4(x - 7)$

50. $y + 9 = 5(x - 3)$

51. $y - 1 = \frac{1}{3}(x + 15)$

52. $y + 10 = \frac{2}{5}(x - 6)$

53. $y - 4 = -\frac{2}{7}(x + 1)$

Simplify each expression. If not possible, write *simplified*.

54. $8x + 3y^2 + 7x - 2y$

55. $4(x - 16) + 6x$

56. $4n - 3m + 9m - n$

57. $6r^2 + 7r$

58. $-2(4g - 5h) - 6g$

59. $9x^2 - 7x + 16y^2$

Study Guide

KeyConcepts

Multiplication and Division Properties of Exponents (Lessons 7-1 and 7-2)

For any nonzero real numbers a and b and any integers m, n, and p, the following are true.

- Product of Powers: $a^m \cdot a^n = a^{m+n}$
- Power of a Power: $(a^m)^n = a^{m \cdot n}$
- Power of a Product: $(ab)^m = a^m b^m$
- Quotient of Powers: $\dfrac{a^m}{a^p} = a^{m-p}$
- Power of a Quotient: $\left(\dfrac{a}{b}\right)^m = \dfrac{a^m}{b^m}$
- Zero Exponent: $a^0 = 1$
- Negative Exponent: $a^{-n} = \dfrac{1}{a^n}$ and $\dfrac{1}{a^{-n}} = a^n$

Rational Exponents (Lesson 7-3)

For any positive real number b and any integers m and $n > 1$, the following are true.

$$b^{\frac{1}{2}} = \sqrt{b} \qquad b^{\frac{1}{n}} = \sqrt[n]{b} \qquad b^{\frac{m}{n}} = \left(\sqrt[n]{b}\right)^m \text{ or } \sqrt[n]{b^m}$$

Scientific Notation (Lesson 7-4)

- A number is in scientific notation if it is in the form $a \times 10^n$, where $1 \le a < 10$.
- To write in standard form:
  - If $n > 0$, move the decimal n places right.
  - If $n < 0$, move the decimal n places left.

Exponential Functions (Lessons 7-5 and 7-6)

- The equation for exponential growth is $y = a(1 + r)^t$, where $r > 0$. The equation for exponential decay is $y = a(1 - r)^t$, where $0 < r < 1$. y is the final amount, a is the initial amount, r is the rate of change, and t is the time in years.

FOLDABLES StudyOrganizer

Be sure the Key Concepts are noted in your Foldable.

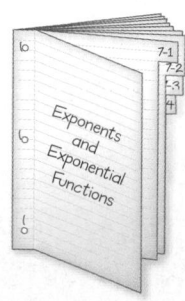

KeyVocabulary

common ratio (p. 438)	monomial (p. 391)
compound interest (p. 433)	negative exponent (p. 400)
constant (p. 391)	nth root (p. 407)
cube root (p. 407)	order of magnitude (p. 401)
exponential decay (p. 424)	rational exponent (p. 406)
exponential equation (p. 409)	recursive formula (p. 445)
exponential function (p. 424)	scientific notation (p. 414)
exponential growth (p. 424)	zero exponent (p. 399)
geometric sequence (p. 438)	

VocabularyCheck

Choose the word or term that best completes each sentence.

1. $7xy^4$ is an example of a(n) _____.

2. The _____ of 95,234 is 10^5.

3. 2 is a(n) _____ of 8.

4. The rules for operations with exponents can be extended to apply to expressions with a(n) _____ such as $7^{\frac{2}{3}}$.

5. A number written in _____ is of the form $a \times 10^n$, where $1 \le a < 10$ and n is an integer.

6. $f(x) = 3^x$ is an example of a(n) _____.

7. $a_1 = 4$ and $a_n = 3a_{n-1} - 12$, if $n \ge 2$, is a(n) _____ for the sequence 4, -8, -20, -32, ….

8. $2^{3x-1} = 16$ is an example of a(n) _____.

9. The equation for _____ is $y = C(1 - r)^t$.

10. If $a^n = b$ for a positive integer n, then a is a(n) _____ of b.

Lesson-by-Lesson Review

7-1 Multiplication Properties of Exponents

Simplify each expression.

11. $x \cdot x^3 \cdot x^5$

12. $(2xy)(-3x^2y^5)$

13. $(-4ab^4)(-5a^5b^2)$

14. $(6x^3y^2)^2$

15. $\left[(2r^3t)^3\right]^2$

16. $(-2u^3)(5u)$

17. $(2x^2)^3(x^3)^3$

18. $\frac{1}{2}(2x^3)^3$

19. GEOMETRY Use the formula $V = \pi r^2 h$ to find the volume of the cylinder.

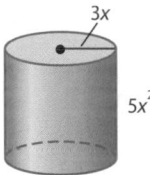

3x

$5x^2$

Example 1

Simplify $(5x^2y^3)(2x^4y)$.

$(5x^2y^3)(2x^4y)$

$\quad = (5 \cdot 2)(x^2 \cdot x^4)(y^3 \cdot y)$ Commutative Property

$\quad = 10x^6y^4$ Product of Powers

Example 2

Simplify $(3a^2b^4)^3$.

$(3a^2b^4)^3 = 3^3(a^2)^3(b^4)^3$ Power of a Product

$\qquad\qquad = 27a^6b^{12}$ Simplify.

7-2 Division Properties of Exponents

Simplify each expression. Assume that no denominator equals zero.

20. $\frac{(3x)^0}{2a}$

21. $\left(\frac{3xy^3}{2z}\right)^3$

22. $\frac{12y^{-4}}{3y^{-5}}$

23. $a^{-3}b^0c^6$

24. $\frac{-15x^7y^8z^4}{-45x^3y^5z^3}$

25. $\frac{(3x^{-1})^{-2}}{(3x^2)^{-2}}$

26. $\left(\frac{6xy^{11}z^9}{48x^6yz^{-7}}\right)^0$

27. $\left(\frac{12}{2}\right)\left(\frac{x}{y^5}\right)\left(\frac{y^4}{x^4}\right)$

28. GEOMETRY The area of a rectangle is $25x^2y^4$ square feet. The width of the rectangle is $5xy$ feet. What is the length of the rectangle?

5xy

Example 3

Simplify $\frac{2k^4m^3}{4k^2m}$. Assume that no denominator equals zero.

$\frac{2k^4m^3}{4k^2m} = \left(\frac{2}{4}\right)\left(\frac{k^4}{k^2}\right)\left(\frac{m^3}{m}\right)$ Group powers with the same base.

$\quad = \left(\frac{1}{2}\right)k^{4-2}\,m^{3-1}$ Quotient of Powers

$\quad = \frac{k^2m^2}{2}$ Simplify.

Example 4

Simplify $\frac{t^4uv^{-2}}{t^{-3}u^7}$. Assume that no denominator equals zero.

$\frac{t^4uv^{-2}}{t^{-3}u^7} = \left(\frac{t^4}{t^{-3}}\right)\left(\frac{u}{u^7}\right)(v^{-2})$ Group the powers with the same base.

$\quad = (t^{4+3})(u^{1-7})(v^{-2})$ Quotient of Powers

$\quad = t^7u^{-6}v^{-2}$ Simplify.

$\quad = \frac{t^7}{u^6v^2}$ Simplify.

7-3 Rational Exponents

Simplify.

29. $\sqrt[3]{343}$

30. $\sqrt[6]{729}$

31. $625^{\frac{1}{4}}$

32. $\left(\frac{8}{27}\right)^{\frac{1}{3}}$

33. $256^{\frac{3}{4}}$

34. $32^{\frac{2}{5}}$

35. $343^{\frac{4}{3}}$

36. $\left(\frac{4}{49}\right)^{\frac{3}{2}}$

Solve each equation.

37. $6^x = 7776$

38. $4^{4x-1} = 32$

Example 5

Simplify $125^{\frac{2}{3}}$.

$$
\begin{aligned}
125^{\frac{2}{3}} &= \left(\sqrt[3]{125}\right)^2 && b^{\frac{m}{n}} = \left(\sqrt[n]{b}\right)^m \\
&= \left(\sqrt[3]{5 \cdot 5 \cdot 5}\right)^2 && 125 = 5^3 \\
&= 5^2 \text{ or } 25 && \text{Simplify.}
\end{aligned}
$$

Example 6

Solve $9^{x-1} = 729$.

$$
\begin{aligned}
9^{x-1} &= 729 && \text{Original equation} \\
9^{x-1} &= 9^3 && \text{Rewrite 729 as } 9^3. \\
x - 1 &= 3 && \text{Power Property of Equality} \\
x &= 4 && \text{Add 1 to each side.}
\end{aligned}
$$

7-4 Scientific Notation

Express each number in scientific notation.

39. 2,300,000

40. 0.0000543

41. ASTRONOMY Earth has a diameter of about 8000 miles. Jupiter has a diameter of about 88,000 miles. Write in scientific notation the ratio of Earth's diameter to Jupiter's diameter.

Example 7

Express 300,000,000 in scientific notation.

Step 1 $300,000,000 \longrightarrow 3.00000000$

Step 2 The decimal point moved 8 places to the left, so $n = 8$.

Step 3 $300,000,000 = 3 \times 10^8$

7-5 Exponential Functions

Graph each function. Find the *y*-intercept, and state the domain and range.

42. $y = 2^x$

43. $y = 3^x + 1$

44. $y = 4^x + 2$

45. $y = 2^x - 3$

46. BIOLOGY The population of bacteria in a petri dish increases according to the model $p = 550(2.7)^{0.008t}$, where t is the number of hours and $t = 0$ corresponds to 1:00 P.M. Use this model to estimate the number of bacteria in the dish at 5:00 P.M.

Example 8

Graph $y = 3^x + 6$. Find the *y*-intercept, and state the domain and range.

x	$3^x + 6$	y
-3	$3^{-3} + 6$	6.04
-2	$3^{-2} + 6$	6.11
-1	$3^{-1} + 6$	6.33
0	$3^0 + 6$	7
1	$3^1 + 6$	9

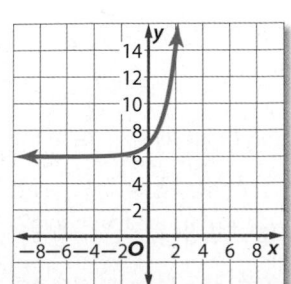

The *y*-intercept is (0, 7). The domain is all real numbers, and the range is all real numbers greater than 6.

7-6 Growth and Decay

47. Find the final value of $2500 invested at an interest rate of 2% compounded monthly for 10 years.

48. COMPUTERS Zita's computer is depreciating at a rate of 3% per year. She bought the computer for $1200.

    a. Write an equation to represent this situation.

    b. What will the computer's value be after 5 years?

Example 9

Find the final value of $2000 invested at an interest rate of 3% compounded quarterly for 8 years.

$$A = P\left(1 + \frac{r}{n}\right)^{nt}$$ Compound interest equation

$$= 2000\left(1 + \frac{0.03}{4}\right)^{4(8)}$$ $P = 2000, r = 0.03,$ $n = 4,$ and $t = 8$

$$\approx \$2540.22$$ Use a calculator.

7-7 Geometric Sequences as Exponential Functions

Find the next three terms in each geometric sequence.

49. $-1, 1, -1, 1, \ldots$

50. $3, 9, 27, \ldots$

51. $256, 128, 64, \ldots$

Write the equation for the nth term of each geometric sequence.

52. $-1, 1, -1, 1, \ldots$

53. $3, 9, 27, \ldots$

54. $256, 128, 64, \ldots$

55. SPORTS A basketball is dropped from a height of 20 feet. It bounces to $\frac{1}{2}$ its height after each bounce. Draw a graph to represent the situation.

Example 10

Find the next three terms in the geometric sequence $2, 6, 18, \ldots$.

| **Step 1** | Find the common ratio. Each number is 3 times the previous number, so $r = 3$. |

| **Step 2** | Multiply each term by the common ratio to find the next three terms. |

$18 \times 3 = 54, 54 \times 3 = 162, 162 \times 3 = 486$

The next three terms are 54, 162, and 486.

Example 11

Write the equation for the nth term of the geometric sequence $-3, 12, -48, \ldots$.

The common ratio is -4. So $r = -4$.

$a_n = a_1 r^{n-1}$ Formula for the nth term

$a_n = -3(-4)^{n-1}$ $a_1 = -3$ and $r = -4$

7-8 Recursive Formulas

Find the first five terms of each sequence.

56. $a_1 = 11, a_n = a_{n-1} - 4, n \geq 2$

57. $a_1 = 3, a_n = 2a_{n-1} + 6, n \geq 2$

Write a recursive formula for each sequence.

58. $2, 7, 12, 17, \ldots$

59. $32, 16, 8, 4, \ldots$

60. $2, 5, 11, 23, \ldots$

Example 12

Write a recursive formula for $3, 1, -1, -3, \ldots$.

| **Step 1** | First subtract each term from the term that follows it. |

$1 - 3 = -2, -1 - 1 = -2, -3 - (-1) = -2$

There is a common difference of -2. The sequence is arithmetic.

| **Step 2** | Use the formula for an arithmetic sequence. |

$a_n = a_{n-1} + d$ Recursive formula

$a_n = a_{n-1} + (-2)$ $d = -2$

| **Step 3** | The first term a_1 is 3, and $n \geq 2$. |

A recursive formula is $a_1 = 3, a_n = a_{n-1} - 2, n \geq 2$.

Simplify each expression.

1. $(x^2)(7x^8)$

2. $(5a^7bc^2)(-6a^2bc^5)$

3. **MULTIPLE CHOICE** Express the volume of the solid as a monomial.

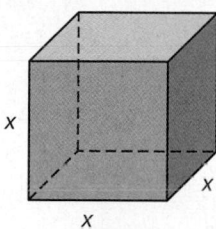

    A x^3 C $6x^3$

    B $6x$ D x^6

Simplify each expression. Assume that no denominator equals 0.

4. $\dfrac{x^6y^8}{x^2}$

5. $\left(\dfrac{2a^4b^3}{c^6}\right)^0$

6. $\dfrac{2xy^{-7}}{8x}$

Simplify.

7. $\sqrt[3]{1000}$ 8. $\sqrt[5]{3125}$

9. $1728^{\frac{1}{3}}$ 10. $\left(\dfrac{16}{81}\right)^{\frac{1}{2}}$

11. $27^{\frac{2}{3}}$ 12. $10{,}000^{\frac{3}{4}}$

13. $27^{\frac{5}{3}}$ 14. $\left(\dfrac{1}{121}\right)^{\frac{3}{2}}$

Solve each equation.

15. $12^x = 1728$

16. $7^{x-1} = 2401$

17. $9^{x-3} = 729$

Express each number in scientific notation.

18. 0.00021

19. $58{,}000$

Express each number in standard form.

20. 2.9×10^{-5}

21. 9.1×10^6

Evaluate each product or quotient. Express the results in scientific notation.

22. $(2.5 \times 10^3)(3 \times 10^4)$

23. $\dfrac{8.8 \times 10^2}{4 \times 10^{-4}}$

24. **ASTRONOMY** The average distance from Mercury to the Sun is 35,980,000 miles. Express this distance in scientific notation.

Graph each function. Find the y-intercept, and state the domain and range.

25. $y = 2(5)^x$

26. $y = -3(11)^x$

27. $y = 3^x + 2$

Find the next three terms in each geometric sequence.

28. $2, -6, 18, \ldots$

29. $1000, 500, 250, \ldots$

30. $32, 8, 2, \ldots$

31. **MULTIPLE CHOICE** Lynne invested $500 into an account with a 6.5% interest rate compounded monthly. How much will Lynne's investment be worth in 10 years?

    F $600.00

    G $938.57

    H $956.09

    J $957.02

32. **INVESTMENTS** Shelly's investment of $3000 has been losing value at a rate of 3% each year. What will her investment be worth in 6 years?

Find the first five terms of each sequence.

33. $a_1 = 18, a_n = a_{n-1} - 4, n \geq 2$

34. $a_1 = -2, a_n = 4a_{n-1} + 5, n \geq 2$

Preparing for Standardized Tests

Using a Scientific or Graphing Calculator

Scientific and graphing calculators are powerful problem-solving tools. There are times when a calculator can be used to make computations faster and easier, such as computations with very large numbers. However, there are times when using a calculator is necessary, like the estimation of irrational numbers.

Strategies for Using a Scientific or Graphing Calculator

Step 1

Familiarize yourself with the various functions of a scientific or graphing calculator as well as when they should be used:

- **Exponents** scientific notation, calculating with large or small numbers

- **Pi** solving circle problems, like circumference and area

- **Square roots** distance on a coordinate plane, Pythagorean theorem

- **Graphs** analyzing paired data in a scatter plot, graphing functions, finding roots of equations

Step 2

Use your scientific or graphing calculator to solve the problem.

- Remember to work as efficiently as possible. Some steps may be done mentally or by hand, while others should be completed using your calculator.

- If time permits, check your answer.

Standardized Test Example

Read the problem. Identify what you need to know. Then use the information in the problem to solve.

The distance from the Sun to Jupiter is approximately 7.786×10^{11} meters. If the speed of light is about 3×10^8 meters per second, how long does it take for light from the Sun to reach Jupiter? Round to the nearest minute.

A about 43 minutes

B about 51 minutes

C about 1876 minutes

D about 2595 minutes

McGraw-Hill Companies, Inc., Mazer Creative Services, photographer

Read the problem carefully. You are given the approximate distance from the Sun to Jupiter as well as the speed of light. Both quantities are given in scientific notation. You are asked to find how many minutes it takes for light from the Sun to reach Jupiter. Use the relationship distance = rate × time to find the amount of time.

$$d = r \times t$$

$$\frac{d}{r} = t$$

To find the amount of time, divide the distance by the rate. Notice, however, that the units for time will be seconds.

$$\frac{7.786 \times 10^{11} \text{ m}}{3 \times 10^8 \text{ m/s}} = t \text{ seconds}$$

Use a scientific calculator to quickly find the quotient. On most scientific calculators, the EE key is used to enter numbers in scientific notation.

KEYSTROKES: (7.786 [2nd] [EE] 11) ÷ (3 [2nd] [EE] 8) [ENTER]

The result is 2595.33333333 seconds. To convert this number to minutes, use your calculator to divide the result by 60. This gives an answer of about 43.2555 minutes. The answer is A.

Exercises

Read each problem. Identify what you need to know. Then use the information in the problem to solve.

1. Since its creation 5 years ago, approximately 2.504×10^7 items have been sold or traded on a popular online website. What is the average daily number of items sold or traded over the 5-year period?

   A about 9640 items per day

   B about 13,720 items per day

   C about 1,025,000 items per day

   D about 5,008,000 items per day

2. Evaluate $\sqrt{ab}$ if $a = 121$ and $b = 23$.

   F about 5.26

   G about 9.90

   H about 12

   J about 52.75

3. The population of the United States is about 3.034×10^8 people. The land area of the country is about 3.54×10^6 square miles. What is the average *population density* (number of people per square mile) of the United States?

   A about 136.3 people per square mile

   B about 112.5 people per square mile

   C about 94.3 people per square mile

   D about 85.7 people per square mile

4. Eleece is making a cover for the marching band's bass drum. The drum has a diameter of 20 inches. Estimate the area of the face of the bass drum.

   F 31.41 square inches

   G 62.83 square inches

   H 78.54 square inches

   J 314.16 square inches

Multiple Choice

Read each question. Then fill in the correct answer on the answer document provided by your teacher or on a sheet of paper.

1. Express the area of the triangle below as a monomial.

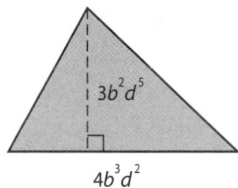

$3b^2d^5$

$4b^3d^2$

A $12b^5d^7$

B $12b^6d^{10}$

C $6b^6d^{10}$

D $6b^5d^7$

2. Simplify the following expression.

$$\left(\frac{2w^2z^5}{3y^4}\right)^3$$

F $\dfrac{2w^5z^8}{3y^7}$

G $\dfrac{8w^6z^{15}}{27y^{12}}$

H $\dfrac{8w^5z^8}{27y^7}$

J $\dfrac{2w^6z^{15}}{3y^{12}}$

3. Which equation of a line is perpendicular to $y = \frac{3}{5}x - 3$?

A $y = -\frac{5}{3}x + 2$ **C** $y = \frac{5}{3}x - 2$

B $y = -\frac{3}{5}x + 2$ **D** $y = \frac{3}{5}x - 2$

Test-TakingTip

Question 2 Use the laws of exponents to simplify the expression. Remember, to find the power of a power, multiply the exponents.

4. Write a recursive formula for the sequence of the number of squares in each figure.

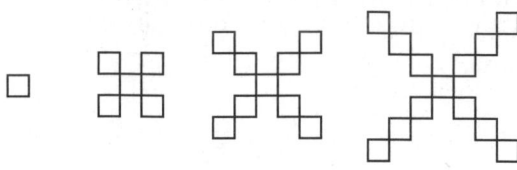

F $a_1 = 1, a_n = 4a_{n-1} - 3, n \geq 1$

G $a_1 = 1, a_n = 4a_{n-1}, n \geq 2$

H $a_1 = 1, a_n = a_{n-1} + 4, n \geq 2$

J $a_1 = 1, a_n = 4a_{n-1} + 4, n \geq 2$

5. Evaluate $(4.2 \times 10^6)(5.7 \times 10^8)$.

A 2.394×10^{15}

B 23.94×10^{14}

C 9.9×10^{14}

D 2.394×10^{48}

6. Which inequality is shown in the graph?

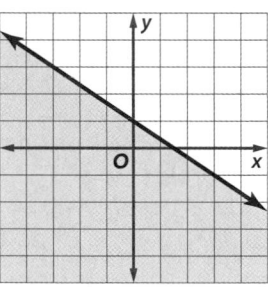

F $y \leq -\frac{2}{3}x - 1$

G $y \leq -\frac{3}{4}x - 1$

H $y \leq -\frac{2}{3}x + 1$

J $y \leq -\frac{3}{4}x + 1$

7. Jaden created a Web site for the Science Olympiad team. The total number of hits the site has received is shown.

Day	Total Hits	Day	Total Hits
3	5	17	27
6	7	21	33
10	12	26	40
13	17	34	55

a. Find an equation for the regression line.

b. Predict the number of total hits that the Web site will have received on day 46.

8. Find the value of x so that the figures have the same area.

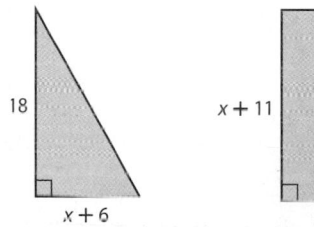

9. What is the solution to the following system of equations? Show your work.

$$\begin{cases} y = 6x - 1 \\ y = 6x + 4 \end{cases}$$

10. GRIDDED RESPONSE At a family fun center, the Wilson and Sanchez families each bought video game tokens and batting cage tokens as shown in the table.

Family	Wilson	Sanchez
Number of Video Game Tokens	25	30
Number of Batting Cage Tokens	8	6
Total Cost	$26.50	$25.50

What is the cost in dollars of a batting cage token at the family fun center?

Record your answers on a sheet of paper. Show your work.

11. The table below shows the distances from the Sun to Mercury, Earth, Mars, and Saturn. Use the data to answer each question.

Planet	Distance from Sun (km)
Mercury	5.79×10^7
Earth	1.50×10^8
Mars	2.28×10^8
Saturn	1.43×10^9

a. Of the planets listed, which one is the closest to the Sun?

b. About how many times as far from the Sun is Mars as Earth?

Need ExtraHelp?

If you missed Question...	1	2	3	4	5	6	7	8	9	10	11
Go to Lesson...	7-1	7-2	4-4	6-6	7-3	5-6	4-6	2-4	6-2	6-4	7-4

Quadratic Expressions and Equations

Then

○ You applied the laws of exponents and explored exponential functions.

Now

○ In this chapter, you will:

- Add, subtract, and multiply polynomials.
- Factor trinomials.
- Factor differences of squares.
- Graph quadratic functions.
- Solve quadratic equations.

Why? ▲

○ **ARCHITECTURE** Quadratic equations can be used to model the shape of architectural structures.

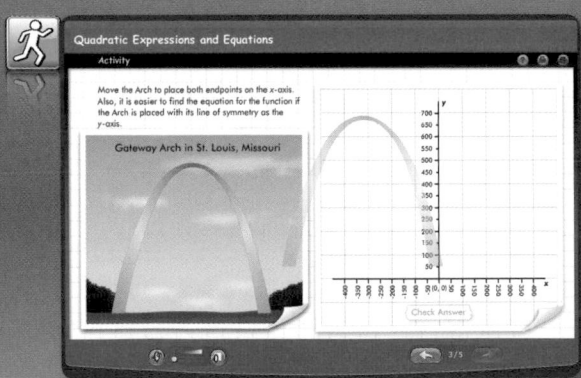

connectED.mcgraw-hill.com   **Your Digital Math Portal**

| Animation | Vocabulary | eGlossary | Personal Tutor | Virtual Manipulatives | Graphing Calculator | Audio | Foldables | Self-Check Practice | Worksheets |

McGraw-Hill Education

Diagnose Readiness | You have two options for checking prerequisite skills.

1 **Textbook Option** Take the Quick Check below. Refer to the Quick Review for help.

QuickCheck	**Quick**Review

QuickCheck

Rewrite each expression using the Distributive Property. Then simplify.

1. $a(a + 5)$

2. $2(3 + x)$

3. $n(n - 3n^2 + 2)$

4. $-6(x^2 - 5x + 6)$

5. **FINANCIAL LITERACY** Five friends will pay $9 per ticket, $3 per drink, and $6 per popcorn at the movies. Write an expression that could be used to determine the cost for them to go to the movies.

Simplify each expression. If not possible, write *simplified*.

6. $3u + 10u$

7. $5a - 2 + 6a$

8. $6m^2 - 8m$

9. $4w^2 + w + 15w^2$

10. $2x^2 + 5 - 11x^2$

11. $8v^3 - 27$

12. $4k^2 + 2k - 2k + 1$

13. $a^2 - 4a - 4a + 16$

14. $6y^2 + 2y - 3y - 1$

15. $9g^2 - 3g - 6g + 2$

Simplify.

16. $b(b^6)$

17. $4n^3(n^2)$

18. $8m(4m^2)$

19. $-5z^4(3z^5)$

20. $5xy(4x^3y)$

21. $(-2a^4c^5)(7ac^4)$

22. **GEOMETRY** A square is $6x^3$ inches on each side. What is the area of the square?

QuickReview

Example 1

Rewrite $6x(-3x - 5x - 5x^2 + x^3)$ using the Distributive Property. Then simplify.

$6x(-3x - 5x - 5x^2 + x^3)$

$= 6x(-3x) + 6x(-5x) + 6x(-5x^2) + 6x(x^3)$

$= -18x^2 - 30x^2 - 30x^3 + 6x^4$

$= -48x^2 - 30x^3 + 6x^4$

Example 2

Simplify $8c + 6 - 4c + 2c^2$.

$8c + 6 - 4c + 2c^2 = 2c^2 + 8c - 4c + 6$

$= 2c^2 + (8 - 4)c + 6$

$= 2c^2 + 4c + 6$

Example 3

Simplify $(-2y^3)(9y^4)$.

$(9y^3)(-2y^4) = (-2 \cdot 9)(y^3 \cdot y^4)$

$= (-2 \cdot 9)(y^{3 + 4})$

$= -18y^7$

2 **Online Option** Take an online self-check Chapter Readiness Quiz at <u>connectED.mcgraw-hill.com</u>.

Get Started on the Chapter

You will learn several new concepts, skills, and vocabulary terms as you study Chapter 8. To get ready, identify important terms and organize your resources. You may wish to refer to Chapter 0 to review prerequisite skills.

FOLDABLES StudyOrganizer

Quadratic Expressions and Equations Make this Foldable to help you organize your Chapter 8 notes about quadratic expressions and equations. Begin with five sheets of grid paper.

1 **Fold** in half along the width. On the first three sheets, cut 5 centimeters along the fold at the ends. On the second two sheets cut in the center, stopping 5 centimeters from the ends.

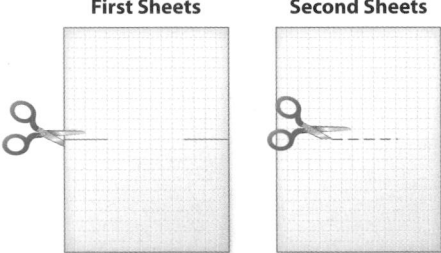

2 **Insert** the first sheets through the second sheets and align the folds. Label the front Chapter 8, Quadratic Expressions and Equations. Label the pages with lesson numbers and the last page with vocabulary.

NewVocabulary

English		Español
polynomial	p. 465	polinomio
binomial	p. 465	binomio
trinomial	p. 465	trinomio
degree of a monomial	p. 465	grado de un monomio
degree of a polynomial	p. 465	grado de un polinomio
standard form of a polynomial	p. 466	forma estándar de polinomio
leading coefficient	p. 466	coeficiente lider
FOIL method	p. 481	método foil
quadratic expression	p. 481	expression cuadrática
factoring	p. 494	factorización
factoring by grouping	p. 495	factorización por agrupamiento
Zero Product Property	p. 496	propiedad del producto de cero
quadratic equation	p. 506	ecuación cuadrática
prime polynomial	p. 512	polinomio primo
difference of two squares	p. 516	diferencia de cuadrados
perfect square trinomial	p. 522	trinomio cuadrado perfecto
Square Root Property	p. 525	Propiedad de la raíz cuadrada

ReviewVocabulary

absolute value valor absoluto the absolute value of any number n is the distance the number is from zero on a number line and is written $|n|$

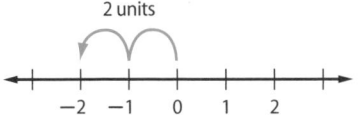

2 units

-2 -1 0   1   2

The absolute value of -2 is 2 because it is 2 units from 0.

perfect square cuadrado perfecto a number with a square root that is a rational number

8-1

Algebra Lab
Adding and Subtracting Polynomials

Algebra tiles can be used to model polynomials. A polynomial is a monomial or the sum of monomials. The diagram below shows the models.

CCSS Common Core State Standards
Content Standards
A.APR.1 Understand that polynomials form a system analogous to the integers, namely, they are closed under the operations of addition, subtraction, and multiplication; add, subtract, and multiply polynomials.

Polynomial Models

- Polynomials are modeled using three types of tiles.

- Each tile has an opposite.

Activity 1   Model Polynomials

Use algebra tiles to model each polynomial.

- $5x$

  To model this polynomial, you will need 5 green x-tiles.

- $-2x^2 + x + 3$

  To model this polynomial, you will need 2 red $-x^2$-tiles, 1 green x-tile, and 3 yellow 1-tiles.

Monomials such as $3x$ and $-2x$ are called *like terms* because they have the same variable to the same power.

Polynomial Models

- Like terms are represented by tiles that have the same shape and size.

- A *zero pair* may be formed by pairing one tile with its opposite. You can remove or add zero pairs without changing the polynomial.

like terms

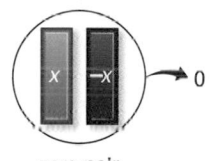
zero pair

Activity 2   Add Polynomials

Use algebra tiles to find $(2x^2 - 3x + 5) + (x^2 + 6x - 4)$.

Step 1

Model each polynomial.

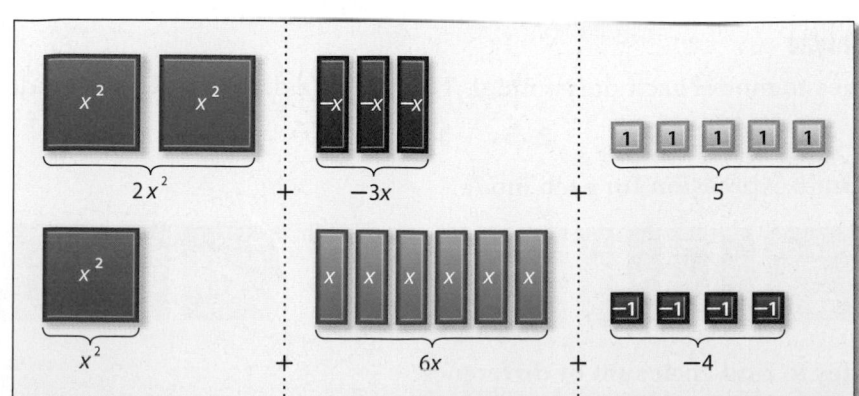

(continued on the next page)

Step 2

Combine like terms and remove zero pairs.

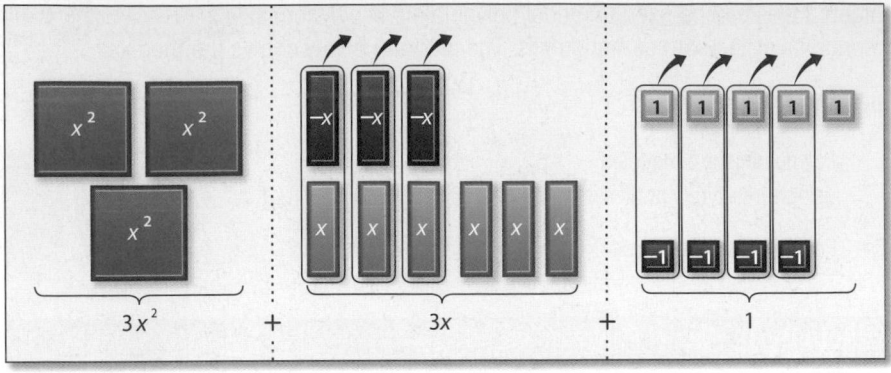

Step 3 Write the polynomial.

$$(2x^2 - 3x + 5) + (x^2 + 6x - 4) = 3x^2 + 3x + 1$$

Activity 3 Subtract Polynomials

Use algebra tiles to find $(4x + 5) - (-3x + 1)$.

Step 1 Model the polynomial $4x + 5$.

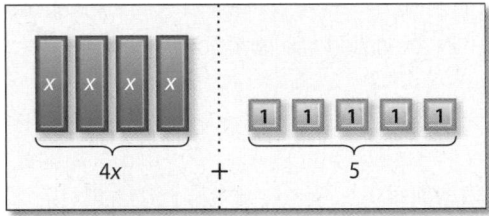

Step 2 To subtract $-3x + 1$, remove 3 red $-x$-tiles and 1 yellow 1-tile. You can remove the 1-tile, but there are no $-x$-tiles. Add 3 zero pairs of x-tiles. Then remove the 3 red $-x$-tiles.

Step 3 Write the polynomial.
$(4x + 5) - (-3x + 1) = 7x + 4$

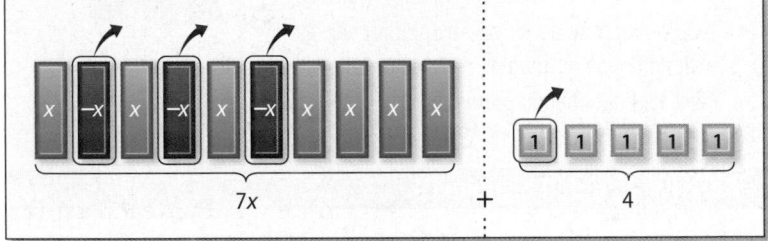

Model and Analyze

Use algebra tiles to model each polynomial. Then draw a diagram of your model.

1. $-2x^2$ **2.** $5x - 4$ **3.** $x^2 - 4x$

Write an algebraic expression for each model.

4.

5.

Use algebra tiles to find each sum or difference.

6. $(x^2 + 5x - 2) + (3x^2 - 2x + 6)$ **7.** $(2x^2 + 8x + 1) - (x^2 - 4x - 2)$ **8.** $(-4x^2 + x) - (x^2 + 5x)$

Adding and Subtracting Polynomials

::Then
- You identified monomials and their characteristics.

::Now
1. Write polynomials in standard form.
2. Add and subtract polynomials.

::Why?
- In 2017, sales of digital audio players are expected to reach record numbers. The sales data can be modeled by the equation $U = -2.7t^2 + 49.4t + 128.7$, where U is the number of units shipped in millions and t is the number of years since 2005.

The expression $-2.7t^2 + 49.4t + 128.7$ is an example of a polynomial. Polynomials can be used to model situations.

NewVocabulary
polynomial
binomial
trinomial
degree of a monomial
degree of a polynomial
standard form of a polynomial
leading coefficient

Common Core State Standards

Content Standards
A.SSE.1a Interpret parts of an expression, such as terms, factors, and coefficients.

A.APR.1 Understand that polynomials form a system analogous to the integers, namely, they are closed under the operations of addition, subtraction, and multiplication; add, subtract, and multiply polynomials.

Mathematical Practices
3 Construct viable arguments and critique the reasoning of others.

---

1 Polynomials in Standard Form A **polynomial** is a monomial or the sum of monomials, each called a *term* of the polynomial. Some polynomials have special names. A **binomial** is the sum of *two* monomials, and a **trinomial** is the sum of *three* monomials.

Monomial	Binomial	Trinomial
$5x$	$2x^2 + 7$	$x^3 - 10x + 1$

The **degree of a monomial** is the sum of the exponents of all its variables. A nonzero constant term has degree 0, and zero has no degree.

The **degree of a polynomial** is the greatest degree of any term in the polynomial. You can find the degree of a polynomial by finding the degree of each term. Polynomials are named based on their degree.

Degree	Name
0	constant
1	linear
2	quadratic
3	cubic
4	quartic
5	quintic
6 or more	6th degree, 7th degree, and so on

Example 1 Identify Polynomials

Determine whether each expression is a polynomial. If it is a polynomial, find the degree and determine whether it is a *monomial, binomial,* or *trinomial*.

Expression	Is it a polynomial?	Degree	Monomial, binomial, or trinomial?
a. $4y - 5xz$	Yes; $4y - 5xz$ is the sum of $4y$ and $-5xz$.	2	binomial
b. -6.5	Yes; -6.5 is a real number.	0	monomial
c. $7a^{-3} + 9b$	No; $7a^{-3} = \frac{7}{a^3}$, which is not a monomial.	—	—
d. $6x^3 + 4x + x + 3$	Yes; $6x^3 + 4x + x + 3 = 6x^3 + 5x + 3$, the sum of three monomials.	3	trinomial

▶ **Guided**Practice

1A. x

1B. $-3y^2 - 2y + 4y - 1$

1C. $5rx + 7tuv$

1D. $10x^{-4} - 8x^a$

The terms of a polynomial can be written in any order. However, polynomials in one variable are usually written in standard form. The **standard form of a polynomial** has the terms in order from greatest to least degree. In this form, the coefficient of the first term is called the **leading coefficient**.

leading coefficient   greatest degree

Standard form: $4x^3 - 5x^2 + 2x + 7$

Example 2 Standard Form of a Polynomial

Write each polynomial in standard form. Identify the leading coefficient.

a. $3x^2 + 4x^5 - 7x$

Find the degree of each term.

Degree: 2   5   1

Polynomial: $3x^2 + 4x^5 - 7x$

The greatest degree is 5. Therefore, the polynomial can be rewritten as $4x^5 + 3x^2 - 7x$, with a leading coefficient of 4.

b. $5y - 9 - 2y^4 - 6y^3$

Find the degree of each term.

Degree: 1   0   4   3

Polynomial: $5y - 9 - 2y^4 - 6y^3$

The greatest degree is 4. Therefore, the polynomial can be rewritten as $-2y^4 - 6y^3 + 5y - 9$, with a leading coefficient of -2.

▶ **Guided**Practice

2A. $8 - 2x^2 + 4x^4 - 3x$

2B. $y + 5y^3 - 2y^2 - 7y^6 + 10$

2 Add and Subtract Polynomials Adding polynomials involves adding like terms. You can group like terms by using a horizontal or vertical format.

Example 3 Add Polynomials

Find each sum.

a. $\left(2x^2 + 5x - 7\right) + \left(3 - 4x^2 + 6x\right)$

Horizontal Method

Group and combine like terms.

$\left(2x^2 + 5x - 7\right) + \left(3 - 4x^2 + 6x\right)$

$= \left[2x^2 + (-4x^2)\right] + \left[5x + 6x\right] + \left[-7 + 3\right]$ Group like terms.

$= -2x^2 + 11x - 4$ Combine like terms.

b. $\left(3y + y^3 - 5\right) + \left(4y^2 - 4y + 2y^3 + 8\right)$

Vertical Method

Align like terms in columns and combine.

$ y^3 + 0y^2 + 3y - 5$ Insert a placeholder to help align the terms.

$\underline{(+)\ 2y^3 + 4y^2 - 4y + 8}$ Align and combine like terms.

$ 3y^3 + 4y^2 - y + 3$

StudyTip

Vertical Method Notice that the polynomials are written in standard form with like terms aligned. Since there is no y^2-term in the first polynomial, $0y^2$ is used as a placeholder.

▶ **Guided**Practice

3A. $\left(5x^2 - 3x + 4\right) + \left(6x - 3x^2 - 3\right)$

3B. $\left(y^4 - 3y + 7\right) + \left(2y^3 + 2y - 2y^4 - 11\right)$

StudyTip

Additive Inverse When finding the additive inverse of a polynomial, you are multiplying every term by −1.

You can subtract a polynomial by adding its additive inverse. To find the additive inverse of a polynomial, write the opposite of each term, as shown.

$$-(3x^2 + 2x - 6) = \underline{-3x^2 - 2x + 6}$$

Additive Inverse

Example 4 Subtract Polynomials

Find each difference.

a. $(3 - 2x + 2x^2) - (4x - 5 + 3x^2)$

   Horizontal Method

   Subtract $4x - 5 + 3x^2$ by adding its additive inverse.

   $(3 - 2x + 2x^2) - (4x - 5 + 3x^2)$

   $= (3 - 2x + 2x^2) + (-4x + 5 - 3x^2)$ The additive inverse of $4x - 5 + 3x^2$ is $-4x + 5 - 3x^2$.

   $= [2x^2 + (-3x^2)] + [(-2x) + (-4x)] + [3 + 5]$ Group like terms.

   $= -x^2 - 6x + 8$ Combine like terms.

b. $(7p + 4p^3 - 8) - (3p^2 + 2 - 9p)$

   Vertical Method

   Align like terms in columns and subtract by adding the additive inverse.

   $$\begin{array}{r} 4p^3 + 0p^2 + 7p - 8 \\ (-)\quad\quad 3p^2 - 9p + 2 \\ \hline \end{array}$$

   Add the opposite.

   $$\begin{array}{r} 4p^3 + 0p^2 + 7p - 8 \\ (+)\quad -3p^2 + 9p - 2 \\ \hline 4p^3 - 3p^2 + 16p - 10 \end{array}$$

▸ **Guided**Practice

4A. $(4x^3 - 3x^2 + 6x - 4) - (-2x^3 + x^2 - 2)$

4B. $(8y - 10 + 5y^2) - (7 - y^3 + 12y)$

Adding or subtracting integers results in an integer, so the set of integers is closed under addition and subtraction. Similarly, adding or subtracting polynomials results in a polynomial, so the set of polynomials is closed under addition and subtraction.

Real-World Example 5 Add and Subtract Polynomials

ELECTRONICS The equations $P = 7m + 137$ and $C = 4m + 78$ represent the number of cell phones P and digital cameras C sold in m months at an electronics store. Write an equation for the total monthly sales T of phones and cameras. Then predict the number of phones and cameras sold in 10 months.

To write an equation that represents the total sales T, add the equations that represent the number of cell phones P and digital cameras C.

$T = 7m + 137 + 4m + 78$

$\quad = 11m + 215$

Substitute 10 for m to predict the number of phones and cameras sold in 10 months.

$T = 11(10) + 215$

$\quad = 110 + 215$ or 325

Therefore, a total of 325 cell phones and digital cameras will be sold in 10 months.

▸ **Guided**Practice

5. Use the information above to write an equation that represents the difference in the monthly sales of cell phones and the monthly sales of digital cameras. Use the equation to predict the difference in monthly sales in 24 months.

Real-World Link

Sales of digital cameras recently increased by 42% in one year. Sales are expected to increase by at least 15% each year as consumers upgrade their cameras.

Source: Big Planet Marketing Company

Digital Vision/Alamy

Example 1   **Determine whether each expression is a polynomial. If it is a polynomial, find the degree and determine whether it is a *monomial*, *binomial*, or *trinomial*.**

1. $7ab + 6b^2 - 2a^3$ **2.** $2y - 5 + 3y^2$

3. $3x^2$ **4.** $\dfrac{4m}{3p}$

5. $5m^2p^3 + 6$ **6.** $5q^{-4} + 6q$

Example 2   **Write each polynomial in standard form. Identify the leading coefficient.**

7. $2x^5 - 12 + 3x$ **8.** $-4d^4 + 1 - d^2$

9. $4z - 2z^2 - 5z^4$ **10.** $2a + 4a^3 - 5a^2 - 1$

Examples 3–4 **Find each sum or difference.**

11. $(6x^3 - 4) + (-2x^3 + 9)$ **12.** $(g^3 - 2g^2 + 5g + 6) - (g^2 + 2g)$

(13) $(4 + 2a^2 - 2a) - (3a^2 - 8a + 7)$ **14.** $(8y - 4y^2) + (3y - 9y^2)$

15. $(-4z^3 - 2z + 8) - (4z^3 + 3z^2 - 5)$ **16.** $(-3d^2 - 8 + 2d) + (4d - 12 + d^2)$

17. $(y + 5) + (2y + 4y^2 - 2)$ **18.** $(3n^3 - 5n + n^2) - (-8n^2 + 3n^3)$

Example 5   **19. CCSS SENSE-MAKING** The total number of students T who traveled for spring break consists of two groups: students who flew to their destinations F and students who drove to their destination D. The number (in thousands) of students who flew and the total number of students who flew or drove can be modeled by the following equations, where n is the number of years since 1995.

$$T = 14n + 21 \qquad F = 8n + 7$$

   a. Write an equation that models the number of students who drove to their destination for this time period.

   b. Predict the number of students who will drive to their destination in 2012.

   c. How many students will drive or fly to their destination in 2015?

Practice and Problem Solving   Extra Practice is on page R8.

Example 1   **Determine whether each expression is a polynomial. If it is a polynomial, find the degree and determine whether it is a *monomial*, *binomial*, or *trinomial*.**

20. $\dfrac{5y^3}{x^2} + 4x$ **21.** 21

22. $c^4 - 2c^2 + 1$ **23.** $d + 3d^c$

24. $a - a^2$ **25.** $5n^3 + nq^3$

Example 2   **Write each polynomial in standard form. Identify the leading coefficient.**

26. $5x^2 - 2 + 3x$ **27.** $8y + 7y^3$

28. $4 - 3c - 5c^2$ **29.** $-y^3 + 3y - 3y^2 + 2$

30. $11t + 2t^2 - 3 + t^5$ **31.** $2 + r - r^3$

32. $\dfrac{1}{2}x - 3x^4 + 7$ **33.** $-9b^2 + 10b - b^6$

Examples 3–4 **Find each sum or difference.**

34. $(2c^2 + 6c + 4) + (5c^2 - 7)$

35 $(2x + 3x^2) - (7 - 8x^2)$

36. $(3c^3 - c + 11) - (c^2 + 2c + 8)$

37. $(z^2 + z) + (z^2 - 11)$

38. $(2x - 2y + 1) - (3y + 4x)$

39. $(4a - 5b^2 + 3) + (6 - 2a + 3b^2)$

40. $(x^2y - 3x^2 + y) + (3y - 2x^2y)$

41. $(-8xy + 3x^2 - 5y) + (4x^2 - 2y + 6xy)$

42. $(5n - 2p^2 + 2np) - (4p^2 + 4n)$

43. $(4rxt - 8r^2x + x^2) - (6rx^2 + 5rxt - 2x^2)$

Example 5

44. PETS From 1999 through 2009, the number of dogs D and the number of cats C (in hundreds) adopted from animal shelters in the United States are modeled by the equations $D = 2n + 3$ and $C = n + 4$, where n is the number of years since 1999.

a. Write a function that models the total number T of dogs and cats adopted in hundreds for this time period.

b. If this trend continues, how many dogs and cats will be adopted in 2013?

Classify each polynomial according to its degree and number of terms.

45. $4x - 3x^2 + 5$

46. $11z^3$

47. $9 + y^4$

48. $3x^3 - 7$

49. $-2x^5 - x^2 + 5x - 8$

50. $10t - 4t^2 + 6t^3$

51. ENROLLMENT In a rapidly growing school system, the numbers (in hundreds) of total students is represented by N and the number of students in Kindergarten through 5th grade is represented by P. The equations $N = 1.25t^2 - t + 7.5$ and $P = 0.7t^2 - 0.95t + 3.8$, models the number of students enrolled from 2000 to 2009, where t is the number of years since 2000.

a. Write an equation modeling the number of students S in grades 6 through 12 enrolled for this time period.

b. How many students were enrolled in grades 6 through 12 in the school system in 2007?

52. CCSS REASONING The perimeter of the triangle can be represented by the expression $3x^2 - 7x + 2$. Write a polynomial that represents the measure of the third side.

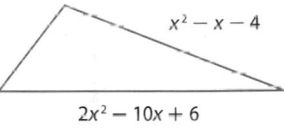

$x^2 - x - 4$

$2x^2 - 10x + 6$

53. GEOMETRY Consider the rectangle.

a. What does $(4x^2 + 2x - 1)(2x^2 - x + 3)$ represent?

b. What does $2(4x^2 + 2x - 1) + 2(2x^2 - x + 3)$ represent?

$4x^2 + 2x - 1$

$2x^2 - x + 3$

Find each sum or difference.

54. $(4x + 2y - 6z) + (5y - 2z + 7x) + (-9z - 2x - 3y)$

55. $(5a^2 - 4) + (a^2 - 2a + 12) + (4a^2 - 6a + 8)$

56. $(3c^2 - 7) + (4c + 7) - (c^2 + 5c - 8)$

57. $(3n^3 + 3n - 10) - (4n^2 - 5n) + (4n^3 - 3n^2 - 9n + 4)$

58. FOOTBALL The National Football League is divided into two conferences, the American A and the National N. From 2002 through 2009, the total attendance T (in thousands) for both conferences and for the American Conference games can be modeled by the following equations, where x is the number of years since 2002.

$T = -0.69x^3 + 55.83x^2 + 643.31x + 10{,}538$ $A = -3.78x^3 + 58.96x^2 + 265.96x + 5257$

Estimate how many people attended National Conference football games in 2009.

59 CAR RENTAL The cost to rent a car for a day is $15 plus $0.15 for each mile driven.

    a. Write a polynomial that represents the cost of renting a car for m miles.

    b. If a car is driven 145 miles, how much would it cost to rent?

    c. If a car is driven 105 miles each day for four days, how much would it cost to rent a car?

    d. If a car is driven 220 miles each day for seven days, how much would it cost to rent a car?

60. 🔄 MULTIPLE REPRESENTATIONS In this problem, you will explore perimeter and area.

    a. Geometric Draw three rectangles that each have a perimeter of 400 feet.

    b. Tabular Record the width and length of each rectangle in a table like the one shown below. Find the area of each rectangle.

Rectangle	Length	Width	Area
1	100 ft		
2	50 ft		
3	75 ft		
4	x ft		

    c. Graphical On a coordinate system, graph the area of rectangle 4 in terms of the length, x. Use the graph to determine the largest area possible.

    d. Analytical Determine the length and width that produce the largest area.

H.O.T. Problems   Use Higher-Order Thinking Skills

61. CCSS CRITIQUE Cheyenne and Sebastian are finding $(2x^2 - x) - (3x + 3x^2 - 2)$. Is either of them correct? Explain your reasoning.

Cheyenne	Sebastian
$(2x^2 - x) - (3x + 3x^2 - 2)$	$(2x^2 - x) - (3x + 3x^2 - 2)$
$= (2x^2 - x) + (-3x + 3x^2 - 2)$	$= (2x^2 - x) + (-3x - 3x^2 - 2)$
$= 5x^2 - 4x - 2$	$= -x^2 - 4x - 2$

62. REASONING Determine whether each of the following statements is *true* or *false*. Explain your reasoning.

    a. A binomial can have a degree of zero.

    b. The order in which polynomials are subtracted does not matter.

63. CHALLENGE Write a polynomial that represents the sum of an odd integer $2n + 1$ and the next two consecutive odd integers.

64. ✍️ WRITING IN MATH Why would you add or subtract equations that represent real-world situations? Explain.

65. WRITING IN MATH Describe how to add and subtract polynomials using both the vertical and horizontal formats.

66. Three consecutive integers can be represented by x, $x + 1$, and $x + 2$. What is the sum of these three integers?

    A $x(x + 1)(x + 2)$ **C** $3x + 3$

    B $x^3 + 3$ **D** $x + 3$

67. SHORT RESPONSE What is the perimeter of a square with sides that measure $2x + 3$ units?

68. Jim cuts a board in the shape of a regular hexagon and pounds in a nail at each vertex, as shown. How many rubber bands will he need to stretch a rubber band across every possible pair of nails?

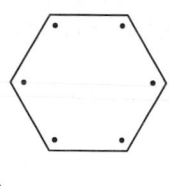

    F 15   **G** 14   **H** 12   **J** 9

69. Which ordered pair is in the solution set of the system of inequalities shown in the graph?

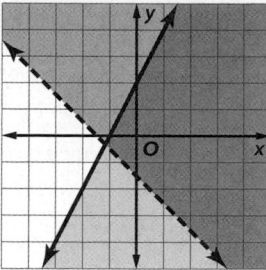

    A $(-3, 0)$ **C** $(5, 0)$

    B $(0, -3)$ **D** $(0, 5)$

Spiral Review

70. COMPUTERS A computer technician charges by the hour to fix and repair computer equipment. The total cost of the technician for one hour is $75, for two hours is $125, for three hours is $175, for four hours is $225, and so on. Write a recursive formula for the sequence. (Lesson 7-8)

Determine whether each sequence is *arithmetic, geometric,* or *neither.* Explain. (Lesson 7-7)

71. $8, -32, 128, -512, \ldots$ **72.** $25, 8, -9, -26, \ldots$ **73.** $1, \dfrac{1}{2}, \dfrac{1}{3}, \dfrac{1}{4}, \ldots$

74. $43, 52, 61, 70, \ldots$ **75.** $-27, -16, -5, 6, \ldots$ **76.** $200, 100, 50, 25, \ldots$

77. JOBS Kimi received an offer for a new job. She wants to compare the offer with her current job. What is total amount of sales that Kimi must get each month to make the same income at either job? (Lesson 6-2)

> New Offer
> $600/mo 2% commission
>
> Current Job
> $1000/mo 1.5% commission

Determine whether each sequence is an arithmetic sequence. If it is, state the common difference. (Lesson 3-5)

78. $24, 16, 8, 0, \ldots$ **79.** $3\dfrac{1}{4}, 6\dfrac{1}{2}, 13, 26, \ldots$ **80.** $7, 6, 5, 4, \ldots$

81. $10, 12, 15, 18, \ldots$ **82.** $-15, -11, -7, -3, \ldots$ **83.** $-0.3, 0.2, 0.7, 1.2, \ldots$

Skills Review

Simplify.

84. $t(t^5)(t^7)$ **85.** $n^3(n^2)(-2n^3)$ **86.** $(5l^5v^2)(10t^3v^4)$ **87.** $(-8u^4z^5)(5uz^4)$

88. $[(3)^2]^3$ **89.** $[(2)^3]^2$ **90.** $(2m^4k^3)^2(-3mk^2)^3$ **91.** $(6xy^2)^2(2x^2y^2z^2)^3$

Multiplying a Polynomial by a Monomial

- You multiplied monomials.

1. Multiply a polynomial by a monomial.

2. Solve equations involving the products of monomials and polynomials.

- Charmaine Brooks is opening a fitness club. She tells the contractor that the length of the fitness room should be three times the width plus 8 feet.

To cover the floor with mats for exercise classes, Ms. Brooks needs to know the area of the floor. So she multiplies the width times the length, $w(3w + 8)$.

 Common Core State Standards

Content Standards
A.APR.1 Understand that polynomials form a system analogous to the integers, namely, they are closed under the operations of addition, subtraction, and multiplication; add, subtract, and multiply polynomials.

Mathematical Practices
5 Use appropriate tools strategically.

1 Polynomial Multiplied by Monomial
To find the product of a polynomial and a monomial, you can use the Distributive Property.

PT

Example 1 Multiply a Polynomial by a Monomial

Find $-3x^2(7x^2 - x + 4)$.

Horizontal Method

$-3x^2(7x^2 - x + 4)$ Original expression

$= -3x^2(7x^2) - (-3x^2)(x) + (-3x^2)(4)$ Distributive Property

$= -21x^4 - (-3x^3) + (-12x^2)$ Multiply.

$= -21x^4 + 3x^3 - 12x^2$ Simplify.

Vertical Method

$$
\begin{array}{r}
7x^2 - x + 4 \\
(\times) \qquad\quad -3x^2 \\
\hline
-21x^4 + 3x^3 - 12x^2
\end{array}
$$

Distributive Property

Multiply.

▶ **Guided**Practice

Find each product.

1A. $5a^2(-4a^2 + 2a - 7)$ **1B.** $-6d^3(3d^4 - 2d^3 - d + 9)$

We can use this same method more than once to simplify large expressions.

PT

Example 2 Simplify Expressions

Simplify $2p(-4p^2 + 5p) - 5(2p^2 + 20)$.

$2p(-4p^2 + 5p) - 5(2p^2 + 20)$ Original expression

$= (2p)(-4p^2) + (2p)(5p) + (-5)(2p^2) + (-5)(20)$ Distributive Property

$= -8p^3 + 10p^2 - 10p^2 - 100$ Multiply.

$= -8p^3 + (10p^2 - 10p^2) - 100$ Commutative and Associative Properties

$= -8p^3 - 100$ Combine like terms.

David Schmidt/Masterfile

GuidedPractice

Simplify each expression.

2A. $3(5x^2 + 2x - 4) - x(7x^2 + 2x - 3)$ **2B.** $15t(10y^3t^5 + 5y^2t) - 2y(yt^2 + 4y^2)$

We can use the Distributive Property to multiply monomials by polynomials and solve real world problems.

Standardized Test Example 3 Write and Evaluate a Polynomial Expression

GRIDDED RESPONSE The theme for a school dance is "Solid Gold." For one decoration, Kana is covering a trapezoid-shaped piece of poster board with metallic gold paper to look like a bar of gold. If the height of the poster board is 18 inches, how much metallic paper will Kana need in square inches?

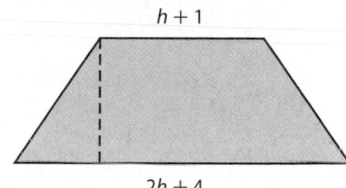

Read the Test Item

The question is asking you to find the area of the trapezoid with a height of h and bases of $h + 1$ and $2h + 4$.

Solve the Test Item

Write an equation to represent the area of the trapezoid.
Let $b_1 = h + 1$, let $b_2 = 2h + 4$ and let $h =$ height of the trapezoid.

$$A = \frac{1}{2}h(b_1 + b_2) \qquad \text{Area of a trapezoid}$$

$$= \frac{1}{2}h[(h + 1) + (2h + 4)] \qquad b_1 = h + 1 \text{ and } b_2 = 2h + 4$$

$$= \frac{1}{2}h(3h + 5) \qquad \text{Add and simplify.}$$

$$= \frac{3}{2}h^2 + \frac{5}{2}h \qquad \text{Distributive Property}$$

$$= \frac{3}{2}(18)^2 + \frac{5}{2}(18) \qquad h = 18$$

$$= 531 \qquad \text{Simplify.}$$

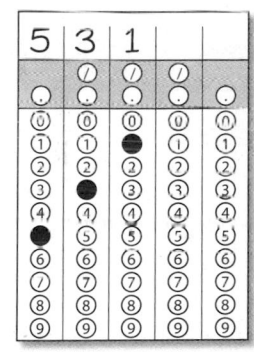

Kana will need 531 square inches of metallic paper.
Grid in your response of 531.

GuidedPractice

3. **SHORT RESPONSE** Kachima is making triangular bandanas for the dogs and cats in her pet club. The base of the bandana is the length of the collar with 4 inches added to each end to tie it on. The height is $\frac{1}{2}$ of the collar length.

A. If Kachima's dog has a collar length of 12 inches, how much fabric does she need in square inches?

B. If Kachima makes a bandana for her friend's cat with a 6-inch collar, how much fabric does Kachima need in square inches?

2 Solve Equations with Polynomial Expressions

We can use the Distributive Property to solve equations that involve the products of monomials and polynomials.

StudyTip

Combining Like Terms When simplifying a long expression, it may be helpful to put a circle around one set of like terms, a rectangle around another set, a triangle around another set, and so on.

Example 4 Equations with Polynomials on Both Sides

Solve $2a(5a - 2) + 3a(2a + 6) + 8 = a(4a + 1) + 2a(6a - 4) + 50$.

$2a(5a - 2) + 3a(2a + 6) + 8 = a(4a + 1) + 2a(6a - 4) + 50$	Original equation
$10a^2 - 4a + 6a^2 + 18a + 8 = 4a^2 + a + 12a^2 - 8a + 50$	Distributive Property
$16a^2 + 14a + 8 = 16a^2 - 7a + 50$	Combine like terms.
$14a + 8 = -7a + 50$	Subtract $16a^2$ from each side.
$21a + 8 = 50$	Add $7a$ to each side.
$21a = 42$	Subtract 8 from each side.
$a = 2$	Divide each side by 21.

CHECK

$2a(5a - 2) + 3a(2a + 6) + 8 = a(4a + 1) + 2a(6a - 4) + 50$

$2(2)[5(2) - 2] + 3(2)[2(2) + 6] + 8 \overset{?}{=} 2[4(2) + 1] + 2(2)[6(2) - 4] + 50$

$4(8) + 6(10) + 8 \overset{?}{=} 2(9) + 4(8) + 50$ Simplify.

$32 + 60 + 8 \overset{?}{=} 18 + 32 + 50$ Multiply.

$100 = 100 \checkmark$ Add and subtract.

Guided Practice

Solve each equation.

4A. $2x(x + 4) + 7 = (x + 8) + 2x(x + 1) + 12$

4B. $d(d + 3) - d(d - 4) = 9d - 16$

Check Your Understanding

○ = Step-by-Step Solutions begin on page R13.

Example 1   Find each product.

1. $5w(-3w^2 + 2w - 4)$ **2.** $6g^2(3g^3 + 4g^2 + 10g - 1)$

3. $4km^2(8km^2 + 2k^2m + 5k)$ **4.** $-3p^4r^3(2p^2r^4 - 6p^6r^3 - 5)$

⑤ $2ab(7a^4b^2 + a^5b - 2a)$ **6.** $c^2d^3(5cd^7 - 3c^3d^2 - 4d^3)$

Example 2   Simplify each expression.

7. $t(4t^2 + 15t + 4) - 4(3t - 1)$ **8.** $x(3x^2 + 4) + 2(7x - 3)$

9. $-2d(d^3c^2 - 4dc^2 + 2d^2c) + c^2(dc^2 - 3d^4)$

10. $-5w^2(8w^2x - 11wx^2) + 6x(9wx^4 - 4w - 3x^2)$

Example 3   **11. GRIDDED RESPONSE** Marlene is buying a new plasma television. The height of the screen of the television is one half the width plus 5 inches. The width is 30 inches. Find the height of the screen in inches.

Example 4   Solve each equation.

12. $-6(11 - 2c) = 7(-2 - 2c)$ **13.** $t(2t + 3) + 20 = 2t(t - 3)$

14. $-2(w + 1) + w = 7 - 4w$ **15.** $3(y - 2) + 2y = 4y + 14$

16. $a(a + 3) + a(a - 6) + 35 = a(a - 5) + a(a + 7)$

17. $n(n - 4) + n(n + 8) = n(n - 13) + n(n + 1) + 16$

Example 1   Find each product.

18. $b(b^2 - 12b + 1)$

19. $f(f^2 + 2f + 25)$

20. $-3m^3(2m^3 - 12m^2 + 2m + 25)$

21. $2j^2(5j^3 - 15j^2 + 2j + 2)$

22. $2pr^2(2pr + 5p^2r - 15p)$

23. $4t^3u(2t^2u^2 - 10tu^4 + 2)$

Example 2   Simplify each expression.

24. $-3(5x^2 + 2x + 9) + x(2x - 3)$

25. $a(-8a^2 + 2a + 4) + 3(6a^2 - 4)$

26. $-4d(5d^2 - 12) + 7(d + 5)$

27. $-9g(-2g + g^2) + 3(g^2 + 4)$

28. $2j(7j^2k^2 + jk^2 + 5k) - 9k(-2j^2k^2 + 2k^2 + 3j)$

29. $4n(2n^3p^2 - 3np^2 + 5n) + 4p(6n^2p - 2np^2 + 3p)$

Example 3   30. **DAMS** A new dam being built has the shape of a trapezoid. The length of the base at the bottom of the dam is 2 times the height. The length of the base at the top of the dam is $\frac{1}{5}$ times the height minus 30 feet.

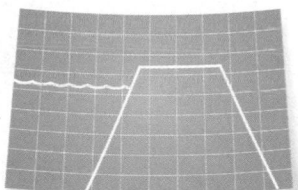

a. Write an expression to find the area of the trapezoidal cross section of the dam.

b. If the height of the dam is 180 feet, find the area of this cross section.

Example 4   Solve each equation.

(31) $7(t^2 + 5t - 9) + t = t(7t - 2) + 13$

32. $w(4w + 6) + 2w = 2(2w^2 + 7w - 3)$

33. $5(4z + 6) - 2(z - 4) = 7z(z + 4) - z(7z - 2) - 48$

34. $9c(c - 11) + 10(5c - 3) = 3c(c + 5) + c(6c - 3) - 30$

35. $2f(5f - 2) - 10(f^2 - 3f + 6) = -8f(f + 4) + 4(2f^2 - 7f)$

36. $2k(-3k + 4) + 6(k^2 + 10) = k(4k + 8) - 2k(2k + 5)$

Simplify each expression.

37. $\frac{2}{3}np^2(30p^2 + 9n^2p - 12)$

38. $\frac{3}{5}r^2t(10r^3 + 5rt^3 + 15t^2)$

39. $-5q^2w^3(4q + 7w) + 4qw^2(7q^2w + 2q) - 3qw(3q^2w^2 + 9)$

40. $-x^2z(2z^2 + 4xz^3) + xz^2(xz + 5x^3z) + x^2z^3(3x^2z + 4xz)$

41. **PARKING** A parking garage charges $30 per month plus $0.50 per daytime hour and $0.25 per hour during nights and weekends. Suppose Trent parks in the garage for 47 hours in January and h of those are night and weekend hours.

a. Find an expression for Trent's January bill.

b. Find the cost if Trent had 12 hours of night and weekend hours.

42. **CCSS MODELING** Che is building a dog house for his new puppy. The upper face of the dog house is a trapezoid. If the height of the trapezoid is 12 inches, find the area of the face of this piece of the dog house.

43 **TENNIS** The tennis club is building a new tennis court with a path around it.

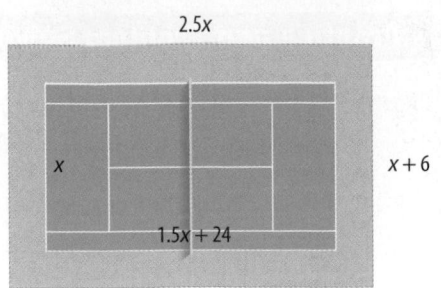

a. Write an expression for the area of the tennis court.

b. Write an expression for the area of the path.

c. If $x = 36$ feet, what is the perimeter of the outside of the path?

44. ⬣ **MULTIPLE REPRESENTATIONS** In this problem, you will investigate the degree of the product of a monomial and a polynomial.

a. **Tabular** Write three monomials of different degrees and three polynomials of different degrees. Determine the degree of each monomial and polynomial. Multiply the monomials by the polynomials. Determine the degree of each product. Record your results in a table like the one shown below.

Monomial	Degree	Polynomial	Degree	Product of Monomial and Polynomial	Degree

b. **Verbal** Make a conjecture about the degree of the product of a monomial and a polynomial. What is the degree of the product of a monomial of degree a and a polynomial of degree b?

45. **ERROR ANALYSIS** Pearl and Ted both worked on this problem. Is either of them correct? Explain your reasoning.

Pearl

$2x^2(3x^2 + 4x + 2)$

$6x^4 + 8x^2 + 4x^2$

$6x^4 + 12x^2$

Ted

$2x^2(3x^2 + 4x + 2)$

$6x^4 + 8x^3 + 4x^2$

46. Ⓒ **PERSEVERANCE** Find p such that $3x^p(4x^{2p+3} + 2x^{3p-2}) = 12x^{12} + 6x^{10}$.

47. **CHALLENGE** Simplify $4x^{-3}y^2(2x^5y^{-4} + 6x^{-7}y^6 - 4x^0y^{-2})$.

48. **REASONING** Is there a value for x that makes the statement $(x + 2)^2 = x^2 + 2^2$ true? If so, find a value for x. Explain your reasoning.

49. **OPEN ENDED** Write a monomial and a polynomial using n as the variable. Find their product.

50. **WRITING IN MATH** Describe the steps to multiply a polynomial by a monomial.

51. Every week a store sells j jeans and t T-shirts. The store makes $8 for each T-shirt and $12 for each pair of jeans. Which of the following expressions represents the total amount of money, in dollars, the store makes every week?

A $8j + 12t$ **C** $20(j + t)$

B $12j + 8t$ **D** $96jt$

52. If $a = 5x + 7y$ and $b = 2y - 3x$, what is $a + b$?

F $2x - 9y$ **H** $2x + 9y$

G $3y + 4x$ **J** $2x - 5y$

53. GEOMETRY A triangle has sides of length 5 inches and 8.5 inches. Which of the following cannot be the length of the third side?

A 3.5 inches

B 4 inches

C 5.5 inches

D 12 inches

54. SHORT RESPONSE Write an equation in which x varies directly as the cube of y and inversely as the square of z.

Find each sum or difference. (Lesson 8-1)

55. $(2x^2 - 7) + (8 - 5x^2)$

56. $(3z^2 + 2z - 1) + (z^2 - 6)$

57. $(2a - 4a^2 + 1) - (5a^2 - 2a - 6)$

58. $(a^3 - 3a^2 + 4) - (4a^2 + 7)$

59. $(2ab - 3a + 4b) + (5a + 4ab)$

60. $(8c^3 - 3c^2 + c - 2) - (3c^3 + 9)$

Write a recursive formula for each sequence. (Lesson 7-8)

61. $16, 2, -12, -26, \ldots$

62. $-5, 3, 11, 19, \ldots$

63. $27, 43, 59, 75, \ldots$

64. $80, -200, 500, -1250, \ldots$

65. $100, 60, 36, 21.6, \ldots$

66. $\frac{1}{16}, \frac{1}{4}, 1, 4, \ldots$

67. TRAVEL In 2003, about 9.5 million people took cruises. Between 2003 and 2008, the number increased by about 740,000 each year. Write the point-slope form of an equation to find the total number of people y taking a cruise for any year x. Estimate the number of people who took a cruise in 2010. (Lesson 4-3)

Write an equation in function notation for each relation. (Lesson 3-6)

68.

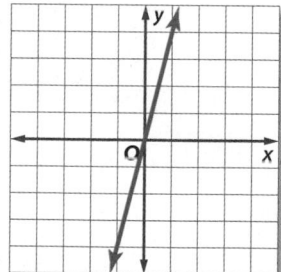

69.

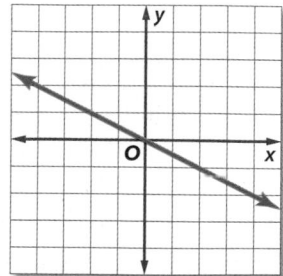

Simplify.

70. $b(b^2)(b^3)$

71. $2y(3y^2)$

72. $-y^4(-2y^3)$

73. $-3z^3(-5z^4 + 2z)$

74. $2m(-4m^4) - 3(-5m^3)$

75. $4p^2(-2p^3) + 2p^4(5p^6)$

8-3

Algebra Lab
Multiplying Polynomials

You can use algebra tiles to find the product of two binomials.

 Common Core State Standards
Content Standards
A.APR.1 Understand that polynomials form a system analogous to the integers, namely, they are closed under the operations of addition, subtraction, and multiplication; add, subtract, and multiply polynomials.

Activity 1 Multiply Binomials

Use algebra tiles to find $(x + 3)(x + 4)$.

The rectangle will have a width of $x + 3$ and a length of $x + 4$. Use algebra tiles to mark off the dimensions on a product mat. Then complete the rectangle with algebra tiles.

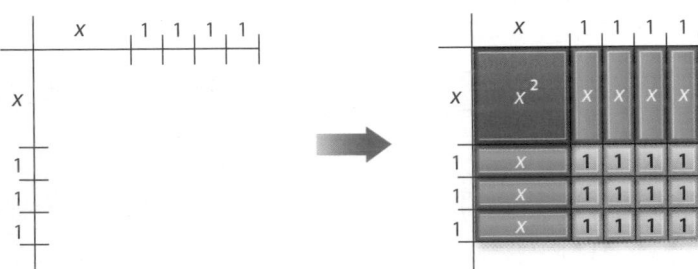

The rectangle consists of 1 blue x^2-tile, 7 green x-tiles, and 12 yellow 1-tiles.
The area of the rectangle is $x^2 + 7x + 12$. So, $(x + 3)(x + 4) = x^2 + 7x + 12$.

Activity 2 Multiply Binomials

Use algebra tiles to find $(x - 2)(x - 5)$.

Step 1 The rectangle will have a width of $x - 2$ and a length of $x - 5$. Use algebra tiles to mark off the dimensions on a product mat. Then begin to make the rectangle with algebra tiles.

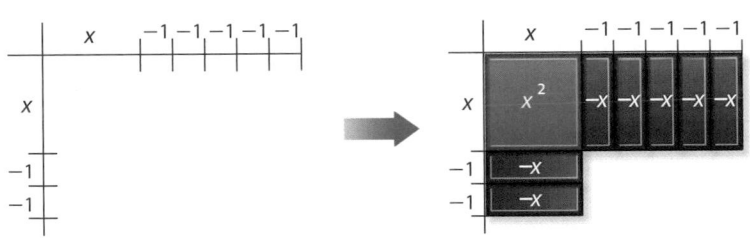

Step 2 Determine whether to use 10 yellow 1-tiles or 10 red -1-tiles to complete the rectangle. The area of each yellow tile is the product of -1 and -1. Fill in the space with 10 yellow 1-tiles to complete the rectangle.

The rectangle consists of 1 blue x^2-tile, 7 red $-x$-tiles, and 10 yellow 1-tiles. The area of the rectangle is $x^2 - 7x + 10$. So, $(x - 2)(x - 5) = x^2 - 7x + 10$.

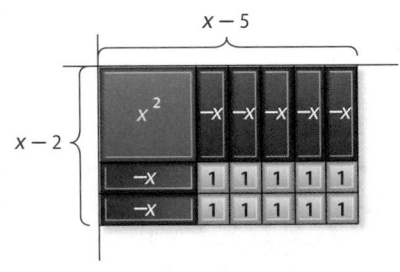

Activity 3 Multiply Binomials

Use algebra tiles to find $(x - 4)(2x + 3)$.

Step 1 The rectangle will have a width of $x - 4$ and a length of $2x + 3$. Use algebra tiles to mark off the dimensions on a product mat. Then begin to make the rectangle with algebra tiles.

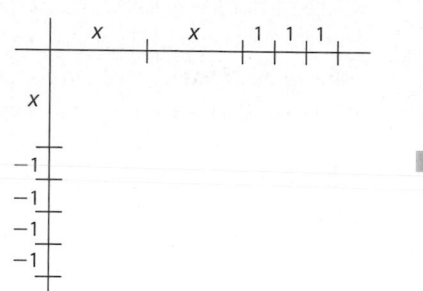

Step 2 Determine what color x-tiles and what color 1-tiles to use to complete the rectangle. The area of each red x-tile is the product of x and -1. The area of each red -1-tile is represented by $1(-1)$ or -1.

Complete the rectangle with 4 red x-tiles and 12 red -1-tiles.

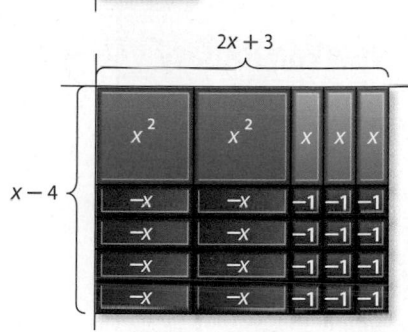

Step 3 Rearrange the tiles to simplify the polynomial you have formed. Notice that a 3 zero pair are formed by three positive and three negative x-tiles.

There are 2 blue x^2-tiles, 5 red $-x$-tiles, and 12 red -1-tiles left. In simplest form, $(x - 4)(2x + 3) = 2x^2 - 5x - 12$.

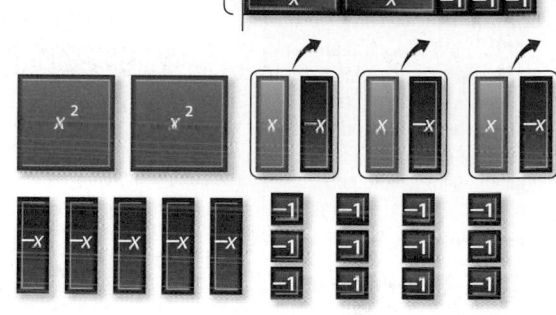

Model and Analyze

Use algebra tiles to find each product.

1. $(x + 1)(x + 4)$
2. $(x - 3)(x - 2)$
3. $(x + 5)(x - 1)$
4. $(x + 2)(2x + 3)$
5. $(x - 1)(2x - 1)$
6. $(x + 4)(2x - 5)$

Is each statement *true* or *false*? Justify your answer with a drawing of algebra tiles.

7. $(x - 4)(x - 2) = x^2 - 6x + 8$
8. $(x + 3)(x + 5) = x^2 + 15$

9. **WRITING IN MATH** You can also use the Distributive Property to find the product of two binomials. The figure at the right shows the model for $(x + 4)(x + 5)$ separated into four parts. Write a sentence or two explaining how this model shows the use of the Distributive Property.

Multiplying Polynomials

∴ Then	∴ Now	∴ Why?
• You multiplied polynomials by monomials.	**1** Multiply binomials by using the FOIL method. **2** Multiply polynomials by using the Distributive Property.	• Bodyboards, which are used to ride waves, are made of foam and are more rectangular than surfboards. A bodyboard's dimensions are determined by the height and skill level of the user. The length of Ann's bodyboard should be Ann's height h minus 32 inches or $h - 32$. The board's width should be half of Ann's height plus 11 inches or $\frac{1}{2}h + 11$. To approximate the area of the bodyboard, you need to find $(h - 32)\left(\frac{1}{2}h + 11\right)$.

 NewVocabulary
FOIL method
quadratic expression

 Common Core State Standards

Content Standards
A.APR.1 Understand that polynomials form a system analogous to the integers, namely, they are closed under the operations of addition, subtraction, and multiplication; add, subtract, and multiply polynomials.

Mathematical Practices
7 Look for and make use of structure.

1 Multiply Binomials To multiply two binomials such as $h - 32$ and $\frac{1}{2}h + 11$, the Distributive Property is used. Binomials can be multiplied horizontally or vertically.

Example 1 The Distributive Property

Find each product.

a. $(2x + 3)(x + 5)$

Vertical Method

Multiply by 5.
$$\begin{array}{r} 2x + 3 \\ (\times)\ x + 5 \\ \hline 10x + 15 \end{array}$$
$5(2x + 3) = 10x + 15$

Multiply by x.
$$\begin{array}{r} 2x + 3 \\ (\times)\ x + 5 \\ \hline 10x + 15 \\ 2x^2 + 3x \\ \hline \end{array}$$
$x(2x + 3) = 2x^2 + 3x$

Combine like terms.
$$\begin{array}{r} 2x + 3 \\ (\times)\ x + 5 \\ \hline 10x + 15 \\ 2x^2 + 3x \\ \hline \end{array}$$
$2x^2 + 13x + 15$

Horizontal Method

$$\begin{aligned}
(2x + 3)(x + 5) &= 2x(x + 5) + 3(x + 5) && \text{Rewrite as the sum of two products.} \\
&= 2x^2 + 10x + 3x + 15 && \text{Distributive Property} \\
&= 2x^2 + 13x + 15 && \text{Combine like terms.}
\end{aligned}$$

b. $(x - 2)(3x + 4)$

Vertical Method

Multiply by 4.
$$\begin{array}{r} x - 2 \\ (\times)\ 3x + 4 \\ \hline 4x - 8 \end{array}$$
$4(x - 2) = 4x - 8$

Multiply by $3x$.
$$\begin{array}{r} x - 2 \\ (\times)\ 3x + 4 \\ \hline 4x - 8 \\ 3x^2 - 6x \\ \hline \end{array}$$
$3x(x - 2) = 3x^2 - 6x$

Combine like terms.
$$\begin{array}{r} x - 2 \\ (\times)\ 3x + 4 \\ \hline 4x - 8 \\ 3x^2 - 6x \\ \hline \end{array}$$
$3x^2 - 2x - 8$

Horizontal Method

$$\begin{aligned}
(x - 2)(3x + 4) &= x(3x + 4) - 2(3x + 4) && \text{Rewrite as the difference of two products.} \\
&= 3x^2 + 4x - 6x - 8 && \text{Distributive Property} \\
&= 3x^2 - 2x - 8 && \text{Combine like terms.}
\end{aligned}$$

Peter Cade/Iconica/Getty Images

▶ **Guided**Practice

1A. $(3m + 4)(m + 5)$ **1B.** $(5y - 2)(y + 8)$

A shortcut version of the Distributive Property for multiplying binomials is called the **FOIL method**.

KeyConcept FOIL Method

| Words | To multiply two binomials, find the sum of the products of **F** the *First* terms, **O** the *Outer* terms, **I** the *Inner* terms, **L** and the *Last* terms. |

Example

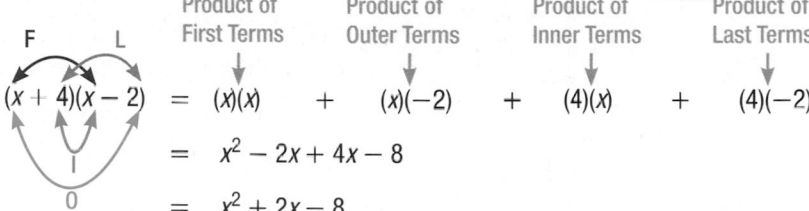

| | Product of First Terms | Product of Outer Terms | Product of Inner Terms | Product of Last Terms |

$$(x + 4)(x - 2) = (x)(x) + (x)(-2) + (4)(x) + (4)(-2)$$
$$= x^2 - 2x + 4x - 8$$
$$= x^2 + 2x - 8$$

ReadingMath

Polynomials as Factors The expression $(x + 4)(x - 2)$ is read *the quantity x plus 4 times the quantity x minus 2.*

Example 2 FOIL Method

Find each product.

a. $(2y - 7)(3y + 5)$

$$(2y - 7)(3y + 5) = (2y)(3y) + (2y)(5) + (-7)(3y) + (-7)(5) \qquad \text{FOIL method}$$
$$= 6y^2 + 10y - 21y - 35 \qquad \text{Multiply.}$$
$$= 6y^2 - 11y - 35 \qquad \text{Combine like terms.}$$

b. $(4a - 5)(2a - 9)$

$$(4a - 5)(2a - 9)$$
$$= (4a)(2a) + (4a)(-9) + (-5)(2a) + (-5)(-9) \qquad \text{FOIL method}$$
$$= 8a^2 - 36a - 10a + 45 \qquad \text{Multiply.}$$
$$= 8a^2 - 46a + 45 \qquad \text{Combine like terms.}$$

▶ **Guided**Practice

2A. $(x + 3)(x - 4)$ **2B.** $(4b - 5)(3b + 2)$

2C. $(2y - 5)(y - 6)$ **2D.** $(5a + 2)(3a - 4)$

Notice that when two linear expressions are multiplied, the result is a quadratic expression. A **quadratic expression** is an expression in one variable with a degree of 2. When three linear expressions are multiplied, the result has a degree of 3.

The FOIL method can be used to find an expression that represents the area of a rectangular object when the lengths of the sides are given as binomials.

Real-World Example 3 FOIL Method

SWIMMING POOL A contractor is building a deck around a rectangular swimming pool. The deck is x feet from every side of the pool. Write an expression for the total area of the pool and deck.

15 ft

20 ft

x

x

Real-World Link

The cost of a swimming pool depends on many factors, including the size of the pool, whether the pool is an above-ground or an in-ground pool, and the material used.

Source: American Dream Homes

Understand We need to find an expression for the total area of the pool and deck.

Plan Find the product of the length and width of the pool with the deck.

Solve Since the deck is the same distance from every side of the pool, the length and width of the pool are $2x$ longer. So, the length can be represented by $2x + 20$ and the width can be represented by $2x + 15$.

$$\begin{aligned} \text{Area} &= \text{length} \cdot \text{width} &&\text{Area of a rectangle}\\ &= (2x + 20)(2x + 15) &&\text{Substitution}\\ &= (2x)(2x) + (2x)(15) + (20)(2x) + (20)(15) &&\text{FOIL Method}\\ &= 4x^2 + 30x + 40x + 300 &&\text{Multiply.}\\ &= 4x^2 + 70x + 300 &&\text{Combine like terms.} \end{aligned}$$

So, the total area of the deck and pool is $4x^2 + 70x + 300$.

Check Choose a value for x. Substitute this value into $(2x + 20)(2x + 15)$ and $4x^2 + 70x + 300$. The result should be the same for both expressions.

▶ **Guided**Practice

3. If the pool is 25 feet long and 20 feet wide, find the area of the pool and deck.

---

2 **Multiply Polynomials** The Distributive Property can also be used to multiply any two polynomials.

Example 4 The Distributive Property

Find each product.

a. $(6x + 5)(2x^2 - 3x - 5)$

$(6x + 5)(2x^2 - 3x - 5)$

$$\begin{aligned} &= 6x(2x^2 - 3x - 5) + 5(2x^2 - 3x - 5) &&\text{Distributive Property}\\ &= 12x^3 - 18x^2 - 30x + 10x^2 - 15x - 25 &&\text{Multiply.}\\ &= 12x^3 - 8x^2 - 45x - 25 &&\text{Combine like terms.} \end{aligned}$$

StudyTip

Multiplying Polynomials If a polynomial with c terms and a polynomial with d terms are multiplied together, there will be $c \cdot d$ terms before simplifying. In Example 4a, there are $2 \cdot 3$ or 6 terms before simplifying.

b. $(2y^2 + 3y - 1)(3y^2 - 5y + 2)$

$(2y^2 + 3y - 1)(3y^2 - 5y + 2)$

$$\begin{aligned} &= 2y^2(3y^2 - 5y + 2) + 3y(3y^2 - 5y + 2) - 1(3y^2 - 5y + 2) &&\text{Distributive Property}\\ &= 6y^4 - 10y^3 + 4y^2 + 9y^3 - 15y^2 + 6y - 3y^2 + 5y - 2 &&\text{Multiply.}\\ &= 6y^4 - y^3 - 14y^2 + 11y - 2 &&\text{Combine like terms.} \end{aligned}$$

▶ **Guided**Practice

4A. $(3x - 5)(2x^2 + 7x - 8)$ **4B.** $(m^2 + 2m - 3)(4m^2 - 7m + 5)$

Examples 1–2 Find each product.

1. $(x + 5)(x + 2)$ **2.** $(y - 2)(y + 4)$ **3.** $(b - 7)(b + 3)$

4. $(4n + 3)(n + 9)$ **5.** $(8h - 1)(2h - 3)$ **6.** $(2a + 9)(5a - 6)$

Example 3

7. FRAME Hugo is designing a frame as shown at the right. The frame has a width of x inches all the way around. Write an expression that represents the total area of the picture and frame.

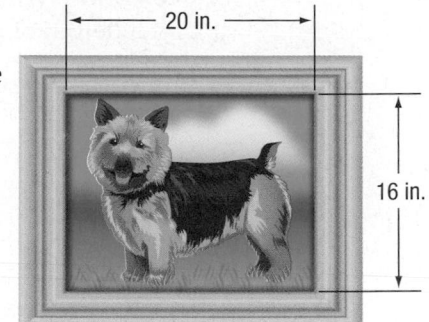

Example 4 Find each product.

8. $(2a - 9)(3a^2 + 4a - 4)$

9. $(4y^2 - 3)(4y^2 + 7y + 2)$

10. $(x^2 - 4x + 5)(5x^2 + 3x - 4)$

11. $(2n^2 + 3n - 6)(5n^2 - 2n - 8)$

Practice and Problem Solving **Extra Practice is on page R8.**

Examples 1–2 Find each product.

12. $(3c - 5)(c + 3)$ **13.** $(g + 10)(2g - 5)$ **14.** $(6a + 5)(5a + 3)$

⑮ $(4x + 1)(6x + 3)$ **16.** $(5y - 4)(3y - 1)$ **17.** $(6d - 5)(4d - 7)$

18. $(3m + 5)(2m + 3)$ **19.** $(7n - 6)(7n - 6)$ **20.** $(12t - 5)(12t + 5)$

21. $(5r + 7)(5r - 7)$ **22.** $(8w + 4x)(5w - 6x)$ **23.** $(11z - 5y)(3z + 2y)$

Example 3

24. GARDEN A walkway surrounds a rectangular garden. The width of the garden is 8 feet, and the length is 6 feet. The width x of the walkway around the garden is the same on every side. Write an expression that represents the total area of the garden and walkway.

Example 4 Find each product.

25. $(2y - 11)(y^2 - 3y + 2)$ **26.** $(4a + 7)(9a^2 + 2a - 7)$

27. $(m^2 - 5m + 4)(m^2 + 7m - 3)$ **28.** $(x^2 + 5x - 1)(5x^2 - 6x + 1)$

29. $(3b^3 - 4b - 7)(2b^2 - b - 9)$ **30.** $(6z^2 - 5z - 2)(3z^3 - 2z - 4)$

Simplify.

31. $(m + 2)[(m^2 + 3m - 6) + (m^2 - 2m + 4)]$

32. $[(t^2 + 3t - 8) - (t^2 - 2t + 6)](t - 4)$

CCSS STRUCTURE Find an expression to represent the area of each shaded region.

33.
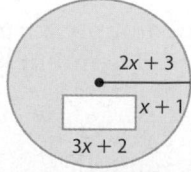

34.
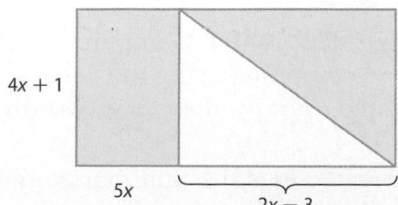

35 **VOLLEYBALL** The dimensions of a sand volleyball court are represented by a width of $6y - 5$ feet and a length of $3y + 4$ feet.

   a. Write an expression that represents the area of the court.

   b. The length of a sand volleyball court is 31 feet. Find the area of the court.

36. GEOMETRY Write an expression for the area of a triangle with a base of $2x + 3$ and a height of $3x - 1$.

Find each product.

37. $(a - 2b)^2$ **38.** $(3c + 4d)^2$ **39.** $(x - 5y)^2$

40. $(2r - 3t)^3$ **41.** $(5g + 2h)^3$ **42.** $(4y + 3z)(4y - 3z)^2$

43. CONSTRUCTION A sandbox kit allows you to build a square sandbox or a rectangular sandbox as shown.

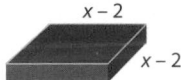
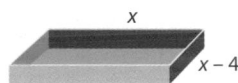

   a. What are the possible values of x? Explain.

   b. Which shape has the greater area?

   c. What is the difference in areas between the two?

44. **MULTIPLE REPRESENTATIONS** In this problem, you will investigate the square of a sum.

   a. Tabular Copy and complete the table for each sum.

Expression	(Expression)²
$x + 5$	
$3y + 1$	
$z + q$	

   b. Verbal Make a conjecture about the terms of the square of a sum.

   c. Symbolic For a sum of the form $a + b$, write an expression for the square of the sum.

H.O.T. Problems   Use Higher-Order Thinking Skills

45. REASONING Determine if the following statement is *sometimes*, *always*, or *never* true. Explain your reasoning.

    The FOIL method can be used to multiply a binomial and a trinomial.

46. CHALLENGE Find $(x^m + x^p)(x^{m-1} - x^{1-p} + x^p)$.

47. OPEN ENDED Write a binomial and a trinomial involving a single variable. Then find their product.

48. CCSS REGULARITY Compare and contrast the procedure used to multiply a trinomial by a binomial using the vertical method with the procedure used to multiply a three-digit number by a two-digit number.

49. WRITING IN MATH Summarize the methods that can be used to multiply polynomials.

50. What is the product of $2x - 5$ and $3x + 4$?

   A $5x - 1$

   B $6x^2 - 7x - 20$

   C $6x^2 - 20$

   D $6x^2 + 7x - 20$

51. Which statement is correct about the symmetry of this design?

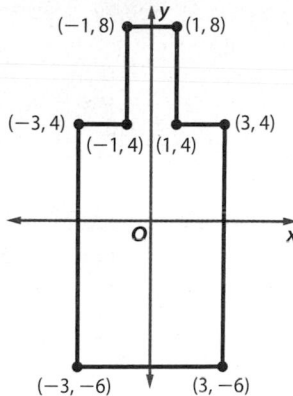

   F The design is symmetrical only about the y-axis.

   G The design is symmetrical only about the x-axis.

   H The design is symmetrical about both the y- and the x-axes.

   J The design has no symmetry.

52. Which point on the number line represents a number that, when cubed, will result in a number greater than itself?

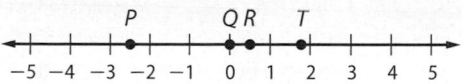

   A P **C** R

   B Q **D** T

53. **SHORT RESPONSE** For a science project, Jodi selected three bean plants of equal height. Then, for five days, she measured their heights in centimeters and plotted the values on the graph below.

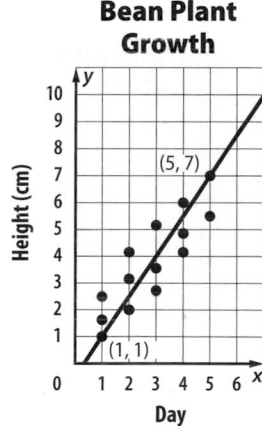

She drew a line of best fit on the graph. What is the slope of the line that she drew?

54. **SAVINGS** Carrie has $6000 to invest. She puts x dollars of this money into a savings account that earns 2% interest per year. She uses the rest of the money to purchase a certificate of deposit that earns 4% interest. Write an equation for the amount of money that Carrie will have in one year. (Lesson 8-2)

Find each sum or difference. (Lesson 8-1)

55. $(7a^2 - 5) + (-3a^2 + 10)$

56. $(8n - 2n^2) + (4n - 6n^2)$

57. $(4 + n^3 + 3n^2) + (2n^3 - 9n^2 + 6)$

58. $(-4u^2 - 9 + 2u) + (6u + 14 + 2u^2)$

59. $(b + 4) + (c + 3b - 2)$

60. $(3a^3 - 6a) - (3a^3 + 5a)$

61. $(-4m^3 - m + 10) - (3m^3 + 3m^2 - 7)$

62. $(3a + 4ab + 3b) - (2b + 5a + 8ab)$

Simplify.

63. $(-2t^4)^3 - 3(-2t^3)^4$ **64.** $(-3h^2)^3 - 2(-h^3)^2$ **65.** $2(-5y^3)^2 + (-3y^3)^3$ **66.** $3(-6n^4)^2 + (-2n^2)^2$

Special Products

::Then	::Now	::Why?
● You multiplied binomials by using the FOIL method.	**1** Find squares of sums and differences. **2** Find the product of a sum and a difference.	● Colby wants to attach a dartboard to a square piece of corkboard. If the radius of the dartboard is $r + 12$, how large does the square corkboard need to be? Colby knows that the diameter of the dartboard is $2(r + 12)$ or $2r + 24$. Each side of the square also measures $2r + 24$. To find how much corkboard is needed, Colby must find the area of the square: $A = (2r + 24)^2$.

 Common Core State Standards

Content Standards
A.APR.1 Understand that polynomials form a system analogous to the integers, namely, they are closed under the operations of addition, subtraction, and multiplication; add, subtract, and multiply polynomials.

Mathematical Practices
8 Look for and express regularity in repeated reasoning.

1 Squares of Sums and Differences Some pairs of binomials, such as squares like $(2r + 24)^2$, have products that follow a specific pattern. Using the pattern can make multiplying easier. The square of a sum, $(a + b)^2$ or $(a + b)(a + b)$, is one of those products.

$$(a + b)^2 = a^2 + ab + ab + b^2$$

KeyConcept Square of a Sum

Words   The square of $a + b$ is the square of a plus twice the product of a and b plus the square of b.

Symbols $(a + b)^2 = (a + b)(a + b)$ **Example** $(x + 4)^2 = (x + 4)(x + 4)$

$\qquad\qquad = a^2 + 2ab + b^2$ $\qquad\qquad\qquad\qquad\qquad = x^2 + 8x + 16$

Example 1 Square of a Sum

Find $(3x + 5)^2$.

$(a + b)^2 = a^2 + 2ab + b^2$ $\qquad$ Square of a sum

$(3x + 5)^2 = (3x)^2 + 2(3x)(5) + 5^2$ $\qquad$ $a = 3x, b = 5$

$\qquad\quad = 9x^2 + 30x + 25$ $\qquad$ Simplify. Use FOIL to check your solution.

▶ **Guided**Practice

Find each product.

1A. $(8c + 3d)^2$ $\qquad\qquad\qquad\qquad\qquad$ **1B.** $(3x + 4y)^2$

There is also a pattern for the *square of a difference*. Write $a - b$ as $a + (-b)$ and square it using the square of a sum pattern.

$$(a - b)^2 = [a + (-b)]^2$$
$$= a^2 + 2(a)(-b) + (-b)^2 \qquad \text{Square of a sum}$$
$$= a^2 - 2ab + b^2 \qquad \text{Simplify.}$$

 KeyConcept Square of a Difference

Words	The square of $a - b$ is the square of a minus twice the product of a and b plus the square of b.

Symbols $(a - b)^2 = (a - b)(a - b)$ **Example** $(x - 3)^2 = (x - 3)(x - 3)$
$$= a^2 - 2ab + b^2 \qquad\qquad\qquad\qquad = x^2 - 6x + 9$$

WatchOut!

CCSS Regularity Remember that $(x - 7)^2$ does not equal $x^2 - 7^2$, or $x^2 - 49$.

$(x - 7)^2$
$= (x - 7)(x - 7)$
$= x^2 - 14x + 49$

Example 2 Square of a Difference

Find $(2x - 5y)^2$.

$$(a - b)^2 = a^2 - 2ab + b^2 \qquad \text{Square of a difference}$$
$$(2x - 5y)^2 = (2x)^2 - 2(2x)(5y) + (5y)^2 \qquad a = 2x \text{ and } b = 5y$$
$$= 4x^2 - 20xy + 25y^2 \qquad \text{Simplify.}$$

GuidedPractice

Find each product.

2A. $(6p - 1)^2$

2B. $(a - 2b)^2$

The product of the square of a sum or the square of a difference is called a *perfect square trinomial*. We can use these to find patterns to solve real-world problems.

Real-World Example 3 Square of a Difference

PHYSICAL SCIENCE Each edge of a cube of aluminum is 4 centimeters less than each edge of a cube of copper. Write an equation to model the surface area of the aluminum cube.

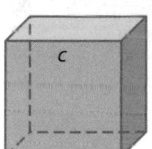
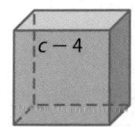

Let $c =$ the length of each edge of the cube of copper. So, each edge of the cube of aluminum is $c - 4$.

$$SA = 6s^2 \qquad \text{Formula for surface area of a cube}$$
$$SA = 6(c - 4)^2 \qquad \text{Replace } s \text{ with } c - 4.$$
$$SA = 6[c^2 - 2(4)(c) + 4^2] \qquad \text{Square of a difference}$$
$$SA = 6(c^2 - 8c + 16) \qquad \text{Simplify.}$$

GuidedPractice

3. GARDENING Alano has a garden that is g feet long and g feet wide. He wants to add 3 feet to the length and the width.

    A. Show how the new area of the garden can be modeled by the square of a binomial.

    B. Find the square of this binomial.

2 Product of a Sum and a Difference

Now we will see what the result is when we multiply a sum and a difference, or $(a + b)(a - b)$. Recall that $a - b$ can be written as $a + (-b)$.

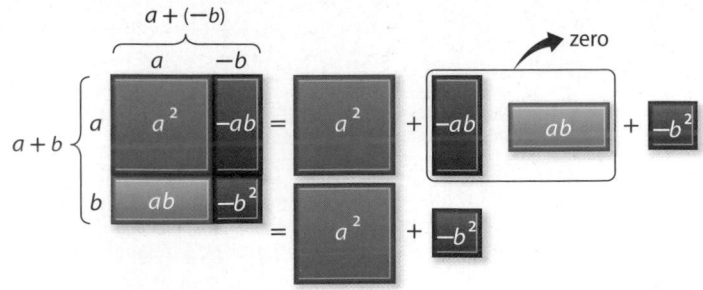

Notice that the middle terms are opposites and add to a zero pair.
So $(a + b)(a - b) = a^2 - ab + ab - b^2 = a^2 - b^2$.

StudyTip

Patterns When using any of these patterns, a and b can be numbers, variables, or expressions with numbers and variables.

KeyConcept Product of a Sum and a Difference

Words The product of $a + b$ and $a - b$ is the square of a minus the square of b.

Symbols
$$(a + b)(a - b) = (a - b)(a + b)$$
$$= a^2 - b^2$$

Example 4 Product of a Sum and a Difference

Find $(2x^2 + 3)(2x^2 - 3)$.

$(a + b)(a - b) = a^2 - b^2$ Product of a sum and a difference

$(2x^2 + 3)(2x^2 - 3) = (2x^2)^2 - (3)^2$ $a = 2x^2$ and $b = 3$

$= 4x^4 - 9$ Simplify.

GuidedPractice

Find each product.

4A. $(3n + 2)(3n - 2)$ **4B.** $(4c - 7d)(4c + 7d)$

Check Your Understanding

= Step-by-Step Solutions begin on page R13.

Examples 1–2 Find each product.

1. $(x + 5)^2$ **2.** $(11 - a)^2$ **3** $(2x + 7y)^2$

4. $(3m - 4)(3m - 4)$ **5.** $(g - 4h)(g - 4h)$ **6.** $(3c + 6d)^2$

Example 3

7. GENETICS The color of a Labrador retriever's fur is genetic. Dark genes D are dominant over yellow genes y. A dog with genes DD or Dy will have dark fur. A dog with genes yy will have yellow fur. Pepper's genes for fur color are Dy, and Ramiro's are yy.

	D	y
D	DD	Dy
y	Dy	yy

   a. Write an expression for the possible fur colors of Pepper's and Ramiro's puppies.

   b. What is the probability that a puppy will have yellow fur?

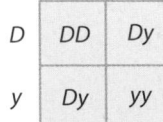

Example 4 Find each product.

    8. $(a - 3)(a + 3)$ **9.** $(x + 5)(x - 5)$

    10. $(6y - 7)(6y + 7)$ **11.** $(9t + 6)(9t - 6)$

Practice and Problem Solving
Extra Practice is on page R8.

Examples 1–2 Find each product.

    12. $(a + 10)(a + 10)$ **13.** $(b - 6)(b - 6)$

    14. $(h + 7)^2$ **15.** $(x + 6)^2$

    16. $(8 - m)^2$ **17.** $(9 - 2y)^2$

    18. $(2b + 3)^2$ **19.** $(5t - 2)^2$

    20. $(8h - 4n)^2$

Example 3 **21. GENETICS** The ability to roll your tongue is inherited genetically from parents if either parent has the dominant trait T. Children of two parents without the trait will not be able to roll their tongues.

    a. Show how the combinations can be modeled by the square of a sum.

    b. Predict the percent of children that will have both dominant genes, one dominant gene, and both recessive genes.

	T	t
T	TT	Tt
t	Tt	tt

Example 4 Find each product.

    22. $(u + 3)(u - 3)$ **23** $(b + 7)(b - 7)$ **24.** $(2 + x)(2 - x)$

    25. $(4 - x)(4 + x)$ **26.** $(2q + 5r)(2q - 5r)$ **27.** $(3a^2 + 7b)(3a^2 - 7b)$

    28. $(5y + 7)^2$ **29.** $(8 - 10a)^2$ **30.** $(10x - 2)(10x + 2)$

    31. $(3t + 12)(3t - 12)$ **32.** $(a + 4b)^2$ **33.** $(3q - 5r)^2$

    34. $(2c - 9d)^2$ **35.** $(g + 5h)^2$ **36.** $(6y - 13)(6y + 13)$

    37. $(3a^4 - b)(3a^4 + b)$ **38.** $(5x^2 - y^2)^2$ **39.** $(8a^2 - 9b^3)(8a^2 + 9b^3)$

    40. $\left(\frac{3}{1}k + 8\right)^2$ **41.** $\left(\frac{2}{5}y - 4\right)^2$ **42.** $(7z^2 + 5y^2)(7z^2 - 5y^2)$

    43. $(2m + 3)(2m - 3)(m + 4)$ **44.** $(r + 2)(r - 5)(r - 2)(r + 5)$

45. CCSS SENSE-MAKING Write a polynomial that represents the area of the figure at the right.

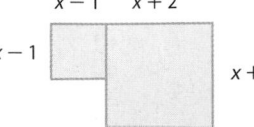

46. FLYING DISKS A flying disk shaped like a circle has a radius of $x + 3$ inches.

    a. Write an expression representing the area of the flying disk.

    b. A hole with a radius of $x - 1$ inches is cut in the center of the disk. Write an expression for the remaining area.

GEOMETRY Find the area of each shaded region.

47.

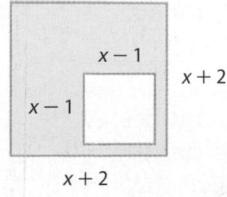

48.

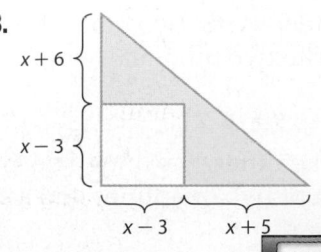

connectED.mcgraw-hill.com   **489**

Find each product.

49. $(c + d)(c + d)(c + d)$ **50.** $(2a - b)^3$ **51.** $(f + g)(f - g)(f + g)$

52. $(k - m)(k + m)(k - m)$ **53.** $(n - p)^2(n + p)$ **54.** $(q + r)^2(q - r)$

55 **WRESTLING** A high school wrestling mat must be a square with 38-foot sides and contain two circles as shown. Suppose the inner circle has a radius of r feet, and the radius of the outer circle is nine feet longer than the inner circle.

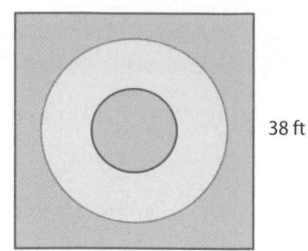

38 ft

a. Write an expression for the area of the larger circle.

b. Write an expression for the area of the portion of the square outside the larger circle.

56. **MULTIPLE REPRESENTATIONS** In this problem, you will investigate a pattern. Begin with a square piece of construction paper. Label each edge of the paper a. In any of the corners, draw a smaller square and label the edges b.

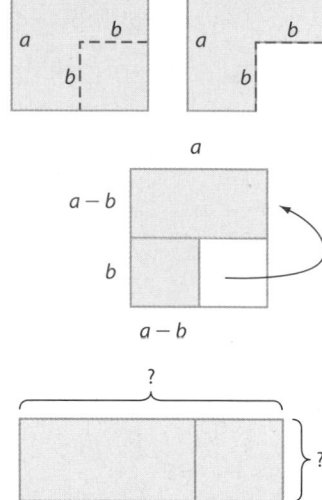

a. Numerical Find the area of each of the squares.

b. Concrete Cut the smaller square out of the corner. What is the area of the shape?

c. Analytical Remove the smaller rectangle on the bottom. Turn it and slide it next to the top rectangle. What is the length of the new arrangement? What is the width? What is the area?

d. Analytical What pattern does this verify?

H.O.T. Problems   Use Higher-Order Thinking Skills

57. **WHICH ONE DOESN'T BELONG?** Which expression does not belong? Explain.

| $(2c - d)(2c - d)$ | $(2c + d)(2c - d)$ | $(2c + d)(2c + d)$ | $(c + d)(c + d)$ |

58. **CCSS STRUCTURE** Does a pattern exist for the cube of a sum $(a + b)^3$?

a. Investigate this question by finding the product $(a + b)(a + b)(a + b)$.

b. Use the pattern you discovered in part **a** to find $(x + 2)^3$.

c. Draw a diagram of a geometric model for $(a + b)^3$.

d. What is the pattern for the cube of a difference, $(a - b)^3$?

59. **REASONING** Find c that makes $25x^2 - 90x + c$ a perfect square trinomial.

60. **OPEN ENDED** Write two binomials with a product that is a binomial. Then write two binomials with a product that is not a binomial.

61. **WRITING IN MATH** Describe how to square the sum of two quantities, square the difference of two quantities, and how to find the product of a sum of two quantities and a difference of two quantities.

62. GRIDDED RESPONSE In the right triangle, $\overline{DB}$ bisects $\angle B$. What is the measure of $\angle ADB$ in degrees?

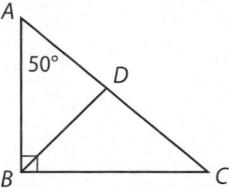

63. What is the product of $(2a - 3)$ and $(2a - 3)$?

  A $4a^2 + 12a + 9$ **C** $4a^2 - 12a - 9$

  B $4a^2 + 9$ **D** $4a^2 - 12a + 9$

64. Myron can drive 4 miles in m minutes. At this rate, how many minutes will it take him to drive 19 miles?

  F $76m$ **H** $\dfrac{4m}{19}$

  G $\dfrac{19m}{4}$ **J** $\dfrac{4}{19m}$

65. What property is illustrated by the equation $2x + 0 = 2x$?

  A Commutative Property of Addition

  B Additive Inverse Property

  C Additive Identity Property

  D Associative Property of Addition

Spiral Review

Find each product. (Lesson 8-3)

66. $(y - 4)(y - 2)$

67. $(2c - 1)(c + 3)$

68. $(d - 9)(d + 5)$

69. $(4h - 3)(2h - 7)$

70. $(3x + 5)(2x + 3)$

71. $(5m + 4)(8m + 3)$

Simplify. (Lesson 8-2)

72. $x(2x - 7) + 5x$

73. $c(c - 8) + 2c(c + 3)$

74. $8y(-3y + 7) - 11y^2$

75. $-2d(5d) - 3d(d + 6)$

76. $5m(2m^3 + m^2 + 8) + 4m$

77. $3p(6p - 4) + 2\left(\frac{1}{2}p^2 - 3p\right)$

Use substitution to solve each system of equations. (Lesson 6-2)

78. $4c = 3d + 3$
   $c = d - 1$

79. $c - 5d = 2$
   $2c + d = 4$

80. $5r - t = 5$
   $-4r + 5t = 17$

81. BIOLOGY Each type of fish thrives in a specific range of temperatures. The best temperatures for sharks range from 18°C to 22°C, inclusive. Write a compound inequality to represent temperatures where sharks will not thrive. (Lesson 5-4)

Write an equation of the line that passes through each pair of points. (Lesson 4-2)

82. $(1, 1), (7, 4)$

83. $(5, 7), (0, 6)$

84. $(5, 1), (8, -2)$

85. COFFEE A coffee store wants to create a mix using two coffees. How many pounds of coffee A should be mixed with 9 pounds of coffee B to get a mixture that can sell for $6.95 per pound? (Lesson 2-9)

Skills Review

Write each polynomial in standard form. Identify the leading coefficient.

86. $2x^2 - x^4 - 8 + x$

87. $-5p^4 + p^2 + 12 + 2p^5$

88. $-10 + a^3 - a + 6a^2$

Determine whether each expression is a polynomial. If it is a polynomial, find the degree and determine whether it is a *monomial*, *binomial*, or *trinomial*. (Lesson 8-1)

1. $3y^2 - 2$

2. $4t^5 + 3t^2 + t$

3. $\dfrac{3x}{5y}$

4. ax^{-3}

5. $3b^2$

6. $2x^{-3} - 4x + 1$

7. **POPULATION** The table shows the population density for Nevada for various years. (Lesson 8-1)

Year	Years Since 1930	People/ Square Mile
1930	0	0.8
1960	30	2.6
1980	50	7.3
1990	60	10.9
2000	70	18.2

a. The population density d of Nevada from 1930 to 2000 can be modeled by $d = 0.005n^2 - 0.127n + 1$, where n represents the number of years since 1930. Identify the type of polynomial for $0.005n^2 - 0.127n + 1$.

b. What is the degree of the polynomial?

c. Predict the population density of Nevada for 2020 and for 2030. Explain your method.

Find each sum or difference. (Lesson 8-1)

8. $(y^2 + 2y + 3) + (y^2 + 3y - 1)$

9. $(3n^3 - 2n + 7) - (n^2 - 2n + 8)$

10. $(5d + d^2) - (4 - 4d^2)$

11. $(x + 4) + (3x + 2x^2 - 7)$

12. $(3a - 3b + 2) - (4a + 5b)$

13. $(8x - y^2 + 3) + (9 - 3x + 2y^2)$

Find each product. (Lesson 8-2)

14. $6y(y^2 + 3y + 1)$

15. $3n(n^2 - 5n + 2)$

16. $d^2(-4 - 3d + 2d^2)$

17. $-2xy(3x^2 + 2xy - 4y^2)$

18. $ab^2(12a + 5b - ab)$

19. $x^2y^4(3xy^2 - x + 2y^2)$

20. **MULTIPLE CHOICE** Simplify $x(4x + 5) + 3(2x^2 - 4x + 1)$. (Lesson 8-2)

    A $10x^2 + 17x + 3$ **C** $2x^2 - 7x + 3$

    B $10x^2 - 7x + 3$ **D** $2x^2 + 17x + 3$

Find each product. (Lesson 8-3)

21. $(x + 2)(x + 5)$

22. $(3b - 2)(b - 4)$

23. $(n - 5)(n + 3)$

24. $(4c - 2)(c + 2)$

25. $(k - 1)(k - 3k^2)$

26. $(8d - 3)(2d^2 + d + 1)$

27. **MANUFACTURING** A company is designing a box for dry pasta in the shape of a rectangular prism. The length is 2 inches more than twice the width, and the height is 3 inches more than the length. Write an expression, in terms of the width, for the volume of the box. (Lesson 8-3)

Find each product. (Lesson 8-4)

28. $(x + 2)^2$

29. $(n - 11)^2$

30. $(4b - 2)^2$

31. $(6c + 3)^2$

32. $(5d - 3)(5d + 3)$

33. $(9k + 1)(9k - 1)$

34. **DISC GOLF** The discs approved for use in disc golf vary in size. (Lesson 8-4)

    Smallest disc   **Largest disc**

a. Write two different expressions for the area of the largest disc.

b. If x is 10.5, what are the areas of the smallest and largest discs?

Algebra Lab
Factoring Using the Distributive Property

When two or more numbers are multiplied, these numbers are *factors* of the product. Sometimes you know the product of binomials and are asked to find the factors. This is called factoring. You can use algebra tiles and a product mat to factor binomials.

CCSS Common Core State Standards
Content Standards
A.SSE.2 Use the structure of an expression to identify ways to rewrite it.

Activity 1 Use Algebra Tiles to Factor $2x - 8$

Step 1 Model $2x - 8$.

Step 2 Arrange the tiles into a rectangle. The total area of the rectangle represents the product, and its length and width represent the factors.

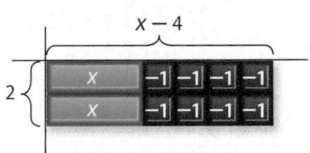

The rectangle has a width of 2 and a length of $x - 4$. Therefore, $2x - 8 = 2(x - 4)$.

Activity 2 Use Algebra Tiles to Factor $x^2 + 3x$

Step 1 Model $x^2 + 3x$.

Step 2 Arrange the tiles into a rectangle.

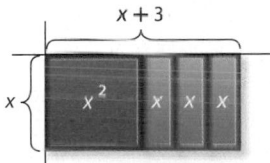

The rectangle has a width of x and a length of $x + 3$. Therefore, $x^2 + 3x = x(x + 3)$.

Model and Analyze

Use algebra tiles to factor each binomial.

1. $4x + 12$ **2.** $4x - 6$ **3.** $3x^2 + 4x$ **4.** $10 - 2x$

Determine whether each binomial can be factored. Justify your answer with a drawing.

5. $6x - 9$ **6.** $5x - 4$ **7.** $4x^2 + 7$ **8.** $x^2 + 3x$

9. **WRITING IN MATH** Write a paragraph that explains how you can use algebra tiles to determine whether a binomial can be factored. Include an example of one binomial that can be factored and one that cannot.

Using the Distributive Property

∴Then	∴Now	∴Why?
● Used the Distributive Property to evaluate expressions.	**1** Use the Distributive Property to factor polynomials. **2** Solve equations of the form $ax^2 + bx = 0$.	● The cost of rent for Mr. Cole's store is determined by the square footage of the space. The area of the store can be modeled by the equation $A = 1.6w^2 + 6w$, where w is the width of the store in feet. We can use factoring and the Zero Product Property to find possible dimensions of the store.

NewVocabulary
factoring
factoring by grouping
Zero Product Property

Common Core State Standards

Content Standards
A.SSE.2 Use the structure of an expression to identify ways to rewrite it.

A.SSE.3a Factor a quadratic expression to reveal the zeros of the function it defines.

Mathematical Practices
2 Reason abstractly and quantitatively.

1 **Use the Distributive Property to Factor** You have used the Distributive Property to multiply a monomial by a polynomial. You can work backward to express a polynomial as the product of a monomial factor and a polynomial factor.

$$1.6w^2 + 6w = 1.6w(w) + 6(w)$$
$$= w(1.6w + 6)$$

So, $w(1.6w + 6)$ is the *factored form* of $1.6w^2 + 6w$. **Factoring** a polynomial involves finding the *completely* factored form.

Example 1 Use the Distributive Property

Use the Distributive Property to factor each polynomial.

a. $27y^2 + 18y$

Find the GCF of each term.

$27y^2 = ③ \cdot ③ \cdot 3 \cdot ⓨ \cdot y$ Factor each term.
$18y = 2 \cdot ③ \cdot ③ \cdot ⓨ$ Circle common factors.
GCF $= 3 \cdot 3 \cdot y$ or $9y$

Write each term as the product of the GCF and the remaining factors. Use the Distributive Property to *factor out* the GCF.

$27y^2 + 18y = 9y(3y) + 9y(2)$ Rewrite each term using the GCF.
$= 9y(3y + 2)$ Distributive Property

b. $-4a^2b - 8ab^2 + 2ab$

$-4a^2b = -1 \cdot ② \cdot 2 \cdot ⓐ \cdot a \cdot ⓑ$ Factor each term.
$-8ab^2 = -1 \cdot ② \cdot 2 \cdot 2 \cdot ⓐ \cdot ⓑ \cdot b$ Circle common factors.
$2ab = ② \cdot ⓐ \cdot ⓑ$
GCF $= 2 \cdot a \cdot b$ or $2ab$

$-4a^2b - 8ab^2 + 2ab = 2ab(-2a) - 2ab(4b) + 2ab(1)$ Rewrite each term using the GCF.
$= 2ab(-2a - 4b + 1)$ Distributive Property

▸ **Guided**Practice

1A. $15w - 3v$ **1B.** $7u^2t^2 + 21ut^2 - ut$

Using the Distributive Property to factor polynomials with four or more terms is called **factoring by grouping** because terms are put into groups and then factored. The Distributive Property is then applied to a common binomial factor.

KeyConcept Factoring by Grouping

Words   A polynomial can be factored by grouping only if all of the following conditions exist.

- There are four or more terms.
- Terms have common factors that can be grouped together.
- There are two common factors that are identical or additive inverses of each other.

Symbols $ax + bx + ay + by = (ax + bx) + (ay + by)$
$$= x(a + b) + y(a + b)$$
$$= (x + y)(a + b)$$

Example 2 Factor by Grouping

Factor $4qr + 8r + 3q + 6$.

$4qr + 8r + 3q + 6$ Original expression
$\quad = (4qr + 8r) + (3q + 6)$ Group terms with common factors.
$\quad = 4r(q + 2) + 3(q + 2)$ Factor the GCF from each group.

Notice that $(q + 2)$ is common in both groups, so it becomes the GCF.
$\quad = (4r + 3)(q + 2)$ Distributive Property

▶ **Guided**Practice

Factor each polynomial.

2A. $rn + 5n - r - 5$ **2B.** $3np + 15p - 4n - 20$

It can be helpful to recognize when binomials are additive inverses of each other. For example $6 - a = -1(a - 6)$.

Example 3 Factor by Grouping with Additive Inverses

Factor $2mk - 12m + 42 - 7k$.

$2mk - 12m + 42 - 7k$
$\quad = (2mk - 12m) + (42 - 7k)$ Group terms with common factors.
$\quad = 2m(k - 6) + 7(6 - k)$ Factor the GCF from each group.
$\quad = 2m(k - 6) + 7[(-1)(k - 6)]$ $6 - k = -1(k - 6)$
$\quad = 2m(k - 6) - 7(k - 6)$ Associative Property
$\quad = (2m - 7)(k - 6)$ Distributive Property

▶ **Guided**Practice

Factor each polynomial.

3A. $c - 2cd + 8d - 4$ **3B.** $3p - 2p^2 - 18p + 27$

2 Solve Equations by Factoring
Some equations can be solved by factoring. Consider the following.

$$3(0) = 0 \qquad 0(2-2) = 0 \qquad -312(0) = 0 \qquad 0(0.25) = 0$$

Notice that in each case, at least one of the factors is 0. These examples are demonstrations of the **Zero Product Property**.

KeyConcept Zero Product Property

Words	If the product of two factors is 0, then at least one of the factors must be 0.
Symbols	For any real numbers a and b, if $ab = 0$, then $a = 0$, $b = 0$, or both a and b equal zero.

Recall that a solution or root of an equation is any value that makes the equation true.

Example 4 Solve Equations

> **Watch**Out!
>
> **Unknown Value** It may be tempting to solve an equation by dividing each side by the variable. However, the variable has an unknown value, so you may be dividing by 0, which is undefined.

Solve each equation. Check your solutions.

a. $(2d + 6)(3d - 15) = 0$

$(2d + 6)(3d - 15) = 0$		Original equation
$2d + 6 = 0 \quad$ or $\quad 3d - 15 = 0$		Zero Product Property
$2d = -6 \qquad\qquad 3d = 15$		Solve each equation.
$d = -3 \qquad\qquad d = 5$		Divide.

The roots are -3 and 5.

CHECK Substitute -3 and 5 for d in the original equation.

$$(2d + 6)(3d - 15) = 0 \qquad\qquad (2d + 6)(3d - 15) = 0$$
$$[2(-3) + 6][3(-3) - 15] \overset{?}{=} 0 \qquad [2(5) + 6][3(5) - 15] \overset{?}{=} 0$$
$$(-6 + 6)(-9 - 15) \overset{?}{=} 0 \qquad\qquad (10 + 6)(15 - 15) \overset{?}{=} 0$$
$$(0)(-24) \overset{?}{=} 0 \qquad\qquad\qquad 16(0) \overset{?}{=} 0$$
$$0 = 0 \checkmark \qquad\qquad\qquad\qquad 0 = 0 \checkmark$$

b. $c^2 = 3c$

$c^2 = 3c$	Original equation
$c^2 - 3c = 0$	Subtract $3c$ from each side to get 0 on one side of the equation.
$c(c - 3) = 0$	Factor by using the GCF to get the form $ab = 0$.
$c = 0 \quad$ or $\quad c - 3 = 0$	Zero Product Property
$c = 3$	Solve each equation.

The roots are 0 and 3. Check by substituting 0 and 3 for c.

▶ **Guided**Practice

4A. $3n(n + 2) = 0$ **4B.** $8b^2 - 40b = 0$ **4C.** $x^2 = -10x$

Real-World Example 5 Use Factoring

AGILITY Penny is a Fox Terrier who competes with her trainer in the agility course. Within the course, Penny must leap over a hurdle. Penny's jump can be modeled by the equation $h = -16t^2 + 20t$, where h is the height of the leap in inches at t seconds. Find the values of t when $h = 0$.

$h = -16t^2 + 20t$	Original equation
$0 = -16t^2 + 20t$	Substitution, $h = 0$
$0 = 4t(-4t + 5)$	Factor by using the GCF.
$4t = 0$ or $-4t + 5 = 0$	Zero Product Property
$t = 0$ $-4t = -5$	Solve each equation.
$t = \dfrac{5}{4}$ or 1.25	Divide each side by -4.

Penny's height is 0 inches at 0 seconds and 1.25 seconds into the jump.

▸ **Guided**Practice

5. KANGAROOS The hop of a kangaroo can be modeled by $h = 24t - 16t^2$ where h represents the height of the hop in meters and t is the time in seconds. Find the values of t when $h = 0$.

Check Your Understanding

$\bigcirc$ = Step-by-Step Solutions begin on page R13.

Example 1 Use the Distributive Property to factor each polynomial.

1. $21b - 15a$

2. $14c^2 + 2c$

3. $10g^2h^2 + 9gh^2 - g^2h$

4. $12jk^2 + 6j^2k + 2j^2k^2$

Examples 2–3 Factor each polynomial.

5 $np + 2n + 8p + 16$

6. $xy - 7x + 7y - 49$

7. $3bc - 2b - 10 + 15c$

8. $9fg - 45f - 7g + 35$

Example 4 Solve each equation. Check your solutions.

9. $3k(k + 10) = 0$

10. $(4m + 2)(3m - 9) = 0$

11. $20p^2 - 15p = 0$

12. $r^2 = 14r$

Example 5 **13. SPIDERS** Jumping spiders can commonly be found in homes and barns throughout the United States. A jumping spider's jump can be modeled by the equation $h = 33.3t - 16t^2$, where t represents the time in seconds and h is the height in feet.

  a. When is the spider's height at 0 feet?

  b. What is the spider's height after 1 second? after 2 seconds?

14. ⒸⒸⓈⓈ **REASONING** At a Fourth of July celebration, a rocket is launched straight up with an initial velocity of 125 feet per second. The height h of the rocket in feet above sea level is modeled by the formula $h = 125t - 16t^2$, where t is the time in seconds after the rocket is launched.

  a. What is the height of the rocket when it returns to the ground?

  b. Let $h = 0$ in the equation and solve for t.

  c. How many seconds will it take for the rocket to return to the ground?

Example 1   Use the Distributive Property to factor each polynomial.

15. $16t - 40y$

16. $30v + 50x$

17. $2k^2 + 4k$

18. $5z^2 + 10z$

19. $4a^2b^2 + 2a^2b - 10ab^2$

20. $5c^2v - 15c^2v^2 + 5c^2v^3$

Examples 2–3 Factor each polynomial.

21. $fg - 5g + 4f - 20$

22. $a^2 - 4a - 24 + 6a$

23. $hj - 2h + 5j - 10$

24. $xy - 2x - 2 + y$

25. $45pq - 27q - 50p + 30$

26. $24ty - 18t + 4y - 3$

27. $3dt - 21d + 35 - 5t$

28. $8r^2 + 12r$

29. $21th - 3t - 35h + 5$

30. $vp + 12v + 8p + 96$

31. $5br - 25b + 2r - 10$

32. $2nu - 8u + 3n - 12$

33. $5gf^2 + g^2f + 15gf$

34. $rp - 9r + 9p - 81$

35. $27cd^2 - 18c^2d^2 + 3cd$

36. $18r^3t^2 + 12r^2t^2 - 6r^2t$

37. $48tu - 90t + 32u - 60$

38. $16gh + 24g - 2h - 3$

Example 4   Solve each equation. Check your solutions.

39. $3b(9b - 27) = 0$

40. $2n(3n + 3) = 0$

41. $(8z + 4)(5z + 10) = 0$

42. $(7x + 3)(2x - 6) = 0$

43. $b^2 = -3b$

44. $a^2 = 4a$

Example 5   **45.** **CCSS** **SENSE-MAKING** Use the drawing at the right.

  a. Write an expression in factored form to represent the area of the blue section.

  b. Write an expression in factored form to represent the area of the region formed by the outer edge.

  c. Write an expression in factored form to represent the yellow region.

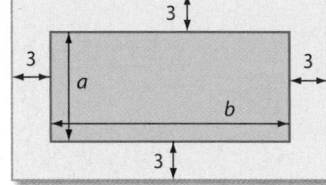

46. **FIREWORKS** A ten-inch fireworks shell is fired from ground level. The height of the shell in feet is given by the formula $h = 263t - 16t^2$, where t is the time in seconds after launch.

  a. Write the expression that represents the height in factored form.

  b. At what time will the height be 0? Is this answer practical? Explain.

  c. What is the height of the shell 8 seconds and 10 seconds after being fired?

  d. At 10 seconds, is the shell rising or falling?

47. **ARCHITECTURE** The frame of a doorway is an arch that can be modeled by the graph of the equation $y = -3x^2 + 12x$, where x and y are measured in feet. On a coordinate plane, the floor is represented by the x-axis.

  a. Make a table of values for the height of the arch if $x = 0, 1, 2, 3,$ and 4 feet.

  b. Plot the points from the table on a coordinate plane and connect the points to form a smooth curve to represent the arch.

  c. How high is the doorway?

48. RIDES Suppose the height of a rider after being dropped can be modeled by $h = -16t^2 - 96t + 160$, where h is the height in feet and t is time in seconds.

    a. Write an expression to represent the height in factored form.

    b. From what height is the rider initially dropped?

    c. At what height will the rider be after 3 seconds of falling? Is this possible? Explain.

49 ARCHERY The height h in feet of an arrow can be modeled by the equation $h = 64t - 16t^2$, where t is time in seconds. Ignoring the height of the archer, how long after the arrow is released does it hit the ground?

50. TENNIS A tennis player hits a tennis ball upward with an initial velocity of 80 feet per second. The height h in feet of the tennis ball can be modeled by the equation $h = 80t - 16t^2$, where t is time in seconds. Ignoring the height of the tennis player, how long does it take the ball to hit the ground?

51. MULTIPLE REPRESENTATIONS In this problem, you will explore the *box method* of factoring. To factor $x^2 + x - 6$, write the first term in the top left-hand corner of the box, and then write the last term in the lower right-hand corner.

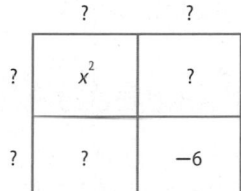

    a. Analytical Determine which two factors have a product of -6 and a sum of 1.

    b. Symbolic Write each factor in an empty square in the box. Include the positive or negative sign and variable.

    c. Analytical Find the factor for each row and column of the box. What are the factors of $x^2 + x - 6$?

    d. Verbal Describe how you would use the box method to factor $x^2 - 3x - 40$.

H.O.T. Problems   Use Higher-Order Thinking Skills

52. CCSS CRITIQUE Hernando and Rachel are solving $2m^2 - 4m$. Is either of them correct? Explain your reasoning.

Hernando	Rachel
$2m^2 = 4m$	$2m^2 = 4m$
$\dfrac{2m^2}{m} = \dfrac{4m^2}{2m}$	$2m^2 - 4m = 0$
$2m = 2$	$2m(m - 2) = 0$
$m = 1$	$2m = 0 \ or \ m - 2 = 0$
	$m = 0 \ or \ 2$

53. CHALLENGE Given the equation $(ax + b)(ax - b) = 0$, solve for x. What do we know about the values of a and b?

54. OPEN ENDED Write a four-term polynomial that can be factored by grouping. Then factor the polynomial.

55. REASONING Given the equation $c = a^2 - ab$, for what values of a and b does $c = 0$?

56. WRITING IN MATH Explain how to solve a quadratic equation by using the Zero Product Property.

57. Which is a factor of $6z^2 - 3z - 2 + 4z$?

   A $2z + 1$ **C** $z + 2$

   B $3z - 2$ **D** $2z - 1$

58. PROBABILITY Hailey has 10 blocks: 2 red, 4 blue, 3 yellow, and 1 green. What is the probability that a randomly chosen block will be either red or yellow?

   F $\frac{3}{10}$ **H** $\frac{1}{2}$

   G $\frac{1}{5}$ **J** $\frac{7}{10}$

59. GRIDDED RESPONSE Cho is making a 140-inch by 160-inch quilt with quilt squares that measure 8 inches on each side. How many will be needed to make the quilt?

60. GEOMETRY The area of the right triangle shown below is $5h$ square centimeters. What is the height of the triangle?

   A 2 cm

   B 5 cm

   C 8 cm

   D 10 cm

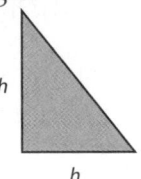

$2h$

h

61. GENETICS Brown genes B are dominant over blue genes b. A person with genes BB or Bb has brown eyes. Someone with genes bb has blue eyes. Elisa has brown eyes with Bb genes, and Bob has blue eyes. Write an expression for the possible eye coloring of Elisa and Bob's children. Determine the probability that their child would have blue eyes. (Lesson 8-4)

Find each product. (Lesson 8-2)

62. $n(n^2 - 4n + 3)$ **63.** $2b(b^2 + b - 5)$

64. $-c(4c^2 + 2c - 2)$ **65.** $-4x(x^3 + x^2 + 2x - 1)$

66. $2ab(4a^2b + 2ab - 2b^2)$ **67.** $-3xy(x^2 + xy + 2y^2)$

Simplify. (Lesson 7-1)

68. $(ab^4)(ab^2)$ **69.** $(p^5r^4)(p^2r)$ **70.** $(-7c^3d^4)(4cd^3)$

71. $(9xy^7)^2$ **72.** $\left[(3^2)^4\right]^2$ **73.** $\left[(4^2)^3\right]^2$

74. BASKETBALL In basketball, a free throw is 1 point and a field goal is either 2 or 3 points. In a season, Tim Duncan of the San Antonio Spurs scored a total of 1342 points. The total number of 2-point field goals and 3-point field goals was 517, and he made 305 of the 455 free throws that he attempted. Find the number of 2-point field goals and 3-point field goals Duncan made that season. (Lesson 6-4)

Solve each inequality. Check your solution. (Lesson 5-3)

75. $3y - 4 > -37$ **76.** $-5q + 9 > 24$ **77.** $-2k + 12 < 30$

78. $5q + 7 \le 3(q + 1)$ **79.** $\frac{z}{4} + 7 \ge -5$ **80.** $8c - (c - 5) > c + 17$

Find each product.

81. $(a + 2)(a + 5)$ **82.** $(d + 4)(d + 10)$ **83.** $(z - 1)(z - 8)$

84. $(c + 9)(c - 3)$ **85.** $(x - 7)(x - 6)$ **86.** $(g - 2)(g + 11)$

Algebra Lab
Factoring Trinomials

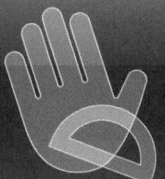

You can use algebra tiles to factor trinomials. If a polynomial represents the area of a rectangle formed by algebra tiles, then the rectangle's length and width are *factors* of the area. If a rectangle cannot be formed to represent the trinomial, then the trinomial is not factorable.

CCSS **Common Core State Standards**
Content Standards
A.SSE.2 Use the structure of an expression to identify ways to rewrite it.

Activity 1 Factor $x^2 + bx + c$

Use algebra tiles to factor $x^2 + 4x + 3$.

Step 1 Model $x^2 + 4x + 3$.

Step 2 Place the x^2-tile at the corner of the product mat. Arrange the 1-tiles into a rectangular array. Because 3 is prime, the 3 tiles can be arranged in a rectangle in one way, a 1-by-3 rectangle.

Step 3 Complete the rectangle with the x-tiles.

The rectangle has a width of $x + 1$ and a length of $x + 3$.

Therefore, $x^2 + 4x + 3 = (x + 1)(x + 3)$.

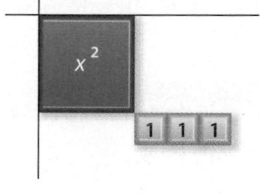

Activity 2 Factor $x^2 + bx + c$

Use algebra tiles to factor $x^2 + 8x + 12$.

Step 1 Model $x^2 + 8x + 12$.

Step 2 Place the x^2-tile at the corner of the product mat. Arrange the 1-tiles into a rectangular array. Since $12 = 3 \times 4$, try a 3-by-4 rectangle. Try to complete the rectangle. Notice that there is an extra x-tile.

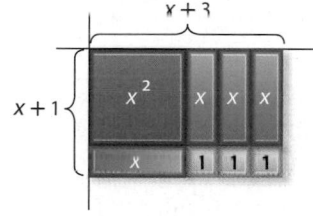

Step 3 Arrange the 1-tiles into a 2-by-6 rectangular array. This time you can complete the rectangle with the x-tiles.

The rectangle has a width of $x + 2$ and a length of $x + 6$.

Therefore, $x^2 + 8x + 12 = (x + 2)(x + 6)$.

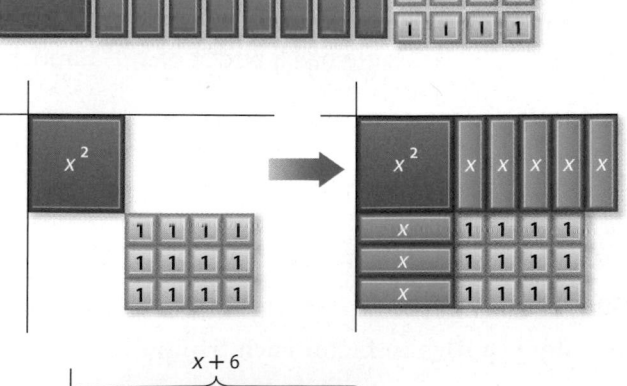

Algebra Lab
Factoring Trinomials *Continued*

Activity 3   Factor $x^2 - bx + c$

Use algebra tiles to factor $x^2 - 5x + 6$.

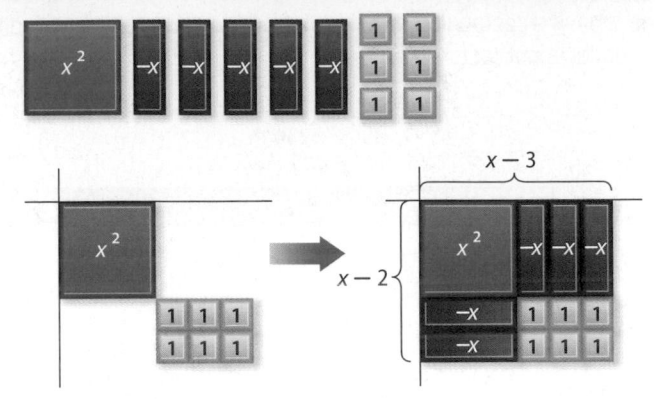

Step 1 Model $x^2 - 5x + 6$.

Step 2 Place the x^2-tile at the corner of the product mat. Arrange the 1-tiles into a 2-by-3 rectangular array as shown.

Step 3 Complete the rectangle with the x-tiles. The rectangle has a width of $x - 2$ and a length of $x - 3$.

Therefore, $x^2 - 5x + 6 = (x - 2)(x - 3)$.

Activity 4   Factor $x^2 - bx - c$

Use algebra tiles to factor $x^2 - 4x - 5$.

Step 1 Model $x^2 - 4x - 5$.

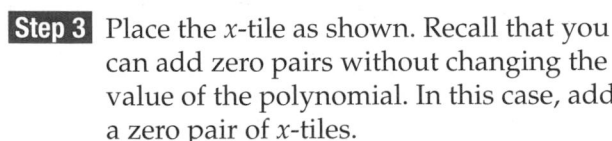

Step 2 Place the x^2-tile at the corner of the product mat. Arrange the 1-tiles into a 1-by-5 rectangular array as shown.

Step 3 Place the x-tile as shown. Recall that you can add zero pairs without changing the value of the polynomial. In this case, add a zero pair of x-tiles.

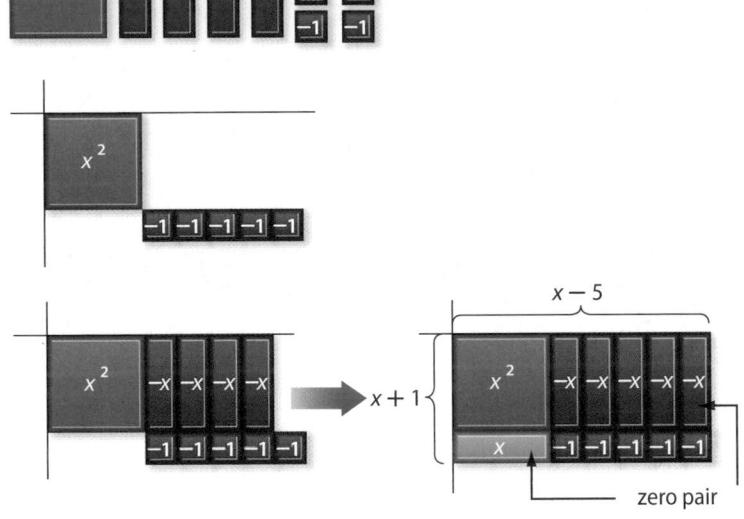

The rectangle has a width of $x + 1$ and a length of $x - 5$.

Therefore, $x^2 - 4x - 5 = (x + 1)(x - 5)$.

Model and Analyze
Use algebra tiles to factor each trinomial.

1. $x^2 + 3x + 2$ **2.** $x^2 + 6x + 8$ **3.** $x^2 + 3x - 4$ **4.** $x^2 - 7x + 12$

5. $x^2 + 7x + 10$ **6.** $x^2 - 2x + 1$ **7.** $x^2 + x - 12$ **8.** $x^2 - 8x + 15$

Tell whether each trinomial can be factored. Justify your answer with a drawing.

9. $x^2 + 3x + 6$ **10.** $x^2 - 5x - 6$ **11.** $x^2 - x - 4$ **12.** $x^2 - 4$

13. WRITING IN MATH How can you use algebra tiles to determine whether a trinomial can be factored?

Solving $x^2 + bx + c = 0$

Then
- You multiplied binomials by using the FOIL method.

Now
1. Factor trinomials of the form $x^2 + bx + c$.
2. Solve equations of the form $x^2 + bx + c = 0$.

Why?
- Diana is having a rectangular in-ground swimming pool installed and she wants to include a 24-foot fence around the pool. The pool requires a space of 36 square feet. What dimensions should the pool have?

To solve this problem, the landscape architect needs to find two numbers that have a product of 36 and a sum of 12, half the perimeter of the pool.

 NewVocabulary
quadratic equation

 Common Core State Standards

Content Standards
A.SSE.3a Factor a quadratic expression to reveal the zeros of the function it defines.

A.REI.4b Solve quadratic equations by inspection (e.g., for $x^2 = 49$), taking square roots, completing the square, the quadratic formula and factoring, as appropriate to the initial form of the equation. Recognize when the quadratic formula gives complex solutions and write them as $a \pm bi$ for real numbers a and b.

Mathematical Practices
7 Look for and make use of structure.
8 Look for and express regularity in repeated reasoning.

1 Factor $x^2 + bx + c$ You have learned how to multiply two binomials using the FOIL method. Each of the binomials was a factor of the product. The pattern for multiplying two binomials can be used to factor certain types of trinomials.

$$
\begin{aligned}
(x + 3)(x + 4) &= x^2 + 4x + 3x + 3 \cdot 4 && \text{Use the FOIL method.} \\
&= x^2 + (4 + 3)x + 3 \cdot 4 && \text{Distributive Property} \\
&= x^2 + 7x + 12 && \text{Simplify.}
\end{aligned}
$$

Notice that the coefficient of the middle term, $7x$, is the sum of 3 and 4, and the last term, 12, is the product of 3 and 4.

Observe the following pattern in this multiplication.

$$
\begin{aligned}
(x + 3)(x + 4) &= x^2 + (4 + 3)x + (3 \cdot 4) \\
(x + m)(x + p) &= x^2 + (p + m)x + mp && \text{Let } 3 = m \text{ and } 4 = p. \\
&= x^2 + \underbrace{(m + p)}_{bx}x + \underbrace{mp}_{c} && \text{Commutative } (+) \\
& \quad\quad x^2 + \quad bx \quad + \quad c && b = m + p \text{ and } c = mp
\end{aligned}
$$

Notice that the coefficient of the middle term is the sum of m and p, and the last term is the product of m and p. This pattern can be used to factor trinomials of the form $x^2 + bx + c$.

> **KeyConcept** Factoring $x^2 + bx + c$
>
> | | |
> |---|---|
> | **Words** | To factor trinomials in the form $x^2 + bx + c$, find two integers, m and p, with a sum of b and a product of c. Then write $x^2 + bx + c$ as $(x + m)(x + p)$. |
> | **Symbols** | $x^2 + bx + c = (x + m)(x + p)$ when $m + p = b$ and $mp = c$. |
> | **Example** | $x^2 + 6x + 8 = (x + 2)(x + 4)$, because $2 + 4 = 6$ and $2 \cdot 4 = 8$. |

When c is positive, its factors have the same signs. Both of the factors are positive or negative based upon the sign of b. If b is positive, the factors are positive. If b is negative, the factors are negative.

Example 1 *b* and *c* are Positive

Factor $x^2 + 9x + 20$.

In this trinomial, $b = 9$ and $c = 20$. Since c is positive and b is positive, you need to find two positive factors with a sum of 9 and a product of 20. Make an organized list of the factors of 20, and look for the pair of factors with a sum of 9.

Factors of 20	Sum of Factors
1, 20	21
2, 10	12
4, 5	9

The correct factors are 4 and 5.

$$x^2 + 9x + 20 = (x + m)(x + p)$$

Write the pattern.

$$= (x + 4)(x + 5)$$

$m = 4$ and $p = 5$

CHECK You can check this result by multiplying the two factors. The product should be equal to the original expression.

$$(x + 4)(x + 5) = x^2 + 5x + 4x + 20 \qquad \text{FOIL Method}$$
$$= x^2 + 9x + 20 \checkmark \qquad \text{Simplify.}$$

▶ **Guided**Practice

Factor each polynomial.

1A. $d^2 + 11d + 24$ **1B.** $9 + 10t + t^2$

When factoring a trinomial in which b is negative and c is positive, use what you know about the product of binomials to narrow the list of possible factors.

Example 2 *b* is Negative and *c* is Positive

Factor $x^2 - 8x + 12$. Confirm your answer using a graphing calculator.

In this trinomial, $b = -8$ and $c = 12$. Since c is positive and b is negative, you need to find two negative factors with a sum of -8 and a product of 12.

Factors of 12	Sum of Factors
$-1, -12$	-13
$-2, -6$	-8
$-3, -4$	-7

The correct factors are -2 and -6.

$$x^2 - 8x + 12 = (x + m)(x + p) \qquad \text{Write the pattern.}$$
$$= (x - 2)(x - 6) \qquad m = -2 \text{ and } p = -6$$

CHECK Graph $y = x^2 - 8x + 12$ and $y = (x - 2)(x - 6)$ on the same screen. Since only one graph appears, the two graphs must coincide. Therefore, the trinomial has been factored correctly. ✓

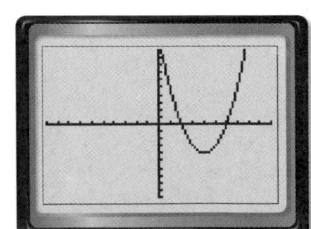

[−10, 10] scl: 1 by [−10, 10] scl: 1

▶ **Guided**Practice

Factor each polynomial.

2A. $21 - 22m + m^2$ **2B.** $w^2 - 11w + 28$

When c is negative, its factors have opposite signs. To determine which factor is positive and which is negative, look at the sign of b. The factor with the greater absolute value has the same sign as b.

Example 3 c is Negative

Factor each polynomial. Confirm your answers using a graphing calculator.

a. $x^2 + 2x - 15$

In this trinomial, $b = 2$ and $c = -15$. Since c is negative, the factors m and p have opposite signs. So either m or p is negative, but not both. Since b is positive, the factor with the greater absolute value is also positive.

List the factors of -15, where one factor of each pair is negative. Look for the pair of factors with a sum of 2.

Factors of -15	Sum of Factors
$-1, 15$	14
$-3, 5$	2

The correct factors are -3 and 5.

$$x^2 + 2x - 15 = (x + m)(x + p)$$
$$= (x - 3)(x + 5)$$

Write the pattern.

$m = -3$ and $p = 5$

CHECK $(x - 3)(x + 5) = x^2 + 5x - 3x - 15$ FOIL Method

$$= x^2 + 2x - 15 \checkmark$$ Simplify.

b. $x^2 - 7x - 18$

In this trinomial, $b = -7$ and $c = -18$. Either m or p is negative, but not both. Since b is negative, the factor with the greater absolute value is also negative.

List the factors of -18, where one factor of each pair is negative. Look for the pair of factors with a sum of -7.

Factors of -18	Sum of Factors
$1, -18$	-17
$2, -9$	-7
$3, -6$	-3

The correct factors are 2 and -9.

$$x^2 - 7x - 18 = (x + m)(x + p)$$
$$= (x + 2)(x - 9)$$

Write the pattern.

$m = 2$ and $p = -9$

CHECK Graph $y = x^2 - 7x - 18$ and $y = (x + 2)(x - 9)$ on the same screen.

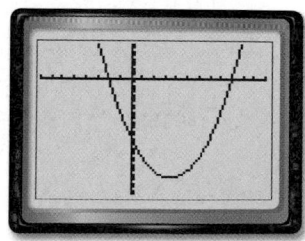

[−10, 15] scl: 1 by [−40, 20] scl: 1

The graphs coincide. Therefore, the trinomial has been factored correctly. $\checkmark$

GuidedPractice

3A. $y^2 + 13y - 48$ **3B.** $r^2 - 2r - 24$

2 Solve Equations by Factoring

A **quadratic equation** can be written in the standard form $ax^2 + bx + c = 0$, where $a \neq 0$. Some equations of the form $x^2 + bx + c = 0$ can be solved by factoring and then using the Zero Product Property.

Example 4 Solve an Equation by Factoring

Solve $x^2 + 6x = 27$. Check your solutions.

$x^2 + 6x = 27$	Original equation
$x^2 + 6x - 27 = 0$	Subtract 27 from each side.
$(x - 3)(x + 9) = 0$	Factor.
$x - 3 = 0 \quad \text{or} \quad x + 9 = 0$	Zero Product Property
$x = 3 \qquad\qquad x = -9$	Solve each equation.

The roots are 3 and -9.

CHECK Substitute 3 and -9 for x in the original equation.

$$x^2 + 6x = 27 \qquad\qquad x^2 + 6x = 27$$
$$(3)^2 + 6(3) \overset{?}{=} 27 \qquad (-9)^2 + 6(-9) \overset{?}{=} 27$$
$$9 + 18 \overset{?}{=} 27 \qquad\qquad 81 - 54 \overset{?}{=} 27$$
$$27 = 27 \checkmark \qquad\qquad 27 = 27 \checkmark$$

GuidedPractice

Solve each equation. Check your solutions.

4A. $z^2 - 3z = 70$

4B. $x^2 + 3x - 18 = 0$

StudyTip

Solving an Equation By Factoring Remember to get 0 on one side of the equation before factoring.

Factoring can be useful when solving real-world problems.

Real-World Example 5 Solve a Problem by Factoring

DESIGN Ling is designing a poster. The top of the poster is 4 inches long and the rest of the poster is 2 inches longer than the width. If the poster requires 616 square inches of poster board, find the width w of the poster.

Understand You want to find the width of the poster.

Plan Since the poster is a rectangle, width · length = area.

Solve Let w = the width of the poster. The length is $w + 2 + 4$ or $w + 6$.

$w(w + 6) = 616$	Write the equation.
$w^2 + 6w = 616$	Multiply.
$w^2 + 6w - 616 = 0$	Subtract 616 from each side.
$(w + 28)(w - 22) = 0$	Factor.
$w + 28 = 0 \quad \text{or} \quad w - 22 = 0$	Zero Product Property
$w = -28 \qquad\qquad w = 22$	Solve each equation.

Since dimensions cannot be negative, the width is 22 inches.

Check If the width is 22 inches, then the area of the poster is $22 \cdot (22 + 6)$ or 616 square inches, which is the amount the poster requires. $\checkmark$

GuidedPractice

5. GEOMETRY The height of a parallelogram is 18 centimeters less than its base. If the area is 175 square centimeters, what is its height?

Real-World Link

A company that produces event signs recommends foamcore boards for event signs that will be used only once. For signs used more than once, use a stronger type of foamcore board.

Source: MegaPrint Inc.

Examples 1–3 Factor each polynomial. Confirm your answers using a graphing calculator.

1. $x^2 + 14x + 24$

2. $y^2 - 7y - 30$

3. $n^2 + 4n - 21$

4. $m^2 - 15m + 50$

Example 4 Solve each equation. Check your solutions.

5. $x^2 - 4x - 21 = 0$

6. $n^2 - 3n + 2 = 0$

7. $x^2 - 15x + 54 = 0$

8. $x^2 + 12x = -32$

9. $x^2 - x - 72 = 0$

10. $x^2 - 10x = -24$

Example 5

11. FRAMING Tina bought a frame for a photo, but the photo is too big for the frame. Tina needs to reduce the width and length of the photo by the same amount. The area of the photo should be reduced to half the original area. If the original photo is 12 inches by 16 inches, what will be the dimensions of the smaller photo?

Practice and Problem Solving Extra Practice is on page R8.

Examples 1–3 Factor each polynomial. Confirm your answers using a graphing calculator.

12. $x^2 + 17x + 42$

13. $y^2 - 17y + 72$

14. $a^2 + 8a - 48$

15. $n^2 - 2n - 35$

16. $44 + 15h + h^2$

17. $40 - 22x + x^2$

18. $-24 - 10x + x^2$

19. $-42 - m + m^2$

Example 4 Solve each equation. Check your solutions.

20. $x^2 - 7x + 12 = 0$

21 $y^2 + y = 20$

22. $x^2 - 6x = 27$

23. $a^2 + 11a = -18$

24. $c^2 + 10c + 9 = 0$

25. $x^2 - 18x = -32$

26. $n^2 - 120 = 7n$

27. $d^2 + 56 = -18d$

28. $y^2 - 90 = 13y$

29. $h^2 + 48 = 16h$

Example 5

30. GEOMETRY A triangle has an area of 36 square feet. If the height of the triangle is 6 feet more than its base, what are its height and base?

31. GEOMETRY A rectangle has an area represented by $x^2 - 4x - 12$ square feet. If the length is $x + 2$ feet, what is the width of the rectangle?

32. SOCCER The width of a high school soccer field is 45 yards shorter than its length.

a. Define a variable, and write an expression for the area of the field.

b. The area of the field is 9000 square yards. Find the dimensions.

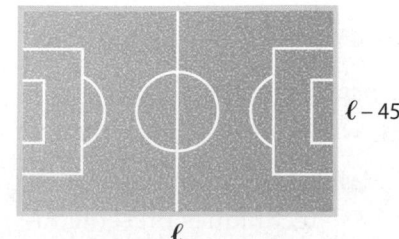

CCSS STRUCTURE Factor each polynomial.

33. $q^2 + 11qr + 18r^2$

34. $x^2 - 14xy - 51y^2$

35. $x^2 - 6xy + 5y^2$

36. $a^2 + 10ab - 39b^2$

37) SWIMMING The length of a rectangular swimming pool is 20 feet greater than its width. The area of the pool is 525 square feet.

   a. Define a variable and write an equation for the area of the pool.

   b. Solve the equation.

   c. Interpret the solutions. Do both solutions make sense? Explain.

GEOMETRY Find an expression for the perimeter of a rectangle with the given area.

38. $A = x^2 + 24x - 81$

39. $A = x^2 + 13x - 90$

40. ⬕ **MULTIPLE REPRESENTATIONS** In this problem, you will explore factoring when the leading coefficient is not 1.

   a. Tabular Copy and complete the table below.

Product of Two Binomials	$ax^2 + mx + px + c$	$ax^2 + bx + c$	$m \times p$	$a \times c$
$(2x + 3)(x + 4)$	$2x^2 + 8x + 3x + 12$	$2x^2 + 11x + 12$	24	24
$(x + 1)(3x + 5)$				
$(2x - 1)(4x + 1)$				
$(3x + 5)(4x - 2)$				

   b. Analytical How are m and p related to a and c?

   c. Analytical How are m and p related to b?

   d. Verbal Describe a process you can use for factoring a polynomial of the form $ax^2 + bx + c$.

H.O.T. Problems   *Use Higher-Order Thinking Skills*

41. ERROR ANALYSIS Jerome and Charles have factored $x^2 + 6x - 16$. Is either of them correct? Explain your reasoning.

> *Jerome*
> $x^2 + 6x - 16 = (x + 2)(x - 8)$

> *Charles*
> $x^2 + 6x - 16 = (x - 2)(x + 8)$

CCSS ARGUMENTS Find all values of k so that each polynomial can be factored using integers.

42. $x^2 + kx - 19$

43. $x^2 + kx + 14$

44. $x^2 - 8x + k, k > 0$

45. $x^2 - 5x + k, k > 0$

46. REASONING For any factorable trinomial, $x^2 + bx + c$, will the absolute value of b *sometimes*, *always*, or *never* be less than the absolute value of c? Explain.

47. OPEN ENDED Give an example of a trinomial that can be factored using the factoring techniques presented in this lesson. Then factor the trinomial.

48. CHALLENGE Factor $(4y - 5)^2 + 3(4y - 5) - 70$.

49. WRITING IN MATH Explain how to factor trinomials of the form $x^2 + bx + c$ and how to determine the signs of the factors of c.

50. Which inequality is shown in the graph below?

A $y \leq -\frac{3}{4}x + 3$

B $y < -\frac{3}{4}x + 3$

C $y > -\frac{3}{4}x + 3$

D $y \geq -\frac{3}{4}x + 3$

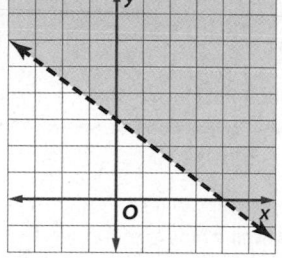

51. SHORT RESPONSE Olivia must earn more than $254 from selling candy bars in order to go on a trip with the National Honor Society. If each candy bar is sold for $1.25, what is the fewest candy bars she must sell?

52. GEOMETRY Which expression represents the length of the rectangle?

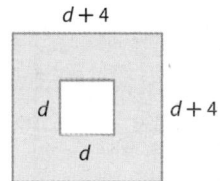

F $x + 5$

G $x + 6$

H $x - 6$

J $x - 5$

53. The difference of 21 and a number n is 6. Which equation shows the relationship?

A $21 - n = 6$ **C** $21n = 6$

B $21 + n = 6$ **D** $6n = -21$

Spiral Review

Factor each polynomial. (Lesson 8-5)

54. $10a^2 + 40a$

55. $11x + 44x^2y$

56. $2m^3p^2 - 16mp^2 + 8mp$

57. $2ax + 6xc + ba + 3bc$

58. $8ac - 2ad + 4bc - bd$

59. $x^2 - xy - xy + y^2$

60. Write a polynomial that represents the area of the shaded region in the figure at the right. (Lesson 8-4)

Use elimination to solve each system of equations. (Lesson 6-3)

61. $-x + y = 9$
$x + 2y = 30$

62. $5a + 2b = 4$
$-5a - b = -7$

63. $2c + d = 12$
$c - d = -3$

64. $6x + 2y = 14$
$5x - 2y = 8$

65. LANDSCAPING Kendrick is planning a circular flower garden with a low fence around the border. He has 38 feet of fence. What is the radius of the largest garden he can make? (*Hint:* $C = 2\pi r$) (Lesson 5-2)

Skills Review

Factor each polynomial.

66. $6mx - 4m + 3rx - 2r$

67. $3ax - 6bx + 8b - 4a$

68. $2d^2g + 2fg + 4d^2h + 4fh$

Solving $ax^2 + bx + c = 0$

·· Then

- You factored trinomials of the form $x^2 + bx + c$.

·· Now

1. Factor trinomials of the form $ax^2 + bx + c$.

2. Solve equations of the form $ax^2 + bx + c = 0$.

·· Why?

- The path of a rider on the amusement park ride shown at the right can be modeled by $16t^2 - 5t + 120$.

  Factoring this expression can help the ride operators determine how long a rider rides on the initial swing.

 NewVocabulary
prime polynomial

 Common Core State Standards

Content Standards
A.SSE.3a Factor a quadratic expression to reveal the zeros of the function it defines.

A.REI.4b Solve quadratic equations by inspection (e.g., for $x^2 = 49$), taking square roots, completing the square, the quadratic formula and factoring, as appropriate to the initial form of the equation. Recognize when the quadratic formula gives complex solutions and write them as $a \pm bi$ for real numbers a and b.

Mathematical Practices
4 Model with mathematics.

1 Factor $ax^2 + bx + c$ In the last lesson, you factored quadratic expressions of the form $ax^2 + bx + c$, where $a = 1$. In this lesson, you will apply the factoring methods to quadratic expressions in which a is not 1.

The dimensions of the rectangle formed by the algebra tiles are the factors of $2x^2 + 5x + 3$. The factors of $2x^2 + 5x + 3$ are $x + 1$ and $2x + 3$.

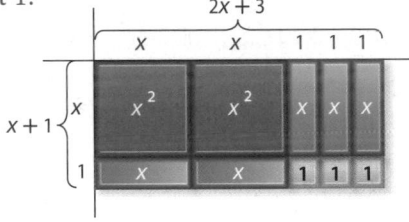

You can also use the method of factoring by grouping to solve this expression.

Step 1 Apply the pattern: $2x^2 + 5x + 3 = 2x^2 + mx + px + 3$.

Step 2 Find two numbers that have a product of $2 \cdot 3$ or 6 and a sum of 5.

Factors of 6	Sum of Factors
1, 6	7
2, 3	5

Step 3 Use grouping to find the factors.

$$2x^2 + 5x + 3 = 2x^2 + mx + px + 3 \qquad \text{Write the pattern.}$$
$$= 2x^2 + 2x + 3x + 3 \qquad m = 2 \text{ and } p = 3$$
$$= (2x^2 + 2x) + (3x + 3) \qquad \text{Group terms with common factors.}$$
$$= 2x(x + 1) + 3(x + 1) \qquad \text{Factor the GCF.}$$
$$= (2x + 3)(x + 1) \qquad x + 1 \text{ is the common factor.}$$

Therefore, $2x^2 + 5x + 3 = (2x + 3)(x + 1)$.

⚙ KeyConcept Factoring $ax^2 + bx + c$

Words
To factor trinomials of the form $ax^2 + bx + c$, find two integers, m and p, with a sum of b and a product of ac. Then write $ax^2 + bx + c$ as $ax^2 + mx + px + c$, and factor by grouping.

Example
$$5x^2 - 13x + 6 = 5x^2 - 10x - 3x + 6 \qquad m = -10 \text{ and } p = -3$$
$$= 5x(x - 2) + (-3)(x - 2)$$
$$= (5x - 3)(x - 2)$$

Example 1 Factor $ax^2 + bx + c$

Factor each trinomial.

a. $7x^2 + 29x + 4$

In this trinomial, $a = 7$, $b = 29$, and $c = 4$. You need to find two numbers with a sum of 29 and a product of $7 \cdot 4$ or 28. Make a list of the factors of 28 and look for the pair of factors with the sum of 29.

Factors of 28	Sum of Factors
1, 28	29

The correct factors are 1 and 28.

$$7x^2 + 29x + 4 = 7x^2 + mx + px + 4 \qquad \text{Write the pattern.}$$
$$= 7x^2 + 1x + 28x + 4 \qquad m = 1 \text{ and } p = 28$$
$$= (7x^2 + 1x) + (28x + 4) \qquad \text{Group terms with common factors.}$$
$$= x(7x + 1) + 4(7x + 1) \qquad \text{Factor the GCF.}$$
$$= (x + 4)(7x + 1) \qquad 7x + 1 \text{ is the common factor.}$$

b. $3x^2 + 15x + 18$

The GCF of the terms $3x^2$, $15x$, and 18 is 3. Factor this first.

$$3x^2 + 15x + 18 = 3(x^2 + 5x + 6) \qquad \text{Distributive Property}$$
$$= 3(x + 3)(x + 2) \qquad \text{Find two factors of 6 with a sum of 5.}$$

> **GuidedPractice**
>
> **1A.** $5x^2 + 13x + 6$ | **1B.** $6x^2 + 22x - 8$

> **StudyTip**
>
> **Greatest Common Factor**
> Always look for a GCF of the terms of a polynomial before you factor.

Sometimes the coefficient of the x-term is negative.

Example 2 Factor $ax^2 - bx + c$

Factor $3x^2 - 17x + 20$.

In this trinomial, $a = 3$, $b = -17$, and $c = 20$. Since b is negative, $m + p$ will be negative. Since c is positive, mp will be positive.

To determine m and p, list the negative factors of ac or 60. The sum of m and p should be -17.

Factors of 60	Sum of Factors
$-2, -30$	-32
$-3, -20$	-23
$-4, -15$	-19
$-5, -12$	-17

The correct factors are -5 and -12.

$$3x^2 - 17x + 20 = 3x^2 - 12x - 5x + 20 \qquad m = -12 \text{ and } p = -5$$
$$= (3x^2 - 12x) + (-5x + 20) \qquad \text{Group terms with common factors.}$$
$$= 3x(x - 4) + (-5)(x - 4) \qquad \text{Factor the GCF.}$$
$$= (3x - 5)(x - 4) \qquad \text{Distributive Property}$$

> **GuidedPractice**
>
> **2A.** $2n^2 - n - 1$ | **2B.** $10y^2 - 35y + 30$

Real-WorldCareer

Urban Planner Urban planners design the layout of an area. They take into consideration the available land and geographical and environmental factors to design an area that benefits the community the most. City planners have a bachelor's degree in planning and almost half have a master's degree.

LWA/Dann Tardif/Blend Images/Getty Images

A polynomial that cannot be written as a product of two polynomials with integral coefficients is called a **prime polynomial**.

PT

Example 3 Determine Whether a Polynomial is Prime

Factor $4x^2 - 3x + 5$, if possible. If the polynomial cannot be factored using integers, write *prime*.

In this trinomial, $a = 4$, $b = -3$, and $c = 5$.
Since b is negative, $m + p$ is negative. Since c is positive, mp is positive. So, m and p are both negative. Next, list the factors of 20. Look for the pair with a sum of -3.

Factors of 20	Sum of Factors
$-20, -1$	-21
$-4, -5$	-9
$-2, -10$	-12

There are no factors with a sum of -3. So the quadratic expression cannot be factored using integers. Therefore, $4x^2 - 3x + 5$ is prime.

▶ **Guided**Practice

Factor each polynomial, if possible. If the polynomial cannot be factored using integers, write *prime*.

3A. $4r^2 - r + 7$

3B. $2x^2 + 3x - 5$

2 **Solve Equations by Factoring** A model for the height of a projectile is given by $h = -16t^2 + vt + h_0$, where h is the height in feet, t is the time in seconds, v is the initial upward velocity in feet per second, and h_0 is the initial height in feet. Equations of the form $ax^2 + bx + c = 0$ can be solved by factoring and by using the Zero Product Property.

PT

● Real-World Example 4 Solve Equations by Factoring

WILDLIFE Suppose a cheetah pouncing on an antelope leaps with an initial upward velocity of 19 feet per second. How long is the cheetah in the air if it lands on the antelope's hind quarter, 3 feet from the ground?

$h = -16t^2 + vt + h_0$	Equation for height
$3 = -16t^2 + 19t + 0$	$h = 3$, $v = 19$, and $h_0 = 0$
$0 = -16t^2 + 19t - 3$	Subtract 3 from each side.
$0 = 16t^2 - 19t + 3$	Multiply each side by -1.
$0 = (16t - 3)(t - 1)$	Factor $16t^2 - 19t + 3$.
$16t - 3 = 0 \quad$ or $\quad t - 1 = 0$	Zero Product Property
$16t = 3 \qquad\qquad t = 1$	Solve each equation.
$t = \dfrac{3}{16}$	

The solutions are $\dfrac{3}{16}$ and 1 seconds. It takes the cheetah $\dfrac{3}{16}$ second to reach a height of 3 feet on his way up. It takes the cheetah 1 second to reach a height of 3 feet on his way down. So, the cheetah is in the air 1 second before he catches the antelope.

▶ **Guided**Practice

4. PHYSICAL SCIENCE A person throws a ball upward from a 506-foot tall building. The ball's height h in feet after t seconds is given by the equation $h = -16t^2 + 48t + 506$. The ball lands on a balcony that is 218 feet above the ground. How many seconds was it in the air?

Real-World Link

Cheetahs are the fastest land animals in the world, reaching speeds of up to 70 mph. It can accelerate from 0 to 40 mph in 3 strides. It takes just seconds for the cheetah to reach the full speed of 70 mph.

Source: Cheetah Conservation Fund

WatchOut!

Keep the -1 Do not forget to carry the -1 that was factored out through the rest of the steps or multiply both sides by -1.

Examples 1–3 **Factor each polynomial, if possible. If the polynomial cannot be factored using integers, write *prime*.**

1. $3x^2 + 17x + 10$ **2.** $2x^2 + 22x + 56$

3. $5x^2 - 3x + 4$ **4.** $3x^2 - 11x - 20$

Example 4 **Solve each equation. Confirm your answers using a graphing calculator.**

5. $2x^2 + 9x + 9 = 0$ **6.** $3x^2 + 17x + 20 = 0$

7. $3x^2 - 10x + 8 = 0$ **8.** $2x^2 - 17x + 30 = 0$

9. **CCSS MODELING** Ken throws the discus at a school meet.

　　a. What is the initial height of the discus?

　　b. After how many seconds does the discus hit the ground?

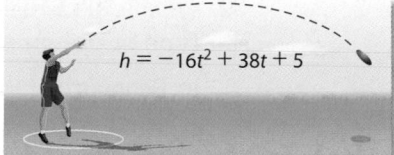
$h = -16t^2 + 38t + 5$

Examples 1–3 **Factor each polynomial, if possible. If the polynomial cannot be factored using integers, write *prime*.**

10. $5x^2 + 34x + 24$ **11** $2x^2 + 19x + 24$ **12.** $4x^2 + 22x + 10$

13. $4x^2 + 38x + 70$ **14.** $2x^2 - 3x - 9$ **15.** $4x^2 - 13x + 10$

16. $2x^2 + 3x + 6$ **17.** $5x^2 + 3x + 4$ **18.** $12x^2 + 69x + 45$

19. $4x^2 - 5x + 7$ **20.** $5x^2 + 23x + 24$ **21.** $3x^2 - 8x + 15$

Example 4 **22.** **SHOT PUT** An athlete throws a shot put with an initial upward velocity of 29 feet per second and from an initial height of 6 feet.

　　a. Write an equation that models the height of the shot put in feet with respect to time in seconds.

　　b. After how many seconds will the shot put hit the ground?

Solve each equation. Confirm your answers using a graphing calculator.

23. $2x^2 + 9x - 18 = 0$ **24.** $4x^2 + 17x + 15 = 0$

25. $-3x^2 + 26x = 16$ **26.** $-2x^2 + 13x = 15$

27. $-3x^2 + 5x = -2$ **28.** $-4x^2 + 19x = -30$

29. **BASKETBALL** When Jerald shoots a free throw, the ball is 6 feet from the floor and has an initial upward velocity of 20 feet per second. The hoop is 10 feet from the floor.

　　a. Use the vertical motion model to determine an equation that models Jerald's free throw.

　　b. How long is the basketball in the air before it reaches the hoop?

　　c. Raymond shoots a free throw that is 5 foot 9 inches from the floor with the same initial upward velocity. Will the ball be in the air more or less time? Explain.

30. **DIVING** Ben dives from a 36-foot platform. The equation $h = -16t^2 + 14t + 36$ models the dive. How long will it take Ben to reach the water?

31 **NUMBER THEORY** Six times the square of a number x plus 11 times the number equals 2. What are possible values of x?

Factor each polynomial, if possible. If the polynomial cannot be factored using integers, write *prime*.

32. $-6x^2 - 23x - 20$ **33.** $-4x^2 - 15x - 14$ **34.** $-5x^2 + 18x + 8$

35. $-6x^2 + 31x - 35$ **36.** $-4x^2 + 5x - 12$ **37.** $-12x^2 + x + 20$

38. URBAN PLANNING The city has commissioned the building of a rectangular park. The area of the park can be expressed as $660x^2 + 524x + 85$. Factor this expression to find binomials with integer coefficients that represent possible dimensions of the park. If $x = 8$, what is a possible perimeter of the park?

39. ⤵ **MULTIPLE REPRESENTATIONS** In this problem, you will explore factoring a special type of polynomial.

    a. Geometric Draw a square and label the sides a. Within this square, draw a smaller square that shares a vertex with the first square. Label the sides b. What are the areas of the two squares?

    b. Geometric Cut and remove the small square. What is the area of the remaining region?

    c. Analytical Draw a diagonal line between the inside corner and outside corner of the figure, and cut along this line to make two congruent pieces. Then rearrange the two pieces to form a rectangle. What are the dimensions?

    d. Analytical Write the area of the rectangle as the product of two binomials.

    e. Verbal Complete this statement: $a^2 - b^2 = \dots$ Why is this statement true?

H.O.T. Problems   Use Higher-Order Thinking Skills

40. **CCSS** **CRITIQUE** Zachary and Samantha are solving $6x^2 - x = 12$. Is either of them correct? Explain your reasoning.

Zachary	Samantha
$6x^2 - x = 12$	$6x^2 - x = 12$
$x(6x - 1) = 12$	$6x^2 - x - 12 = 0$
$x = 12$ or $6x - 1 = 12$	$(2x - 3)(3x + 4) = 0$
$6x = 13$	$2x - 3 = 0$ or $3x + 4 = 0$
$x = \dfrac{13}{6}$	$x = \dfrac{3}{2}$ $x = -\dfrac{4}{3}$

41. REASONING A square has an area of $9x^2 + 30xy + 25y^2$ square inches. The dimensions are binomials with positive integer coefficients. What is the perimeter of the square? Explain.

42. CHALLENGE Find all values of k so that $2x^2 + kx + 12$ can be factored as two binomials using integers.

43. 🖹 **WRITING IN MATH** What should you consider when solving a quadratic equation that models a real-world situation?

44. WRITING IN MATH Explain how to determine which values should be chosen for m and p when factoring a polynomial of the form $ax^2 + bx + c$.

45. Gridded Response Savannah has two sisters. One sister is 8 years older than her and the other sister is 2 years younger than her. The product of Savannah's sisters' ages is 56. How old is Savannah?

46. What is the product of $\frac{2}{3}a^3b^5$ and $\frac{3}{5}a^5b^2$?

A $\frac{2}{5}a^8b^7$

B $\frac{2}{5}a^2b^3$

C $\frac{2}{5}a^8b^3$

D $\frac{2}{5}a^2b^7$

47. What is the solution set of $x^2 + 2x - 24 = 0$?

F $\{-4, 6\}$ **H** $\{-3, 8\}$

G $\{3, -8\}$ **J** $\{4, -6\}$

48. Which is the solution set of $x \geq -2$?

A
−6−5−4−3−2−1 0 1 2 3 4

B
−6−5−4−3−2−1 0 1 2 3 4

C
−6−5−4−3−2−1 0 1 2 3 4

D
−6−5−4−3−2−1 0 1 2 3 4

Spiral Review

Factor each polynomial. (Lesson 8-6)

49. $x^2 - 9x + 14$

50. $n^2 - 8n + 15$

51. $x^2 - 5x - 24$

52. $z^2 + 15z + 36$

53. $r^2 + 3r - 40$

54. $v^2 + 16v + 63$

Solve each equation. Check your solutions. (Lesson 8-5)

55. $a(a - 9) = 0$

56. $(2y + 6)(y - 1) = 0$

57. $10x^2 - 20x = 0$

58. $8b^2 - 12b = 0$

59. $15a^2 = 60a$

60. $33x^2 = -22x$

61. ART A painter has 32 units of yellow dye and 54 units of blue dye to make two shades of green. The units needed to make a gallon of light green and a gallon of dark green are shown. Make a graph showing the numbers of gallons of the two greens she can make, and list three possible solutions. (Lesson 6-6)

Color	Units of Yellow Dye	Units of Blue Dye
light green	4	1
dark green	1	6

Solve each compound inequality. Then graph the solution set. (Lesson 5-4)

62. $k + 2 > 12$ and $k + 2 \leq 18$

63. $d - 4 > 3$ or $d - 4 \leq 1$

64. $3 < 2x - 3 < 15$

65. $3t - 7 \geq 5$ and $2t + 6 \leq 12$

66. $h - 10 < -21$ or $h + 3 < 2$

67. $4 < 2y - 2 < 10$

68. FINANCIAL LITERACY A home security company provides security systems for $5 per week, plus an installation fee. The total cost for installation and 12 weeks of service is $210. Write the point-slope form of an equation to find the total fee y for any number of weeks x. What is the installation fee? (Lesson 4-3)

Skills Review

Find the principal square root of each number.

69. 16

70. 36

71. 64

72. 81

73. 121

74. 100

Differences of Squares

:·Then

- You factored trinomials into two binomials.

:·Now

1. Factor binomials that are the difference of squares.

2. Use the difference of squares to solve equations.

:·Why?

- Computer graphics designers use a combination of art and mathematics skills to design images and videos. They use equations to form shapes and lines on computers. Factoring can help to determine the dimensions and shapes of the figures.

 **NewVocabulary**
difference of two squares

 Common Core State Standards

Content Standards
A.SSE.3a Factor a quadratic expression to reveal the zeros of the function it defines.

A.REI.4b Solve quadratic equations by inspection (e.g., for $x^2 = 49$), taking square roots, completing the square, the quadratic formula and factoring, as appropriate to the initial form of the equation. Recognize when the quadratic formula gives complex solutions and write them as $a \pm bi$ for real numbers a and b.

Mathematical Practices
1 Make sense of problems and persevere in solving them.

1 Factor Differences of Squares You have previously learned about the product of the sum and difference of two quantities. This resulting product is referred to as the **difference of two squares**. So, the factored form of the difference of squares is called the product of the sum and difference of the two quantities.

KeyConcept Difference of Squares

Symbols $\quad a^2 - b^2 = (a + b)(a - b)$ or $(a - b)(a + b)$

Examples $\quad x^2 - 25 = (x + 5)(x - 5)$ or $(x - 5)(x + 5)$

$\qquad\qquad t^2 - 64 = (t + 8)(t - 8)$ or $(t - 8)(t + 8)$

Example 1 Factor Differences of Squares

Factor each polynomial.

a. $16h^2 - 9a^2$

$16h^2 - 9a^2 = (4h)^2 - (3a)^2$ Write in the form of $a^2 - b^2$.

$\qquad\qquad = (4h + 3a)(4h - 3a)$ Factor the difference of squares.

b. $121 - 4b^2$

$121 - 4b^2 = (11)^2 - (2b)^2$ Write in the form of $a^2 - b^2$.

$\qquad\qquad = (11 - 2b)(11 + 2b)$ Factor the difference of squares.

c. $27g^3 - 3g$

Because the terms have a common factor, factor out the GCF first. Then proceed with other factoring techniques.

$27g^3 - 3g = 3g(9g^2 - 1)$ Factor out the GCF of $3g$.

$\qquad\qquad = 3g[(3g)^2 - (1)^2]$ Write in the form $a^2 - b^2$.

$\qquad\qquad = 3g(3g - 1)(3g + 1)$ Factor the difference of squares.

▸ **Guided**Practice

1A. $81 - c^2$ **1B.** $64g^2 - h^2$

1C. $9x^3 - 4x$ **1D.** $-4y^3 + 9y$

To factor a polynomial completely, a technique may need to be applied more than once. This also applies to the difference of squares pattern.

Example 2 Apply a Technique More than Once

Factor each polynomial.

a. $b^4 - 16$

$$b^4 - 16 = (b^2)^2 - (4)^2 \qquad \text{Write } b^4 - 16 \text{ in } a^2 - b^2 \text{ form.}$$
$$= (b^2 + 4)(b^2 - 4) \qquad \text{Factor the difference of squares.}$$
$$= (b^2 + 4)(b^2 - 2^2) \qquad b^2 - 4 \text{ is also a difference of squares.}$$
$$= (b^2 + 4)(b + 2)(b - 2) \qquad \text{Factor the difference of squares.}$$

> **Watch**Out!
>
> **Sum of Squares** The sum of squares, $a^2 + b^2$, does not factor into $(a + b)(a + b)$. The sum of squares is a prime polynomial and cannot be factored.

b. $625 - x^4$

$$625 - x^4 = (25)^2 - (x^2)^2 \qquad \text{Write } 625 - x^4 \text{ in } a^2 - b^2 \text{ form.}$$
$$= (25 + x^2)(25 - x^2) \qquad \text{Factor the difference of squares.}$$
$$= (25 + x^2)(5^2 - x^2) \qquad \text{Write } 25 - x^2 \text{ in } a^2 - b^2 \text{ form.}$$
$$= (25 + x^2)(5 - x)(5 + x) \qquad \text{Factor the difference of squares.}$$

▶ **Guided**Practice

2A. $y^4 - 1$ **2B.** $4a^4 - b^4$

2C. $81 - x^4$ **2D.** $16y^4 - 1$

Sometimes more than one factoring technique needs to be applied to ensure that a polynomial is factored completely.

Example 3 Apply Different Techniques

Factor each polynomial.

a. $5x^5 - 45x$

$$5x^5 - 45x = 5x(x^4 - 9) \qquad \text{Factor out GCF.}$$
$$= 5x[(x^2)^2 - (3)^2] \qquad \text{Write } x^4 - 9 \text{ in the form } a^2 - b^2.$$
$$= 5x(x^2 - 3)(x^2 + 3) \qquad \text{Factor the difference of squares.}$$

$x^2 - 3$ is not a difference of squares because 3 is not a perfect square.

b. $7x^3 + 21x^2 - 7x - 21$

$$7x^3 + 21x^2 - 7x - 21 \qquad \text{Original expression}$$
$$= 7(x^3 + 3x^2 - x - 3) \qquad \text{Factor out GCF.}$$
$$= 7[(x^3 + 3x^2) - (x + 3)] \qquad \text{Group terms with common factors.}$$
$$= 7[x^2(x + 3) - 1(x + 3)] \qquad \text{Factor each grouping.}$$
$$= 7(x + 3)(x^2 - 1) \qquad x + 3 \text{ is the common factor.}$$
$$= 7(x + 3)(x + 1)(x - 1) \qquad \text{Factor the difference of squares.}$$

▶ **Guided**Practice

3A. $2y^4 - 50$ **3B.** $6x^4 - 96$

3C. $2m^3 + m^2 - 50m - 25$ **3D.** $r^3 + 6r^2 + 11r + 66$

2 Solve Equations by Factoring

After factoring, you can apply the Zero Product Property to an equation that is written as the product of factors set equal to 0.

Test-TakingTip

 Sense-Making

Another method that can be used to solve this equation is to substitute each answer choice into the equation.

Standardized Test Example 4 Solve an Equation by Factoring

In the equation $y = x^2 - \frac{9}{16}$, which is a value of x when $y = 0$?

A $-\frac{9}{4}$ **B** 0 **C** $\frac{3}{4}$ **D** $\frac{9}{4}$

Read the Test Item

Replace y with 0 and then solve.

Solve the Test Item

$y = x^2 - \frac{9}{16}$ Original equation

$0 = x^2 - \frac{9}{16}$ Replace y with 0.

$0 = x^2 - \left(\frac{3}{4}\right)^2$ Write in the form $a^2 - b^2$.

$0 = \left(x + \frac{3}{4}\right)\left(x - \frac{3}{4}\right)$ Factor the difference of squares.

$0 = x + \frac{3}{4}$ or $0 = x - \frac{3}{4}$ Zero Product Property

$x = -\frac{3}{4}$ $x = \frac{3}{4}$ The correct answer is C.

▶ **GuidedPractice**

4. Which are the solutions of $18x^3 = 50x$?

F $0, \frac{5}{3}$ **G** $-\frac{5}{3}, \frac{5}{3}$ **H** $-\frac{5}{3}, \frac{5}{3}, 0$ **J** $-\frac{5}{3}, \frac{5}{3}, 1$

Check Your Understanding

◯ = **Step-by-Step Solutions begin on page R13.**

Examples 1–3 Factor each polynomial.

1. $x^2 - 9$

2. $4a^2 - 25$

3. $9m^2 - 144$

4. $2p^3 - 162p$

5. $u^4 - 81$

6. $2d^4 - 32f^4$

7 $20r^4 - 45n^4$

8. $256n^4 - c^4$

9. $2c^3 + 3c^2 - 2c - 3$

10. $f^3 - 4f^2 - 9f + 36$

11. $3t^3 + 2t^2 - 48t - 32$

12. $w^3 - 3w^2 - 9w + 27$

Example 4 **EXTENDED RESPONSE** During an accident, skid marks may result from sudden breaking. The formula $\frac{1}{24}s^2 = d$ approximates a vehicle's speed s in miles per hour given the length d in feet of the skid marks on dry concrete.

13. If skid marks on dry concrete are 54 feet long, how fast was the car traveling when the brakes were applied?

14. If the skid marks on dry concrete are 150 feet long, how fast was the car traveling when the brakes were applied?

Examples 1–3 Factor each polynomial.

15. $q^2 - 121$

16. $r^4 - k^4$

17. $6n^4 - 6$

18. $w^4 - 625$

19. $r^2 - 9t^2$

20. $2c^2 - 32d^2$

21. $h^3 - 100h$

22. $h^4 - 256$

23. $2x^3 - x^2 - 162x + 81$

24. $x^2 - 4y^2$

25. $7h^4 - 7p^4$

26. $3c^3 + 2c^2 - 147c - 98$

27. $6k^2h^4 - 54k^4$

28. $5a^3 - 20a$

29. $f^3 + 2f^2 - 64f - 128$

30. $3r^3 - 192r$

31. $10q^3 - 1210q$

32. $3xn^4 - 27x^3$

33. $p^3r^5 - p^3r$

34. $8c^3 - 8c$

35. $r^3 - 5r^2 - 100r + 500$

36. $3t^3 - 7t^2 - 3t + 7$

37. $a^2 - 49$

38. $4m^3 + 9m^2 - 36m - 81$

39. $3m^4 + 243$

40. $3x^3 + x^2 - 75x - 25$

41. $12a^3 + 2a^2 - 192a - 32$

42. $x^4 + 6x^3 - 36x^2 - 216x$

43. $15m^3 + 12m^2 - 375m - 300$

Example 4

44. GEOMETRY The drawing at the right is a square with a square cut out of it.

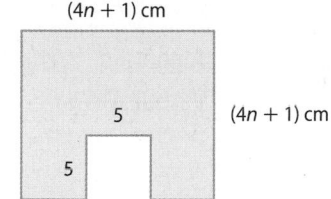

(4n + 1) cm

5

(4n + 1) cm

5

a. Write an expression that represents the area of the shaded region.

b. Find the dimensions of a rectangle with the same area as the shaded region in the drawing. Assume that the dimensions of the rectangle must be represented by binomials with integral coefficients.

45. DECORATIONS An arch decorated with balloons was used to decorate the gym for the spring dance. The shape of the arch can be modeled by the equation $y = -0.5x^2 + 4.5x$, where x and y are measured in feet and the x-axis represents the floor.

a. Write the expression that represents the height of the arch in factored form.

b. How far apart are the two points where the arch touches the floor?

c. Graph this equation on your calculator. What is the highest point of the arch?

46. CCSS SENSE-MAKING Zelda is building a deck in her backyard. The plans for the deck show that it is to be 24 feet by 24 feet. Zelda wants to reduce one dimension by a number of feet and increase the other dimension by the same number of feet. If the area of the reduced deck is 512 square feet, what are the dimensions of the deck?

47 SALES The sales of a particular CD can be modeled by the equation $S = -25m^2 + 125m$, where S is the number of CDs sold in thousands, and m is the number of months that it is on the market.

a. In what month should the music store expect the CD to stop selling?

b. In what month will CD sales peak?

c. How many copies will the CD sell at its peak?

Solve each equation by factoring. Confirm your answers using a graphing calculator.

48. $36w^2 = 121$

49 $100 = 25x^2$

50. $64x^2 - 1 = 0$

51. $4y^2 - \dfrac{9}{16} = 0$

52. $\dfrac{1}{4}b^2 = 16$

53. $81 - \dfrac{1}{25}x^2 = 0$

54. $9d^2 - 81 = 0$

55. $4a^2 = \dfrac{9}{64}$

56. ⚙ **MULTIPLE REPRESENTATIONS** In this problem, you will investigate perfect square trinomials.

   a. Tabular Copy and complete the table below by factoring each polynomial. Then write the first and last terms of the given polynomials as perfect squares.

   b. Analytical Write the middle term of each polynomial using the square roots of the perfect squares of the first and last terms.

Polynomial	Factored Polynomial	First Term	Last Term	Middle Term
$4x^2 + 12x + 9$	$(2x + 3)(2x + 3)$	$4x^2 = (2x)^2$	$9 = 3^2$	
$9x^2 - 24x + 16$				
$4x^2 - 20x + 25$				
$16x^2 + 24x + 9$				
$25x^2 + 20x + 4$				

   c. Algebraic Write the pattern for a perfect square trinomial.

   d. Verbal What conditions must be met for a trinomial to be classified as a perfect square trinomial?

H.O.T. Problems   *Use Higher-Order Thinking Skills*

57. ERROR ANALYSIS Elizabeth and Lorenzo are factoring an expression. Is either of them correct? Explain your reasoning.

> **Elizabeth**
> $16x^4 - 25y^2 =$
> $(4x - 5y)(4x + 5y)$

> **Lorenzo**
> $16x^4 - 25y^2 =$
> $(4x^2 - 5y)(4x^2 + 5y)$

58. CHALLENGE Factor and simplify $9 - (k + 3)^2$, a difference of squares.

59. ⒸⒸⓈⓈ **PERSEVERANCE** Factor $x^{16} - 81$.

60. REASONING Write and factor a binomial that is the difference of two perfect squares and that has a greatest common factor of $5mk$.

61. REASONING Determine whether the following statement is *true* or *false*. Give an example or counterexample to justify your answer.

   All binomials that have a perfect square in each of the two terms can be factored.

62. OPEN ENDED Write a binomial in which the difference of squares pattern must be repeated to factor it completely. Then factor the binomial.

63. WRITING IN MATH Describe why the difference of squares pattern has no middle term with a variable.

64. One of the roots of $2x^2 + 13x = 24$ is -8. What is the other root?

A $-\dfrac{3}{2}$ C $\dfrac{2}{3}$

B $-\dfrac{2}{3}$ D $\dfrac{3}{2}$

65. Which of the following is the sum of both solutions of the equation $x^2 + 3x = 54$?

F -21 H 3

G -3 J 21

66. What are the x-intercepts of the graph of $y = -3x^2 + 7x + 20$?

A $\dfrac{5}{3}, -4$ C $-\dfrac{5}{3}, 4$

B $-\dfrac{5}{3}, -4$ D $\dfrac{5}{3}, 4$

67. EXTENDED RESPONSE Two cars leave Cleveland at the same time from different parts of the city and both drive to Cincinnati. The distance in miles of the cars from the center of Cleveland can be represented by the two equations below, where t represents the time in hours.

Car A: $65t + 15$ Car B: $60t + 25$

a. Which car is faster? Explain.

b. Find an expression that models the distance between the two cars.

c. How far apart are the cars after $2\frac{1}{2}$ hours?

Factor each trinomial, if possible. If the trinomial cannot be factored using integers, write *prime*. (Lesson 8-7)

68. $5x^2 - 17x + 14$

69. $5a^2 - 3a + 15$

70. $10x^2 - 20xy + 10y^2$

Solve each equation. Check your solutions. (Lesson 8-6)

71. $n^2 - 9n = -18$

72. $10 + a^2 = -7a$

73. $22x - x^2 = 96$

74. SAVINGS Victoria and Trey each want to buy a scooter. In how many weeks will Victoria and Trey have saved the same amount of money, and how much will each of them have saved? (Lesson 6-2)

Trey $8 per week $18 so far

Victoria $5 per week $25 so far

Solve each inequality. Graph the solution set on a number line. (Lesson 5-1)

75. $t + 14 \geq 18$

76. $d + 5 \leq 7$

77. $-5 + k > -1$

78. $5 < 3 + g$

79. $2 \leq -1 + m$

80. $2y > -8 + y$

81. FITNESS Silvia is beginning an exercise program that calls for 20 minutes of walking each day for the first week. Each week thereafter, she has to increase her daily walking for a week by 7 minutes. In which week will she first walk over an hour a day? (Lesson 3-5)

Find each product.

82. $(x - 6)^2$

83. $(x - 2)(x - 2)$

84. $(x + 3)(x + 3)$

85. $(2x - 5)^2$

86. $(6x - 1)^2$

87. $(4x + 5)(4x + 5)$

Perfect Squares

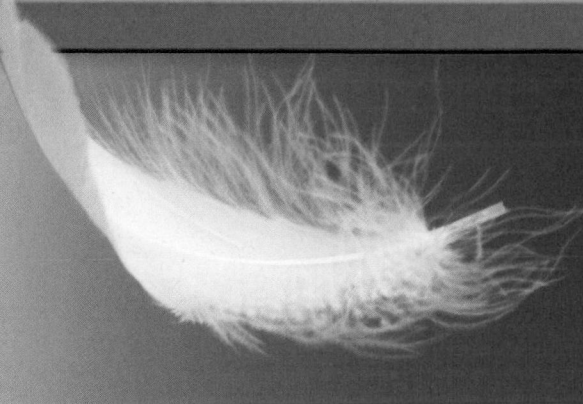

:·Then	:·Now	:·Why?
● You found the product of a sum and difference.	**1** Factor perfect square trinomials. **2** Solve equations involving perfect squares.	● In a vacuum, a feather and a piano would fall at the same speed, or velocity. To find about how long it takes an object to hit the ground if it is dropped from an initial height of h_0 feet above ground, you would need to solve the equation $0 = -16t^2 + h_0$, where t is time in seconds after the object is dropped.

NewVocabulary
perfect square trinomial

Common Core State Standards

Content Standards
A.SSE.3a Factor a quadratic expression to reveal the zeros of the function it defines.

A.REI.1 Explain each step in solving a simple equation as following from the equality of numbers asserted at the previous step, starting from the assumption that the original equation has a solution. Construct a viable argument to justify a solution method.

Mathematical Practices
6 Attend to precision.

1 Factor Perfect Square Trinomials You have learned the patterns for the products of the binomials $(a + b)^2$ and $(a - b)^2$. Recall that these are special products that follow specific patterns.

$$(a + b)^2 = (a + b)(a + b)$$
$$= a^2 + ab + ab + b^2$$
$$= a^2 + 2ab + b^2$$

$$(a - b)^2 = (a - b)(a - b)$$
$$= a^2 - ab - ab + b^2$$
$$= a^2 - 2ab + b^2$$

These products are called **perfect square trinomials**, because they are the squares of binomials. The above patterns can help you factor perfect square trinomials.

For a trinomial to be factorable as a perfect square, the first and last terms must be perfect squares and the middle term must be two times the square roots of the first and last terms.

The trinomial $16x^2 + 24x + 9$ is a perfect square trinomial, as illustrated below.

$$16x^2 + 24x + 9$$

Is the first term a perfect square? Yes, because $16x^2 = (4x)^2$.	Is the middle term twice the product of the square roots of the first and last terms? Yes, because $24x = 2(4x)(3)$.	Is the last term a perfect square? Yes, because $9 = 3^2$.

◆ KeyConcept Factoring Perfect Square Trinomials

Symbols $a^2 + 2ab + b^2 = (a + b)(a + b) = (a + b)^2$

$a^2 - 2ab + b^2 = (a - b)(a - b) = (a - b)^2$

Examples $x^2 + 8x + 16 = (x + 4)(x + 4)$ or $(x + 4)^2$

$x^2 - 6x + 9 = (x - 3)(x - 3)$ or $(x - 3)^2$

Example 1 Recognize and Factor Perfect Square Trinomials

Determine whether each trinomial is a perfect square trinomial. Write *yes* or *no*. If so, factor it.

a. $4y^2 + 12y + 9$

① Is the first term a perfect square? Yes, $4y^2 = (2y)^2$.

② Is the last term a perfect square? Yes, $9 = 3^2$.

③ Is the middle term equal to $2(2y)(3)$? Yes, $12y = 2(2y)(3)$

Since all three conditions are satisfied, $4y^2 + 12y + 9$ is a perfect square trinomial.

$$4y^2 + 12y + 9 = (2y)^2 + 2(2y)(3) + 3^2 \qquad \text{Write as } a^2 + 2ab + b^2.$$
$$= (2y + 3)^2 \qquad \text{Factor using the pattern.}$$

b. $9x^2 - 6x + 4$

① Is the first term a perfect square? Yes, $9x^2 = (3x)^2$.

② Is the last term a perfect square? Yes, $4 = 2^2$.

③ Is the middle term equal to $-2(3x)(2)$? No, $-6x \neq -2(3x)(2)$.

Since the middle term does not satisfy the required condition, $9x^2 - 6x + 4$ is not a perfect square trinomial.

GuidedPractice

1A. $9y^2 + 24y + 16$

1B. $2a^2 + 10a + 25$

A polynomial is completely factored when it is written as a product of prime polynomials. More than one method might be needed to factor a polynomial completely. When completely factoring a polynomial, the Concept Summary can help you decide where to start.

Remember, if the polynomial does not fit any pattern or cannot be factored, the polynomial is prime.

ConceptSummary Factoring Methods

Steps	Number of Terms	Examples
Step 1 Factor out the GCF.	any	$4x^3 + 2x^2 - 6x = 2x(2x^2 + x - 3)$
Step 2 Check for a difference of squares or a perfect square trinomial.	2 or 3	$9x^2 - 16 = (3x + 4)(3x - 4)$ $16x^2 + 24x + 9 = (4x + 3)^2$
Step 3 Apply the factoring patterns for $x^2 + bx + c$ or $ax^2 + bx + c$ (general trinomials), or factor by grouping.	3 or 4	$x^2 - 8x + 12 = (x - 2)(x - 6)$ $2x^2 + 13x + 6 = (2x + 1)(x + 6)$ $12y^2 + 9y + 8y + 6$ $\quad = (12y^2 + 9y) + (8y + 6)$ $\quad = 3y(4y + 3) + 2(4y + 3)$ $\quad = (4y + 3)(3y + 2)$

Example 2 Factor Completely

Factor each polynomial, if possible. If the polynomial cannot be factored, write *prime*.

a. $5x^2 - 80$

  Step 1 The GCF of $5x^2$ and -80 is 5, so factor it out.

  Step 2 Since there are two terms, check for a difference of squares.

$$5x^2 - 80 = 5(x^2 - 16) \qquad \text{5 is the GCF of the terms.}$$
$$= 5(x^2 - 4^2) \qquad x^2 = x \cdot x \text{ and } 16 = 4 \cdot 4$$
$$= 5(x - 4)(x + 4) \qquad \text{Factor the difference of squares.}$$

b. $9x^2 - 6x - 35$

  Step 1 The GCF of $9x^2$, $-6x$, and -35 is 1.

  Step 2 Since 35 is not a perfect square, this is not a perfect square trinomial.

  Step 3 Factor using the pattern $ax^2 + bx + c$. Are there two numbers with a product of $9(-35)$ or -315 and a sum of -6? Yes, the product of 15 and -21 is -315, and the sum is -6.

$$9x^2 - 6x - 35 = 9x^2 + mx + px - 35 \qquad \text{Write the pattern.}$$
$$= 9x^2 + 15x - 21x - 35 \qquad m = 15 \text{ and } n = -21$$
$$= (9x^2 + 15x) + (-21x - 35) \qquad \text{Group terms with common factors.}$$
$$= 3x(3x + 5) - 7(3x + 5) \qquad \text{Factor out the GCF from each grouping.}$$
$$= (3x + 5)(3x - 7) \qquad 3x + 5 \text{ is the common factor.}$$

▶ **Guided**Practice

2A. $2x^2 - 32$ **2B.** $12x^2 + 5x - 25$

> **Study**Tip
>
> **Check Your Answer** You can check your answer by:
> - Using the FOIL method.
> - Using the Distributive Property.
> - Graphing the original expression and factored expression and comparing the graphs.
>
> If the product of the factors does not match the original expression exactly, the answer is incorrect.

2 Solve Equations with Perfect Squares When solving equations involving repeated factors, it is only necessary to set one of the repeated factors equal to zero.

Example 3 Solve Equations with Repeated Factors

Solve $9x^2 - 48x = -64$.

$$9x^2 - 48x = -64 \qquad \text{Original equation}$$
$$9x^2 - 48x + 64 = 0 \qquad \text{Add 64 to each side.}$$
$$(3x)^2 - 2(3x)(8) + (8)^2 = 0 \qquad \text{Recognize } 9x^2 - 48x + 64 \text{ as a perfect square trinomial.}$$
$$(3x - 8)^2 = 0 \qquad \text{Factor the perfect square trinomial.}$$
$$(3x - 8)(3x - 8) = 0 \qquad \text{Write } (3x - 8)^2 \text{ as two factors.}$$
$$3x - 8 = 0 \qquad \text{Set the repeated factor equal to zero.}$$
$$3x = 8 \qquad \text{Add 8 to each side.}$$
$$x = \frac{8}{3} \qquad \text{Divide each side by 3.}$$

▶ **Guided**Practice

Solve each equation. Check your solutions.

3A. $a^2 + 12a + 36 = 0$ **3B.** $y^2 - \frac{4}{3}y + \frac{4}{9} = 0$

You have solved equations like $x^2 - 16 = 0$ by factoring. You can also use the definition of a square root to solve the equation.

$$x^2 - 16 = 0 \qquad \text{Original equation}$$
$$x^2 = 16 \qquad \text{Add 16 to each side.}$$
$$x = \pm\sqrt{16} \qquad \text{Take the square root of each side.}$$

Remember that there are two square roots of 16, namely 4 and -4. Therefore, the solution set is $\{-4, 4\}$. You can express this as $\{\pm 4\}$.

KeyConcept Square Root Property

Words	To solve a quadratic equation in the form $x^2 = n$, take the square root of each side.
Symbols	For any number $n \geq 0$, if $x^2 = n$, then $x = \pm\sqrt{n}$.
Example	$x^2 = 25$
	$x = \pm\sqrt{25}$ or ± 5

In the equation $x^2 = n$, if n is not a perfect square, you need to approximate the square root. Use a calculator to find an approximation. If n is a perfect square, you will have an exact answer.

Example 4 Use the Square Root Property

Solve each equation. Check your solutions.

a. $(y - 6)^2 = 81$

$$(y - 6)^2 = 81 \qquad \text{Original equation}$$
$$y - 6 = \pm\sqrt{81} \qquad \text{Square Root Property}$$
$$y - 6 = \pm 9 \qquad 81 = 9 \cdot 9$$
$$y = 6 \pm 9 \qquad \text{Add 6 to each side.}$$
$$y = 6 + 9 \quad \text{or} \quad y = 6 - 9 \qquad \text{Separate into two equations.}$$
$$= 15 \qquad\qquad = -3 \qquad \text{Simplify.}$$

The roots are 15 and -3.   Check in the original equation.

b. $(x + 6)^2 = 12$

$$(x + 6)^2 = 12 \qquad \text{Original equation}$$
$$x + 6 = \pm\sqrt{12} \qquad \text{Square Root Property}$$
$$x = -6 \pm\sqrt{12} \qquad \text{Subtract 6 from each side.}$$

The roots are $-6 \pm\sqrt{12}$ or $-6 + \sqrt{12}$ and $-6 - \sqrt{12}$.

Using a calculator, $-6 + \sqrt{12} \approx -2.54$ and $-6 - \sqrt{12} \approx -9.46$.

GuidedPractice

4A. $(a - 10)^2 = 121$ **4B.** $(z + 3)^2 = 26$

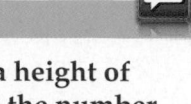

Math History Link

Galileo Galilei (1564–1642) Galileo was the first person to prove that objects of different weights fall at the same velocity by dropping two objects of different weights from the top of the Leaning Tower of Pisa in 1589.

Real-World Example 5 Solve an Equation

PHYSICAL SCIENCE During an experiment, a ball is dropped from a height of 205 feet. The formula $h = -16t^2 + h_0$ can be used to approximate the number of seconds t it takes for the ball to reach height h from an initial height of h_0 in feet. Find the time it takes the ball to reach the ground.

At ground level, $h = 0$ and the initial height is 205, so $h_0 = 205$.

$h = -16t^2 + h_0$	Original Formula
$0 = -16t^2 + 205$	Replace h with 0 and h_0 with 205.
$-205 = -16t^2$	Subtract 205 from each side.
$12.8125 = t^2$	Divide each side by -16.
$\pm 3.6 \approx t$	Use the Square Root Property.

Since a negative number does not make sense in this situation, the solution is 3.6. It takes about 3.6 seconds for the ball to reach the ground.

▶ **Guided**Practice

5. Find the time it takes a ball to reach the ground if it is dropped from a bridge that is half as high as the one described above.

Check Your Understanding

 = Step-by-Step Solutions begin on page R13.

Example 1 Determine whether each trinomial is a perfect square trinomial. Write *yes* or *no*. If so, factor it.

1. $25x^2 + 60x + 36$ **2.** $6x^2 + 30x + 36$

Example 2 Factor each polynomial, if possible. If the polynomial cannot be factored, write *prime*.

3. $2x^2 - x - 28$ **4.** $6x^2 - 34x + 48$

5. $4x^2 + 64$ **6.** $4x^2 + 9x - 16$

Examples 3–4 Solve each equation. Confirm your answers using a graphing calculator.

7. $4x^2 = 36$ **8.** $25a^2 - 40a = -16$

9. $64y^2 - 48y + 18 = 9$ **10.** $(z + 5)^2 = 47$

Example 5 **11.** **CCSS** **REASONING** While painting his bedroom, Nick drops his paintbrush off his ladder from a height of 6 feet. Use the formula $h = -16t^2 + h_0$ to approximate the number of seconds it takes for the paintbrush to hit the floor.

Practice and Problem Solving

Extra Practice is on page R8.

Example 1 Determine whether each trinomial is a perfect square trinomial. Write *yes* or *no*. If so, factor it.

12. $4x^2 - 42x + 110$ **13.** $16x^2 - 56x + 49$

14. $81x^2 - 90x + 25$ **15** $x^2 + 26x + 168$

Example 2

Factor each polynomial, if possible. If the polynomial cannot be factored, write *prime*.

16. $24d^2 + 39d - 18$

17. $8x^2 + 10x - 21$

18. $2b^2 + 12b - 24$

19. $8y^2 - 200z^2$

20. $16a^2 - 121b^2$

21. $12m^3 - 22m^2 - 70m$

22. $8c^2 - 88c + 242$

23. $12x^2 - 84x + 147$

24. $w^4 - w^2$

25. $12p^3 - 3p$

26. $16q^3 - 48q^2 + 36q$

27. $4t^3 + 10t^2 - 84t$

28. $x^3 + 2x^2y - 4x - 8y$

29. $2a^2b^2 - 2a^2 - 2ab^3 + 2ab$

30. $2r^3 - r^2 - 72r + 36$

31. $3k^3 - 24k^2 + 48k$

32. $4c^4d - 10c^3d + 4c^2d^3 - 10cd^3$

33. $g^2 + 2g - 3h^2 + 4h$

Solve each equation. Confirm your answers using a graphing calculator.

34. $4m^2 - 24m + 36 = 0$

35 $(y - 4)^2 = 7$

36. $a^2 + \frac{10}{7}a + \frac{25}{49} = 0$

37. $x^2 - \frac{3}{2}x + \frac{9}{16} = 0$

38. $x^2 + 8x + 16 = 25$

39. $5x^2 - 60x = -180$

40. $4x^2 = 80x - 400$

41. $9 - 54x = -81x^2$

42. $4c^2 + 4c + 1 = 15$

43. $x^2 - 16x + 64 = 6$

44. PHYSICAL SCIENCE For an experiment in physics class, a water balloon is dropped from the window of the school building. The window is 40 feet high. How long does it take until the balloon hits the ground? Round to the nearest hundredth.

45. SCREENS The area A in square feet of a projected picture on a movie screen can be modeled by the equation $A = 0.25d^2$, where d represents the distance from a projector to a movie screen. At what distance will the projected picture have an area of 100 square feet?

Example 5 **46. GEOMETRY** The area of a square is represented by $9x^2 - 42x + 49$. Find the length of each side.

47. GEOMETRY The area of a square is represented by $16x^2 + 40x + 25$. Find the length of each side.

48. GEOMETRY The volume of a rectangular prism is represented by the expression $8y^3 + 40y^2 + 50y$. Find the possible dimensions of the prism if the dimensions are represented by polynomials with integer coefficients.

49 POOLS Ichiro wants to buy an above-ground swimming pool for his yard. Model A is 42 inches deep and holds 1750 cubic feet of water. The length of the rectangular pool is 5 feet more than the width.

a. What is the surface area of the water?

b. What are the dimensions of the pool?

c. Model B pool holds twice as much water as Model A. What are some possible dimensions for this pool?

d. Model C has length and width that are both twice as long as Model A, but the height is the same. What is the ratio of the volume of Model A to Model C?

50. GEOMETRY Use the rectangular prism at the right.

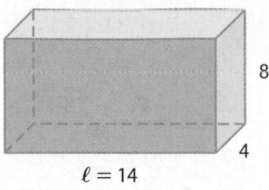

a. Write an expression for the height and width of the prism in terms of the length, ℓ.

b. Write a polynomial for the volume of the prism in terms of the length.

51. CCSS PRECISION A zoo has an aquarium shaped like a rectangular prism. It has a volume of 180 cubic feet. The height of the aquarium is 9 feet taller than the width, and the length is 4 feet shorter than the width. What are the dimensions of the aquarium?

52 ELECTION For the student council elections, Franco is building the voting box shown with a volume of 96 cubic inches. What are the dimensions of the voting box?

H.O.T. Problems   Use Higher-Order Thinking Skills

53. ERROR ANALYSIS Debbie and Adriano are factoring the expression $x^8 - x^4$ completely. Is either of them correct? Explain your reasoning.

Debbie	Adriano
$x^8 - x^4 = x^4(x^2 + 1)(x^2 - 1)$	$x^8 - x^4 = x^4(x^2 + 1)(x - 1)(x + 1)$

54. CHALLENGE Factor $x^{n+6} + x^{n+2} + x^n$ completely.

55. OPEN ENDED Write a perfect square trinomial equation in which the coefficient of the middle term is negative and the last term is a fraction. Solve the equation.

56. REASONING A counterexample is a specific case in which a statement is false. Find a counterexample to the following statement.

A polynomial equation of degree three always has three real solutions.

57. CCSS REGULARITY Explain how to factor a polynomial completely.

58. WHICH ONE DOESN'T BELONG? Identify the trinomial that does not belong. Explain.

| $4x^2 - 36x + 81$ | $25x^2 + 10x + 1$ | $4x^2 + 10x + 4$ | $9x^2 - 24x + 16$ |

59. OPEN ENDED Write a binomial that can be factored using the difference of two squares twice. Set your binomial equal to zero and solve the equation.

60. WRITING IN MATH Explain how to determine whether a trinomial is a perfect square trinomial.

61. What is the solution set for the equation $(x - 3)^2 = 25$?

    A $\{-8, 2\}$ **C** $\{4, 14\}$

    B $\{-2, 8\}$ **D** $\{-4, 14\}$

62. SHORT RESPONSE Write an equation in slope-intercept form for the graph shown below.

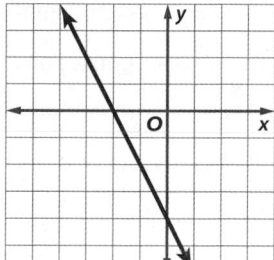

63. At an amphitheater, the price of 2 lawn seats and 2 pavilion seats is $120. The price of 3 lawn seats and 4 pavilion seats is $225. How much do lawn and pavilion seats cost?

    F $20 and $41.25

    G $10 and $50

    H $15 and $45

    J $30 and $30

64. GEOMETRY The circumference of a circle is $\frac{6\pi}{5}$ units. What is the area of the circle?

    A $\frac{9\pi}{25}$ units^2 **B** $\frac{3\pi}{5}$ units^2

    C $\frac{6\pi}{5}$ units^2 **D** $\frac{12\pi}{5}$ units^2

Factor each polynomial, if possible. If the polynomial cannot be factored, write *prime*. (Lesson 8-8)

65. $x^2 - 16$

66. $4x^2 - 81y^2$

67. $1 - 100p^2$

68. $3a^2 - 20$

69. $25n^2 - 1$

70. $36 - 9c^2$

Solve each equation. Confirm your answers using a graphing calculator. (Lesson 8-7)

71. $4x^2 - 8x - 32 = 0$

72. $6x^2 - 48x + 90 = 0$

73. $14x^2 + 14x = 28$

74. $2x^2 - 10x = 48$

75. $5x^2 - 25x = -30$

76. $8x^2 - 16x - 192$

SOUND The intensity of sound can be measured in watts per square meter. The table gives the watts per square meter for some common sounds. (Lesson 7-2)

77. How many times more intense is the sound from busy street traffic than sound from normal conversation?

78. Which sound is 10,000 times as loud as a busy street traffic?

79. How does the intensity of a whisper compare to that of normal conversation?

Watts Per Square Meter	Common Sounds
10^{-11}	rustling leaves
10^{-10}	whisper
10^{-6}	normal conversation
10^{-5}	busy street traffic
10^{-4}	vacuum cleaner
10^{-1}	front rows of rock concert
10^1	threshold of pain
10^2	military jet takeoff

Find the slope of the line that passes through each pair of points.

80. $(5, 7), (-2, -3)$

81. $(2, -1), (5, -3)$

82. $(-4, -1), (-3, -3)$

83. $(-3, -4), (5, -1)$

84. $(-2, 3), (8, 3)$

85. $(-5, 4), (-5, -1)$

Study Guide

KeyConcepts

Operations with Polynomials (Lessons 8-1 through 8-4)

- To add or subtract polynomials, add or subtract like terms.
- To multiply polynomials, use the Distributive Property.
- Special products: $(a + b)^2 = a^2 + 2ab + b^2$
  $(a - b)^2 = a^2 - 2ab + b^2$
  $(a + b)(a - b) = a^2 - b^2$

Factoring Using the Distributive Property (Lesson 8-5)

- Using the Distributive Property to factor polynomials with four or more terms is called factoring by grouping.
  $ax + bx + ay + by = x(a + b) + y(a + b)$
  $= (a + b)(x + y)$

Solving Quadratic Equations by Factoring
(Lessons 8-6 through 8-8)

- To factor $x^2 + bx + c$, find m and p with a sum of b and a product of c. Then write $x^2 + bx + c$ as $(x + m)(x + p)$.

- To factor $ax^2 + bx + c$, find m and p with a sum of b and a product of ac. Then write as $ax^2 + mx + px + c$ and factor by grouping.

- $a^2 - b^2 = (a - b)(a + b)$

Perfect Squares and Factoring (Lesson 8-9)

- For a trinomial to be a perfect square, the first and last terms must be perfect squares, and the middle term must be twice the product of the square roots of the first and last terms.

- For any number $n \geq 0$, if $x^2 = n$, then $x = \pm\sqrt{n}$.

FOLDABLES StudyOrganizer

Be sure the Key Concepts are noted in your Foldable.

Lesson 8-3

KeyVocabulary

binomial (p. 465)

degree of a monomial (p. 465)

degree of a polynomial (p. 465)

difference of two squares (p. 516)

factoring (p. 494)

factoring by grouping (p. 495)

FOIL method (p. 481)

leading coefficient (p. 466)

perfect square trinomial (p. 522)

polynomial (p. 465)

prime polynomial (p. 512)

quadratic equation (p. 506)

quadratic expression (p. 481)

Square Root Property (p. 525)

standard form of a polynomial (p. 466)

trinomial (p. 465)

Zero Product Property (p. 496)

VocabularyCheck

State whether each sentence is *true* or *false*. If *false*, replace the underlined phrase or expression to make a true sentence.

1. $x^2 + 5x + 6$ is an example of a prime polynomial.

2. $(x + 5)(x - 5)$ is the factorization of a difference of squares.

3. $4x^2 - 2x + 7$ is a polynomial of degree $\underline{2}$.

4. $(x + 5)(x - 2)$ is the factored form of $x^2 - 3x - 10$.

5. Expressions with four or more unlike terms can sometimes be factored by grouping.

6. The Zero Product Property states that if $\underline{ab = 1}$, then $\underline{a}$ or $\underline{b}$ is 1.

7. $x^2 - 12x + 36$ is an example of a perfect square trinomial.

8. The leading coefficient of $1 + 6a + 9a^2$ is $\underline{1}$.

9. $x^2 - 16$ is an example of a perfect square trinomial.

10. The FOIL method is used to multiply two trinomials.

Lesson-by-Lesson Review

8-1 **Adding and Subtracting Polynomials**

Write each polynomial in standard form.

11. $x + 2 + 3x^2$ **12.** $1 - x^4$

13. $2 + 3x + x^2$ **14.** $3x^5 - 2 + 6x - 2x^2 + x^3$

Find each sum or difference.

15. $(x^3 + 2) + (-3x^3 - 5)$

16. $a^2 + 5a - 3 - (2a^2 - 4a + 3)$

17. $(4x - 3x^2 + 5) + (2x^2 - 5x + 1)$

18. PICTURE FRAMES Jean is framing a painting that is a rectangle. What is the perimeter of the frame?

$5x + 3$

$2x^2 - 3x + 1$

Example 1

Write $3 - x^2 + 4x$ in standard form.

Step 1 Find the degree of each term.

3:	degree 0
$-x^2$:	degree 2
4x:	degree 1

Step 2 Write the terms in descending order of degree.

$3 - x^2 + 4x = -x^2 + 4x + 3$

Example 2

Find $(8r^2 + 3r) - (10r^2 - 5)$.

$(8r^2 + 3r) - (10r^2 - 5)$

$= (8r^2 + 3r) + (-10r^2 + 5)$ Use the additive inverse.

$= (8r^2 - 10r^2) + 3r + 5$ Group like terms.

$= -2r^2 + 3r + 5$ Add like terms.

8-2 **Multiplying a Polynomial by a Monomial**

Solve each equation.

19. $x^2(x + 2) = x(x^2 + 2x + 1)$

20. $2x(x + 3) - 2(x^2 + 3)$

21. $2(4w + w^2) - 6 = 2w(w - 4) + 10$

22. GEOMETRY Find the area of the rectangle.

$3x$

$x^2 + x - 7$

Example 3

Solve $m(2m - 5) + m = 2m(m - 6) + 16$.

$m(2m - 5) + m = 2m(m - 6) + 16$

$2m^2 - 5m + m = 2m^2 - 12m + 16$

$2m^2 - 4m = 2m^2 - 12m + 16$

$4m = -12m + 16$

$8m = 16$

$m = 2$

8-3 **Multiplying Polynomials**

Find each product.

23. $(x - 3)(x + 7)$ **24.** $(3a - 2)(6a + 5)$

25. $(3r - 7t)(2r + 5t)$ **26.** $(2x + 5)(5x + 2)$

27. PARKING LOT The parking lot shown is to be paved. What is the area to be paved?

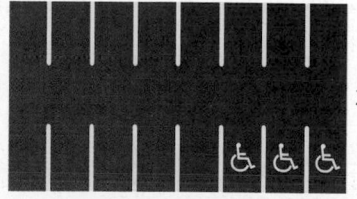

$2x + 3$

$5x - 4$

Example 4

Find $(6x - 5)(x + 4)$.

$(6x - 5)(x + 4)$

          F   O   I   L

$= (6x)(x) + (6x)(4) + (-5)(x) + (-5)(4)$

$= 6x^2 + 24x - 5x - 20$ Multiply.

$= 6x^2 + 19x - 20$ Combine like terms.

8-4 Special Products

Find each product.

28. $(x + 5)(x - 5)$

29. $(3x - 2)^2$

30. $(5x + 4)^2$

31. $(2x - 3)(2x + 3)$

32. $(2r + 5t)^2$

33. $(3m - 2)(3m + 2)$

34. GEOMETRY Write an expression to represent the area of the shaded region.

2x + 5

x + 2

x − 2

2x − 5

Example 5

Find $(x - 7)^2$.

$(a - b)^2 = a^2 - 2ab + b^2$ Square of a Difference

$(x - 7)^2 = x^2 - 2(x)(7) + (-7)^2$ $a = x$ and $b = 7$

$\qquad\quad = x^2 - 14x + 49$ Simplify.

Example 6

Find $(5a - 4)(5a + 4)$.

$(a + b)(a - b) = a^2 - b^2$ Product of a Sum and Difference

$(5a - 4)(5a + 4) = (5a)^2 - (4)^2$ $a = 5a$ and $b = 4$

$\qquad\qquad\qquad = 25a^2 - 16$ Simplify.

8-5 Using the Distributive Property

Use the Distributive Property to factor each polynomial.

35. $12x + 24y$

36. $14x^2y - 21xy + 35xy^2$

37. $8xy - 16x^3y + 10y$

38. $a^2 - 4ac + ab - 4bc$

39. $2x^2 - 3xz - 2xy + 3yz$

40. $24am - 9an + 40bm - 15bn$

Solve each equation. Check your solutions.

41. $x(3x - 6) = 0$

42. $6x^2 = 12x$

43. $x^2 = 3x$

44. $3x^2 = 5x$

45. GEOMETRY The area of the rectangle shown is $x^3 - 2x^2 + 5x$ square units. What is the length?

x

Example 7

Factor $12y^2 + 9y + 8y + 6$.

$12y^2 + 9y + 8y + 6$

$= (12y^2 + 9y) + (8y + 6)$ Group terms with common factors.

$= 3y(4y + 3) + 2(4y + 3)$ Factor the GCF from each group.

$= (4y + 3)(3y + 2)$ Distributive Property

Example 8

Solve $x^2 - 6x = 0$. Check your solutions.

Write the equation so that it is of the form $ab = 0$.

$x^2 - 6x = 0$ Original equation

$x(x - 6) = 0$ Factor by using the GCF.

$x = 0$ or $x - 6 = 0$ Zero Product Property

$\qquad\qquad\quad x = 6$ Solve.

The roots are 0 and 6. Check by substituting 0 and 6 for x in the original equation.

8-6 Solving $x^2 + bx + c = 0$

Factor each trinomial. Confirm your answers using a graphing calculator.

46. $x^2 - 8x + 15$ **47.** $x^2 + 9x + 20$

48. $x^2 - 5x - 6$ **49.** $x^2 + 3x - 18$

Solve each equation. Check your solutions.

50. $x^2 + 5x - 50 = 0$

51. $x^2 - 6x + 8 = 0$

52. $x^2 + 12x + 32 = 0$

53. $x^2 - 2x - 48 = 0$

54. $x^2 + 11x + 10 = 0$

55. ART An artist is working on a painting that is 3 inches longer than it is wide. The area of the painting is 154 square inches. What is the length of the painting?

Example 9

Factor $x^2 + 10x + 21$

$b = 10$ and $c = 21$, so $m + p$ is positive and mp is positive. Therefore, m and p must both be positive. List the positive factors of 21, and look for the pair of factors with a sum of 10.

Factors of 21	Sum of 10
1, 21	22
3, 7	10

The correct factors are 3 and 7.

$x^2 + 10x + 21 = (x + m)(x + p)$ Write the pattern.

$\qquad\qquad\qquad = (x + 3)(x + 7)$ $m = 3$ and $p = 7$

8-7 Solving $ax^2 + bx + c = 0$

Factor each trinomial, if possible. If the trinomial cannot be factored, write *prime*.

56. $12x^2 + 22x - 14$

57. $2y^2 - 9y + 3$

58. $3x^2 - 6x - 45$

59. $2a^2 + 13a - 24$

Solve each equation. Confirm your answers using a graphing calculator.

60. $40x^2 + 2x = 24$

61. $2x^2 - 3x - 20 = 0$

62. $-16t^2 + 36t - 8 = 0$

63. $6x^2 - 7x - 5 = 0$

64. GEOMETRY The area of the rectangle shown is $6x^2 + 11x - 7$ square units. What is the width of the rectangle?

2x − 1

Example 10

Factor $12a^2 + 17a + 6$

$a - 12$, $b = 17$, and $c = 6$. Since b is positive, $m + p$ is positive. Since c is positive, mp is positive. So, m and p are both positive. List the factors of 12(6) or 72, where both factors are positive.

Factors of 72	Sum of 17
1, 72	73
2, 36	38
3, 24	27
4, 18	22
6, 12	18
8, 9	17

The correct factors are 8 and 9.

$12a^2 + 17a + 6 = 12a^2 + ma + pa + 6$

$\qquad\qquad\qquad\quad = 12a^2 + 8a + 9a + 6$

$\qquad\qquad\qquad\quad = (12a^2 + 8a) + (9a + 6)$

$\qquad\qquad\qquad\quad = 4a(3a + 2) + 3(3a + 2)$

$\qquad\qquad\qquad\quad = (3a + 2)(4a + 3)$

So, $12a^2 + 17a + 6 = (3a + 2)(4a + 3)$.

8-8 Differences of Squares

Factor each polynomial.

65. $y^2 - 81$

66. $64 - 25x^2$

67. $16a^2 - 21b^2$

68. $3x^2 - 3$

Solve each equation by factoring. Confirm your answers using a graphing calculator.

69. $a^2 - 25 = 0$

70. $9x^2 - 25 = 0$

71. $81 - y^2 = 0$

72. $x^2 - 5 = 20$

73. EROSION A boulder falls down a mountain into water 64 feet below. The distance d that the boulder falls in t seconds is given by the equation $d = 16t^2$. How long does it take the boulder to hit the water?

Example 11

Solve $x^2 - 4 = 12$ by factoring.

$x^2 - 4 = 12$	Original equation
$x^2 - 16 = 0$	Subtract 12 from each side.
$x^2 - (4)^2 = 0$	$16 = 4^2$
$(x + 4)(x - 4) = 0$	Factor the difference of squares.
$x + 4 = 0 \quad \text{or} \quad x - 4 = 0$	Zero Product Property
$x = -4 \qquad\qquad x = 4$	Solve each equation.

The solutions are -4 and 4.

8-9 Perfect Squares

Factor each polynomial, if possible. If the polynomial cannot be factored write *prime*.

74. $x^2 + 12x + 36$

75. $x^2 + 5x + 25$

76. $9y^2 - 12y + 4$

77. $4 - 28a + 49a^2$

78. $x^4 - 1$

79. $x^4 - 16x^2$

Solve each equation. Confirm your answers using a graphing calculator.

80. $(x - 5)^2 = 121$

81. $4c^2 + 4c + 1 = 9$

82. $4y^2 = 64$

83. $16d^2 + 40d + 25 = 9$

84. LANDSCAPING A sidewalk of equal width is being built around a square yard. What is the width of the sidewalk?

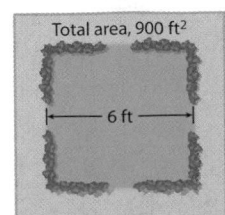

Total area, 900 ft²

6 ft

Example 12

Solve $(x - 9)^2 = 144$.

$(x - 9)^2 = 144$	Original equation
$x - 9 = \pm\sqrt{144}$	Square Root Property
$x - 9 = \pm 12$	$144 = 12 \cdot 12$
$x = 9 \pm 12$	Add 9 to each side.
$x = 9 + 12 \ \text{or} \ x = 9 - 12$	Zero Product Property
$x = 21 \qquad\qquad x = -3$	Solve.

CHECK

$(x - 9)^2 = 144$
$(21 - 9)^2 \stackrel{?}{=} 144$
$(12)^2 \stackrel{?}{=} 144$
$\qquad 144 = 144 \ ✓$

$(x - 9)^2 = 144$
$(-3 - 9)^2 \stackrel{?}{=} 144$
$(-12)^2 \stackrel{?}{=} 144$
$\qquad 144 = 144 \ ✓$

Find each sum or difference.

1. $(x + 5) + (x^2 - 3x + 7)$

2. $(7m - 8n^2 + 3n) - (-2n^2 + 4m - 3n)$

3. **MULTIPLE CHOICE** Antonia is carpeting two of the rooms in her house. The dimensions are shown. Which expression represents the total area to be carpeted?

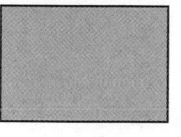 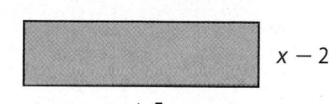

x $x - 2$

$x + 5$

$x + 3$

A $x^2 + 3x$

B $2x^2 + 6x - 10$

C $x^2 + 3x - 5$

D $8x + 12$

Find each product.

4. $a(a^2 + 2a - 10)$

5. $(2a - 5)(3a + 5)$

6. $(x - 3)(x^2 + 5x - 6)$

7. $(x + 3)^2$

8. $(2b - 5)(2b + 5)$

9. **FINANCIAL LITERACY** Suppose you invest $4000 in a 2-year certificate of deposit (CD).

a. If the interest rate is 5% per year, the expression $4000(1 + 0.05)^2$ can be evaluated to find the total amount of money after two years. Explain the numbers in this expression.

b. Find the amount at the end of two years.

c. Suppose you invest $10,000 in a CD for 4 years at an annual rate of 6.25%. What is the total amount of money you will have after 4 years?

10. **MULTIPLE CHOICE** The area of the rectangle shown below is $2x^2 - x - 15$ square units. What is the width of the rectangle?

F $x - 5$

G $x + 3$

H $x - 3$

J $2x - 3$

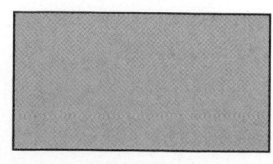

$2x + 5$

Solve each equation.

11. $5(t^2 - 3t + 2) = t(5t - 2)$

12. $3x(x + 2) = 3(x^2 - 2)$

Factor each polynomial.

13. $5xy - 10x$

14. $7ab + 14ab^2 + 21a^2b$

15. $4x^2 + 8x + x + 2$

16. $10a^2 - 50a - a + 5$

Solve each equation. Confirm your answers using a graphing calculator.

17. $y(y - 14) = 0$

18. $3x(x + 6) = 0$

19. $a^2 = 12a$

20. **MULTIPLE CHOICE** Chantel is carpeting a room that has an area of $x^2 - 100$ square feet. If the width of the room is $x - 10$ feet, what is the length of the room?

A $x - 10$ ft

B $x + 10$ ft

C $x - 100$ ft

D 10 ft

Factor each trinomial.

21. $x^2 + 7x + 6$

22. $x^2 - 3x - 28$

23. $10x^2 - x - 3$

24. $15x^2 + 7x - 2$

25. $x^2 - 25$

26. $4x^2 - 81$

27. $9x^2 - 12x + 4$

28. $16x^2 + 40x + 25$

Solve each equation. Confirm your answers using a graphing calculator.

29. $x^2 - 4x = 21$

30. $x^2 - 2x - 24 = 0$

31. $6x^2 - 5x - 6 = 0$

32. $2x^2 - 13x + 20 = 0$

33. **MULTIPLE CHOICE** Which choice is a factor of $x^4 - 1$ when it is factored completely?

F $x^2 - 1$

G $x - 1$

H x

J 1

Solve Multi-Step Problems

Some problems that you will encounter on standardized tests require you to solve multiple parts in order to come up with the final solution. Use this lesson to practice these types of problems.

Strategies for Solving Multi-Step Problems

Step 1

Read the problem statement carefully.

Ask yourself:

- What am I being asked to solve? What information is given?

- Are there any intermediate steps that need to be completed before I can solve the problem?

Step 2

Organize your approach.

- List the steps you will need to complete in order to solve the problem.

- Remember that there may be more than one possible way to solve the problem.

Step 3

Solve and check.

- Work as efficiently as possible to complete each step and solve.

- If time permits, check your answer.

Standardized Test Example

Read the problem. Identify what you need to know. Then use the information in the problem to solve.

A florist has 80 roses, 50 tulips, and 20 lilies that he wants to use to create bouquets. He wants to create the maximum number of bouquets possible and use all of the flowers. Each bouquet should have the same number of each type of flower. How many roses will be in each bouquet?

A 4 roses

C 10 roses

B 8 roses

D 15 roses

@Palladium/age Fotostock

Read the problem carefully. You are given the number of roses, tulips, and lilies and told that bouquets will be made using the same number of flowers in each. You need to find the number of roses that will be in each bouquet.

Step 1 Find the GCF of the number of roses, tulips, and lilies.

Step 2 Use the GCF to determine how many bouquets will be made.

Step 3 Divide the total number of roses by the number of bouquets.

Step 1 Write the prime factorization of each number of flowers to find the GCF.

$$80 = \mathbf{2} \cdot 2 \cdot 2 \cdot 2 \cdot \mathbf{5}$$

$$50 = \mathbf{2} \cdot \mathbf{5} \cdot 5$$

$$20 = \mathbf{2} \cdot 2 \cdot \mathbf{5}$$

$$\text{GCF} = 2 \cdot 5 = 10$$

Step 2 The GCF of the number of roses, tulips, and lilies tells you how many bouquets can be made because each bouquet will contain the same number of flowers. So, the florist can make a total of 10 bouquets.

Step 3 Divide the number of roses by the number of bouquets to find the number of roses in each bouquet.

$$\frac{80}{10} = 8$$

So, there will be 8 roses in each bouquet. The answer is B.

Exercises

Read each problem. Identify what you need to know. Then use the information in the problem to solve.

1. Which of the following values is not a solution to $x^3 - 3x^2 - 25x + 75 = 0$?

   A $x = 5$ **C** $x = -3$

   B $x = 3$ **D** $x = -5$

2. There are 12 teachers, 90 students, and 36 parent volunteers going on a field trip. Mrs. Bartholomew wants to divide everyone into equal groups with the same number of teachers, students, and parents in each group. If she makes as many groups as possible, how many students will be in each group?

   F 6   **H** 12

   G 9   **J** 15

3. What is the area of the square?

   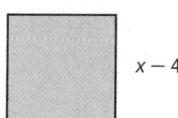

   $x - 4$

   A $x^2 + 16$

   B $4x - 16$

   C $x^2 - 8x - 16$

   D $x^2 - 8x + 16$

4. Students are selling magazines to raise money for a field trip. They make $2.75 for each magazine they sell. If they want to raise $600, what is the least amount of magazines they need to sell?

   F 121   **H** 202

   G 177   **J** 219

Multiple Choice

Read each question. Then fill in the correct answer on the answer document provided by your teacher or on a sheet of paper.

1. Each year a local country club sponsors a tennis tournament. Play starts with 256 participants. During each round, half of the players are eliminated. How many players remain after 6 rounds?

   A 128

   B 64

   C 16

   D 4

2. Evaluate $\dfrac{5^6 - 5^5}{4}$.

   F 5^6

   G 5^5

   H $\dfrac{5}{4}$

   J $\dfrac{25}{4}$

3. Factor $mn + 5m - 3n - 15$.

   A $(mn - 3)(5)$

   B $(n - 3)(m + 5)$

   C $(m - 5)(n + 3)$

   D $(m - 3)(n + 5)$

4. Which of the following is a solution to $x^2 + 6x - 112 = 0$?

   F -14

   G -8

   H 6

   J 12

5. Which of the following polynomials is prime?

   A $5x^2 + 34x + 24$

   B $4x^2 + 22x + 10$

   C $4x^2 + 38x + 70$

   D $5x^2 + 3x + 4$

6. Which of the following is not a factor of the polynomial $45a^2 - 80b^2$?

   F 5

   G $3a - 4b$

   H $2a - 5b$

   J $3a + 4b$

7. A rectangular gift box has dimensions that can be represented as shown in the figure. The volume of the box is $56w$ cubic inches. Which of the following is *not* a dimension of the box?

   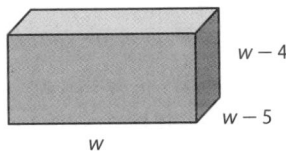

   A 6 in.

   B 7 in.

   C 8 in.

   D 12 in.

8. Factor the polynomial $y^2 - 9y + 20$.

   F $(y - 2)(y - 10)$

   G $(y - 4)(y - 5)$

   H $(y - 2)(y - 7)$

   J $(y - 5)(y + 2)$

9. Which of the following numbers is less than zero?

   A 1.03×10^{-21}

   B 7.5×10^2

   C 8.21543×10^{10}

   D none of the above

Test-TakingTip

Question 4 If time permits, be sure to check your answer. Substitute it into the equation to see if you get a true number sentence.

Record your answers on the answer sheet provided by your teacher or on a sheet of paper.

10. **GRIDDED RESPONSE** Mr. Branson bought a total of 9 tickets to the zoo. He bought children tickets at the rate of $6.50 and adult tickets for $9.25 each. If he spent $69.50 altogether, how many adult tickets did Mr. Branson purchase?

11. What is the domain of the following relation? $\{(2, -1), (4, 3), (7, 6)\}$

12. Ken just added 15 more songs to his digital media player, making the total number of songs more than 84. Draw a number line that represents the original number of songs he had on his digital media player.

13. Carlos bought a rare painting in 1995 for $14,200. By 2003, the painting was worth $17,120. Assuming that a linear relationship exists, write an equation in slope-intercept form that represents the value V of the painting after t years.

14. The equation $h = -16t^2 + 40t + 3$ models the height h in feet of a soccer ball after t seconds. What is the height of the ball after 2 seconds?

15. Marcel spent $24.50 on peanuts and walnuts for a party. He bought 1.5 pounds more peanuts than walnuts. How many pounds of peanuts and walnuts did he buy?

Product	Price per pound
Peanuts p	$3.80
Cashews c	$6.90
Walnuts w	$5.60

16. **GRIDDED RESPONSE** The amount of money that Humberto earns varies directly as the number of hours that he works as shown in the graph. How much money will he earn for working 40 hours next week? Express your answer in dollars.

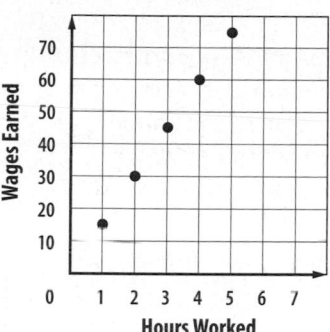

Record your answers on a sheet of paper. Show your work.

17. The height in feet of a model rocket t seconds after being launched into the air is given by the function $h(t) = -16t^2 + 200t$.

    a. Write the expression that shows the height of the rocket in factored form.

    b. At what time(s) is the height of the rocket equal to zero feet above the ground? Describe the real world meaning of your answer.

    c. What is the greatest height reached by the model rocket? When does this occur?

Need ExtraHelp?

If you missed Question...	1	2	3	4	5	6	7	8	9	10	11	12	13	14	15	16	17
Go to Lesson...	7-7	7-2	8-5	8-6	8-3	8-7	8-9	8-6	7-4	6-5	1-6	5-1	4-2	8-7	2-9	3-4	8-5

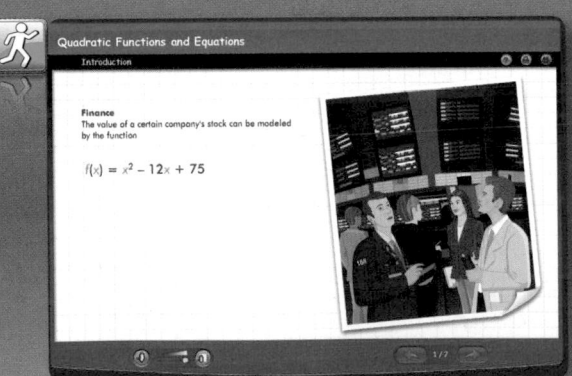

··Then

○ You solved quadratic equations by factoring and by using the Square Root Property.

··Now

○ In this chapter, you will:

- Solve quadratic equations by graphing, completing the square, and using the Quadratic Formula.

- Analyze functions with successive differences and ratios.

- Identify and graph special functions.

··Why? ▲

○ **FINANCE** The value of a certain company's stock can be modeled by the function $f(x) = x^2 - 12x + 75$. By graphing this quadratic function, we can make an educated guess as to how the stock will perform in the near future.

Quadratic Functions and Equations
Introduction

Finance
The value of a certain company's stock can be modeled by the function

$f(x) = x^2 - 12x + 75$

1/7

ᗩ connectED.mcgraw-hill.com **Your Digital Math Portal**

Animation   Vocabulary   eGlossary   Personal Tutor   Virtual Manipulatives   Graphing Calculator   Audio   Foldables   Self-Check Practice   Worksheets

Diagnose Readiness | You have two options for checking prerequisite skills.

 Textbook Option Take the Quick Check below. Refer to the Quick Review for help.

| **Quick**Check | **Quick**Review |

QuickCheck

Use a table of values to graph each equation.

1. $y = x + 3$ **2.** $y = 2x + 2$

3. $y = -2x - 3$ **4.** $y = 0.5x - 1$

5. $4x - 3y = 12$ **6.** $3y = 6 + 9x$

7. SAVINGS Jack has $100 to buy a game system. He plans to save $10 each week. Graph an equation to show the total amount T Jack will have in w weeks.

QuickReview

Example 1

Use a table of values to graph $y = 3x + 1$.

x	$y = 3x + 1$	y
-1	$3(-1) + 1$	-2
0	$3(0) + 1$	1
1	$3(1) + 1$	4
2	$3(2) + 1$	7

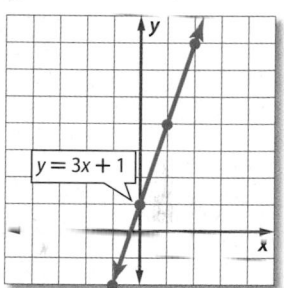

Determine whether each trinomial is a perfect square trinomial. Write *yes* or *no*. If so, factor it.

8. $a^2 + 12a + 36$ **9.** $x^2 + 5x + 25$

10. $x^2 - 12x + 32$ **11.** $x^2 + 20x + 100$

12. $4x^2 + 28x + 49$ **13.** $k^2 - 16k + 64$

14. $a^2 - 22a + 121$ **15.** $5t^2 - 12t + 25$

Example 2

Determine whether $x^2 - 10x + 25$ is a perfect square trinomial. Write *yes* or *no*. If so, factor it.

1. Is the first term a perfect square? yes

2. Is the last term a perfect square? yes

3. Is the middle term equal to $-2(1x)(5)$? yes

$x^2 - 10x + 25 = (x - 5)^2$

Evaluate each expression if $a = -2$, $b = -1$, $c = 0$, and $d = 2.5$.

16. $|a - 3|$ **17.** $|2a + 1|$

18. $|4 - b|$ **19.** $\left|\frac{1}{2}b - 2\right|$

20. $|12 - 4c|$ **21.** $|2c - 3| + 1$

22. $|4d - 6|$ **23.** $|3d - 2| - 8$

Example 3

Evaluate $|2x + 1| - 7$ if $x = -1$.

$$\begin{aligned}
|2x + 1| - 7 &= |2(-1) + 1| - 7 & x = -1 \\
&= |-2 + 1| - 7 & \text{Multiply.} \\
&= |-1| - 7 & \text{Add.} \\
&= 1 - 7 & |-1| = 1 \\
&= -6 & \text{Subtract.}
\end{aligned}$$

---

2 Online Option Take an online self-check Chapter Readiness Quiz at connectED.mcgraw-hill.com.

You will learn several new concepts, skills, and vocabulary terms as you study Chapter 9. To get ready, identify important terms and organize your resources. You may wish to refer to Chapter 0 to review prerequisite skills.

FOLDABLES StudyOrganizer

Quadratic Functions and Equations Make this Foldable to help you organize your Chapter 9 notes about quadratic functions. Begin with a sheet of notebook paper.

1 **Fold** the sheet of paper along the length so that the edge of the paper aligns with the margin rule on the paper.

2 **Fold** the sheet twice widthwise to form four sections.

3 **Unfold** the sheet, and cut along the folds on the front flap only.

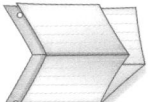

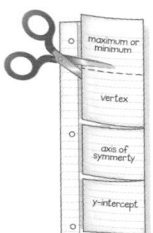

4 **Label** each section as shown.

NewVocabulary

English		Español
quadratic function	p. 543	función cuadrática
parabola	p. 543	parábola
axis of symmetry	p. 543	eje de simetría
vertex	p. 543	vértice
minimum	p. 543	mínimo
maximum	p. 543	máximo
double root	p. 556	doble raíz
transformation	p. 564	transformación
completing the square	p. 574	completar el cuadrado
Quadratic Formula	p. 583	Formula cuadrática
discriminant	p. 586	discriminante
step function	p. 598	función etapa
greatest integer function	p. 598	función del máximo entero
absolute value function	p. 599	función del valor absoluto

ReviewVocabulary

domain dominio all the possible values of the independent variable, x

leading coefficient coeficiente delantero the coefficient of the first term of a polynomial written in standard form

range rango all the possible values of the dependent variable, y

In the function represented by the table, the domain is {0, 2, 4, 6}, and the range is {3, 5, 7, 9}.

x	y
0	3
2	5
4	7
6	9

Graphing Quadratic Functions

∴ Then	∴ Now	∴ Why?
● You graphed linear and exponential functions.	● **1** Analyze the characteristics of graphs of quadratic functions. **2** Graph quadratic functions.	● The Innovention Fountain in Epcot's Futureworld in Orlando, Florida, is an elaborate display of water, light, and music. The sprayers shoot water in shapes that can be modeled by quadratic equations. You can use graphs of these equations to show the path of the water.

NewVocabulary
quadratic function
standard form
parabola
axis of symmetry
vertex
minimum
maximum

Common Core State Standards

Content Standards
F.IF.4 For a function that models a relationship between two quantities, interpret key features of graphs and tables in terms of the quantities, and sketch graphs showing key features given a verbal description of the relationship.

F.IF.7a Graph linear and quadratic functions and show intercepts, maxima, and minima.

Mathematical Practices
2 Reason abstractly and quantitatively.

1 **Characteristics of Quadratic Functions** **Quadratic functions** are nonlinear and can be written in the form $f(x) = ax^2 + bx + c$, where $a \neq 0$. This form is called the **standard form** of a quadratic function.

The shape of the graph of a quadratic function is called a **parabola**. Parabolas are symmetric about a central line called the **axis of symmetry**. The axis of symmetry intersects a parabola at only one point, called the **vertex**.

KeyConcept Quadratic Functions

Parent Function:	$f(x) = x^2$
Standard Form:	$f(x) = ax^2 + bx + c$
Type of Graph:	parabola
Axis of Symmetry:	$x = -\dfrac{b}{2a}$
y-intercept:	c

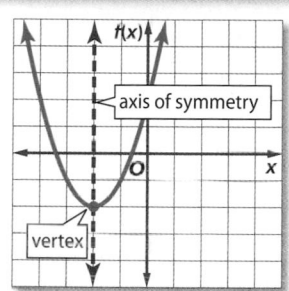

When $a > 0$, the graph of $y = ax^2 + bx + c$ opens upward. The lowest point on the graph is the **minimum**. When $a < 0$, the graph opens downward. The highest point is the **maximum**. The maximum or minimum is the vertex.

Example 1 Graph a Parabola

Use a table of values to graph $y = 3x^2 + 6x - 4$. State the domain and range.

x	y
1	5
0	−4
−1	−7
−2	−4
−3	5

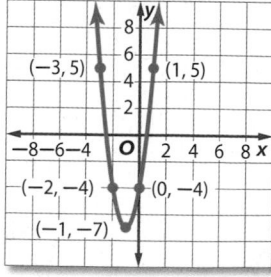

Graph the ordered pairs, and connect them to create a smooth curve. The parabola extends to infinity. The domain is all real numbers. The range is $\{y \mid y \geq -7\}$, because −7 is the minimum.

▶ **Guided**Practice

1. Use a table of values to graph $y = x^2 + 3$. State the domain and range.

Recall that figures with symmetry are those in which each half of the figure matches exactly.

A parabola is symmetric about the axis of symmetry. Every point on the parabola to the left of the axis of symmetry has a corresponding point on the other half. The function is increasing on one side of the axis of symmetry and decreasing on the other side.

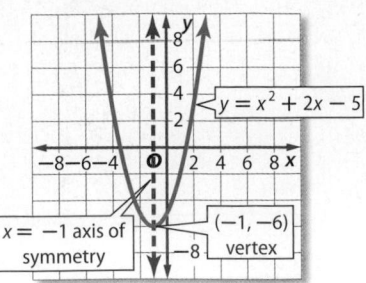

When identifying characteristics from a graph, it is often easiest to locate the vertex first. It is either the maximum or minimum point of the graph.

Example 2 Identify Characteristics from Graphs

Find the vertex, the equation of the axis of symmetry, and the y-intercept of each graph.

a.

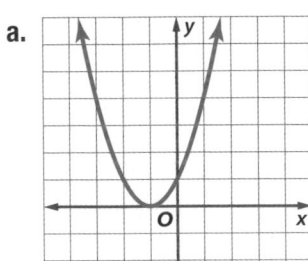

Step 1 Find the vertex.
Because the parabola opens upward, the vertex is located at the minimum point of the parabola. It is located at $(-1, 0)$.

Step 2 Find the axis of symmetry.
The axis of symmetry is the line that goes through the vertex and divides the parabola into congruent halves. It is located at $x = -1$.

Step 3 Find the y-intercept.
The y-intercept is the point where the graph intersects the y-axis. It is located at $(0, 1)$, so the y-intercept is 1.

b.

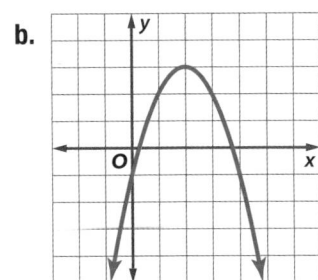

Step 1 Find the vertex.
The parabola opens downward, so the vertex is located at its maximum point, $(2, 3)$.

Step 2 Find the axis of symmetry.
The axis of symmetry is located at $x = 2$.

Step 3 Find the y-intercept.
The y-intercept is where the parabola crosses the y-axis. It is located at $(0, -1)$, so the y-intercept is -1.

▶ **Guided**Practice

2A.

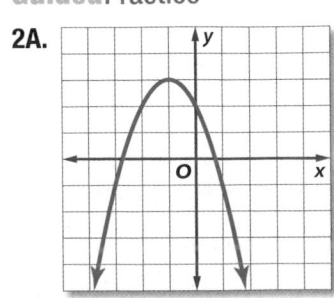

2B.

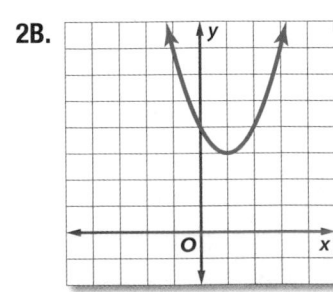

Example 3 Identify Characteristics from Functions

Find the vertex, the equation of the axis of symmetry, and the y-intercept of each function.

a. $y = 2x^2 + 4x - 3$

$x = -\dfrac{b}{2a}$ Formula for the equation of the axis of symmetry

$x = -\dfrac{4}{2 \cdot 2}$ $a = 2$ and $b = 4$

$x = -1$ Simplify.

The equation for the axis of symmetry is $x = -1$.

To find the vertex, use the value you found for the axis of symmetry as the x-coordinate of the vertex. Find the y-coordinate using the original equation.

$y = 2x^2 + 4x - 3$ Original equation

$ = 2(-1)^2 + 4(-1) - 3$ $x = -1$

$ = -5$ Simplify.

The vertex is at $(-1, -5)$.

The y-intercept always occurs at $(0, c)$. So, the y-intercept is -3.

b. $y = -x^2 + 6x + 4$

$x = -\dfrac{b}{2a}$ Formula for the equation of the axis of symmetry

$x = -\dfrac{6}{2(-1)}$ $a = -1$ and $b = 6$

$x = 3$ Simplify.

The equation of the axis of symmetry is $x = 3$.

$y = -x^2 + 6x + 4$ Original equation

$ = -(3)^2 + 6(3) + 4$ $x = 3$

$ = 13$ Simplify.

The vertex is at $(3, 13)$.

The y-intercept is 4.

▶ **Guided Practice**

3A. $y = -3x^2 + 6x - 5$ **3B.** $y = 2x^2 + 2x + 2$

Next you will learn how to identify whether the vertex is a maximum or a minimum.

◆ KeyConcept Maximum and Minimum Values

Words

The graph of $f(x) = ax^2 + bx + c$, where $a \neq 0$:
- opens upward and has a minimum value when $a > 0$, and
- opens downward and has a maximum value when $a < 0$.
- The range of a quadratic function is all real numbers greater than or equal to the minimum, or all real numbers less than or equal to the maximum.

Examples

a is positive. a is negative.

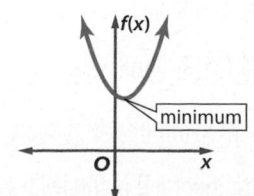

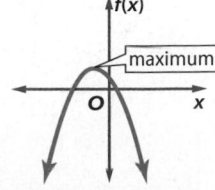

Example 4 Maximum and Minimum Values

Consider $f(x) = -2x^2 - 4x + 6$.

a. Determine whether the function has a *maximum* or *minimum* value.

For $f(x) = -2x^2 - 4x + 6$, $a = -2$, $b = -4$, and $c = 6$.

Because a is negative the graph opens down, so the function has a maximum value.

b. State the maximum or minimum value of the function.

The maximum value is the y-coordinate of the vertex.

The x-coordinate of the vertex is $\dfrac{-b}{2a}$ or $\dfrac{4}{2(-2)}$ or -1.

$f(x) = -2x^2 - 4x + 6$	Original function
$f(-1) = -2(-1)^2 - 4(-1) + 6$	$x = -1$
$f(-1) = 8$	Simplify.

The maximum value is 8.

c. State the domain and range of the function.

The domain is all real numbers. The range is all real numbers less than or equal to the maximum value, or $\{y \mid y \le 8\}$.

GuidedPractice

Consider $g(x) = 2x^2 - 4x - 1$.

4A. Determine whether the function has a *maximum* or *minimum* value.

4B. State the maximum or minimum value.

4C. State the domain and range of the function.

2 **Graph Quadratic Functions** You have learned how to find several important characteristics of quadratic functions.

KeyConcept Graph Quadratic Functions

Step 1 Find the equation of the axis of symmetry.

Step 2 Find the vertex, and determine whether it is a maximum or minimum.

Step 3 Find the y-intercept.

Step 4 Use symmetry to find additional points on the graph, if necessary.

Step 5 Connect the points with a smooth curve.

Example 5 Graph Quadratic Functions

Graph $f(x) = x^2 + 4x + 3$.

Step 1 Find the equation of the axis of symmetry.

$$x = \frac{-b}{2a} \qquad \text{Formula for the equation of the axis of symmetry}$$

$$x = \frac{-4}{2 \cdot 1} \quad \text{or} \quad -2 \qquad a = 1 \text{ and } b = 4$$

Step 2 Find the vertex, and determine whether it is a maximum or minimum.

$$f(x) = x^2 + 4x + 3 \qquad \text{Original equation}$$
$$= (-2)^2 + 4(-2) + 3 \qquad x = -2$$
$$= -1 \qquad \text{Simplify.}$$

The vertex lies at $(-2, -1)$. Because a is positive the graph opens up, and the vertex is a minimum.

Step 3 Find the y-intercept.

$$f(x) = x^2 + 4x + 3 \qquad \text{Original equation}$$
$$= (0)^2 + 4(0) + 3 \qquad x = 0$$
$$= 3 \qquad \text{The } y\text{-intercept is 3.}$$

Step 4 The axis of symmetry divides the parabola into two equal parts. So if there is a point on one side, there is a corresponding point on the other side that is the same distance from the axis of symmetry and has the same y-value.

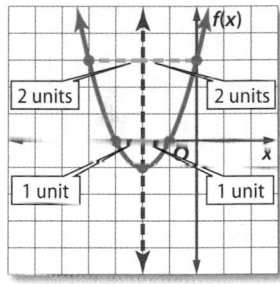

Step 5 Connect the points with a smooth curve.

GuidedPractice **Graph each function.**

5A. $f(x) = -2x^2 + 2x - 1$ **5B.** $f(x) = 3x^2 - 6x + 2$

There are general differences between linear, exponential, and quadratic functions.

	Linear Functions	Exponential Functions	Quadratic Functions
Equation	$y = mx + b$	$y = ab^x, a \neq 0, b > 0, b \neq 1$	$y = ax^2 + bx + c, a \neq 0$
Degree	1	x	2
Graph	line	curve	parabola
Increasing / Decreasing	$m > 0$: y is increasing on the entire domain. $m < 0$: y is decreasing on the entire domain.	$a > 0, b > 1$ or $a < 0$, $0 < b < 1$: y is increasing on the entire domain. $a > 0, 0 < b < 1$ or $a < 0$, $b > 1$: y is decreasing on the entire domain.	$a > 0$: y is decreasing to the left of the axis of symmetry and increasing on the right. $a < 0$: y is increasing to the left of the axis of symmetry and decreasing on the right.
End Behavior	$m > 0$: as x increases, y increases; as x decreases, y decreases. $m < 0$: as x increases, y decreases; as x decreases, y increases	$b > 1$: as x decreases, y approaches 0; $a > 0$, as x increases, y increases; $a < 0$, as x increases, y decreases. $0 < b < 1$: as x increases, y approaches 0; $a > 0$, as x decreases, y increases; $a < 0$, as x decreases, y decreases.	$a > 0$: as x increases, y increases; as x decreases, y increases. $a < 0$: as x increases, y decreases; as x decreases, y decreases

You have used what you know about quadratic functions, parabolas, and symmetry to create graphs. You can analyze these graphs to solve real-world problems.

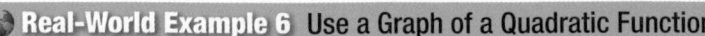

Real-World Example 6 Use a Graph of a Quadratic Function

SCHOOL SPIRIT The cheerleaders at Lake High School launch T-shirts into the crowd every time the Lakers score a touchdown. The height of the T-shirt can be modeled by the function $h(x) = -16x^2 + 48x + 6$, where $h(x)$ represents the height in feet of the T-shirt after x seconds.

a. Graph the function.

$$x = -\frac{b}{2a}$$ Equation of the axis of symmetry

$$x = -\frac{48}{2(-16)} \text{ or } \frac{3}{2}$$ $a = -16$ and $b = 48$

The equation of the axis of symmetry is $x = \frac{3}{2}$. Thus, the x-coordinate for the vertex is $\frac{3}{2}$.

$$y = -16x^2 + 48x + 6$$ Original equation

$$= -16\left(\frac{3}{2}\right)^2 + 48\left(\frac{3}{2}\right) + 6$$ $x = \frac{3}{2}$

$$= -16\left(\frac{9}{4}\right) + 48\left(\frac{3}{2}\right) + 6$$ $\left(\frac{3}{2}\right)^2 = \frac{9}{4}$

$$= -36 + 72 + 6 \text{ or } 42$$ Simplify.

The vertex is at $\left(\frac{3}{2}, 42\right)$.

Let's find another point. Choose an x-value of 0 and substitute. Our new point is at $(0, 6)$. The point paired with it on the other side of the axis of symmetry is $(3, 6)$.

Repeat this and choose an x-value of 1 to get $(1, 38)$ and its corresponding point $(2, 38)$. Connect these points and create a smooth curve.

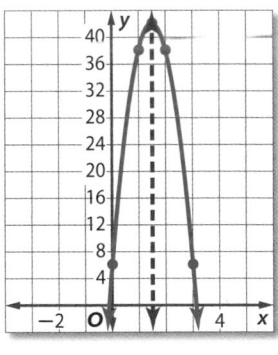

b. At what height was the T-shirt launched?

The T-shirt is launched when time equals 0, or at the y-intercept.

So, the T-shirt was launched 6 feet from the ground.

c. What is the maximum height of the T-shirt? When was the maximum height reached?

The maximum height of the T-shirt occurs at the vertex.

So the T-shirt reaches a maximum height of 42 feet. The time was $\frac{3}{2}$ or 1.5 seconds after launch.

▶ **Guided**Practice

6. TRACK Emilio is competing in the javelin throw. The height of the javelin can be modeled by the equation $y = -16x^2 + 64x + 6$, where y represents the height in feet of the javelin after x seconds.

A. Graph the path of the javelin.

B. At what height is the javelin thrown?

C. What is the maximum height of the javelin?

Example 1 Use a table of values to graph each equation. State the domain and range.

1. $y = 2x^2 + 4x - 6$ 2. $y = x^2 + 2x - 1$

3. $y = x^2 - 6x - 3$ 4. $y = 3x^2 - 6x - 5$

Example 2 Find the vertex, the equation of the axis of symmetry, and the y-intercept of each graph.

5.
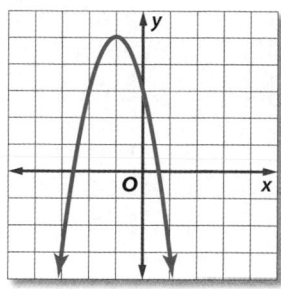

6.
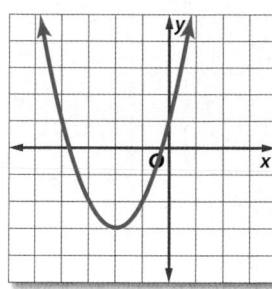

7.
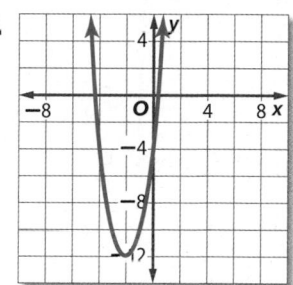

8.
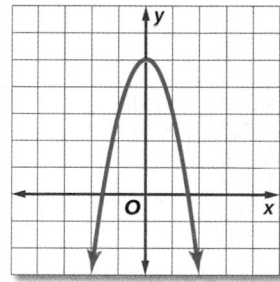

Example 3 Find the vertex, the equation of the axis of symmetry, and the y-intercept of the graph of each function.

9. $y = -3x^2 + 6x - 1$ 10. $y = -x^2 + 2x + 1$

11. $y = x^2 - 4x + 5$ 12. $y = 4x^2 - 8x + 9$

Example 4 Consider each function.

a. Determine whether the function has *maximum* or *minimum* value.

b. State the maximum or minimum value.

c. What are the domain and range of the function?

13 $y = -x^2 + 4x - 3$ 14. $y = -x^2 - 2x + 2$

15. $y = -3x^2 + 6x + 3$ 16. $y = -2x^2 + 8x - 6$

Example 5 Graph each function.

17. $f(x) = -3x^2 + 6x + 3$ 18. $f(x) = -2x^2 + 4x + 1$

19. $f(x) = 2x^2 - 8x - 4$ 20. $f(x) = 3x^2 - 6x - 1$

Example 6 21. **CCSS REASONING** A juggler is tossing a ball into the air. The height of the ball in feet can be modeled by the equation $y = -16x^2 + 16x + 5$, where y represents the height of the ball at x seconds.

a. Graph this equation.

b. At what height is the ball thrown?

c. What is the maximum height of the ball?

Example 1   Use a table of values to graph each equation. State the domain and range.

22. $y = x^2 + 4x + 6$ **23.** $y = 2x^2 + 4x + 7$ **24.** $y = 2x^2 - 8x - 5$

25. $y = 3x^2 + 12x + 5$ **26.** $y = 3x^2 - 6x - 2$ **27.** $y = x^2 - 2x - 1$

Example 2   Find the vertex, the equation of the axis of symmetry, and the y-intercept of each graph.

28.

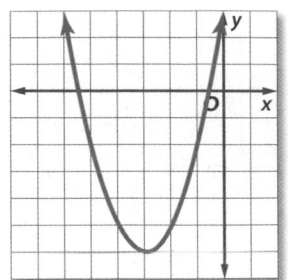

29.

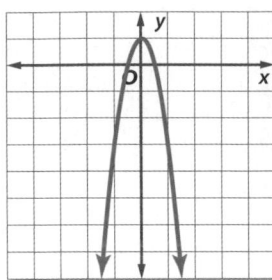

30.

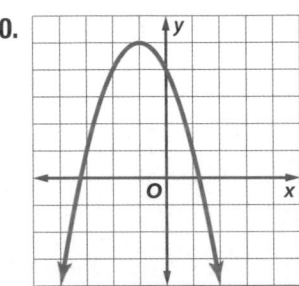

31.

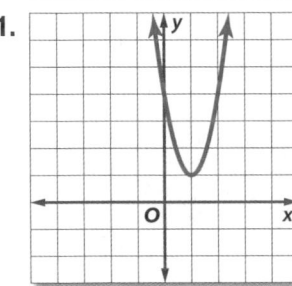

32.

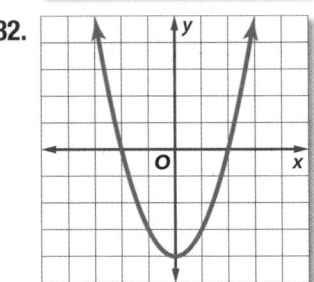

33.

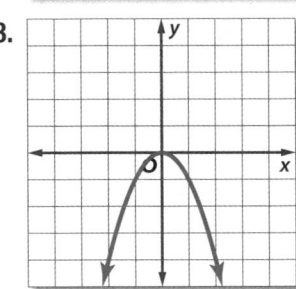

Example 3   Find the vertex, the equation of the axis of symmetry, and the y-intercept of each function.

34. $y = x^2 + 8x + 10$ **35** $y = 2x^2 + 12x + 10$ **36.** $y = -3x^2 - 6x + 7$

37. $y = -x^2 - 6x - 5$ **38.** $y = 5x^2 + 20x + 10$ **39.** $y = 7x^2 - 28x + 14$

40. $y = 2x^2 - 12x + 6$ **41.** $y = -3x^2 + 6x - 18$ **42.** $y = -x^2 + 10x - 13$

Example 4   Consider each function.

   a. Determine whether the function has a *maximum* or *minimum* value.

   b. State the maximum or minimum value.

   c. What are the domain and range of the function?

   43. $y = -2x^2 - 8x + 1$ **44.** $y = x^2 + 4x - 5$ **45.** $y = 3x^2 + 18x - 21$

   46. $y = -2x^2 - 16x + 18$ **47.** $y = -x^2 - 14x - 16$ **48.** $y = 4x^2 + 40x + 44$

   49. $y = -x^2 - 6x - 5$ **50.** $y = 2x^2 + 4x + 6$ **51.** $y = -3x^2 - 12x - 9$

Example 5   Graph each function.

   52. $y = -3x^2 + 6x - 4$ **53.** $y = -2x^2 - 4x - 3$ **54.** $y = -2x^2 - 8x + 2$

   55. $y = x^2 + 6x - 6$ **56.** $y = x^2 - 2x + 2$ **57.** $y = 3x^2 - 12x + 5$

Example 6

58. BOATING Miranda has her boat docked on the west side of Casper Point. She is boating over to Casper Marina, which is located strictly east of where her boat is docked. The equation $d = -16t^2 + 66t$ models the distance she travels north of her starting point, where d is the number of feet and t is the time traveled in minutes.

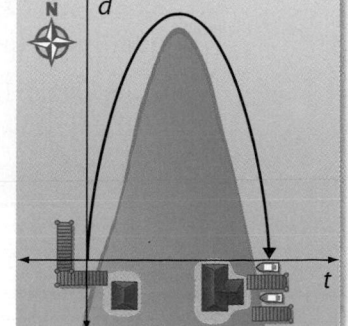

   a. Graph this equation.

   b. What is the maximum number of feet north that she traveled?

   c. How long did it take her to reach Casper Marina?

GRAPHING CALCULATOR Graph each equation. Use the TRACE feature to find the vertex on the graph. Round to the nearest thousandth if necessary.

59. $y = 4x^2 + 10x + 6$ **60.** $y = 8x^2 - 8x + 8$

61. $y = -5x^2 - 3x - 8$ **62.** $y = -7x^2 + 12x - 10$

63. GOLF The average amateur golfer can hit a ball with an initial upward velocity of 31.3 meters per second. The height can be modeled by the equation $h = -4.9t^2 + 31.3t$, where h is the height of the ball, in meters, after t seconds.

   a. Graph this equation. What do the portions of the graph where $h > 0$ represent in the context of the situation? What does the end behavior of the graph represent?

   b. At what height is the ball hit?

   c. What is the maximum height of the ball?

   d. How long did it take for the ball to hit the ground?

   e. State a reasonable range and domain for this situation.

64. FUNDRAISING The marching band is selling poinsettias to buy new uniforms. Last year the band charged $5 each, and they sold 150. They want to increase the price this year, and they expect to lose 10 sales for each $1 increase. The sales revenue R, in dollars, generated by selling the poinsettias is predicted by the function $R = (5 + p)(150 - 10p)$, where p is the number of $1 price increases.

   a. Write the function in standard form.

   b. Find the maximum value of the function.

   c. At what price should the poinsettias be sold to generate the most sales revenue? Explain your reasoning.

65 FOOTBALL A football is kicked up from ground level at an initial upward velocity of 90 feet per second. The equation $h = -16t^2 + 90t$ gives the height h of the football after t seconds.

   a. What is the height of the ball after one second?

   b. When is the ball 126 feet high?

   c. When is the height of the ball 0 feet? What do these points represent in the context of the situation?

66. CCSS STRUCTURE Let $f(x) = x^2 - 9$.

   a. What is the domain of $f(x)$?

   b. What is the range of $f(x)$?

   c. For what values of x is $f(x)$ negative?

   d. When x is a real number, what are the domain and range of $f(x) = \sqrt{x^2 - 9}$?

67. ⚙ **MULTIPLE REPRESENTATIONS** In this problem, you will investigate solving quadratic equations using tables.

  a. Algebraic Determine the related function for each equation. Copy and complete the first two columns of the table below.

Equation	Related Function	Zeros	*y*-Values
$x^2 - x = 12$			
$x^2 + 8x = 9$			
$x^2 = 14x - 24$			
$x^2 + 16x = -28$			

  b. Graphical Graph each related function with a graphing calculator.

  c. Analytical The number of zeros is equal to the degree of the related function. Use the table feature on your calculator to determine the zeros of each related function. Record the zeros in the table above. Also record the values of the function one unit less than and one unit more than each zero.

  d. Verbal Examine the function values for *x*-values just before and just after a zero. What happens to the sign of the function value before and after a zero?

H.O.T. Problems   Use Higher-Order Thinking Skills

68. OPEN ENDED Write a quadratic function for which the graph has an axis of symmetry of $x = -\dfrac{3}{8}$. Summarize your steps.

69. ERROR ANALYSIS Jade thinks that the parabolas represented by the graph and the description have the same axis of symmetry. Chase disagrees. Who is correct? Explain your reasoning.

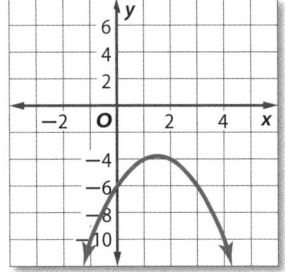

> *a parabola that opens downward,*
>
> *passing through (0, 6) and having a vertex at (2, 2)*

70. CHALLENGE Using the axis of symmetry, the *y*-intercept, and one *x*-intercept, write an equation for the graph shown.

71. CCSS **STRUCTURE** The graph of a quadratic function has a vertex at (2, 0). One point on the graph is (5, 9). Find another point on the graph. Explain how you found it.

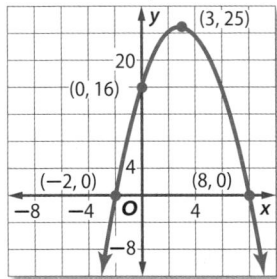

72. OPEN ENDED Describe a real-world situation that involves a quadratic equation. Explain what the vertex represents.

73. REASONING Provide a counterexample that is a specific case to show that the following statement is false. *The vertex of a parabola is always the minimum of the graph.*

74. WRITING IN MATH Use tables and graphs to compare and contrast an exponential function $f(x) = ab^x + c$, where $a \neq 0$, $b > 0$, and $b \neq 1$, a quadratic function $g(x) = ax^2 + c$, and a linear function $h(x) = ax + c$. Include intercepts, portions of the graph where the functions are increasing, decreasing, positive, or negative, relative maxima and minima, symmetries, and end behavior. Which function eventually exceeds the others?

75. Which of the following is an equation for the line that passes through $(2, -5)$ and is perpendicular to $2x + 4y = 8$?

A $y = 2x + 10$ **C** $y = 2x - 9$

B $y = -\frac{1}{2}x - 4$ **D** $y = -2x - 1$

76. GEOMETRY The area of the circle is 36π square units. If the radius is doubled, what is the area of the new circle?

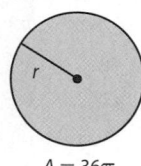

$A = 36\pi$

F 1296π units^2 **H** 72π units^2

G 144π units^2 **J** 9π units^2

77. What is the range of the function $f(x) = -4x^2 - \frac{1}{2}$?

A $\left\{\text{all integers less than or equal to } \frac{1}{2}\right\}$

B {all nonnegative integers}

C {all real numbers}

D $\left\{\text{all real numbers less than or equal to } -\frac{1}{2}\right\}$

78. SHORT RESPONSE Dylan delivers newspapers for extra money. He starts delivering the newspapers at 3:15 P.M. and finishes at 5:05 P.M. How long does it take Dylan to complete his route?

Determine whether each trinomial is a perfect square trinomial. Write *yes* or *no*. If so, factor it. (Lesson 8-9)

79. $4x^2 + 4x + 1$ **80.** $4x^2 - 20x + 25$ **81.** $9x^2 + 8x + 16$

Factor each polynomial if possible. If the polynomial cannot be factored, write *prime*. (Lesson 8-8)

82. $n^2 - 16$ **83.** $x^2 + 25$ **84.** $9 - 4a^2$

Find each product. (Lesson 8-3)

85. $(b - 7)(b + 3)$ **86.** $(c - 6)(c - 5)$ **87.** $(2x - 1)(x + 9)$

88. MULTIPLE BIRTHS The number of quadruplet births Q in the United States in recent years can be modeled by $Q = -0.5t^3 + 11.7t^2 - 21.5t + 218.6$, where t represents the number of years since 2002. What is the expected number of quadruplet births in the United States in 2017? (Lesson 8-1)

Use elimination to solve each system of equations. (Lesson 6-4)

89. $2x + y = 5$ **90.** $4x - 3y = 12$ **91.** $2x - 3y = 2$

     $3x - 2y = 4$ $x + 2y = 14$ $5x + 4y = 28$

92. HEALTH About 20% of the time you sleep is spent in rapid eye movement (REM), which is associated with dreaming. If an adult sleeps 7 to 8 hours, how much time is spent in REM sleep? (Lesson 5-4)

Find the *x*-intercept of the graph of each equation.

93. $x + 2y = 10$ **94.** $2x - 3y = 12$ **95.** $3x - y = -18$

9-1

Algebra Lab
Rate of Change of a Quadratic Function

CCSS Common Core State Standards
Content Standards
F.IF.6 Calculate and interpret the average rate of change of a function (presented symbolically or as a table) over a specified interval. Estimate the rate of change from a graph.

A model rocket is launched from the ground with an upward velocity of 144 feet per second. The function $y = -16x^2 + 144x$ models the height y of the rocket in feet after x seconds. Using this function, we can investigate the rate of change of a quadratic function.

Activity

Step 1 Copy the table below.

x	0	0.5	1.0	1.5	...	9.0
y	0					
Rate of Change	−					

Step 2 Find the value of y for each value of x from 0 through 9.

Step 3 Graph the ordered pairs (x, y) on grid paper. Connect the points with a smooth curve. Notice that the function *increases* when $0 < x < 4.5$ and *decreases* when $4.5 < x < 9$.

Step 4 Recall that the *rate of change* is the change in y divided by the change in x. Find the rate of change for each half second interval of x and y.

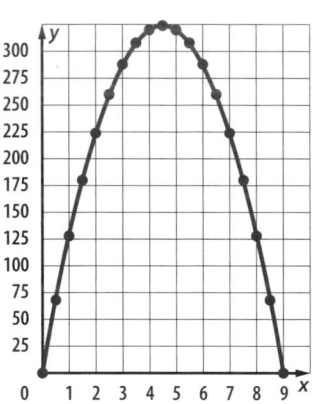

Exercises

Use the quadratic function $y = x^2$.

1. Make a table, similar to the one in the Activity, for the function using $x = -4, -3, -2, -1, 0, 1, 2, 3,$ and 4. Find the values of y for each x-value.

2. Graph the ordered pairs on grid paper. Connect the points with a smooth curve. Describe where the function is increasing and where it is decreasing.

3. Find the rate of change for each column starting with $x = -3$. Compare the rates of change when the function is increasing and when it is decreasing.

4. **CHALLENGE** If an object is dropped from 100 feet in the air and air resistance is ignored, the object will fall at a rate that can be modeled by the function $f(x) = -16x^2 + 100$, where $f(x)$ represents the object's height in feet after x seconds. Make a table like that in Exercise 1, selecting appropriate values for x. Fill in the x-values, the y-values, and rates of change. Compare the rates of change. Describe any patterns that you see.

Solving Quadratic Equations by Graphing

:: Then	:: Now	:: Why?
• You solved quadratic equations by factoring.	**1** Solve quadratic equations by graphing. **2** Estimate solutions of quadratic equations by graphing.	• Dorton Arena at the state fairgrounds in Raleigh, North Carolina, has a shape created by two intersecting parabolas. The shape of one of the parabolas can be modeled by $y = -x^2 + 127x$, where x is the width of the parabola in feet, and y is the length of the parabola in feet. The x-intercepts of the graph of this function can be used to find the distance between the points where the parabola meets the ground.

NewVocabulary
double root

Common Core State Standards

Content Standards
A.REI.4b Solve quadratic equations by inspection (e.g., for $x^2 = 49$), taking square roots, completing the square, the quadratic formula and factoring, as appropriate to the initial form of the equation. Recognize when the quadratic formula gives complex solutions and write them as $a \pm bi$ for real numbers a and b.

F.IF.7a Graph linear and quadratic functions and show intercepts, maxima, and minima.

Mathematical Practices
3 Construct viable arguments and critique the reasoning of others.
6 Attend to precision

1 **Solve by Graphing** A quadratic equation can be written in the standard form $ax^2 + bx + c = 0$, where $a \neq 0$. To write a quadratic function as an equation, replace y or $f(x)$ with 0. Recall that the solutions or roots of an equation can be identified by finding the x-intercepts of the related graph. Quadratic equations may have two, one, or no real solutions.

KeyConcept Solutions of Quadratic Equations

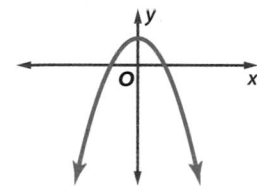
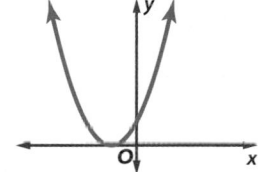
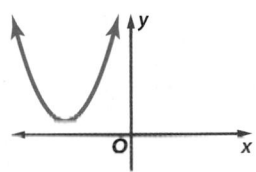

| *two* unique real solutions | *one* unique real solution | *no* real solutions |

Example 1 Two Roots

Solve $x^2 - 2x - 8 = 0$ by graphing.

Graph the related function $f(x) = x^2 - 2x - 8$.

The x-intercepts of the graph appear to be at -2 and 4, so the solutions are -2 and 4.

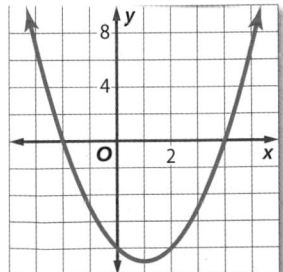

CHECK Check each solution in the original equation.

$$x^2 - 2x - 8 = 0 \qquad \text{Original equation} \qquad x^2 - 2x - 8 = 0$$
$$(-2)^2 - 2(-2) - 8 \stackrel{?}{=} 0 \qquad x = -2 \text{ or } x = 4 \qquad (4)^2 - 2(4) - 8 \stackrel{?}{=} 0$$
$$0 = 0 \checkmark \qquad \text{Simplify.} \qquad 0 = 0 \checkmark$$

> **GuidedPractice** Solve each equation by graphing.

1A. $-x^2 - 3x + 18 = 0$ **1B.** $x^2 - 4x + 3 = 0$

The solutions in Example 1 were two distinct numbers. Sometimes the two roots are the same number, called a **double root**.

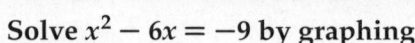

Example 2 Double Root

Solve $x^2 - 6x = -9$ by graphing.

Step 1 Rewrite the equation in standard form.

$$x^2 - 6x = -9 \qquad \text{Original equation}$$
$$x^2 - 6x + 9 = 0 \qquad \text{Add 9 to each side.}$$

Step 2 Graph the related function
$f(x) = x^2 - 6x + 9.$

Step 3 Locate the x-intercepts of the graph. Notice that the vertex of the parabola is the only x-intercept. Therefore, there is only one solution, 3.

CHECK Solve by factoring.

$$x^2 - 6x + 9 = 0 \qquad \text{Original equation}$$
$$(x - 3)(x - 3) = 0 \qquad \text{Factor.}$$
$$x - 3 = 0 \quad \text{or} \quad x - 3 = 0 \qquad \text{Zero Product Property}$$
$$x = 3 \qquad\qquad x = 3 \qquad \text{Add 3 to each side.}$$

The only solution is 3.

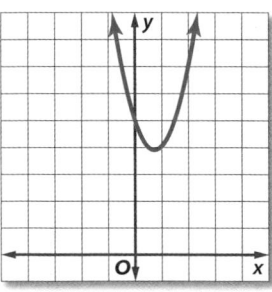

> **Watch**Out!
>
> **CCSS** Precision Solutions found from the graph of an equation may appear to be exact. Check them in the original equation to be sure.

▶ **Guided**Practice

Solve each equation by graphing.

2A. $x^2 + 25 = 10x$

2B. $x^2 = -8x - 16$

Sometimes the roots are not real numbers. Quadratic equations with solutions that are not real numbers lead us to extend the number system to allow for solutions of these equations. These numbers are called *complex numbers*. You will study complex numbers in Algebra 2.

Example 3 No Real Roots

Solve $2x^2 - 3x + 5 = 0$ by graphing.

Step 1 Rewrite the equation in standard form.
This equation is written in standard form.

Step 2 Graph the related function
$f(x) = 2x^2 - 3x + 5.$

Step 3 Locate the x-intercepts of the graph. This graph has no x-intercepts. Therefore, this equation has no real number solutions. The solution set is ∅.

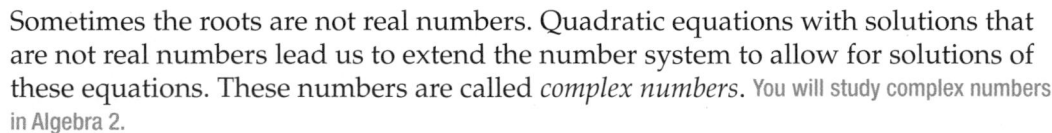

▶ **Guided**Practice

Solve each equation by graphing.

3A. $-x^2 - 3x = 5$

3B. $-2x^2 - 8 = 6x$

2 Estimate Solutions

The real roots found thus far have been integers. However, the roots of quadratic equations are usually not integers. In these cases, use estimation to approximate the roots of the equation.

Example 4 Approximate Roots with a Table

Solve $x^2 + 6x + 6 = 0$ by graphing. If integral roots cannot be found, estimate the roots to the nearest tenth.

Graph the related function $f(x) = x^2 + 6x + 6$.

The x-intercepts are located between −5 and −4 and between −2 and −1.

Make a table using an increment of 0.1 for the x-values located between −5 and −4 and between −2 and −1.

Look for a change in the signs of the function values. The function value that is closest to zero is the best approximation for a zero of the function.

x	−4.9	−4.8	−4.7	−4.6	−4.5	−4.4	−4.3	−4.2	−4.1
y	0.61	0.24	−0.11	−0.44	−0.75	−1.04	−1.31	−1.56	−1.79

x	−1.9	−1.8	−1.7	−1.6	−1.5	−1.4	−1.3	−1.2	−1.1
y	−1.79	−1.56	−1.31	−1.04	−0.75	−0.44	−0.11	0.24	0.61

For each table, the function value that is closest to zero when the sign changes is −0.11. Thus, the roots are approximately −4.7 and −1.3.

▶ **Guided**Practice

4. Solve $2x^2 + 6x - 3 = 0$ by graphing. If integral roots cannot be found, estimate the roots to the nearest tenth.

Approximating the x-intercepts of graphs is helpful for real-world applications.

● Real-World Example 5 Approximate Roots with a Calculator

SOCCER A goalie kicks a soccer ball with an upward velocity of 65 feet per second, and her foot meets the ball 1 foot off the ground. The quadratic function $h = -16t^2 + 65t + 1$ represents the height of the ball h in feet after t seconds. Approximately how long is the ball in the air?

You need to find the roots of the equation $-16t^2 + 65t + 1 = 0$. Use a graphing calculator to graph the related function $f(x) = -16t^2 + 65t + 1$.

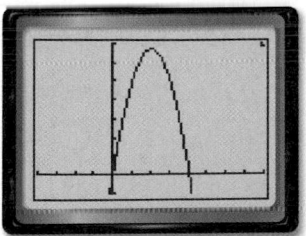

[−4, 7] scl: 1 by [−10, 70] scl: 10

The positive x-intercept of the graph is approximately 4. Therefore, the ball is in the air for approximately 4 seconds.

▶ **Guided**Practice

5. If the goalie kicks the soccer ball with an upward velocity of 55 feet per second and his foot meets the ball 2 feet off the ground, approximately how long is the ball in the air?

Examples 1–3 Solve each equation by graphing.

1. $x^2 + 3x - 10 = 0$

2. $2x^2 - 8x = 0$

3. $x^2 + 4x = -4$

4. $x^2 + 12 = -8x$

Example 4 Solve each equation by graphing. If integral roots cannot be found, estimate the roots to the nearest tenth.

5. $-x^2 - 5x + 1 = 0$

6. $-9 = x^2$

7. $x^2 = 25$

8. $x^2 - 8x = -9$

Example 5 **9. SCIENCE FAIR** Ricky built a model rocket. Its flight can be modeled by the equation shown, where h is the height of the rocket in feet after t seconds. About how long was Ricky's rocket in the air?

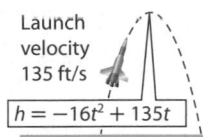

Launch velocity 135 ft/s

$h = -16t^2 + 135t$

Practice and Problem Solving

Extra Practice is on page R9.

Examples 1–3 Solve each equation by graphing.

10. $x^2 + 7x + 14 = 0$

11. $x^2 + 2x - 24 = 0$

12. $x^2 - 16x + 64 = 0$

13. $x^2 - 5x + 12 = 0$

14. $x^2 + 14x = -49$

15. $x^2 = 2x - 1$

16. $x^2 - 10x = -16$

17. $-2x^2 - 8x = 13$

18. $2x^2 - 16x = -30$

19. $2x^2 = -24x - 72$

20. $-3x^2 + 2x = 15$

21. $x^2 = -2x + 80$

Example 4 Solve each equation by graphing. If integral roots cannot be found, estimate the roots to the nearest tenth.

22. $x^2 + 2x - 9 = 0$

23. $x^2 - 4x = 20$

24. $x^2 + 3x = 18$

25. $2x^2 - 9x = -8$

26. $3x^2 = -2x + 7$

27. $5x = 25 - x^2$

Example 5 **28. SOFTBALL** The equation $h = -16t^2 + 47t + 3$ models the height h, in feet, of a ball that Sofia hits after t seconds. How long is the ball in the air?

29. RIDES The Terror Tower launches riders straight up and returns straight down. The equation $h = -16t^2 + 122t$ models the height h, in feet, of the riders from their starting position after t seconds. How long is it until the riders return to the bottom?

Use factoring to determine how many times the graph of each function intersects the x-axis. Identify each zero.

30. $y = x^2 - 8x + 16$

31. $y = x^2 + 4x + 4$

32. $y = x^2 + 2x - 24$

33. $y = x^2 + 12x + 32$

34. NUMBER THEORY Use a quadratic equation to find two numbers that have a sum of 9 and a product of 20.

35. NUMBER THEORY Use a quadratic equation to find two numbers that have a sum of 1 and a product of −12.

36. **CCSS MODELING** The height of a golf ball in the air can be modeled by the equation $h = -16t^2 + 76t$, where h is the height in feet of the ball after t seconds.

   a. How long was the ball in the air?

   b. What is the ball's maximum height?

   c. When will the ball reach its maximum height?

37 SKIING Stefanie is in a freestyle aerial competition. The equation $h = -16t^2 + 30t + 10$ models Stefanie's height h, in feet, t seconds after leaving the ramp.

a. How long is Stefanie in the air?

b. When will Stefanie reach a height of 15 feet?

c. To earn bonus points in the competition, you must reach a height of 20 feet. Will Stefanie earn bonus points?

38. MULTIPLE REPRESENTATIONS In this problem, you will explore how to further interpret the relationship between quadratic functions and graphs.

a. **Graphical** Graph $y = x^2$.

b. **Analytical** Name the vertex and two other points on the graph.

c. **Graphical** Graph $y = x^2 + 2$, $y = x^2 + 4$, and $y = x^2 + 6$ on the same coordinate plane as the previous graph.

d. **Analytical** Name the vertex and two points from each of these graphs that have the same x-coordinates as the first graph.

e. **Analytical** What conclusion can you draw from this?

GRAPHING CALCULATOR Solve each equation by graphing.

39. $x^3 - 3x^2 - 6x + 8 = 0$

40. $x^3 - 8x^2 + 15x = 0$

H.O.T. Problems   Use Higher-Order Thinking Skills

41. CCSS CRITIQUE Iku and Zachary are finding the number of real zeros of the function graphed at the right. Iku says that the function has no real zeros because there are no x-intercepts. Zachary says that the function has one real zero because the graph has a y-intercept. Is either of them correct? Explain your reasoning.

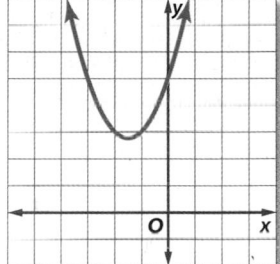

42. OPEN ENDED Describe a real-world situation in which a thrown object travels in the air. Write an equation that models the height of the object with respect to time, and determine how long the object travels in the air.

43. REASONING The graph shown is that of a *quadratic inequality*. Analyze the graph, and determine whether the y-value of a solution of the inequality is *sometimes*, *always*, or *never* greater than 2. Explain.

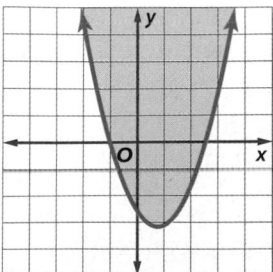

44. CHALLENGE Write a quadratic equation that has the roots described.

a. one double root

b. one rational (nonintegral) root and one integral root

c. two distinct integral roots that are additive opposites.

45. CHALLENGE Find the roots of $x^2 = 2.25$ without using a calculator. Explain your strategy.

46. WRITING IN MATH Explain how to approximate the roots of a quadratic equation when the roots are not integers.

47. Adrahan earned 50 out of 80 points on a test. What percentage did Adrahan score on the test?

   A 62.5%   **C** 6.25%

   B 16%   **D** 1.6%

48. Ernesto needs to loosen a bolt. He needs a wrench that is smaller than a $\frac{7}{8}$-inch wrench, but larger than a $\frac{3}{4}$-inch wrench. Which of the following sizes should Ernesto use?

   F $\frac{3}{8}$ inch   **H** $\frac{13}{16}$ inch

   G $\frac{5}{8}$ inch   **J** $\frac{15}{16}$ inch

49. EXTENDED RESPONSE Two boats leave a dock. One boat travels 4 miles east and then 5 miles north. The second boat travels 12 miles south and 9 miles west. Draw a diagram that represents the paths traveled by the boats. How far apart are the boats in miles?

50. The formula $s = \frac{1}{2}at^2$ represents the distance s in meters that a free-falling object will fall on a planet or moon in a given time t in seconds. Solve the formula for a, the acceleration due to gravity.

   A $a = \frac{1}{2}t^2 - s$ **C** $a = s - \frac{1}{2}t^2$

   B $a = 2s - t^2$ **D** $a = \frac{2s}{t^2}$

Spiral Review

Write the equation of the axis of symmetry, and find the coordinates of the vertex of the graph of each function. Identify the vertex as a maximum or minimum. Then graph the function. (Lesson 9-1)

51. $y = 3x^2$

52. $y = -4x^2 - 5$

53. $y = -x^2 + 4x - 7$

54. $y = x^2 - 6x - 8$

55. $y = 3x^2 + 2x + 1$

56. $y = -4x^2 - 8x + 5$

Solve each equation. Check the solutions. (Lesson 8-9)

57. $2x^2 = 32$

58. $(x - 4)^2 = 25$

59. $4x^2 - 4x + 1 = 16$

60. $2x^2 + 16x = -32$

61. $(x + 3)^2 = 5$

62. $4x^2 - 12x = -9$

Find each sum or difference. (Lesson 8-1)

63. $(3n^2 - 3) + (4 + 4n^2)$

64. $(2d^2 - 7d - 3) - (4d^2 + 7)$

65. $(2b^3 - 4b^2 + 4) - (3b^4 + 5b^2 - 9)$

66. $(8 - 4h^2 + 6h^4) + (5h^2 - 3 + 2h^3)$

67. GEOMETRY Supplementary angles are two angles with measures that have a sum of 180°. For the supplementary angles in the figure, the measure of the larger angle is 24° greater than the measure of the smaller angle. Write and solve a system of equations to find these measures. (Lesson 6-5)

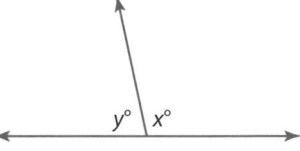

Write an equation in point-slope form for the line that passes through each point with the given slope. (Lesson 4-3)

68. $(2, 5), m = 3$

69. $(-3, 6), m = -7$

70. $(-1, -2), m = -\frac{1}{2}$

Skills Review

Graph each function.

71. $y = x^2 + 5$

72. $y = x^2 - 8$

73. $y = 2x^2 - 7$

74. $y = -x^2 + 2$

75. $y = -0.5x^2 - 3$

76. $y = (-x)^2 + 1$

Graphing Technology Lab
Quadratic Inequalities

Recall that the graph of a linear inequality consists of the boundary and the shaded half plane. The solution set of the inequality lies in the shaded region of the graph. Graphing quadratic inequalities is similar to graphing linear inequalities.

Activity 1 Shade Inside a Parabola

Graph $y \geq x^2 - 5x + 4$ in a standard viewing window.

First, clear all functions from the **Y=** list.

To graph $y \geq x^2 - 5x + 4$, enter the equation in the **Y=** list. Then use the left arrow to select =. Press [ENTER] until shading above the line is selected.

KEYSTROKES: [◄] [◄] [ENTER] [ENTER] [►] [►] [X,T,θ,n] [x²] [−] 5 [X,T,θ,n] [+] 4 [ZOOM] 6

[−10, 10] scl: 1 by [−10, 10] scl: 1

All ordered pairs for which y is *greater than or equal* to $x^2 - 5x + 4$ lie *above or on* the line and are solutions.

A similar procedure will be used to graph an inequality in which the shading is outside of the parabola.

Activity 2 Shade Outside a Parabola

Graph $y - 4 \leq x^2 - 5x$ in a standard viewing window.

First, clear the graph that is displayed.

KEYSTROKES: [Y=] [CLEAR]

Then rewrite $y - 4 \leq x^2 - 5x$ as $y \leq x^2 - 5x + 4$, and graph it.

KEYSTROKES: [◄] [◄] [ENTER] [ENTER] [ENTER] [►] [►] [X,T,θ,n] [x²] [−] 5 [X,T,θ,n] [+] 4 [GRAPH]

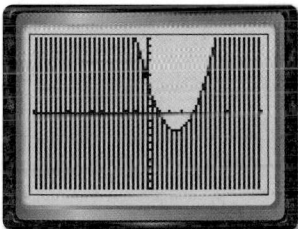

[−10, 10] scl: 1 by [−10, 10] scl: 1

All ordered pairs for which y is *less than or equal* to $x^2 - 5x + 4$ lie *below or on* the line and are solutions.

Exercises

1. Compare and contrast the two graphs shown above.

2. Graph $y - 2x + 6 \geq 5x^2$ in the standard viewing window. Name three solutions of the inequality.

3. Graph $y - 6x \leq -x^2 - 3$ in the standard viewing window. Name three solutions of the inequality.

Graphing Technology Lab
Family of Quadratic Functions

You have studied the effects of changing parameters in the equations of linear and exponential functions. You can use a graphing calculator to analyze how changing the parameters of the equation of a quadratic function affects the graphs in the family of quadratic functions.

CCSS Common Core State Standards
Content Standards
F.IF.7a Graph linear and quadratic functions and show intercepts, maxima, and minima.
F.BF.3 Identify the effect on the graph of replacing $f(x)$ by $f(x) + k$, $kf(x)$, $f(kx)$, and $f(x + k)$ for specific values of k (both positive and negative); find the value of k given the graphs. Experiment with cases and illustrate an explanation of the effects on the graph using technology.

PT

Activity 1 Change k in $y = a(x - h)^2 + k$

Graph the set of equations on the same screen in the standard viewing window. Describe any similarities and differences among the graphs.

$y = x^2, y = x^2 + 2, y = x^2 - 4$

Enter the equations in the **Y =** list and graph in the standard viewing window. Use the **ZOOM** feature to investigate the key features of the graphs.

The graphs have the same shape, and all open up. The vertex of each graph is on the y-axis, which is the axis of symmetry.

However, the graphs have different vertical positions. The graph of $y = x^2 + 2$ is shifted up 2 units. The graph of $y = x^2 - 4$ is shifted down 4 units.

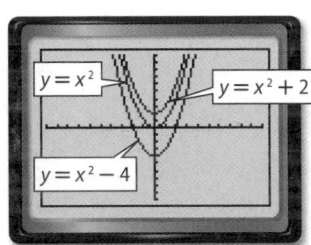

[−10, 10] scl: 1 by [−10, 10] scl: 1

Changing the value of h in $y = a(x - h)^2 + k$ affects the graphs in a different way than changing k.

Activity 2 Change h in $y = a(x - h)^2 + k$

Graph the set of equations on the same screen in the standard viewing window. Describe any similarities and differences among the graphs.

$y = x^2, y = (x + 2)^2, y = (x - 4)^2$

The graphs have the same shape, and all open up. The vertex of each graph is on the x-axis.

However, the graphs have different horizontal positions. Each has a different axis of symmetry. The graph of $y = (x + 2)^2$ is shifted to the left 2 units. The graph of $y = (x - 4)^2$ is shifted to the right 4 units.

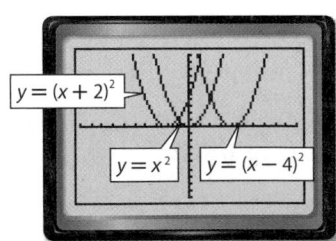

[−10, 10] scl: 1 by [−10, 10] scl: 1

It appears that changing the values of h and k in $y = a(x - h)^2 + k$ moves the graph vertically or horizontally. How does changing the value of a affect the graphs?

Activity 3 Change a in $y = a(x - h)^2 + k$

Graph each set of equations on the same screen in the standard viewing window. Describe any similarities and differences among the graphs.

a. $y = x^2$, $y = 2x^2$, $y = \frac{1}{3}x^2$

The graphs have the same vertex, they have the same axis of symmetry, and all open up.

However, the graphs have different widths. The graph of $y = 2x^2$ is narrower than the graph of $y = x^2$. The graph of $y = \frac{1}{3}x^2$ is wider than the graph of $y = x^2$.

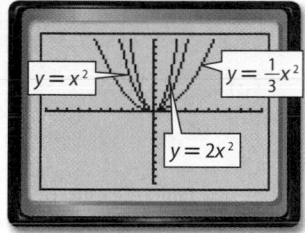

[−10, 10] scl: 1 by [−10, 10] scl: 1

b. $y = x^2$, $y = -\frac{1}{3}x^2$, $y = -2x^2$

The graphs have the same vertex and the same axis of symmetry.

However, the graphs of $y = -\frac{1}{3}x^2$ and $y = -2x^2$ open down. Also the graph of $y = -2x^2$ is narrower than the graph of $y = x^2$. The graph of $y = -\frac{1}{3}x^2$ is wider than the graph of $y = x^2$.

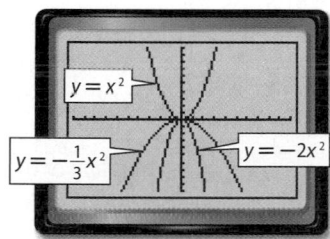

[−10, 10] scl: 1 by [−10, 10] scl: 1

Model and Analyze

How does each parameter affect the graph of $y = a(x - h)^2 + k$? Give examples.

1. k **2.** h **3.** a

Examine each pair of equations and predict the similarities and differences in their graphs. Use a graphing calculator to confirm your predictions. Write a sentence or two comparing the two graphs.

4. $y = x^2$, $y = x^2 + 3$ **5.** $y = \frac{1}{2}x^2$, $y = 3x^2$

6. $y = x^2$, $y = (x - 5)^2$ **7.** $y = 3x^2$, $y = -3x^2$

8. $y = x^2$, $y = -4x^2$ **9.** $y = x^2 - 1$, $y = x^2 + 2$

10. $y = \frac{1}{2}x^2 + 3$, $y = -2x^2$ **11.** $y = x^2 - 4$, $y = (x - 4)^2$

Transformations of Quadratic Functions

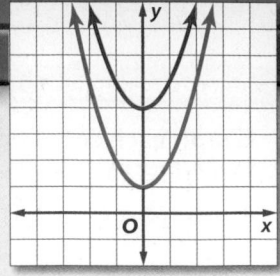

::Then	::Now	::Why?
● You graphed quadratic functions by using the vertex and axis of symmetry.	**1** Apply translations to quadratic functions. **2** Apply dilations and reflections to quadratic functions.	● The graphs of the parabolas shown at the right are the same size and shape, but notice that the vertex of the red parabola is higher on the y-axis than the vertex of the blue parabola. Shifting a parabola up and down is an example of a transformation.

 NewVocabulary
transformation
translation
dilation
reflection
vertex form

Common Core State Standards

Content Standards
A.SSE.3b Complete the square in a quadratic expression to reveal the maximum or minimum value of the function it defines.

F.IF.7a Graph linear and quadratic functions and show intercepts, maxima, and minima.

Mathematical Practices
1 Make sense of problems and persevere in solving them.
8 Look for and express regularity in repeated reasoning.

1 **Translations** A **transformation** changes the position or size of a figure. One transformation, a **translation**, moves a figure up, down, left, or right. When a constant k is added to or subtracted from the parent function, the graph of the resulting function $f(x) \pm k$ is the graph of the parent function translated up or down.

The parent function of the family of quadratics is $f(x) = x^2$. All other quadratic functions have graphs that are transformations of the graph of $f(x) = x^2$.

KeyConcept Vertical Translations

The graph of $f(x) = x^2 + k$ is the graph of $f(x) = x^2$ translated vertically.

If $k > 0$, the graph of $f(x) = x^2$ is translated $|k|$ units **up**.

If $k < 0$, the graph of $f(x) = x^2$ is translated $|k|$ units **down**.

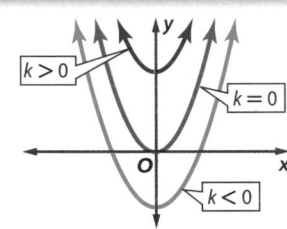

Example 1 Describe and Graph Translations

Describe how the graph of each function is related to the graph of $f(x) = x^2$.

a. $h(x) = x^2 + 3$

$k = 3$ and $3 > 0$
$h(x)$ is a translation of the graph of $f(x) = x^2$ up 3 units.

b. $g(x) = x^2 - 4$

$k = -4$ and $-4 < 0$
$g(x)$ is a translation of the graph of $f(x) = x^2$ down 4 units.

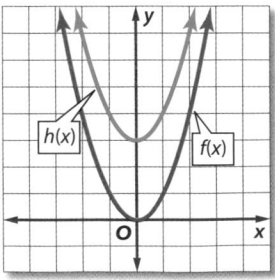

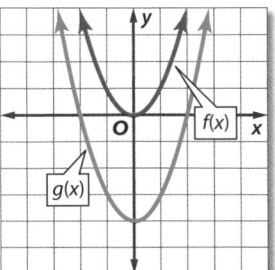

GuidedPractice

1A. $f(x) = x^2 - 7$ **1B.** $g(x) = 5 + x^2$ **1C.** $h(x) = -5 + x^2$ **1D.** $f(x) = x^2 + 1$

A quadratic graph can be translated horizontally by subtracting an h term from x.

KeyConcept Horizontal Translations

The graph of $g(x) = (x - h)^2$ is the graph of $f(x) = x^2$ translated horizontally.

If $h > 0$, the graph of $f(x) = x^2$ is translated h units to the **right**.

If $h < 0$, the graph of $f(x) = x^2$ is translated $|h|$ units to the **left**.

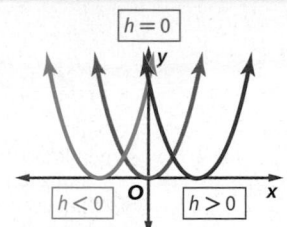

Example 2 Horizontal Translations

Describe how the graph of each function is related to the graph of $f(x) = x^2$.

a. $g(x) = (x - 2)^2$

$k = 0, h = 2$ and $2 > 0$
$g(x)$ is a translation of the graph of $f(x) = x^2$ to the right 2 units.

b. $g(x) = (x + 1)^2$

$k = 0, h = -1$ and $-1 < 0$
$g(x)$ is a translation of the graph of $f(x) = x^2$ to the left 1 unit.

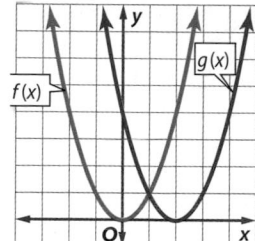

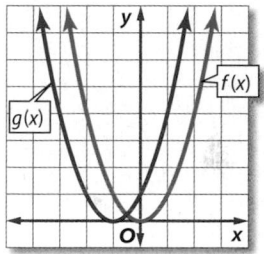

▶ **Guided**Practice

2A. $g(x) = (x - 3)^2$

2B. $g(x) = (x + 2)^2$

A quadratic graph can be translated both horizontally and vertically.

Example 3 Horizontal and Vertical Translations

Describe how the graph of each function is related to the graph of $f(x) = x^2$.

a. $g(x) = (x - 3)^2 + 2$

$k = 2, h = 3$ and $3 > 0$
$g(x)$ is a translation of the graph of $f(x) = x^2$ to the right 3 units and up 2 units.

b. $g(x) = (x + 3)^2 - 1$

$k = -1, h = -3$ and $-3 < 0$
$g(x)$ is a translation of the graph of $f(x) = x^2$ to the left 3 units and down 1 unit.

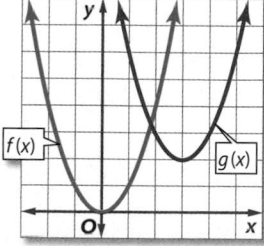

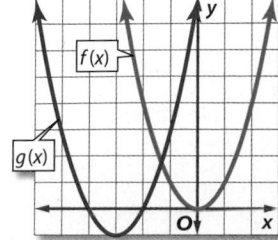

▶ **Guided**Practice

3A. $g(x) = (x + 2)^2 + 3$

3B. $g(x) = (x - 4)^2 - 4$

2 Dilations and Reflections Another type of transformation is a dilation. A **dilation** makes the graph narrower than the parent graph or wider than the parent graph. When the parent function $f(x) = x^2$ is multiplied by a constant a, the graph of the resulting function $f(x) = ax^2$ is either stretched or compressed vertically.

KeyConcept Dilations

The graph of $g(x) = ax^2$ is the graph of $f(x) = x^2$ stretched or compressed vertically.

If $|a| > 1$, the graph of $f(x) = x^2$ is stretched vertically.

If $0 < |a| < 1$, the graph of $f(x) = x^2$ is compressed vertically.

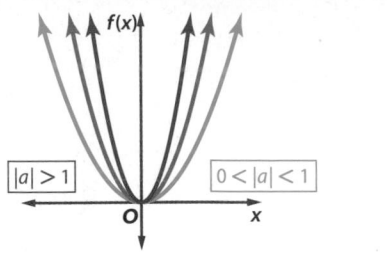

PT

Example 4 Describe and Graph Dilations

StudyTip

CCSS Sense-Making When the graph of a quadratic function is stretched vertically, the shape of the graph is narrower than that of the parent function. When it is compressed vertically, the graph is wider than the parent function.

Describe how the graph of each function is related to the graph of $f(x) = x^2$.

a. $h(x) = \frac{1}{2}x^2$

$a = \frac{1}{2}$ and $0 < \frac{1}{2} < 1$

$h(x)$ is a dilation of the graph of $f(x) = x^2$ that is compressed vertically.

b. $g(x) = 3x^2 + 2$

$a = 3$ and $3 > 1$, $k = 2$ and $2 > 0$

$g(x)$ is a dilation of the graph of $f(x) = x^2$ that is stretched vertically and translated up 2 units.

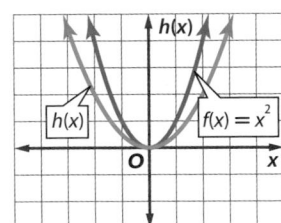
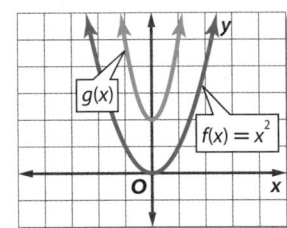

GuidedPractice

4A. $j(x) = 2x^2$ **4B.** $h(x) = 5x^2 - 2$ **4C.** $g(x) = \frac{1}{3}x^2 + 2$

A **reflection** flips a figure across a line.

KeyConcept Reflections

The graph of $-f(x)$ is the reflection of the graph of $f(x) = x^2$ across the x-axis.

The graph of $f(-x)$ is the reflection of the graph of $f(x) = x^2$ across the y-axis.

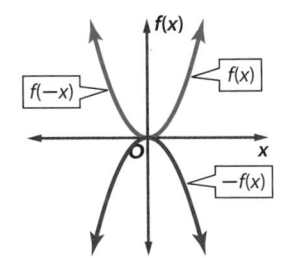

StudyTip

Reflection A reflection of $f(x) = x^2$ across the y-axis results in the same function, because $f(-x) = (-x)^2 = x^2$.

WatchOut!

Transformations The graph of $f(x) = -ax^2$ can result in two transformations of the graph of $f(x) = x^2$: a reflection across the x-axis if $a > 0$ and either a compression or expansion depending on the absolute value of a.

Example 5 Describe and Graph Reflections

Describe how the graph of each function is related to the graph of $f(x) = x^2$.

a. $g(x) = -2x^2 - 3$

- $a = -2$, $-2 < 0$, and $|-2| > 1$, so there is a reflection across the x-axis and the graph is vertically stretched.
- $k = -3$ and $-3 < 0$, so there is a translation down 3 units.

b. $h(x) = -4(x + 2)^2 + 1$

- $a = -4$, $-4 < 0$, and $|-4| > 1$, so there is a reflection across the x-axis and the graph is vertically stretched.
- $h = -2$ and $-2 < 0$, so there is a translation 2 units to the left.
- $k = 1$ and $1 > 0$, so there is a translation up 1 unit.

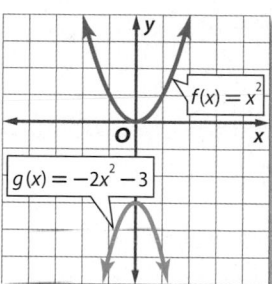

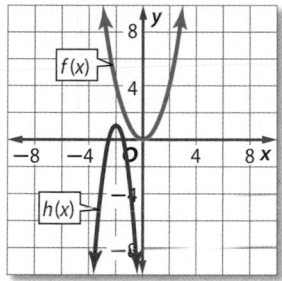

▶ **Guided**Practice

5A. $h(x) = 2(-x)^2 - 9$ **5B.** $g(x) = \frac{1}{5}x^2 + 3$ **5C.** $j(x) = -2(x - 1)^2 - 2$

You can use what you know about the characteristics of graphs of quadratic equations to match an equation with a graph.

Standardized Test Example 6 Identify an Equation for a Graph

Which is an equation for the function shown in the graph?

A $y = \frac{1}{2}x^2 - 5$ **C** $y = -\frac{1}{2}x^2 + 5$

B $y = -2x^2 - 5$ **D** $y = 2x^2 + 5$

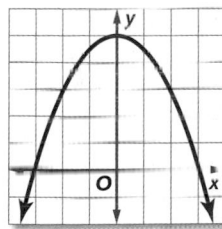

Read the Test Item

You are given a graph. You need to find its equation.

Solve the Test Item

The graph opens downward, so the graph of $y = x^2$ has been reflected across the x-axis. The leading coefficient should be negative, so eliminate choices A and D.

The parabola is translated up 5 units, so $k = 5$. Look at the equations. Only choices C and D have $k = 5$. The answer is C.

▶ **Guided**Practice

6. Which is the graph of $y = -3x^2 + 1$?

F
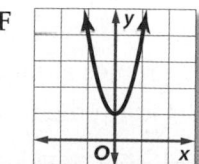

G
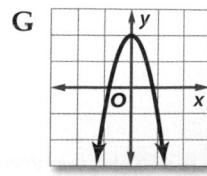

H
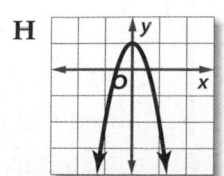

J
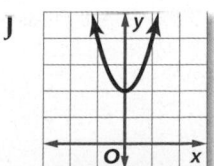

A quadratic function written in the form $f(x) = a(x - h)^2 + k$ is said to be in **vertex form**. Transformations of the parent graph are easily found from an equation in vertex form.

ConceptSummary Transformations of Quadratic Functions

$$f(x) = a(x - h)^2 + k$$

h, Horizontal Translation
h units to the right if h is positive
$|h|$ units to the left if h is negative

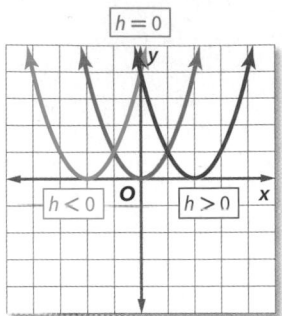

k, Vertical Translation
k units up if k is positive
$|k|$ units down if k is negative

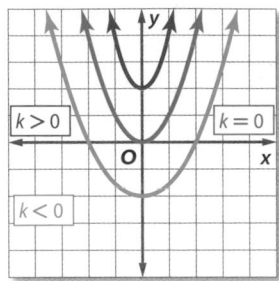

a, Reflection
If $a > 0$, the graph opens up.
If $a < 0$, the graph opens down.

a, Dilation
If $|a| > 1$, the graph is stretched vertically. If $0 < |a| < 1$, the graph is compressed vertically.

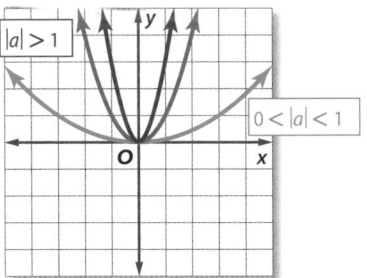

Real-World Example 7 Transformations with a Calculator

FIREWORKS During a firework show, the height h in meters of a specific rocket after t seconds can be modeled by $h(t) = -4.6(t - 3)^2 + 75$. Graph the function. How is it related to the graph of $f(x) = x^2$?

Four separate transformations are occurring.

The negative sign of the coefficient of x^2 causes a reflection across the x-axis. A dilation occurs, which stretches the graph vertically. There are also translations up 75 units and to the right 3 units.

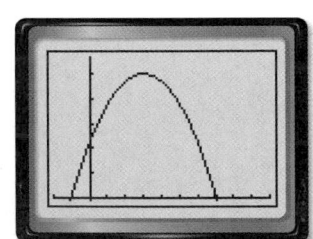

[−2, 10] scl: 1 by [−2, 85] scl: 15

▶ **Guided**Practice

7. **MONUMENTS** The St. Louis Arch resembles a quadratic and can be modeled by $h(x) = -\frac{2}{315}x^2 + 630$. Graph the function. How is it related to the graph of $f(x) = x^2$?

Examples
1–5, 7

Describe how the graph of each function is related to the graph of $f(x) = x^2$.

1. $g(x) = x^2 - 11$

2. $h(x) = \frac{1}{2}(x-2)^2$

3. $h(x) = -x^2 + 8$

4. $g(x) = x^2 + 6$

5. $g(x) = -4(x+3)^2$

6. $h(x) = -x^2 - 2$

Example 6

7. MULTIPLE CHOICE Which is an equation for the function shown in the graph?

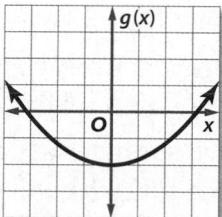

A $g(x) = \frac{1}{5}x^2 + 2$

C $g(x) = \frac{1}{5}x^2 - 2$

B $g(x) = -5x^2 - 2$

D $g(x) = -\frac{1}{5}x^2 - 2$

Practice and Problem Solving   Extra Practice is on page R9.

Examples
1–5, 7

Describe how the graph of each function is related to the graph of $f(x) = x^2$.

8. $g(x) = -10 + x^2$

9 $h(x) = -7 - x^2$

10. $g(x) = 2(x-3)^2 + 8$

11. $h(x) = 6 + \frac{2}{3}x^2$

12. $g(x) = -5 - \frac{4}{3}x^2$

13. $h(x) = 3 + \frac{5}{2}x^2$

14. $g(x) = 0.25x^2 - 1.1$

15. $h(x) = 1.35(x+1)^2 + 2.6$

16. $g(x) = \frac{3}{4}x^2 + \frac{5}{6}$

17. $h(x) = 1.01x^2 - 6.5$

Example 6

Match each equation to its graph.

A
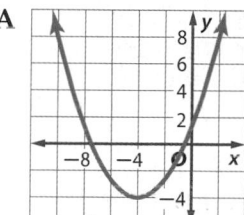

B
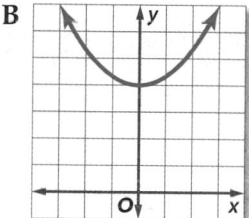

C
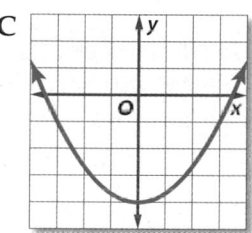

D
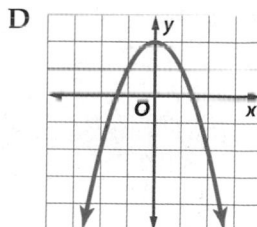

E
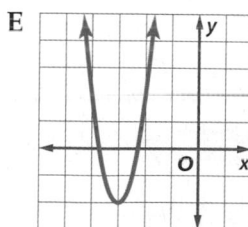

F
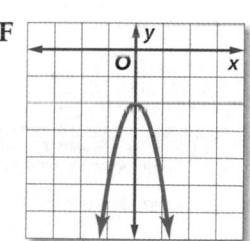

18. $y = \frac{1}{3}x^2 - 4$

19. $y = \frac{1}{3}(x+4)^2 - 4$

20. $y = \frac{1}{3}x^2 + 4$

21. $y = -3x^2 - 2$

22. $y = -x^2 + 2$

23. $y = (2x+6)^2 + 2$

24. SQUIRRELS A squirrel 12 feet above the ground drops an acorn from a tree. The function $h = -16t^2 + 12$ models the height of the acorn above the ground in feet after t seconds. Graph the function, and compare this graph to the graph of its parent function.

CCSS REGULARITY List the functions in order from the most stretched vertically to the least stretched vertically graph.

25. $g(x) = 2x^2, h(x) = \frac{1}{2}x^2$

26. $g(x) = -3x^2, h(x) = \frac{2}{3}x^2$

27. $g(x) = -4x^2, h(x) = 6x^2, f(x) = 0.3x^2$

28. $g(x) = -x^2, h(x) = \frac{5}{3}x^2, f(x) = -4.5x^2$

29 **ROCKS** A rock falls from a cliff 300 feet above the ground. At the same time, another rock falls from a cliff 700 feet above the ground.

    a. Write functions that model the height h of each rock after t seconds.

    b. If the rocks fall at the same time, how much sooner will the first rock reach the ground?

30. **SPRINKLERS** The path of water from a sprinkler can be modeled by quadratic functions. The following functions model paths for three different sprinklers.

    Sprinkler A: $y = -0.35x^2 + 3.5$ Sprinkler B: $y = -0.21x^2 + 1.7$
    Sprinkler C: $y = -0.08x^2 + 2.4$

    a. Which sprinkler will send water the farthest? Explain.

    b. Which sprinkler will send water the highest? Explain.

    c. Which sprinkler will produce the narrowest path? Explain.

31. **GOLF** The path of a drive can be modeled by a quadratic function where $g(x)$ is the vertical distance in yards of the ball from the ground and x is the horizontal distance in yards.

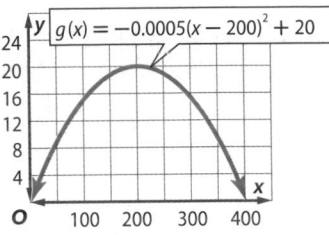

    a. How can you obtain $g(x)$ from the graph of $f(x) = x^2$.

    b. A second golfer hits a ball from the red tee, which is 30 yards closer to the hole. What function $h(x)$ can be used to describe the second golfer's shot?

Describe the transformations to obtain the graph of $g(x)$ from the graph of $f(x)$.

32. $f(x) = x^2 + 3$
    $g(x) = x^2 - 2$

33. $f(x) = x^2 - 4$
    $g(x) = (x - 2)^2 + 7$

34. $f(x) = -6x^2$
    $g(x) = -3x^2$

35. **COMBINING FUNCTIONS** An engineer created a self-refueling generator that burns fuel according to the function $g(t) = -t^2 + 10t + 200$, where t represents the time in hours and $g(t)$ represents the number of gallons remaining.

    a. How long will it take for the generator to run out of fuel?

    b. The engine self-refuels at a rate of 40 gallons per hour. Write a linear function $h(t)$ to represent the refueling of the generator.

    c. Find $T(t) = g(t) + h(t)$. What does this new function represent?

    d. Will the generator run out of fuel? If so, when?

H.O.T. Problems   Use Higher-Order Thinking Skills

36. **REASONING** Are the following statements *sometimes*, *always*, or *never* true? Explain.

    a. The graph of $y = x^2 + k$ has its vertex at the origin.

    b. The graphs of $y = ax^2$ and its reflection over the x-axis are the same width.

    c. The graph of $y = x^2 + k$, where $k \geq 0$, and the graph of a quadratic with vertex at $(0, -3)$ have the same maximum or minimum point.

37. **CHALLENGE** Write a function of the form $y = ax^2 + k$ with a graph that passes through the points $(-2, 3)$ and $(4, 15)$.

38. **CCSS ARGUMENTS** Determine whether all quadratic functions that are reflected across the y-axis produce the same graph. Explain your answer.

39. **OPEN ENDED** Write a quadratic function that opens downward and is wider than the parent graph.

40. **WRITING IN MATH** Describe how the values of a and k affect the graphical and tabular representations for the functions $y = ax^2$, $y = x^2 + k$, and $y = ax^2 + k$.

41. SHORT RESPONSE A tutor charges a flat fee of $55 and $30 for each hour of work. Write a function that represents the total charge C, in terms of the number of hours h worked.

42. Which *best* describes the graph of $y = 2x^2$?

A a line with a y-intercept of 2 and an x-intercept at the origin

B a parabola with a minimum point at $(0, 0)$ and that is wider than the graph of $y = x^2$

C a parabola with a maximum point at $(0, 0)$ and that is narrower than the graph of $y = x^2$

D a parabola with a minimum point at $(0, 0)$ and that is narrower than the graph of $y = x^2$

43. Candace is 5 feet tall. If 1 inch is about 2.54 centimeters, how tall is Candace to the nearest centimeter?

F 13 cm   **H** 123 cm
G 26 cm   **J** 152 cm

44. While in England, Imani spent 49.60 British pounds on a pair of jeans. If this is equivalent to $100 in U.S. currency, how many British pounds would Imani have spent on a sweater that cost $60?

A 2976 pounds
B 29.76 pounds
C 19.84 pounds
D 8.26 pounds

Spiral Review

Solve each equation by graphing. (Lesson 9-2)

45. $x^2 + 6 = 0$

46. $x^2 - 10x = -24$

47. $x^2 + 5x + 4 = 0$

48. $2x^2 - x = 3$

49. $2x^2 - x = 15$

50. $12x^2 = -11x + 15$

Find the vertex, the equation of the axis of symmetry, and the y-intercept of each graph. (Lesson 9-1)

51.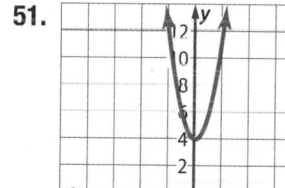

52.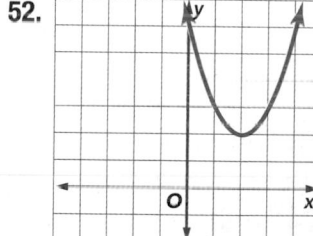

53.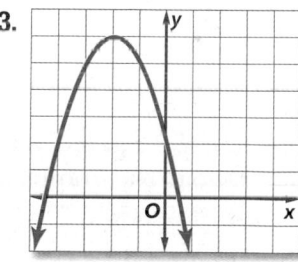

54. CLASS TRIP Mr. Wong's American History class will take taxis from their hotel in Washington, D.C., to the Lincoln Memorial. The fare is $2.75 for the first mile and $1.25 for each additional mile. If the distance is m miles and t taxis are needed, write an expression for the cost to transport the group. (Lesson 8-2)

Solve each inequality. Check your solution. (Lesson 5-3)

55. $-3t + 6 \leq -3$

56. $59 > -5 - 8f$

57. $-2 - \dfrac{d}{5} < 23$

Skills Review

Determine whether each trinomial is a perfect square trinomial. If so, factor it.

58. $16x^2 - 24x + 9$

59. $9x^2 + 6x + 1$

60. $25x^2 - 60x + 36$

61. $x^2 - 8x + 81$

62. $36x^2 - 84x + 49$

63. $4x^2 - 3x + 9$

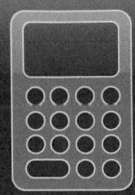

Graphing Technology Lab
Systems of Linear and Quadratic Equations

You can use a graphing calculator to solve systems involving linear and quadratic equations.

CCSS Common Core State Standards
Content Standards
A.REI.7 Solve a simple system consisting of a linear equation and a quadratic equation in two variables algebraically and graphically.
F.IF.7a Graph linear and quadratic functions and show intercepts, maxima, and minima.

Activity 1   Solve a System of Equations Graphically

Use a graphing calculator to solve the system of equations.

$$y = x^2 - x - 6$$
$$y = x - 3$$

Step 1 Enter each equation in the Y= list.

KEYSTROKES: $\boxed{X,T,\theta,n}$ $\boxed{x^2}$ $\boxed{-}$ $\boxed{X,T,\theta,n}$ $\boxed{-}$
6 $\boxed{ENTER}$ $\boxed{X,T,\theta,n}$ $\boxed{-}$ 3

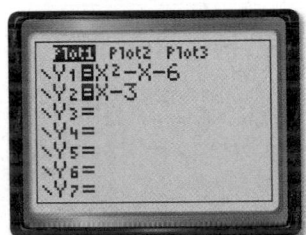

Step 2 Graph the system.   KEYSTROKES: $\boxed{GRAPH}$

The graphs intersect at two points. So, there are two solutions.

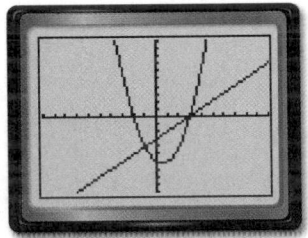

[−10, 10] scl: 1 by [−10, 10] scl: 1

Step 3 Find the intersection on the left by using the **CALC** menu.

KEYSTROKES: $\boxed{2nd}$ [CALC] 5 $\boxed{ENTER}$ $\boxed{ENTER}$

Use the arrow keys to move the cursor close to the intersection on the left. Press $\boxed{ENTER}$ again.

The graphs intersect at $(-1, -4)$.

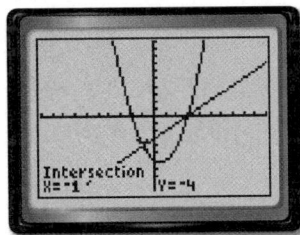

[−10, 10] scl: 1 by [−10, 10] scl: 1

Step 4 Repeat Step 3 but move the cursor to the other intersection. The graphs intersect at $(3, 0)$.

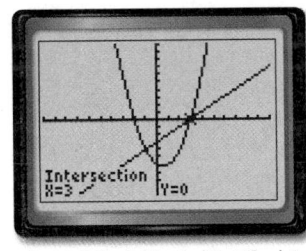

[−10, 10] scl: 1 by [−10, 10] scl: 1

You can use a graphing calculator to verify solutions of systems found algebraically.

Activity 2   Check Solutions Graphically

Solve the system of equations algebraically. Use a graphing calculator to check your solutions.

$$y = 2x - 6$$
$$y = x^2 - 8x + 19$$

Step 1 Set the expressions equal to each other, and solve for x.

$x^2 - 8x + 19 = 2x - 6$	Substitute $x^2 - 8x + 19$ for y.
$x^2 - 10x + 25 = 0$	Simplify.
$(x - 5)^2 = 0$	Factor.
$x = 5$	Solve for x.

Step 2 Substitute the x-value into either equation to find the y-value: $y = 2(5) - 6$ or 4.

Step 3 Graph the system and find the point(s) of intersection as in Activity 1.

The graphs intersect at $(5, 4)$. Thus, the solution of the system of equations is $(5, 4)$.

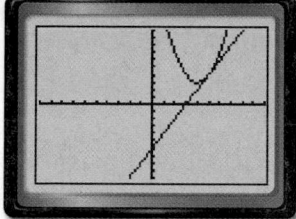

[−10, 10] scl: 1 by [−10, 10] scl: 1

You can solve a quadratic equation graphically by writing each side of the equation as a separate function. The x-coordinate of the point(s) of intersection will be the solution of the equation, since at that point(s) the original equations are true.

Activity 3 Use a System to Solve an Equation

Use a system of equations to solve $x^2 - 3x + 1 = \frac{11}{4}x - 6$.

Step 1 Write as a system of equations.
$$y = x^2 - 3x + 1$$
$$y = \frac{11}{4}x - 6$$

Step 2 Enter the equations into the graphing calculator, and graph them.

Step 3 Use the [CALC] menu to find the two points of intersection.

The graphs intersect at $(1.75, -1.1875)$ and $(4, 5)$. Thus, the solutions of $x^2 - 3x + 1 = \frac{11}{4}x - 6$ are 1.75 and 4.

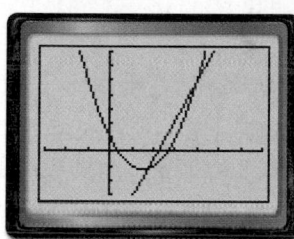

[−3, 7] scl: 1 by [−3, 7] scl: 1

Exercises

Use a graphing calculator to solve each system of equations.

1. $y = x^2$
$y = 2x$

2. $y = -2x^2 + 7x - 2$
$y = 3 - 4x$

3. $y = -x^2 + 4$
$y = \frac{1}{2}x + 5$

Solve each system of equations algebraically. Use a graphing calculator to check your solutions.

4. $y = x^2 + 7x + 12$
$y = 2x + 8$

5. $y = x^2 - x - 20$
$y = 3x + 12$

6. $y = 3x^2 - x - 2$
$y = -2x + 2$

Use a system of equations to solve each equation.

7. $x^2 = -2x - 1$

8. $\frac{1}{2}x^2 - 4 = 3x + 4$

9. $x^2 + 5x + 5 = -x - 8$

CHALLENGE Use a graphing calculator to solve other types of systems.

10. $y = x^2 + 3x - 5$
$y = -x^2$

11. $y = \frac{3}{4}x$
$x^2 + y^2 = 1$ (*Hint:* Enter as two functions, $y = \sqrt{1 - x^2}$ and $y = -\sqrt{1 - x^2}$.)

Solving Quadratic Equations by Completing the Square

∴Then	∴Now	∴Why?

- You solved quadratic equations by using the square root property.

- **1** Complete the square to write perfect square trinomials.

- **2** Solve quadratic equations by completing the square.

- In competitions, skateboarders may launch themselves from a half pipe into the air to perform tricks. The equation $h = -16t^2 + 20t + 12$ can be used to model their height, in feet, after t seconds.

  To find how long a skateboarder is in the air if he is 25 feet above the half pipe, you can solve $25 = -16t^2 + 20t + 12$ by using a method called completing the square.

 NewVocabulary
completing the square

 Common Core State Standards

Content Standards
A.REI.4 Solve quadratic equations in one variable.

a. Use the method of completing the square to transform any quadratic equation in x into an equation of the form $(x - p)^2 = q$ that has the same solutions. Derive the quadratic formula from this form.

b. Solve quadratic equations by inspection (e.g., for $x^2 = 49$), taking square roots, completing the square, the quadratic formula and factoring, as appropriate to the initial form of the equation. Recognize when the quadratic formula gives complex solutions and write them as $a \pm bi$ for real numbers a and b.

F.IF.8a Use the process of factoring and completing the square in a quadratic function to show zeros, extreme values, and symmetry of the graph, and interpret these in terms of a context.

Mathematical Practices
4 Model with mathematics.

1 Complete the Square You have previously solved equations by taking the square root of each side. This method worked only because the expression on the left-hand side was a perfect square. In perfect square trinomials in which the leading coefficient is 1, there is a relationship between the **coefficient of the x-term** and the **constant term**.

$$(x + 5)^2 = x^2 + 2(5)(x) + 5^2$$
$$= x^2 + 10x + 25$$

Notice that $\left(\dfrac{10}{2}\right)^2 = 25$. To get the constant term, divide the coefficient of the x-term by 2 and square the result. Any quadratic expression in the form $x^2 + bx$ can be made into a perfect square by using a method called **completing the square**.

KeyConcept Completing the Square

Words	To complete the square for any quadratic expression of the form $x^2 + bx$, follow the steps below.
	Step 1 Find one half of b, the coefficient of x.
	Step 2 Square the result in Step 1.
	Step 3 Add the result of Step 2 to $x^2 + bx$.
Symbols	$x^2 + bx + \left(\dfrac{b}{2}\right)^2 = \left(x + \dfrac{b}{2}\right)^2$

Example 1 Complete the Square

Find the value of c that makes $x^2 + 4x + c$ a perfect square trinomial.

Method 1 Use algebra tiles.

Arrange the tiles for $x^2 + 4x$ so that the two sides of the figure are congruent.

To make the figure a square, add 4 positive 1-tiles.

Method 2 Use complete the square algorithm.

Step 1 Find $\frac{1}{2}$ of 4.	$\frac{4}{2} = 2$
Step 2 Square the result in Step 1.	$2^2 = 4$
Step 3 Add the result of Step 2 to $x^2 + 4x$.	$x^2 + 4x + 4$

Thus, $c = 4$. Notice that $x^2 + 4x + 4 = (x + 2)^2$.

▶ **Guided**Practice

1. Find the value of c that makes $r^2 - 8r + c$ a perfect square trinomial.

2 **Solve Equations by Completing the Square** You can complete the square to solve quadratic equations. First, you must isolate the x^2- and bx-terms.

Example 2 Solve an Equation by Completing the Square

Solve $x^2 - 6x + 12 = 19$ by completing the square.

$x^2 - 6x + 12 = 19$	Original equation
$x^2 - 6x = 7$	Subtract 12 from each side.
$x^2 - 6x + 9 = 7 + 9$	Since $\left(\frac{-6}{2}\right)^2 = 9$, add 9 to each side.
$(x - 3)^2 = 16$	Factor $x^2 - 6x + 9$.
$x - 3 = \pm 4$	Take the square root of each side.
$x = 3 \pm 4$	Add 3 to each side.

$x = 3 + 4$ or $x = 3 - 4$	Separate the solutions.
$= 7$ $= -1$	The solutions are 7 and -1.

▶ **Guided**Practice

2. Solve $x^2 - 12x + 3 = 8$ by completing the square.

To solve a quadratic equation in which the leading coefficient is not 1, divide each term by the coefficient. Then isolate the x^2- and x-terms and complete the square.

Example 3 Equation with $a \neq 1$

Solve $-2x^2 + 8x - 18 = 0$ by completing the square.

$-2x^2 + 8x - 18 = 0$	Original equation
$\dfrac{-2x^2 + 8x - 18}{-2} = \dfrac{0}{-2}$	Divide each side by -2.
$x^2 - 4x + 9 = 0$	Simplify.
$x^2 - 4x = -9$	Subtract 9 from each side.
$x^2 - 4x + 4 = -9 + 4$	Since $\left(\frac{-4}{2}\right)^2 = 4$, add 4 to each side.
$(x - 2)^2 = -5$	Factor $x^2 - 4x + 4$.

No real number has a negative square. So, this equation has no real solutions.

▶ **Guided**Practice

3. Solve $3x^2 - 9x - 3 = 21$ by completing the square.

Real-World Example 4 Solve a Problem by Completing the Square

JERSEYS The senior class at Bay High School buys jerseys to wear to the football games. The cost of the jerseys can be modeled by the equation $C = 0.1x^2 + 2.4x + 25$, where C is the amount it costs to buy x jerseys. How many jerseys can they purchase for $430?

The seniors have $430, so set the equation equal to 430 and complete the square.

$0.1x^2 + 2.4x + 25 = 430$	Original equation
$\dfrac{0.1x^2 + 2.4x + 25}{0.1} = \dfrac{430}{0.1}$	Divide each side by 0.1.
$x^2 + 24x + 250 = 4300$	Simplify.
$x^2 + 24x + 250 - \mathbf{250} = 4300 - \mathbf{250}$	Subtract 250 from each side.
$x^2 + 24x = 4050$	Simplify.
$x^2 + 24x + \mathbf{144} = 4050 + \mathbf{144}$	Since $\left(\dfrac{24}{2}\right)^2 = 144$, add 144 to each side.
$x^2 + 24x + 144 = 4194$	Simplify.
$(x + 12)^2 = 4194$	Factor $x^2 + 24x + 144$.
$x + 12 = \pm\sqrt{4194}$	Take the square root of each side.
$x = -12 \pm\sqrt{4194}$	Subtract 12 from each side.

Use a calculator to approximate each value of x.

$x = -12 + \sqrt{4194}$ or $x = -12 - \sqrt{4194}$ Separate the solutions.

≈ 52.8 ≈ -76.8 Evaluate.

Since you cannot buy a negative number of jerseys, the negative solution is not reasonable. The seniors can afford to buy 52 jerseys.

GuidedPractice

4. If the senior class were able to raise $620, how many jerseys could they buy?

Real-WorldLink

The oldest public high school rivalry takes place between Wellesley High School and Needham Heights High School in Massachusetts. The first football game between them took place on Thanksgiving morning in 1882 in Needham.

Source: USA Football

Check Your Understanding

 = Step-by-Step Solutions begin on page R13.

Example 1   Find the value of c that makes each trinomial a perfect square.

1 $x^2 - 18x + c$ **2.** $x^2 + 22x + c$

3. $x^2 + 9x + c$ **4.** $x^2 - 7x + c$

Examples 2–3 Solve each equation by completing the square. Round to the nearest tenth if necessary.

5. $x^2 + 4x = 6$ **6.** $x^2 - 8x = -9$

7. $4x^2 + 9x - 1 = 0$ **8.** $-2x^2 + 10x + 22 = 4$

Example 4   **9.** **MODELING** Collin is building a deck on the back of his family's house. He has enough lumber for the deck to be 144 square feet. The length should be 10 feet more than its width. What should the dimensions of the deck be?

Example 1   Find the value of c that makes each trinomial a perfect square.

10. $x^2 + 26x + c$ **11.** $x^2 - 24x + c$ **12.** $x^2 - 19x + c$

13. $x^2 + 17x + c$ **14.** $x^2 + 5x + c$ **15.** $x^2 - 13x + c$

16. $x^2 - 22x + c$ **17.** $x^2 - 15x + c$ **18.** $x^2 + 24x + c$

Examples 2–3 Solve each equation by completing the square. Round to the nearest tenth if necessary.

19 $x^2 + 6x - 16 = 0$ **20.** $x^2 - 2x - 14 = 0$

21. $x^2 - 8x - 1 = 8$ **22.** $x^2 + 3x + 21 = 22$

23. $x^2 - 11x + 3 = 5$ **24.** $5x^2 - 10x = 23$

25. $2x^2 - 2x + 7 = 5$ **26.** $3x^2 + 12x + 81 = 15$

27. $4x^2 + 6x = 12$ **28.** $4x^2 + 5 = 10x$

29. $-2x^2 + 10x = -14$ **30.** $-3x^2 - 12 = 14x$

Example 4   **31. FINANCIAL LITERACY** The price p in dollars for a particular stock can be modeled by the quadratic equation $p = 3.5t - 0.05t^2$, where t represents the number of days after the stock is purchased. When is the stock worth $60?

GEOMETRY Find the value of x for each figure. Round to the nearest tenth if necessary.

32. $A = 45$ in^2

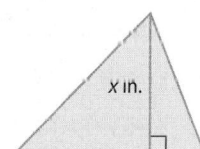

x in.

$(x + 8)$ in.

33. $A = 110$ ft^2

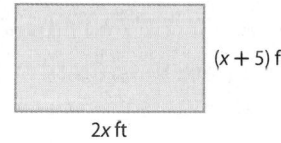

$(x + 5)$ ft

$2x$ ft

34. NUMBER THEORY The product of two consecutive even integers is 224. Find the integers.

35. CCSS PRECISION The product of two consecutive negative odd integers is 483. Find the integers.

36. GEOMETRY Find the area of the triangle below.

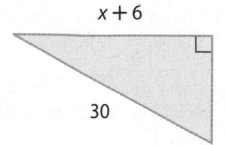

$x + 6$

x

30

Solve each equation by completing the square. Round to the nearest tenth if necessary.

37. $0.2x^2 - 0.2x - 0.4 = 0$ **38.** $0.5x^2 = 2x - 0.3$

39. $2x^2 - \dfrac{11}{5}x = -\dfrac{3}{10}$ **40.** $\dfrac{2}{3}x^2 - \dfrac{4}{3}x = \dfrac{5}{6}$

41. $\dfrac{1}{4}x^2 + 2x = \dfrac{3}{8}$ **42.** $\dfrac{2}{5}x^2 + 2x = \dfrac{1}{5}$

43 **ASTRONOMY** The height of an object t seconds after it is dropped is given by the equation $h = -\frac{1}{2}gt^2 + h_0$, where h_0 is the initial height and g is the acceleration due to gravity. The acceleration due to gravity near the surface of Mars is 3.73 m/s^2, while on Earth it is 9.8 m/s^2. Suppose an object is dropped from an initial height of 120 meters above the surface of each planet.

a. On which planet would the object reach the ground first?

b. How long would it take the object to reach the ground on each planet? Round each answer to the nearest tenth.

c. Do the times that it takes the object to reach the ground seem reasonable? Explain your reasoning.

44. Find all values of c that make $x^2 + cx + 100$ a perfect square trinomial.

45. Find all values of c that make $x^2 + cx + 225$ a perfect square trinomial.

46. PAINTING Before she begins painting a picture, Donna stretches her canvas over a wood frame. The frame has a length of 60 inches and a width of 4 inches. She has enough canvas to cover 480 square inches. Donna decides to increase the dimensions of the frame. If the increase in the length is 10 times the increase in the width, what will the dimensions of the frame be?

47. ⟳ **MULTIPLE REPRESENTATIONS** In this problem, you will investigate a property of quadratic equations.

a. Tabular Copy the table shown and complete the second column.

b. Algebraic Set each trinomial equal to zero, and solve the equation by completing the square. Complete the last column of the table with the number of roots of each equation.

Trinomial	$b^2 - 4ac$	Number of Roots
$x^2 - 8x + 16$	0	1
$2x^2 - 11x + 3$		
$3x^2 + 6x + 9$		
$x^2 - 2x + 7$		
$x^2 + 10x + 25$		
$x^2 + 3x - 12$		

c. Verbal Compare the number of roots of each equation to the result in the $b^2 - 4ac$ column. Is there a relationship between these values? If so, describe it.

d. Analytical Predict how many solutions $2x^2 - 9x + 15 = 0$ will have. Verify your prediction by solving the equation.

H.O.T. Problems *Use Higher-Order Thinking Skills*

48. **CCSS** **PERSEVERANCE** Given $y = ax^2 + bx + c$ with $a \neq 0$, derive the equation for the axis of symmetry by completing the square and rewriting the equation in the form $y = a(x - h)^2 + k$.

49. REASONING Determine the number of solutions $x^2 + bx = c$ has if $c < -\left(\frac{b}{2}\right)^2$. Explain.

50. WHICH ONE DOESN'T BELONG? Identify the expression that does not belong with the other three. Explain your reasoning.

$$n^2 - n + \frac{1}{4} \qquad n^2 + n + \frac{1}{4} \qquad n^2 - \frac{2}{3}n + \frac{1}{9} \qquad n^2 + \frac{1}{3}n + \frac{1}{9}$$

51. OPEN ENDED Write a quadratic equation for which the only solution is 4.

52. WRITING IN MATH Compare and contrast the following strategies for solving $x^2 - 5x - 7 = 0$: completing the square, graphing, and factoring.

53. The length of a rectangle is 3 times its width. The area of the rectangle is 75 square feet. Find the length of the rectangle in feet.

A 25   **B** 15   **C** 10   **D** 5

54. PROBABILITY At a festival, winners of a game draw a token for a prize. There is one token for each prize. The prizes include 9 movie passes, 8 stuffed animals, 5 hats, 10 jump ropes, and 4 glow necklaces. What is the probability that the first person to draw a token will win a movie pass?

F $\frac{1}{36}$ **G** $\frac{1}{9}$ **H** $\frac{9}{61}$ **J** $\frac{1}{4}$

55. GRIDDED RESPONSE The population of a town can be modeled by $P = 22{,}000 + 125t$, where P represents the population and t represents the number of years from 2000. How many years after 2000 will the population be 26,000?

56. Percy delivers pizzas for Pizza King. He is paid $6 an hour plus $2.50 for each pizza he delivers. Percy earned $280 last week. If he worked a total of 30 hours, how many pizzas did he deliver?

A 250 pizzas

B 184 pizzas

C 40 pizzas

D 34 pizzas

Spiral Review

Describe how the graph of each function is related to the graph of $f(x) = x^2$.
(Lesson 9-3)

57. $g(x) = -12 + x^2$

58. $h(x) = (x + 2)^2$

59. $g(x) = 2x^2 + 5$

60. $h(x) = \frac{2}{3}(x - 6)^2$

61. $g(x) = 6 + \frac{4}{3}x^2$

62. $h(x) = -1 - \frac{3}{2}x^2$

63. RIDES A popular amusement park ride whisks riders to the top of a 250-foot tower and drops them. A function for the height of a rider is $h = -16t^2 + 250$, where h is the height and t is the time in seconds. The ride stops the descent of the rider 40 feet above the ground. Write an equation that models the drop of the rider. How long does it take to fall from 250 feet to 40 feet? (Lesson 9-2)

Simplify. Assume that no denominator is equal to zero. (Lesson 7-2)

64. $\dfrac{a^6}{a^3}$

65. $\dfrac{4^7}{4^5}$

66. $\dfrac{c^3 d^4}{cd^7}$

67. $\left(\dfrac{4h^{-2}g}{2g^5}\right)^0$

68. $\dfrac{5q^{-2}t^6}{10q^2 t^{-4}}$

69. $b^3(m^{-3})(b^{-6})$

Solve each open sentence. (Lesson 5-5)

70. $|y - 2| > 7$

71. $|z + 5| < 3$

72. $|2b + 7| \le -6$

73. $|3 - 2y| \ge 8$

74. $|9 - 4m| < -1$

75. $|5c - 2| \le 13$

Skills Review

Evaluate $\sqrt{b^2 - 4ac}$ for each set of values. Round to the nearest tenth if necessary.

76. $a - 2, b - -5, c - 2$

77. $a = 1, b - 12, c - 11$

78. $a = -9, b = 10, c = -1$

79. $a = 1, b = 7, c = -3$

80. $a = 2, b = -4, c = -6$

81. $a = 3, b = 1, c = 2$

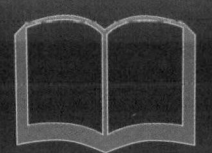

EXTEND

9-4

Algebra Lab
Finding the Maximum or Minimum Value

In Lesson 9-3, we learned about the vertex form of the equation of a quadratic function. You will now learn how to write equations in vertex form and use them to identify key characteristics of the graphs of quadratic functions.

CCSS **Common Core State Standards**
Content Standards
A.SSE.3b Complete the square in a quadratic expression to reveal the maximum or minimum value of the function it defines.
F.IF.8a Use the process of factoring and completing the square in a quadratic function to show zeros, extreme values, and symmetry of the graph, and interpret these in terms of a context.

Activity 1   Find a Minimum

Write $y = x^2 + 4x - 10$ in vertex form. Identify the axis of symmetry, extrema, and zeros. Then graph the function.

Step 1 Complete the square to write the function in vertex form.

$y = x^2 + 4x - 10$	Original function
$y + 10 = x^2 + 4x$	Add 10 to each side.
$y + 10 + 4 = x^2 + 4x + 4$	Since $\left(\frac{4}{2}\right)^2 = 4$, add 4 to each side.
$y + 14 = (x + 2)^2$	Factor $x^2 + 4x + 4$.
$y = (x + 2)^2 - 14$	Subtract 14 from each side to write in vertex form.

Step 2 Identify the axis of symmetry and extrema based on the equation in vertex form. The vertex is at (h, k) or $(-2, -14)$. Since there is no negative sign before the x^2-term, the parabola opens up and has a minimum at $(-2, -14)$. The equation of the axis of symmetry is $x = -2$.

Step 3 Solve for x to find the zeros.

$(x + 2)^2 - 14 = 0$	Vertex form, $y = 0$
$(x + 2)^2 = 14$	Add 14 to each side.
$x + 2 = \pm\sqrt{14}$	Take square root of each side.
$x \approx -5.74$ or 1.74	Subtract 2 from each side.

The zeros are approximately -5.74 and 1.74.

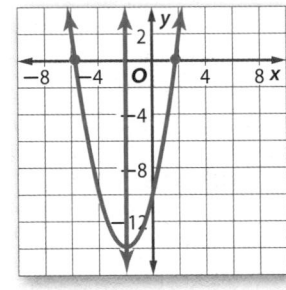

Step 4 Use the key features to graph the function.

There may be a negative coefficient before the quadratic term. When this is the case, the parabola will open down and have a maximum.

Activity 2   Find a Maximum

Write $y = -x^2 + 6x - 5$ in vertex form. Identify the axis of symmetry, extrema, and zeros. Then graph the function.

Step 1 Complete the square to write the equation of the function in vertex form.

$y = -x^2 + 6x - 5$	Original function
$y + 5 = -x^2 + 6x$	Add 5 to each side.
$y + 5 = -(x^2 - 6x)$	Factor out -1.
$y + 5 - 9 = -(x^2 - 6x + 9)$	Since $\left(\frac{6}{2}\right)^2 = 9$, add -9 to each side.
$y - 4 = -(x - 3)^2$	Factor $x^2 - 6x + 9$.
$y = -(x - 3)^2 + 4$	Add 4 to each side to write in vertex form.

Step 2 Identify the axis of symmetry and extrema based on the equation in vertex form. The vertex is at (h, k) or $(3, 4)$. Since there is a negative sign before the x^2-term, the parabola opens down and has a maximum at $(3, 4)$. The equation of the axis of symmetry is $x = 3$.

Step 3 Solve for x to find the zeros.

$$0 = -(x - 3)^2 + 4 \qquad \text{Vertex form, } y = 0$$
$$(x - 3)^2 = 4 \qquad \text{Add } (x - 3)^2 \text{ to each side.}$$
$$x - 3 = \pm 2 \qquad \text{Take the square root of each side.}$$
$$x = 5 \text{ or } 1 \qquad \text{Add 3 to each.}$$

Step 4 Use the key features to graph the function.

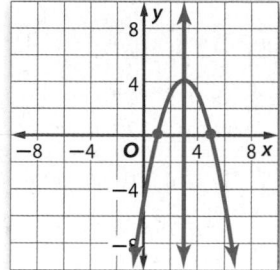

Analyze the Results

1. Why do you need to complete the square to write the equation of a quadratic function in vertex form?

Write each function in vertex form. Identify the axis of symmetry, extrema, and zeros. Then graph the function.

2. $y = x^2 + 6x$

3. $y = x^2 - 8x + 6$

4. $y = x^2 + 2x - 12$

5. $y = x^2 + 6x + 8$

6. $y = x^2 - 4x + 3$

7. $y = x^2 - 2.4x - 2.2$

8. $y = -4x^2 + 16x - 11$

9. $y = 3x^2 - 12x + 5$

10. $y = -x^2 + 6x - 5$

Activity 3   Use Extrema in the Real World

DIVING Alexis jumps from a diving platform upward and outward before diving into the pool. The function $h = -9.8t^2 + 4.9t + 10$, where h is the height of the diver in meters above the pool after t seconds approximates Alexis's dive. Graph the function, then find the maximum height that she reaches and the equation of the axis of symmetry.

Step 1 Graph the function.

Step 2 Complete the square to write the eqution of the function in vertex form.
$$h = -9.8t^2 + 4.9t + 10$$
$$h = -9.8(t - 0.25)^2 + 10.6125$$

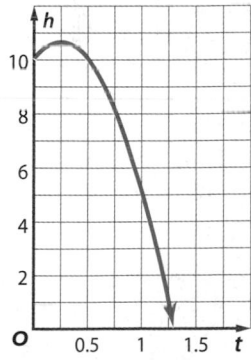

Step 3 The vertex is at $(0.25, 10.6125)$, so the maximum height is 10.6125 meters. The equation of the axis of symmetry is $x = 0.25$.

Exercise

11. **SOFTBALL** Jenna throws a ball in the air. The function $h = -16t^2 + 40t + 5$, where h is the height in feet and t represents the time in seconds, approximates Jenna's throw. Graph the function, then find the maximum height of the ball and the equation of the axis of symmetry. When does the ball hit the ground?

Use a table of values to graph each equation. State the domain and range. (Lesson 9-1)

1. $y = x^2 + 3x + 1$

2. $y = 2x^2 - 4x + 3$

3. $y = -x^2 - 3x - 3$

4. $y = -3x^2 - x + 1$

Consider $y = x^2 - 5x + 4$. (Lesson 9-1)

5. Write the equation of the axis of symmetry.

6. Find the coordinates of the vertex. Is it a maximum or minimum point?

7. Graph the function.

8. **SOCCER** A soccer ball is kicked from ground level with an initial upward velocity of 90 feet per second. The equation $h = -16t^2 + 90t$ gives the height h of the ball after t seconds. (Lesson 9-1)

   a. What is the height of the ball after one second?

   b. How many seconds will it take for the ball to reach its maximum height?

   c. When is the height of the ball 0 feet? What do these points represent in this situation?

Solve each equation by graphing. If integral roots cannot be found, estimate the roots to the nearest tenth. (Lesson 9-2)

9. $x^2 + 5x + 6 = 0$

10. $x^2 + 8 = -6x$

11. $-x^2 + 3x - 1 = 0$

12. $x^2 = 12$

13. **BASEBALL** Juan hits a baseball. The equation $h = -16t^2 + 120t$ models the height h, in feet, of the ball after t seconds. How long is the ball in the air? (Lesson 9-2)

14. **CONSTRUCTION** Christopher is repairing the roof on a shed. He accidentally dropped a box of nails from a height of 14 feet. This is represented by the equation $h = -16t^2 + 14$, where h is the height in feet and t is the time in seconds. Describe how the graph is related to $h = t^2$. (Lesson 9-3)

15. **PARTIES** Della's parents are throwing a Sweet 16 party for her. At 10:00, a ball will slide 25 feet down a pole and light up. A function that models the drop is $h = -t^2 + 5t + 25$, where h is height in feet of the ball after t seconds. How many seconds will it take for the ball to reach the bottom of the pole? (Lesson 9-2)

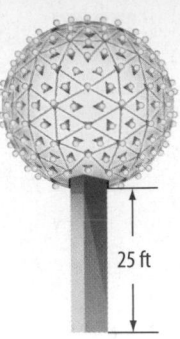

25 ft

Describe how the graph of each function is related to the graph of $f(x) = x^2$. (Lesson 9-3)

16. $g(x) = x^2 + 3$

17. $h(x) = 2x^2$

18. $g(x) = x^2 - 6$

19. $h(x) = \frac{1}{5}x^2$

20. $g(x) = -x^2 + 1$

21. $h(x) = -\frac{5}{8}x^2$

22. **MULTIPLE CHOICE** Which is an equation for the function shown in the graph? (Lesson 9-3)

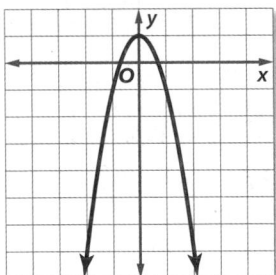

   A $y = -2x^2$

   B $y = 2x^2 + 1$

   C $y = x^2 - 1$

   D $y = -2x^2 + 1$

Solve each equation by completing the square. Round to the nearest tenth. (Lesson 9-4)

23. $x^2 + 4x + 2 = 0$

24. $x^2 - 2x - 10 = 0$

25. $2x^2 + 4x - 5 = 7$

Solving Quadratic Equations by Using the Quadratic Formula

- You solved quadratic equations by completing the square.

1. Solve quadratic equations by using the Quadratic Formula.

2. Use the discriminant to determine the number of solutions of a quadratic equation.

- For adult women, the normal systolic blood pressure P in millimeters of mercury (mm Hg) can be modeled by $P = 0.01a^2 + 0.05a + 107$, where a is age in years. This equation can be used to approximate the age of a woman with a certain systolic blood pressure. However, it would be difficult to solve by factoring, graphing, or completing the square.

NewVocabulary
Quadratic Formula
discriminant

Common Core State Standards

Content Standards
A.REI.4 Solve quadratic equations in one variable.

a. Use the method of completing the square to transform any quadratic equation in x into an equation of the form $(x - p)^2 = q$ that has the same solutions. Derive the quadratic formula from this form.

b. Solve quadratic equations by inspection (e.g., for $x^2 = 49$), taking square roots, completing the square, the quadratic formula and factoring, as appropriate to the initial form of the equation. Recognize when the quadratic formula gives complex solutions and write them as $a \pm bi$ for real numbers a and b.

Mathematical Practices
6 Attend to precision.

1 **Quadratic Formula** Completing the square of the quadratic equation $ax^2 + bx + c = 0$ produces a formula that allows you to find the solutions of *any* quadratic equation. This formula is called the **Quadratic Formula**.

KeyConcept The Quadratic Formula

The solutions of a quadratic equation $ax^2 + bx + c = 0$, where $a \neq 0$, are given by the Quadratic Formula.

$$x = \frac{-b \pm \sqrt{b^2 - 4ac}}{2a}$$

You will derive this formula in Lesson 10-2.

PT

Example 1 Use the Quadratic Formula

Solve $x^2 - 12x = -20$ **by using the Quadratic Formula.**

Step 1 Rewrite the equation in standard form.

$$x^2 - 12x = -20 \quad \text{Original equation}$$
$$x^2 - 12x + 20 = 0 \quad \text{Add 20 to each side.}$$

Step 2 Apply the Quadratic Formula.

$$x = \frac{-b \pm \sqrt{b^2 - 4ac}}{2a} \quad \text{Quadratic Formula}$$

$$= \frac{-(-12) \pm \sqrt{(-12)^2 - 4(1)(20)}}{2(1)} \quad a = 1, b = -12, \text{ and } c = 20$$

$$= \frac{12 \pm \sqrt{144 - 80}}{2} \quad \text{Multiply.}$$

$$= \frac{12 \pm \sqrt{64}}{2} \text{ or } \frac{12 \pm 8}{2} \quad \text{Subtract and take the square root.}$$

$$x = \frac{12 - 8}{2} \text{ or } x = \frac{12 + 8}{2} \quad \text{Separate the solutions.}$$

$$= 2 \qquad\qquad = 10 \qquad \text{The solutions are 2 and 10.}$$

▶ **Guided**Practice

1. Solve $2x^2 + 9x = 18$ by using the Quadratic Formula.

The solutions of quadratic equations are not always integers.

Example 2 Use the Quadratic Formula

Solve each equation by using the Quadratic Formula. Round to the nearest tenth if necessary.

a. $3x^2 + 5x - 12 = 0$

For this equation, $a = 3$, $b = 5$, and $c = -12$.

$$x = \frac{-b \pm \sqrt{b^2 - 4ac}}{2a} \qquad \text{Quadratic Formula}$$

$$= \frac{-(5) \pm \sqrt{(5)^2 - 4(3)(-12)}}{2(3)} \qquad a = 3, b = 5, \text{ and } c = -12$$

$$= \frac{-5 \pm \sqrt{25 + 144}}{6} \qquad \text{Multiply.}$$

$$= \frac{-5 \pm \sqrt{169}}{6} \text{ or } \frac{-5 \pm 13}{6} \qquad \text{Add and simplify.}$$

$$x = \frac{-5 - 13}{6} \text{ or } x = \frac{-5 + 13}{6} \qquad \text{Separate the solutions.}$$

$$= -3 \qquad\qquad = \frac{4}{3} \qquad \text{Simplify.}$$

The solutions are -3 and $\frac{4}{3}$.

b. $10x^2 - 5x = 25$

Step 1 Rewrite the equation in standard form.

$$10x^2 - 5x = 25 \qquad \text{Original equation}$$

$$10x^2 - 5x - 25 = 0 \qquad \text{Subtract 25 from each side.}$$

Step 2 Apply the Quadratic Formula.

$$x = \frac{-b \pm \sqrt{b^2 - 4ac}}{2a} \qquad \text{Quadratic Formula}$$

$$= \frac{-(-5) \pm \sqrt{(-5)^2 - 4(10)(-25)}}{2(10)} \qquad a = 10, b = -5, \text{ and } c = -25$$

$$= \frac{5 \pm \sqrt{25 + 1000}}{20} \qquad \text{Multiply.}$$

$$= \frac{5 \pm \sqrt{1025}}{20} \qquad \text{Add.}$$

$$= \frac{5 - \sqrt{1025}}{20} \text{ or } \frac{5 + \sqrt{1025}}{20} \qquad \text{Separate the solutions.}$$

$$\approx -1.4 \qquad\qquad \approx 1.9 \qquad \text{Simplify.}$$

The solutions are about -1.4 and 1.9.

StudyTip

CCSS Precision In Example 2, the number $\sqrt{1025}$ is irrational, so the calculator can only give you an approximation of its value. So, the exact answer in Example 2 is $\frac{5 \pm \sqrt{1025}}{20}$. The numbers -1.4 and 1.9 are approximations.

GuidedPractice

2A. $4x^2 - 24x + 35 = 0$

2B. $3x^2 - 2x - 9 = 0$

You can solve quadratic equations by using one of many equivalent methods. No one way is always best.

Example 3 Solve Quadratic Equations Using Different Methods

Solve $x^2 - 4x = 12$.

Method 1 Graphing

Rewrite the equation in standard form.

$x^2 - 4x = 12$ Original equation

$x^2 - 4x - 12 = 0$ Subtract 12 from each side.

Graph the related function $f(x) = x^2 - 4x - 12$. Locate the x-intercepts of the graph.

The solutions are **−2** and **6**.

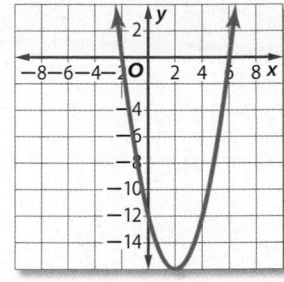

Method 2 Factoring

$x^2 - 4x = 12$ Original equation

$x^2 - 4x - 12 = 0$ Subtract 12 from each side.

$(x - 6)(x + 2) = 0$ Factor.

$x - 6 = 0$ or $x + 2 = 0$ Zero Product Property

$x = 6$ $x = -2$ Solve for x.

Method 3 Completing the Square

The equation is in the correct form to complete the square, since the leading coefficient is 1 and the x^2 and x terms are isolated.

$x^2 - 4x = 12$ Original equation

$x^2 - 4x + 4 = 12 + 4$ Since $\left(\frac{-4}{2}\right)^2 = 4$, add 4 to each side.

$(x - 2)^2 = 16$ Factor $x^2 - 4x + 4$.

$x - 2 = \pm 4$ Take the square root of each side.

$x = 2 \pm 4$ Add 2 to each side.

$x = 2 + 4$ or $x = 2 - 4$ Separate the solutions.

$= 6$ $= -2$ Simplify.

Method 4 Quadratic Formula

From Method 1, the standard form of the equation is $x^2 - 4x - 12 = 0$.

$x = \dfrac{-b \pm \sqrt{b^2 - 4ac}}{2a}$ Quadratic Formula

$= \dfrac{-(-4) \pm \sqrt{(-4)^2 - 4(1)(-12)}}{2(1)}$ $a = 1$, $b = -4$, and $c = -12$

$= \dfrac{4 \pm \sqrt{16 + 48}}{2}$ Multiply.

$= \dfrac{4 \pm \sqrt{64}}{2}$ or $\dfrac{4 \pm 8}{2}$ Add and simplify.

$x = \dfrac{4 - 8}{2}$ or $x = \dfrac{4 + 8}{2}$ Separate the solutions.

$= -2$ $= 6$ Simplify.

▶ **Guided**Practice

Solve each equation.

3A. $2x^2 - 17x + 8 = 0$ **3B.** $4x^2 - 4x - 11 = 0$

ConceptSummary Solving Quadratic Equations

Method	When to Use
Factoring	Use when the constant term is 0 or if the factors are easily determined. Not all equations are factorable.
Graphing	Use when an approximate solution is sufficient.
Using Square Roots	Use when an equation can be written in the form $x^2 = n$. Can only be used if the equation has no x-term.
Completing the Square	Can be used for any equation $ax^2 + bx + c = 0$, but is simplest to apply when b is even and $a = 1$.
Quadratic Formula	Can be used for any equation $ax^2 + bx + c = 0$.

2 The Discriminant In the Quadratic Formula, the expression under the radical sign, $b^2 - 4ac$, is called the **discriminant**. The discriminant can be used to determine the number of real solutions of a quadratic equation.

StudyTip

Discriminant Recall that when the left side of the standard form of an equation is a perfect square trinomial, there is only one solution. Therefore, the discriminant of a perfect square trinomial will always be zero.

🔄 KeyConcept Using the Discriminant

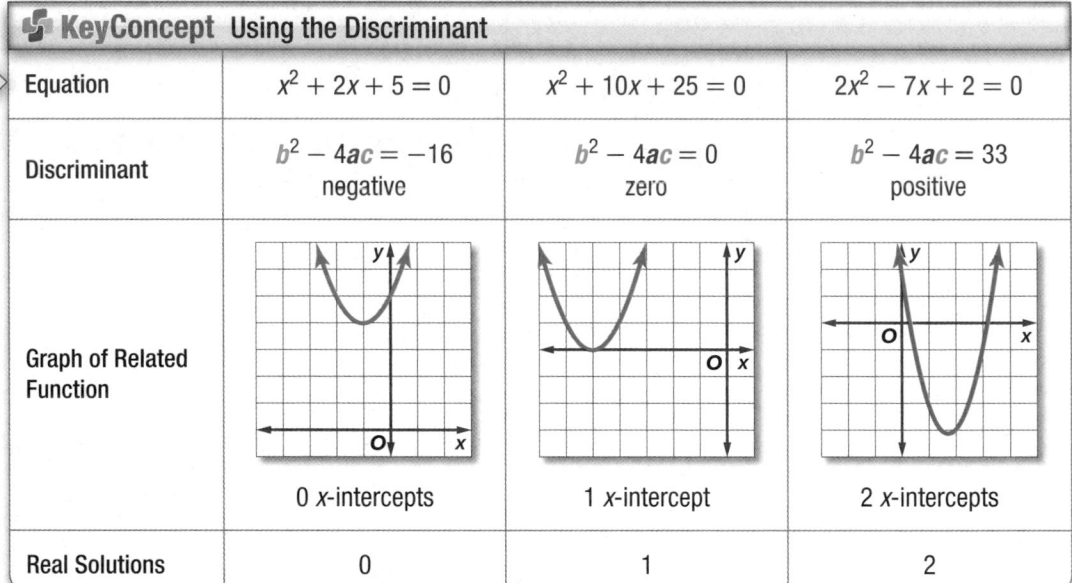

Equation	$x^2 + 2x + 5 = 0$	$x^2 + 10x + 25 = 0$	$2x^2 - 7x + 2 = 0$
Discriminant	$b^2 - 4ac = -16$ negative	$b^2 - 4ac = 0$ zero	$b^2 - 4ac = 33$ positive
Graph of Related Function	0 x-intercepts	1 x-intercept	2 x-intercepts
Real Solutions	0	1	2

Example 4 Use the Discriminant

State the value of the discriminant of $4x^2 + 5x = -3$. Then determine the number of real solutions of the equation.

Step 1 Rewrite in standard form. $4x^2 - 5x = -3 \longrightarrow 4x^2 - 5x + 3 = 0$

Step 2 Find the discriminant.

$$b^2 - 4ac = (-5)^2 - 4(4)(3) \qquad a = 4, b = -5, \text{ and } c = 3$$

$$= -23 \qquad \text{Simplify.}$$

Since the discriminant is negative, the equation has no real solutions.

▶ **Guided**Practice

4A. $2x^2 + 11x + 15 = 0$ **4B.** $9x^2 - 30x + 25 = 0$

Examples 1–2 Solve each equation by using the Quadratic Formula. Round to the nearest tenth if necessary.

1. $x^2 - 2x - 15 = 0$
2. $x^2 - 10x + 16 = 0$
3. $x^2 - 8x = -10$
4. $x^2 + 3x = 12$
5. $10x^2 - 31x + 15 = 0$
6. $5x^2 + 5 = -13x$

Example 3 Solve each equation. State which method you used.

7. $2x^2 + 11x - 6 = 0$
8. $2x^2 - 3x - 6 = 0$
9. $9x^2 = 25$
10. $x^2 - 9x = -19$

Example 4 State the value of the discriminant for each equation. Then determine the number of real solutions of the equation.

11. $x^2 - 9x + 21 = 0$
12. $2x^2 - 11x + 10 = 0$
13. $9x^2 + 24x = -16$
14. $3x^2 - x = 8$

15. **TRAMPOLINE** Eva springs from a trampoline to dunk a basketball. Her height h in feet can be modeled by the equation $h = -16t^2 + 22.3t + 2$, where t is time in seconds. Use the discriminant to determine if Eva will reach a height of 10 feet. Explain.

Practice and Problem Solving   Extra Practice is on page R9.

Examples 1–2 Solve each equation by using the Quadratic Formula. Round to the nearest tenth if necessary.

16. $4x^2 + 5x - 6 = 0$
17. $x^2 + 16 = 0$
18. $6x^2 - 12x + 1 = 0$
19. $5x^2 - 8x = 6$
20. $2x^2 - 5x = 7$
21. $5x^2 + 21x = -18$
22. $81x^2 = 9$
23. $8x^2 + 12x = 8$
24. $4x^2 = -16x - 16$
25. $10x^2 = -7x + 6$
26. $-3x^2 = 8x - 12$
27. $2x^2 = 12x - 18$

28. **AMUSEMENT PARKS** The Demon Drop at Cedar Point in Ohio takes riders to the top of a tower and drops them 60 feet. A function that approximates this ride is $h = -16t^2 + 64t - 60$, where h is the height in feet and t is the time in seconds. About how many seconds does it take for riders to drop 60 feet?

Example 3 Solve each equation. State which method you used.

29. $2x^2 - 8x = 12$
30. $3x^2 - 24x = -36$
31. $x^2 - 3x = 10$
32. $4x^2 + 100 = 0$
33. $x^2 = -7x - 5$
34. $12 - 12x = -3x^2$

Example 4 State the value of the discriminant for each equation. Then determine the number of real solutions of the equation.

35. $0.2x^2 - 1.5x + 2.9 = 0$
36. $2x^2 - 5x + 20 = 0$
37. $x^2 - \frac{4}{5}x = 3$
38. $0.5x^2 - 2x = -2$
39. $2.25x^2 - 3x = -1$
40. $2x^2 = \frac{5}{2}x + \frac{3}{2}$

41. **CCSS MODELING** The percent of U.S. households with high-speed Internet h can be estimated by $h = -0.2n^2 + 7.2n + 1.5$, where n is the number of years since 1990.

   a. Use the Quadratic Formula to determine when 20% of the population will have high-speed Internet.

   b. Is a quadratic equation a good model for this information? Explain.

42. TRAFFIC The equation $d = 0.05v^2 + 1.1v$ models the distance d in feet it takes a car traveling at a speed of v miles per hour to come to a complete stop. If Hannah's car stopped after 250 feet on a highway with a speed limit of 65 miles per hour, was she speeding? Explain your reasoning.

Without graphing, determine the number of x-intercepts of the graph of the related function for each equation.

43. $4.25x + 3 = -3x^2$ **44.** $x^2 + \frac{2}{25} = \frac{3}{5}x$ **45.** $0.25x^2 + x = -1$

Solve each equation by using the Quadratic Formula. Round to the nearest tenth if necessary.

46. $-2x^2 - 7x = -1.5$ **47.** $2.3x^2 - 1.4x = 6.8$ **48.** $x^2 - 2x = 5$

49) POSTER Bartolo is making a poster for the dance. He wants to cover three fourths of the area with text.

    a. Write an equation for the area of the section with text.

    b. Solve the equation by using the Quadratic Formula.

    c. What should be the margins of the poster?

50. ⬚ **MULTIPLE REPRESENTATIONS** In this problem, you will investigate writing a quadratic equation with given roots. If p is a root of $0 = ax^2 + bx + c$, then $(x - p)$ is a factor of $ax^2 + bx + c$.

    a. Tabular Copy and complete the first two columns of the table.

    b. Algebraic Multiply the factors to write each equation with integral coefficients. Use the equations to complete the last column of the table. Write each equation.

    c. Analytical How could you write an equation with three roots? Test your conjecture by writing an equation with roots 1, 2, and 3. Is the equation quadratic? Explain.

Roots	Factors	Equation
2, 5	$(x - 2), (x - 5)$	$(x - 2)(x - 5) = 0$ $x^2 - 7x + 10 = 0$
1, 9		
$-1, 3$		
0, 6		
$\frac{1}{2}, 7$		
$-\frac{2}{3}, 4$		

H.O.T. Problems   Use Higher-Order Thinking Skills

51. CHALLENGE Find all values of k such that $2x^2 - 3x + 5k = 0$ has two solutions.

52. REASONING Use factoring techniques to determine the number of real zeros of $f(x) = x^2 - 8x + 16$. Compare this method to using the discriminant.

CCSS STRUCTURE Determine whether there are *two*, *one*, or *no* real solutions of each equation.

53. The graph of the related quadratic function does not have an x-intercept.

54. The graph of the related quadratic function touches but does not cross the x-axis.

55. The graph of the related quadratic function intersects the x-axis twice.

56. Both a and b are greater than 0 and c is less than 0 in a quadratic equation.

57. ✏️ **WRITING IN MATH** Why can the discriminant be used to confirm the number of real solutions of a quadratic equation?

58. WRITING IN MATH Describe the advantages and disadvantages of each method of solving quadratic equations. Why are the methods equivalent? Which method do you prefer, and why?

59. If n is an even integer, which expression represents the product of three consecutive even integers?

 A $n(n + 1)(n + 2)$
 B $(n + 1)(n + 2)(n + 3)$
 C $3n + 2$
 D $n(n + 2)(n + 4)$

60. SHORT RESPONSE The triangle shown is an isosceles triangle. What is the value of x?

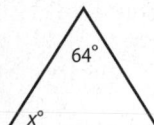

61. Which statement best describes the graph of $x = 5$?

 F It is parallel to the x-axis.
 G It is parallel to the y-axis.
 H It passes through the point (2, 5).
 J It has a y-intercept of 5.

62. What are the solutions of the quadratic equation $6h^2 + 6h = 72$?

 A 3 or -4 **C** no solution
 B -3 or 4   **D** 12 or -48

Solve each equation by completing the square. Round to the nearest tenth if necessary. (Lesson 9-4)

63. $6x^2 - 17x + 12 = 0$

64. $x^2 - 9x = -12$

65. $4x^2 = 20x - 25$

Describe the transformations needed to obtain the graph of $g(x)$ from the graph of $f(x)$. (Lesson 9-3)

66. $f(x) = 4x^2$
   $g(x) = 2x^2$

67. $f(x) = x^2 + 5$
   $g(x) = x^2 - 1$

68. $f(x) = x^2 - 6$
   $g(x) = x^2 + 3$

Determine whether each graph shows a *positive*, a *negative*, or *no* correlation. If there is a positive or negative correlation, describe its meaning in the situation. (Lesson 4-5)

69.

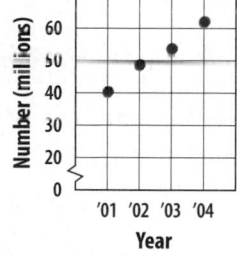

70.

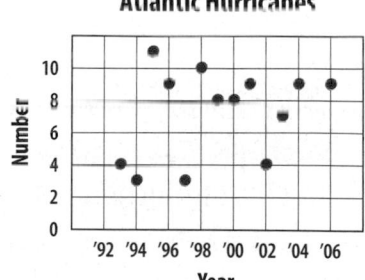

71. ENTERTAINMENT Coach Washington wants to take her softball team out for pizza and soft drinks after the last game of the season. A large pizza costs $12 and a pitcher of a soft drink costs $3. She does not want to spend more than $60. Write an inequality that represents this situation and graph the solution set. (Lesson 5-6)

Determine whether each sequence is *arithmetic*, *geometric*, or *neither*. Explain.

72. 20, 25, 30, …

73. 1000, 950, 900, …

74. 200, 350, 650, …

75. 6, 18, 54, …

76. 2, 4, 16, …

77. 8, -4, 2 …

Analyzing Functions with Successive Differences

:: Then	:: Now	:: Why?
● You graphed linear, quadratic, and exponential functions.	**1** Identify linear, quadratic, and exponential functions from given data. **2** Write equations that model data.	● Every year the golf team sells candy to raise money for charity. By knowing what type of function models the sales of the candy, they can determine the best price of the candy.

Common Core State Standards

Content Standards
F.IF.6 Calculate and interpret the average rate of change of a function (presented symbolically or as a table) over a specified interval. Estimate the rate of change from a graph.

F.LE.1 Distinguish between situations that can be modeled with linear functions and with exponential functions.

a. Prove that linear functions grow by equal differences over equal intervals, and that exponential functions grow by equal factors over equal intervals.

b. Recognize situations in which one quantity changes at a constant rate per unit interval relative to another.

c. Recognize situations in which a quantity grows or decays by a constant percent rate per unit interval relative to another.

Mathematical Practices
7 Look for and make use of structure.

1 **Identify Functions** You can use linear functions, quadratic functions, and exponential functions to model data. The general forms of the equations and a graph of each function type are listed below.

ConceptSummary Linear and Nonlinear Functions

Linear Function	Quadratic Function	Exponential Function
$y = mx + b$	$y = ax^2 + bx + c$	$y = ab^x$, when $b > 0$

Example 1 Choose a Model Using Graphs

Graph each set of ordered pairs. Determine whether the ordered pairs represent a *linear* function, a *quadratic* function, or an *exponential* function.

a. {(−2, 5), (−1, 2), (0, 1), (1, 2), (2, 5)}

The ordered pairs appear to represent a quadratic function.

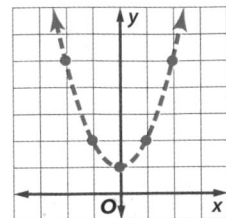

b. $\left\{\left(-2, \frac{1}{4}\right), \left(-1, \frac{1}{2}\right), (0, 1), (1, 2), (2, 4)\right\}$

The ordered pairs appear to represent an exponential function.

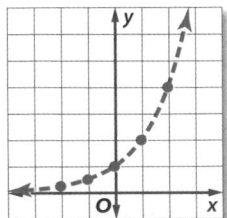

▶ **Guided** Practice

1A. (−2, −3), (−1, −1), (0, 1), (1, 3)

1B. (−1, 0.25), (0, 1), (1, 4), (2, 16)

Another way to determine which model best describes data is to use patterns. The differences of successive y-values are called *first differences*. The differences of successive first differences are called *second differences*.

- If the differences of successive y-values are all equal, the data represent a linear function.

- If the second differences are all equal, but the first differences are not equal, the data represent a quadratic function.

- If the ratios of successive y-values are all equal and $r \neq 1$, the data represent an exponential function.

WatchOut!

x-Values Before you check for successive differences or ratios, make sure the *x*-values are increasing by the same amount.

Example 2 Choose a Model Using Differences or Ratios

Look for a pattern in each table of values to determine which kind of model best describes the data.

a.

x	−2	−1	0	1	2
y	−8	−3	2	7	12

First differences:

$$\begin{array}{ccccc} -8 & -3 & 2 & 7 & 12 \\ & 5 & 5 & 5 & 5 \end{array}$$

Since the first differences are all equal, the table of values represents a linear function.

b.

x	−1	0	1	2	3
y	8	4	2	1	0.5

First differences:

$$\begin{array}{ccccc} 8 & 4 & 2 & 1 & 0.5 \\ & -4 & -2 & -1 & -0.5 \end{array}$$

The first differences are not all equal. So, the table of values does not represent a linear function. Find the second differences and compare.

First differences: −4   −2   −1   −0.5

Second differences: 2   1   0.5

The second differences are not all equal. So, the table of values does not represent a quadratic function. Find the ratios of the y-values and compare.

$$\begin{array}{ccccc} 8 & 4 & 2 & 1 & 0.5 \end{array}$$

Ratios: $\dfrac{4}{8} = \dfrac{1}{2}$ $\dfrac{2}{4} = \dfrac{1}{2}$ $\dfrac{1}{2}$ $\dfrac{0.5}{1} = \dfrac{1}{2}$

The ratios of successive y-values are equal. Therefore, the table of values can be modeled by an exponential function.

Guided Practice

2A.

x	−3	−2	−1	0	1
y	−3	−7	−9	−9	−7

2B.

x	−2	−1	0	1	2
y	−18	−13	−8	−3	2

2 **Write Equations** Once you find the model that best describes the data, you can write an equation for the function. For a quadratic function in this lesson, the equation will have the form $y = ax^2$.

Example 3 Write an Equation

Determine which kind of model best describes the data. Then write an equation for the function that models the data.

x	−4	−3	−2	−1	0
y	32	18	8	2	0

Step 1 Determine which model fits the data.

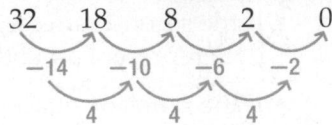

First differences:

Second differences:

Since the second differences are equal, a quadratic function models the data.

Step 2 Write an equation for the function that models the data.

The equation has the form $y = ax^2$. Find the value of a by choosing one of the ordered pairs from the table of values. Let's use $(−1, 2)$.

$y = ax^2$ Equation for quadratic function

$2 = a(−1)^2$ $x = −1$ and $y = 2$

$2 = a$ An equation that models the data is $y = 2x^2$.

WatchOut!

Finding *a* In Example 3, the point (0, 0) cannot be used to find the value of *a*. You will have to divide each side by 0, giving you an undefined value for *a*.

▶ **Guided**Practice

3A.

x	−2	−1	0	1	2
y	11	7	3	−1	−5

3B.

x	−3	−2	−1	0	1
y	0.375	0.75	1.5	3	6

🌐 **Real-World Example 4** Write an Equation for a Real-World Situation

BOOK CLUB The table shows the number of book club members for four consecutive years. Determine which model best represents the data. Then write a function that models the data.

Understand We need to find a model for the data, and then write a function.

Time (years)	0	1	2	3	4
Members	5	10	20	40	80

Plan Find a pattern using successive differences or ratios. Then use the general form of the equation to write a function.

Solve The constant ratio is 2. This is the value of the base. An exponential function of the form $y = ab^x$ models the data.

$y = ab^x$ Equation for exponential function

$5 = a(2)^0$ $x = 0, y = 5$, and $b = 2$

$5 = a$ The equation that models the data is $y = 5 \cdot 2^x$.

Check You used (0, 5) to write the function. Verify that every other ordered pair satisfies the equation.

Real-World Link

A poll by the National Education Association found that 87% of all teens polled found reading relaxing, 85% viewed reading as rewarding, and 79% found reading exciting.

Source: *American Demographics*

▶ **Guided**Practice

4. ADVERTISING The table shows the cost of placing an ad in a newspaper. Determine a model that best represents the data and write a function that models the data.

No. of Lines	5	6	7	8
Total Cost ($)	14.50	16.60	18.70	20.80

Example 1 Graph each set of ordered pairs. Determine whether the ordered pairs represent a *linear* function, a *quadratic* function, or an *exponential* function.

1. $(-2, 8), (-1, 5), (0, 2), (1, -1)$

2. $(-3, 7), (-2, 3), (-1, 1), (0, 1), (1, 3)$

3. $(-3, 8), (-2, 4), (-1, 2), (0, 1), (1, 0.5)$

4. $(0, 2), (1, 2.5), (2, 3), (3, 3.5)$

Example 2 Look for a pattern in each table of values to determine which kind of model best describes the data.

5.

x	0	1	2	3	4
y	5	8	17	32	53

6.

x	−3	−2	−1	0
y	−6.75	−7.5	−8.25	−9

7.

x	−1	0	1	2	3
y	3	6	12	24	48

8.

x	3	4	5	6	7
y	−1.5	0	2.5	6	10.5

Example 3 Determine which kind of model best describes the data. Then write an equation for the function that models the data.

9.

x	−1	0	1	2	3
y	1	3	9	27	81

10.

x	−5	−4	−3	−2	−1
y	125	80	45	20	5

11.

x	−3	−2	−1	0	1
y	1	1.5	2	2.5	3

12.

x	−1	0	1	2
y	−1.25	−1	−0.75	−0.5

Example 4 **13. PLANTS** The table shows the height of a plant for four consecutive weeks. Determine which kind of function best models the height. Then write a function that models the data.

Week	0	1	2	3	4
Height (in.)	3	3.5	4	4.5	5

Practice and Problem Solving

Extra Practice is on page R9.

Example 1 Graph each set of ordered pairs. Determine whether the ordered pairs represent a *linear* function, a *quadratic* function, or an *exponential* function.

14. $(-1, 1), (0, -2), (1, -3), (2, -2), (3, 1)$

15. $(1, 2.75), (2, 2.5), (3, 2.25), (4, 2)$

16. $(-3, 0.25), (-2, 0.5), (-1, 1), (0, 2)$

17. $(-3, -11), (-2, -5), (-1, -3), (0, -5)$

18. $(-2, 6), (-1, 1), (0, -4), (1, -9)$

19. $(-1, 8), (0, 2), (1, 0.5), (2, 0.125)$

Examples 2–3 Look for a pattern in each table of values to determine which kind of model best describes the data. Then write an equation for the function that models the data.

20.

x	−3	−2	−1	0
y	−8.8	−8.6	−8.4	−8.2

21

x	−2	−1	0	1	2
y	10	2.5	0	2.5	10

22.

x	−1	0	1	2	3
y	0.75	3	12	48	192

23.

x	−2	−1	0	1	2
y	0.008	0.04	0.2	1	5

24.

x	0	1	2	3	4
y	0	4.2	16.8	37.8	67.2

25.

x	−3	−2	−1	0	1
y	14.75	9.75	4.75	−0.25	−5.25

Example 4

26. WEB SITES A company tracked the number of visitors to its Web site over 4 days. Determine which kind of model best represents the number of visitors to the Web site with respect to time. Then write a function that models the data.

Day	0	1	2	3	4
Visitors (in thousands)	0	0.9	3.6	8.1	14.4

27. CALLING The cost of an international call depends on the length of the call. The table shows the cost for up to 6 minutes.

Length of call (min)	1	2	3	4	5	6
Cost ($)	0.12	0.24	0.36	0.48	0.60	0.72

a. Graph the data and determine which kind of function best models the data.

b. Write an equation for the function that models the data.

c. Use your equation to determine how much a 10-minute call would cost.

28. DEPRECIATION The value of a car depreciates over time. The table shows the value of a car over a period of time.

Year	0	1	2	3	4
Value ($)	18,500	15,910	13,682.60	11,767.04	10,119.65

a. Determine which kind of function best models the data.

b. Write an equation for the function that models the data.

c. Use your equation to determine how much the car is worth after 7 years.

29. BACTERIA A scientist estimates that a bacteria culture with an initial population of 12 will triple every hour.

a. Make a table to show the bacteria population for the first 4 hours.

b. Which kind of model best represents the data?

c. Write a function that models the data.

d. How many bacteria will there be after 8 hours?

30. PRINTING A printing company charges the fees shown to print flyers. Write a function that models the total cost of the flyers, and determine how much 30 flyers would cost.

Quick 2 U Printing
Set Up Fee $25
15¢ each flyer

H.O.T. Problems Use Higher-Order Thinking Skills

31. CHALLENGE Write a function that has constant second differences, first differences that are not constant, a y-intercept of -5, and contains the point $(2, 3)$.

32. CCSS ARGUMENTS What type of function will have constant third differences but not constant second differences? Explain.

33. OPEN ENDED Write a linear function that has a constant first difference of 4.

34. PROOF Write a paragraph proof to show that linear functions grow by equal differences over equal intervals, and exponential functions grow by equal factors over equal intervals. (*Hint:* Let $y = ax$ represent a linear function and let $y = a^x$ represent an exponential function.)

35. WRITING IN MATH How can you determine whether a given set of data should be modeled by a *linear* function, a *quadratic* function, or an *exponential* function?

36. SHORT RESPONSE Write an equation that models the data in the table.

x	0	1	2	3	4
y	3	6	12	24	48

37. What is the equation of the line below?

A $y = \frac{2}{5}x + 2$

B $y = \frac{2}{5}x - 2$

C $y = \frac{5}{2}x + 2$

D $y = \frac{5}{2}x - 2$

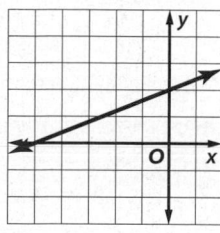

38. The point $(r, -4)$ lies on a line with an equation of $2x + 3y = -8$. Find the value of r.

F -10 H 2

G 0 J 8

39. GEOMETRY The rectangle has an area of 220 square feet. Find the length ℓ.

A 8 feet

B 10 feet

C 22 feet

D 34 feet

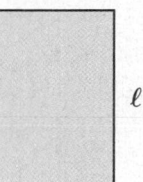

$\ell + 12$

ℓ

Spiral Review

Solve each equation by using the Quadratic Formula. Round to the nearest tenth if necessary. (Lesson 9-5)

40. $6x^2 - 3x - 30 = 0$

41. $4x^2 + 18x = 10$

42. $2x^2 + 6x = 7$

Solve each equation by taking the square root of each side. Round to the nearest tenth if necessary. (Lesson 9-4)

43. $x^2 = 25$

44. $x^2 + 6x + 9 = 16$

45. $x^2 - 14x + 49 = 15$

46. INVESTMENTS Joey's investment of $2500 has been decreasing in value at a rate of 1.5% each year. What will his investment be worth in 5 years? (Lesson 7-7)

Write an equation for the nth term of each geometric sequence, and find the seventh term of each sequence. (Lesson 7-8)

47. $1, 2, 4, 8, \ldots$

48. $-20, -10, -5, \ldots$

49. $4, -12, 36, \ldots$

50. $99, -33, 11, \ldots$

51. $22, 44, 88, \ldots$

52. $\frac{2}{3}, \frac{1}{3}, \frac{1}{6}, \ldots$

53. CANOE RENTAL To rent a canoe, you must pay a daily rate plus $10 per hour. Ilia and her friends rented a canoe for 3 hours and paid $45. Write a linear equation for the cost C of renting the canoe for h hours, and determine how much it cost to rent the canoe for 8 hours. (Lesson 4-2)

Determine whether each equation is a linear equation. If so, write the equation in standard form. (Lesson 3-1)

54. $3x = 5y$

55. $6 - y = 2x$

56. $6xy + 3x = 4$

57. $y + 5 = 0$

58. $7y = 2x + 5x$

59. $y = 4x^2 - 1$

Skills Review

Evaluate each expression if $x = -3$, $y = -1$, and $z = 4$.

60. $|x - 4|$

61. $|2y + 1|$

62. $|4 - z|$

63. $\left|\frac{1}{2}x + 2\right|$

64. $|12 - 4z|$

65. $|2y - 3| - 6$

Graphing Technology Lab
Curve Fitting

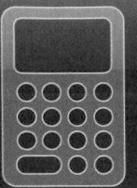

If there is a constant increase or decrease in data values, there is a linear trend. If the values are increasing or decreasing more and more rapidly, there may be a quadratic or exponential trend.

CCSS **Common Core State Standards**
Content Standards
F.LE.2 Construct linear and exponential functions, including arithmetic and geometric sequences, given a graph, a description of a relationship, or two input-output pairs (include reading these from a table).
S.ID.6a Fit a function to the data; use functions fitted to data to solve problems in the context of the data. Use given functions or choose a function suggested by the context. Emphasize linear, quadratic, and exponential models.

Linear Trend

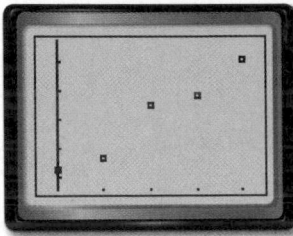

[0, 5] scl: 1 by [0, 6] scl: 1

Quadratic Trend

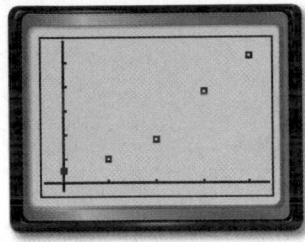

[0, 5] scl: 1 by [0, 6] scl: 1

Exponential Trend

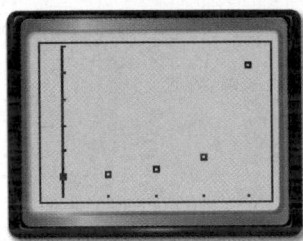

[0, 5] scl: 1 by [0, 6] scl: 1

With a graphing calculator, you can find the appropriate regression equation.

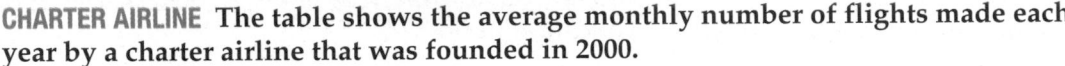

Activity

CHARTER AIRLINE The table shows the average monthly number of flights made each year by a charter airline that was founded in 2000.

Year	2000	2001	2002	2003	2004	2005	2006	2007
Flights	17	20	24	28	33	38	44	50

Step 1 Make a scatter plot.

• Enter the number of years since 2000 in **L1** and the number of flights in **L2**.

KEYSTROKES: *Review entering a list on page 255.*

• Use **STAT PLOT** to graph the scatter plot.

KEYSTROKES: *Review statistical plots on page 256.*

Use ZOOM 9 to graph.

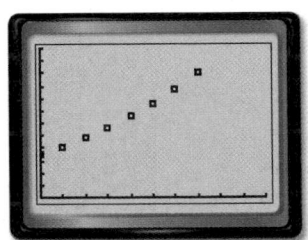

[0, 10] scl: 1 by [0, 60] scl: 5

From the scatter plot we can see that the data may have either a quadratic trend or an exponential trend.

Step 2 Find the regression equation.

We will check both trends by examining their regression equations.

• Select **DiagnosticOn** from the **CATALOG**.

• Select **QuadReg** on the **STAT** menu.

KEYSTROKES: STAT ▶ 5 ENTER ENTER

The equation is in the form $y = ax^2 + bx + c$.

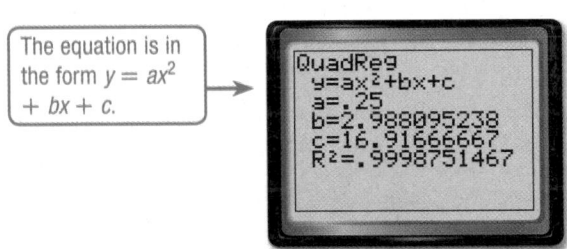

The equation is about $y = 0.25x^2 + 3x + 17$.

R^2 is the **coefficient of determination**. The closer R^2 is to 1, the better the model. To acquire the exponential equation select **ExpReg** on the **STAT** menu. To choose a quadratic or exponential model, fit both and use the one with the R^2 value closer to 1.

Step 3 Graph the quadratic regression equation.

- Copy the equation to the **Y=** list and graph.

KEYSTROKES: Y= VARS 5 ▶
▶ 1 ZOOM 9

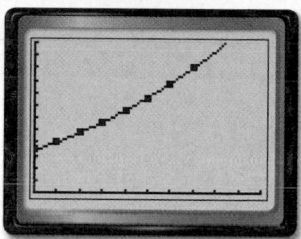

[0, 10] scl: 1 by [0, 60] scl: 5

Step 4 Predict using the equation.

If this trend continues, we can use the graph of our equation to predict the monthly number of flights the airline will make in a specific year. Let's check the year 2020. First adjust the window.

KEYSTROKES: 2nd CALC 1 At $x =$ enter 20 ENTER.

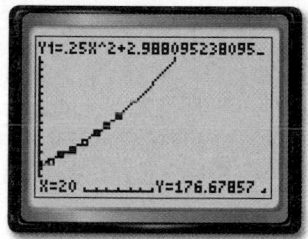

[0, 25] scl: 1 by [0, 200] scl: 5

There will be approximately 177 flights per month if this trend continues.

Exercises

Plot each set of data points. Determine whether to use a *linear*, *quadratic* or *exponential* regression equation. State the coefficient of determination.

1.

x	y
1	30
2	40
3	50
4	55
5	50
6	40

2.

x	y
0.0	12.1
0.1	9.6
0.2	6.3
0.3	5.5
0.4	4.8
0.5	1.9

3.

x	y
0	1.1
2	3.3
4	2.9
6	5.6
8	11.9
10	19.8

4.

x	y
1	1.67
5	2.59
9	4.37
13	6.12
17	5.48
21	3.12

5. BAKING Alyssa baked a cake and is waiting for it to cool so she can ice it. The table shows the temperature of the cake every 5 minutes after Alyssa took it out of the oven.

a. Make a scatter plot of the data.

b. Which regression equation has an R^2 value closest to 1? Is this the equation that best fits the context of the problem? Explain your reasoning.

c. Find an appropriate regression equation, and state the coefficient of determination. What is the domain and range?

d. Alyssa will ice the cake when it reaches room temperature (70°F). Use the regression equation to predict when she can ice her cake.

Time (min)	Temperature (°F)
0	350
5	244
10	178
15	137
20	112
25	96
30	89

Special Functions

Onoky/Getty Images

:: Then

- You identified and graphed linear, exponential, and quadratic functions.

:: Now

1. Identify and graph step functions.

2. Identify and graph absolute value and piecewise-defined functions.

:: Why?

- Kim is ordering books online. The site charges for shipping based on the amount of the order. If the order is less than $10, shipping costs $3. If the order is at least $10 but less than $20, it will cost $5 to ship it.

 NewVocabulary

step function
piecewise-linear function
greatest integer function
absolute value function
piecewise-defined function

 Common Core State Standards

Content Standards

F.IF.4 For a function that models a relationship between two quantities, interpret key features of graphs and tables in terms of the quantities, and sketch graphs showing key features given a verbal description of the relationship.

F.IF.7b Graph square root, cube root, and piecewise-defined functions, including step functions and absolute value functions.

Mathematical Practices
4 Model with mathematics.

1 Step Functions The graph of a **step function** is a series of line segments. Because each part of a step function is linear, this type of function is called a **piecewise-linear function**. One example of a step function is the **greatest integer function**, written as $f(x) = [\![x]\!]$, where $f(x)$ is the greatest integer not greater than x. For example, $[\![6.8]\!] = 6$ because 6 is the greatest integer that is not greater than 6.8.

KeyConcept Greatest Integer Function

Parent function:	$f(x) = [\![x]\!]$
Type of graph:	disjointed line segments
Domain:	all real numbers
Range:	all integers

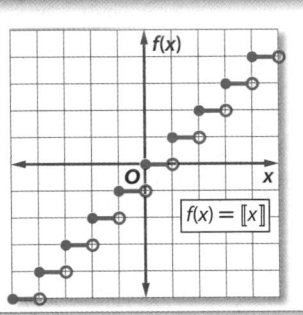

Example 1 Greatest Integer Function

Graph $f(x) = [\![x + 2]\!]$. State the domain and range.

First, make a table. Select a few values between integers. On the graph, dots represent included points. Circles represent points not included.

x	$x + 2$	$[\![x + 2]\!]$
0	2	2
0.25	2.25	2
0.5	2.5	2
1	3	3
1.25	3.25	3
1.5	3.5	3
2	4	4
2.25	4.25	4

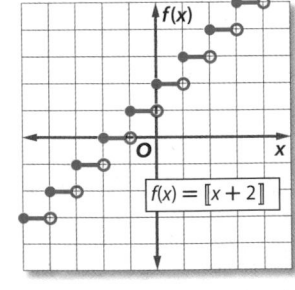

Note that this is the graph of $f(x) = [\![x]\!]$ shifted 2 units to the left.

Because the dots and circles overlap, the domain is all real numbers. The range is all integers. Notice that the graph has no symmetry and no maximum or minimum values. As x increases, $f(x)$ increases, and as x decreases, $f(x)$ decreases.

▶ **Guided**Practice

1. Graph $g(x) = 2[\![x]\!]$. State the domain and range.

Step functions can be used to represent many real-world situations involving money.

Real-World Link

North Americans are the most likely to have cell phones with 93.2% of the population owning phones.

Source: IT Facts

Real-World Example 2 Step Function

CELL PHONE PLANS Cell phone companies charge by the minute, not by the second. A cell phone company charges $0.45 per minute or any fraction thereof for exceeding the number of minutes allotted on each plan. Draw a graph that represents this situation.

The total cost for the extra minutes will be a multiple of $0.45, and the graph will be a step function. If the time is greater than 0 but less than or equal to 1 minute, the charge will be $0.45. If the time is greater than 2 but is less than or equal to 3 minutes, you will be charged for 3 minutes or $1.35.

x	f(x)
$0 < x \leq 1$	0.45
$1 < x \leq 2$	0.90
$2 < x \leq 3$	1.35
$3 < x \leq 4$	1.80
$4 < x \leq 5$	2.25
$5 < x \leq 6$	2.70
$6 < x \leq 7$	3.15

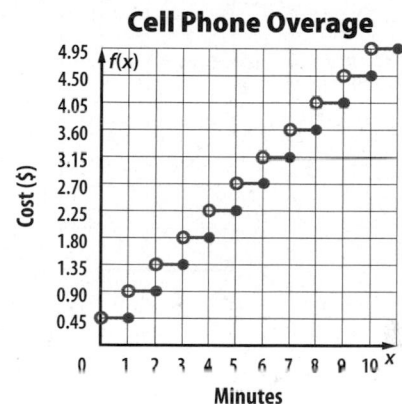

GuidedPractice

2. PARKING A garage charges $4 for the first hour and $1 for each additional hour. Draw a graph that represents this situation.

ReviewVocabulary

absolute value the distance a number is from zero on a number line; written | n |

2 Absolute Value Functions

Another type of piecewise-linear function is the **absolute value function**. Recall that the absolute value of a number is always nonnegative. So in the absolute value parent function, written as $f(x) = |x|$, all of the values of the range are nonnegative.

KeyConcept Absolute Value Function

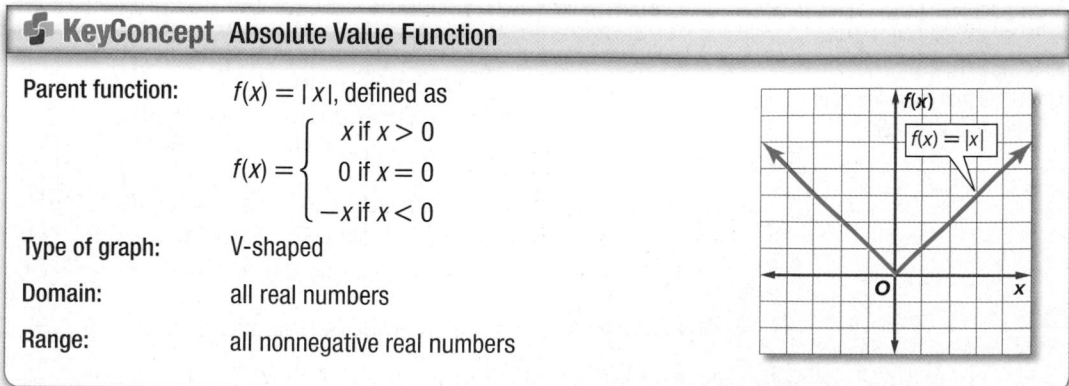

Parent function:	$f(x) = |x|$, defined as $f(x) = \begin{cases} x & \text{if } x > 0 \\ 0 & \text{if } x = 0 \\ -x & \text{if } x < 0 \end{cases}$
Type of graph:	V-shaped
Domain:	all real numbers
Range:	all nonnegative real numbers

The absolute value function is called a **piecewise-defined function** because it is defined using two or more expressions.

Example 3 Absolute Value Function

Graph $f(x) = |x - 4|$. State the domain and range.

Since $f(x)$ cannot be negative, the minimum point of the graph is where $f(x) = 0$.

$f(x) = |x - 4|$	Original function
$0 = x - 4$	Replace $f(x)$ with 0 and $|x - 4|$ with $x - 4$.
$4 = x$	Add 4 to each side.

Next make a table of values. Include values for $x > 4$ and $x < 4$.

$f(x) = |x - 4|$	
x	$f(x)$
-2	6
0	4
2	2
4	0
5	1
6	2
7	3
8	4

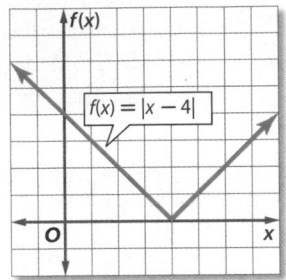

The domain is all real numbers. The range is all real numbers greater than or equal to 0. Note that this is the graph of $f(x) = |x|$ shifted 4 units to the right. Notice that the graph is symmetric about the line $x = 4$, and the minimum value of the function is 0 at $x = 4$. As x increases, $f(x)$ increases, and as x decreases, $f(x)$ increases.

GuidedPractice

3. Graph $f(x) = |2x + 1|$. State the domain and range.

Math HistoryLink

Florence Nightingale David (1909–1993) A renowned statistician born in Ivington, England, Florence Nightingale David received the first Elizabeth L. Scott Award for her "efforts in opening the door to women in statistics; … for research contributions to … statistical methods …; and her spirit as a lecturer and as a role model."

Not all piecewise-defined functions are absolute value functions. Step functions are also piecewise-defined functions. In fact, all piecewise-linear functions are piecewise-defined.

Example 4 Piecewise-Defined Function

Graph $f(x) = \begin{cases} -2x & \text{if } x > 1 \\ x + 3 & \text{if } x \leq 1 \end{cases}$. State the domain and range.

Graph the first expression. Create a table of values for when $x > 1$, $f(x) = -2x$ and draw the graph. Since x is not equal to 1, place a circle at $(1, -2)$.

Next, graph the second expression. Create a table of values for when $x \leq 1$, $f(x) = x + 3$ and draw the graph. If $x = 1$, then $f(x) = 4$; place a dot at $(1, 4)$.

The domain is all real numbers. The range is $y \leq 4$.

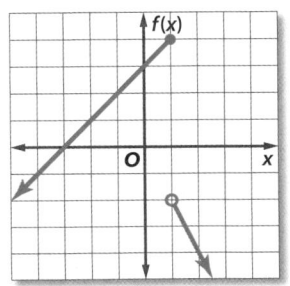

StudyTip

Piecewise Functions
To graph a piecewise-defined function, graph each "piece" separately. There should be a dot or line that contains each member of the domain.

GuidedPractice

4. Graph $f(x) = \begin{cases} 2x + 1 & \text{if } x > 0 \\ 3 & \text{if } x \leq 0 \end{cases}$. State the domain and range.

ConceptSummary Special Functions

Step Function	Absolute Value Function	Piecewise-Defined Function
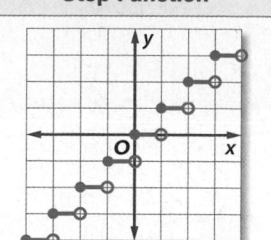	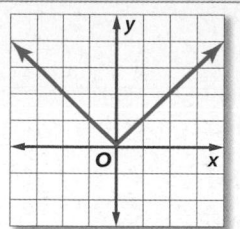	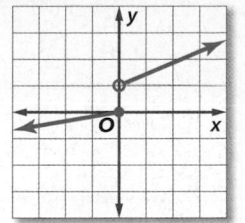

Check Your Understanding

○ = Step-by-Step Solutions begin on page R13.

Example 1   **Graph each function. State the domain and range.**

1. $f(x) = \frac{1}{2}[\![x]\!]$

2. $g(x) = -[\![x]\!]$

3. $h(x) = [\![2x]\!]$

Order Total ($)	Shipping Cost ($)
0–15	3.99
15.01–30	5.99
30.01–50	6.99
50.01–75	7.99
75.01–100	8.99
Over $100	9.99

Example 2   **4. SHIPPING** Elan is ordering a gift for his dad online.
The table shows the shipping rates. Graph the
step function.

Examples 3–4 **Graph each function. State the domain and range.**

5. $f(x) = |x - 3|$

6. $g(x) = |2x + 4|$

7. $f(x) = \begin{cases} 2x - 1 \text{ if } x > -1 \\ -x \text{ if } x \leq -1 \end{cases}$

8. $g(x) = \begin{cases} -3x - 2 \text{ if } x > -2 \\ -x + 1 \text{ if } x \leq -2 \end{cases}$

Practice and Problem Solving

Extra Practice is on page R9.

Example 1   **Graph each function. State the domain and range.**

9 $f(x) = 3[\![x]\!]$

10. $f(x) = [\![-x]\!]$

11. $g(x) = -2[\![x]\!]$

12. $g(x) = [\![x]\!] + 3$

13. $h(x) = [\![x]\!] - 1$

14. $h(x) = \frac{1}{2}[\![x]\!] + 1$

Example 2   **15. CAB FARES** Lauren wants to take a taxi from a hotel to a friend's house. The rate
is $3 plus $1.50 per mile after the first mile. Every fraction of a mile is rounded
up to the next mile.

a. Draw a graph to represent the cost of using a taxi cab.

b. What is the cost if the trip is 8.5 miles long?

16. **CCSS MODELING** The United States Postal Service increases the rate of postage
periodically. The table shows the cost to mail a letter weighing 1 ounce or less from
1995 through 2009. Draw a step graph to represent the data.

Year	1995	1999	2001	2002	2006	2007	2008	2009
Cost ($)	0.32	0.33	0.34	0.37	0.39	0.41	0.42	0.44

Examples 3–4 Graph each function. State the domain and range.

17. $f(x) = |2x - 1|$ 　　　　　　**18.** $f(x) = |x + 5|$

19. $g(x) = |-3x - 5|$ 　　　　　**20.** $g(x) = |-x - 3|$

21. $f(x) = \left|\frac{1}{2}x - 2\right|$ 　　　　　**22.** $f(x) = \left|\frac{1}{3}x + 2\right|$

23. $g(x) = |x + 2| + 3$ 　　　　　**24.** $g(x) = |2x - 3| + 1$

25. $f(x) = \begin{cases} \frac{1}{2}x - 1 \text{ if } x > 3 \\ -2x + 3 \text{ if } x \le 3 \end{cases}$ 　　**26.** $f(x) = \begin{cases} 2x - 5 \text{ if } x > 1 \\ 4x - 3 \text{ if } x \le 1 \end{cases}$

27. $f(x) = \begin{cases} 2x + 3 \text{ if } x \ge -3 \\ -\frac{1}{3}x + 1 \text{ if } x < -3 \end{cases}$ 　　**28.** $f(x) = \begin{cases} 3x + 4 \text{ if } x \ge 1 \\ x + 3 \text{ if } x < 1 \end{cases}$

29. $f(x) = \begin{cases} 3x + 2 \text{ if } x > -1 \\ -\frac{1}{2}x - 3 \text{ if } x \le -1 \end{cases}$ 　　**30.** $f(x) = \begin{cases} 2x + 1 \text{ if } x < -2 \\ -3x - 1 \text{ if } x \ge -2 \end{cases}$

Determine the domain and range of each function.

 31.
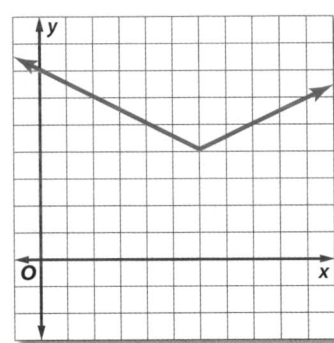

32.
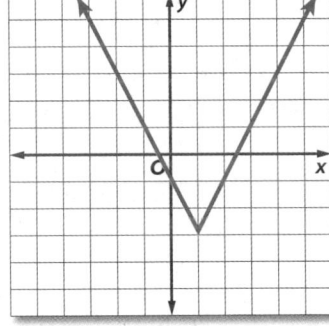

33.
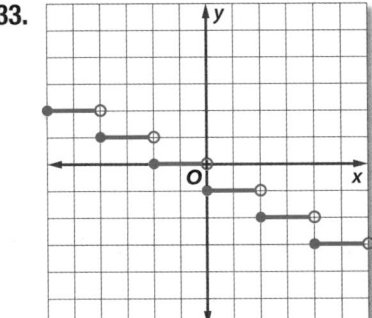

34.
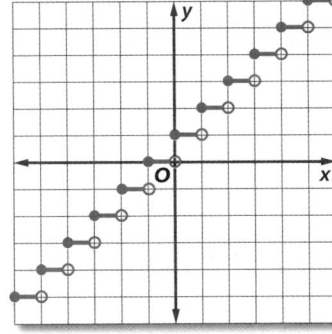

35.
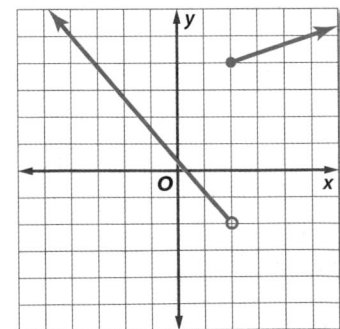

36.
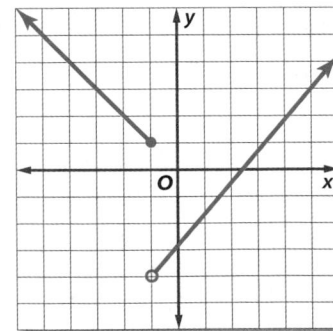

37. BOATING According to Boat Minnesota, the maximum number of people that can safely ride in a boat is determined by the boat's length and width. The table shows some guidelines for the length of a boat that is 6 feet wide. Graph this relation.

Length of Boat (ft)	18–19	20–22	23–24
Number of People	7	8	9

For Exercises 38–41, match each graph to one of the following equations.

A	B	C	D
$y = 2x - 1$	$y = [\![2x]\!] - 1$	$y = |2x| - 1$	$y = \begin{cases} 2x + 1 \text{ if } x > 0 \\ -2x + 1 \text{ if } x \leq 0 \end{cases}$

38.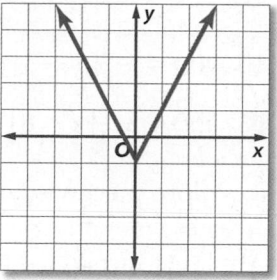

39.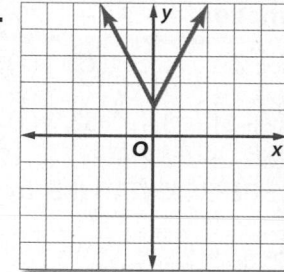

40.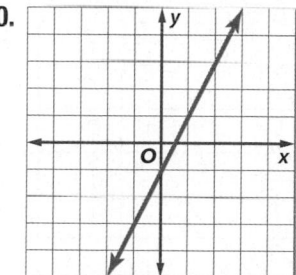

41.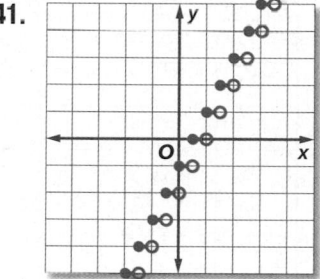

42. **CAR LEASE** As part of Marcus' leasing agreement, he will be charged $0.20 per mile for each mile over 12,000. Any fraction of a mile is rounded up to the next mile. Make a step graph to represent the cost of going over the mileage.

43. **BASEBALL** A baseball team is ordering T-shirts with the team logo on the front and the players' names on the back. For an order of 10 shirts or less, a graphic design store charges $10 to set up the artwork plus $10 per shirt, $4 each for the team logo, and $2 to print the last name for an order of 10 shirts or less. For orders of 11–20 shirts, a 5% discount is given. For orders of more than 20 shirts, a 10% discount is given.

   a. Organize the information into a table. Include a column showing the total order price for each size order.

   b. Write an equation representing the total price for an order of x shirts.

   c. Graph the piecewise relation.

44. Consider the function $f(x) = |2x + 3|$.

   a. Make a table of values where x is all integers from -5 to 5, inclusive.

   b. Plot the points on a coordinate grid.

   c. Graph the function.

45. Consider the function $f(x) = |2x| + 3$.

   a. Make a table of values where x is all integers from -5 to 5, inclusive.

   b. Plot the points on a coordinate grid.

   c. Graph the function.

   d. Describe how this graph is different from the graph in Exercise 44.

46. DANCE A local studio owner will teach up to four students by herself. Her instructors can teach up to 5 students each. Draw a step function graph that best describes the number of instructors needed for the different numbers of students.

47. THEATERS A community theater will only perform a show if there are at least 250 pre-sale ticket requests. Additional performances will be added for each 250 requests after that. Draw a step function graph that best describes this situation.

Graph each function.

48. $f(x) = \frac{1}{2}|x| + 2$

49 $g(x) = \frac{1}{3}|x| + 4$

50. $h(x) = -2|x - 3| + 2$

51. $f(x) = -4|x + 2| - 3$

52. $g(x) = -\frac{2}{3}|x + 6| - 1$

53. $h(x) = -\frac{3}{4}|x - 8| + 1$

54. 🔁 **MULTIPLE REPRESENTATIONS** In this problem, you will explore piecewise linear functions.

a. Tabular Copy and complete the table of values for $f(x) = |[\![x]\!]|$ and $g(x) = [\![|x|]\!]$.

x	$[\![x]\!]$	$f(x) = |[\![x]\!]|$	$|x|$	$g(x) = [\![|x|]\!]$
−3	−3	3	3	3
−2.5				
−2				
0				
0.5				
1				
1.5				

b. Graphical Graph each function on a coordinate plane.

c. Analytical Compare and contrast the graphs of $f(x)$ and $g(x)$.

H.O.T. Problems   Use Higher-Order Thinking Skills

55. REASONING Does the piecewise relation $y = \begin{cases} -2x + 4 \text{ if } x \geq 2 \\ -\frac{1}{2}x - 1 \text{ if } x \leq 4 \end{cases}$ represent a function? Why or why not?

CCSS SENSE-MAKING Refer to the graph.

56. Write an absolute value function that represents the graph.

57. Write a piecewise function to represent the graph.

58. What are the domain and range?

59. WRITING IN MATH Compare and contrast the graphs of absolute value, step, and piecewise-defined functions with the graphs of quadratic and exponential functions. Discuss the domains, ranges, maxima, minima, and symmetry.

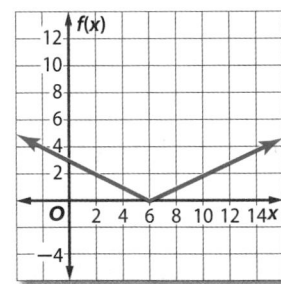

60. CHALLENGE A bicyclist travels up and down a hill with a vertical cross section that can be modeled by $y = -\frac{1}{4}|x - 400| + 100$, where x and y are measured in feet.

a. If $0 \leq x \leq 800$, find the slope for the uphill portion of the trip and downhill portion of the trip.

b. Graph this function. What are the domain and range?

61. Which equation represents a line that is perpendicular to the graph and passes through the point at (2, 0)?

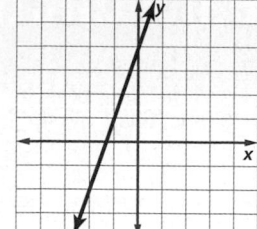

A $y = 3x - 6$

B $y = -3x + 6$

C $y = -\frac{1}{3}x + \frac{2}{3}$

D $y = \frac{1}{3}x - \frac{2}{3}$

62. A giant tortoise travels at a rate of 0.17 mile per hour. Which equation models the time t it would take the giant tortoise to travel 0.8 mile?

F $t = \dfrac{0.8}{0.17}$

H $t = \dfrac{0.17}{0.8}$

G $t = (0.17)(0.8)$

J $0.8 = \dfrac{0.17}{t}$

63. GEOMETRY If $\triangle JKL$ is similar to $\triangle JNM$ what is the value a?

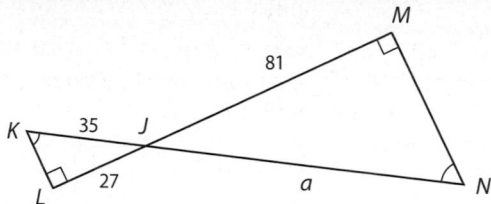

A 62.5

B 105

C 125

D 155.5

64. GRIDDED RESPONSE What is the difference in the value of $2.1(x + 3.2)$, when $x = 5$ and when $x = 3$?

Look for a pattern in each table of values to determine which model best describes the data. (Lesson 9-6)

65.

x	0	1	2	3	4
y	1	3	5	7	9

66.

x	−2	−1	0	1	2
y	5	2	1	2	5

67.

x	−1	0	1	2	3
y	1	2	4	8	16

68.

x	5	6	7	8	9
y	−2.5	−1.5	1.5	6.5	13.5

69. TESTS Determine whether the graph at the right shows a *positive*, *negative*, or *no* correlation. If there is a correlation, describe its meaning. (Lesson 4-5)

Suppose y varies directly as x. (Lesson 3-4)

70. If $y = 2.5$ when $x = 0.5$, find y when $x = 20$.

71. If $y = -6.6$ when $x = 9.9$, find y when $x = 6.6$.

72. If $y = 2.6$ when $x = 0.25$, find y when $x = 1.125$.

73. If $y = 6$ when $x = 0.6$, find x when $y = 12$.

Test Scores

Test Scores vs *Study Time (min)*

Evaluate each expression. If necessary, round to the nearest hundredth.

74. $\sqrt{9}$

75. $\sqrt{12}$

76. $\sqrt{4.5}$

77. $3\sqrt{16}$

78. $2\sqrt{10}$

79. $\sqrt{5} - 2$

9-7

Graphing Technology Lab
Piecewise-Linear Functions

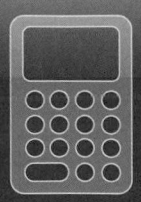

You can use a graphing calculator to graph and analyze various piecewise functions, including greatest integer functions and absolute value functions.

CCSS **Common Core State Standards**
Content Standards
F.IF.7b Graph square root, cube root, and piecewise-defined functions, including step functions and absolute value functions.

Activity 1   Greatest Integer Functions

Graph $f(x) = [\![x]\!]$ **in the standard viewing window.**

The calculator may need to be changed to dot mode for the function to graph correctly. Press $\boxed{\text{MODE}}$ then use the arrow and $\boxed{\text{ENTER}}$ keys to select **DOT**.

Enter the equation in the **Y=** list. Then graph the equation.

KEYSTROKES: $\boxed{\text{Y=}}$ $\boxed{\text{MATH}}$ $\boxed{\blacktriangleright}$ 5 $\boxed{\text{X,T,}\theta,n}$ $\boxed{)}$ $\boxed{\text{ZOOM}}$ 6

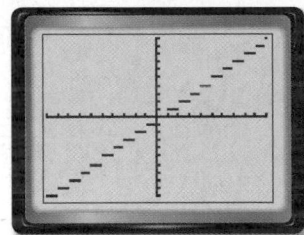

[−10, 10] scl: 1 by [−10, 10] scl: 1

1A. How does the graph of $f(x) = [\![x]\!]$ compare to the graph of $f(x) = x$?

1B. What are the domain and range of the function $f(x) = [\![x]\!]$? Explain.

The graphs of piecewise functions are affected by changes in parameters.

Activity 2   Absolute Value Functions

Graph $y = |x| - 3$ **and** $y = |x| + 1$ **in the standard viewing window.**

Enter the equations in the **Y=** list. Then graph.

KEYSTROKES: $\boxed{\text{Y=}}$ $\boxed{\text{MATH}}$ $\boxed{\blacktriangleright}$ 1 $\boxed{\text{X,T,}\theta,n}$ $\boxed{)}$ $\boxed{-}$ 3 $\boxed{\text{ENTER}}$ $\boxed{\text{MATH}}$ $\boxed{\blacktriangleright}$ 1

$\boxed{\text{X,T,}\theta,n}$ $\boxed{)}$ $\boxed{+}$ 1 $\boxed{\text{ZOOM}}$ 6

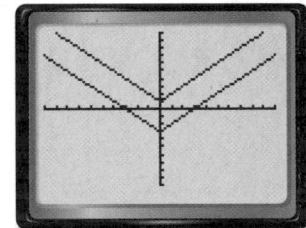

[−10, 10] scl: 1 by [−10, 10] scl: 1

2A. Compare and contrast the graphs to the graph of $y = |x|$.

2B. How does the value of k affect the graph of $y = |x| + k$?

Analyze the Results

1. A parking garage charges $4 for every hour or fraction of an hour. Is this situation modeled by a *linear* function or a *step* function? Explain your reasoning.

2. A maintenance technician is testing an elevator system. The technician starts the elevator at the fifth floor. It is sent to the ground floor, then back to the fifth floor. Assume the elevator travels at a constant rate. Should the height of the elevator be modeled by a step function or an absolute value function? Explain.

Because the points on a graph are solutions of its equation, the x-coordinates of points where $y = f(x)$ and $y = g(x)$ intersect are solutions of $f(x) = g(x)$. For example, the solution of $5x - 2 = |x|$ is the intersection of the graphs of $y = 5x - 2$ and $y = |x|$. Write each equation as a system of equations, and then use a graph or a table to solve it.

3. $5x - 2 = |x|$ **4.** $2|x - 2| = x - 1$ **5.** $|4x + 2| = -|x| + 3$

Study Guide

KeyConcepts

Graphing Quadratic Functions (Lesson 9-1)

- A quadratic function can be described by an equation of the form $y = ax^2 + bx + c$, where $a \neq 0$.
- The axis of symmetry for the graph of $y = ax^2 + bx + c$, where $a \neq 0$, is $x = -\dfrac{b}{2a}$.

Solving Quadratic Equations (Lessons 9-2, 9-4, and 9-5)

- Quadratic equations can be solved by graphing. The solutions are the x-intercepts or zeros of the related quadratic function.
- Quadratic equations can be solved by completing the square. To complete the square for $x^2 + bx$, find $\dfrac{1}{2}$ of b, square this result, and then add the result to $x^2 + bx$.
- Quadratic equations can be solved by using the Quadratic Formula, $x = \dfrac{-b \pm \sqrt{b^2 - 4ac}}{2a}$.

Transformations of Quadratic Functions (Lesson 9-3)

- $f(x) = x^2 + c$ translates the graph up or down.
- $f(x) = ax^2$ compresses or expands the graph vertically.

Special Functions (Lesson 9-7)

- The greatest integer function is written as $f(x) = [\![x]\!]$, where $f(x)$ is the greatest integer not greater than x.
- The absolute value function is written as $f(x) = |x|$, where $f(x)$ is the distance from x to 0 on a number line.

FOLDABLES StudyOrganizer

Be sure the Key Concepts are noted in your Foldable.

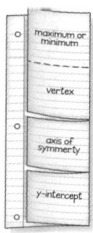

KeyVocabulary

absolute value function (p. 599)

axis of symmetry (p. 543)

completing the square (p. 574)

dilation (p. 566)

discriminant (p. 586)

double root (p. 556)

greatest integer function (p. 598)

maximum (p. 543)

minimum (p. 543)

parabola (p. 543)

piecewise-defined function (p. 599)

piecewise-linear function (p. 598)

Quadratic Formula (p. 583)

quadratic function (p. 543)

reflection (p. 566)

standard form (p. 543)

step function (p. 598)

transformation (p. 564)

translation (p. 564)

vertex (p. 543)

VocabularyCheck

State whether each sentence is **true** or **false**. If **false**, replace the underlined term to make a true sentence.

1. The <u>axis of symmetry</u> of a quadratic function can be found by using the equation $x = -\dfrac{b}{2a}$.

2. The <u>vertex</u> is the maximum or minimum point of a parabola.

3. The graph of a quadratic function is a <u>straight line</u>.

4. The graph of a quadratic function has a <u>maximum</u> if the coefficient of the x^2-term is positive.

5. A quadratic equation with a graph that has two x-intercepts has <u>one</u> real root.

6. The expression $b^2 - 4ac$ is called the <u>discriminant</u>.

7. A function that is defined differently for different parts of its domain is called a <u>piecewise-defined function</u>.

8. The <u>range</u> of the greatest integer function is the set of all real numbers.

9. The solutions of a quadratic equation are called <u>roots</u>.

10. The graph of the parent function is <u>translated down</u> to form the graph of $f(x) = x^2 + 5$.

Lesson-by-Lesson Review

9-1 Graphing Quadratic Functions

Consider each equation.

a. Determine whether the function has a *maximum* or *minimum* value.

b. State the maximum or minimum value.

c. What are the domain and range of the function?

11. $y = x^2 - 4x + 4$

12. $y = -x^2 + 3x$

13. $y = x^2 - 2x - 3$

14. $y = -x^2 + 2$.

15. ROCKET A toy rocket is launched with an upward velocity of 32 feet per second. The equation $h = -16t^2 + 32t$ gives the height of the ball t seconds after it is launched.

a. Determine whether the function has a *maximum* or *minimum* value.

b. State the maximum or minimum value.

c. State a reasonable domain and range for this situation.

Example 1

Consider $f(x) = x^2 + 6x + 5$.

a. Determine whether the function has a *maximum* or *minimum* value.

For $f(x) = x^2 + 6x + 5$, $a = 1$, $b = 6$, and $c = 5$.

Because a is positive, the graph opens up, so the function has a minimum value.

b. State the *maximum* or *minimum* value of the function.

The minimum value is the y-coordinate of the vertex.

The x-coordinate of the vertex is $\frac{-b}{2a}$ or $\frac{-6}{2(1)}$ or -3.

$f(x) = x^2 + 6x + 5$ Original function

$f(-3) = (-3)^2 + 6(-3) + 5$ $x = -3$

$f(-3) = -4$ Simplify.

The minimum value is -4.

c. State the domain and range of the function.

The domain is all real numbers. The range is all real numbers greater than or equal to the minimum value, or $\{y \mid y \geq -4\}$.

9-2 Solving Quadratic Equations by Graphing

Solve each equation by graphing. If integral roots cannot be found, estimate the roots to the nearest tenth.

16. $x^2 - 3x - 4 = 0$

17. $-x^2 + 6x - 9 = 0$

18. $x^2 - x - 12 = 0$

19. $x^2 + 4x - 3 = 0$

20. $x^2 - 10x = -21$

21. $6x^2 - 13x = 15$

22. NUMBER THEORY Find two numbers that have a sum of 2 and a product of -15.

Example 2

Solve $x^2 - x - 6 = 0$ by graphing.

Graph the related function $f(x) = x^2 - x - 6$.

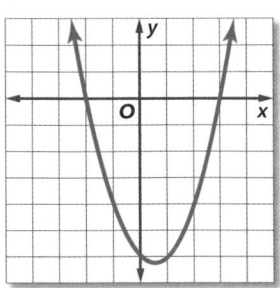

The x-intercepts of the graph appear to be at -2 and 3, so the solutions are -2 and 3.

9-3 Transformations of Quadratic Functions

Describe how the graph of each function is related to the graph of $f(x) = x^2$.

23. $f(x) = x^2 + 8$

24. $f(x) = x^2 - 3$

25. $f(x) = 2x^2$

26. $f(x) = 4x^2 - 18$

27. $f(x) = \frac{1}{3}x^2$

28. $f(x) = \frac{1}{4}x^2$

29. Write an equation for the function shown in the graph.

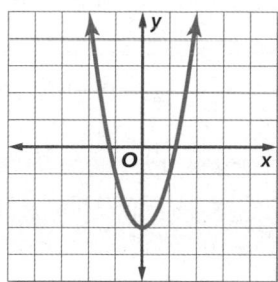

30. PHYSICS A ball is dropped off a cliff that is 100 feet high. The function $h = -16t^2 + 100$ models the height h of the ball after t seconds. Compare the graph of this function to the graph of $h = t^2$.

Example 3

Describe how the graph of $f(x) = x^2 - 2$ is related to the graph of $f(x) = x^2$.

The graph of $f(x) = x^2 + c$ represents a translation up or down of the parent graph.

Since $c = -2$, the translation is down.

So, the graph is translated down 2 units from the parent function.

Example 4

Write an equation for the function shown in the graph.

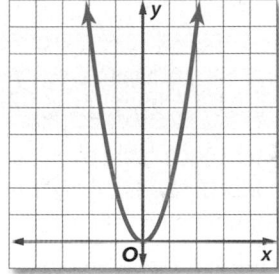

Since the graph opens upward, the leading coefficient must be positive. The parabola has not been translated up or down, so $c = 0$. Since the graph is stretched vertically, it must be of the form of $f(x) = ax^2$ where $a > 1$. The equation for the function is $y = 2x^2$.

9-4 Solving Quadratic Equations by Completing the Square

Solve each equation by completing the square. Round to the nearest tenth if necessary.

31. $x^2 + 6x + 9 = 16$

32. $-a^2 - 10a + 25 = 25$

33. $y^2 - 8y + 16 = 36$

34. $y^2 - 6y + 2 = 0$

35. $n^2 - 7n = 5$

36. $-3x^2 + 4 = 0$

37. NUMBER THEORY Find two numbers that have a sum of -2 and a product of -48.

Example 5

Solve $x^2 - 16x + 32 = 0$ by completing the square. Round to the nearest tenth if necessary.

Isolate the x^2- and x-terms. Then complete the square and solve.

$x^2 - 16x + 32 = 0$	Original equation
$x^2 - 16x = -32$	Isolate the x^2- and x-terms.
$x^2 - 16x + 64 = -32 + 64$	Complete the square.
$(x - 8)^2 = 32$	Factor.
$x - 8 = \pm\sqrt{32}$	Take the square root.
$x = 8 \pm \sqrt{32}$	Add 8 to each side.
$x = 8 \pm 4\sqrt{2}$	Simplify.

The solutions are about 2.3 and 13.7.

9-5 Solving Quadratic Equations by Using the Quadratic Formula

Solve each equation by using the Quadratic Formula. Round to the nearest tenth if necessary.

38. $x^2 - 8x = 20$

39. $21x^2 + 5x - 7 = 0$

40. $d^2 - 5d + 6 = 0$

41. $2f^2 + 7f - 15 = 0$

42. $2h^2 + 8h + 3 = 3$

43. $4x^2 + 4x = 15$

44. GEOMETRY The area of a square can be quadrupled by increasing the side length and width by 4 inches. What is the side length?

Example 6

Solve $x^2 + 10x + 9 = 0$ by using the Quadratic Formula.

$$x = \frac{-b \pm \sqrt{b^2 - 4ac}}{2a} \quad \text{Quadratic Formula}$$

$$= \frac{-10 \pm \sqrt{10^2 - 4(1)(9)}}{2(1)} \quad a = 1, b = 10, c = 9$$

$$= \frac{-10 \pm \sqrt{64}}{2} \quad \text{Simplify.}$$

$$x = \frac{-10 + 8}{2} \quad \text{or} \quad x = \frac{-10 - 8}{2} \quad \text{Separate the solutions.}$$

$$= -1 \qquad\qquad = -9 \quad \text{Simplify.}$$

9-6 Analyzing Functions with Successive Differences

Look for a pattern in each table of values to determine which kind of model best describes the data. Then write an equation for the function that models the data.

45.

x	0	1	2	3	4
y	0	3	12	27	48

46.

x	0	1	2	3	4
y	1	2	4	8	16

47.

x	0	1	2	3	4
y	0	−1	−4	−9	−16

Example 7

Determine the model that best describes the data. Then write an equation for the function that models the data.

x	0	1	2	3	4
y	3	4	5	6	7

Step 1 First differences:

A linear function models the data.

Step 2 The slope is 1 and the y-intercept is 3, so the equation is $y = x + 3$.

9-7 Special Functions

Graph each function. State the domain and range.

48. $f(x) = [\![x]\!]$

49. $f(x) = [\![2x]\!]$

50. $f(x) = |x|$

51. $f(x) = |2x - 2|$

52. $f(x) = \begin{cases} x - 2 & \text{if } x < 1 \\ 3x & \text{if } x \geq 1 \end{cases}$

53. $f(x) = \begin{cases} 2x - 3 & \text{if } x \leq 2 \\ x + 1 & \text{if } x > 2 \end{cases}$

Example 8

Graph $f(x) = |x + 3|$. State the domain and range.

x	f(x)
−5	2
−4	1
−3	0
−2	1
−1	2

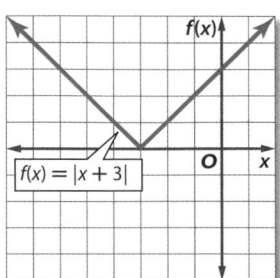

The domain is all real numbers, and the range is $f(x) \geq 0$.

Use a table of values to graph the following functions. State the domain and range.

1. $y = x^2 + 2x + 5$

2. $y = 2x^2 - 3x + 1$

Consider $y = x^2 - 7x + 6$.

3. Determine whether the function has a *maximum* or *minimum* value.

4. State the maximum or minimum value.

5. What are the domain and range?

Solve each equation by graphing. If integral roots cannot be found, estimate the roots to the nearest tenth.

6. $x^2 + 7x + 10 = 0$

7. $x^2 - 5 = -3x$

Describe how the graph of each function is related to the graph of $f(x) = x^2$.

8. $g(x) = x^2 - 5$

9. $g(x) = -3x^2$

10. $h(x) = \frac{1}{2}x^2 + 4$

11. **MULTIPLE CHOICE** Which is an equation for the function shown in the graph?

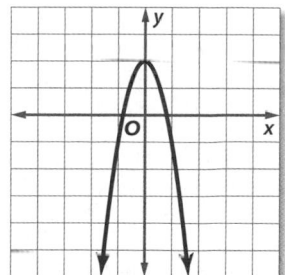

   A $y = -3x^2$

   B $y = 3x^2 + 1$

   C $y = x^2 + 2$

   D $y = -3x^2 + 2$

Solve each equation by completing the square.

12. $x^2 + 2x + 5 = 0$

13. $x^2 - x - 6 = 0$

14. $2x^2 - 36 = -6x$

Solve each equation by using the Quadratic Formula. Round to the nearest tenth if necessary.

15. $x^2 - x - 30 = 0$

16. $x^2 - 10x = -15$

17. $2x^2 + x - 15 = 0$

18. **BASEBALL** Elias hits a baseball into the air. The equation $h = -16t^2 + 60t + 3$ models the height h in feet of the ball after t seconds. How long is the ball in the air?

19. Graph $\{(-2, 4), (-1, 1), (0, 0), (1, 1), (2, 4)\}$. Determine whether the ordered pairs represent a *linear function*, a *quadratic function*, or an *exponential function*.

20. Look for a pattern in the table to determine which kind of model best describes the data.

x	0	1	2	3	4
y	1	3	5	7	9

21. **CAR CLUB** The table shows the number of car club members for four consecutive years after it began.

Time (years)	0	1	2	3	4
Members	10	20	40	80	160

   a. Determine which model best represents the data.

   b. Write a function that models the data.

   c. Predict the number of car club members after 6 years.

Graph each function.

22. $f(x) = |x - 1|$

23. $f(x) = -|2x|$

24. $f(x) = [\![x]\!]$

25. $f(x) = \begin{cases} 2x - 1 \text{ if } x < 2 \\ x - 3 \text{ if } x \geq 2 \end{cases}$

26. Determine the domain and range of the function graphed below.

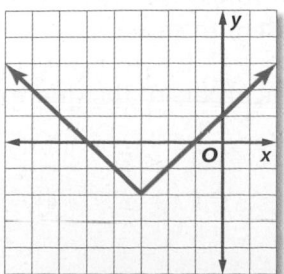

Preparing for Standardized Tests

Use a Formula

A *formula* is an equation that shows a relationship among certain quantities. Many standardized test problems will require using a formula to solve them.

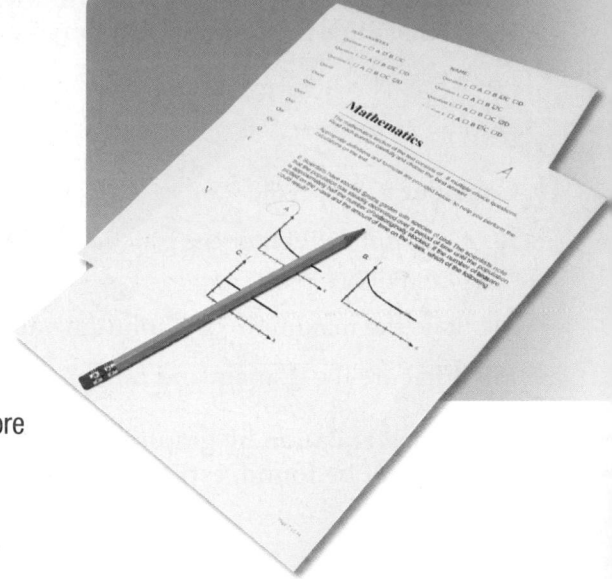

Strategies for Using a Formula

Step 1

Become familiar with common formulas and their uses. You may or may not be given access to a formula sheet to use during the test.

- **If given a formula sheet,** be sure to practice with the formulas on it before taking the test so you know how to apply them.

- **If not given a formula sheet,** study and practice with common formulas such as perimeter, area, and volume formulas, the Distance Formula, the Pythagorean Theorem, the Midpoint Formula, the Quadratic Formula, and others.

Step 2

Choose a formula and solve.

- **Ask Yourself:** What quantities are given in the problem statement?
- **Ask Yourself:** What quantities am I looking for?
- **Ask Yourself:** Is there a formula I know that relates these quantities?
- **Write:** Write the formula out that you have chosen each time.
- **Solve:** Substitute known quantities into the formula and solve for the unknown quantity.
- **Check:** Check your answer if time permits.

Standardized Test Example

Read the problem. Identify what you need to know. Then use the information in the problem to solve.

Find the exact roots of the quadratic equation $-2x^2 + 6x + 5 = 0$.

A $\dfrac{3 \pm \sqrt{17}}{4}$ **C** $\dfrac{3 \pm \sqrt{19}}{2}$

B $\dfrac{4 \pm \sqrt{17}}{3}$ **D** $\dfrac{3 \pm \sqrt{19}}{4}$

Read the problem carefully. You are given a quadratic equation and asked to find the exact roots of the equation. Use the **Quadratic Formula** to find the roots.

$$-2x^2 + 6x + 5 = 0$$ Original equation

$$a = -2, b = 6, c = 5$$ Identify the coefficients of the equation.

$$x = \frac{-b \pm \sqrt{b^2 - 4ac}}{2a}$$ Quadratic Formula

$$= \frac{-(6) \pm \sqrt{(6)^2 - 4(-2)(5)}}{2(-2)}$$ $a = -2, b = 6,$ and $c = 5$

$$= \frac{-6 \pm \sqrt{36 - (-40)}}{-4}$$ Simplify.

$$= \frac{-6 \pm \sqrt{76}}{-4}$$ Subtract.

$$= \frac{-6 \pm 2\sqrt{19}}{-4}$$ $\sqrt{76} = \sqrt{4 \cdot 19}$ or $2\sqrt{19}$.

$$= \frac{-2(3 \pm \sqrt{19})}{-2(2)}$$ Factor out -2 from the numerator and denominator.

$$= \frac{3 \pm \sqrt{19}}{2}$$ Simplify.

The roots of the equation are $\dfrac{3 + \sqrt{19}}{2}$ and $\dfrac{3 - \sqrt{19}}{2}$. The correct answer is C.

Exercises

Read each problem. Identify what you need to know. Then use the information in the problem to solve.

1. Find the exact roots of the quadratic equation $x^2 + 5x - 12 = 0$.

   A $\dfrac{-5 \pm \sqrt{73}}{2}$

   B $\dfrac{4 \pm \sqrt{61}}{3}$

   C $\dfrac{-3 \pm \sqrt{73}}{4}$

   D $\dfrac{-1 \pm \sqrt{61}}{2}$

2. The area of a triangle in which the length of the base is 4 centimeters greater than twice the height is 80 square centimeters. What is the length of the base of the triangle?

   F -10

   G 8

   H 16

   J 20

3. Find the volume of the figure below.

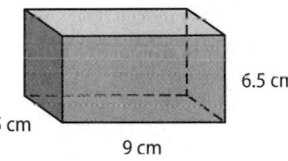

6.5 cm
5 cm
9 cm

   A 18.5 cm^3

   B 91 cm^3

   C 272 cm^3

   D 292.5 cm^3

4. Myron is traveling 263.5 miles at an average rate of 62 miles per hour. How long will it take Myron to complete his trip?

   F 4 h 10 min

   G 4 h 15 min

   H 5 h 10 min

   J 5 h 25 min

Multiple Choice

Read each question. Then fill in the correct answer on the answer document provided by your teacher or on a sheet of paper.

1. What is the vertex of the parabola graphed below?

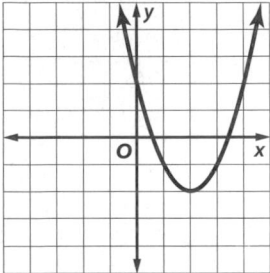

   A (2, 0)

   B (0, 2)

   C (−2, 2)

   D (2, −2)

2. Write an equation in slope-intercept form with a slope of $\frac{9}{10}$ and y-intercept of 3.

   F $y = 3x + \frac{9}{10}$

   G $y = \frac{9}{10}x + 3$

   H $y = \frac{9}{10}x - 3$

   J $y = 3x - \frac{9}{10}$

3. Use the Quadratic Formula to find the exact solutions of the equation $2x^2 - 6x + 3 = 0$.

   A $\frac{3 \pm \sqrt{3}}{2}$

   B $\frac{3 \pm \sqrt{2}}{4}$

   C $\frac{2 \pm \sqrt{5}}{3}$

   D $\frac{5 \pm \sqrt{2}}{2}$

4. Write an expression for the area of the rectangle below.

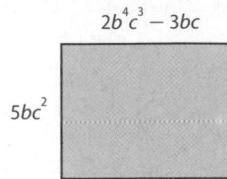

$2b^4c^3 - 3bc$

$5bc^2$

   F $10b^5c^5 - 3bc$

   G $10b^5c^5 - 15b^2c^3$

   H $2b^5c^5 - 3b^2c^3$

   J $10b^4c^6 - 15bc^2$

5. Solve the quadratic equation below by graphing.

$$x^2 - 2x - 15 = 0$$

   A −1, 4

   B −3, 5

   C 3, −5

   D ∅

6. Jason is playing games at a family fun center. So far he has won 38 prize tickets. How many more tickets would he need to win to place him in the gold prize category?

Number of Tickets	Prize Category
1–20	bronze
21–40	silver
41–60	gold
61–80	platinum

   F $2 \le t \le 22$

   G $3 \le t \le 22$

   H $1 \le t \le 20$

   J $3 \le t \le 20$

Test-TakingTip

Question 5 If permitted, you can use a graphing calculator to quickly graph an equation and find its roots.

Record your answers on the answer sheet provided by your teacher or on a sheet of paper.

7. **GRIDDED RESPONSE** Misty purchased a car several years ago for $21,459. The value of the car depreciated at a rate of 15% annually. What was the value of the car after 5 years? Round your answer to the nearest whole dollar.

8. Use the graph of the quadratic equation shown below to answer each question.

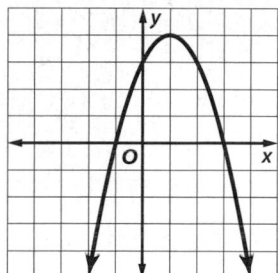

   a. What is the vertex?

   b. What is the y-intercept?

   c. What is the axis of symmetry?

   d. What are the roots of the corresponding quadratic equation?

9. The cost of 5 notebooks and 3 pens is $9.75. The cost of 4 notebooks and 6 pens is $10.50.

   a. Write a system of equations to model the situation.

   b. Solve the system of equations. How much does each item cost?

10. The table shows the total cost of renting a canoe for n hours.

Number of Hours (n)	Rental Cost (C)
1	$15
2	$20
3	$25
4	$30

   a. Write a function to represent the situation.

   b. How much would it cost to rent the canoe for 7 hours?

Extended Response

Record your answers on a sheet of paper. Show your work.

11. Use the equation and its graph to answer each question.

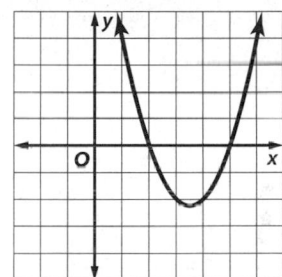

   a. Factor $x^2 - 7x + 10$.

   b. What are the solutions of $x^2 - 7x + 10 = 0$?

   c. What do you notice about the graph of the quadratic equation and where it crosses the x-axis? How do these values compare to the solutions of $x^2 - 7x + 10 = 0$? Explain.

Need ExtraHelp?

If you missed Question...	1	2	3	4	5	6	7	8	9	10	11
Go to Lesson...	9-1	4-2	9-5	8-2	9-2	5-1	7-6	9-1	6-4	3-5	8-6

CHAPTER 10

Radical Functions and Geometry

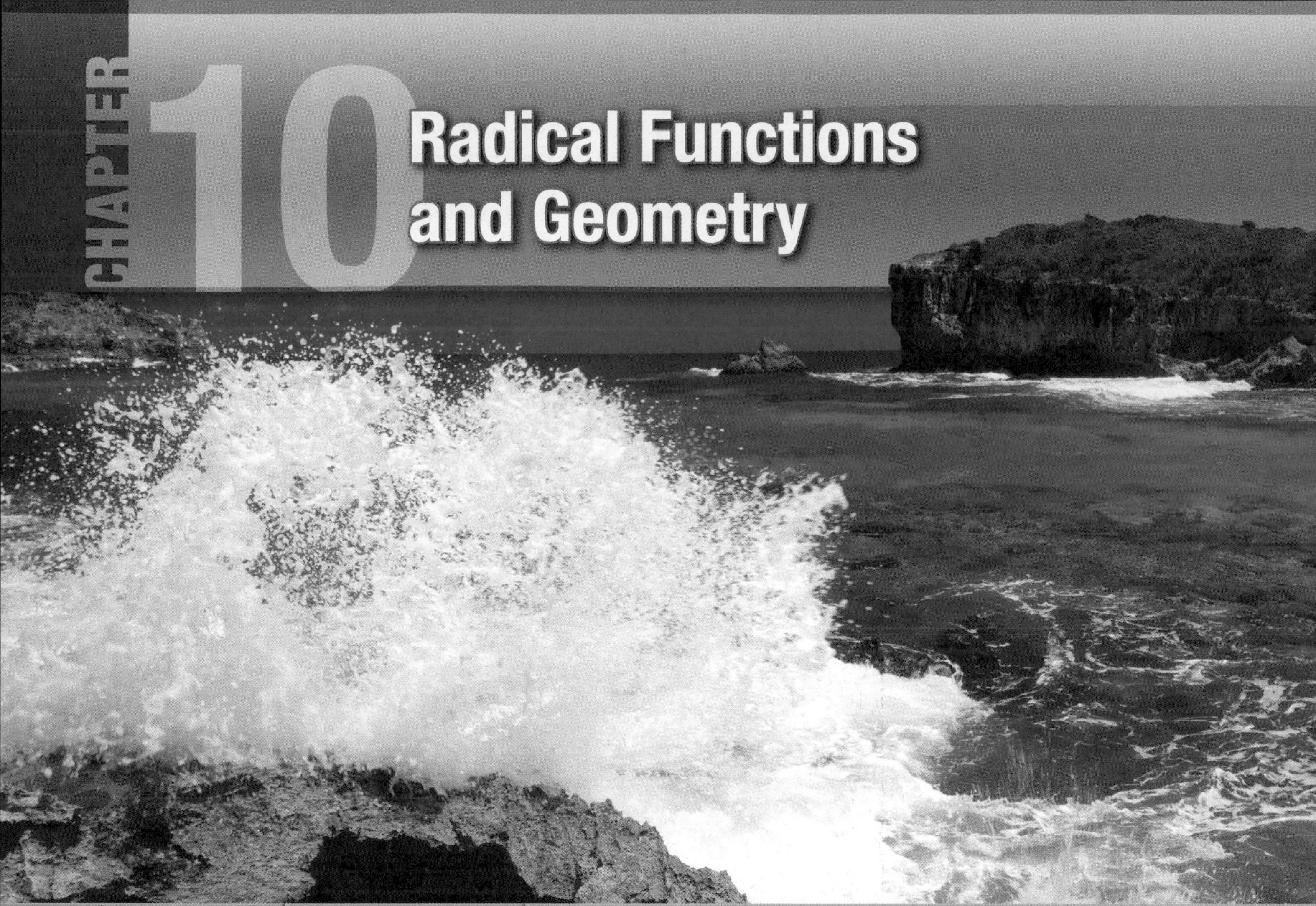

Then

You solved quadratic and exponential equations.

Now

In this chapter, you will:

- Graph and transform radical functions.
- Simplify, add, subtract, and multiply radical expressions.
- Solve radical equations.
- Use the Pythagorean Theorem.
- Find trigonometric ratios.

Why? ▲

OCEANS Tsunamis, or large waves, are generated by undersea earthquakes. A radical equation can be used to find the speed of a tsunami in meters per second or the depth of the ocean in meters.

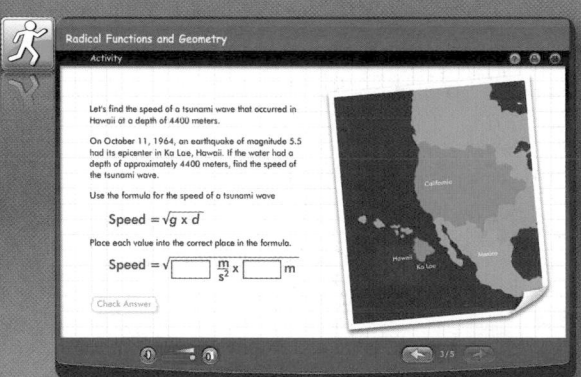

Radical Functions and Geometry
Activity
Let's find the speed of a tsunami wave that occurred in Hawaii at a depth of 4400 meters.
On October 11, 1964, an earthquake of magnitude 5.5 had its epicenter in Ka Lae, Hawaii. If the water had a depth of approximately 4400 meters, find the speed of the tsunami wave.
Use the formula for the speed of a tsunami wave
Speed = √g x d
Place each value into the correct place in the formula.
Speed = √ m/s² x m
Check Answer
California
Hawaii
Ka Lae
Mexico
3/5

connectED.mcgraw-hill.com **Your Digital Math Portal**

Animation | Vocabulary | eGlossary | Personal Tutor | Virtual Manipulatives | Graphing Calculator | Audio | Foldables | Self-Check Practice | Worksheets

Jean-Pierre Pieuchot/Photodisc/Getty Images

Get Ready for the Chapter

Diagnose Readiness | You have two options for checking prerequisite skills.

1 Textbook Option
Take the Quick Check below. Refer to the Quick Review for help.

QuickCheck	QuickReview

Find each square root. If necessary, round to the nearest hundredth.

1. $\sqrt{82}$ **2.** $\sqrt{26}$

3. $\sqrt{15}$ **4.** $\sqrt{99}$

5. SANDBOX Isaac is making a square sandbox with an area of 100 square feet. How long is a side of the sandbox?

Example 1

Find the square root of $\sqrt{50}$. If necessary, round to the nearest hundredth.

$\sqrt{50} = 7.071067812....$ Use a calculator.

To the nearest hundredth, $\sqrt{50} = 7.07$.

Simplify each expression.

6. $(21x + 15y)$ $(9x$ $4y)$

7. $13x - 5y + 2y$

8. $(10a - 5b) + (6a + 5b)$

9. $6m + 5n + 4 - 3m - 2n + 6$

10. $x + y - 3x - 4y + 2x - 8y$

Example 2

Simplify $3x + 7y - 4x - 8y$.

$3x + 7y - 4x - 8y$

$\quad = (3x - 4x) + (7y - 8y)$ Combine like terms.

$\quad = -x - y$ Simplify.

Solve each equation.

11. $2x^2 - 4x = 0$ **12.** $6x^2 - 5x - 4 = 0$

13. $x^2 - 7x + 10 = 0$ **14.** $2x^2 + 7x - 5 = -1$

15. GEOMETRY The area of the rectangle is 90 square feet. Find x.

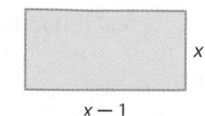

x

$x - 1$

Example 3

Solve $x^2 - 5x + 6 = 0$.

$x^2 - 5x + 6 = 0$ Original equation

$(x - 3)(x - 2) = 0$ Factor.

$x - 3 = 0$ or $x - 2 = 0$ Zero Product Property

$x = 3$ $x = 2$ Solve each equation.

---

2 Online Option
Take an online self-check Chapter Readiness Quiz at connectED.mcgraw-hill.com.

Get Started on the Chapter

You will learn several new concepts, skills, and vocabulary terms as you study Chapter 10. To get ready, identify important terms and organize your resources. You may wish to refer to Chapter 0 to review prerequisite skills.

FOLDABLES StudyOrganizer

Radical Functions and Geometry Make this Foldable to help you organize your Chapter 10 notes about radical functions and geometry. Begin with four sheets of grid paper.

1 **Fold** in half along the width.

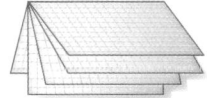

2 **Staple** along the fold.

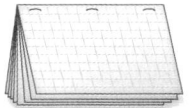

3 **Turn** the fold to the left and write the title of the chapter on the front. On each left-hand page of the booklet, write the title of a lesson from the chapter.

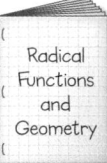

Radical
Functions
and
Geometry

NewVocabulary

English		Español
radicand	p. 621	radicando
radical function	p. 621	función radicales
conjugate	p. 630	conjugado
radical equations	p. 642	ecuaciones radicales
hypotenuse	p. 648	hipotenusa
legs	p. 648	catetos
converse	p. 649	recíproco
midpoint	p. 654	punto medio
sine	p. 656	seno
cosine	p. 656	coseno
tangent	p. 656	tangente
trigonometry	p. 656	trigonometría
inverse cosine	p. 658	coseno inverso
inverse sine	p. 658	seno inverso
inverse tangent	p. 658	tangente inverse

ReviewVocabulary

FOIL method metodo FOIL to multiply two binomials, find the sum of the products of the First terms, Outer terms, Inner terms, and Last terms

perfect square cuadrado perfecto a number with a square root that is a rational number

proportion proporcion an equation of the form $\frac{a}{b} = \frac{c}{d}$, $b \neq 0$, $d \neq 0$ stating that two ratios are equivalent

$$\frac{a}{b} \diagup\!\!\!\!\diagdown \frac{c}{d}$$
$$ad = bc$$

Algebra Lab
Inverse Functions

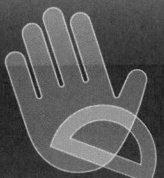

You have discovered that every nonhorizontal linear function has an inverse function. You have learned how to find the inverse of any function by exchanging the coordinates for a set of ordered pairs. In the following activity, we will exchange coordinates to find the inverse of a quadratic function and determine whether the inverse is a function.

CCSS **Common Core State Standards**
Content Standards
F.BF.4a Solve an equation of the form $f(x) = c$ for a simple function f that has an inverse and write an expression for the inverse.

Activity 1   Exchange Coordinates

Find the inverse of $y = x^2$ by exchanging the coordinates. Is the inverse a function?

Step 1 Make a table of values for $y = x^2$ using x from -3 to 3.

x	-3	-2	-1	0	1	2	3
y	9	4	1	0	1	4	9

Step 2 Write the coordinates as a set of ordered pairs.

$\{(-3, 9), (-2, 4), (-1, 1), (0, 0), (1, 1), (2, 4), (3, 9)\}$

Step 3 Exchange the x- and y-coordinates in each ordered pair to form the inverse.

$\{(9, -3), (4, -2), \mathbf{(1, -1)}, (0, 0), \mathbf{(1, 1)}, (4, 2), (9, 3)\}$

Step 4 Examine the set of ordered pairs and determine if it would be a function. This set of ordered pairs would not be a function because each x-value is not paired with a unique y-value. For example, there are two y-values when $x = 1$.

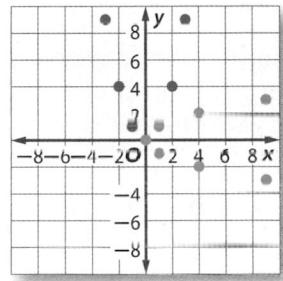

You have also learned how to find the inverse of a linear function algebraically. In the next activity, you will find the inverse of the quadratic function from Activity 1.

Activity 2   Use Algebra

Find the inverse of $y = x^2$ algebraically. Check by graphing the function, its inverse, and the line $y = x$.

Step 1 Find the inverse algebraically.

$y = x^2$ Original function

$x = y^2$ Interchange x and y.

$\pm\sqrt{x} = \sqrt{y^2}$ Take the square root of each side.

$\pm\sqrt{x} = y$ Simplify.

The inverse of $y = x^2$ is $y = \pm\sqrt{x}$.

(continued on the next page)

Algebra Lab
Inverse Functions Continued

Step 2 On a coordinate plane, plot and connect the sets of points from Steps 2 and 3 of Activity 1 with a smooth curve to graph $y = x^2$ and its inverse. Graph the line $y = x$.

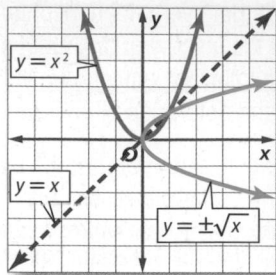

Step 3 The graph of $y = \pm\sqrt{x}$ does not pass the vertical line test for a function. The inverse is not a function.

Many functions like $y = x^2$ have inverse relations that are not functions. It is often possible to limit the domains of these functions so that their inverses will be functions.

Activity 3 Restricted Domains

Restrict the domain of $y = x^2$ so that its inverse is a function.

Notice from Activity 2 that the graph of $y = x^2$ is symmetric about the y-axis. If we restrict the domain of $y = x^2$ to either $x \geq 0$ or $x \leq 0$, we are left with half of the graph.

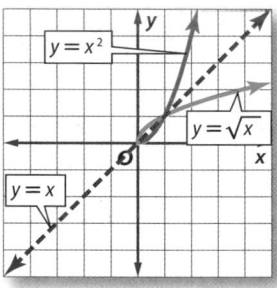

For $x \geq 0$, the graph of $y = x^2$ is now the portion of the parabola to the right of the y-axis. Its inverse is its reflection across the line $y = x$, which is the top portion of the graph of $y = \pm\sqrt{x}$.

Since each x-value of this reflection is paired with a unique y-value, the inverse is now a function.

Exercises

Write a set of ordered pairs for the inverse of each function by making a table of values for x from 3 to 3 and exchanging the coordinates. Is the inverse a function?

1. $y = x^2 - 3$ 2. $y = (x - 1)^2$

3. $y = 2x^2$ 4. $y = 3x^2 - 2$

Find the inverse of each function algebraically. Is the inverse a function?

5. $y = x^2 + 2$ 6. $y = (x - 1)^2$

7. $y = (x + 3)^2 - 4$ 8. $y = 4x^2 + 2$

Name a restricted domain for each function for which its inverse would be a function.

9. $y = x^2 - 1$ 10. $y = (x + 2)^2$

11. $y = (x - 2)^2 + 1$ 12. $y = 3x^2 - 1$

:: Then	:: Now	:: Why?
● You graphed and analyzed linear, exponential, and quadratic functions.	**1** Graph and analyze dilations of radical functions. **2** Graph and analyze reflections and translations of radical functions.	● Scientists use sounds of whales to track their movements. The distance to a whale can be found by relating time to the speed of sound in water. The speed of sound in water can be described by the *square root function* $c = \sqrt{\dfrac{E}{d}}$, where E represents the bulk modulus elasticity of the water and d represents the density of the water.

 NewVocabulary
square root function
radical function
radicand

 Common Core State Standards

Content Standards

F.IF.4 For a function that models a relationship between two quantities, interpret key features of graphs and tables in terms of the quantities, and sketch graphs showing key features given a verbal description of the relationship.

F.IF.7b Graph square root, cube root, and piecewise-defined functions, including step functions and absolute value functions.

Mathematical Practices
6 Attend to precision.

1 **Dilations of Radical Functions** A **square root function** contains the square root of a variable. Square root functions are a type of **radical function**. The expression under the radical sign is called the **radicand**. For a square root to be a real number, the radicand cannot be negative. Values that make the radicand negative are not included in the domain.

KeyConcept Square Root Function

Parent Function:	$f(x) = \sqrt{x}$
Type of Graph:	curve
Domain:	$\{x \mid x \geq 0\}$
Range:	$\{y \mid y \geq 0\}$

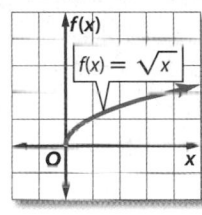

Example 1 Dilation of the Square Root Function

Graph $f(x) = 2\sqrt{x}$. State the domain and range.

Step 1 Make a table.

x	0	0.5	1	2	3	4
f(x)	0	≈1.4	2	≈2.8	≈3.5	4

The domain is $\{x \mid x \geq 0\}$, and the range is $\{y \mid y \geq 0\}$. Notice that the graph is increasing on the entire domain, the minimum value is 0, and there is no symmetry.

Step 2 Plot points. Draw a smooth curve.

▶ **Guided**Practice

1A. $g(x) = 4\sqrt{x}$ **1B.** $h(x) = 6\sqrt{x}$

 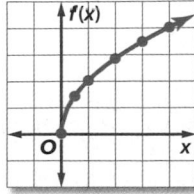

2 Reflections and Translations of Radical Functions
Recall that when the value of a is negative in the quadratic function $f(x) = ax^2$, the graph of the parent function is reflected across the x-axis.

StudyTip

Graphing Radical Functions
Choose perfect squares for x-values that will result in coordinates that are easy to plot.

KeyConcept Graphing $y = a\sqrt{x + h} + k$

Step 1 Draw the graph of $y = a\sqrt{x}$. The graph starts at the origin and passes through $(1, a)$. If $a > 0$, the graph is in quadrant I. If $a < 0$, the graph is reflected across the x-axis and is in quadrant IV.

Step 2 Translate the graph k units up if $k > 0$ and $|k|$ units down if $k < 0$.

Step 3 Translate the graph h units left if $h > 0$ and $|h|$ units right if $h < 0$.

Example 2 Reflection of the Square Root Function

Graph $y = -3\sqrt{x}$. Compare to the parent graph. State the domain and range.

Make a table of values. Then plot the points on a coordinate system and draw a smooth curve that connects them.

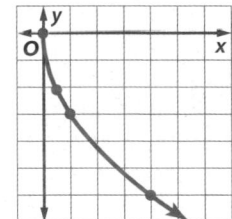

x	0	0.5	1	4
y	0	≈−2.1	−3	−6

Notice that the graph is in the 4th quadrant. It is obtained by stretching the graph of $y = \sqrt{x}$ vertically and then reflecting across the x-axis. The domain is $\{x \mid x \geq 0\}$, and the range is $\{y \mid y \leq 0\}$.

▸ **GuidedPractice**

2A. $y = -2\sqrt{x}$ **2B.** $y = -4\sqrt{x}$

StudyTip

Translating Radical Functions
If $h > 0$, a radical function $f(x) = \sqrt{x - h}$ is a horizontal translation h units to the right.
$f(x) = \sqrt{x + h}$ is a horizontal translation h units to the left.

Example 3 Translation of the Square Root Function

Graph each function. Compare to the parent graph. State the domain and range.

a. $g(x) = \sqrt{x} + 1$

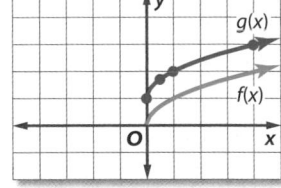

x	0	0.5	1	4	9
y	0	≈1.7	2	3	4

Notice that the values of $g(x)$ are 1 greater than those of $f(x) = \sqrt{x}$. This is a vertical translation 1 unit up from the parent function. The domain is $\{x \mid x \geq 0\}$, and the range is $\{y \mid y \geq 1\}$.

b. $h(x) = \sqrt{x - 2}$

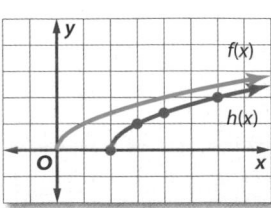

x	2	3	4	6
y	0	1	≈1.4	2

This is a horizontal translation 2 units to the right of the parent function. The domain is $\{x \mid x \geq 2\}$, and the range is $\{y \mid y \geq 0\}$.

3A. $g(x) = \sqrt{x} - 4$

3B. $h(x) = \sqrt{x + 3}$

Physical phenomena such as motion can be modeled by radical functions. Often these functions are transformations of the parent square root function.

Real-World Example 4 Analyze a Radical Function

BRIDGES The Golden Gate Bridge is about 67 meters above the water. The velocity v of a freely falling object that has fallen h meters is given by $v = \sqrt{2gh}$, where g is the constant 9.8 meters per second squared. Graph the function. If an object is dropped from the bridge, what is its velocity when it hits the water?

Use a graphing calculator to graph the function.
To find the velocity of the object, substitute 67 meters for h.

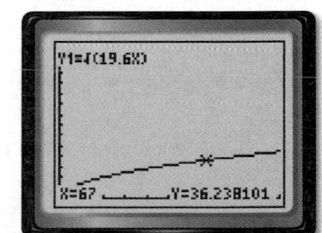

$v = \sqrt{2gh}$ Original function

$ = \sqrt{2(9.8)(67)}$ $g = 9.8$ and $h = 67$

$ = \sqrt{1313.2}$ Simplify.

$ \approx 36.2 \text{ m/s}$ Use a calculator.

The velocity of the object is about 36.2 meters per second after dropping 67 meters.

Real-WorldLink

Approximately 39 million cars cross the Golden Gate Bridge in San Francisco each year.

Source: San Francisco Convention and Visitors Bureau

GuidedPractice

4. Use the graph above to estimate the initial height of an object if it is moving at 20 meters per second when it hits the water.

Transformations such as reflections, translations, and dilations can be combined in one equation.

Example 5 Transformations of the Square Root Function

Graph $y = -2\sqrt{x} + 1$, and compare to the parent graph. State the domain and range.

x	0	1	4	9
y	1	−1	−3	−5

This graph is the result of a vertical stretch of the graph of $y = \sqrt{x}$ followed by a reflection across the x-axis, and then a translation 1 unit up. The domain is $\{x \mid x \geq 0\}$, and the range is $\{y \mid y \leq 1\}$.

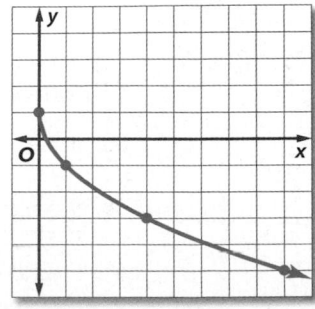

GuidedPractice

5A. $y = \frac{1}{2}\sqrt{x} - 1$

5B. $y = -2\sqrt{x - 1}$

Examples 1–3 Graph each function. Compare to the parent graph. State the domain and range.

1. $y = 3\sqrt{x}$

2. $y = -5\sqrt{x}$

3. $y = \frac{1}{3}\sqrt{x}$

4. $y = -\frac{1}{2}\sqrt{x}$

5. $y = \sqrt{x} + 3$

6. $y = \sqrt{x} - 2$

7. $y = \sqrt{x + 2}$

8. $y = \sqrt{x - 3}$

Example 4   **9. FREE FALL** The time t, in seconds, that it takes an object to fall a distance d, in feet, is given by the function $t = \frac{1}{4}\sqrt{d}$ (assuming zero air resistance). Graph the function, and state the domain and range.

Example 5   Graph each function, and compare to the parent graph. State the domain and range.

10. $y = \frac{1}{2}\sqrt{x} + 2$

11. $y = -\frac{1}{4}\sqrt{x} - 1$

12. $y = -2\sqrt{x + 1}$

13. $y = 3\sqrt{x - 2}$

Practice and Problem Solving   Extra Practice is on page R10.

Examples 1–3 Graph each function. Compare to the parent graph. State the domain and range.

14. $y = 5\sqrt{x}$ **15** $y = \frac{1}{2}\sqrt{x}$ **16.** $y = -\frac{1}{3}\sqrt{x}$ **17.** $y = 7\sqrt{x}$

18. $y = -\frac{1}{4}\sqrt{x}$ **19.** $y = -\sqrt{x}$ **20.** $y = -\frac{1}{5}\sqrt{x}$ **21.** $y = -7\sqrt{x}$

22. $y = \sqrt{x} + 2$ **23.** $y = \sqrt{x} + 4$ **24.** $y = \sqrt{x} - 1$

25. $y = \sqrt{x} - 3$ **26.** $y = \sqrt{x} + 1.5$ **27.** $y = \sqrt{x} - 2.5$

28. $y = \sqrt{x + 4}$ **29.** $y = \sqrt{x - 4}$ **30.** $y = \sqrt{x + 1}$

31. $y = \sqrt{x - 0.5}$ **32.** $y = \sqrt{x + 5}$ **33.** $y = \sqrt{x - 1.5}$

Example 4   **34. GEOMETRY** The perimeter of a square is given by the function $P = 4\sqrt{A}$, where A is the area of the square.

   a. Graph the function.

   b. Determine the perimeter of a square with an area of 225 m².

   c. When will the perimeter and the area be the same value?

Example 5   Graph each function, and compare to the parent graph. State the domain and range.

35. $y = -2\sqrt{x} + 2$ **36.** $y = -3\sqrt{x} - 3$ **37.** $y = \frac{1}{2}\sqrt{x + 2}$

38. $y = -\sqrt{x - 1}$ **39.** $y = \frac{1}{4}\sqrt{x - 1} + 2$ **40.** $y = \frac{1}{2}\sqrt{x - 2} + 1$

41. ENERGY An object has kinetic energy when it is in motion. The velocity in meters per second of an object of mass m kilograms with an energy of E joules is given by the function $v = \sqrt{\dfrac{2E}{m}}$. Use a graphing calculator to graph the function that represents the velocity of a basketball with a mass of 0.6 kilogram.

42. **GEOMETRY** The radius of a circle is given by $r = \sqrt{\frac{A}{\pi}}$, where A is the area of the circle.

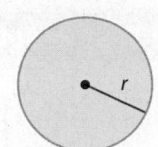

a. Graph the function.

b. Use a graphing calculator to determine the radius of a circle that has an area of 27 in².

43. **SPEED OF SOUND** The speed of sound in air is determined by the temperature of the air. The speed c in meters per second is given by $c = 331.5 \sqrt{1 + \frac{t}{273.15}}$, where t is the temperature of the air in degrees Celsius.

a. Use a graphing calculator to graph the function.

b. How fast does sound travel when the temperature is 55°C?

c. How is the speed of sound affected when the temperature increases to 65°C?

44. **MULTIPLE REPRESENTATIONS** In this problem, you will explore the relationship between the graphs of square root functions and parabolas.

a. **Graphical** Graph $y = x^2$ on a coordinate system.

b. **Algebraic** Write a piecewise-defined function to describe the graph of $y^2 = x$ in each quadrant.

c. **Graphical** On the same coordinate system, graph $y = \sqrt{x}$ and $y = -\sqrt{x}$.

d. **Graphical** On the same coordinate system, graph $y = x$. Plot the points (2, 4), (4, 2), and (1, 1).

e. **Analytical** Compare the graph of the parabola to the graphs of the square root functions.

H.O.T. Problems   Use Higher-Order Thinking Skills

CHALLENGE Determine whether each statement is *true* or *false*. Provide an example or counterexample to support your answer.

45. Numbers in the domain of a radical function will always be nonnegative.

46. Numbers in the range of a radical function will always be nonnegative.

47. **WRITING IN MATH** Why are there limitations on the domain and range of square root functions?

48. **CCSS TOOLS** Write a radical function with a domain of all real numbers greater than or equal to 2 and a range of all real numbers less than or equal to 5.

49. **WHICH DOES NOT BELONG?** Identify the equation that does not belong. Explain.

| $y = 3\sqrt{x}$ | $y = 0.7\sqrt{x}$ | $y = \sqrt{x} + 3$ | $y = \dfrac{\sqrt{x}}{6}$ |

50. **OPEN ENDED** Write a function that is a reflection, translation, and a dilation of the parent graph $y = \sqrt{x}$.

51. **REASONING** If the range of the function $y = a\sqrt{x}$ is $\{y \mid y \le 0\}$, what can you conclude about the value of a? Explain your reasoning.

52. **WRITING IN MATH** Compare and contrast the graphs of $f(x) = \sqrt{x} + 2$ and $g(x) = \sqrt{x + 2}$.

53.

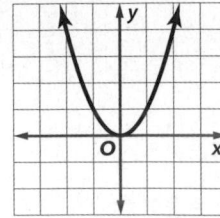

Which function *best* represents the graph?

A $y = x^2$ **C** $y = \sqrt{x}$

B $y = 2^x$ **D** $y = x$

54. The statement "$x < 10$ and $3x - 2 \geq 7$" is true when x is equal to what?

F 0 **H** 8

G 2 **J** 12

55. Which of the following is the equation of a line parallel to $y = -\frac{1}{2}x + 3$ and passing through $(-2, -1)$?

A $y = \frac{1}{2}x$ **C** $y = -\frac{1}{2}x + 2$

B $y = 2x + 3$ **D** $y = -\frac{1}{2}x - 2$

56. SHORT RESPONSE A landscaper needs to mulch 6 rectangular flower beds that are 8 feet by 4 feet and 4 circular flower beds each with a radius of 3 feet. One bag of mulch covers 25 square feet. How many bags of mulch are needed to cover the flower beds?

Spiral Review

Graph each function. (Lesson 9-7)

57. $f(x) = |3x + 2|$

58. $f(x) = \begin{cases} x - 2 \text{ if } x > -1 \\ x + 3 \text{ if } x \leq -1 \end{cases}$

59. $f(x) = [\![x + 1]\!]$

60. $f(x) = \left| \frac{1}{4}x - 1 \right|$

Graph each set of ordered pairs. Determine whether the ordered pairs represent a *linear* function, a *quadratic* function, or an *exponential* function. (Lesson 9-6)

61. $\{(-2, 5), (-1, 3), (0, 1), (1, -1), (2, -3)\}$

62. $\{(0, 0), (1, 3), (2, 4), (3, 3), (4, 0)\}$

63. $\left\{ \left(-2, \frac{1}{4} \right), (0, 1), (1, 2), (2, 4), (3, 8) \right\}$

64. $\{(-3, 1), (-2, -5), (-1, -7), (0, -5), (1, 1)\}$

65. HEALTH Aida exercises every day by walking and jogging at least 3 miles. Aida walks at a rate of 4 miles per hour and jogs at a rate of 8 miles per hour. Suppose she has at most one half-hour to exercise today. (Lesson 6-6)

    a. Draw a graph showing the possible amounts of time she can spend walking and jogging today.

    b. List three possible solutions.

66. NUTRITION Determine whether the graph shows a *positive*, *negative*, or *no* correlation. If there is a positive or negative correlation, describe its meaning in the situation. (Lesson 4-5)

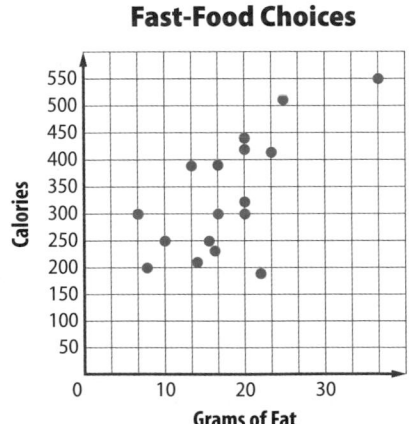

Fast-Food Choices

Skills Review

Factor each monomial completely.

67. $28n^3$

68. $-33a^2b$

69. $150rt$

70. $-378nq^2r^2$

71. $225a^3b^2c$

72. $-160x^2y^4$

10-1 Graphing Technology Lab
Graphing Square Root Functions

For a square root to be a real number, the radicand cannot be negative. When graphing a radical function, determine when the radicand would be negative and exclude those values from the domain.

 Common Core State Standards
Content Standards
F.IF.7b Graph square root, cube root, and piecewise-defined functions, including step functions and absolute value functions.
Mathematical Practices
5 Use appropriate tools strategically.

Activity 1 Parent Function

Graph $y = \sqrt{x}$.

Enter the equation in the **Y=** list, and graph in the standard viewing window.

KEYSTROKES: Y= 2nd [$\sqrt{}$] X,T,θ,n) ZOOM 6

1A. Examine the graph. What is the domain of the function?

1B. What is the range of the function?

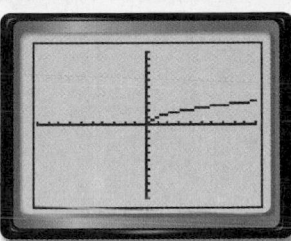

[−10, 10] scl: 1 by [−10, 10] scl: 1

Activity 2 Translation of Parent Function

Graph $y = \sqrt{x - 2}$.

Enter the equation in the **Y=** list, and graph in the standard viewing window.

KEYSTROKES: Y= 2nd [$\sqrt{}$] X,T,θ,n − 2) ZOOM 6

2A. What are the domain and range of the function?

2B. How does the graph of $y = \sqrt{x - 2}$ compare to the graph of the parent function $y = \sqrt{x}$?

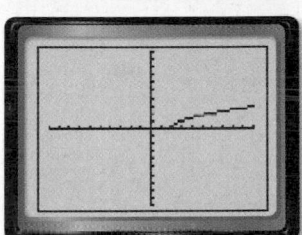

[−10, 10] scl: 1 by [−10, 10] scl: 1

Exercises

Graph each equation, and sketch the graph on your paper. State the domain and range. Describe how the graph differs from that of the parent function $y = \sqrt{x}$.

1. $y = \sqrt{x - 1}$ **2.** $y = \sqrt{x + 3}$ **3.** $y = \sqrt{x} - 2$ **4.** $y = \sqrt{-x}$

5. $y = -\sqrt{x}$ **6.** $y = \sqrt{2x}$ **7.** $y = \sqrt{2 - x}$ **8.** $y = \sqrt{x - 3} + 2$

Solve each equation for y. Does the equation represent a function? Explain your reasoning.

9. $x = y^2$

10. $x^2 + y^2 = 4$

11. $x^2 + y^2 = 2$

Write a function with a graph that translates $y = \sqrt{x}$ in each way.

12. shifted 4 units to the left

13. shifted up 7 units

14. shifted down 6 units

15. shifted 5 units to the right and up 3 units

Simplifying Radical Expressions

:: Then	:: Now	:: Why?
● You simplified radicals.	**1** Simplify radical expressions by using the Product Property of Square Roots.	● The Sunshine Skyway Bridge across Florida's Tampa Bay is supported by 21 steel cables, each 9 inches in diameter.
	2 Simplify radical expressions by using the Quotient Property of Square Roots.	To find the diameter a steel cable should have to support a given weight, you can use the equation $d = \sqrt{\frac{w}{8}}$, where d is the diameter of the cable in inches and w is the weight in tons.

NewVocabulary
radical expression
rationalizing the
denominator
conjugate

Common Core State Standards

Content Standards
A.REI.4a Use the method of completing the square to transform any quadratic equation in x into an equation of the form $(x - p)^2 = q$ that has the same solutions. Derive the quadratic formula from this form.

Mathematical Practices
7 Look for and make use of structure.
8 Look for and express regularity in repeated reasoning.

1 Product Property of Square Roots A **radical expression** contains a radical, such as a square root. Recall the expression under the radical sign is called the radicand. A radicand is in simplest form if the following three conditions are true.

• No radicands have perfect square factors other than 1.

• No radicands contain fractions.

• No radicals appear in the denominator of a fraction.

The following property can be used to simplify square roots.

> **KeyConcept** Product Property of Square Roots
>
> | Words | For any nonnegative real numbers a and b, the square root of ab is equal to the square root of a times the square root of b. |
> |---|---|
> | Symbols | $\sqrt{ab} = \sqrt{a} \cdot \sqrt{b}$, if $a \geq 0$ and $b \geq 0$ |
> | Examples | $\sqrt{4 \cdot 9} = \sqrt{36}$ or 6 $\qquad$ $\sqrt{4 \cdot 9} = \sqrt{4} \cdot \sqrt{9} = 2 \cdot 3$ or 6 |

Example 1 Simplify Square Roots

Simplify $\sqrt{80}$.

$\sqrt{80} = \sqrt{2 \cdot 2 \cdot 2 \cdot 2 \cdot 5}$ $\qquad$ Prime factorization of 80

$\phantom{\sqrt{80}} = \sqrt{2^2} \cdot \sqrt{2^2} \cdot \sqrt{5}$ $\qquad$ Product Property of Square Roots

$\phantom{\sqrt{80}} = 2 \cdot 2 \cdot \sqrt{5}$ or $4\sqrt{5}$ $\qquad$ Simplify.

▶ **Guided**Practice

1A. $\sqrt{54}$ $\qquad\qquad\qquad\qquad$ **1B.** $\sqrt{180}$

Eyecon Images/Alamy

Example 2 Multiply Square Roots

Simplify $\sqrt{2} \cdot \sqrt{14}$.

$\sqrt{2} \cdot \sqrt{14} = \sqrt{2} \cdot \sqrt{2} \cdot \sqrt{7}$ Product Property of Square Roots

$\phantom{\sqrt{2} \cdot \sqrt{14}} = \sqrt{2^2} \cdot \sqrt{7}$ or $2\sqrt{7}$ Product Property of Square Roots

▶ **Guided**Practice

2A. $\sqrt{5} \cdot \sqrt{10}$ **2B.** $\sqrt{6} \cdot \sqrt{8}$

Consider the expression $\sqrt{x^2}$. It may seem that $x = \sqrt{x^2}$, but when finding the principal square root of an expression containing variables, you have to be sure that the result is not negative. Consider $x = -3$.

$$\sqrt{x^2} \stackrel{?}{=} x$$

$$\sqrt{(-3)^2} \stackrel{?}{=} -3 \qquad \text{Replace } x \text{ with } -3.$$

$$\sqrt{9} \stackrel{?}{=} -3 \qquad (-3)^2 = 9$$

$$3 \neq -3 \qquad \sqrt{9} = 3$$

Notice in this case, if the right hand side of the equation were $|x|$, the equation would be true. For expressions where the exponent of the variable inside a radical is even and the simplified exponent is odd, you must use absolute value.

$$\sqrt{x^2} = |x| \qquad \sqrt{x^3} = x\sqrt{x} \qquad \sqrt{x^4} = x^2 \qquad \sqrt{x^6} = |x^3|$$

Example 3 Simplify a Square Root with Variables

Simplify $\sqrt{90x^3y^4z^5}$.

$\sqrt{90x^3y^4z^5} = \sqrt{2 \cdot 3^2 \cdot 5 \cdot x^3 \cdot y^4 \cdot z^5}$ Prime factorization

$\phantom{\sqrt{90x^3y^4z^5}} = \sqrt{2} \cdot \sqrt{3^2} \cdot \sqrt{5} \cdot \sqrt{x^2} \cdot \sqrt{x} \cdot \sqrt{y^4} \cdot \sqrt{z^4} \cdot \sqrt{z}$ Product Property

$\phantom{\sqrt{90x^3y^4z^5}} = \sqrt{2} \cdot 3 \cdot \sqrt{5} \cdot x \cdot \sqrt{x} \cdot y^2 \cdot z^2 \cdot \sqrt{z}$ Simplify.

$\phantom{\sqrt{90x^3y^4z^5}} = 3y^2z^2x\sqrt{10xz}$ Simplify.

▶ **Guided**Practice

3A. $\sqrt{32r^2k^4t^5}$ **3B.** $\sqrt{56xy^{10}z^5}$

2 Quotient Property of Square Roots To divide square roots and simplify radical expressions, you can use the Quotient Property of Square Roots.

ReadingMath

Fractions in the Radicand
The expression $\sqrt{\frac{a}{b}}$ is read *the square root of a over b,* or *the square root of the quantity of a over b.*

🔑 KeyConcept Quotient Property of Square Roots

Words   For any real numbers a and b, where $a \geq 0$ and $b > 0$, the square root of $\frac{a}{b}$ is equal to the square root of a divided by the square root of b.

Symbols $\sqrt{\dfrac{a}{b}} = \dfrac{\sqrt{a}}{\sqrt{b}}$

You can use the properties of square roots to **rationalize the denominator** of a fraction with a radical. This involves multiplying the numerator and denominator by a factor that eliminates radicals in the denominator.

Standardized Test Example 4 Rationalize a Denominator

Which expression is equivalent to $\sqrt{\dfrac{35}{15}}$?

A $\dfrac{5\sqrt{21}}{15}$ B $\dfrac{\sqrt{21}}{3}$ C $\dfrac{\sqrt{525}}{15}$ D $\dfrac{\sqrt{35}}{15}$

Read the Test Item The radical expression needs to be simplified.

Solve the Test Item

$\sqrt{\dfrac{35}{15}} = \sqrt{\dfrac{7}{3}}$ Reduce $\dfrac{35}{15}$ to $\dfrac{7}{3}$.

$= \dfrac{\sqrt{7}}{\sqrt{3}}$ Quotient Property of Square Roots

$= \dfrac{\sqrt{7}}{\sqrt{3}} \cdot \dfrac{\sqrt{3}}{\sqrt{3}}$ Multiply by $\dfrac{\sqrt{3}}{\sqrt{3}}$.

$= \dfrac{\sqrt{21}}{3}$ Product Property of Square Roots

The correct choice is B.

▶ **Guided**Practice

4. Simplify $\dfrac{\sqrt{6y}}{\sqrt{12}}$.

F $\dfrac{\sqrt{y}}{2}$ G $\dfrac{\sqrt{y}}{4}$ H $\dfrac{\sqrt{2y}}{2}$ J $\dfrac{\sqrt{2y}}{4}$

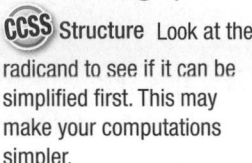

Test-Taking Tip

CCSS Structure Look at the radicand to see if it can be simplified first. This may make your computations simpler.

Binomials of the form $a\sqrt{b} + c\sqrt{d}$ and $a\sqrt{b} - c\sqrt{d}$, where a, b, c, and d are rational numbers, are called **conjugates**. For example, $2 + \sqrt{7}$ and $2 - \sqrt{7}$ are conjugates. The product of two conjugates is a rational number and can be found using the pattern for the difference of squares.

Example 5 Use Conjugates to Rationalize a Denominator

Simplify $\dfrac{3}{5 + \sqrt{2}}$.

$\dfrac{3}{5 + \sqrt{2}} = \dfrac{3}{5 + \sqrt{2}} \cdot \dfrac{5 - \sqrt{2}}{5 - \sqrt{2}}$ The conjugate of $5 + \sqrt{2}$ is $5 - \sqrt{2}$.

$= \dfrac{3(5 - \sqrt{2})}{5^2 - (\sqrt{2})^2}$ $(a - b)(a + b) = a^2 - b^2$

$= \dfrac{15 - 3\sqrt{2}}{25 - 2}$ or $\dfrac{15 - 3\sqrt{2}}{23}$ $(\sqrt{2})^2 = 2$

▶ **Guided**Practice Simplify each expression.

5A. $\dfrac{3}{2 + \sqrt{2}}$ 5B. $\dfrac{7}{3 - \sqrt{7}}$

Examples 1–3 Simplify each expression.

1. $\sqrt{24}$

2. $3\sqrt{16}$

3. $2\sqrt{25}$

4. $\sqrt{10} \cdot \sqrt{14}$

5. $\sqrt{3} \cdot \sqrt{18}$

6. $3\sqrt{10} \cdot 4\sqrt{10}$

7. $\sqrt{60x^4y^7}$

8. $\sqrt{88m^3p^2r^5}$

9. $\sqrt{99ab^5c^2}$

Example 4 **10. MULTIPLE CHOICE** Which expression is equivalent to $\sqrt{\dfrac{45}{10}}$?

    A $\dfrac{5\sqrt{2}}{10}$ **B** $\dfrac{\sqrt{45}}{10}$ **C** $\dfrac{\sqrt{50}}{10}$ **D** $\dfrac{3\sqrt{2}}{2}$

Example 5 Simplify each expression.

11. $\dfrac{3}{3 + \sqrt{5}}$

12. $\dfrac{5}{2 - \sqrt{6}}$

13. $\dfrac{2}{1 - \sqrt{10}}$

14. $\dfrac{1}{4 + \sqrt{12}}$

15. $\dfrac{4}{6 - \sqrt{7}}$

16. $\dfrac{6}{5 + \sqrt{11}}$

Practice and Problem Solving   Extra Practice is on page R10.

Examples 1–3 Simplify each expression.

17. $\sqrt{52}$

18. $\sqrt{56}$

19. $\sqrt{72}$

20. $3\sqrt{18}$

21. $\sqrt{243}$

22. $\sqrt{245}$

23. $\sqrt{5} \cdot \sqrt{10}$

24. $\sqrt{10} \cdot \sqrt{20}$

25. $3\sqrt{8} \cdot 2\sqrt{7}$

26. $4\sqrt{2} \cdot 5\sqrt{8}$

27. $3\sqrt{25t^2}$

28. $5\sqrt{81q^5}$

29. $\sqrt{28a^2b^3}$

30. $\sqrt{75qr^3}$

31. $7\sqrt{63m^3p}$

32. $4\sqrt{66g^2h^4}$

33. $\sqrt{2ab^2} \cdot \sqrt{10a^5b}$

34. $\sqrt{4c^3d^3} \cdot \sqrt{8c^3d}$

35 ROLLER COASTER Starting from a stationary position, the velocity v of a roller coaster in feet per second at the bottom of a hill can be approximated by $v = \sqrt{64h}$, where h is the height of the hill in feet.

  a. Simplify the equation.

  b. Determine the velocity of a roller coaster at the bottom of a 134-foot hill.

36. CCSS PRECISION When fighting a fire, the velocity v of water being pumped into the air is modeled by the function $v = \sqrt{2hg}$, where h represents the maximum height of the water and g represents the acceleration due to gravity (32 ft/s²).

  a. Solve the function for h.

  b. The Hollowville Fire Department needs a pump that will propel water 80 feet into the air. Will a pump advertised to project water with a velocity of 70 feet per second meet their needs? Explain.

  c. The Jackson Fire Department must purchase a pump that will propel water 90 feet into the air. Will a pump that is advertised to project water with a velocity of 77 feet per second meet the fire department's need? Explain.

Examples 4–5 Simplify each expression.

(37) $\sqrt{\dfrac{32}{t^4}}$

38. $\sqrt{\dfrac{27}{m^5}}$

39. $\dfrac{\sqrt{68ac^3}}{\sqrt{27a^2}}$

40. $\dfrac{\sqrt{h^3}}{\sqrt{8}}$

41. $\sqrt{\dfrac{3}{16}} \cdot \sqrt{\dfrac{9}{5}}$

42. $\sqrt{\dfrac{7}{2}} \cdot \sqrt{\dfrac{5}{3}}$

43. $\dfrac{7}{5 + \sqrt{3}}$

44. $\dfrac{9}{6 - \sqrt{8}}$

45. $\dfrac{3\sqrt{3}}{-2 + \sqrt{6}}$

46. $\dfrac{3}{\sqrt{7} - \sqrt{2}}$

47. $\dfrac{5}{\sqrt{6} + \sqrt{3}}$

48. $\dfrac{2\sqrt{5}}{2\sqrt{7} + 3\sqrt{3}}$

49. ELECTRICITY The amount of current in amperes I that an appliance uses can be calculated using the formula $I = \sqrt{\dfrac{P}{R}}$, where P is the power in watts and R is the resistance in ohms.

    a. Simplify the formula.

    b. How much current does an appliance use if the power used is 75 watts and the resistance is 5 ohms?

50. KINETIC ENERGY The speed v of a ball can be determined by the equation $v = \sqrt{\dfrac{2k}{m}}$, where k is the kinetic energy and m is the mass of the ball.

    a. Simplify the formula if the mass of the ball is 3 kilograms.

    b. If the ball is traveling 7 meters per second, what is the kinetic energy of the ball in Joules?

51. SUBMARINES The greatest distance d in miles that a lookout can see on a clear day is modeled by the formula shown. Determine how high the submarine would have to raise its periscope to see a ship, if the submarine is the given distances away from the ship.

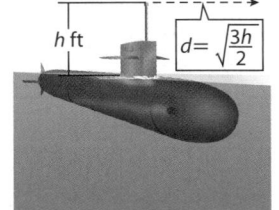

Distance	3	6	9	12	15
Height					

H.O.T. Problems   Use Higher-Order Thinking Skills

52. CCSS STRUCTURE Explain how to solve $\dfrac{\sqrt{3} + 2}{x} = \dfrac{\sqrt{3} - 1}{\sqrt{3}}$.

53. CHALLENGE Simplify each expression.

    a. $\sqrt[3]{27}$ **b.** $\sqrt[3]{40}$ **c.** $\sqrt[3]{750}$

54. REASONING Marge takes a number, subtracts 4, multiplies by 4, takes the square root, and takes the reciprocal to get $\dfrac{1}{2}$. What number did she start with? Write a formula to describe the process.

55. OPEN ENDED Write two binomials of the form $a\sqrt{b} + c\sqrt{f}$ and $a\sqrt{b} - c\sqrt{f}$. Then find their product.

56. CHALLENGE Use the Quotient Property of Square Roots to derive the Quadratic Formula by solving the quadratic equation $ax^2 + bx + c = 0$. (*Hint*: Begin by completing the square.)

57. WRITING IN MATH Summarize how to write a radical expression in simplest form.

58. Jerry's electric bill is $23 less than his natural gas bill. The two bills are a total of $109. Which of the following equations can be used to find the amount of his natural gas bill?

   A $g + g = 109$ **C** $g - 23 = 109$

   B $23 + 2g = 109$ **D** $2g - 23 = 109$

59. Solve $a^2 - 2a + 1 = 25$.

   F $-4, -6$ **H** $-4, 6$

   G $4, -6$ **J** $4, 6$

60. The expression $\sqrt{160x^2y^5}$ is equivalent to which of the following?

   A $16|x|y^2\sqrt{10y}$ **C** $4|x|y^2\sqrt{10y}$

   B $|x|y^2\sqrt{160y}$ **D** $10|x|y^2\sqrt{4y}$

61. GRIDDED RESPONSE Miki earns $10 an hour and 10% commission on sales. If Miki worked 38 hours and had a total sales of $1275 last week, how much did she make?

Graph each function. Compare to the parent graph. State the domain and range. (Lesson 10-1)

62. $y = 2\sqrt{x} - 1$ **63.** $y = \frac{1}{2}\sqrt{x}$ **64.** $y = 2\sqrt{x + 2}$

65. $y = -\sqrt{x + 1}$ **66.** $y = -3\sqrt{x - 3}$ **67.** $y = -2\sqrt{x} + 1$

Determine the domain and range for each function. (Lesson 9-7)

68. $f(x) = |2x - 5|$ **69.** $h(x) = [\![x - 1]\!]$ **70.** $g(x) = \begin{cases} -3x + 4 \text{ if } x > 2 \\ x - 1 \text{ if } x \le 2 \end{cases}$

Solve each equation by using the Quadratic Formula. Round to the nearest tenth if necessary. (Lesson 9-5)

71. $x^2 - 25 = 0$ **72.** $r^2 + 25 = 0$ **73.** $4w^2 + 100 = 40w$

74. $2r^2 + r - 14 = 0$ **75.** $5v^2 - 7v = 1$ **76.** $11z^2 - z = 3$

Factor each polynomial, if possible. If the polynomial cannot be factored, write prime. (Lesson 8-8)

77. $n^2 - 81$ **78.** $4 - 9a^2$ **79.** $2x^5 - 98x^3$

80. $32x^4 - 2y^4$ **81.** $4l^2 - 27$ **82.** $x^3 - 3x^2 - 9x + 27$

83. POPULATION The country of Latvia has been experiencing a 1.1% annual decrease in population. In 2009, its population was 2,261,294. If the trend continues, predict Latvia's population in 2019. (Lesson 7-6)

84. TOMATOES There are more than 10,000 varieties of tomatoes. One seed company produces seed packages for 200 varieties of tomatoes. For how many varieties do they not provide seeds? (Lesson 5-1)

Write the prime factorization of each number.

85. 24   **86.** 88   **87.** 180

88. 31   **89.** 60   **90.** 90

Algebra Lab
Rational and Irrational Numbers

A set is **closed** under an operation if for any numbers in the set, the result of the operation is also in the set. A set may be closed under one operation and not closed under another.

CCSS Common Core State Standards
Content Standards
N.RN.3 Explain why the sum or product of two rational numbers is rational; that the sum of a rational number and an irrational number is irrational; and that the product of a nonzero rational number and an irrational number is irrational.
Mathematical Practices
7 Look for and make use of structure.

Activity 1   Closure of Rational Numbers and Irrational Numbers

Are the sets of rational and irrational numbers closed under multiplication? under addition?

Step 1 To determine if each set is closed under multiplication, examine several products of two rational factors and then two irrational factors.

Rational: $5 \times 2 = 10$; $-3 \times 4 = -12$; $3.7 \times 0.5 = 1.85$; $\frac{3}{4} \times \frac{2}{3} = \frac{1}{2}$

Irrational: $\pi \times \sqrt{2} = \sqrt{2}\pi$; $\sqrt{3} \times \sqrt{7} = \sqrt{21}$; $\sqrt{5} \times \sqrt{5} = 5$

The product of each pair of rational numbers is rational. However, the products of pairs of irrational numbers are both irrational and rational. Thus, it appears that the set of rational numbers is closed under multiplication, but the set of irrational numbers is not.

Step 2 Repeat this process for addition.

Rational: $3 + 8 = 11$; $-4 + 7 = 3$; $3.7 + 5.82 = 9.52$; $\frac{2}{5} + \frac{1}{4} = \frac{13}{20}$

Irrational: $\sqrt{3} + \pi = \sqrt{3} + \pi$; $3\sqrt{5} + 6\sqrt{5} = 9\sqrt{5}$; $\sqrt{12} + \sqrt{50} = 2\sqrt{3} + 5\sqrt{2}$

The sum of each pair of rational numbers is rational, and the sum of each pair of irrational numbers is irrational. Both sets are closed under addition.

Activity 2   Rational and Irrational Numbers

What kind of numbers are the product and sum of a rational and irrational number?

Step 1 Examine the products of several pairs of rational and irrational numbers.

$3 \times \sqrt{8} = 6\sqrt{2}$; $\frac{3}{4} \times \sqrt{2} = \frac{3\sqrt{2}}{4}$; $1 \times \sqrt{7} = \sqrt{7}$; $0 \times \sqrt{5} = 0$

The product is rational only when the rational factor is 0. The product of each nonzero rational number and irrational number is irrational.

Step 2 Find the sums of several pairs of a rational and irrational number.

$5 + \sqrt{3} = 5 + \sqrt{3}$; $\frac{2}{3} + \sqrt{5} = \frac{2 + 3\sqrt{5}}{3}$; $-4 + \sqrt{6} = -1(4 - \sqrt{6})$

The sum of each rational and irrational number is irrational.

Analyze the Results

1. What kinds of numbers are the difference of two unique rational numbers, two unique irrational numbers, and a rational and an irrational number?

2. Is the quotient of every rational and irrational number always another rational or irrational number? If not, provide a counterexample.

3. **CHALLENGE** Recall that rational numbers are numbers that can be written in the form $\frac{a}{b}$, where a and b are integers and $b \neq 0$. Using $\frac{a}{b}$ and $\frac{c}{d}$ show that the sum and product of two rational numbers must always be a rational number.

10-3 Operations with Radical Expressions

:: Then	:: Now	:: Why?

- You simplified radical expressions.

- **1** Add and subtract radical expressions.

- **2** Multiply radical expressions.

- Conchita is going to run in her neighborhood to get ready for the soccer season. She plans to run the course that she has laid out three times each day.

  How far does Conchita have to run to complete the course that she laid out?

  How far does she run every day?

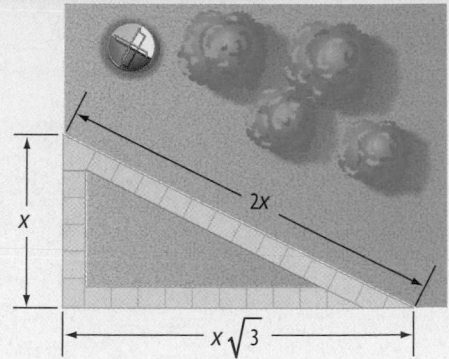

Common Core State Standards

Content Standards
N.RN.2 Rewrite expressions involving radicals and rational exponents using the properties of exponents.

Mathematical Practices
2 Reason abstractly and quantitatively.

1 **Add or Subtract Radical Expressions** To add or subtract radical expressions, the radicands must be alike in the same way that monomial terms must be alike to add or subtract.

Monomials	Radical Expressions

$$4a + 2a = (4 + 2)a$$
$$= 6a$$

$$4\sqrt{5} + 2\sqrt{5} = (4 + 2)\sqrt{5}$$
$$= 6\sqrt{5}$$

$$9b - 2b = (9 - 2)b$$
$$= 7b$$

$$9\sqrt{3} - 2\sqrt{3} = (9 - 2)\sqrt{3}$$
$$= 7\sqrt{3}$$

Notice that when adding and subtracting radical expressions, the radicand does not change. This is the same as when adding or subtracting monomials.

Example 1 Add and Subtract Expressions with Like Radicands

Simplify each expression.

a. $5\sqrt{2} + 7\sqrt{2} - 6\sqrt{2}$

$$5\sqrt{2} + 7\sqrt{2} - 6\sqrt{2} = (5 + 7 - 6)\sqrt{2} \qquad \text{Distributive Property}$$
$$= 6\sqrt{2} \qquad \text{Simplify.}$$

b. $10\sqrt{7} + 5\sqrt{11} + 4\sqrt{7} - 6\sqrt{11}$

$$10\sqrt{7} + 5\sqrt{11} + 4\sqrt{7} - 6\sqrt{11} = (10 + 4)\sqrt{7} + (5 - 6)\sqrt{11} \qquad \text{Distributive Property}$$
$$= 14\sqrt{7} - \sqrt{11} \qquad \text{Simplify.}$$

▶ **Guided**Practice

1A. $3\sqrt{2} - 5\sqrt{2} + 4\sqrt{2}$

1B. $6\sqrt{11} + 2\sqrt{11} - 9\sqrt{11}$

1C. $15\sqrt{3} - 14\sqrt{5} + 6\sqrt{5} - 11\sqrt{3}$

1D. $4\sqrt{3} + 3\sqrt{7} - 6\sqrt{3} + 3\sqrt{7}$

Not all radical expressions have like radicands. Simplifying the expressions may make it possible to have like radicands so that they can be added or subtracted.

Example 2 Add and Subtract Expressions with Unlike Radicands

Simplify $2\sqrt{18} + 2\sqrt{32} + \sqrt{72}$.

$$2\sqrt{18} + 2\sqrt{32} + \sqrt{72} = 2(\sqrt{3^2} \cdot \sqrt{2}) + 2(\sqrt{4^2} \cdot \sqrt{2}) + (\sqrt{6^2} \cdot \sqrt{2}) \qquad \text{Product Property}$$

$$= 2(3\sqrt{2}) + 2(4\sqrt{2}) + (6\sqrt{2}) \qquad \text{Simplify.}$$

$$= 6\sqrt{2} + 8\sqrt{2} + 6\sqrt{2} \qquad \text{Multiply.}$$

$$= 20\sqrt{2} \qquad \text{Simplify.}$$

▶ **Guided**Practice

2A. $4\sqrt{54} + 2\sqrt{24}$

2B. $4\sqrt{12} - 6\sqrt{48}$

2C. $3\sqrt{45} + \sqrt{20} - \sqrt{245}$

2D. $\sqrt{24} - \sqrt{54} + \sqrt{96}$

2 Multiply Radical Expressions Multiplying radical expressions is similar to multiplying monomial algebraic expressions. Let $x \geq 0$.

Monomials	Radical Expressions
$(2x)(3x) = 2 \cdot 3 \cdot x \cdot x$	$(2\sqrt{x})(3\sqrt{x}) = 2 \cdot 3 \cdot \sqrt{x} \cdot \sqrt{x}$
$= 6x^2$	$= 6x$

You can also apply the Distributive Property to radical expressions.

Example 3 Multiply Radical Expressions

Simplify each expression.

a. $3\sqrt{2} \cdot 2\sqrt{6}$

$$3\sqrt{2} \cdot 2\sqrt{6} = (3 \cdot 2)(\sqrt{2} \cdot \sqrt{6}) \qquad \text{Associative Property}$$

$$= 6(\sqrt{12}) \qquad \text{Multiply.}$$

$$= 6(2\sqrt{3}) \qquad \text{Simplify.}$$

$$= 12\sqrt{3} \qquad \text{Multiply.}$$

b. $3\sqrt{5}(2\sqrt{5} + 5\sqrt{3})$

$$3\sqrt{5}(2\sqrt{5} + 5\sqrt{3}) = (3\sqrt{5} \cdot 2\sqrt{5}) + (3\sqrt{5} \cdot 5\sqrt{3}) \qquad \text{Distributive Property}$$

$$= [(3 \cdot 2)(\sqrt{5} \cdot \sqrt{5})] + [(3 \cdot 5)(\sqrt{5} \cdot \sqrt{3})] \qquad \text{Associative Property}$$

$$= [6(\sqrt{25})] + [15(\sqrt{15})] \qquad \text{Multiply.}$$

$$= [6(5)] + [15(\sqrt{15})] \qquad \text{Simplify.}$$

$$= 30 + 15\sqrt{15} \qquad \text{Multiply.}$$

▶ **Guided**Practice

3A. $2\sqrt{6} \cdot 7\sqrt{3}$

3B. $9\sqrt{5} \cdot 11\sqrt{15}$

3C. $3\sqrt{2}(4\sqrt{3} + 6\sqrt{2})$

3D. $5\sqrt{3}(3\sqrt{2} - \sqrt{3})$

You can also multiply radical expressions with more than one term in each factor. This is similar to multiplying two algebraic binomials with variables.

Real-World Example 4 Multiply Radical Expressions

GEOMETRY Find the area of the rectangle in simplest form.

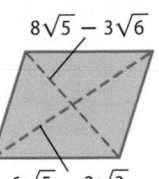

$$A = (5\sqrt{2} - \sqrt{3})(\sqrt{5} + 4\sqrt{3}) \qquad A = \ell \cdot w$$

First Terms   Outer Terms   Inner Terms   Last Terms

$$= \overbrace{(5\sqrt{2})(\sqrt{5})} + \overbrace{(5\sqrt{2})(4\sqrt{3})} + \overbrace{(-\sqrt{3})(\sqrt{5})} + \overbrace{(\sqrt{3})(4\sqrt{3})}$$

$$= 5\sqrt{10} + 20\sqrt{6} - \sqrt{15} - 4\sqrt{9} \qquad \text{Multiply.}$$

$$= 5\sqrt{10} + 20\sqrt{6} - \sqrt{15} - 12 \qquad \text{Simplify.}$$

> **ReviewVocabulary**
> **FOIL Method** Multiply two binomials by finding the sum of the products of the First terms, the Outer terms, the Inner terms, and the Last terms.

▶ **Guided**Practice

4. GEOMETRY The area A of a rhombus can be found using the equation $A = \frac{1}{2}d_1d_2$, where d_1 and d_2 are the lengths of the diagonals. What is the area of the rhombus at the right?

ConceptSummary Operations with Radical Expressions

Operation	Symbols	Example
addition, $b \geq 0$	$a\sqrt{b} + c\sqrt{b} = (a + c)\sqrt{b}$ like radicands	$4\sqrt{3} + 6\sqrt{3} = (4 + 6)\sqrt{3}$ $= 10\sqrt{3}$
subtraction, $b \geq 0$	$a\sqrt{b} - c\sqrt{b} = (a - c)\sqrt{b}$ like radicands	$12\sqrt{5} - 8\sqrt{5} = (12 - 8)\sqrt{5}$ $= 4\sqrt{5}$
multiplication, $b \geq 0, g \geq 0$	$a\sqrt{b}(f\sqrt{g}) = af\sqrt{bg}$ Radicands do not have to be like radicands.	$3\sqrt{2}(5\sqrt{7}) = (3 \cdot 5)(\sqrt{2 \cdot 7})$ $= 15\sqrt{14}$

Check Your Understanding

◯ = Step-by-Step Solutions begin on page R13.

Examples 1–3 Simplify each expression.

1. $3\sqrt{5} + 6\sqrt{5}$

2. $8\sqrt{3} + 5\sqrt{3}$

3. $\sqrt{7} - 6\sqrt{7}$

4. $10\sqrt{2} - 6\sqrt{2}$

5. $4\sqrt{5} + 2\sqrt{20}$

6. $\sqrt{12} - \sqrt{3}$

7. $\sqrt{8} + \sqrt{12} + \sqrt{18}$

8. $\sqrt{27} + 2\sqrt{3} - \sqrt{12}$

9. $9\sqrt{2}(4\sqrt{6})$

10. $4\sqrt{3}(8\sqrt{3})$

11. $\sqrt{3}(\sqrt{7} + 3\sqrt{2})$

12. $\sqrt{5}(\sqrt{2} + 4\sqrt{2})$

Example 4 **13. GEOMETRY** The area A of a triangle can be found by using the formula $A = \frac{1}{2}bh$, where b represents the base and h is the height. What is the area of the triangle at the right?

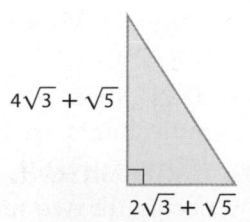

Examples 1–3 Simplify each expression.

14. $7\sqrt{5} + 4\sqrt{5}$

15. $2\sqrt{6} + 9\sqrt{6}$

16. $3\sqrt{5} - 2\sqrt{20}$

17. $3\sqrt{50} - 3\sqrt{32}$

18. $7\sqrt{3} - 2\sqrt{2} + 3\sqrt{2} + 5\sqrt{3}$

19. $\sqrt{5}(\sqrt{2} + 4\sqrt{2})$

20. $\sqrt{6}(2\sqrt{10} + 3\sqrt{2})$

21. $4\sqrt{5}(3\sqrt{5} + 8\sqrt{2})$

22. $5\sqrt{3}(6\sqrt{10} - 6\sqrt{3})$

23. $(\sqrt{3} - \sqrt{2})(\sqrt{15} + \sqrt{12})$

24. $(3\sqrt{11} + 3\sqrt{15})(3\sqrt{3} - 2\sqrt{2})$

25. $(5\sqrt{2} + 3\sqrt{5})(2\sqrt{10} - 5)$

Example 4 **26. GEOMETRY** Find the perimeter and area of a rectangle with a width of $2\sqrt{7} - 2\sqrt{5}$ and a length of $3\sqrt{7} + 3\sqrt{5}$.

Simplify each expression.

27. $\sqrt{\dfrac{1}{5}} - \sqrt{5}$

28. $\sqrt{\dfrac{2}{3}} + \sqrt{6}$

29. $2\sqrt{\dfrac{1}{2}} + 2\sqrt{2} - \sqrt{8}$

30. $8\sqrt{\dfrac{5}{4}} + 3\sqrt{20} - 10\sqrt{\dfrac{1}{5}}$ **31.** $(3 - \sqrt{5})^2$

32. $(\sqrt{2} + \sqrt{3})^2$

33 **ROLLER COASTERS** The velocity v in feet per second of a roller coaster at the bottom of a hill is related to the vertical drop h in feet and the velocity v_0 of the coaster at the top of the hill by the formula $v_0 = \sqrt{v^2 - 64h}$.

a. What velocity must a coaster have at the top of a 225-foot hill to achieve a velocity of 120 feet per second at the bottom?

b. Explain why $v_0 = v - 8\sqrt{h}$ is not equivalent to the formula given.

34. FINANCIAL LITERACY Tadi invests \$225 in a savings account. In two years, Tadi has \$232 in his account. You can use the formula $r = \sqrt{\dfrac{v_2}{v_0}} - 1$ to find the average annual interest rate r that the account has earned. The initial investment is v_0, and v_2 is the amount in two years. What was the average annual interest rate that Tadi's account earned?

35. ELECTRICITY Electricians can calculate the electrical current in amps A by using the formula $A = \dfrac{\sqrt{w}}{\sqrt{r}}$, where w is the power in watts and r the resistance in ohms. How much electrical current is running through a microwave oven that has 850 watts of power and 5 ohms of resistance? Write the number of amps in simplest radical form, and then estimate the amount of current to the nearest tenth.

H.O.T. Problems Use Higher-Order Thinking Skills

36. CHALLENGE Determine whether the following statement is *true* or *false*. Provide a proof or counterexample to support your answer.

$$x + y > \sqrt{x^2 + y^2} \text{ when } x > 0 \text{ and } y > 0$$

37. CCSS ARGUMENTS Make a conjecture about the sum of a rational number and an irrational number. Is the sum *rational* or *irrational*? Is the product of a nonzero rational number and an irrational number *rational* or *irrational*? Explain your reasoning.

38. OPEN ENDED Write an equation that shows a sum of two radicals with different radicands. Explain how you could combine these terms.

39. WRITING IN MATH Describe step by step how to multiply two radical expressions, each with two terms. Write an example to demonstrate your description.

40. SHORT RESPONSE The population of a town is 13,000 and is increasing by about 250 people per year. This can be represented by the equation $p = 13{,}000 + 250y$, where y is the number of years from now and p represents the population. In how many years will the population of the town be 14,500?

41. GEOMETRY Which expression represents the sum of the lengths of the 12 edges on this rectangular solid?

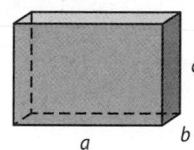

    A $2(a + b + c)$
    B $3(a + b + c)$
    C $4(a + b + c)$
    D $12(a + b + c)$

42. Evaluate $\sqrt{n - 9}$ and $\sqrt{n} - \sqrt{9}$ for $n = 25$.

    F $4; 4$ **G** $4; 2$
    H $2; 4$ **J** $2; 2$

43. The current I in a simple electrical circuit is given by the formula $I = \dfrac{V}{R}$, where V is the voltage and R is the resistance of the circuit. If the voltage remains unchanged, what effect will doubling the resistance of the circuit have on the current?

    A The current will remain the same.
    B The current will double its previous value.
    C The current will be half its previous value.
    D The current will be two units more than its previous value.

Simplify. (Lesson 10-2)

44. $\sqrt{18}$ **45.** $\sqrt{24}$ **46.** $\sqrt{60}$

47. $\sqrt{50a^3b^5}$ **48.** $\sqrt{169x^4y^7}$ **49.** $\sqrt{63c^3d^4f^5}$

Graph each function. Compare to the parent graph. State the domain and range. (Lesson 10-1)

50. $y = 2\sqrt{x}$ **51.** $y = -3\sqrt{x}$ **52.** $y = \sqrt{x + 1}$

53. $y = \sqrt{x - 4}$ **54.** $y = \sqrt{x} + 3$ **55.** $y = \sqrt{x} - 2$

Factor each trinomial. (Lesson 8-5)

56. $x^2 + 12x + 27$ **57.** $y^2 + 13y + 30$ **58.** $p^2 - 17p + 72$

59. $x^2 + 6x - 7$ **60.** $y^2 - y - 42$ **61.** $-72 + 6w + w^2$

62. FINANCIAL LITERACY Determine the value of an investment if \$400 is invested at an interest rate of 7.25% compounded quarterly for 7 years. (Lesson 7-6)

Solve each equation. Round each solution to the nearest tenth, if necessary.

63. $-4c - 1.2 = 0.8$ **64.** $-2.6q - 33.7 = 84.1$ **65.** $0.3m + 4 = 9.6$

66. $-10 - \dfrac{n}{5} = 6$ **67.** $\dfrac{-4h - (-5)}{-7} = 13$ **68.** $3.6t + 6 - 2.5t = 8$

10-3

Algebra Lab
Simplifying nth Root Expressions

The inverse of raising a number to the nth power is finding the nth root of a number. The **index** of a radical expression indicates to what root the value under the radicand is being taken. The fourth root of a number is indicated with an index of 4. When simplifying a radical expression in which there is a variable with an exponent in the radicand, divide the exponent by the index.

CCSS **Common Core State Standards**
Content Standards
N.RN.2 Rewrite expressions involving radicals and rational exponents using the properties of exponents.

$$13 \div 5 = 2 \text{ R } 3 \longrightarrow \boxed{\text{index}} \rightarrow \sqrt[5]{x^{13}} = x^2 \cdot \sqrt[5]{x^3} \leftarrow \boxed{\text{remainder}}$$

with $\boxed{\text{quotient}}$ pointing to the exponent.

Example 1 Simplify Expressions

Simplify each expression.

a. $\sqrt[3]{x^7}$

$\sqrt[3]{x^7} = x^2 \sqrt[3]{x}$ $7 \div 3 = 2 \text{ R } 1$

b. $\sqrt[5]{32x^9}$

$\sqrt[5]{32x^9} = \sqrt[5]{32} \cdot \sqrt[5]{x^9}$ Multiplication Property

$= 2x\sqrt[5]{x^4}$ $9 \div 5 = 1 \text{ R } 4$

The properties of square roots (and nth roots) also apply when the radicand contains fractions. Separate the numerator and denominator and then simplify them individually.

Example 2 Simplify Expressions with Fractions

Simplify $\sqrt[3]{\dfrac{27x^5}{8y^3}}$.

$\sqrt[3]{\dfrac{27x^5}{8y^3}} = \dfrac{\sqrt[3]{27}}{\sqrt[3]{8}} \cdot \dfrac{\sqrt[3]{x^5}}{\sqrt[3]{y^3}}$ Multiplication Property of Radicals

$= \dfrac{3}{2} \cdot \dfrac{x\sqrt[3]{x^2}}{y}$ Simplify.

$= \dfrac{3x\sqrt[3]{x^2}}{2y}$ Multiplication Property of Radicals

---

The indices *and* the radicands must be alike in order to add or subtract radical expressions.

Example 3 Combine Like Terms

Simplify $8\sqrt[4]{\dfrac{4}{3}} + \sqrt[4]{\dfrac{5}{4}} - 3\sqrt[4]{\dfrac{4}{3}} + \sqrt[3]{\dfrac{4}{3}}$.

Combine the expressions with identical indices and radicands. Then simplify.

$8\sqrt[4]{\dfrac{4}{3}} + \sqrt[4]{\dfrac{5}{4}} - 3\sqrt[4]{\dfrac{4}{3}} + \sqrt[3]{\dfrac{4}{3}} = (8 - 3)\sqrt[4]{\dfrac{4}{3}} + \sqrt[4]{\dfrac{5}{4}} + \sqrt[3]{\dfrac{4}{3}}$ Associative Property

$= 5\sqrt[4]{\dfrac{4}{3}} + \sqrt[4]{\dfrac{5}{4}} + \sqrt[3]{\dfrac{4}{3}}$ Simplify.

When multiplying radical expressions, ensure that the indices are the same. Then multiply the radicands and simplify if possible. Once none of the remaining terms can be combined or simplified, the expression is considered simplified.

Example 4 Simplify Expressions with Products

Simplify $5\sqrt[4]{6} \cdot 2\sqrt[4]{12} \cdot \sqrt[3]{10}$.

Multiply the radicands with identical indexes.

$$5\sqrt[4]{6} \cdot 2\sqrt[4]{12} \cdot \sqrt[3]{10} = (5 \cdot 2)(\sqrt[4]{6} \cdot \sqrt[4]{12}) \cdot \sqrt[3]{10} \qquad \text{Associative Property}$$
$$= 10 \cdot (\sqrt[4]{6} \cdot \sqrt[4]{12}) \cdot \sqrt[3]{10} \qquad \text{Multiply.}$$
$$= 10\sqrt[4]{72}\sqrt[3]{10} \qquad \text{Multiply.}$$

The properties of radical expressions still hold when variables are in the radicand.

Example 5 Simplify Expressions with Several Operations

Simplify $6\sqrt[4]{x} \cdot \sqrt[4]{x^3} + 3(\sqrt[3]{x} + 2\sqrt[3]{x})$.

Follow the order of operations and the properties of radical expressions.

$$6\sqrt[4]{x} \cdot \sqrt[4]{x^3} + 3(\sqrt[3]{x} + 2\sqrt[3]{x}) = 6\sqrt[4]{x} \cdot \sqrt[4]{x^3} + 3(3\sqrt[3]{x}) \qquad \text{Add like terms.}$$
$$= 6\sqrt[4]{x \cdot x^3} + 3(3\sqrt[3]{x}) \qquad \text{Associative Property}$$
$$= 6\sqrt[4]{x^4} + 9\sqrt[3]{x} \qquad \text{Multiply.}$$
$$= 6x + 9\sqrt[3]{x} \qquad \text{Simplify.}$$

Exercises

Simplify each expression.

1. $\sqrt[3]{c^6}$

2. $\sqrt[3]{16d^9}$

3. $\sqrt[3]{9} \cdot \sqrt[3]{6} \cdot \sqrt[3]{3}$

4. $\sqrt[3]{\dfrac{8a^4}{125b^7}}$

5. $\sqrt[5]{\dfrac{32x^4}{5y^6z^5}}$

6. $\sqrt[4]{\dfrac{3}{2}} + 5\sqrt[4]{\dfrac{3}{2}} - 2\sqrt[4]{\dfrac{2}{3}}$

7. $3\sqrt[4]{6} \cdot 4\sqrt[3]{6} \cdot 5\sqrt[4]{8}$

8. $3\sqrt[4]{x^2} + 2\sqrt[4]{x} \cdot 4\sqrt[4]{x}$

9. $\sqrt[5]{a} \cdot 2\sqrt[5]{a^3} - 2(\sqrt[5]{a} + 4\sqrt[5]{a})$

10. $\sqrt[4]{\dfrac{x}{4}} + 5\sqrt[4]{\dfrac{x}{4}} - 2\sqrt[4]{\dfrac{2x}{3}}$

11. $\sqrt[4]{\dfrac{8a^2}{15b^3}} \cdot 3\sqrt[4]{\dfrac{2a^3}{27b}}$

12. $\sqrt[4]{\dfrac{16x^3}{81y^5}} + 3\sqrt[4]{\dfrac{x^3}{y}} + \sqrt[3]{\dfrac{16x}{y^8}}$

Think About It

13. Provide an example in which two radical expressions with *unlike* radicands can be combined by addition.

14. Provide an example in which two radical expressions with identical indices and with like variables in the radicand *cannot* be combined by addition.

Radical Equations

- You added, subtracted, and multiplied radical expressions.

1 Solve radical equations.

2 Solve radical equations with extraneous solutions.

- The waterline length of a sailboat is the length of the line made by the water's edge when the boat is full. A sailboat's hull speed is the fastest speed that it can travel.

  You can estimate hull speed h by using the formula $h = 1.34\sqrt{\ell}$, where ℓ is the length of the sailboat's waterline.

 NewVocabulary
radical equations
extraneous solutions

 Common Core State Standards

Content Standards
N.RN.2 Rewrite expressions involving radicals and rational exponents using the properties of exponents.

A.CED.2 Create equations in two or more variables to represent relationships between quantities; graph equations on coordinate axes with labels and scales.

Mathematical Practices
3 Construct viable arguments and critique the reasoning of others.
4 Model with mathematics.

1 **Radical Equations** Equations that contain variables in the radicand, like $h = 1.34\sqrt{\ell}$, are called **radical equations**. To solve, isolate the desired variable on one side of the equation first. Then square each side of the equation to eliminate the radical.

> **KeyConcept** Power Property of Equality
>
> | Words | If you square both sides of a true equation, the resulting equation is still true. |
> |---|---|
> | Symbols | If $a = b$, then $a^2 = b^2$. |
> | Examples | If $\sqrt{x} = 4$, then $(\sqrt{x})^2 = 4^2$. |

Real-World Example 1 Variable as a Radicand

SAILING Idris and Sebastian are sailing in a friend's sailboat. They measure the hull speed at 9 nautical miles per hour. Find the length of the sailboat's waterline. Round to the nearest foot.

Understand You know how fast the boat will travel and that it relates to the length.

Plan The boat travels at 9 nautical miles per hour. The formula for hull speed is $h = 1.34\sqrt{\ell}$.

Solve

$h = 1.34\sqrt{\ell}$ Formula for hull speed

$9 = 1.34\sqrt{\ell}$ Substitute 9 for h.

$\dfrac{9}{1.34} = \dfrac{1.34\sqrt{\ell}}{1.34}$ Divide each side by 1.34.

$6.72 \approx \sqrt{\ell}$ Simplify.

$(6.72)^2 \approx (\sqrt{\ell})^2$ Square each side of the equation.

$45.16 \approx \ell$ Simplify.

The sailboat's waterline length is about 45 feet.

Check Check by substituting the estimate into the original formula.

$h = 1.34\sqrt{\ell}$ Formula for hull speed

$9 \stackrel{?}{=} 1.34\sqrt{45}$ $h = 9$ and $\ell = 45$

$9 \approx 8.98899327$ ✓   Multiply.

David Sanger/The Image Bank/Getty Images

> **Guided**Practice

1. **DRIVING** The equation $v = \sqrt{2.5r}$ represents the maximum velocity that a car can travel safely on an unbanked curve when v is the maximum velocity in miles per hour and r is the radius of the turn in feet. If a road is designed for a maximum speed of 65 miles per hour, what is the radius of the turn?

To solve a radical equation, isolate the radical first. Then square both sides of the equation.

Example 2 Expression as a Radicand

Solve $\sqrt{a + 5} + 7 = 12$.

$\sqrt{a + 5} + 7 = 12$	Original equation
$\sqrt{a + 5} = 5$	Subtract 7 from each side.
$\left(\sqrt{a + 5}\right)^2 = 5^2$	Square each side.
$a + 5 = 25$	Simplify.
$a = 20$	Subtract 5 from each side.

> **Guided**Practice

Solve each equation.

2A. $\sqrt{c - 3} - 2 = 4$ **2B.** $4 + \sqrt{h + 1} = 14$

WatchOut!

Squaring Each Side
Remember that when you square each side of the equation, you must square the entire side of the equation, even if there is more than one term on the side.

---

2 **Extraneous Solutions** Squaring each side of an equation sometimes produces a solution that is not a solution of the original equation. These are called **extraneous solutions**. Therefore, you must check all solutions in the original equation.

Example 3 Variable on Each Side

Solve $\sqrt{k + 1} = k - 1$. Check your solution.

$\sqrt{k + 1} = k - 1$	Original equation
$\left(\sqrt{k + 1}\right)^2 = (k - 1)^2$	Square each side.
$k + 1 = k^2 - 2k + 1$	Simplify.
$0 = k^2 - 3k$	Subtract k and 1 from each side.
$0 = k(k - 3)$	Factor.
$k = 0$ or $k - 3 = 0$	Zero Product Property
$k = 3$	Solve.

CHECK

$\sqrt{k + 1} = k - 1$	Original equation	$\sqrt{k + 1} = k - 1$	Original equation
$\sqrt{0 + 1} \stackrel{?}{=} 0 - 1$	$k = 0$	$\sqrt{3 + 1} \stackrel{?}{=} 3 - 1$	$k = 3$
$\sqrt{1} \stackrel{?}{=} -1$	Simplify.	$\sqrt{4} \stackrel{?}{=} 2$	Simplify.
$1 \neq -1$ ✗	False	$2 = 2$ ✓	True

Since 0 does not satisfy the original equation, 3 is the only solution.

> **Guided**Practice

Solve each equation. Check your solution.

3A. $\sqrt{t + 5} = t + 3$ **3B.** $x - 3 = \sqrt{x - 1}$

StudyTip

Extraneous Solutions
When checking solutions for extraneous solutions, we are only interested in principal roots.

Example 1 **1. GEOMETRY** The surface area of a basketball is x square inches. What is the radius of the basketball if the formula for the surface area of a sphere is $SA = 4\pi r^2$?

Examples 2–3 Solve each equation. Check your solution.

2. $\sqrt{10h} + 1 = 21$ **3.** $\sqrt{7r + 2} + 3 = 7$ **4.** $5 + \sqrt{g - 3} = 6$

5. $\sqrt{3x - 5} = x - 5$ **6.** $\sqrt{2n + 3} = n$ **7.** $\sqrt{a - 2} + 4 = a$

Practice and Problem Solving   Extra Practice is on page R10.

Example 1   **8. EXERCISE** Suppose the function $S = \pi\sqrt{\dfrac{9.8\ell}{1.6}}$, where S represents speed in meters per second and ℓ is the leg length of a person in meters, can approximate the maximum speed that a person can run.

    a. What is the maximum running speed of a person with a leg length of 1.1 meters to the nearest tenth of a meter?

    b. What is the leg length of a person with a running speed of 6.7 meters per second to the nearest tenth of a meter?

    c. As leg length increases, does maximum speed increase or decrease? Explain.

Examples 2–3 Solve each equation. Check your solution.

⑨ $\sqrt{a} + 11 = 21$ **10.** $\sqrt{t} - 4 = 7$ **11.** $\sqrt{n - 3} = 6$

12. $\sqrt{c + 10} = 4$ **13.** $\sqrt{h - 5} = 2\sqrt{3}$ **14.** $\sqrt{k + 7} = 3\sqrt{2}$

15. $y = \sqrt{12 - y}$ **16.** $\sqrt{u + 6} = u$ **17.** $\sqrt{r + 3} = r - 3$

18. $\sqrt{1 - 2t} = 1 + t$ **19.** $5\sqrt{a - 3} + 4 = 14$ **20.** $2\sqrt{x - 11} - 8 = 4$

21. RIDES The amount of time t, in seconds, that it takes a simple pendulum to complete a full swing is called the *period*. It is given by $t = 2\pi\sqrt{\dfrac{\ell}{32}}$, where ℓ is the length of the pendulum, in feet.

    a. The Giant Swing completes a period in about 8 seconds. About how long is the pendulum's arm? Round to the nearest foot.

    b. Does increasing the length of the pendulum increase or decrease the period? Explain.

Solve each equation. Check your solution.

22. $\sqrt{6a - 6} = a + 1$ **23.** $\sqrt{x^2 + 9x + 15} = x + 5$ **24.** $6\sqrt{\dfrac{5k}{4}} - 3 = 0$

25. $\sqrt{\dfrac{5y}{6}} - 10 = 4$ **26.** $\sqrt{2a^2 - 121} = a$ **27.** $\sqrt{5x^2 - 9} = 2x$

28. CCSS REASONING The formula for the slant height c of a cone is $c = \sqrt{h^2 + r^2}$, where h is the height of the cone and r is the radius of its base. Find the height of the cone if the slant height is 4 units and the radius is 2 units. Round to the nearest tenth.

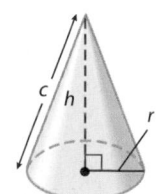

29.  **MULTIPLE REPRESENTATIONS** Consider $\sqrt{2x - 7} = x - 7$.

    a. Graphical Clear the **Y=** list. Enter the left side of the equation as $Y_1 = \sqrt{2x - 7}$. Enter the right side of the equation as $Y_2 = x - 7$. Press $\boxed{\text{GRAPH}}$.

    b. Graphical Sketch what is shown on the screen.

    c. Analytical Use the **intersect** feature on the **CALC** menu to find the point of intersection.

    d. Analytical Solve the radical equation algebraically. How does your solution compare to the solution from the graph?

30. **PACKAGING** A cylindrical container of chocolate drink mix has a volume of 162 cubic inches. The radius r of the container can be found by using the formula $r = \sqrt{\dfrac{V}{\pi h}}$, where V is the volume of the container and h is the height.

    a. If the radius is 2.5 inches, find the height of the container. Round to the nearest hundredth.

    b. If the height of the container is 10 inches, find the radius. Round to the nearest hundredth.

H.O.T. Problems   Use Higher-Order Thinking Skills

31. **CCSS CRITIQUE** Jada and Fina solved $\sqrt{6 - b} = \sqrt{b + 10}$. Is either of them correct? Explain.

Jada	Fina
$\sqrt{6 - b} = \sqrt{b + 10}$	$\sqrt{6 - b} = \sqrt{b + 10}$
$(\sqrt{6 - b})^2 = (\sqrt{b + 10})^2$	$(\sqrt{6 - b})^2 = (\sqrt{b + 10})^2$
$6 - b = b + 10$	$6 - b = b + 10$
$-2b = 4$	$2b = 4$
$b = -2$	$b = 2$
Check $\sqrt{6 - (-2)} \overset{?}{=} \sqrt{(-2) + 10}$	Check $\sqrt{6 - (2)} \overset{?}{=} \sqrt{(2) + 10}$
$\sqrt{8} = \sqrt{8}$ ✓	$\sqrt{4} \neq \sqrt{12}$ ✗
	no solution

32. **REASONING** Which equation has the same solution set as $\sqrt{4} = \sqrt{x + 2}$? Explain.

    A. $\sqrt{4} = \sqrt{x} + \sqrt{2}$ **B.** $4 = x + 2$ **C.** $2 - \sqrt{2} = \sqrt{x}$

33. **REASONING** Explain how solving $5 = \sqrt{x} + 1$ is different from solving $5 = \sqrt{x + 1}$.

34. **OPEN ENDED** Write a radical equation with a variable on each side. Then solve the equation.

35. **REASONING** Is the following equation *sometimes, always* or *never* true? Explain.
$$\sqrt{(x - 2)^2} = x - 2$$

36. **CHALLENGE** Solve $\sqrt{x + 9} = \sqrt{3} + \sqrt{x}$.

37. **WRITING IN MATH** Write some general rules about how to solve radical equations. Demonstrate your rules by solving a radical equation.

38. SHORT RESPONSE Zack needs to drill a hole at A, B, C, D, and E on circle P.

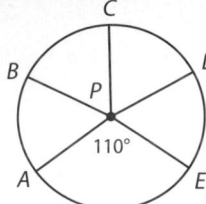

If Zack drills holes so that $m\angle APE = 110°$ and the other four angles are congruent, what is $m\angle CPD$?

39. Which expression is undefined when $w = 3$?

A $\dfrac{w - 3}{w + 1}$

C $\dfrac{w + 1}{w^2 - 3w}$

B $\dfrac{w^2 - 3w}{3w}$

D $\dfrac{3w}{3w^2}$

40. What is the slope of a line that is parallel to the line?

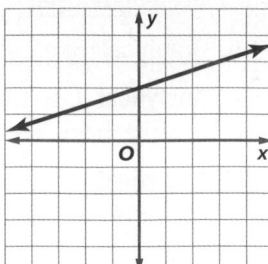

F -3

H $\dfrac{1}{3}$

G $-\dfrac{1}{3}$

J 3

41. What are the solutions of $\sqrt{x + 3} - 1 = x - 4$?

A $1, 6$

C 1

B $-1, -6$

D 6

42. ELECTRICITY The voltage V required for a circuit is given by $V = \sqrt{PR}$, where P is the power in watts and R is the resistance in ohms. How many more volts are needed to light a 100-watt light bulb than a 75-watt light bulb if the resistance of both is 110 ohms? (Lesson 10-3)

Simplify each expression. (Lesson 10-2)

43. $\sqrt{6} \cdot \sqrt{8}$

44. $\sqrt{3} \cdot \sqrt{6}$

45. $7\sqrt{3} \cdot 2\sqrt{6}$

46. $\sqrt{\dfrac{27}{a^2}}$

47. $\sqrt{\dfrac{5c^5}{4d^5}}$

48. $\dfrac{\sqrt{9x^3 y}}{\sqrt{16x^2 y^2}}$

49. PHYSICAL SCIENCE A projectile is shot straight up from ground level. Its height h, in feet, after t seconds is given by $h = 96t - 16t^2$. Find the value(s) of t when h is 96 feet. (Lesson 9-5)

Factor each trinomial, if possible. If the trinomial cannot be factored using integers, write *prime*. (Lesson 8-7)

50. $2x^2 + 7x + 5$

51. $6p^2 + 5p - 6$

52. $5d^2 + 6d - 8$

53. $8k^2 - 19k + 9$

54. $9g^2 - 12g + 4$

55. $2a^2 - 9a - 18$

Determine whether each expression is a monomial. Write *yes* or *no*. Explain. (Lesson 7-1)

56. 12

57. $4x^3$

58. $a - 2b$

59. $4n + 5p$

60. $\dfrac{x}{y^2}$

61. $\dfrac{1}{5}abc^{14}$

Simplify.

62. 9^2

63. 10^6

64. 4^5

65. $(8v)^2$

66. $\left(\dfrac{w^3}{9}\right)^2$

67. $\left(10y^2\right)^3$

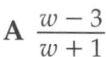

Graph each function. Compare to the parent graph. State the domain and range. (Lesson 10-1)

1. $y = 2\sqrt{x}$

2. $y = -4\sqrt{x}$

3. $y = \frac{1}{2}\sqrt{x}$

4. $y = \sqrt{x} - 3$

5. $y = \sqrt{x-1}$

6. $y = 2\sqrt{x-2}$

7. **MULTIPLE CHOICE** The length of the side of a square is given by the function $s = \sqrt{A}$, where A is the area of the square. What is the length of the side of a square that has an area of 121 square inches? (Lesson 10-1)

A 121 inches   C 11 inches

B 44 inches   D 10 inches

Simplify each expression. (Lesson 10-2)

8. $2\sqrt{25}$

9. $\sqrt{12} \cdot \sqrt{8}$

10. $\sqrt{72xy^9z^0}$

11. $\dfrac{3}{1+\sqrt{5}}$

12. $\dfrac{1}{5\sqrt{7}}$

13. **SATELLITES** A satellite is launched into orbit 200 kilometers above Earth. The orbital velocity of a satellite is given by the formula $v = \sqrt{\dfrac{Gm_E}{r}}$. v is velocity in meters per second, G is a given constant, m_E is the mass of Earth, and r is the radius of the satellite's orbit in meters. (Lesson 10-2)

    a. The radius of Earth is 6,380,000 meters. What is the radius of the satellite's orbit in meters?

    b. The mass of Earth is 5.97×10^{24} kilograms, and the constant G is 6.67×10^{-11} N $\cdot \dfrac{m^2}{kg^2}$ where N is in Newtons. Use the formula to find the orbital velocity of the satellite in meters per second.

14. **MULTIPLE CHOICE** Which expression is equivalent to $\sqrt{\dfrac{16}{32}}$? (Lesson 10-2)

F $\dfrac{1}{2}$

G $\dfrac{\sqrt{2}}{2}$

H 2

J 4

Simplify each expression. (Lesson 10-3)

15. $3\sqrt{2} + 5\sqrt{2}$

16. $\sqrt{11} - 3\sqrt{11}$

17. $6\sqrt{2} + 4\sqrt{50}$

18. $\sqrt{27} - \sqrt{48}$

19. $4\sqrt{3}(2\sqrt{6})$

20. $3\sqrt{20}(2\sqrt{5})$

21. $(\sqrt{5} + \sqrt{7})(\sqrt{20} + \sqrt{3})$

22. **GEOMETRY** Find the area of the rectangle. (Lesson 10-3)

$6\sqrt{10}$

$3\sqrt{2}$

Solve each equation. Check your solution. (Lesson 10-4)

23. $\sqrt{5x} - 1 = 4$

24. $\sqrt{a-2} = 6$

25. $\sqrt{15-x} = 4$

26. $\sqrt{3x^2 - 32} = x$

27. $\sqrt{2x-1} = 2x - 7$

28. $\sqrt{x+1} + 2 = 4$

29. **GEOMETRY** The lateral surface area S of a cone can be found by using the formula $S = \pi r\sqrt{r^2 + h^2}$, where r is the radius of the base and h is the height of the cone. Find the height of the cone. (Lesson 10-4)

$S = 121$ in^2

h

3 in.

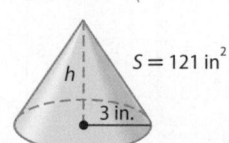

10-5 The Pythagorean Theorem

∷ Then	∷ Now	∷ Why?
● You solved quadratic equations by using the Square Root Property.	**1** Solve problems by using the Pythagorean Theorem. **2** Determine whether a triangle is a right triangle.	● Televisions are measured along the diagonal of the screen. If the height and width of the screen is known, the Pythagorean Theorem can be used to find the measure of the diagonal.

 NewVocabulary
hypotenuse
legs
converse
Pythagorean triple

 Common Core State Standards

Mathematical Practices
1 Make sense of problems and persevere in solving them.

1 **The Pythagorean Theorem** In a right triangle, the side opposite the right angle, called the **hypotenuse**, is always the longest. The other two sides are the **legs**.

> **⬧ KeyConcept** The Pythagorean Theorem
>
> | Words | If a triangle is a right triangle, then the square of the length of the hypotenuse is equal to the sum of the squares of the lengths of the legs. |
> |---|---|
> | Symbols | $c^2 = a^2 + b^2$ |
>
>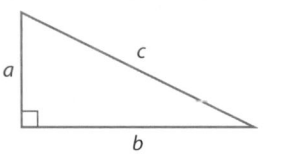

Example 1 Find the Length of a Side

Find each missing length. If necessary, round to the nearest hundredth.

a.

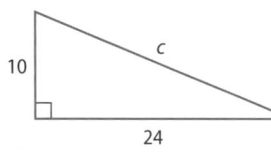

$c^2 = a^2 + b^2$	Pythagorean Theorem
$c^2 = 10^2 + 24^2$	$a = 10$ and $b = 24$
$c^2 = 100 + 576$	Evaluate squares.
$c^2 = 676$	Simplify.
$c = \pm\sqrt{676}$	Take the square root of each side.
$c = \pm 26$	$(\pm 26)^2 = 676$

Length cannot be negative, so the missing length is 26 units.

b.

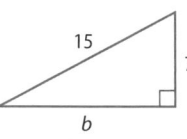

$c^2 = a^2 + b^2$	Pythagorean Theorem
$15^2 = 7^2 + b^2$	$a = 7$ and $c = 15$
$225 = 49 + b^2$	Evaluate squares.
$176 = b^2$	Subtract 49 from each side.
$\pm\sqrt{176} = b$	Take the square root of each side.
$\pm 13.27 \approx b$	Use a calculator to evaluate $\sqrt{176}$.

The missing length is 13.27 units.

▶ **Guided**Practice

1A.

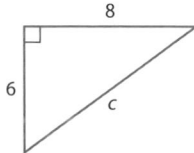

1B.

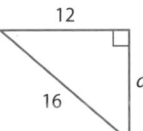

Real-World Link

A keelboat is a sailboat with a weighted keel, a vertical fin at the bottom of the boat. Keels are 20 to 30 inches in length.

Source: United States Sailing Association

Real-World Example 2 Find the Length of a Side

SAILING The sail of a keelboat forms a right triangle as shown. Find the height of the sail.

$20^2 = h^2 + 10^2$	Pythagorean Theorem
$400 = h^2 + 100$	Evaluate squares.
$300 = h^2$	Subtract 100 from each side.
$\pm 17.32 \approx h$	Take the square root of each side.
$17.32 \approx h$	Use the positive value.

20 ft
h
10 ft

The sail is approximately 17.32 feet high.

GuidedPractice

2. Suppose the longest side of the sail is 30 feet long and the shortest side is 14 feet long. Find the height of the sail.

2 Right Triangles If you exchange the phrases after *if* and *then* of an if-then statement, the result is the **converse** of the statement. The converse of the Pythagorean Theorem can be used to determine whether a triangle is a right triangle.

KeyConcept Converse of the Pythagorean Theorem

If a triangle has side lengths a, b, and c such that $c^2 = a^2 + b^2$, then the triangle is a right triangle. If $c^2 \neq a^2 + b^2$, then the triangle is not a right triangle.

A **Pythagorean triple** is a group of three counting numbers that satisfy the equation $c^2 = a^2 + b^2$, where c is the greatest number. Examples include (3, 4, 5) and (5, 12, 13). Multiples of Pythagorean triples also satisfy the converse of the Pythagorean Theorem, so (6, 8, 10) is also a Pythagorean triple.

Example 3 Check for Right Triangles

Determine whether 9, 12, and 16 can be the lengths of the sides of a right triangle.

Since the measure of the longest side is 16, let $c = 16$, $a = 9$, and $b = 12$.

$c^2 = a^2 + b^2$	Pythagorean Theorem
$16^2 \overset{?}{=} 9^2 + 12^2$	$a = 9$, $b = 12$, and $c = 16$
$256 \overset{?}{=} 81 + 144$	Evaluate squares.
$256 \neq 225$	Add.

Since $c^2 \neq a^2 + b^2$, segments with these measures cannot form a right triangle.

GuidedPractice

Determine whether each set of measures can be the lengths of the sides of a right triangle.

3A. 30, 40, 50　　　　　　　　　　　　　**3B.** 6, 12, 18

Check Your Understanding   ⬤ = Step-by-Step Solutions begin on page R13.

Example 1   **Find each missing length. If necessary, round to the nearest hundredth.**

1.
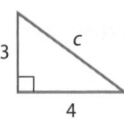

2.
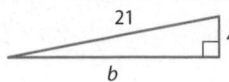

3.
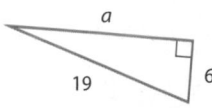

4.
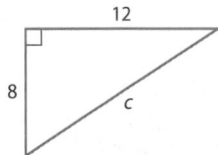

Example 2   **5. BASEBALL** A baseball diamond is a square. The distance between consecutive bases is 90 feet.

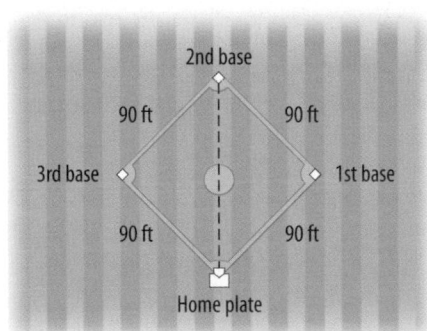

   a. How far does a catcher have to throw the ball from home plate to second base?

   b. How far does a third baseman throw the ball to the first baseman from a point in the baseline 15 feet from third to second base?

   c. A base runner going from first to second base is 100 feet from home plate. How far is the runner from second base?

Example 3   **Determine whether each set of measures can be the lengths of the sides of a right triangle.**

   6. 8, 12, 16   **7.** 28, 45, 53

   8. 7, 24, 25   **9.** 15, 25, 45

Practice and Problem Solving   Extra Practice is on page R10.

Example 1   **Find each missing length. If necessary, round to the nearest hundredth.**

10.
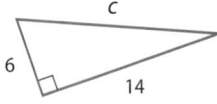

11
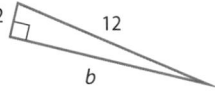

12.
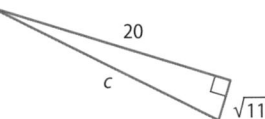

13.
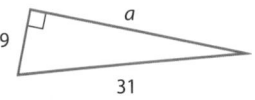

14.
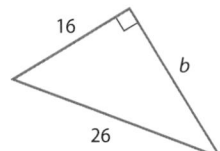

15.
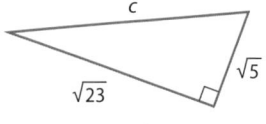

16.
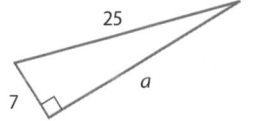

17.
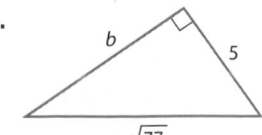

18.
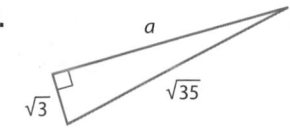

Example 2

19 **TELEVISION** Larry is buying an entertainment stand for his television. The diagonal of his television is 42 inches. The space for the television measures 30 inches by 36 inches. Will Larry's television fit? Explain.

Example 3 **Determine whether each set of measures can be the lengths of the sides of a right triangle. Then determine whether they form a Pythagorean triple.**

20. 9, 40, 41

21. $3, 2\sqrt{10}, \sqrt{41}$

22. $4, \sqrt{26}, 12$

23. $\sqrt{5}, 7, 14$

24. 8, 31.5, 32.5

25. $\sqrt{65}, 6\sqrt{2}, \sqrt{97}$

26. 18, 24, 30

27. 36, 77, 85

28. 17, 33, 98

29. **GEOMETRY** Refer to the triangle at the right.

   a. What is a?

   b. Find the area of the triangle.

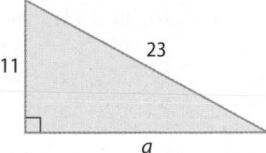

30. **GARDENING** Khaliah wants to plant flowers in a triangular plot. She has three lengths of plastic garden edging that measure 8 feet, 15 feet, and 17 feet. Determine whether these pieces form a right triangle. Explain.

31. **LADDER** Mr. Takeo is locked out of his house. The only open window is on the second floor. There is a bush along the edge of the house, so he places the neighbor's ladder 10 feet from the house. To the nearest foot, what length of ladder does he need to reach the window?

CCSS TOOLS **Find the length of the hypotenuse. Round to the nearest hundredth.**

32.

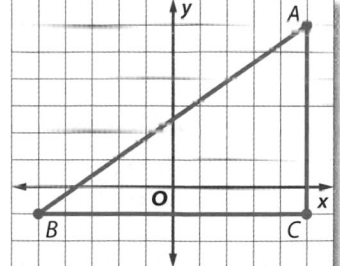

33.

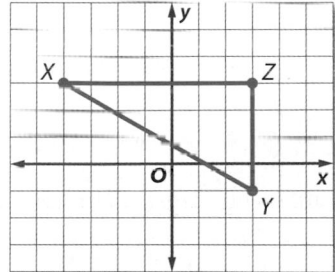

34. **DOLLHOUSE** Alonso is building a dollhouse for his sister's birthday. The roof is 24 inches across and the slanted side is 16 inches long as shown. Find the height of the roof to the nearest tenth of an inch.

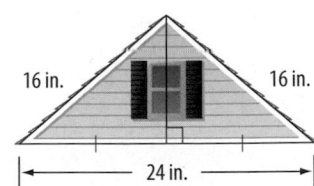

35. **GEOMETRY** Each side of a cube is 5 inches long. Find the length of a diagonal of the cube.

36. **TOWN SQUARES** The largest town square in the world is Tiananmen Square in Beijing, China, covering 98 acres.

   a. One square mile is 640 acres. Assuming that Tiananmen Square is a square, how many feet long is a side to the nearest foot?

   b. To the nearest foot, what is the diagonal distance across Tiananmen Square?

37. TRUCKS Violeta needs to construct a ramp to roll a cart of moving boxes from her garage into the back of her truck. How long does the ramp have to be?

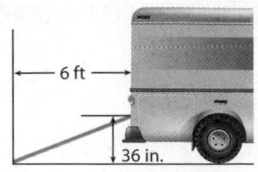
6 ft
36 in.

If c is the measure of the hypotenuse of a right triangle, find each missing measure. If necessary, round to the nearest hundredth.

38. $a = x, b = x + 41, c = 85$

39 $a = 8, b = x, c = x + 2$

40. $a = 12, b = x - 2, c = x$

41. $a = x, b = x + 7, c = 97$

42. $a = x - 47, b = x, c = x + 2$

43. $a = x - 32, b = x - 1, c = x$

44. GEOMETRY A right triangle has one leg that is 8 inches shorter than the other leg. The hypotenuse is 30 inches long. Find the length of each leg.

45. ✦ MULTIPLE REPRESENTATIONS In this problem, you will derive a method for finding the midpoint and length of a segment on the coordinate plane.

a. Graphical Use a graph to find the lengths of the segments between $(3, 2)$ and $(8, 2)$ and between $(4, 1)$ and $(4, 9)$. Then find the midpoint of each segment.

b. Logical Use what you learned in part **a** to write expressions for the lengths of the segments between (x_1, y) and (x_2, y) and between (x, y_1) and (x, y_2). What would be the midpoint of each segment?

c. Analytical Based on your results from part **b**, find the midpoint of the segment with endpoints at (x_1, y_1), and (x_2, y_2).

d. Analytical Use the Pythagorean Theorem to write an expression for the distance between (x_1, y_1), and (x_2, y_2).

H.O.T. Problems   Use Higher-Order Thinking Skills

46. ERROR ANALYSIS Wyatt and Dario are determining whether 36, 77, and 85 form a Pythagorean triple. Is either of them correct? Explain your reasoning.

Wyatt	Dario
$36^2 + 77^2 \stackrel{?}{=} 85^2$	$36^2 + 85^2 \stackrel{?}{=} 77^2$
$1296 + 5929 \stackrel{?}{=} 7225$	$1296 + 7725 \stackrel{?}{=} 5929$
$7225 = 7225$	$9021 \neq 5929$
yes	no

47. CCSS PERSEVERANCE Find the value of x in the figure.

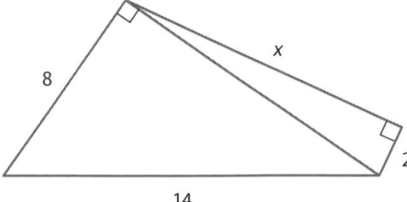
8
x
14
2

48. REASONING Provide a counterexample that is a specific case to show that the statement is false. *Any two right triangles with the same hypotenuse have the same area.*

49. OPEN ENDED Draw a right triangle that has a hypotenuse of $\sqrt{72}$ units.

50. WRITING IN MATH Explain how to determine whether segments in three lengths could form a right triangle.

51. GEOMETRY Find the missing length.

A -17

B $-\sqrt{161}$

C $\sqrt{161}$

D 17

52. What is a solution of this equation?

$$x + 1 = \sqrt{x + 1}$$

F $0, 3$

G 3

H 0

J no solutions

53. SHORT RESPONSE A plumber charges $40 for the first hour of each house call plus $8 for each additional half hour. If the plumber works for 4 hours, how much does he charge?

54. Find the next term in the geometric sequence $4, 3, \frac{9}{4}, \frac{27}{16}, \ldots$.

A $\frac{64}{81}$　　B $\frac{81}{64}$　　C $\frac{4}{3}$　　D $\frac{243}{64}$

Spiral Review

Solve each equation. Check your solution. (Lesson 10-4)

55. $\sqrt{x} = 16$

56. $\sqrt{4x} = 64$

57. $\sqrt{10x} = 10$

58. $\sqrt{8x} + 1 = 65$

59. $\sqrt{x + 1} + 2 = 4$

60. $\sqrt{x - 15} = 3 - \sqrt{x}$

Simplify each expression. (Lesson 10-3)

61. $2\sqrt{3} + 5\sqrt{3}$

62. $4\sqrt{5} - 2\sqrt{5}$

63. $6\sqrt{7} + 2\sqrt{28}$

64. $\sqrt{18} - 4\sqrt{2}$

65. $3\sqrt{5} - 5\sqrt{3} + 9\sqrt{5}$

66. $4\sqrt{3} + 6\sqrt{12}$

Describe how the graph of each function is related to the graph of $f(x) = x^2$. (Lesson 9-3)

67. $g(x) = x^2 - 8$

68. $h(x) = \frac{1}{4}x^2$

69. $h(x) = -x^2 + 5$

70. $g(x) = (x + 10)^2$

71. $g(x) = -2x^2$

72. $h(x) = -x^2 - \frac{4}{3}$

73. ROCK CLIMBING While rock climbing, Damaris launches a grappling hook from a height of 6 feet with an initial upward velocity of 56 feet per second. The hook just misses the stone ledge that she wants to scale. As it falls, the hook anchors on a ledge 30 feet above the ground. How long was the hook in the air? (Lesson 8-7)

Find each product. (Lesson 8-3)

74. $(b + 8)(b + 2)$

75. $(x - 4)(x - 9)$

76. $(y + 4)(y - 8)$

77. $(p + 2)(p - 10)$

78. $(2w - 5)(w + 7)$

79. $(8d + 3)(5d + 2)$

80. BUSINESS The amount of money spent at West Outlet Mall continues to increase. The total $T(x)$ in millions of dollars can be estimated by the function $T(x) = 12(1.12)^x$, where x is the number of years after it opened in 2005. Find the amount of sales in 2015, 2016, and 2017. (Lesson 7-5)

Skills Review

Solve each proportion.

81. $\frac{x}{5} = \frac{12}{3}$

82. $\frac{12}{x} = \frac{3}{4}$

83. $\frac{5}{4} = \frac{10}{x}$

84. $\frac{3}{5} = \frac{12}{x + 8}$

10-5 Algebra Lab
Distance on the Coordinate Plane

The Pythagorean Theorem can be extended to develop a formula for finding the distance between two points on a coordinate plane.

Activity Find Distance

Find the distance between the points at $P(4, 3)$ and $Q(1, -2)$.

Step 1 Graph the points and $\overline{PQ}$.

Step 2 Draw a vertical segment down from P and a horizontal segment to the right from Q. Label the the point of intersection R. Notice that $\triangle PQR$ is a right triangle.

Step 3 Use the Pythagorean Theorem.

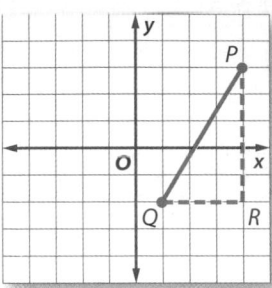

$c^2 = a^2 + b^2$	Pythagorean Theorem
$PQ^2 = 3^2 + 5^2$	$a = 3$ and $c = 5$
$PQ^2 = 9 + 25$	Evaluate squares.
$PQ^2 = 34$	Subtract 49 from each side.
$PQ = \pm\sqrt{34}$	Take the square root of each side.
$PQ \approx 5.83$	Use a calculator to evaluate $\sqrt{34}$. Use the positive value.

The distance between two points can be generalized as follows.

🔑 KeyConcept The Distance Formula

Words The distance d between any two points with coordinates (x_1, y_1) and (x_2, y_2) is given by the following formula.

$$d = \sqrt{(x_2 - x_1)^2 + (y_2 - y_1)^2}$$

Model

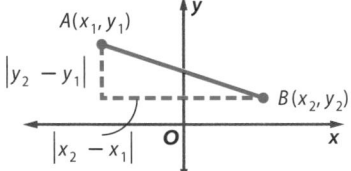

Model and Analyze

Find the distance between the points with the given coordinates.

1. $(6, -2), (12, 8)$ **2.** $(4, 8), (-3, -6)$ **3.** $(3, 0), (6, -2)$

4. $(-2, -4), (-5, -3)$ **5.** $(5, 1), (0, 4)$ **6.** $(-5, 2), (4, -2)$

The Midpoint Formula states that the midpoint M of a line segment with endpoints at (x_1, y_1) and (x_2, y_2) is given by $M = \left(\dfrac{x_1 + x_2}{2}, \dfrac{y_1 + y_2}{2}\right)$. Find the coordinates of the midpoint of the segment with the given endpoints.

7. $(5, -10), (5, 8)$ **8.** $(2, -2), (6, 2)$ **9.** $(5, 0), (0, 3)$

10. $(-4, 1), (3, -1)$ **11.** $(3, -17), (2, -8)$ **12.** $(-2, 2), (4, 10)$

13. The point that is equidistant from both endpoints is the **midpoint** of a segment. Use the Midpoint Formula to find the midpoint of the segment with endpoints at $(-16, -7)$ and $(-4, -3)$. Then use the Distance Formula to verify that the midpoint you found is correct. (*Hint:* The midpoint must be on the segment.)

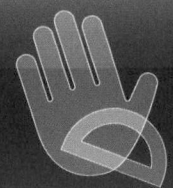

10-6

Algebra Lab
Investigating
Trigonometric Ratios

Recall that *similar triangles* have the same shape, but not necessarily the same size. You can use similar triangles to investigate the ratios of the lengths of sides of right triangles.

Activity   Collect the Data

Step 1 Use a ruler and grid paper to draw several right triangles with legs in a ratio of 5:8. Include right triangles with the side lengths listed in the table below and several more right triangles similar to these three. Label the vertices of each triangle as *A*, *B*, and *C*, where *C* is at the right angle, *B* is opposite the longest leg, and *A* is opposite the shortest leg.

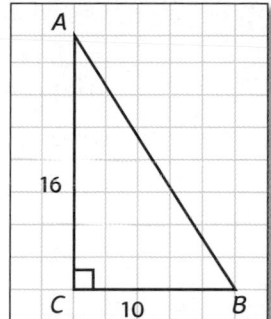

Step 2 Copy the table below. Complete the first three columns by measuring the hypotenuse (side $\overline{AB}$) in each right triangle you created and recording its length to the nearest tenth.

Step 3 Calculate and record the ratios in the middle two columns. Round to the nearest hundredth.

Step 4 Use a protractor to carefully measure angles *A* and *B* to the nearest degree in each right triangle. Record the angle measures in the table.

Side Lengths			Ratios		Angle Measures		
side *BC*	side *AC*	side *AB*	$\dfrac{BC}{AC}$	$\dfrac{BC}{AB}$	angle *A*	angle *B*	angle *C*
2.5	4						90°
5	8						90°
10	16						90°
							90°
							90°
							90°

Analyze the Results

1. Examine the measures and ratios in the table. What do you notice? Write a sentence or two to describe any patterns you see.

Make a Conjecture

2. For any right triangle similar to the ones you have drawn here, what will be the value of the ratio of the length of the shortest leg to the length of the longest leg?

3. If you draw a right triangle and calculate the ratio of the length of the shortest leg to the length of the hypotenuse to be approximately 0.53, what will be the measure of the larger acute angle in the right triangle?

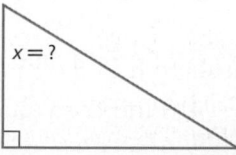

x = ?

10-6 Trigonometric Ratios

·· Then
- You used the Pythagorean Theorem

·· Now
1. Find trigonometric ratios of angles.
2. Use trigonometry to solve triangles.

·· Why?
- If a road has a percent grade of 8%, this means the road rises or falls 8 feet over a horizontal distance of 100 feet. Trigonometric ratios can be used to determine the angle that the road rises or falls.

 NewVocabulary
trigonometry
trigonometric ratio
sine
cosine
tangent
solving the triangle
inverse sine
inverse cosine
inverse tangent

 Common Core State Standards

Mathematical Practices
5 Use appropriate tools strategically.

1 Trigonometric Ratios

Trigonometry is the study of relationships among the angles and sides of triangles. A **trigonometric ratio** is a ratio that compares the side lengths of two sides of a right triangle. The three most common trigonometric ratios, **sine**, **cosine**, and **tangent**, are described below.

KeyConcept Trigonometric Ratios

Words	Symbols	Model
sine of $\angle A = \dfrac{\text{leg opposite } \angle A}{\text{hypotenuse}}$	$\sin A = \dfrac{a}{c}$	
cosine of $\angle A = \dfrac{\text{leg adjacent to } \angle A}{\text{hypotenuse}}$	$\cos A = \dfrac{b}{c}$	
tangent of $\angle A = \dfrac{\text{leg opposite } \angle A}{\text{Leg adjacent to } \angle A}$	$\tan A = \dfrac{a}{b}$	

Opposite, adjacent, and hypotenuse are abbreviated *opp*, *adj*, and *hyp*, respectively.

Example 1 Find Sine, Cosine, and Tangent Ratios

Find the values of the three trigonometric ratios for angle A.

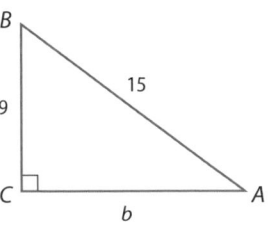

Step 1 Use the Pythagorean Theorem to find b.

$a^2 + b^2 = c^2$ Pythagorean Theorem

$9^2 + b^2 = 15^2$ $a = 9$ and $c = 15$

$81 + b^2 = 225$ Simplify.

$b^2 = 144$ Subtract 81 from each side.

$b = 12$ Take the square root of each side.

Step 2 Use the side lengths to write the trigonometric ratios.

$\sin A = \dfrac{\text{opp}}{\text{hyp}} = \dfrac{9}{15} = \dfrac{3}{5}$ $\cos A = \dfrac{\text{adj}}{\text{hyp}} = \dfrac{12}{15} = \dfrac{4}{5}$ $\tan A = \dfrac{\text{opp}}{\text{adj}} = \dfrac{9}{12} = \dfrac{3}{4}$

▶ **GuidedPractice**

1. Find the values of the three trigonometric ratios for angle B.

Masterfile

WatchOut!

CCSS Tools Make sure your graphing calculator is in degree mode.

Example 2 Use a Calculator to Evaluate Expressions

Use a calculator to find cos 42° to the nearest ten-thousandth.

KEYSTROKES: [COS] 42 [ )] [ENTER]

Rounded to the nearest ten-thousandth, $\cos 42° \approx 0.7431$.

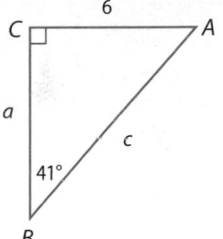

▶ **Guided**Practice

2A. $\sin 31°$ **2B.** $\tan 76°$ **2C.** $\cos 55°$

---

2 **Use Trigonometric Ratios** When you find all unknown measures of the sides and angles of a right triangle, you are **solving the triangle**. You can find the missing measures if you know the measure of two sides of the triangle or the measure of one side and the measure of one acute angle.

Example 3 Solve a Triangle

Solve the right triangle. Round each side length to the nearest tenth.

StudyTip

Remembering Trigonometric Ratios SOH–CAH–TOA can be used to help you remember the ratios for sine, cosine, and tangent. Each letter represents a word.

$\sin A = \dfrac{\text{opp}}{\text{hyp}}$

$\cos A = \dfrac{\text{adj}}{\text{hyp}}$

$\tan A = \dfrac{\text{opp}}{\text{adj}}$

Step 1 Find the measure of $\angle A$. $180° - (90° + 41°) = 49°$ The measure of $\angle A = 49°$.

Step 2 Find a. Since you are given the measure of the side opposite $\angle B$ and are finding the measure of the side adjacent to $\angle B$, use the tangent ratio.

$\tan 41° = \dfrac{6}{a}$ Definition of tangent

$a \tan 41° = 6$ Multiply each side by a.

$a = \dfrac{6}{\tan 41°}$ or about 6.9   Divide each side by tan 41°. Use a calculator.

So the measure of a or $\overline{BC}$ is about 6.9.

Step 3 Find c. Since you are given the measure of the side opposite $\angle B$ and are finding the measure of the hypotenuse, use the sine ratio.

$\sin 41° = \dfrac{6}{c}$ Definition of sine

$c \sin 41° = 6$ Multiply each side by c.

$c = \dfrac{6}{\sin 41°}$ or about 9.1   Divide each side by sin 41°. Use a calculator.

So the measure of c or $\overline{AB}$ is about 9.1.

▶ **Guided**Practice

3A.

3B.

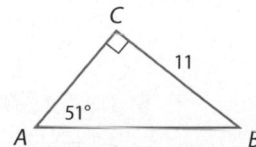

Real-World Example 4 Find a Missing Side Length

EXERCISE A trainer sets the incline on a treadmill to 10°. The walking surface of the treadmill is 5 feet long. About how many inches is the end of the treadmill from the floor?

$$\sin 10° = \frac{h}{5}$$ Definition of sine

$$5 \cdot \sin 10° = h$$ Multiply each side by 5.

$$0.87 \approx h$$ Use a calculator.

The value of *h* is in feet. Multiply 0.87 by 12 to convert feet to inches. The trainer raised the treadmill about 10.4 inches.

▶ **Guided**Practice

4. SKATEBOARDING The angle that a skateboarding ramp forms with the ground is 25° and the height of the ramp is 6 feet. Determine the length of the ramp.

A trigonometric function has a rule given by a trigonometric ratio. If you know the sine, cosine, or tangent of an acute angle, you can use the *inverse* of the trigonometric function to find the measure of the angle.

KeyConcept Inverse Trigonometric Functions

Words	If ∠A is an acute angle and the sine of A is x, then the **inverse sine** of x is the measure of ∠A.
Symbols	If sin A = x, then sin⁻¹ x = m∠A.
Words	If ∠A is an acute angle and the cosine of A is x, then the **inverse cosine** of x is the measure of ∠A.
Symbols	If cos A = x, then cos⁻¹ x = m∠A.
Words	If ∠A is an acute angle and the tangent of A is x, then the **inverse tangent** of x is the measure of ∠A.
Symbols	If tan A = x, then tan⁻¹ x = m∠A.

Example 5 Find a Missing Angle Measure

Find *m*∠Y to the nearest degree.

You know the measure of the side adjacent to ∠Y and the measure of the hypotenuse. Use the cosine ratio.

$$\cos Y = \frac{8}{19}$$ Definition of cosine

Use a calculator and the **[COS⁻¹]** function to find the measure of the angle.

KEYSTROKES: [2nd] [COS⁻¹] 8 [÷] 19 [ )] [ENTER] 65.098937   So, *m*∠Y = 65°.

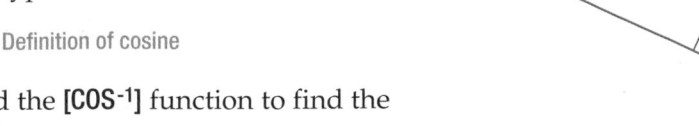

▶ **Guided**Practice

5. Find *m*∠X to the nearest degree if *XY* = 14 and *YZ* = 5.

Masterfile

Example 1   Find the values of the three trigonometric ratios for angle A.

1.
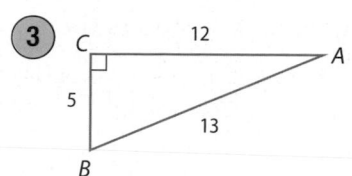

2.

3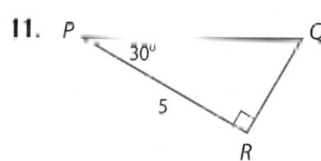

4.

Example 2   **CCSS TOOLS** Use a calculator to find the value of each trigonometric ratio to the nearest ten-thousandth.

5. $\sin 37°$ **6.** $\cos 23°$ **7.** $\tan 14°$ **8.** $\cos 82°$

Example 3   Solve each right triangle. Round each side length to the nearest tenth.

9.

10.

11.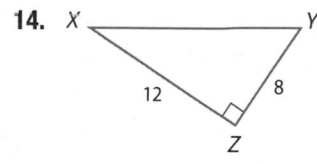

12.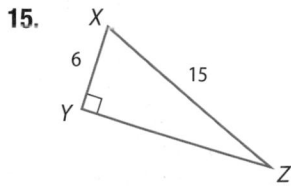

Example 4   **13. SNOWBOARDING** A hill used for snowboarding has a vertical drop of 3500 feet. The angle the run makes with the ground is 18°. Estimate the length of r.

3500 ft

Example 5   Find $m\angle X$ for each right triangle to the nearest degree.

14.

15.

16.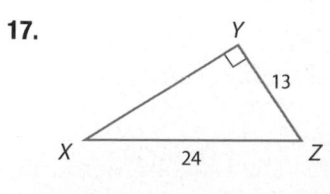

17.

Example 1 Find the values of the three trigonometric ratios for angle *B*.

18.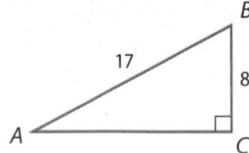

19.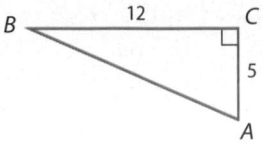

20.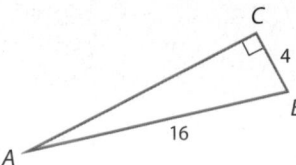

Example 2 **CCSS TOOLS** Use a calculator to find the value of each trigonometric ratio to the nearest ten-thousandth.

21. tan 2°

22. sin 89°

23. cos 44°

24. tan 45°

25. sin 73°

26. cos 90°

27. sin 30°

28. tan 60°

Example 3 Solve each right triangle. Round each side length to the nearest tenth.

29.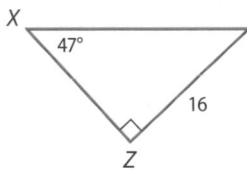

30.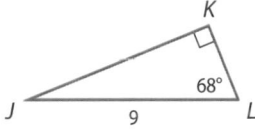

31.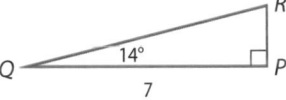

32.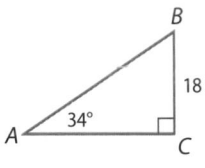

33.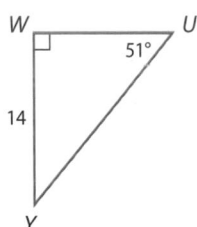

34.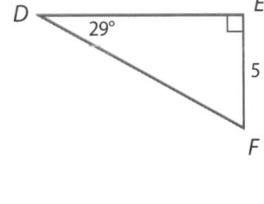

Example 4 **35. ESCALATORS** At a local mall, an escalator is 110 feet long. The angle the escalator makes with the ground is 29°. Find the height reached by the escalator.

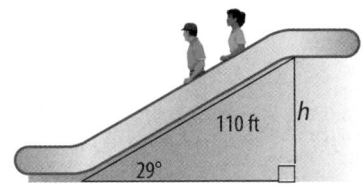

Example 5 Find *m∠J* for each right triangle to the nearest degree.

36.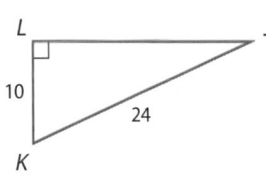

37.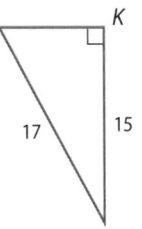

38.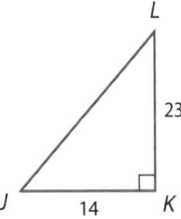

39.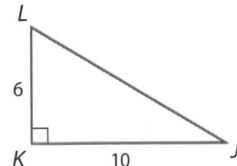

40.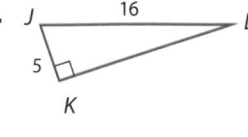

41.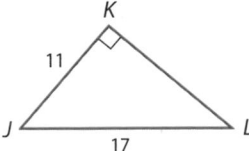

42. MONUMENTS The Lincoln Memorial building measures 204 feet long, 134 feet wide, and 99 feet tall. Chloe is looking at the top of the monument at an angle of 55°. How far away is she standing from the monument?

43 **AIRPLANES** Ella looks down at a city from an airplane window. The airplane is 5000 feet in the air, and she looks down at an angle of 8°. Determine the horizontal distance to the city.

44. FORESTS A forest ranger estimates the height of a tree is about 175 feet. If the forest ranger is standing 100 feet from the base of the tree, what is the measure of the angle formed by the ranger and the top of the tree?

Suppose ∠A is an acute angle of right triangle ABC.

45. Find $\sin A$ and $\tan A$ if $\cos A = \frac{3}{4}$.

46. Find $\tan A$ and $\cos A$ if $\sin A = \frac{2}{7}$.

47. Find $\cos A$ and $\tan A$ if $\sin A = \frac{1}{4}$.

48. Find $\sin A$ and $\cos A$ if $\tan A = \frac{5}{3}$.

49. SUBMARINES A submarine descends into the ocean at an angle of 10° below the water line and travels 3 miles diagonally. How far beneath the surface of the water has the submarine reached?

50. ⟳ **MULTIPLE REPRESENTATIONS** In this problem, you will explore a relationship between the sine and cosine functions.

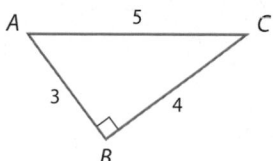 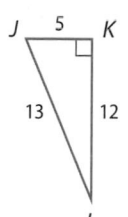 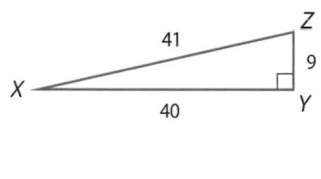

a. Tabular Copy and complete the table using the triangles shown above.

Triangle	Trigonometric Ratios		$\sin^2$	$\cos^2$	$\sin^2 + \cos^2 =$
ABC	$\sin A =$	$\cos A =$	$\sin^2 A =$	$\cos^2 A =$	
	$\sin C =$	$\cos C =$	$\sin^2 C =$	$\cos^2 C =$	
JKL	$\sin J =$	$\cos J =$	$\sin^2 J =$	$\cos^2 J =$	
	$\sin L =$	$\cos L =$	$\sin^2 L =$	$\cos^2 L =$	
XYZ	$\sin X =$	$\cos X =$	$\sin^2 X =$	$\cos^2 X =$	
	$\sin Z =$	$\cos Z =$	$\sin^2 Z =$	$\cos^2 Z =$	

b. Verbal Make a conjecture about the sum of the squares of the sine and cosine functions of an acute angle in a right triangle.

H.O.T. Problems   Use Higher-Order Thinking Skills

51. CHALLENGE Find a and c in the triangle shown.

52. REASONING Use the definitions of the sine and cosine ratios to define the tangent ratio.

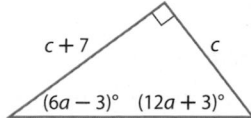

53. 🖉 **WRITING IN MATH** How can triangles be used to solve problems?

54. **CCSS ARGUMENTS** The sine and cosine of an acute angle in a right triangle are equal. What can you conclude about the triangle?

55. WRITING IN MATH Explain how to use trigonometric ratios to find the missing length of a side of a right triangle given the measure of one acute angle and the length of one side.

56. Which graph below represents the solution set for $-2 \leq x \leq 4$?

A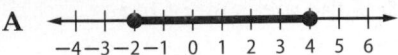

B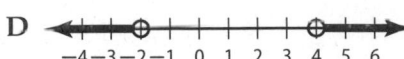

C

D

57. PROBABILITY Suppose one chip is chosen from a bin with the chips shown. To the nearest tenth, what is the probability that a green chip is chosen?

Color	Number
yellow	7
blue	9
orange	3
green	5
red	6

F 0.2   **H** 0.6

G 0.5   **J** 0.8

58. In the graph, for what value(s) of x is $y = 0$?

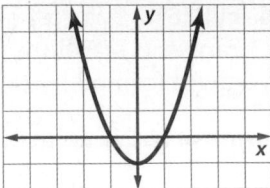

A 0   **C** 1

B -1 **D** 1 and -1

59. EXTENDED RESPONSE A 16-foot ladder is placed against the side of a house so that the bottom of the ladder is 8 feet from the base of the house.

a. If the bottom of the ladder is moved closer to the base of the house, does the height reached by the ladder increase or decrease?

b. What conclusion can you make about the distance between the bottom of the ladder and the base of the house and the height reached by the ladder?

c. How high does the ladder reach if the ladder is 3 feet from the base of the house?

If c is the measure of the hypotenuse of a right triangle, find each missing measure. If necessary, round to the nearest hundredth. (Lesson 10-5)

60. $a = 16, b = 63, c = ?$

61. $b = 3, a = \sqrt{112}, c = ?$

62. $c = 14, a = 9, b = ?$

63. $a = 6, b = 3, c = ?$

64. $b = \sqrt{77}, c = 12, a = ?$

65. $a = 4, b = \sqrt{11}, c = ?$

66. AVIATION The relationship between a plane's length L in feet and the pounds P its wings can lift is described by $L = \sqrt{kP}$, where k is the constant of proportionality. Find k for this plane to the nearest hundredth. (Lesson 10-4)

Weight 870,000 lb

232 ft

67. FINANCIAL LITERACY A salesperson is paid $32,000 a year plus 5% of the amount in sales made. What is the amount of sales needed to have an annual income greater than $45,000? (Lesson 5-3)

Solve each proportion.

68. $\frac{8}{9} = \frac{6}{z}$

69. $\frac{p}{6} = \frac{4}{3}$

70. $\frac{0.3}{r} = \frac{0.9}{1.7}$

71. $\frac{0.6}{1.1} = \frac{y}{8.47}$

Study Guide

KeyConcepts

Square Root Functions (Lesson 10-1)

- A square root function contains the square root of a variable.
- The parent function of the family of square root functions is $f(x) = \sqrt{x}$.

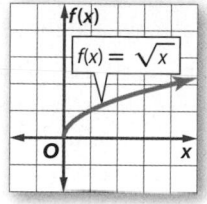

Simplifying Radical Expressions (Lesson 10-2)

- A radical expression is in simplest form when
  - no radicands have perfect square factors other than 1,
  - no radicals contain fractions,
  - and no radicals appear in the denominator of a fraction.

Operations with Radical Expressions and Equations (Lessons 10-3 and 10-4)

- Radical expressions with like radicals can be added or subtracted.
- Use the FOIL method to multiply radical expressions.

Pythagorean Theorem and Trigonometric Ratios (Lessons 10-5 and 10-6)

Pythagorean Theorem $c^2 = a^2 + b^2$

$\sin A = \dfrac{a}{c}$

$\cos A = \dfrac{b}{c}$

$\tan A = \dfrac{a}{b}$

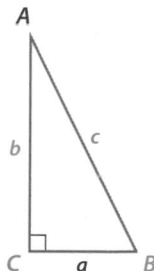

FOLDABLES StudyOrganizer

Be sure the Key Concepts are noted in your Foldable.

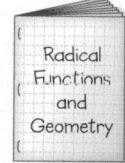

Radical Functions and Geometry

KeyVocabulary

conjugate (p. 630)

converse (p. 649)

cosine (p. 656)

Distance Formula (p. 654)

extraneous solution (p. 653)

hypotenuse (p. 648)

inverse cosine (p. 658)

inverse sine (p. 658)

inverse tangent (p. 658)

legs (p. 648)

midpoint (p. 654)

Pythagorean triple (p. 649)

radical equation (p. 642)

radical expression (p. 628)

radical function (p. 621)

radicand (p. 621)

rationalizing the denominator (p. 630)

sine (p. 656)

solving the triangle (p. 657)

square root function (p. 621)

tangent (p. 656)

trigonometric ratio (p. 656)

trigonometry (p. 656)

VocabularyCheck

State whether each sentence is *true* or *false*. If *false*, replace the underlined word, phrase, expression, or number to make a true sentence.

1. A triangle with sides having measures of <u>3, 4, and 6</u> is a right triangle.

2. The expressions <u>$12\sqrt{4}$</u> and $\sqrt{288}$ are equivalent.

3. The expressions $2 + \sqrt{5}$ and <u>$2 - \sqrt{5}$</u> are conjugates.

4. In the expression $-5\sqrt{2}$, the radicand is <u>2</u>.

5. The <u>shortest</u> side of a right triangle is the hypotenuse.

6. The cosine of an angle is found by dividing the measure of the side <u>opposite</u> the angle by the hypotenuse.

7. The domain of the function $y = \sqrt{x}$ is <u>$\{x \mid x \leq 0\}$</u>.

8. After the first step in solving $\sqrt{2x + 4} = x + 5$, you would have <u>$2x + 4 = x^2 + 10x + 25$</u>

9. The converse of the Pythagorean Theorem is <u>true</u>.

10. The range of the function $y = \sqrt{x}$ is <u>$\{y \mid y > 0\}$</u>.

Lesson-by-Lesson Review

10-1 Square Root Functions

Graph each function. Compare to the parent graph. State the domain and range.

11. $y = \sqrt{x} - 3$

12. $y = \sqrt{x} + 2$

13. $y = -5\sqrt{x}$

14. $y = \sqrt{x} - 6$

15. $y = \sqrt{x-1}$

16. $y = \sqrt{x} + 5$

17. GEOMETRY The function $s = \sqrt{A}$ can be used to find the length of a side of a square given its area. Use this function to determine the length of a side of a square with an area of 90 square inches. Round to the nearest tenth if necessary.

Example 1

Graph $y = -3\sqrt{x}$. Compare to the parent graph. State the domain and range.

Make a table. Choose nonnegative values for x.

x	0	1	2	3	4
y	0	-3	≈ -4.2	≈ -5.2	-6

Plot points and draw a smooth curve.

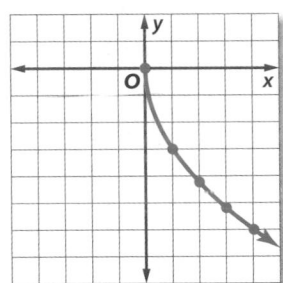

The graph of $y = \sqrt{x}$ is stretched vertically and is reflected across the x-axis.

The domain is $\{x \mid x \geq 0\}$.

The range is $\{y \mid y \leq 0\}$.

10-2 Simplifying Radical Expressions

Simplify.

18. $\sqrt{36x^2y^7}$ **19.** $\sqrt{20ab^3}$

20. $\sqrt{3} \cdot \sqrt{6}$ **21.** $2\sqrt{3} \cdot 3\sqrt{12}$

22. $(4 - \sqrt{5})^2$ **23.** $(1 + \sqrt{2})^2$

24. $\sqrt{\dfrac{50}{a^2}}$ **25.** $\sqrt{\dfrac{2}{5}} \cdot \sqrt{\dfrac{3}{4}}$

26. $\dfrac{3}{2 - \sqrt{5}}$ **27.** $\dfrac{5}{\sqrt{7} + 6}$

28. WEATHER To estimate how long a thunderstorm will last, use $t = \sqrt{\dfrac{d^3}{216}}$, where t is the time in hours and d is the diameter of the storm in miles. A storm is 10 miles in diameter. How long will it last?

Example 2

Simplify $\dfrac{2}{4 + \sqrt{3}}$.

$\dfrac{2}{4 + \sqrt{3}}$ Original expression

$= \dfrac{2}{4 + \sqrt{3}} \cdot \dfrac{4 - \sqrt{3}}{4 - \sqrt{3}}$ Rationalize the denominator.

$= \dfrac{2(4) - 2\sqrt{3}}{4^2 - (\sqrt{3})^2}$ $(a - b)(a + b) = a^2 - b^2$

$= \dfrac{8 - 2\sqrt{3}}{16 - 3}$ $(\sqrt{3})^2 = 3$

$= \dfrac{8 - 2\sqrt{3}}{13}$ Simplify.

10-3 Operations with Radical Expressions

Simplify each expression.

29. $\sqrt{6} - \sqrt{54} + 3\sqrt{12} + 5\sqrt{3}$

30. $2\sqrt{6} - \sqrt{48}$

31. $4\sqrt{3x} - 3\sqrt{3x} + 3\sqrt{3x}$

32. $\sqrt{50} + \sqrt{75}$

33. $\sqrt{2}(5 + 3\sqrt{3})$

34. $(2\sqrt{3} - \sqrt{5})(\sqrt{10} + 4\sqrt{6})$

35. $(6\sqrt{5} + 2)(4\sqrt{2} + \sqrt{3})$

36. MOTION The velocity of a dropped object when it hits the ground can be found using $v = \sqrt{2gd}$, where v is the velocity in feet per second, g is the acceleration due to gravity, and d is the distance in feet the object drops. Find the speed of a penny when it hits the ground, after being dropped from 984 feet. Use 32 feet per second squared for g.

Example 3

Simplify $2\sqrt{6} - \sqrt{24}$.

$$
\begin{aligned}
2\sqrt{6} - \sqrt{24} &= 2\sqrt{6} - \sqrt{4 \cdot 6} && \text{Product Property} \\
&= 2\sqrt{6} - 2\sqrt{6} && \text{Simplify.} \\
&= 0 && \text{Simplify.}
\end{aligned}
$$

Example 4

Simplify $(\sqrt{3} - \sqrt{2})(\sqrt{3} + 2\sqrt{2})$.

$$
\begin{aligned}
&(\sqrt{3} - \sqrt{2})(\sqrt{3} + 2\sqrt{2}) \\
&= (\sqrt{3})(\sqrt{3}) + (\sqrt{3})(2\sqrt{2}) + (-\sqrt{2})(\sqrt{3}) + (\sqrt{2})(2\sqrt{2}) \\
&= 3 + 2\sqrt{6} - \sqrt{6} + 4 \\
&= 7 + \sqrt{6}
\end{aligned}
$$

10-4 Radical Equations

Solve each equation. Check your solution.

37. $10 + 2\sqrt{x} = 0$

38. $\sqrt{5 - 4x} - 6 = 7$

39. $\sqrt{a + 4} = 6$

40. $\sqrt{3x} = 2$

41. $\sqrt{x + 4} = x - 8$

42. $\sqrt{3x - 14} + x = 6$

43. FREE FALL Assuming no air resistance, the time t in seconds that it takes an object to fall h feet can be determined by $t = \dfrac{\sqrt{h}}{4}$. If a skydiver jumps from an airplane and free falls for 10 seconds before opening the parachute, how many feet does she free fall?

Example 5

Solve $\sqrt{7x + 4} - 18 = 5$.

$$
\begin{aligned}
\sqrt{7x + 4} - 18 &= 5 && \text{Original equation} \\
\sqrt{7x + 4} &= 23 && \text{Add 18 to each side.} \\
\left(\sqrt{7x + 4}\right)^2 &= 23^2 && \text{Square each side.} \\
7x + 4 &= 529 && \text{Simplify.} \\
7x &= 525 && \text{Subtract 4 from each side.} \\
x &= 75 && \text{Divide each side by 7.}
\end{aligned}
$$

CHECK
$$
\begin{aligned}
\sqrt{7x + 4} - 18 &= 5 && \text{Original equation} \\
\sqrt{7(75) + 4} - 18 &\stackrel{?}{=} 5 && x = 75 \\
\sqrt{525 + 4} - 18 &\stackrel{?}{=} 5 && \text{Multiply.} \\
\sqrt{529} - 18 &\stackrel{?}{=} 5 && \text{Add.} \\
23 - 18 &\stackrel{?}{=} 5 && \text{Simplify.} \\
5 &= 5 \checkmark && \text{True.}
\end{aligned}
$$

10-5 The Pythagorean Theorem

Determine whether each set of measures can be the lengths of the sides of a right triangle.

44. 6, 8, 10   **45.** 3, 4, 5

46. 12, 16, 21   **47.** 10, 12, 15

48. 2, 3, 4   **49.** 7, 24, 25

50. 5, 12, 13   **51.** 15, 19, 23

52. LADDER A ladder is leaning on a building. The base of the ladder is 10 feet from the building, and the ladder reaches up 15 feet on the building. How long is the ladder?

Example 6

Determine whether the set of measures 12, 16, and 20 can be the lengths of the sides of a right triangle.

$$a^2 + b^2 = c^2 \qquad \text{Pythagorean Theorem}$$

$$12^2 + 16^2 \stackrel{?}{=} 20^2 \qquad a = 12, b = 16, \text{ and } c = 20$$

$$144 + 256 \stackrel{?}{=} 400 \qquad \text{Multiply.}$$

$$400 = 400 \checkmark \qquad \text{Add.}$$

The measures can be the lengths of the sides of a right triangle.

10-6 Trigonometric Ratios

Find the values of the three trigonometric ratios for angle *A*.

53.
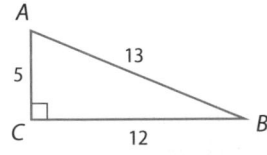

54.
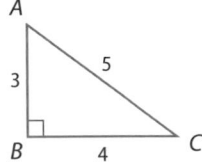

55. RAMPS How long is the ramp?

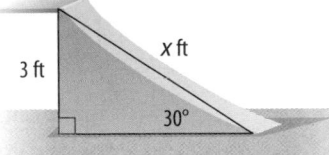

Example 7

Find the values of the three trigonometric ratios for angle *A*.

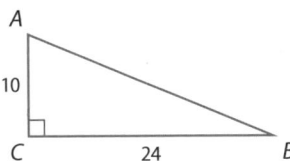

Find the hypotenuse: $c^2 = 10^2 + 24^2$, so $c = 26$.

$$\sin A = \frac{\text{leg opposite } \angle A}{\text{hypotenuse}} = \frac{24}{26} = \frac{12}{13}$$

$$\cos A = \frac{\text{leg adjacent } \angle A}{\text{hypotenuse}} = \frac{10}{26} = \frac{5}{13}$$

$$\tan A = \frac{\text{leg opposite } \angle A}{\text{leg adjacent } \angle A} = \frac{24}{10} = \frac{12}{5}$$

Graph each function, and compare to the parent graph. State the domain and range.

1. $y = -\sqrt{x}$

2. $y = \frac{1}{4}\sqrt{x}$

3. $y = \sqrt{x} + 5$

4. $y = \sqrt{x + 4}$

5. **MULTIPLE CHOICE** The length of the side of a square is given by the function $s = \sqrt{A}$, where A is the area of the square. What is the perimeter of a square that has an area of 64 square inches?

A 64 inches

B 8 inches

C 32 inches

D 16 inches

Simplify each expression.

6. $5\sqrt{36}$

7. $\dfrac{3}{1 - \sqrt{2}}$

8. $2\sqrt{3} + 7\sqrt{3}$

9. $3\sqrt{6}(5\sqrt{2})$

10. **MULTIPLE CHOICE** Find the area of the rectangle.

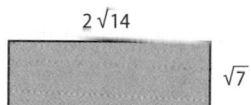

$2\sqrt{14}$

$\sqrt{7}$

F $7\sqrt{2}$

G 14

H $14\sqrt{2}$

J $98\sqrt{2}$

Solve each equation. Check your solution.

11. $\sqrt{10x} = 20$

12. $\sqrt{4x - 3} = 6 - x$

13. **PACKAGING** A cylindrical container of chocolate drink mix has a volume of about 162 in³. The radius of the container can be found by using the formula $r = \sqrt{\dfrac{V}{\pi h}}$, where r is the radius and h is the height. If the height is 8.25 inches, find the radius of the container.

Find each missing length. If necessary, round to the nearest tenth.

14.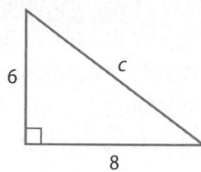

6

c

8

15.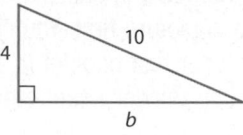

4

10

b

16. **DELIVERY** Ben and Amado are delivering a freezer. The bank in front of the house is the same height as the back of the truck. They set up their ramp as shown. What is the length of the slanted part of the ramp to the nearest foot?

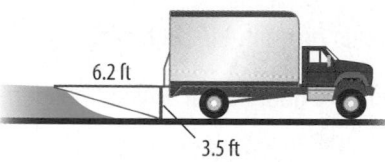

6.2 ft

3.5 ft

17. Find the values of the three trigonometric ratios for angle A.

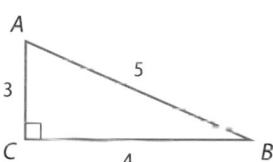

A

3

5

C

4

B

18. Find $m\angle X$ to the nearest degree.

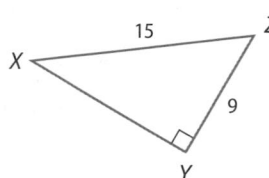

15

Z

X

9

Y

19. **LIGHTHOUSE** How tall is the lighthouse?

79°

buoy

25 ft

Preparing for Standardized Tests

Draw a Picture

Sometimes it is easier to visualize how to solve a problem if you draw a picture first. You can sketch your picture on scrap paper or in your test booklet (if allowed). Be careful not make any marks on your answer sheet other than your answers.

Strategies for Drawing a Picture

Step 1

Read the problem statement carefully.

Ask yourself:

- What am I being asked to solve?

- What information is given in the problem?

- What is the unknown quantity for which I need to solve?

Step 2

Sketch and label your picture.

- Draw your picture as clearly and accurately as possible.

- Label the picture carefully. Be sure to include all of the information given in the problem statement.

Step 3

Solve the problem.

- Use your picture to help you model the problem situation with an equation. Then solve the equation.

- Check your answer to make sure it is reasonable.

Standardized Test Example

Read the problem. Identify what you need to know. Then use the information in the problem to solve. Show your work.

An 18-foot ladder is leaning against a building. For stability, the base of the ladder must be 36 inches away from the wall. How far up the wall does the ladder reach?

Read the problem statement carefully. You know the height of the ladder leaning against the building and you know that the base of the ladder must be 36 inches away from the wall. You need to find how far up the wall the ladder reaches.

Scoring Rubric	
Criteria	Score
Full Credit: The answer is correct and a full explanation is provided that shows each step.	2
Partial Credit: • The answer is correct, but the explanation is incomplete. • The answer is incorrect, but the explanation is correct.	1
No Credit: Either an answer is not provided or the answer does not make sense.	0

Example of a 2-point response:

First convert all measurements to feet.

36 inches = 3 feet

Use a right triangle to find how high the ladder reaches. Draw and label a triangle to represent the situation.

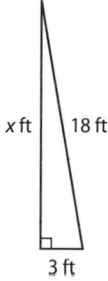

x ft   18 ft

3 ft

You know the measures of a leg and the hypotenuse, and need to know the length of the other leg. So you can use the Pythagorean Theorem.

$$c^2 = a^2 + b^2$$
$$18^2 = 3^2 + b^2$$
$$324 = 9 + b^2$$
$$315 = b^2$$
$$\pm 315 = b$$
$$17.7 \approx b$$

The ladder reaches about 17.7 feet or about 17 feet 9 inches.

Exercises

Read each problem. Identify what you need to know. Then use the information in the problem to solve. Show your work.

1. A building casts a 15-foot shadow, while a billboard casts a 4.5-foot shadow. If the billboard is 26 feet high, what is the height of the building? Round to the nearest tenth if necessary.

2. A space shuttle is directed toward the Moon, but drifts 1.2° from its intended course. The distance from Earth to the Moon is about 240,000 miles. If the pilot doesn't get the shuttle back on course, how far will the shuttle have drifted from its intended landing position?

Multiple Choice

Read each question. Then fill in the correct answer on the answer document provided by your teacher or a sheet of paper.

1. What is the equation of the square root function graphed below?

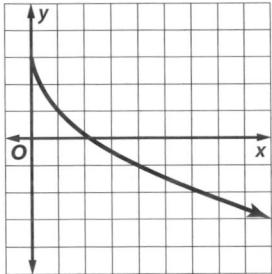

A $y = -2\sqrt{x} + 1$

B $y = -2\sqrt{x} + 3$

C $y = 2\sqrt{x} + 3$

D $y = 2\sqrt{x} + 1$

2. Simplify $\dfrac{1}{4 + \sqrt{2}}$.

F $\dfrac{4 + \sqrt{2}}{14}$

G $\dfrac{2 - \sqrt{2}}{7}$

H $\dfrac{4 - \sqrt{2}}{14}$

J $\dfrac{2 + \sqrt{2}}{7}$

3. What is the area of the triangle below?

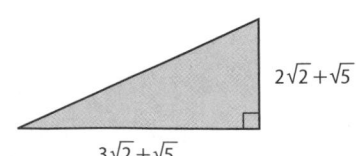

A $3\sqrt{2} + 10\sqrt{5}$

B $17 + 5\sqrt{10}$

C $12\sqrt{2} + 8\sqrt{5}$

D $8.5 + 2.5\sqrt{10}$

4. The formula for the slant height c of a cone is $c = \sqrt{h^2 + r^2}$, where h is the height of the cone and r is the radius of its base. What is the radius of the cone below? Round to the nearest tenth.

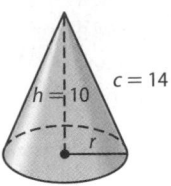

F 4.9   **H** 9.8

G 6.3   **J** 10.2

5. Which of the following sets of measures could not be the sides of a right triangle?

A (12, 16, 24)   **C** (24, 45, 51)

B (10, 24, 26)   **D** (18, 24, 30)

6. Which of the following is an equation of the line perpendicular to $4x - 2y = 6$ and passing through $(4, -4)$?

F $y = -\dfrac{3}{4}x + 3$

G $y = -\dfrac{3}{4}x - 1$

H $y = -\dfrac{1}{2}x - 4$

J $y = -\dfrac{1}{2}x - 2$

7. The scale on a map shows that 1.5 centimeters is equivalent to 40 miles. If the distance on the map between two cities is 8 centimeters, about how many miles apart are the cities?

A 178 miles

B 213 miles

C 224 miles

D 275 miles

Test-TakingTip

Question 4 Substitute for c and h in the formula. Then solve for r.

Record your answers on the answer sheet provided by your teacher or on a sheet of paper.

8. GRIDDED RESPONSE How many times does the graph of $y = x^2 - 4x + 10$ cross the x-axis?

9. Factor $2x^4 - 32$ completely.

10. GRIDDED RESPONSE In football, a field goal is worth 3 points, and the extra point after a touchdown is worth 1 point. During the 2006 season, John Kasay of the Carolina Panthers scored a total of 100 points for his team by making a total of 52 field goals and extra points. How many field goals did he make?

11. Shannon bought a satellite radio and a subscription to satellite radio. What is the total cost for his first year of service?

Item	Cost
radio	$39.99
subscription	$11.99 per month

12. GRIDDED RESPONSE The distance required for a car to stop is directly proportional to the square of its velocity. If a car can stop in 242 meters at 22 kilometers per hour, how many meters are needed to stop at 30 kilometers per hour?

13. The highest point in Kentucky is at an elevation of 4145 feet above sea level. The lowest point in the state is at an elevation of 257 feet above sea level. Write an inequality to describe the possible elevations in Kentucky.

14. Simplify the expression below. Show your work.
$$\left(\frac{-2r^{-2}q^5t^2}{5r^4q^2t^{-3}} \right)^{-2}$$

15. GRIDDED RESPONSE For the first home basketball game, 652 tickets were sold for a total revenue of $5216. If each ticket costs the same, how much is the cost per ticket? State your answer in dollars.

Record your answers on a sheet of paper. Show your work.

16. Karen is making a map of her hometown using a coordinate grid. The scale of her map is 1 unit = 2.5 miles.

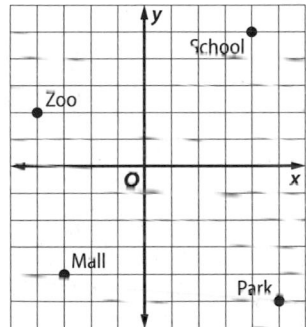

a. Use the Pythagorean Theorem to find the actual distance between Karen's school and the park. Round to the nearest tenth of a mile if necessary.

b. Suppose Karen's house is located at (0.5, 0.5). What is farthest from her house, the zoo, the park, the school, or the mall?

Need ExtraHelp?															
If you missed Question...	1	2	3	4	5	6	7	8	9	10	11	12	13	14	15	16
Go to Lesson...	10-1	10-2	10-3	10-4	10-5	4-4	2-6	9-5	8-8	6-4	1-4	3-6	5-4	7-2	2-2	10-5

Rational Functions and Equations

Then

○ You simplified expressions involving monomials and polynomials.

Now

○ In this chapter, you will:

- Identify and graph inverse variations.

- Identify excluded values of rational functions.

- Multiply, divide, and add rational expressions.

- Divide polynomials.

- Solve rational equations.

Why? ▲

○ **HOCKEY** The time it will take for a puck hit from the blue line to reach the goal line is given by the rational expression $\frac{64}{x}$, where x is the speed of the puck in feet per seconds. If a player hits the puck at 100 miles per hour, the puck will reach the goal line in 0.34 second.

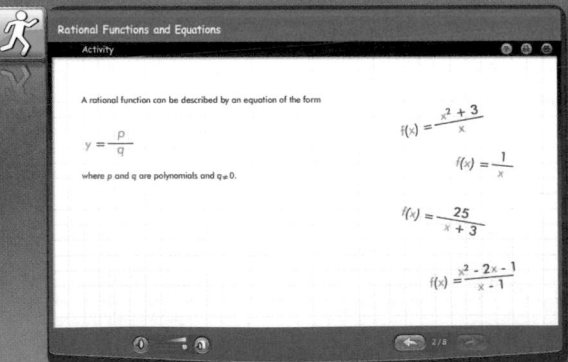

 **connectED.mcgraw-hill.com**   **Your Digital Math Portal**

Animation	Vocabulary	eGlossary	Personal Tutor	Virtual Manipulatives	Graphing Calculator	Audio	Foldables	Self-Check Practice	Worksheets

Get Ready for the Chapter

Diagnose Readiness | **You have two options for checking prerequisite skills.**

1 Textbook Option Take the Quick Check below. Refer to the Quick Review for help.

QuickCheck

Solve each proportion.

1. $\frac{y}{3} = \frac{8}{9}$
2. $\frac{5}{12} = \frac{x}{36}$
3. $\frac{7}{2} = \frac{y}{3}$
4. $\frac{5}{x} = \frac{10}{4}$
5. **DRAWING** Rosie is making a scale drawing. She is using the scale 1 inch = 3 feet. How many inches will represent 10 feet?

Factor each polynomial, if possible. If the polynomial cannot be factored using integers, write *prime*.

6. $4a - 12$
7. $2x^2 - 4x$
8. $6xy + 15x$
9. $2c^2d - 4c^2d^2$
10. $d^2 - 12d + 27$
11. $m^2 - m - 132$
12. $3x^2 + 2x - 1$
13. $6x^2 - 5x - 4$
14. **AREA** The area of a rectangle is $x^2 + 5x + 6$. What binomial expressions could represent the side lengths of the rectangle?

$A = x^2 + 5x + 6$

QuickReview

Example 1

Solve $\frac{3}{5} = \frac{x}{12}$.

$\frac{3}{5} = \frac{x}{12}$	Original equation
$3 \cdot 12 = 5 \cdot x$	Cross products
$36 = 5x$	Simplify.
$\frac{36}{5} = \frac{5x}{5}$	Divide each side by 5.
$\frac{36}{5} = x$	Simplify.

Example 2

Factor $45x^2 + 27x$.

Find the GCF of the terms.

$45x^2 = 3 \cdot 3 \cdot 5 \cdot x \cdot x$

$27x = 3 \cdot 3 \cdot 3 \cdot x$

GCF $= 3 \cdot 3 \cdot x$ or $9x$

$$45x^2 + 27x = 9x(5x) + 9x(3)$$
$$= 9x(5x + 3)$$

2 Online Option Take an online self-check Chapter Readiness Quiz at connectED.mcgraw-hill.com.

Get Started on the Chapter

You will learn several new concepts, skills, and vocabulary terms as you study Chapter 11. To get ready, identify important terms and organize your resources. You may wish to refer to Chapter 0 to review prerequisite skills.

FOLDABLES StudyOrganizer

Rational Functions and Equations Make this Foldable to help you organize your Chapter 11 notes about rational functions and equations. Begin with 3 sheets of notebook paper.

1 **Take** one sheet of paper and fold in half along the width. Cut 1 inch slits on each side of the paper.

2 **Stack** the two sheets of paper and fold in half along the width. Cut a slit through the center stopping 1 inch from each side.

3 **Insert** the first sheet through the second sheets and align the folds to form a booklet. Label the cover with the chapter title.

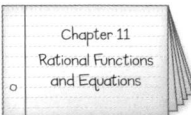

Chapter 11
Rational Functions
and Equations

NewVocabulary

English		Español
inverse variation	p. 676	variación inversa
product rule	p. 677	regla del producto
excluded value	p. 684	valores excluidos
rational function	p. 684	función racional
asymptote	p. 685	asíntota
rational expression	p. 690	expresión racional
least common multiple (LCM)	p. 713	mínimo común múltiplo (mcm)
least common denominator (LCD)	p. 714	mínimo común denominador (mcd)
complex fraction	p. 720	fracción compleja
mixed expression	p. 720	expresión mixta
rational equation	p. 726	ecuacion racional
extraneous solutions	p. 727	soluciones extrañas
work problems	p. 728	problemas de trabajo
rate problems	p. 729	problemas de tasas

ReviewVocabulary

direct variation variación directa an equation of the form $y = kx$, where $k \neq 0$

Quotient of Powers cociente de potencia $\dfrac{a^m}{a^n} = a^{m-n}$

$$\frac{x^5}{x^3} = \frac{x \cdot x \cdot x \cdot x \cdot x}{x \cdot x \cdot x} = x \cdot x \text{ or } x^2$$

$$\frac{x^5}{x^3} = x^{5-3} \text{ or } x^2$$

Zero Product Property propiedad del producto de cero if the product of two factors is 0, then at least one of the factors must be 0

11-1

Graphing Technology Lab
Inverse Variation

You can use a data collection device to investigate the relationship between volume and pressure.

Set Up The Lab

- Connect a syringe to the gas pressure sensor. Then connect the data collection device to both the sensor and the calculator as shown.

- Start the collection program and select the sensor.

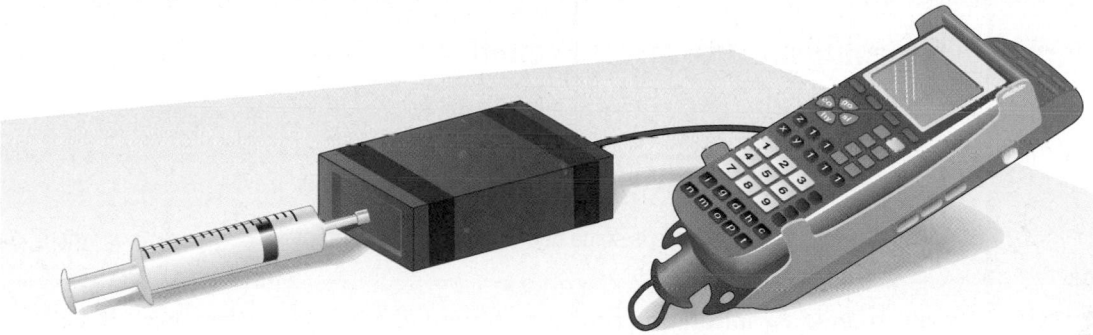

Activity   Collect Data

Step 1 Open the valve between the atmosphere and the syringe. Set the inside ring of the syringe to 20 mL and close the valve. This ensures that the amount of air inside the syringe will be constant throughout the experiment.

Step 2 Press the plunger of the syringe to the 5 mL mark. Wait for the pressure gauge to stop changing, then take the data reading. Enter 5 as the volume in the calculator. The pressure is measured in atmospheres (atm).

Step 3 Repeat step 2, pressing the plunger to 7.5 mL, 10.0 mL, 12.5 mL, 15.0 mL, 17.5 mL, and 20.0 mL. Record the volume from each data reading.

Step 4 After taking the last data reading, use **STAT PLOT** to create a line graph.

Exercises

1. Does the pressure vary directly as the volume? Explain.

2. As the volume changes from 10 to 20 mL, what happens to the pressure?

3. Predict what the pressure of the gas in the syringe would be if the volume increased to 40 mL.

4. Add a column to the data table to find the product of the volume and the pressure for each data reading. What pattern do you observe?

5. **MAKE A CONJECTURE** The relationship between the pressure and volume of a gas is called Boyle's Law. Write an equation relating the volume v in milliliters and pressure p in atmospheres in your experiment. Compare your conjecture to those of two classmates. Formulate mathematical questions about their conjectures.

Inverse Variation

:: Then	:: Now	:: Why?
● You solved problems involving direct variation.	**1** Identify and use inverse variations. **2** Graph inverse variations.	● The time it takes a runner to finish a race is inversely proportional to the average pace of the runner. The runner's time decreases as the pace of the runner increases. So, these quantities are *inversely proportional*.

 NewVocabulary

inverse variation
product rule

 Common Core State Standards

Mathematical Practices
1 Make sense of problems and persevere in solving them.

1 **Identify and Use Inverse Variations** An **inverse variation** can be represented by the equation $y = \frac{k}{x}$ or $xy = k$.

> ⚘ **KeyConcept** Inverse Variation
>
> y varies inversely as x if there is some nonzero constant k such that $y = \frac{k}{x}$ or $xy = k$, where $x, y \neq 0$.

In an inverse variation, the product of two values remains constant. Recall that a relationship of the form $y = kx$ is a *direct variation*. The constant k is called the *constant of variation* or the *constant of proportionality*.

> **Example 1** Identify Inverse and Direct Variations
>
> Determine whether each table or equation represents an *inverse* or a *direct* variation. Explain.
>
> **a.**
>
> | x | y |
> |---|---|
> | 1 | 16 |
> | 2 | 8 |
> | 4 | 4 |
>
> In an inverse variation, xy equals a constant k. Find xy for each ordered pair in the table.
>
> $1 \cdot 16 = 16 \quad 2 \cdot 8 = 16 \quad 4 \cdot 4 = 16$
>
> The product is constant, so the table represents an inverse variation.
>
> **b.**
>
> | x | y |
> |---|---|
> | 1 | 3 |
> | 2 | 6 |
> | 3 | 9 |
>
> Notice that xy is not constant. So, the table does not represent an indirect variation.
>
> $3 = k(1) \qquad 6 - k(2) \qquad 9 = k(3)$
> $3 = k \qquad\quad 3 = k \qquad\quad 3 = k$
>
> The table of values represents the direct variation $y = 3x$.
>
> **c.** $x = 2y$
>
> The equation can be written as $y = \frac{1}{2}x$. Therefore, it represents a direct variation.
>
> **d.** $2xy = 10$
>
> $2xy = 10$ Write the equation.
> $xy = 5$ Divide each side by 2.
>
> The equation represents an inverse variation.
>
> ▶ **Guided**Practice
>
> **1A.**
>
> | x | 1 | 2 | 5 |
> |---|---|---|---|
> | y | 10 | 5 | 2 |
>
> **1B.** $-2x = y$

Stockbyte/Getty Images

You can use $xy = k$ to write an inverse variation equation that relates x and y.

Example 2 Write an Inverse Variation

Assume that y varies inversely as x. If $y = 18$ when $x = 2$, write an inverse variation equation that relates x and y.

$$xy = k \qquad \text{Inverse variation equation}$$
$$2(18) = k \qquad x = 2 \text{ and } y = 18$$
$$36 = k \qquad \text{Simplify.}$$

The constant of variation is 36. So, an equation that relates x and y is $xy = 36$ or $y = \dfrac{36}{x}$.

▶ **Guided**Practice

2. Assume that y varies inversely as x. If $y = 5$ when $x = -4$, write an inverse variation equation that relates x and y.

If (x_1, y_1) and (x_2, y_2) are solutions of an inverse variation, then $x_1y_1 = k$ and $x_2y_2 = k$.

$$x_1y_1 = k \text{ and } x_2y_2 = k$$
$$x_1y_1 = x_2y_2 \qquad \text{Substitute } x_2y_2 \text{ for } k.$$

The equation $x_1y_1 = x_2y_2$ is called the **product rule** for inverse variations.

▣ KeyConcept Product Rule for Inverse Variations

Words If (x_1, y_1) and (x_2, y_2) are solutions of an inverse variation, then the products x_1y_1 and x_2y_2 are equal.

Symbols $x_1y_1 = x_2y_2$ or $\dfrac{x_1}{x_2} = \dfrac{y_2}{y_1}$

Example 3 Solve for *x* or *y*

Assume that y varies inversely as x. If $y = 3$ when $x = 12$, find x when $y = 4$.

$$x_1y_1 = x_2y_2 \qquad \text{Product rule for inverse variations}$$
$$12 \cdot 3 = x_2 \cdot 4 \qquad x_1 = 12, y_1 = 3, \text{ and } y_2 = 4$$
$$36 = x_2 \cdot 4 \qquad \text{Simplify.}$$
$$\frac{36}{4} = x_2 \qquad \text{Divide each side by 4.}$$
$$9 = x_2 \qquad \text{Simplify.}$$

So, when $y = 4$, $x = 9$.

▶ **Guided**Practice

3. If y varies inversely as x and $y = 4$ when $x = -8$, find y when $x = -4$.

The product rule for inverse variations can be used to write an equation to solve real-world problems.

Real-World Example 4 Use Inverse Variations

PHYSICS The acceleration a of a hockey puck is inversely proportional to its mass m. Suppose a hockey puck with a mass of 164 grams is hit so that it accelerates 122 m/s². Find the acceleration of a 158-gram hockey puck if the same amount of force is applied.

Make a table to organize the information.

Let $m_1 = 164$, $a_1 = 122$, and $m_2 = 164$. Solve for a_2.

Puck	Mass	Acceleration
1	164 g	122 m/s²
2	158 g	a_2

$\quad m_1 a_1 = m_2 a_2$ Use the product rule to write an equation.

$164 \cdot 122 = 158 a_2$ $m_1 = 164$, $a_1 = 122$, and $m_2 = 158$

$\quad 20{,}008 = 158 a_2$ Simplify.

$\quad\quad 126.6 \approx a_2$ Divide each side by 158 and simplify.

The 158-gram puck has an acceleration of approximately 126.6 m/s².

GuidedPractice

4. RACING Manuel runs an average of 8 miles per hour and finishes a race in 0.39 hour. Dyani finished the race in 0.35 hour. What was her average pace?

2 Graph Inverse Variations
The graph of an inverse variation is not a straight line like the graph of a direct variation.

Example 5 Graph an Inverse Variation

Graph an inverse variation equation in which $y = 8$ when $x = 3$.

Step 1 Write an inverse variation equation.

$\quad xy = k$ Inverse variation equation

$\quad 3(8) = k$ $x = 3$, $y = 8$

$\quad\quad 24 = k$ Simplify.

The inverse variation equation is $xy = 24$ or $y = \dfrac{24}{x}$.

Step 2 Choose values for x and y that have a product of 24.

Step 3 Plot each point and draw a smooth curve that connects the points.

x	y
−12	−2
−8	−3
−4	−6
−2	−12
0	undefined
2	12
3	8
6	4
12	2

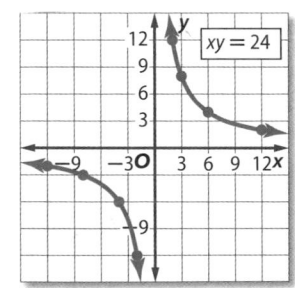

Notice that since y is undefined when $x = 0$, there is no point on the graph when $x = 0$. This graph is called a *hyperbola*.

GuidedPractice

5. Graph an inverse variation equation in which $y = 16$ when $x = 4$.

ConceptSummary Direct and Inverse Variations

Direct Variation	Inverse Variation

Direct Variation

$y = kx$ ($k > 0$)

$y = kx$ ($k < 0$)

- $y = kx$
- y varies directly as x.
- The ratio $\dfrac{y}{x}$ is a constant.

Inverse Variation

$y = \dfrac{k}{x}$ ($k > 0$)

$y = \dfrac{k}{x}$ ($k < 0$)

- $y = \dfrac{k}{x}$
- y varies inversely as x.
- The product xy is a constant.

Check Your Understanding

= Step-by-Step Solutions begin on page R13.

Example 1 Determine whether each table or equation represents an *inverse* or a *direct* variation. Explain.

1.

x	1	4	8	12
y	2	8	16	24

2.

x	1	2	3	4
y	24	12	8	6

3. $xy = 4$

4. $y = \dfrac{x}{10}$

Examples 2, 5 Assume that y varies inversely as x. Write an inverse variation equation that relates x and y. Then graph the equation.

5. $y = 8$ when $x = 6$

6. $y = 2$ when $x = 5$

7. $y = 3$ when $x = -10$

8. $y = -1$ when $x = -12$

Example 3 Solve. Assume that y varies inversely as x.

9. If $y = 8$ when $x = 4$, find x when $y = 2$.

10. If $y = 7$ when $x = 6$, find y when $x = -21$.

11. If $y = -5$ when $x = 9$, find y when $x = 6$.

Example 4 **12. RACING** The time it takes to complete a go-cart race course is inversely proportional to the average speed of the go-cart. One rider has an average speed of 73.3 feet per second and completes the course in 30 seconds. Another rider completes the course in 25 seconds. What was the average speed of the second rider?

13. OPTOMETRY When a person does not have clear vision, an optometrist can prescribe lenses to correct the condition. The power P of a lens, in a unit called diopters, is equal to 1 divided by the focal length f, in meters, of the lens.

  a. Graph the inverse variation $P = \dfrac{1}{f}$.

  b. Find the powers of lenses with focal lengths $+0.2$ to -0.4 meters.

Example 1   Determine whether each table or equation represents an *inverse* or a *direct* variation. Explain.

14.
x	y
1	30
2	15
5	6
6	5

15.
x	y
2	−6
3	−9
4	−12
5	−15

16.
x	y
−4	−2
−2	−1
2	1
4	2

17.
x	y
−5	8
−2	20
4	−10
8	−5

18. $5x - y = 0$ **19.** $xy = \dfrac{1}{4}$ **20.** $x = 14y$ **21.** $\dfrac{y}{x} = 9$

Examples 2, 5 Assume that y varies inversely as x. Write an inverse variation equation that relates x and y. Then graph the equation.

22. $y = 2$ when $x = 20$ **23.** $y = 18$ when $x = 4$ **24.** $y = -6$ when $x = -3$

25. $y = -4$ when $x = -3$ **26.** $y = -4$ when $x = 16$ **27.** $y = 12$ when $x = -9$

Example 3   Solve. Assume that y varies inversely as x.

28. If $y = 12$ when $x = 3$, find x when $y = 6$.

29. If $y = 5$ when $x = 6$, find x when $y = 2$.

30. If $y = 4$ when $x = 14$, find x when $y = -5$.

31. If $y = 9$ when $x = 9$, find y when $x = -27$.

32. If $y = 15$ when $x = -2$, find y when $x = 3$.

33. If $y = -8$ when $x = -12$, find y when $x = 10$.

Example 4   **34. EARTH SCIENCE** The water level in a river varies inversely with air temperature. When the air temperature was 90° Fahrenheit, the water level was 11 feet. If the air temperature was 110° Fahrenheit, what was the level of water in the river?

35 **MUSIC** When under equal tension, the frequency of a vibrating string in a piano varies inversely with the string length. If a string that is 420 millimeters in length vibrates at a frequency of 523 cycles a second, at what frequency will a 707-millimeter string vibrate?

Determine whether each situation is an example of an *inverse* or a *direct* variation. Justify your reasoning.

36. The drama club can afford to purchase 10 wigs at $2 each or 5 wigs at $4 each.

37. The Spring family buys several lemonades for $1.50 each.

38. Nicole earns $14 for babysitting 2 hours, and $21 for babysitting 3 hours.

39. Thirty video game tokens are divided evenly among a group of friends.

Determine whether each table or graph represents an *inverse* or a *direct* variation. Explain.

40.
x	y
5	1
8	1.6
11	2.2

41.
x	y
−3	−7
−2	−10.5
4	5.25

42.
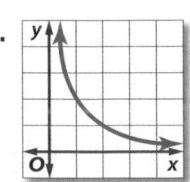

43.
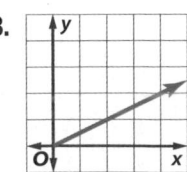

44. PHYSICAL SCIENCE When two people are balanced on a seesaw, their distances from the center of the seesaw are inversely proportional to their weights. If a 118-pound person sits 1.8 meters from the center of the seesaw, how far should a 125-pound person sit from the center to balance the seesaw?

Solve. Assume that y varies inversely as x.

45 If $y = 9.2$ when $x = 6$, find x when $y = 3$.

46. If $y = 3.8$ when $x = 1.5$, find x when $y = 0.3$.

47. If $y = \frac{1}{5}$ when $x = -20$, find y when $x = -\frac{8}{5}$.

48. If $y = -6.3$ when $x = \frac{2}{3}$, find y when $x = 8$.

49. SWIMMING Logan and Brianna each bought a pool membership. Their average cost per day is inversely proportional to the number of days that they go to the pool. Logan went to the pool 25 days for an average cost per day of $5.60. Brianna went to the pool 35 days. What was her average cost per day?

50. PHYSICAL SCIENCE The amount of force required to do a certain amount of work in moving an object is inversely proportional to the distance that the object is moved. Suppose 90 N of force is required to move an object 10 feet. Find the force needed to move another object 15 feet if the same amount of work is done.

51. DRIVING Lina must practice driving 40 hours with a parent or guardian before she is allowed to take the test to get her driver's license. She plans to practice the same number of hours each week.

   a. Let h represent the number of hours per week that she practices driving. Make a table showing the number of weeks w that she will need to practice for the following values of h: 1, 2, 4, 5, 8, and 10.

   b. Describe how the number of weeks changes as the number of hours per week increases.

   c. Write and graph an equation that shows the relationship between h and w.

H.O.T. Problems   Use Higher-Order Thinking Skills

52. CCSS CRITIQUE Christian and Trevor found an equation such that x and y vary inversely, and $y = 10$ when $x = 5$. Is either of them correct? Explain.

Christian	Trevor
$k = \dfrac{y}{x}$	$k = xy$
$= \dfrac{10}{2}$ or 5	$= (5)(10)$ or 50
$y = 5x$	$y = \dfrac{50}{x}$

53. CHALLENGE Suppose f varies inversely with g, and g varies inversely with h. What is the relationship between f and h?

54. REASONING Does $xy = -k$ represent an inverse variation when $k \neq 0$? Explain.

55. OPEN ENDED Give a real-world situation or phenomena that can be modeled by an inverse variation equation. Use the correct terminology to describe your example and explain why this situation is an inverse variation.

56. WRITING IN MATH Compare and contrast direct and inverse variation. Include a description of the relationship between slope and the graphs of a direct and inverse variation.

57. Given a constant force, the acceleration of an object varies inversely with its mass. Assume that a constant force is acting on an object with a mass of 6 pounds resulting in an acceleration of 10 ft/s². The same force acts on another object with a mass of 12 pounds. What would be the resulting acceleration?

A 4 ft/s²   **C** 6 ft/s²
B 5 ft/s²   **D** 7 ft/s²

58. Fiona had an average of 56% on her first seven tests. What would she have to make on her eighth test to average 60% on 8 tests?

F 82%   **H** 98%
G 88%   **J** 100%

59. Anthony takes a picture of a 1-meter snake beside a brick wall. When he develops the pictures, the 1-meter snake is 2 centimeters long and the wall is 4.5 centimeters high. What was the actual height of the brick wall?

A 2.25 cm
B 22.5 cm
C 225 cm
D 2250 cm

60. SHORT RESPONSE Find the area of the rectangle.

$(3 + x)$ cm
$(12 + x)$ cm

For each triangle, find sin A, cos A, and tan A to the nearest ten-thousandth. (Lesson 10-6)

61.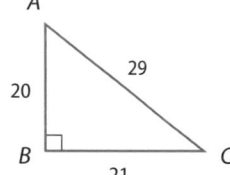

62.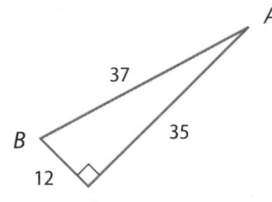

63.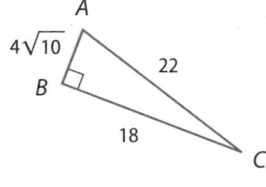

64. CRAFTS Jane is making a stained glass window using several triangular pieces of glass like the one shown. What is the length of the third side? (Lesson 10-5)

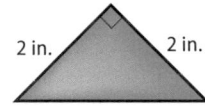

Solve each equation. (Lesson 10-4)

65. $\sqrt{10c} + 2 = 5$

66. $\sqrt{9h + 19} = 9$

67. $\sqrt{7k + 2} + 2 = 5$

68. $\sqrt{5r - 1} = r - 5$

69. $6 + \sqrt{2x + 11} = -x$

70. $4 + \sqrt{4t - 4} = t$

Simplify. Assume that no denominator is equal to zero.

71. $\dfrac{7^8}{7^6}$

72. $\dfrac{x^8 y^{12}}{x^2 y^7}$

73. $\dfrac{5pq^7}{10p^6 q^3}$

74. $\left(\dfrac{2c^3 d}{7z^2}\right)^3$

75. $\left(\dfrac{4a^2 b}{2c^3}\right)^2$

76. $y^0 (y^5)(y^{-9})$

77. $\dfrac{(4m^{-3} n^5)^0}{mn}$

78. $\dfrac{(3x^2 y^5)^0}{(21x^5 y^2)^0}$

Graphing Technology Lab
Family of Rational Functions

You can use a graphing calculator to analyze how changing the parameters a and b in $y = \dfrac{a}{x - b} + c$ affects the graphs in the family of rational functions.

Activity   Change Parameters

Graph each set of equations on the same screen in the standard viewing window. Describe any similarities and differences among the graphs.

a. $y = \dfrac{1}{x}$, $y = \dfrac{1}{x} + 2$, $y = \dfrac{1}{x} - 4$

Enter the equations in the Y= list and graph in the standard viewing window.

The graphs have the same shape. Each graph approaches the y-axis on both sides. However, the graphs have different vertical positions.

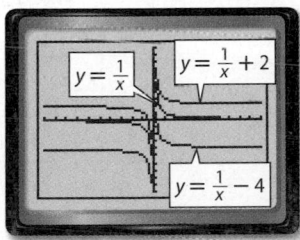

[−10, 10] scl: 1 by [−10, 10] scl: 1

b. $y = \dfrac{1}{x}$, $y = \dfrac{1}{x + 2}$, $y = \dfrac{1}{x - 4}$

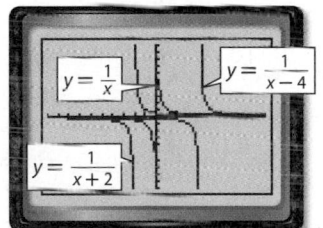

[−10, 10] scl: 1 by [−10, 10] scl: 1

The graphs have the same shape, and all approach the x-axis from both sides. However, the graphs have different horizontal positions.

c. $y = \dfrac{1}{x}$, $y = \dfrac{2}{x}$, $y = \dfrac{4}{x}$

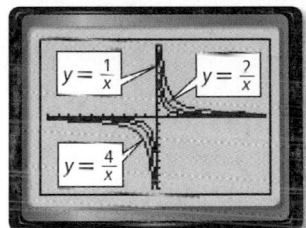

[−10, 10] scl: 1 by [−10, 10] scl: 1

The graphs all approach the x-axis and the y-axis from both sides. However, the graphs have different shapes.

Model and Analyze

1. How do a, b, and c affect the graph of $y = \dfrac{a}{x - b} + c$? Give examples.

Examine each pair of equations and predict the similarities and differences in their graphs. Use a graphing calculator to confirm your predictions. Write a sentence or two comparing the two graphs.

2. $y = \dfrac{1}{x}$, $y = \dfrac{1}{x} + 2$

3. $y = \dfrac{1}{x}$, $y = \dfrac{1}{x + 5}$

4. $y = \dfrac{1}{x}$, $y = \dfrac{3}{x}$

Rational Functions

: · Then

- You wrote inverse variation equations.

: · Now

1. Identify excluded values.
2. Identify and use asymptotes to graph rational functions.

: · Why?

- Trina is reading a 300-page book. The average number of pages she reads each day y is given by $y = \frac{300}{x}$, where x is the number of days that she reads.

 NewVocabulary
rational function
excluded value
asymptote

 Common Core State Standards

Content Standards
A.CED.2 Create equations in two or more variables to represent relationships between quantities; graph equations on coordinate axes with labels and scales.

Mathematical Practices
3 Construct viable arguments and critique the reasoning of others.
7 Look for and make use of structure.

1 **Identify Excluded Values** The function $y = \frac{300}{x}$ is an example of a **rational function**. This function is nonlinear.

KeyConcept Rational Functions

Words	Graph
A rational function can be described by an equation of the form $y = \frac{p}{q}$, where p and q are polynomials and $q \neq 0$.	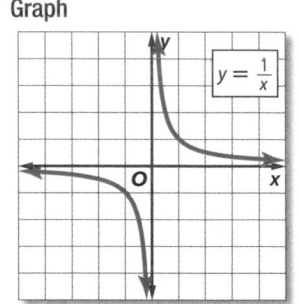 $y = \frac{1}{x}$

Parent function: $f(x) = \frac{1}{x}$

Type of graph: hyperbola

Domain: $\{x \mid x \neq 0\}$

Range: $\{y \mid y \neq 0\}$

Since division by zero is undefined, any value of a variable that results in a denominator of zero in a rational function is excluded from the domain of the function. These are called **excluded values** for the rational function.

Example 1 Find Excluded Values

State the excluded value for each function.

a. $y = -\frac{2}{x}$

The denominator cannot equal 0. So, the excluded value is $x = 0$.

b. $y = \frac{2}{x + 1}$

$x + 1 = 0$ 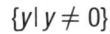 Set the denominator equal to 0.

$x = -1$

The excluded value is $x = -1$.

c. $y = \frac{5}{4x - 8}$

$4x - 8 = 0$

$4x = 8$

$x = 2$

The excluded value is $x = 2$.

GuidedPractice

1A. $y = \frac{5}{2x}$

1B. $y = \frac{x}{x - 7}$

1C. $y = \frac{4}{3x + 9}$

Depending on the real-world situation, in addition to excluding x-values that make a denominator zero from the domain of a rational function, additional values might have to be excluded from the domain as well.

Real-World Example 2 Graph Real-Life Rational Functions

BALLOONS If there are x people in the basket of a hot air balloon, the function $y = \frac{20}{x}$ represents the number of square feet y per person. Graph this function.

Since the number of people cannot be zero or less, it is reasonable to exclude negative values and only use positive values for x.

Number of People x	2	4	5	10
Square Feet per Person y	10	5	4	2

Notice that as x increases y approaches 0. This is reasonable since as the number of people increases, the space per person gets closer to 0.

▶ **Guided**Practice

2. GEOMETRY A rectangle has an area of 18 square inches. The function $\ell = \frac{18}{w}$ shows the relationship between the length and width. Graph the function.

2 **Identify and Use Asymptotes** In Example 2, an excluded value is $x = 0$. Notice that the graph approaches the vertical line $x = 0$, but never touches it.

The graph also approaches but never touches the horizontal line $y = 0$. The lines $x = 0$ and $y = 0$ are called *asymptotes*. An **asymptote** is a line that the graph of a function approaches.

🔑 KeyConcept Asymptotes

Words   A rational function in the form $y = \frac{a}{x - b} + c$, $a \neq 0$, has a vertical asymptote at the x-value that makes the denominator equal zero, $x = b$. It has a horizontal asymptote at $y = c$.

Model

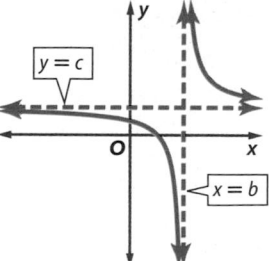

Example

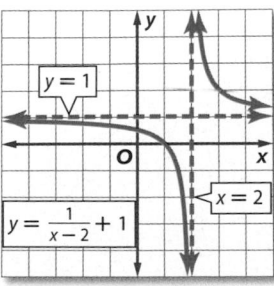

The domain of $y = \frac{a}{x - b} + c$ is all real numbers except $x = b$. The range is all real numbers except $y = c$. Rational functions cannot be traced with a pencil that never leaves the paper, so choose x-values on both sides of the vertical asymptote to graph both portions of the function.

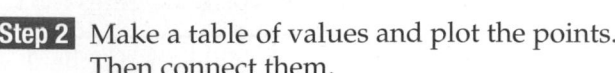

Identify the asymptotes of each function. Then graph the function.

a. $y = \frac{2}{x} - 4$

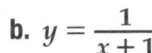 Identify and graph the asymptotes using dashed lines.

vertical asymptote: $x = 0$
horizontal asymptote: $y = -4$

Step 2 Make a table of values and plot the points. Then connect them.

x	−2	−1	1	2
y	−5	−6	−2	−3

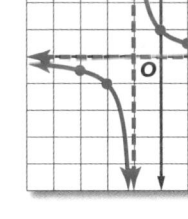

b. $y = \frac{1}{x + 1}$

Step 1 To find the vertical asymptote, find the excluded value.

$x + 1 = 0$ Set the denominator equal to 0.

$x = -1$ Subtract 1 from each side.

vertical asymptote: $x = -1$
horizontal asymptote: $y = 0$

Step 2

x	−3	−2	0	1
y	−0.5	−1	1	0.5

▶ **Guided**Practice

3A. $y = -\frac{6}{x}$ **3B.** $y = \frac{1}{x - 3}$ **3C.** $y = \frac{2}{x + 2} + 1$

Four types of nonlinear functions are shown below.

ConceptSummary **Families of Functions**

Quadratic	Exponential	Radical	Rational
Parent function: $y = x^2$	Parent function: varies	Parent function: $y = \sqrt{x}$	Parent function: $y = \frac{1}{x}$
General form: $y = ax^2 + bx + c$	General form: $y = ab^x$	General form: $y = \sqrt{x - b} + c$	General form: $y = \frac{a}{x - b} + c$

Example 1 State the excluded value for each function.

1. $y = \dfrac{5}{x}$ **2.** $y = \dfrac{1}{x+3}$ **3.** $y = \dfrac{x+2}{x-1}$ **4.** $y = \dfrac{x}{2x-8}$

Example 2 **5. PARTY PLANNING** The cost of decorations for a party is $32. This is split among a group of friends. The amount each person pays y is given by $y = \dfrac{32}{x}$, where x is the number of people. Graph the function.

Example 3 Identify the asymptotes of each function. Then graph the function.

6. $y = \dfrac{2}{x}$ **7.** $y = \dfrac{3}{x} - 1$ **8.** $y = \dfrac{1}{x-2}$

9. $y = \dfrac{-4}{x+2}$ **10.** $y = \dfrac{3}{x-1} + 2$ **11.** $y = \dfrac{2}{x+1} - 5$

Practice and Problem Solving Extra Practice is on page R11.

Example 1 State the excluded value for each function.

12. $y = \dfrac{-1}{x}$ **13.** $y = \dfrac{8}{x-8}$ **14.** $y = \dfrac{x}{x+2}$ **15.** $y = \dfrac{4}{x+6}$

16. $y = \dfrac{x+1}{x-3}$ **17.** $y = \dfrac{2x+5}{x+5}$ **18.** $y = \dfrac{7}{5x-10}$ **19.** $y = \dfrac{x}{2x+14}$

Example 2 **20. ANTELOPES** A pronghorn antelope can run 40 miles without stopping. The average speed is given by $y = \dfrac{40}{x}$, where x is the time it takes to run the distance.

a. Graph $y = \dfrac{40}{x}$.

b. Describe the asymptotes.

21. CYCLING A cyclist rides 10 miles each morning. Her average speed y is given by $y = \dfrac{10}{x}$, where x is the time it takes her to ride 10 miles. Graph the function.

Example 3 Identify the asymptotes of each function. Then graph the function.

22. $y = \dfrac{5}{x}$ **(23)** $y = \dfrac{-3}{x}$ **24.** $y = \dfrac{2}{x} + 3$

25. $y = \dfrac{1}{x} - 2$ **26.** $y = \dfrac{1}{x+3}$ **27.** $y = \dfrac{1}{x-2}$

28. $y = \dfrac{-2}{x+1}$ **29.** $y = \dfrac{4}{x-1}$ **30.** $y = \dfrac{1}{x-2} + 1$

31. $y = \dfrac{3}{x-1} - 2$ **32.** $y = \dfrac{2}{x+1} - 4$ **33.** $y = \dfrac{-1}{x+4} + 3$

34. READING Refer to the application at the beginning of the lesson.

a. Graph the function. Interpret key features of the graph in terms of the situation.

b. Choose a point on the graph, and describe what it means in the context of the situation.

35. CCSS STRUCTURE The graph shows a translation of the graph of $y = \dfrac{1}{x}$.

a. Describe the asymptotes.

b. Write a possible function for the graph.

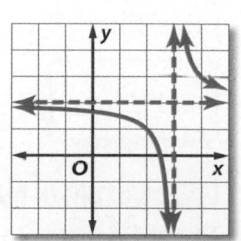

36. BIRDS A long-tailed jaeger is a sea bird that can migrate 5000 miles or more each year. The average rate in miles per hour r can be given by the function $r = \dfrac{5000}{t}$, where t is the time in hours. Use the function to determine the average rate of the bird if it spends 250 hours flying.

37. CLASS TRIP The freshmen class is going to a science museum. As part of the trip, each person is also contributing an equal amount of money to name a star.

Welcome to
The Museum

Admission $8.50
As a special memory
of your visit,
name a star for $95

  a. Write a verbal description for the cost per person.

  b. Write an equation to represent the total cost y per person if p people go to the museum.

  c. Use a graphing calculator to graph the equation. Interpret key features of the graph in terms of the situation.

  d. Estimate the number of people needed for the total cost of the trip to be about $15.

Graph each function. Identify the asymptotes.

38. $y = \dfrac{4x + 3}{2x - 4}$ **39.** $y = \dfrac{x^2}{x^2 - 1}$ **40.** $y = \dfrac{x}{x^2 - 9}$

41 GEOMETRY The equation $h = \dfrac{2(64)}{b_1 + 8}$ represents the height h of a trapezoid with an area of 64 square units. The trapezoid has two opposite sides that are parallel and h units apart; one is b_1 units long and another is 8 units long.

  a. Describe a reasonable domain and range for the function.

  b. Graph the function in the first quadrant.

  c. Use the graph to estimate the value of h when $b_1 = 10$.

H.O.T. Problems   Use Higher-Order Thinking Skills

42. CHALLENGE Graph $y = \dfrac{1}{x^2 - 4}$. State the domain and the range of the function.

43. REASONING Without graphing, describe the transformation that takes place between the graph of $y = \dfrac{1}{x}$ and the graph of $y = \dfrac{1}{x + 5} - 2$.

44. OPEN ENDED Write a rational function if the asymptotes of the graph are at $x = 3$ and $y = 1$. Explain how you found the function.

45. CCSS ARGUMENTS Is the following statement *true* or *false*? If false, give a counterexample.

    The graph of a rational function will have at least one intercept.

46. WHICH ONE DOESN'T BELONG Identify the function that does not belong with the other three. Explain your reasoning.

$$y = \dfrac{4}{x} \qquad y = \dfrac{6}{x + 1} \qquad y = \dfrac{8}{x} + 1 \qquad y = \dfrac{10}{2x}$$

47. WRITING IN MATH How are the properties of a rational function reflected in its graph?

48. Simplify $\dfrac{2a^2d}{3bc} \cdot \dfrac{9b^2c}{16ad^2}$.

A $\dfrac{abd}{c}$ **C** $\dfrac{6a}{4bd}$

B $\dfrac{ab}{d}$ **D** $\dfrac{3ab}{8d}$

49. SHORT RESPONSE One day Lola ran 100 meters in 15 seconds, 200 meters in 45 seconds, and 300 meters over low hurdles in one and a half minutes. How many more seconds did it take her to run 300 meters over low hurdles than the 200-meter dash?

50. Scott and Ian started a T-shirt printing business. The total start-up costs were $450. It costs $5.50 to print one T-shirt. Write a rational function $A(x)$ for the average cost of producing x T-shirts.

F $A(x) = \dfrac{450 + 5.5x}{x}$ **H** $A(x) = 450x + 5.5$

G $A(x) = \dfrac{450}{x} + 5.5$ **J** $A(x) = 450 + 5.5x$

51. GEOMETRY Which of the following is a quadrilateral with exactly one pair of parallel sides?

A parallelogram   **C** square

B rectangle   **D** trapezoid

Spiral Review

52. TRAVEL The Brooks family can drive to the beach, which is 220 miles away, in 4 hours if they drive 55 miles per hour. Kendra says that they would save at least a half an hour if they were to drive 65 miles per hour. Is Kendra correct? Explain. (Lesson 11-1)

Use a calculator to find the measure of each angle to the nearest degree. (Lesson 10-6)

53. $\sin C = 0.9781$ **54.** $\tan H = 0.6473$ **55.** $\cos K = 0.7658$

56. $\tan Y = 3.6541$ **57.** $\cos U = 0.5000$ **58.** $\sin N = 0.3832$

If c is the measure of the hypotenuse of a right triangle, find each missing measure. If necessary round to the nearest hundredth. (Lesson 10-5)

59. $a = 15, b = 60, c = ?$ **60.** $a = 17, c = 35, b = ?$ **61.** $a = \sqrt{110}, b = 1, c = ?$

62. $a = \sqrt{17}, b = \sqrt{12}, c = ?$ **63.** $a = 6, c = 11, b = ?$ **64.** $a = 9, b = 6, c = ?$

65. SIGHT The formula $d = \sqrt{\dfrac{3h}{2}}$ represents the distance d in miles that a person h feet high can see. Irene is standing on a cliff that is 310 feet above sea level. How far can Irene see from the cliff? Write a simplified radical expression and a decimal approximation. (Lesson 10-2)

310 ft

Skills Review

Factor each trinomial.

66. $x^2 + 11x + 24$ **67.** $w^2 + 13w - 48$ **68.** $p^2 - 2p - 35$ **69.** $72 + 27a + a^2$

70. $c^2 + 12c + 35$ **71.** $d^2 - 7d + 10$ **72.** $g^2 - 19g + 60$ **73.** $n^2 + 3n - 54$

74. $5x^2 + 27x + 10$ **75.** $24b^2 - 14b - 3$ **76.** $12a^2 - 13a - 35$ **77.** $6x^2 - 14x - 12$

Simplifying Rational Expressions

:·Then	:·Now	:·Why?
● You simplified expressions involving the quotient of monomials.	**1** Identify values excluded from the domain of a rational expression. **2** Simplify rational expressions.	● Big-O is a "hubless" Ferris wheel in Tokyo, Japan. The *centripetal force*, or the force acting toward the center, is given by $\frac{mv^2}{r}$, where m is the mass of the Ferris wheel, v is the velocity, and r is the radius.

 NewVocabulary
rational expression

 **Common Core State Standards**

Mathematical Practice
7 Look for and make use of structure.

1 **Identify Excluded Values** The expression $\frac{mv^2}{r}$ is an example of a rational expression. A **rational expression** is an algebraic fraction whose numerator and denominator are polynomials. Since division by zero is undefined, the polynomial in the denominator cannot be 0.

Example 1 Find Excluded Values

State the excluded values for each rational expression.

a. $\dfrac{-8}{r^2 - 36}$

Exclude the values for which $r^2 - 36 = 0$.

$r^2 - 36 = 0$	The denominator cannot be zero.
$(r - 6)(r + 6) = 0$	Factor.
$r - 6 = 0 \quad$ or $\quad r + 6 = 0$	Zero Product Property
$r = 6 \qquad\qquad r = -6$	Therefore, r cannot equal 6 or -6.

b. $\dfrac{n^2}{n^2 + 4n - 5}$

Exclude the values for which $n^2 + 4n - 5 = 0$.

$n^2 + 4n - 5 = 0$	The denominator cannot be zero.
$(n - 1)(n + 5) = 0$	Factor.
$n - 1 = 0 \quad$ or $\quad n + 5 = 0$	Zero Product Property
$n = 1 \qquad\qquad n = -5$	Therefore, n cannot equal 1 or -5.

▶ **GuidedPractice**

1A. $\dfrac{5x}{x^2 - 81}$

1B. $\dfrac{3a - 2}{a^2 + 6a + 8}$

Real-World Example 2 Use Rational Expressions

GEOMETRY Find the height of a cylinder that has a volume of 821 cubic inches and a radius of 7 inches. Round to the nearest tenth.

Understand You have a rational expression with two variables, V and r.

Plan Substitute 821 for V and 7 for r and simplify.

Solve $\dfrac{V}{\pi r^2} = \dfrac{821}{\pi(7)^2}$ Replace V with 821 and r with 7.

≈ 5.3 The height of the cylinder is about 5.3 inches.

Check Use estimation to determine whether the answer is reasonable.

$\dfrac{800}{3(50)} \approx 5$ ✓ The solution is reasonable.

GuidedPractice

2. Find the height of the cylinder that has a volume of 710 cubic inches and a diameter of 18 inches.

2 Simplify Expressions A rational expression is in simplest form when the numerator and denominator have no common factors except 1. To simplify a rational expression, divide out any common factors of the numerator and denominator.

KeyConcept Simplifying Rational Expressions

Words Let a, b, and c, be polynomials with $a \neq 0$ and $c \neq 0$.

Symbols $\dfrac{ba}{ca} = \dfrac{b \cdot a}{c \cdot a} = \dfrac{b}{c}$

Example $\dfrac{3x - 9}{4x - 12} = \dfrac{3(x - 3)}{4(x - 3)} = \dfrac{3}{4}$

Standardized Test Example 3 Use GCF to Simplify an Expression

Which expression is equivalent to $\dfrac{(-3x^2)(4x^5)}{9x^6}$?

A $\dfrac{4}{3}x$ **B** $\dfrac{4}{3x}$ **C** $-\dfrac{4}{3x}$ **D** $-\dfrac{4}{3}x$

Read the Test Item The expression represents the product of two monomials and the division of that product by another monomial.

Solve the Test Item

Step 1 Factor the numerator and denominator, using their GCF.

$\dfrac{(3x^6)(-4x)}{(3x^6)(3)}$

Step 2 Simplify. The correct answer is D.

$\dfrac{(\overset{1}{\cancel{3x^6}})(-4x)}{\underset{1}{\cancel{(3x^6)}}(3)}$ or $-\dfrac{4}{3}x$

GuidedPractice

3. Which expression is equivalent to $\dfrac{16c^2b^4}{8c^3b}$?

F $\dfrac{2b^3}{c}$ **G** $\dfrac{b^3}{2c}$ **H** $\dfrac{1}{2b^3c}$ **J** $2b^3c$

You can use the same procedure to simplify a rational expression in which the numerator and denominator are polynomials.

Example 4 Simplify Rational Expressions

Simplify $\dfrac{2r + 18}{r^2 + 8r - 9}$. State the excluded values of r.

$\dfrac{2r + 18}{r^2 + 8r - 9} = \dfrac{2(r + 9)}{(r + 9)(r - 1)}$ Factor.

$= \dfrac{2(\overset{1}{\cancel{r + 9}})}{\cancel{(r + 9)}(r - 1)}$ or $\dfrac{2}{r - 1}$ Divide the numerator and denominator by the GCF, $r + 9$.

Exclude the values for which $r^2 + 8r - 9$ equals 0.

$r^2 + 8r - 9 = 0$ The denominator cannot equal zero.

$(r + 9)(r - 1) = 0$ Factor.

$r = -9$ or $r = 1$ Zero Product Property

So, $r \neq -9$ and $r \neq 1$.

StudyTip

CCSS Structure Determine the excluded values using the original expression rather than the simplified expression.

▶ **Guided**Practice

Simplify each rational expression. State the excluded values of the variables.

4A. $\dfrac{n + 3}{n^2 + 10n + 21}$ **4B.** $\dfrac{y^2 + 9y - 10}{2y + 20}$

When simplifying rational expressions, look for binomials that are opposites. For example, $5 - x$ and $x - 5$ are opposites because $5 - x = -1(x - 5)$. So, you can write $\dfrac{x - 5}{5 - x}$ as $\dfrac{x - 5}{-1(x - 5)}$.

Example 5 Recognize Opposites

Simplify $\dfrac{36 - t^2}{5t - 30}$. State the excluded values of t.

$\dfrac{36 - t^2}{5t - 30} = \dfrac{(6 - t)(6 + t)}{5(t - 6)}$ Factor.

$= \dfrac{-1(t - 6)(6 + t)}{5(t - 6)}$ Rewrite $6 - t$ as $-1(t - 6)$.

$= \dfrac{-1(\overset{1}{\cancel{t - 6}})(6 + t)}{5\cancel{(t - 6)}}$ or $-\dfrac{6 + t}{5}$ Divide out the common factor, $t - 6$.

Exclude the values for which $5t - 30$ equals 0.

$5t - 30 = 0$ The denominator cannot equal zero.

$5t = 30$ Add 30 to each side.

$t = 6$ Divide each side by 5.

So, $t \neq 6$.

▶ **Guided**Practice

Simplify each expression. State the excluded values of x.

5A. $\dfrac{12x + 36}{x^2 - x - 12}$ **5B.** $\dfrac{x^2 - 2x - 35}{x^2 - 9x + 14}$

Recall that to find the zeros of a quadratic function, you need to find the values of x when $f(x) = 0$. The zeros of a rational function are found in the same way.

Example 6 Rational Functions

Find the zeros of $f(x) = \dfrac{x^2 + 3x - 18}{x - 3}$.

$$f(x) = \dfrac{x^2 + 3x - 18}{x - 3} \qquad \text{Original function}$$

$$0 = \dfrac{x^2 + 3x - 18}{x - 3} \qquad f(x) = 0$$

$$0 = \dfrac{(x + 6)(x - 3)}{x - 3} \qquad \text{Factor.}$$

$$0 = \dfrac{(x + 6)(x \overset{1}{\cancel{- 3}})}{\underset{1}{\cancel{x - 3}}} \qquad \text{Divide out common factors.}$$

$$0 = x + 6 \qquad \text{Simplify.}$$

When $x = -6$, the numerator becomes 0, so $f(x) = 0$. Therefore, the zero of the function is -6.

▶ **Guided**Practice

Find the zeros of each function.

6A. $f(x) = \dfrac{x^2 + 2x - 15}{x + 1}$

6B. $f(x) = \dfrac{x^2 + 6x + 8}{x^2 + x - 2}$

Check Your Understanding

 = Step-by-Step Solutions begin on page R13.

Example 1 **State the excluded values for each rational expression.**

    1. $\dfrac{8}{x^2 - 16}$

    2. $\dfrac{3m}{m^2 - 6m + 5}$

Example 2 **3. PHYSICAL SCIENCE** A 0.16-kilogram ball attached to a string is being spun in a circle 7.26 meters per second. The expression $\dfrac{mv^2}{r}$, where m is the mass of the ball, v is the velocity, and r is the radius, can be used to find the force that keeps the ball spinning in a circle. If the circle has a radius of 0.5 meter, find the force that must be exerted to keep the ball spinning. Round to the nearest tenth.

Examples 3–5 **Simplify each expression. State the excluded values of the variables.**

    4. $\dfrac{28ab^3}{16a^2b}$

    5 $\dfrac{(-3r)(10r^4)}{6r^5}$

    6. $\dfrac{5d + 15}{d^2 - d - 12}$

    7. $\dfrac{x^2 + 11x + 28}{x + 4}$

    8. $\dfrac{2r - 12}{r^2 - 36}$

    9. $\dfrac{3y - 27}{81 - y^2}$

Example 6 **Find the zeros of each function.**

    10. $f(x) = \dfrac{x^2 - x - 12}{x - 2}$

    11. $f(x) = \dfrac{x^2 - x - 6}{x^2 + 8x + 12}$

Example 1   State the excluded values for each rational expression.

12. $\dfrac{-n}{n^2 - 49}$

13. $\dfrac{5x + 1}{x^2 - 1}$

14. $\dfrac{12a}{a^2 - 3a - 10}$

15. $\dfrac{k^2 - 4}{k^2 + 5k - 24}$

Example 2

16. **GEOMETRY** The volume of a rectangular prism is $3x^3 + 34x^2 + 72x - 64$. If the height is $x + 4$, what is the area of the base of the prism?

17. **GEOMETRY** Use the circle at the right to write the ratio $\dfrac{\text{circumference}}{\text{area}}$. Then simplify. State the excluded value of the variable.

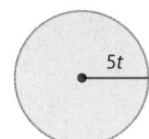

5t

Examples 3–5 Simplify each expression. State the excluded values of the variables.

18. $\dfrac{15x^4y^2}{40x^3y^3}$

19. $\dfrac{32n^2p}{2n^4p}$

20. $\dfrac{(4t^3)(2t)}{20t^2}$

21. $\dfrac{(7c^2)(-6c^3)}{21c^4}$

22. $\dfrac{4x - 24}{x^2 - 12x + 36}$

23. $\dfrac{a^2 + 3a}{a^2 - 3a - 18}$

24. $\dfrac{n^2 + 7n - 18}{n - 2}$

25. $\dfrac{x^2 + 4x - 32}{x + 8}$

26. $\dfrac{x^2 - 25}{x^2 + 5x}$

27. $\dfrac{2p^2 - 14p}{p^2 - 49}$

28. $\dfrac{2x - 10}{25 - x^2}$

29. $\dfrac{64 - c^2}{c^2 - 7c - 8}$

Example 6   Find the zeros of each function.

30. $f(x) = \dfrac{x^2 - x - 12}{x^2 + 2x - 35}$

31 $f(x) = \dfrac{x^2 + 3x - 4}{x^2 + 9x + 20}$

32. $f(x) = \dfrac{2x^2 + 11x - 40}{2x + 5}$

33. $f(x) = \dfrac{3x^2 - 18x + 24}{x - 6}$

34. $f(x) = \dfrac{x^3 + x^2 - 6x}{x - 1}$

35. $f(x) = \dfrac{x^3 - 4x^2 - 12x}{x + 2}$

36. **PYRAMIDS** The perimeter of the base of the Pyramid of the Sun is 4π times the height. The perimeter of the base of the Great Pyramid of Giza is 2π times the height. Write and simplify each ratio comparing the base perimeters.

Pyramid	Height (ft)
Pyramid of the Sun (Mexico)	233.5
Great Pyramid (Egypt)	481.4

   a. Pyramid of the Sun to the Great Pyramid

   b. Great Pyramid to the Pyramid of the Sun

37. **CCSS REASONING** George Ferris built the first Ferris wheel for the World's Columbian Exposition in 1893. It had a diameter of 250 feet.

   a. To find the speed traveled by a car located on the wheel, you can find the circumference of a circle and divide by the time it takes for one rotation. Write a rational expression for the speed of a car rotating in time t.

   b. Suppose the first Ferris wheel rotated once every 5 minutes. What was the speed of a car on the circumference in feet per minute?

   Simplify each expression. State the excluded values of the variables.

38. $\dfrac{3a^2b^4 + 9a^3b - 6a^5b}{3a^2b}$

39. $\dfrac{8x^5 - 10xy^2}{2xy^3}$

40. $\dfrac{x + 5}{3x^2 + 14x - 5}$

41 **PACKAGING** To minimize packaging expenses, a company uses packages that have the least surface area to volume ratio. For each figure, write a ratio comparing the surface area to the volume. Then simplify. State the excluded values of the variables.

a.

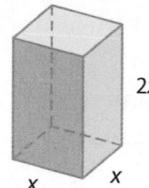

2x

x   x

b.

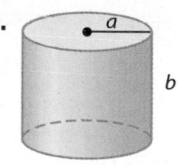

a

b

42. HISTORY The diagram shows how a lever may have been used to move blocks.

a. The mechanical advantage of a lever is $\dfrac{L_A}{L_R}$, where L_A is the length of the effort arm and L_R is the length of the resistance arm. Find the mechanical advantage of the lever shown.

b. The force placed on the rock is the product of the mechanical advantage and the force applied to the end of the lever. If the Egyptian worker can apply a force of 180 pounds, what is the greatest weight he can lift with the lever?

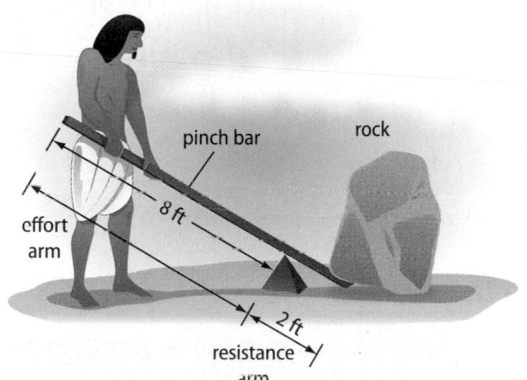

pinch bar   rock

8 ft

effort arm

2 ft

resistance arm

c. To lift a 535-pound rock using a 7-foot lever with the fulcrum 2 feet from the rock, how much force will have to be used?

H.O.T. Problems   Use Higher-Order Thinking Skills

43. ERROR ANALYSIS Colleen and Sanson examined $\dfrac{12x + 36}{x^2 - x - 12}$ and found the excluded value(s). Is either of them correct? Explain.

Colleen	Sanson
$\dfrac{12x + 36}{x^2 - x - 12} - \dfrac{12(x + 3)}{(x - 4)(x + 3)}$ The excluded values are 4 and -3.	$\dfrac{12x + 36}{x^2 - x - 12} = \dfrac{12(x + 3)}{(x - 4)(x + 3)}$ $= \dfrac{12\cancel{(x + 3)}}{(x - 4)\cancel{(x + 3)}}$ $= \dfrac{12}{x - 4}$ The excluded value is 4.

44. **CCSS** **PRECISION** Compare and contrast the key features of the graphs of $y = x - 2$ and $y = \dfrac{x^2 + 5x - 14}{x + 7}$.

45. REASONING Explain why every polynomial is also a rational expression.

46. OPEN ENDED Write a rational expression with excluded values -2 and 2. Explain how you found the expression.

47. REASONING Is $\dfrac{2x^2 - 4x}{x - 2}$ in simplest form? Justify your answer.

48. WRITING IN MATH List the steps you would use to simplify $\dfrac{x^2 + x - 20}{x + 5}$. State the excluded value.

49. Simplify $\dfrac{2x + 4}{2}$.

   A $x + 1$

   B x

   C $x + 2$

   D $\dfrac{x}{2}$

51. GEOMETRY What is the name of the figure?

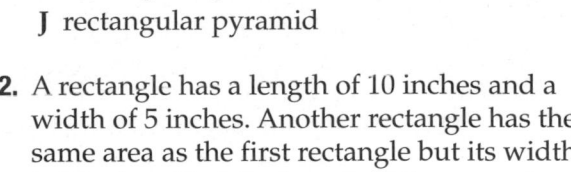

   F triangular pyramid

   G triangular prism

   H rectangular prism

   J rectangular pyramid

50. SHORT RESPONSE Shiro is buying a car for $5800. He can pay the full amount in cash, or he can pay $1000 down and $230 a month for 24 months. How much more would he pay for the car on the second plan?

52. A rectangle has a length of 10 inches and a width of 5 inches. Another rectangle has the same area as the first rectangle but its width is 2 inches. Find the length of the second rectangle.

   A 60 in.   **C** 30 in.

   B 45 in.   **D** 25 in.

Spiral Review

State the excluded value for each function. (Lesson 11-2)

53. $y = \dfrac{6}{x}$

54. $y = \dfrac{2}{x - 5}$

55. $y = \dfrac{x - 4}{x - 3}$

56. $y = \dfrac{3x}{2x + 6}$

Solve. Assume that y varies inversely as x. (Lesson 11-1)

57. If $y = 10$ when $x = 4$, find x when $y = 2$.

58. If $y = 12$ when $x = 3$, find x when $y = 6$.

59. If $y = -5$ when $x = 3$, find x when $y = -3$.

60. If $y = 21$ when $x = -6$, find x when $y = 7$.

61. CRAFTS Melinda is working on a quilt using the pattern shown. She has several right triangular pieces of material with two sides that measure 6 inches. What is the length of the third side? (Lesson 10-5)

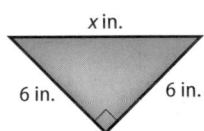

Simplify. (Lesson 10-2)

62. $\sqrt{20}$

63. $\sqrt{18}$

64. $\sqrt{2} \cdot \sqrt{8}$

65. $2\sqrt{32}$

66. $\sqrt{5} \cdot \sqrt{6}$

67. $\sqrt{40a^2}$

68. $\sqrt{\dfrac{t}{8}}$

69. $\sqrt{\dfrac{2}{7}} \cdot \sqrt{\dfrac{7}{3}}$

70. FINANCIAL LITERACY Determine the amount of an investment if $250 is invested at an interest rate of 7.3% compounded quarterly for 40 years. (Lesson 9-3)

Skills Review

Find the greatest common factor for each set of monomials.

71. $2x, 8x^2$

72. $3y^2, 7y^3$

73. $7g, 10h$

74. $21c^2d^3, 14cd^2$

75. $9qt^2, 18q^2t^2, 27qt$

76. $10ab, 25a^2b^2, 30a^2b$

Graphing Technology Lab
Simplifying Rational Expressions

When simplifying rational expressions, you can use a graphing calculator to support your answer. If the graphs of the original expression and the simplified expression overlap, they are equivalent. You can also use the graphs to see excluded values.

Activity   Simplify a Rational Expression

Simplify $\dfrac{x^2 - 16}{x^2 + 8x + 16}$.

Step 1 Factor the numerator and denominator.

$$\frac{x^2 - 16}{x^2 + 8x + 16} = \frac{(x - 4)(x + 4)}{(x + 4)(x + 4)}$$

$$= \frac{(x - 4)}{(x + 4)}$$

When $x = -4$, $x + 4 = 0$. Therefore, x cannot equal -4 because you cannot divide by zero.

Step 2 Graph the original expression.

- Set the calculator to Dot mode.
- Enter $\dfrac{x^2 - 16}{x^2 + 8x + 16}$ as **Y₁** and graph in the standard viewing window.

KEYSTROKES:

```
MODE ▼ ▼ ▼ ▼ ▶
ENTER Y= ( X,T,θ,n x²
− 16 ) ÷ ( X,T,θ,n
x² + 8 X,T,θ,n + 16 )
ZOOM 6
```

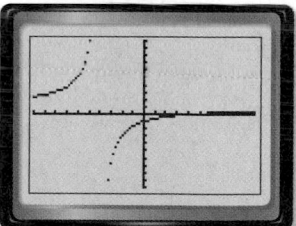

[−10, 10] scl: 1 by [−10, 10] scl: 1

Step 3 Graph the simplified expression.

Enter $\dfrac{(x - 4)}{(x + 4)}$ as **Y₂** and graph.

KEYSTROKES: Y= ▼ (X,T,θ,n − 4)
÷ (X,T,θ,n + 4)
GRAPH

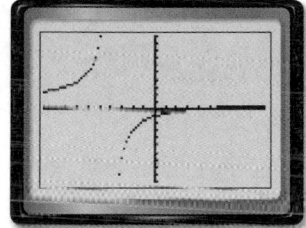

[−10, 10] scl: 1 by [−10, 10] scl: 1

Since the graphs overlap, the two expressions are equivalent.

Exercises

Simplify each expression. Then verify your answer graphically. Name the excluded values.

1. $\dfrac{5x + 15}{x^2 + 10x + 21}$

2. $\dfrac{x^2 - 8x + 12}{x^2 + 7x - 18}$

3. $\dfrac{2x^2 + 6x + 4}{3x^2 + 9x + 6}$

4. a. Simplify $\dfrac{3x - 8}{6x^2 - 16x}$.

   b. How can you use the **TABLE** function to verify that the original expression and the simplified expression are equivalent?

   c. How does the **TABLE** function show you that an x-value is an excluded value?

Multiplying and Dividing Rational Expressions

:: Then	:: Now	:: Why?
● You multiplied polynomials and divided monomials.	**1** Multiply rational expressions. **2** Divide rational expressions.	● A recent survey showed 10- to 17-year olds talk or text on their cell phones an average of 3.75 hours per day during the summer. The expression below can be used to find the average number of minutes youth talk or text on their phones during summer, approximately 90 days.

$$90 \; \cancel{\text{days}} \cdot \frac{3.75 \; \cancel{\text{hours}}}{\cancel{\text{day}}} \cdot \frac{60 \text{ minutes}}{1 \; \cancel{\text{hour}}} = 20{,}250 \text{ minutes}$$

CCSS Common Core State Standards

Mathematical Practices
1 Make sense of problems and persevere in solving them.

1 Multiply Rational Expressions
To multiply fractions, you multiply numerators and multiply denominators. Use this same method to multiply rational expressions.

> **KeyConcept** Multiplying Rational Expressions
>
> **Words** Let a, b, c, and d be polynomials with $b \neq 0$ and $d \neq 0$. Then, $\dfrac{a}{b} \cdot \dfrac{c}{d} = \dfrac{ac}{bd}$.
>
> **Example** $\dfrac{x}{2x-3} \cdot \dfrac{4x^2}{5} = \dfrac{4x^3}{5(2x-3)}$

Example 1 Multiply Expressions Involving Monomials

Find each product.

a. $\dfrac{r^2 x}{9t^3} \cdot \dfrac{3t^4}{r}$

Divide by the common factors before multiplying.

$$\dfrac{r^2 x}{9t^3} \cdot \dfrac{3t^4}{r} = \dfrac{\cancel{r^2}^{\,r} x}{\cancel{9t^3}_{3\;1}} \cdot \dfrac{\cancel{3t^4}^{1\;t}}{\cancel{r}_{1}}$$ Divide by the common factors 3, r, and t^3.

$$= \dfrac{rxt}{3}$$ Simplify.

b. $\dfrac{a+4}{a^2} \cdot \dfrac{a}{a^2 + 2a - 8}$

$$\dfrac{a+4}{a^2} \cdot \dfrac{a}{a^2 + 2a - 8} = \dfrac{a+4}{a^2} \cdot \dfrac{a}{(a+4)(a-2)}$$ Factor the denominator.

$$= \dfrac{\cancel{a+4}^{1}}{\cancel{a^2}_{a}} \cdot \dfrac{\cancel{a}^{1}}{\cancel{(a+4)}_{1}(a-2)}$$ The GCF is $a(a+4)$.

$$= \dfrac{1}{a(a-2)} \text{ or } \dfrac{1}{a^2 - 2a}$$ Simplify.

▶ **Guided Practice**

1A. $\dfrac{3x}{16x^2} \cdot \dfrac{8x^2}{3}$

1B. $\dfrac{x+3}{x} \cdot \dfrac{5}{x^2 + 7x + 12}$

1C. $\dfrac{y^2 - 3y - 4}{y+5} \cdot \dfrac{y+5}{y^2 - 4y}$

When you multiply fractions that involve units of measure, you can divide by the units in the same way that you divide by variables. Recall that this process is called *dimensional analysis*. You can use dimensional analysis to convert units of measure within a system and between systems.

Real-World Example 2 Dimensional Analysis

SKI RACING Ann Proctor won the 2007 World Waterski Racing Championship race in her category when she finished the 88-kilometer course in 51.23 minutes. What was her average speed in miles per hour? (*Hint*: 1 km ≈ 0.62 mi)

$$\frac{88 \text{ km}}{51.23 \text{ min}} \cdot \frac{0.62 \text{ mi}}{1 \text{ km}} \cdot \frac{60 \text{ min}}{1 \text{ h}} = \frac{88 \cancel{\text{ km}}}{51.23 \cancel{\text{ min}}} \cdot \frac{0.62 \text{ mi}}{1 \cancel{\text{ km}}} \cdot \frac{60 \cancel{\text{ min}}}{1 \text{ h}}$$

$$= \frac{88 \cdot 0.62 \text{ mi} \cdot 60}{51.23 \cdot 1 \cdot 1 \text{ h}} \qquad \text{Simplify.}$$

$$= \frac{3273.6 \text{ mi}}{51.23 \text{ h}} \qquad \text{Multiply.}$$

$$\approx \frac{63.9 \text{ mi}}{\text{h}} \qquad \text{Divide the numerator and the denominator by 51.23.}$$

Her average speed was 63.9 miles per hour.

▶ **Guided**Practice

2. SKI RACING What was Ann Proctor's speed in feet per second?

2 Divide Rational Expressions

To divide by a fraction, you multiply by the reciprocal. You can use this same method to divide by a rational expression.

KeyConcept Dividing Rational Expressions

Symbols Let a, b, c, and d be polynomials with $b \neq 0$, $c \neq 0$, and $d \neq 0$. Then, $\dfrac{a}{b} \div \dfrac{c}{d} = \dfrac{a}{b} \cdot \dfrac{d}{c} = \dfrac{ad}{bc}$.

Example $\dfrac{x-3}{x} \div \dfrac{2x^2}{5} = \dfrac{x-3}{x} \cdot \dfrac{5}{2x^2} = \dfrac{5(x-3)}{2x^3}$

Example 3 Divide by a Rational Expression

Find $\dfrac{4}{15n^3} \div \dfrac{12}{25n}$.

$$\frac{4}{15n^3} \div \frac{12}{25n} = \frac{4}{15n^3} \cdot \frac{25n}{12} \qquad \text{Multiply by } \tfrac{25n}{12}, \text{ the reciprocal of } \tfrac{12}{25n}.$$

$$= \frac{\overset{1}{\cancel{4}}}{\underset{3n^2}{\cancel{15n^3}}} \cdot \frac{\overset{5}{\cancel{25n}}}{\underset{3}{\cancel{12}}} \qquad \text{Divide by common factors 4, 5, and } n.$$

$$= \frac{5}{9n^2} \qquad \text{Simplify.}$$

▶ **Guided**Practice

Find each quotient.

3A. $\dfrac{15y^2}{4x} \div \dfrac{5y}{8x^3}$

3B. $\dfrac{12a^2}{5b} \div \dfrac{25a}{6b^2}$

Example 4 Divide by Rational Expressions and Polynomials

Find each quotient.

a. $\dfrac{2x + 6}{x^2} \div (x + 3)$

$$\dfrac{2x + 6}{x^2} \div (x + 3) = \dfrac{2x + 6}{x^2} \div \dfrac{x + 3}{1}$$ Write the binomial as a fraction.

$$= \dfrac{2x + 6}{x^2} \cdot \dfrac{1}{x + 3}$$ Multiply by the reciprocal of $x + 3$.

$$= \dfrac{2(x + 3)}{x^2} \cdot \dfrac{1}{x + 3}$$ Factor $4x + 6$.

$$= \dfrac{2(\overset{1}{\cancel{x + 3}})}{x^2} \cdot \dfrac{1}{\underset{1}{\cancel{x + 3}}} \text{ or } \dfrac{2}{x^2}$$ Divide out the common factor and simplify.

b. $\dfrac{a - 2}{4a + 4} \div \dfrac{a + 5}{a + 1}$

$$\dfrac{a - 2}{4a + 4} \div \dfrac{a + 5}{a + 1} = \dfrac{a - 2}{4a + 4} \cdot \dfrac{a + 1}{a + 5}$$ Multiply by the reciprocal.

$$= \dfrac{a - 2}{4(a + 1)} \cdot \dfrac{\overset{1}{\cancel{a + 1}}}{a + 5}$$ Factor $4a + 4$.

$$= \dfrac{a - 2}{4(a + 5)}$$ The GCF is $a + 1$ and simplify.

▶ **Guided**Practice

4A. $\dfrac{4d - 8}{2d - 6} \div \dfrac{2d - 4}{d - 4}$ **4B.** $\dfrac{b + 4}{3b + 2} \div \dfrac{3b + 12}{b + 1}$

Sometimes you must factor a quadratic expression before you can simplify the quotient of rational expressions.

Example 5 Expression Involving Polynomials

Find $\dfrac{y - 3}{y^2 - 10y + 16} \div \dfrac{y^2 - 9}{y - 8}$.

$$\dfrac{y - 3}{y^2 - 10y + 16} \div \dfrac{y^2 - 9}{y - 8}$$

$$= \dfrac{y - 3}{y^2 - 10y + 16} \cdot \dfrac{y - 8}{y^2 - 9}$$ Multiply by the reciprocal, $\dfrac{y - 8}{y^2 - 9}$.

$$= \dfrac{y - 3}{(y - 2)(y - 8)} \cdot \dfrac{y - 8}{(y - 3)(y + 3)}$$ Factor $y^2 - 10y + 16$ and $y^2 - 9$.

$$= \dfrac{\overset{1}{\cancel{y - 3}}}{(y - 2)\underset{1}{\cancel{(y - 8)}}} \cdot \dfrac{\overset{1}{\cancel{y - 8}}}{\underset{1}{\cancel{(y - 3)}}(y + 3)}$$ The GCF is $(y - 3)(y - 8)$.

$$= \dfrac{1}{(y - 2)(y + 3)}$$ Simplify.

▶ **Guided**Practice

Find each quotient.

5A. $\dfrac{p^2 - 4}{5p} \div \dfrac{p - 2}{p + q}$ **5B.** $\dfrac{q^2 + 3q + 2}{12} \div \dfrac{q + 1}{q^2 + 4}$

Check Your Understanding ◯ = Step-by-Step Solutions begin on page R13.

Example 1 Find each product.

1. $\dfrac{2x^3}{7x} \cdot \dfrac{14}{x}$

2. $\dfrac{3ab}{4c^4} \cdot \dfrac{16c^2}{9b}$

3. $\dfrac{t^2}{(t-5)(t+5)} \cdot \dfrac{t+5}{6t}$

4. $\dfrac{8}{r+1} \cdot \dfrac{r^2-1}{2}$

Example 2

5. **CCSS PRECISION** The slowest land mammal is the three-toed sloth. It travels 0.07 mile per hour on the ground. What is this speed in feet per minute?

6. **EXERCISE** One hour of moderate inline skating burns approximately 330 Calories. If Nelia plans to do inline skating for 3 hours a week, how many Calories will she burn in a year from the skating?

Examples 3–5 Find each quotient.

7. $\dfrac{8}{3x^2} \div \dfrac{4}{x}$

8. $\dfrac{c^5}{2} \div \dfrac{c^3}{6d^2}$

9. $\dfrac{b^2+6b+5}{6b+6} \div (b+5)$

10. $\dfrac{2x+8}{x+3} \div \dfrac{x+4}{x^2+6x+9}$

Practice and Problem Solving Extra Practice is on page R11.

Example 1 Find each product.

11. $\dfrac{10n^2}{4} \cdot \dfrac{2}{n}$

12. $\dfrac{12c^3}{21b} \cdot \dfrac{14b^2}{6c}$

13. $\dfrac{x^5y}{2z^3} \cdot \dfrac{18z^4}{xy}$

14. $\dfrac{5c^3d}{c^4d} \cdot \dfrac{f^2d^3c}{10cf^4}$

15. $\dfrac{9}{t-2} \cdot \dfrac{(t+2)(t-2)}{3}$

16. $\dfrac{(a+4)(a-5)}{a^2} \cdot \dfrac{6a}{a+4}$

17. $\dfrac{(k+6)(k-1)}{k+2} \cdot \dfrac{(k+1)(k+2)}{(k+1)(k-1)}$

18. $\dfrac{(r-8)(r+3)}{r} \cdot \dfrac{2r}{(r+8)(r+3)}$

19. $\dfrac{n^2+n-2}{n+2} \cdot \dfrac{4n}{n-1}$

20. $\dfrac{y^2-1}{y^2-49} \cdot \dfrac{y-7}{y+1}$

Example 2

21. **FINANCIAL LITERACY** A scarf bought in Italy cost 18 Euros. The exchange rate at the time was 1 U.S. dollar = 0.73 Euro.
   a. How much did the scarf cost in U.S. dollars?
   b. If the exchange rate at the time was 1 Canadian dollar = 0.69 Euro, how much did the scarf cost in Canadian dollars?

22. **ROLLER COASTERS** A roller coaster has 6 trains. Each train has 3 cars, and each car seats 4 people. Write and simplify an expression including units to find the total number of people that can ride the roller coaster at one time.

Examples 3–5 Find each quotient.

23. $\dfrac{x^5}{y} \div \dfrac{x}{y^2}$

24. $\dfrac{3r^4}{k^2} \div \dfrac{18r^3}{k}$

25. $\dfrac{21b^3}{4c^2} \div \dfrac{7}{6c^2}$

26. $\dfrac{f^4g^2h}{x^2y} \div f^3g$

27. $\dfrac{6b-12}{b+5} \div (12b+18)$

28. $\dfrac{k+3}{k+2} \div \dfrac{k}{5k+10}$

29. $\dfrac{5x^2}{x^2-5x+4} \div \dfrac{10x}{x-1}$

30. $\dfrac{n^2+7n+12}{16n^2} \div \dfrac{n+3}{2n}$

31. $\dfrac{r+2}{r+1} \div \dfrac{4}{r^2+3r+2}$

32. $\dfrac{3a}{a^2+2a+1} \div \dfrac{a-1}{a+1}$

33. BEARS A grizzly bear runs 110 feet in 5 seconds. What is the average speed of the bear in miles per hour?

34. SEWING The fabric that Megan wants to buy for a costume she is making costs $7.50 per yard. To the nearest tenth, how many meters can she buy with $24? (*Hint*: 1 inch = 2.54 centimeters)

35. TRAVEL A plane flies 1250 kilometers at an average speed of 540 miles per hour.

    a. Write a product to find the number of hours it took for the plane to make the trip. Include the units. (*Hint*: 1 mile = 1.609 kilometers).

    b. Find the number of hours it took for the plane to make the trip. Round to the nearest tenth.

36. VOLUNTEERING Tyree is passing out cups of orange drink from a 3.5-gallon cooler. If each cup of orange drink is 4.25 ounces, about how many cups can he hand out? (*Hint*: There are 128 ounces in a gallon.)

37. LAND Louisiana loses about 30 square miles of land each year to coastal erosion, hurricanes, and other natural and human causes. Approximately how many square yards of land are lost per month? (*Hint*: Use 1 square mile = 3,097,600 square yards.)

38. GEOMETRY Write an expression to represent the length of the rectangle.

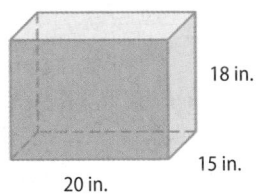

$A = x^2 + 2x - 24$ $\dfrac{x-4}{x-3}$

Convert each rate. Round to the nearest tenth.

39. 46 feet per second to miles per hour

40. 29.5 meters per second to kilometers per hour

41. 28 milliliters per second to cups per minute. (*Hint*: 1 liter ≈ 0.908 quart)

42. 32.4 meters per second to miles per hour. (*Hint*: 1 mile ≈ 1.609 kilometers)

43. LIFE SCIENCE A human heart pumps about a cup of blood each time it beats. On average, a person's heart beats about 70 times a minute. Write and simplify an expression to find how many gallons of blood are pumped per hour.

44. GEOMETRY Refer to the prism at the right.

    a. Find the volume in cubic inches.

    b. Use the ratio $\dfrac{1 \text{ foot}^3}{1728 \text{ inches}^3}$ to write a multiplication expression to convert the volume to cubic feet. Then convert the volume.

18 in.

20 in.

15 in.

CCSS PRECISION Find each product. Describe what the final answer represents.

45. $\dfrac{\$9.80}{1 \text{ hour}} \cdot \dfrac{15 \text{ hours}}{1 \text{ week}} \cdot \dfrac{52 \text{ weeks}}{1 \text{ year}}$

46. $\dfrac{\$2.85}{1 \text{ gallon of gasoline}} \cdot \dfrac{15 \text{ gallons of gasoline}}{1 \text{ fill-up}} \cdot \dfrac{3 \text{ fill-ups}}{1 \text{ month}} \cdot \dfrac{1 \text{ month}}{30 \text{ days}}$

47. $\dfrac{32 \text{ meters}}{1 \text{ second}} \cdot \dfrac{60 \text{ seconds}}{1 \text{ minute}} \cdot \dfrac{60 \text{ minutes}}{1 \text{ hour}} \cdot \dfrac{1 \text{ kilometer}}{1000 \text{ meters}} \cdot \dfrac{1 \text{ mile}}{1.609 \text{ kilometers}}$

48. $\dfrac{\$32,000}{1 \text{ year}} \cdot \dfrac{1 \text{ year}}{52 \text{ weeks}} \cdot \dfrac{1 \text{ week}}{40 \text{ hours}}$

49. SPACE The highest speed at which any spacecraft has ever escaped from Earth is 35,800 miles per hour by the *New Horizons* probe, which was launched in 2006. Convert this speed to feet per second. Round to the nearest tenth.

50. ELECTRICITY Simplify the expression below to find the cost of running a 3500-watt air conditioner for one week.

$$3500 \text{ watts} \cdot \frac{1 \text{ kilowatt}}{1000 \text{ watts}} \cdot \frac{168 \text{ hours}}{1 \text{ week}} \cdot \frac{10 \text{ cents}}{1 \text{ kilowatt} \cdot \text{hours}} \cdot \frac{1 \text{ dollar}}{100 \text{ cents}}$$

51. AMUSEMENT PARKS In a ride, riders stand along the wall of a circular room with a radius of 3.1 meters. The room completes 27 rotations per minute.

a. Write an expression for the number of meters the room moves per second.

b. Simplify the expression you wrote in Part **a** and describe what it means.

52. AQUARIUMS An aquarium is a rectangular prism 30 inches long, 15 inches wide, and 18 inches high.

a. Sketch and label a diagram of the aquarium. Then find its volume in cubic inches.

b. Describe how to use the ratio $\dfrac{1 \text{ ft}^3}{1728 \text{ in}^3}$ to find the volume of the tank in cubic feet. Then find the volume. Round to the nearest tenth.

c. Pure water weighs 62 pounds per cubic foot. How much would the water in the tank weigh if the tank were filled?

d. A saltwater aquarium requires 2 pounds of minerals for each cubic foot of water. Find the percent concentration by weight of the saltwater by dividing the weight of the minerals by the sum of the weights of the minerals and the water.

e. Kim accidentally added twice the required minerals. Find the percent concentration by weight. Is it twice the recommended concentration? Explain.

H.O.T. Problems Use Higher-Order Thinking Skills

53. ERROR ANALYSIS Mei and Tamika are finding $\dfrac{2x+6}{x+5} \div \dfrac{2}{x+5}$. Is either of them correct? Explain.

54. REASONING Find the missing term in $\underline{\ \ ?\ \ } \div \dfrac{10x^3}{21} = \dfrac{3}{2x}$. Justify your answer.

55. CCSS PERSEVERANCE Find $\dfrac{x^2 - 3x - 10}{x^2 + 2x - 35} \cdot \dfrac{x^2 + 4x - 21}{x^2 + 9x + 14}$. Write in simplest form.

56. WRITING IN MATH Give an example and describe how you could use dimensional analysis to solve a real-world problem involving rational expressions.

57. OPEN ENDED Give an example of a real-world situation that could be modeled by the quotient of two rational expressions. Provide an example of this quotient.

58. WRITE A QUESTION A classmate found that the product of two rational expressions is $\dfrac{9x - 3}{(x+3)(3x+1)}$. Write a question to help her find the excluded values.

59. WRITING IN MATH Describe how to use dimensional analysis to find the number of hours in one year.

60. GEOMETRY The perimeter of a rectangle is 30 inches. Its area is 54 square inches. Find the length of the longest side.

A 6 inches

B 9 inches

C 12 inches

D 30 inches

61. Find $\dfrac{c^2 - c - 6}{2c - 10} \div \dfrac{2c + 4}{3c - 15}$.

F $\dfrac{3(c - 3)}{4}$

H $\dfrac{4(c - 3)}{3}$

G $\dfrac{c + 5}{c - 3}$

J $\dfrac{c - 3}{c - 5}$

62. EXTENDED RESPONSE The weekly salaries of six employees at a fast food restaurant are $140, $220, $90, $180, $140, $200.

a. What is the mean of the six salaries?

b. What is the median of the six salaries?

c. What is the mode of the six salaries?

63. Tito has three times as many CDs as Dasan. Dasan has two thirds as many CDs as Brant. Brant has 27 CDs. How many CDs does Tito have?

A 54

C 27

B 32

D 18

Spiral Review

Simplify each expression. State the excluded values of the variables. (Lesson 11-3)

64. $\dfrac{20x^2y}{25xy}$

65. $\dfrac{14g^3h^2}{42gh^3}$

66. $\dfrac{64qt}{16q^2t^3}$

67. $\dfrac{y^2 + 10y + 16}{y + 2}$

68. $\dfrac{p^2 - 9}{p^2 - 5p + 6}$

69. $\dfrac{z^2 + z - 2}{z^2 - 3z + 2}$

Identify the asymptotes of each function. (Lesson 11-2)

70. $y = \dfrac{2}{x}$

71. $y = \dfrac{3}{x} + 5$

72. $y = \dfrac{1}{x - 5} - 4$

73. $y = \dfrac{1}{x + 3}$

74. $y = \dfrac{-1}{x + 6} + 7$

75. $y = \dfrac{2}{x - 8} - 3$

76. FORESTRY The number of board feet B that a log will yield can be estimated by using the formula $B = \dfrac{L}{16}(D^2 - 8D + 16)$, where D is the diameter in inches and L is the log length in feet. For logs that are 16 feet long, what diameter will yield approximately 256 board feet? (Lesson 8-9)

Find the degree of each polynomial. (Lesson 8-1)

77. 2

78. $-3a$

79. $5x^2 + 3x$

80. $d^4 - 6c^2$

81. $2x^3 - 4z + 8xz$

82. $3d^4 + 5d^3 - 4c^2 + 1$

83. DRIVING Tires should be kept within 2 pounds per square inch (psi) of the manufacturer's recommended tire pressure. If the recommendation for a tire is 30 psi, what is the range of acceptable pressures? (Lesson 5-5)

Skills Review

Factor each polynomial.

84. $x^2 - 18x - 40$

85. $x^2 - 5x + 6$

86. $x^2 - 2x - 24$

87. $3x^2 + 7x - 20$

88. $2x^2 + x - 15$

89. $8x^2 - 4x - 40$

1. Determine whether the table represents an inverse variation. Explain. (Lesson 11-1)

x	y
2	8
4	4
8	2
16	1

Assume that y varies inversely as x. Write an inverse variation equation that relates x and y. (Lesson 11-1)

2. $y = 5$ when $x = 10$

3. $y = -2$ when $x = 12$

Solve. Assume that y varies inversely as x. (Lesson 11-1)

4. If $y = 6$ when $x = 3$, find x when $y = 5$.

5. If $y = 3$ when $x = 2$, find y when $x = 4$.

State the excluded value for each function. (Lesson 11-2)

6. $y = \dfrac{2}{x}$

7. $y = \dfrac{1}{x - 6}$

Identify the asymptotes of each function. (Lesson 11-2)

8. $y = \dfrac{3}{2x + 4}$

9. $y = \dfrac{2}{x - 4}$

10. **MULTIPLE CHOICE** Jorge has $x^2 + 8x + 15$ square yards of carpet. He wants to carpet rooms that have areas of $x^2 + 5x + 6$ square yards. Write and simplify an expression to show how many rooms he can carpet. (Lesson 11-3)

   A $\dfrac{x + 5}{x + 3}$

   B $\dfrac{x + 5}{x + 2}$

   C $\dfrac{x + 3}{x + 2}$

   D $\dfrac{x + 5}{x + 6}$

Simplify each expression. State the excluded values of the variables. (Lesson 11-3)

11. $\dfrac{16x^2 y^3}{8xy}$

12. $\dfrac{z - 5}{z^2 - 7z + 10}$

13. $\dfrac{3x - 15}{x^2 - 25}$

Find each product. (Lesson 11-4)

14. $\dfrac{(x + 5)(x - 3)}{x^3} \cdot \dfrac{5x}{x - 3}$

15. $\dfrac{a^2 + 2a + 1}{a + 1} \cdot \dfrac{a - 1}{a^2 - 1}$

16. $\dfrac{m}{m^2 + 3m + 2} \cdot \dfrac{m + 2}{m^2}$

17. **MULTIPLE CHOICE** Find the area of the rectangle. (Lesson 11-4)

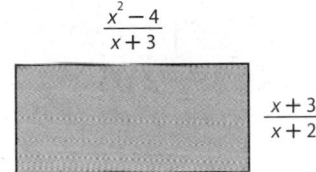

$\dfrac{x^2 - 4}{x + 3}$

$\dfrac{x + 3}{x + 2}$

   F $\dfrac{x + 2}{x - 2}$

   G $\dfrac{x + 3}{x - 2}$

   H 1

   J $x - 2$

Find each quotient. (Lesson 11-4)

18. $\dfrac{x^4}{y^2} \div \dfrac{x}{y}$

19. $\dfrac{x + 3}{2x + 6} \div \dfrac{3x - 6}{4x - 8}$

20. $\dfrac{x^2 + 7x + 12}{x^2 - 25} \div \dfrac{x^2 - 9}{2x + 10}$

21. **ANIMALS** American quarter horses excel at sprinting short distances. A quarter horse named Evening Snow holds the world record for the quarter mile with a time of 20.94 seconds. What is this speed in miles per hour and in meters per second? (Lesson 11-4)

Dividing Polynomials

:·Then

:·Now

:·Why?

● You divided rational expressions.

1 Divide a polynomial by a monomial.

2 Divide a polynomial by a binomial.

● The equation below describes the distance d a horse travels when its initial velocity is 4 m/s, its final velocity is v m/s, and its acceleration is a m/s^2.

$$d = \frac{v^2 - 4^2}{2a}$$

There are different ways to simplify the expression.

Keep as one fraction.

$$\frac{v^2 - 4^2}{2a} = \frac{v^2 - 16}{2a}$$

Divide each term by 2a.

$$\frac{v^2 - 4^2}{2a} = \frac{v^2}{2a} - \frac{4^2}{2a}$$
$$= \frac{v^2}{2a} - \frac{8}{a}$$

 Common Core State Standards

Mathematical Practices
8 Look for and express regularity in repeated reasoning.

1 **Divide Polynomials by Monomials** To divide a polynomial by a monomial, divide each term of the polynomial by the monomial.

Example 1 Divide Polynomials by Monomials

Find each quotient.

a. $\left(2x^2 + 16x\right) \div 2x$

$$\left(2x^2 + 16x\right) \div 2x = \frac{2x^2 + 16x}{2x} \qquad \text{Write as a fraction.}$$

$$= \frac{2x^2}{2x} + \frac{16x}{2x} \qquad \text{Divide each term by 2}x.$$

$$= \frac{\overset{x}{\cancel{2x^2}}}{\underset{1}{\cancel{2x}}} + \frac{\overset{8}{\cancel{16x}}}{\underset{1}{\cancel{2x}}} \qquad \text{Divide out common factors.}$$

$$= x + 8 \qquad \text{Simplify.}$$

b. $\left(b^2 + 12b - 14\right) \div 3b$

$$\left(b^2 + 12b - 14\right) \div 3b = \frac{b^2 + 12b - 14}{3b} \qquad \text{Write as a fraction.}$$

$$= \frac{b^2}{3b} + \frac{12b}{3b} - \frac{14}{3b} \qquad \text{Divide each term by 3}b.$$

$$= \frac{\overset{b}{\cancel{b^2}}}{\underset{3}{\cancel{3b}}} + \frac{\overset{4}{\cancel{12b}}}{\underset{1}{\cancel{3b}}} - \frac{14}{3b} \qquad \text{Divide out common factors.}$$

$$= \frac{b}{3} + 4 - \frac{14}{3b} \qquad \text{Simplify.}$$

▶ **Guided**Practice

1A. $\left(3q^3 - 6q\right) \div 3q$

1B. $\left(4t^5 - 5t^2 - 12\right) \div 2t^2$

1C. $\left(4r^6 + 3r^4 - 2r^2\right) \div 2r$

1D. $\left(6w^3 - 3w\right) \div 4w^2$

2 Divide Polynomials by Binomials
You can also divide polynomials by binomials. When a polynomial can be factored and common factors can be divided out, write the division as a rational expression and simplify.

Example 2 Divide a Polynomial by a Binomial

Find $(h^2 + 9h + 18) \div (h + 6)$.

$$(h^2 + 9h + 18) \div (h + 6) = \frac{h^2 + 9h + 18}{h + 6}$$ Write as a rational expression.

$$= \frac{(h + 3)(h + 6)}{h + 6}$$ Factor the numerator.

$$= \frac{(h + 3)\cancel{(h + 6)}^{1}}{\cancel{h + 6}_{1}}$$ Divide out common factors.

$$= h + 3$$ Simplify.

▶ **Guided** Practice Find each quotient.

2A. $(b^2 - 2b - 15) \div (b + 3)$ **2B.** $(x^2 + 11x + 24) \div (x + 8)$

If the polynomial cannot be factored or if there are no common factors by which to divide, you must use long division.

Example 3 Use Long Division

Find $\left(y^2 + 4y + 12\right) \div (y + 3)$ by using long division.

Step 1 Divide the first term of the dividend, y^2, by the first term of the divisor, y.

$$\begin{array}{r} y \\ y + 3 \overline{)y^2 + 4y + 12} \\ \underline{(-)\, y^2 + 3y} \\ 1y + 12 \end{array}$$

$y^2 \div y = y$

Multiply y and $y + 3$

Subtract. Bring down the 12.

Step 2 Divide the first term of the partial dividend, $1y$, by the first term of the divisor, y.

$$\begin{array}{r} y + 1 \\ y + 3 \overline{)y^2 + 4y + 12} \\ \underline{(-)\, y^2 + 3y} \\ 1y + 12 \\ \underline{(-)\, y + 3} \\ 9 \end{array}$$

Subtract. Bring down the 12.

Multiply 1 and $y + 3$.

Subtract.

So, $\left(y^2 + 4y + 12\right) \div (y + 3)$ is $y + 1$ with a remainder of 9. This answer can be written as $y + 1 + \dfrac{9}{y + 3}$.

▶ **Guided** Practice Find each quotient.

3A. $(3x^2 + 9x - 15) \div (x + 5)$ **3B.** $\left(n^2 + 6n + 2\right) \div (n - 2)$

Real-World Example 4 Divide Polynomials to Solve a Problem

PARTIES The expression $5x + 250$ represents the cost of renting a picnic shelter and food for x people. The total cost is divided evenly among all the people except for the two who bought decorations. Find $(5x + 250) \div (x - 2)$ to determine how much each person pays.

$$
\begin{array}{r}
5 \\
x - 2 \overline{)5x + 250} \\
\underline{(-)\ 5x - 10} \\
260
\end{array}
$$

So, $5 + \dfrac{260}{x-2}$ represents the amount each person pays.

▶ **Guided**Practice

4. GEOMETRY The area of a rectangle is $(2x^2 + 10x - 1)$ square units, and the width is $(x + 1)$ units. What is the length?

When a dividend is written in standard form and a power is missing, add a term of that power with a coefficient of zero.

Example 5 Insert Missing Terms

Find $(c^3 + 5c - 6) \div (c - 1)$.

$$
\begin{array}{r}
c^2 + c + 6 \\
c - 1 \overline{)c^3 + 0c^2 + 5c - 6} \\
\underline{(-)\ c^3 - c^2} \\
c^2 + 5c \\
\underline{(-)\ c^2 - c} \\
6c - 6 \\
\underline{(-)\ 6c - 6} \\
0
\end{array}
$$

Insert a c^2-term that has a coefficient of 0.
Multiply c^2 and $c - 1$.
Subtract. Bring down the $5c$.
Multiply c and $c - 1$.
Subtract. Bring down the -6.
Multiply 6 and $c - 1$.
Subtract.

So, $(c^3 + 5c - 6) \div (c - 1) = c^2 + c + 6$.

▶ **Guided**Practice **Find each quotient.**

5A. $(2r^3 + 2r^2 - 4) \div (r - 1)$

5B. $(x^4 + 2x^3 + 6x - 10) \div (x + 2)$

Check Your Understanding

 = Step-by-Step Solutions begin on page R13.

Examples 1–2 Find each quotient.

1 $(8a^2 + 20a) \div 4a$

2. $(4z^3 + 1) \div 2z$

3. $(12n^3 - 6n^2 + 15) \div 6n$

4. $(t^2 + 5t + 4) \div (t + 4)$

5. $(x^2 + 3x - 28) \div (x + 7)$

6. $(x^2 + x - 20) \div (x - 4)$

Example 4

7. CHEMISTRY The number of beakers that can be filled with $50 + x$ milliliters of a solution is given by $(400 + 3x) \div (50 + x)$. How many beakers can be filled.

Examples 3–5 Find each quotient. Use long division.

8. $(n^2 + 3n + 10) \div (n - 1)$

9. $(4y^2 + 8y + 3) \div (y + 2)$

10. $(4h^3 + 6h^2 - 3) \div (2h + 3)$

11. $(9n^3 - 13n + 8) \div (3n - 1)$

Examples 1–2 Find each quotient.

12. $(14x^2 + 7x) \div 7x$

13. $(a^3 + 4a^2 - 18a) \div a$

14. $(5q^3 + q) \div q$

15. $(6n^2 - 12n + 3) \div 3n$

16. $(8k^2 - 6) \div 2k$

17. $(9m^2 + 5m) \div 6m$

18. $(a^2 + a - 12) \div (a - 3)$

19. $(x^2 - 6x - 16) \div (x + 2)$

20. $(r^2 - 12r + 11) \div (r - 1)$

21 $(k^2 - 5k - 24) \div (k - 8)$

22. $(y^2 - 36) \div (y^2 + 6y)$

23. $(a^3 - 4a^2) \div (a - 4)$

24. $(c^3 - 9c) \div (c - 3)$

25. $(4t^2 - 1) \div (2t + 1)$

26. $(6x^3 + 15x^2 - 60x + 39) \div (3x^2)$

27. $(2h^3 + 8h^2 - 3h - 12) \div (h + 4)$

Example 4

28. GEOMETRY The area of a rectangle is $(x^3 - 4x^2)$ square units, and the width is $(x - 4)$ units. What is the length?

29. MANUFACTURING The expression $-n^2 + 18n + 850$ represents the number of baseball caps produced by n workers. Find $(-n^2 + 18n + 850) \div n$ to write an expression for average number of caps produced per person.

Examples 3, 5 Find each quotient. Use long division.

30. $(b^2 + 3b - 9) \div (b + 5)$

31. $(a^2 + 4a + 3) \div (a - 1)$

32. $(2y^2 - 3y + 1) \div (y - 2)$

33. $(4n^2 - 3n + 6) \div (n - 2)$

34. $(p^3 - 4p^2 + 9) \div (p - 1)$

35. $(t^3 - 2t - 4) \div (t + 4)$

36. $(6x^3 + 5x^2 + 9) \div (2x + 3)$

37. $(8c^3 + 6c - 5) \div (4c - 2)$

38. GEOMETRY The volume of a prism with a triangular base is $10w^3 + 23w^2 + 5w - 2$. The height of the prism is $2w + 1$, and the height of the triangle is $5w - 1$. What is the measure of the base of the triangle? (*Hint:* $V = Bh$)

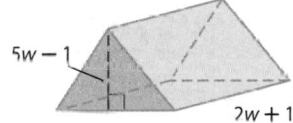

$5w - 1$

$2w + 1$

Use long division to find the expression that represents the missing length.

39.

$A = x^2 - 3x - 18$?

$x - 6$

40.

$A = 4x^2 + 16x + 16$ $2x + 4$

?

41. Determine the quotient when $x^3 + 11x + 14$ is divided by $x + 2$.

42. What is $14y^5 + 21y^4 - 6y^3 - 9y^2 + 32y + 48$ divided by $2y + 3$?

43. CCSS STRUCTURE Consider $f(x) = \dfrac{3x + 4}{x - 1}$.

   a. Rewrite the function as a quotient plus a remainder. Then graph the quotient, ignoring the remainder.

   b. Graph the original function using a graphing calculator.

   c. How are the graphs of the function and quotient related?

   d. What happens to the graph near the excluded value of x?

44. ROAD TRIP The first Ski Club van has been on the road for 20 minutes, and the second van has been on the road for 35 minutes.

    a. Write an expression for the amount of time that each van has spent on the road after an additional t minutes.

    b. Write a ratio for the first van's time on the road to the second van's time on the road and use long division to rewrite this ratio as an expression. Then find the ratio of the first van's time on the road to the second van's time on the road after 60 minutes, 200 minutes.

45 BOILING POINT The temperature at which water boils decreases by about 0.9°F for every 500 feet above sea level. The boiling point at sea level is 212°F.

    a. Write an equation for the temperature T at which water boils x feet above sea level.

    b. Mount Whitney, the tallest point in California, is 14,494 feet above sea level. At approximately what temperature does water boil on Mount Whitney?

46. ⚙ MULTIPLE REPRESENTATIONS In this problem, you will use picture models to help divide expressions.

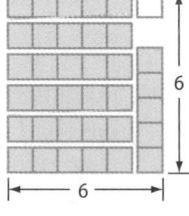

    a. Analytical The first figure models $6^2 \div 7$. Notice that the square is divided into seven equal parts. What are the quotient and the remainder? What division problem does the second figure model?

    b. Concrete Draw figures for $3^2 \div 4$ and $2^2 \div 3$.

    c. Verbal Do you observe a pattern in the previous exercises? Express this pattern algebraically.

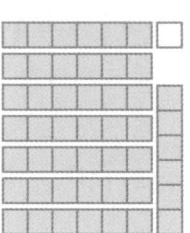

    d. Analytical Use long division to find $x^2 \div (x + 1)$. Does this result match your expression from part **c**?

H.O.T. Problems   Use Higher-Order Thinking Skills

47. ERROR ANALYSIS Alvin and Andrea are dividing $c^3 + 6c - 4$ by $c + 2$. Is either of them correct? Explain your reasoning.

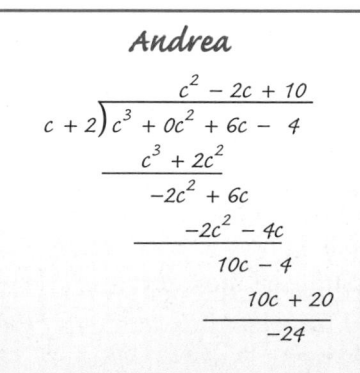

48. ⟨CCSS⟩ REGULARITY The quotient of two polynomials is $4x^2 - x - 7 + \dfrac{11x + 15}{x^2 + x + 2}$. What are the polynomials?

49. OPEN ENDED Write a division problem involving polynomials that you would solve by using long division. Explain your answer.

50. WRITING IN MATH Describe the steps to find $(w^2 - 2w - 30) \div (w + 7)$.

51. Simplify $\dfrac{21x^3 - 35x^2}{7x}$.

  A $3x^2 - 5x$ **C** $3x - 5$

  B $4x^2 - 6x$ **D** $5x - 3$

52. EXTENDED RESPONSE The box shown is designed to hold rice.

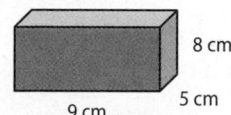

8 cm

5 cm

9 cm

  a. What is the volume of the box?

  b. What is the area of the label on the box, if the label covers all surfaces?

53. Simplify $\dfrac{x^2 + 7x + 12}{x^2 + 5x + 6}$.

  F $x + 4$ **H** $x + 2$

  G $\dfrac{x + 4}{x + 2}$ **J** $\dfrac{x + 2}{x + 4}$

54. Susana bought cards at 6 for $10. She decorated them and sold them at 4 for $10. She made $60 in profit. How many cards did she buy and sell if she had none left?

  A 25   **C** 60

  B 53   **D** 72

Spiral Review

Find each product. (Lesson 11-4)

55. $\dfrac{3x^3}{8x} \cdot \dfrac{16}{x}$

56. $\dfrac{3ad}{4c^4} \cdot \dfrac{8c^2}{6d}$

57. $\dfrac{t^2}{(t - 4)(t + 4)} \cdot \dfrac{t - 4}{6t}$

58. $\dfrac{10}{r - 2} \cdot \dfrac{r^2 - 4}{2}$

Find the zeros of each function. (Lesson 11-3)

59. $f(x) = \dfrac{x + 2}{x^2 - 6x + 8}$

60. $f(x) = \dfrac{x^2 - 3x - 4}{x^2 - x - 12}$

61. $f(x) = \dfrac{x^2 + 6x + 9}{x^2 - 9}$

62. SHADOWS A flagpole casts a shadow that is 10 feet long when the Sun is at an elevation of 68°. How tall is the flagpole? (Lesson 10-6)

Solve each equation. Check your solution. (Lesson 10-4)

63. $\sqrt{h} - 9$

64. $\sqrt{x + 3} = -5$

65. $3 + 5\sqrt{n} = 18$

66. $\sqrt{x - 5} = 2\sqrt{6}$

Solve each equation by using the Quadratic Formula. Round to the nearest tenth if necessary. (Lesson 9-5)

67. $v^2 + 12v + 20 = 0$

68. $3t^2 - 7t - 20 = 0$

69. $5y^2 - y - 4 = 0$

70. $2x^2 + 98 = 28x$

71. $2n^2 - 7n - 3 = 0$

72. $2w^2 = -(7w + 3)$

73. THEATER A backdrop for a play uses a series of thin metal arches attached to the stage floor. For each arch the height y, in feet, is modeled by the equation $y = -x^2 + 6x$, where x is the distance, in feet, across the bottom of the arch. (Lesson 9-2)

  a. Graph the related function and determine the width of the arch at the floor.

  b. What is the height at the top of the arch?

Skills Review

Find each sum.

74. $(3a^2 + 2a - 12) + (8a + 7 - 2a^2)$

75. $(2c^3 + 3cd - d^2) + (-5cd - 2c^3 + 2d^2)$

Adding and Subtracting Rational Expressions

::Then

- You added and subtracted polynomials.

::Now

1. Add and subtract rational expressions with like denominators.
2. Add and subtract rational expressions with unlike denominators.

::Why?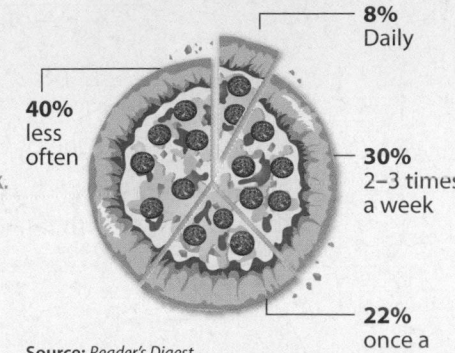

- A survey asked families how often they eat takeout. To determine the fraction of those surveyed who eat takeout more than once a week, you can add. Remember that percents can be written as fractions with denominators of 100.

| 2–3 times a week | plus | daily | equals | more than once a week. |

$$\frac{30}{100} + \frac{8}{100} = \frac{38}{100}$$

Thus, $\frac{38}{100}$ or 38% eat takeout more than once a week.

How Many Times a Week Families Eat Takeout

- 8% Daily
- 40% less often
- 30% 2–3 times a week
- 22% once a week

Source: *Reader's Digest*

 NewVocabulary
least common multiple (LCM)
least common denominator (LCD)

 Common Core State Standards

Mathematical Practices
6 Attend to precision.

1 Add and Subtract Rational Expressions with Like Denominators

To add or subtract rational expressions that have the same denominator, add or subtract the numerators and write the sum or difference over the common denominator.

> **KeyConcept** Add or Subtract Rational Expressions with Like Denominators
>
> Let a, b, and c be polynomials with $c \neq 0$.
>
> $$\frac{a}{c} + \frac{b}{c} = \frac{a+b}{c} \qquad\qquad \frac{a}{c} - \frac{b}{c} = \frac{a-b}{c}$$

Example 1 Add Rational Expressions with Like Denominators

Find $\dfrac{5n}{n+3} + \dfrac{15}{n+3}$.

$$\frac{5n}{n+3} + \frac{15}{n+3} = \frac{5n+15}{n+3} \qquad \text{The common denominator is } n+3.$$

$$= \frac{5(n+3)}{n+3} \qquad \text{Factor the numerator.}$$

$$= \frac{5(\cancel{n+3})^1}{\cancel{n+3}_1} \qquad \text{Divide by the common factor, } n+3.$$

$$= \frac{5}{1} \text{ or } 5 \qquad \text{Simplify.}$$

▸ **Guided**Practice

Find each sum.

1A. $\dfrac{8c}{6} + \dfrac{5c}{6}$

1B. $\dfrac{4t}{5xy} + \dfrac{7}{5xy}$

1C. $\dfrac{3y}{3+y} + \dfrac{y^2}{3+y}$

Example 2 Subtract Rational Expressions with Like Denominators

Find $\dfrac{3m-5}{m+4} - \dfrac{4m+2}{m+4}$.

$$\dfrac{3m-5}{m+4} - \dfrac{4m+2}{m+4} = \dfrac{(3m-5)-(4m+2)}{m+4}$$ The common denominator is $m+4$.

$$= \dfrac{(3m-5)+[-(4m+2)]}{m+4}$$ The additive inverse of $(4m+2)$ is $-(4m+2)$.

$$= \dfrac{3m-5-4m-2}{m+4}$$ Distributive Property

$$= \dfrac{-m-7}{m+4}$$ Simplify.

StudyTip

Checking Answers You can check whether you have simplified a rational expression correctly by substituting values, but this does not guarantee that the expressions are always equal. If the results are different, check for an error.

▶ **Guided**Practice

Find each difference.

2A. $\dfrac{2h+4}{h+1} - \dfrac{5+h}{h+1}$ **2B.** $\dfrac{17h+4}{15h-5} - \dfrac{2h-6}{15h-5}$

You can sometimes use additive inverses to form like denominators.

Example 3 Inverse Denominators

Find $\dfrac{3n}{n-4} + \dfrac{6n}{4-n}$.

$$\dfrac{3n}{n-4} + \dfrac{6n}{4-n} = \dfrac{3n}{n-4} + \dfrac{6n}{-(n-4)}$$ Rewrite $4-n$ as $-(n-4)$.

$$= \dfrac{3n}{n-4} - \dfrac{6n}{n-4}$$ Rewrite so the denominators are the same.

$$= \dfrac{3n-6n}{n-4} \text{ or } -\dfrac{3n}{n-4}$$ Subtract the numerators and simplify.

▶ **Guided**Practice

Find each sum or difference.

3A. $\dfrac{t^2}{t-3} + \dfrac{3}{3-t}$ **3B.** $\dfrac{2p}{p-1} - \dfrac{2p}{1-p}$

2 Add and Subtract with Unlike Denominators The **least common multiple (LCM)** is the least number that is a multiple of two or more numbers or polynomials.

Example 4 LCMs of Polynomials

Find the LCM of each pair of polynomials.

a. $6x$ and $4x^3$

Step 1 Find the prime factors of each expression.

$$6x = 2 \cdot 3 \cdot x \qquad\qquad 4x^3 = 2 \cdot 2 \cdot x \cdot x \cdot x$$

Step 2 Use each prime factor, 2, 3, and x, the greatest number of times it appears in either of the factorizations.

$$6x = 2 \cdot 3 \cdot x \qquad\qquad 4x^3 = 2 \cdot 2 \cdot x \cdot x \cdot x$$

$$\text{LCM} = 2 \cdot 2 \cdot 3 \cdot x \cdot x \cdot x \text{ or } 12x^3$$

b. $n^2 + 5n + 4$ and $(n + 1)^2$

$n^2 + 5n + 4 = (n + 1)(n + 4)$ Factor each expression.

$(n + 1)^2 = (n + 1)(n + 1)$

$(n + 1)$ is a factor twice in the second expression. $(n + 4)$ is a factor once.

$\text{LCM} = (n + 1)(n + 1)(n + 4)$ or $(n + 1)^2(n + 4)$

> **Guided**Practice
>
> **4A.** $8m^2t$ and $12m^2t^3$ **4B.** $x^2 - 2x - 8$ and $x^2 - 5x - 14$

To add or subtract fractions with unlike denominators, you need to rename the fractions using the least common multiple of the denominators, called the **least common denominator (LCD)**.

KeyConcept   Add or Subtract Rational Expressions with Unlike Denominators

Step 1   Find the LCD.

Step 2   Write each rational expression as an equivalent expression with the LCD as the denominator.

Step 3   Add or subtract the numerators and write the result over the common denominator.

Step 4   Simplify if possible.

Example 5   Add Rational Expressions with Unlike Denominators

Find $\dfrac{3t + 2}{t^2 - 2t - 3} + \dfrac{t + 1}{t - 3}$.

Find the LCD. Since $t^2 - 2t - 3 = (t - 3)(t + 1)$, the LCD is $(t - 3)(t + 1)$.

$\dfrac{3t + 2}{t^2 - 2t - 3} + \dfrac{t + 1}{t - 3} = \dfrac{3t + 2}{(t - 3)(t + 1)} + \dfrac{t + 1}{t - 3}$ Factor $t^2 - 2t - 3$.

$= \dfrac{3t + 2}{(t - 3)(t + 1)} + \dfrac{t + 1}{t - 3}\left(\dfrac{t + 1}{t + 1}\right)$ Write $\dfrac{t + 1}{t - 3}$ using the LCD.

$= \dfrac{3t + 2}{(t - 3)(t + 1)} + \dfrac{t^2 + 2t + 1}{(t - 3)(t + 1)}$ Simplify.

$= \dfrac{3t + 2 + t^2 + 2t + 1}{(t - 3)(t + 1)}$ Add the numerators.

$= \dfrac{t^2 + 5t + 3}{(t - 3)(t + 1)}$ Simplify.

> **Guided**Practice
>
> **5A.** $\dfrac{4d^2}{d} + \dfrac{d + 2}{d^2}$ 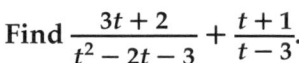   **5B.** $\dfrac{b + 3}{b} + \dfrac{b - 5}{b + 1}$

The formula $\text{time} = \frac{\text{distance}}{\text{rate}}$ is helpful in solving real-world applications.

Real-World Example 6 Add Rational Expressions

HANG GLIDING For the first 5000 meters, a hang glider travels at a rate of x meters per minute. Then, due to a stronger wind, it travels 6000 meters at a speed that is 3 times as fast.

a. Write an expression to represent how much time the hang glider is flying.

Understand For the first 5000 meters, the hang glider's speed is x. For the last 6000 meters, the hang glider's speed is $3x$.

Plan Use the formula $d = r \times t$ or $t = \frac{d}{r}$ to represent the time t of each section of the hang glider's trip, with rate r and distance d.

Solve Time to fly 5000 meters: $\frac{d}{r} = \frac{5000}{x}$ $d = 5000, r = x$

Time to fly 6000 meters: $\frac{d}{r} = \frac{6000}{3x}$ $d = 6000, r = 3x$

Total flying time: $\frac{5000}{x} + \frac{6000}{3x}$

$\frac{5000}{x} + \frac{6000}{3x} = \frac{5000}{x}\left(\frac{3}{3}\right) + \frac{6000}{3x}$ The LCD is $3x$.

$= \frac{15,000}{3x} + \frac{6000}{3x}$ Multiply.

$= \frac{\overset{7000}{\cancel{21,000}}}{\underset{1}{\cancel{3x}}}$ or $\frac{7000}{x}$ Simplify.

Check $\frac{5000}{x} + \frac{6000}{3x} = \frac{5000}{1} + \frac{6000}{3(1)}$ Let $x = 1$ in the original expression.

$= 5000 + 2000$ or 7000 Simplify.

$\frac{7000}{x} = \frac{7000}{1}$ or 7000 Let $x = 1$ in the answer expression. Simplify.

Since the expressions have the same value for $x = 1$, the answer is reasonable. ✓

b. If the hang glider is flying at a rate of 600 meters per minute for the first 5000 meters, find the total amount of time that the hang glider is flying.

$\frac{7000}{x} = \frac{7000}{600}$ Substitute 600 for x in the expression.

≈ 11.7 Simplify.

So, the hang glider is flying for approximately 11.7 minutes.

c. If the hang glider flew for approximately 15 minutes, find the rate the hang glider flew for the first 5000 meters.

$\frac{7000}{x} = 15$ Set the expression equal to 15.

$7000 = 15x$ Multiply each side by x.

$466.7 \approx x$ Divide each side by 15 and simplify.

The hang glider was flying at a rate of 466.7 meters per minute.

▶ **Guided**Practice

6. TRAINS A train travels 5 miles from Lynbrook to Long Beach and then back. The train travels about 1.2 times as fast returning from Long Beach. If r is the train's speed from Lynbrook to Long Beach, write and simplify an expression for the total time of the round trip.

To subtract rational expressions with unlike denominators, rename the expressions using the LCD. Then subtract the numerators.

Example 7 Subtract Rational Expressions with Unlike Denominators

Find $\dfrac{5}{x} - \dfrac{2x+1}{4x}$.

$$\dfrac{5}{x} - \dfrac{2x+1}{4x} = \dfrac{5}{x}\left(\dfrac{4}{4}\right) - \dfrac{2x+1}{4x} \qquad \text{Write } \dfrac{5}{x} \text{ using the LCD, } 4x.$$

$$= \dfrac{20}{4x} - \dfrac{2x+1}{4x} \qquad \text{Simplify.}$$

$$= \dfrac{20 - (2x+1)}{4x} \qquad \text{Subtract the numerators.}$$

$$= \dfrac{20 - 2x - 1}{4x} \text{ or } \dfrac{19 - 2x}{4x} \qquad \text{Simplify.}$$

StudyTip

Simplifying Answers
When simplifying a rational expression, you can leave the denominator in factored form, or multiply the terms.

▶ **Guided**Practice

Find each difference.

7A. $\dfrac{6}{t+3} - \dfrac{7}{t}$

7B. $\dfrac{y}{y-3} - \dfrac{2}{y^2+y-12}$

Check Your Understanding

◯ = Step-by-Step Solutions begin on page R13.

Examples 1–3 Find each sum or difference.

1. $\dfrac{3}{7n} + \dfrac{2}{7n}$

2. $\dfrac{x+8}{2} + \dfrac{x}{2}$

3. $\dfrac{14r}{9-r} - \dfrac{2r}{r-9}$

4. $\dfrac{7}{5t} - \dfrac{3+t}{5t}$

Example 4 Find the LCM of each pair of polynomials.

5. $3t, 8t^2$

6. $5m + 15, 2m + 6$

7. $(x^2 - 8x + 7), (x^2 + x - 2)$

Examples 5–7 Find each sum or difference.

8. $\dfrac{6}{n^4} + \dfrac{2}{n^2}$

9. $\dfrac{3}{4x} + \dfrac{2}{5y}$

10. $\dfrac{4}{5n} - \dfrac{1}{10n^3}$

11. $\dfrac{8}{3c} - \dfrac{-5}{6d}$

12. $\dfrac{a}{a+4} + \dfrac{6}{a+2}$

13. $\dfrac{x}{x-3} - \dfrac{3}{x+2}$

Example 6 **14. EXERCISE** Joseph walks 10 times around the track at a rate of x laps per hour. He runs 8 times around the track at a rate of $3x$ laps per hour. Write and simplify an expression for the total time it takes him to go around the track 18 times.

Practice and Problem Solving

Extra Practice is on page R11.

Examples 1–3 Find each sum or difference.

15 $\dfrac{a}{4} + \dfrac{3a}{4}$

16. $\dfrac{1}{6m} + \dfrac{5m}{6m}$

17. $\dfrac{5y}{6} - \dfrac{y}{6}$

18. $\dfrac{11}{4r} - \dfrac{-1}{4r}$

19. $\dfrac{8b}{ab} + \dfrac{3a}{ab}$

20. $\dfrac{t+2}{3} + \dfrac{t+5}{3}$

21. $\dfrac{3c-7}{2c-1} + \dfrac{2c+1}{1-2c}$

22. $\dfrac{15x}{33x-9} + \dfrac{3}{9-33x}$

23. $\dfrac{n+6}{10} - \dfrac{n+1}{10}$

24. $\dfrac{5x+2}{2x+5} - \dfrac{x-8}{2x+5}$

25. $\dfrac{w+2}{8w} - \dfrac{2w-3}{8w}$

26. $\dfrac{3a+1}{a-1} - \dfrac{a+4}{a-1}$

Example 4   Find the LCM of each pair of polynomials.

27. x^3y, x^2y^2 **28.** $5ab, 10b$ **29.** $(3r - 1), (r + 2)$

30. $2n - 10, 4n - 20$ **31.** $(x^2 + 9x + 18), x + 3$ **32.** $(k^2 - 2k - 8), (k + 2)^2$

Examples 5, 7 Find each sum or difference.

33. $\dfrac{5}{4x} + \dfrac{1}{10x}$ **34.** $\dfrac{6}{r} + \dfrac{2}{r^2}$ **35.** $\dfrac{3}{2a} + \dfrac{1}{5b}$

36. $\dfrac{6g}{g + 5} - \dfrac{g - 2}{2g}$ **37.** $\dfrac{7}{4k + 8} - \dfrac{k}{k + 2}$ **38.** $\dfrac{5}{2d + 2} - \dfrac{d}{d + 5}$

39. $\dfrac{-2}{7r} + \dfrac{4}{t}$ **40.** $\dfrac{n}{n - 2} + \dfrac{n}{n + 1}$ **41.** $\dfrac{d}{d + 5} + \dfrac{7}{d - 1}$

42. $\dfrac{4}{a} - \dfrac{1}{3a}$ **43.** $\dfrac{6}{5t^2} - \dfrac{2}{3t}$ **44.** $\dfrac{7}{4r} - \dfrac{3}{t}$

45. $\dfrac{w - 3}{w^2 - w - 20} + \dfrac{w}{w + 4}$ **46.** $\dfrac{n}{2n + 10} + \dfrac{1}{n^2 - 25}$

47. $\dfrac{2x}{x^2 + 8x + 15} - \dfrac{x + 3}{x + 5}$ **48.** $\dfrac{r - 3}{r^2 + 6r + 9} - \dfrac{r - 9}{r^2 - 9}$

Example 6

49. TRAVEL Grace walks to her friend's house 2 miles away and then jogs back home. Her jogging speed is 2.5 times her walking speed w.

   a. Write and simplify an expression to represent the amount of time Grace spends going to and coming from her friend's house.

   b. If Grace walks about 3.5 miles per hour, how many minutes did she spend going to and from her friend's house?

50. BOATS A boat travels 3 miles downstream and then goes 6 miles upstream. The speed of a boat going downstream is $x + 2$ miles per hour and going upstream is $x - 2$ miles per hour, where x is the speed of the boat in still water.

   a. Write and simplify an expression to represent the total time it takes the boat to travel 3 miles downstream and 6 miles upstream.

   b. If the rate of the boat in still water x is 4 miles per hour, how long did it take the boat to travel the 9 miles?

51. SCHOOL On Saturday, Mr. Kim graded 2 geometry tests and all his algebra tests a. On Sunday, he graded 7 more geometry tests. Write an expression to represent the fraction of tests he graded if he had 18 more geometry tests than algebra tests.

52. CCSS REASONING A total of 1248 people attended the school play. The same number x attended each of the two Sunday performances. There were twice as many people at the Saturday performance than at both Sunday performances. Write an expression to represent the fraction of people who attended on Saturday.

Find each sum or difference.

53. $\dfrac{x + 5}{x^2 - 4} - \dfrac{3}{x^2 - 4}$ **54.** $\dfrac{18y}{9y + 2} - \dfrac{-4}{-2 - 9y}$

55. $\dfrac{k^2 - 26}{k - 5} - \dfrac{1}{5 - k}$ **56.** $\dfrac{8}{c - 1} + \dfrac{c}{1 - c}$

57. $\dfrac{2}{x - 1} + \dfrac{3}{x + 1} - \dfrac{4x - 2}{x^2 - 1}$ **58.** $\dfrac{x^2 - x - 12}{x^2 - 11x + 30} - \dfrac{x - 4}{18 - x}$

59. $\dfrac{a^2 - 5a}{3a - 18} - \dfrac{7a - 36}{3a - 18}$ **60.** $\dfrac{8n - 3}{n^2 + 8n + 12} - \dfrac{5n - 9}{n^2 + 8n + 12}$

61. $\dfrac{x^2 - 16}{x^3} + \dfrac{x^3 + 1}{x^4}$ **62.** $\dfrac{x}{7x - 3} + \dfrac{x + 2}{15x + 30}$

63. $\dfrac{5x}{3x^2 + 19x - 14} - \dfrac{1}{9x^2 - 12x + 4}$ **64.** $\dfrac{2x + 7}{x^2 - y^2} + \dfrac{-5}{x^2 - 2xy + y^2}$

65. TRIATHLONS In a sprint triathlon, athletes swim 400 meters, bike 20 kilometers, and run 5 kilometers. An athlete bikes 12 times as fast as she swims and runs 5 times as fast as she swims.

   a. Simplify $\dfrac{400}{x} + \dfrac{20{,}000}{12x} + \dfrac{5000}{5x}$, an expression that represents the time it takes the athlete to complete the sprint triathlon.

   b. If the athlete swims 40 meters per minute, find the total time it takes her to complete the triathlon.

GEOMETRY Write an expression for the perimeter of each figure.

66.

67.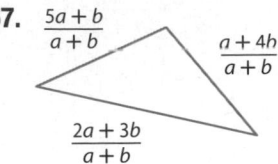

68.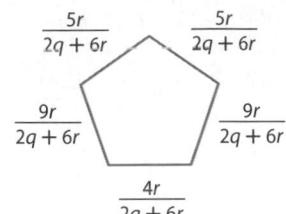

69 **BIKES** Marina rides her bike at an average rate of 10 miles per hour. On one day, she rides 9 miles and then rides around a large loop x miles long. On the second day, she rides 5 miles and then rides around the loop three times.

   a. Write an expression to represent the total time she spent riding her bike on those two days. Then simplify the expression.

   b. If the loop is 2 miles long, how long did Marina ride on those two days?

70. TRAVEL The Showalter family drives 80 miles to a college football game. On the trip home, their average speed is about 3 miles per hour slower.

   a. Let x represent the average speed of the car on the way to the game. Write and simplify an expression to represent the total time it took driving to the game and then back home.

   b. If their average speed on the way to the game was 68 miles per hour, how long did it take the Showalter family to drive to the game and back? Round to the nearest tenth.

71. CCSS PERSEVERANCE Find $\left(\dfrac{4}{7y-2} + \dfrac{7y}{2-7y}\right)\left(\dfrac{y+5}{6} - \dfrac{y+3}{6}\right)$.

72. WRITING IN MATH Describe in words the steps you use to find the LCD in an addition or subtraction of rational expressions with unlike denominators.

73. CHALLENGE Is the following statement *sometimes*, *always*, or *never* true? Explain.

$$\frac{a}{x} + \frac{b}{y} = \frac{ay + bx}{xy}; x \neq 0, y \neq 0$$

74. OPEN ENDED Describe a real-life situation that could be expressed by adding two rational expressions that are fractions. Explain what the denominator and numerator represent in both expressions.

75. WRITING IN MATH Describe how to add rational expressions with denominators that are additive inverses.

76. SHORT RESPONSE An object is launched upwards with an initial velocity of 19.6 meters per second from a 58.8-meter-tall platform. The equation for the object's height h, in meters, at time t seconds after launch is $h(t) = -4.9t^2 + 19.6t + 58.8$. How long after the launch does the object strike the ground?

77. Simplify $\frac{2}{5} + \frac{3}{25} + \frac{1}{10}$.

  A $\frac{2}{5}$ **C** $\frac{31}{50}$

  B $\frac{3}{5}$ **D** $\frac{5}{3}$

78. STATISTICS Courtney has grades of 84, 65, and 76 on three math tests. What grade must she earn on the next test to have an average of exactly 80 for the four tests?

  F 80   **H** 92

  G 84   **J** 95

79. Simplify $\frac{2}{x} + \frac{3}{x^2} + \frac{1}{2x}$.

  A $\frac{3x + 2}{x^2}$ **C** $\frac{5x + 6}{2x^2}$

  B $\frac{6}{2x^2}$ **D** $\frac{6 + x}{x^2}$

Find each quotient. (Lesson 11-5)

80. $(6x^2 + 10x) \div 2x$

81. $(15y^3 + 14y) \div 3y$

82. $(10a^3 - 20a^2 + 5a) \div 5a$

Convert each rate. Round to the nearest tenth if necessary. (Lesson 11-4)

83. 23 feet per second to miles per hour

84. 118 milliliters per second to quarts per hour (*Hint*: 1 liter ≈ 1.06 quarts)

Find the length of the missing side. If necessary, round to the nearest hundredth. (Lesson 10-5)

85.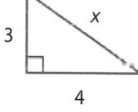

86.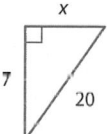

87.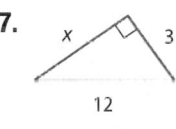

88. AMUSEMENT RIDE The height h in feet of a car above the exit ramp of a free-fall ride can be modeled by $h(t) = -16t^2 + s$, where t is the time in seconds after the car drops, and s is the starting height of the car in feet. If the designer wants the free fall to last 3 seconds, what should be the starting height in feet? (Lesson 8-9)

Express each number in scientific notation. (Lesson 7-4)

89. 12,300   **90.** 0.0000375   **91.** 1,255,000

92. FINANCIAL LITERACY Ruben has $13 to order pizza. The pizza costs $7.50 plus $1.25 per topping. He plans to tip 15% of the total cost. Write and solve an inequality to find out how many toppings he can order. (Lesson 5-3)

Find each quotient.

93. $\frac{12}{3x^2} \div \frac{6}{x}$ **94.** $\frac{g^4}{2} \div \frac{g^3}{8d^2}$ **95.** $\frac{4y - 8}{y + 1} \div (y - 2)$

Mixed Expressions and Complex Fractions

∴Then	∴Now	∴Why?
● You simplified rational expressions.	**1** Simplify mixed expressions. **2** Simplify complex fractions.	● A Top Fuel dragster can cover $\frac{1}{4}$ mile in $4\frac{2}{5}$ seconds. The average speed in miles per second can be described by the expression below. It is called a complex fraction. $$\frac{\frac{1}{4}\text{ mile}}{4\frac{2}{5}\text{ seconds}}$$

NewVocabulary

mixed expression
complex fraction

Common Core State Standards

Mathematical Practices
4 Model with mathematics.

1 Simplify Mixed Expressions An expression like $2 + \frac{4}{x+1}$ is called a **mixed expression** because it contains the sum of a monomial, 2, and a rational expression, $\frac{4}{x+1}$. You can use the LCD to change a mixed expression to a rational expression.

Example 1 Change Mixed Expressions to Rational Expressions

Write $2 + \frac{4}{x-1}$ as a rational expression.

$$2 + \frac{4}{x-1} = \frac{2(x-1)}{x-1} + \frac{4}{x-1} \qquad \text{The LCD is } x-1.$$

$$= \frac{2(x-1)+4}{x-1} \qquad \text{Add the numerators.}$$

$$= \frac{2x-2+4}{x-1} \qquad \text{Distributive Property}$$

$$= \frac{2x+2}{x-1} \qquad \text{Simplify.}$$

▶ **Guided**Practice

Write each mixed expression as a rational expression.

1A. $2 + \frac{5}{x}$ **1B.** $\frac{6y}{4y+8} + 5y$

2 Simplify Complex Fractions A **complex fraction** has one or more fractions in the numerator or denominator. You can simplify by using division.

numerical complex fraction	algebraic complex fraction
$$\frac{\frac{2}{3}}{\frac{5}{8}} = \frac{2}{3} \div \frac{5}{8}$$ $$= \frac{2}{3} \times \frac{8}{5}$$ $$= \frac{16}{15}$$	$$\frac{\frac{a}{b}}{\frac{c}{d}} = \frac{a}{b} \div \frac{c}{d}$$ $$= \frac{a}{b} \times \frac{d}{c}$$ $$= \frac{ad}{bc}$$

To simplify a complex fraction, write it as a division expression. Then find the reciprocal of the second expression and multiply.

Real-World Example 2 Use Complex Fractions to Solve Problems

RACING Refer to the application at the beginning of the lesson. Find the average speed of the Top Fuel dragster in miles per minute.

Real-World Link

A Jr. Dragster is a half-scale verson of a Top Fuel dragster. This car, which can go $\frac{1}{8}$ mile in $7\frac{9}{10}$ seconds, is designed to be driven by kids ages 8–17 in the NHRA Jr. Drag Racing League.

Source: NHRA

$$\frac{\frac{1}{4}\text{ mile}}{4\frac{2}{5}\text{ seconds}} = \frac{\frac{1}{4}\text{ mile}}{4\frac{2}{5}\text{ seconds}} \times \frac{60\text{ seconds}}{1\text{ minute}}$$ Convert seconds to minutes. Divide by common units.

$$= \frac{\frac{1}{4} \times 60}{4\frac{2}{5}}$$ Simplify.

$$= \frac{\frac{60}{4}}{\frac{22}{5}}$$ Express each term as an improper fraction.

$$= \frac{\overset{15}{\cancel{60}} \times 5}{\underset{1}{\cancel{4}} \times 22}$$ Use the rule $\frac{\frac{a}{b}}{\frac{c}{d}} = \frac{ad}{bc}$.

$$= \frac{75}{22}\text{ or }3\frac{9}{22}$$ Simplify.

So, the average speed of the Top Fuel dragster is $3\frac{9}{22}$ miles per minute.

▸ **Guided**Practice

2. RACING Refer to the information about the Jr. Dragster at the left. What is the average speed of the car in feet per second?

To simplify complex fractions, you can either use the rule as in Example 2, or you can rewrite the fraction as a division expression, as shown below.

Example 3 Complex Fractions Involving Monomials

Simplify $\dfrac{\frac{8t^2}{v}}{\frac{4t}{v^3}}$.

$$\frac{\frac{8t^2}{v}}{\frac{4t}{v^3}} = \frac{8t^2}{v} \div \frac{4t}{v^3}$$ Write as a division expression.

$$= \frac{8t^2}{v} \times \frac{v^3}{4t}$$ To divide, multiply by the reciprocal.

$$= \frac{\overset{2t}{\cancel{8t^2}}}{\underset{1}{\cancel{v}}} \times \frac{\overset{v^2}{\cancel{v^3}}}{\underset{1}{\cancel{4t}}}\text{ or }2tv^2$$ Divide by the common factors $4t$ and v and simplify.

▸ **Guided**Practice

Simplify each expression.

3A. $\dfrac{\frac{g^3h}{b}}{\frac{gh^3}{b^2}}$

3B. $\dfrac{\frac{-24m^3t^5}{p^2h}}{\frac{16pm^2}{t^4h}}$

Complex fractions may also involve polynomials.

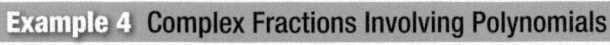

Example 4 Complex Fractions Involving Polynomials

Simplify each expression.

a. $\dfrac{\dfrac{2}{y+3}}{\dfrac{5}{y^2-9}}$

$\dfrac{\dfrac{2}{y+3}}{\dfrac{5}{y^2-9}} = \dfrac{2}{y+3} \div \dfrac{5}{y^2-9}$ Write as a division expression.

$= \dfrac{2}{y+3} \times \dfrac{y^2-9}{5}$ To divide, multiply by the reciprocal.

$= \dfrac{2}{y+3} \times \dfrac{(y-3)(y+3)}{5}$ Factor y^2-9.

$= \dfrac{2}{\cancel{y+3}\,_1} \times \dfrac{(y-3)\cancel{(y+3)}^{\,1}}{5}$ Divide by the GCF, $y+3$.

$= \dfrac{2(y-3)}{5}$ Simplify.

b. $\dfrac{\dfrac{n^2+7n-18}{n^2-2n+1}}{\dfrac{n^2-81}{n-1}}$

$\dfrac{\dfrac{n^2+7n-18}{n^2-2n+1}}{\dfrac{n^2-81}{n-1}} = \dfrac{n^2+7n-18}{n^2-2n+1} \div \dfrac{n^2-81}{n-1}$ Write as a division expression.

$= \dfrac{n^2+7n-18}{n^2-2n+1} \times \dfrac{n-1}{n^2-81}$ Multiply by the reciprocal.

$= \dfrac{(n-2)(n+9)}{(n-1)(n-1)} \times \dfrac{n-1}{(n-9)(n+9)}$ Factor the polynomials.

$= \dfrac{(n-2)\cancel{(n+9)}^{\,1}}{(n-1)\cancel{(n-1)}_1} \times \dfrac{\cancel{n-1}^{\,1}}{(n-9)\cancel{(n+9)}_1}$ Divide out the common factors.

$= \dfrac{n-2}{(n-1)(n-9)}$ Simplify.

Guided Practice

4A. $\dfrac{\dfrac{a+7}{4}}{\dfrac{a^2-49}{10}}$

4B. $\dfrac{\dfrac{x+4}{x-1}}{\dfrac{x^2+6x+8}{2x-2}}$

4C. $\dfrac{\dfrac{c-d}{j+p}}{\dfrac{c^2-d^2}{j^2-p^2}}$

4D. $\dfrac{\dfrac{n^2+4n-21}{n^2-9n+18}}{\dfrac{n^2+3n-28}{n^2-10n+24}}$

Real-World Career

Lab Technician
Lab technicians work with scientists, running experiments, conducting research projects, and running routine diagnostic samples. Lab technicians in any field need at least a two-year associate degree.

Study Tip

Factoring When simplifying fractions involving polynomials, factor the numerator and the denominator of each expression if possible.

Example 1 Write each mixed expression as a rational expression.

1. $\dfrac{2}{n} + 4$

2. $r + \dfrac{1}{3r}$

3. $6 + \dfrac{5}{t + 1}$

4. $\dfrac{x + 7}{2x} - 5x$

Example 2 **5. ROWING** Rico rowed a canoe $2\frac{1}{2}$ miles in $\frac{1}{3}$ hour.

   a. Write an expression to represent his speed in miles per hour.

   b. Simplify the expression to find his average speed.

Examples 3–4 Simplify each expression.

6. $\dfrac{2\frac{1}{3}}{1\frac{2}{5}}$

7. $\dfrac{\frac{4}{5}}{6\frac{2}{3}}$

8. $\dfrac{\frac{a^2}{b^3}}{\frac{b^5}{a}}$

9. $\dfrac{\frac{y^4}{x^2}}{\frac{xy^2}{2x^2}}$

10. $\dfrac{\frac{6}{x-2}}{\frac{3}{x^2-x-2}}$

11. $\dfrac{\frac{r+s}{x^2-y^2}}{\frac{(r+s)^2}{x-y}}$

12. $\dfrac{\frac{2+q}{q^2-4}}{\frac{q+4}{q^2-6q+8}}$

13. $\dfrac{\frac{p+3}{p^2+p-6}}{\frac{p^2+4p+3}{p^2+6p+9}}$

Practice and Problem Solving

Extra Practice is on page R11.

Example 1 Write each mixed expression as a rational expression.

14. $10 + \dfrac{6}{f}$

(15) $p - \dfrac{7}{2p}$

16. $5a - \dfrac{2a}{b}$

17. $3h + \dfrac{1 + h}{h}$

18. $t + \dfrac{v + w}{v - w}$

19. $n^2 + \dfrac{n - 1}{n + 4}$

20. $(k + 2) + \dfrac{k - 1}{k - 2}$

21. $(d - 6) + \dfrac{d + 1}{d - 7}$

22. $\dfrac{h - 3}{h + 5} - (h + 2)$

Example 2 **23. READING** Ebony reads $6\frac{3}{4}$ pages of a book in 9 minutes. What is her average reading rate in pages per minute?

24. HORSES A thoroughbred can run $\frac{1}{2}$ mile in about $\frac{3}{4}$ minute. What is the horse's speed in miles per hour?

Examples 3–4 Simplify each expression.

25. $\dfrac{2\frac{2}{9}}{3\frac{1}{3}}$

26. $\dfrac{5\frac{3}{5}}{2\frac{1}{7}}$

27. $\dfrac{\frac{g^2}{h}}{\frac{g^5}{h^2}}$

28. $\dfrac{\frac{5n^4}{p^3}}{\frac{6n}{5p}}$

29. $\dfrac{\frac{2}{a}}{\frac{1}{a+6}}$

30. $\dfrac{\frac{t+5}{9}}{\frac{t^2-t-30}{12}}$

31. $\dfrac{\frac{j^2-16}{j^2+10j+16}}{\frac{15}{j+8}}$

32. $\dfrac{\frac{x-3}{x^2+3x+2}}{\frac{x^2-9}{x+1}}$

33. CCSS MODELING The Centralville High School Cooking Club has $12\frac{1}{2}$ pounds of flour with which to make tortillas. There are $3\frac{3}{4}$ cups of flour in a pound, and it takes about $\frac{1}{3}$ cup of flour per tortilla. How many tortillas can they make?

34. SCOOTER The speed v of an object spinning in a circle equals the circumference of the circle divided by the time T it takes the object to complete one revolution.

    a. Use the variables v, r (the radius of the circle), and T to write a formula describing the speed of a spinning object.

    b. A scooter has tires with a radius of $3\frac{1}{2}$ inches. The tires make one revolution every $\frac{1}{10}$ second. Find the speed in miles per hour. Round to the nearest tenth.

35 SCIENCE The *density* of an object equals $\frac{m}{V}$, where m is the mass of the object and V is the volume. The densities of four metals are shown in the table. Identify the metal of each ball described below. (*Hint*: The volume of a sphere is $V = \frac{4}{3}\pi r^3$.)

Metal	Density (kg/m³)
copper	8900
gold	19,300
iron	7800
lead	11,300

    a. A metal ball has a mass of 15.6 kilograms and a radius of 0.0748 meter.

    b. A metal ball has a mass of 285.3 kilograms and a radius of 0.1819 meter.

36. SIRENS As an ambulance approaches, the siren sounds different than if it were sitting still. If the ambulance is moving toward you at v miles per hour and blowing the siren at a frequency of f, then you hear the siren as if it were blowing at a frequency h. This can be described by the equation $h = \dfrac{f}{1 - \frac{v}{s}}$, where s is the speed of sound, approximately 760 miles per hour.

    a. Simplify the complex fraction in the formula.

    b. Suppose a siren blows at 45 cycles per minute and is moving toward you at 65 miles per hour. Find the frequency of the siren as you hear it.

Simplify each expression.

37. $15 - \dfrac{17x + 5}{5x + 10}$

38. $\dfrac{\frac{b}{b+3} + 2}{b^2 - 2b - 8}$

39. $\dfrac{1 + \frac{2c^2 - 6c - 10}{c + 7}}{2c + 1}$

40. $\dfrac{y - \frac{12}{y - 4}}{y - \frac{18}{y - 3}}$

41. $\dfrac{\frac{x^2 - 4x - 32}{x + 1}}{\frac{x^2 + 6x + 8}{x^2 - 1}}$

42. $\dfrac{\frac{r^2 - 9r}{r^2 + 7r + 10}}{\frac{r^2 + 5r}{r^2 + r - 2}}$

H.O.T. Problems   Use Higher-Order Thinking Skills

43. REASONING Describe the first step to simplify the expression below.

$$\dfrac{\left(\frac{y}{x} - \frac{x}{y}\right)}{\frac{x + y}{xy}}$$

44. REASONING Is $\dfrac{n}{1 - \frac{5}{p}} + \dfrac{n}{\frac{5}{p} - 1}$ *sometimes*, *always*, or *never* equal to 0? Explain.

45. CCSS PERSEVERANCE Simplify the rational expression below.

$$\dfrac{\frac{1}{t - 1} + \frac{1}{t + 1}}{\frac{1}{t} - \frac{1}{t^2}}$$

46. OPEN ENDED Write a complex fraction that, when simplified, results in $\frac{1}{x}$.

47. WRITING IN MATH Explain how complex fractions can be used to solve a problem involving distance, rate, and time. Give an example.

48. A number is between 44 squared and 45 squared. 5 squared is one of its factors, and it is a multiple of 13. Find the number.

    A 1950

    B 2000

    C 2025

    D 2050

49. SHORT RESPONSE Bernard is reading a 445-page book. He has already read 157 pages. If he reads 24 pages a day, how long will it take him to finish the book?

50. GEOMETRY Angela wanted a round rug to fit her room that is 16 feet wide. The rug should just meet the edges. What is the area of the rug rounded to the nearest tenth?

    F 50.3 ft^2 **H** 152.2 ft^2

    G 100.5 ft^2 **J** 201.1 ft^2

51. Simplify $7x + \dfrac{10}{2xy}$.

    A $\dfrac{7x + 10}{2xy}$ **C** $\dfrac{17x}{2xy}$

    B $\dfrac{7x^2y + 5}{xy}$ **D** $\dfrac{7xy + 5}{x^2y}$

Find each sum or difference. (Lesson 11-6)

52. $\dfrac{6}{7x} - \dfrac{5 + x}{7x}$

53. $\dfrac{4}{d - 1} + \dfrac{d}{1 - d}$

54. $\dfrac{3q + 2}{2q + 1} + \dfrac{q - 5}{2q + 1}$

55. $\dfrac{2}{5m} - \dfrac{1}{15m^3}$

56. $\dfrac{10}{3g} - \dfrac{-3}{4h}$

57. $\dfrac{b}{b + 3} + \dfrac{6}{b - 2}$

Find each quotient. Use long division. (Lesson 11-5)

58. $(x^2 - 2x - 30) \div (x + 7)$

59. $(a^2 + 4u - 22) \div (a - 3)$

60. $(3q^2 + 20q + 11) \div (q + 6)$

61. $(3y^3 + 8y^2 + y - 7) \div (y + 2)$

62. $(6t^3 - 9t^2 + 6) \div (2t - 3)$

63. $(9h^3 + 5h - 8) \div (3h - 2)$

64. GEOMETRY A rectangle has a base of 8 meters and a height of 14 meters. What is the length of the diagonal? (Lesson 10-5)

Graph each function. Determine the domain and range. (Lesson 10-1)

65. $y = 2\sqrt{x}$

66. $y = -3\sqrt{x}$

67. $y = \dfrac{1}{4}\sqrt{x}$

Factor each polynomial. If the polynomial cannot be factored, write *prime*. (Lesson 8-8)

68. $x^2 - 81$

69. $a^2 - 121$

70. $n^2 + 100$

71. $-25 + 4y^2$

72. $p^4 - 16$

73. $4t^4 - 4$

74. PARKS A youth group traveling in two vans visited Mammoth Cave in Kentucky. The number of people in each van and the total cost of the cave are shown. Find the adult price and the student price of the tour. (Lesson 6-3)

Van	Number of Adults	Number of Students	Total Cost
A	2	5	$77
B	2	7	$95

Solve each equation.

75. $6x = 24$

76. $5y - 1 = 19$

77. $2t + 7 = 21$

78. $\dfrac{p}{3} = -4.2$

79. $\dfrac{2m + 1}{4} = -5.5$

80. $\dfrac{3}{4}g = \dfrac{1}{2}$

Rational Equations

- You solved proportions.

1 Solve rational equations.

2 Use rational equations to solve problems.

- Oceanic species of dolphins can swim 5 miles per hour faster than coastal species of dolphins. An oceanic dolphin can swim 3 miles in the same time that it takes a coastal dolphin to swim 2 miles.

Dolphins			
Species	Distance	Rate	Time
coastal	2 miles	x mph	t hours
oceanic	3 miles	x + 5 mph	t hours

Since time $= \dfrac{\text{distance}}{\text{rate}}$, the equation below represents this situation.

Time an oceanic dolphin swims 3 miles   equals   time a coastal dolphin swims 2 miles.

$$\text{distance} \rightarrow \underset{\text{rate} \rightarrow}{\dfrac{3}{x+5}} = \dfrac{2}{x} \begin{array}{l}\leftarrow \text{distance} \\ \leftarrow \text{rate}\end{array}$$

NewVocabulary
rational equation
extraneous solution
work problem
rate problem

Common Core State Standards

Content Standards
A.CED.2 Create equations in two or more variables to represent relationships between quantities; graph equations on coordinate axes with labels and scales.

Mathematical Practices
2 Reason abstractly and quantitatively.
4 Model with mathematics.

1 **Solve Rational Equations** A **rational equation** contains one or more rational expressions. When a rational equation is a proportion, you can use cross products to solve it.

Real-World Example 1   Use Cross Products to Solve Equations

DOLPHINS Refer to the information above. Solve $\dfrac{3}{x+5} = \dfrac{2}{x}$ to find the speed of a coastal dolphin. Check the solution.

$\dfrac{3}{x+5} = \dfrac{2}{x}$	Original equation
$3x = 2(x+5)$	Find the cross products.
$3x = 2x + 10$	Distributive Property
$x = 10$	Subtract 2x from each side.

So, a coastal dolphin can swim 10 miles per hour.

CHECK

$\dfrac{3}{x+5} = \dfrac{2}{x}$	Original equation
$\dfrac{3}{10+5} \stackrel{?}{=} \dfrac{2}{10}$	Replace x with 10.
$\dfrac{3}{15} \stackrel{?}{=} \dfrac{1}{5}$	Simplify.
$\dfrac{1}{5} = \dfrac{1}{5}$ ✓	Simplify.

▶ **Guided**Practice

Solve each equation. Check the solution.

1A. $\dfrac{7}{y-3} = \dfrac{3}{y+1}$ **1B.** $\dfrac{13}{10} = \dfrac{2f+0.2}{7}$

Another method that can be used to solve any rational equation is to find the LCD of all the fractions in the equation. Then multiply each side of the equation by the LCD to eliminate the fractions.

Example 2 Use the LCD to Solve Rational Equations

Solve $\dfrac{4}{y} + \dfrac{5y}{y+1} = 5$. **Check the solution.**

Step 1 Find the LCD.

The LCD of $\dfrac{4}{y}$ and $\dfrac{5y}{y+1}$ is $y(y+1)$.

Step 2 Multiply each side of the equation by the LCD.

$$\dfrac{4}{y} + \dfrac{5y}{y+1} = 5 \qquad \text{Original equation}$$

$$y(y+1)\left(\dfrac{4}{y} + \dfrac{5y}{y+1}\right) = y(y+1)(5) \qquad \begin{array}{l}\text{Multiply each side by}\\ \text{the LCD, } y(y+1).\end{array}$$

$$\left(\dfrac{\overset{1}{\cancel{y}}(y+1)}{1}\cdot\dfrac{4}{\cancel{y}}\right) + \left(\dfrac{y\overset{1}{\cancel{(y+1)}}}{1}\cdot\dfrac{5y}{\cancel{y+1}}\right) = y(y+1)(5) \qquad \text{Distributive Property}$$

$$(y+1)4 + y(5y) = y(y+1)(5) \qquad \text{Simplify.}$$

$$4y + 4 + 5y^2 = 5y^2 + 5y \qquad \text{Multiply.}$$

$$4y + 4 + 5y^2 - 5y^2 = 5y^2 - 5y^2 + 5y \qquad \begin{array}{l}\text{Subtract } 5y^2 \text{ from each}\\ \text{side.}\end{array}$$

$$4y + 4 = 5y \qquad \text{Simplify.}$$

$$4y - 4y + 4 = 5y - 4y \qquad \begin{array}{l}\text{Subtract } 4y \text{ from each}\\ \text{side.}\end{array}$$

$$4 = y \qquad \text{Simplify.}$$

CHECK $\dfrac{4}{y} + \dfrac{5y}{y+1} = 5 \qquad$ Original equation

$\dfrac{4}{4} + \dfrac{5(4)}{4+1} \overset{?}{=} 5 \qquad$ Replace y with 4.

$1 + 4 \overset{?}{=} 5 \qquad$ Simplify.

$5 = 5 \checkmark \qquad$ Simplify.

StudyTip

Solutions It is important to check the solutions of rational equations to be sure that they satisfy the original equation.

> **Guided**Practice

Solve each equation. Check your solutions.

2A. $\dfrac{2b-5}{b-2}$ $2 = \dfrac{3}{b+2}$

2B. $1 + \dfrac{1}{c+2} = \dfrac{28}{c^2 + 2c}$

2C. $\dfrac{y+2}{y-2} - \dfrac{2}{y+2} = -\dfrac{7}{3}$

2D. $\dfrac{n}{3n+6} - \dfrac{n}{5n+10} = \dfrac{2}{5}$

VocabularyLink

extraneous
Everyday Use
irrelevant or unimportant

extraneous solution
Math Use a result that is not a solution of the original equation

Recall that any value of a variable that makes the denominator of a rational expression zero must be excluded from the domain.

In the same way, when a solution of a rational equation results in a zero in the denominator, that solution must be excluded. Such solutions are called **extraneous solutions**.

$$\dfrac{4+x}{x-5} + \dfrac{1}{x} = \dfrac{2}{x+1} \qquad \text{5, 0, and } -1 \text{ cannot be solutions.}$$

Example 3 Extraneous Solutions

Solve $\dfrac{2n}{n-5} + \dfrac{4n-30}{n-5} = 5$. State any extraneous solutions.

$$\dfrac{2n}{n-5} + \dfrac{4n-30}{n-5} = 5 \qquad \text{Original equation}$$

$$(n-5)\left(\dfrac{2n}{n-5} + \dfrac{4n-30}{n-5}\right) = (n-5)5 \qquad \text{Multiply each side by the LCD, } n-5.$$

$$\left(\dfrac{\cancel{n-5}^{1}}{1} \cdot \dfrac{2n}{\cancel{n-5}_{1}}\right) + \left(\dfrac{\cancel{n-5}^{1}}{1} \cdot \dfrac{4n-30}{\cancel{n-5}_{1}}\right) = (n-5)5 \qquad \text{Distributive Property}$$

$$2n + 4n - 30 = 5n - 25 \qquad \text{Simplify.}$$

$$6n - 30 = 5n - 25 \qquad \text{Add like terms.}$$

$$6n - 5n - 30 = 5n - 5n - 25 \qquad \text{Subtract } 5n \text{ from each side.}$$

$$n - 30 = -25 \qquad \text{Simplify.}$$

$$n - 30 + 30 = -25 + 30 \qquad \text{Add 30 to each side.}$$

$$n = 5 \qquad \text{Simplify.}$$

Since $n = 5$ results in a zero in the denominator of the original equation, it is an extraneous solution. So, the equation has no solution.

GuidedPractice

3. Solve $\dfrac{n^2 - 3n}{n^2 - 4} - \dfrac{10}{n^2 - 4} = 2$. State any extraneous solutions.

StudyTip

Solutions It is possible to get both a valid solution and an extraneous solution when solving a rational equation.

2 Use Rational Equations to Solve Problems You can use rational equations to solve **work problems**, or problems involving work rates.

Real-World Example 4 Work Problem

JOBS At his part-time job at the zoo, Ping can clean the bird area in 2 hours. Natalie can clean the same area in 1 hour and 15 minutes. How long would it take them if they worked together?

Understand It takes Ping 2 hours to complete the job and Natalie $1\frac{1}{4}$ hours.

You need to find the rate that each person works and the total time t that it will take them if they work together.

Plan Find the fraction of the job that each person can do in an hour.

Ping's rate $\longrightarrow \dfrac{1 \text{ job}}{2 \text{ hours}} = \dfrac{1}{2}$ job per hour

Natalie's rate $\longrightarrow \dfrac{1 \text{ job}}{1\frac{1}{4} \text{ hours}}$ or $\dfrac{1 \text{ job}}{\frac{5}{4} \text{ hours}} = \dfrac{4}{5}$ job per hour

Since rate · time = fraction of job done, multiply each rate by the time t to represent the amount of the job done by each person.

StudyTip

CCSS Reasoning When solving work problems, remember that each term should represent the portion of a job completed in one unit of time.

Solve

| Fraction of job Ping completes | plus | fraction of job Natalie completes | equals | 1 job. |

$$\frac{1}{2}t \qquad + \qquad \frac{4}{5}t \qquad = \qquad 1$$

$$10\left(\frac{1}{2}t + \frac{4}{5}t\right) = 10(1) \qquad \text{Multiply each side by the LCD, 10.}$$

$$10\left(\frac{1}{2}t\right) + 10\left(\frac{4}{5}t\right) = 10 \qquad \text{Distributive Property}$$

$$5t + 8t = 10 \qquad \text{Simplify.}$$

$$t = \frac{10}{13} \qquad \text{Add like terms and divide each side by 13.}$$

So, it would take them $\frac{10}{13}$ hour or about 46 minutes to complete the job if they work together.

Check In $\frac{10}{13}$ hour, Ping would complete $\frac{1}{2} \cdot \frac{10}{13}$ or $\frac{5}{13}$ of the job and Natalie would complete $\frac{4}{5} \cdot \frac{10}{13}$ or $\frac{8}{13}$ of the job. Together, they complete $\frac{5}{13} + \frac{8}{13}$ or 1 whole job. So, the answer is reasonable. ✓

▶ **Guided**Practice

4. RAKING Jenna can rake the leaves in 2 hours. It takes her brother Benjamin 3 hours. How long would it take them if they worked together?

Rational equations can also be used to solve **rate problems**.

🔵 **Real-World Example 5** Rate Problem

AIRPLANES An airplane takes off and flies an average of 480 miles per hour. Another plane leaves 15 minutes later and flies to the same city traveling 560 miles per hour. How long will it take the second plane to pass the first plane?

Record the information that you know in a table.

Plane	Distance	Rate	Time
1	d miles	480 mi/h	t hours
2	d miles	560 mi/h	$t - \frac{1}{4}$ hours

◀ Plane 2 took off 15 minutes, or $\frac{1}{4}$ hour, after Plane 1

Since both planes will have traveled the same distance when Plane 2 passes Plane 1, you can write the following equation.

Distance for Plane 1 = Distance for Plane 2

$$480 \cdot t = 560 \cdot \left(t - \frac{1}{4}\right) \qquad \text{distance = rate} \cdot \text{time}$$

$$480t = (560 \cdot t) - \left(560 \cdot \frac{1}{4}\right) \qquad \text{Distributive Property}$$

$$480t = 560t - 140 \qquad \text{Simplify.}$$

$$-80t = -140 \qquad \text{Subtract } 560t \text{ from each side.}$$

$$t = 1.75 \qquad \text{Divide each side by } -80.$$

So, the second plane passes the first plane after 1.75 hours.

Real-World Link

The longest nonstop commercial flight was 13,422 miles from Hong Kong Airport in China to London Heathrow in the United Kingdom. It took 22 hours and 42 minutes.

Source: Guinness Book of World Records

▶ **Guided**Practice

5. Lenora leaves the house walking at 3 miles per hour. After 10 minutes, her mother leaves the house riding a bicycle at 10 miles per hour. In how many minutes will Lenora's mother catch her?

Examples 1–3 Solve each equation. State any extraneous solutions.

1. $\dfrac{2}{x+1} = \dfrac{4}{x}$

2. $\dfrac{t+3}{5} = \dfrac{2t+3}{9}$

3. $\dfrac{a+3}{a} - \dfrac{6}{5a} = \dfrac{1}{a}$

4. $4 - \dfrac{p}{p-1} = \dfrac{2}{p-1}$

5. $\dfrac{2t}{t+1} + \dfrac{4}{t-1} = 2$

6. $\dfrac{x+3}{x^2-1} - \dfrac{2x}{x-1} = 1$

Example 4

7. **WEEDING** Maurice can weed the garden in 45 minutes. Olinda can weed the garden in 50 minutes. How long would it take them to weed the garden if they work together?

Example 5

8. **LANDSCAPING** Hunter is filling a 3.5-gallon bucket to water plants at a faucet that flows at a rate of 1.75 gallons a minute. If he were to add a hose that flows at a rate of 1.45 gallons per minute, how many minutes would it take him to fill the bucket? Round to the nearest tenth.

Practice and Problem Solving   **Extra Practice is on page R11.**

Examples 1–3 Solve each equation. State any extraneous solutions.

9 $\dfrac{8}{n} = \dfrac{3}{n-5}$

10. $\dfrac{6}{t+2} = \dfrac{4}{t}$

11. $\dfrac{3g+2}{12} = \dfrac{g}{2}$

12. $\dfrac{5h}{4} + \dfrac{1}{2} = \dfrac{3h}{8}$

13. $\dfrac{2}{3w} = \dfrac{2}{15} + \dfrac{12}{5w}$

14. $\dfrac{c-4}{c+1} - \dfrac{c}{c-1}$

15. $\dfrac{x-1}{x+1} - \dfrac{2x}{x-1} = -1$

16. $\dfrac{y+4}{y-2} + \dfrac{6}{y-2} = \dfrac{1}{y+3}$

17. $\dfrac{a}{a+3} + \dfrac{a^2}{a+3} = 2$

18. $\dfrac{12}{a+3} + \dfrac{6}{a^2-9} = \dfrac{8}{a+3}$

19. $\dfrac{3n}{n-1} + \dfrac{6n-9}{n-1} = 6$

20. $\dfrac{n^2-n-6}{n^2-n} - \dfrac{n-5}{n-1} = \dfrac{n-3}{n^2-n}$

Example 4

21. **PAINTING** It takes Noah 3 hours to paint one side of a fence. It takes Gilberto 5 hours. How long would it take them if they worked together?

22. **DISHWASHING** Ron works as a dishwasher and can wash 500 plates in two hours and 15 minutes. Chris can finish the 500 plates in 3 hours. About how long would it take them to finish all of the plates if they work together?

Example 5

23. **ICE** A hotel has two ice machines in its kitchen. How many hours would it take both machines to make 60 pounds of ice? Round to the nearest tenth.

30 lb/day   25 lb/day

24. **CYCLING** Two cyclists travel in opposite directions around a 5.6-mile circular trail. They start at the same time. The first cyclist completes the trail in 22 minutes and the second in 28 minutes. At what time do they pass each other?

GRAPHING CALCULATOR For each function, a) describe the shape of the graph, b) use factoring to simplify the function, and c) find the zeros of the function.

25. $f(x) = \dfrac{x^2 - x - 30}{x - 6}$

26. $f(x) = \dfrac{x^3 + x^2 - 2x}{x + 2}$

27. $f(x) = \dfrac{x^3 + 6x^2 + 12x}{x}$

28. **CCSS REASONING** Morgan can paint a standard-sized house in about 5 days. For his latest job, Morgan hires two assistants. At what rate must these assistants work for Morgan to meet a deadline of two days?

29. AIRPLANES Headwinds push against a plane and reduce its total speed, while tailwinds push on a plane and increase its total speed. Let w equal the speed of the wind, r equal the speed set by the pilot, and s equal the total speed.

   a. Write an equation for the total speed with a headwind and an equation for the total speed with a tailwind.

   b. Use the rate formula to write an equation for the distance traveled by a plane with a headwind and another equation for the distance traveled by a plane with a tailwind. Then solve each equation for time instead of distance.

30. MIXTURES A pitcher of fruit juice has 3 pints of pineapple juice and 2 pints of orange juice. Erin wants to add more orange juice so that the fruit juice mixture is 60% orange juice. Let x equal the pints of orange juice that she needs to add.

   a. Copy and complete the table below.

Juice	Pints of Orange Juice	Total Pints of Juice	Percent of Orange Juice
original mixture		5	
final mixture	2 + x		0.6

   b. Write and solve an equation to find the pints of orange juice to add.

31 DORMITORIES The number of hours h it takes to clean a dormitory varies inversely with the number of people cleaning it c and directly with the number of people living there p.

   a. Write an equation showing how h, c, and p are related. (*Hint*: Include the constant k.)

   b. It takes 8 hours for 5 people to clean the dormitory when there are 100 people there. How long will it take to clean the dormitory if there are 10 people cleaning and the number of people living in the dorm stays the same?

Solve each equation. State any extraneous solutions.

32. $\dfrac{4b + 2}{b^2 - 3b} + \dfrac{b + 2}{b} = \dfrac{b - 1}{b}$

33. $\dfrac{x^2 - x - 3}{x - 1} + \dfrac{x^3 + 2x^2}{x - 1} = 3$

34. $\dfrac{y^2 + 5y - 6}{y^3 - 2y^2} - \dfrac{5}{y} - \dfrac{6}{y^3 - 2y^2}$

35. $\dfrac{x - \frac{6}{5}}{x} - \dfrac{x}{x - 5} \dfrac{10\frac{1}{2}}{} = \dfrac{x + 21}{x^2 - 5x}$

H.O.T. Problems   Use Higher-Order Thinking Skills

36. CHALLENGE Solve $\dfrac{2x}{x - 2} + \dfrac{x^2 + 3x}{(x + 1)(x - 2)} = \dfrac{2}{(x + 1)(x - 2)}$.

37. REASONING How is an excluded value of a rational expression related to an extraneous solution of a corresponding rational equation? Explain.

38. 📝 **WRITING IN MATH** Why should you check solutions of rational equations?

39. CCSS ARGUMENTS Find a counterexample for the following statement.

     The solution of a rational equation can never be zero.

40. WRITING IN MATH Describe the steps for solving a rational equation that is not a proportion.

41. It takes Cheng 4 hours to build a fence. If he hires Odell to help him, they can do the job in 3 hours. If Odell built the same fence alone, how long would it take him?

   A $1\frac{5}{7}$ hours   **C** 8 hours

   B $3\frac{2}{3}$ hours   **D** 12 hours

42. In the 1000-meter race, Zoe finished 35 meters ahead of Taryn and 53 meters ahead of Evan. How far was Taryn ahead of Evan?

   F 18 m   **G** 35 m   **H** 53 m   **J** 88 m

43. Twenty gallons of lemonade were poured into two containers of different sizes. Express the amount of lemonade poured into the smaller container in terms of g, the amount poured into the larger container.

   A $g + 20$ **C** $g - 20$

   B $20 + g$ **D** $20 - g$

44. GRIDDED RESPONSE The gym has 2-kilogram and 5-kilogram disks for weight lifting. They have fourteen disks in all. The total weight of the 2-kilogram disks is the same as the total weight of the 5-kilogram disks. How many 2-kilogram disks are there?

Spiral Review

Simplify each expression. (Lesson 11-7)

45. $\dfrac{\frac{c^2}{d}}{\frac{c^3}{d^2}}$

46. $\dfrac{\frac{5g^3}{h^2}}{\frac{6g}{5h}}$

47. $\dfrac{\frac{2}{h}}{\frac{4}{b-3}}$

48. $\dfrac{\frac{q-2}{9}}{\frac{q^2-6q+8}{12}}$

Find the LCM of each pair of polynomials. (Lesson 11-6)

49. $2h, 4h^2$ **50.** $5c^2, 12c^3$ **51.** $x - 4, x + 2$ **52.** $p - 7, 2(p - 14)$

Look for a pattern in each table of values to determine which kind of model best describes the data. (Lesson 9-6)

53.

x	0	1	2	3	4
y	4	5	6	7	8

54.

x	1	2	3	4	5
y	2	4	8	16	32

55.

x	−3	−2	−1	0	1
y	14	9	6	5	6

56.

x	3	4	5	6	7
y	3	5	7	9	11

57. GENETICS Brown genes B are dominant over blue genes b. A person with genes BB or Bb has brown eyes. Someone with genes bb has blue eyes. Mrs. Dunn has brown eyes with genes Bb, and Mr. Dunn has blue eyes. Write an expression for the possible eye coloring of their children. Then find the probability that a child would have blue eyes. (Lesson 8-4)

Solve each inequality. Check your solution. (Lesson 5-2)

58. $\dfrac{b}{10} \le 5$ **59.** $-7 > -\dfrac{r}{7}$ **60.** $\dfrac{5}{8}y \ge -15$

Skills Review

Determine the probability of each event if you randomly select a marble from a bag containing 9 red marbles, 6 blue marbles, and 5 yellow marbles.

61. P(blue)   **62.** P(red)   **63.** P(not yellow)

11-8 Graphing Technology Lab
Solving Rational Equations

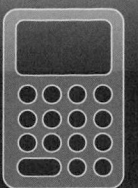

You can use TI-Nspire Technology to solve rational equations by graphing, by using tables, and by using a computer algebra system (CAS).

To solve by graphing, graph both sides of the equation and locate the point(s) of intersection.

CCSS Common Core State Standards
Content Standards
A.REI.11 Explain why the x-coordinates of the points where the graphs of the equations $y = f(x)$ and $y = g(x)$ intersect are the solutions of the equation $f(x) = g(x)$; find the solutions approximately, e.g., using technology to graph the functions, make tables of values, or find successive approximations. Include cases where $f(x)$ and/or $g(x)$ are linear, polynomial, rational, absolute value, exponential, and logarithmic functions.

Mathematical Practices
5 Use appropriate tools strategically.

Activity 1   Solve a Rational Equation by Graphing

Solve $\dfrac{5}{x+2} = \dfrac{3}{x}$ by graphing.

Step 1 Add a new **Graphs** page.

Step 2 Use the **Window Settings** option from the **Window/Zoom** menu to adjust the window to -20 to 20 for both x and y. Set both scales to 2.

Step 3 Enter $\dfrac{5}{x+2}$ into **f1(x)** and $\dfrac{3}{x}$ into **f2(x)**.

Step 4 Change the thickness of the graph of **f1(x)** by selecting the graph of **f1(x)** and the **ctrl menu Attributes** option.

Step 5 Use the **Intersection Point(s)** tool from the **Points & Lines** menu to find the intersection of the two graphs. Select the graph of **f1(x)** enter and then the graph of **f2(x)** enter.

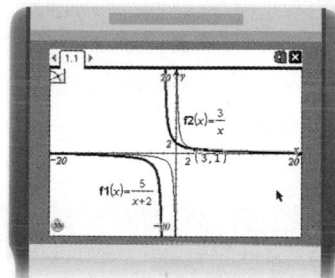

$[-20, 20]$ scl: 2 by $[-20, 20]$ scl: 2

The graphs intersect at $(3, 1)$. This means that $\dfrac{5}{x+2}$ and $\dfrac{3}{x}$ both equal 1 when $x = 3$. Thus, the solution of $\dfrac{5}{x+2} = \dfrac{3}{x}$ is $x = 3$.

Exercises

Use a graphing calculator to solve each equation.

1. $\dfrac{5}{x} + \dfrac{4}{x} = 10$

2. $\dfrac{12}{x} + \dfrac{3}{4} = \dfrac{3}{2}$

3. $\dfrac{6}{x} + \dfrac{3}{2x} = 12$

4. $\dfrac{4}{x} + \dfrac{3}{4x} = \dfrac{1}{8}$

5. $\dfrac{4}{x} + \dfrac{x-2}{2x} = x$

6. $\dfrac{3}{3x-2} + \dfrac{5}{x} = 0$

7. $\dfrac{2x+1}{2} + \dfrac{3}{2x} = \dfrac{2}{x}$

8. $\dfrac{x}{x+2} + x = \dfrac{5x+8}{x+2}$

9. $\dfrac{1}{2x} + \dfrac{5}{x} = \dfrac{3}{x-1}$

10. $\dfrac{4x-3}{x-2} + \dfrac{2x+5}{x-2} = 6$

Activity 2 Solve a Rational Equation by Using a Table

Solve $\dfrac{2x+1}{3} = \dfrac{x+2}{2}$ using a table.

Step 1 Add a new **Lists & Spreadsheet** page.

Step 2 Label column A as x. Enter values from -4 to 4 in cells A1 to A9.

Step 3 In column B in the formula row, enter the left side of the rational equation, with parenthesis around the binomials. In column C in the formula row, enter the right side of the rational equation, with parenthesis around the binomials. Specify **Variable Reference** when prompted.

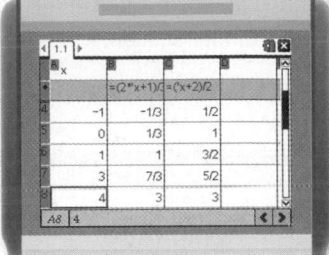

Scroll until you see where the values in Columns B and C are equal. This occurs at $x = 4$. Therefore the solution of $\dfrac{2x+1}{3} = \dfrac{x+2}{2}$ is 4.

You can also use a computer algebra system (CAS) to solve rational equations.

Activity 3 Solve a Rational Equation by Using a CAS

Solve $\dfrac{x-3}{x} - \dfrac{x-4}{x-2} = \dfrac{1}{x}$ using a CAS.

Step1 Add a new **Calculator** page.

Step 2 To solve, select the **Solve** tool from the **Algebra** menu. Enter the left side of the equation with parenthesis around the binomials. Enter = and the right side of the equation. Then type a comma, followed by x, and then **enter**.

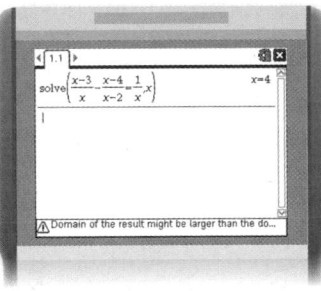

The solution of 4 is displayed.

Exercises

Use a table or CAS to solve each equation.

11. $\dfrac{2}{x} + \dfrac{2+x}{2} = \dfrac{x+3}{2}$

12. $\dfrac{4}{x-2} = -\dfrac{1}{x+3}$

13. $\dfrac{3}{x+2} + \dfrac{4}{x-1} = 0$

14. $\dfrac{1}{x+1} + \dfrac{2}{x-1} = 0$

15. $\dfrac{2}{x+4} + \dfrac{4}{x-1} = 0$

16. $\dfrac{1}{x-2} + \dfrac{x+2}{4} = 2x$

17. $\dfrac{2x}{x+3} + \dfrac{x+1}{2} = x$

18. $\dfrac{2}{x-3} + \dfrac{3}{x-2} = \dfrac{4}{x}$

19. $\dfrac{x^2}{x+1} + \dfrac{x}{x-1} = x$

Study Guide

KeyConcepts

Inverse Variation (Lesson 11-1)

- You can use $\dfrac{x_1}{x_2} = \dfrac{y_2}{y_1}$ to solve problems involving inverse variation.

Rational Functions (Lesson 11-2)

- Excluded values are values of a variable that result in a denominator of zero.

- If vertical asymptotes occur, it will be at excluded values.

Rational Expressions (Lessons 11-3 and 11-4)

- Multiplying rational expressions is similar to multiplying rational numbers.

- Divide rational expressions by multiplying by the reciprocal of the divisor.

Dividing Polynomials (Lesson 11-5)

- To divide a polynomial by a monomial, divide each term of the polynomial by the monomial.

Adding and Subtracting Rational Expressions
(Lesson 11-6)

- Rewrite rational expressions with unlike denominators using the least common denominator (LCD). Then add or subtract.

Complex Fractions (Lesson 11-7)

- Simplify complex fractions by writing them as division problems.

Solving Rational Equations (Lesson 11-8)

- Use cross products to solve rational equations with a single fraction on each side of the equals sign.

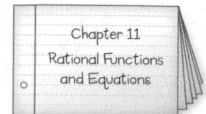 **StudyOrganizer**

Be sure the Key Concepts are noted in your Foldable.

Chapter 11
Rational Functions
and Equations

KeyVocabulary

asymptote (p. 685)

complex fraction (p. 720)

excluded value (p. 684)

extraneous solution (p. 727)

inverse variation (p. 676)

least common denominator
(LCD) (p. 714)

least common multiple
(LCM) (p. 713)

mixed expression (p. 720)

product rule (p. 677)

rate problems (p. 729)

rational equation (p. 726)

rational expression (p. 690)

rational function (p. 684)

work problems (p. 728)

VocabularyCheck

State whether each sentence is *true* or *false*. If *false*, replace the underlined word, phrase, expression, or number to make a true sentence.

1. The least common multiple for $x^2 - 25$ and $x - 5$ is $\underline{x - 5}$.

2. If the product of two variables is a nonzero constant, the relationship is an <u>inverse variation</u>.

3. If the line $x = a$ is a vertical <u>asymptote</u> of a rational function, then a is an excluded value.

4. A rational expression is a fraction in which the numerator and denominator are <u>fractions</u>.

5. The excluded values for $\dfrac{x}{x^2 + 5x + 6}$ are <u>−2 and −3</u>.

6. The equation $\dfrac{3x}{x - 2} = \dfrac{6}{x - 2}$ has an extraneous solution, <u>2</u>.

7. A <u>rational expression</u> has one or more fractions in the numerator and denominator.

8. The expression $\dfrac{\frac{1}{2}}{\frac{3}{4}}$ can be simplified to $\underline{\dfrac{2}{3}}$.

9. A <u>direct variation</u> can be represented by an equation of the form $k = xy$, where k is a nonzero constant.

10. The rational function $y = \dfrac{2}{x - 1} + 3$ has a horizontal asymptote at $\underline{y = 3}$.

Lesson-by-Lesson Review

11-1 Inverse Variation

Solve. Assume that y varies inversely as x.

11. If $y = 4$ when $x = 1$, find x when $y = 12$

12. If $y = -1$ when $x = -3$, find y when $x = -9$

13. If $y = 1.5$ when $x = 6$, find y when $x = -16$

14. PHYSICS A 135-pound person sits 5 feet from the center of a seesaw. How far from the center should a 108-pound person sit to balance the seesaw?

Example 1

If y varies inversely as x and $y = 28$ when $x = 42$, find y when $x = 56$.

Let $x_1 = 42$, $x_2 = 56$, and $y_1 = 28$. Solve for y_2.

$$\frac{x_1}{x_2} = \frac{y_2}{y_1} \qquad \text{Proportion for inverse variation}$$

$$\frac{42}{56} = \frac{y_2}{28} \qquad \text{Substitution}$$

$$1176 = 56y_2 \qquad \text{Cross multiply.}$$

$$21 = y_2$$

Thus, $y = 21$ when $x = 56$.

11-2 Rational Functions

State the excluded value for each function.

15. $y = \dfrac{1}{x-3}$

16. $y = \dfrac{2}{2x-5}$

17. $y = \dfrac{3}{3x-6}$

18. $y = \dfrac{-1}{2x+8}$

19. PIZZA PARTY Katelyn ordered pizza and soda for her study group for \$38. The cost per person y is given by $y = \dfrac{38}{x}$, where x is the number of people in the study group. Graph the function and describe the asymptotes.

Example 2

State the excluded value for the function $y = \dfrac{1}{4x+16}$.

Set the denominator equal to zero.

$$4x + 16 = 0$$

$$4x + 16 - 16 = 0 - 16 \qquad \text{Subtract 16 from each side.}$$

$$4x = -16 \qquad \text{Simplify.}$$

$$x = -4 \qquad \text{Divide each side by 4.}$$

11-3 Simplifying Rational Expressions

Simplify each expression.

20. $\dfrac{2xy^2}{16xyz}$

21. $\dfrac{x+4}{x^2+12x+32}$

22. $\dfrac{x^2+10x+21}{x^3+x^2-42x}$

23. $\dfrac{y^2-25}{y^2+3y-10}$

24. $\dfrac{3x^3}{3x^3+6x^2}$

25. $\dfrac{4y^2}{8y^4+16y^3}$

State the excluded values for each function.

26. $y = \dfrac{x}{x^2+9x+18}$

27. $y = \dfrac{10}{6x^2+7x-3}$

Example 3

Simplify $\dfrac{a^2-7a+12}{a^2-13a+36}$.

Factor and simplify.

$$\frac{a^2-7a+12}{a^2-13a+36} = \frac{(a-3)(a-4)}{(a-9)(a-4)} \qquad \text{Factor.}$$

$$= \frac{a-3}{a-9} \qquad \text{Simplify.}$$

11-4 Multiplying and Dividing Rational Expressions

Find each product or quotient.

28. $\dfrac{6x^2y^4}{12} \cdot \dfrac{3x^3y^2}{xy}$

29. $\dfrac{3x-6}{x^2-9} \cdot \dfrac{x+3}{x^2-2x}$

30. $\dfrac{x^2}{x+4} \div \dfrac{3x}{x^2-16}$

31. $\dfrac{3b-12}{b+4} \div \left(b^2-6b+8\right)$

32. $\dfrac{2a^2+7a-15}{a+5} \div \dfrac{9a^2-4}{3a+2}$

33. GEOMETRY Find the area of the rectangle shown. Write the answer in simplest form.

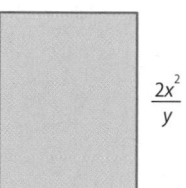

$\dfrac{2x^2}{y}$

$\dfrac{y^2}{2x}$

Example 4

Find $\dfrac{7b^2}{9} \cdot \dfrac{6a^2}{b}$.

$\dfrac{7b^2}{9} \cdot \dfrac{6a^2}{b} = \dfrac{42a^2b^2}{9b}$ Multiply.

$\phantom{\dfrac{7b^2}{9} \cdot \dfrac{6a^2}{b}} = \dfrac{14a^2b}{3}$ Simplify.

Example 5

Find $\dfrac{x^2-25}{x^2-9} \div \dfrac{x+5}{x-3}$.

$\dfrac{x^2-25}{x^2-9} \div \dfrac{x+5}{x-3} = \dfrac{(x+5)(x-5)}{(x+3)(x-3)} \div \dfrac{x+5}{x-3}$ Factor.

$ = \dfrac{\cancel{(x+5)}(x-5)}{(x+3)\cancel{(x-3)}} \cdot \dfrac{\cancel{x-3}}{\cancel{x+5}}$ Multiply by the reciprocal.

$ = \dfrac{x-5}{x+3}$ Simplify.

11-5 Dividing Polynomials

Find each quotient.

34. $\left(x^3-2x^2-22x+21\right) \div (x-3)$

35. $\left(x^3+7x^2+10x-6\right) : (x+3)$

36. $\left(5x^2y^2-10x^2y+5xy\right) \div 5xy$

37. $\left(48y^2+8y+7\right) \div (12y-1)$

38. GEOMETRY The area of a rectangle is $x^2+7x+13$. If the length is $(x+4)$, what is the width of the rectangle?

Example 6

Find $(4x^2+17x-1) \div (4x+1)$.

$$\require{enclose}\begin{array}{r} x+4 \\ 4x+1 \enclose{longdiv}{4x^2+17x-1} \\ \underline{4x^2+x} \\ 16x-1 \\ \underline{16x+4} \\ -5 \end{array}$$

Multiply x and $4x+1$.

Subtract, bring down -1.

Multiply 4 and $4x+1$.

Subtract.

The quotient is $x+4 - \dfrac{5}{4x+1}$.

11-6 Adding and Subtracting Rational Expressions

Find each sum or difference.

39. $\dfrac{5a}{b} - \dfrac{2a}{b}$ **40.** $\dfrac{-3}{2n-3} + \dfrac{2n}{2n-3}$

41. $\dfrac{3}{y+1} - \dfrac{y}{y-3}$ **42.** $\dfrac{1}{x+1} + \dfrac{3}{x-2}$

43. DESIGN Miguel is decorating a model of a room that is $\dfrac{2x}{x+4}$ feet long and $\dfrac{8}{x+4}$ feet wide. What is the perimeter of the room?

Example 7

Find $\dfrac{x^2}{x+1} + \dfrac{2x+1}{x+1}$.

$\dfrac{x^2}{x+1} + \dfrac{2x+1}{x+1} = \dfrac{x^2+2x+1}{x+1}$ Add the numerators.

$ = \dfrac{(x+1)(x+1)}{x+1}$ Factor.

$ = x+1$ Simplify.

11-7 Mixed Expressions and Complex Fractions

Simplify each expression.

44. $\dfrac{\frac{a^2 b^4}{c}}{\frac{a^3 b}{c^2}}$

45. $\dfrac{x - \frac{35}{x+2}}{x + \frac{42}{x+13}}$

46. $\dfrac{\frac{x^2 - 25}{x+2}}{\frac{x-5}{x^2 - 4}}$

47. $\dfrac{y + 9 - \frac{6}{y+4}}{y + 4 + \frac{2}{y+1}}$

48. FABRICS Donna makes sets of three different sized tablecloths to sell. A small one takes one-half yard of fabric, a medium one takes five-eighths yard, and a large one takes one and one-quarter yards.

　　a. How many yards of fabric does she need to make a tablecloth of each size?

　　b. If Donna can make a tablecloth of each size in 38 minutes, what is the average rate of fabric she uses in yards per hour?

Example 8

Simplify $\dfrac{\frac{x+3}{6}}{\frac{x^2 - 2x - 15}{x}}$.

Write as a division expression.

$$\dfrac{\frac{x+3}{6}}{\frac{x^2 - 2x - 15}{x}} = \frac{x+3}{6} \div \frac{x^2 - 2x - 15}{x}$$

$$= \frac{x+3}{6} \cdot \frac{x}{x^2 - 2x - 15}$$

$$= \frac{x+3}{6} \cdot \frac{x}{(x+3)(x-5)}$$

$$= \frac{x}{6(x-5)}$$

11-8 Rational Equations

Solve each equation. State any extraneous solutions.

49. $\dfrac{5n}{6} + \dfrac{1}{n-2} = \dfrac{n+1}{3(n-2)}$

50. $\dfrac{4x}{3} + \dfrac{7}{2} = \dfrac{7x}{12} - 14$

51. $\dfrac{11}{2x} + \dfrac{2}{4x} = \dfrac{1}{4}$

52. $\dfrac{1}{x+4} - \dfrac{1}{x-1} = \dfrac{2}{x^2 + 3x - 4}$

53. $\dfrac{1}{n-2} = \dfrac{n}{8}$

54. PAINTING Anne can paint a room in 6 hours. Oljay can paint a room in 4 hours. How long will it take them to paint the room working together?

Example 9

Solve $\dfrac{3}{x^2 + 3x} + \dfrac{x+2}{x+3} = \dfrac{1}{x}$.

$$\frac{3}{x^2 + 3x} + \frac{x+2}{x+3} = \frac{1}{x}$$

$$x(x+3)\left(\frac{3}{x(x+3)}\right) + x(x+3)\left(\frac{x+2}{x+3}\right) = x(x+3)\left(\frac{1}{x}\right)$$

$$3 + x(x+2) = 1(x+3)$$

$$3 + x^2 + 2x = x + 3$$

$$x^2 + x = 0$$

$$x(x+1) = 0$$

$$x = 0 \text{ or } x = -1$$

The solution is -1, and there is an extraneous solution of 0.

Practice Test

Determine whether each table represents an inverse variation. Explain.

1.

x	y
2	10
4	12
8	14

2.

x	y
2	2
4	1
8	$\frac{1}{2}$

Find each product or quotient.

3. $\dfrac{(x+6)(x-2)}{x^3} \cdot \dfrac{7x^2}{x-3}$

4. $\dfrac{(x+3)}{y^2} \div \dfrac{x^2-9}{y}$

Solve. Assume that y varies inversely as x.

5. If $y = 3$ when $x = 9$, find x when $y = 1$.

6. If $y = 2$ when $x = 0.5$, find y when $x = 3$.

Simplify each expression. State the excluded values of the variables.

7. $\dfrac{z-6}{z^2-3z-18}$

8. $\dfrac{4x-28}{x^2-49}$

9. MULTIPLE CHOICE The area of a rectangle is $x^2 + 5x + 6$ square feet. If the width is $x + 2$, what is the length of the rectangle?

$x + 2$

A $x + 2$

B $x + 3$

C 1

D 3

Simplify each expression.

10. $\dfrac{2\frac{1}{3}}{3\frac{1}{2}}$

11. $\dfrac{\frac{x^2-25}{x-2}}{\frac{x-5}{x-2}}$

12. $\dfrac{\frac{a-4}{a^2+6a+8}}{\frac{a^2-3a-4}{a^2-a-6}}$

13. $\dfrac{\frac{y^2+10y+24}{y^2-9}}{\frac{3y^2+17y-6}{2y^2-11y+15}}$

Find each quotient.

14. $(2x^2 + 10x) \div 2x$

15. $(4x^2 - 8x + 5) \div (2x + 1)$

16. $(3x^2 - 14x - 3) \div (x - 5)$

Assume that y varies inversely as x. Write an inverse variation equation that relates x and y.

17. $y = 2$ when $x = 8$

18. $y = -3$ when $x = 1$

Find each sum or difference.

19. $\dfrac{15}{x} - \dfrac{6}{x}$

20. $\dfrac{t-5}{t-6} + \dfrac{t+8}{t-6}$

21. $\dfrac{1}{x-6} + \dfrac{3}{x-2}$

22. $\dfrac{2x^2+8x+5}{x^2-2x-24} - \dfrac{x}{x-6}$

State the excluded value or values for each function.

23. $y = \dfrac{6}{x-1}$

24. $y = \dfrac{5}{x^2-5x-24}$

Identify the asymptotes of each function.

25. $y = \dfrac{2}{(x-4)(x+2)}$

26. $y = \dfrac{4}{x^2+3x-28} + 2$

27. MULTIPLE CHOICE Lee can shovel the driveway in 3 hours, and Susan can shovel the driveway in 2 hours. How long will it take them working together?

F 6 hours

G 5 hours

H $\frac{3}{2}$ hours

J $\frac{6}{5}$ hours

28. PAINTING Sydney can paint a 60-square foot wall in 40 minutes. Working with her friend Cleveland, the two of them can paint the wall in 25 minutes. How long would it take Cleveland to do the job himself?

Preparing for Standardized Tests

Model with an Equation

In order to successfully solve some standardized test questions, you will need to be able to write equations to model different situations. Use this lesson to practice solving these types of problems.

Strategies for Modeling with Equations

Step 1

Read the problem statement carefully.

Ask yourself:

- What am I being asked to solve?

- What information is given in the problem?

- What is the unknown quantity that I need to find?

Step 2

Translate the problem statement into an equation.

- Assign a variable to the unknown quantity.

- Write the word sentence as a mathematical number sentence.

- Look for keywords such as *is, is the same as, is equal to,* or *is identical to* that indicate where to place the equals sign.

Step 3

Solve the equation.

- Solve for the unknown in the equation.

- Check your answer to be sure it is reasonable and that it answers the question in the problem statement.

Standardized Test Example

Read the problem. Identify what you need to know. Then use the information in the problem to solve.

It takes Craig 75 minutes to paint a small room. If Delsin can paint the same room in 60 minutes, how long would it take them to paint the room if they work together? Round to the nearest tenth.

A about 33.3 minutes

C about 45.1 minutes

B about 38.4 minutes

D about 50.3 minutes

Read the problem carefully. You know how long it takes Craig and Delsin to paint a room individually. Model the situation with an equation to find how long it would take them to paint the room if they work together.

Find the rate that each person works when painting individually.

Craig's rate: $\dfrac{1 \text{ job}}{75 \text{ minutes}} = \dfrac{1}{75}$ job per minute

Delsin's rate: $\dfrac{1 \text{ job}}{60 \text{ minutes}} = \dfrac{1}{60}$ job per minute

Let t represent the number of minutes it would take them to complete the job working together. Multiply each rate by the time t to represent the portion of the job done by each painter. Add these expressions and set them equal to 1 job. Then solve for t.

Portion that Craig completes	plus	portion that Delsin completes	equals	1 job.
$\dfrac{1}{75}t$	$+$	$\dfrac{1}{60}t$	$=$	1

Solve for t:

$\dfrac{1}{75}t + \dfrac{1}{60}t = 1$	Original equation
$300\left(\dfrac{1}{75}t + \dfrac{1}{60}t\right) = 300(1)$	Multiply each side by the LCD, 300.
$4t + 5t = 300$	Simplify.
$9t = 300$	Combine like terms
$t \approx 33.3$	Divide each side by 9.

So, it would take Craig and Delsin about 33.3 minutes to paint the room working together. The correct answer is A.

Exercises

Read each problem. Identify what you need to know. Then use the information in the problem to solve.

1. Hana can finish a puzzle in 6 hours, while Eric can finish one in 5 hours. How long would it take them to finish a puzzle together? Round to the nearest tenth.

   A about 1.8 hours

   B about 2.4 hours

   C about 2.5 hours

   D about 2.7 hours

2. Roberto wants to print 500 flyers for his landscaping business. His printer can complete the job in 35 minutes, and his brother's printer can print them in 45 minutes. How long would it take to print the flyers using both printers? Round to the nearest whole minute.

   F about 15 minutes

   G about 18 minutes

   H about 20 minutes

   J about 23 minutes

Multiple Choice

Read each problem. Then fill in the correct answer on the answer document provided by your teacher or on a sheet of paper.

1. What is the inverse variation equation for the numbers shown in the table?

x	y
−8	16
−4	32
2	−64
8	−16
16	−8

A $y = -2x$

B $y = 8x$

C $xy = 24$

D $xy = -128$

2. Suppose a square has a side length given by the expression $\frac{x+5}{8x}$. What is the perimeter of the square?

F $\frac{4x + 20}{5x}$

G $\frac{2x + 10}{x}$

H $\frac{x + 5}{4x}$

J $\frac{x + 5}{2x}$

3. Find the length of the hypotenuse if the legs of a right triangle are 2 centimeters and 10 centimeters.

A 8.1 cm

B 8.5 cm

C 9.6 cm

D 10.2 cm

Test-Taking Tip

Question 1 Sometimes you can eliminate answer choices as unreasonable because they do not fit the requirements of the problem. Choices A and B show direct variation equations, so they can be eliminated.

4. In 1990, the population of a country was about 3.66 million people. By 2010, this number had grown to about 4.04 million people. What was the annual rate of change in population from 1990 to 2010?

F about 15,000 people per year

G about 19,000 people per year

H about 24,000 people per year

J about 38,000 people per year

5. Ricky's Rentals rented 12 more bicycles than scooters last weekend for a total revenue of $2,125. How many scooters were rented?

Item	Rental Fee
Bicycle	$20
Scooter	$45

A 26

B 29

C 37

D 41

6. The table shows the relationship between calories and fat grams contained in orders of french fries from various restaurants.

Calories	Fat Grams
240	14
280	15
310	16
260	12
340	16
350	18
300	13

Assuming the data can best be described by a linear model, how many fat grams would be expected to be contained in a 315-calorie order of french fries?

F 12 fat grams

G 13 fat grams

H 16 fat grams

J 8051 fat grams

Short Response/Gridded Response

Record your answers on the answer sheet provided by your teacher or on a sheet of paper.

7. Suppose the first term of a geometric sequence is 3 and the fourth term is 192.

  a. What is the common ratio of the sequence?

  b. Write an equation that can be used to find the nth term of the sequence.

  c. What is the sixth term of the sequence?

8. **GRIDDED RESPONSE** Peggy is having a cement walkway installed around the perimeter of her swimming pool with the dimensions shown below. If $x = 3$ find the area, in square feet, of the pool and walkway.

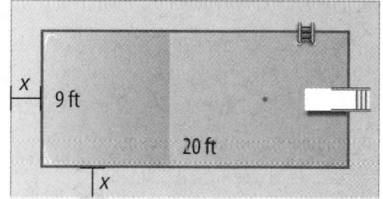

9. Use the equation $y = 2(4 + x)$ to answer each question.

  a. Complete the following table for the different values of x.

  b. Plot the points from the table on a coordinate grid. What do you notice about the points?

x	y
1	
2	
3	
4	
5	
6	

10. Jason received a $50 gift certificate for his birthday. He wants to buy a DVD and a poster from a media store. (Assume that sales tax is included in the prices.) Write and solve a linear inequality to show how much he would have left to spend after making these purchases.

Weekend Blowout Sale
- ★ All DVDs only **$14.95**
- ★ All CDs only **$11.25**
- ★ All posters only **$10.99**

11. Simplify the complex fraction. Show your work.

$$\frac{\dfrac{5}{x-3}}{\dfrac{x-6}{x^2-x-6}}$$

Extended Response

Record your answers on a sheet of paper. Show your work.

12. Carl's father is building a tool chest that is shaped like a rectangular prism. He wants the tool chest to have a surface area of 62 square feet. The height of the chest will be 1 foot shorter than the width. The length will be 3 feet longer than the height.

  a. Sketch a model to represent the problem.

  b. Write a polynomial that represents the surface area of the tool chest.

  c. What are the dimensions of the tool chest?

Need ExtraHelp?

If you missed Question...	1	2	3	4	5	6	7	8	9	10	11	12
Go to Lesson...	11-1	11-6	10-5	3-3	6-2	4-5	7-7	8-3	1-4	5-1	11-7	8-9

CHAPTER 12

Statistics and Probability

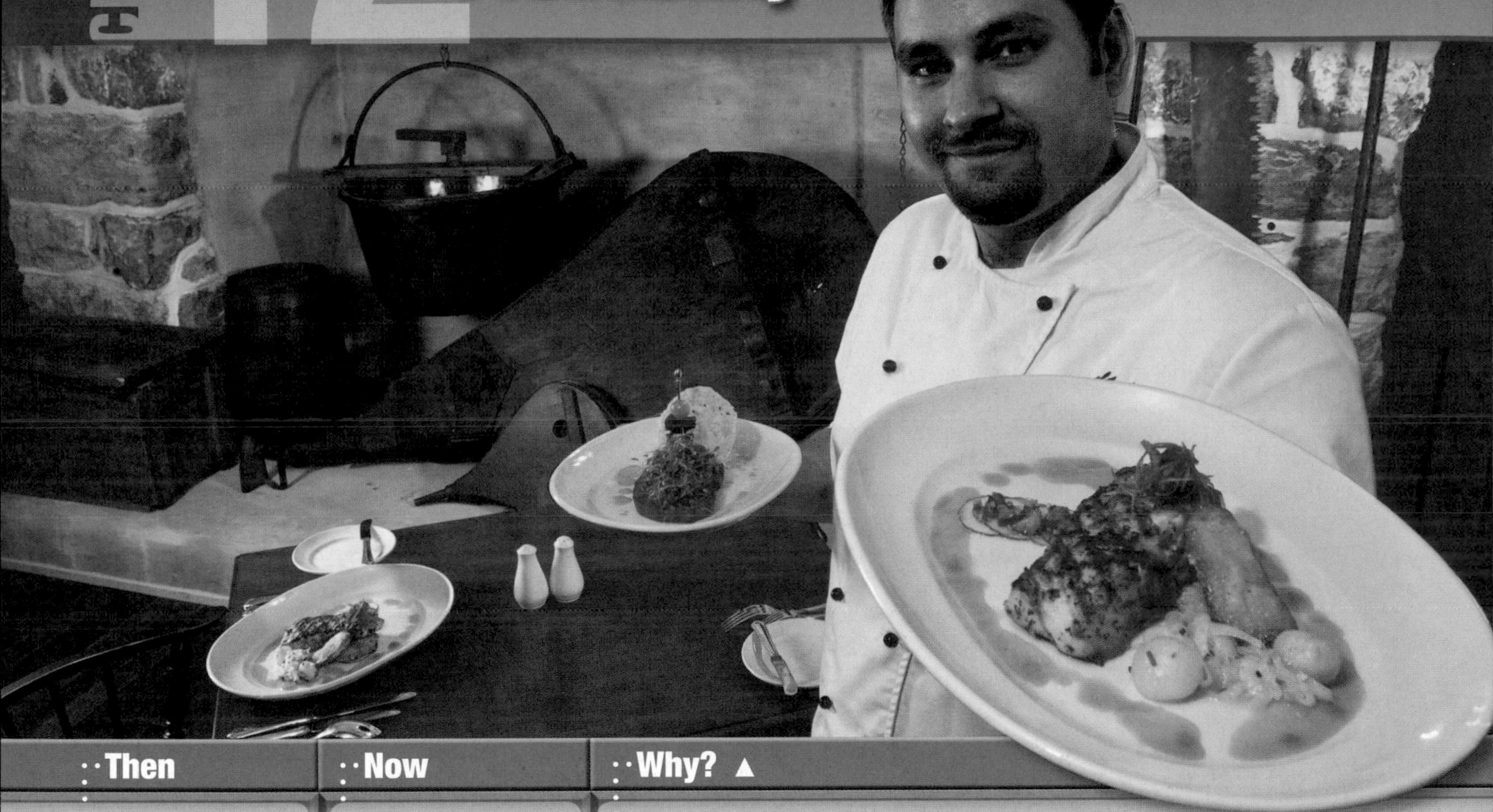

Then

- You calculated simple probability.

Now

- In this chapter, you will:
  - Design surveys and evaluate results.
  - Use permutations and combinations.
  - Find probabilities of compound events.
  - Design and use simulations.

Why? ▲

- **RESTAURANTS** A restaurant may ask their customers to complete a survey about their visit. The survey data can be analyzed using statistical methods. The restaurant staff can learn more about their customers and how to improve their experiences in the restaurant.

connectED.mcgraw-hill.com **Your Digital Math Portal**

Animation | Vocabulary | eGlossary | Personal Tutor | Virtual Manipulatives | Graphing Calculator | Audio | Foldables | Self-Check Practice | Worksheets

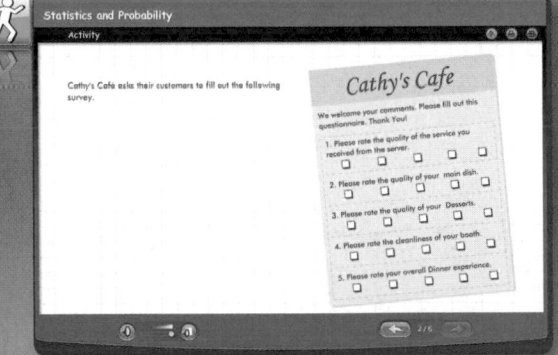

Richard Nowitz/National Geographic/Getty Images

Get Ready for the Chapter

Diagnose Readiness | You have two options for checking prerequisite skills.

1 **Textbook Option** Take the Quick Check below. Refer to the Quick Review for help.

QuickCheck	**Quick**Review

QuickCheck

Determine the probability of each event if you randomly select a cube from a bag containing 6 red cubes, 4 yellow cubes, 3 blue cubes, and 1 green cube.

1. P(red) 2. P(blue)

3. P(not red) 4. P(white)

5. Jim rolls a die with 6 sides. What is the probability of rolling a 5?

6. Malika spins a spinner that is divided into 8 equal sections. Each section is a different color, including blue. What is the probability the spinner lands on the blue section?

QuickReview

Example 1

Determine the probability of selecting a green cube if you randomly select a cube from a bag containing 6 red cubes, 4 yellow cubes, and 1 green cube.

There is 1 green cube and a total of 11 cubes in the bag.

$$\frac{1}{11} = \frac{\text{number of green cubes}}{\text{total number of cubes}}$$

The probability of selecting a green cube is $\frac{1}{11}$.

Find each product.

7. $\dfrac{5}{4} \cdot \dfrac{2}{3}$ 8. $\dfrac{4}{19} \cdot \dfrac{7}{20}$

9. $\dfrac{4}{32} \cdot \dfrac{7}{32}$ 10. $\dfrac{5}{12} \cdot \dfrac{6}{11}$

11. $\dfrac{56}{100} \cdot \dfrac{24}{100}$ 12. $\dfrac{9}{34} \cdot \dfrac{17}{27}$

Example 2

Find $\dfrac{4}{5} \cdot \dfrac{3}{4}$.

$\dfrac{4}{5} \cdot \dfrac{3}{4} = \dfrac{4 \cdot 3}{5 \cdot 4}$ Multiply the numerators and the denominators.

$= \dfrac{12}{20}$ Simplify.

$= \dfrac{3}{5}$ Rename in simplest form.

Write each fraction as a percent. Round to the nearest tenth.

13. $\dfrac{14}{17}$ 14. $\dfrac{7}{8}$

15. $\dfrac{107}{125}$ 16. $\dfrac{625}{1024}$

17. **SHOPPERS** At the mall, 700 of the 2000 people shopping were under the age of 21. What percent of the shoppers were under 21?

Example 3

Write the fraction $\dfrac{33}{80}$ as a percent. Round to the nearest tenth.

$\dfrac{33}{80} \approx 0.413$ Simplify and round.

$0.413 \cdot 100 = 41.3$ Multiply the decimal by 100.

$\dfrac{33}{80}$ written as a percent is about 41.3%.

---

2 **Online Option** Take an online self-check Chapter Readiness Quiz at connectED.mcgraw-hill.com.

Get Started on the Chapter

You will learn several new concepts, skills, and vocabulary terms as you study Chapter 12. To get ready, identify important terms and organize your resources. You may wish to refer to Chapter 0 to review prerequisite skills.

FOLDABLES StudyOrganizer

Statistics and Probability Make this Foldable to help you organize your Chapter 12 notes about Statistics and Probability. Begin with 8 sheets of $8\frac{1}{2}$" by 11" paper.

1 **Fold** each sheet of paper in half. Cut 1 inch from the end to the fold. Then cut 1 inch along the fold.

2 **Label** 7 of the 8 sheets with the lesson number and title.

3 **Label** the inside of each sheet with Definitions and Examples.

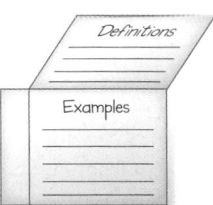

4 **Stack** the sheets. Staple along the left side. Write the title of the chapter on the first page.

NewVocabulary

English		Español
population	p. 747	población
sample	p. 747	muestra
bias	p. 748	tendencia
observational study	p. 749	estudio de observación
experiment	p. 749	experimento
statistic	p. 757	estadística
parameter	p. 757	parámetro
standard deviation	p. 759	desviación estándar
distribution	p. 764	distribución
symmetric distribution	p. 764	distribución simétrica
theoretical probability	p. 780	probabilidad teórica
experimental probability	p. 780	probabilidad experimental
simulation	p. 781	simulación
permutation	p. 786	permutación
combination	p. 787	combinación
compound event	p. 793	evento compuesto
independent events	p. 793	eventos independientes
dependent events	p. 794	eventos dependientes
mutually exclusive	p. 795	mutuamente exclusivos
random variable	p. 803	variable aleatoria
probability distribution	p. 804	distribución de probabilidad
expected value	p. 805	valor esperado

ReviewVocabulary

probability probilidad the ratio of favorable outcomes to the total possible outcomes

sample space espacio muestral the list of all possible outcomes

Then
- You displayed results from studies.

Now
1 Classify and analyze samples.
2 Classify and analyze studies.

Why?
- A high school principal is trying to determine whether the school should change its mascot and decides to survey some of the students.

NewVocabulary
population
sample
simple random sample
systematic sample
self-selected sample
convenience sample
stratified sample
bias
survey
observational study
experiment

1 Sampling A **population** consists of all of the members of a group of interest. Since it may be impractical to examine every member of a population, a **sample** or subset is sometimes selected to represent the population. The sample can then be analyzed to draw conclusions about the entire population. For instance, in the example above, the population is the entire student body. The sample is the students who participate in the survey.

In a **simple random sample**, each member of the population has an equal chance of being selected as part of the sample. The principal can generate a simple random sample by randomly drawing student ID numbers from a hat that contains the ID number of every student.

Some other types of samples are described below.

In a **systematic sample**, members are selected according to a specified interval from a random starting point, such as selecting every third student.

In a **self-selected sample**, members volunteer to be included in the sample.

freshmen   sophomores

juniors   seniors

In a **convenience sample**, members that are readily available or easy to reach are selected, such as the students on a particular bus.

In a **stratified sample**, the population is first divided into similar, nonoverlapping groups. Members are then randomly selected from each group.

Factors that can influence the type of sample used include cost, time, and the availability of willing participants.

Jeff Minton/Stone/Getty Images

Real-World Example 1 Classify a Random Sample

ZOOS Animals in a zoo are divided by species. Then two animals are selected at random from each group to have their blood tested.

a. Identify the sample, and suggest a population from which it was selected.

Sample: the two animals selected from each species
Population: all of the animals in the zoo

b. Classify the sample as *simple, systematic, self-selected, convenience*, or *stratified*. Explain your reasoning.

This is a stratified sample. The animals were divided into categories before there was a random selection.

GuidedPractice

Identify each sample, and suggest a population from which it was selected. Then classify the sample as *simple, systematic, self-selected, convenience*, or *stratified*. Explain your reasoning.

1A. CONTESTS Refer to the information at the left. After the competition, contestants were asked to complete an online survey about possible future categories.

1B. FOOD At a popular diner, the manager checks the quality of the burgers every 20 minutes, starting at a randomly selected time.

Real-World Link

The Jamaican Jerk Festival is held annually in Pembroke Pines, Florida. Cooks may enter a competition where each entrant must prepare any three dishes from the following categories: jerked pork, jerked chicken, jerked seafood, or other.

Source: Jamaican Jerk Festival

Sample data are often used to estimate a characteristic of a population. Therefore, a sample should be selected so that it is representative of the entire population. Also, the larger the sample size, or the more samples taken, the more closely it approximates the population.

A **bias** is an error that results in a misrepresentation of a population. If a sample favors one conclusion over another, the sample is biased and the data are invalid.

Example 2 Biased and Unbiased Samples

Identify each sample as *biased* or *unbiased*. Explain your reasoning.

a. MUSIC Every fifth person coming into a grocery store is asked to name their favorite radio station.

The sample is unbiased because the participants are randomly selected. The fact that the sample is selected at a grocery store has no bearing on the conclusion because many different kinds of people shop at grocery stores.

b. MUSIC Every fifth person at the Country Music Showcase is asked to name their favorite radio station.

The sample is biased because the participants are selected at a country music show, and therefore people may be more likely to select a country music station.

GuidedPractice

2A. POLITICS A journalist surveys 15 student members of the Young Republicans to determine the overall political opinions of the student body.

2B. SHOES A shoe store conducts a study of which shoes are most popular. The store surveys every third girl and boy that enters the store.

2 **Studies** After a sample is selected, information can be collected using one of the following study types.

KeyConcept Study Types

Type	Definition	Example
survey	Data are collected from responses given by a sample regarding their characteristics, behaviors, or opinions.	To determine whether the student body is happy with the spring dance theme, the dance committee asks a sample of students for their opinion.
observational study	Members of a sample are measured or observed without being affected by the study.	A gaming company watches a group of teens play a selection of video games and notes the ones they play the most.
experiment	The sample is divided into two groups: • an *experimental group* that undergoes a change, and • a *control group* that does not undergo the change. The effect on the experimental group is then compared to the control group.	A teacher administers a paper-and-pencil test to one of two Algebra classes. He administers a computer-based test that covers the same material to the other class. The teacher compares the scores and the completion times of the two classes.

Factors that can influence the type of study conducted are cost, time, and the objective of the study.

Example 3 Classify Study Techniques

Determine whether each situation describes a *survey*, an *observational study*, or an *experiment*. Explain your reasoning.

a. CHARITY A local charity is interested in finding out whether residents would use a curbside pick-up service for donations. They distribute 30 questionnaires to people living in the neighborhood.

This is a survey. The data are gathered from responses given by members of the sample.

b. ADVERTISING A company shows five different commercials that advertise the same product to a group of students. The company records the students' reactions to each.

This is an observational study. The data are gathered from observing the students.

▶ **GuidedPractice**

Determine whether each situation describes a *survey*, an *observational study*, or an *experiment*. Explain your reasoning.

3A. RESEARCH Scientists study the behavior of two groups of rats to determine their reaction to sugar.

Group 1 — Food with sugar   Group 2 — Food with no sugar

3B. RECREATION The city council wants to build a community recreational center. They call 1000 random citizens asking if they would consider paying a fee to join if the center is built.

The design of a survey can introduce bias. Survey questions should not:

- be confusing,
- encourage the members of the sample to answer a certain way,
- cause a strong reaction, or
- address more than one issue at a time.

Example 4 Biased and Unbiased Survey Questions

Identify each survey question as *biased* or *unbiased*. If biased, explain your reasoning.

a. Is your favorite ice cream flavor plain vanilla or delicious chocolate?

This question is biased. It limits the responses to vanilla or chocolate and encourages the participant to answer a certain way by using the adjectives *plain* and *delicious*.

b. What type(s) of reading material(s) do you enjoy?

This question is unbiased. It does not encourage participants to answer a certain way, and it is clearly stated.

▶ **Guided**Practice

Identify each survey question as *biased* or *unbiased*. If biased, explain your reasoning.

4A. Do you like animals, and would you ever consider having a dog, a cat, or a hamster as a pet?

4B. How much time do you spend working on school work each night?

Bias can also be introduced in the design of an experiment. To avoid this, members of the control group and the experimental group should be randomly selected. Also, the only difference between the two groups should be the change being observed.

● Real-World Example 5 Biased and Unbiased Experimental Designs

TRAINER A dog trainer wants to test training speed with a new method. She selects three terriers as the control group and teaches a trick using her normal method. She selects three greyhounds as the experimental group and teaches the same trick using the new method. She then compares the training times. Identify the experiment as *biased* or *unbiased*. If biased, explain your reasoning.

This experiment is biased. A terrier may be more apt to learn tricks than a greyhound, or vice versa. The bias could be corrected by using a random sample consisting entirely of terriers or greyhounds.

▶ **Guided**Practice

5. TRACK A track coach wants to test whether new shoes that claim to increase speed will produce faster times for his runners. He randomly selects five runners as the control group and records their times throughout the season using their normal shoes. He randomly selects five runners as the experimental group and records their times throughout the season using the new shoes. At the end of the season, he compares any improvements made by each group. Identify the experiment as *biased* or *unbiased*. If biased, explain your reasoning.

Real-World Career

Dog Trainer Dog trainers may seek professional certification by passing an examination after compiling 300 hours experience in dog training, acquiring a high school diploma, and obtaining three letters of reference.

Source: Certification Council for Professional Dog Trainers

Example 1 Identify each sample, and suggest a population from which it was selected. Then classify the sample as *simple, systematic, self-selected, convenience,* or *stratified.* Explain your reasoning.

  1. SHOWER At a bridal shower, a sticker was placed on the bottom of three random plates. The guests who receive the stickered plates will win a prize.

  2. BOOK CLUB Mr. Peterson surveys the students in his English classes to gauge the student body's interest in forming a book club.

Example 2 Identify each sample as *biased* or *unbiased.* Explain your reasoning.

  3. ELECTION A group of students stands at the door of the school and asks every tenth student who they would vote for in the upcoming class elections and why.

  4. SHOPPING Every fifteenth shopper at a clothing store is asked what they would want most for their birthday.

Example 3 Determine whether each situation describes a *survey,* an *observational study,* or an *experiment.* Explain your reasoning.

  5. TELEVISION A television network wants to conduct a cartoon marathon. To choose the episodes, they mail a questionnaire to people selected at random throughout the country.

  6. FOOD A frozen food company is considering creating frozen meals with tofu instead of meat. At a testing, they randomly give half of a group of 100 people the meals with meat and the other half the same meals with tofu and ask the people how they like the meals.

Example 4 Identify each survey question as *biased* or *unbiased.* If biased, explain your reasoning.

  7 What are you planning to do over summer vacation?

  8. Do you think we should serve mouth-watering steak or chicken?

  9. Don't you think Suzanne should be the class president?

  10. What type of music do you listen to?

Example 5 Identify the experiment as *biased* or *unbiased.* If biased, explain your reasoning.

  11. POOLS A national pool manufacturer wants to determine if a new advertising strategy will increase sales. The company continues to use the normal advertising strategy in its stores located in Indiana, Pennsylvania, and Kentucky. The company uses the new advertising strategy in Florida, Louisiana, and Georgia. The company then compares the sales.

  12. EDUCATION A school district wants to determine if having school in session year-round will improve the performance of students. They select one of their schools to be in session year-round and compare the test scores of those students with the test scores of the other students in the district.

Check out our beautiful pools that can be used year-round!

Example 1   Identify each sample, and suggest a population from which it was selected. Then classify the sample as *simple, systematic, self-selected, convenience,* or *stratified.* Explain your reasoning.

13. **SPORTS CARDS** Greg divides his baseball cards by teams. Then he randomly selects four cards from each team and records the players' RBIs.

14. **CARS** The service manager at a car dealership inspects every fifth car to make sure that cars are detailed after being serviced.

15. **RAFFLE** The students who attended a prom committee meeting were each given a raffle ticket for a drawing of five prizes.

16. **MUSIC** A music store asks its customers to submit suggestions for local bands that should play on Friday nights.

Example 2   Identify each sample as *biased* or *unbiased.* Explain your reasoning.

17. **ACTORS** A random sample of ten people is asked to name their favorite actor.

18. **BASKETBALL** Every fifth athlete at a basketball camp is asked to name their favorite brand of basketball shoe.

19. **TELEVISION** Every tenth person entering a gas station is asked to name their favorite television program.

20. **MUSIC** Every fifth person entering a play is asked to name their favorite style of music.

Example 3   Determine whether each situation describes a *survey,* an *observational study,* or an *experiment.* Explain your reasoning.

21. **PARTIES** Federico is throwing a party for one of his friends. He is trying to decide on a theme. He sends a piece of paper in each invitation, asking guests questions to get their opinions.

22. **VOLUNTEER** Jaime finds 50 students, half of whom volunteer at a homeless shelter, and compares their grade point averages, extracurricular activities, and involvement in school clubs.

(23) **SOCCER** A researcher organizes a soccer game in hot weather. One team wears short-sleeved shirts, while the other team wears long sleeves.

24. **SALONS** A salon emails customers, asking them to rate their experience during their last appointment.

Example 4   Identify each survey question as *biased* or *unbiased.* If biased, explain your reasoning.

25. What outdoor activities do you enjoy?

26. Do you think the comedian's stupid antics are funny?

27. Do you like to listen to music, read a book, or watch movies?

28. What is your favorite Web site?

Example 5   Identify the experiment as *biased* or *unbiased.* If biased, explain your reasoning.

29. **EMPLOYMENT** The management of a company hopes to increase the morale of their employees. They select some employees at random and move them into two identical office buildings. They build a recreation room in one of the buildings for employees to use. They then compare the morale of the employees in each building.

30. **CONCERTS** The manager of a band wants to see if a light show will improve their concerts. For the final date of the band's tour in their hometown, they perform while doing the light show. The manager then compares the reviews of this concert with the reviews of the rest of the tour.

Identify the observational study as *biased* or *unbiased*. If biased, explain your reasoning.

31. **GOLF** A golf club manufacturer watches a group of randomly selected golfers test a prototype club and notes their performances and reactions.

32. **MOVIES** A production company sets up a test audience of friends and family of the production crew to view the new movie and notes their reactions.

Determine whether each situation calls for a *survey*, an *experiment*, or an *observational study*. Explain your reasoning.

33. **GYMS** A gym owner wants to test whether changing the color of the walls improves member satisfaction.

34. **GAMING** A manufacturer invites twenty randomly selected teens to try out a new gaming system and notes their reactions as they play.

35. **SCHOOLS** A student asks 100 randomly selected neighbors if they think the school should build a new football field.

36. **SHOES** A shoe company's Web site allows customers to design their own shoes. This program keeps a count of styles and colors chosen by customers.

    a. Identify the sample. From what population was the sample selected?

    b. State the method of data collection.

    c. Tell whether the sample is *biased* or *unbiased*. Explain.

    d. If unbiased, classify the sample as *simple, stratified, systematic, self-selected,* or *convenience*.

H.O.T. Problems   Use Higher-Order Thinking Skills

37. **ERROR ANALYSIS** Amy and Esteban are describing one way to increase the accuracy of a survey. Is either of them correct? Explain your reasoning.

> **Amy**
> The survey should include as many people in the population as possible.

> **Esteban**
> The sample for the survey should be chosen randomly. Several random samples should be taken.

38. **CCSS CRITIQUE** Consider the following survey proposal.

    Question: How do students feel about the new dress code?
    Method: Take a simple random sample from each of the four classes. Use this sample to conduct the survey.
    Discuss the strengths and weaknesses of this survey.

39. **REASONING** Charlie wants to determine who the most popular athlete is. He conducts three different surveys. For the first survey, he asks 20 random students at school. For the second survey, he asks 50 random people at the mall. For the third survey, he asks 150 random people at a concert. The most popular athlete was different for each survey. Which survey do you think would most likely represent the population? Explain your reasoning.

40. **OPEN ENDED** Design and conduct a simple experiment.

41. **WRITING IN MATH** Why are accurate studies important to companies?

42. GRIDDED RESPONSE The first stage of a rocket burns 28 seconds longer than the second stage. If the total burning time for both stages is 152 seconds, how many seconds does the first stage burn?

43. Ms. Brinkman invested $30,000; part at 5%, and part at 8%. The total interest on the investment was $2100 after one year. How much did she invest at 8%?

  A $10,000   **C** $20,000

  B $15,000   **D** $25,000

44. A pair of $25 jeans is on sale for 15% off. What is the sale price?

  F $21.25   **H** $23.25

  G $22.25   **J** $24.25

45. GEOMETRY A piece of wire 42 centimeters long is bent into the shape of a rectangle with a width that is twice its length. Find the dimensions of the rectangle.

  A 5 cm, 12 cm   **C** 9 cm, 16 cm

  B 7 cm, 14 cm   **D** 11 cm, 18 cm

Spiral Review

Solve each equation. State any extraneous solutions. (Lesson 11-8)

46. $\dfrac{3}{c} = \dfrac{2}{c+2}$

47. $\dfrac{4}{f} = \dfrac{2}{f-3}$

48. $\dfrac{j}{j+2} = \dfrac{j-6}{j-2}$

49. $\dfrac{h-2}{h} = \dfrac{h-2}{h-5}$

50. $\dfrac{3m}{4} + \dfrac{1}{3} = \dfrac{3m+4}{6}$

51. $\dfrac{6}{5} + \dfrac{4p}{3} = \dfrac{8p}{5}$

52. $\dfrac{r-2}{r+2} - \dfrac{3r}{r-2} = -2$

53. $\dfrac{t-3}{t+3} - \dfrac{2t}{t-3} = -1$

54. $\dfrac{4v}{2v+3} - \dfrac{2v}{2v-3} = 1$

55. SPORTS When air is pumped into a ball, the pressure required can be computed by

using the formula $P = \dfrac{3412.94}{\frac{4\pi r^3}{3}}$, where P represents the pressure in pounds per

square inch (psi), and r is the radius of the ball in inches. (Lesson 11-7)

  a. Simplify the complex fraction.

  b. Suppose the air pressure inside the ball is 8 psi. Approximate the radius of the ball to the nearest hundredth.

56. ROLLER COASTERS Suppose a roller coaster climbs 208 feet higher than its starting point, while making a horizontal advance of 360 feet. When it comes down, it makes a horizontal advance of 44 feet. (Lesson 10-5)

  a. How far will it travel to get to the top of the ride?

  b. How far will it travel on the downhill track?

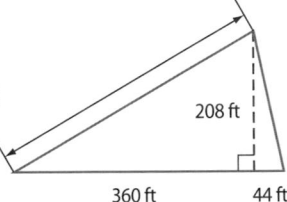

Skills Check

57. PHYSICAL SCIENCE Mr. Blackwell's students recorded the height of an object above the ground at several intervals after it was dropped from a height of 5 meters.

Time (s)	0	0.2	0.4	0.6	0.8	1
Height (cm)	500	480	422	324	186	10

Draw a graph showing the relationship between the height of the falling object and time.

Algebra Lab
Evaluating Published Data

An article in the *International Business Times* asserts that "the rate of decline in teen smoking in the United States has slowed down." This claim is supported by survey results from a study by the Centers for Disease Control and Prevention (CDC). This survey was conducted over an 18-year period from 1991 to 2009 and included a minimum of 10,904 to a maximum of 16,410 students in 110 to 159 high schools. The study reported cigarette use among high school students as shown in the following table. Current use was defined as smoking a cigarette at least one day in the last 30, while frequent use was defined as smoking at least 20 days in the last 30.

CCSS Common Core State Standards
Mathematical Practices
3 Construct viable arguments and critique the reasoning of others.

Year	1991	1993	1995	1997	1999	2001	2003	2005	2007	2009
Current Use (%)	27.5	30.5	34.8	36.4	34.8	28.5	21.9	23.0	20.0	19.5
Frequent Use (%)	12.7	13.8	16.1	16.7	16.8	13.8	9.7	9.4	8.1	7.3

The report contained the following graph comparing the numbers of current high school smokers and adults who consider themselves smokers.

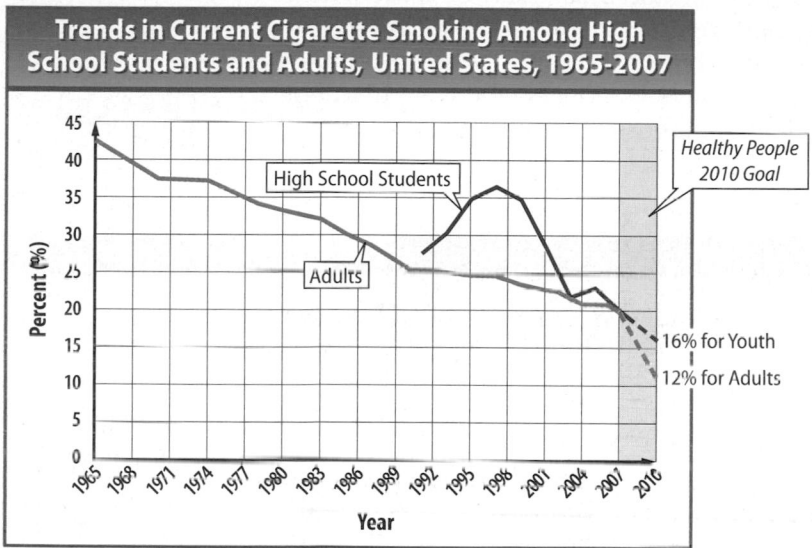

Trends in Current Cigarette Smoking Among High School Students and Adults, United States, 1965-2007

Activity 1 Evaluate the Design of a Study

Evaluate the design of this study and the source of the data.

a. What might be a problem that could occur when conducting this survey?

b. What biases could occur from their choice of a sample group?

a. A survey about activities that are deemed unacceptable might not get truthful responses from all students who complete the survey. Many students may refuse to complete the survey if they think the data might be used to punish them.

b. The study does not identify what types of schools were used in the survey. The sample may not represent all ethnic and economic groups that are found throughout the United States.

(continued on the next page)

Bias may be introduced when data are collected or by how the results are analyzed or reported.

Activity 2 Evaluate an Analysis and Display

For this study, evaluate how the data are analyzed and displayed.

a. Is there a problem with the conclusion made in the article that the rate of decline in teen smoking is slowing? If so, what is it?

b. The graph appears to show that percentage of high school smokers is greater than the percentage of adult smokers. Is this accurate? Explain.

a. The assertion generalizes all teens, while the survey was taken from high school students. Teens who have graduated or dropped out of school were not included and may have vastly different smoking habits than the sample group.

b. The line labeled *Adults* represents the percent of adults who classify themselves as current cigarette smokers. These adults would most likely be those who smoke on a daily basis. The line labeled *High School Students* is for those students who have smoked at least once in the last 30 days. A better comparison with adult smokers would be those high school students who were frequent smokers. These percents would be less than those for adults.

Activity 3 Compare Data

On a graphing calculator, make a double line graph comparing the rates of change of the current use and frequent use of cigarettes by high school students.

Step 1 First enter the data into L1, L2, and L3. Press [STAT] 1 and enter the years from 1991 to 2009 in **L1**, then the percents of current use in **L2**, and the percents for frequent use in **L3**.

Step 2 Use the **Stat Plot** function to produce a double line graph. Use **Plot1** to graph the line for the current use and **Plot2** to graph the line for the frequent use.

Step 3 Next, adjust the window settings and graph the double line graph.

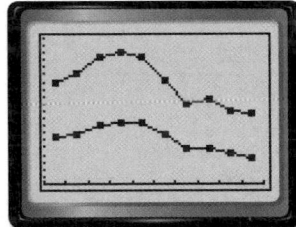

[1990, 2010] scl: 2 by [0, 40] scl: 2

Analyze the Results

1. Would a graph of frequent smokers imply a conclusion similar to the one implied by the graph of current smokers for all high school students? Explain.

2. What might be a reason that the CDC would use the percents of current use rather than frequent use?

3. List five things to consider whenever you are evaluating published data in the media.

4. Find a report in the media and evaluate the published data using your list from Exercise 3.

Statistics and Parameters

Halfdark/fStop/Getty Images

∷ Then

- You analyzed data collection techniques.

∷ Now

- **1** Identify sample statistics and population parameters.

- **2** Analyze data sets using statistics.

∷ Why?

- At the start of every class period for one week, each of Mr. Day's algebra students randomly draws 9 pennies from a jar of 1000 pennies. Each student calculates the mean age of the random sample of pennies drawn and then returns the pennies to the jar.

  How does the mean age for 9 pennies compare to the mean age of all 1000 pennies?

 NewVocabulary
statistical inference
statistic
parameter
mean absolute deviation (MAD)
standard deviation
variance

 Common Core State Standards

Content Standards
S.ID.2 Use statistics appropriate to the shape of the data distribution to compare center (median, mean) and spread (interquartile range, standard deviation) of two or more different data sets.

Mathematical Practices
2 Reason abstractly and quantitatively.
6 Attend to precision.

1 **Statistics and Parameters** The statistics of a sample are used to draw conclusions about the entire population. This is called **statistical inference**. In the scenario above, each student takes a random sample of pennies from the jar. The jar of 1000 pennies represents the population.

A **statistic** is a measure that describes a characteristic of a sample. A **parameter** is a measure that describes a characteristic of a population. Parameters are fixed values that can be determined by the entire population, but are typically estimated based on the statistics of a carefully chosen random sample. A statistic can and usually will vary from sample to sample. A parameter will not change, for it represents the entire population.

Example 1 Statistics and Parameters

Identify the sample and the population for each situation. Then describe the sample statistic and the population parameter.

a. **At a local university, a random sample of 40 scholarship applicants is selected. The mean grade-point average of the 40 applicants is calculated.**

Sample: the group of 40 scholarship applicants
Population: all applicants
Sample statistic: mean grade-point average of the sample
Population parameter: mean grade-point average of all applicants

b. **A stratified random sample of registered nurses is selected from all hospitals in a three-county area, and the median salary is calculated.**

Sample: randomly selected registered nurses from hospitals in the three-county area
Population: all nurses at the hospitals in the same region
Sample statistic: median salary of nurses in the sample
Population parameter: median salary of all nurses in all hospitals in a three-county area.

▸ **Guided**Practice

1. **CEREAL** Starting with a randomly selected box of cereal from the manufacturing line, every 50th box of cereal is removed and weighed. The mode weight of a day's sample is calculated.

2 Statistical Analysis

As shown in Lesson 0-12, univariate data can be represented by measures of central tendency, such as the mean, median, and mode. Univariate data can also be represented by measures of variation that assess the variability of the data. Some examples are the range, quartiles, interquartile range, mean absolute deviation, and standard deviation.

The **mean absolute deviation (MAD)** is the average of the absolute values of the differences between the mean and each value in the data set. The mean absolute deviation is used to predict errors and judge how well the mean represents the data.

KeyConcept Mean Absolute Deviation

Step 1 Find the mean, $\bar{x}$.

Step 2 Find the absolute value of the difference between each data value x_n and the mean, $|\bar{x} - x_n|$.

Step 3 Find the sum of all of the values in Step 2.

Step 4 Divide the sum by the number of values in the set of data n.

Formula $\text{MAD} = \dfrac{|\bar{x} - x_1| + |\bar{x} - x_2| + \ldots + |\bar{x} - x_n|}{n}$

StudyTip

Ellipsis The ellipsis in the formula for MAD denotes "and so on". All of the terms between the second term and the last term are implied to save space.

Example 2 Mean Absolute Deviation

MARKETING Each person who visited the Comic Book Shoppe's Web site was asked to enter the number of comic books they buy each month. They received the following responses in one day: {2, 2, 3, 4, 14}. Find and interpret the mean absolute deviation.

Step 1 Find the mean.

$$\bar{x} = \frac{2 + 2 + 3 + 4 + 14}{5} \text{ or } 5$$

Step 2 Find the absolute values of the differences.

$x_1 = 2: |\bar{x} - x_1| = |5 - 2| \text{ or } 3$ $x_2 = 2: |\bar{x} - x_2| = |5 - 2| \text{ or } 3$

$x_3 = 3: |\bar{x} - x_3| = |5 - 3| \text{ or } 2$ $x_4 = 4: |\bar{x} - x_4| = |5 - 4| \text{ or } 1$

$x_5 = 14: |\bar{x} - x_5| = |5 - 14| \text{ or } 9$

Step 3 Find the sum.

$$3 + 3 + 2 + 1 + 9 = 18$$

Step 4 Find the mean absolute deviation.

$\text{MAD} = \dfrac{|\bar{x} - x_1| + |\bar{x} - x_2| + \ldots + |\bar{x} - x_n|}{n}$ Formula for Mean Absolute Deviation

$= \dfrac{18}{5} \text{ or } 3.6$ The sum is 18 and $n = 5$.

A mean absolute deviation of 3.6 indicates that the data, on average, are 3.6 units away from the mean. This value is significantly influenced by the outlier 14. Without the outlier, the data set would have a mean of 2.75 and a mean absolute deviation of 0.75.

▶ **Guided**Practice

2. DANCES The prom committee kept a count of how many tickets it sold each day during lunch: {12, 32, 36, 41, 22, 47, 51, 33, 37, 49}. Find and interpret the mean absolute deviation of these data.

In a set of data, the **standard deviation** shows how the data deviate from the mean. A low standard deviation indicates that the data tend to be very close to the mean, while a high standard deviation indicates that the data are spread out over a larger range of values.

The standard deviation is represented by the lowercase Greek letter sigma, σ. The **variance** σ^2 of the data is the square of the standard deviation.

StudyTip

Symbols The mean of a sample and the mean of a population are calculated the same way. $\bar{x}$ refers to the mean of a sample and μ refers to the mean of a population. In this text, $\bar{x}$ will refer to both.

KeyConcept Standard Deviation

Step 1 Find the mean, $\bar{x}$.

Step 2 Find the square of the difference between each data value x_n and the mean, $(\bar{x} - x_n)^2$.

Step 3 Find the sum of all of the values in Step 2.

Step 4 Divide the sum by the number of values in the set of data n. This value is the variance.

Step 5 Take the square root of the variance.

Formula $\sigma = \sqrt{\dfrac{(\bar{x} - x_1)^2 + (\bar{x} - x_2)^2 + \ldots + (\bar{x} - x_n)^2}{n}}$

Example 3 Variance and Standard Deviation

ELECTRONICS Ed surveys his classmates to find out how many electronic gadgets each person has in their home. Find and interpret the standard deviation of the data set.

$$\{9, 10, 11, 6, 9, 11, 9, 8, 11, 8, 7, 9, 11, 11, 5\}$$

Step 1 Find the mean.

$$\bar{x} = \frac{9 + 10 + 11 + 6 + 9 + 11 + 9 + 8 + 11 + 8 + 7 + 9 + 11 + 11 + 5}{15} \text{ or } 9$$

Step 2 Find the square of the differences, $(\bar{x} - x_n)^2$.

$(9 - 9)^2 = 0$	$(9 - 10)^2 = 1$	$(9 - 11)^2 = 4$	$(9 - 6)^2 = 9$	$(9 - 9)^2 = 0$
$(9 - 11)^2 = 4$	$(9 - 9)^2 = 0$	$(9 - 8)^2 = 1$	$(9 - 11)^2 = 4$	$(9 - 8)^2 = 1$
$(9 - 7)^2 = 4$	$(9 - 9)^2 = 0$	$(9 - 11)^2 = 4$	$(9 - 11)^2 = 4$	$(9 - 5)^2 = 16$

Step 3 Find the sum.

$$0 + 1 + 4 + 9 + 0 + 4 + 0 + 1 + 4 + 1 + 4 + 0 + 4 + 4 + 16 = 52$$

Step 4 Find the variance.

$$\sigma^2 = \frac{(\bar{x} - x_1)^2 + (\bar{x} - x_2)^2 + \ldots + (\bar{x} - x_n)^2}{n} \qquad \text{Formula for Variance}$$

$$= \frac{52}{15} \text{ or about } 3.47 \qquad \text{The sum is 52 and } n = 15.$$

Step 5 Find the standard deviation.

$$\sigma = \sqrt{\sigma^2} \qquad \text{Square Root of the Variance}$$

$$\approx \sqrt{3.47} \text{ or about } 1.86$$

A standard deviation of 1.86 is small compared to the mean of 9. This suggests that most of the data values are relatively close to the mean.

▶ **GuidedPractice**

3. DIET Caleb tracked his Calorie intake for a week. Find and interpret the standard deviation of his Calorie intake.

1950, 2000, 2100, 2000, 1900, 2100, 2000

The mean and standard deviation can be used to compare two different sets of data.

StudyTip

Symbols The standard deviation of a sample s and the standard deviation of a population σ are calculated in different ways. In this text, you will calculate the standard deviation of a population.

Example 4 Compare Two Sets of Data

Miguel plays golf at Table Rock and Blackhawk golf courses. Compare the means and standard deviations of each set of Miguel's scores.

Table Rock				
81	78	79	82	80
80	79	83	81	80

Blackhawk				
84	79	86	78	77
88	85	79	87	86

Use a graphing calculator to find the mean and standard deviation. Clear all lists. Then press STAT ENTER, and enter each data value into **L1**. To view the statistics, press STAT ▶ 1 ENTER.

Table Rock

Blackhawk

Miguel's mean score at Table Rock is 80.3 with a standard deviation of about 1.4. His mean score at Blackhawk is 82.9 with a standard deviation of about 4.0. Therefore, he tends to score lower at Table Rock. The greater standard deviation at Blackhawk indicates that there is greater variability to his scores at that course, but he is more consistent at Table Rock.

▶ **Guided**Practice

4. SWIMMING Anna is considering two different lineups for her 4 × 100 relay team. Below are the times in minutes recorded for each lineup. Compare the means and standard deviations of each set of data.

Lineup A				
4.25	4.31	4.19	4.40	4.23
4.18	4.71	4.56	4.32	4.39

Lineup B				
4.47	4.68	4.25	4.41	4.49
4.18	4.27	4.69	4.32	4.44

Check Your Understanding

 = Step-by-Step Solutions begin on page R13.

Example 1

1. **BOOKS** A stratified random sample of 1000 college students in the United States is surveyed about how much money they spend on books per year. Identify the sample and the population. Then describe the sample statistic and the population parameter.

Example 2

2. **AMUSEMENT PARKS** An amusement park manager kept track of how many bags of cotton candy they sold each hour on a Saturday: {16, 24, 15, 17, 22, 16, 18, 24, 17, 13, 25, 21}. Find and interpret the mean absolute deviation.

Example 3

3 **PART-TIME JOBS** Ms. Johnson asks all of the members of the girls' tennis team to find the number of hours each week they work at part-time jobs: {10, 12, 0, 6, 9, 15, 12, 10, 11, 20}. Find and interpret the standard deviation of the data set.

Example 4

4. **CCSS MODELING** Mr. Jones recorded the number of pull-ups done by his students. Compare the means and standard deviations of each group.
Boys: {5, 16, 3, 8, 4, 12, 2, 15, 0, 1, 9, 3} Girls: {2, 4, 0, 3, 5, 4, 6, 1, 3, 8, 3, 4}

Example 1 Identify the sample and the population for each situation. Then describe the sample statistic and the population parameter.

5. **POLITICS** A random sample of 1003 Mercy County voters is asked if they would vote for the incumbent for governor. The percent responding *yes* is calculated.

6. **ACTIVITIES** A stratified random sample of high school students from each school in the county was polled about the time spent each week on extracurricular activities.

7. **MONEY** A stratified random sample of 2500 high school students across the country was asked how much money they spent each month.

Example 2 8. **DVDS** A math teacher asked all of his students to count the number of DVDs they owned. Find and interpret the mean absolute deviation.

Number of DVDs					
26	39	5	82	12	14
0	3	15	19	41	6
2	0	11	1	19	29

9. **SWIMMING** The owner of a public swimming pool tracked the daily attendance. Find and interpret the mean absolute deviation.

Daily Attendance					
86	45	91	104	95	88
111	85	79	102	166	103
89	94	79	103	88	84

Example 3 10. **CCSS REASONING** Samantha wants to see if she is getting a fair wage for babysitting at $8.50 per hour. She takes a survey of her friends to see what they charge per hour. The results are {$8.00, $8.50, $9.00, $7.50, $15.00, $8.25, $8.75}. Find and interpret the standard deviation of the data.

11. **ARCHERY** Carla participates in competitive archery. Each competition allows a maximum of 90 points. Carla's results for the last 8 competitions are {76, 78, 81, 75, 80, 80, 76, 77}. Find and interpret the standard deviation of the data.

Example 4 12. **BASKETBALL** The coach of the Wildcats basketball team is comparing the number of fouls called against his team with the number called against their rivals, the Trojans. He records the number of fouls called against each team for each game of the season. Compare the means and standard deviations of each set of data.

Wildcats			
15	12	13	9
11	12	14	12
8	16	9	9
11	13	12	14

Trojans			
9	10	14	13
7	8	10	10
9	7	11	9
12	11	13	8

13. **MOVIE RATINGS** Two movies were rated by the same group of students. Ratings were from 1 to 10, with 10 being the best.

a. Compare the means and standard deviations of each set of data.

b. Provide an argument for why Movie A would be preferred. Movie B?

Movie A			
7	8	7	6
8	6	7	8
6	8	8	6
7	7	8	8

Movie B			
9	5	10	6
3	10	9	4
8	3	9	9
2	8	10	3

14. PENNIES Mr. Day has another jar of pennies on his desk. There are 30 pennies in this jar. Theo chooses 5 pennies from the jar. Lola chooses 10 pennies, and Peter chooses 20 pennies. Pennies are chosen and replaced.

a. Theo's pennies are {1974, 1975, 1981, 1999, 1992}. Find the mean absolute deviation.

b. Lola's pennies are {2004, 1999, 2004, 2005, 1991, 2003, 2005, 2000, 2001, 1998}. Find the mean absolute deviation.

c. Peter's pennies are {2007, 2005, 1975, 2003, 2005, 1997, 1992, 1994, 1991, 1992, 2000, 1999, 2005, 1982, 2005, 2004, 1998, 2001, 2002, 2006}. Find the mean absolute deviation.

Years of Pennies in Jar					
2001	1990	2000	1982	1991	1975
2007	1981	2005	2007	2003	2005
1997	1974	1992	1994	1991	1992
2000	1995	1999	2005	2006	2005
2004	2004	1998	2001	2002	2006

d. Find the mean absolute deviation for all of the pennies in the jar. Which sample most accurately reflected the population mean? Explain.

15. RUNNING The results of a 5K race are published in a local paper. Over a thousand people participated, but only the times of the top 15 finishers are listed.

15th Annual 5K Road Race					
Place	Time (min:s)	Place	Time (min:s)	Place	Time (min:s)
1	15:56	6	16:34	11	17:14
2	16:06	7	16:41	12	17:46
3	16:11	8	16:54	13	17:56
4	16:21	9	17:00	14	17:57
5	16:26	10	17:03	15	18:03

a. Find the mean and standard deviation of the top 15 running times. (*Hint*: Convert each time to seconds.)

b. Identify the sample and population.

c. Analyze the sample. Classify the data as *quantitative* or *qualitative*. Can a statistical analysis of the sample be applied to the population? Explain your reasoning.

H.O.T. Problems   Use Higher-Order Thinking Skills

16. CCSS CRITIQUE Jennifer and Megan are determining one way to decrease the size of the standard deviation of a set of data. Is either of them correct? Explain.

> *Jennifer*
> Remove the outliers
> from the data set.

> *Megan*
> Add data values to the data
> set that are equal to the mean.

17. REASONING Determine whether the statement *Two random samples taken from the same population will have the same mean and standard deviation* is *sometimes*, *always*, or *never* true. Explain.

18. OPEN ENDED Describe a situation in which it would be useful to use a sample mean to help estimate a population mean. How could you collect a random sample?

19. CHALLENGE Write a set of data with a standard deviation that is equal to the mean absolute deviation.

WRITING IN MATH Compare and contrast each of the following.

20. statistics and parameters

21. standard deviation and mean absolute deviation

22. Melina bought a shirt that was marked 20% off. She paid $15.75. What was the original price?

   A $16.69   C $18.69

   B $17.69   D $19.69

23. **SHORT RESPONSE** A group of student ambassadors visited the Capitol building. Twenty students met with the local representative. This was 16% of the students. How many student ambassadors were there altogether?

24. The tallest 7 trees in a park have heights in meters of 19, 24, 17, 26, 24, 20, and 18. Find the mean absolute deviation of their heights.

   F 3.0   H 3.4

   G 3.2   J 21

25. It takes 3 hours for a boat to travel 27 miles upstream. The same boat can travel 30 miles downstream in 2 hours. Find the speed of the boat.

   A 3 mph   C 12 mph

   B 5 mph   D 14 mph

Spiral Review

Identify each sample as *biased* or *unbiased*. Explain your reasoning. (Lesson 12-1)

26. **SHOPPING** Every tenth person walking into the mall is asked to name their favorite store.

27. **MUSIC** Every fifth person at a rock concert is asked to name their favorite radio station.

Simplify each expression. (Lesson 11-3)

28. $\dfrac{x^2 - 8x + 15}{x^2 + 3x - 18}$

29. $\dfrac{x^2 - x - 12}{x^2 - 6x + 8}$

30. $\dfrac{x^2 - x - 30}{x^2 - 4x - 12}$

31. **SOCCER** The number of members of the local soccer association has increased by 6% every year. As of the beginning of 2010, there were 880 members. (Lesson 7-6)

   a. Write an equation for the number of members of the association t years after 2010.

   b. If this trend continues, predict how many members the association will have in 2020.

32. **GEOMETRY** If the side length of a cube is s, the volume is represented by s^3, and the surface area is represented by $6s^2$. (Lessons 7-1 and 7-2)

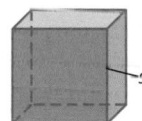

   a. Are the expressions for volume and surface area monomials? Explain.

   b. If the side of a cube measures 3 feet, find the volume and surface area.

   c. Find a side length s such that the volume and surface area have the same measure.

   d. The volume of a cylinder can be found by multiplying the radius squared times the height times π, or $V = \pi r^2 h$. Suppose you have two cylinders. Each measure of the second is twice the measure of the first, so $V = \pi(2r)^2(2h)$. What is the ratio of the volume of the first cylinder to the second cylinder?

Skills Review

Find the range, median, lower quartile, and upper quartile for each set of data.

33. {15, 23, 46, 36, 15, 19}

34. {55, 57, 39, 72, 46, 53, 81}

35. {21, 25, 19, 18, 22, 16, 27}

36. {52, 29, 72, 64, 33, 49, 51, 68}

37. {8, 12, 9, 11, 11, 10, 14, 18}

38. {133, 119, 147, 94, 141, 106, 118, 149}

12-3 Distributions of Data

·· Then	·· Now	·· Why?
● You calculated measures of central tendency and variation.	**1** Describe the shape of a distribution. **2** Use the shapes of distributions to select appropriate statistics.	● While training for the 100-meter dash, Sarah pulled a muscle in her lower back. After being cleared for practice, she continued to train. Sarah's median time was about 12.34 seconds, but her average time dropped to about 12.53 seconds.

 NewVocabulary
distribution
negatively skewed
    distribution
symmetric distribution
positively skewed
    distribution

 Common Core State Standards

Content Standards
S.ID.2 Use statistics appropriate to the shape of the data distribution to compare center (median, mean) and spread (interquartile range, standard deviation) of two or more different data sets.

S.ID.3 Interpret differences in shape, center, and spread in the context of the data sets, accounting for possible effects of extreme data points (outliers).

Mathematical Practices
5 Use appropriate tools strategically.

1 Describing Distributions A **distribution** of data shows the observed or theoretical frequency of each possible data value. Recall that a histogram is a type of bar graph used to display data that have been organized into equal intervals. A histogram is useful when viewing the overall distribution of the data within a set over its range. You can see the shape of the distribution by drawing a curve over the histogram.

KeyConcept Symmetric and Skewed Distributions

Negatively Skewed Distribution	Symmetric Distribution	Positively Skewed Distribution
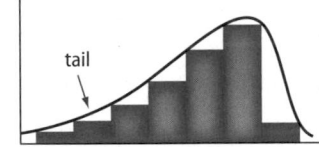	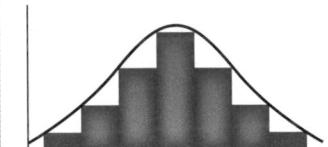	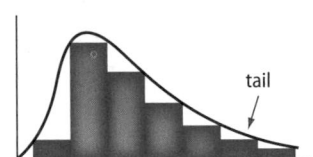
The majority of the data are on the right.	The data are evenly distributed.	The majority of the data are on the left.

Example 1 Distribution Using a Histogram

Use a graphing calculator to construct a histogram for the data, and use it to describe the shape of the distribution.

25, 22, 31, 25, 26, 35, 18, 39, 22, 32, 34, 26, 42, 23, 40, 36, 18, 30
26, 30, 37, 23, 19, 33, 24, 29, 39, 21, 43, 25, 34, 24, 26, 30, 21, 22

First, press [STAT] [ENTER] and enter each data value.
Then, press [2nd] [STAT PLOT] [ENTER] [ENTER] and choose ⬚. Press [ZOOM] [ZoomStat] to adjust the window.

The graph is high on the left and has a tail on the right. Therefore, the distribution is positively skewed.

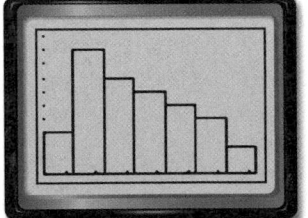

[17, 45] scl: 4 by [0, 10] scl: 1

▶ **Guided**Practice

1. Use a graphing calculator to construct a histogram for the data, and use it to describe the shape of the distribution.

8, 11, 15, 25, 21, 26, 20, 12, 32, 20, 31, 14, 19, 27, 22, 21, 14, 8
6, 23, 18, 16, 28, 25, 16, 20, 29, 24, 17, 35, 20, 27, 10, 16, 22, 12

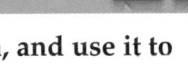

A box-and-whisker plot can also be used to identify the shape of a distribution. Recall from Lesson 0-13 that a box-and-whisker plot displays the spread of a data set by dividing it into four quartiles. The data from Example 1 are displayed below.

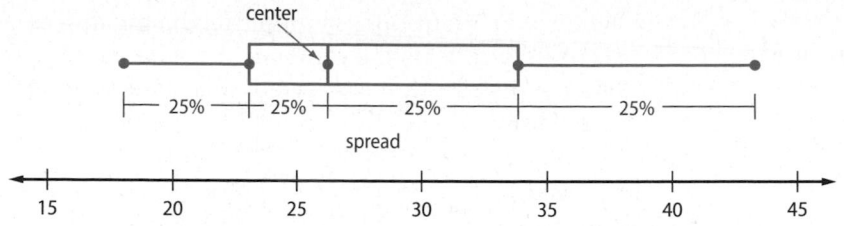

Notice that the left whisker is shorter than the right whisker, and that the line representing the median is closer to the left whisker. This represents a peak on the left and a tail to the right.

KeyConcept Symmetric and Skewed Box-and-Whisker Plots

Negatively Skewed	**Symmetric**	**Positively Skewed**
50% 50%	50% 50%	50% 50%
The left whisker is longer than the right. The median is closer to the shorter whisker.	The whiskers are the same length. The median is in the center of the data.	The right whisker is longer than the left. The median is closer to the shorter whisker.

Example 2 Distribution Using a Box-and-Whisker Plot

Use a graphing calculator to construct a box-and-whisker plot for the data, and use it to determine the shape of the distribution.

9, 17, 15, 10, 16, 2, 17, 19, 10, 18, 14, 8, 20, 20, 3, 21, 12, 11
5, 26, 15, 28, 12, 5, 27, 26, 15, 53, 12, 7, 22, 11, 8, 16, 22, 15

Enter the data as **L1**. Press 2nd [STAT PLOT] ENTER ENTER and choose ⊡⋯. Adjust the window to the dimensions shown.

The lengths of the whiskers are approximately equal, and the median is in the middle of the data. This indicates that the data are equally distributed to the left and right of the median. Thus, the distribution is symmetric.

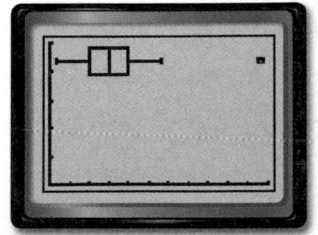

[0, 55] scl: 5 by [0, 5] scl: 1

StudyTip

Outliers In Example 2, notice that the outlier does not affect the shape of the distribution.

> **Guided**Practice

2. Use a graphing calculator to construct a box-and-whisker plot for the data, and use it to describe the shape of the distribution.

40, 50, 35, 48, 43, 31, 52, 42, 54, 38, 50, 46, 49, 43, 40, 50, 32, 53
51, 43, 47, 41, 49, 50, 34, 54, 51, 44, 54, 39, 47, 35, 51, 44, 48, 37

2 Analyzing Distributions You have learned that data can be described using statistics. The mean and median describe the center. The standard deviation and quartiles describe the spread. You can use the shape of the distribution to choose the most appropriate statistics that describe the center and spread of a set of data.

When a distribution is symmetric, the mean accurately reflects the center of the data. However, when a distribution is skewed, this statistic is not as reliable.

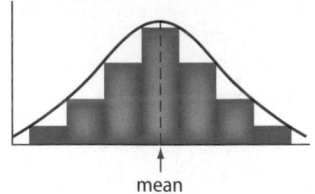

In Lesson 0-12, you discovered that outliers can have a strong effect on the mean of a data set, while the median is less affected. So, when a distribution is skewed, the mean lies away from the majority of the data toward the tail. The median is less affected and stays near the majority of the data.

Negatively Skewed Distribution

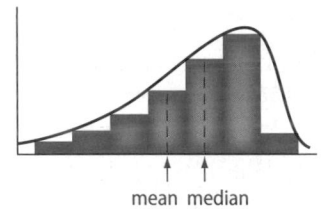

Positively Skewed Distribution

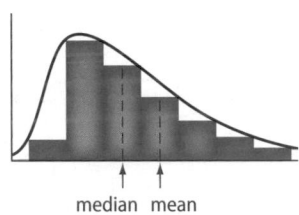

When choosing appropriate statistics to represent a set of data, first determine the shape of the distribution.

- If the distribution is relatively symmetric, the mean and standard deviation can be used.
- If the distribution is skewed or has outliers, use the five-number summary.

Example 3 Choose Appropriate Statistics

Describe the center and spread of the data using either the mean and standard deviation or the five-number summary. Justify your choice by constructing a histogram for the data.

21, 28, 16, 30, 25, 34, 21, 47, 18, 36, 24, 28, 30, 15, 33, 24, 32, 22
27, 38, 23, 29, 15, 27, 33, 19, 34, 29, 23, 26, 19, 30, 25, 13, 20, 25

Use a graphing calculator to create a histogram. The graph is high in the middle and low on the left and right. Therefore, the distribution is symmetric.

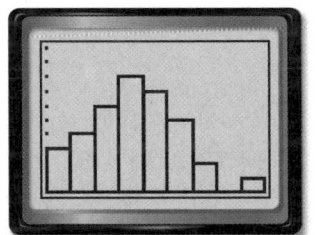

[12, 48] scl: 4 by [0, 10] scl: 1

The distribution is symmetric, so use the mean and standard deviation to describe the center and spread. Press STAT ▶ ENTER ENTER.

The mean $\bar{x}$ is about 26.1 with standard deviation σ of about 7.1.

TechnologyTip

CCSS Tools On a graphing calculator, each bar is called a *bin*. The width of each bin can be adjusted by pressing WINDOW and changing Xscl. View the histogram using different bin widths and compare the results to determine the appropriate bin width.

3. Describe the center and spread of the data using either the mean and standard deviation or the five-number summary. Justify your choice by creating a histogram for the data.

19, 2, 25, 14, 24, 20, 27, 30, 14, 25, 19, 32, 21, 31, 25, 16, 24, 22
29, 6, 26, 32, 17, 26, 24, 26, 32, 10, 28, 19, 26, 24, 11, 23, 19, 8

A box-and-whisker plot is helpful when viewing a skewed distribution since it is constructed using the five-number summary.

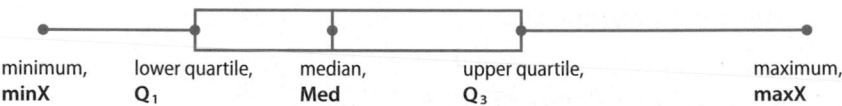

| minimum, minX | lower quartile, Q_1 | median, Med | upper quartile, Q_3 | maximum, maxX |

Real-World Example 4 Choose Appropriate Statistics

COMMUNITY SERVICE The number of community service hours each of Ms. Tucci's students completed is shown. Describe the center and spread of the data using either the mean and standard deviation or the five-number summary. Justify your choice by constructing a box-and-whisker plot for the data.

Community Service Hours												
6	13	8	7	19	12	2	19	11	22	7	33	13
3	8	10	5	25	16	6	14	7	20	10	30	

Use a graphing calculator to create a box-and-whisker plot. The right whisker is longer than the left and the median is closer to the left whisker. Therefore, the distribution is positively skewed.

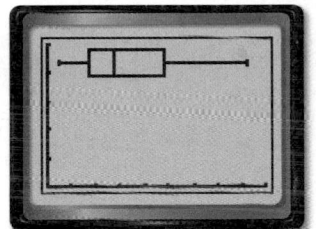

[0, 36] scl: 4 by [0, 5] scl: 1

The distribution is positively skewed, so use the five-number summary. The range is 33 − 2 or 31. The median number of hours completed is 11, and half of the students completed between 7 and 19 hours.

▶ GuidedPractice

4. **FUNDRAISER** The money raised per student in Mr. Bulanda's 5th period class is shown. Describe the center and spread of the data using either the mean and standard deviation or the five-number summary. Justify your choice by creating a box-and-whisker plot for the data.

Money Raised per Student (dollars)									
41	27	52	18	42	32	16	95	27	65
36	45	5	34	50	15	62	38	57	20
38	21	33	58	25	42	31	8	40	28

Paul Burns/Photodisc/Getty Images

Examples 1–2 Use a graphing calculator to construct a histogram and a box-and-whisker plot for the data. Then describe the shape of the distribution.

1. 80, 84, 68, 64, 57, 88, 61, 72, 76, 80, 83, 77, 78, 82, 65, 70, 83, 78
73, 79, 70, 62, 69, 66, 79, 80, 86, 82, 73, 75, 71, 81, 74, 83, 77, 73

2. 30, 24, 35, 84, 60, 42, 29, 16, 68, 47, 22, 74, 34, 21, 48, 91, 66, 51
33, 29, 18, 31, 54, 75, 23, 45, 25, 32, 57, 40, 23, 32, 47, 67, 62, 23

Example 3 Describe the center and spread of the data using either the mean and standard deviation or the five-number summary. Justify your choice by constructing a histogram for the data.

3. 58, 66, 52, 75, 60, 56, 78, 63, 59, 54, 60, 67, 72, 80, 68, 88, 55, 60
59, 61, 82, 70, 67, 60, 58, 86, 74, 61, 92, 76, 58, 62, 66, 74, 69, 64

Example 4 **4. PRESENTATIONS** The length of the students' presentations in Ms. Monroe's 2nd period class are shown. Describe the center and spread of the data using either the mean and standard deviation or the five-number summary. Justify your choice by constructing a box-and-whisker plot for the data.

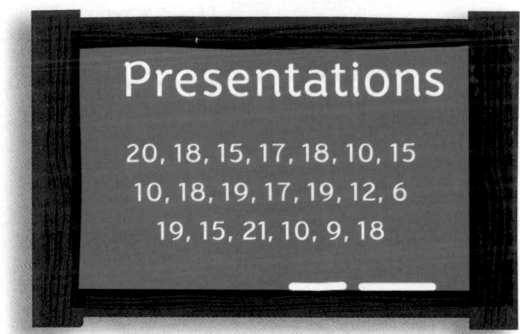

Presentations

20, 18, 15, 17, 18, 10, 15
10, 18, 19, 17, 19, 12, 6
19, 15, 21, 10, 9, 18

Practice and Problem Solving

Extra Practice is on page R12.

Examples 1–2 Use a graphing calculator to construct a histogram and a box-and-whisker plot for the data. Then describe the shape of the distribution.

5. 55, 65, 70, 73, 25, 36, 33, 47, 52, 54, 55, 60, 45, 39, 48, 55, 46, 38
50, 54, 63, 31, 49, 54, 68, 35, 27, 45, 53, 62, 47, 41, 50, 76, 67, 49

6. 42, 48, 51, 39, 47, 50, 48, 51, 54, 46, 49, 36, 50, 55, 51, 43, 46, 37
50, 52, 43, 40, 33, 51, 45, 53, 44, 40, 52, 54, 48, 51, 47, 43, 50, 46

Example 3 Describe the center and spread of the data using either the mean and standard deviation or the five-number summary. Justify your choice by constructing a histogram for the data.

7 32, 44, 50, 49, 21, 12, 27, 41, 48, 30, 50, 23, 37, 16, 49, 53, 33, 25
35, 40, 48, 39, 50, 24, 15, 29, 37, 50, 36, 43, 49, 44, 46, 27, 42, 47

8. 82, 86, 74, 90, 70, 81, 89, 88, 75, 72, 69, 91, 96, 82, 80, 78, 74, 94
85, 77, 80, 67, 76, 84, 80, 83, 88, 92, 87, 79, 84, 96, 85, 73, 82, 83

Example 4 **9. WEATHER** The daily low temperatures for New Carlisle over a 30-day period are shown. Describe the center and spread of the data using either the mean and standard deviation or the five-number summary. Justify your choice by constructing a box-and-whisker plot for the data.

Temperature (°F)														
48	50	55	53	57	53	44	61	57	49	51	58	46	54	57
50	55	47	57	48	58	53	49	56	59	52	48	55	53	51

10. **TRACK** Refer to the beginning of the lesson. Sarah's 100-meter dash times are shown.

   a. Use a graphing calculator to create a box-and-whisker plot. Describe the center and spread of the data.

   b. Sarah's slowest time prior to pulling a muscle was 12.50 seconds. Use a graphing calculator to create a box-and-whisker plot that *does not* include the times that she ran after pulling the muscle. Then describe the center and spread of the new data set.

100-meter dash (seconds)				
12.20	12.35	13.60	12.24	12.72
12.18	12.06	12.41	12.28	13.06
12.87	12.04	12.38	12.20	13.12
12.30	13.27	12.93	12.16	12.02
12.50	12.14	11.97	12.24	13.09
12.46	12.33	13.57	11.96	13.34

   c. What effect does removing the times recorded after Sarah pulled a muscle have on the shape of the distribution and on how you should describe the center and spread?

11. **MENU** The prices for entrees at a restaurant are shown.

   a. Use a graphing calculator to create a box-and-whisker plot. Describe the center and spread of the data.

   b. The owner of the restaurant decides to eliminate all entrees that cost more than $15. Use a graphing calculator to create a box-and-whisker plot that reflects this change. Then describe the center and spread of the new data set.

Entree Prices ($)				
9.00	11.25	16.50	9.50	13.00
18.50	7.75	11.50	13.75	9.75
8.00	16.50	12.50	10.25	17.75
13.00	10.75	16.75	8.50	11.50

H.O.T. Problems   Use Higher-Order Thinking Skills

CHALLENGE Identify the box-and-whisker plot that corresponds to each of the following histograms.

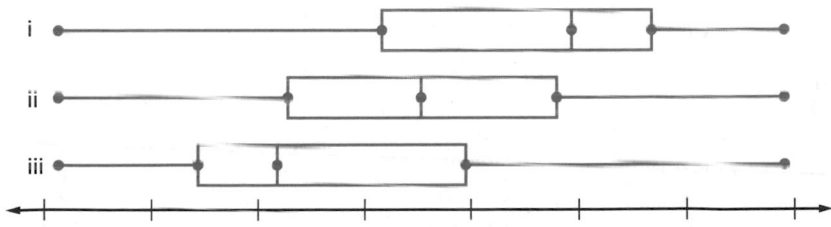

12.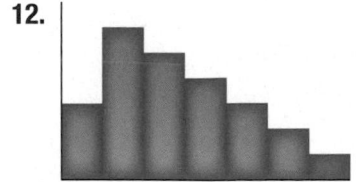

13.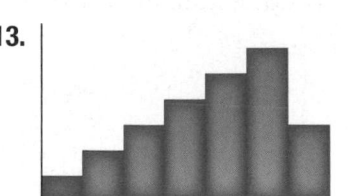

14.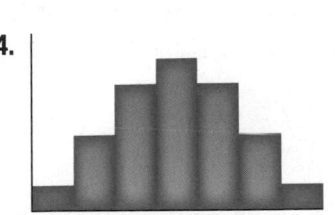

15. **CCSS ARGUMENTS** Research and write a definition for a *bimodal distribution*. How can the measures of center and spread of a bimodal distribution be described?

16. **OPEN ENDED** Give an example of a set of real-world data with a distribution that is symmetric and one with a distribution that is not symmetric.

17. **WRITING IN MATH** Explain why the mean and standard deviation are used to describe the center and spread of a symmetrical distribution and the five-number summary is used to describe the center and spread of a skewed distribution.

18. At the county fair, 1000 tickets were sold. Adult tickets cost $8.50, children's tickets cost $4.50, and a total of $7300 was collected. How many children's tickets were sold?

   A 700   **C** 400

   B 600   **D** 300

19. Edward has 20 dimes and nickels, which together total $1.40. How many nickels does he have?

   F 12   **H** 8

   G 10   **J** 6

20. If 4.5 kilometers is about 2.8 miles, about how many miles is 6.1 kilometers?

   A 3.2 miles   **C** 3.8 miles

   B 3.6 miles   **D** 4.0 miles

21. EXTENDED RESPONSE Three times the width of a certain rectangle exceeds twice its length by three inches, and four times its length is twelve more than its perimeter.

   a. Translate the sentences into equations.

   b. Find the dimensions of the rectangle.

   c. What is the area of the rectangle?

Spiral Review

Identify the sample and the population for each situation. Then describe the sample statistic and the population parameter. (Lesson 12-2)

22. AMUSEMENT PARK A systematic sample of 250 guests is asked how much money they spent on concessions inside the park. The median amount of money is calculated.

23. PROM A random sample of 100 high school seniors at North Boyton High School is surveyed, and the mean amount of money spent on prom by a senior is calculated.

Identify each survey question as *biased* or *unbiased*. If biased, explain your reasoning. (Lesson 12-1)

24. What do you like the most about reality television shows, and which one is your favorite?

25. Are you planning on seeing the school play?

26. Don't you agree that the school should renovate the library?

27. GARDENING Trey planted a triangular garden. Write an expression for the perimeter of the triangle. (Lesson 11-6)

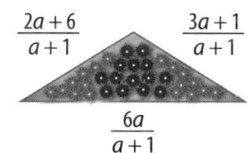

$$\frac{2a+6}{a+1} \qquad \frac{3a+1}{a+1}$$

$$\frac{6a}{a+1}$$

Find the inverse of each function. (Lesson 4-7)

28. $f(x) = 2x - 14$

29. $f(x) = 17 - 5x$

30. $f(x) = \frac{1}{4}x + 3$

31. $f(x) = -\frac{1}{7}x - 1$

32. $f(x) = \frac{2}{3}x + 6$

33. $f(x) = 12 - \frac{3}{5}x$

Skills Review

A bowl contains 3 red chips, 6 green chips, 5 yellow chips, and 8 orange chips. A chip is drawn randomly. Find each probability.

34. red

35. orange

36. yellow or green

37. not orange

38. not green

39. red or orange

Comparing Sets of Data

- You calculated measures of central tendency and variation.

- **1** Determine the effect that transformations of data have on measures of central tendency and variation.

  2 Compare data using measures of central tendency and variation.

- Tom gets paid hourly to do landscaping work. Because he is such a good employee, Tom is planning to ask his boss for a bonus. Tom's initial pay for a month is shown. He is trying to decide whether he should ask for an extra $5 per day or a 10% increase in his daily wages.

Tom's Pay ($)		
44	52	50
40	48	46
44	52	54
58	42	52
54	50	52
42	52	46
56	48	44
50	42	

 NewVocabulary
linear transformation

 Common Core State Standards

Content Standards
S.ID.2 Use statistics appropriate to the shape of the data distribution to compare center (median, mean) and spread (interquartile range, standard deviation) of two or more different data sets.

S.ID.3 Interpret differences in shape, center, and spread in the context of the data sets, accounting for possible effects of extreme data points (outliers).

Mathematical Practices
1 Make sense of problems and persevere in solving them.

1 Transformations of Data To see the effect that an extra $5 per day would have on Tom's daily pay, we can find the new daily pay values and compare the measures of center and variation for the two sets of data. The new data can be found by performing a *linear transformation*. A **linear transformation** is an operation performed on a data set that can be written as a linear function. Tom's daily pay after the $5 bonus can be found using $y = 5 + x$, where x represents his original daily pay and y represents his daily pay after the bonus.

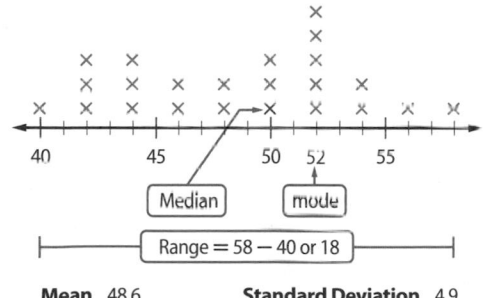

Tom's Earnings Before Extra $5

Median mode

Range = 58 − 40 or 18

Mean 48.6   **Standard Deviation** 4.9

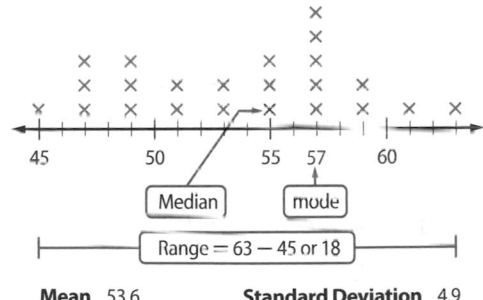

Tom's Earnings With Extra $5

Median mode

Range = 63 − 45 or 18

Mean 53.6   **Standard Deviation** 4.9

Notice that each value was translated 5 units to the right. Thus, the mean, median, and mode increased by 5. Since the new minimum and maximum values also increased by 5, the range remained the same. The standard deviation is unchanged because the amount by which each value deviates from the mean stayed the same.

These results occur when any positive or negative number is added to every value in a set of data.

KeyConcept Transformations Using Addition

If a real number k is added to every value in a set of data, then:

- the mean, median, and mode of the new data set can be found by adding k to the mean, median, and mode of the original data set, and

- the range and standard deviation will not change.

Example 1 Transformation Using Addition

Find the mean, median, mode, range, and standard deviation of the data set obtained after adding 7 to each value.

$$13, 5, 8, 12, 7, 4, 5, 8, 14, 11, 13, 8$$

Method 1 Find the mean, median, mode, range, and standard deviation of the original data set.

| Mean | 9 | Mode | 8 | Standard Deviation | 3.3 |
| Median | 8 | Range | 10 | | |

Add 7 to the mean, median, and mode. The range and standard deviation are unchanged.

| Mean | 16 | Mode | 15 | Standard Deviation | 3.3 |
| Median | 15 | Range | 10 | | |

Method 2 Add 7 to each data value.

$$20, 12, 15, 19, 14, 11, 12, 15, 21, 18, 20, 15$$

Find the mean, median, mode, range, and standard deviation of the new data set.

| Mean | 16 | Mode | 15 | Standard Deviation | 3.3 |
| Median | 15 | Range | 10 | | |

▶ **Guided**Practice

1. Find the mean, median, mode, range, and standard deviation of the data set obtained after adding −4 to each value.

$$27, 41, 15, 36, 26, 40, 53, 38, 37, 24, 45, 26$$

To see the effect that a daily increase of 10% has on the data set, we can multiply each value by 1.10 and recalculate the measures of center and variation.

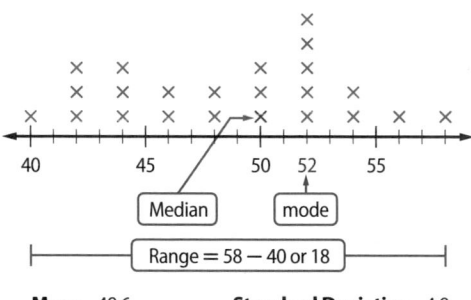

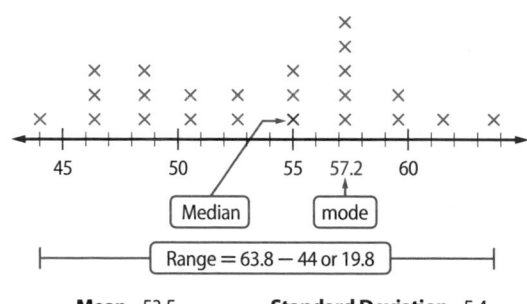

Notice that each value did not increase by the same amount, but did increase by a factor of 1.10. Thus, the mean, median, and mode increased by a factor of 1.10. Since each value was increased by a constant percent and not by a constant amount, the range and standard deviation both changed, also increasing by a factor of 1.10.

KeyConcept Transformations Using Multiplication

If every value in a set of data is multiplied by a constant k, $k > 0$, then the mean, median, mode, range, and standard deviation of the new data set can be found by multiplying each original statistic by k.

Since the medians for both bonuses are equal and the means are approximately equal, Tom should ask for the bonus that he thinks he has the best chance of receiving.

Example 2 Transformation Using Multiplication

Find the mean, median, mode, range, and standard deviation of the data set obtained after multiplying each value by 3.

21, 12, 15, 18, 16, 10, 12, 19, 17, 18, 12, 22

Find the mean, median, mode, range, and standard deviation of the original data set.

| Mean | 16 | Mode | 12 | Standard Deviation | 3.7 |
| Median | 16.5 | Range | 12 | | |

Multiply the mean, median, mode, range, and standard deviation by 3.

| Mean | 48 | Mode | 36 | Standard Deviation | 11.1 |
| Median | 49.5 | Range | 36 | | |

> **Guided**Practice

2. Find the mean, median, mode, range, and standard deviation of the data set obtained after multiplying each value by 0.8.

63, 47, 54, 60, 55, 46, 51, 60, 58, 50, 56, 60

2 Comparing Distributions

Recall that when choosing appropriate statistics to represent data, you should first analyze the shape of the distribution. The same is true when comparing distributions.

- Use the mean and standard deviation to compare two symmetric distributions.

- Use the five-number summaries to compare two skewed distributions or a symmetric distribution and a skewed distribution.

Example 3 Compare Data Using Histograms

QUIZ SCORES Robert and Elaine's quiz scores for the first semester of Algebra 1 are shown below.

Robert's Quiz Scores	Elaine's Quiz Scores
85, 95, 70, 87, 78, 82, 84, 84, 85, 99, 88, 74, 75, 89, 79, 80, 92, 91, 96, 81	89, 76, 87, 86, 92, 77, 78, 83, 83, 82, 81, 82, 84, 85, 85, 86, 89, 93, 77, 85

a. Use a graphing calculator to construct a histogram for each set of data. Then describe the shape of each distribution.

Enter Robert's quiz scores as **L1** and Elaine's quiz scores as **L2**.

TechnologyTip

Histograms To create a histogram for a set of data in L2, press `2nd` [STAT PLOT] `ENTER` `ENTER`, choose ⏹, and enter L2 for **Xlist**.

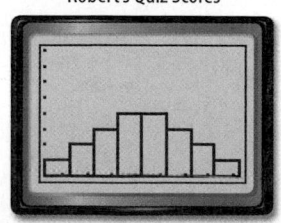

Robert's Quiz Scores

[69, 101] scl: 4 by [0, 8] scl: 1

Elaine's Quiz Scores

[69, 101] scl: 4 by [0, 8] scl: 1

Both distributions are high in the middle and low on the left and right. Therefore, both distributions are symmetric.

b. Compare the data sets using either the means and standard deviations or the five-number summaries. Justify your choice.

Both distributions are symmetric, so use the means and standard deviations to describe the centers and spreads.

Robert's Quiz Scores

Elaine's Quiz Scores

The means for the students' quiz scores are approximately equal, but Robert's quiz scores have a much higher standard deviation than Elaine's quiz scores. This means that Elaine's quiz scores are generally closer to her mean than Robert's quiz scores are to his mean.

▶ **Guided**Practice

COMMUTE The students in two of Mr. Martin's classes found the average number of minutes that they each spent traveling to school each day.

3A. Use a graphing calculator to construct a histogram for each set of data. Then describe the shape of each distribution.

3B. Compare the data sets using either the means and standard deviations or the five-number summaries. Justify your choice.

2nd Period (minutes)
8, 4, 18, 7, 13, 26, 12, 6, 20, 5, 9, 24, 8, 16, 31, 13, 17, 10, 8, 22, 12, 25, 13, 11, 18, 12, 16, 22, 25, 33

7th Period (minutes)
21, 4, 20, 13, 22, 6, 10, 23, 13, 25, 14, 16, 19, 21, 19, 8, 20, 18, 9, 14, 21, 17, 19, 22, 4, 19, 21, 26

Box-and-whisker plots are useful for comparisons of data because they can be displayed on the same screen.

> 🌐 **Real-World Example 4** Compare Data Using Box-and-Whisker Plots

FOOTBALL Kurt's total rushing yards per game for his junior and senior seasons are shown.

Junior Season (yards)					
16	20	72	4	25	18
34	10	42	17	56	12

Senior Season (yards)					
77	54	109	60	156	72
39	83	73	101	46	80

a. Use a graphing calculator to construct a box-and-whisker plot for each set of data. Then describe the shape of each distribution.

Enter Kurt's rushing yards from his junior season as **L1** and his rushing yards from his senior season as **L2**. Graph both box-and-whisker plots on the same screen by graphing **L1** as **Plot1** and **L2** as **Plot2**.

For Kurt's junior season, the right whisker is longer than the left, and the median is closer to the left whisker. The distribution is positively skewed.

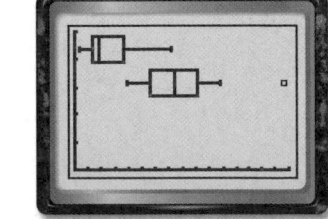

[0, 160] scl: 10 by [0, 5] scl: 1

For Kurt's senior season, the lengths of the whiskers are approximately equal, and the median is in the middle of the data. The distribution is symmetric.

©Thinkstock/Corbis

b. Compare the data sets using either the means and standard deviations or the five-number summaries. Justify your choice.

One distribution is symmetric and the other is skewed, so use the five-number summaries to compare the data.

The upper quartile for Kurt's junior season was 38, while the minimum for his senior season was 39. This means that Kurt rushed for more yards in every game during his senior season than 75% of the games during his junior season.

The maximum for Kurt's junior season was 72, while his median for his senior season was 75. This means that in half of his games during his senior year, he rushed for more yards than in any game during his junior season. Overall, we can conclude that Kurt rushed for many more yards during his senior season than during his junior season.

▶ **Guided**Practice

BASKETBALL The points Vanessa scored per game during her junior and senior seasons are shown.

4A. Use a graphing calculator to construct a histogram for each set of data. Then describe the shape of each distribution.

4B. Compare the data sets using either the means and standard deviations or the five-number summaries. Justify your choice.

Junior Season (points)	Senior Season (points)
10, 12, 6, 10, 13, 8, 12, 3, 21, 14, 7, 0, 15, 6, 16, 8, 17, 3, 17, 2	10, 32, 3, 22, 20, 30, 26, 24, 5, 22, 28, 32, 26, 21, 6, 20, 24, 18, 12, 25

Check Your Understanding ⬤ **= Step-by-Step Solutions begin on page R13.**

Example 1 **Find the mean, median, mode, range, and standard deviation of each data set that is obtained after adding the given constant to each value.**

1. 10, 13, 9, 8, 15, 8, 13, 12, 7, 8, 11, 12; + (−7)   **2.** 38, 36, 37, 42, 31, 44, 37, 45, 29, 42, 30, 42; + 23

Example 2 **Find the mean, median, mode, range, and standard deviation of each data set that is obtained after multiplying each value by the given constant.**

3 6, 10, 3, 7, 4, 9, 3, 8, 5, 11, 2, 1; × 3   **4.** 42, 39, 45, 44, 37, 42, 38, 37, 41, 49, 42, 36; × 0.5

Example 3 **5. TRACK** Mark and Kyle's long jump distances are shown.

Kyle's Distances (ft)	Mark's Distances (ft)
17.2, 18.28, 18.56, 17.28, 17.36, 18.08, 17.43, 17.71, 17.46, 18.26, 17.51, 17.58, 17.41, 18.21, 17.34, 17.63, 17.55, 17.26, 17.18, 17.78, 17.51, 17.83, 17.92, 18.04, 17.91	18.88, 19.24, 17.63, 18.69, 17.74, 19.18, 17.92, 18.96, 18.19, 18.21, 18.46, 17.47, 18.49, 17.86, 18.93, 18.73, 18.34, 18.67, 18.56, 18.79, 18.47, 18.84, 18.87, 17.94, 18.7

a. Use a graphing calculator to construct a histogram for each set of data. Then describe the shape of each distribution.

b. Compare the data sets using either the means and standard deviations or the five-number summaries. Justify your choice.

Example 4

6. TIPS Miguel and Stephanie are servers at a restaurant. The tips that they earned to the nearest dollar over the past 15 workdays are shown.

Miguel's Tips ($)
14, 68, 52, 21, 63, 32, 43, 35, 70, 37, 42, 16, 47, 38, 48

Stephanie's Tips ($)
34, 52, 43, 39, 41, 50, 46, 36, 37, 47, 39, 49, 44, 36, 50

a. Use a graphing calculator to construct a box-and-whisker plot for each set of data. Then describe the shape of each distribution.

b. Compare the data sets using either the means and standard deviations or the five-number summaries. Justify your choice.

Practice and Problem Solving

Extra Practice is on page R12.

Example 1 Find the mean, median, mode, range, and standard deviation of each data set that is obtained after adding the given constant to each value.

7. 52, 53, 49, 61, 57, 52, 48, 60, 50, 47; $+ 8$ **8.** 101, 99, 97, 88, 92, 100, 97, 89, 94, 90; $+ (-13)$

9. 27, 21, 34, 42, 20, 19, 18, 26, 25, 33; $+ (-4)$ **10.** 72, 56, 71, 63, 68, 59, 77, 74, 76, 66; $+ 16$

Example 2 Find the mean, median, mode, range, and standard deviation of each data set that is obtained after multiplying each value by the given constant.

11. 11, 7, 3, 13, 16, 8, 3, 11, 17, 3; $\times 4$ **12.** 64, 42, 58, 40, 61, 67, 58, 52, 51, 49; $\times 0.2$

13. 33, 37, 38, 29, 35, 37, 27, 40, 28, 31; $\times 0.8$ **14.** 1, 5, 4, 2, 1, 3, 6, 2, 5, 1; $\times 6.5$

Example 3 **15. BOOKS** The page counts for the books that the students chose are shown.

1st Period
388, 439, 206, 438, 413, 253, 311, 427, 258, 511, 283, 578, 291, 358, 297, 303, 325, 506, 331, 482, 343, 372, 456, 267, 484, 227

6th Period
357, 294, 506, 392, 296, 467, 308, 319, 485, 333, 352, 405, 359, 451, 378, 490, 379, 401, 409, 421, 341, 438, 297, 440, 500, 312, 502

a. Use a graphing calculator to construct a histogram for each set of data. Then describe the shape of each distribution.

b. Compare the data sets using either the means and standard deviations or the five-number summaries. Justify your choice.

16. TELEVISIONS The prices for a sample of televisions are shown.

The Electronics Superstore
46, 25, 62, 45, 30, 43, 40, 46, 33, 53, 35, 38, 39, 40, 52, 42, 44, 48, 50, 35, 32, 55, 28, 58

Game Central
53, 49, 26, 61, 40, 50, 42, 35, 45, 48, 31, 48, 33, 50, 35, 55, 38, 50, 42, 53, 44, 54, 48, 58

a. Use a graphing calculator to construct a histogram for each set of data. Then describe the shape of each distribution.

b. Compare the data sets using either the means and standard deviations or the five-number summaries. Justify your choice.

Example 4 **17 BRAINTEASERS** The time that it took Leon and Cassie to complete puzzles is shown.

Leon's Times (minutes)
4.5, 1.8, 3.2, 5.1, 2.0, 2.6, 4.8, 2.4, 2.2, 2.8, 1.8, 2.2, 3.9, 2.3, 3.3, 2.4

Cassie's Times (minutes)
2.3, 5.8, 4.8, 3.3, 5.2, 4.6, 3.6, 5.7, 3.8, 4.2, 5.0, 4.3, 5.5, 4.9, 2.4, 5.2

a. Use a graphing calculator to construct a box-and-whisker plot for each set of data. Then describe the shape of each distribution.

b. Compare the data sets using either the means and standard deviations or the five-number summaries. Justify your choice.

18. DANCE The total amount of money that a sample of students spent to attend the homecoming dance is shown.

Boys (dollars)
114, 98, 131, 83, 91, 64, 94, 77, 96, 105, 72, 108, 87, 112, 58, 126

Girls (dollars)
124, 74, 105, 133, 85, 162, 90, 109, 94, 102, 98, 171, 138, 89, 154, 76

a. Use a graphing calculator to construct a box-and-whisker plot for each set of data. Then describe the shape of each distribution.

b. Compare the data sets using either the means and standard deviations or the five-number summaries. Justify your choice.

19. LANDSCAPING Refer to the beginning of the lesson. Rhonda, another employee that works with Tom, earned the following over the past month.

a. Find the mean, median, mode, range, and standard deviation of Rhonda's earnings.

b. A $5 bonus had been added to each of Rhonda's daily earnings. Find the mean, median, mode, range, and standard deviation of Rhonda's earnings before the $5 bonus.

Rhonda's Pay ($)		
45	55	53
47	53	54
44	56	59
63	47	53
60	57	62
44	50	45
60	53	49
62	47	

20. SHOPPING The items Lorenzo purchased are shown.

a. Find the mean, median, mode, range, and standard deviation of the prices.

b. A 7% sales tax was added to the price of each item. Find the mean, median, mode, range, and standard deviation of the items without the sales tax.

Baseball hat	$14.98
Jeans	$24.61
T-shirt	$12.84
T-shirt	$16.05
Backpack	$42.80
Folders	$2.14
Sweatshirt	$19.26

H.O.T. Problems   Use Higher-Order Thinking Skills

21 CHALLENGE A salesperson has 15 SUVs priced between $33,000 and $37,000 and 5 luxury cars priced between $44,000 and $48,000. The average price for all of the vehicles is $39,250. The salesperson decides to reduce the prices of the SUVs by $2000 per vehicle. What is the new average price for all of the vehicles?

22. REASONING If every value in a set of data is multiplied by a constant k, $k < 0$, then how can the mean, median, mode, range, and standard deviation of the new data set be found?

23. WRITING IN MATH Compare and contrast the benefits of displaying data using histograms and box-and-whisker plots.

24. CCSS REGULARITY If k is added to every value in a set of data, and then each resulting value is multiplied by a constant m, $m > 0$, how can the mean, median, mode, range, and standard deviation of the new data set be found? Explain your reasoning.

25. WRITING IN MATH Explain why the mean and standard deviation are used to compare the center and spread of two symmetrical distributions and the five-number summary is used to compare the center and spread of two skewed distributions or a symmetric distribution and a skewed distribution.

26. A store manager recorded the number of customers each day for a week: {46, 57, 63, 78, 91, 110, 101}. Find the mean absolute deviation.

A 16.8   **C** 19.4

B 18.1   **D** 22.7

27. SHORT RESPONSE Solve the right triangle. Round each side length to the nearest tenth.

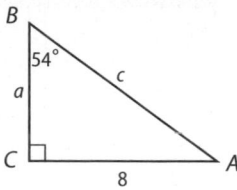

28. A research company divides a group of volunteers by age, and then randomly selects volunteers from each group to complete a survey. What type of sample is this?

F simple   **H** self-selected

G systematic   **J** stratified

29. Which set of measures can be the measures of the sides of a right triangle?

A 6, 7, 9

B 9, 12, 19

C 12, 15, 17

D 14, 48, 50

Spiral Review

30. Use a graphing calculator to construct a histogram for the data, and use it to describe the shape of the distribution. (Lesson 12-3)

23, 45, 50, 22, 37, 24, 36, 46, 24, 52, 25, 42, 25, 26, 54, 47, 27, 55
63, 28, 29, 30, 45, 31, 55, 43, 32, 34, 30, 23, 30, 35, 27, 35, 38, 40

31. SUBSCRIPTIONS Ms. Wilson's students are selling magazine subscriptions. Her students recorded the total number of subscriptions they each sold: {8, 12, 10, 7, 4, 3, 0, 4, 9, 0, 5, 3, 23, 6, 2}. Find and interpret the standard deviation of the data set. (Lesson 12-2)

Find the value of x for each figure. Round to the nearest tenth if necessary. (Lesson 9-4)

32. $A = 45$ in^2

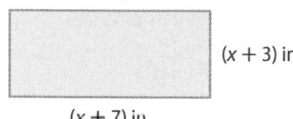

33. $A = 20$ ft^2

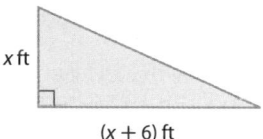

34. $A = 42$ m^2

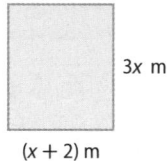

Factor each polynomial. (Lesson 8-6)

35. $x^2 - 4x - 21$ **36.** $11x + x^2 + 30$ **37.** $32 + x^2 - 12x$

38. $-36 - 9x + x^2$ **39.** $x^2 + 12x + 20$ **40.** $-x + x^2 - 42$

41. MANUFACTURING A company is designing a box for dry pasta in the shape of a rectangular prism. The length is 2 inches more than twice the width, and the height is 3 inches more than the length. Write an expression for the volume of the box. (Lesson 8-3)

Skills Review

Find the degree of each polynomial.

42. $2x^2 + 5y - 21$ **43.** $16xy^3 - 17x^2y - 16z^3$ **44.** $3ac^3d + 14a^2$

45. 18   **46.** $3a^2b^3 + 11ab^2c$ **47.** $7x + 11$

Identify each sample, and suggest a population from which it was selected. Then classify the sample as *simple, systematic, self-selected, convenience,* or *stratified.* Explain your reasoning. (Lesson 12-1)

1. **CEREAL** A cereal company invites 100 random children and parents to test a new cereal and records the reactions.

2. **SCHOOL LUNCH** A school is creating a new lunch menu. They send out a questionnaire to all students with odd homeroom numbers to determine what items should be on the new menu.

3. **MASCOTS** The cheerleaders send out a flyer with pictures of possible options for the new mascot to all the girls in the school. The girls mark their favorite mascot and send it back. The new mascot is chosen from the survey.

Identify each sample as *biased* or *unbiased.* Explain your reasoning. (Lesson 12-1)

4. **ART** Every fifth person leaving the art museum is asked to name their favorite piece.

5. **SHOPPING** Each person leaving the Earring Pagoda is asked to name their favorite store in the mall.

6. **FOOTBALL** Every 10th student leaving the student union at Ohio State is asked to name their favorite college football team.

Identify the sample and the population for each situation. Then describe the sample statistic and the population parameter. (Lesson 12-2)

7. **DINING** At a restaurant, a random sample of 15 diners is selected. The amount of money spent on each meal is recorded.

8. **POOLS** A random sample of 25 children at a community pool is asked if they visit the pool at least once each week. The percent responding *yes* is calculated.

9. **PLAY AREA** Ian listed the ages of the children playing at the play area at the mall. Find and interpret the standard deviation of the data set. (Lesson 12-2)

$$\{2, 3, 2, 2, 4, 2, 3, 2, 8, 3, 4, 2\}$$

10. **MULTIPLE CHOICE** Several friends are chipping in to buy a gift for their teacher. Indigo is keeping track of how much each friend spends. Find the mean absolute deviation. (Lesson 12-2)

$$\{\$10, \$5, \$3, \$6, \$7, \$8\}$$

A $1.83   C $2.40

B $2.22   D $6.50

11. Use a graphing calculator to construct a histogram for the data, and use it to describe the shape of the distribution. (Lesson 12-3)

$$19, 36, 26, 36, 40, 31, 30, 33, 23, 38, 23, 46$$

12. Describe the center and spread of the data using either the mean and standard deviation or the five-number summary. Justify your choice by constructing a box-and-whisker plot for the data. (Lesson 12-3)

$$9, 11, 2, 6, 8, 10, 6, 3, 10, 11, 9, 8, 3,$$
$$8, 5, 11, 14, 6, 8, 6, 11, 5, 9, 10, 8$$

13. **MULTIPLE CHOICE** Which pair of box-and-whisker plots depicts two positively skewed sets of data in which 75% of one set of data is larger than 75% of the other set of data? (Lesson 12-4)

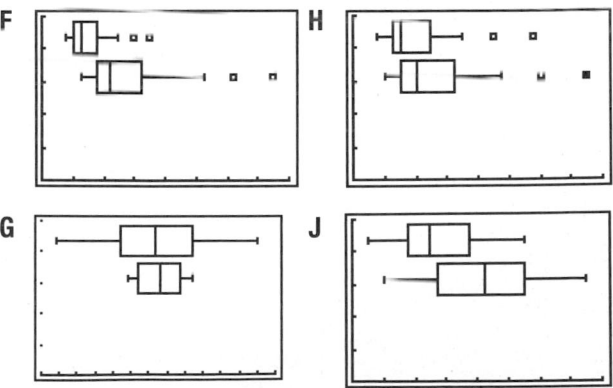

Find the mean, median, mode, range, and standard deviation of each data set that is obtained after adding the given constant to each value. (Lesson 12-4)

14. $6, 9, 0, 15, 9, 14, 11, 13, 9, 5, 8, 6; +(-3)$

15. $19, 22, 10, 17, 26, 24, 12, 22, 18, 17; +8$

Simulations

Then
- You calculated simple probability.

Now
1. Calculate experimental probabilities.
2. Design simulations and summarize data from simulations.

Why?
- Alex has been practicing his penalty kicks. He expects to be able to make at least 66% of his penalty kicks. To test this, he takes 50 penalty kicks, of which he makes 35.

NewVocabulary
theoretical probability
experimental probability
relative frequency
simulation
probability model

Common Core State Standards

Mathematical Practices
4 Model with mathematics.

1 Experimental Probability A **theoretical probability** is the ratio of the number of favorable outcomes to the total number of outcomes. The theoretical probability that a coin lands heads up is $\frac{1}{2}$ or 50%. The theoretical probability suggests that 5 of 10 tosses will most likely result in the coin landing heads up.

We can toss a coin 10 times and record the outcomes. The **experimental probability**, or **relative frequency**, is determined using data from experiments. It is the ratio of the number of times an outcome occurs to the total number of events or trials. In this case, the experimental probability of tossing the coin and having it land heads up is $\frac{3}{10}$ or 30%.

Outcome	Frequency
heads	3
tails	7

Standardized Test Example 1 Find Experimental Probability

A die is rolled 50 times and the results are recorded. Find the experimental probability of rolling at least a 5.

A $\frac{6}{50}$ B $\frac{7}{50}$ C $\frac{13}{50}$ D $\frac{37}{50}$

Outcome	Frequency
1	6
2	10
3	14
4	7
5	6
6	7

Read the Test Item
We are asked to find the probability of rolling at least a 5. Therefore, we need to consider rolling a 5 and rolling a 6.

Solve the Test Item

$P(\text{at least a 5}) = \dfrac{\text{number of 5s and 6s rolled}}{\text{total number of rolls}}$

$= \dfrac{6+7}{50}$ or $\dfrac{13}{50}$

The experimental probability of rolling at least a 5 is $\frac{13}{50}$ or 26%.

The correct answer is C.

Guided Practice

1. Students were asked how they travel to school each morning. Find the experimental probability of randomly selecting a student who does not ride in a car or bus.

F $\frac{9}{47}$ G $\frac{17}{47}$ H $\frac{21}{47}$ J $\frac{38}{47}$

Mode	Frequency
bike	3
bus	21
car	17
walk	6

Notice that the experimental probability of rolling a 3 was $\frac{7}{25}$ or 28%, but the theoretical probability is $\frac{1}{6}$ or about 16.7%. This is because theoretical probability is what *is expected to* happen and experimental probability is what *actually* happens.

2 Simulations

An experiment that would be difficult or impractical to perform can be modeled by a **simulation**. In a simulation, a **probability model** is used to recreate a situation so that the experimental probability of an outcome can be found. A probability model is a mathematical model used to represent the theoretical probability of the outcomes in an experiment. Coins, dice, random number generators, or other objects can be used as probability models.

To design a simulation, use the following steps.

KeyConcept Designing a Simulation

Step 1 Determine each possible outcome and its theoretical probability.

Step 2 Describe an appropriate probability model for the situation that accurately represents the theoretical probability of each outcome.

Step 3 Define what a trial is for the situation, and state the number of trials to be conducted.

Real-World Example 2 Design a Simulation

QUALITY CONTROL Eloy inspects bike frames as they come through the assembly line. From previous observations, he expects to find a weld defect in one out of every ten frames and a design defect in one out of every twenty frames that he inspects. Design a simulation using Eloy's expectation of defects. Assume that a frame can only have one of the defects.

Step 1 There are three possible outcomes: weld defect, design defect, and no defects. Use Eloy's expectation of defects to calculate the theoretical probability of each outcome.

Possible Outcomes	Theoretical Probability
Weld defect	10%
Design defect	5%
No defects	85%

Step 2 We can use the random number generator on a graphing calculator. Assign the integers 0–19 to accurately represent the probability data.

Outcome	Represented by
Weld defect	0, 1
Design defect	2
No defects	3–19

Step 3 A trial will represent selecting a frame at random. The simulation can consist of any number of trials. We will use 40.

GuidedPractice

2. FREE AGENT A survey asked team executives for which team they thought free agent Bonny Solomon would choose to play next season. The results are shown. Design a simulation that can be used to estimate the probability that Bonny chooses one of these three teams.

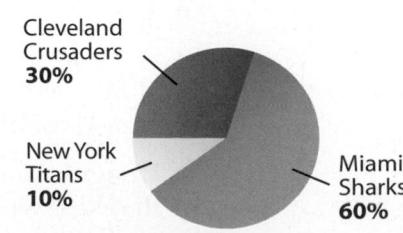

Cleveland Crusaders **30%**

New York Titans **10%**

Miami Sharks **60%**

Real-WorldCareer

Assemblers Assemblers may work as part of a team, where all members are capable of performing each task. In the automobile manufacturing industry, the mean wage for a team assembler in 2009 was $49,360.

Source: Bureau of Labor Statistics

StudyTip

CCSS Modeling Simulations should involve data that are easier to obtain than the actual data you are modeling. Another probability model for Example 2 would be to place 20 marbles in a bag letting 2 red marbles represent a weld defect, 1 orange marble represent a design defect, and 17 yellow marbles represent no defects.

Monty Rakusen/Cultura/Getty Images

After designing a simulation, conduct it and analyze the results.

Example 3 Conduct and Evaluate a Simulation

QUALITY CONTROL Refer to the simulation in Example 2. Conduct the simulation and report the results.

Press [MATH] [◄] and select **[randInt (]**. Then press 0 [,] 19 [,] 40 [)] [ENTER]. Use the left and right arrow buttons to view the results. Make a frequency table and record the results.

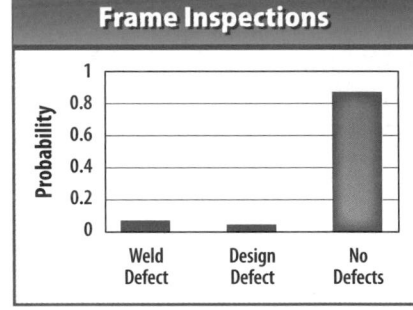

Outcome	Tally	Frequency
Weld defect	|||	3
Design defect	||	2
No defects	ℍℍ ℍℍ ℍℍ ℍℍ ℍℍ ℍℍ ℍℍ	35
Total		40

Calculate the experimental probability of finding each type of defect.

Weld defect $\frac{3}{40}$ or 0.075   Design defect $\frac{2}{40}$ or 0.05

No defects $\frac{35}{40}$ or 0.875

The experimental probabilities that a frame will have a weld defect, a design defect, or no defects in this case are 7.5%, 5%, and 87.5%, respectively. Notice that the probabilities are close to, and in the case of a design defect, the same as, the theoretical probabilities.

Make a bar graph of these results.

 GuidedPractice

3. FREE AGENT Conduct the simulation in Guided Practice 2. Then report the results.

Frame Inspections

Math History Link

Ada Byron Lovelace (1815–1852) Lovelace is known for generating a sequence of numbers using an early model of a computer. She is credited with being the first computer programmer. In 1980, the U.S. Department of Defense named a programming language after her.

TechnologyTip

Random Number Generator After running the random number generator, you can store the results as a list by pressing [STO►] and entering L1.

Check Your Understanding

○ = Step-by-Step Solutions begin on page R13.

Example 1

1 **MULTIPLE CHOICE** A movie theater employee surveyed a sample of customers as they exited the theater. Find the experimental probability of randomly selecting a customer who is older than 12 but younger than 46.

A $\frac{24}{127}$ C $\frac{99}{127}$

B $\frac{157}{254}$ D $\frac{213}{254}$

Age	Frequency
0–7	13
8–12	28
13–17	48
18–23	42
24–31	27
32–45	40
46–64	33
65+	23

Examples 2–3 **2. FOOTBALL** Rico is the kicker on the football team. Last season, he made 94% of his extra points.

a. Design a simulation that can be used to estimate the probability that Rico will make his next extra point.

b. Conduct the simulation, and report the results.

Example 1 **3** **CARDS** Javier is drawing a card from a standard deck of cards, recording the suit, and then replacing the card in the deck. The table below shows his results.

Suit	clubs	diamonds	hearts	spades
Frequency	7	4	5	9

a. Find the experimental probability of drawing a heart.

b. Find the experimental probability of drawing a black card.

c. Javier repeated his test. The results of the second test are shown below. Combine the results from the second test with the results from the first test, and then find the experimental probability of drawing a spade.

Suit	clubs	diamonds	hearts	spades
Frequency	5	8	6	6

4. **MUSIC** Shannon's digital media player has a large collection of songs. She randomly toggles through the songs and then records the genre. The graph shows her results.

a. Find the experimental probability of selecting a country song.

b. Find the experimental probability of selecting a song that is not rock.

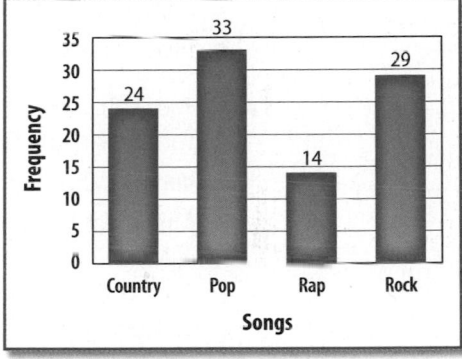

Examples 2–3 5. **BATTING AVERAGE** In a computer baseball game, a player has a batting average of .300. That is, he gets a hit 300 out of 1000, or 30%, of the times he is at bat.

a. Design a simulation that can be used to estimate the probability that the player will get a hit at his next at bat.

b. Conduct the simulation, and report the results.

6. **JEANS** Julie examines the stitching on pairs of jeans that are produced at a manufacturing plant. She expects to find defects in 1 out of every 20 pairs.

a. Design a simulation that can be used to estimate the probability that the next pair of jeans that Julie examines has a defect.

b. Conduct the simulation, and report the results.

7. **FOOD** For a promotion, the concession stands at a football stadium are giving away free items. For every tenth customer, a wheel is spun to choose the customer's prize. Each prize is equally likely.

a. Design a simulation that can be used to estimate the probability that the next spin is one of the five prizes.

b. Conduct the simulation, and report the results.

8. **CCSS MODELING** For its twentieth anniversary, a store randomly gives each customer a prize from the following choices: a free music download, a free game download, a free bag of popcorn, or a free DVD. The chances of winning each prize are equal.

a. Design a simulation that can be used to estimate the probability that the next prize given is one of the four prizes.

b. Conduct the simulation, and report the results.

9 **GAMES** Games at the fair require the majority of players to lose in order for game owners to make a profit. New games are tested to make sure they have sufficient difficulty. The results of three test groups are listed in the table. The owners want a maximum of 33% of players to win. There were 50 participants in each test group.

a. What is the experimental probability that the participant was a winner in the second group?

b. What is the experimental probability of winning for all three groups?

Result	Group 1	Group 2	Group 3
Winners	13	15	19
Losers	37	35	31

c. **DECISION MAKING** Should this game be used? Explain your reasoning.

10. **TEST** Jack forgot to study for his multiple-choice science quiz and is going to guess for each question. There are 20 questions, each with 4 possible answers.

a. Design a simulation that can be used to estimate the number of questions that Jack answers correctly.

b. Conduct the experiment from part **a** five times, and complete the table.

c. How many should Jack expect to answer correctly?

Simulation	Number of Correct Answers
1	
2	
3	
4	
5	

11. 🔁 **MULTIPLE REPRESENTATIONS** In this problem, you will explore the effect the number of trials has on the experimental probability of an event.

a. **Verbal** What is the probability of rolling a 1 on a die?

b. **Analytical** Design a simulation that can be used to estimate the probability that the next number rolled on a die is a 1.

c. **Analytical** Conduct the simulation from part **b** for 10, 20, 50, and 100 trials, and complete the table.

d. **Analytical** As the number of trials increases, what is happening to the experimental probability?

e. **Verbal** The *Law of Large Numbers* states that as the number of trials increases, the experimental probability gets closer to the theoretical probability. If you continued the simulation, each time increasing the number of trials, what would you expect the experimental probability for rolling a 1 to approach?

Rolling a 1		
Trials	Frequency	Experimental Probability
10		
20		
50		
100		

H.O.T. Problems   Use Higher-Order Thinking Skills

12. **CCSS ARGUMENTS** The experimental probability of heads when a coin is tossed 15 times is *sometimes*, *never*, or *always* equal to the theoretical probability. Explain your reasoning.

13. **OPEN ENDED** Describe a situation at your school that could be represented by a simulation. Then design the simulation.

14. **CHALLENGE** *True* or *false*: If the theoretical probability of an event is 1, the experimental probability of the event cannot be 0. Explain your reasoning.

15. **REASONING** Jeremy tosses a coin several times and finds that the experimental probability for it landing heads up is 25%. Should Jeremy be concerned about the fairness of the coin? Explain your reasoning.

16. ✏️ **WRITING IN MATH** What should you consider when using the results of a simulation to make a prediction?

17. GEOMETRY Suppose a covered water tank in the shape of a right circular cylinder is thirty feet long and eight feet in diameter. What is the surface area of the cylinder?

A 272π ft^2 **C** 224π ft^2

B 248π ft^2 **D** 153π ft^2

18. SHORT RESPONSE Two consecutive numbers have a sum of 91. What are the numbers?

19. Solve $\dfrac{2x}{x-2} + \dfrac{8}{x} = 6$.

F 1   **H** 2

G 1, 4   **J** 2, 6

20. Mr. Bahn has $20,000 to invest. He invests part at 6% and the rest at 7%. He earns $1280 in interest within a year. How much did he invest at 7%?

A $12,000   **C** $9950

B $11,275   **D** $8000

Find the mean, median, mode, range, and standard deviation of each data set that is obtained after adding the given constant to each value. (Lesson 12-4)

21. 12, 16, 4, 8, 7, 11, 9, 4; $+ 5$

22. 1, 4, 3, 9, 12, 6, 7, 3; $+ 12$

23. 18, 12, 8, 13, 7, 15, 8, 6; $+ (-3)$

24. DANCE RECITAL The number of dance students in each act of a dance recital is shown. Describe the center and spread of the data using either the mean and standard deviation or the five-number summary. Justify your choice by constructing a box-and-whisker plot for the data. (Lesson 12-3)

Number of Dance Students														
13	15	1	20	14	4	18	2	17	10	22	1	22	15	17
21	10	18	14	18	2	10	20	15	18	4	19	12	16	5

25. PARTIES Student Council is planning a party for the school volunteers. There are five 66-ounce unopened bottles of soda left from a recent dance. When poured over ice, $5\frac{1}{2}$ ounces of soda fills a cup. How many servings of soda do they have? (Lesson 11-7)

Write an inverse variation equation that relates x and y. Assume that y varies inversely as x. Then solve. (Lesson 11-1)

26. If $y = 8.5$ when $x = -1$, find x when $y = -1$.

27. If $y = 8$ when $x = 1.55$, find x when $y = -0.62$.

28. If $y = 6.4$ when $x = 4.4$, find x when $y = 3.2$.

29. DELIVERY Ben and Amado are delivering a freezer. The bank in front of the house is the same height as the back of the truck. They set up their ramp as shown. What is the length of the slanted part of the ramp to the nearest tenth of a foot? (Lesson 10-5)

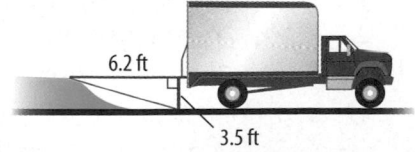

6.2 ft

3.5 ft

Write each fraction as a percent rounded to the nearest whole number.

30. $\dfrac{26}{58}$

31. $\dfrac{55}{125}$

32. $\dfrac{14}{128}$

33. $\dfrac{82}{110}$

34. $\dfrac{76}{124}$

35. $\dfrac{23}{86}$

Permutations and Combinations

Then
- You used the Fundamental Counting Principle.

Now
1. Use permutations.
2. Use combinations.

Why?
- Angie's coach told her that she would bat sixth in the softball game. When a coach decides on the team's lineup before a game, the order in which she fills in the names is important because it determines the order in which the players will bat.

NewVocabulary
permutation
factorial
combination

 Common Core State Standards

Mathematical Practices
8 Look for and express regularity in repeated reasoning.

1 Permutations When the objects in a sample space are arranged so that order is important, each possible arrangement is called a **permutation**. Three permutations of batting orders for the first four batters are shown below.

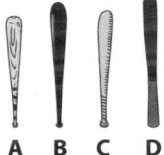
A B C D

A D C B

D B C A

Use the Fundamental Counting Principle to find the total number of permutations.

number of permutations		choices for 1st batter		choices for 2nd batter		choices for 3rd batter		choices for 4th batter
Permutations =		4	$\cdot$	3	$\cdot$	2	$\cdot$	1

$$= 24$$

There are 24 ways to arrange the first four batters. The expression used to find the number of arrangements, $4 \cdot 3 \cdot 2 \cdot 1$, can be written as $4!$, which is read *4 factorial*.

KeyConcept Factorial

Words	The **factorial** of a positive integer n is the product of the positive integers less than or equal to n. $0!$ is defined to be 1.
Symbols	$n! = n \cdot (n - 1) \cdot (n - 2) \cdot \ldots \cdot 1$, where $n \geq 1$

PT

Example 1 Permutations Using Factorials

TRAVEL A travel agency is planning a European vacation package. How many ways can the agency arrange the 5 cities along the tour?

There are five choices for the first city, four choices for the second city, and so on.

Number of ways to arrange the cities = $5!$
$$= 5 \cdot 4 \cdot 3 \cdot 2 \cdot 1 \text{ or } 120$$

There are 120 ways to arrange the cities.

GuidedPractice

1. **MOVIES** Lloyd and five friends go to a movie. In how many different ways can they sit together in a row of 6 empty seats?

Suppose Angie's coach had 5 players in mind for the first 3 spots in the lineup. How many different batting orders could she make?

<table>
<tr><td align="center">choices for
1st batter</td><td></td><td align="center">choices for
2nd batter</td><td></td><td align="center">choices for
3rd batter</td></tr>
<tr><td align="center">5</td><td align="center">·</td><td align="center">4</td><td align="center">·</td><td align="center">3</td><td align="center">= 60 permutations</td></tr>
</table>

Once the 3 spots are filled, there are **5** − 3 or **2** players left over. Notice that $5 \cdot 4 \cdot 3$ can also be written as $= \frac{5 \cdot 4 \cdot 3 \cdot 2 \cdot 1}{2 \cdot 1} = \frac{5!}{2!}$ or $\frac{5!}{(5-3)!}$. This result is generalized below.

> ### KeyConcept Permutation Formula
>
> **Words**   The number of permutations of n objects taken r at a time is the quotient of $n!$ and $(n − r)!$.
>
> **Symbols** $P(n, r) = \dfrac{n!}{(n - r)!}$

Example 2 Use the Permutation Formula

LIBRARY The librarian is placing 6 of 10 magazines in a school showcase. How many ways can she arrange the magazines in the case?

$P(n, r) = \dfrac{n!}{(n - r)!}$ Permutation Formula

$P(10, 6) = \dfrac{10!}{(10 - 6)!}$ $n = 10$ and $r = 6$

$= \dfrac{10!}{4!}$ Simplify.

$= \dfrac{10 \cdot 9 \cdot 8 \cdot 7 \cdot 6 \cdot 5 \cdot \cancel{4} \cdot \cancel{3} \cdot \cancel{2} \cdot \cancel{1}}{\cancel{4} \cdot \cancel{3} \cdot \cancel{2} \cdot \cancel{1}}$ Divide by common factors.

$= 151{,}200$ Simplify.

There are 151,200 ways for the librarian to arrange the magazines.

▶ **Guided**Practice

2. **TROPHIES** James wants to place 5 of his 8 trophies on the fireplace. How many ways can he arrange the trophies?

2 Combinations Recall that the coach had 60 possible lineups when considering 5 players for 3 spots. What if order didn't matter and he only cares *if* players are in the lineup? A selection of objects in which order is *not* important is called a **combination**. From how many combinations of 3 players can the coach choose?

Angie Beth Cora

Cora Angie Beth

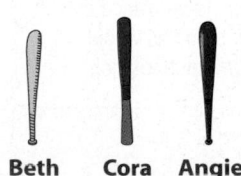

Beth Cora Angie

This same combination of players can be arranged 3! or 6 different ways. So 6 out of the 60 possible lineups include Angie, Beth, and Cora. The same will be true of every other combination. Therefore, out of the 60 possible lineups, there are $\frac{60}{6}$ or 10 different combinations of players.

Notice that we divided the number of permutations by the number of possible arrangements for every combination of elements $r!$ to obtain the number of combinations.

KeyConcept Combination Formula

Words	The number of combinations of n objects taken r at a time is the quotient of $n!$ and $(n - r)!r!$.
Symbols	$C(n, r) = \dfrac{n!}{(n - r)!r!}$

Example 3 Use the Combination Formula

RETAIL Marques works part-time at a local department store. His manager asked him to make a display using 5 different styles of shirts from 8 available styles. How many ways can Marques choose the shirts?

$$C(n, r) = \frac{n!}{(n - r)!r!} \qquad \text{Combination Formula}$$

$$C(8, 5) = \frac{8!}{(8 - 5)!5!} \qquad n = 8 \text{ and } r = 5$$

$$= \frac{8!}{3!5!} \qquad \text{Simplify.}$$

$$= \frac{8 \cdot 7 \cdot 6 \cdot \cancel{5} \cdot \cancel{4} \cdot \cancel{3} \cdot \cancel{2} \cdot \cancel{1}}{3 \cdot 2 \cdot 1 \cdot \cancel{5} \cdot \cancel{4} \cdot \cancel{3} \cdot \cancel{2} \cdot \cancel{1}} \qquad \text{Divide by common factors.}$$

$$= \frac{336}{6} \text{ or } 56 \qquad \text{Simplify.}$$

GuidedPractice

3. **SPRING DANCE** A group of four students is selecting corsages and boutonnières to wear to the spring dance. They can choose from 4 roses, 6 carnations, and 8 tulips. If all of the flowers are different, in how many ways can 4 flowers be chosen?

Part of the difficulty in learning permutations and combinations is distinguishing between them. Remember that order matters with permutations.

Example 4 Identifying Permutations and Combinations

Identify each situation as a *permutation* or a *combination*.

a. A playlist of songs on a digital media player is being played so that songs do not repeat. In how many ways can the songs be played?

Since the question is really about the different orders of songs, this is a permutation.

b. Amy orders an ice cream cone with three different flavors. There are 26 possible flavors. How many different sets of three flavors can she choose?

Since the order in which the flavors are selected is not important, this is a combination.

GuidedPractice

4A. A teacher uses a random number generator to create a seating chart for 20 students.

4B. Ten athletes enter a race. The top three finishers move on to the next round.

We can use permutations or combinations to find the probability of an event.

Example 5 Probability with Permutations and Combinations

SPANISH CLUB The Spanish club is forming two teams for the International Festival. Rebekah and Lydia are among the nine who volunteered to lead a team. If the advisor assigns positions at random, what is the probability that Rebekah is chosen to lead the food team and Lydia is chosen to lead the tickets team?

Step 1 Find the total number of outcomes.
Since we care about specific positions, this is a permutation. Find the number of permutations of 9 people taken 2 at a time.

$$P(n, r) = \frac{n!}{(n-r)!}$$ Permutation Formula

$$P(9, 2) = \frac{9!}{(9-2)!} \text{ or } 72$$ $n = 9$ and $r = 2$

There are 72 possible outcomes

Step 2 Find the successes.
Of the 72 permutations, only one has Rebekah leading the food team and Lydia leading the tickets team.

Step 3 Find the probability.

$$P = \frac{\text{number of successes}}{\text{number of possible outcomes}} = \frac{1}{72}$$ Probability Formula

The probability of Rebekah leading the food team and Lydia leading the tickets team is $\frac{1}{72}$.

> **StudyTip**
> Probability Find the probability of an event by dividing the number of favorable outcomes by the number of possible outcomes.

▶ **Guided**Practice

5. STUDENT COUNCIL Four of the twelve members of the student council are being randomly selected to hand out flyers at the mall. What is the probability that Hasina, Amy, Bonny, and Tim are all selected?

Check Your Understanding

 — Step-by-Step Solutions begin on page R13.

Example 1   **1. CHARITY** A youth charity group is holding a raffle and wants to display a picture of the 6 prizes on a flyer. How many ways can they arrange the prizes in a row?

Examples 2–4 Identify each situation as a *permutation* or a *combination*.

     2. choosing 3 different pizza toppings from a list of 12

     3. choosing team captains for a football team

     4. choosing the first-, second-, and third-place winner of an art competition

Evaluate each expression.

     5. $P(7, 2)$ **6.** $P(9, 3)$ **7** $C(6, 4)$ **8.** $C(5, 2)$

     9. MENUS There are 14 toppings listed in a pizza menu. How many different six-topping pizzas are possible?

Example 5   **10. CCSS MODELING** Students are given 5-digit passwords for their accounts on the school's computer system. If no numbers can repeat, what is the probability that a student's password is 93152?

Example 1   **11. SCIENCE FAIR** There are 8 finalists in a science fair competition. How many ways can they stand on the stage?

Evaluate each expression.

12. $P(6, 6)$ **13.** $P(5, 1)$ **14.** $P(4, 1)$ **15.** $P(7, 3)$

16. $C(7, 6)$ **17.** $C(5, 3)$ **18.** $C(5, 5)$ **19.** $C(3, 0)$

Examples 2–4   **Identify each situation as a *permutation* or a *combination*.**

20. selecting 5 books to read from a list of 8

21. an arrangement of the letters in the word *probability*

22. a list of students by class ranking

23. a playlist of songs on a digital media player

24. selecting 4 different ingredients out of 8 for a salad

25. CCSS TOOLS Abigail works at the jewelry store in the mall. Her manger asks her to place 3 of the 12 birthstone necklaces in the front display case. How many ways can she arrange the necklaces in the display case?

26. RECYCLING Juana is setting two recycling bins at the end of her driveway for pick-up. She has four bins from which to choose. How many ways can she pick the bins to set out?

Example 5   **27 MUSIC** What is the probability that the first 8 songs that are played on Kenneth's playlist are country songs?

Exercise Playlist

Country	8 songs
Rock	6 songs
Rap	4 songs

28. AMUSEMENT PARKS Tino is entering an amusement park with 5 of his friends. At the gate they must go through a turnstile one at a time. How many ways can Tino and his friends go through the turnstile?

29. PAGEANTS The Teen Miss USA pageant has 51 delegates. If the judges choose Teen Miss USA and four runners-up, how many ways can they be chosen?

30. BASKETBALL A coach has to select 5 of 12 players on the basketball team to start the game. How many different groups of players could be selected as starters?

31. ICE CREAM How many ways can a customer choose 3 flavors of ice cream at The Dairy Barn?

32. GAMES Tonisha is playing a board game in which you make words to score points. There are 12 letters left in the box, and she must choose 4. She cannot see the letters.
   a. Suppose the 12 letters are all different. How many ways can she choose 4 of the 12?
   b. She chooses *A*, *T*, *R*, and *E*. How many different arrangements of three letters can she make from these letters?
   c. How many of the three-letter arrangements are words? List them.

THE DAIRY BARN

5 varieties of chocolate
4 varieties of candy-flavor
6 varieties of berry-flavor

33. DANCE At the spring dance, Christy and 7 of her friends sit on one side of a table. How many ways can Christy and her friends fill the 8 empty seats?

34. HORSEBACK RIDING Trish and Charliqua entered a horseback riding camp with 22 other people. Six riders are randomly selected to work with the head instructor. How many different groups of people can be placed with the head instructor?

35 BOWLING Chris and Kelly entered a bowling tournament.

Bowling Tournament!
32 Entries
2 Players Per Lane
Positions are Chosen by Drawing

   a. What is the probability that Chris and Kelly are selected to bowl against each other in *any* lane?

   b. What is the probability that Chris and Kelly are selected to bowl against each other in the *last* lane?

36. SECURITY Banks lock an account after three incorrect PIN entries. Suppose a bank that uses four-digit PINs in which the digits cannot repeat allowed unlimited incorrect entries. What is the maximum time it would take a hacker using a computer program that can enter 100 different codes per second to enter the correct PIN?

37. DECISION MAKING Westerville High School is putting on a play. In all, 4 freshmen, 5 sophomores, 6 juniors, and 8 seniors tried out for the 12 open spots.

   a. How many ways can the 12 spots be chosen?

   b. If the students are chosen randomly, what is the probability that at least one senior will be chosen?

   c. What probability model can we use to randomly select the first spot?

   d. How does this model change when selecting the second spot?

38. COMPOUND PROBABILITY There are 7 red marbles, 8 purple marbles, and 6 green marbles in a bag. Fifteen marbles are randomly selected. Use the following steps to determine the probability of selecting 5 of each color.

   a. Find the number of possible outcomes.

   b. Find the number of successes for selecting 5 of 7 red marbles, for selecting 5 of 8 purple marbles, and for selecting 5 of 6 green marbles.

   c. Use the Fundamental Counting Principle to find the total number of successes.

   d. Determine the probability of selecting 5 of each color.

H.O.T. Problems   Use Higher-Order Thinking Skills

39. ERROR ANALYSIS Sydney and Ming are determining how many 4-person committees are possible if 10 people are available. Is either of them correct? Explain.

                Sydney   *Ming*

$$_{10}P_4 = \frac{10!}{(10-4)!} \qquad\qquad _{10}C_4 = \frac{10!}{(10-4)!4!}$$

$$= 5040 \qquad\qquad\qquad\qquad\qquad = 210$$

40. CCSS PERSEVERANCE Seven identical mathematics books and 4 identical science books are to be stored on one shelf. How many different ways can the books be arranged?

41. WHICH ONE DOESN'T BELONG? Determine which situation does not belong. Explain.

choosing 5 players on a quiz team	choosing 10 colored marbles from a bag
choosing 4 horses from 6 to run a race	ranking students in a senior class

42. REASONING Determine whether the statement $_nP_r = {}_nC_r$ is *sometimes*, *always*, or *never* true. Explain your reasoning.

43. WRITING IN MATH Write a situation in which 3 of 8 objects are selected and order is not important.

44. In how many ways can 3 of 8 different flowers be planted along one side of a road?

A 342 B 338 C 336 D 328

45. If Jack can eat 21 hard-boiled eggs in 15 minutes, how many can he eat in 25 minutes if he continues eating at the same pace?

F 18 G 35 H 36 J 37

46. Shante has 30 coins, quarters and dimes, that total $5.70. How many quarters does she have?

A 12   C 18
B 15   D 20

47. SHORT RESPONSE There are 3 red candies in a bag of 20 candies. What is the probability of selecting a red candy?

Spiral Review

Vacationers were asked how many evenings they spent eating out during their trip. Find each experimental probability. (Lesson 12-5)

Number of Evenings	Frequency
0	14
1	39
2	28
3	21
4	10
5+	13

48. a vacationer ate out at least once

49. a vacationer ate out less than three times

50. Find the mean, median, mode, range, and standard deviation of the data set obtained after multiplying each value by 1.3. (Lesson 12-4)

$$26, 15, 19, 31, 47, 44, 38, 26, 28, 19$$

51. PET CARE Kendra takes care of pets while their owners are out of town. One week she has three dogs that all eat the same kind of dog food. How many bags of food should Kendra buy for one week? (Lesson 11-6)

Max
12 days/bag

Miles
15 days/bag

Stormy
16 days/bag

Find each product. (Lesson 11-4)

52. $\dfrac{8}{x^2} \cdot \dfrac{x^4}{4x}$

53. $\dfrac{10r^3}{6n^3} \cdot \dfrac{42n^2}{35r^3}$

54. $\dfrac{10y^3z^2}{6wx^3} \cdot \dfrac{12w^2x^2}{25y^2z^4}$

55. $\dfrac{(n-1)(n+1)}{(n+1)} \cdot \dfrac{(n-4)}{(n-1)(n+4)}$

56. $\dfrac{(x-8)}{(x+8)(x-3)} \cdot \dfrac{(x+4)(x-3)}{(x-8)}$

57. $\dfrac{3a^2b}{2gh} \cdot \dfrac{24g^2h}{15ab^2}$

58. COOKING The formula $t = \dfrac{40(25 + 1.85a)}{50 - 1.85a}$ relates the time t in minutes that it takes to cook an average-size potato in an oven at an altitude of a thousands of feet. (Lesson 11-3)

a. What is the value of a for an altitude of 4500 feet?

b. Calculate the time it takes to cook a potato at an altitude of 3500 feet and at 7000 feet. How do your cooking times compare at these two altitudes?

Skills Review

Ten red tiles, 12 blue tiles, 8 green tiles, 4 yellow tiles, 10 red tiles, and 12 black tiles are placed in a bag and selected at random. Find each probability.

59. P(blue)

60. P(red)

61. P(black or yellow)

62. P(green or red)

63. P(not blue)

64. P(not green)

Probability of Compound Events

- You calculated simple probability.

1 Find probabilities of independent and dependent events.

2 Find probabilities of mutually exclusive events.

- Evita is flying from Cleveland to Honolulu. The airline reports that the flight from Cleveland to Honolulu has a 40% on-time record. The airline also reported that they lose luggage 5% of the time. What is the probability that both the flight will be on time and Evita's luggage will arrive?

 NewVocabulary
compound event
joint probability
independent events
dependent events
mutually exclusive events

 **Common Core State Standards**

Mathematical Practices
5 Use appropriate tools strategically.

1 Independent and Dependent Events Recall that one event, like flying from Cleveland to Honolulu, is called a *simple event*. A **compound event** is made up of two or more simple events. So, the probability that the flight will be on time and the luggage arrives is an example of a compound event. The probability of compound events is called **joint probability**. The plane being on time may not affect whether luggage is lost. These two events are called **independent events** because the outcome of one event does not affect the outcome of the other.

> **KeyConcept** Probability of Independent Events
>
> Words   If two events, A and B, are independent, then the probability of both events occurring is the product of the probability of A and the probability of B.
>
> Model
>
>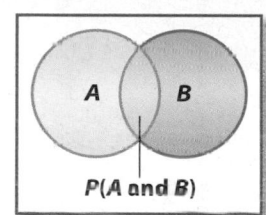
>
> $P(A \text{ and } B)$
>
> Symbols $P(A \text{ and } B) = P(A) \cdot P(B)$

> **Real-World Example 1** Independent Events
>
> **MARBLES** A bag contains 6 black marbles, 9 blue marbles, 4 yellow marbles, and 2 green marbles. A marble is selected, replaced, and a second marble is selected. Find the probability of selecting a black marble, then a yellow marble.
>
> First marble: $P(\text{black}) = \dfrac{6}{21}$ ← number of black marbles / total number of marbles
>
> Second marble: $P(\text{yellow}) = \dfrac{4}{21}$ ← number of yellow marbles / total number of marbles
>
> $P(\text{black, yellow}) = P(\text{black}) \cdot P(\text{yellow})$ Probability of independent events
>
> $\qquad\qquad\qquad = \dfrac{6}{21} \cdot \dfrac{4}{21} \text{ or } \dfrac{24}{441}$ Substitution
>
> The probability is $\dfrac{24}{441}$ or about 5.4%.
>
> **GuidedPractice**   Find each probability.
>
> **1A.** $P(\text{blue, green})$ **1B.** $P(\text{not black, blue})$

When the outcome of one event affects the outcome of another event, they are **dependent events**. In Example 1, if the marble was not placed back in the bag, then drawing the two marbles would have been dependent events. The probability of drawing the second marble depends on what marble was drawn first.

KeyConcept Probability of Dependent Events

Words	If two events, A and B, are dependent, then the probability of both events occurring is the product of the probability of A and the probability of B after A occurs.
Symbols	$P(A \text{ and } B) = P(A) \cdot P(B \text{ following } A)$

Recall that the complement of a set is the set of all objects that do *not* belong to the given set. In a standard deck of cards, the complement of drawing a heart is drawing a diamond, club, or spade. So, the probability of drawing a heart is $\frac{13}{52}$, and the probability of not drawing a heart is $\frac{52-13}{52}$ or $\frac{39}{52}$.

The sum of the probabilities for any two complementary events is 1.

Real-World Example 2 Dependent Events

CARDS Cynthia randomly draws three cards from a standard deck one at a time without replacement. Find the probability that the cards are drawn in the given order.

a. $P(\text{diamond, spade, diamond})$

First card: $P(\text{diamond}) = \frac{13}{52}$ or $\frac{1}{4}$ ← number of diamonds / total number of cards

Second card: $P(\text{spade}) = \frac{13}{51}$ ← number of spades / number of cards remaining

Third card: $P(\text{diamond}) = \frac{12}{50}$ or $\frac{6}{25}$ ← number of diamonds remaining / number of cards remaining

$P(\text{diamond, spade, diamond}) = P(\text{diamond}) \cdot P(\text{spade}) \cdot P(\text{diamond})$

$$= \frac{1}{4} \cdot \frac{13}{51} \cdot \frac{6}{25} \text{ or } \frac{13}{850} \qquad \text{Substitution}$$

The probability is $\frac{13}{850}$ or about 1.5%.

b. $P(\text{four, four, not a jack})$

After Cynthia draws the first two fours from the deck of 52 cards, there are 50 cards left. Since neither of these cards are jacks, there are still four jacks left in the deck. So, there are $52 - 2 - 4$ or 46 cards that are not jacks.

$P(\text{four, four, not a jack}) = P(\text{four}) \cdot P(\text{four}) \cdot P(\text{not a jack})$

$$= \frac{4}{52} \cdot \frac{3}{51} \cdot \frac{46}{50}$$

$$= \frac{552}{132,600} \text{ or } \frac{23}{5525}$$

The probability is $\frac{23}{5525}$ or about 0.4%.

GuidedPractice

Find each probability.

2A. $P(\text{two, five, not a five})$

2B. $P(\text{heart, not a heart, heart})$

Problem-SolvingTip

CCSS Tools Acting out the situation can help you understand what the question is asking. Use a deck of cards to represent the situation described in the problem.

2 Mutually Exclusive Events
Events that cannot occur at the same time are called **mutually exclusive events**. Suppose you wanted to find the probability of drawing a heart or a diamond. Since a card cannot be both a heart and a diamond, the events are mutually exclusive.

StudyTip

and and *or* While probabilities involving *and* deal with independent and dependent events, probabilities involving *or* deal with mutually exclusive and non-mutually exclusive events.

KeyConcept Probability of Mutually Exclusive Events

Words If two events, A and B, are mutually exclusive, then the probability that either A or B occurs is the sum of their probabilities.

Symbols $P(A \text{ or } B) = P(A) + P(B)$

Model

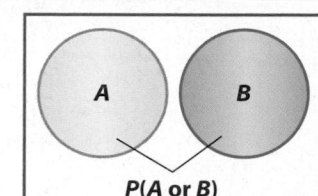

$P(A \text{ or } B)$

Real-World Example 3 Mutually Exclusive Events

A die is being rolled. Find each probability.

a. $P(3 \text{ or } 5)$

Since a die cannot show both a 3 and a 5 at the same time, these events are mutually exclusive.

$P(\text{rolling a } 3) = \dfrac{1}{6}$ ← number of sides with a 3 / total number of sides

$P(\text{rolling a } 5) = \dfrac{1}{6}$ ← number of sides with a 5 / total number of sides

StudyTip

Alternative Method In Example 3a, you could have placed the number of possible outcomes over the total number of outcomes.
$\dfrac{1+1}{6} = \dfrac{2}{6}$ or $\dfrac{1}{3}$

$P(3 \text{ or } 5) = P(\text{rolling a } 3) + P(\text{rolling a } 5)$ Probability of mutually exclusive events

$\qquad\qquad = \dfrac{1}{6} + \dfrac{1}{6}$ Substitution

$\qquad\qquad = \dfrac{2}{6} \text{ or } \dfrac{1}{3}$ Add.

The probability of rolling a 3 or a 5 is $\dfrac{1}{3}$ or about 33%.

b. $P(\text{at least } 4)$

Rolling at least a 4 means you can roll either a 4, 5, or a 6. So, you need to find the probability of rolling a 4, 5, or a 6.

$P(\text{rolling a } 4) = \dfrac{1}{6}$ ← number of sides with a 4 / total number of sides

$P(\text{rolling a } 5) = \dfrac{1}{6}$ ← number of sides with a 5 / total number of sides

$P(\text{rolling a } 6) = \dfrac{1}{6}$ ← number of sides with a 6 / total number of sides

$P(\text{at least } 4) = P(\text{rolling a } 4) + P(\text{rolling a } 5) + P(\text{rolling a } 6)$ Mutually exclusive events

$\qquad\qquad = \dfrac{1}{6} + \dfrac{1}{6} + \dfrac{1}{6}$ Substitution

$\qquad\qquad = \dfrac{3}{6} \text{ or } \dfrac{1}{2}$ Add.

The probability of rolling at least a 4 is $\dfrac{1}{2}$ or about 50%.

▶ **Guided**Practice

3A. $P(\text{less than } 3)$ 　　　　　　　　　　　**3B.** $P(\text{even})$

Suppose you want to find the probability of randomly drawing a 2 or a diamond from a standard deck of cards. Since it is possible to draw a card that is both a 2 and a diamond, these events are not mutually exclusive.

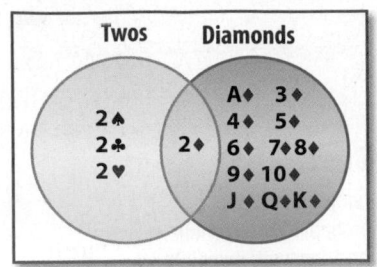

Twos   Diamonds

$P(2)$ $P(\text{diamond})$ $P(2, \text{diamond})$

$\frac{4}{52}$ $\frac{13}{52}$ $\frac{1}{52}$

In the first two fractions above, the probability of drawing the two of diamonds is counted twice, once for a two and once for a diamond. To find the correct probability, subtract $P(2 \text{ of diamonds})$ from the sum of the first two probabilities.

$$P(2 \text{ or a diamond}) = P(2) + P(\text{diamond}) - P(2 \text{ of diamonds})$$

$$= \frac{4}{52} + \frac{13}{52} - \frac{1}{52}$$

$$= \frac{16}{52} \text{ or } \frac{4}{13} \qquad \text{The probability is } \frac{4}{13} \text{ or about 31\%.}$$

⚙ KeyConcept Probability of Events that are Not Mutually Exclusive

Words   If two events, *A* and *B*, are not mutually exclusive, then the probability that either *A* or *B* occurs is the sum of their probabilities decreased by the probability of both occurring.

Model

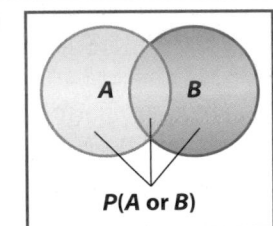

P(A or B)

Symbols $P(A \text{ or } B) = P(A) + P(B) - P(A \text{ and } B)$

◐ Real-World Example 4 Events that are Not Mutually Exclusive

STUDENT ATHLETES Of 240 girls, 176 are on the Honor Roll, 48 play sports, and 36 are on the Honor Roll and play sports. What is the probability that a randomly selected student plays sports or is on the Honor Roll?

Since some students play sports and are on the Honor Roll, the events are not mutually exclusive.

$$P(\text{sports}) = \frac{48}{240} \qquad P(\text{Honor Roll}) = \frac{176}{240} \qquad P(\text{sports and Honor Roll}) = \frac{36}{240}$$

$$P(\text{sports or Honor Roll}) = P(\text{sports}) + P(\text{HR}) - P(\text{sports and HR})$$

$$= \frac{48}{240} + \frac{176}{240} - \frac{36}{240} \qquad \text{Substitution}$$

$$= \frac{188}{240} \text{ or } \frac{47}{60} \qquad\qquad \text{Simplify.}$$

The probability is $\frac{47}{60}$ or about 78%.

▶ **GuidedPractice**

4. PETS Out of 5200 households surveyed, 2107 had a dog, 807 had a cat, and 303 had both a dog and a cat. What is the probability that a randomly selected household has a dog or a cat?

Examples 1–2 Determine whether the events are *independent* or *dependent*. Then find the probability.

1. **BABYSITTING** A toy bin contains 12 toys, 8 stuffed animals, and 3 board games. Marsha randomly chooses 2 items for the child she is babysitting. What is the probability that she chose 2 stuffed animals as the first two choices?

2. **FRUIT** A basket contains 6 apples, 5 bananas, 4 oranges, and 5 peaches. Drew randomly chooses one piece of fruit, eats it, and chooses another. What is the probability that he chose a banana and then an apple?

3. **MONEY** Nakos has 4 quarters, 3 dimes, and 2 nickels in his pocket. Nakos randomly picks two coins out of his pocket. What is the probability that he did not choose a dime either time, if he replaced the first coin before choosing a second coin?

4. **BOOKS** Joanna needs a book to prop up a table leg. She randomly selects a book, puts it back on the shelf, and selects another book. What is the probability that Joanna selected two math books?

Examples 3–4 A card is drawn from a standard deck of playing cards. Determine whether the events are *mutually exclusive* or *not mutually exclusive*. Then find the probability.

5. P(two or queen)

6. P(diamond or heart)

7. P(seven or club)

8. P(spade or ace)

Examples 1–2 Determine whether the events are *independent* or *dependent*. Then find the probability.

9. **COINS** If a coin is tossed 4 times, what is the probability of getting tails all 4 times?

10. **DICE** A die is rolled twice. What is the probability of rolling two different numbers?

11. **CANDY** A box of chocolates contains 10 milk chocolates, 8 dark chocolates, and 6 white chocolates. Sung randomly chooses a chocolate, eats it, and then randomly chooses another. What is the probability that Sung chose a milk chocolate and then a white chocolate?

12. **DICE** A die is rolled twice. What is the probability of rolling the same numbers?

13. **PETS** Chuck and Rashid went to a pet store to buy dog food. They chose from 10 brands of dry food, 6 brands of canned food, and 3 brands of pet snacks. What is the probability that both chose dry food, if Chuck randomly chose first and liked the first brand he picked up?

14. **CCSS MODELING** A rental agency has 12 white sedans, 8 gray sedans, 6 red sedans, and 3 green sedans for rent. Mr. Escobar rents a sedan, returns it because the radio is broken, and gets another sedan. Assuming the returned sedan remains in circulation, what is the probability that Mr. Escobar was given a green sedan and then a gray sedan?

Determine whether the events are *mutually exclusive* or *not mutually exclusive*. Then find the probability.

15. **BOWLING** Cindy's bowling records indicate that for any frame, the probability that she will bowl a strike is 30%, a spare 45%, and neither 25%. What is the probability that she will bowl either a spare or a strike for any given frame?

16. **SPORTS CARDS** Dario owns 145 baseball cards, 102 football cards, and 48 basketball cards. What is the probability that he randomly selects a baseball or a football card?

17. **SCHOLARSHIPS** 3000 essays were received for a $5000 college scholarship. 2865 essays were the required length, 2577 of the applicants had the minimum required grade-point average, and 2486 had the required length and minimum grade-point average. What is the probability that an essay selected at random will have the required length or the required grade-point average?

18. **KITTENS** Ruby's cat had 8 kittens. The litter included 2 orange females, 3 mixed-color females, 1 orange male, and 2 mixed-color males. Ruby wants to keep one kitten. What is the probability that she randomly chooses a kitten that is female or orange?

CHIPS **A restaurant serves red, blue, and yellow tortilla chips. The bowl of chips Gabriel receives has 10 red chips, 8 blue chips, and 12 yellow chips. After Gabriel chooses a chip, he eats it. Find each probability.**

19. P(red, blue)

20. P(blue, yellow)

21. P(yellow, not blue)

22. P(red, not yellow)

23 **SOCKS** Damon has 14 white socks, 6 black socks, and 4 blue socks in his drawer. If he chooses two socks at random, what is the probability that the first two socks are white?

CCSS **TOOLS** **Cards are being randomly drawn from a standard deck of cards. Find each probability.**

24. P(heart or spade)   **25.** P(spade or club)   **26.** P(queen or heart)

27. P(jack or spade)   **28.** P(five or prime number)   **29.** P(ace or black)

30. **CANDY** A bag contains 10 red, 6 green, 7 yellow, and 5 orange jelly beans. What is the probability of randomly choosing a red jelly bean, replacing, randomly choosing another red jelly bean, replacing, and then randomly choosing an orange jelly bean?

31. **SPORTS** The extracurricular activities in which the senior class at Valley View High School participate are shown in the Venn diagram.

a. How many students are in the senior class?

b. How many students participate in athletics?

c. If a student is randomly chosen, what is the probability that the student participates in athletics or drama?

d. If a student is randomly chosen, what is the probability that the student participates in only drama and band?

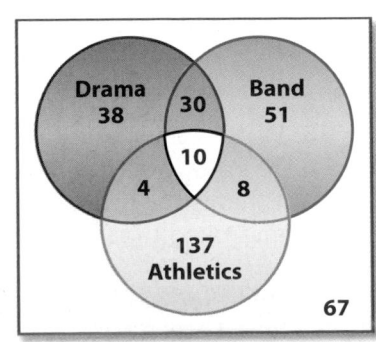

32. TILES Kirsten and José are playing a game. Kirsten places tiles numbered 1 to 50 in a bag. José selects a tile at random. If he selects a prime number or a number greater than 40, then he wins the game. What is the probability that José will win on his first turn?

33 ⟳ **MULTIPLE REPRESENTATIONS** In this problem, you will explore **conditional probability**, which is the probability that event *B* occurs given that event *A* has already occurred. It is calculated by dividing the probability of the occurrence of both events by the probability of the occurrence of the first event. The notation for conditional probability is $P(B \mid A)$, read *the probability of B, given A*.

a. Graphical Draw a Venn diagram to illustrate $P(A \text{ and } B)$.

b. Verbal Tell how to find $P(B \mid A)$ given the Venn diagram.

c. Analytical A jar contains 12 marbles, of which 8 marbles are red and 4 marbles are green. If marbles are chosen without replacement, find $P(\text{red})$ and $P(\text{red, green})$.

d. Analytical Using the probabilities from part **c** and the Venn diagram in part **a**, determine the probability of choosing a green marble on the second selection, given that the first marble selected was red.

e. Analytical Write a formula for finding a conditional probability.

f. Analytical Use the definition from part **e** to answer the following: At a basketball game, 80% of the fans cheered for the home team. In the same crowd, 20% of the fans were waving banners and cheering for the home team. What is the probability that a fan waved a banner given that the fan cheered for the home team?

H.O.T. Problems   Use Higher-Order Thinking Skills

34. ERROR ANALYSIS George and Aliyah are determining the probability of randomly choosing a blue or red marble from a bag of 8 blue marbles, 6 red marbles, 8 yellow marbles, and 4 white marbles. Is either of them correct? Explain.

George

$P(\text{blue or red}) = P(\text{blue}) \cdot P(\text{red})$

$= \dfrac{8}{26} \cdot \dfrac{6}{26}$

$= \dfrac{48}{676}$

about 7%

Aliyah

$P(\text{blue or red}) = P(\text{blue}) + P(\text{red})$

$= \dfrac{8}{26} + \dfrac{6}{26}$

$= \dfrac{14}{26}$

about 54%

35. CHALLENGE In some cases, if one bulb in a string of holiday lights fails to work, the whole string will not light. If each bulb in a set has a 99.5% chance of working, what is the maximum number of lights that can be strung together with at least a 90% chance that the whole string will light?

36. CCSS REGULARITY Suppose there are three events *A*, *B*, and *C* that are not mutually exclusive. List all of the probabilities you would need to consider in order to calculate $P(A \text{ or } B \text{ or } C)$. Then write the formula you would use to calculate it.

37. OPEN ENDED Describe a situation in your life that involves dependent and independent events. Explain why the events are dependent or independent.

38. WRITING IN MATH Explain why the subtraction occurs when finding the probability of two events that are not mutually exclusive.

39. In how many ways can a committee of 4 be selected from a group of 12 people?

A 48

B 483

C 495

D 11,880

40. A total of 925 tickets were sold for $5925. If adult tickets cost $7.50 and children's tickets cost $3.00, how many adult tickets were sold?

F 700   H 325

G 600   J 225

41. SHORT RESPONSE A circular swimming pool with a diameter of 28 feet has a deck of uniform width built around it. If the area of the deck is 60π square feet, find its width.

42. The probability of heads landing up when you flip a coin is $\frac{1}{2}$. What is the probability of getting tails if you flip it again?

A $\frac{1}{4}$ C $\frac{1}{2}$

B $\frac{1}{3}$ D $\frac{3}{4}$

43. SHOPPING The Millers have twelve grandchildren, 5 boys and 7 girls. For their anniversary, the grandchildren decided to pool their money and have three of them shop for the entire group. (Lesson 12-6)

  a. Does this situation represent a *combination* or *permutation*?

  b. How many ways are there to choose the three?

  c. What is the probability that all three will be girls?

44. HAIR STYLIST Tia is a hair stylist. Last week, 70% of her clients who called to make appointments made an appointment for a basic haircut. (Lesson 12-5)

  a. Design a simulation that can be used to estimate the probability that the next client that makes an appointment will schedule a basic haircut.

  b. Conduct the simulation, and report the results.

Solve each equation. State any extraneous solutions. (Lesson 11-8)

45. $\dfrac{4}{a} = \dfrac{3}{a-2}$

46. $\dfrac{3}{x} = \dfrac{1}{x-2}$

47. $\dfrac{x}{x+1} = \dfrac{x-6}{x-1}$

48. $\dfrac{2n}{3} + \dfrac{1}{2} = \dfrac{2n-3}{6}$

49. COOKING Hannah was making candy using a two-quart pan. As she stirred the mixture, she noticed that the pan was about $\frac{2}{3}$ full. If each piece of candy has a volume of about $\frac{3}{4}$ ounce, approximately how many pieces of candy will Hannah make? (*Hint:* There are 32 ounces in a quart.) (Lesson 11-3)

50. GEOMETRY A rectangle has a width of $3\sqrt{5}$ centimeters and a length of $4\sqrt{10}$ centimeters. Find the area of the rectangle. Write as a simplified radical expression. (Lesson 10-2)

Solve each equation. Check your solution.

51. $\sqrt{-3a} = 6$

52. $\sqrt{a} = 100$

53. $\sqrt{-k} = 4$

54. $5\sqrt{2} = \sqrt{x}$

55. $3\sqrt{7} = \sqrt{-y}$

56. $3\sqrt{4a} - 2 = 10$

Algebra Lab
Two-Way Frequency Tables

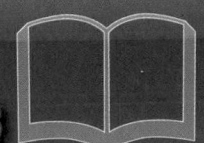

Joana sent out a survey to the freshmen and sophomores, asking if they were planning on attending the dance. One way of organizing her responses is to use a two-way frequency table. A **two-way frequency table** or *contingency table* is used to show the frequencies of data from a survey or experiment classified according to two categories, with the rows indicating one category and the columns indicating the other.

CCSS **Common Core State Standards**
Content Standards
S.ID.5 Summarize categorical data for two categories in two-way frequency tables. Interpret relative frequencies in the context of the data (including joint, marginal, and conditional relative frequencies). Recognize possible associations and trends in the data.

For Joana's survey, the two categories are *class* and *attendance*. These categories can be split into subcategories: *freshman* and *sophomore* for *class*, and *attending* and *not attending* for *attendance*.

Class	Attending	Not Attending	Totals
Freshman		subcategories	
Sophomore			
Totals			

Activity 1 Two-Way Frequency Table

DANCE **Sixty-six freshmen responded to the survey, with 32 saying that they would be attending. Of the 84 sophomores that responded, 46 said they would attend. Organize the data in a two-way table.**

Step 1 Find the values for every combination of subcategories. One combination is freshmen / not attending. Since 32 of 66 freshmen are attending, 66 − 32 or 34 freshmen are *not* attending. These combinations are called **joint frequencies.**

Step 2 Place every combination in the corresponding cell.

Step 3 Find the totals of each subcategory and place them in their corresponding cell. These values are called **marginal frequencies.**

Class	Attending	Not Attending	Totals
Freshman	32	34	66
Sophomore	46	38	84
Totals	78	72	150

joint frequencies

marginal frequencies

Step 4 Find the sum of each set of marginal frequencies. These two sums should be equal. Place the value in the bottom right corner.

marginal frequencies

Analyze the Results

1. How many students responded to the survey?

2. How many of the students that were surveyed are attending the dance?

3. How many of the surveyed sophomores are not attending the dance?

4. What does each of the joint frequencies represent?

5. What does each of the marginal frequencies represent?

6. **WORK** Heather sent out a survey asking who was working during the holiday. Of the 50 boys who responded, 34 said *yes*. Of the 45 girls who responded, 21 said *no*. Create a two-way frequency table of the results.

7. **SOCCER** Pamela asked if anyone would be interested in a co-ed soccer team. Of the 28 boys who responded, 18 said that they would play and 4 were undecided. Of the 22 girls who responded, 6 said they did not want to play and 3 were undecided. Create a two-way frequency table of the results.

A **relative frequency** is the ratio of the number of observations in a category to the total number of observations. Relative frequencies are also probabilities. To create a relative frequency two-way table, divide each of the values by the total number of observations and replace them with their corresponding decimals or percents.

Class	Attending	Not Attending	Totals
Freshman	$\frac{32}{150} \approx 21.3\%$	22.7%	44%
Sophomore	30.7%	25.3%	56%
Totals	52%	48%	100%

A **conditional relative frequency** is the ratio of the joint frequency to the marginal frequency. For example, given that a student is a freshman, what is the conditional relative frequency that he or she is going to the dance? In other words, what is the probability that a freshman is going to the dance?

Activity 2 Two-Way Conditional Relative Frequency Table

DANCE Joana wants to determine the conditional relative frequencies (or probabilities) given the fact that she knows the class of the respondents.

Step 1 Refer to the table in Activity 1. A total of 66 freshmen responded, and 32 said *yes*. Therefore, the conditional relative frequency that a respondent said *yes* given that the respondent is a freshman is $\frac{32}{66}$.

Step 2 Place every conditional relative frequency in the corresponding cell.

Step 3 The conditional relative frequencies for each row should sum to 100%.

Conditional Relative Frequencies by Class			
Class	Attending	Not Attending	Totals
Freshman	$\frac{32}{66} \approx 48\%$	$\frac{34}{66} \approx 52\%$	100%
Sophomore	$\frac{46}{84} \approx 55\%$	$\frac{38}{84} \approx 45\%$	100%

Analyze the Results

8. Given that a respondent was a sophomore, what is the probability that he or she said *no*?

9. What does each of the conditional relative frequencies represent?

10. Why do you think that the columns do not sum to 100%?

11. Create a two-way conditional relative frequency table for the category *attendance*.

12. Given that a respondent was not attending, what is the probability that he or she is a freshman?

13. **ACTIVITIES** The managers, staff, and assistants were given three options for the holiday activity: a potluck, a dinner at a restaurant, and a gift exchange. Five of the 11 managers want a dinner, while 3 want a potluck. Eleven of the 45 staff members want a gift exchange, while 18 want a dinner. Ten of the 32 assistants want a dinner, while 8 of them want a gift exchange.

   a. Create a two-way frequency table.

   b. Convert the two-way frequency table into a relative frequency table.

   c. Create two conditional relative frequency tables: one for the activities and one for the employees.

<table>
<tr><th>··Then</th><th>··Now</th><th>··Why?</th></tr>
</table>

- You found probabilities of events.

1 Find probabilities by using random variables.

2 Find the expected value of a probability distribution.

- A gaming software company with five online games on the market is interested in how many games each of their customers play. They surveyed 1000 randomly chosen customers. The results of the survey are shown.

Number of Computer Games	Number of Customers
1	130
2	110
3	150
4	500
5	110

NewVocabulary
random variable
discrete random variable
probability distribution
probability graph
expected value

Common Core State Standards

Mathematical Practices
4 Model with mathematics.

1 **Random Variables and Probability** A variable with a value that is the numerical outcome of a random event is called a **random variable**. A random variable with a countable number of possibilities is a **discrete random variable**. We can let the random variable G represent the number of different games. So, G can equal 1, 2, 3, 4, or 5.

Example 1 Random Variables

A graduation supply company offers five items that can be purchased for graduation: a diploma frame, graduation picture, cap and gown, senior key ring, and class pin. The school takes a poll of the seniors to see how many of these items each senior is buying. The results are shown.

Number of Items Being Purchased	Number of Seniors
0	12
1	122
2	134
3	115
4	145
5	97

a. **Find the probability that a randomly chosen senior is buying 3 items.**

Let X represent the number of items being purchased. There is only one outcome in which 3 items are being purchased, and there is a total of 625 seniors.

$P(X = 3) = \dfrac{3 \text{ items being purchased}}{\text{seniors surveyed}}$ $P(X = n)$ is the probability of X occurring n times.

$= \dfrac{115}{625}$ or $\dfrac{23}{125}$

The probability is $\dfrac{23}{125}$ or 18.4%.

b. **Find the probability that a randomly chosen senior buys at least 4 items.**

There are 145 + 97 or 242 seniors who are purchasing at least 4 items.

$P(X \geq 4) = \dfrac{242}{625}$ The probability is $\dfrac{242}{625}$ or about 38.7%.

GuidedPractice

GRADES After an algebra test, there are 7 students with As, 9 with Bs, 11 with Cs, 3 with Ds, and 2 with Fs.

1A. Find the probability that a randomly chosen student has a C.

1B. Find the probability that a randomly chosen student has at least a B.

2 **Probability Distributions** A **probability distribution** is the probability of every possible value of the random variable. A **probability graph** is a bar graph that displays a probability distribution.

> **KeyConcept** Properties of Probability Distributions
>
> • The probability of each value of X is greater than or equal to 0 and is less than or equal to 1.
>
> • The sum of the probabilities of all values of X is 1.

Real-World Example 2 Probability Distribution

PIZZA The table shows the probability distribution of the number of times a customer orders pizza each month.

Pizzas Ordered Per Month	
X = Number of Pizzas	Probability
0	0.10
1	0.12
2	0.44
3	0.24
4+	0.10

a. Show that the distribution is valid.

• For each value of X, the probability is greater than or equal to 0 and less than or equal to 1.

• The sum of the probabilities, $0.10 + 0.12 + 0.44 + 0.24 + 0.10$, is 1.

b. What is the probability that a customer orders pizza fewer than three times per month?

The probability of a compound event is the sum of the probabilities of each individual event. The probability of a customer ordering fewer than 3 times per month is the sum of the probability of ordering 2 times per month plus the probability of ordering one time per month.

$P(X < 3) = P(X = 2) + P(X = 1) + P(X = 0)$ Sum of individual probabilities

$\qquad\qquad = 0.44 + 0.12 + 0.10$ $P(X = 2) = 0.44$, $P(X = 1) = 0.12$, and $P(X = 0) = 0.10$

$\qquad\qquad = 0.66$ Add.

c. Make a probability graph of the data.

Use the data from the probability distribution table to draw a bar graph. Remember to label each axis and give the graph a title.

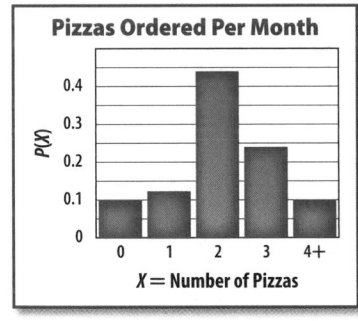

Pizzas Ordered Per Month

▶ **GuidedPractice**

The table shows the probability distribution of adults who play golf by age range.

2A. Show that the distribution is valid.

2B. What is the probability that an adult golfer is 35 years old or older?

2C. Make a probability graph of the data.

Golfers By Age	
A = Ages	Probability
18–24	0.13
25–34	0.18
35–44	0.21
45–54	0.19
55–64	0.12
65+	0.17

The **expected value** $E(X)$ of a discrete random variable is the weighted average of the variable. The "weight" applied to each value is its theoretical probability. The expected value tells you what average value to expect after many trials.

Real-World Example 3 Expected Value

GAMES A candy bar manufacturer is holding a contest. The potential prizes and the probability of winning each prize are shown in the table.

Prize	gift certificate	year supply of candy bars	trip	new car
Prize Value	$50	$250	$4000	$25,000
Probability	1 in 500	1 in 2000	1 in 2,000,000	1 in 20,000,000

a. Create a probability distribution.

Find the probability associated with each prize. Note that the probability of winning $0 is found by subtracting the probability of winning something from 1.

X	P(X)
$50	0.002
$250	0.0005
$4000	0.0000005
$25,000	0.00000005
$0	0.99749945

b. Calculate the expected value.

$$E(X) = [X_1 \cdot P(X_1)] + [X_2 \cdot P(X_2)] + \ldots + [X_n \cdot P(X_n)]$$

$$= 50(0.002) + 250(0.0005) + 4000(0.0000005) + 25,000(0.00000005)$$
$$+ 0(0.997\ldots)$$

$$= 0.1 + 0.125 + 0.002 + 0.00125 + 0 \text{ or } 0.22825$$

The expected value is 0.22825 or about $0.23.

c. Interpret your results.

The expected value of 0.22825 means that one candy bar purchase can be expected to win about $0.23.

> **Guided**Practice

3. CONTEST Wendy entered a contest with the following prize values.

Prize Value	$100	$500	$5000	$50,000
Probability	1 in 1000	1 in 5000	1 in 100,000	1 in 1,000,000

A. Create a probability distribution.

B. Calculate the expected value.

C. Interpret your results.

Real-World Link

Sweepstakes The odds of winning the grand prize in some sweepstakes can be 1 in over 500 million.

StudyTip

Probability of Winning $0 The probability of winning $0 is included in order for the sum of the probabilities in the probability distribution to equal 1.

Example 1   1. **GPS** A car dealership surveys 10,000 of its
customers with a GPS system in their vehicles to ask
how often they have used the system within the past
year. The results are shown in the table.

    a. Find the probability that a randomly chosen
customer will have used the GPS system more than
20 times.

    b. Find the probability that a randomly chosen
customer will have used the GPS system no more
than 10 times.

Customers Using the GPS System	
Uses	**Customers**
0	1382
1–5	2350
6–10	2010
11–15	1863
16–20	1925
21+	470

Example 2   2. **JEANS** A fashion boutique ordered jeans with different
numbers of stripes down the outside seams. The table
shows the probability distribution of the number of each
type of jean that was sold in a particular week.

    a. Show that the distribution is valid.

    b. What is the probability that a randomly chosen pair of
jeans has fewer than 3 stripes?

    c. Make a probability graph of the data.

Types of Jeans Sold	
Number of Stripes	**Probability**
0	0.15
1	0.19
2	0.26
3	0.22
4	0.18

Example 3   3. **CCSS ARGUMENTS** The producers of a game show
provided the probability of winning the prizes for one
of the games.

    a. Create a probability distribution.

    b. Calculate the expected value.

    c. Interpret your results.

Prize Value	Probability
$1000	1 in 80
$5000	1 in 200
$25,000	1 in 1000
$100,000	1 in 10,000

Practice and Problem Solving   Extra Practice is on page R12.

Example 1   4. **HOME THEATER** An electronics store sells
components and speakers for home theaters
individually. The store surveyed its home
theater customers to see how many of the
10 components they bought. The results
are shown in the table.

    a. Find the probability that a randomly chosen customer
bought 5 or 6 components.

    b. Find the probability that a randomly chosen customer
bought fewer than 5 components.

Home Theater Purchases	
Components	**Customers**
0–2	26
3–4	42
5–6	33
7–8	24
9–10	40

    5. **FOOD DRIVE** Ms. Valdez's biology class held a food
drive that lasted four days. The class kept track of the
number of packages donated each day.

    a. Find the probability that a randomly chosen
product was donated on the fourth day.

    b. Find the probability that a randomly chosen
product was donated on the first or second day.

Food Drive Donations Count	
Day	**Packages**
1	36
2	22
3	12
4	45

Example 2

6. **MUSIC** A Web site conducted a survey on the number of different formats on which teens have music. The table shows a probability distribution of the results.

   a. Show that the distribution is valid.

   b. What is the probability that a student randomly chosen will have music on 2 or more formats?

   c. Make a probability graph of the data.

Formats	Probability
1	0.35
2	0.31
3	0.19
4	0.11
5	0.02
6+	0.02

7. **GRADES** Mr. Rockwell's Algebra class took a chapter test last week. The table shows the probability distribution of the results.

   a. Show that the distribution is valid.

   b. What is the probability that a student chosen at random will have no higher than a 3?

   c. Make a probability graph of the data.

Score	Probability
4	0.29
3	0.43
2	0.17
1	0.11
0	0

Example 3

8. **CCSS ARGUMENTS** Kylie entered a drawing at the county fair. The table shows the value and probability of winning each prize.

   a. Create a probability distribution.

   b. Calculate the expected value.

   c. Interpret your results.

Prize Value	Probability
$20	1 in 50
$50	1 in 100
$100	1 in 250
$250	1 in 1000

9. **CONTESTS** Nikia entered a contest where each ticket cost $1. The table shows the value and probability of each prize.

   a. Create a probability distribution.

   b. Calculate the expected value.

   c. Interpret your results.

Prize Value	Probability
$200	1 in 500
$1000	1 in 5000
$5000	1 in 25,000
$25,000	1 in 100,000

10. **MARKETING** A retail marketing group conducted a survey on teen shopping habits and asked the teens for the number of stores they visited to complete their holiday shopping. The table shows the probability distribution of the results.

Number of Stores	0–2	3–5	6–8	9–11	12+
Probability	0.35	0.32	0.17	0.11	0.05

   a. Show that the distribution is valid.

   b. What is the probability that a shopper chosen at random will shop at more than 5 stores but fewer than 12?

   c. Make a probability graph of the data.

11. **DECISION MAKING** An auto insurance company uses many variables to calculate each driver's six-month payment. The table at the right shows the probability of a specific driver getting into an accident that will cost the company $10,000.

   a. Taking all of the probabilities into account, what is the probability of the driver having an accident?

   b. What is the company's expected payout for this driver?

   c. What should the company charge the driver for a six-month policy? Explain your reasoning.

Variable	Probability
driving record	1 in 50
vehicle type	1 in 500
age	1 in 250
gender	1 in 200
residence	1 in 1000

12. DECISION MAKING Amber is investing in stocks for her math class. After analyzing five stocks, she has determined the following probabilities.

Stock	Cost Per Share	Probability of $20 Gain	Probability of No Gain	Probability of $30 Loss
A	$100	47%	25%	28%
B	$200	40%	40%	20%
C	$200	50%	20%	30%
D	$200	60%	0%	40%
E	$300	30%	60%	10%

a. Calculate the expected gain or loss for one share of each stock.

b. If Amber is allowed to spend $1000, what combination of stocks should she purchase to ensure the greatest expected value? Explain your reasoning.

13. ⬧ **MULTIPLE REPRESENTATIONS** In this problem, you will investigate probability distributions and simulations.

a. **Tabular** Construct a relative-frequency table showing the theoretical probability for the sum obtained from rolling a die and spinning the spinner at the right.

b. **Graphical** Make a probability graph of the data.

c. **Analytical** Calculate the expected value of one spin.

d. **Concrete** Design a simulation for 50 trials. Explain your reasoning. Conduct the simulation and tally your results.

e. **Graphical** Make a probability graph of the data in the simulation. Compare and contrast the two graphs.

H.O.T. Problems

14. CHALLENGE What is wrong with the probability distribution shown? Explain your reasoning.

15. REASONING Suppose two dice are rolled twelve times. Which sum is most likely to occur? Make a table to show the probability distribution. Then make a probability graph to confirm your answer.

16. **CCSS STRUCTURE** Explain why the sum of the probabilities in a probability distribution should always be 1. Include an example.

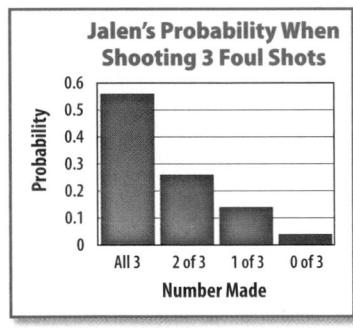

17. REASONING Determine whether the following statement is *true* or *false*. Explain your reasoning.
   Discrete random variables can take on an infinite number of values.

18. OPEN ENDED Write a real-world problem in which you could find a probability distribution. Create a probability graph for your data.

19. REASONING Determine whether the following statement is *true* or *false*. Explain your reasoning.
   The expected value of a random variable is the value for the random variable most likely to occur.

20. WRITING IN MATH Write a real-world story in which you are the owner of a business. Explain how you could use a probability distribution to help you make a business decision.

21. A coin is flipped and a die is rolled. What is the probability of the coin landing heads up and rolling a 3?

A $\frac{1}{12}$ C $\frac{1}{8}$

B $\frac{1}{9}$ D $\frac{2}{3}$

22. SHORT RESPONSE How many different ways can the letters P, Q, R, S be arranged?

23. Suppose there are 10 tickets in a box for a drawing numbered as follows: 1, 2, 2, 3, 4, 4, 6, 6, 9, and 9. A single ticket is randomly chosen from the box. What is the probability of drawing a ticket with a number less than 10?

F $\frac{1}{5}$ G $\frac{3}{10}$ H 1   J 0

24. GEOMETRY The height of a triangle is five inches less than the length of its base. If the area of the triangle is 52 square inches, find the base and the height.

A 15 in., 9 in.   C 13 in., 8 in.

B 11 in., 7 in.   D 17 in., 11 in.

25. PET TOYS Johnda is looking for 2 puppy toys at the pet store. They have a box of clearance items that contains 6 balls, 5 tug toys, 8 rawhide chews, and 4 chew toys. If Johnda reaches in the box and pulls out two items, what is the probability that she will pull out two tug toys? (Lesson 12-7)

26. GAMES For a certain game, each player rolls four dice at the same time. (Lesson 12-6)

   a. Do the outcomes of rolling the four dice represent a permutation or combination? Explain.

   b. How many outcomes are possible?

   c. What is the probability that four dice show the same number on a single roll?

Find each sum. (Lesson 11-6)

27. $\dfrac{4}{a^2} + \dfrac{6}{a}$

28. $\dfrac{3}{b^3} + \dfrac{7}{b^2}$

29. $\dfrac{4}{d+6} + \dfrac{5}{d-5}$

30. $\dfrac{f}{f+5} + \dfrac{4}{f-4}$

31. $\dfrac{8h}{h+6} + \dfrac{h}{h-3}$

32. $\dfrac{7k}{k-3} + \dfrac{k}{k+2}$

Find the values of the three trigonometric ratios for angle A. (Lesson 10-6)

33.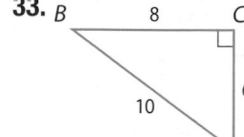

34.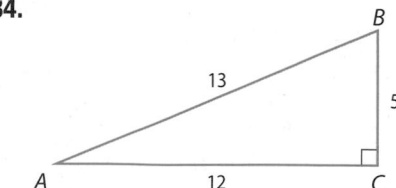

35.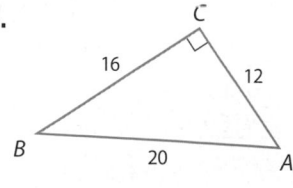

Simplify each expression. (Lesson 10-2)

36. $\sqrt{\dfrac{50}{x^4}}$

37. $\dfrac{\sqrt{t^3}}{\sqrt{18}}$

38. $\sqrt{\dfrac{15}{14}} \cdot \sqrt{\dfrac{21}{10}}$

39. $\dfrac{6}{3-\sqrt{5}}$

40. $\dfrac{3}{\sqrt{7}+\sqrt{6}}$

41. $\dfrac{\sqrt{2}}{\sqrt{8}-\sqrt{6}}$

Graphing Technology Lab
The Normal Curve

When there are a large number of values in a data set, the frequency distribution tends to cluster around the mean of the set in a distribution (or shape) called a **normal distribution**. The graph of a normal distribution is called a **normal curve**. Since the shape of the graph resembles a bell, the graph is also called a *bell curve*.

Data sets that have a normal distribution include reaction times of drivers that are the same age, achievement test scores, and the heights of people that are the same age.

CCSS Common Core State Standards
Content Standards
S.ID.2 Use statistics appropriate to the shape of the data distribution to compare center (median, mean) and spread (interquartile range, standard deviation) of two or more different data sets.

Mathematical Practices
2 Reason abstractly and quantitatively.

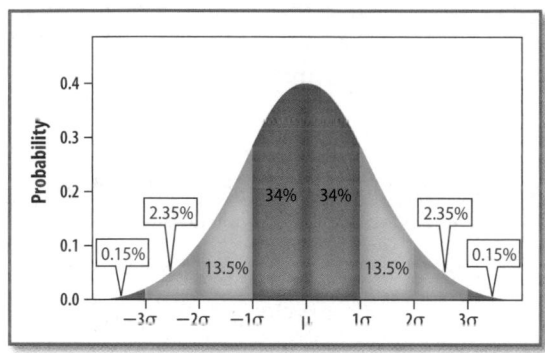

You can use a graphing calculator to graph and analyze a normal distribution if the mean and standard deviation of the data are known.

Activity 1   Graph a Normal Distribution

HEIGHT The mean height of 15-year-old boys in the city where Isaac lives is 67 inches, with a standard deviation of 2.8 inches. Use a normal distribution to represent these data.

Step 1 Set the viewing window. WINDOW

- Xmin = 67 $-$ 3 $\times$ 2.8 ENTER **58.6**
- Xmax = 67 $+$ 3 $\times$ 2.8 ENTER **75.4**
- Xscl = 2.8 ENTER
- Ymin = 0 ENTER
- Ymax = 1 $\div$ (2 $\times$ 2.8) ENTER **.17857142...**
- Yscale = 1 ENTER

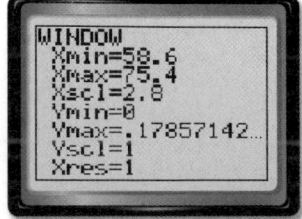

Step 2 By entering the mean and standard deviation into the calculator, we can graph the corresponding normal curve. Enter the values using the following keystrokes.

KEYSTROKES: Y= 2nd [DISTR] ENTER
X,T,θ,*n* , 67 , 2.8
) GRAPH

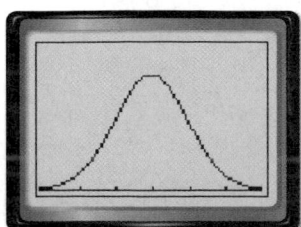

[58.6, 75.4] scl: 2.8 by [0, 0.17857142] scl: 1

The probability of a range of values is the area under the curve.

Activity 2 Analyze a Normal Distribution

Use the graph to answer questions about the data. What is the probability that Isaac will be at most 67 inches tall when he is 15?

The sum of all the y-values up to $x = 67$ would give us the probability that Isaac's height will be less than or equal to 67 inches. This is also the area under the curve. We will shade the area under the curve from negative infinity to 67 inches and find the area of the shaded portion of the graph.

Step 1 Use the **ShadeNorm** function.

KEYSTROKES: [2nd] [DISTR] [▶] [ENTER]

Step 2 Shade the graph.

Next enter the lowest value, highest value, mean, and standard deviation.

On the TI-84 Plus, -1×10^{99} represents negative infinity.

KEYSTROKES: [(–)] 1 [2nd] [EE] 99 [,] 67 [,] 67 [,] 2.8 [)] [ENTER]

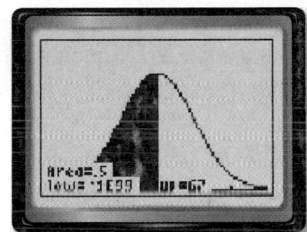

[58.6, 75.4] scl: 2.8 by [0, 0.17857142] scl: 1

The area is given as 0.5. The probability that Isaac will be 67 inches tall is 0.5 or 50%. Since the mean value is 67, we expect the probability to be 50%.

Exercises

1. What is the probability that Isaac will be at least 6 feet tall when he is 15?

2. What is the probability that Isaac will be between 65 and 68 inches?

3. The **z-score** represents the number of standard deviations that a given data value is from the mean. The z-score for a data value X is given by $z = \dfrac{X - \mu}{\sigma}$, where μ is the mean and σ is the standard deviation. Find and interpret the z-score of a height of 73 inches.

4. Find and interpret the z-score of a height of 61 inches.

Extension

Refer to the curve at the right.

5. Compare this curve to the normal curve in Activity 1.

6. Describe where an outlier of the data set would be graphed on this curve.

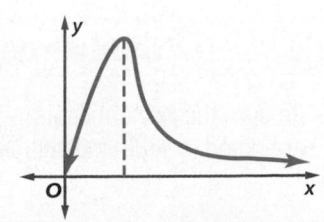

Study Guide

KeyConcepts

Samples and Studies (Lesson 12-1)

- Some types of samples are simple random, systematic, convenience, self-selected, and stratified.
- Three study types are the survey, observational study, and experiment.

Statistics and Parameters (Lesson 12-2)

- The mean absolute deviation is used to predict errors and judge how well the mean represents the data.
- A low standard deviation indicates that the data tend to be very close to the mean, while a high standard deviation indicates that the data are spread out over a larger range.

Distributions of Data and Comparing Sets of Data
(Lessons 12-3 and 12-4)

- In a negatively skewed distribution, the majority of the data are on the right. In a positively skewed distribution, the majority of the data are on the left. In a symmetric distribution, the data are evenly distributed.

Simulations (Lesson 12-5)

- Simulations are used to perform experiments that would be difficult or impossible to perform in real life.

Permutations and Combinations and Probability of Compound Events (Lessons 12-6 and 12-7)

- A selection of objects in which order is important is called a permutation. A selection of objects in which order is not important is called a combination.
- For independent events, the outcome of one does not affect the outcome of the other.

Probability Distributions (Lesson 12-8)

- For each value of X, $0 \leq P(X) \leq 1$. The sum of the probabilities of each value of X is 1.
- The expected value $E(X)$ of a discrete random variable of a probability distribution is its weighted average.

FOLDABLES StudyOrganizer

Be sure the Key Concepts are noted in your Foldable.

Statistics and Probability

KeyVocabulary

bias (p. 748)

combination (p. 787)

compound event (p. 793)

convenience sample (p. 747)

dependent events (p. 794)

discrete random variable (p. 803)

distribution (p. 764)

expected value $E(X)$ (p. 805)

experiment (p. 749)

experimental probability (p. 780)

independent events (p. 793)

linear transformation (p. 771)

mean absolute deviation (MAD) (p. 758)

mutually exclusive events (p. 795)

observational study (p. 749)

parameter (p. 757)

permutation (p. 786)

population (p. 747)

probability distribution (p. 804)

probability graph (p. 804)

random variable (p. 803)

sample (p. 747)

self-selected sample (p. 747)

simple random sample (p. 747)

simulation (p. 781)

standard deviation (p. 759)

statistic (p. 757)

statistical inference (p. 757)

stratified sample (p. 747)

survey (p. 749)

systematic sample (p. 747)

theoretical probability (p. 780)

variance (p. 759)

VocabularyCheck

Choose the term that best completes each sentence.

1. An arrangement in which order is important is called a (combination, permutation).

2. A (parameter, statistic) is a measure that describes the characteristic of a sample.

3. A (sample, population) consists of all of the members of a group.

4. (Experimental probability, Theoretical probability) is the ratio of the number of favorable outcomes to the total number of outcomes.

5. A variable with a value that is the numerical outcome of a random event is called a (discrete random variable, random variable).

Lesson-by-Lesson Review

12-1 Samples and Studies

Identify the sample as *biased* or *unbiased*. Explain your reasoning.

6. GOVERNMENT To determine whether voters support a new trade agreement, 5 people from the list of registered voters in each state are selected at random.

Determine whether each situation describes a *survey*, an *observational study*, or an *experiment*. Explain your reasoning.

7. SCHOOL DANCE The homecoming dance committee sends out a questionnaire to all the girls in the school to decide on a theme for the homecoming dance.

8. MILKSHAKE Mary wants to test her milkshake recipe using honey instead of sugar. She randomly gives half of her 6 friends milkshakes sweetened with honey and the other half the same milkshakes sweetened with sugar. Then she asks them how they like the milkshakes

Example 1

MUSIC A randomly selected group of people who listen to a country music radio station is asked to name their favorite type of music. Identify the sample as *biased* or *unbiased*. Explain your reasoning.

The sample is biased. People listening to a country music radio station are likely to vote for country music.

Example 2

BOOK COVER An artist is trying to choose a cover for a children's book. She sends out a flyer with the two covers to all of the students at one school. She asks them to check the favorite cover. Determine whether this situation describes a *survey*, an *observational study*, or an *experiment*. Explain your reasoning.

This is a survey. The data are gathered from the responses given by members in the sample.

12-2 Statistics and Parameters

9. SHOVELING Ben shovels driveways to raise money. The number of driveways he shovels each day is {2, 4, 3, 5, 3}. Find and interpret the mean absolute deviation.

10. CANDY BARS Luci is keeping track of the number of candy bars each member of the drill team sold. The results are {20, 25, 30, 50, 40, 60, 20, 10, 42}. Find and interpret the mean absolute deviation.

11. FOOD A fast food company polls a random sample of its day and night customers to find how many times a month they eat out. Compare the means and standard deviations of each data set.

Day Customers	Night Customers
10, 3, 12, 15, 7, 8, 4, 12, 9, 14, 12, 9	15, 12, 13, 9, 11, 12, 14, 12, 8, 16, 9, 9

Example 3

GIFTS Joshua is collecting money from his family for a Mother's Day gift. He keeps track of how much each person has donated: {10, 5, 20, 15, 10}. Find and interpret the mean absolute deviation.

Step 1 Find the mean: $\bar{x} = \dfrac{10 + 5 + 20 + 15 + 10}{5}$ or 12.

Step 2 Find the absolute values of the differences.

$x_1 = 10: |12 - 10|$ or 2 $\qquad x_2 = 5: |12 - 5|$ or 7

$x_3 = 20: |12 - 20|$ or 8 $\qquad x_4 = 15: |12 - 15|$ or 3

$x_5 = 10: |12 - 10|$ or 2

Step 3 Find the sum: $2 + 7 + 8 + 3 + 2 = 22$.

Step 4 Find the mean absolute deviation.

$$\text{MAD} = \frac{|\bar{x} - x_1| + |\bar{x} - x_2| + \dots + |\bar{x} - x_n|}{n}$$
$$= \frac{22}{5} \text{ or } 4.4$$

A mean absolute deviation of 4.4 indicates that the data, on average, are 4.4 units away from the mean.

12-3 Distributions of Data

Use a graphing calculator to construct a histogram for the data. Then describe the shape of the distribution.

12. 55, 62, 32, 56, 31, 59, 19, 61, 8, 48, 41, 69, 32, 63, 48, 60, 43, 66, 71, 70, 49, 56, 21, 67

13. 4, 19, 62, 28, 26, 59, 33, 39, 36, 72, 46, 48, 49, 44, 72, 76, 55, 53, 55, 62, 66, 69, 71, 74

14. **MILK** A grocery store manager tracked the amount of milk in gallons sold each day. Describe the center and spread of the data using either the mean and standard deviation or the five-number summary. Justify your choice by constructing a box-and-whisker plot for the data.

Gallons of Milk Sold Per Day					
383	296	354	288	195	372
421	367	411	355	296	321
403	357	432	229	180	266

Example 4

DRIVING TESTS Several driving test results are shown. Describe the center and spread of the data using either the mean and standard deviation or the five-number summary. Justify your choice by constructing a box-and-whisker plot for the data.

Driving Test Scores					
80	95	100	95	95	100
100	90	75	60	90	80

Use a graphing calculator to create a box-and-whisker plot.

The left whisker is longer than the right and the median is closer to the right whisker. Therefore, the distribution is negatively skewed.

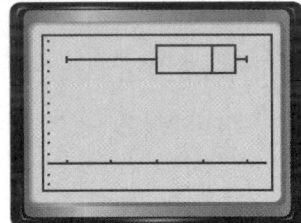

[56, 104] scl: 10 by [−2, 12] scl: 1

Use the five-number summary. The range is 40. The median score is 92.5, and half of the drivers scored between 80 and 97.5.

12-4 Comparing Sets of Data

Find the mean, median, mode, range, and standard deviation of each data set that is obtained after adding the given constant to each value.

15. 27, 21, 34, 42, 20, 19, 18, 26, 25, 33; +(−4)

16. 72, 56, 71, 63, 68, 59, 77, 74, 76, 66; +16

17. **SCHOOL** Principal Andrews tracked the number of disciplinary actions given by Ms. Miller and Ms. Anderson to their students each week.

Ms. Miller	Ms. Anderson
9, 16, 12, 11, 12, 9, 10, 14, 13, 10, 9, 10, 11, 9, 12, 10, 11, 12	7, 1, 0, 4, 2, 1, 6, 2, 2, 1, 4, 3, 0, 7, 0, 2, 5, 0

a. Use a graphing calculator to construct a histogram for each set of data. Then describe the shape of each distribution.

b. Compare the data sets using either the means and standard deviations or the five-number summaries. Justify your choice.

Example 5

Find the mean, median, mode, range, and standard deviation of the data set obtained after adding 6 to each value.

12, 15, 11, 12, 14, 16, 15, 12, 10, 13

Find the mean, median, mode, range, and standard deviation of the original data set.

Mean 13   Mode 12   Standard Deviation 1.8

Median 12.5   Range 6

Add 6 to the mean, median, and mode. The range and standard deviation are unchanged.

Mean 19   Mode 18   Standard Deviation 1.8

Median 18.5   Range 6

12-5 Simulations

18. GAMES While watching a game at a carnival where participants guess which of three shells is covering a ball, Jeremy tallies the following results.

Ball Location	left	middle	right
Frequency	16	18	33

    a. Find the experimental probability of the ball being under the right shell.

    b. Find the experimental probability of the ball not being under the middle shell.

19. SCHOOL BUS Christy has determined that the school bus is late 60% of the time.

    a. Design a simulation that can be used to estimate the probability that the school bus is late today.

    b. Conduct the simulation, and report the results.

Example 6

GROUPS Before the random drawing of groups, Dawn has determined that she is 20% likely to get placed in the same group as Sherry. Design a simulation that can be used to estimate the probability of Dawn and Sherry being in the same group.

Step 1 There are two possible outcomes.

Possible Outcomes	Theoretical Probability
Grouped together	20%
Not grouped together	80%

Step 2 We can use the random number generator on a graphing calculator. Assign the integers 1–5 to accurately represent the probability data.

Outcome	Represented by
Grouped together	1
Not grouped together	2–5

Step 3 A trial will represent one drawing of groups. The simulation can consist of any number of trials. We will use 20.

12-6 Permutations and Combinations

Identify each situation as a *permutation* or a *combination*.

20. selecting 3 different toppings for pizza from a list of 15

21. an arrangement of textbooks on a bookshelf

22. a list of teams participating in a tournament

23. a ranking of students by scores

24. CLASS PHOTO The Spanish teacher at South High School wants to arrange 7 students who traveled to Mexico for a yearbook photo. In how many ways can the students be arranged?

25. GOLF BALLS A golf bag contains 5 white golf balls, 6 yellow golf balls, and 4 orange golf balls. Two balls are pulled from the bag at random. What is the probability that both balls are orange golf balls?

Example 7

ALBUM COVERS An artist is trying to choose five covers for children's books. There are 10 different covers to choose from. How many ways can the artist choose the covers?

$$C(n, r) = \frac{n!}{(n - r)! r!} \qquad \text{Combination Formula}$$

$$C(10, 5) = \frac{10!}{(10 - 5)! 5!} \qquad n = 10 \text{ and } r = 5$$

$$= \frac{10!}{5! 5!} \qquad \text{Simplify.}$$

$$= \frac{10 \cdot 9 \cdot 8 \cdot 7 \cdot 6 \cdot \cancel{5} \cdot \cancel{4} \cdot \cancel{3} \cdot \cancel{2} \cdot \cancel{1}}{5 \cdot 4 \cdot 3 \cdot 2 \cdot 1 \cdot \cancel{5} \cdot \cancel{4} \cdot \cancel{3} \cdot \cancel{2} \cdot \cancel{1}}$$

$$= \frac{30,240}{120} \text{ or } 252 \qquad \text{Simplify.}$$

There are 252 ways for the aritist to choose 5 covers.

12-7 Probability of Compound Events

26. MUSIC Tracie is playing a mix playlist with 6 classic rock, 8 pop, and 4 dance songs.

a. If she selects random play and the songs can repeat, what is the probability that she hears 2 classic rock songs and then a dance song?

b. If the songs cannot repeat, what is the probability that she hears 3 dance songs in a row?

A box contains 8 red chips, 6 blue chips, and 12 white chips. Chips are randomly drawn from the box and are not replaced. Find each probability.

27. P(red, white, blue)   **28.** P(red, red, red)

29. P(red, white, white)   **30.** P(blue, blue)

One card is randomly drawn from a standard deck of 52 cards. Find each probability.

31. P(heart or red)

32. P(10 or spade)

Example 8

Determine if rolling a die three times is an *independent* or *dependent* event. Then find the probability of rolling a 5 all three times.

These events are independent because the outcome of one does not affect the outcome of the others.

The probability of rolling a 5 is $\frac{1}{6}$.

$P(5, 5, 5) = P(5) \cdot P(5) \cdot P(5)$

$= \frac{1}{6} \cdot \frac{1}{6} \cdot \frac{1}{6}$ or $\frac{1}{216}$

Example 9

A bag of colored paper clips contains 30 red clips, 22 blue clips, and 22 green clips. Three clips are drawn randomly from the bag and are not replaced. Find P(blue, red, green).

First clip: P(blue) $= \frac{22}{74}$ Second clip: P(red) $= \frac{30}{73}$

Third clip: P(green) $= \frac{22}{72}$

P(blue, red, green) $= \frac{22}{74} \cdot \frac{30}{73} \cdot \frac{22}{72}$ or $\frac{605}{16,206}$

12-8 Probability Distributions

A local cable provider asked its subscribers how many television sets they had in their homes. The results of their survey are shown in the probability distribution.

X = Number of Televisions	Probability
1	0.18
2	0.36
3	0.34
4	0.08
5+	0.04

33. Show that the probability distribution is valid.

34. If a household is selected at random, what is the probability that it has fewer than 4 televisions?

Example 10

The table shows the probability distribution for the number of extracurricular activities in which students at Midpark High School participate.

X = Number of Activities	Probability
0	0.04
1	0.12
2	0.37
3	0.30
4+	0.17

a. Show that the distribution is valid.

For each X, the probability is greater than or equal to 0 and less than or equal to 1.

$0.04 + 0.12 + 0.37 + 0.30 + 0.17 = 1$

b. What is the probability that a randomly chosen student participates in 1 to 3 activities?

$P(1 \leq X \leq 3) = P(X = 1) + P(X = 2) + P(X = 3)$

$= 0.12 + 0.37 + 0.30 = 0.79$ or 79%

1. **CHOCOLATE** Rico is selling candy. If Marisa randomly selects two candy bars to purchase, what is the probability that she buys a milk chocolate bar followed by a caramel bar?

RICO'S CHOCOLATES

8 Milk Chocolate Bars
6 Caramel Bars
5 Dark Chocolate Bars
6 Peanut Butter Bars

2. A die is rolled 200 times. What is the experimental probability of rolling less than 3?

Outcome	Frequency
1	30
2	26
3	44
4	30
5	22
6	40

3. **MULTIPLE CHOICE** Use a graphing calculator to construct a histogram for the data, and use it to describe the shape of the distribution.

16, 18, 14, 31, 19, 18, 10, 29,
12, 12, 28, 19, 17, 26, 15, 20

A positively skewed   C symmetric

B negatively skewed   D none of the above

Find the mean, median, mode, range, and standard deviation of each data set that is obtained after multiplying each value by the given constant.

4. 9, 17, 31, 21, 17, 25, 13, 9, 12, 9; × 3

5. 16, 14, 23, 41, 38, 29, 18, 13, 16; × 0.25

Identify each sample as *biased* or *unbiased*. Explain your reasoning.

6. **NEWSPAPERS** A survey is sent to all people who subscribe to the *Dispatch* to determine what newspaper people prefer to read.

7. **SHOPPING** Each person leaving the Maxtowne Mall is asked to name their favorite clothing store in the mall.

8. **SALES** Nate is keeping track of how much people spent at the school bookstore in one day. Find and interpret the mean absolute deviation for the data: 1, 1, 2, 3, 4, 5, 12.

9. **PIZZA** How many ways can 3 different toppings be chosen from a list of 10 toppings?

10. **EDUCATION** Kristin surveys 200 people in her school to determine how many nights per week students do homework. The results are shown in the table.

Number of Nights	Number of Students
0	10
1	30
2	50
3	90
4	10
5 or more	10

a. Find the probability that a randomly chosen student will have studied more than 4 nights.

b. Find the probability that a randomly chosen student will have studied no more than 3 nights.

11. **MULTIPLE CHOICE** The second graders are divided into boys and girls. Then 2 girls and 2 boys are chosen at random to represent the class at the Pride Assembly. Which of the following best describes the sample?

F simple   H systematic

G stratified   J self-selected

Identify each situation as a *permutation* or a *combination*.

12. a student's daily class schedule

13. a list of teachers' names at school

14. gold, silver, and bronze medalists

15. **RAFFLE** Carmen is considering paying $1 for a raffle ticket. What is the expected value of this ticket?

Prize Value	$10	$50	$500
Probability	1 in 100	1 in 500	1 in 5000

Preparing for Standardized Tests

Organize Data

Sometimes you may be given a set of data that you need to analyze in order to solve problems on a standardized test. Use this lesson to practice organizing data to help you solve problems.

Strategies for Organizing Data

Step 1

When you are given a problem statement containing data, consider:

- **making a list** of the data.
- **using a table** to organize the data.
- **using a data display** (such as a bar graph, Venn diagram, circle graph, line graph, or box-and-whisker plot) to organize the data.

Step 2

Organize the data.

- Create your table, list, or data display.
- If possible, fill in any missing values that can be found by intermediate computations.

Step 3

Analyze the data to solve the problem.

- Reread the problem statement to determine what you are being asked to solve.
- Use the properties of algebra to work with the organized data and solve the problem.
- If time permits, go back and check your answer.

Standardized Test Example

Read the problem. Identify what you need to know. Then use the information in the problem to solve. Show your work.

Of the 24 students in a music class, 10 play the flute, 14 play the piano, and 13 play the guitar. Two students play the flute only, 5 the piano only, and 7 the guitar only. One student plays the flute and the guitar but not the piano. Two students play the piano and guitar but not the flute. Three students play all the instruments. If a student is selected at random, what is the probability that he or she plays the piano and flute, but not the guitar?

Scoring Rubric	
Criteria	Score
Full Credit: The answer is correct and a full explanation is provided that shows each step.	2
Partial Credit: • The answer is correct, but the explanation is incomplete. • The answer is incorrect, but the explanation is correct.	1
No Credit: Either an answer is not provided or the answer does not make sense.	0

Read the problem carefully. The data is difficult to analyze as it is presented. Use a Venn diagram to organize the data and solve the problem.

Example of a 2-point response:

Use a Venn diagram to organize the data. Fill in all of the information given in the problem statement. There are 14 students who play the piano, so $14 - 5 - 2 - 3$ or 4 students play the piano and the flute, but not the guitar. Find the probability.

$P(\text{piano and flute}) = \dfrac{4}{24}$ or $\dfrac{1}{6}$

So, the probability that a randomly selected student plays the piano and flute but not the guitar is $\dfrac{1}{6}$.

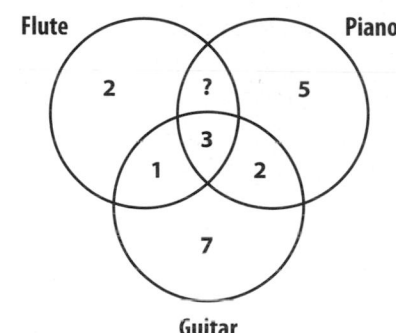

Exercises

Read the problem. Identify what you need to know. Then use the information in the problem to solve. Show your work.

1. There are 40 students, 9 camp counselors, and 5 teachers at Camp Kern. Each person is assigned to one activity this afternoon. There are 9 students going hiking and 17 students going horseback riding. Of the camp counselors, 2 will supervise the hike and 3 will help with the canoe trip. There are 2 teachers helping with the canoe trip and 2 going horseback riding. Suppose a person is selected at random during the afternoon activities. What is the probability that the one selected is a student on the canoe trip or a camp counselor on a horse? Express your answer as a fraction.

2. The table shows the number of coins in a piggy bank.

Coin	Number
Penny	16
Nickel	18
Dime	20
Quarter	10

   a. Find the probability that a randomly selected coin will be a dime.

   b. Find the probability that a randomly selected coin will be either a nickel or a quarter.

3. It takes Craig 40 minutes to mow his family's lawn. His brother Jacob can do the same job in 50 minutes. How long would it take them to mow the lawn together? Round your answer to the nearest tenth of a minute.

Multiple Choice

Read each question. Then fill in the correct answer on the answer document provided by your teacher or on a sheet of paper.

1. What are the excluded values of the variable in the expression $\dfrac{x^2 - x - 12}{x^2 - x - 2}$?

   A $-1, 2$

   B $-2, 2$

   C $-2, 1$

   D $-3, 4$

2. The table shows the number of Calories in twelve different snacks. Find the mean absolute deviation.

Number of Calories in Snacks			
122	91	149	121
64	138	342	72
179	105	99	114

   F 46

   G 43

   H 1.5

   J 0.8

3. Which of the following is *not* a factor of $x^4 - 6x^2 - 27$?

   A $x^2 + 3$

   B $x - 3$

   C $x + 3$

   D $x^2 - 3$

4. Eduardo has 20 CDs. He wants to choose 3 of them at random to take on a road trip. How many different ways can he do this if the order is *not* important?

   F 60

   G 84

   H 1140

   J 6840

5. Which of the following does *not* accurately describe the graph $y = -2x^2 + 4$?

   A The parabola is symmetric about the y-axis.

   B The parabola opens downward.

   C The parabola has the origin as its vertex.

   D The parabola crosses the x-axis in two different places.

6. The highest point in North Carolina is Mt. Mitchell at an elevation of 2,037 meters above sea level. Suppose the position of a hiker is given by the function $p(t) = -2.5t + 2,037$, where t is the number of minutes. Which of the following is the best interpretation of the slope of the function?

   F The hiker's initial position was 2,037 feet below sea level.

   G The hiker's initial position was 2,037 feet above sea level.

   H The hiker is descending at a rate of 2.5 meters per minute.

   J The hiker is ascending at a rate of 2.5 meters per minute

7. Jorge has made 39 out of 52 free throw attempts this season. What is the experimental probability that he makes a free throw?

   A 54%

   B 68%

   C 75%

   D 79%

8. The graph of which equation passes through the points $(-1, -3)$ and $(-2, 3)$?

   F $y = -6x - 9$

   G $y = -\dfrac{1}{4}x + 3$

   H $y = 4x - 5$

   J $y = \dfrac{2}{3}x + 1$

9. At a museum, each child admission costs $5.75 and each adult costs $8.25. How much does it cost a family that consists of 2 adults and 4 children?

   A $34.50

   B $39.50

   C $44.50

   D $49.50

> **Test-TakingTip**
>
> Question 4 Since order is not important, you are looking for the number of combinations of CDs that can be chosen.

Short Response/Gridded Response

Record your answers on the answer sheet provided by your teacher or on a sheet of paper.

10. GRIDDED RESPONSE Suppose Colleen spins the spinner below 80 times and records the results in a frequency table. How many times should she expect to spin a vowel?

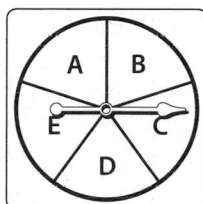

11. What is the value of $\sin B$? Express your answer as a fraction.

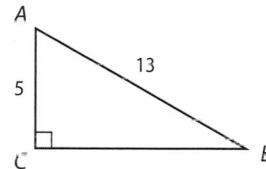

12. Graph $f(x) \geq |x - 2|$ on a coordinate grid.

13. GRIDDED RESPONSE Find the standard deviation of the set of data below to the nearest tenth.

14	11	9	6
10	16	15	13
9	12	19	10

14. Larissa has 5 peanut butter cookies, 7 chocolate chip cookies, 4 sugar cookies, and 9 oatmeal raisin cookies in a jar. If she picks two cookies at random without replacing them, what is the probability that she will choose a peanut butter cookie then a sugar cookie? Express your answer as a fraction.

15. Write an expression that describes the area in square units of a triangle with a height of $4c^3d^2$ and a base of $3cd^4$.

16. Casey made 84 field goals during the basketball season for a total of 183 points. Each field goal was worth either 2 or 3 points. How many 2-point and 3-point field goals did Casey make during the season?

17. GRIDDED RESPONSE The booster club pays $180 to rent a concession stand at a football game. They purchase cans of soda for $0.25 and sell them at the game for $1.15. How many cans of soda must they sell to break even?

Extended Response

Record your answers on a sheet of paper. Show your work.

18. To predict whether or not an issue on a ballot will pass or fail, a committee randomly calls 250 houses with area codes that are inside the voting district and asks the opinions of registered voters. Based on these efforts, the committee determines that 71% (±2.5%) of the voting population supports the issue. The committee concludes that the issue will pass.

a. Identify the sample.

b. Describe the population.

c. What method of data collection did the committee use: survey, experiment, or observational survey? Explain.

d. Is the sample *biased* or *unbiased*. Explain.

e. If unbiased, classify the sample as *simple, systematic, self-selected, convenience,* or *stratified.* Explain.

Need ExtraHelp?																		
If you missed Question...	1	2	3	4	5	6	7	8	9	10	11	12	13	14	15	16	17	18
Go to Lesson...	11-3	12-2	8-9	12-6	9-3	3-3	12-5	4-2	1-3	12-5	10-6	9-7	12-2	12-7	7-1	6-5	2-4	12-1

Student Handbook

This **Student Handbook** can
help you answer these questions.

What if I Need More Practice?

Extra Practice   R1

The **Extra Practice** section provides additional
problems for each lesson so you have ample
opportunity to practice new skills.

What if I Need to Check a Homework Answer?

Selected Answers and Solutions   R13

The answers to odd-numbered problems are included
in **Selected Answers and Solutions**.

What if I Forget a Vocabulary Word?

Glossary/Glosario   R97

The **English-Spanish Glossary** provides definitions
and page numbers of important or difficult words
used throughout the textbook.

What if I Need to Find Something Quickly?

Index   R118

The **Index** alphabetically lists the subjects covered
throughout the entire textbook and the pages on
which each subject can be found.

What if I Forget a Formula?

Formulas and Measures,   **Inside Back Cover**
Symbols and Properties

Inside the back cover of your math book is a list of
Formulas and Symbols that are used in the book.

Extra Practice

Write an algebraic expression for each verbal expression. (Lesson 1-1)

1. 6 times a number m

2. a number t less twelve

Evaluate each expression if $m = 2$, $t = 6$, and $z = 5$. (Lesson 1-2)

3. $2(t - z) + \frac{14}{m}$

4. $(m + 2z)^2 + 12tz$

5. SPORTS Adam mows lawns at an average rate of 40 minutes per lawn. Write and evaluate an expression to find the number of hours Adam spent mowing last weekend. (Lesson 1-2)

Lawns Per Day	
Friday	3
Saturday	11
Sunday	4

Evaluate each expression. Name the property used in each step. (Lesson 1-3)

6. $14\left(5 - \frac{1}{5} \cdot 25\right) + 2 \div (4 \cdot 1)$

7. $3(14 + 8 + 6) - 1 \cdot 18$

8. PETS Rosa takes two dogs and one cat to the veterinarian during her vacation. (Lesson 1-3)

Pet	Board	Bath
dog	$25/day	$12
cat	$15/day	$0

a. Use the table to find the total cost of boarding both dogs and the cat for 5 days.

b. If Rosa has the veterinarian give all of the pets a bath while she is on vacation, what is the new total cost?

Use the Distributive Property to rewrite each expression. Then evaluate. (Lesson 1-4)

9. $14(102)$

10. $5\frac{1}{6}(30)$

11. ARTS Logan sells handmade wooden products. He charges $25 per bowl, $14 per picture frame, and $30 per jewelry box. On Friday, he sells 3 bowls, 6 picture frames, and 2 jewelry boxes. On Saturday, he sells 6 bowls, 14 picture frames, and 3 jewelry boxes. Write and evaluate an expression for his total sales. (Lesson 1-4)

Find the solution set for each equation if the replacement sets are $m = \{0, 2, 4, 5, 8\}$ and $n = \{-2, 0, 2, 7, 9\}$. (Lesson 1-5)

12. $4(m - 2) = 8$

13. $6(6 - n) = 24$

Identify the independent and dependent variables for each relation. (Lesson 1-6)

14. Increasing the amount of fertilizer put on a plant increases the rate at which it grows.

15. Pam babysits to save money. The more kids she babysits, the more money she makes.

16. INCOME Jonathan draws the graph at the right to describe his income throughout his career. Describe what is happening in the graph. (Lesson 1-6)

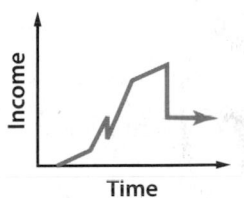

Determine whether the relation is a function. Explain. (Lesson 1-7)

17. $\{(2, 2), (-5, 2), (6, 6), (9, 4), (4, 9)\}$

18. $\{(0, 2), (1, 7), (0, -6), (4, 8), (-3, -1)\}$

19. Identify the function graphed as *linear* or *nonlinear*. Then estimate and interpret the intercepts of the graph, any symmetry, where the function is positive, negative, increasing, and decreasing, the x-coordinate of any relative extrema, and the end behavior of the graph. (Lesson 1-8)

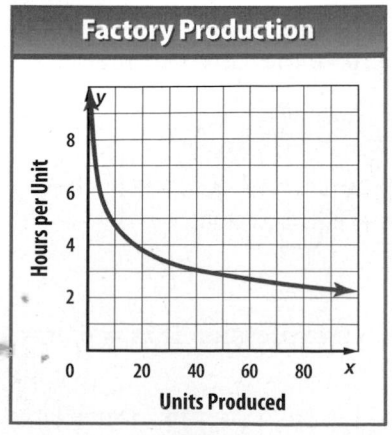

Translate each sentence into a formula. (Lesson 2-1)

1. The number of seconds is 60 times the number of minutes.

2. A gallon contains 8 pints.

3. **THEATER** A theater has 1400 seats. Write and use an equation to find the number of rows if each row has 35 seats. (Lesson 2-1)

Solve each equation. Check your solution.
(Lesson 2-2)

4. $-18t + 26 = -19$

5. $\frac{1}{2}b + \frac{3}{4} = \frac{1}{6}$

Write an equation and solve each problem.
(Lesson 2-3)

6. Find three consecutive even integers with a sum of 78.

7. Eight more than half a number is negative two.

Solve each equation. Check your solution.
(Lesson 2-3)

8. $-3w + 16 = 14.5$

9. $\frac{4}{7} - \frac{k}{3} = \frac{1}{2}$

10. **ZOO** The Martin and Smith families went to the zoo. What is the cost of an adult ticket if they spent $75 total? (Lesson 2-3)

Family	Adults
Martin	2
Smith	3

Solve each equation. Check your solution.
(Lesson 2-4)

11. $16x - 8 = 21 - x$

12. $\frac{x}{2} + \frac{1}{5} = 3x$

13. **GEOMETRY** Find the value of x so that the figures have the same perimeter. (Lesson 2-4)

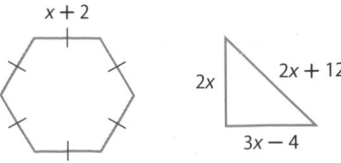

Evaluate each expression if $m = 6$, $n = 15$, and $p = \frac{1}{2}$. (Lesson 2-5)

14. $|m - 12| - 3p$

15. $18p + |n - m|$

Write an equation involving absolute value for each graph. (Lesson 2-5)

16.

17.
$$\begin{array}{c}\end{array}$$

Solve each proportion. If necessary, round to the nearest hundredth. (Lesson 2-6)

18. $\frac{1.7}{n} = \frac{16}{30}$

19. $\frac{418}{83} = \frac{b}{7}$

20. $\frac{30}{y} = \frac{75}{135}$

21. **ART** Neil is enlarging a photo to hang on the wall. To keep the pictures proportional, what length is the unknown side? (Lesson 2-6)

 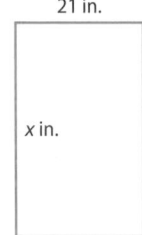

Find the discounted price of each item. (Lesson 2-7)

22. sofa: $575
    discount: 20%

23. cell phone: $80
    discount: 12%

24. **CAMPING** Ty's backpack weighs 38.6 pounds. Roger's backpack weighs 15% more. How much does Roger's backpack weigh? (Lesson 2-7)

Solve each equation or formula for the variable indicated. (Lesson 2-8)

25. $\frac{2b - a}{c} = -\frac{1}{2}d$, for a

26. $y = w(h - 5)$, for h

27. $3x - 4z = 7 - xy$, for x

28. **WEIGHT** A watermelon weighs 6.3 pounds. One pound is approximately 0.454 kilogram. How many kilograms does the watermelon weigh? (Lesson 2-8)

29. **PAINT** Jeff painted 400 square feet in 50 minutes. He then painted 700 square feet in 75 minutes. What is Jeff's average painting speed? (Lesson 2-9)

30. **CHEMISTRY** Sarah has 65 milliliters of a 30% solution. How many millimeters of 75% solution should she add to obtain the required 35% solution? (Lesson 2-9)

Determine whether each equation is a linear equation. Write *yes* or *no*. If yes, write the equation in standard form. (Lesson 3-1)

1. $-6xy + 2y = 3x$

2. $\frac{1}{2}y - 7 = 3x$

Graph each equation by making a table. (Lesson 3-1)

3. $2y - x = 5$

4. $x + y = 6$

Solve each equation. (Lesson 3-2)

5. $-5x + 14 = 7x - 28$

6. $-\frac{1}{2}x + 8 = 6x - 12 - \frac{13}{2}x$

7. NURSERY The function $b = 100 - 2.5f$ represents the remaining balance of store credit Louie has at Blooms Nursery. Find the zero and explain what it means in this situation. (Lesson 3-2)

Determine whether each function is linear. Write *yes* or *no*. Explain. (Lesson 3-3)

8.

x	−2	0	2	4	6
y	0	6	12	18	24

9.

x	7	4	1	−2	−5
y	14	2	12	4	10

10. WEATHER Refer to the graph. (Lesson 3-3)

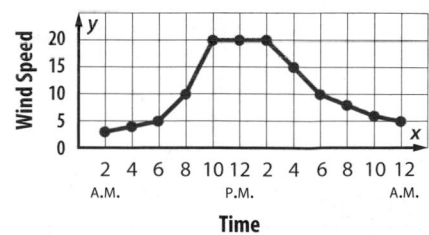

a. Find the rate of change in wind speed between 6 A.M. and 8 A.M.

b. Is there a greater change in wind speed during the day? If so, when does it occur?

c. The meteorologist says a storm came through at some point during the day. When do you think this may have happened? Explain your reasoning.

Name the constant of variation for each equation. Then find the slope of the line that passes through each pair of points. (Lesson 3-4)

11.

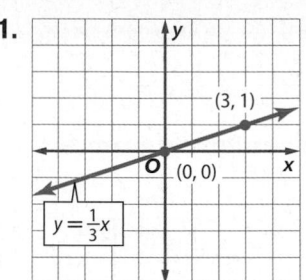

12.

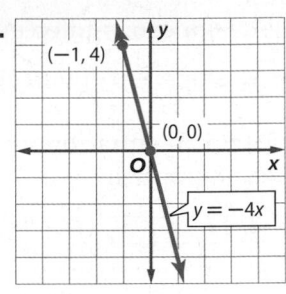

Suppose y varies directly as x. Write a direct variation equation that relates x and y. Then solve. (Lesson 3-4)

13. If $y = -6$ when $x = 9$, find y when $x = -3$.

14. If $y = -7$ when $x = -1$, find x when $y = 0$.

Determine whether each sequence is an arithmetic sequence. Write *yes* or *no*. Explain. (Lesson 3-5)

15. $-2, 2, -4, 4, -6, 6 \ldots$

16. $-6, -3, 0, 3, 6 \ldots$

Write an equation for the nth term of each arithmetic sequence. Then graph the first five terms of the sequence. (Lesson 3-5)

17. $3, 3.5, 4, 4.5\ldots$

18. $1, -1.5, -4, -6.5\ldots$

19. NEWSPAPER The table shows the number of newspapers Daniel delivers. (Lesson 3-6)

Blocks	Papers
5	40
6	48
7	56
8	64

a. Graph the data.

b. Write an equation to describe the relationship.

c. Find the number of papers delivered if he has 11 blocks.

Write an equation in function notation for each relation. (Lesson 3-6)

20.

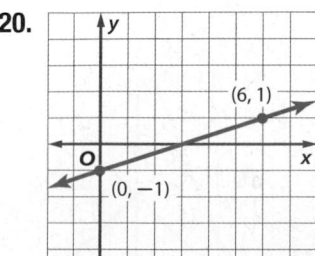

21.

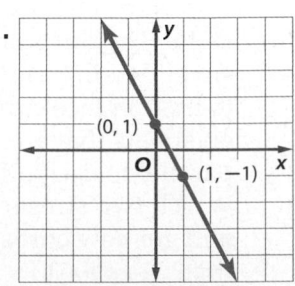

Extra Practice

Write an equation of a line in slope-intercept form with the given slope and y-intercept. Then graph the equation. (Lesson 4-1)

1. slope: 6, y-intercept: -4

2. slope: $\frac{1}{3}$, y-intercept: 3

Write an equation in slope-intercept form for each graph shown. (Lesson 4-1)

3.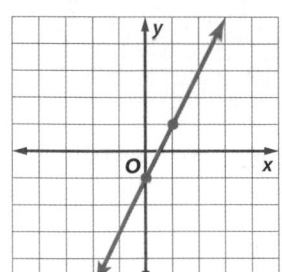

4.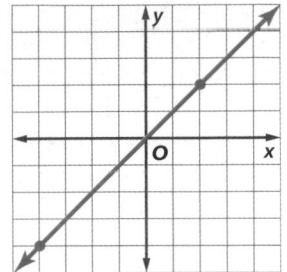

5. **PARTY** Mr. Ramirez paid $60 for 15 specialty balloons for his son's birthday. (Lesson 4-2)

  a. Write an equation in slope-intercept form to find the cost C of b specialty balloons.

  b. How much would 20 of the balloons cost?

Write an equation in slope-intercept form of the line that passes through each pair of points. (Lesson 4-2)

6. $(2, 0)$, $\left(4, \frac{1}{2}\right)$

7. $(-5, -2)$, $(5, 2)$

Write each equation in standard form. (Lesson 4-3)

8. $\frac{1}{2}(y + 2) = 6x$

9. $7y - 13 = 2x + 5$

Write each equation in slope-intercept form. (Lesson 4-3)

10. $2y - 8 = \frac{1}{2}(x + 3)$

11. $y - \frac{1}{2} = 3x - \frac{3}{4}$

Determine whether the graphs of the following equations are *parallel* or *perpendicular*. Explain. (Lesson 4-4)

12. $y = -2x + 6$, $2y = x - 3$

13. $y = 3x - 2$, $-3y = x + 6$

14. **MAPS** The director of street repairs wants to first replace curbs on streets that are parallel to each other. Which two streets will get new curbs first? (Lesson 4-4)

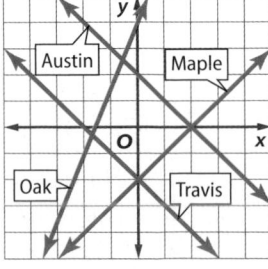

Write an equation in slope-intercept form for the line that passes through the given point and is parallel to the graph of the equation. (Lesson 4-4)

15. $(-1, 6)$, $y = \frac{1}{4}x - 4$

16. $(5, 7)$, $y = -x + 5$

Determine whether each graph shows a *positive*, *negative*, or *no* correlation. If there is a positive or negative correlation, describe its meaning in the situation. (Lesson 4-5)

17.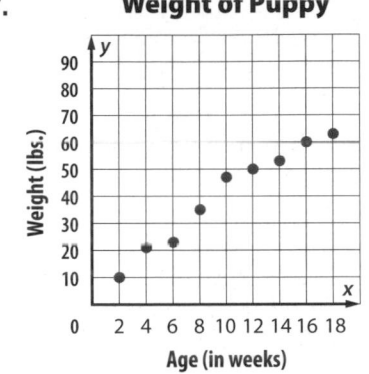

18.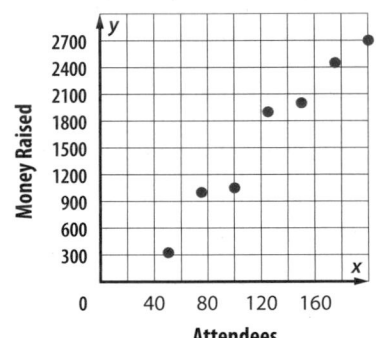

Write an equation of the regression line for the data in each table. Then find the correlation coefficient. (Lesson 4-6)

19. The table shows the numbers of plants in each garden and turnips produced.

Plants	2	3	4	5	6	7
Turnips	12	19	24	32	34	45

20. The table shows the amount of shrimp caught each day and the retail price of the shrimp.

Pounds (1000s)	48	52	60	65	73
Cost/lb ($)	3.50	3.42	3.35	3.15	3.08

Find the inverse of each function. (Lesson 4-7)

21. $f(x) = -3x + 8$

22. $f(x) = \frac{1}{2}x + 7$

Solve each inequality. Then graph the solution set on a number line. (Lesson 5-1)

1. $6t \le 3$

2. $14 > k + 2$

3. $16 \le 4n$

4. $6c < 5c + 3$

Define a variable, write an inequality, and solve each problem. Check your solution. (Lesson 5-1)

5. The sum of six times a number and three is less than the product of seven and a number.

6. Three times a number is greater than or equal to the sum of twice a number and 12.

Solve each inequality. Graph the solution on a number line. (Lesson 5-2)

7. $\frac{c}{8} > \frac{1}{4}$

8. $-3 \le 4m$

Define a variable, write an inequality, and solve each problem. Then interpret your solution. (Lesson 5-2)

9. READING Thomas has a 432-page book to read in 12 days. At least how many pages must he read per day to finish the book on time?

10. DOGS Laura has a maximum of 91 minutes to walk 7 dogs. How much time can she spend walking each dog?

Solve each inequality. Graph the solution on a number line. (Lesson 5-3)

11. $1.2x + 6 < 4.6x - 3$

12. $4\left(3g + \frac{1}{2}\right) \le -6(3 + 2g)$

13. PIZZA Sam orders 3 large pizzas. Each pizza costs $12, and each topping costs $0.50. Sam has $38 to spend. Write and solve an inequality to find the greatest number of toppings Sam can afford. (Lesson 5-3)

Toppings

mushrooms olives
jalapeños pineapple
tomatoes
Canadian bacon onions

Solve each compound inequality. Then graph the solution set. (Lesson 5-4)

14. $6w + 3 < 9$ or $\frac{1}{2}w \ge 2$

15. $16 \ge 3x - 5$ and $3x - 2 > 2(x - 1)$

16. COUPON Victor has a coupon that is valid only for juice sold in containers between 16 and 32 ounces. (Lesson 5-4)

a. Write a compound inequality that describes acceptable juice container sizes.

b. Graph the inequality.

Solve each inequality. Then graph the solution set. (Lesson 5-5)

17. $\left|\frac{1}{2}z + 6\right| \le 4$

18. $|3p + 3| > 1$

19. SHOP Kristi is shopping online for gifts for her friends. (Lesson 5-5)

a. If prices ranged from $1.50 above and below the average CD price, find the range of prices.

Average Prices	
CD	$15.50
Book	$19
Shirt	$32

b. Prices for the book varied $2.25 from the average. Write the range of average book prices.

c. Graph the solution set for shirt prices if they varied $6 below to $4 above the average.

Graph each inequality. (Lesson 5-6)

20. $3(x + y) \ge 6$

21. $\frac{1}{2}y < 2(-1 - x)$

Use a graph to solve each inequality. (Lesson 5-6)

22. $5x + 3 > -2$

23. $y - 8 \le -3$

24. DOG WASH It costs Pups and Suds $975 a week to operate their business. (Lesson 5-6)

Prices

Small/Med Dog $20
Large Dog $30

a. The prices that they charge are shown. Write an inequality to describe how many of each type of dog they need to service to make a profit.

b. How many dogs must they wash to make a profit each week?

Use the graph below to determine whether each system is *consistent* or *inconsistent* and if it is *independent* or *dependent*. (Lesson 6-1)

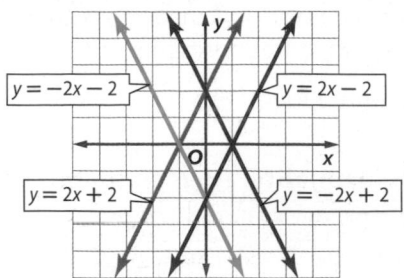

1. $y = 2x + 2$
$y = -2x - 2$

2. $y = -2x + 2$
$y = -2x - 2$

3. **DANCES** Mario and Tanesha are inflating balloons for the school dance. Mario has 12 balloons inflated and is inflating additional balloons at a rate of 3 balloons per minute. Tanesha has 16 balloons inflated and is inflating additional balloons at a rate of 2 balloons per minute. (Lesson 6-1)

   a. Write a system of equations to represent the situation.

   b. Graph each equation.

   c. How long will it take Mario to have more balloons filled than Tanesha?

Use substitution to solve each system of equations. (Lesson 6-2)

4. $x = -y + 3$
$3y + 2x = 10$

5. $-x + 2y = 6$
$4y - 2x = 11$

6. $y - 7x = 2$
$2x + 3 = 5y$

7. $-2y = x + 3$
$\frac{3}{2} = -\frac{1}{2}x - y$

8. **FRUIT** Sarah and Toni each bought fruit for a fundraiser. If Toni spent \$4.30 and Sarah spent \$2.80, how much does each type of fruit cost? (Lesson 6-3)

Girl	Apples	Oranges
Toni	6	5
Sarah	6	2

Use elimination to solve each system of equations. (Lesson 6-3)

9. $2m + 3n = 16$
$-3m - 3n = -4$

10. $-5k + 4j = 8$
$-5k - 6j = -12$

11. The difference of three times a number and a second number is two. The sum of the two numbers is fourteen. What are the two numbers? (Lesson 6-3)

Use elimination to solve each system of equations. (Lesson 6-4)

12. $1.6x + 2.2y = 5.4$
$-3.2x + 4y = -2.4$

13. $2x + 5y = -8$
$4x - 2y = 0$

14. **CARNIVALS** Scott and Isaac went to the school carnival. Use the table shown to determine how many tickets a ride and game each cost. (Lesson 6-4)

Rider	Rides	Games	Tickets
Scott	5	4	19
Isaac	7	2	23

15. **SAVINGS** Caleb made \$105 mowing lawns and walking dogs, charging the rates shown. If he mowed half as many lawns as dogs walked, how many lawns did he mow and how many dogs did he walk? (Lesson 6-5)

\$7.50 per dog   **\$20 per lawn**

Determine the best method to solve each system of equations. Then solve the system. (Lesson 6-5)

16. $4x + 2y = 12$
$-y - 4x = 2$

17. $y + 3x = 11$
$-2x + 3y = 11$

18. **BABYSITTING** Kelsey and Emma babysit after school to earn extra money. Kelsey made \$52 by charging \$10 per hour and \$4 per child. Emma made \$67.50 by charging \$15 per hour and \$2.50 per child. (Lesson 6-5)

   a. Write a system of equations to represent the situation.

   b. How many hours and how many children did each babysit?

Solve each system of inequalities by graphing. (Lesson 6-6)

19. $0.5x - y \geq 3$
$x + y < 3$

20. $y < 2x - 1$
$y > 4(1 + 0.5x)$

Simplify each expression. (Lesson 7-1)

1. $(2xy^2)^3(2x^2y^3z)$ **2.** $(2ab^2c^3)(3ad^2)^2$

GEOMETRY **Express the area of each triangle as a monomial.** (Lesson 7-1)

3.

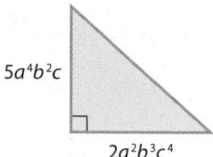

$5a^4b^2c$

$2a^2b^3c^4$

4.

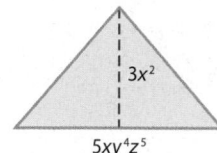

$3x^2$

$5xy^4z^5$

Simplify each expression. Assume that no denominator equals zero. (Lesson 7-2)

5. $\dfrac{3m^5n^3p^4}{5m^6n^4p^2q^5}$ **6.** $\left(\dfrac{x^2yz^3}{2xy^3z^3}\right)^3$

7. $\dfrac{2a^{-1}b^{-2}c^{-3}}{7a^2b^3c^4d^{-5}}$ **8.** $\left(\dfrac{4x^{-2}y^4z^5}{5x^5y^{-3}z^{-2}}\right)^{-1}$

Write each expression in radical form, or write each radical in exponential form. (Lesson 7-3)

9. $13^{\frac{1}{3}}$ **10.** $(7k)^{\frac{1}{2}}$

11. $\sqrt{17a}$ **12.** $3\sqrt{2xyz^2}$

Simplify. (Lesson 7-3)

13. $\sqrt[4]{\dfrac{81}{625}}$ **14.** $\sqrt[5]{0.00001}$

15. $4096^{\frac{1}{3}}$ **16.** $\left(\dfrac{125}{343}\right)^{\frac{4}{3}}$

Solve each equation. (Lesson 7-3)

17. $81^x = \dfrac{1}{3}$ **18.** $3^{4x} = 3^{x+1}$

Express each number in scientific notation.
(Lesson 7-4)

19. 22,100,000,000   **20.** 0.000000003088

Evaluate each product. Express the results in both scientific notation and standard form.
(Lesson 7-4)

21. $(6.2 \times 10^3)(1.77 \times 10^6)$

22. $(4.08 \times 10^{-4})^2$

Graph each equation. Find the y-intercept, and state the domain and range. (Lesson 7-5)

23. $f(x) = -3^x - 1$ **24.** $f(x) = \left(\dfrac{1}{2}\right)^x + 3$

25. Determine whether the set of data shown below displays exponential behavior. Write *yes* or *no*. Explain why or why not. (Lesson 7-5)

x	-1	0	1	2	3	4
y	3	1	$\frac{1}{3}$	$\frac{1}{9}$	$\frac{1}{27}$	$\frac{1}{81}$

26. **POPULATION** A neighborhood had 4518 residents in 2006. The number of residents has been declining by 3.5% each year. How many residents will there be in 2012? (Lesson 7-6)

27. **MONEY** Sarah put $3000 in an investment that gets 6.2% compounded quarterly for 8 years. What will her investment be worth at the end of the 8 years? (Lesson 7-6)

28. **SOCCER** The Westside Soccer League has 186 players. They expect a 7.5% increase in players for at least the next 4 years. How many players will they have at that point? (Lesson 7-6)

Determine whether each sequence is *arithmetic*, *geometric*, or *neither*. Explain. (Lesson 7-7)

29. $\dfrac{1}{2}, -\dfrac{1}{4}, 0, \dfrac{1}{4}, \dfrac{1}{2} \ldots$ **30.** $100, 90, 85, 75, 60\ldots$

Find the next three terms in each geometric sequence. (Lesson 7-7)

31. $48, -96, 192, -384, 768\ldots$

32. $150, 75, 37.5, 18.75, 9.375\ldots$

Find the first five terms of each sequence. (Lesson 7-8)

33. $a_1 = 5, a_n = 3.5a_{n-1} + 1, n \geq 2$

34. $a_1 = 12, a_n = -\dfrac{1}{2}a_{n-1} + \dfrac{5}{2}, n \geq 2$

Write a recursive formula for each sequence.
(Lesson 7-8)

35. $7, 16, 43, 124, \ldots$ **36.** $729, 243, 81, 27, \ldots$

Extra Practice

Extra Practice

Find each sum or difference. (Lesson 8-1)

1. $(7g^3 + 2g^2 - 12) - (-2g^3 - 4g)$

2. $(-3h^2 + 3h - 6) + (5h^2 - 3h - 10)$

Simplify each expression. (Lesson 8-2)

3. $-\frac{1}{2}n^3p^2(5np^3 - 3n^2p^2 + 8n)$

4. $6j^2(-3j + 3k^2) - 2k^2(2j + 10j^2)$

Solve each equation. (Lesson 8-2)

5. $-4(b + 3) + b(b - 3) = -b(6 - b) + 2(b - 3)$

6. $3(a - 3) + a(a - 1) + 12 = a(a - 2) + 3(a - 2) + 4$

Find each product. (Lesson 8-3)

7. $(-3t - 16)(5t + 2)$ **8.** $\left(4p + \frac{1}{2}\right)\left(\frac{1}{2}p + 4\right)$

9. SIDEWALKS Reynoldsville is repairing sidewalks. If the sidewalk is the same width around a city block, write an expression for the area of the block and the sidewalk. (Lesson 8-3)

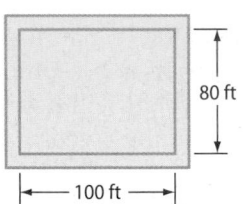

80 ft

100 ft

Find each product. (Lesson 8-4)

10. $\left(\frac{1}{2}m + 3\right)^2$ **11.** $(2n - 6)(2n + 6)$

12. $(5a - 4)^2$ **13.** $(x - 2y)(x + 2y)$

Use the Distributive Property to factor each polynomial. (Lesson 8-5)

14. $4m^3n^2 + 16m^2n^3 - 8m^3n^4$

15. $12j^4k^4 + 36j^3k^2 - 3j^2k^5$

Factor each polynomial. (Lesson 8-5)

16. $x^2 - 4x + 3xy - 12y$

17. $4a - 10ab + 6b - 15b^2$

18. HEIGHT The height h of a ball bounced off the ground after t seconds is modeled by the equation $h = -16t^2 + 28.8t$. (Lesson 8-5)

   a. What is the height of the ball at 1.5 seconds?

   b. How many seconds before the ball hits the ground again?

Factor each polynomial. (Lesson 8-6)

19. $t^2 + 2t - 15$ **20.** $d^2 - 3d - 28$

21. $m^2 + 5m - 14$ **22.** $x^2 - 4x - 45$

Solve each equation. Check your solution. (Lesson 8-6)

23. $h^2 + 3h - 4 = 0$ **24.** $a^2 + 9a + 18 = 0$

25. $x^2 - x - 6 = 0$ **26.** $y^2 + 2y - 15 = 0$

Factor each polynomial, if possible. If the polynomial cannot be factored using integers, write _prime_. (Lesson 8-7)

27. $6x^2 + 21x - 90$ **28.** $3x^2 - 11x - 42$

29. $6x^2 - 13x - 5$ **30.** $5y^2 - 3y + 11$

31. PRIZES A machine is used to throw T-shirts into the crowd at the Hornets basketball games. (Lesson 8-7)

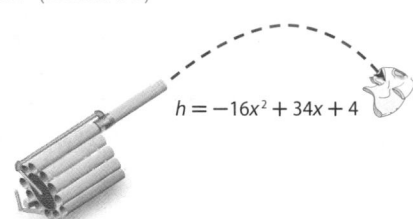

$h = -16x^2 + 34x + 4$

   a. What is the initial height of the T-shirt?

   b. If the T-shirt is caught after 2 seconds, what is the height?

Factor each polynomial. (Lesson 8-8)

32. $\frac{1}{2}t^2 - 162$ **33.** $25d^3 - 49d$

34. $196t^2u^3 - 144u^3$ **35.** $169a^4b^6 - 121c^8$

36. $4g^2 - 1296h^2$ **37.** $18a^3 + 27a^2 - 50a - 75$

38. FRUIT An apple fell 25 feet from a tree. The formula $h = -16t^2 + 25$ can be used to approximate the number of seconds it will take the apple to hit the ground. (Lesson 8-8)

   a. How long will it take the apple to hit the ground?

   b. If you catch it at 4 feet, how long did the apple drop?

Determine whether each trinomial is a perfect square trinomial. Write _yes_ or _no_. If so, factor it. (Lesson 8-9)

39. $64x^2 - 32x + 4$ **40.** $4a^2 - 12a + 16$

41. $12y^2 - 36y + 27$ **42.** $75b^3 - 60ab^2 + 12a^2b$

Find the vertex, the equation of the axis of symmetry, and the y-intercept of the graph of each equation. (Lesson 9-1)

1. $y = 4x^2 + 8x - 5$ **2.** $y = -2x^2 + 8x + 5$

3. $y = x^2 - 8x + 9$ **4.** $y = 4x^2 + 16x - 6$

5. KICKBALL A kickball is kicked in the air. The equation $h = -16t^2 + 60t$ gives the height h of the ball in feet after t seconds. (Lesson 9-1)

    a. What is the height of the ball after one second?

    b. When will the ball reach its maximum height?

    c. When will the ball hit the ground?

Solve each equation by graphing. (Lesson 9-2)

6. $-2x^2 - 2x + 4 = 0$ **7.** $x^2 = -2x + 3$

8. MARBLES Jason shot a marble straight up using a slingshot. The equation $h = -16t^2 + 42t + 5.5$ models the height h, in feet, of the marble after t seconds. After how long will the marble hit the ground? (Lesson 9-2)

Describe how the graph of each function is related to the graph of $f(x) = x^2$. (Lesson 9-3)

9. $g(x) = -x^2 - 4$ **10.** $h(x) = 7x^2 + 2$

Match each equation to its graph. (Lesson 9-3)

A 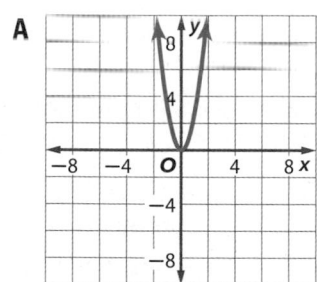   **B**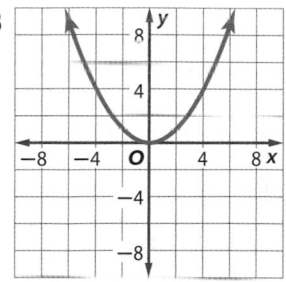

C 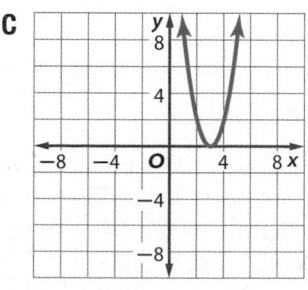   **D**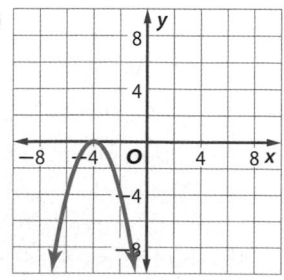

11. $y = 3x^2$ **12.** $y = \frac{1}{4}x^2$

13. $y = -(x + 4)^2$ **14.** $y = 2(x - 3)^2$

15. BIOLOGY The number of cells in a Petri dish can be modeled by the quadratic equation $n = 6t^2 - 4.5t + 74$, where t is the number of hours the cells have been in the dish. When will there be 200 cells in the Petri dish? (Lesson 9-4)

Solve each equation by completing the square. Round to the nearest tenth if necessary. (Lesson 9-4)

16. $x^2 + 4x - 8 = 5$

17. $3x^2 + 5x = 18$

18. Find the value of x in the figure if the area is 36 square inches. (Lesson 9-4)

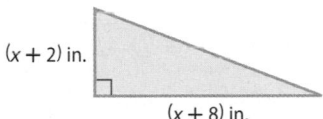

$(x + 2)$ in.

$(x + 8)$ in.

Solve each equation by using the Quadratic Formula. Round to the nearest tenth if necessary. (Lesson 9-5)

19. $3x^2 + 10x = 15$

20. $\frac{1}{2}x^2 - 8x + 6 = 0$

State the value of the discriminant. Then determine the number of solutions of the equation. (Lesson 9-5)

21. $4x^2 - 12x = -9$

22. $3x^2 + 8 = 9x$

Look for a pattern in each table of values to determine which kind of model best describes the data. (Lesson 9-6)

23.

x	2	3	4	5	6
y	$\frac{9}{4}$	$\frac{27}{8}$	$\frac{81}{16}$	$\frac{243}{32}$	$\frac{729}{64}$

24.

x	−2	−1	0	1	2
y	−13	−6.25	0	5.75	11

Graph each function. State the domain and range. (Lesson 9-7)

25. $f(x) = |x + 5|$

26. $f(x) = 2[\![x]\!]$

Graph each function. Compare to the parent graph. State the domain and range. (Lesson 10-1)

1. $y = 4\sqrt{x+2}$ **2.** $y = -3\sqrt{x}$

3. $y = -2\sqrt{x-2}$ **4.** $y = \sqrt{x+2} + 2$

5. PENDULUMS Find the period T of a pendulum, the time in seconds it takes to swing from one side to the other and back, if the length of the pendulum ℓ is 50 meters and $T = 2\pi\sqrt{\dfrac{\ell}{g}}$, where g is the gravitational constant, 9.8 meters per second squared. (Lesson 10-1)

Simplify each expression. (Lesson 10-2)

6. $\sqrt{24} \cdot 3\sqrt{14}$ **7.** $\sqrt{45x^4y^3z^6}$

8. $\dfrac{7}{6 - \sqrt{10}}$ **9.** $\dfrac{3}{8 + \sqrt{14}}$

Simplify each expression. (Lesson 10-3)

10. $6\sqrt{18} + 3\sqrt{2}$ **11.** $\sqrt{6}(\sqrt{24} + 3\sqrt{3})$

12. $4\sqrt{7}(3\sqrt{63})$ **13.** $\sqrt{12} + 5\sqrt{48} - 2\sqrt{3}$

14. $\sqrt{\dfrac{1}{3}} - \sqrt{3}$ **15.** $(2\sqrt{3} - 2\sqrt{5})(2\sqrt{15} - 4)$

16. PARKS Crandall Lake has a swimming area shaped like a trapezoid. The area can be found using the formula $A = \frac{1}{2}h(b_1 + b_2)$, where h represents the height and b_1 and b_2 are the lengths of the two bases. What is the area available for swimming? (Lesson 10-3)

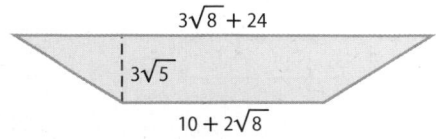

Solve each equation. Check your solution.
(Lesson 10-4)

17. $\sqrt{24 - n} = \dfrac{n}{2}$ **18.** $\sqrt{b + 6} = 3\sqrt{15}$

19. $f = 2\sqrt{3f + 6}$ **20.** $\sqrt{6 - m} = m + 6$

21. $\sqrt{t} = \sqrt{4t - 6}$ **22.** $5 + \sqrt{17 - m} = m$

Find each missing length. If necessary, round to the nearest hundredth. (Lesson 10-5)

23.

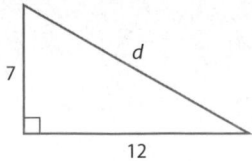

24.

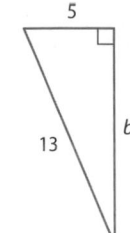

25. BAKING Jasmen is cutting triangles from dough to bake for her math class. If she wants the base to be 1.5 inches and the height to be 3 inches, how long is the hypotenuse? (Lesson 10-5)

Use a calculator to find the value of each trigonometric ratio to the nearest ten-thousandth.
(Lesson 10-6)

26. $\cos 52°$ **27.** $\tan 28°$

28. $\sin 17°$ **29.** $\cos 75°$

30. $\tan 55°$ **31.** $\sin 65°$

Solve each right triangle. Round each side length to the nearest tenth. (Lesson 10-6)

32.

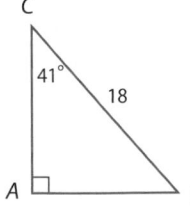

33.

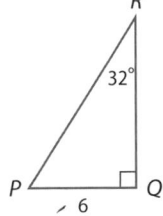

34. ROLLERBLADING The path for rollerblading in the park has a vertical rise of 35 feet. The angle the rise makes with the path is 75°. How long is the incline? (Lesson 10-6)

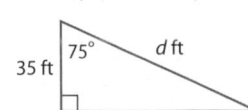

Determine whether each table or equation represents an *inverse* or *direct* variation. Explain. (Lesson 11-1)

1.

x	1	2	3	4	5
y	−3	−6	−9	−12	−15

2. $y = -\dfrac{2}{x}$

3.

x	−2	−$\frac{1}{2}$	$\frac{1}{2}$	2
y	−3	−12	12	3

State the excluded value for each function. (Lesson 11-2)

4. $y = \dfrac{2x - 3}{x - 5}$

5. $y = \dfrac{x + 3}{3x - 3}$

Identify the asymptotes of each function. Then graph the function. (Lesson 11-2)

6. $y = \dfrac{3}{2x - 2}$

7. $y = \dfrac{2}{3x} - 4$

8. FUNDRAISER The Sophomore class has committed to walking 200 miles for a fundraiser. The equation $y = \dfrac{200}{x}$ models the number of miles each person will walk depending on the number of volunteers. Graph the function. (Lesson 11-2)

State the excluded value(s) for each rational expression. (Lesson 11-3)

9. $\dfrac{6t}{2t^2 - 18}$

10. $\dfrac{11}{2x^2 - x - 28}$

11. $\dfrac{2x + 1}{x - 4}$

12. $\dfrac{7y^2}{2y^2 - 5y - 12}$

Find the zeros of each function. (Lesson 11-3)

13. $f(x) = \dfrac{x^2 - 2x - 8}{2x^2 + 8x + 6}$

14. $f(x) = \dfrac{x^2 + 7x + 10}{x + 5}$

15. $f(x) = \dfrac{x^3 + x^2 - 2x}{x^2 - 2x + 1}$

Find each quotient. (Lesson 11-4)

16. $\dfrac{3a^3}{6b^2} \div \dfrac{9a^6}{4b^3}$

17. $\dfrac{2t - 1}{t + 2} \div \dfrac{6t^2 - 21t + 9}{3t^2 + 2t - 8}$

18. $\dfrac{4x^2yz}{5z^3} \div \dfrac{2yz^4}{7xy^2z^3}$

19. $\dfrac{4g^2 + 4g - 8}{6h^2 + 6h - 12} \div \dfrac{5g + 10}{5h - 5}$

20. BOOKS John has read 6 books recently. If he reads 40 words per minute, each page has approximately 500 words, and each book has an average of 225 pages, how many hours did John spend reading? (Lesson 11-4)

21. AREA The area A of a rectangle is given by ℓw. Find the length of the unknown side. (Lesson 11-5)

$A = 2x^2 + x - 3$

$2x + 3$

Find each quotient. (Lesson 11-5)

22. $\dfrac{m^2 - m - 6}{m} \div (m - 3)$

23. $(2g^2 - 3g - 2) \div (g - 2)$

Find each sum or difference. (Lesson 11-6)

24. $\dfrac{7}{g} - \dfrac{2}{3g}$

25. $\dfrac{3n}{2n + 4} + \dfrac{n - 2}{2n + 4}$

26. $\dfrac{-6}{k + 2} + \dfrac{k}{k - 2}$

27. $\dfrac{2p - 3}{p^2 \ 5p + 6} - \dfrac{5}{p^2 - 9}$

28. TRAINING Miguel is training for a triathlon. He runs 5 miles at x miles per hour and 7 miles at $1.5x$ miles per hour. Write and simplify an expression for the time it takes him to run the 12 miles. (Lesson 11-6)

Simplify each expression. (Lesson 11-7)

29. $\dfrac{4\frac{2}{5}}{2\frac{5}{9}}$

30. $\dfrac{\frac{3}{2a + 2}}{\frac{1 + a}{4a + 8}}$

Solve each equation. State any extraneous solutions. (Lesson 11-8)

31. $\dfrac{a}{a - 1} - 1 = \dfrac{a}{2}$

32. $\dfrac{c}{c + 2} + c = \dfrac{5c + 8}{c + 2}$

33. $\dfrac{4}{w - 2} = \dfrac{-1}{w + 3}$

34. $\dfrac{1}{y + 2} + \dfrac{1}{y - 2} = \dfrac{3}{y^2 - 4}$

35. RECYCLING Jamie and Adam volunteer at a recycle sorting center. Jamie sorts 3 bins in 2 hours. Adam sorts 3 bins in 3 hours. If they work together, how long would it take them to sort the 3 bins? (Lesson 11-8)

Extra Practice

Determine whether each situation calls for a *survey*, an *experiment*, or an *observational study*. Explain your reasoning. (Lesson 12-1)

1. **AUTO REPAIR** An auto repair company sends each customer a letter asking them to rate their experience during their last service appointment.

2. **ICE CREAM** An ice cream manufacturer is testing a new formula for their chocolate cherry flavor. They randomly give half of a group of 150 people the original formula, and the other half gets the new formula. Both are asked how they like their ice cream.

3. **GAS** A random sample of 1000 car owners in Los Angeles were surveyed about the price they last paid for a gallon of gasoline. Identify the sample and the population for the situation. Then describe the sample statistic and the population parameter. (Lesson 12-2)

4. **BOWLING** Tina's results for 8 bowling games are {110, 123, 147, 119, 153, 142, 113, 143}. Find and interpret the standard deviation of the data. (Lesson 12-2)

For Exercises 5 and 6, use these data.
{12, 18, 21, 18, 19, 18, 16, 23, 20, 15, 17, 18}

5. Describe the center and spread of the data using either the mean and standard deviation or the five-number summary. Justify your choice by constructing a histogram. (Lesson 12-3)

6. Find the mean, median, mode, range, and standard deviation of the data after multiplying each value by 3. (Lesson 12-4)

7. **BASKETBALL** Keisha has a 75% free throw average. Describe how to simulate her next 10 free throws. (Lesson 12-5)

8. **FLIGHTS** Three airlines have the departure results shown. (Lesson 12-5)

Status	A	B	C
on-time	21	19	24
late	6	3	9

   a. What is the experimental probability of each airline having an on-time departure?

   b. What is the experimental probability for an on-time departure for all the airlines?

Evaluate each expression. (Lesson 12-6)

9. $P(8, 5)$ 10. $P(7, 3)$

11. $C(8, 5)$ 12. $C(7, 3)$

13. **DOG SHOW** There are 12 dogs in the finals for best in show. If 4 dogs will get ribbons, how many ways can the ribbons be awarded? (Lesson 12-6)

14. **SOCKS** A drawer contains 4 blue socks, 8 red socks, and 12 white socks. What is the probability of drawing two white socks in successive draws? (Lesson 12-7)

A card is drawn from a standard deck of playing cards. Determine whether the events are *mutually exclusive* or *not mutually exclusive*. Then find the probability. (Lesson 12-7)

15. P(eight or king)   16. P(heart or club)

17. P(two or red card)   18. P(even number or ace)

19. **SWEATERS** A local pet store sells five different sizes of dog sweaters. The table shows the probability distribution of each size sold in a month. (Lesson 12-8)

Dog Sweater Sizes	
Size	Probability
S	0.38
M	0.29
L	0.13
XL	0.11
XL+	0.09

   a. Show that the distribution is valid.

   b. What is the probability that a randomly chosen purchase is XL or larger?

   c. Find the probability that a customer purchased a medium sweater.

20. **MOVIES** A movie theater surveyed all attendees to a movie on Sunday afternoon to determine the average age. The table shows the results. (Lesson 12-8)

   a. Which category will have the highest probability? Explain.

   b. Find the probability of a person being between 21 and 40 years of age.

   c. Find the probability of someone being over 40.

Movie Attendee Age	
Age	Customers
0–10	11
11–20	115
21–30	82
31–40	15
41–50	37
51+	28

Selected Answers and Solutions

Lesson 0-1

1. estimate; about 700 mi   **3.** estimate; about 7 times   **5.** exact; $98.75

Lesson 0-2

1. integers, rationals   **3.** irrationals   **5.** irrationals
7. rationals   **9.** rationals   **11.** irrationals
13. $-\frac{6}{5}, -\frac{3}{5}, \frac{3}{4},$ and $\frac{7}{5}$

$-\frac{6}{5}$ $-\frac{3}{5}$ $\frac{3}{4}$ $\frac{7}{5}$

$\leftarrow$|+|+|+●|+|●|+|+|+|+|●|+|●|+|+|$\rightarrow$
−2 −1.6 −1.2 −0.8 −0.4 0 0.4 0.8 1.2 1.6 2

15. $2\frac{1}{4}, 2.\overline{3}, \sqrt{7},$ and $\sqrt{8}$

$-2\frac{1}{4}$ 2.3333... $\sqrt{7}$ $\sqrt{8}$

$\leftarrow$|+|+●|●|+|+|●|+|●|+|+|$\rightarrow$
2.0 2.1 2.2 2.3 2.4 2.5 2.6 2.7 2.8 2.9 3.0

17. $-3\frac{3}{4}, -3.5, -\sqrt{10},$ and $-\frac{15}{5}$

$-3\frac{3}{4}$ -3.5 $-\sqrt{10}$ $-\frac{15}{5}$

$\leftarrow$|+|●|+|●|+|●|+|●|+|$\rightarrow$
−4.0 −3.8 −3.6 −3.4 −3.2 −3.0 −2.8

19. $\frac{5}{9}$ **21.** $\frac{13}{99}$ **23.** −5   **25.** ±6   **27.** +1.2   **29.** $\frac{4}{7}$
31. $\frac{5}{18}$ **33.** 16   **35.** 26

Lesson 0-3

1. 5   **3.** −27   **5.** −22   **7.** −32   **9.** 22   **11.** 5   **13.** 8
15. −9   **17.** −115   **19.** 17°   **21.** $150   **23.** $125

Lesson 0-4

1. <   **3.** <   **5.** =   **7.** 3.06, $3\frac{1}{6}, 3\frac{3}{4}$, 3.8   **9.** −0.5, −$\frac{1}{9}$,
$\frac{1}{10}$, 0.11   **11.** $\frac{3}{5}$ **13.** $\frac{1}{16}$ **15.** 1   **17.** $2\frac{2}{3}$ **19.** $\frac{1}{9}$
21. $\frac{1}{6}$ **23.** $\frac{17}{30}$ **25.** $\frac{1}{4}$ **27.** −36.9   **29.** −19.33
31. 153.8   **33.** 93.3   **35.** $-\frac{5}{6}$ **37.** $\frac{9}{20}$ **39.** $\frac{2}{3}$ **41.** $\frac{3}{10}$

Lesson 0-5

1. 0.85   **3.** −7.05   **5.** 60   **7.** −4.8   **9.** −1.52
11. $\frac{6}{35}$ **13.** $\frac{2}{33}$ **15.** $\frac{21}{4}$ or $5\frac{1}{4}$ **17.** $-\frac{1}{2}$ **19.** $-\frac{1}{8}$
21. $\frac{10}{11}$ **23.** $\frac{5}{2}$ or $2\frac{1}{2}$ **25.** $\frac{7}{6}$ or $1\frac{1}{6}$ **27.** $-\frac{23}{14}$ or $-1\frac{9}{14}$
29. $-\frac{3}{16}$ **31.** 2   **33.** 3   **35.** $-\frac{3}{10}$ **37.** $\frac{9}{2}$ or $4\frac{1}{2}$
39. $\frac{11}{20}$ **41.** $\frac{5}{18}$ **43.** 3 slices   **45.** 34 uniforms
47. 6 ribbons

Lesson 0-6

1. $\frac{1}{20}$ **3.** $\frac{11}{100}$ **5.** $\frac{39}{50}$ **7.** $\frac{3}{500}$ **9.** 14   **11.** 40%

13. 160   **15.** 9.5   **17.** 48   **19.** 0.25%   **21.** 24.5
23. 150%   **25.** 90%   **27.** 5%   **29a.** 20 g
29b. 2350 mg   **29c.** 44%   **31.** 6 animals

Lesson 0-7

1. 20 m   **3.** 90 in.   **5.** 32 in.   **7.** 29 ft   **9.** 25.0 in.
11. 31.4 in.   **13.** 23.2 m   **15.** 848.2 in.   **17.** 13.4 cm
19. 10.3 ft

Lesson 0-8

1. 6 cm^2 **3.** 120 m^2 **5.** 81 ft^2 **7.** 9 ft^2 **9.** 14.1 in^2
11. 12.6 ft^2 **13.** 50.3 cm^2 **15.** 201.1 in^2 **17.** 7 ft
19. 20.5 units^2 **21.** 22.1 cm^2 **23.** 4.0 cm^2

Lesson 0-9

1. 30 cm^3 **3.** 48 yd^3 **5.** 1404 ft^3 **7.** 20 m^3
9. 27 m^3 **11.** 2070 in^3 **13.** 1 ft   **15.** 4 cm
17. 2770.9 in^3 **19a.** 128 ft^3 **19b.** 80 ft^3 **19c.** 5 ft 4 in.

Lesson 0-10

1. 68 in^2 **3.** 220 mm^2 **5.** 37 ft^2 **7.** 48 m^2
9. 216 in^2 **11.** 480.7 in^2 **13.** 24 m^2 **15.** 77 ft^2
17. 40.8 in^2

Lesson 0-11

1. $\frac{4}{15}$ **3.** $\frac{1}{2}$ **5.** $\frac{5}{6}$ **7.** $\frac{1}{2}$ **9.** $\frac{2}{3}$ **11.** 20   **13.** 12 codes
15. $\frac{11}{24}$ **17.** 1:5   **19.** 13:11   **21.** 16 orders

Lesson 0-12

1. 5 students; 4 students; 3 students; 10 students
3. 54.75 mph; 54 mph; 53 mph; 8 mph   **5.** ≈2.8; 2.75; 2; 4   **7.** 128   **9.** $309; $311; $312; $314; $399
11. 2 books; 5 books; 10 books; 17 books; 18 books
13. 16 years old; 19 years old; 21 years old; 24 years old; 45 years old   **15.** ≈138.3 mi, 101.5 mi; no outliers   **17.** ≈0.286, 0.296; 0.201; ≈0.295, 0.300; mean

Lesson 0-13

1.

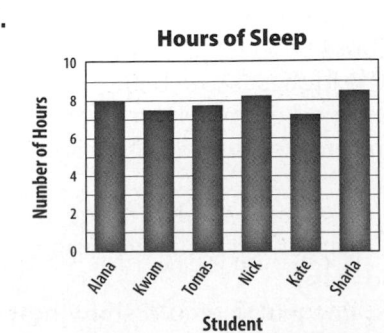

Hours of Sleep

3.

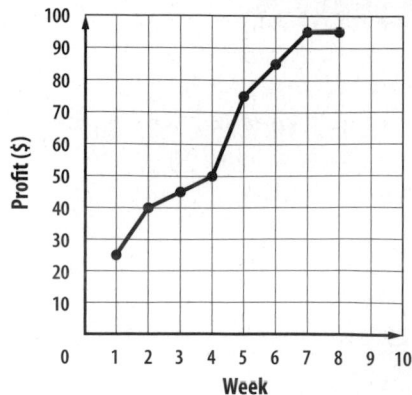

Lawn Care Profits

5.

Stem	Leaf
1	8 8
2	1 3 6 6 6 8
3	0 1 1 2 3 4
4	7

Key: 1|8 = 18

Removing 47 leaves Q_1 the same, changes Q_2 to 27 and Q_3 to 31.

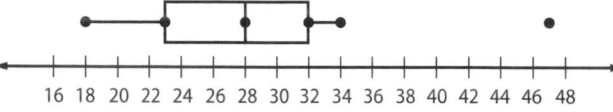

7.

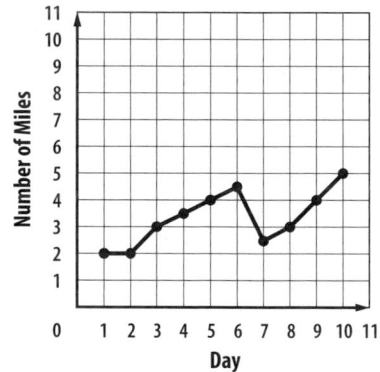

Miles Jogged

9a.

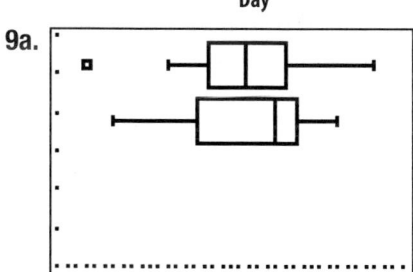

9b. Most of the data for third period are spread fairly evenly from about 80 to 89, with the lowest score being 67 and the highest score being 99. Most of the data for sixth period are between 79 and 91, with the lowest score for the class being 70 and the highest score being 95.

9c. The sixth period class has a smaller range, a higher median, and a larger interquartile range than the third period class.

11. Sample answer: a line graph would show how the cost of a seat changes during those years.

13a.

Stem	Leaf
11	9
12	4 6 9
13	0 0 3 5 6 7 8
14	0

Key: 11|9 = 119

13b.

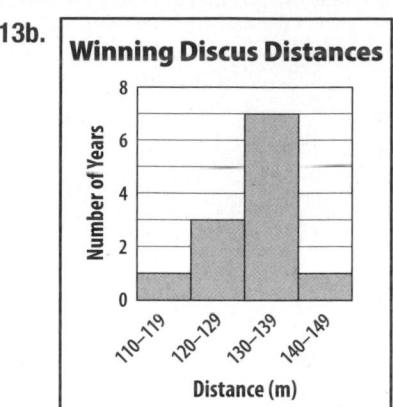

Winning Discus Distances

13c. Sample answer: The histogram shows frequencies, while the stem and leaf shows all data points.

13d. Sample answer: The winning distance increased by 16 meters from 2000 to 2010. If this continues, in 2030 the winning distance will be 32 meters more than in 2010, or 172 meters. It is unreasonable to expect that every year girls will be able to throw farther and farther, at some time the distance will level off.

CHAPTER 1
Expressions, Equations, and Functions

Get Ready

1. $\frac{2}{3}$ **3.** 3 **5.** *simplest form* **7.** 19 **9.** $\frac{8}{11}$
11. 8.2 cm **13.** 20 m **15.** 34.02 **17.** 1.9 **19.** 0.56

Lesson 1-1

1. Sample answer: the product of 2 and m
3. Sample answer: a squared minus 18 times b
5. $6 - t$ **7.** $1 - \frac{r}{7}$ **9.** $n^3 + 5$ **11.** Sample answer: four times a number q **13.** Sample answer: 15 plus r **15.** Sample answer: 3 times x squared
(17) Sample answer: 6 more than the product of 2 and a
19. $7 + x$ **21.** $5n$ **23.** $\frac{f}{10}$ **25.** $3n + 16$ **27.** $k^2 - 11$
29. $\pi r^2 h$ **31.** Sample answer: twenty-five plus six times a number squared **33.** Sample answer: three times a number raised to the fifth power divided by two
(35) a. Words: $\frac{3}{4}$ of the number of dreams
Expression: $\frac{3}{4} \cdot d$
The expression is $\frac{3}{4}d$.

b. $\frac{3}{4}(28) = 21$ dreams

37a.

10^2	$\cdot$	10^1	$= 10 \cdot 10 \cdot 10$	$=$	10^3
10^2	$\cdot$	10^2	$= 10 \cdot 10 \cdot 10 \cdot 10$	$=$	10^4
10^2	$\cdot$	10^3	$= 10 \cdot 10 \cdot 10 \cdot 10 \cdot 10$	$=$	10^5
10^2	$\cdot$	10^4	$= 10 \cdot 10 \cdot 10 \cdot 10 \cdot 10 \cdot 10$	$=$	10^6

37b. $10^2 \cdot 10^x = 10^{(2+x)}$

37c. The exponent of the product of two powers is the sum of the exponents of the powers with the same base. **39.** Sample answer: x is the number of minutes it takes to walk between my house and school. $2x + 15$ represents the amount of time in minutes I spend walking each day since I walk to and from school and I take my dog on a 15 minute walk. **41.** 6 **43.** D **45.** $\frac{3l}{36}$

47.

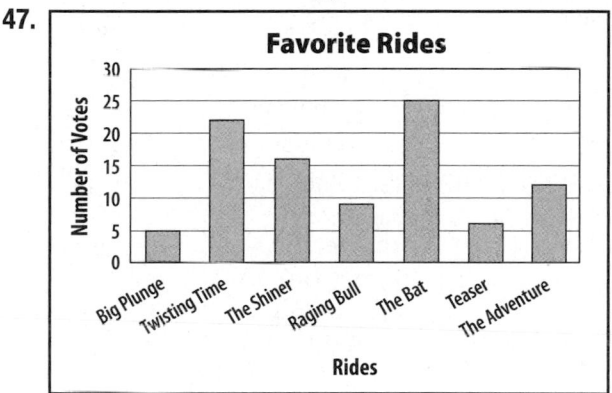

Favorite Rides

49. mean = 5.6; median = 6.5; mode = 7
51. mean = 15.25; median = 15.5; mode = 24
53. $\frac{21}{55}$ **55.** $\frac{20}{9}$ **57.** 1.46 **59.** 24.61 **61.** 21.16

Lesson 1-2

1. 81 **3.** 243
5 $5 \cdot 5 - 1 \cdot 3 = 25 - 3$
$\qquad\qquad\qquad = 22$
7. 28 **9.** 12 **11.** 20 **13.** $20 + 3 \times 4.95$; $34.85
15. 49 **17.** 64 **19.** 14 **21.** 36 **23.** 14
25. 142 **27.** 36 **29.** 3 **31.** 1
33 $(2t + 3g) \div 4 = (2(11) + 3(2)) \div 4$
$\qquad\qquad\qquad\quad = (22 + 6) \div 4$
$\qquad\qquad\qquad\quad = (28) \div 4$
$\qquad\qquad\qquad\quad = 7$
35. 149 **37.** $3344 - 148 = 3196$ **39.** 16 **41.** 729
43. 177 **45.** 324 **47.** 29 **49.** 4080 **51.** $\frac{97}{31}$ **53.** 0
55. $28(7) + 12(9.75) + 30(7) + 15(9.75)$; $669.25
57 **a.**

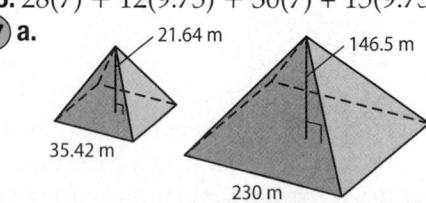

21.64 m
146.5 m
35.42 m
230 m

b. Words: one third times 230 squared times 146.5 minus one third times 35.42 squared times 21.64

c. Expression: $\frac{1}{3}(230)^2(146.5) - \frac{1}{3}(35.42)^2(21.64)$
$\qquad\qquad \approx 2583283.33 - 9049.68$
$\qquad\qquad \approx 2574233.656 \text{ m}^3$

59. Curtis; Tara subtracted $10 - 9$ before multiplying 4 by 10. **61.** Sample answer: $5 + 4 - 3 - 2 - 1$
63. Sample answer: Area of a trapezoid: $\frac{1}{2}h(b_1 + b_2)$; according to the order of operations, you have to add the lengths of the bases together first and then multiply by the height and by $\frac{1}{2}$. **65.** A **67A.** G
67B. Sample answers given.

a.

c.

d.

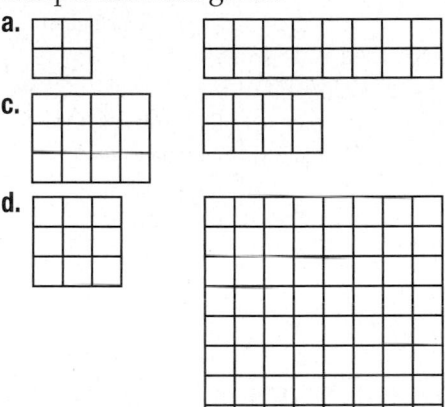

69. 14 minus 9 times c **71.** the difference of 4 and v divided by w **73.** 9π units^2 **75.** $12b$ units^2
77. 2.57 **79.** 13.192 **81.** $\frac{2}{3}$

Lesson 1-3

1. $(1 \div 5)5 \cdot 14$
$\quad = \frac{1}{5} \cdot 5 \cdot 14$ Substitution
$\quad = (1) \cdot 14$ Multiplicative Inverse
$\quad = 14$ Multiplicative Identity
3. $5(14 - 5) + 6(3 + 7) = 5(9) + 6(10)$ Substitution
$\qquad\qquad\qquad\qquad\quad = 45 + 60$ Substitution
$\qquad\qquad\qquad\qquad\quad = 105$ Substitution
5. $23 + 42 + 37$
$\quad = 23 + 37 + 42$ Commutative (+)
$\quad = (23 + 37) + 42$ Associative (+)
$\quad = 60 + 42$ Substitution
$\quad = 102$ Substitution
7. $3 \cdot 7 \cdot 10 \cdot 2$
$\quad = 3 \cdot 2 \cdot 7 \cdot 10$ Commutative (×)
$\quad = (3 \cdot 2) \cdot (7 \cdot 10)$ Associative (×)
$\quad = 6 \cdot 70$ Substitution
$\quad = 420$ Substitution
9 $3(22 - 3 \cdot 7) = 3(22 - 21)$ Substitution
$\qquad\qquad\qquad = 3(1)$ Substitution
$\qquad\qquad\qquad = 3$ Multiplicative Identity
11. $\frac{3}{4}[4 \div (7 - 4)]$
$\quad = \frac{3}{4}[4 \div 3]$ Substitution
$\quad = \frac{3}{4} \times \frac{4}{3}$ Substitution
$\quad = 1$ Multiplicative Inverse
13. $2(3 \cdot 2 - 5) + 3 \cdot \frac{1}{3}$
$\quad = 2(6 - 5) + 3 \cdot \frac{1}{3}$ Substitution

$$= 2(1) + 3 \cdot \frac{1}{3}$$ Substitution

$$= 2 + 3 \cdot \frac{1}{3}$$ Multiplicative Identity

$$= 2 + 1$$ Multiplicative Inverse

$$= 3$$ Substitution

15. $2 \cdot \frac{22}{7} \cdot 14^2 + 2 \cdot \frac{22}{7} \cdot 14 \cdot 7$

   $= 2 \cdot \frac{22}{7} \cdot 196 + 2 \cdot \frac{22}{7} \cdot 14 \cdot 7$ Substitution

   $= \frac{44}{7} \cdot 196 + \frac{44}{7} \cdot 14 \cdot 7$ Substitution

   $= 1232 + 616$ Substitution

   $= 1848$ Substitution

   The surface area is 1848 in^2.

17. $25 + 14 + 15 + 36$

   $= 25 + 15 + 14 + 36$ Commutative(+)

   $= (25 + 15) + (14 + 36)$ Associative(+)

   $= 40 + 50$ Substitution

   $= 90$ Substitution

19. $3\frac{2}{3} + 4 + 5\frac{1}{3} = 3\frac{2}{3} + 5\frac{1}{3} + 4$ Commutative (+)

   $= \left(3\frac{2}{3} + 5\frac{1}{3}\right) + 4$ Associative (+)

   $= 9 + 4$ Substitution

   $= 13$ Substitution

21. $4.3 + 2.4 + 3.6 + 9.7$

   $= 4.3 + 9.7 + 2.4 + 3.6$ Commutative (+)

   $= (4.3 + 9.7) + (2.4 + 3.6)$ Associative (+)

   $= 14 + 6$ Substitution

   $= 20$ Substitution

23. $12 \cdot 2 \cdot 6 \cdot 5 = 12 \cdot 6 \cdot 2 \cdot 5$ Commutative (×)

   $= (12 \cdot 6) \cdot (2 \cdot 5)$ Associative (×)

   $= 72 \cdot 10$ Substitution

   $= 720$ Substitution

25. $0.2 \cdot 4.6 \cdot 5 = (0.2 \cdot 4.6) \cdot 5$ Associative (×)

   $= 0.92 \cdot 5$ Substitution

   $= 4.6$ Substitution

27. $1\frac{5}{6} \cdot 24 \cdot 3\frac{1}{11} = 1\frac{5}{6}\left(24 \cdot 3\frac{1}{11}\right)$ Associative (×)

   $= 1\frac{5}{6}\left(\frac{24}{1} \cdot \frac{34}{11}\right)$

   $= 1\frac{5}{6} \cdot \frac{816}{11}$ Substitution

   $= \frac{8976}{66}$ Substitution

   $= 136$ Substitution

29a. Sample answer: $2(10.95) + 3(7.5) + 2(5) + 5(18.99)$; $2(10.95 + 5) + 3(7.5) + 5(18.99)$ **29b.** $149.35

31 $4(-1) + 9(4) - 2(6) = -4 + 36 - 12$

   $= 32 - 12$

   $= 20$

33. -18 **35.** 192 **37.** Additive Identity; $35 + 0 = 35$

39. 0; Additive Identity   **41.** 7; Reflexive Property

43. 3; Multiplicative Identity   **45.** 2; Commutative Property   **47.** 3; Multiplicative Inverse   **49.** $108

51. 88 units   **53** **a.**

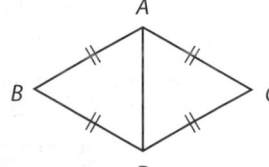

b. $\overline{AD} \cong \overline{AD}$ by the Reflexive Property. The Transitive Property shows that if $\overline{AB} \cong \overline{AC}$ and $\overline{AC} \cong \overline{DC}$, then $\overline{AB} \cong \overline{DC}$ and if $\overline{AB} \cong \overline{BD}$ and $\overline{AB} \cong \overline{AC}$, then $\overline{BD} \cong \overline{AC}$.

c. Since the sides are all congruent, each side has a length x. So, $P = x + x + x + x$.

55. Sample answer: You cannot divide by 0.

57. Sometimes; when a number is subtracted from itself then it holds but otherwise it does not.

59. $(2j)k = 2(jk)$; The other three equations illustrate the Commutative Property of Addition or Multiplication. This equation represents the Associative Property of Multiplication.   **61.** D

63. C   **65.** 14   **67.** 6   **69.** 26 ft; 40 ft^2

71. about 64.7 %   **73.** $\frac{23}{2}$ **75.** $\frac{6}{35}$ **77.** $\frac{6}{11}$ **79.** 6

Lesson 1-4

1. $25(12 + 15)$; $675   **3.** $\left(6 + \frac{1}{9}\right)9$; 55   **5.** $g(5) + (-9)(5)$; $5g - 45$ **7.** simplified

9. $4(2x + 6)$

   $= 4(2x) + 4(6)$ Distributive Property

   $= 8x + 24$ Multiply.

11 $4(5 + 3 + 4) = 4(8 + 4)$

   $= 4(12)$

   $= 48$ activities

13. $6(4) + 6(5)$; 54   **15.** $6(6) - 6(1)$; 30   **17.** $14(8) - 14(5)$; 42   **19.** $4(7) - 4(2)$; 20   **21.** $7(500 - 3)$; 3479

23. $36\left(3 + \frac{1}{4}\right)$; 117   **25.** $2(x) + 2(4)$; $2x + 8$

27. $4(8) + (-3m)(8)$; $32 - 24m$ **29.** $18r$ **31.** $2m + 7$

33. $34 - 68n$ **35.** $13m + 5p$ **37.** $4fg + 17g$

39. $7(a^2 + b) - 4(a^2 + b)$

   $= 7a^2 + 7b - 4a^2 - 4b$ Substitution

   $= 7a^2 - 4a^2 + 7b - 4b$ Commutative (+)

   $= (7 - 4)a^2 + (7 - 4)b$ Distributive Prop.

   $= 3a^2 + 3b$ Substitution

41 A hexagon has six sides so an expression for the perimeter is $6(3x + 5)$.

   $6(3x + 5) = 6(3x) + 6(5)$

             $= 18x + 30$ units

43. $14m + 11g$ **45.** $12k^3 + 12k$ **47.** $19x + 8$

49. $9 - 54b$ **51.** $12c - 6cd^2 + 6d$ **53.** $7y^3 + y^4$

55a. $2(x + 3)$

55b.

Area	Factored form
$2x + 6$	$2(x + 3)$
$3x + 3$	$3(x + 1)$
$3x - 12$	$3(x - 4)$
$5x + 10$	$5(x + 2)$

55c. Divide each term of the expression by the same number. Then write the expression as a product.

57. It should be considered a property of both. Both operations are used in $a(b + c) = ab + ac$.   **59.** Sample answer: You can use the Distributive Property to calculate quickly by expressing any number as a sum or difference of a more convenient number. Answers should include the following: Both methods result in the correct answer. In one method you multiply then add, and in the other you add then multiply. **61.** G **63.** about $\frac{1}{3}$ or 33%

65. $0.24 \cdot 8 \cdot 7.05 = (0.24 \cdot 8) \cdot 7.05$ Associative (×)
$$= 1.92 \cdot 7.05 \quad \text{Substitution}$$
$$= 13.536 \quad \text{Substitution}$$

67. $\frac{4[6(30) + 3(20)]}{60}$; 16 hours   **69.** 21:48

71. 384 in^2 **73.** 15   **75.** 60   **77.** 192

Lesson 1-5

1. {13}   **3.** {12}   **5.** B   **7.** −68   **9.** all real numbers
11. {12}   **13.** {5}   **15.** {16}   **17.** {3}   **19.** 14   **21.** 2
23. 2   **25.** 5   **27.** no solution   **29.** all real numbers
(31) $(2^4 - 3 \cdot 5)q + 13 = (2 \cdot 9 - 4^2)q + \left(\frac{3 \cdot 4}{12} - 1\right)$
$$(16 - 15)q + 13 = (18 - 16)q + (1 - 1)$$
$$1q + 13 = 2q + 0$$
$$1q + 13 = 2q$$
$$13 = 2q - 1q$$
$$13 = 1q$$
$$q = 13$$

33. 41 students

(35) Words: the number of calories equals 2836 plus 3091
Expression: $C = 2836 + 3091$ Solve: $C = 5927$

37.

x	3x − 2	y
−2	3(−2) − 2	−8
−1	3(−1) − 2	−5
0	3(0) − 2	−2
1	3(1) − 2	1
2	3(2) − 2	4

39. 20   **41.** 66   **43.** 5   **45.** $c - 15$ **47a.** $5 = \frac{1000}{r}$; 20

47b.

Intial Pressure p$_1$ (mm Hg)	Final Pressure p$_2$ (mm Hg)	Resistance r (mm Hg/L/min)	Blood Flow Rate F (L/min)
100	0	20	5
100	0	30	≈ 3.33
165	5	40	4
90	30	10	12

49. solution   **51.** not a solution   **53.** solution
55. solution
57.

x	3x + 5	y
−2	3(−2) + 5	−1
−1	3(−1) + 5	2
0	3(0) + 5	5
1	3(1) + 5	8
2	3(2) + 5	11

59.

x	$\frac{1}{2}x + 2$	y
−2	$\frac{1}{2}(-2) + 2$	1
−1	$\frac{1}{2}(-1) + 2$	1.5
0	$\frac{1}{2}(0) + 2$	2
1	$\frac{1}{2}(1) + 2$	2.5
2	$\frac{1}{2}(2) + 2$	3

61a.

61b. perimeter of rectangle = $2(2 + w) + 2w$ or $4 + 4w$; perimeter of triangle = $2(w + 1) + 12 = 2w + 14$.
61c. $4 + 4w = 2w + 14$; $w = 5$ in.

(63) b.

Layers	1	2	3	4	5	6	7
Cubes	4	8	12	16	20	24	28

   c. From the table, we can tell that each layer adds 4 more cubes. Notice $8 - 4 = 4$; $12 - 8 = 4$; $16 - 12 = 4$; $20 - 16 = 4$; $24 - 20 = 4$; $28 - 24 = 4$

   d. The number of cubes is 4 times the number of layers, or $c = 4L$.

65. Sample answer: $3x + 12 = 3(x + 4)$ **67.** Tom; Li-Cheng added $6 + 4$ instead of dividing 6 by 8. She did not follow the order of operations.
69. Sample answer: $3x - 2 = 23$ **71.** C   **73.** G
75. $30(500 + 750)$ **77.** $p = \frac{1}{12}$; Multiplicative

Inverse   **79.** 1040 in^3 **81.** $\frac{3}{20}$ **83.** estimate; 10 gal

85. 6.74   **87.** 1.65   **89.** $\frac{29}{28}$

Lesson 1-6

1.

x	y
4	3
−2	2
5	−6

Domain   Range

$D = \{-2, 4, 5\}$; $R = \{-6, 2, 3\}$
3. I: the temperature of the compound; D: the pressure of the compound   **5.** I: number of concert tickets, D: cost of tickets   **7.** The track team starts by running or walking, and then stops for a short period of time, then continues at the same pace. Finally, they run or walk at a slower pace.

9.

x	y
0	0
-3	2
6	4
-1	1

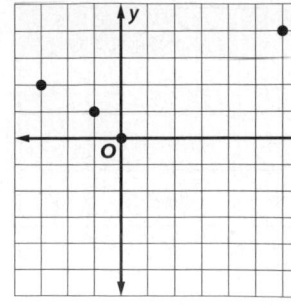

Domain   Range

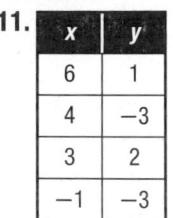

D = {0, -3, 6, -1};
R = {0, 2, 4, 1}

11.

x	y
6	1
4	-3
3	2
-1	-3

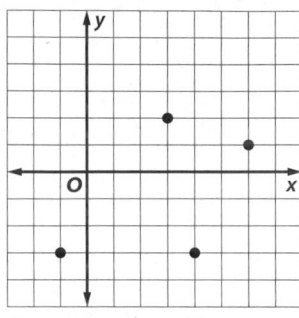

Domain   Range

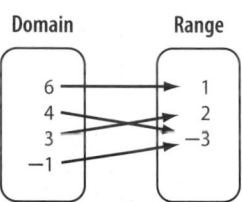

D = {6, 4, 3, -1};
R = {1, -3, 2}

13.

x	y
6	7
3	-2
8	8
-6	2
2	-6

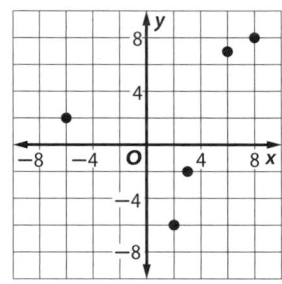

Domain   Range

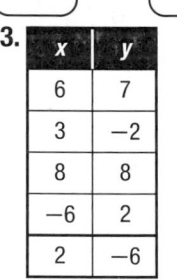

D = {-6, 2, 3, 6, 8};
R = {-6, -2, 2, 7, 8}

(15) The number of students who attend is the independent variable because it does not depend on the amount of food there will be. The amount of food is the dependent variable because it depends on the number of students who attend.

17. The bungee jumper starts at the maximum height, and then jumps. After the initial jump, the jumper bounces up and down until coming to a rest.

(19) Use the graph to determine what is happening to the value of the baseball card. The values are continually increasing.

21. (1, 5); The dog walker earns $5 for walking 1 dog. **23.** I: number of dogs walked; D: amount earned **25.** (5, 6); In the year 2005, sales were about $6 million. **27.** {(1, 2.50), (2, 4.50), (5, 10.50), (8, 16.50)}; D = {1, 2, 5, 8}; R = {2.50, 4.50, 10.50, 16.50} **29.** {(4, -1), (8, 9), (-2, -6), (7, -3)}
31. {(4, -2), (-1, 3), (-2,-1), (1, 4)}
33. Sample answer: **35.** Sample answer:

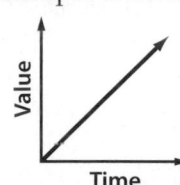   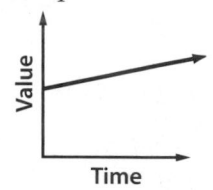

(37) a.

b	$w = 2\left(\frac{b}{3}\right)$	w
100	$w = 2\left(\frac{100}{3}\right)$	66.7
105	$w = 2\left(\frac{105}{3}\right)$	70
110	$w = 2\left(\frac{110}{3}\right)$	73.3
115	$w = 2\left(\frac{115}{3}\right)$	76.7
120	$w = 2\left(\frac{120}{3}\right)$	80
125	$w = 2\left(\frac{125}{3}\right)$	83.3
130	$w = 2\left(\frac{130}{3}\right)$	86.7

b. The independent variable is the body weight b. The dependent variable is the water weight w.

c. The domain is the set of b values. D = {100, 105, 110, 115, 120, 125, 130}. The range is the set of all w values. R = {66.7, 70, 73.3, 76.7, 80, 83.3, 86.7}

Water Weight Per Body Weight

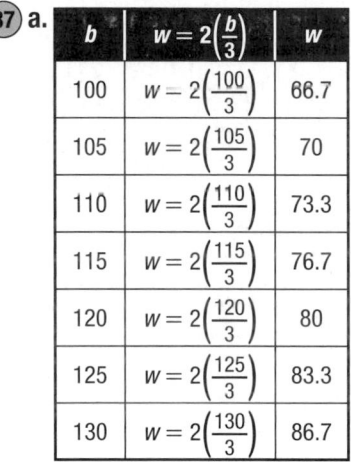

d. Graph the following ordered pairs: (66.7, 100), (70, 105), (73.3, 110), (76.7, 115), (80, 120), (83.3, 125), (86.7, 130).

Body Weight Per Water Weight

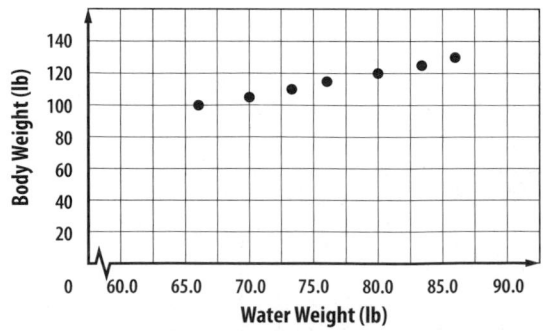

This graph shows what a person's body weight would be based on their water weight.

41. Reversing the coordinates gives (1, 0), (3, 1), (5, 2), and (7, 3).

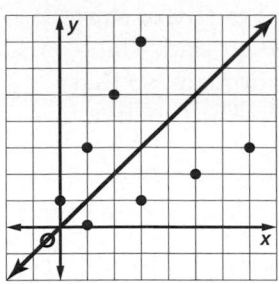 Each point in the original relation is the same distance from the line as the corresponding points of the reverse relation. The graphs are symmetric about the line $y = x$.

43. B   **45.** $(-1, -3)$ **47.** 2   **49.** 3   **51.** $\frac{1}{8}$
53. 50.27 cm   **55.** 64   **57.** 6.25   **59.** 49

Lesson 1-7

1. Yes; for each input there is exactly one output.
3. No; the domain value 2 is paired with 2 and -4.   **5.** No; when $x = 0$, $y = 1$ and $y = 6$.
7. Yes; its graph passes the vertical line test.
9a. {(0, 48,560), (1, 48,710), (2, 48,948), (3, 49,091)}
9b.

School Enrollment

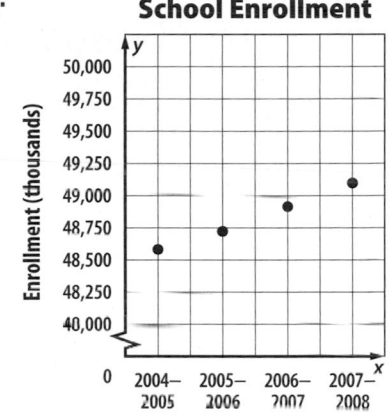

9c. The domain is the school year and the range is the enrollment.
11. $f(-3) = 6(-3) + 7$
$= -18 + 7$
$= -11$
13. $6r - 5$ **15.** $a^2 + 5$ **17.** $6q + 13$ **19.** $b^2 - 4$
21. No; the domain value 4 is paired with both 5 and 6.   **23.** Yes; for each input there is exactly one output.   **25.** Yes; the graph passes the vertical line test.   **27.** yes   **29.** yes   **31.** yes   **33.** -1 **35.** 14
37. -4 **39.** $-8y - 3$ **41.** $-2c + 7$ **43.** $-10d - 15$
45. a. Create a table using the rule given.

t	$0.8t + 72$	$f(t)$
0	0.8(0) + 72	72
10	0.8(10) + 72	80
20	0.8(20) + 72	88
30	0.8(30) + 72	96
40	0.8(40) + 72	104
50	0.8(50) + 72	112

Plot the ordered pairs on a coordinate plane.

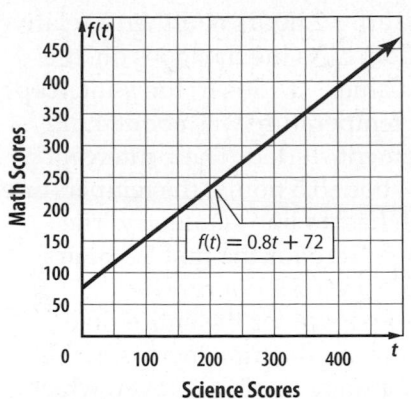

When the science score is 0, the math score is 72. For each point the science score increases, the math score increases by 0.8 points.

b. $308 = 0.8t + 72$
$236 = 0.8t$
$295 = t$

c. The domain is the set of science scores, the range is the set of math scores.

47. The graph represents a function because each x-value is paired with only one y-value.
49. Sample answer: {(−2, 3), (0, 3), (2, 5)}

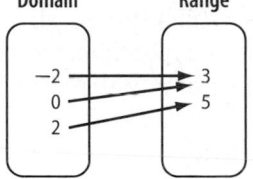

51. $f(g + 3.5) = -4.3g - 17.05$ **53.** Sample answer: $f(x) = 3x + 2$ **55.** Sample answer: You can determine whether each element of the domain is paired with exactly one element of the range. For example, if given a graph, you could use the vertical line test; if a vertical line intersects the graph more than once, then the relation that the graph represents is not a function.   **57.** J
59. her first game   **61.** $\frac{13}{2}$ **63.** 4(1.99) + 10(0.25) + 4(1.85) = 17.86, so the cost is $17.86.   **65.** sample answer: two thirds times x **67.** 38.016 cm^3
69. 288,000 mm^3 **71.** -1 **73.** 40   **75.** 65

Lesson 1-8

1. Nonlinear; the y-intercept is 0, so there is no change in the stock value at the opening bell. The x-intercepts are 0, about 3.2, and about 4.5, so there is no change in the stock value after 0 hours, after about 3.2 hours, and after about 4.5 hours, respectively, after the opening bell. The graph has no line symmetry. The stock went up in value for the first 3.2 hours, then dropped below the starting value from about 3.2 hours until 4.5 hours, and finally went up again after 4.5 hours. The stock value starts the day increasing in value for the first 2 hours, then it goes down in value from 2 hours until 4 hours, and after 4 hours it goes up in value for the remainder of the day. The stock had a relative high value after 2 hours and then a relative low value after 4 hours. As the day goes on, the stock increases in value. **3.** Linear; the *y*-intercept is about 45, so the temperature was about 45°F when the measurement started. The *x*-intercept is about 5.5, so after about 5.5 hours, the temperature was 0°F. The graph has no line symmetry. The temperature is above zero for the first 5.5 hours, and then below zero after 5.5 hours. The temperature is going down for the entire time. There are no extrema. As the time increases, the temperature will continue to drop forever, which is not very likely.

5 Since the graph is a curve and not a line, the graph is nonlinear. The graph intersects the *y*-axis at about (0, 20), so the *y*-intercept of the graph is about 20. This means that the purchase price of the vehicle was about $20,000. The graph approaches but never intersects the *x* axis, so the graph has no *x*-intercept. This means that the value of the vehicle will never reach 0. There is no line over which the graph can be folded so that both halves match exactly. Therefore the graph has no line symmetry. The graph lies entirely above the *x*-axis, therefore the function is positive for all values of *x*. This means that the value of the vehicle is always positive. When viewed from left to right, the graph goes down for all values of *x*, so the function is decreasing for all values of *x*, which means that the value of the vehicle is always decreasing. There are no relatively high or low points on the graph, so the function has no extrema. As you move right, the graph goes down, so the end behavior of the graph is that as *x* increases, *y* decreases. This means that as the number of years since the car was purchased increases, the value of the vehicle decreases.

7. Nonlinear; the *y*-intercept is about 100. This means that the Web site had 100 hits before the time began. There is no *x*-intercept. The function is positive for all values of *x*. This means that the Web site has never experienced a time of inactivity. The function is increasing for all values of *x*, with no relative maxima or minima. As *x* increases, *y* increases, which means that the upward trend in the number of hits is expected to continue.
9. Nonlinear; the *x*- and *y*-intercepts are both 0, which means that a pendulum with no length cannot complete a swing. The function is positive and increasing for all values of *x*. Also, as *x* increases, *y* increases. The function has no relative minima or maxima. This means that as the pendulum gets longer, the time it takes for it to

complete one full swing increases. **11.** Sample answer: The function has a *y*-intercept of 0 and an *x*-intercept of 0, indicating that the plant started with no height as a seed in the ground. The function is increasing over its domain, so that plant was always getting taller. The function has no relative extrema.

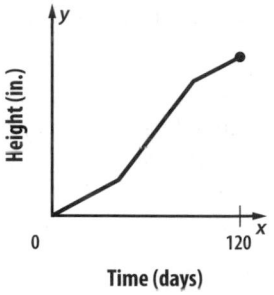
Time (days)

13. Sample answer: The function has a *y*-intercept of 27, indicating that the initial balance of the loan was $27,000. The *x*-intercept of 4 indicates that the loan was paid off after 4 years. The function is decreasing over its entire domain, indicating that the amount owed on the loan was always decreasing. The function has no relative extrema.

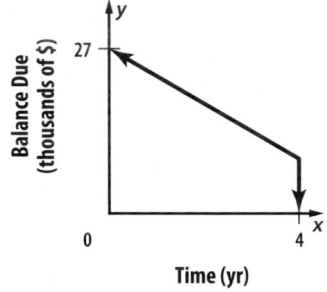
Time (yr)

15 Plot the *x*-intercepts at (−2, 0) and (2, 0) and the *y*-intercept at (0, −4). Since the graph is nonlinear and decreasing for *x* < 0, draw a smooth curve starting somewhere to the left and above (−2, 0) that moves down through (−2, 0) to (0, −4). Since the graph has a relative minimum at *x* = 0 and is increasing for *x* > 0, turn at the point (0, −4) and draw a smooth curve moving up as you move right, through (2, 0) and continuing to the upper right portion of the graph. Sample graph:

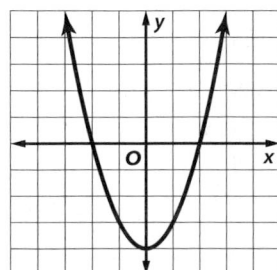

17. Sample graph:

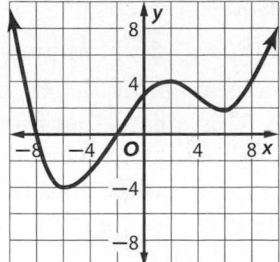

19. As x increases or decreases, y approaches 0.
21. The graph has a relative maximum at about $x = 2$ and a relative minimum at about $x = 4.5$. This means that the weekly gasoline price spiked around week 2 at a high of about \$3.50/gal and dipped around week 5 to a low of about \$1.50/gal.

Average Weekly Gasoline Price

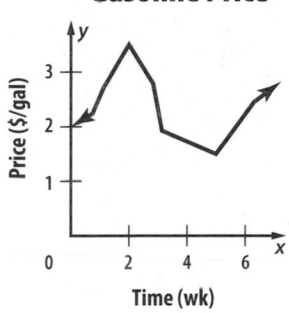

23. C   **25.** A   **27.** yes   **29.** yes   **31.** $d^2 + 3d$
33. $3(z - 2x)$ **35.** 49   **37.** 17.64

Chapter 1   Study Guide and Review

1. true   **3.** false; not in simplest form   **5.** true
7. false; multiplicative identity   **9.** the product of 3 and x squared   **11.** $x + 9$ **13.** $4x - 5$ **15.** 216
17. $2.50 + 3.25g$ **19.** 18   **21.** 2   **23.** 3   **25.** 5
27. $2.75(3) + 4.25(2)$; \$16.75

29. $[5 \div (8 - 6)]$
$= [5 \div 2]\frac{2}{5}$ Substitution
$= \frac{5}{2} \cdot \frac{2}{5}$ Substitution
$= 1$ Multiplicative Inverse

31. $2 \cdot \frac{1}{2} + 4(4 \cdot 2 - 7)$
$= 2 \cdot \frac{1}{2} + 4(8 - 7)$ Substitution
$= 2 \cdot \frac{1}{2} + 4(1)$ Substitution
$= 1 + 4(1)$ Multiplicative Inverse
$= 1 + 4$ Multiplicative Identity
$= 5$ Substitution

33. $7\frac{2}{5} + 5 + 2\frac{3}{5}$
$= 7\frac{2}{5} + 2\frac{3}{5} + 5$ Commutative (+)
$= \left(7\frac{2}{5} + 2\frac{3}{5}\right) + 5$ Associative (+)
$= 10 + 5$ Substitution
$= 15$ Substitution

35. $5.3 + 2.8 + 3.7 + 6.2$
$= 5.3 + 3.7 + 2.8 + 6.2$ Commutative (+)
$= (5.3 + 3.7) + (2.8 + 6.2)$ Associative (+)
$= 9 + 9$ Substitution
$= 18$ Substitution

37. $(2)6 + (3)6$; 30   **39.** $8(6) - 8(2)$; 32   **41.** $-2(5) - (-2)(3)$; -4 **43.** $3(x) + 3(2)$; $3x + 6$ **45.** $6(d) - 6(3)$; $6d - 18$ **47.** $(9y)(-3) - (6)(-3)$; $-27y + 18$
49. $4(3 + 5 + 4)$; 48   **51.** {7}   **53.** {9}   **55.** {5}   **57.** 9

59.

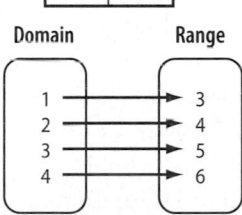

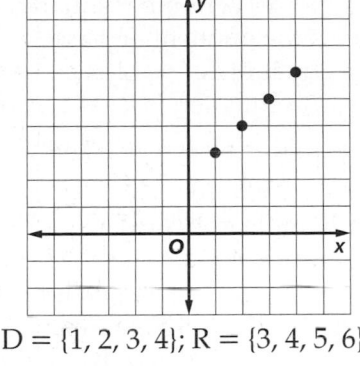

$D = \{1, 2, 3, 4\}$; $R = \{3, 4, 5, 6\}$

61.

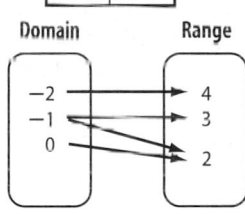

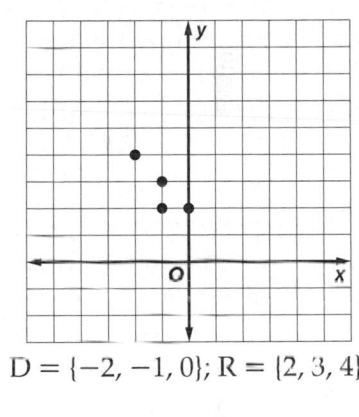

$D = \{-2, -1, 0\}$; $R = \{2, 3, 4\}$

63. $\{(-2, -2), (0, -3), (2, -2), (2, 0), (4, -1)\}$
65. function   **67.** not a function   **69.** 1   **71.** 13
73. $9p^2 - 3$ **75.** Nonlinear; the graph intersects the y-axis at about $(0, 56)$, so the y-intercept is about 56. This means that about 56,000 U.S. patents were granted in 1980. The graph has no symmetry. The graph does not intersect the x-axis, so there is no x-intercept. This means that in no year were 0 patents granted. The function is positive for all values of x, so the number of patents will always have a positive value. The function is increasing for all values of x. The y-intercept is a relative minimum, so the number of patents granted was at its lowest in 1980. As x increases, y increases. As x decreases, y decreases.

Get Ready

1. $3n - 4$ **3.** $2b - 11$ **5.** 2   **7.** 11   **9.** 11
11. $28.40   **13.** 20%   **15.** 21%

Lesson 2-1

1. $15 - 3r = 6$
(3) Words: A number squared plus 12 is the same as the quotient of p and 4.
Equation: $n^2 + 12 = p \div 4$
The equation is $n^2 + 12 = p \div 4$.
5. $8 + 3k = 5k - 3$ **7.** $\frac{25}{t} + 6 = 2t + 1$ **9.** $1900 + 30w = 2500$; 20   **11.** $P = 5s$ **13.** $4\pi r^2 = S$
15. Sample answer: The product of seven and m minus q is equal to 23.   **17.** Sample answer: Three times the sum of g and eight is the same as 4 times h minus 10.   **19.** Sample answer: A team of gymnasts competed in a regional meet. Each member of the team won 3 medals. There were a total of 45 medals won by the team. How many team members were there?   **21.** $f - 5g = 25 - f$
23. $4(14 + c) = a^2$ **25.** $3 \cdot 10 = 12f$; $2\frac{1}{2}$ flats
(27) Words: C is five ninths times the difference of F and 32.
Equation: $C = \frac{5}{9} \cdot (F - 32)$
The equation is $C = \frac{5}{9}(F - 32)$.
29. $I = prt$ **31.** Sample answer: Four times m is equal to fifty-two.   **33.** Sample answer: Fifteen less than the square of r equals the sum of t and nineteen.   **35.** Sample answer: One third minus four fifths of z is four thirds of y cubed.
37. Sample answer: Ashley has a credit card that charges 12% interest on the principal balance. If Ashley's payment was $224, what was the principal balance on the credit card?   **39.** Sample answer: Fred was teaching his friends a new card game. Each player gets 5 cards, and 7 cards are placed in the center of the table. Since there are 52 cards in a deck, find how many players can play the game.   **41.** C   **43.** D
(45) Words: the number of tent stakes + packets of drink mix + bottles of water = 17
$d = 3t$
$w = t + 2$
$$t + d + w = 17$$
$$t + 3t + (t + 2) = 17$$
$$5t + 2 = 17$$
$$5t + 2 - 2 = 17 - 2$$
$$5t = 15$$
$$t = 3$$
She brought 3 tent stakes.
47. Sample answer: My favorite television show

has 30 new episodes each year. So far eight have aired. How many new episodes are left?
49. $\ell = \dfrac{P - 2w}{2}$ **51.** C   **53.** 180 m   **55.** Nonlinear; the graph intersects the y-axis at about (0, 0.8), so the y-intercept is about 0.8. This means that the population of Phoenix was about 800,000 in 1980. The graph has no symmetry. The graph does not intersect the x-axis, so there is no x-intercept. This means that the population will always have a positive value. The function is positive for all values of x. The function is increasing for all values of x. The y-intercept is a relative minimum, so the population was at its lowest in 1980. As x increases, y increases. As x decreases, y decreases.
57a. independent: number of sides; dependent: interior angle sum   **57b.** Domain: all integers greater than or equal to 3; Range: all positive integer multiples of 180   **57c.** Discrete; sample answer: There cannot be a polygon with 3.5 sides, so the function cannot be continuous.
59. 1,000,000   **61.** 125

Lesson 2-2

1. 28   **3.** $\frac{5}{6}$ **5.** 9   **7.** -4.1 **9.** $-3\frac{1}{4}$ **11.** 16
13. $\frac{10}{9}$ or $1\frac{1}{9}$ **15.** $-\frac{4}{7}$ **17.** $22.75   **19.** 116   **21.** 22
23. -11
(25)
$$-16 - (-t) = -45$$
$$-16 + 16 - (-t) = -45 + 16$$
$$t = -29$$
Check: $-16 + (-29) = -45$
$$-45 = -45 \text{ Yes}$$
27. -32 **29.** -7 **31.** $1\frac{1}{8}$ **33.** $1\frac{2}{7}$ **35.** -708 **37.** 33
39. -2 **41.** $-1\frac{1}{9}$ **43.** $24.9 = 8.1 + t$; 16.8 hours
45. -77 **47.** $\frac{16}{3}$ **49.** -10 **51.** $-\frac{10}{7}$ or $-1\frac{3}{7}$ **53.** 18
55. 225   **57.** $\frac{2}{3} = -8n$; $n = -\frac{1}{12}$ **59.** $\frac{4}{5} = \frac{10}{16}n$; $n = \frac{32}{25}$
(61) Words: Four and four fifths times a number is one and one fifth.
Equation: $4\frac{4}{5} \cdot n = 1\frac{1}{5}$
The equation is $4\frac{4}{5}n = 1\frac{1}{5}$.
Solve:
$$4\tfrac{4}{5}n = 1\tfrac{1}{5}$$
$$\tfrac{24}{5}n = \tfrac{6}{5}$$
$$5\left(\tfrac{24}{5}n\right) = 5\left(\tfrac{6}{5}\right)$$
$$24n = 6$$
$$n = \tfrac{6}{24}$$
$$n = \tfrac{1}{4}$$
63. $555 = 139 + p$; 416   **65.** $180 = t + 154$; 26 s
67. $1.6 - m = 0.8$; $0.8 million

69 Words: 45 million fewer than 57 million is the number who have blogs.
Equation: $57 - 45$
Solve: $57 - 45 = 12$ million

71a. $350 + m = 1000$; $650   **71b.** $350 + 225 + m = 1000$; $425   **71c.** $6t = 1000$; 167   **73.** Sample answer: $12 + n = 25$; subtract 12 from each side or add -12 to each side.

75a. Sometimes; $0 + 0 = 0$ but $2 + 2 \neq 2$.
75b. Always; this is the Addition Identity Property.   **77a.** $x = \frac{12}{a}$ **77b.** $x = 15 - a$
77c. $x = a - 5$ **77d.** $x = 10a$ **79.** C   **81.** F
83. $2r + 3k = 13$ **85.** $m^2 - p^3 = 16$
87. $12(5 + 8 + 2)$; 180 hours

Lesson 2-3

1

$$3m + 4 = -11$$
$$3m + 4 - 4 = -11 - 4$$
$$3m = -15$$
$$m = -5$$

Check: $3m + 4 = -11$
$3(-5) + 4 = -11$
$-15 + 4 = -11$
$-11 = -11$ Yes

3. -55 **5.** 61   **7.** $12 - 2n = -34$; 23
9. $n + (n + 2) + (n + 4) = 75$; 23, 25, 27   **11.** -5
13. -5 **15.** 70   **17.** 27   **19.** 16   **21.** -61

23 Equation: $49.99 + 0.15m = 100$
$$49.99 - 49.99 + 0.15m = 100 - 49.99$$
$$0.15m = 50.01$$
$$m \approx 333$$

So, he can use the phone for $650 + 333$ or 983 minutes.

25. $17 = 6x - 13$; $n = 5$ **27.** $n + (n + 2) + (n + 4) = 141$; 45, 47, 49   **29.** $n + (n + 1) + (n + 2) + (n + 3) = -142$; $-37, -36, -35, -34$ **31.** $-7\frac{3}{5}$ **33.** -72
35. 108   **37.** $\frac{4}{5}$ **39.** $\frac{33}{14}$ **41.** $7\frac{1}{4}$ yr or 7 yr 3 mo

43

$$3.7q + 26.2 = 111.67$$
$$3.7q + 26.2 - 26.2 = 111.67 - 26.2$$
$$3.7q = 85.47$$
$$q = 23.1$$

45. 31.6   **47.** -3.5 **49.** 5
51a. $5x + 275 = x(6 + 15 + 9)$; 11 visits

51b.

Visits	Cost for Members	Cost for Nonmembers
3	290	90
6	305	180
9	320	270
12	335	360
15	350	450

51c.

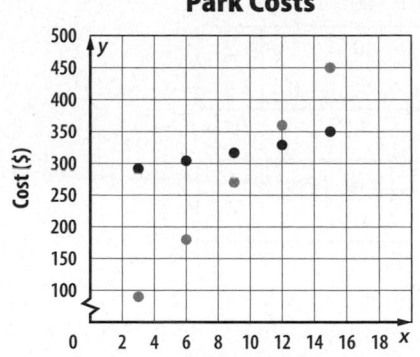

Park Costs

Both functions are linear. If a person is going to visit the park fewer than 11 times, it will be cheaper to be a nonmember.   **53.** Sample answer: A pair of designer jeans costs $60. This is $40 more than twice the cost of a T-shirt. How much is the T-shirt? The T-shirt costs $10.   **55a.** No; for there to be a solution there must be a number for which $a + 4 = a + 5$.   **55b.** Yes; for $b = 0$, $\frac{1 + b}{1 - b} = \frac{1 + 0}{1 - 0}$ or 1.
55c. No; $c - 5 = 5 - c$ when $c = 5$. However, $\frac{c - 5}{5 - c}$ is undefined for $c = 5$ since the fraction represents division by 0.   **57.** Sample answer: In order to solve the equation $4k + 20 = 236$, you would first subtract 20 from each side and then divide each side by 4.
59. 84   **61.** B   **63.** 1379   **65.** Three times a number h is increased by 7 to equal 20.   **67.** Three multiplied by a number p is the same as the difference of 8 times p and r.   **69.** The product of $\frac{1}{2}$ and v is equal to the product of $\frac{2}{3}$ and v plus 4.
71. 0; Additive Identity   **73.** 4; Additive Inverse
75. 53   **77.** 1000

Lesson 2-4

1. 4   **3.** -7 **5.** no solution   **7.** all numbers
9. A   **11.** 4
13

$$6 + 3t = 8t - 14$$
$$6 + 3t - 3t = 8t - 3t - 14$$
$$6 = 5t - 14$$
$$6 + 14 = 5t - 14 + 14$$
$$20 = 5t$$
$$4 = t$$

Check: $6 + 3t = 8t - 14$
$6 + 3(4) = 8(4) - 14$
$6 + 12 = 32 - 14$
$18 = 18$ Yes

15. $2\frac{2}{5}$ **17.** 6   **19.** -5 **21.** 1   **23.** $-4, -2$
25. no solution   **27.** all numbers   **29.** -25 **31.** 15
33. 3   **35.** -2 **37.** $-2, 0$
39 Equation: $1500 + 0.80x = 1.59x$
Solve: $1500 + 0.80x = 1.59x$
$$1500 + 0.80x - 0.80x = 1.59x - 0.80x$$
$$1500 = 0.79x$$
$$1899 \approx x$$

41a. Sample answer: $y = 2x + 4$

x	−2	−1	0	1	2
y	0	2	4	6	8

$y = -x - 2$

x	−2	−1	0	1	2
y	0	−1	−2	−3	−4

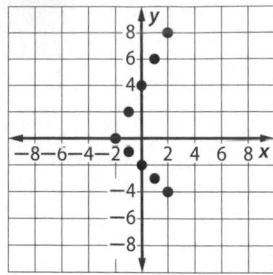

41b. −2   **41c.** Sample answer: The solution in part **b** is the x-coordinate for the point of intersection on the graph.

43. Sample answer: $2x + 1 = \frac{3}{2}x - 2$; First I chose $\frac{3}{2}$ as the fractional coefficient. Then I chose 2 for the coefficient for the variable on the other side of the equation. After substituting −6 in for x on both sides, 1 must be added to the left and 2 must be subtracted from the right to balance the equation.   **45a.** Incorrect; the 2 must be distributed over both g and 5; 6.   **45b.** correct   **45c.** Incorrect; to eliminate −6z on the right side of the equals sign, 6z must be added to each side of the equation; 1.
47. Sample answer: If the equation has variables on both sides of the equation, you must first add or subtract one of the terms from both sides of the equation so that the variable is left on only one side of the equation. Then, solving the equation uses the same steps.   **49.** J   **51.** A   **53.** $-2\frac{2}{3}$
55. −15   **57.** −15   **59.** $34   **61.** 2; Multiplicative Identity   **63.** $\frac{2}{3}$; Additive Identity   **65.** 7; Substitution   **67.** $5(m + k) = 7k$ **69.** 5
71. −24   **73.** 11

Lesson 2-5

1. 15   **3.** −4
5. {4, −2};
7. {−6, −2}
9. ∅
(11) Find the point that is the same distance from −2 and 4. This is the midpoint between −2 and 4, which is 1. The distance from −2 to 1 is 3 units. The distance from 4 to 1 is 3 units. So, an equation is $|x - 1| = 3$

(13) $|2x + z| + 2y = |2(2.1) + (−4.2)| + 2(3)$
$= |4.2 + (−4.2)| + 6$
$= |0| + 6$
$= 0 + 6$
$= 6$
15. −7.4   **17.** 8.4   **19.** −9.6   **21.** 0.4
23. {−11, −9};
25. {7, −3};
27. ∅
29. {0, 6}
31. 11% to 19%   **33.** $|x| = 4$ **35.** $|x − 1| = 4$
37. {−24, 16}
39. $\left\{3, -\frac{9}{5}\right\}$
41. no solution
43a. $|x − 52| = 2$; {50, 54}   **43b.** $|x − 54| = 1$; {52, 54}
43c. 203 and 214 seconds   **45a.** 47 to 53 mph
45b. Sample answer: The speedometer was calibrated more accurately than the speedometer in part **a.**
47. $|x| = 1\frac{1}{2}$ **49.** $\left|x - \frac{1}{4}\right| = \frac{1}{4}$ **51.** $\left|x + \frac{1}{3}\right| = 1$
(53) a. Let h be the number of people who can clearly hear voices.
   Equation: $|h − 20{,}000| = 1000$
   b. $|h − 20{,}000| = 1000$
   $h − 20{,}000 = 1000$ or $h − 20{,}000 = −1000$
   $h − 20{,}000 + 20{,}000 = 1000 + 20{,}000$ or
   $h − 20{,}000 + 20{,}000 = −1000 + 20{,}000$
   $h = 21{,}000$ or $h = 19{,}000$
   c. To find the range, find $21{,}000 − 19{,}000 = 2000$
(55) a. Each school could answer every question correctly, earning 50 points. They could also answer every question incorrectly, earning −50 points.
   b. Let m = the score on the math section. The maximum distance between m and the initial score of 160 is 50, so $|m − 160| = 50$.
   c. Every tenth value between -50 and 50 is possible. Let c = correct, n = incorrect, and u = unanswered. Here are some combinations of answers. $50 = 5c$, $40 = 4c + 1u$, $30 = 3c + 2u$, $20 = 2c + 3u$, $10 = 2c + 1n + 2u$, $0 = 2c + 2n + u$, $-10 = 1n + 4u$, $-20 = 2n + 3u$, $-30 = 3n + 2u$, $-40 = 4n + u$, $-50 = 5u$.
57. Sometimes; when $x = −1$, the value is 0.
59. Sometimes; when c is a negative value the inequality is true.   **61.** An absolute value

represents a distance from zero on a number line. A distance can never be a negative number.

63. Wesley; the absolute value of a number cannot be a negative number.

65. D   **67.** A   **69.** $\frac{1}{2}n + 16 = \frac{2}{3}n - 4$; 120   **71.** 10 in.

73. $\frac{2}{5}n = -24$; -60 **75.** $12 = \frac{1}{5}n$; 60

Lesson 2-6

1. no

3 $\frac{1.4}{2.1}$ is written in simplified form. $\frac{2.8}{4.4} = \frac{1.4}{2.2}$. Since the fractions are not equal, the ratios are not equivalent.

5. 5   **7.** about 253.3 min or 4 hours 13.3 min

9. yes   **11.** no   **13.** yes   **15.** 40   **17.** 29.25   **19.** 9.8

21. 1.32   **23.** 0.84

25 $\frac{t}{0.3} = \frac{1.7}{0.9}$
$0.9t = 1.7(0.3)$
$0.9t = 0.51$
$t \approx 0.57$

27. 6   **29.** 11   **31.** 156 mi   **33.** about $262.59   **35.** 18

37. 0.8   **39.** 11   **41.** 130 students

43 **a.** Write each ratio.

for 2003, $\frac{\text{indoor theaters}}{\text{total theaters}} = \frac{35{,}361}{35{,}995}$

for 2004, $\frac{\text{indoor theaters}}{\text{total theaters}} = \frac{36{,}012}{36{,}653}$

for 2005, $\frac{\text{indoor theaters}}{\text{total theaters}} = \frac{37{,}092}{37{,}740}$

for 2006, $\frac{\text{indoor theaters}}{\text{total theaters}} = \frac{37{,}776}{38{,}425}$

for 2007, $\frac{\text{indoor theaters}}{\text{total theaters}} = \frac{38{,}159}{38{,}794}$

for 2008, $\frac{\text{indoor theaters}}{\text{total theaters}} = \frac{38{,}201}{38{,}834}$

for 2009, $\frac{\text{indoor theaters}}{\text{total theaters}} = \frac{38{,}605}{39{,}233}$

b. None of the ratios form a proportion.

45a.
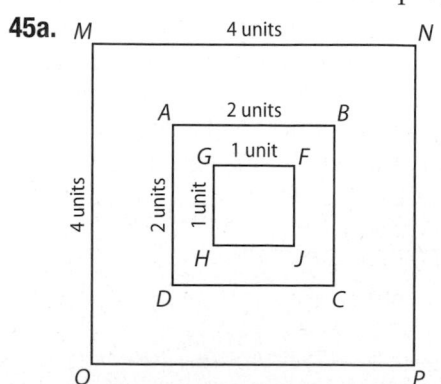

45b.

ABCD		MNPQ		FGHJ	
Side length	2	Side length	4	Side length	1
Perimeter	8	Perimeter	16	Perimeter	4

45c. If the length of a side is increased by a factor,

---

the perimeter is also increased by that factor. If the length of the sides are decreased by a factor, the perimeter is also decreased by the same factor.

47. Ratios and rates each compare two numbers by using division. However, rates compare two measurements that involve different units of measure.   **49.** If the tank is about $\frac{9}{16}$ full, he has about $\frac{9}{16} \times 10$ or $5\frac{5}{8}$ gal of gas left. At 32 miles per gallon, he will be able to travel $32 \times 5\frac{5}{8}$ or 180 miles. Since Atlanta is 200 miles away, he will run out of gas about 20 miles before reaching the city if he doesn't stop to get gas.   **51.** C   **53.** G   **55.** ∅

57. {10, −7}   **59.** 30 years   **61.** −7   **63.** −48   **65.** 13

67. 5.5   **69.** 3.5

Lesson 2-7

1 It is an increase.
$125 - 78 = 47$
$47 \div 78 \approx 0.60.$
The percent of increase is about 60%.

3. inc.; 33%   **5.** 146 mi   **7.** $38.42   **9.** $53.07

11. $17.21   **13.** $22.10

15 It is a decrease.
$16 - 10 = 6$
$6 \div 16 \approx 0.38.$
The percent of decrease is about 38%.

17. dec.; 77%   **19.** inc.; 127%   **21.** inc.; 90%

23. $12,400   **25.** $47.48   **27.** $27.31   **29.** $10.66

31. $76.49   **33.** $16.42   **35.** $11.99   **37.** $48.04

39. about 20.7% increase   **41a.** First girl's dress - $15; Second girl's dress = $25.50   **41b.** the second girl by $0.50

43 Find the percent of increase or decrease for each grocery item. Milk had the biggest increase with a 38.7% increase.

45. Sample answer: A CD is on sale for $9.99. If tax is 6.5%, what will the CD cost?   **47.** Xavier; Maddie divided by the new amount instead of the original amount.   **49.** Sample answer: Retail stores use percents of decrease when the prices of items are discounted in a sale; salary increases are usually given as a percent of increase. To find the percent of change, subtract the original from the new amount. Then write a proportion, comparing the change to the original amount. The answer should be written as a percent.   **51.** $72   **53.** C

55. 12   **57.** 4   **59.** 5.6   **61.** 3   **63.** −6   **65.** −7

67. $0.99x + 1.29y$ **69.** Sample answer: Six more than twice a number f equals nineteen.   **71.** Sample answer: The product of three and a number a when added to 5 is equal to the difference of 27 and two times a.   **73.** Sample answer: The fourth power of a number d increased by sixty-four is three times that number d to the third power plus seventy-seven.

(1)
$$5a + c = -8a$$
$$5a - 5a + c = -8a - 5a$$
$$c = -13a$$
$$-\frac{c}{13} = a$$

3. $k = -7n - m$ **5a.** $h = \dfrac{V}{\pi r^2}$ **5b.** 8 in.
7. about 0.43875 ft

(9)
$$x = b - cd$$
$$x - b = -cd$$
$$\frac{x - b}{-d} = c$$

11. $m = \dfrac{-n + p}{10}$ **13.** $v = \dfrac{9}{5}(z - w)$ **15.** $f = \dfrac{6g - 10}{d}$

17a. $v_f = at + v_i$ **17b.** 10 ft/s^2 **19.** 49.8 L

21. $t = \dfrac{w - 11v}{31}$ **23.** $c = \dfrac{-13 + f}{10 - d}$ **25.** 1.0 mm/s

27. 3.9 km/s **29.** $t - 7 = r + 6; t = r + 13$

31. $\dfrac{9}{10}g = 7 + \dfrac{2}{3}k; k = \dfrac{3}{2}\left[\dfrac{9}{10}g - 7\right]$

(33) $S = 2w(\ell + h) + 2lh$
$$214 = 2(6)(7 + h) + 2(7)h$$
$$214 = 12(7 + h) + 14h$$
$$214 = 84 + 12h + 14h$$
$$130 = 26h$$
$$5 = h$$
So, 5 inches.

35. about 364 in^3 **37.** Sandrea; she performed each step correctly; Fernando omitted the negative sign from $-5b$. **39a.** $x = \dfrac{y - 1}{yn - 1}$ **39b.** $y = -\dfrac{1}{3}x$ **41.** D

43. 15 **45.** \$101.76 **47.** \$46.33 **49.** \$56.95
51. 1.67 **53.** 5.14 **55.** $50(7.50) + 90(5.00)$; \$825
57. -0.5 **59.** -1.5 **61.** 2

(1)

	Weight	Price	Total Price
Soup	10	0.15	0.15(10)
Salad	x	0.20	0.20x
Total			3.30

$$0.15(10) + 0.20x = 3.30$$
$$1.50 + 0.20x = 3.30$$
$$0.20x = 1.80$$
$$x = 9$$
She bought 9 oz of salad.

3. 10 mph **5.** 2 hours

(7) a.

	Amount	Percent	Total
Metallic Balloons	b	\$2.00	2.00b
Bunches of helium Balloons	$b - 36$	\$3.50	3.50($b - 36$)

b. $2.00b + 3.50(b - 36) = 281$

c. $2.00b + 3.50(b - 36) = 281$
$$2b + 3.5b - 126 = 281$$
$$5.5b = 281 + 126$$
$$5.5b = 407$$
$$b = 74$$
There were 74 metallic balloons sold.

d. $b - 36 = 74 - 36$
$$= 38$$
There were 38 bunches of balloons sold.

9. about 16.67 gal **11.** about 22.2 mph **13.** $1\dfrac{1}{7}$ hours or 1 h 8 min 34 s **15.** 10 gal **17.** 10.89 mph

(19) a. $D = rt$
$$= 65(6)$$
$$= 390 \quad \text{He could drive 390 miles.}$$

b. $D = rt$
$$\frac{D}{r} = t$$
$$\frac{625}{65} = t$$
$$9.62 \approx t \quad \text{It will take about 9.62 hours.}$$

21. 33 mi **23.** Sample answer: For a 50% solution being added to a 100% solution to produce a 75% resulting solution, the quantity of each must be the same. **25.** Sample answer: How many grams of salt must be added to 36 grams of a 15% salt solution to obtain a 50% salt solution? **27.** B
29. C **31.** $\dfrac{-5 + b}{2b}$ **33.** $\dfrac{A}{2\pi r} - r$ **35.** Sample answer: The quotient of n and -6 is the same as the sum of two times n and one. **37.** Sample answer: The sum of three and twice x squared is equal to twenty-one. **39.** (4, 25); Sample answer: If four cars are washed, \$25 is earned. **41.** \$583.50
43. -2 **45.** -7 **47.** 24

Chapter 2   Study Guide and Review

1. false, variable **3.** true **5.** false, ratio **7.** false, decrease **9.** $5x + 3 = 15$ **11.** $\dfrac{1}{2}m^3 = 4m - 9$
13. h squared minus five times h plus six is equal to zero. **15.** width: 8 ft, length: 19 ft **17.** -5 **19.** 2.1
21. 6 **23.** 14 **25.** 6 **27.** -11 **29.** 17 **31.** 2
33. 38.1 **35.** 19, 21, 23 **37.** 3 **39.** -2 **41.** 2
43. -8 **45.** 21 **47.** 28 **49.** -144 **51.** $\{-5, 17\}$

53. $\{-27, 63\}$

55. yes **57.** 20 **59.** 12 **61.** increase, 25%
63. decrease, 17% **65.** \$52.19 **67.** \$55.20
69. \$33.75 **71.** $y = \dfrac{9 - 3x}{2}$ **73.** $m = \dfrac{15 - 9n}{-5}$
75. $y = \dfrac{5}{2}(m - n)$ **77.** $h = \dfrac{2A}{a + b}$ **79.** 52 mph

Get Ready

1.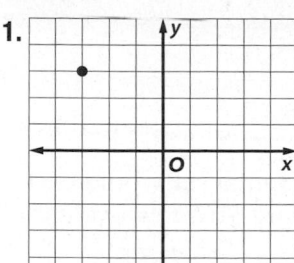

3.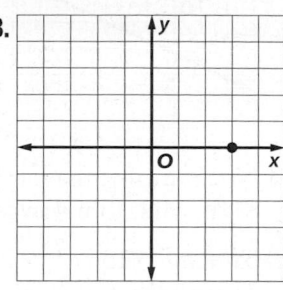

5.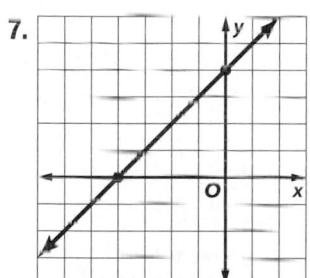

7. $(3, -1)$ **9.** $(3, 2)$
11. $(5, 0)$ **13.** $y = -3x + 1$
15. $y = \frac{5}{2}x - 6$
17. $y = -10x + 6$
19. $\frac{1}{4}$ **21.** 0 **23.** about
13.5 million

Lesson 3-1

1. yes; $x - y = -5$ **3.** yes; $y = 1$ **5.** $25, -4$; The
x-intercept 25 means that after 25 minutes, the
temperature is $0°F$. The y-intercept -4 means that
at time 0, the temperature is $-4°F$.

7.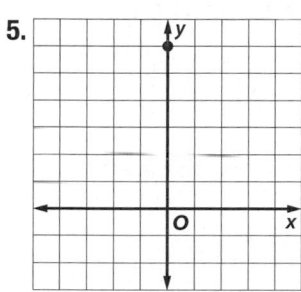

9.

x	$y = 2 - \frac{x}{2}$	y	(x, y)
-4	$y = 2 - \frac{(-4)}{2}$	4	$(-4, 4)$
-2	$y = 2 - \frac{(-2)}{2}$	3	$(-2, 3)$
0	$y = 2 - \frac{0}{2}$	2	$(0, 2)$
2	$y = 2 - \frac{2}{2}$	1	$(2, 1)$
4	$y = 2 - \frac{4}{2}$	0	$(4, 0)$

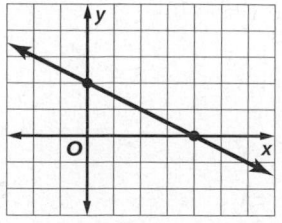

11.

x	$y = 3$	y	(x, y)
-2	$y = 3$	3	$(-2, 3)$
-1	$y = 3$	3	$(-1, 3)$
0	$y = 3$	3	$(0, 3)$
1	$y = 3$	3	$(1, 3)$
2	$y = 3$	3	$(2, 3)$

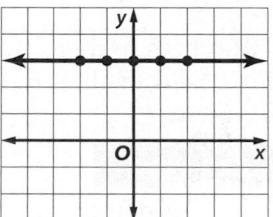

13 $5x + y^2 = 25$
Since the y term is squared, this equation
cannot be written in the form $Ax + By = C$, so it
is not a linear equation.

15. no   **17.** yes;
$4x + y = 0$ **19.** $3, 4$
21. $6, 20$; The x-intercept
represents the number
of seconds that it takes
the eagle to land. The
y-intercept represents the
initial height of the eagle.

23.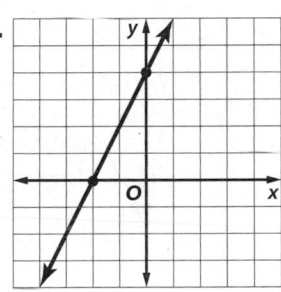

25.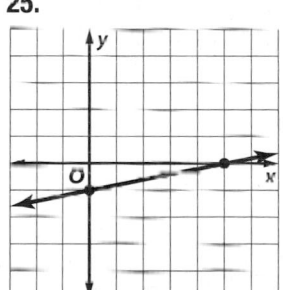

27.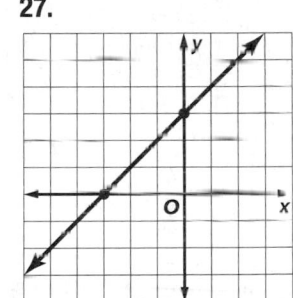

29.

x	y
-2	0
-2	1
-2	2

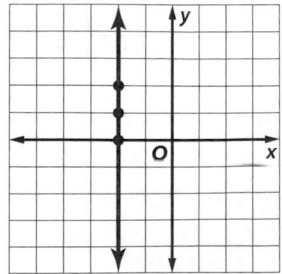

31.

x	y
-1	8
0	0
1	-8

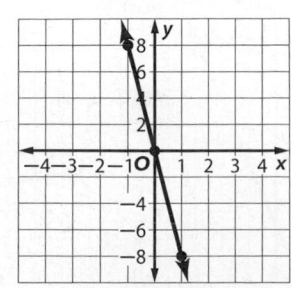

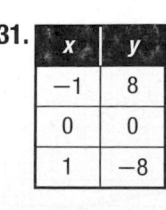

33.

x	y
0	8
1	7
2	6

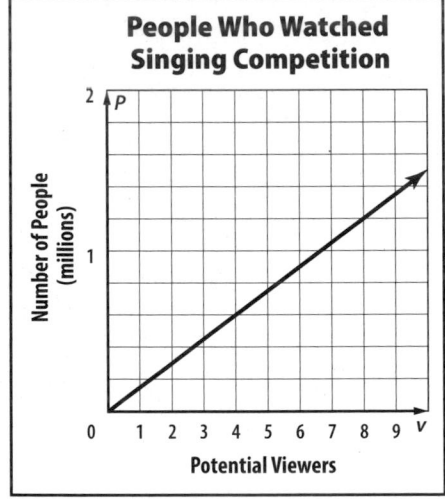

35 a. The domain is all real numbers so there are infinitely many solutions. Select values from the domain and make a table.

v	p = 0.15v	p	(v, p)
0	p = 0.15(0)	0	(0, 0)
2	p = 0.15(2)	0.3	(2, 0.3)
4	p = 0.15(4)	0.6	(4, 0.6)
6	p = 0.15(6)	0.9	(6, 0.9)
8	p = 0.15(8)	1.2	(8, 1.2)
10	p = 0.15(10)	1.5	(10, 1.5)

b. Create ordered pairs and graph them.

People Who Watched Singing Competition

(Number of People (millions) vs Potential Viewers)

c. Using the graph, when there are 14 million potential viewers, there will about 2.1 million people who watch.

d. A negative does not make sense because you cannot have a negative number of viewers.

37. yes; $3x - 4y = 60$ **39.** yes; $3a = 2$
41. yes; $9m - 8n = -60$

43. 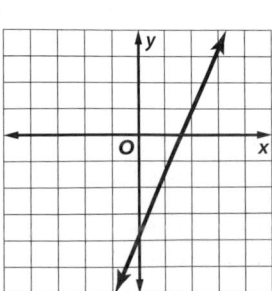   **45.**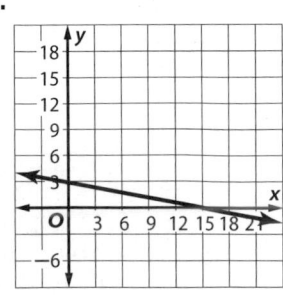

47.

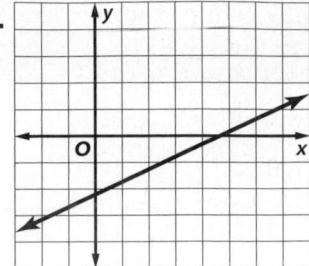

49. No; Sample answer: The rental car would cost $176. Mrs. Johnson has only $160 to spend.

51 $5x + 3y = 15$
To find the x-intercept, let $y = 0$.
$$5x + 3y = 15$$
$$5x + 3(0) = 15$$
$$5x = 15$$
$$x = 3$$
The x-intercept is 3. This means the graph intersects the x-axis at (3, 0).
To find the y-intercept, let $x = 0$.
$$5x + 3y = 15$$
$$5(0) + 3y = 15$$
$$3y = 15$$
$$y = 5$$
The y-intercept is 5. This means the graph crosses the y-axis at (0, 5).

53. $2\frac{1}{2}; -1\frac{2}{3}$ **55.** $12; -3$

57a.

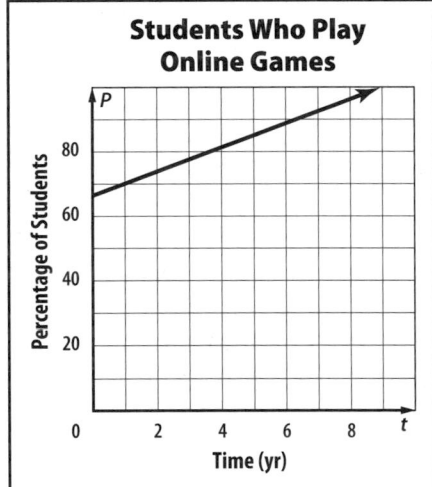

Students Who Play Online Games

(Percentage of Students vs Time (yr))

57b. 96%

59.

Perimeter of a Square	
Side Length	Perimeter
1	4
2	8
3	12
4	16

Sample answer: Yes; we used the formula $P = 4s$, which is linear.

Area of a Square	
Side Length	Area
1	1
2	4
3	9
4	16

Sample answer: No; we used the formula $A = s^2$, which is not linear.

Volume of a Cube	
Side Length	Volume
1	1
2	8
3	27
4	64

Sample answer: No; we used the formula $V = s^3$, which is not linear.

61. Sample answer: $y = 8$; horizontal line
63. Sample answer: $x - y = 0$; line through $(0, 0)$
65. D **67.** \$30 **69.** 270 rolls of solid wrap, 210 rolls of print wrap **71.** $g = \dfrac{5 + m}{2 + h}$ **73.** $z = \dfrac{c - b}{2}$
75. $-\dfrac{23}{14}$ **77.** -56

Lesson 3-2

1. 3 **3.** $\dfrac{1}{2}$ **5.** no solution **7.** no solution
9. Tyrone must deliver 40 newspapers for the papers in his bag to weigh 0 pounds. **11.** -3
13. no solution **15.** $-\dfrac{10}{7}$ or $-1\dfrac{3}{7}$

17
$$5x - 5 = 5x + 2$$
$$5x - 5 + 5 = 5x + 2 + 5$$
$$5x = 5x + 7$$
$$5x - 5x = 5x - 5x + 7$$
$$0 = 7$$
There are no solutions.

19. no solution **21.** no solution **23.** 100; She can download a total of 100 songs before the gift card is completely used. **25.** -8 **27.** $\dfrac{10}{3}$ or $3\dfrac{1}{3}$
29. $-\dfrac{34}{13}$ or $-2\dfrac{8}{13}$ **31.** $\dfrac{17}{25}$ **33.** $\dfrac{15}{8}$ or $1\dfrac{7}{8}$ **35.** 3
37. 4:00 P.M.

39. -3

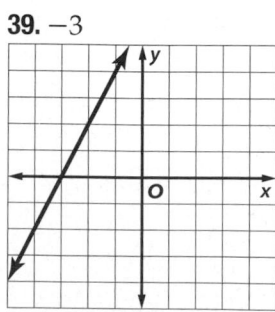

41. -2

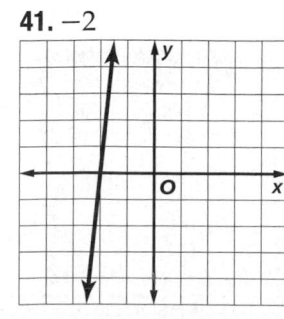

43. $\dfrac{9}{8}$ or $1\dfrac{1}{8}$

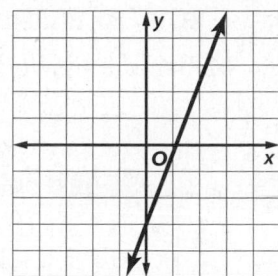

45 **a.** Sample answers given:

Number of Songs Downloaded	Total Cost (\$)	Total Cost / Number Songs Downloaded
2	4	$\dfrac{4}{2} = 2$
4	8	$\dfrac{8}{4} = 2$
6	12	$\dfrac{12}{6} = 2$
8	16	$\dfrac{16}{8} = 2$
10	20	$\dfrac{20}{10} = 2$

b. As the number of songs downloaded increases by 2, the cost increases by 4.
c. The value of the total cost divided by the number of songs downloaded represents the cost per song. It costs \$2 per song.
47. 3 **49.** Sample answer: $3 + 4x = 0$; $y = 3 + 4x$ or $f(x) = 3 + 4x$ **51.** A **53.** B **55.** $-5, 10$ **57.** $7, -2$
59. $3m + 2n = \dfrac{4}{p}$ **61.** $\dfrac{5}{2}$ **63.** $-\dfrac{1}{2}$ **65.** $\dfrac{2}{3}$ **67.** 11

Lesson 3-3

1. $\dfrac{4}{3}$ **3a.** 1.035; There was an average increase in ticket price of \$1.035 per year. **3b.** Sample answer: 1998–2000; A steeper segment means a greater rate of change. **3c.** Sample answer: 1998–2000; Ticket prices show a sharp increase. **5.** No; the rate of change is not constant. **7.** -1
9. $\dfrac{7}{9}$ **11.** 0 **13.** -8
15 rate of change $= \dfrac{\text{change in } y}{\text{change in } x}$
$$= \dfrac{9 - 15}{2 - 1}$$
$$= \dfrac{-6}{1}$$
$$= -6$$
17. $\dfrac{1}{2}$ **19a.** Sample answer: $P = -1221t + 19{,}820$
19b. The car value depreciates by \$1221 each year.
19c. \$11,273 **21.** No; the y-values do not increase at a constant rate. **23.** Yes; both the x-values and the y-values increase at a constant rate.
25 $m = \dfrac{y_2 - y_1}{x_2 - x_1}$
$$= \dfrac{1 - (-2)}{1 - 8}$$
$$= \dfrac{1 + 2}{1 - 8}$$
$$= -\dfrac{3}{7}$$

27. undefined **29.** $\frac{5}{17}$ **31.** 0 **33.** undefined

35. $\frac{10}{3}$ **37.** $\frac{3}{4}$ **39.** 6 **41.** Sample answer: about -1

43. $\frac{15}{4}$ **45.** $-\frac{2}{3}$

47 **a.** Plot the ordered pairs on a coordinate plane. Connect the points with a line.

Michael Redd's PPG

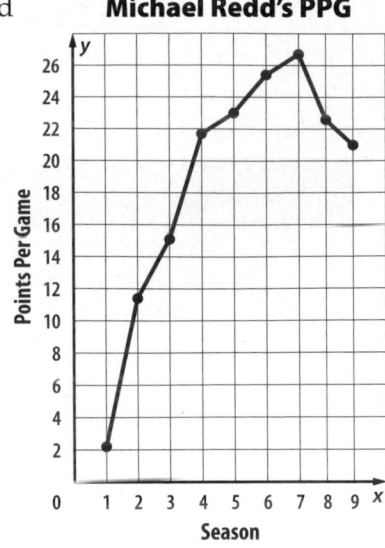

b. The steepest line is between season 1 and season 2, so that is when the PPG increased the most.

c. The rate of change was much more dramatic or steeper in the first four seasons, the rate of change leveled off the next three seasons and is negative and steeper the last two seasons.

49. The rate of change is $2\frac{1}{4}$ inches of growth per week. **51.** Sample answer: Slope can be used to describe a rate of change. Rate of change is a ratio that describes how much one quantity changes with respect to a change in another quantity. The slope of a line is also a ratio and it is the ratio of the change in the y-coordinates to the change in the x-coordinates. **53.** A **55.** \$4 **57.** -6 **59.** 4

61. $-1, 2$ **63.** 12 **65.** $\frac{5}{16}$ **67.** 5

Lesson 3-4

1. $-\frac{4}{5}; -\frac{4}{5}$

3.

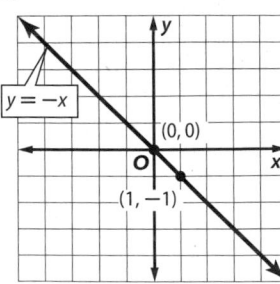

5.

7. $y = \frac{5}{4}x$; 40

9a. $y = \frac{12}{5}x$;

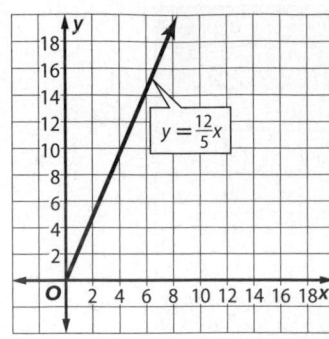

9b. 40

11 The constant of variation is -5.

$$m = \frac{5 - 0}{-1 - 0}$$

$$= \frac{5}{-1}$$

$$= -5$$

13. $-\frac{1}{5}; -\frac{1}{5}$ **15.** $-12; -12$

17. 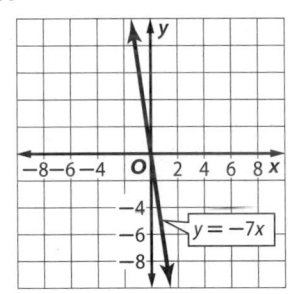 **19.**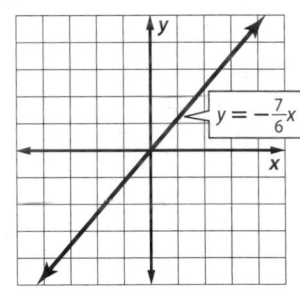

21. 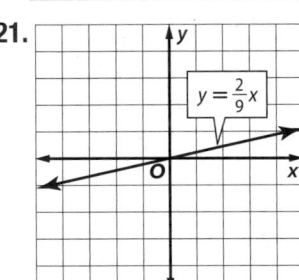 **23.**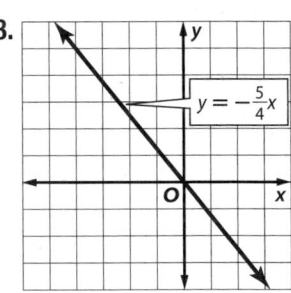

25 $y = kx$

$22 = k(8)$

$\frac{11}{4} = k$

Therefore, the direct variation equation is

$y = \frac{11}{4}x$.

$y = \frac{11}{4}x$

$y = \frac{11}{4}(-16)$ Substitute -16 for x.

$y = 11(-4)$

$y = -44$

Therefore, $y = -44$ when $x = -16$.

27. $y = 14x$; $1\frac{1}{7}$ **29a.** $y = 1800x$ **29b.** 7 yr 6 mo

31. $y = 20x$; $\frac{5}{4}$ **33.** $y = -3.75x$; -30

35. dark green **37.** lime green

39 Since the songs are \$0.99 each, the equation is $T = 0.99d$. The graph of the equation runs through the origin with a slope of 0.99.

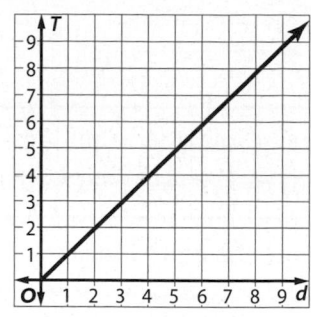

41a.

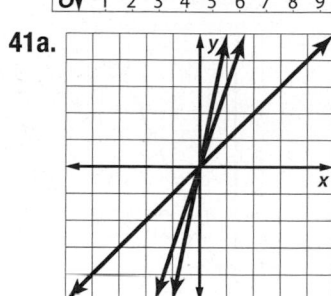

41b. Sample answer: The constant of variation, slope, and rate of change of a graph all have the same value. **41c.** Sample answer: Find the absolute value of k in each equation. The one with the greater value of $|k|$ has the steeper graph.
43. $C = 9.95n$ **45.** $z = \frac{1}{9}x$; It is the only equation that is a direct variation. **47.** Sample answer: $y = 0.50x$ represents the cost of x apples.

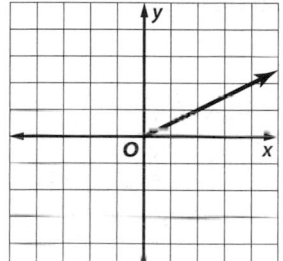

The rate of change, 0.50, is the cost per apple.

49. Neither; the slope is constant, but it is k.
51. A **53.** D **55.** 6.5; There was an average increase of 6.5 channels per year.
57. -7 **59.** -4

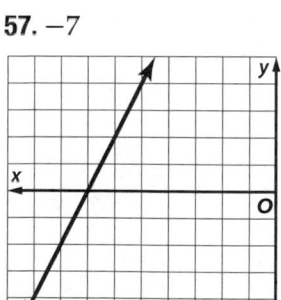   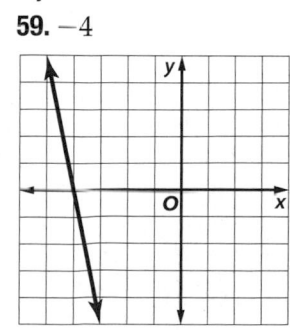

61. 12

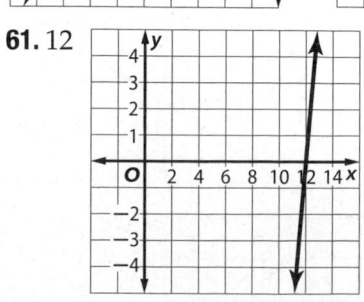

63. 12 **65.** -2
67. -28 **69.** -12
71. -6 **73.** -12

1. No; there is no common difference. **3.** 0, -3, -6
5. $a_n = 17 - 2n$

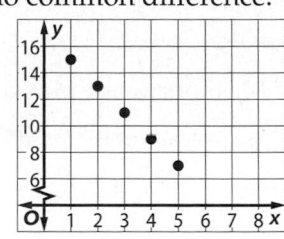

7. $a(n) = 55n + 525$

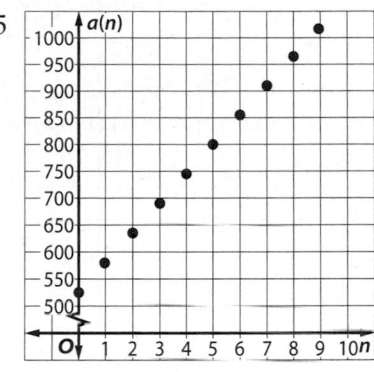

9. No; there is no common difference. **11.** Yes; the common difference is 2.6. **13.** 30, 36, 42

(15) Step 1: Find the common difference by subtracting successive terms.

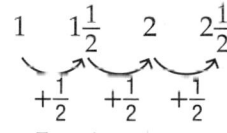

$-\frac{1}{2}$, 0, $\frac{1}{2}$, 1, … The common difference is $\frac{1}{2}$.
$+\frac{1}{2}$ $+\frac{1}{2}$ $+\frac{1}{2}$

Step 2: Add $\frac{1}{2}$ to the last term of the sequence to get the next term.

1   $1\frac{1}{2}$ 2 $2\frac{1}{2}$ The next three terms are $1\frac{1}{2}$, 2, $2\frac{1}{2}$.
$+\frac{1}{2}$ $+\frac{1}{2}$ $+\frac{1}{2}$

17. $3\frac{7}{12}$, $4\frac{1}{3}$, $5\frac{1}{12}$

19. $a_n = 5n - 7$ **21.** $a_n = 0.25n - 1$

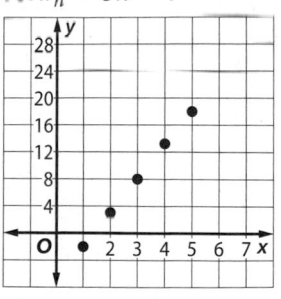   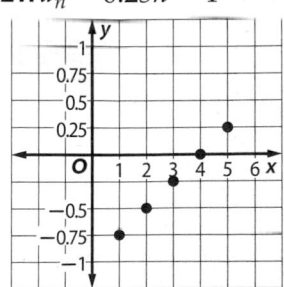

23a. $f(n) = 0.80n$

23b.

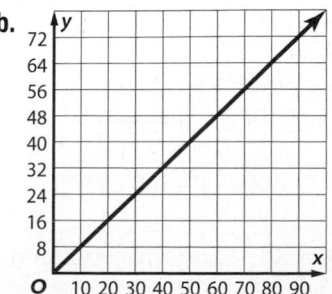

$D = \{10, 20, 30, 40, …\}$

25 The ordered pairs are (10, 7.50), (15, 8.75), (20, 10), (25, 11.25). So, the rate of change is $\frac{8.75 - 7.50}{15 - 10} = \frac{1.25}{5} = 0.25$.

The cost is \$0.25 per word plus a flat fee.
$f(n) = 0.25n + b$
$7.50 = 0.25(10) + b$
$7.50 = 2.50 + b$
$5 = b$

So, the equation is $f(n) = 0.25n + 5$.

27. 77 **29.** 25,646 **31a.** $A_n = 2.5 + 0.5n$
31b. week 15 **31c.** Sample answer: No; eventually the number of miles ran per day will become unrealistic. **33.** -1 **35a.** Yes; there is a common difference.; x; $5x + 1$, $6x + 1$, $7x + 1$.
35b. No; there is no common difference.
37. 8 yr **39.** H **41.** 3, 3 **43.** $-\frac{3}{7}$ **45.** 3 **47.** 2
49. Sample answer: $453{,}000 - d = 369{,}000$; 84,000
51–55.

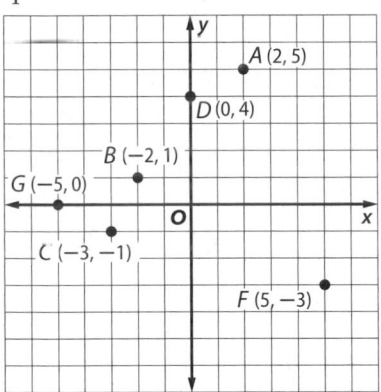

Lesson 3-6

1a.

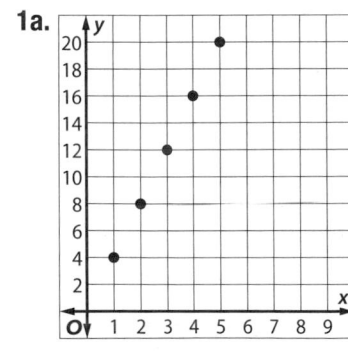

1b. $y = 4x$ **1c.** The perimeter is 4 times the length of the side.
3. $f(x) = -x + 3$

5 Select points from the graph and place them in a table.

x	0	1	2	3
y	0	2	4	6

The difference between the x-values is 1, while the difference between the y-values is 2. This suggests that $y = 2x$.
This works for each ordered pair, so the equation is $f(x) = 2x$.

7. $f(x) = 3x - 2$ **9.** $f(n) = 3n - 3$; nonproportional; the function does not describe a direct variation.

11. $y = 2.25x + 2.50$

13 **a.** Sample answer:

Number of T-shirts ordered	5	10	15	20	25
Cost (\$)	13	23	33	43	53

b. In functional notation, the equation is $C(t) = 2t + 3$.

c. The equation goes through (0, 3) and has a slope of 2.

d. The relationship is nonproportional because $\frac{13}{5} \neq \frac{23}{10} \neq \frac{33}{15} \neq \frac{43}{20} \neq \frac{53}{25}$.

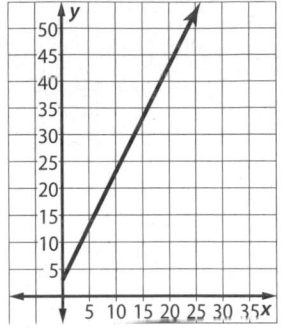

15. Sample answer: 4, 7, 10, 13; add a common difference of 3; $a_n = 3n + 1$. **17.** $f(n) = 3n + 2$ is the related function for the arithmetic sequence 5, 8, 11, 14, …, but it is not proportional. The line through (1, 5) and (2, 8) does not pass through (0, 0). **19.** D

21. H **23.** 13, 53, 63 **25.** $\frac{5}{4}, \frac{11}{8}, \frac{3}{2}$ **27.** $y = 7x$; -12 **29a.** $V = \frac{1}{3}\pi r^2 h$ **29b.** about 3142 cm³
31. $y = 3x - 5$

33.

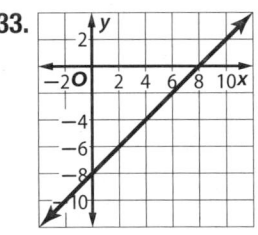

35.

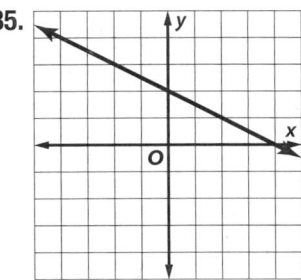

Chapter 3 Study Guide and Review

1. true **3.** false; common difference **5.** true
7. false; The slope of $y = 5$ is 0. **9.** true
11. $-8, 6$

13.

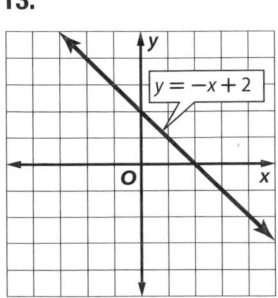

15.

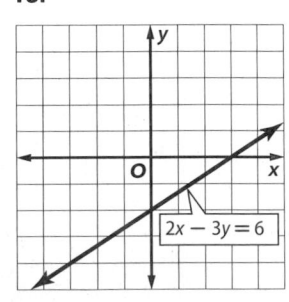

17a.

t	0	1	2	3	4	5
d	0	1.6	3.2	4.8	6.4	8

Speed of Sound

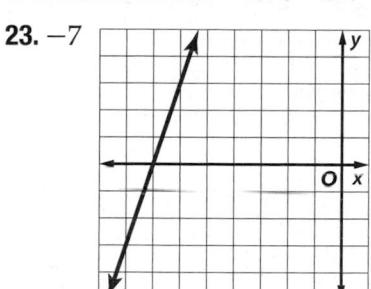

17b. about 11 km   **19.** 6   **21.** $-\dfrac{1}{2}$

23. -7

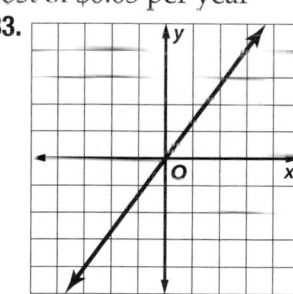

25. 9

27. 3   **29.** $-\dfrac{1}{2}$ **31.** -0.05; an average decrease in cost of $0.05 per year

33.

35. $y = 7.5x$; $y = 60$
37. $y = -x$; $y = -7$
39. 26, 31, 36
41. $a_n = 5n + 1$
43. $a_n = 4820n$; 15 s

45a.

45b. $f(x) = 1.25x$
45c. $7.50

CHAPTER 4
Equations of Linear Functions

Get Ready

1. 13   **3.** 14   **5.** $282.50   **7.** $x = 3 + 2y$

---

9. $x = \dfrac{3}{4}y + 3$ **11.** $(4, 2)$ **13.** $(2, -4)$ **15.** $(-3, -3)$

Lesson 4-1

(1) $y = mx + b$
$y = 2x + 4$
To graph, plot the y-intercept $(0, 4)$. Then use the slope of 2 to move up 2 and right 1 from the y-intercept to find the next point. Connect the points with a straight line.

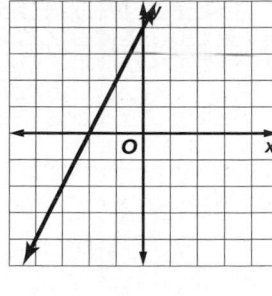

3. $y = \dfrac{3}{4}x - 1$ **5.**

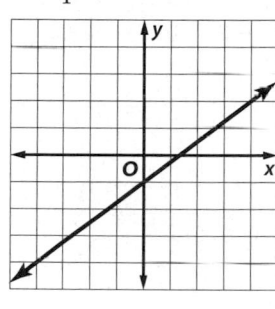

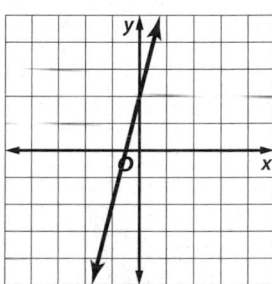

7.

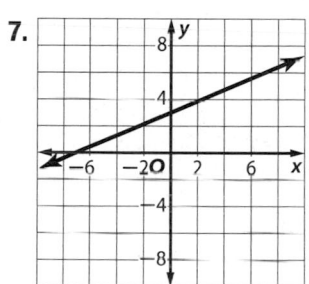

9.

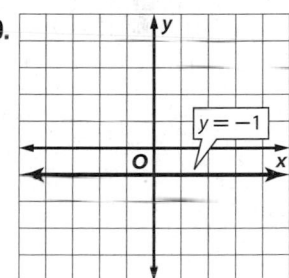

11. $y = \dfrac{2}{3}x + 2$ **13.** not possible   **15a.** $S = 10w + 75$

15b.

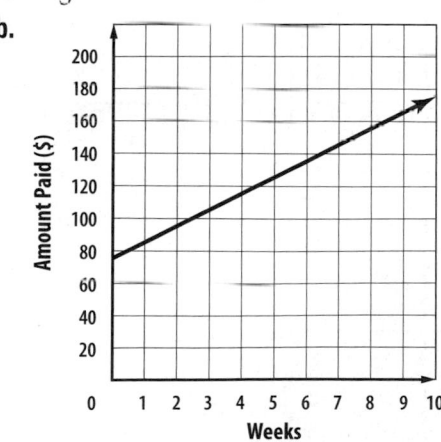

15c. $155

(17) $y = mx + b$
$y = 5x + 8$
To graph, plot the y-intercept $(0, 8)$. Then use the slope of 5 to move up 5 and right 1 from the y-intercept to find the next point. Connect the points with a straight line.

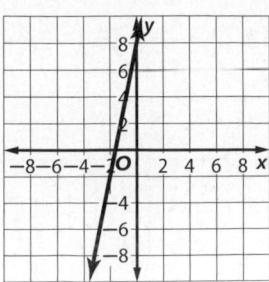

19. $y = -4x + 6$ **21.** $y = 3x - 4$

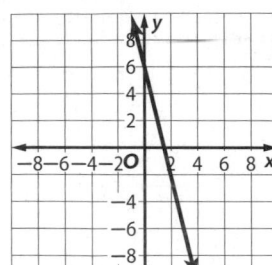 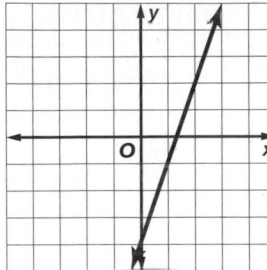

23. 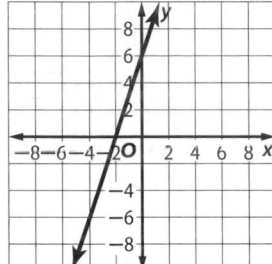   **25.**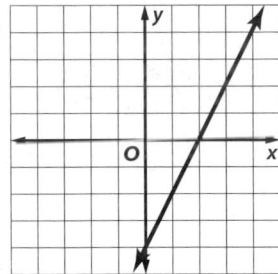

27. 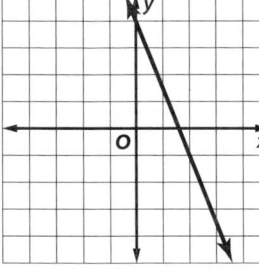   **29.**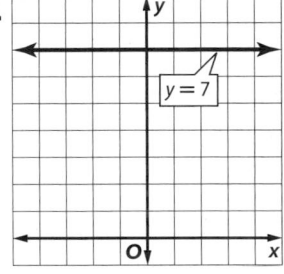

31. 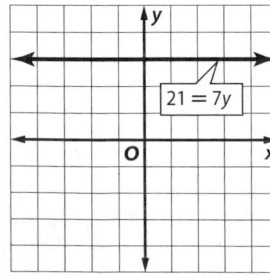   **33.** $y = -\dfrac{3}{5}x + 4$
35. $y = \dfrac{1}{2}x - 3$

37 a. Words: the population is 1267 plus 123 per year.
Equation: $P = 1267 + 123t$

b. Graph the equation by plotting the y-intercept of
(0, 1267). Then use the slope of 123 to move 123
up and 1 right.

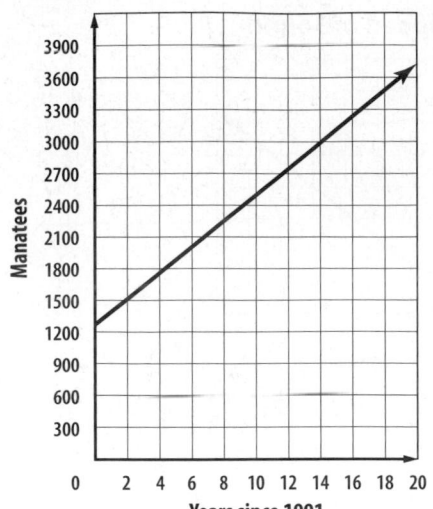

c. $P = 1267 + 123t$
$P = 1267 + 123(15)$
$P = 1267 + 1845$
$P = 3112$ manatees

39. $y = \dfrac{2}{3}x - 5$ **41.** $y = -\dfrac{3}{7}x + 2$ **43.** $y = 5$

45. 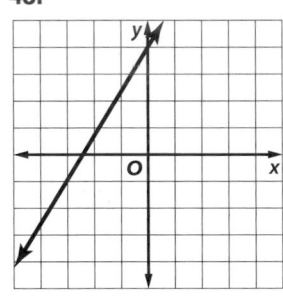   **47.**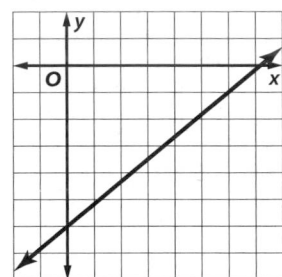

49.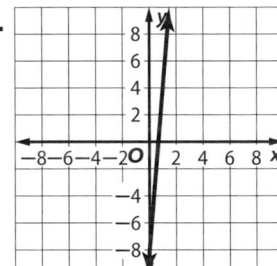

51a. $T = 157c + 218$
51b. \$5242   **53.** $y = 0.5x$
$+ 7.5$ **55.** $y = -1.5x -$
0.25 **57.** $y = 3x$
59a. $C = 45m + 145$
59b. the cost per month
to maintain the
membership

59c. the start up fee   **59d.** \$1225
61a. $S = 9125t + 3305$ **61b.** 12 yr   **63.** No; because
a vertical line has no slope, it cannot be written in
slope-intercept form.   **65.** Sample answer: Assume
that the coefficient of y is not 0. We would first
have to rewrite the equation in slope-intercept
form. The rate of change is also the slope, so, the
coefficient for the x-variable is the rate of change.
Assume that the coefficient of y is not 0.
67. B   **69.** C   **71.** $a_n = 4n - 1$; nonproportional,
does not contain (0, 0)   **73.** $a_n = 3n - 3$;
nonproportional, does not contain (0, 0)
75a. \$25,500   **75b.** \$142,500   **77.** $y = -4x; -5$
79. $y = 0.8x; -7.5$ **81.** $-\dfrac{2}{5}$ **83.** 0

1. $y = 3x - 12$ **3.** $y = -x + 6$ **5.** $y = -3x + 9$
7. $y = 5x + 8$ **9a.** $C = 35p + 75$ **9b.** $600

11 $y = mx + b$
$4 = -1(-1) + b$
$4 = 1 + b$
$3 = b$
So, the equation is $y = -x + 3$.
13. $y = 8x - 55$ **15.** $y = 2x + 2$ **17.** $y = -x + 3$
19. $y = 7x - 16$ **21.** $y = 2x$ **23a.** $y = 0.2x + 0.4$
23b. 4.4 million **25.** $y = \frac{1}{2}x$ **27.** $y = -\frac{3}{4}x + 8\frac{1}{2}$
29. $y = \frac{2}{7}x - 2\frac{4}{7}$ **31a.** $G = 6.4t + 49.7$

31b.

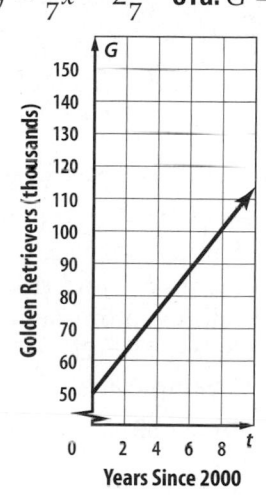

Golden Retrievers (thousands)
Years Since 2000

31c. 158,500
33a. $2.75
33b. $35.40

35 First, find the slope.
$m = \dfrac{y_2 - y_1}{x_2 - x_1}$
$= \dfrac{5 - (-3)}{2 - 5}$
$= \dfrac{5 + 3}{2 - 5}$
$= \dfrac{8}{-3}$
$= -\dfrac{8}{3}$

Next, use the slope-intercept formula.
$y = mx + b$
$5 = -\dfrac{8}{3}(2) + b$
$5 = -\dfrac{16}{3} + b$
$\dfrac{31}{3} = b$
$10\dfrac{1}{3} = b$
So, the equation is $y = -2\frac{2}{3}x + 10\frac{1}{3}$.

37. $y = -x - \dfrac{7}{12}$ **39.** Yes; substituting 6 and -2 for x and y, respectively, results in an equation that is true. **41.** B; x represents the number of raffle tickets sold, y represents the total amount of money in the treasury. **43a.** 605.2 **43b.** 2032; In that year, the waste would be 0 tons. After that, the waste would be a negative amount, which is impossible.

45 **a.** $C = 52t + b$
$275 = 52(5) + b$
$275 = 260 + b$
$15 = b$
So, the equation is $C = 52t + 15$.

b.

Number of tickets	3	4	6	7
Cost ($)	171	223	327	379

c. Graph the equation by graphing the y-intercept $(0, 15)$ and use the slope of 52 to find the next point.

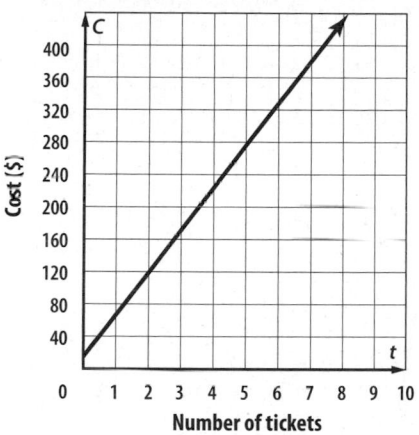

Cost ($)
Number of tickets

Eight tickets would be $431.

47. Jacinta; Teresa switched the x- and y-coordinates on the point that she entered in step 3.
49a. $y = -\dfrac{A}{B}x + \dfrac{C}{B}$ **49b.** slope $= -\dfrac{A}{B}$
49c. y-intercept $= \dfrac{C}{B}$ **49d.** no, $B \neq 0$ **51.** Sample answer: If the problem is about something that could suddenly change, such as weather or prices, the graph could suddenly spike up. You need a constant rate of change to produce a linear graph.
53. D **55.** B

57.

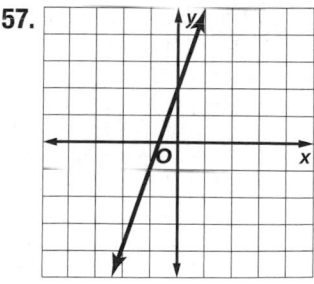

59.

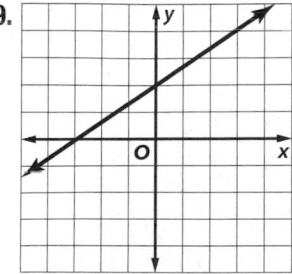

61.

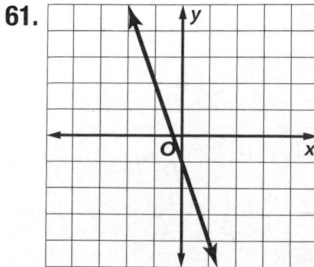

63. $f(x) = -2x$

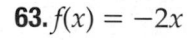

65a.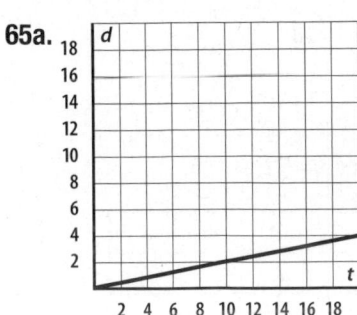

65b. about 14 seconds
67. −22 **69.** −207
71. 1.5 **73.** 7
75. −1 **77.** 1

Lesson 4-3

1 $y - y^2 = m(x - x^2)$
$(y - 5) = -6(x - (-2))$
$(y - 5) = -6(x + 2)$
Plot the point $(-2, 5)$ and use the slope of -6 to find the next point.

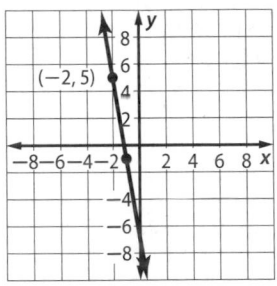

3. $y - 3 = -\frac{1}{2}(x - 4)$

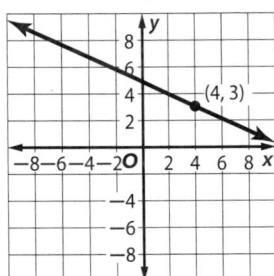

5. $5x + y = -22$ **7.** $y = 4x + 34$ **9.** $y = x + 13$
11. $y - 3 = 7(x - 5)$ **13.** $y + 3 = -1(x + 6)$

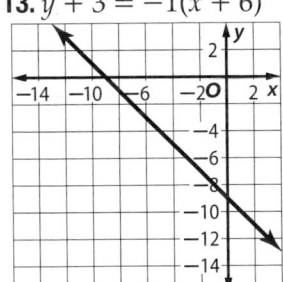

15. $y - 11 = \frac{4}{3}(x + 2)$

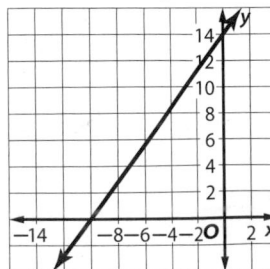

17. $y + 9 = -\frac{7}{5}(x + 2)$
19. $2x - y = 6$ **21.** $6x + y = -45$ **23.** $9x - 10y = 43$
25. $x + 6y = -7$ **27.** $y = -2x + 20$ **29.** $y = -6x - 47$ **31.** $y = \frac{1}{6}x - \frac{8}{3}$
33. $y = -\frac{2}{3}x - 5$

35 The slope is −5 and the point is (6, 4). So, the equation is $y - 4 = -5(x - 6)$. Replace x with 2 to determine how many were rented the second week.
$y - 4 = -5(2 - 6)$
$y - 4 = -5(-4)$
$y - 4 = 20$
$\quad\quad y = 24$
So, 24 copies were rented.
37. $x + y = 6$ **39.** $5x + 4y = 20$
41. $y + 1 = \frac{3}{2}(x + 4)$

43 $y + \frac{3}{5} = x - \frac{2}{5}$
$y + \frac{3}{5} - \frac{3}{5} = x - \frac{2}{5} - \frac{3}{5}$
$y = x - \frac{5}{5}$
$y = x - 1$

45. $y = \frac{5}{6}x$ **47.** $y - 4 = \frac{4}{7}(x + 9)$; $y = \frac{4}{7}x + \frac{64}{7}$; $4x - 7y = -64$ **49.** $y + 4 = 3(x + 1)$; The slope-intercept form is not $y = 3x + 2$. **51.** Sample answer: Jocari spent \$14 to go to an amusement park and ride ponies. The price she paid included admission. The 5 pony rides cost \$2 each; $y - 14 = 2(x - 5)$, $-2x + y = 4$, $y = 2x + 4$. **53.** Sample answer:
$y - g = \frac{j - g}{h - f}(x - f)$ **55.** B **57.** J **59.** $y = x - 2$
61. $y = -2x + 1$ **63.** $y = -2$ **65.** $y = -2x + 6$
67. $y = \frac{1}{2}x + 3$ **69.** $y = 3$ **71.** Yes; there are only 364 seats. **73.** $a = \frac{v - r}{t}$ **75.** $b = \frac{-t + 5}{4}$

Lesson 4-4

1. $y = \frac{1}{2}x + 2\frac{1}{2}$ **3.** Slope of $\overline{AC} = \frac{1 - 7}{-2 - 5}$ or $\frac{6}{7}$; slope of $\overline{BD} = \frac{-3 - 4}{3 - (-3)}$ or $-\frac{7}{6}$; the paths are perpendicular.

5 Graph each line on a coordinate plane.
$y = -2x$ and $4y = 2x + 4$ are perpendicular to $2y = x$; $2y = x$ and $4y = 2x + 4$ are parallel.
7. $y = 2x + 7$ **9.** $y = \frac{3}{2}x$ **11.** $y = x - 5$
13. $y = -5x + 2$ **15.** $y = -\frac{3}{4}x + 1\frac{1}{2}$ **17.** Yes; the line containing $\overline{AD}$ and the line containing $\overline{BC}$ have the same slope, $\frac{1}{3}$. Therefore one pair of sides is parallel. The slope of $\overline{AB}$ is undefined and the slope of $\overline{CD}$ is $-\frac{5}{3}$. **19.** Yes; the slopes are −6 and $\frac{1}{6}$. **21.** $2x - 8y = -24$ and $4x + y = -2$ are perpendicular; $2x - 8y = -24$ and $x - 4y = 4$ are parallel.

23 The slope of the given line is −2. So, the slope of a line perpendicular is $\frac{1}{2}$.
$y = mx + b$
$-2 = \frac{1}{2}(-3) + b$

$$-2 = -\frac{3}{2} + b$$
$$-\frac{1}{2} = b$$

The equation is $y = \frac{1}{2}x - \frac{1}{2}$.

25. $y = -3x - 7$ **27.** $y = -\frac{1}{5}x + 8\frac{3}{5}$ **29.** $y = 2x + 16$

31. $y = -\frac{1}{5}x - \frac{3}{25}$ **33.** neither **35.** perpendicular

37. neither **39.** $y = 7x$

41 Find the slopes.

$$m = \frac{-1 - 6}{4 - 2} \qquad m = \frac{12 - 10}{14 - 7}$$
$$= \frac{-7}{2} \qquad\qquad = \frac{2}{7}$$
$$= -\frac{7}{2}$$

Since the slopes are opposite reciprocals, the objects are perpendicular.

43a.

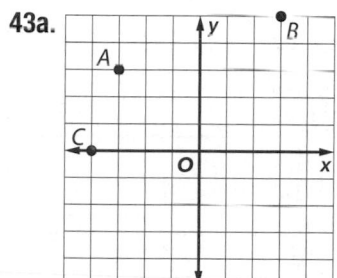

43b. Sample answer: (2, 2); $\overline{AB}$ and $\overline{CD}$ both have slope $\frac{1}{3}$, and $\overline{AC}$ and $\overline{BD}$ both have slope 3.

43c. Two; sample answer: Move C to $(-2, 0)$ and move D to $(4, 2)$. Moving C changes the slope of AC to -3. This is the opposite reciprocal of the slope $\overline{AC}$, $\frac{1}{3}$. Moving D also changes the slope of $\overline{BD}$ so $\overline{BD}$ is perpendicular to $\overline{AB}$ and $\overline{CD}$ and it is parallel to $\overline{AC}$. **45.** Sample answer: Parallel lines: similarities: The domain and range are all real numbers, the functions are both either increasing or decreasing on the entire domain, the end behavior is the same; differences: x- and y-intercepts are different. Perpendicular lines: similarities: The domain and range are all real numbers; differences: One function is increasing and the other is decreasing on the entire domain, as x decreases, y increases for one function and decreases for the other and as x increases, y increases for one function and decreases for the other. **47.** Carmen is correct; she correctly determined the slope of the perpendicular line.
49. A **51.** B **53.** $-4x + y = 5$ **55.** $5x + y = -8$
57. $5x - 6y = 14$ **59a.** $C = 10h + 15$ **59b.** $95
61. $y = -5x - 21$ **63.** $y = 2x - 1$ **65.** $y = -5x - 6$
67. simplified **69a.** $25(5) + 10(8.5) + 35(5) + 12(8.5)$
69b. $487

71.

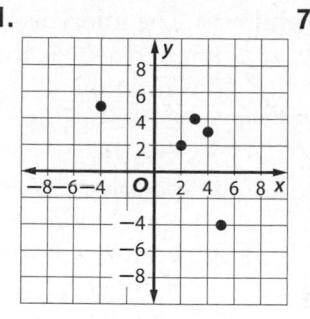

$D = \{-7, 4, -2, -3\};$
$R = \{4, 3, 2, -4, 5\}$

73.

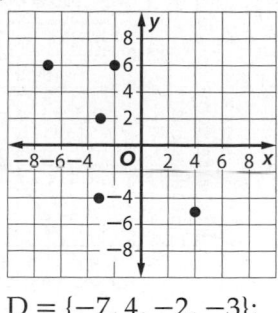

$D = \{-7, 4, -2, -3\};$
$R = \{6, -4, -5, 2\}$

Lesson 4-5

1. Positive; the longer you practice free throws, the more free throws you will make.

3a.

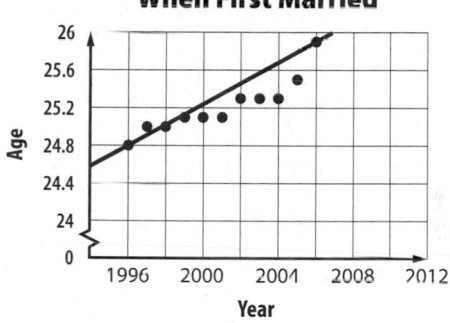

Positive; independent variable is year and dependent variable is median age of females when they were first married. **3b.** See above graph.
3c. Sample answer: Using (1996, 24.8) and (2006, 25.9) and rounding, $y = 0.11x - 194.8$ **3d.** Sample answer: 27.0 **3e.** Yes, according to the equation, the median age would be 31.4, which is likely.

5 As the height increases, the percentage decreases. The graph shows a slight negative correlation. This correlation means that the taller a player is, the lower their percentage of 3-point shots made is.

7 There is no pattern to the graph, so there is no correlation between the speed of a vehicle and the miles per gallon.

9a. $y = -648.5x + 74,447.5$ **9b.** $61,478 **9c.** No; the average attendance will fluctuate with other variables such as how good the team is that year.

11 a. The independent variable is the duration of the eruptions and the dependent variable is the interval of the eruptions. This is because the duration of the eruptions is not affected by the interval.

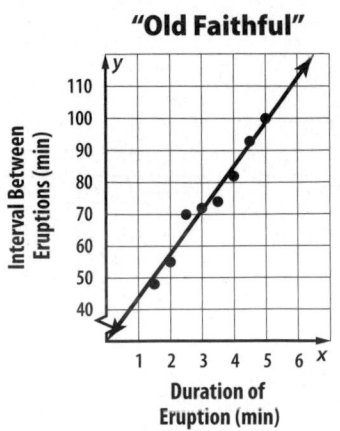

"Old Faithful"

Interval Between Eruptions (min)

Duration of Eruption (min)

The interval increases as the duration increases, so there is a positive correlation between the independent and dependent variables.

b. Use (2, 55) and (4, 82)

$$m = \frac{y_2 - y_1}{x_2 - x_1}$$

$$= \frac{82 - 55}{4 - 2}$$

$$-\frac{27}{2}$$

$$= 13.5$$

So, the slope is 13.5.

$y = 13.5x + b$	$y = 13.5x + 28$
$55 = 13.5(2) + b$	$y = 13.5(7.5) + 28$
$55 = 27 + b$	$y = 101.25 + 28$
$28 = b$	$y = 129.25$ min

c. Sample answer: The duration of an eruption is not dependent on the previous interval. Only the interval can be predicted by the length of the eruption.

13. Sample answer: The salary of an individual and the years of experience that they have; this would be a positive correlation because the more experience an individual has, the higher the salary would probably be. **15.** Neither; line g has the same number of points above the line and below the line. Line f is close to 2 of the points; but for the rest of the data, there are 3 points above and 3 points below the line. **17.** Sample answer: You can visualize a line to determine whether the data has a positive or negative correlation. The following graph shows the ages and heights of people. To predict a person's age given his or her height, write a linear equation for the line of fit. Then substitute the person's height and solve for the corresponding age. You can use the pattern in the scatter plot to make decisions.

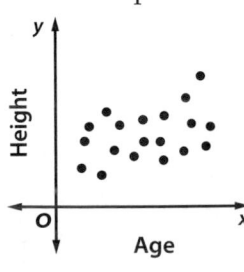

Height

Age

19. F **21.** 22 days
23. neither
25. perpendicular
27. $2x + y = 1$
29. $x - 2y = 12$
31. $2x + 5y = 26$

33.

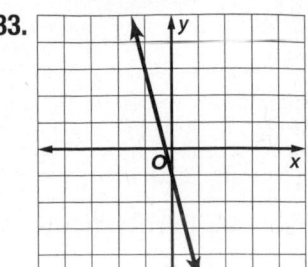

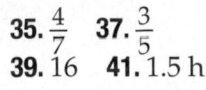

35. $\frac{4}{7}$ **37.** $\frac{3}{5}$
39. 16 **41.** 1.5 h

43.

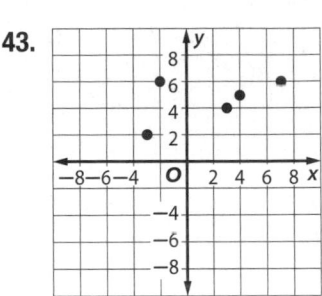

$D = \{7, 3, 4, -2, -3\};$
$R = \{6, 4, 5, 2\}$

Lesson 4-6

1a. $y = 1.18x + 11; 0.7181$
1b. The residuals appear to be randomly scattered, so the regression line fits the data reasonably well.

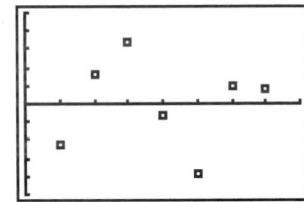

[0, 8] scl: 1.5 by [−5, 5] scl: 1.5

3a. $y = -271.88x + 554.48$
3b. $78.69

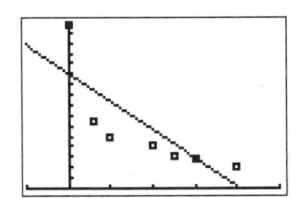

[−0.5, 2.5] scl: 1 by [0, 785] scl: 10

5 Step 1: Enter the data by pressing STAT and selecting the Edit option. Let the year 2000 be represented by 0. Enter the years since 2000 into List 1 (L1). These will represent the x-values. Enter the number of auditions into List 2 (L2). These will represent the y-values.
Step 2: Perform the regression by pressing STAT and selecting the CALC option. Scroll down to LinReg($ax + b$) and press ENTER .
The equation is $y = 3.54x + 5.54$.
The correlation coefficient is 0.9007.

7 a. Enter the data using 0 for 1975. Use med-med to find $y = 601.44x + 1236.13$.
b. $2003 - 1975 = 28$. Substitute 28 into the equation in **a** to get 18,076.45. There were about 18,076 entrants in 2003.

9a. $y = 0.095x - 94.58$

9b.

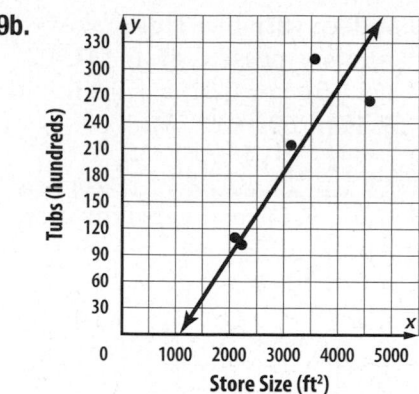

9c. about 48 tubs; about 380 tubs

11a. $y = 0.0326x + 1.598$

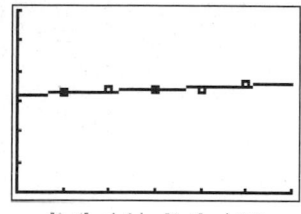

[0, 6] scl: 1 by [0, 3] scl: 0.5

11b. The regression line is a good fit as the residuals appear to almost be on the line.

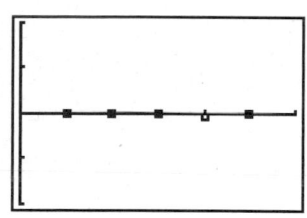

[0, 6] scl: 1 by [−2, 2] scl: 1

11c. about 2.12 million people

13a. $y = 87,390.5x + 4,018,431$ **13b.** about 5,591,460

15.

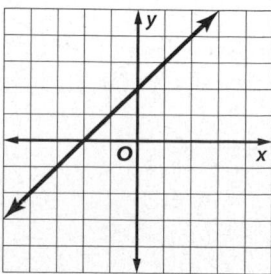

Sample answer: Men: $y = -2.92x + 95.92$; women: $y = -7x + 106$; women's scores have a steeper slope than men's.

19. C   **21.** H   **23a.** negative correlation   **23b.** $3600
23c. No; according to the line of fit, the cost would be $0.   **25.** $3x - y = 1$ **27.** $2x + y = 8$ **29.** $2x - 3y = -21$ **31.** $\frac{4}{7}$ **33.** $\frac{3}{5}$ **35.** 3   **37.** $a^2 - a + 1$

39.

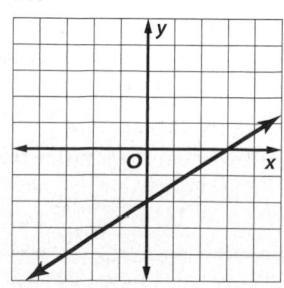

41.

1. $\{(-15, 4), (-18, -8), (-16.5, -2), (-15.25, 3)\}$

3.

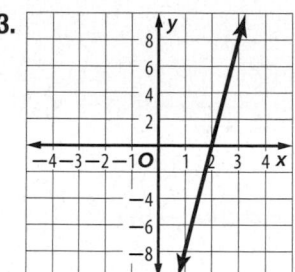

5. $f^{-1}(x) = -\frac{1}{2}x + \frac{7}{2}$ **7a.** $C^{-1}(x) = \frac{1}{70}x - \frac{60}{7}$

7b. x is Dwayne's total cost, and $C^{-1}(x)$ is the number of games Dwayne attended.   **7c.** 5

(9) Exchange the coordinates of the ordered pairs.
$(-4, -49) \rightarrow (-49, -4)$ $(8, 35) \rightarrow (35, 8)$
$(-1, -28) \rightarrow (-28, -1)$ $(4, 7) \rightarrow (7, 4)$
The inverse is $\{(-49, -4), (35, 8), (-28, -1), (7, 4)\}$.

11. $\{(7.4, -3), (4, -1), (0.6, 1), (-2.8, 3), (-6.2, 5)\}$

13.

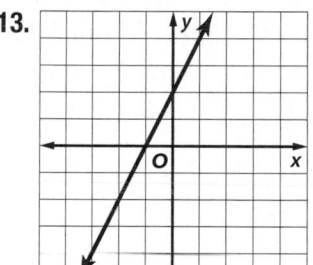

(15)
$$f(x) = 17 - \frac{1}{3}x$$
$$y = 17 - \frac{1}{3}x$$
$$x = 17 - \frac{1}{3}y$$
$$x - 17 = -\frac{1}{3}y$$
$$-3(x - 17) = y$$
$$-3x + 51 = y$$
The inverse of $f(x)$ is $f^{-1}(x) = -3x + 51$.

17. $f^{-1}(x) = -\frac{1}{6}x + 2$ **19.** $f^{-1}(x) = -\frac{3}{4}x - 12$

21a. $C^{-1}(x) = \frac{1}{35}x - \frac{2}{7}$ **21b.** x is the total amount collected from the Fosters, and $C^{-1}(x)$ is the number of times Chuck mowed the Fosters' lawn.

21c. 22   **23.** $f^{-1}(x) = 15 - 5x$ **25.** $f^{-1}(x) = \frac{3}{2}x + 12$

27. $f^{-1}(x) = 3x - 3$ **29.** B   **31.** A

(33) If the graph of $f(x)$ contains the points $(-3, 6)$ and $(6, 12)$, then the graph of $f^{-1}(x)$ contains the points $(6, -3)$ and $(12, 6)$. Find the slope of the line that passes through these points.
$$m = \frac{y_2 - y_1}{x_2 - x_1}$$
$$= \frac{6 - (-3)}{12 - 6}$$
$$= \frac{9}{6} \text{ or } \frac{3}{2}$$
Choose $(12, 6)$ and find the y-intercept of the line.

$$y = mx + b$$
$$6 = \frac{3}{2}(12) + b$$
$$6 = 18 + b$$
$$-12 = b$$

The line that passes through $(6, -3)$ and $(12, 6)$ is $y = \frac{3}{2}x - 12$. An equation for $f^{-1}(x)$ is $f^{-1}(x) = \frac{3}{2}x - 12$

35. $f^{-1}(x) = \frac{1}{4}x + \frac{3}{4}$ **37a.** $A(x) = 8(x - 3)$ or $A(x) = 8x - 24$

37b.

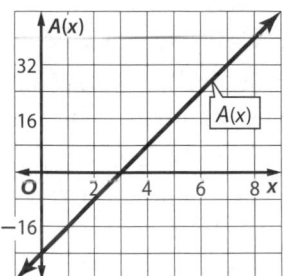

Sample answer: The domain represents possible values of x. The range represents the area of the rectangle and must be positive. This means that the domain of $A(x)$ is all real numbers greater than 3, and the range of $A(x)$ is all positive real numbers. **37c.** $A^{-1}(x) = \frac{1}{8}x + 3$; x is the area of the rectangle and $A^{-1}(x)$ is the value of x in the expression for the length of the side of the rectangle $x - 3$.

37d.

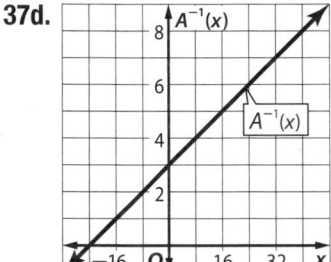

Sample answer: The domain represents the area of the rectangle and must be positive. The range represents possible values for x in the expression $x - 3$. This means that the domain of $A^{-1}(x)$ is all positive real numbers, and the range of $A^{-1}(x)$ is all real numbers greater than 3. **37e.** Sample answer: The domain of $A(x)$ is the range of $A^{-1}(x)$, and the range of $A(x)$ is the domain of $A^{-1}(x)$.
39. $a = 2$; $b = 14$ **41.** Sometimes; sample answer: $f(x)$ and $g(x)$ do not need to be inverse functions for $f(a) = b$ and $g(b) = a$. For example, if $f(x) = 2x + 10$, then $f(2) = 14$ and if $g(x) = x - 12$, then $g(14) = 2$, but $f(x)$ and $g(x)$ are not inverse functions. However, if $f(x)$ and $g(x)$ are inverse functions, then $f(a) = b$ and $g(b) = a$. **43.** Sample answer: A situation may require substituting values for the dependent variable into a function. By finding the inverse of the function, the dependent variable

becomes the independent variable. This makes the substitution an easier process. **45.** F **47.** 4.2
49. $y = 8.235x - 17.365$ **51.** $y = 0.325x + 0.89$
53. 100 **55.** 11.7 **57.** 171 **59.** -77 **61.** 100

Chapter 4   Study Guide and Review

1. true **3.** true **5.** true **7.** false, inverse function
9. false, slope-intercept form
11. $y = -2x - 9$ **13.** $y = -\frac{5}{8}x - 2$

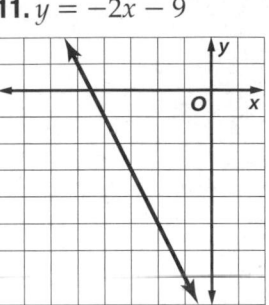   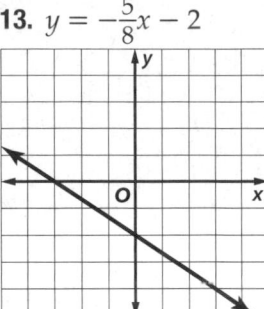

15.   **17.**

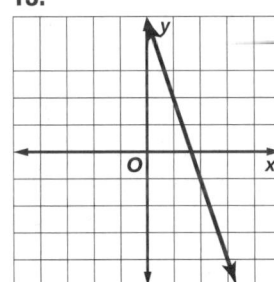   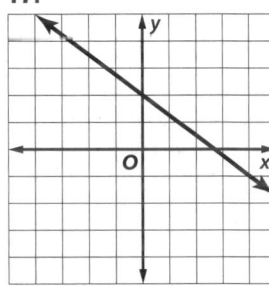

19. $y = 3x - 1$ **21.** $y = \frac{2}{5}x + \frac{1}{5}$ **23.** $y = x - 3$
25. $y = \frac{1}{2}x + \frac{7}{2}$ **27.** $y = 60x + 450$ **29.** $y - 1 = -3(x + 2)$ **31.** $5x - y = 7$ **33.** $x - 2y = 11$
35. $y = 3x - 13$ **37.** $y = 5x + 2$ **39.** $y = x + 3$
41. $y = -2x - 7$ **43.** $y = -\frac{1}{3}x + \frac{14}{3}$
45. $y = -3x - 13$ **47.** positive **49.** $y = 5.36x + 11$; 65
51. $\{(3.5, 7), (8, 6.2), (2.7, -4), (1.4, -12)\}$
53. $\{(2.7, -4), (3.8, -1), (4.1, 0), (7.2, 3)\}$
55. $f^{-1}(x) = \frac{11}{5}x - 22$ **57.** $f^{-1}(x) = -\frac{1}{4}x - 3$
59. $f^{-1}(x) = -\frac{3}{2}x + \frac{3}{8}$

CHAPTER 5
Linear Inequalities

Get Ready

1. -10 **3.** 24.6 **5.** -11 **7.** 21 **9.** 5 **11.** -4
13. $\{-29, 7\}$ **15.** 34%, 30%

Lesson 5-1

1. $\{x \mid x > 10\}$

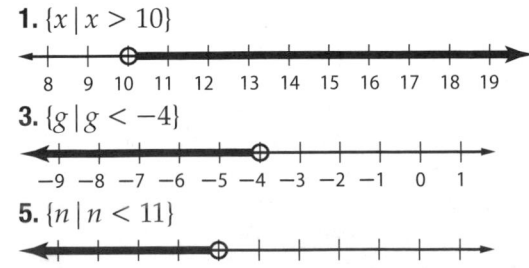

3. $\{g \mid g < -4\}$

5. $\{n \mid n < 11\}$

7. $\{r \mid r > 6\}$

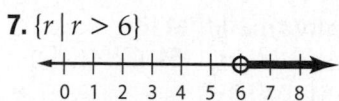

9. Sample answer: Let n = the number, $2n + 4 \geq n + 10; \{n \mid n \geq 6\}$.   **11.** no more than 92 ft

(13) $\quad p - 6 \geq 3$
$p - 6 + 6 \geq 3 + 6$
$\qquad p \geq 9 \quad \{p \mid p \geq 9\}$
Place a closed circle on 9 and an arrow to the right.

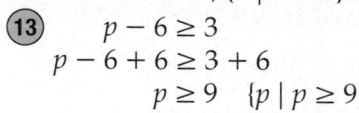

15. $\{t \mid t > -5\}$

17. $\{r \mid r < -5\}$

19. $\{q \mid q \leq 7\}$

21. $\{h \mid h < 30\}$

23. $\{c \mid c < -27\}$

25. $\{z \mid z \leq 4\}$

27. $\{y \mid y \leq -6\}$

29. $\{a \mid a > -9\}$

31. Sample answer: Let n = the number, $2n + 5 \leq n - 3; \{n \mid n \leq -8\}$.   **33.** Sample answer: Let n = the number, $6n - 8 < 5n + 21; \{n \mid n < 29\}$.   **35.** Sample answer: Let n = the number of online teens that do not use the Internet at school in millions; $n > 21 - 16; \{n \mid n > 5\}$, at least 5 million teens use the Internet but not at school.   **37.** Sample answer: Let t = the original water temperature; $t + 4 < 81; \{t \mid t < 77\}$; the water temperature was originally less than 77°.

(39) Let x represent the amount left on the card.
$x + 32 + 26 < 75$
$\qquad x + 58 < 75$
$x + 58 - 58 < 75 - 58$
$\qquad\qquad x < 17$
She will have no more than $17 left on the card.

41. $\{c \mid c \geq 3.7\}$

43. $\left\{ k \mid k > -\dfrac{5}{12} \right\}$

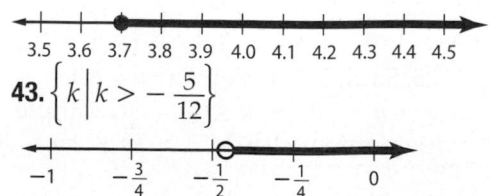

45a. 12 lb   18 lb

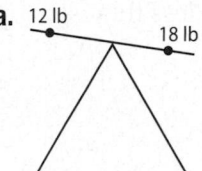

45b. 12 lb < 18 lb

45c.

	12	<	18
$2x$	24	<	36
$3x$	36	<	54
$4x$	48	<	72
$\frac{1}{2}x$	6	<	9
$\frac{1}{3}x$	4	<	6
$\frac{1}{4}x$	3	<	$4\frac{1}{2}$

45d. If a true inequality is multiplied by a positive number, the resulting inequality is also true. If a true inequality is divided by a positive number, the resulting inequality is also true.

47. 10   **49.** 3   **51.** 26   **53.** $c < a < d < b$ **55.** Solving linear inequalities is similar to solving linear equations. You must isolate the variable on one side of the inequality. To graph, if the problem is a less than or a greater than inequality, an open circle is used. Otherwise a dot is used. If the variable is on the left hand side of the inequality, and the inequality sign is less than (or less than or equal to), the graph extends to the left; otherwise it extends to the right.   **57.** C   **59.** B   **61.** $f(x)^{-1} = \frac{1}{7}x + 4$ **63.** $f(x)^{-1} = -3x - 24$ **65.** $y = -x - 2$ **67.** $y = -2x - 1$ **69.** blue   **71.** 25   **73.** $y = 7x$; $210   **75.** -30 **77.** $\frac{1}{10}$ **79.** 16   **81.** $-\frac{1}{9}$

Lesson 5-2

1. Let d = the number of DVDs sold; $15d > 5500$; $d > 366.67$; the band sold at least 367 DVDs.
3. $\{r \mid r \geq 8\}$

5. $\{h \mid h < -10\}$

7. $\{v \mid v > -12\}$

9. $\{z \mid z \geq -8\}$

11. Let p = the number of pay periods for which Rodrigo will need to save; $25p \geq 560$; $p \geq 22.4$; Rodrigo will need to save for 23 weeks.

(13) $\quad \dfrac{1}{2}a < 20$
$2\left(\dfrac{1}{2}a\right) < 2(20)$
$\qquad a < 40 \quad \{a \mid a < 40\}$

Check by substituting values less than 40.

15. $\{d \mid d \geq 68\}$

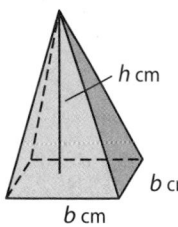

60 62 64 66 68 70 72 74 76

17. $\{f \mid f < 432\}$

416 418 420 422 424 426 428 430 432 434

19. $\{j \mid j \leq -16\}$

−20 −18 −16 −14 −12 −10 −8 −6 −4 −2 0 2

21. $\{p \mid p \leq 16\}$

0 2 4 6 8 10 12 14 16 18

23. $\{y \mid y > -16\}$

−18 −16 −14 −12 −10 −8 −6 −4 −2 0

25. $\{v \mid v < 12\}$

0 2 4 6 8 10 12 14 16 18

27. $\left\{b \mid b \leq -\dfrac{3}{4}\right\}$

−5 −4 −3 −2 −1 0 1 2 3 4 5

29. $\left\{f \mid f < -\dfrac{5}{7}\right\}$

−5 −4 −3 −2 −1 0 1 2 3 4 5

31. no more than 4   **33.** no more than 32 people

35. b   **37.** d

39 $\dfrac{2}{3}x < 42$

$3\left(\dfrac{2}{3}x\right) < 3(42)$

$2x < 126$

$x < 63$ fewer than 63 employees

41a.   **41b.** $h = \dfrac{216}{b^2}$

41c.

b	1	3	6	9	12
h	216	24	6	$\dfrac{8}{3}$	$\dfrac{3}{2}$

41d. $b < h$ when $0 < b < 6$; $b > h$ when $h < 6$.

43a. $x > -\dfrac{5}{a}$ **43b.** $x \geq 8a$ **43c.** $x \leq -\dfrac{6}{a}$

45. Sometimes; the statement is true when $a > 0$ and $b < 0$.   **47.** Sample answer: The same processes are used when solving linear inequalities and equations that involve addition, subtraction, multiplication, or division by a positive number. However, when a linear inequality is multiplied or divided by a negative number, the inequality

symbol must change directions so that the inequality remains true.   **49.** 10 in.   **51.** C

53. $\left\{y \mid y \geq 21\right\}$

0 2 4 6 8 10 12 14 16 18 20 22

55. $f^{-1}(x) = -\dfrac{1}{6}x + 3$ **57.** $f^{-1}(x) = \dfrac{1}{4}x + \dfrac{5}{4}$

59. 2 hours   **61.** $\{1, 7\}$ **63.** 2   **65.** $\dfrac{33}{8}$ **67.** 3

Lesson 5-3

1. $4n + 60 \leq 800$; $n \leq 185$; at most 185 lb per person

3 $6h - 10 \geq 32$

$6h - 10 + 10 \geq 32 + 10$

$6h \geq 42$

$h \geq 7$

$\{h \mid h \geq 7\}$

0 2 4 6 8 10 12 14 18

5. $\{x \mid x < -12\}$

−20 −18 −16 −14 −12 −10 −8 −6 −4 −2 0 2 4

7. Sample answer: Let n = the number; $4n - 6 > 8 + 2n$; $\{n \mid n > 7\}$.   **9.** $\{v \mid v \geq 0\}$

−10 −8 −6 −4 −2 0 2 4 6 8 10

11. ∅

−10 −8 −6 −4 −2 0 2 4 6 8 10

13 $21 > 15 + 2a$

$21 - 15 > 2a$

$6 > 2a$

$3 > a$ $\{a \mid a < 3\}$

−10 −8 −6 −4 −2 0 2 4 6 8 10

15. $\{w \mid w > 56\}$

46 48 50 52 54 56 58 60 62 64

17. $\{w \mid w < -3\}$

−10 −8 −6 −4 −2 0 2 4 6 8 10

19. $\left\{p \mid p > -\dfrac{24}{5}\right\}$

−10 −8 −6 −4 −2 0 2 4 6 8 10

21. $\{h \mid h < -15\}$

−20 −18 −16 −14 −12 −10 −8 −6 −4 −2 0 2

23. Sample answer: Let n = the number; $\dfrac{2}{3}n + 6 \geq 22$; $\{n \mid n \geq 24\}$.   **25.** Sample answer: Let n = the number; $8n - 27 \leq -n + 18$; $\{n \mid n \leq 5\}$.   **27.** Sample answer: Let n = the number; $3(n + 7) > 5n - 13$; $\{n \mid n < 17\}$

29. $\left\{n \mid n > -\dfrac{1}{3}\right\}$

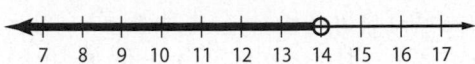

31. ∅

33. $\{t \mid t \geq -1\}$

35. Sample answer: Let s = the amount of sales made, $35{,}000 + 0.08s > 65{,}000$; $\{s \mid s > 375{,}000\}$; the sales must be more than \$375,000.

37.
$$6(m - 3) > 5(2m + 4) \quad \text{Original inequality}$$
$$6m - 18 > 10m + 20 \quad \text{Distributive Property}$$
$$6m - 18 - 6m > 10m + 20 - 6m \quad \text{Subtract } 6m \text{ from each side.}$$
$$-18 > 4m + 20 \quad \text{Simplify.}$$
$$-18 - 20 > 4m + 20 - 20 \quad \text{Subtract 20 from each side.}$$
$$-38 > 4m \quad \text{Simplify.}$$
$$\dfrac{-38}{4} > \dfrac{4m}{4} \quad \text{Divide each side by 4.}$$
$$-9.5 > m \quad \text{Simplify.}$$
$$\{m \mid m < -9.5\}$$

39a. $5t + 565 \geq 1500$; $t \geq 187$

39b.

41. a. Words: temperature can be greater than 104
$$t > 104$$
b. $F > 104$
$$\dfrac{9}{5}C + 32 > 104$$
$$\dfrac{9}{5}C > 72$$
$$5\left(\dfrac{9}{5}C > 72\right)$$
$$9C > 360$$
$$C > 40$$

43. 1, 3, 5, 7; 3, 5, 7, 9; 5, 7, 9, 11; 7, 9, 11, 13

45. $\left\{x \mid x \geq \dfrac{1}{2}\right\}$ **47.** $\{m \mid m \geq 18\}$ **49.** $\{x \mid x \leq 8\}$

51. $\{x \mid x > -6\}$ **53.** $\{x \mid x \geq 1.5\}$ **55.** Add $3p$ and 2 to each side. The inequality becomes $9 \geq 3p$. Then divide each side by 3 to get $3 \geq p$.

57a. $\left\{x \mid x \geq -\dfrac{9}{2a}\right\}$ **57b.** $\left\{x \mid x > \dfrac{2}{1+a}\right\}$

57c. $\{x \mid x < 6a\}$ **59.** Sample answer: The solution set for an inequality that results in a false statement is the empty set, as in $12 < -15$. The solution set for an inequality in which any value of x results in a true statement is all real numbers, as in $12 \leq 12$.

61. G **63.** D **65.** $\{b \mid b > -4\}$
67. $\{h \mid h < 14\}$

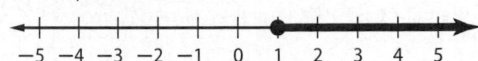
69. $\{m \mid m \geq 1\}$

71. 4
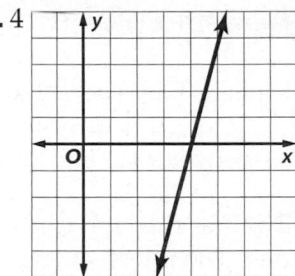

73. about 70.1 million
75. 8 **77.** $12(29.95 + 4)$ or $12(29.95) + 12(4)$; \$407.40
79.
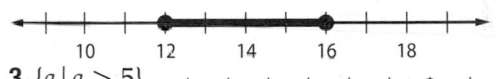
81.

83.

Lesson 5-4

1. $\{p \mid 12 \leq p \leq 16\}$

3. $\{a \mid a > 5\}$

5. $11 \text{ psi} \leq x \leq 56 \text{ psi}$

7.
$$n + 2 \leq 5 \qquad\qquad n + 6 \geq -6$$
$$n + 2 - 2 \leq -5 - 2 \quad \text{and} \quad n + 6 - 6 \geq -6 - 6$$
$$n \leq -7 \qquad\qquad n \geq -12$$
The solution set is $\{n \mid -12 \leq n \leq -7\}$.

9. $\{t \mid t \geq 1 \text{ or } t < -1\}$

11. $\{c \mid -1 \leq c < 2\}$

13. $\{m \mid m \text{ is a real number.}\}$

15. $\{y \mid y < -3\}$

17. Sample answer: Let x = the smaller of two consecutive odd numbers, then $8 \leq 2x + 2 \leq 24$; $3 \leq x \leq 11$; 3, 5; 5, 7; 7, 9; 9, 11; 11, 13

19. The graph shows $x > -3$ and $x \leq 2$, so the inequality is $-3 < x \leq 2$.
21. $x < -4$ or $x > -3$ **23.** $x \leq -3$ or $x > 0$
25. $\left\{a \mid -3 < a \leq \dfrac{1}{2}\right\}$

27. $\{n \mid n < -3 \text{ or } n > -3\}$

29. Sample answer: Let n = the number; $5 \le n - 8 \le 14$; $\{n \mid 13 \le n \le 22\}$. **31.** $-5n > 35$ or $-5n < 10$; $\{n \mid n < -7 \text{ or } n > -2\}$ **33.** $t < 75$ or $t > 90$

35. Sample answer: Let t = the temperature; $23 \le t \le 33$.

37 a. Category 3 has wind speeds between 111 and 130: $111 \le x \le 130$

Category 4 has wind speeds between 131 and 155: $131 \le x \le 155$

b. The union is all values between 111 and 155: $111 \le x \le 155$.

The intersection is all values they have in common, which is none. The intersection is the empty set, $\varnothing$.

39. Neither; Chloe did not add 5 to 3, and Jonas did not add 5 to 7. **41.** Sample answer: $x \le 2$ or $x \ge 4$ **43.** Sample answer: The speed at which a roller coaster runs while staying on the track could represent a compound inequality that is an intersection. **45.** H **47.** B **49.** at least 22 subscriptions **51.** 5.6 **53.** 3.94 **55.** 3.28 **57.** Yes; for each input there is exactly one output. **59.** No; the domain value -4 is paired with both -1 and 11.

61. $5 + (4 - 2^2)$
$= 5 + (4 - 4)$
$= 5 + 0$
$= 5$

63. $2(4 \cdot 9 - 3) + 5 \cdot \dfrac{1}{5}$
$= 2(36 - 3) + 5 \cdot \dfrac{1}{5}$
$= 2(33) + 5 \cdot \dfrac{1}{5}$
$= 66 + 5 \cdot \dfrac{1}{5}$
$= 66 + 1$
$= 67$

65. 3 **67.** $12\frac{2}{3}$ **69.** 30 **71.** 10

Lesson 5-5

1. $\{a \mid 2 < a < 8\}$

3. $\varnothing$

5. $\{n \mid n \le -8 \text{ or } n \ge -2\}$

7. $\{m \mid 70.10 \le m \le 71.60\}$

9 $|r + 1| \le 2$

Case 1:		Case 2:
$r + 1$ is positive.	and	$r + 1$ is negative.
$r + 1 \le 2$	and	$r + 1 \ge -2$
$r \le 1$	and	$r \ge -3$

The solution set is $\{r \mid -3 \le r \le 1\}$.

11. $\{h \mid -3 < h < 5\}$

13. $\varnothing$

15. $\{k \mid k < 1 \text{ or } k > 7\}$

17. $\{p \mid p \le -3 \text{ or } p \ge 2\}$

19. $\{c \mid c \text{ is a real number.}\}$

21. $\{n \mid n \le -5\frac{1}{4} \text{ or } n \ge 3\frac{3}{4}\}$

23. $\{h \mid -5\frac{2}{3} < h < 5\}$

25. $\varnothing$

27. $\{r \mid -2 < r < \frac{2}{3}\}$

29. $\{h \mid -1.5 < h < 4.5\}$

31a. $\{t \mid t < 32 \text{ or } t > 212\}$

31b.

31c. $|t - 122| > 90$ **33.** $|x + 1| \le 4$

35. $|x - 5.5| > 4.5$

37 $|g - 52| \le 5$

Case 1:		Case 2:
$g - 52$ is positive.	and	$g - 52$ is negative.
$g - 52 \le 5$	and	$g - 52 \ge -5$
$g \le 57$	and	$g \ge 47$

The solution is $\{g \mid 47 \le g \le 57\}$.

39. $|t - 38| \le 1.5$ **41.** $|c - 55| \le 3$ **43.** No; Sample answer: Lucita forgot to change the direction of the inequality sign for the negative case of the absolute value. **45.** Sample answer: If $t = 0$, then the absolute value is equal to 0, not greater than 0.

47. Sample answer: When an absolute value is on the left and the inequality symbol is $<$ or $\le$, the compound sentence uses *and*, and if the inequality symbol is $>$ or $\ge$, the compound sentence used *or*. To solve, if $|x| < n$, then set up and solve the inequalities $x < n$ and $x > -n$, and if $|x| > n$, then set up and solve the inequalities $x > n$ or $x < -n$.

49. J **51.** B **53.** $\{t \mid 5 \le t \le 6\}$

55. Sample answer: $6 + 22w \le 87$; up to 3 withdrawals **57.** 18 **59.** -20

61.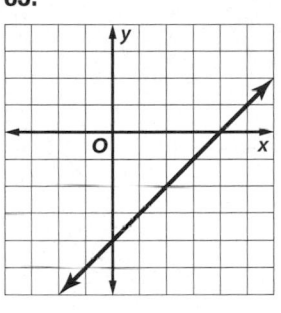

63.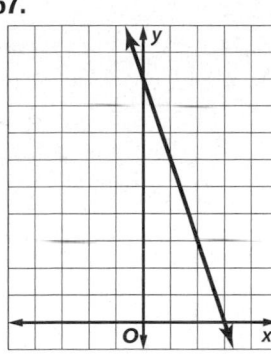

65.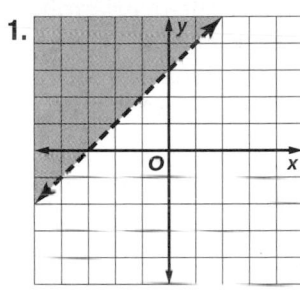

67.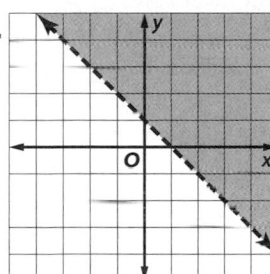

Lesson 5-6

1.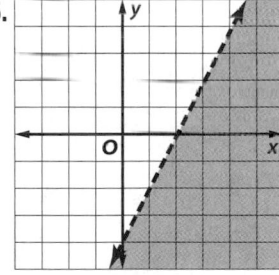

3.

5.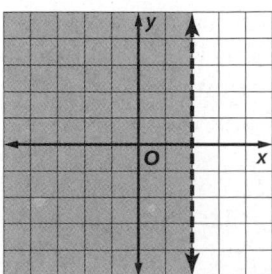

7. $x < 2$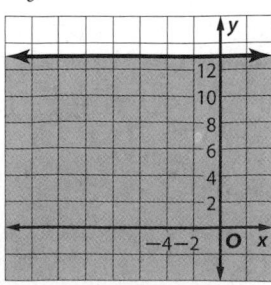

9. $y \leq 13$

11a. $115x + 685y > 2300$ **11b.** Sample answer:
1 skim board and 4 surfboards

13.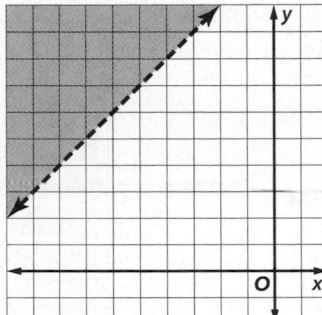

15.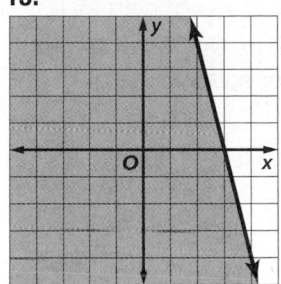

17.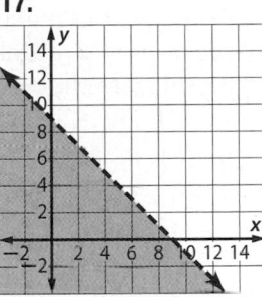

19.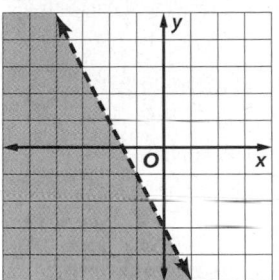

21.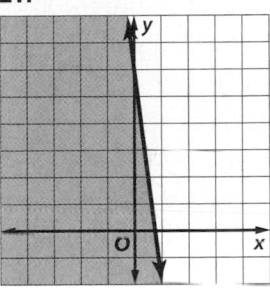

23.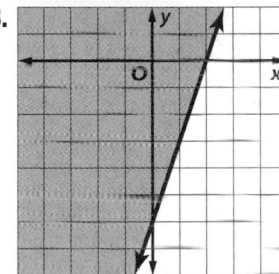

25. $x > 2$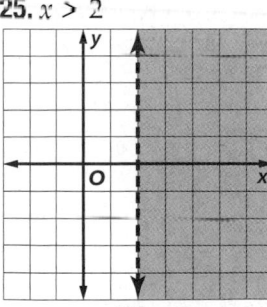

27. $y \leq 5$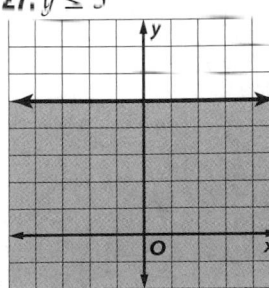

29. $x > -\dfrac{19}{14}$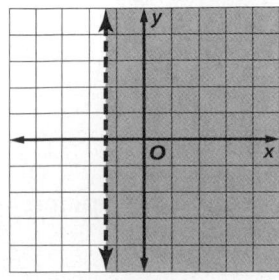

31. $x < -\dfrac{2}{3}$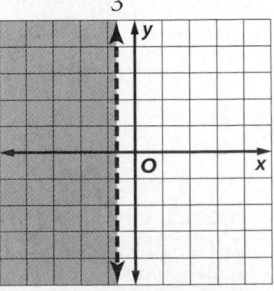

33. $x \le -\frac{2}{3}$

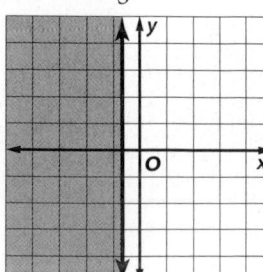

35. $x > \frac{3}{7}$

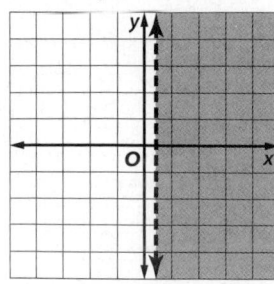

37 **a.** They need to make more than $2000. Let x represent the number of hot dogs sold and let y represent the number of sodas sold.
$1x + 1.25y \ge 2000$

b. Step 1: First, solve for y in terms of x.
$$x + 1.25y \ge 2000$$
$$1.25y \ge -x + 2000$$
$$y \ge 0.80x + 2000$$
Then, graph $y = 0.80x + 2000$. Because the inequality involves $\ge$, graph the boundary with a solid line.

Step 2: Select a test point in either half-plane. A simple choice is $(0, 0)$.

$x + 1.25y \ge 2000$ Original inequality
$0 + 1.25(0) \overset{?}{\ge} 2000$ Substitution
$0 \not\ge 2000$ false

Step 3: Since this statement is not true, the half-plane containing the origin is not the solution. Shade the other half-plane.

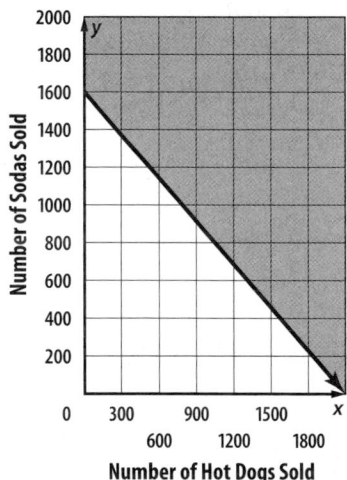

c. Sample answer: (400, 1600), (200, 1500), (300, 1400), (400, 1300), (1000, 1000)

d. Sample points should be in the shaded region of the graph in part **b**.

39 $x < -4$
Step 1: Graph $x = 4$ with a dotted line since it involves $<$.
Step 2: Choose a test point in either half-plane. A simple choice is $(0, 0)$.
$0 < -4$

R46 | Selected Answers

Step 3: Since this statement is not true, the half-plane containing the origin is not the solution. Shade the other half-plane.

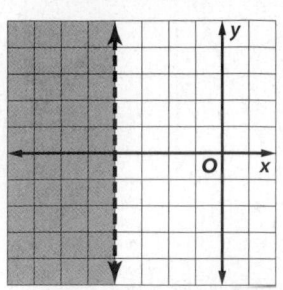

(2, 1): $2 < -4$ no
(−3, 0): $-3 < -4$ no
(0, −3): $0 < -4$ no
(−5, −5): $-5 < -4$ yes
(−4, 2): $4 < -4$ no

41.

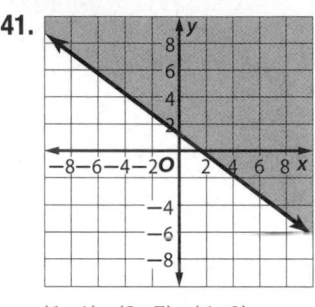

(1, 1), (2, 5), (6, 0)

43.

(7, 5), (5, 3), (2, −5)

45a. $y \le 3x - 4$; $y \le -x + 4$

45b.

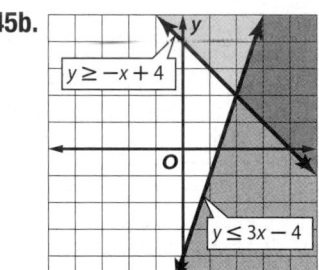

45c. The overlapping region represents the solutions that make both A and B true.

45d.

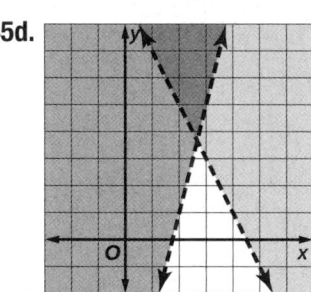

47. Sample answer: $y \le -x + 1$

49. Sample answer: The inequality $y > 10x + 45$ represents the cost of a monthly smartphone data plan with a flat rate of $45 for the first 2 GB of data used, plus $10 per each additional GB of data used. Both the domain and range are nonnegative real numbers because the GB used and the total cost cannot be negative.

51. B   **53.** F

55. $\{y \mid y > 6 \text{ or } y < -2\}$ **57.** $\varnothing$ **59.** $\{p \mid 4 < p < 10\}$

61. $y = 8x - 11$ **63.** $y = -\frac{3}{2}x - 17$ **65.** $r = \frac{w - sm}{10}$

Chapter 5   Study Guide and Review

1. false; more   **3.** false; intersection   **5.** true
7. true   **9.** true

11. $\{w \mid w > 13\}$ (number line: 5 6 7 8 9 10 11 12 13 14 15, open circle at 13, shaded right)

13. $\{h \mid h < -5\}$ (number line: −7−6−5−4−3−2−1 0 1 2 3, open circle at −5, shaded left)

15. $\{p \mid p \le -2\}$ (number line: −5−4−3−2−1 0 1 2 3 4 5, closed circle at −2, shaded left)

17. no more than 9

19. $\{g \mid g \ge -20\}$ (number line: −30 −20 −10 0 10, closed circle at −20, shaded right)

21. $\{w \mid w \le 11\}$ (number line: −20 −10 0 10 20, closed circle at 11, shaded left)

23. $\{t \mid t < -72\}$ (number line: −100 −80 −60 −40 −20 0, open circle at −72, shaded left)

25. $\{h \mid h < 7\}$ (number line: −10−8 −6 −4 −2 0 2 4 6 8 10, open circle at 7, shaded left)

27. $\{x \mid x \le -5\}$ (number line: −10−8 −6 −4 −2 0 2 4 6 8 10, closed circle at −5, shaded left)

29. Sample answer: Let x = the number; $4x - 6 < -2$; $\{x \mid x < 1\}$.

31. $\{m \mid 2 < m < 9\}$ (number line: 0 1 2 3 4 5 6 7 8 9 10 11, open circles at 2 and 9, shaded between)

33. $\{x \mid x \le 3 \text{ or } x > 6\}$ (number line: 0 1 2 3 4 5 6 7 8 9 10, closed circle at 3 shaded left, open circle at 6 shaded right)

35. $\{x \mid -5 < x < 13\}$ (number line: −6−4−2 0 2 4 6 8 10 12 14, open circles at −5 and 13, shaded between)

37. $\{c \mid -7 \le c \le 4\}$ (number line: −10−8−6−4−2 0 2 4 6 8 10, closed circles at −7 and 4, shaded between)

39. $\left\{d \mid -\dfrac{7}{3} \le d \le 3\right\}$ (number line: −5−4−3−2−1 0 1 2 3 4 5, closed circles, shaded between)

41. $\{t \mid t < -13 \text{ or } t > 7\}$ (number line: −14−12−10−8 −6 −4 −2 0 2 4 6 8, open circles at −13 and 7, shaded outward)

43. $\{m \mid \ \ 20 \le m \le -18\}$ (number line: −24−22−20−18−16−14−12−10 −8 −6 −4 −2, closed circles, shaded between)

45.

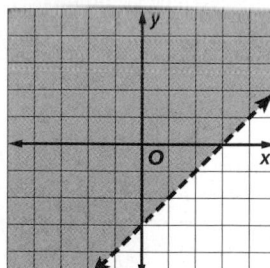

47.

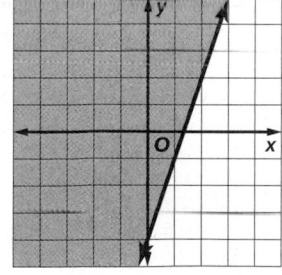

49.

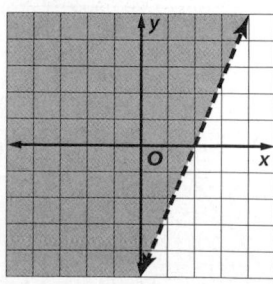

51. $(1, 2), (3, -2)$

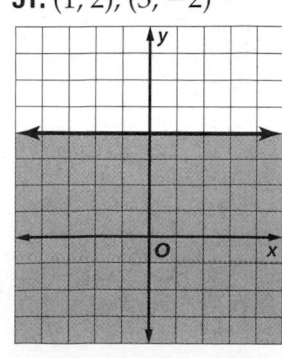

53. $2x + 3y \le 24$

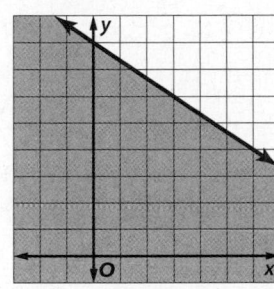

Get Ready

1. $(4, 0)$ **3.** $(-2, -3)$ **5.** $(-1, -1)$ **7.** $x = 6 - 2y$

9. $m = 2n + 6$ **11.** $\ell = \dfrac{P - 2w}{2}$ **13.** $b = \dfrac{2A}{h}$

Lesson 6-1

1. consistent and independent   **3.** inconsistent

5. consistent and independent

7.

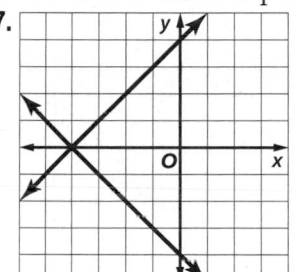

1 solution, $(-4, 0)$

9a. Alberto: $y = 20x + 35$; Ashanti: $y = 10x + 85$

9b.

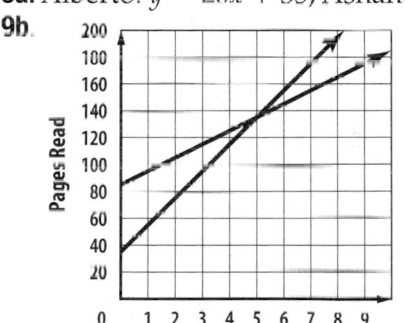

9c. $(5, 135)$; Alberto will have read more after 5 days.

11. consistent and independent

13 Since these two graphs intersect at one point, there is exactly one solution. Therefore, the system is consistent and independent.

15. consistent and independent

17. 1 solution; $\left(-\dfrac{5}{6}, -\dfrac{4}{3}\right)$ **19.** infinitely many

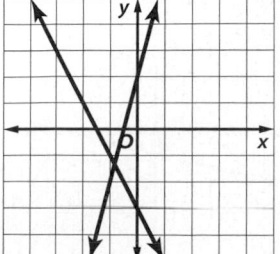

21. 1 solution; (5, −1)

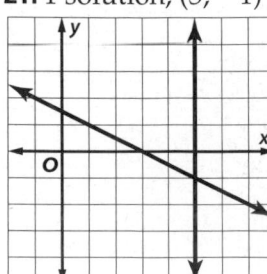

23. no solution

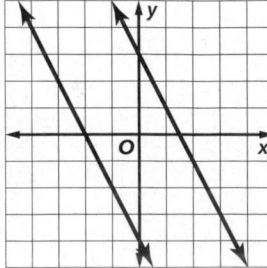

25a. Akira: $y = 30x + 22$; Jen: $y = 20x + 53$

25b.

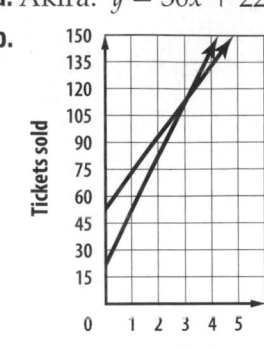

25c. (3.1, 115); After about 3 days Akira will have sold more tickets.

27 Graph the two equations on the same coordinate plane. The graphs appear to intersect at (−4, −2). Check by substituting into the equations.

$y = \frac{1}{2}x$

$-2 = \frac{1}{2}(-4)$

$-2 = -2$ Yes

$y = x + 2$

$-2 = -4 + 2$

$-2 = -2$ Yes

So, the solution is (−4, −2).

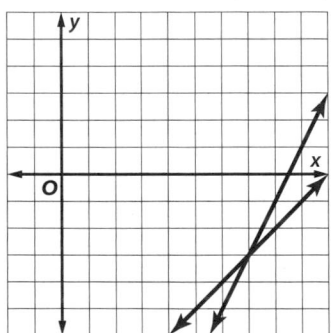

29.

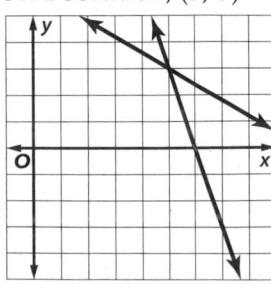

1 solution, (7, −3)

31. 1 solution, (5, 3)

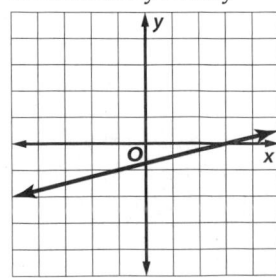

33. infinitely many

35. no solution

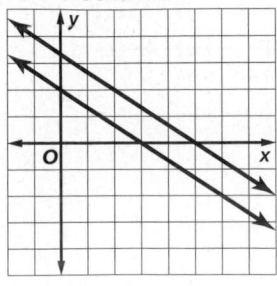

37. no solution

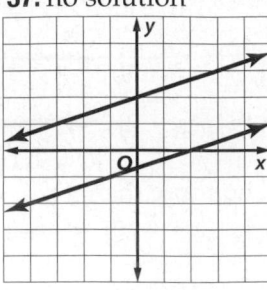

39. no solution

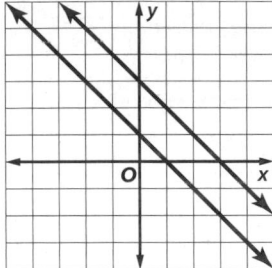

41. 1 solution, (3, −3)

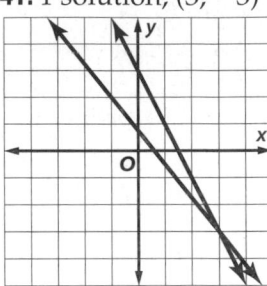

43.

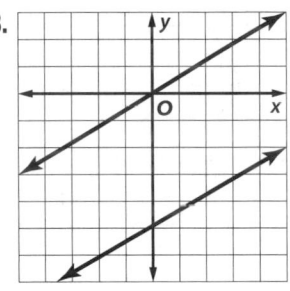

no solution

45 a. Lookatme:
Words: Started at 2.5 million and rose 13.1 million each year.
Expression: $2.5 + 13.1x$; $y = 13.1x + 2.5$
Buyourstuff:
Words: Started at 59 million and dropped by 2 million each year.
Expression: $59 - 2x$; $y = -2x + 59$

b.

Years Since 2005	Lookatme Vistors (mln)	Buyourstuff Vistors (mln)
0	2.5	59
1	15.6	57
2	28.7	55
3	41.8	53
4	54.9	51

c. Plot each equation on the same coordinate plane.

d. The graphs appear to intersect when $x = 3$, so they are about the same 3 years after 2009, or 2012.

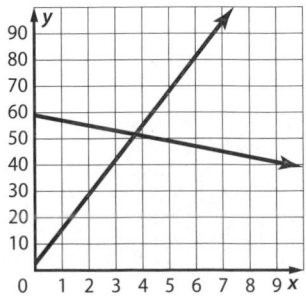

Selected Answers and Solutions

e. The domain, or input values will be all values greater than 0 since negative values do not make sense for years. So, D = $\{x \mid x \geq 0\}$ The range, or output values will be all values greater than 0 since negative values do not make sense for the number of visitors. So, R = $\{y \mid y \geq 0\}$.

47. Francisca; if the item is less than $100, then $10 off is better. If the item is more than $100, then the 10% is better. **49.** If the equations are linear and have more than one common solution, they must be consistent and dependent, which means that they have an infinite number of solutions in common. **51.** Sample answers: $y = 5x + 3$; $y = -5x - 3$; $2y = 10x - 6$ **53.** 14,745,600,000 bacteria **55.** H

57. 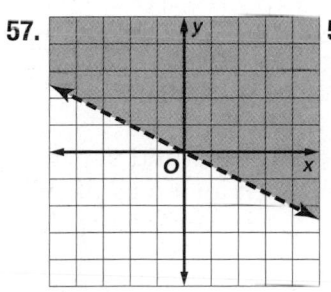 **59.**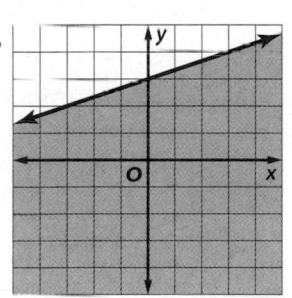

61.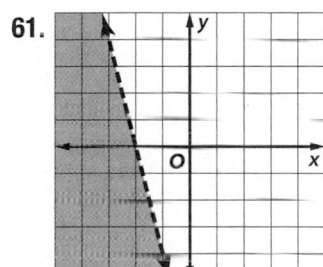

63. 1475 to 1525 books **65.** $y = -3x - 8$
67. $y = \frac{1}{2}x - 3$ **69.** 22 **71.** 92 **73.** −16 **75.** 7

Lesson 6-2

1. (5, 10) **3.** (2, 0) **5.** infinitely many
7a. $x = m\angle X$, $y = m\angle Y$; $x + y = 180$, $x = 24 + y$
7b. $x = 102°$, $y = 78°$

9 Step 1: One equation is already solved for y.
$y = 4x + 5$
$2x + y = 17$
Step 2: Substitute $4x + 5$ for y in the second equation.
$$2x + y = 17$$
$$2x + 4x + 5 = 17$$
$$6x + 5 = 17$$
$$6x = 12$$
$$x = 2$$
Step 3: Substitute 2 for x in either equation to find y.

$y = 4x + 5$
$y = 4(2) + 5$
$y = 8 + 5$
$y = 13$
The solution is (2, 13).

11. (−3, −11) **13.** (−1, 0) **15.** infinitely many **17.** (2, 3) **19.** no solution **21.** (2, 0) **23a.** Let $x =$ number of years since 2000, and let $y =$ the number of nurses; supply, $y = 5599.9x + 1,890,000$; demand, $y = 40,520.7x + 2,000,000$ **23b.** during 1996

25 **a.** Men: 1:51:39 = 60 + 51 = 111, then round up because the number of seconds is greater than 30. So, 1:51:39 rounds to 112.
1:44:51 = 60 + 44 = 104, then round up because the number of seconds is greater than 30. So, 1:44:51 rounds to 105.
Women: 1:54:33 = 60 + 54 = 114, then round up because the number of seconds is greater than 30. So, 1:54:31 rounds to 115.
1:59:14 = 60 + 59 = 119, then round down because the number of seconds is less than 30. So, 1:59:14 rounds to 119.

b. The y-intercept is (0, 112). Find the rate of change.
$$m = \frac{112 - 110}{0 - 5}$$
$$= \frac{2}{-5}$$
$$= -0.4$$
So, the equation is $y = -0.8x + 112$.
The y-intercept is (0, 115). Find the rate of change.
$$m = \frac{118 - 115}{5 - 0}$$
$$= \frac{3}{5}$$
$$- 0.6$$
So, the equation is $y = 0.4x + 115$.

c. Never; If you graph the two equations, the graphs do not cross in the positive values of x. Negative values will not make sense in terms of the word problem.

27. Neither; Guillermo substituted incorrectly for b. Cara solved correctly for b but misinterpreted the pounds of apples bought. **29.** Sample answer: The solutions found by each of these methods should be the same. However, it may be necessary to estimate using a graph. So, when a precise solution is needed, you should use substitution. **31.** An equation containing a variable with a coefficient of 1 can easily be solved for the variable. That expression can then be substituted into the second equation for the variable. **33.** 5/6 **35.** C

37. one solution; $(1, -5)$ **39.** infinitely many solutions

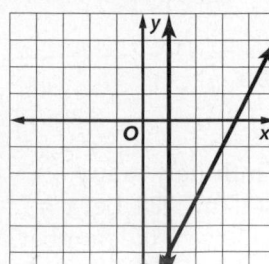 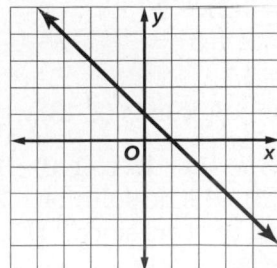

41. $v \geq -2$ **43.** $q \leq -40$ **45.** $t \geq 3$

47. $55b + 15$ **49.** $11h^2 + 12h$

Lesson 6-3

1. $(2, 3)$

(3) Step 1: The like terms are already aligned.
$$7f + 3g = -6$$
$$7f - 2g = -31$$
Step 2: Subtract the equations.
$$\begin{array}{r} 7f + 3g = -6 \\ (-)\,7f - 2g = -31 \\ \hline 5g = 25 \\ g = 5 \end{array}$$
Step 3: Substitute 5 for g in either equation to find f.
$$7f + 3(5) = -6$$
$$7f + 15 = -6$$
$$7f = -21$$
$$f = -3$$
The solution is $(-3, 5)$.

5. 6, 18 **7.** $(-3, 4)$ **9.** $(-3, 1)$ **11.** $(4, -2)$
13. $(8, -7)$ **15.** $(4, 7)$ **17.** $(4, 1.5)$ **19.** 5, 17

(21)

Three times a number	minus	another number	is	-3.
$3x$	$-$	y	$=$	-3

The first number	plus	the second number	is	11.
x	$+$	y	$=$	11

Steps 1 and 2: Write the equations vertically and add.
$$\begin{array}{r} 3x - y = -3 \\ x + y = 11 \\ \hline 4x = 8 \\ x = 2 \end{array}$$
Step 3: Substitute 2 for x in either equation to find y.
$$x + y = 11$$
$$2 + y = 11$$
$$y = 9$$
The numbers are 2 and 9.

23. adult, \$5.95; children, \$3.95 **25.** $(2, -1)$

27. $\left(-\dfrac{5}{6}, 3\right)$ **29.** $\left(2\dfrac{7}{9}, 13\dfrac{1}{3}\right)$ **31a.** $x + y = 66$;
$x = 30 + y$ **31b.** $(48, 18)$ **31c.** There are 48 teams that are not from the U.S. and 18 teams that are from the U.S.

31d.

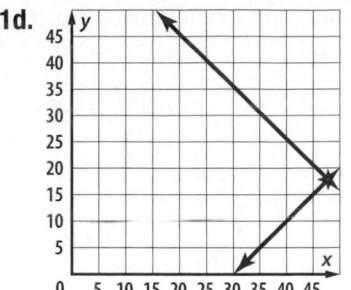

(33) **a.** One way to get 15 points is to use 4 pennies and 3 paper clips.
$$4(3) + 3 = 12 + 3$$
$$= 15$$

b. The total number of objects is 9.
$$p + c = 9$$
Pennies are worth 3 points each and paper clips are worth 1 point each for a total of 15 points.
$$3p + c = 15$$
Solve:
$$\begin{array}{r} p + c = 9 \\ (-)\,3p + c = 15 \\ \hline -2p = -6 \\ p = 3 \end{array}$$
Substitute 3 for p in either equation to find c,
$$p + c = 9$$
$$3 + c = 9$$
$$c = 6$$
So, $p = 3$ and $c = 6$.

c.

p	$c = 9 - p$	$3p + c$
0	9	$3(0) + 9 = 9$
1	8	$3(1) + 8 = 11$
2	7	$3(2) + 7 = 13$
3	6	$3(3) + 6 = 15$
4	5	$3(4) + 5 = 17$
5	4	$3(5) + 4 = 19$

d. Yes; since the pennies are 3 points each, 3 of them makes 9 points. Add the 6 points from 6 paper clips and you get 15 points.

35. The result of the statement is false, so there is no solution. **37.** Sample answer: $-x + y = 5$; I used the solution to create another equation with the coefficient of the x-term being the opposite of its corresponding coefficient. **39.** Sample answer: It would be most beneficial when one variable has either the same or opposite coefficients in each of the equations. **41.** A **43.** B **45.** (15, 5) **47.** (3, 11) **49.** (−2, 2) **51.** Yes; each pair of opposite sides have the same or an undefined slope, so they are parallel. **53.** −5 **55.** −20 **57.** $11w^2 - 9w$ **59.** $-2y - 35$

Lesson 6-4

1. (3, 2)

3 Eliminate y:

$(4x + 2y = -14)(-3)$ $-12x - 6y = 42$
$(5x + 3y = -17)(2)$ $\underline{10x + 6y = -34}$
                          $-2x = 8$
                          $x = -4$

Now, substitute −4 for x in either equation to find the value of y.

$4x + 2y = -14$
$4(-4) + 2y = -14$
$-16 + 2y = -14$
$2y = 2$
$y = 1$

The solution is (−4, 1).

5. 6 mph **7.** (−1, 3) **9.** (−3, 4) **11.** (−2, 3)
13. (3, 5) **15.** (1, −5) **17.** (0, 1)

19

Seven times a number	plus	three times another number	equals	−1.
$7x$	+	$3y$	=	−1

The sum of the two numbers is −3.

$x + y = -3$
$7x + 3y = -1$ $7x + 3y = -1$
$(x + y = -3)(-3)$ $\underline{-3x - 3y = 9}$
                      $4x = 8$
                      $x = 2$

Now, substitute 2 for x in either equation to find y.

$x + y = -3$
$2 + y = -3$
$y = -5$ The two numbers are 2 and −5.

21. (2.5, 3.25) **23.** $\left(3, \frac{1}{2}\right)$ **25a.** $240n + 360s = 3000$ **25b.** $90n + 120s = 1050$ **25c.** (5, 5); 5 nurses and 5 support staff were placed.

27 **a.** Let x be the cost of a batting token and y be the cost of the miniature golf games.
For the first group, the equation is $16x + 3y = 30$.
For the second group, the equation is $22x + 5y = 43$.
b. Solve.

$(16x + 3y = 30)(5)$ $80x + 15y = 150$
$(22x + 5y = 43)(-3)$ $\underline{-66x - 15y = -129}$
                         $14x = 21$
                         $x = 1.5$

Now, substitute 1.5 for x in either equation to find y.

$16x + 3y = 30$
$16(1.5) + 3y = 30$
$24 + 3y = 30$
$3y = 6$
$y = 2$

A batting token costs $1.50 and a game of miniature golf costs $2.

29. One of the equations will be a multiple of the other. **31.** Sample answer: $2x + 3y = 6$, $4x + 9y = 5$
33. Sample answer: It is more helpful to use substitution when one of the variables has a coefficient of 1 or if a coefficient can be reduced to 1 without turning other coefficients into fractions. Otherwise, elimination is more helpful because it will avoid the use of fractions when solving the system. **35.** G **37.** D **39.** (−1, −1) **41.** (9, 3)
43. (0, 6)

45. $m \le 13$ and $m \ge -3$

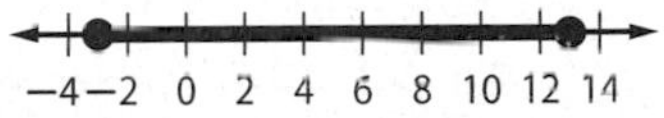

47. $w > 1$ or $w < -10$

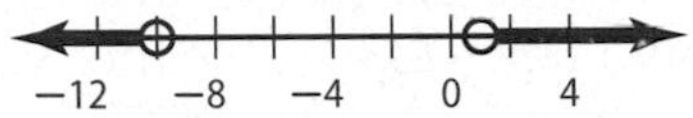

49. $A = \frac{1}{2}bh$ **51.** $V = \ell wh$ **53.** $A = \pi r^2$

Lesson 6-5

1. elim ($\times$); (2, −5) **3.** elim (+); $\left(-\frac{1}{3}, 1\right)$
5a. $4t + 3j = 181$; $t + 2j = 94$ **5b.** substitution
5c. Each T-shirt cost $16 and each pair of jeans cost $39 **7.** subst.; (2, −2) **9.** elim. (−); $\left(1, -\frac{1}{2}\right)$

11 $-5x + 4y = 7$
$-5x - 3y = -14$
Since there are no coefficients of 1, elimination is the best method.

$(-5x + 4y = 7)(-1)$ $5x - 4y = -7$
$-5x - 3y = -14$ $\underline{-5x - 3y = -14}$
                         $-7y = -21$
                         $y = 3$

Now substitute 3 for y in either equation to find x,

$-5x + 4y = 7$
$-5x + 4(3) = 7$
$-5x + 12 = 7$
$-5x = -5$
$x = 1$

The solution is (1, 3).

13. $m + t = 40$ and $m = 3t - 4$; 29 movies, 11 television shows **15.** 880 books; If they sell this number, then their income and expenses both equal $35,200.

17 **a.** Let x be the cost per pound of the aluminum cans and y be the cost per pound of the newspapers.
For Mara, the equation is $9x + 26y = 3.77$.
For Ling, the equation is $9x + 114y = 4.65$.

b. Elimination is the best method for solving these equations.

$(9x + 26y = 3.77)(-1)$ $\quad$ $-9x - 26y = -3.77$
$9x + 114y = 4.65$ $\qquad\qquad\underline{9x + 114y = 4.65}$
$\qquad\qquad\qquad\qquad\qquad\qquad 88y = 0.88$
$\qquad\qquad\qquad\qquad\qquad\qquady = 0.01$

Now substitute 0.01 for y in either equation to find x.

$9x + 26y = 3.77$
$9x + 26(0.01) = 3.77$
$9x + 0.26 = 3.77$
$9x = 3.51$
$x = 0.39$

The aluminum cans are $0.39 per pound. This solution is reasonable.

19a. $1.15 **19b.** $9.15 **21.** Sample answer: $x + y = 12$ and $3x + 2y = 29$, where x represents the cost of a student ticket for the basketball game and y represents the cost of an adult ticket; substitution could be used to solve the system; (5, 7) means the cost of a student ticket is $5 and the cost of an adult ticket is $7.

23. Graphing: (2, 5)

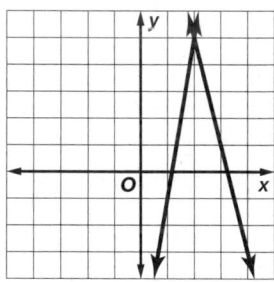

elimination by addition:
$4x + y = 13$
$\underline{6x - y = 7}$
$10x = 20$
$x = 2$
$4(2) + y = 13$
$y = 5$

substitution:
$y = -4x + 13$
$6x - (-4x + 13) = 7$
$6x + 4x - 13 = 7$
$10x = 20$
$x = 2$
$4(2) + y = 13$
$y = 5$

25. The third system; this system is the only one that is not a system of linear equations. **27.** A
29. 10 ft **31.** (0, 3) **33.** (2, 1)

35.

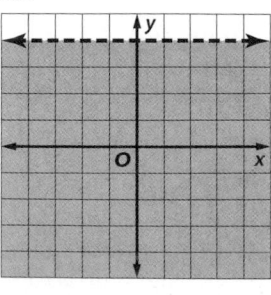

37.

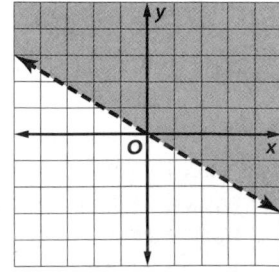

39. -12.31 **41.** 6.6 **43.** -93.19

1.

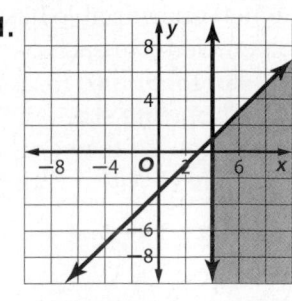

3.

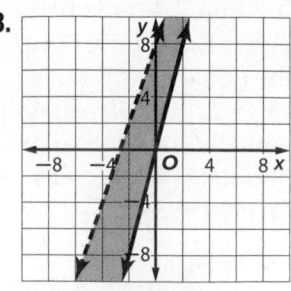

5.

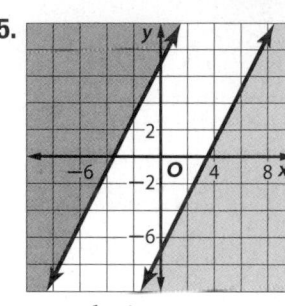

7.

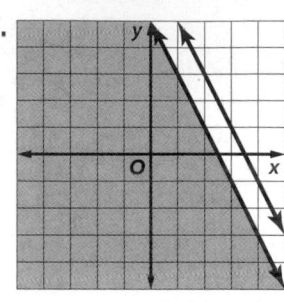

no solution

9a. Let h = the height of the driver in inches and w = the weight of the driver in pounds; $h < 79$ and $w < 295$.

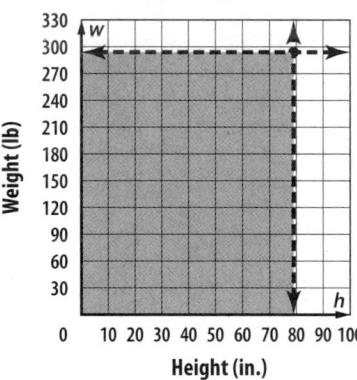

9b. Sample answer: 72 in. and 220 lb **9c.** Yes, the point falls in the overlapping region.

11 Graph both inequalities on the same coordinate plane.
$y \geq 0$ has a solid line.
$y \leq x - 5$ has a solid line.
The solution is the intersection of the shading.

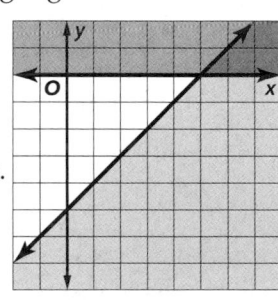

13.

15.

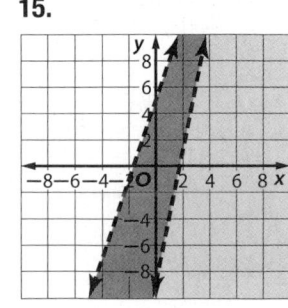

17.

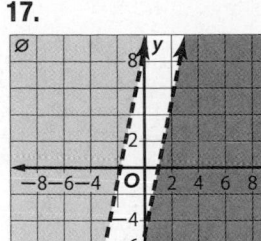

no solution

19.

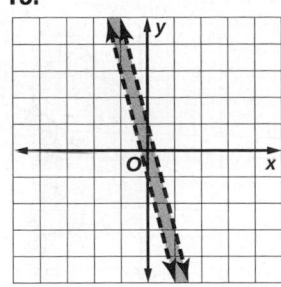

21.

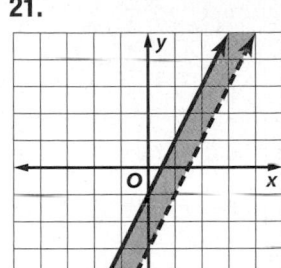

23.

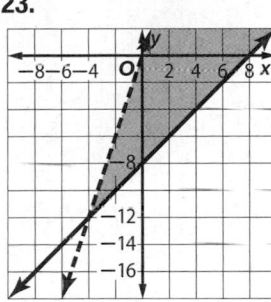

25a. Let f = square footage and let p = price; 1000 $\leq f \leq$ 17,000 and 10,000 $\leq p \leq$ 150,000

Ice Rink Resurfacers

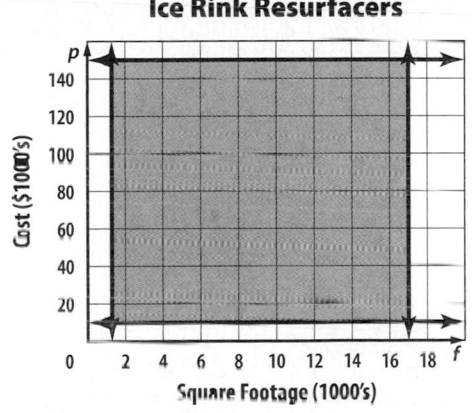

25b. Sample answer: an ice resurfacer for a rink of 5000 ft² and a price of $20,000   **25c.** Yes; the point satisfies each inequality.

27.

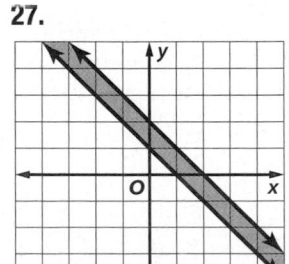

29.

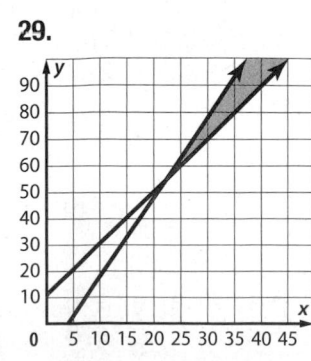

31.

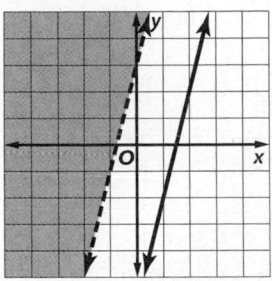

33.

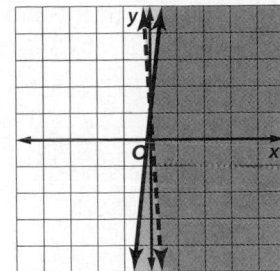

35.

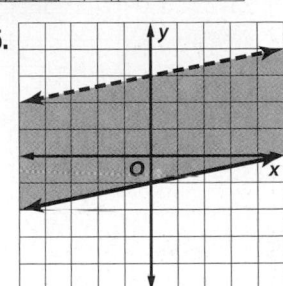

37 a. Let x be the number of hours she works for a photographer and y be the number of hours she works coaching.
$x + y \leq 20$
$15x + 10y \geq 90$
b. Graph both inequalities on the same grid.
$x + y \leq 20$ and $15x + 10y \geq 90$ have solid lines.
The solution is the intersection of the shading.

Earnings

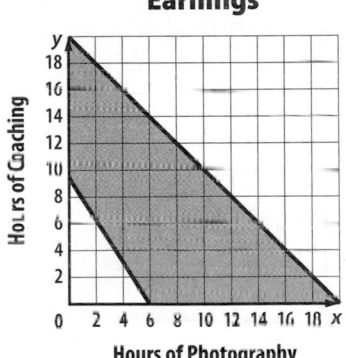

c. Two ordered pairs that are in the shaded area are (6, 10) and (8, 10). This means she could work for the photographer for 6 hours and coach for 10 or work for the photographer for 8 hours and coach for 10.

d. (2, 2) is not a solution because it does not fall in the shaded region. She would not earn enough money.

39. Sometimes; sample answer: $y > 3$, $y < -3$ will have no solution, but $y > -3$, $y < 3$ will have solutions.   **41.** Sample answer: $3x - y < -4$
43. Sample answer: The yellow region represents the beats per minute below the target heart rate. The blue region represents the beats per minute above the target heart rate. The green region represents the beats per minute within the target heart rate. Shading

in different colors clearly shows the overlapping solution set of the system of inequalities.   **45.** D
47. A   **49.** (4, 3)   **51.** (4, −3)

53.

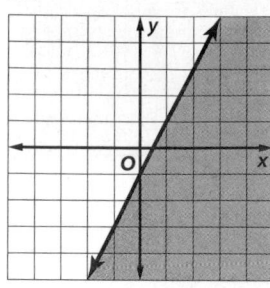

55.

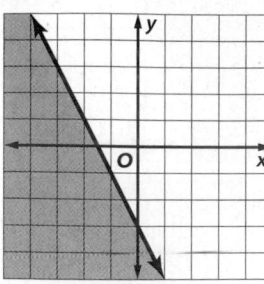

57. 16

Chapter 6   Study Guide and Review

1. true   **3.** false; dependent   **5.** true   **7.** false; system of inequalities

9. one; (3, 2)

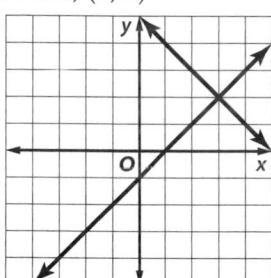

11. one; (0, 2)

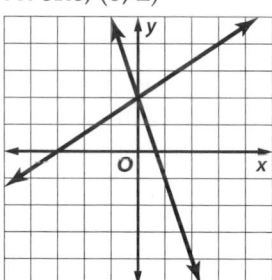

13.

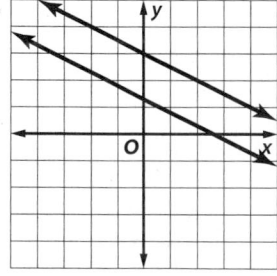

no solution

15. Sample answer: Let x be one number and y the other number; $x + y = 14$; $x − y = 4$; 9 and 5
17. (2, −10)   **19.** (2, −6)
21. (−3, 4)   **23.** (9, 4)
25. (4, −2)   **27.** $\left(\frac{1}{2}, 6\right)$
29. (−3, 5)   **31.** Sample

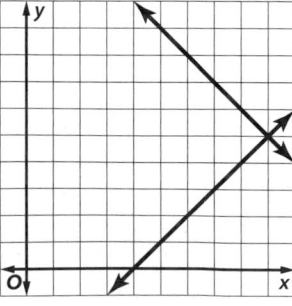

answer: Let f be the first type of card and let c be the second type of card; $f + c = 24, f + 3c = 50$; 11 $1 cards and 13 $3 cards.   **33.** (5, 7)   **35.** (2, 5)   **37.** (6, −1)
39. (1, −2)   **41.** (2, −6)   **43.** (24, −4)   **45.** (−2, 1)
47. (2, 5)   **49.** Sample answer: Let d represent the dimes and let q represent the quarters; $d + q = 25$, $0.10d + 0.25q = 4$; 15 dimes, 10 quarters

51.

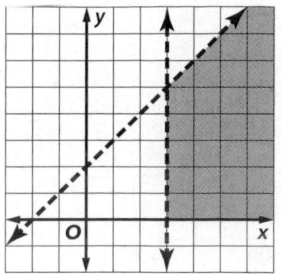

53.

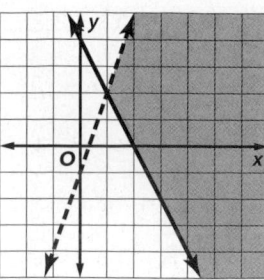

55.

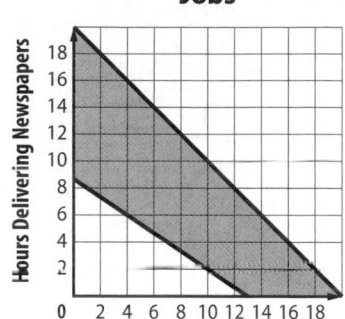

Jobs

CHAPTER 7
Exponents and Exponential Functions

Get Ready

1. 4^5 **3.** 6^2 **5.** b^6 **7.** $\left(\frac{1}{3}\right)^8$ or $\frac{1}{3^8}$ **9.** 4π m^2
11. 24 in^2 **13.** 25   **15.** −64   **17.** $\frac{1}{16}$

Lesson 7-1

1. Yes; constants are monomials.   **3.** No; there is a variable in the denominator.   **5.** Yes; this is a product of a number and variables.   **7.** k^4

⑨ $2q^2(9q^4) = (2 \cdot 9)(q^2 \cdot q^4)$
$= 18q^{2+4}$
$= 18q^6$

11. 3^8 or 6561   **13.** $16a^8b^{18}c^2$ **15.** $81p^{20}t^{24}$
17. $800x^8y^{12}z^4$ **19.** $−18g^7h^3j^{10}$ **21.** Yes; constants are monomials.   **23.** No; there is addition and more than one term.   **25.** Yes; this can be written as the product of a number and a variable.

㉗ $(q^2)(2q^4) = 2(q^2 \cdot q^4)$
$= 2q^{2+4}$
$= 2q^6$

29. $9w^8x^{12}$ **31.** $7b^{14}c^8d^6$ **33.** $j^{20}k^{28}$ **35.** 2^8 or 256
37. $4096r^{12}t^6$ **39.** $20c^5d^5$ **41.** $16a^{21}$ **43.** $512g^{27}h^{18}$
45. $294p^{27}r^{19}$ **47.** $30a^5b^7c^6$ **49.** $0.25x^6$ **51.** $−\frac{27}{64}c^3$
53. $−9x^3y^9$ **55.** $2,985,984r^{28}w^{32}$ **57a.** $0.12c$
57b. $280   **59.** $15x^7$

61 **a.** $V = \pi r^2 h$
$= \pi(2p^3)^2(4p^3)$
$= \pi(2^2)(p^3)^2(4p^3)$
$= \pi(4)(p^6)(4p^3)$
$= \pi(4 \cdot 4)(p^6 \cdot p^3)$
$= \pi(16)(p^{6+3})$
$= 16\pi p^9$

b.

radius	height	Volume
$4p$	p^7	$16\pi p^9$
$4p^2$	p^5	$16\pi p^9$
$2p^3$	$4p^3$	$16\pi p^9$
$2p^4$	$4p$	$16\pi p^9$
$2p$	$4p^7$	$16\pi p^9$

c. If the height of the container is doubled, the volume of the container is doubled. So, the volume is $32\pi p^9$.

63a.

Power	3^4	3^3	3^3	3^1	3^0	3^{-1}	3^{-3}	3^{-3}	3^{-4}
Value	81	27	9	3	1	$\frac{1}{3}$	$\frac{1}{9}$	$\frac{1}{27}$	$\frac{1}{81}$

63b. 1 and $\frac{1}{5}$ **63c.** $\frac{1}{a^n}$ **63d.** Any nonzero number raised to the zero power is 1.

65a.

Equation	Related Expression	Power of x	Linear or Nonlinear
$y = x$	x	1	linear
$y = x^2$	x^2	2	nonlinear
$y = x^3$	x^3	3	nonlinear

65b.

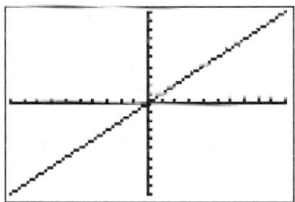

[−10, 10] scl: 1 by [−10, 10] scl: 1

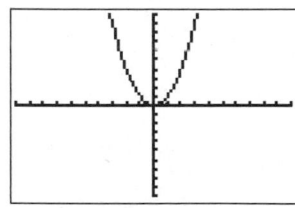

[−10, 10] scl: 1 by [−10, 10] scl: 1

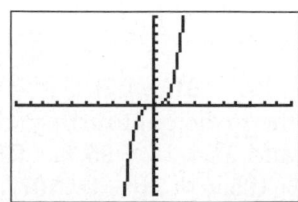

[−10, 10] scl: 1 by [−10, 10] scl: 1

65c. See chart for 65a. **65d.** If the power of x is 1, the equation or its related expression is linear. Otherwise, it is nonlinear. **67.** Sample answer: The area of a circle or $A = \pi r^2$, where r is the radius, can be used to find the area of any circle. The area of a rectangle or $A = w \times \ell$, where w is the width and ℓ is the length, can be used to find the area of any rectangle.
69. F **71.** The x-intercept does not change.

73.

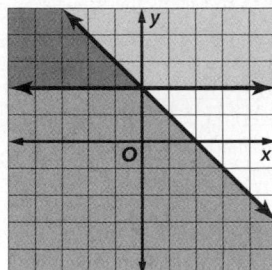

75.

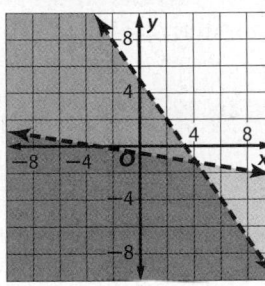

77. $\{p \mid 28 \leq p \leq 32\}$ **79.** 8 **81.** -7.05 **83.** 13

Lesson 7-2

1. $t^3 u^3$

3 $\dfrac{m^6 r^5 p^3}{m^5 r^2 p^3} = \left(\dfrac{m^6}{m^5}\right)\left(\dfrac{r^5}{r^2}\right)\left(\dfrac{p^3}{p^3}\right)$
$= m^{6-5} r^{5-2} p^{3-3}$
$= m^1 r^3 p^0$
$= mr^3$

5. ghm **7.** xyz **9.** $\dfrac{4a^6 b^{10}}{9}$ **11.** $\dfrac{32c^{15} d^{25}}{3125 g^{10}}$ **13.** 1

15. $\dfrac{g^2 h^4}{f^3}$ **17.** $\dfrac{a^5 c^{13}}{3b^9}$ **19.** $m^2 p$ **21.** $\dfrac{r^4 p^2}{4m^3 t^4}$ **23.** $\dfrac{9x^2 y^8}{25z^4}$

25. $\dfrac{p^6 t^{21}}{1000}$ **27.** $a^2 b^7 c$

29 $\left(\dfrac{2r^3 t^6}{5u^9}\right)^4 = \dfrac{2^4 (r^3)^4 (t^6)^4}{5^4 (u^9)^4}$
$= \dfrac{16r^{12} t^{24}}{625 u^{36}}$

31. 1 **33.** $\dfrac{p^4 r^2}{t^3}$ **35.** $\dfrac{-f}{4}$ **37.** $k^2 mp^2$ **39.** $\dfrac{3t^7}{u^6 v^2}$
41. $\dfrac{r^3}{t^2 x^{10}}$ **43.** 10^6; 10^8; about 10^2 or 100 times as many users as hosts **45.** $-\dfrac{w^9}{3}$
47. $1600k^{13}$ **49.** $\dfrac{5q}{r^6 t^3}$ **51.** $\dfrac{4g^{12}}{h^4}$ **53.** $\dfrac{4x^8 y^4}{z^6}$
55. $\dfrac{16z^2}{y^8}$ **57.** 100

59 **a.** the probability is $\frac{1}{6}$ multiplied d times, or $\left(\dfrac{1}{6}\right)^d$.

b. $\left(\dfrac{1}{6}\right)^d = (6^{-1})^d$
$= 6^{-d}$

61. Sometimes; sample answer: The equation is true when $x = 0$, $y = 2$, and $z = 3$, but it is false when $x = 1$, $y = 2$, and $z = 3$.

63. $\dfrac{1}{x^n} = \dfrac{x^0}{x^n} = x^{0-n} = x^{-n}$ **65.** The Quotient of Powers Property is used when dividing two powers with the same base. The exponents are subtracted. The Power of a Quotient Property is used to find the power of a quotient. You find the power of the numerator and the power of the denominator.   **67.** J   **69.** B

71.

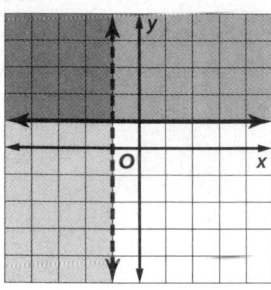

73.

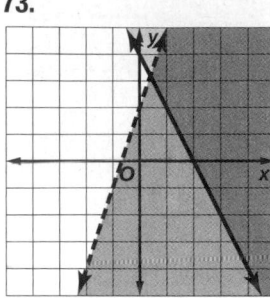

75. $h > 5$ **77.** $u \leq 35$ **79.** $n \geq -2$ **81.** 87   **83.** 121
85. 10,000   **87.** 125   **89.** 4096

Lesson 7-3

1. $\sqrt{12}$ **3.** $33^{\frac{1}{2}}$ **5.** 8   **7.** 7   **9.** 49

11. $216^{\frac{4}{3}} = \left(\sqrt[3]{216}\right)^4 = \left(\sqrt[3]{6 \cdot 6 \cdot 6}\right)^4 = 6^4$ or 1296

13. 4   **15.** 5.5   **17.** $\sqrt{15}$ **19.** $4\sqrt{k}$ **21.** $26^{\frac{1}{2}}$
23. $2(ab)^{\frac{1}{2}}$ **25.** 2   **27.** 6   **29.** 0.1   **31.** 11   **33.** 15
35. $\dfrac{1}{3}$ **37.** 4   **39.** 243   **41.** 625   **43.** $\dfrac{27}{1000}$ **45.** 5
47. $\dfrac{1}{2}$ **49.** $\dfrac{3}{2}$ **51.** 8   **53.** 8

55.

$4^{3x} = 512$	Original equation
$(2^2)^{3x} = 2^9$	Write the expressions with a common base, 2.
$2^{6x} = 2^9$	Power of a Power Property
$6x = 9$	Power Property of Equality
$x = \dfrac{3}{2}$	Divide each side by 6.

57. 4 ft   **59.** $\sqrt[3]{17}$ **61.** $7\sqrt[3]{b}$ **63.** $29^{\frac{1}{3}}$ **65.** $2a^{\frac{1}{3}}$
67. 0.3   **69.** a **71.** 16   **73.** $\dfrac{1}{3}$ **75.** $\dfrac{1}{27}$ **77.** $\dfrac{1}{\sqrt{k}}$
79. 12   **81.** -5 **83.** $-\dfrac{3}{2}$ **85a.** 440 Hz

85b. A below middle C, the 37th note

87.

$r = 0.62V^{\frac{1}{3}}$	Original equation
$3.65 = 0.62V^{\frac{1}{3}}$	$r = \dfrac{7.3}{2}$ or 3.65
$\dfrac{3.65}{0.62} \approx V^{\frac{1}{3}}$	Divide each side by 0.62.
$\left(\dfrac{3.65}{0.62}\right)^{3 \cdot \frac{1}{3}} \approx V^{\frac{1}{3}}$	$\dfrac{3.65}{0.62} = \left(\dfrac{3.65}{0.62}\right)^{3 \cdot \frac{1}{3}}$
$\left(\dfrac{3.65}{0.62}\right)^3 \approx V$	Power Property of Equality
$204.0 \approx V$	Simplify.

$r = 0.62V^{\frac{1}{3}}$	Original equation
$3.8 = 0.62V^{\frac{1}{3}}$	$r = \dfrac{7.6}{2}$ or 3.8
$\dfrac{3.8}{0.62} \approx V^{\frac{1}{3}}$	Divide each side by 0.62.
$\left(\dfrac{3.8}{0.62}\right)^{3 \cdot \frac{1}{3}} \approx V^{\frac{1}{3}}$	$\dfrac{3.8}{0.62} = \left(\dfrac{3.8}{0.62}\right)^{3 \cdot \frac{1}{3}}$
$\left(\dfrac{3.8}{0.62}\right)^3 \approx V$	Power Property of Equality
$230.2 \approx V$	Simplify.

So the volume of a size 3 ball is 204.0 to 230.2 in^3.

$r = 0.62V^{\frac{1}{3}}$	Original equation
$4.0 = 0.62V^{\frac{1}{3}}$	$r = \dfrac{8.0}{2}$ or 4.0
$\dfrac{4.0}{0.62} \approx V^{\frac{1}{3}}$	Divide each side by 0.62.
$\left(\dfrac{4.0}{0.62}\right)^{3 \cdot \frac{1}{3}} \approx V^{\frac{1}{3}}$	$\dfrac{4.0}{0.62} = \left(\dfrac{4.0}{0.62}\right)^{3 \cdot \frac{1}{3}}$
$\left(\dfrac{4.0}{0.62}\right)^3 \approx V$	Power Property of Equality
$268.5 \approx V$	Simplify.

$r = 0.62V^{\frac{1}{3}}$	Original equation
$4.15 = 0.62V^{\frac{1}{3}}$	$r = \dfrac{8.3}{2}$ or 4.15
$\dfrac{4.15}{0.62} \approx V^{\frac{1}{3}}$	Divide each side by 0.62.
$\left(\dfrac{4.15}{0.62}\right)^{3 \cdot \frac{1}{3}} \approx V^{\frac{1}{3}}$	$\dfrac{4.15}{0.62} = \left(\dfrac{4.15}{0.62}\right)^{3 \cdot \frac{1}{3}}$
$\left(\dfrac{4.15}{0.62}\right)^3 \approx V$	Power Property of Equality
$299.9 \approx V$	Simplify.

So the volume of a size 4 ball is 268.5 to 299.9 in^3.

$r = 0.62V^{\frac{1}{3}}$	Original equation
$4.3 = 0.62V^{\frac{1}{3}}$	$r = \dfrac{8.6}{2}$ or 4.3
$\dfrac{4.3}{0.62} \approx V^{\frac{1}{3}}$	Divide each side by 0.62.
$\left(\dfrac{4.3}{0.62}\right)^{3 \cdot \frac{1}{3}} \approx V^{\frac{1}{3}}$	$\dfrac{4.3}{0.62} = \left(\dfrac{4.3}{0.62}\right)^{3 \cdot \frac{1}{3}}$
$\left(\dfrac{4.3}{0.62}\right)^3 \approx V$	Power Property of Equality
$333.6 \approx V$	Simplify.

$r = 0.62V^{\frac{1}{3}}$	Original equation
$4.5 = 0.62V^{\frac{1}{3}}$	$r = \dfrac{9.0}{2}$ or 4.5
$\dfrac{4.5}{0.62} \approx V^{\frac{1}{3}}$	Divide each side by 0.62.
$\left(\dfrac{4.5}{0.62}\right)^{3 \cdot \frac{1}{3}} \approx V^{\frac{1}{3}}$	$\dfrac{4.5}{0.62} = \left(\dfrac{4.5}{0.62}\right)^{3 \cdot \frac{1}{3}}$
$\left(\dfrac{4.5}{0.62}\right)^3 \approx V$	Power Property of Equality
$382.4 \approx V$	Simplify.

So the volume of a size 5 ball is 333.6 to 382.4 in^3.

89. Sample answer: $2^{\frac{1}{2}}$ and $4^{\frac{1}{4}}$ **91.** $-1, 0, 1$
93. Sample answer: 2 is the principal fourth root of 16 because 2 is positive and $2^4 = 16$.   **95.** G   **97.** C
99. c^4d^6 **101.** b^2 **103.** 1   **105.** $y = 3x - 1$ **107.** $y = -2x - 12$ **109.** $y = \dfrac{2}{3}x + 7$ **111.** 1000   **113.** 0.1

1. 1.85×10^8 **3.** 5.64×10^{-4} **5.** 1.3×10^{10}
7. 19,800,000   **9.** 0.00000003405   **11.** 1.74×10^{15};
1,740,000,000,000,000   **13.** 4.7138×10^{-2}; 0.047138
15. 4.5×10^3; 4500   **17.** 8.5×10^{-13};
0.00000000000085   **19a.** 0.01, 0.000001
19b. 1×10^{-2}, 1×10^{-6} **19c.** 0.00000000001;
1×10^{-11}

㉑ 58,600,000
Step 1: 58,600,000 ⟶ 5.8600000
Step 2: The decimal point moved 7 places to the left, so $n = 7$.
Step 3: $58,600,000 = 5.8600000 \times 10^7$
Step 4: 5.86×10^7
23. 1.3×10^{-6} **25.** 7.09×10^{-10} **27.** 6.5×10^9
29. 94,000,000   **31.** 0.0005   **33.** 0.00000622
35. 11,000,000   **37.** 8×10^7; 80,000,000

㊴ $(6.5 \times 10^7)(7.2 \times 10^{-2}) = (6.5 \times 7.2)(10^7 \times 10^{-2})$
$\qquad = 46.8 \times 10^5$
$\qquad = (4.68 \times 10^1) \times 10^5$
$\qquad = 4.68 \times 10^6$
$\qquad = 4,680,000$
41. 2.2×10^7; 22,000,000   **43.** 1.96×10^{12};
1,960,000,000,000   **45.** 6.89×10^5; 689,000
47. 9×10^{-4}; 0.0009   **49.** 5×10^{-6}; 0.000005
51. 5.184×10^{15}; 5,184,000,000,000,000   **53.** 3.969×10^{-9}; 0.000000003969   **55.** 2.74185×10^5; 274,815
57. 6.1×10^{-8}; 0.000000061   **59.** 1.7889×10^{-6};
0.0000017889   **61.** 4.7008×10^3; 4700.8   **63.** 3×10^5

65.

Time	Kilometers Traveled
1 day	2.592×10^{10}
1 week	1.8144×10^{11}
1 month	7.776×10^{11}
1 year	9.4608×10^{12}

㊲ $(6.623 \times 10^9) \div (1.483 \times 10^8)$
$= (6.623 \div 1.483) \times (10^9 \div 10^8)$
$\approx 4.47 \times 10^1$
There are about 44.7 persons per square kilometer.
69a. corn: 9.29×10^7, 92,900,000; soybeans: 6.41×10^7, 64,100,000; cotton: 1.11×10^7, 11,100,000
69b. about 1.4493×10^0; 1.4493   **69c.** about 8.3694×10^0; 8.3694   **71.** Pete is correct; Syreeta moved the decimal point in the wrong direction.   **73.** Always; if the numbers are $a \times 10^m$ and $b \times 10^n$ in scientific notation, then $1 \le a < 10$ and $1 \le b < 10$. So $1 \le ab < 100$.   **75.** Sample answer: Divide the numbers to the left of the × symbols. Then divide the powers of 10. If necessary, rewrite the results in scientific notation. To convert that to standard form, check to see if the exponent is positive or negative. If positive, move the decimal point to the right, and if negative, to the left. The number of places to

move the decimal point is the absolute value of the exponent. Fill in with zeros as needed.
77. H   **79.** B   **81.** 8^3 or 512   **83.** $r^6 t^5$ **85.** $\dfrac{25d^6 g^4}{9h^8}$
87. 10^5 **89.** 4   **91.** 0   **93.** −75

1.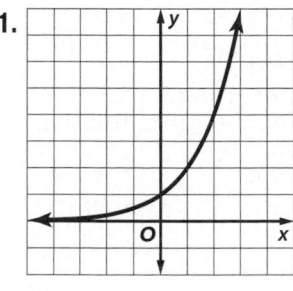
1;
D = {all real numbers};
R = {$y \mid y > 0$}

3.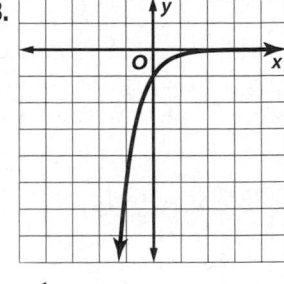
−1;
D = {all real numbers};
R = {$y \mid y < 0$}

5.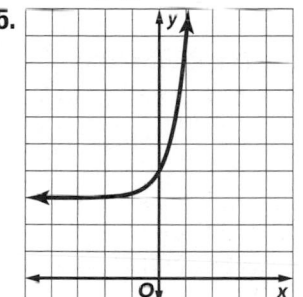

4;
D = {all real numbers};
R = {$y \mid y > 3$}

7a. D = {$d \mid d \ge 0$}, the number of days is greater than or equal to 0; R = {$y \mid y \ge 100$}, the number of fruit flies is greater than or equal to 100.   **7b.** about 198 fruit flies   **9.** Yes; the domain values are at regular intervals, and the range values have a common factor of 4.

11.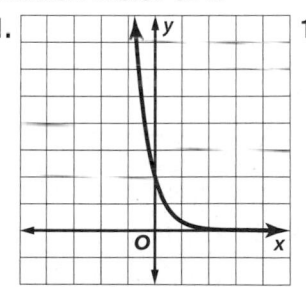
2;
D = {all real numbers};
R = {$y \mid y > 0$}

13.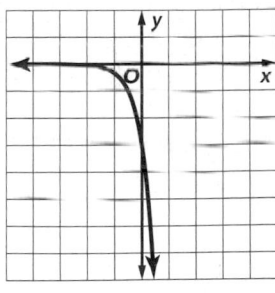
−3;
D = {all real numbers};
R = {$y \mid y < 0$}

15.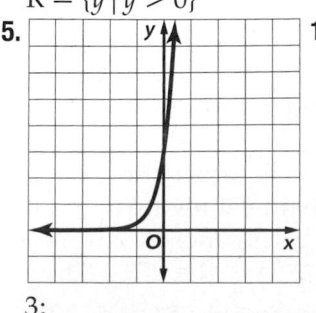
3;
D = {all real numbers};
R = {$y \mid y > 0$}

17.
The y-intercept is −3.5;
D = {all real numbers};
R = {$y \mid y > -4$}

19.

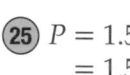

The y-intercept is 3;
D = {all real numbers};
R = $\{y \mid y < 5\}$

21. No; the domain values are at regular intervals of 4.

$2 \times (-2) = 4$
$-4 \times (-2) = 8$
$8 \times (-2) = -16$
$-16 \times (-2) = 32$

The range values differ by the common factor of -2. The range values do not have a positive common factor.

23. Yes; the domain values are at regular intervals, and the range values have a common factor of 2.

25. $P = 1.5^x$
$= 1.5^4$
≈ 5.06 or 506%

This enlargement is about 506% bigger than the original.

27. exponential **29.** linear **31.** neither
33. about 198 students **35.** a vertical stretch by a factor of 3 **37.** a translation down 3 units
39. a vertical stretch by a factor of 5 and a reflection over the x-axis. **41.** $f(x) = 3(2)^x$ **43.** Sample answer: The number of teams competing in a basketball tournament can be represented by $y = 2^x$, where the number of teams competing is y and the number of rounds is x.

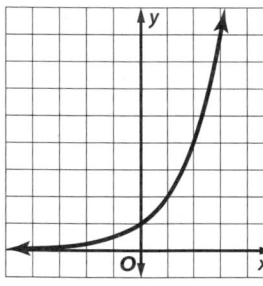

The y-intercept of the graph is 1. The graph increases quickly for $x > 0$. With an exponential model, each team that joins the tournament will play all of the other teams. If the scenario were modeled with a linear function, each team that joined would play a fixed number of teams.

45. Sample answer: First, look for a pattern by making sure that the domain values are at regular intervals and the range values differ by a common factor. **47.** B **49.** A **51.** 2.52×10^2, 252 **53.** 7
55. $\frac{1}{2}$ **57.** 7776 **59.** 32 km^2/min^2 **61.** $(-5, 20)$
63. 9, 11, 13 **65.** 16.5, 19, 21.5 **67.** $\frac{7}{2}, \frac{17}{4}, 5$

1. about \$37,734.73 **3a.** $y = 2200(0.98)^t$ **3b.** about 1624 **5.** about 92,095,349

7. $A = P\left(1 + \frac{r}{n}\right)^{nt}$
$= 6600\left(1 + \frac{0.045}{12}\right)^{12(4)}$
$= 6600(1.00375)^{48}$
≈ 7898.97 or about \$7898.97

9. Sample answer: No; she will have about \$199.94 in the account in 4 years.

11. $y = a(1 + r)^t$
$= 530,000(1 - 0.009)^5$
$= 530,000(0.991)^5$
$\approx 506,575$

13a. $I = 247,900(1.014)^t$ **13b.** about \$288,864
15a. $w(t) = 20,500(0.995)^t$ **15b.** $p(t) = 3000t$
15c. about 15.2 hr **15d.** $C(t) = 3000t + 20,500(0.995)^t$; The function represents the number of gallons of water remaining during the times in which more water is being pumped into the pool. **17.** about 9.2 yr **19.** Sample answer: Exponential models can grow without bound, which is usually not the case of the situation that is being modeled. For instance, a population cannot grow without bound due to space and food constraints. Therefore, when using a model, the situation that is being modeled should be carefully considered when used to make decisions. **21.** C **23.** D

25.

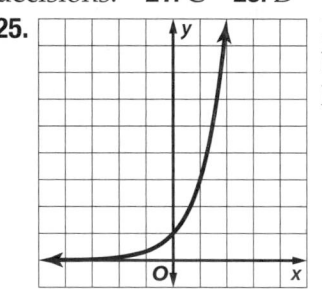

1;
D = {all real numbers},
R = $\{y \mid y > 0\}$

27.

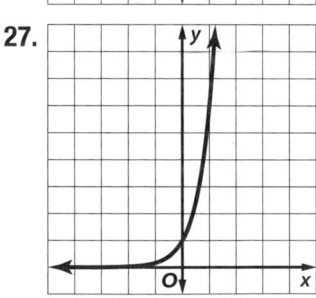

1;
D = {all real numbers},
R = $\{y \mid y > 0\}$ **29.** 3.01×10^{10}; 30,100,000,000
31. 1.21×10^{-4}; 0.000121
33. 1.9154×10^0; 1.9154
35. parallel **37.** neither
39. \$14.77 **41.** \$37.45

43.

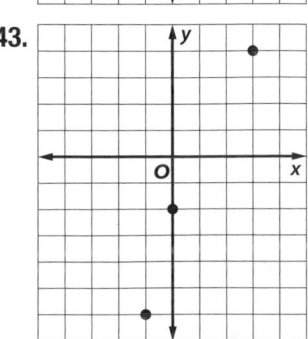

1. Geometric; the common ratio is $\frac{1}{5}$.

3. Arithmetic; the common difference is 3.

5. 160, 320, 640 **7.** $-\frac{1}{16}, -\frac{1}{64}, -\frac{1}{256}$ **9.** $a_n =$
$-6 \cdot (4)^{n-1}; -1536$ **11.** $a_n = 72 \cdot \left(\frac{2}{3}\right)^{n-1}; \frac{4096}{2187}$

13.

Experiment

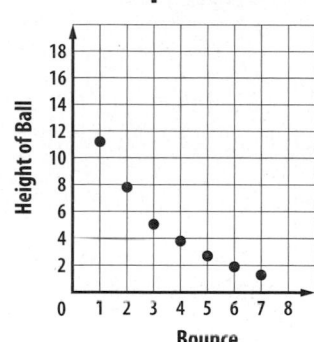

15. Arithmetic; the common difference is 10.

17. Geometric; the common ratio is $\frac{1}{2}$.

19. Neither; there is no common ratio or difference.

21 Step 1: Find the common ratio.
$36 \times \frac{1}{3} = 12$
$12 \times \frac{1}{3} = 4$
The common ratio is $\frac{1}{3}$.
Step 2: Multiply each term by the common ratio to find the next three terms.
$4\left(\frac{1}{3}\right) = \frac{4}{3}$
$\left(\frac{4}{3}\right)\left(\frac{1}{3}\right) = \frac{4}{9}$
$\left(\frac{4}{9}\right)\left(\frac{1}{3}\right) = \frac{4}{27}$
So, the next three terms are $\frac{4}{3}, \frac{4}{9}$, and $\frac{4}{27}$.

23. $\frac{25}{4}, \frac{25}{16}, \frac{25}{64}$ **25.** $-2, \frac{1}{4}, -\frac{1}{32}$ **27.** 134,217,728

29. $-1,572,864$ **31.** 19,683

33 **a.** $1 \times 2 = 2$
$2 \times 2 = 4$
The common ratio is 2, so the second option forms a geometric sequence.

b. first option: She will receive $30(9) = \$270$
second option: In the 9th week, she will get
$a_n = a_1 r^{n-1}$.
$a_n = 1(2)^{9-1}$
$= 1(2)^8$
$= 256$
In the 8th week, she will get $a_n = a_1 r^{n-1}$.
$a_n = 1(2)^{8-1}$
$= 1(2)^7$
$= 128$
Over the nine weeks she will earn: $1 + 2 + 4 + 8 + 16 + 32 + 64 + 128 + 256$ or $\$511$.
She should choose the second option.

35. $9; \frac{1}{3}$

37a.

Richter Number (x)	Increase in Magnitude (y)	Rate of Change (slope)
1	1	—
2	10	9
3	100	90
4	1,000	900
5	10,000	9000

37b.

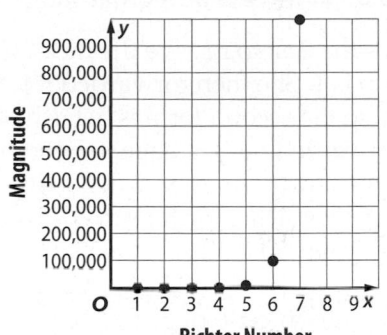

Richter Number

37c. The graph appears to be exponential. The rate of change between any two points does not match any others. **37d.** $1 \cdot (10)^{x-1} = y$ **39.** Neither; Haro calculated the exponent incorrectly. Matthew did not calculate $(-2)^8$ correctly. **41.** Sample answer: When graphed, the terms of a geometric sequence lie on a curve that can be represented by an exponential function. They are different in that the domain of a geometric sequence is the set of natural numbers, while the domain of an exponential function is all real numbers. Thus, geometric sequences are discrete, while exponential functions are continuous. **43.** B

45. 15 dimes and 20 quarters **47.** 162, 486, 1458

49. $\frac{1}{16}, -\frac{1}{32}, \frac{1}{64}$ **51.** 0.1296, 0.07776, 0.046656

53.

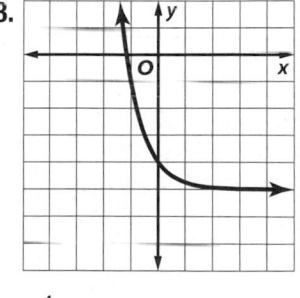

$-4;$
$D = \{\text{all real numbers}\};$
$R = \{y \mid y > -5\}$

55.

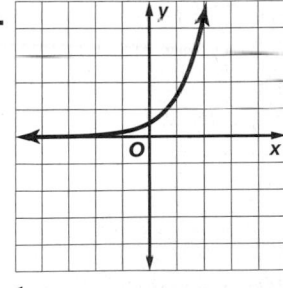

$\frac{1}{2};$
$D = \{\text{all real numbers}\};$
$R = \{y \mid y > 0\}$

57. at least $\$3747$ **59.** $y = -3x - \frac{2}{3}$ **61.** $y = \frac{1}{2}x - 9$

63. $y = -6x - 7$ **65.** $11a - 2$ **67.** $19w^2 + w$

69. $64t - 96$

1. 16, 13, 10, 7, 4 **3.** $a_1 = 1, a_n = a_{n-1} + 5, n \geq 2$

5a. $a_1 = 10, a_n = 0.6a_{n-1}, n \geq 2$

5b. $a_n = 10(0.6)^{n-1}$

7 $a_n = 5n + 8$ is an explicit formula for an arithmetic sequence with $d = 5$ and $a_1 = 5(1) + 8$ or 13. Therefore, $a_1 = 13$, $a_n = a_{n-1} + 5$, $n \geq 2$.
9. $a_n = 22(4)^{n-1}$ **11.** 48, −16, 16, 0, 8   **13.** 12, 15, 24, 51, 132   **15.** $\frac{1}{2}$, 2, $\frac{7}{2}$, 5, $\frac{13}{2}$ **17.** $a_1 = 27$, $a_n = a_{n-1} + 14$, $n \geq 2$ **19.** $a_1 = 100$, $a_n = 0.8a_{n-1}$, $n \geq 2$ **21.** $a_1 = 81$, $a_n = \frac{1}{3}a_{n-1}$, $n \geq 2$ **23.** $a_1 = 3$, $a_n = 4a_{n-1}$, $n \geq 2$ **25.** $a_n = 38\left(\frac{1}{2}\right)^{n-1}$

27 **a.** Barbara was the first to receive the message, so the first term is 1. She then forwarded it to 5 of her friends, so the second term is 5. Each of her 5 friends forwarded the message to 5 more friends, so the third term is 5 · 5 or 25. This pattern continues. The fourth term is 25 · 5 or 125, and the fifth term is 125 · 5 or 625. Therefore, the first five terms are 1, 5, 25, 125, 625.
b. There is a common ratio of 5. The sequence is geometric.
$a_n = r \cdot a_{n-1}$
$a_n = 5a_{n-1}$
The first term a_1 is 1, and $n \geq 2$. So, $a_1 = 1$, $a_n = 5a_{n-1}$, $n \geq 2$.
c. We found that $a_5 = 625$.
$a_6 = 5a_{6-1}$
$\quad = 5a_5$
$\quad = 5(625)$ or 3125
$a_7 = 5a_{7-1}$
$\quad = 5a_6$
$\quad = 5(3125)$ or 15,625
$a_8 = 5a_{8-1}$
$\quad = 5a_7$
$\quad = 5(15,625)$ or 78,125

29a. $a_1 = 10$, $a_n = 1.1a_{n-2}$, $n \geq 2$ **29b.** 16.1 ft
31. Both; sample answer: The sequence can be written as the recursive formula $a_1 = 2$, $a_n = (-1)a_{n-1}$, $n \geq 2$. The sequence can also be written as the explicit formula $a_n = 2(-1)^{n-1}$.
33. False; sample answer: A recursive formula for the sequence 1, 2, 3, … can be written as $a_1 = 1$, $a_n = a_{n-1} + 1$, $n \geq 2$ or as $a_1 = 1$, $a_2 = 2$, $a_n = a_{n-2} + 2$, $n \geq 3$.   **35.** Sample answer: In an explicit formula, the nth term a_n is given as a function of n. In a recursive formula, the nth term a_n is found by performing operations to one or more of the terms that precede it.   **37.** G   **39.** F   **41.** −54, 81, −121.5
43. −64, 32, −16   **45.** 1500; 7500; 37,500
47. adults: \$14; children: \$10   **49.** $4x - y = 16$
51. $x - 3y = -18$ **53.** $2x + 7y = 26$ **55.** $10x - 64$
57. simplified   **59.** simplified

Chapter 7   Study Guide and Review

1. monomial   **3.** cube root   **5.** scientific notation
7. recursive formula   **9.** exponential decay   **11.** x^9

13. $20a^6b^6$ **15.** $64r^{18}t^6$ **17.** $8x^{15}$ **19.** $45\pi x^4$
21. $\frac{27x^3y^9}{8z^3}$ **23.** $\frac{c^6}{a^3}$ **25.** x^6 **27.** $\frac{6}{yx^3}$ **29.** 7   **31.** 5
33. 64   **35.** 2401   **37.** 5   **39.** 2.3×10^6
41. about 9.1×10^{-2}

43. y-intercept 2; D = {all real numbers}; R = $\{y \mid y > 1\}$

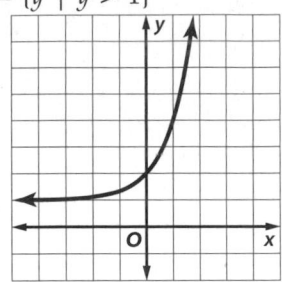

45. y-intercept −2; D = {all real numbers}; R = $\{y \mid y > -3\}$

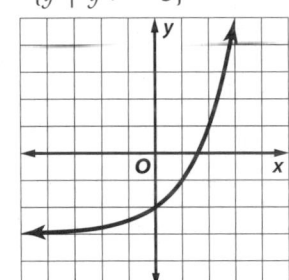

47. \$3053.00   **49.** −1, 1, −1   **51.** 32, 16, 8
53. $a_n = 3(3)^{n-1}$

55.

Basketball Rebound

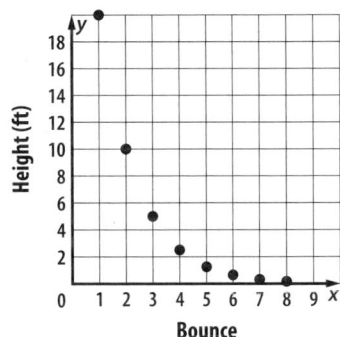

57. 3, 12, 30, 66, 138   **59.** $a_1 = 32$, $a_n = 0.5a_{n-1}$, $n \geq 2$

CHAPTER 8
Quadratic Expressions and Equations

Chapter 8   Get Ready

1. $a(a) + a(5)$; $a^2 + 5a$ **3.** $n(n) + n(-3n^2) + n(2)$; $n^2 - 3n^3 + 2n$ **5.** $5(9 + 3 + 6)$; \$90   **7.** $11a - 2$
9. $19w^2 + w$ **11.** simplified   **13.** $a^2 - 8a + 16$
15. $9g^2 - 9g + 2$ **17.** $4n^5$ **19.** $-15z^9$ **21.** $-14a^5c^9$

Lesson 8-1

1. yes; 3; trinomial   **3.** yes; 2; monomial   **5.** yes; 5; binomial   **7.** $2x^5 + 3x - 12$; 2   **9.** $-5z^4 - 2z^2 + 4z$; −5   **11.** $4x^3 + 5$

13 $(4 + 2a^2 - 2a) - (3a^2 - 8a + 7) = (2a^2 - 2a + 4)$
$- (3a^2 - 8a + 7) = (2a^2 - 3a^2) + (-2a - (-8a)) + (4 - 7) = -a^2 + 6a - 3$
15. $-8z^3 - 3z^2 - 2z + 13$ **17.** $4y^2 + 3y + 3$
19a. $D(n) = 6n + 14$ **19b.** 116,000 students
19c. 301,000 students **21.** yes; 0; monomial
23. No; the exponent is a variable. **25.** yes; 4;
binomial **27.** $7y^3 + 8y$; 7 **29.** $-y^3 - 3y^2 + 3y + 2$;
-1 **31.** $-r^3 + r + 2$; -1 **33.** $-b^6 - 9b^2 + 10b$; -1

35 $(2x + 3x^2) - (7 - 8x^2) = (2x + 3x^2) + (-7 + 8x^2)$
$= [3x^2 + 8x^2] + 2x + (-7)$
$= 11x^2 + 2x - 7$
37. $2z^2 + z - 11$ **39.** $-2b^2 + 2a + 9$
41. $7x^2 - 2xy - 7y$ **43.** $3x^2 - rxt - 8r^2x - 6rx^2$
45. quadratic trinomial **47.** quartic binomial
49. quintic polynomial **51a.** $s = 0.55t^2 - 0.05t + 3.7$
51b. 3030 **53a.** the area of the rectangle
53b. the perimeter of the rectangle
55. $10a^2 - 8a + 16$ **57.** $7n^3 - 7n^2 - n - 6$

59 **a.** Words: \$15 plus \$0.15 per mile
Expression: $15 + 0.15m$
The expression is $15 + 0.15m$.
b. $15 + 0.15m = 15 + 0.15(145)$
$= 15 + 21.75$
$= 36.75$
The cost is \$36.75.

61. Neither; neither of them found the additive
inverse correctly. All terms should have been
multiplied by -1. **63.** $6n + 9$ **65.** Sample answer:
To add polynomials in a horizontal format,
combine like terms. For the vertical format, write
the polynomials in standard form, align like terms
in columns, and combine like terms. To subtract
polynomials in a horizontal format you find the
additive inverse of the polynomial you are
subtracting, and then combine like terms. For the
vertical format, you write the polynomials in
standard form, align like terms in columns, and
subtract by adding the additive inverse.
67. $8x + 12$ units **69.** C **71.** Geometric; the common
ratio is -4. **73.** Neither; there is no common ratio
or difference. **75.** Arithmetic; the common
difference is 11. **77.** \$80,000 **79.** no **81.** no
83. yes; 0.5 **85.** $-2n^8$ **87.** $-40u^5z^9$ **89.** 64
91. $288x^8y^{10}z^6$

Lesson 8-2

1. $-15w^3 + 10w^2 - 20w$ **3.** $32k^2m^4 + 8k^3m^3 + 20k^2m^2$
5 $2ab(7a^4b^2 + a^5b - 2a) = 2ab(7a^4b^2) + 2ab(a^5b) + 2ab(-2a)$
$= 14a^5b^3 + 2a^6b^2 + (-4a^2b)$
$= 14a^5b^3 + 2a^6b^2 - 4a^2b$

7. $4t^3 + 15t^2 - 8t + 4$ **9.** $-5d^4c^2 + 8d^2c^2 - 4d^3c + dc^4$ **11.** 20 **13.** $-\dfrac{20}{9}$ **15.** 20 **17.** 1 **19.** $f^3 + 2f^2 + 25f$ **21.** $10j^5 - 30j^4 + 4j^3 + 4j^2$ **23.** $8t^5u^3 - 40t^4u^5 + 8t^3u$ **25.** $-8a^3 + 20a^2 + 4a - 12$ **27.** $-9g^3 + 21g^2 + 12$ **29.** $8n^4p^2 + 12n^2p^2 + 20n^2 - 8np^3 + 12p^2$
31 $7(t^2 + 5t - 9) + t = t(7t - 2) + 13$
$7t^2 + 35t - 63 + t = 7t^2 - 2t + 13$
$7t^2 + 36t - 63 = 7t^2 - 2t + 13$
$36t - 63 = -2t + 13$
$38t = 76$
$t = 2$
33. $\dfrac{43}{6}$ **35.** $\dfrac{30}{43}$ **37.** $20np^4 + 6n^3p^3 - 8np^2$
39. $-q^3w^3 - 35q^2w^4 + 8q^2w^2 - 27qw$ **41a.** $53.50 - 0.25h$ **41b.** \$50.50
43 **a.** $A = \ell w$
$= (1.5x + 24)x$
$= 1.5x^2 + 24x$
b. $x(x - 9) = x^2 - 9x$
c. $2(2.5x) = 2(2.5)(36)$
$= 180$ ft
$2(x + 6) = 2(36 + 6)$
$= 2(42)$
$= 84$ ft
Perimeter $= 180 + 84$ or 264 ft
45. Ted; Pearl used the Distributive Property
incorrectly. **47.** $8x^2y^{-2} + 24x^{-10}y^8 - 16x^{-3}$
49. Sample answer: $3n, 4n + 1; 12n^2 + 3n$ **51.** B
53. A **55.** $-3x^2 + 1$ **57.** $-9a^2 + 4a + 7$ **59.** $6ab + 2a + 4b$ **61.** $a_1 = 16, a_n = a_{n-1} - 14, n \ge 2$
63. $a_1 = 27, a_n = a_{n-1} + 16, n \ge 2$ **65.** $a_1 = 100,$
$a_n = 0.6a_{n-1}, n \ge 2$ **67.** y $9,500,000 = 740,000(x - 2003); 14,680,000$ people **69.** $f(x) = -0.5x$ **71.** $6y^3$ **73.** $15z^7 - 6z^4$ **75.** $-8p^5 + 10p^{10}$

Lesson 8-3

1. $x^2 + 7x + 10$ **3.** $b^2 - 4b - 21$ **5.** $16h^2 - 26h + 3$
7. $4x^2 + 72x + 320$ **9.** $16y^4 + 28y^3 - 4y^2 - 21y - 6$
11. $10n^4 + 11n^3 - 52n^2 - 12n + 48$
13. $2g^2 + 15g - 50$
15 $(4x + 1)(6x + 3) = 4x(6x) + 4x(3) + 1(6x) + 1(3)$
$= 24x^2 + 12x + 6x + 3$
$= 24x^2 + 18x + 3$
17. $24d^2 - 62d + 35$ **19.** $49n^2 - 84n + 36$
21. $25r^2 - 49$ **23.** $33z^2 + 7yz - 10y^2$ **25.** $2y^3 - 17y^2 + 37y - 22$ **27.** $m^4 + 2m^3 - 34m^2 + 43m - 12$
29. $6b^5 - 3b^4 - 35b^3 - 10b^2 + 43b + 63$ **31.** $2m^3 + 5m^2 - 4$ **33.** $4\pi x^2 + 12\pi x + 9\pi - 3x^2 - 5x - 2$
35 **a.** $A = \ell w$
$= (3y + 4)(6y - 5)$
$= 3y(6y) + 3y(-5) + 4(6y) + 4(-5)$
$= 18y^2 - 15y + 24y - 20$
$= 18y^2 + 9y - 20$
b. $3y + 4 = 31$
$3y = 27$
$y = 9$

So, the width is $6y - 5 = 6(9) - 5$
$$= 54 - 5$$
$$= 49$$

$A = \ell w$
$$= (31)(49)$$
$$= 1519 \text{ ft}^2$$

37. $a^2 - 4ab + 4b^2$ **39.** $x^2 - 10xy + 25y^2$
41. $125g^3 + 150g^2h + 60gh^2 + 8h^3$ **43a.** $x > 4$; If
$x = 4$, the width of the rectangular sandbox would
be zero and if $x < 4$ the width of the rectangular
sandbox would be negative. **43b.** square **43c.** 4 ft^2
45. Always; by grouping two adjacent terms, a
trinomial can be written as a binomial, the sum of
two quantities, and apply the FOIL method. For
example, $(2x + 3)(x^2 + 5x + 7) = (2x + 3)[x^2 + (5x + 7)] = 2x(x^2) + 2x(5x + 7) + 3(x^2) + 3(5x + 7)$.
Then use the Distributive Property and simplify.
47. Sample answer: $x - 1, x^2 - x - 1.$ $(x - 1)(x^2 - x - 1) = x^3 - 2x^2 + 1$ **49.** The Distributive Property
can be used with a vertical or horizontal format by
distributing, multiplying, and combining like
terms. The FOIL method is used with a horizontal
format. You multiply the first, outer, inner, and last
terms of the binomials and then combine like
terms. A rectangular method can also be used by
writing the terms of the polynomials along the top
and left side of a rectangle and then multiplying
the terms and combining like terms.
51. F **53.** $\frac{3}{2}$ **55.** $4a^2 + 5$ **57.** $3n^3 - 6n^2 + 10$
59. $4b + c + 2$ **61.** $-7m^3 - 3m^2 - m + 17$
63. $-56t^{12}$ **65.** $50y^6 - 27y^9$

Lesson 8-4

1. $x^2 + 10x + 25$

③ $(2x + 7y)^2 = (2x)^2 + 2(2x)(7y) + (7y)^2$
$$= 4x^2 + 28xy + 49y^2$$
5. $g^2 - 8gh + 16h^2$ **7a.** $0.5Dy + 0.5y^2$ **7b.** 50%
9. $x^2 - 25$ **11.** $81t^2 - 36$ **13.** $b^2 - 12b + 36$
15. $x^2 + 12x + 36$ **17.** $81 - 36y + 4y^2$ **19.** $25t^2 - 20t + 4$ **21a.** $(T + t)^2 = T^2 + 2Tt + t^2$
21b. TT: 25%; Tt: 50%; tt: 25%

㉓ $(b + 7)(b - 7) = b^2 - (7)^2$
$$= b^2 - 49$$
25. $16 - x^2$ **27.** $9a^4 - 49b^2$ **29.** $64 - 160a + 100a^2$ **31.** $9t^2 - 144$ **33.** $9q^2 - 30qr + 25r^2$
35. $g^2 + 10gh + 25h^2$ **37.** $9a^8 - b^2$ **39.** $64a^4 - 81b^6$
41. $\frac{4}{25}y^2 - \frac{16}{5}y + 16$ **43.** $4m^3 + 16m^2 - 9m - 36$
45. $2x^2 + 2x + 5$ **47.** $6x + 3$ **49.** $c^3 + 3c^2d + 3cd^2 + d^3$ **51.** $f^3 + f^2g - fg^2 - g^3$
53. $n^3 - n^2p - np^2 + p^3$

㊺ a. $A = 3.14(r + 9)^2$
$$= 3.14(r^2 + 18r + 81)$$
$$\approx (3.14r^2 + 56.52r + 254.34) \text{ ft}^2$$

b. $38^2 - (3.14r^2 + 56.52r + 254.34)$
$$= 1444 - 3.14r^2 - 56.52r - 254.34$$
$$\approx (1189.66 - 3.14r^2 - 56.52r) \text{ ft}^2$$
57. Sample answer: $(2c + d)(2c - d)$; The product of
these binomials is a difference of two squares and
does not have a middle term. The other three do.
59. 81 **61.** Sample answer: To find the square of a
sum, apply the FOIL method or apply the pattern.
The square of the sum of two quantities is the first
quantity squared plus two times the product of the
two quantities plus the second quantity squared.
The square of the difference of two quantities is
the first quantity squared minus two times the
product of the two quantities plus the second
quantity squared. The product of the sum and
difference of two quantities is the square of the
first quantity minus the square of the second
quantity. **63.** D **65.** C **67.** $2c^2 + 5c - 3$
69. $8h^2 - 34h + 21$ **71.** $40m^2 + 47m + 12$
73. $3c^2 - 2c$ **75.** $-13d^2 - 18d$ **77.** $19p^2 - 18p$
79. $(2, 0)$ **81.** $t < 18$ or $t > 22$ **83.** $y = \frac{1}{5}x + 6$
85. 15 lb **87.** $2p^5 - 5p^4 + p^2 + 12; 2$

Lesson 8-5

1. $3(7b - 5a)$ **3.** $gh(10gh + 9h - g)$
⑤ $np + 2n + 8p + 16 = (np + 2n) + (8p + 16)$
$$= n(p + 2) + 8(p + 2)$$
$$= (n + 8)(p + 2)$$
7. $(b + 5)(3c - 2)$ **9.** $0, -10$ **11.** $0, \frac{3}{4}$
13a. $0, 2.08125$ **13b.** 17.3 ft, 2.6 ft **15.** $8(2t - 5y)$
17. $2k(k + 2)$ **19.** $2ab(2ab + a - 5b)$
㉑ $fg - 5g + 4f - 20 = (fg - 5g) + (4f - 20)$
$$= g(f - 5) + 4(f - 5)$$
$$= (g + 4)(f - 5)$$
23. $(h + 5)(j - 2)$ **25.** $(9q - 10)(5p - 3)$
27. $(3d - 5)(t - 7)$ **29.** $(3t - 5)(7h - 1)$
31. $(r - 5)(5b + 2)$ **33.** $gf(5f + g + 15)$
35. $3cd(9d - 6cd + 1)$ **37.** $2(8u - 15)(3t + 2)$
39. $0, 3$ **41.** $-\frac{1}{2}, -2$ **43.** $0, -3$ **45a.** ab
45b. $(a + 6)(b + 6)$ **45c.** $6(a + b + 6)$

47a.

x	y
0	0
1	9
2	12
3	9
4	0

47b.

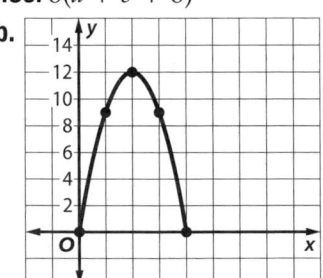

47c. 12 ft
㊾ $h = 64t - 16t^2$
$$0 = 64t - 16t^2$$
$$0 = 16t(4 - t)$$
$16t = 0$ or $4 - t = 0$
$t = 0$ or $4 = t$
So, the arrow hits the ground after 4 seconds.

51a. 3 and -2

51b.

x^2	$+3x$
$-2x$	-6

51c.

	x	$+3$
x	x^2	$+3x$
-2	$-2x$	-6

$(x + 3)(x - 2)$

51d. Sample answer: Place x^2 in the top left-hand corner and place -40 in the lower right-hand corner. Then determine which two factors have a product of -40 and a sum of -3. Then place these factors in the box. Then find the factor of each row and column. The factors will be listed on the very top and far left of the box.   **53.** Since the solutions are $-\frac{b}{a}$ and $\frac{b}{a}$, $a \neq 0$ and b is any real number.
55. Sample answer: $a = 0$ or $a = b$ for any real values of a and b.   **57.** D   **59.** 350   **61.** $0.5Bb + 0.5b^2$; $\frac{1}{2}$
63. $2b^3 + 2b^2 - 10b$ **65.** $-4x^4 - 4x^3 - 8x^2 + 4x^2$
67. $-3x^3y - 3x^2y^2 - 6xy^3$ **69.** p^7r^5 **71.** $81x^2y^{14}$
73. 16,777,216   **75.** $\{y \mid y > -11\}$ **77.** $\{k \mid k > -9\}$
79. $\{z \mid z \geq -48\}$

Lesson 8-6

1. $(x + 2)(x + 12)$ **3.** $(n + 7)(n - 3)$ **5.** $-3, 7$
7. 6, 9   **9.** $-8, 9$ **11.** 8 in. by 12 in.   **13.** $(y - 9)$
$(y - 8)$ **15.** $(n - 7)(n + 5)$ **17.** $(x - 2)(x - 20)$
19. $(m + 6)(m - 7)$

(21)
$$y^2 + y = 20$$
$$y^2 + y - 20 = 0$$
$$(y + 5)(y - 4) = 0$$
$$y + 5 = 0 \quad \text{or} \quad y - 4 = 0$$
$$y = -5 \qquad\qquad y = 4$$

23. $-2, -9$ **25.** 2, 16   **27.** $-4, -14$ **29.** 4, 12
31. $(x - 6)$ ft   **33.** $(q + 2r)(q + 9r)$ **35.** $(x - y)(x - 5y)$
(37) a. Sample answer: Let w represent the width of the swimming pool. So, the length of the pool is $w + 20$.
$$A = \ell w$$
$$= (w + 20)w$$
$$525 = (w + 20)w$$
b.
$$525 = (w + 20)w$$
$$525 = w^2 + 20w$$
$$0 = w^2 + 20w - 525$$
$$0 = (w + 35)(w - 15)$$
$$w + 35 = 0 \quad \text{or} \quad w - 15 = 0$$
$$w = -35 \qquad\qquad w = 15$$

c. The solution of 15 means that the width is 15 feet and the length is 35 feet. The solution -35 does not make sense because length cannot be negative.
39. $4x - 26$ **41.** Charles; Jerome's answer once multiplied is $x^2 - 6x - 16$. The middle term should be positive.   **43.** $-15, -9, 9, 15$ **45.** 4, 6
47. Sample answer: $x^2 + 19x - 20$; $(x - 1)(x + 20)$

49. Sample answer: Find factors m and n such that $m + n = b$ and $mn = c$. If b and c are positive, then m and n are positive. If b is negative and c is positive, then m and n are negative. When c is negative, m and n have different signs and the factor with the greater absolute value has the same sign as b.   **51.** 204   **53.** A   **55.** $11x(1 + 4xy)$
57. $(2x + b)(a + 3c)$ **59.** $(x - y)(x - y)$
61. $(4, 13)$ **63.** $(3, 6)$ **65.** about 6 ft
67. $(3x - 4)(a - 2b)$

Lesson 8-7

1. $(3x + 2)(x + 5)$ **3.** prime   **5.** $-\frac{3}{2}, -3$ **7.** $\frac{4}{3}, 2$
9a. 5 ft   **9b.** 2.5 seconds
(11) $2x^2 + 19x + 24$

factors of 48	sum of 19
3, 16	19

$$2x^2 + 3x + 16x + 24 = (2x^2 + 3x) + (16x + 24)$$
$$= x(2x + 3) + 8(2x + 3)$$
$$= (x + 8)(2x + 3)$$

13. $2(2x + 5)(x + 7)$ **15.** $(4x - 5)(x - 2)$ **17.** prime
19. prime   **21.** prime   **23.** $\frac{3}{2}, -6$ **25.** $\frac{2}{3}, 8$
27. $-\frac{1}{3}, 2$ **29a.** $10 = -16t^2 + 20t + 6$ **29b.** 1 sec
29c. Less; sample answer: It starts closer to the ground so the shot will not have as far to fall.

(31) Words: 6 times the square of a number plus
11 times the number equals 2
Equation: $6 \cdot x^2 + 11 \cdot x = 2$
$$6x^2 + 11x - 2$$
$$6x^2 + 11x - 2 = 0$$

factors of -12	sum of 11
$-1, 12$	11

$$6x^2 + 11x - 2 = 0$$
$$6x^2 - 1x + 12x - 2 = 0$$
$$(6x^2 - 1x) + (12x - 2) = 0$$
$$x(6x - 1) + 2(6x - 1) = 0$$
$$(x + 2)(6x - 1) = 0$$
$$x + 2 = 0 \quad \text{or} \quad 6x - 1 = 0$$
$$x = -2 \qquad\qquad 6x = 1$$
$$\qquad\qquad\qquad x = \frac{1}{6}$$

The numbers are -2 and $\frac{1}{6}$.
33. $-(x + 2)(4x + 7)$ **35.** $-(2x - 7)(3x - 5)$
37. $-(3x - 4)(4x + 5)$ **39a.** a^2 and b^2 **39b.** $a^2 - b^2$
39c. width: $a - b$, length: $a + b$ **39d.** $(a - b)(a + b)$
39e. $(a - b)(a + b)$; the figure with area $a^2 - b^2$ and the rectangle with area $(a - b)(a + b)$ have the same area, so $a^2 - b^2 = (a - b)(a + b)$.   **41.** $(12x + 20y)$ in.; The area of the square equals $(3x + 5y)(3x + 5y)$ in^2, so the length of one side is $(3x + 5y)$ in. The perimeter is $4(3x + 5y)$ or $(12x + 20y)$ in.
43. Sample answer: A quadratic equation may have zero, one, or two solutions. If there are two solutions, you must consider the context of the

situation to determine whether one or both solutions answer the given question. **45.** 6 **47.** J
49. $(x - 2)(x - 7)$ **51.** $(x + 3)(x - 8)$
53. $(r + 8)(r - 5)$ **55.** 0, 9 **57.** 0, 2 **59.** 0, 4

61.

Green Paint

Sample answers: 2 light, 8 dark; 6 light, 8 dark; 7 light, 4 dark

63. $\{d \mid d \le 5 \text{ or } d > 7\}$

65. ø

67. $\{y \mid 3 < y < 6\}$

69. 4 **71.** 8 **73.** 11

Lesson 8-8

1. $(x + 3)(x - 3)$ **3.** $9(m + 4)(m - 4)$
5. $(u^2 + 9)(u + 3)(u - 3)$

7 $20r^4 - 45n^4 = 5(4r^4 - 9n^4)$
$\qquad = 5\left((2r^2)^2 - (3n^2)^2\right)$
$\qquad = 5(2r^2 + 3n^2)(2r^2 - 3n^2)$

9. $(c + 1)(c - 1)(2c + 3)$ **11.** $(t + 4)(t - 4)(3t + 2)$
13. 36 mph **15.** $(q + 11)(q - 11)$
17. $6(n^2 + 1)(n + 1)(n - 1)$ **19.** $(r + 3t)(r - 3t)$
21. $h(h + 10)(h - 10)$ **23.** $(x + 9)(x - 9)(2x - 1)$
25. $7(h^2 + p^2)(h + p)(h - p)$
27. $6k^2(h^2 + 3k)(h^2 - 3k)$ **29.** $(f + 8)(f - 8)(f + 2)$
31. $10q(q + 11)(q - 11)$
33. $p^3r(r + 1)(r - 1)(r^2 + 1)$
35. $(r + 10)(r - 10)(r - 5)$ **37.** $(a + 7)(a - 7)$
39. $3(m^4 + 81)$ **41.** $2(a + 4)(a - 4)(6a + 1)$
43. $3(m + 5)(m - 5)(5m + 4)$ **45a.** $-0.5x(x - 9)$
45b. 9 ft **45c.** 10.125 ft

47 **a.** $S = -25m^2 + 125m$
$\qquad 0 = -25m^2 + 125m$
$\qquad 0 = -25m(m - 5)$
$\qquad 0 = -25m \quad \text{or} \quad 0 = m - 5$
$\qquad 0 = m \qquad\qquad 5 = m$
So, they will stop selling in month 5.
b. The peak will occur halfway between 0 and 5, or 2.5.

c. The peak amount is $S = -25(2.5)^2 + 125(2.5)$
$\qquad\qquad\qquad\quad = -156.25 + 312.5$
$\qquad\qquad\qquad\quad = 156.25$
The peak is 156,250 copies.

49 $100 = 25x^2$
$\quad 0 = 25x^2 - 100$
$\quad 0 = 25(x^2 - 4)$
$\quad 0 = 25(x + 2)(x - 2)$
$x + 2 = 0 \qquad \text{or} \qquad x - 2 = 0$
$\quad x = -2 \qquad\qquad\qquad x = 2$

51. $\frac{3}{8}, -\frac{3}{8}$ **53.** $-45, 45$ **55.** $\frac{3}{16}, -\frac{3}{16}$
57. Lorenzo; sample answer: Checking Elizabeth's answer gives us $16x^2 - 25y^2$. The exponent on x in the final product should be 4. **59.** $(x^4 - 3)(x^4 + 3) \cdot (x^8 + 9)$ **61.** false; $a^2 + b^2$ **63.** When the difference of squares pattern is multiplied together using the FOIL method, the outer and inner terms are opposites of each other. When these terms are added together, the sum is zero. **65.** G **67a.** Car A, because it is traveling at 65 mph, and Car B is traveling at 60 mph. **67b.** $5t - 10$ **67c.** 2.5 mi
69. prime **71.** $\{3, 6\}$ **73.** $\{6, 16\}$
75. $\{t \mid t \ge 4\}$

77. $\{k \mid k > 4\}$

79. $\{m \mid m \ge 3\}$

81. the seventh week **83.** $x^2 - 4x + 4$
85. $4x^2 - 20x + 25$ **87.** $16x^2 + 40x + 25$

Lesson 8-9

1. yes; $(5x + 6)^2$ **3.** $(x - 4)(2x + 7)$ **5.** $4(x^2 + 16)$
7. ± 3 **9.** $\frac{3}{8}$ **11.** 0.6 second **13.** yes; $(4x - 7)^2$

15 Since the last term is not a perfect square, the trinomial is not a perfect square trinomial

17. prime **19.** $8(y - 5z)(y + 5z)$
21. $2m(2m - 7)(3m + 5)$ **23.** $3(2x - 7)^2$
25. $3p(2p + 1)(2p - 1)$ **27.** $2t(t + 6)(2t - 7)$
29. $2a(a - b)(b + 1)(b - 1)$ **31.** $3k(k - 4)(k - 4)$
33. prime

35 $(y - 4)^2 = 7$
$\quad y - 4 = \pm\sqrt{7}$
$\qquad\quad y = 4 \pm \sqrt{7}$

37. $\frac{3}{4}$ **39.** 6 **41.** $\frac{1}{3}$ **43.** $8 \pm \sqrt{6}$ **45.** 20 ft
47. $|4x + 5|$

49 **a.** 42 in. = 3.5 ft., width = w, length = $w + 5$, height = 3.5

Area of the surface = $\ell \cdot w$ or $(w + 5)(w)$
$V = \ell \cdot w \cdot h$
$1750 = (w + 5)(w)(3.5)$
$500 = (w + 5)(w)$
Area of the surface is 500 ft^2

b. $500 = (w + 5)(w)$
$= w^2 + 5w$
$0 = w^2 + 5w - 500$
$= (w + 25)(w - 20)$
$0 = w + 25$ or $0 = w - 20$
$-25 = w$ $\quad\quad$ $20 = w$
The dimensions are 20 ft by 25 ft by 42 in.

c. Because the volume is doubled, we can double any one of the dimensions. 40 ft by 25 ft by 42 in., 20 ft by 50 ft by 42 in., or 20 ft by 25 ft by 84 in.

d. Because 2 of the dimensions are doubled the volume is increased by a factor of 4. The ratio is 1:4

51. 6 ft wide by 2 ft long by 15 ft high
53. Adriano; sample answer: Debbie did not factor the expression completely. **55.** Sample answer: $x^2 - 3x + \frac{9}{4} = 0; \left\{\frac{3}{2}\right\}$ **57.** First look for a GCF in all the terms and factor the GCF out of all the terms. Then, if the polynomial has two terms, check if the terms are the differences of squares and factor if so. If the polynomial has three terms, check if the trinomial will factor into two binomial factors or if it is a perfect square trinomial and factor if so. If the polynomial has four or more terms, factor by grouping. If the polynomial does not have a GCF and cannot be factored, the polynomial is a prime polynomial. **59.** Sample answer: $x^4 - 1; 1, -1$
61. B **63.** H
65. $(x - 4)(x + 4)$ **67.** $(1 - 10p)(1 + 10p)$
69. $(5n - 1)(5n + 1)$ **71.** $\{-2, 4\}$ **73.** $\{-2, 1\}$
75. $\{2, 3\}$ **77.** 10^1 or 10 **79.** $\frac{1}{10,000}$ **81.** $-\frac{2}{3}$
83. $\frac{3}{8}$ **85.** undefined

Chapter 8   Study Guide and Review

1. false; sample answer: $x^2 + 5x + 7$ **3.** true
5. true **7.** true **9.** false; difference of squares
11. $3x^2 + x + 2$ **13.** $x^2 + 3x + 2$ **15.** $-2x^3 - 3$
17. $-x^2 - x + 6$ **19.** 0 **21.** 1 **23.** $x^2 + 4x - 21$
25. $6r^2 + rt - 35t^2$ **27.** $10x^2 + 7x - 12$
29. $9x^2 - 12x + 4$ **31.** $4x^2 - 9$ **33.** $9m^2 - 4$
35. $12(x + 2y)$ **37.** $2y(4x - 8x^3 + 5)$
39. $(2x - 3z)(x - y)$ **41.** 0, 2 **43.** 0, 3
45. $x^2 - 2x + 5$ **47.** $(x + 5)(x + 4)$
49. $(x + 6)(x - 3)$ **51.** 2, 4 **53.** $-6, 8$ **55.** 14 in.
57. prime **59.** $(2a - 3)(a + 8)$ **61.** $4, -\frac{5}{2}$ **63.** $\frac{5}{3}, -\frac{1}{2}$
65. $(y + 9)(y - 9)$ **67.** prime **69.** $5, -5$ **71.** $-9, 9$
73. 2 seconds **75.** prime **77.** $(2 - 7a)^2$
79. $x^2(x + 4)(x - 4)$ **81.** $1, -2$ **83.** $-2, -\frac{1}{2}$

Chapter 9   Get Ready

1. 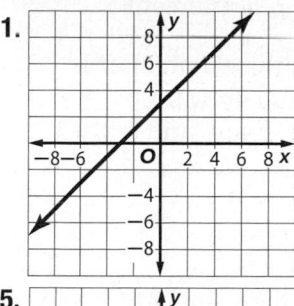 **3.**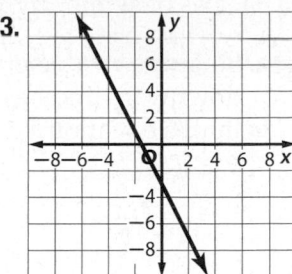

5.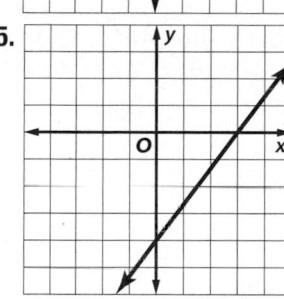

7. **Savings**

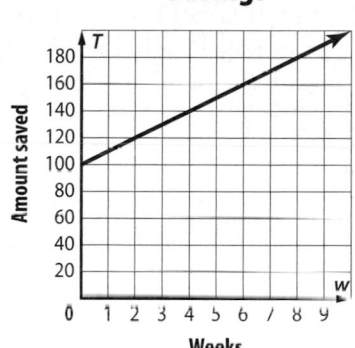

9. no **11.** yes; $(x + 10)^2$ **13.** yes; $(k - 8)^2$ **15.** no
17. 3 **19.** $2\frac{1}{2}$
21. 4 **23.** -2.5

Lesson 9-1

1.

x	y
-3	0
2	-6
-1	-8
0	-6
1	0
2	10

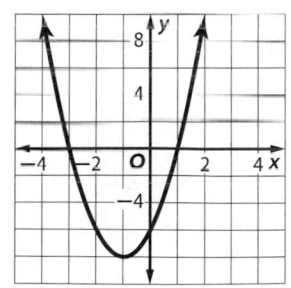

D = {all real numbers}; R = $\{y \mid y \geq -8\}$

3.

x	y
-1	4
0	-3
1	-8
2	-11
3	-12
4	-11
5	-8
6	-3
7	4

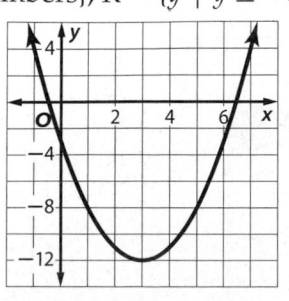

D = {all real numbers}; R = $\{y \mid y \geq -12\}$

Selected Answers and Solutions

5. vertex $(-1, 5)$, axis of symmetry $x = -1$, y-intercept 3 **7.** vertex $(-2, -12)$, axis of symmetry $x = -2$, y-intercept -4 **9.** vertex $(1, 2)$, axis of symmetry $x = 1$, y-intercept -1 **11.** vertex $(2, 1)$, axis of symmetry $x = 2$, y-intercept 5

13 a. Since the a value is -1, the graph opens downward and has a maximum.

b. In this equation $a = -1$, $b = 4$, and $c = -3$.
$$x = \frac{-b}{2a}$$
$$= \frac{-4}{2(-1)}$$
$$= 2$$
To find the vertex, use the value you found for the x-coordinate of the vertex. To find the y-coordinate, substitute the value for x in the original equation.
$$y = -x^2 + 4x - 3$$
$$= -(2)^2 + 4(2) - 3$$
$$= -4 + 8 - 3$$
$$= 1$$
The maximum is at $(2, 1)$.

c. The domain of the function is $D = \{x \mid x$ is all real numbers$\}$. The range is $R = \{y \mid y \leq 1\}$.

15a. maximum **15b.** 6 **15c.** $D = \{$all real numbers$\}$; $R = \{y \mid y \leq 6\}$

17.

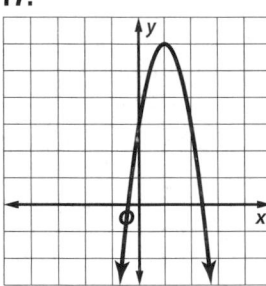

19.

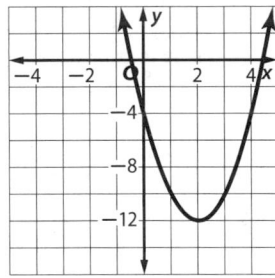

21a.

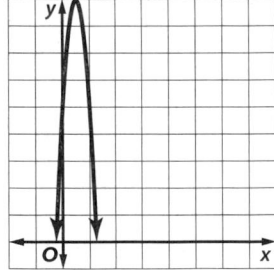

21b. 5 ft **21c.** 9 ft

23.

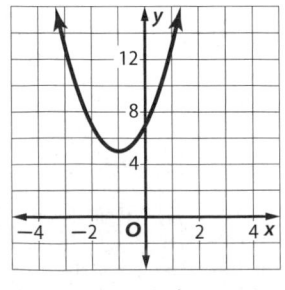

x	y
−3	13
−2	7
−1	5
0	7
1	13

$D = \{$all real numbers$\}$; $R = \{y \mid y \geq 5\}$

25.

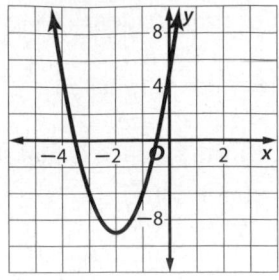

x	y
0	5
−1	−4
−2	−7
−3	−4
−4	5

$D = \{$all real numbers$\}$; $R = \{y \mid y \geq -7\}$

27.

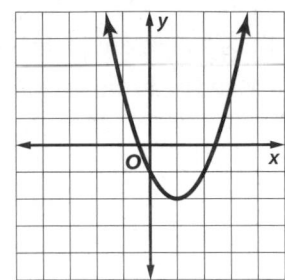

x	y
3	2
2	−1
1	−2
0	−1
−1	2

$D = \{$all real numbers$\}$; $R - \{y \mid y \geq -2\}$

29. vertex $(0, 1)$, axis of symmetry $x = 0$, y-intercept 1 **31.** vertex $(1, 1)$, axis of symmetry $x = 1$, y-intercept 4 **33.** vertex $(0, 0)$, axis of symmetry $x = 0$, y-intercept 0

35 In this equation $a = 2$, $b = 12$, and $c = 10$.
$$x = \frac{-b}{2a} \qquad = \frac{-12}{2(2)}$$
The equation for the axis of symmetry is $x = -3$.
To find the vertex, use the value you found for the axis of symmetry as the x-coordinate of the vertex. To find the y-coordinate, substitute the value for x in the original equation.
$$y = 2x^2 + 12x + 10$$
$$= 2(-3)^2 + 12(-3) + 10$$
$$= -8$$
The vertex is at $(-3, -8)$.
The y-intercept occurs at $(0, c)$. So, in this case, the y-intercept is 10.

37. vertex $(-3, 4)$, axis of symmetry $x = -3$, y-intercept -5 **39.** vertex $(2, -14)$, axis of symmetry $x = 2$, y-intercept 14 **41.** vertex $(1, -15)$, axis of symmetry $x = 1$, y-intercept -18
43a. maximum **43b.** 9 **43c.** $D = \{$all real numbers$\}$, $R = \{y \mid y \leq 9\}$ **45a.** minimum
45b. -48 **45c.** $D = \{$all real numbers$\}$, $R = \{y \mid y \geq -48\}$ **47a.** maximum **47b.** 33 **47c.** $D = \{$all real numbers$\}$, $R = \{y \mid y \leq 33\}$ **49a.** maximum
49b. 4 **49c.** $D = \{$all real numbers$\}$, $R = \{y \mid y \leq 4\}$
51a. maximum **51b.** 3
51c. $D = \{$all real numbers$\}$, $R = \{y \mid y \leq 3\}$

53.

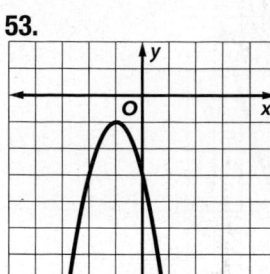

55.

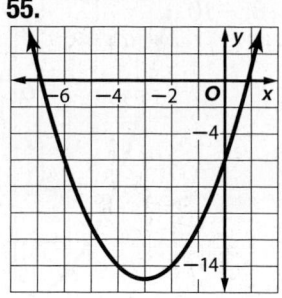

57.

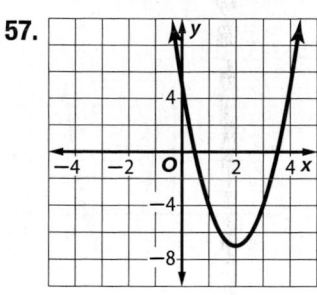

59.

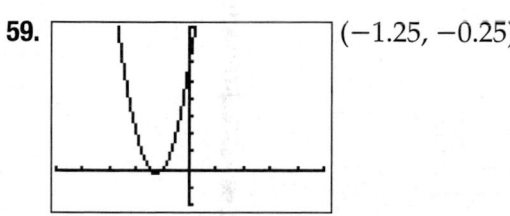

$(-1.25, -0.25)$

$[-5, 5]$ scl:1 by $[-5, 5]$ scl:1

61.

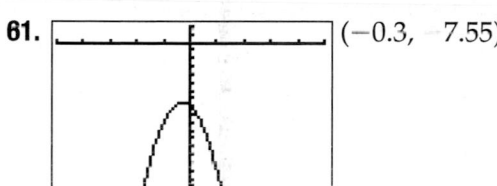

$(-0.3, -7.55)$

$[-5, 5]$ scl: 1 by $[-20, 2]$ scl: 2

63a.

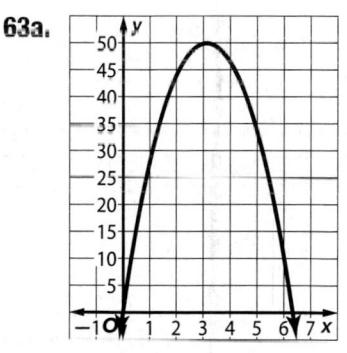

Where $h > 0$, the ball is above the ground. The height of the ball decreases as more time passes.

63b. 0 m **63c.** ≈50.0 m **63d.** ≈6.4 seconds

63e. D = $\{x \mid 0 \le x \le 6.4\}$; R = $\{y \mid 0 \le y \le 50.0\}$

65 a. $h = -16t^2 + 90t$
$= -16(1)^2 + 90(1)$
$= 74$ ft

b. $126 = -16t^2 - 90t$
$0 = -16t^2 - 90t - 126$
$0 = (t - 3)(-16t + 42)$
$t = 3$ and $t = 2.625$

c. $h = -16t^2 + 90t$
$0 = -16t^2 + 90t$

$0 = -16t(t - 5)$
$t = 0$ and $t = 5.625$

These represent the time that the ball leaves the ground initially and the time it returns to the ground.

67 a.

Equation	Related Function	Zeros	y-Values
$x^2 - x = 12$	$y = x^2 - x - 12$	$-3, 4$	$-3: 8, -6; 4: -6, 8$
$x^2 + 8x = 9$	$y = x^2 + 8x - 9$	$-9, 1$	$-9: 11, -9; 1: -9, 11$
$x^2 = 14x - 24$	$y = x^2 - 14x + 24$	$2, 12$	$2: 11, -9; 12: -9, 11$
$x^2 + 16x = -28$	$y = x^2 + 16x + 28$	$-14, -2$	$-14: 13, -11; -2: -11, 13$

b. Use a graphing calculator to graph.

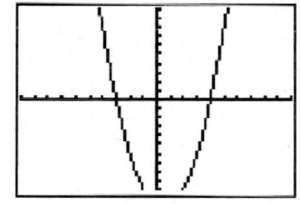

$y = x^2 - x - 12$

$[-10, 10]$ scl: 1 by $[-10, 10]$ scl: 1

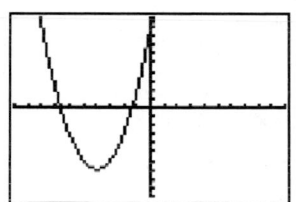

$y = x^2 + 8x - 9$

$[-15, 15]$ scl: 2 by $[-30, 30]$ scl: 5

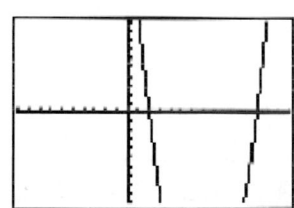

$y = x^2 - 14x + 24$

$[-10, 15]$ scl: 1 by $[-10, 10]$ scl: 1

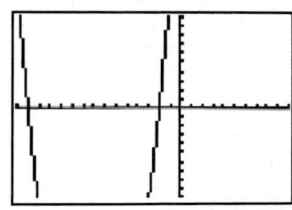

$y = x^2 + 16x + 28$

$[-15, 10]$ scl: 1 by $[-10, 10]$ scl: 1

c. Use the table function on the calculator to identify the zeros. The zeros are the values where y is 0. Also list the y-values that are to the left and right of the zeros.

d. The function values have opposite signs just before and just after the zeros.

69. Chase; the lines of symmetry are $x = 2$ and $x = 1.5$. **75.** C **77.** D **79.** yes; $(2x + 1)^2$ **81.** no

83. prime **85.** $b^2 - 4b - 21$ **87.** $2x^2 + 17x - 9$
89. $(2, 1)$ **91.** $(4, 2)$ **93.** 10 **95.** −6

Lesson 9-2

1. 2, −5

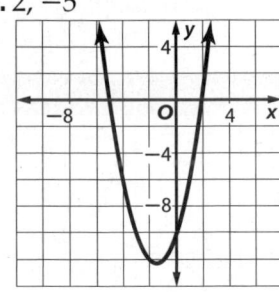

3. −2

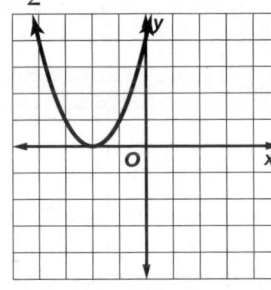

5. −5.2, 0.2

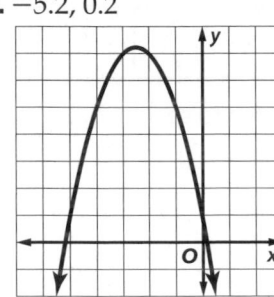

7. 5, −5

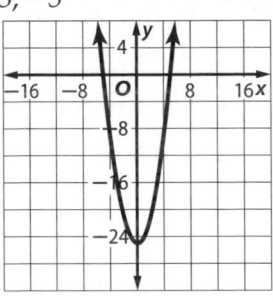

9. about 8.4 seconds

11 Step 1: Graph the related function of $f(x) = x^2 + 2x - 24 = 0$

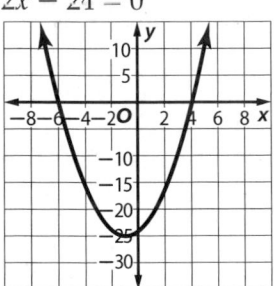

Step 2: The x-intercepts appear to be at −6 and 4, so the solutions are −6 and 4.

13. ∅

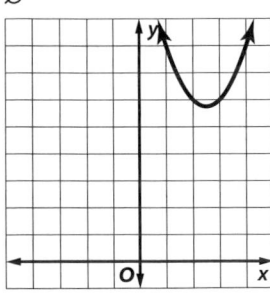

15. 1

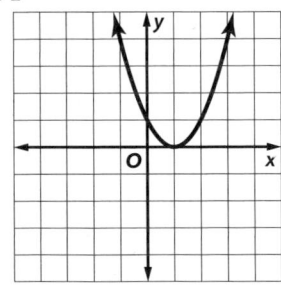

17. ∅

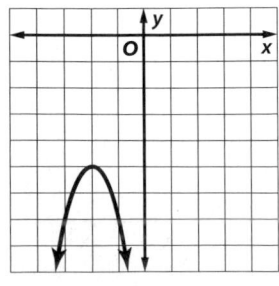

19. −6

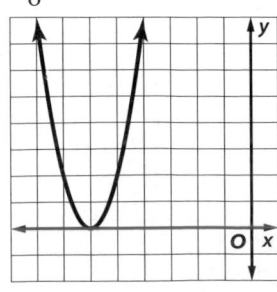

21. 8, −10

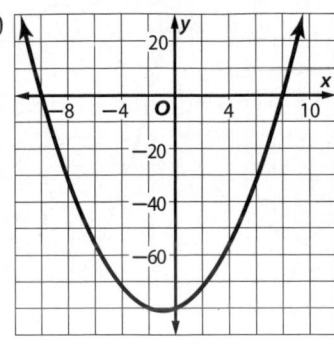

23. 6.9, −2.9 **25.** 3.3, 1.2

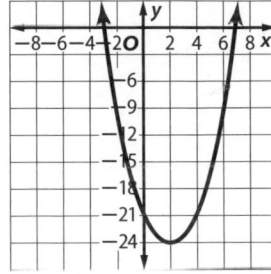

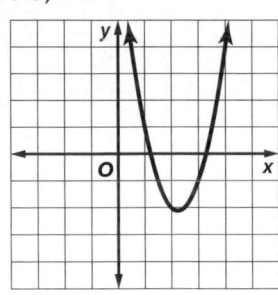

27. 3.1, −8.1

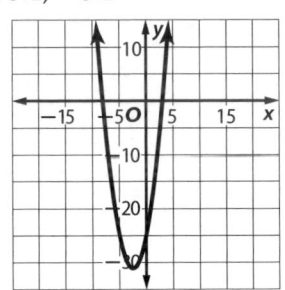

29. about 7.6 seconds **31.** 1; −2
33. 2; −4, −8 **35.** −3, 4

37 **a.** $h = -16t^2 + 30t + 10$
$0 = -16t^2 + 30t + 10$
Graph the equation and find the x-intercepts.

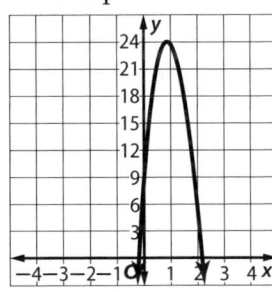

The positive x-intercept appears to be at 2.2, so she is in the air 2.2 seconds.

b. From the graph, she appears to hit a height of 15 feet at 0.2 seconds and 1.7 seconds.

c. $x = \dfrac{-b}{2a}$

$= \dfrac{-30}{2(-16)}$

$h = (-16)(0.9375)^2 + 30(0.9375) + 10$

≈ 24 feet

Her maximum height is about 24 feet, so she gets the bonus points.

39. $(x + 2)(x - 1)(x - 4)$ **41.** Iku; sample answer: The zeros of a quadratic function are the x-intercepts of the graph. Since the graph does not cross the x-axis, there are no x-intercepts and no real zeros.   **43.** Sometimes; for (1, 3), the y-value is greater than 2, but for (1, −1), it is less than 2.   **45.** 1.5 and −1.5; Sample answer: Make a table of values for x from −2.0 to 2.0. Use increments of 0.1.   **47.** A

49.

Boat 1
5 mi
Dock
4 mi
12 mi
Boat 2
9 mi

about 21.4 mi

51. $x = 0$; (0, 0); min

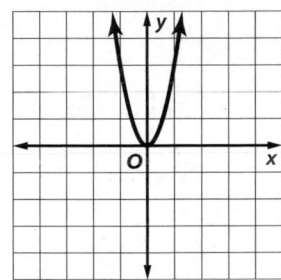

53. $x = 2$; (2, −3); max

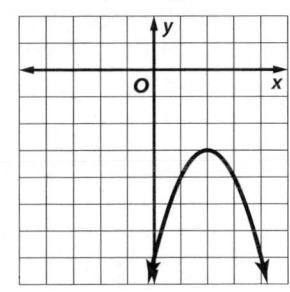

55. $x = -\dfrac{1}{3}$; $\left(-\dfrac{1}{3}, \dfrac{2}{3}\right)$; min

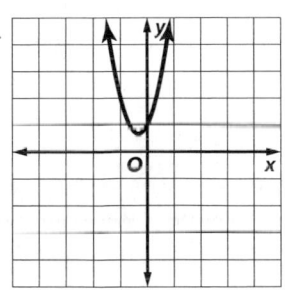

57. −4, 4   **59.** $-\dfrac{3}{2}, \dfrac{5}{2}$ **61.** $-3 \pm \sqrt{5}$ **63.** $7n^2 + 1$
65. $-3b^4 + 2b^3 - 9b^2 + 13$ **67.** $x + y = 180$; $x = y + 24$; 102°, 78°   **69.** $y - 6 = -7(x + 3)$

71.

73.

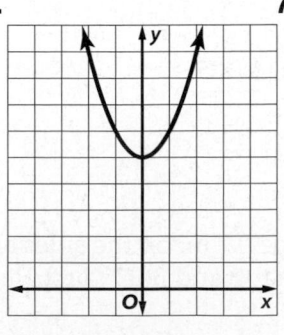

75.

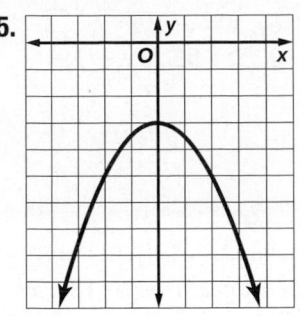

Lesson 9-3

1. translated down   **3.** reflected across the x-axis, translated up   **5.** reflected across the x-axis, translated left and stretched vertically   **7.** C
9 The function can be written as $f(x) = ax^2 + c$ where $a = -1$ and $c = -7$. Since $-7 < 0$ and $-1 < 0$ the graph of $y = -x^2 - 7$ translates the graph of $y = x^2$ down 7 units and reflects it across the x-axis.
11. compressed vertically, translated up
13. stretched vertically, translated up   **15.** translated left and up and stretched vertically   **17.** stretched vertically, translated down   **19.** A   **21.** F   **23.** E
25. $g(x)$, $h(x)$ **27.** $h(x)$, $g(x)$, $f(x)$
29 a. The two equations are $h = -16t^2 + 300$ and $h = -16t^2 + 700$.
b. $0 = -16t^2 + 300$, $t \approx 4.3$; $0 = -16t^2 + 700$, $t \approx 6.6$; $6.6 - 4.3 \approx 2.3$ seconds.
31a. The graph of $g(x)$ is the graph of $f(x)$ translated 200 yards right, compressed vertically, reflected in the x-axis, and translated up 20 yards.
31b. $h(x) = 0.0005(x - 230)^2 + 20$ **33.** Translate the graph of $f(x)$ up 11 units and to the right 5 units.
35a. 20 h   **35b.** $h(t) = 40t$ **35c.** $T(t) = -t^2 + 50t + 200$; the fuel in the tank after t hours with refueling
35d. Yes; after about 53 hours 43 minutes.
37. $y = x^2 - 1$ **39.** Sample answer: $f(x) = -\dfrac{1}{2}x^2$
41. $C = 55 + 30h$ **43.** J
45. ∅

47. 1, −1

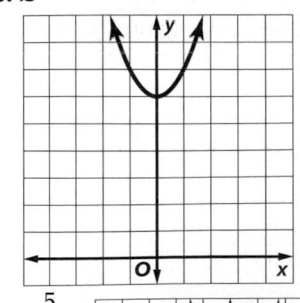

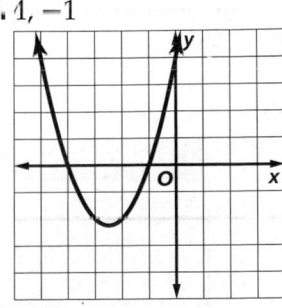

49. $-\dfrac{5}{2}$, 3

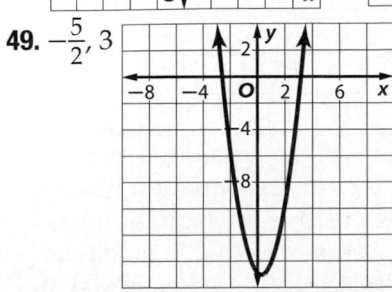

51. $(0, 4)$; $x = 0$; 4   **53.** $(-2, 6)$; $x = -2$; 2
55. $\{t \mid t \geq 3\}$ **57.** $\{d \mid d > -125\}$ **59.** $(3x + 1)^2$
61. no   **63.** no

Lesson 9-4

① Step 1: Find $\frac{1}{2}$ of $-18 = -9$
Step 2: Square the result in step 1: $(-9)^2 = 81$
Step 3: Add the result of step 2 to $x^2 - 18x$: $x^2 - 18x + 81$
Thus, $c = 81$.
3. $\frac{81}{4}$ **5.** $-5.2, 1.2$ **7.** $-2.4, 0.1$ **9.** 8 ft by 18 ft
11. 144   **13.** $\frac{289}{4}$ **15.** $\frac{169}{4}$ **17.** $\frac{225}{4}$
⑲ $x^2 + 6x - 16 = 0$
$x^2 + 6x = 16$
$x^2 + 6x + 9 = 16 + 9$
$(x + 3)^2 = 25$
$x + 3 = \pm 5$
$x = -3 \pm 5$
The solutions are 2 and -8.
21. $-1, 9$ **23.** $-0.2, 11.2$ **25.** $\varnothing$ **27.** $-2.6, 1.1$
29. $-1.1, 6.1$ **31.** on the 30th and 40th days after purchase   **33.** 5.3   **35.** -21 and -23 **37.** $-1, 2$
39. $0.2, 0.9$ **41.** $-8.2, 0.2$
㊸ a. The object on Earth will reach the ground first because it is falling at a faster rate.
b. Mars: $0 = -1.855t^2 + 120$
$-120 = -1.855t^2$
$64.69 \approx t^2$
$\pm 8.0 \approx t$
So, $t \approx 8.0$ seconds
Earth: $0 = -4.9t^2 + 120$
$-120 = -4.9t^2$
$24.49 = t^2$
$\pm 4.9 \approx t$
So, $t = 4.9$ seconds.
c. Sample answer: Yes, the acceleration due to gravity is much greater on Earth than on Mars, so the time to reach the ground should be much less.
45. -30 and 30
47a–b.

Trinomial	$b^2 - 4ac$	Number of Roots
$x^2 - 8x + 16$	0	1
$2x^2 - 11x + 3$	97	2
$3x^2 + 6x + 9$	-72	0
$x^2 - 2x + 7$	-24	0
$x^2 + 10x + 25$	0	1
$x^2 + 3x - 12$	57	2

47c. If $b^2 - 4ac$ is negative, the equation has no real solutions. If $b^2 - 4ac$ is zero, the equation has one solution. If $b^2 - 4ac$ is positive, the equation has 2 solutions.   **47d.** 0 because $b^2 - 4ac$ is negative. The equation has no real solutions because taking the square root of a negative number does not produce a real number.   **49.** None; sample answer: If you add $\left(\frac{b}{2}\right)^2$ to each side of the equation and each side of the inequality, you get $x^2 + bx + \left(\frac{b}{2}\right)^2 = c + \left(\frac{b}{2}\right)^2$ and $c + \left(\frac{b}{2}\right)^2 < 0$. Since the left side of the last equation is a perfect square, it cannot equal the negative number $c + \left(\frac{b}{2}\right)^2$. So, there are no real solutions.   **51.** Sample answer: $x^2 - 8x + 16 = 0$
53. B   **55.** 32   **57.** translated down 12 units
59. expanded vertically, translated up 5 units
61. expanded vertically, translated up 6 units
63. $40 = -16t^2 + 250$; about 3.6 s   **65.** 16   **67.** 1
69. $\frac{1}{m^3 b^3}$ **71.** $\{z \mid -8 < z < -2\}$ **73.** $\{y \mid y \geq 5.5$ or $y \leq -2.5\}$ **75.** $\{c \mid -2.2 \leq c \leq 3\}$ **77.** ± 10 **79.** ± 7.8
81. not a real number

Lesson 9-5

1. $-3, 5$ **3.** $6.4, 1.6$ **5.** $0.6, 2.5$ **7.** $-6, \frac{1}{2}$
9. $\pm \frac{5}{3}$ **11.** -3; no real solutions
13. 0; one real solution   **15.** The discriminant is -14.91, so the equation has no real solutions. Thus, Eva will not reach a height of 10 feet.
⑰ $x^2 + 16 = 0$
For this equation, $a = 1$, $b = 0$, and $c = 16$
$$x = \frac{-b \pm \sqrt{b^2 - 4ac}}{2a}$$
$$= \frac{0 \pm \sqrt{(0)^2 - 4(1)(16)}}{2(1)}$$
$$= \frac{\sqrt{-64}}{2}$$
So, there is no real solution. The solution can be written $\varnothing$.
19. $2.2, -0.6$ **21.** $-3, -\frac{6}{5}$ **23.** $0.5, -2$ **25.** $0.5, -1.2$ **27.** 3   **29.** $-1.2, 5.2$ **31.** $-2, 5$ **33.** $-6.2, -0.8$ **35.** -0.07; no real solution   **37.** 12.64; two real solutions   **39.** 0; one real solution   **41a.** in 1993 and 2023   **41b.** Sample answer: No; the parabola has a maximum at about 66, meaning only 66% of the population would ever have high-speed Internet.   **43.** 0   **45.** 1   **47.** $-1.4, 2.1$
㊾ a. $(20 - 2x)(25 - 7x) = 375$
b. $500 - 50x - 140x + 14x^2 = 375$
$14x^2 - 190x + 125 = 0$
$$x = \frac{-b \pm \sqrt{b^2 - 4ac}}{2a}$$
$$= \frac{190 \pm \sqrt{(190)^2 - 4(14)(125)}}{2(14)}$$
$$= \frac{190 \pm \sqrt{29,100}}{28}$$
≈ 0.7 and 12.9
c. The margins should be 0.7 in. on the sides and $4(0.7)$ or 2.8 in. on the top and $3(0.7)$ or 2.1 in. on the bottom.

51. $k < \frac{9}{40}$ **53.** none **55.** two **57.** Sample answer: If the discriminant is positive, the Quadratic Formula will result in two real solutions because you are adding and subtracting the square root of a positive number in the numerator of the expression. If the discriminant is zero, there will be one real solution because you are adding and subtracting the square root of zero. If the discriminant is negative, there will be no real solutions because you are adding and subtracting the square root of a negative number in the numerator of the expression.
59. D **61.** G **63.** $\frac{4}{3}, \frac{3}{2}$ **65.** $\frac{5}{2}$ **67.** Translate down 6. **69.** Positive; as time goes on, more people use electronic tax returns.

71. $12x + 3y \le 60$

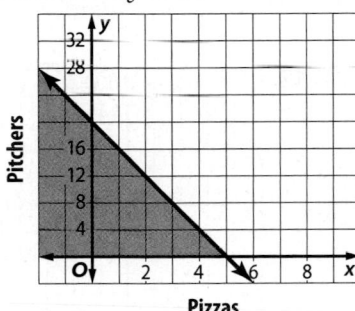

Pitchers / Pizzas

73. Arithmetic; the common difference is -50.
75. Geometric; the common ratio is 3.
77. Geometric; the common ratio is $-\frac{1}{2}$.

Lesson 9-6

1.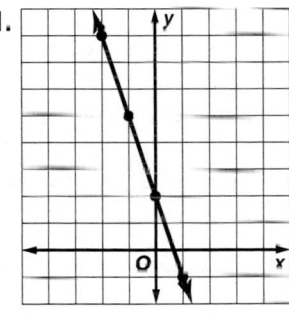
linear

3. exponential

5. quadratic **7.** exponential **9.** exponential; $y = 3 \times 3^x$ **11.** linear; $y = \frac{1}{2}x + \frac{5}{2}$ **13.** linear: $y = 0.5x + 3$

15.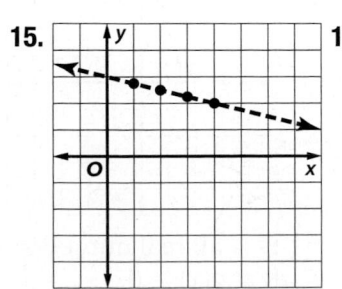
linear

17.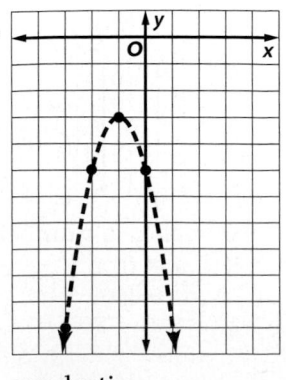
quadratic

19. 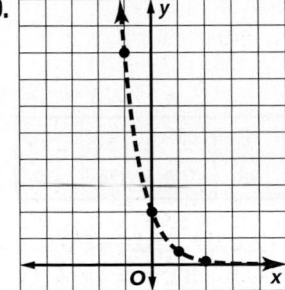 exponential

21 Look for a pattern in the y-values. Start with comparing first differences.

10   2.5   0   2.5   10
  −7.5 −2.5 2.5 7.5

The first differences are not all equal. So, the table of values does not represent a linear function. Find the second differences and compare.

−7.5   −2.5   2.5   7.5
    +5   +5   +5

The second differences are all equal, so the table of values represents a quadratic function. Write an equation for the function that models the data.
The equation has the form $y = ax^2$. Find the value of a by choosing one of the ordered pairs from the table of values. Let's use (2, 10).

$y = ax^2$
$10 = a(2)^2$
$10 = 4a$
$\frac{5}{2} = a$
$2.5 = a$

An equation that models the data is $y = 2.5x^2$.
23. exponential; $y = 0.2 \cdot 5^x$
25. linear; $y = -5x - 0.25$

27 **a.** Graph the ordered pairs on a coordinate plane.

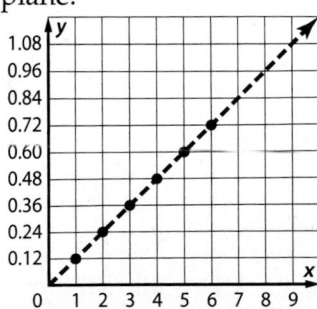

The graph appears to be linear.
b. Look at the first differences of the y-values.

0.12   0.24   0.36   0.48   0.60   0.72
   +0.12 +0.12 +0.12 +0.12 +0.12

The common difference is 0.12.
The equation is $y = 0.12x$.

Selected Answers and Solutions

c. $y = 0.12x$
    $= 0.12(10)$
    $= \$1.20$

29a.

Time (hour)	0	1	2	3	4
Amount of Bacteria	12	36	108	324	972

29b. exponential **29c.** $b = 12 \cdot 3^t$ **29d.** 78,732
31. Sample answer: $y = 2x^2 - 5$ **33.** $y = 4x + 1$
35. Sample answer: The data can be graphed to determine which function best models the data. You can also find the differences in ratios of the y-values. If the first differences are constant, the data can be modeled by a linear function. If the second differences are constant but the first differences are not, the data can be modeled by a quadratic function. If the ratios are constant, then the data can be modeled by an exponential function.
37. A **39.** B **41.** $-5, 0.5$ **43.** ± 5 **45.** $-3.1, 10.9$
47. $a_n = 1(2)^{n-1}$; 64 **49.** $a_n = 4(-3)^{n-1}$; 2916
51. $a_n = 22(2)^{n-1}$; 1408 **53.** $C = 10h + 15$; \$95
55. yes; $2x + y = 6$ **57.** yes; $y = -5$ **59.** no

61. 1 **63.** $\frac{1}{2}$ **65.** -1

Lesson 9-7

1.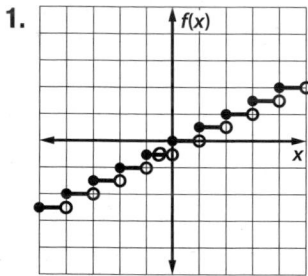
D = all real numbers;
R = all integers
multiples of 0.5

3.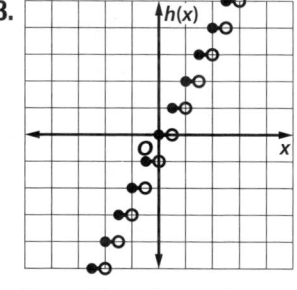
D = all real numbers;
R = all integers

5.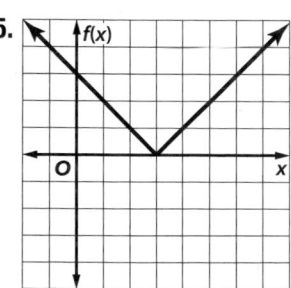
D = all real numbers;
R = $f(x) \geq 0$

7.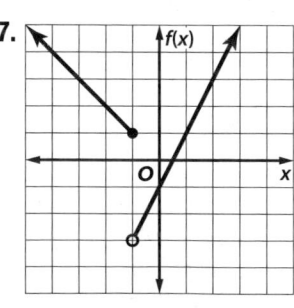
D = all real numbers;
R = $f(x) > -3$

9 $f(x) = 3[\![x]\!]$
Create a table.

x	3[[x]]	f(x)
0	3[[0]]	0
0.25	3[[0.25]]	0
0.5	3[[0.5]]	0
1	3[[1]]	3
1.25	3[[1.25]]	3
1.5	3[[1.5]]	3
2	3[[2]]	6
2.25	3[[2.25]]	6
2.5	3[[2.5]]	6
3	3[[3]]	9

Graph on a coordinate grid.

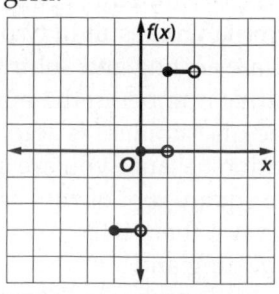

D = all real numbers;
R = all integer multiples of 3

11.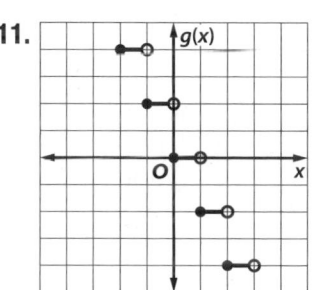
D = all real numbers;
R = all even integers

13.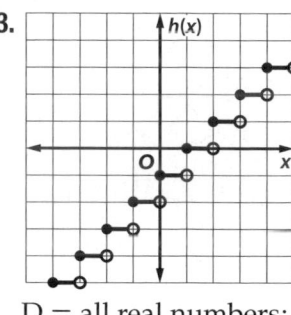
D = all real numbers;
R = all integers

15a.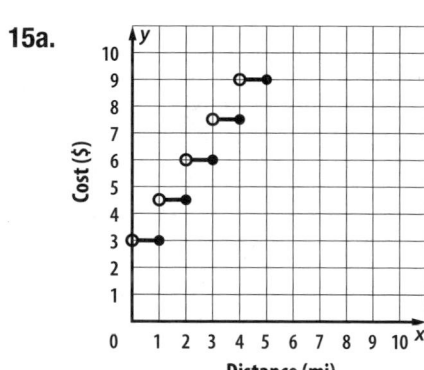

15b. \$16.50

17.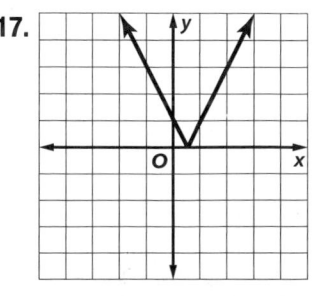
D = all real numbers;
R = $y \geq 0$

19.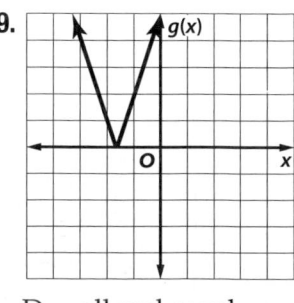
D = all real numbers;
R = $g(x) \geq 0$

21.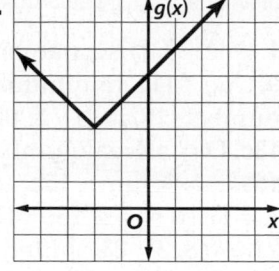

D = all real numbers;
R = f(x) ≥ 0

23.

D = all real numbers;
R = g(x) ≥ 3

25.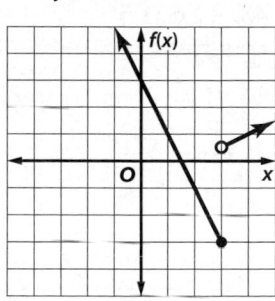

D = all real numbers;
R = all real numbers

27.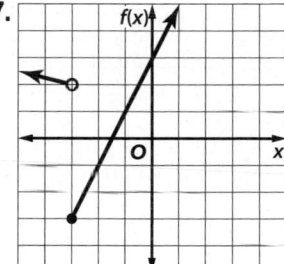

D = all real numbers;
R = f(x) ≥ −3

29.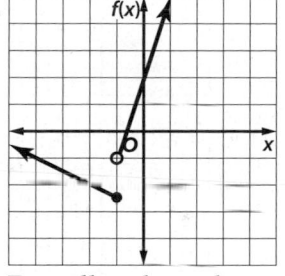

D = all real numbers;
R − f(x) > −2.5

31. The graph extends infinitely to the left and to the right, so the domain is all real numbers. The range is all real numbers greater than or equal to 4 because the line increases from both sides of (6, 4).

33. D − all real numbers; R = all integers

35. D = all real numbers; R = y > −2

37.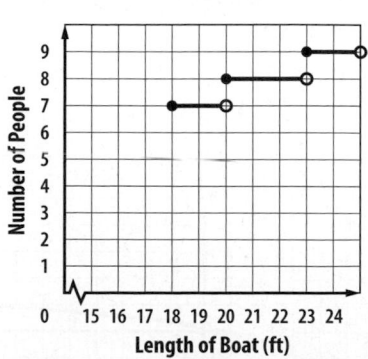

Length of Boat (ft)

39. D   **41.** B

43. a.

Number of Orders	Total Price
1 ≤ x ≤ 10	10 + (10 + 4 + 2)x = 10 + 16x
10 < x ≤ 20	(10 + 16x)(0.95) = 9.5 + 15.20x
x > 20	(10 + 16x)(0.90) = 9 + 14.40x

b. $y = \begin{cases} 10 + 16x \text{ if } 1 \le x \le 10 \\ 9.50 + 15.20x \text{ if } 10 < x \le 20 \\ 9 + 14.40x \text{ if } x > 20 \end{cases}$

c.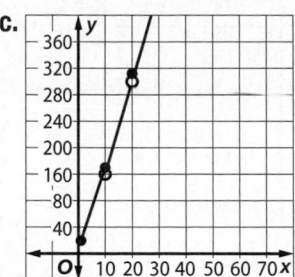

45a.

x	−5	−4	−3	−2	−1	0	1	2	3	4	5
f(x)	13	11	9	7	5	3	5	7	9	11	13

45b.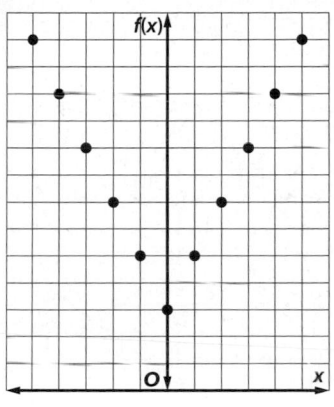

45c.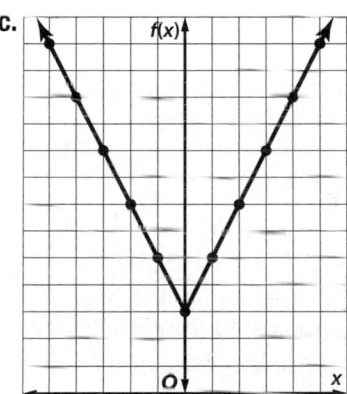

45d. The graph is shifted 1.5 units to the right and 3 units up.

47.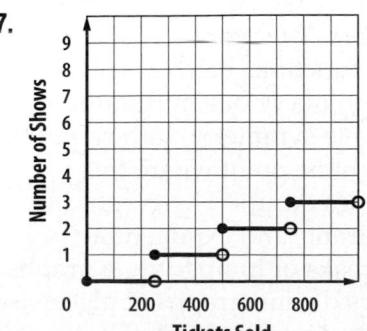

Tickets Sold

49. $g(x) = \frac{1}{3}|x| + 4$
Set up a table.

x	$\frac{1}{3}\lvert x\rvert + 4$	$g(x)$
-6	$\frac{1}{3}\lvert -6\rvert + 4$	6
-3	$\frac{1}{3}\lvert -3\rvert + 4$	5
0	$\frac{1}{3}\lvert 0\rvert + 4$	4
3	$\frac{1}{3}\lvert 3\rvert + 4$	5
6	$\frac{1}{3}\lvert 6\rvert + 4$	6

Plot on a coordinate grid.

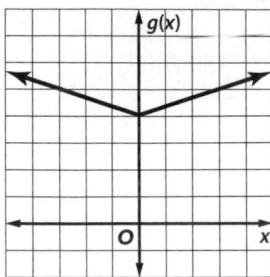

51. 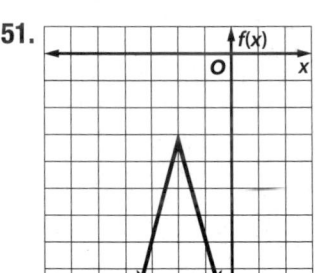 **53.**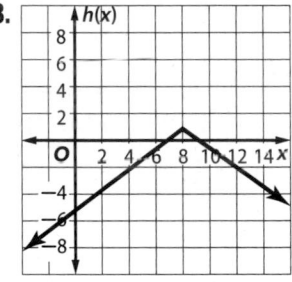

55. No; the pieces of the graph overlap vertically, so the graph fails the vertical line test.

57. $f(x) = \begin{cases} \frac{1}{2}x - 3 & \text{if } x > 6 \\ -\frac{1}{2}x - 1 & \text{if } x \le 6 \end{cases}$

59. Sample answer: The domain of absolute value, step, quadratic, and exponential functions is all real numbers, while some piecewise functions may not be defined for all real numbers. The range of absolute value, step, quadratic, and exponential functions is limited to a portion of the real numbers, but the range of a piecewise-defined function can be all real numbers. The graphs of absolute value and quadratic functions have either one maximum and no minima or one minimum and no maxima, and both have symmetry with respect to a vertical line through the point where this maximum or minimum occurs. The graphs of absolute value, quadratic, and exponential functions have no breaks or jumps, while graphs of step functions always do and graphs of piecewise-defined functions sometimes do. **61.** C
63. B **65.** linear **67.** exponential **69.** Positive; it means the more you study, the better your test score. **71.** -4.4 **73.** 1.2 **75.** 3.46 **77.** 12 **79.** 0.24

1. true **3.** false; parabola **5.** false; two **7.** true
9. true **11a.** minimum **11b.** 0 **11c.** D = all real numbers; R = $y\,|\,y \ge 0$} **13a.** minimum **13b.** -4
13c. D = all real numbers; R = $\{y\,|\,y \ge -4\}$
15a. maximum **15b.** 16 **15c.** D = $\{t\,|\,0 \le t \le 2\}$; R = $\{h\,|\,0 \le h \le 16\}$ **17.** 3 **19.** $-4.6, 0.6$
21. $-0.8, 3$ **23.** shifted up 8 units **25.** vertical stretch **27.** vertical compression **29.** $y = 2x^2 - 3$
31. $1, -7$ **33.** $10, -2$ **35.** $-0.7, 7.7$ **37.** $-8, 6$
39. $-0.7, 0.5$ **41.** $-5, 1.5$ **43.** $-2.5, 1.5$
45. quadratic; $y = 3x^2$ **47.** quadratic; $y = -x^2$

49. D = all real numbers; **51.** D = all real numbers; R = all integers   R = $f(x) \ge 0$

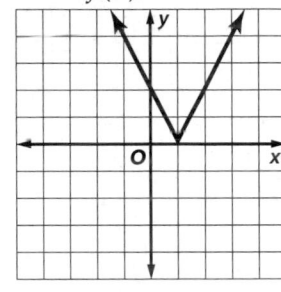

53. D = all real numbers; R = $f(x) \le -1$ or $f(x) > 3$

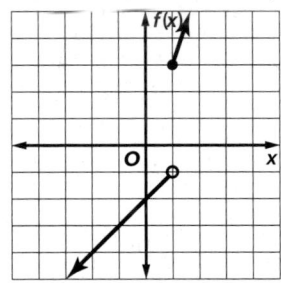

Radical Functions and Geometry

Get Ready

1. 9.06 **3.** 3.87 **5.** 10 ft **7.** $13x - 3y$ **9.** $3m + 3n + 10$ **11.** $0, 2$ **13.** $2, 5$ **15.** 10

Lesson 10-1

1. 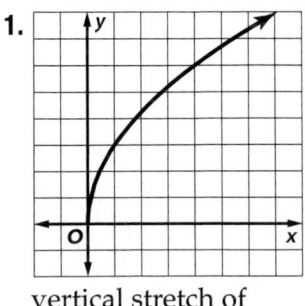 **3.**

vertical stretch of $y = \sqrt{x}$; D = $\{x\,|\,x \ge 0\}$, R = $\{y\,|\,y \ge 0\}$

vertical compression of $y = \sqrt{x}$; D = $\{x\,|\,x \ge 0\}$, R = $\{y\,|\,y \ge 0\}$

5.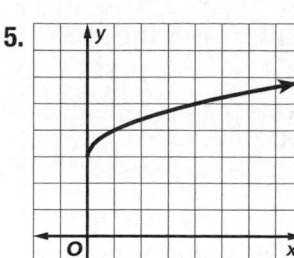

translated up 3;
$D = \{x \mid x \geq 0\}$,
$R = \{y \mid y \geq 3\}$

7.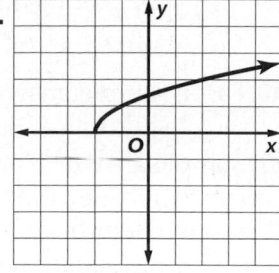

translated left 2;
$D = \{x \mid x \geq -2\}$,
$R = \{y \mid y \geq 0\}$

9.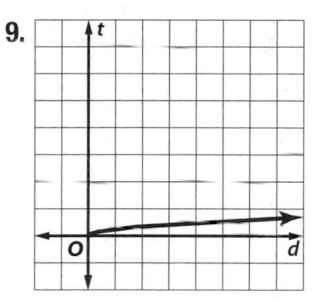

$D = \{d \mid d \geq 0\}$,
$R = \{t \mid t \geq 0\}$

11.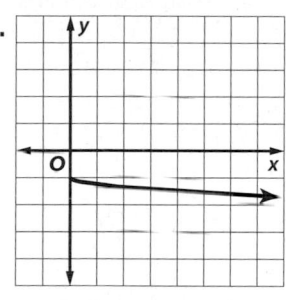

vertical compression
of $\sqrt{x}$, reflected across
the x-axis and
translated down 1;
$D - \{x \mid x \geq 0\}$,
$R = \{y \mid y \leq -1\}$

13.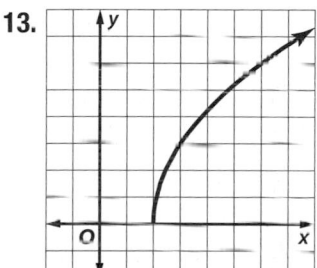

translated right 2 and
vertical stretch of $\sqrt{x}$;
$D = \{x \mid x \geq 2\}$,
$R = \{y \mid y \geq 0\}$

15. Step 1: Make a table.
Choose nonnegative
values for x.

x	y
0	0
0.5	≈0.35
1	0.5
2	0.71
3	0.87
4	1

Step 2: Plot the points and draw a smooth curve.

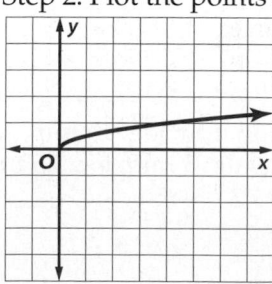

The graph is a vertical compression of $\sqrt{x}$.

The domain is $\{x \mid x \geq 0\}$. The range is $\{y \mid y \geq 0\}$.

17.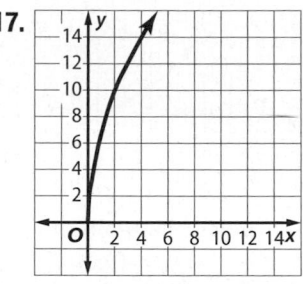

vertical stretch of $\sqrt{x}$;
$D = \{x \mid x \geq 0\}$,
$R = \{y \mid y \geq 0\}$

19.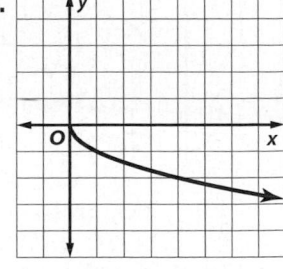

reflected across the
x-axis;
$D = \{x \mid x \geq 0\}$,
$R = \{y \mid y \leq 0\}$

21.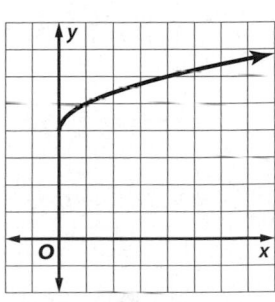

vertical stretch of $\sqrt{x}$
and reflected across the
x-axis;
$D = \{x \mid x \geq 0\}$,
$R = \{y \mid y \leq 0\}$

23.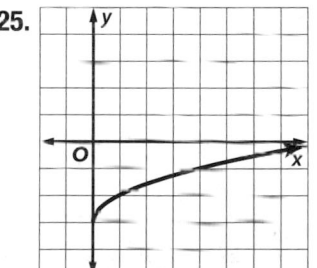

translated up 4;
$D = \{x \mid x \geq 0\}$,
$R = \{y \mid y \geq 4\}$

25.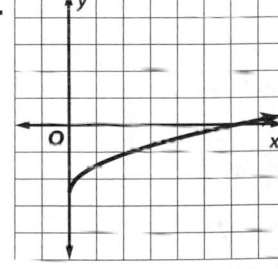

translated down 3;
$D = \{x \mid x \geq 0\}$,
$R = \{y \mid y \geq -3\}$

27.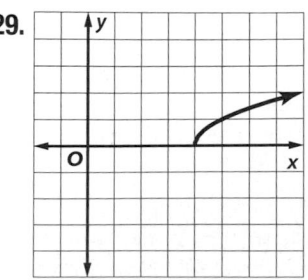

translated down 2.5;
$D = \{x \mid x \geq 0\}$,
$R = \{y \mid y \geq -2.5\}$

29.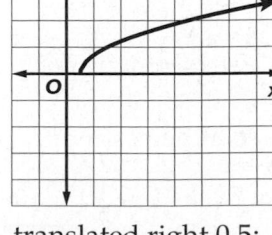

translated right 4;
$D = \{x \mid x \geq 4\}$,
$R = \{y \mid y \geq 0\}$

31.

translated right 0.5;
$D = \{x \mid x \geq 0.5\}$,
$R = \{y \mid y \geq 0\}$

33.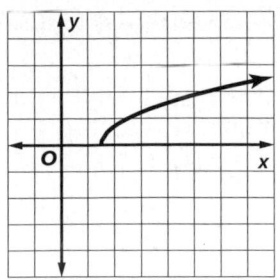

translated right 1.5;
D = {x | x ≥ 1.5},
R = {y | y ≥ 0}

35.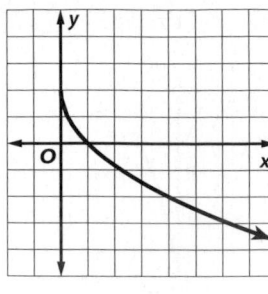

vertical stretch of
y = √x, reflected
across the x-axis, and
translated up 2;
D = {x | x ≥ 0},
R = {y | y ≤ 2}

37.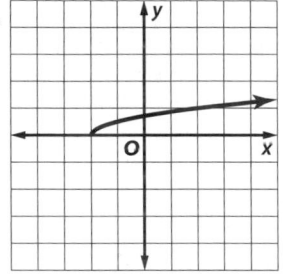

vertical compression
of y = √x and
translated left 2;
D = {x | x ≥ -2},
R = {y | y ≥ 0}

39.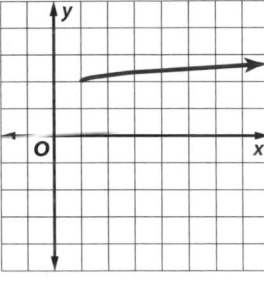

vertical compression
of y = √x and
translated up 2 and
right 1; D = {x | x ≥ 1},
R = {y | y ≥ 2}

41.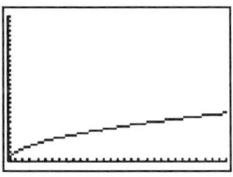

[0, 28] scl: 1 by [0, 28] scl: 1

43 a. Use a graphing calculator to graph the equation.

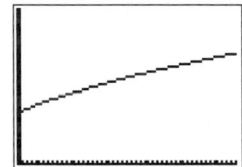

[0, 1000] scl: 20 by [0, 1000] scl: 10

b. $c = 331.5\sqrt{1 + \dfrac{t}{273.15}}$

$= 331.5\sqrt{1 + \dfrac{55}{273.15}}$

$= 331.5\sqrt{1.201}$

≈ 363.3 m/s

c. When t = 65°C:

$c = 331.5\sqrt{1 + \dfrac{t}{273.15}}$

$= 331.5\sqrt{1 + \dfrac{65}{273.15}}$

$= 331.5\sqrt{1.238}$

≈ 368.8 m/s

So, this increase of 10°C increases the speed by about 5.5 m/s.

45. False; sample answer: The domain of $y = \sqrt{x} + 3$ includes −1, −2, and −3.

47. Sample answer: The domain is limited because square roots of negative numbers are imaginary; therefore the radicand must be nonnegative. Since the principal square root of a nonnegative number is a nonnegative number, the range will be nonnegative.

49. $y = \sqrt{x} + 3$; it is a translation of $y = \sqrt{x}$; the other equations represent vertical stretches or compressions. **51.** The value of *a* is negative. For the function to have negative y-values, the value of *a* must be negative.

53. A **55.** D

57.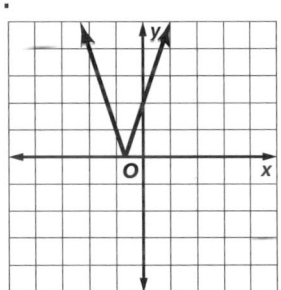

59.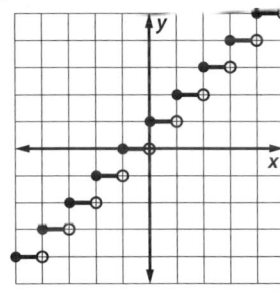

61.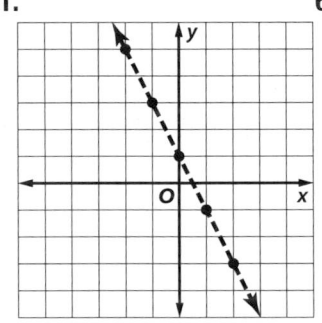

linear

63.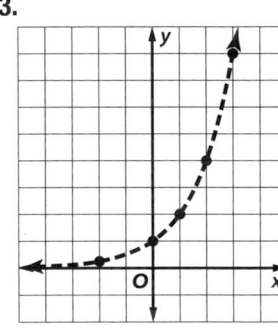

exponential

65a.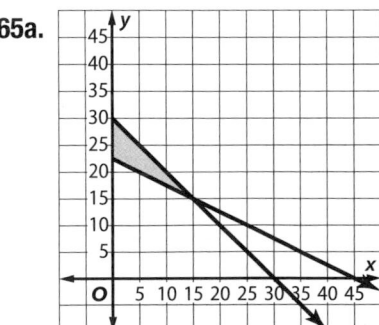

65b. Sample answer: walk: 15 min, jog: 15 min; walk: 10 min, jog: 20 min; walk: 5 min, jog: 25 min

67. 2 · 2 · 7 · n · n · n **69.** 2 · 3 · 5 · 5 · r · t

71. 3 · 3 · 5 · 5 · a · a · a · b · b · c

Lesson 10-2

1. $2\sqrt{6}$ **3.** 10 **5.** $3\sqrt{6}$ **7.** $2x^2y^3\sqrt{15y}$

9. $3b^2|c|\sqrt{11ab}$ **11.** $\dfrac{9-3\sqrt{5}}{4}$ **13.** $\dfrac{2+2\sqrt{10}}{-9}$

15. $\dfrac{24+4\sqrt{7}}{29}$ **17.** $2\sqrt{13}$ **19.** $6\sqrt{2}$ **21.** $9\sqrt{3}$

23. $5\sqrt{2}$ **25.** $12\sqrt{14}$ **27.** $15|t|$ **29.** $2|a|b\sqrt{7b}$

31. $21m\sqrt{7mp}$ **33.** $2a^3b\sqrt{5b}$

35. **a.** $v = \sqrt{64h}$
$= \sqrt{8 \cdot 8h}$
$= \sqrt{8^2} \cdot \sqrt{h}$
$= 8\sqrt{h}$
b. $v = 8\sqrt{h}$
$= 8\sqrt{134}$
≈ 92.6 ft/s

37. $\sqrt{\dfrac{32}{t^4}} = \dfrac{\sqrt{32}}{\sqrt{t^4}}$
$= \dfrac{\sqrt{16 \cdot 2}}{t^2}$
$= \dfrac{\sqrt{16} \cdot \sqrt{2}}{t^2}$
$= \dfrac{4\sqrt{2}}{t^2}$

39. $\dfrac{2c\sqrt{51ac}}{9a}$ **41.** $\dfrac{3\sqrt{15}}{20}$ **43.** $\dfrac{35-7\sqrt{3}}{22}$

45. $\dfrac{6\sqrt{3}+9\sqrt{2}}{2}$ **47.** $\dfrac{5\sqrt{6}-5\sqrt{3}}{3}$ **49a.** $I = \dfrac{\sqrt{PR}}{R}$

49b. 3.9 amps

51.

Distance	3	6	9	12	15
Height	6	24	4	96	150

53. $\pm\dfrac{\sqrt{3}}{3}$ **55.** Sample answer: $1+\sqrt{2}$ and $1-\sqrt{2}$;
$(1+\sqrt{2})(1-\sqrt{2}) = 1 - 2 = -1$ **57.** No radicals can appear in the denominator of a fraction. So, rationalize the denominator to get rid of the radicand in the denominator. Then check if any of the radicands have perfect square factors other than 1. If so, simplify. **59.** H **61.** 507.50

63.
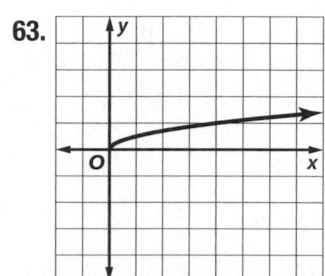
vertical compression of $y = \sqrt{x}$; $D = \{x \mid x \geq 0\}$, $R = \{y \mid y \geq 0\}$

65.
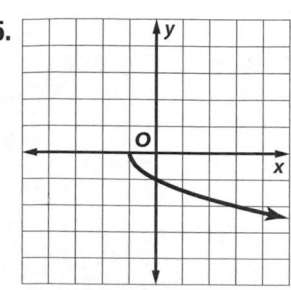
reflected across the x-axis and translated left 1 unit; $D = \{x \mid x \geq -1\}$, $R = \{y \mid y \leq 0\}$

67.
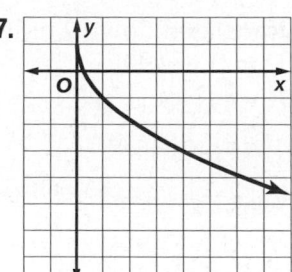
stretched vertically, reflected across the x-axis, and translated up 1 unit; $D = \{x \mid x \geq 0\}$, $R = \{y \mid y \leq 1\}$

69. $D = \{$all real numbers$\}$; $R = \{y \mid y$ is an integer$\}$
71. $-5, 5$ **73.** 5 **75.** $-0.1, 1.5$ **77.** $(n-9)(n+9)$
79. $2x^3(x+7)(x-7)$ **81.** prime **83.** about 2,172,453
85. $2^3 \cdot 3$ **87.** $2^2 \cdot 3^2 \cdot 5$ **89.** $2^2 \cdot 3 \cdot 5$

Lesson 10-3

1. $3\sqrt{5} + 6\sqrt{5} = (3+6)\sqrt{5}$
$= 9\sqrt{5}$
3. $-5\sqrt{7}$ **5.** $8\sqrt{5}$ **7.** $5\sqrt{2} + 2\sqrt{3}$ **9.** $72\sqrt{3}$
11. $\sqrt{21} + 3\sqrt{6}$ **13.** $14.5 + 3\sqrt{15}$ **15.** $11\sqrt{6}$
17. $3\sqrt{2}$ **19.** $5\sqrt{10}$ **21.** $60 + 32\sqrt{10}$ **23.** $3\sqrt{5} + 6 - \sqrt{30} - 2\sqrt{6}$ **25.** $5\sqrt{5} + 5\sqrt{2}$ **27.** $-\dfrac{4\sqrt{5}}{5}$
29. $\sqrt{2}$ **31.** $14 - 6\sqrt{5}$

33. **a.** $v_0 = \sqrt{v^2 - 64h}$
$= \sqrt{(120)^2 - 64(225)}$
$= \sqrt{0}$
$= 0$ ft/s
b. Sample answer: In the formula, we are taking the square root of the difference, not the square root of each term.

35. $\sqrt{170}$; about 13 amps **37.** Irrational; irrational; no rational number could be added to or multiplied by an irrational number so that the result is rational. **39.** Sample answer: You can use the FOIL method. You multiply the first terms within the parentheses. Then you multiply the outer terms within the parentheses. Then you would multiply the inner terms within the parentheses. And, then you would multiply the last terms within each parentheses. Combine any like terms and simplify any radicals. Sample answer: $(\sqrt{2} + \sqrt{3})(\sqrt{5} + \sqrt{7}) = \sqrt{10} + \sqrt{14} + \sqrt{15} + \sqrt{21}$. **41.** C **43.** C **45.** $2\sqrt{6}$
47. $5ab^2\sqrt{2ab}$ **49.** $3cd^2f^2\sqrt{7cf}$

51.
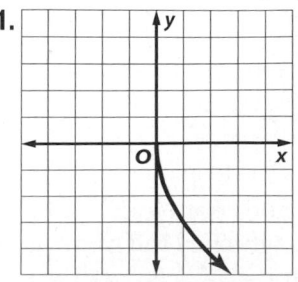
stretched vertically and reflected across the x-axis; $D = \{x \mid x \geq 0\}$, $R = \{y \mid y \leq 0\}$

53.
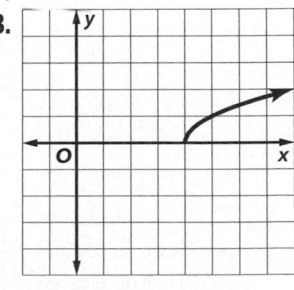
translated right 4 units; $D = \{x \mid x \geq 4\}$, $R = \{y \mid y \geq 0\}$

55.

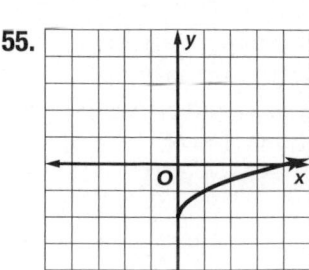

translated down 2 units;
$D = \{x \mid x \geq 0\}$,
$R = \{y \mid y \geq -2\}$

57. $(y + 10)(y + 3)$ **59.** $(x - 1)(x + 7)$
61. $(w + 12)(w - 6)$ **63.** -0.5 **65.** 18.7 **67.** 24

Lesson 10-4

1. $r = \dfrac{\sqrt{\pi x}}{2\pi}$ **3.** 2 **5.** 10 **7.** 6

9 $\sqrt{a} + 11 = 21$
$\sqrt{a} = 10$
$(\sqrt{a})^2 = (10)^2$
$a = 100$

11. 39 **13.** 17 **15.** 3 **17.** 6 **19.** 7 **21a.** 52 ft
21b. Increases; sample answer: If the length is longer, the quotient and square root will be a greater number than before. **23.** no solution
25. 235.2 **27.** 3

29 **a.**

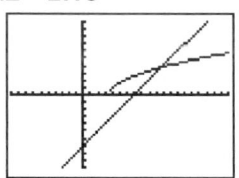

[−10, 20] scl: 1 by [−10, 10] scl: 1

c.

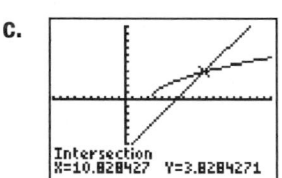

Intersection
X=10.828427 Y=3.8284271

[−10, 20] scl: 1 by [−10, 10] scl: 1

d. $\sqrt{2x - 7} = x - 7$
$(\sqrt{2x - 7})^2 = (x - 7)^2$
$2x - 7 = x^2 - 14x + 49$
$0 = x^2 - 16x + 56$
$x = \dfrac{-b \pm \sqrt{b^2 - 4ac}}{2a}$
$= \dfrac{16 \pm \sqrt{(16)^2 - 4(1)(56)}}{2(1)}$
$= \dfrac{16 \pm \sqrt{32}}{2}$
≈ 10.83 and 5.17

When checking, 5.17 does not work, so the answer is about 10.83 which is the same as we got using the calculator.

31. Jada; Fina had the wrong sign for $2b$ in the fourth step. **33.** Sample answer: In the first equation, you have to isolate the radical first by subtracting 1 from each side. Then square each side to find the value of x. In the second equation, the radical is already isolated, so square each side to start. Then subtract 1 from each side to solve for x. **35.** Sometimes; the equation is true for $x \geq 2$, but false for $x < 2$. **37.** Sample answer: Use addition/subtraction to isolate the term containing the radical on one side. Multiply/divide to change the coefficient of the radical to ± 1. Square each side of the equation to eliminate the radical. Solve the resulting equation using the most appropriate method. Check the answers for any extraneous solutions. **39.** C **41.** D **43.** $4\sqrt{3}$ **45.** $42\sqrt{2}$
47. $\dfrac{c^2\sqrt{5cd}}{2|d^3|}$ **49.** about 1.3 s and 4.7 s
51. $(2p + 3)(3p - 2)$ **53.** Sample answer: Triangles can be used to represent objects or situations with a triangular shape, and trigonometry or the Pythagorean Theorem can be used to find unknown values. **55.** $(2a + 3)(a - 6)$ **57.** Yes; $4x^3$ is the product of a number and three variables.
59. No; $4n + 5p$ shows addition, not multiplication alone of numbers and variables. **61.** Yes; $\frac{1}{5}abc^{14}$ is the product of a number, $\frac{1}{5}$, and several variables.
63. $1,000,000$ **65.** $64v^2$ **67.** $1000y^6$

Lesson 10-5

1. 5 **3.** 18.03 **5a.** about 127 ft **5b.** about 117 ft
5c. about 46 ft **7.** yes **9.** no
11 $a^2 + b^2 = c^2$
$(2)^2 + b^2 = 12^2$
$4 + b^2 = 144$
$b^2 = 140$
$b \approx 11.83$
13. 29.66 **15.** 5.29 **17.** 7.21
19 $a^2 + b^2 = c^2$
$(30)^2 + (36)^2 = c^2$
$900 + 1296 = c^2$
$2196 = c^2$
$46.9 \approx c$
The diagonal of the TV stand is about 46.9 inches which is larger than 42 inches, so the TV will fit.
21. no; no **23.** no; no **25.** no, no **27.** yes; yes
29a. about 20.20 units **29b.** 111.1 units^2 **31.** a 30-ft ladder **33.** 8.06 **35.** $5\sqrt{3}$ in. or about 8.66 in.
37. about 6.7 ft
39 $a^2 + b^2 = c^2$
$(8)^2 + x^2 = (x + 2)^2$
$64 + x^2 = x^2 + 4x + 4$
$64 = 4x + 4$
$60 = 4x$
$15 = x$
So, $b = 15$ and $c = 15 + 2$, or 17
41. $a = 65$; $b = 72$ **43.** $a = 9$; $b = 40$; $c = 41$
45a. 5 units, 8 units; $\left(\frac{5}{2}, 2\right)$, $(4, 5)$

45b. $|x_2 - x_1|$ units, $|y_2 - y_1|$ units; $\left(\dfrac{x_1 + x_2}{2}, y\right), \left(x, \dfrac{y_1 + y_2}{2}\right)$ **45c.** $\left(\dfrac{x_1 + x_2}{2}, \dfrac{y_1 + y_2}{2}\right)$

45d. $\sqrt{(x_2 - x_1)^2 + (y_2 - y_1)^2}$ units

47. $8\sqrt{2}$

49. Sample answer:

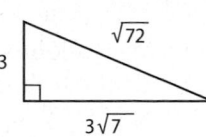

51. C **53.** $88 **55.** 256 **57.** 10 **59.** 3 **61.** $7\sqrt{3}$
63. $10\sqrt{7}$ **65.** $12\sqrt{5} - 5\sqrt{3}$ **67.** translated down 8
69. reflected across the x-axis, translated up 5
71. reflected across the x-axis, stretched vertically
73. 3 seconds **75.** $x^2 - 13x + 36$ **77.** $p^2 - 8p - 20$
79. $40d^2 + 31d + 6$ **81.** 20 **83.** 8

Lesson 10-6

1. $\sin A = \dfrac{24}{25}$; $\cos A = \dfrac{7}{25}$; $\tan A = \dfrac{24}{7}$

③ $\sin A = \dfrac{\text{opposite}}{\text{hypotenuse}} = \dfrac{5}{13}$

$\cos A = \dfrac{\text{adjacent}}{\text{hypotenuse}} = \dfrac{12}{13}$

$\tan A = \dfrac{\text{opposite}}{\text{adjacent}} = \dfrac{5}{12}$

5. 0.6018 **7.** 0.2493 **9.** $m\angle X \approx 51°$; $XY \approx 4.4$;
$YZ \approx 5.4$ **11.** $m\angle Q \approx 60°$; $RQ \approx 2.9$; $PQ \approx 5.8$
13. about 11,326.2 ft **15.** 66° **17.** 33° **19.** $\sin B = \dfrac{5}{13}$; $\cos B = \dfrac{12}{13}$; $\tan B = \dfrac{5}{12}$ **21.** 0.0349 **23.** 0.7193
25. 0.9563 **27.** 0.5

㉙ Step 1: Find the measure of $\angle Y$.
$180° - (90° + 47°) = 43°$
Step 2: Find $\overline{XY}$ or z. Since you are given the measure of the side opposite $\angle X$, and are finding the measure of the hypotenuse, use the sine ratio.
$\sin 47° = \dfrac{16}{z}$
$z \sin 47° = 16$
$z = \dfrac{16}{\sin 47°}$
$z \approx 21.9$ or $XY \approx 21.9$
Step 3: Find $\overline{XZ}$ or y. Since you are given the measure of the side opposite $\angle X$, and are finding the measure of the side adjacent to $\angle X$, use the tangent ratio.
$\tan 47° = \dfrac{16}{y}$
$y \tan 47° = 16$
$y = \dfrac{16}{\tan 47°}$
$y \approx 14.9$ or $XZ \approx 14.9$
31. $m\angle R = 76°$; $QR \approx 7.2$; $PR \approx 1.7$ **33.** $m\angle Y = 39°$;
$WU \approx 11.3$; $UY \approx 18.0$ **35.** about 53 ft **37.** 62°
39. 31° **41.** 50°

㊸ $\tan 8° = \dfrac{5000}{x}$

$x \tan 8° = 5000$
$x = \dfrac{5000}{\tan 8°}$
$x \approx 35{,}577$ ft

45. $\sin A = \dfrac{\sqrt{7}}{4}$; $\tan A = \dfrac{\sqrt{7}}{3}$ **47.** $\cos A = \dfrac{\sqrt{15}}{4}$;
$\tan A = \dfrac{\sqrt{15}}{15}$ **49.** about 0.5 mi **51.** $a = 5$; $c = 7.3$
53. Sample answer: Triangles can be used to represent objects or situations with a triangular shape, and trigonometry or the Pythagorean Theorem can be used to find unknown values.
55. Use the acute angle given and the measure of the known side to set up one of the trigonometric ratios. The sine ratio uses the opposite side and hypotenuse of the triangle. The cosine ratio uses the adjacent side and hypotenuse of the triangle. The tangent ratio uses the opposite and adjacent sides of the triangle. Set up the ratio and solve for the unknown measure. **57.** F **59a.** increase
59b. Sample answer: The sum of their squares is 16^2 or 256. **59c.** about 15.7 ft **61.** 11 **63.** 6.71
65. 5.20 **67.** The amount of sales must be more than $260,000. **69.** 8 **71.** 4.62

Chapter 10 Study Guide and Review

1. false; Sample answer: 3, 4, and 5 **3.** true
5. false; longest **7.** false; $\{x \mid x \geq 0\}$ **9.** true

11.

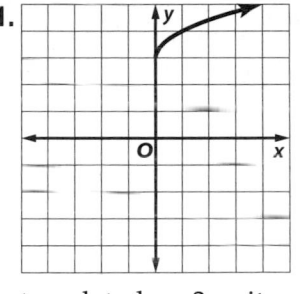

translated up 3 units;
$D = \{x \mid x \geq 0\}$,
$R = \{y \mid y \geq 3\}$

13.

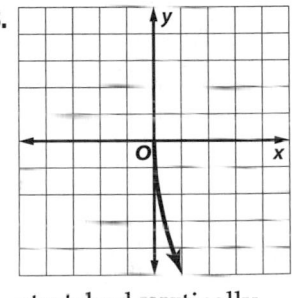

stretched vertically
and reflected across
the x-axis;
$D = \{x \mid x \geq 0\}$,
$R = \{y \mid y \leq 0\}$

15.

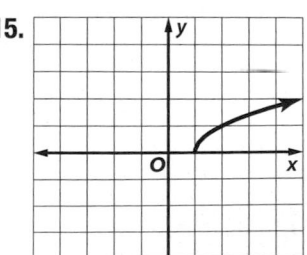

translated right 1 unit;
$D = \{x \mid x \geq 1\}$,
$R = \{y \mid y \geq 0\}$

17. 9.5 in.

19. $2|b|\sqrt{5ab}$

21. 36

23. $2\sqrt{2} + 3$

25. $\dfrac{\sqrt{30}}{10}$ **27.** $\dfrac{5\sqrt{7} - 30}{-29}$ **29.** $-2\sqrt{6} + 11\sqrt{3}$
31. $4\sqrt{3x}$ **33.** $5\sqrt{2} + 3\sqrt{6}$ **35.** $24\sqrt{10} + 8\sqrt{2} + 6\sqrt{15} + 2\sqrt{3}$

37. no solution **39.** 32 **41.** 12 **43.** 1600 ft **45.** yes
47. no **49.** yes **51.** no **53.** $\cos A = \frac{5}{13}$, $\sin A = \frac{12}{13}$, $\tan A = \frac{12}{5}$ **55.** 6 ft

CHAPTER 11
Rational Functions and Equations

Chapter 11 Get Ready

1. $\frac{8}{3}$ **3.** $\frac{21}{2}$ **5.** $\frac{10}{3}$ or $3\frac{1}{3}$ inches **7.** $2x(x-2)$
9. $2c^2d(1-2d)$ **11.** $(m-12)(m+11)$
13. $(3x-4)(2x+1)$

Lesson 11-1

1. Direct; the data in the table can be represented by the equation $y = 2x$. **3.** Inverse; $xy = 4$.

5. 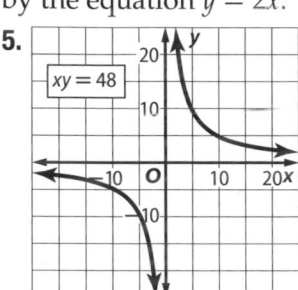 **7.**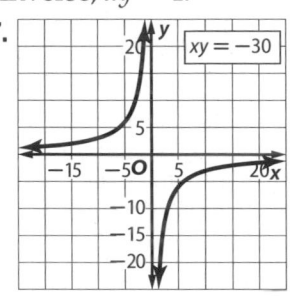

9.
$$x_1y_1 = x_2y_2$$
$$(4)(8) = x_2(2)$$
$$32 = 2x_2$$
$$16 = x_2$$

11. -7.5

13a.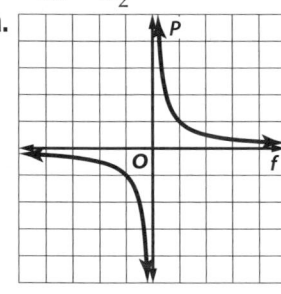

13b. 5 to -2.5 diopters
15. Direct; $y = -3x$
17. Inverse; $xy = -40$
19. Inverse; $xy = \frac{1}{4}$
21. Direct; $y = 9x$

23. $xy = 72$ 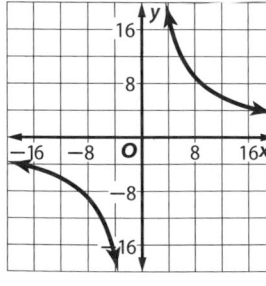 **25.** $xy = 12$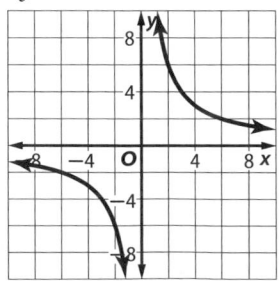

27. 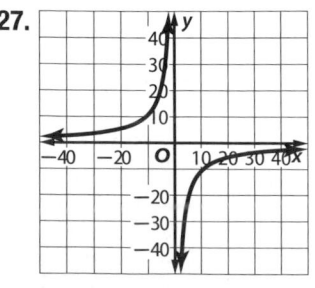 $xy = -108$

29. 15 **31.** -3 **33.** 9.6
35.
$$x_1y_1 = x_2y_2$$
$$(420)(523) = (707)y_2$$
$$219{,}660 = 707y_2$$
$$311 \approx y_2$$
approximately 311 cycles per second
37. Direct; the number of lemonades times the cost per lemonade equals the total cost. So the ratio $\frac{\text{total cost}}{\text{number of lemonades}}$ is a constant \$1.50.
39. Inverse; the number of friends times the number of tokens per person equals the constant 30. **41.** Inverse; $xy = 21$ **43.** Direct; $y = \frac{1}{2}x$
45.
$$x_1y_1 = x_2y_2$$
$$(6)(9.2) = x_2(3)$$
$$55.2 = 3x_2$$
$$18.4 = x_2$$
47. 2.5 **49.** \$4

51a.

Hours per Week h	Number of Weeks w
1	40
2	20
4	10
5	8
8	5
10	4

51b. The number of weeks decreases.

51c. $hw = 40$ or $w = \frac{40}{h}$

Driving

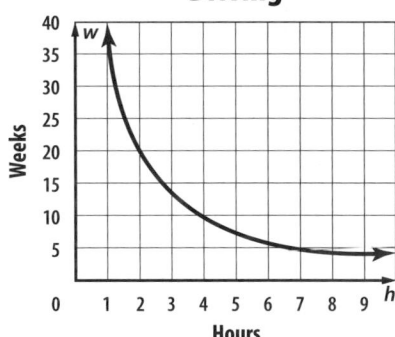

53. direct variation **55.** Sample answer: Newton's Law of Gravitational Force is an example of an inverse variation that models real-world situations. The gravitational force exerted on two objects is inversely proportional to the square of the distances between the two objects. The force exerted on the two objects, times the square of the distance between the two objects, is equal to the gravitational constant times the masses of the two objects. **57.** B **59.** C **61.** $\sin A = 0.7241$, $\cos A = 0.6897$, $\tan A = 1.05$ **63.** $\sin A = 0.8182$, $\cos A = 0.5750$, $\tan A = 1.4230$ **65.** $\frac{9}{10}$ **67.** 1 **69.** no solution **71.** 7^2 or 49 **73.** $\frac{q^4}{2p^5}$ **75.** $\frac{4a^4b^2}{c^6}$ **77.** $\frac{1}{mn}$

1. $x = 0$ **3.** $x = 1$

5.

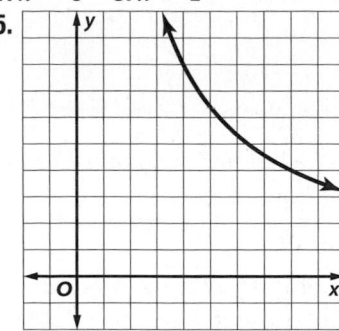

7. $x = 0; y = -1$

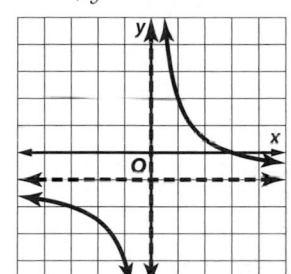

9. $x = -2; y = 0$

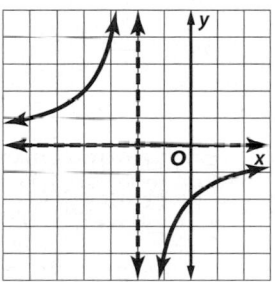

11. $x = -1; y = -5$

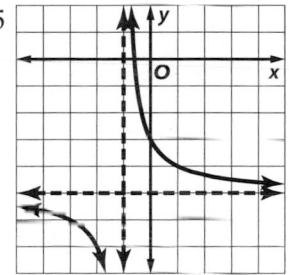

13. $x = 8$ **15.** $x = -6$ **17.** $x = -5$ **19.** $x = -7$

21.

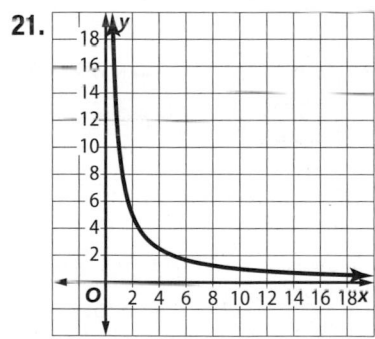

23 Step 1: Identify and graph the asymptotes using dashed lines.

vertical asymptote: $x = 0$
horizontal asymptote: $y = 0$

Step 2: Make a table of values and plot the points. Then connect them.

x	−2	−1	1	2
y	$\frac{3}{2}$	3	−3	$-\frac{3}{2}$

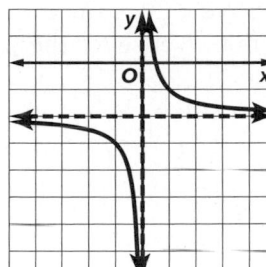

25. $x = 0; y = -2$ **27.** $x = 2; y = 0$

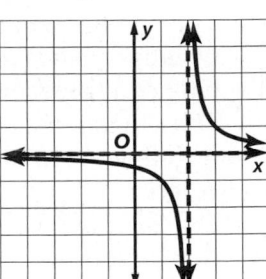
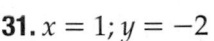
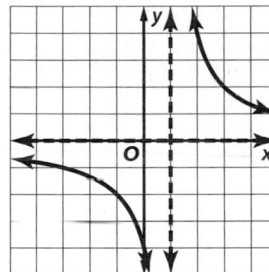
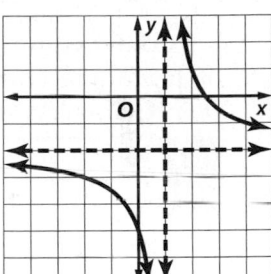

29. $x = 1; y = 0$ **31.** $x = 1; y = -2$

33. $x = -4; y = 3$

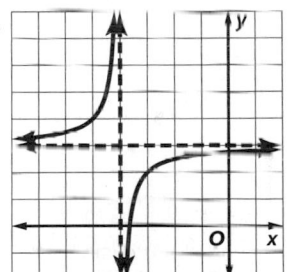

35a. $x = 3$ and $y = 2$ **35b.** $y = \dfrac{1}{x - 3} + 2$

37a. Sample answer: The total cost of the trip equals the cost of a ticket plus the cost of the star-naming package divided by the number of people.   **37b.** $y = \dfrac{95}{p} + 8.50$

37c.

[0, 50] scl: 5 by [0, 50] scl: 5

Sample answer: The end behavior indicates that as the number of people increases, the cost per person approaches 0. Since there is no x-intercept, cost per person will never be 0.

37d. Sample answer: 15 people

39. 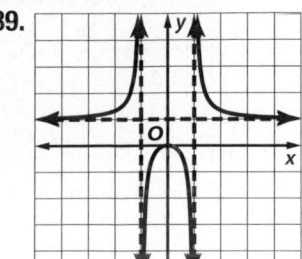 $x = -1, x = 1; y = 1$

41 **a.** The domain will be positive values since negative values do not make sense for the base of a trapezoid.
The range must also be positive values since negative values do not make sense for the height.
D = {all positive real numbers}
R = {all positive real numbers}

b. Step 1: Identify and graph the asymptotes using dashed lines.
vertical asymptote: $b_1 = -8$
horizontal asymptote: $h = 0$
Step 2: Make a table of values and plot the points. Then connect them.

b_1	2	4	6	8
h	12.8	10.7	9.1	8

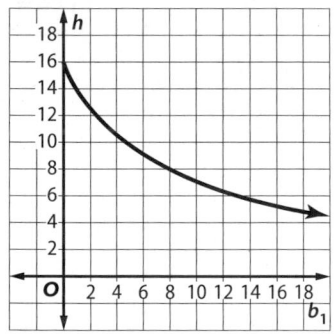

c. When $b_1 = 10$, h will be about 7 units.

43. The graph of $y = \dfrac{1}{x+5} - 2$ is the graph of $y = \dfrac{1}{x}$ translated 5 units to the left and 2 units down.
45. False; sample answer: The graph of $y = \dfrac{1}{x}$ has no x- or y-intercepts. **47.** Sample answer: Vertical asymptotes occur at values that make the denominator 0; horizontal asymptotes occur at $y = c$ for any rational function of the form $y = \dfrac{a}{x-b} + c$. **49.** 45 seconds **51.** D **53.** 78°
55. 40° **57.** 60° **59.** 61.85 **61.** 10.54 **63.** 9.22
65. $\sqrt{465}$ or about 21.56 mi **67.** $(w+16)(w-3)$
69. $(3+a)(24+a)$ **71.** $(d-2)(d-5)$ **73.** $(n+9)(n-6)$
75. $(4b-3)(6b+1)$ **77.** $2(x-3)(3x+2)$

1. $4, -4$ **3.** 16.9 units of force

5 $\dfrac{(-3r)(10r^4)}{6r^5} = \dfrac{-30r^5}{6r^5}$
$= -5r^0$
$= -5$
The excluded value is when $6r^5 = 0$, or when $r = 0$.
7. $x + 7; -4$ **9.** $-\dfrac{3}{9+y}; -9, 9$ **11.** 3 **13.** $1, -1$

15. $3, -8$ **17.** $\dfrac{2\pi(5t)}{\pi(5t)^2} = \dfrac{2}{5t}; 0$ **19.** $\dfrac{16}{n^2}; n \neq 0, p \neq 0$
21. $-2c; 0$ **23.** $\dfrac{a}{a-6}; 6, -3$ **25.** $x - 4; -8$

27. $\dfrac{2p}{p+7}; 7, -7$ **29.** $\dfrac{-(8+c)}{c+1}; 8, -1$

31 $f(x) = \dfrac{x^2 + 3x - 4}{x^2 + 9x + 20}$

$0 = \dfrac{x^2 + 3x - 4}{x^2 + 9x + 20}$ $\quad f(x) = 0$

$0 = \dfrac{(x+4)(x-1)}{(x+4)(x+5)}$ $\quad$ Factor

$0 = \dfrac{\cancel{(x+4)}(x-1)}{\cancel{(x+4)}(x+5)}$ $\quad$ Divide out common factors

$0 = \dfrac{x-1}{x+5}$ $\quad$ Simplify.

When $x = 1$, the numerator becomes 0, which makes $f(x) = 0$. Therefore, the zero of the function is 1.

33. $2, 4$ **35.** $0, 6$ **37a.** $\dfrac{250\pi}{t}$ **37b.** about 157 ft/min

39. $\dfrac{4x^4 - 5y^2}{y^3}; x, y \neq 0$

41 **a.** Surface area: $2(2x^2) + 2(x^2) + 2(2x^2)$
$= 4x^2 + 2x^2 + 4x^2$
$= 10x^2$
Volume: $2x(x)(x) = 2x^3$
So, the ratio is $\dfrac{10x^2}{2x^3} = \dfrac{5}{x}$.
The excluded value is when $2x^3 = 0$, or when $x = 0$.

b. Surface area: $2\pi a^2 + 2\pi ab$
Volume: $\pi a^2 b$
So, the ratio is $\dfrac{2\pi a^2 + 2\pi ab}{\pi a^2 b} = \dfrac{2\pi a^2}{\pi a^2 b} + \dfrac{2\pi ab}{\pi a^2 b}$
$= \dfrac{2a + 2b}{ab}$
The excluded value is when $\pi a^2 b = 0$, or when $a = 0$ or $b = 0$.

43. No; Colleen did not show the simplified expression, and Sanson used the simplified expression to find the excluded value. **45.** Every polynomial P can be written as $\dfrac{P}{1}$, where the numerator and denominator are polynomials; hence every polynomial is also a rational expression. **47.** No; the numerator and denominator have $x - 2$ as a common factor.
49. C **51.** G **53.** 0 **55.** 3 **57.** 20 **59.** 5
61. 8.4 in. **63.** $3\sqrt{2}$ **65.** $8\sqrt{2}$ **67.** $2|a|\sqrt{10}$

69. $\dfrac{\sqrt{6}}{3}$ **71.** $2x$ **73.** 1 **75.** $9qt$

1. $4x$ **3.** $\dfrac{t}{6(t-5)}$ **5.** 6.16 ft/min   **7.** $\dfrac{2}{3x}$ **9.** $\dfrac{1}{6}$

(11) $\dfrac{10n^2}{4} \cdot \dfrac{2}{n} = \dfrac{5n}{1} \cdot \dfrac{1}{1}$
$= 5n$

13. $9x^4z$ **15.** $3(t+2)$ **17.** $k+6$ **19.** $4n$
21a. about \$24.66   **21b.** about \$26.09   **23.** x^4y
25. $\dfrac{9b^3}{2}$ **27.** $\dfrac{b-2}{(b+5)(2b+3)}$ **29.** $\dfrac{x}{2(x-4)}$ **31.** $\dfrac{(r+2)^2}{4}$
33. 15 mi/h   **35a.** $\dfrac{1250\text{ km}}{1} \cdot \dfrac{1\text{ mi}}{1.609\text{ km}} \cdot \dfrac{1\text{ hr}}{540\text{ mi}}$
35b. 1.4 h   **37.** 7,744,000 yd²/mo   **39.** about
31.4 mi/h   **41.** 6.1 c/min   **43.** $\dfrac{1\text{ cup}}{1\text{ beat}} \cdot \dfrac{70\text{ beats}}{1\text{ minute}} \cdot$
$\dfrac{1\text{ gallon}}{16\text{ cups}} \cdot \dfrac{60\text{ minutes}}{1\text{ hour}} = 262.5$ gal/h

(45) $\dfrac{\$9.80}{1\text{ hour}} \cdot \dfrac{15\text{ hours}}{\text{week}} \cdot \dfrac{52\text{ weeks}}{\text{year}} = \dfrac{\$7644}{\text{year}}$
This answer represents earning per year.
47. about 71.6 mi/h; converting 32 meters per
second to miles per hour   **49.** about 52,506.7 ft/s

(51) **a.** $\dfrac{27\text{ rotations}}{1\text{ minute}} \cdot \dfrac{1\text{ minute}}{60\text{ seconds}} \cdot \dfrac{2\pi(3.1)\text{ meters}}{\text{rotation}}$

b. $\dfrac{27\text{ rotations}}{1\text{ minute}} \cdot \dfrac{1\text{ minute}}{60\text{ seconds}} \cdot \dfrac{2\pi(3.1)\text{ meters}}{\text{rotation}}$
$\approx \dfrac{8.8\text{ meters}}{\text{second}}$
This means that the room moves about
8.8 meters per second.

53. Neither; Tamika did not multiply by the
reciprocal, and Mei incorrectly factored out the
2 in $2x+6$.   **55.** $\dfrac{x-3}{x+7}$ **57.** Sample answer: The
height of a cylinder when you know an expression
for the volume and radius; $\dfrac{V}{\pi r^2} = h$; $V = \pi(x^3 - 6x^2$
$+ 9x)$, $r = (x-3)$.   **59.** Sample answer: Write ratios
comparing the number of days in one year and the
number of hours in one day. Then multiply the
ratios: $\dfrac{365\text{ days}}{1\text{ year}} \cdot \dfrac{24\text{ hours}}{1\text{ day}}$; 8760 hours.

61. F   **63.** A   **65.** $\dfrac{g^2}{3h}$; $g \neq 0, h \neq 0$ **67.** $y + 8$; -2
69. $\dfrac{z+2}{z-2}$; 2, 1   **71.** $x = 0, y = 5$ **73.** $x = -3, y = 0$
75. $x = 8, y = -3$ **77.** 0   **79.** 2   **81.** 3   **83.** $\{p \mid 28 \leq$
$p \leq 32\}$ **85.** $(x-2)(x-3)$ **87.** $(x+4)(3x-5)$
89. $4(x+2)(2x-5)$

Lesson 11-5

(1) $(8a^2 + 20a) \div 4a = \dfrac{8a^2}{4a} + \dfrac{20a}{4a}$
$= 2a + 5$

3. $2n^2 - n + \dfrac{5}{2n}$ **5.** $x - 4$ **7.** $3 + \dfrac{250}{x+50}$
9. $4y + \dfrac{3}{y+2}$ **11.** $3n^2 + n - 4 + \dfrac{4}{3n-1}$ **13.** $a^2 +$
$4a - 18$ **15.** $2n - 4 + \dfrac{1}{n}$ **17.** $\dfrac{3}{2}m + \dfrac{5}{6}$ **19.** $x - 8$

(21) $(k^2 - 5k - 24) \div (k - 8) = \dfrac{(k^2 - 5k - 24)}{(k-8)}$
$= \dfrac{(k-8)(k+3)}{k-8}$
$= k + 3$

23. a^2 **25.** $2t - 1$ **27.** $2h^2 - 3$ **29.** $-n + 18 + \dfrac{850}{n}$
31. $a + 5 + \dfrac{8}{a-1}$ **33.** $4n + 5 + \dfrac{16}{n-2}$ **35.** $t^2 - 4t +$
$14 - \dfrac{60}{t+4}$ **37.** $2c^2 + c + 2 - \dfrac{1}{4c-2}$ **39.** $x + 3$
41. $x^2 - 2x + 15 - \dfrac{16}{x+2}$

43a.

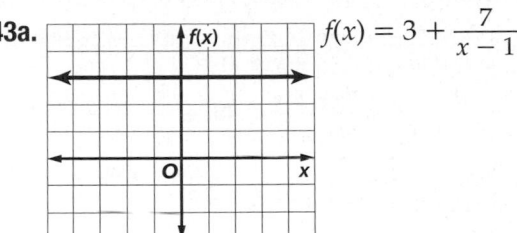

$f(x) = 3 + \dfrac{7}{x-1}$

43b.

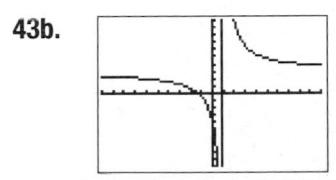

[−10, 10] scl: 1 by [−10, 10] scl:1

43c. The graph of the quotient ignoring the
remainder is an asymptote of the graph of the
function.   **43d.** As x approaches 1 from the left, y
approaches negative infinity. As x approaches 1
from the right, y approaches positive infinity.

(45) **a.** Words: 212° decreases by 0.9 for every
500 feet above sea level.
Equation: $T = 212 - 0.9\left(\dfrac{x}{500}\right)$

b. $212 - 0.9\left(\dfrac{x}{500}\right) = 212 - 0.9\left(\dfrac{14,494}{500}\right)$
$= 212 - 26.1$
$= 185.9°$ F

47. Andrea; Alvin did not take into account the
missing term.   **49.** Sample answer: $(a^2 + 4a - 22)$
$\div (a - 3)$; The polynomial $a^2 + 4a - 22$ is prime, so
the problem can be solved by using long division.
51. A   **53.** G   **55.** $6x$ **57.** $\dfrac{t}{6(t+4)}$ **59.** -2
61. no zero   **63.** 81   **65.** 9   **67.** $-10, -2$ **69.** $-0.8, 1$
71. $-0.4, 3.9$ **73a.**

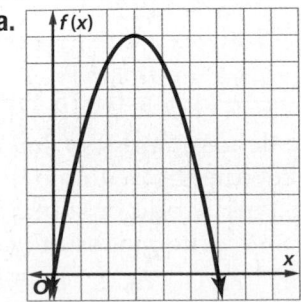

6 ft

73b. 9 ft   **75.** $-2cd + d^2$

1. $\dfrac{5}{7n}$ **3.** $\dfrac{16r}{9-r}$ **5.** $24t^2$ **7.** $(x-7)(x-1)(x+2)$

9. $\dfrac{15y+8x}{20xy}$ **11.** $\dfrac{16d+5c}{6cd}$ **13.** $\dfrac{x^2-x+9}{x^2-x-6}$

(15) $\dfrac{a}{4}+\dfrac{3a}{4}=\dfrac{a+3a}{4}$
$$=\dfrac{4a}{4}$$
$$=a$$

17. $\dfrac{2y}{3}$ **19.** $\dfrac{8b+3a}{ab}$ **21.** $\dfrac{c-8}{2c-1}$ **23.** $\dfrac{1}{2}$ **25.** $\dfrac{-w+5}{8w}$

27. x^3y^2 **29.** $(3r-1)(r+2)$ **31.** $(x+6)(x+3)$

33. $\dfrac{27}{20x}$ **35.** $\dfrac{15b+2a}{10ab}$ **37.** $\dfrac{7-4k}{4(k+2)}$ **39.** $\dfrac{-2t+28r}{7rt}$

41. $\dfrac{d^2+6d+35}{(d+5)(d-1)}$ **43.** $\dfrac{18-10t}{15t^2}$ **45.** $\dfrac{w^2-4w-3}{(w+4)(w-5)}$

47. $\dfrac{-x^2-4x-9}{(x+3)(x+5)}$ **49a.** $\dfrac{2}{w}+\dfrac{2}{2.5w};\dfrac{7}{2.5w}$ **49b.** 48 min

51. $\dfrac{a+9}{2a+18}$ or $\dfrac{1}{2}$

(53) $\dfrac{x+5}{x^2-4}-\dfrac{3}{x^2-4}=\dfrac{x+5-3}{x^2-4}$
$$=\dfrac{x+2}{(x+2)(x-2)}$$
$$=\dfrac{1}{x-2}$$

55. $k+5$ **57.** $\dfrac{1}{x-1}$ **59.** $\dfrac{a-6}{3}$ **61.** $\dfrac{2x^3-16x+1}{x^4}$

63. $\dfrac{15x^2-11x-7}{9x^3+51x^2-80x+28}$ **65a.** $\dfrac{9200}{3x}$ **65b.** about 76 min 40 seconds **67.** 8

(69) a. $t=D/r$
Day 1: $t=\dfrac{9+x}{10}$
Day 2: $t=\dfrac{5+3x}{10}$
So, the total for the 2 days was
$$\dfrac{9+x}{10}+\dfrac{5+3x}{10}=\dfrac{9+x+5+3x}{10}$$
$$=\dfrac{14+4x}{10}$$
$$=\dfrac{7+2x}{5}$$
b. $\dfrac{7+2x}{5}=\dfrac{7+2(2)}{5}$
$$=\dfrac{7+4}{5}$$
$$=\dfrac{11}{5}\text{ or }2.2$$
Marina rode her bike 2.2 hours on those 2 days.

71. $\dfrac{4-7y}{21y-6}$ **73.** always; $\dfrac{a}{x}+\dfrac{b}{y}=\dfrac{a}{x}\cdot\dfrac{y}{y}+\dfrac{b}{y}\cdot\dfrac{x}{x}=$
$\dfrac{ay}{xy}+\dfrac{bx}{yx}=\dfrac{ay+bx}{xy},x,y\neq0$ **75.** First, factor -1 out of one of the denominators so that it is like the other. Then rewrite the denominator without parentheses. Finally, add or subtract the numerators and write the result over the like denominator. **77.** C **79.** C **81.** $5y^2+\dfrac{14}{3}$

83. about 15.7 mi/h **85.** 5 **87.** 11.62 **89.** 1.23×10^4

91. 1.255×10^6 **93.** $\dfrac{2}{3x}$ **95.** $\dfrac{4}{y+1}$

1. $\dfrac{2+4n}{n}$ **3.** $\dfrac{6t+11}{t+1}$ **5a.** $\dfrac{2\frac{1}{2}\text{ mi}}{\frac{1}{3}\text{ h}}$ **5b.** $\dfrac{15}{2}$ or $7\frac{1}{2}$ mi/h

7. $\dfrac{3}{25}$ **9.** $\dfrac{2y^2}{x}$ **11.** $\dfrac{1}{(x+y)(r+s)}$ **13.** $\dfrac{p+3}{p^2-p-2}$

(15) $\dfrac{p-7}{2p}=\dfrac{p(2p)}{2p}-\dfrac{7}{2p}$
$$=\dfrac{2p^2}{2p}-\dfrac{7}{2p}$$
$$=\dfrac{2p^2-7}{2p}$$

17. $\dfrac{3h^2+h+1}{h}$ **19.** $\dfrac{n^3+4n^2+n-1}{n+4}$

21. $\dfrac{d^2-12d+43}{d-7}$ **23.** $\dfrac{3}{4}$ page/min **25.** $\dfrac{2}{3}$ **27.** $\dfrac{h}{g^3}$

29. $\dfrac{2a+12}{a}$ **31.** $\dfrac{(j-4)(j+4)}{15(j+2)}$ **33.** about 140

(35) $D=\dfrac{m}{V};V=\dfrac{4}{3}\pi r^3$
$$D=\dfrac{m}{\frac{4}{3}\pi r^3}$$
$$D=\dfrac{3m}{4\pi r^3}$$
a. $D=\dfrac{3(15.6)}{4\pi(0.0748)^3}\approx8898.794862$ or 8900
The metal is copper.
b. $D=\dfrac{3(285.3)}{4\pi(0.1819)^3}\approx11{,}316.57654$ or 11,300
The metal is lead.

37. $\dfrac{58x+145}{5x+10}$ **39.** $\dfrac{c-3}{c+7}$ **41.** $\dfrac{(x-8)(x-1)}{x+2}$

43. Find the lowest common denominator for the fractions in the numerator and simplify to $\dfrac{y^2-x^2}{xy}$.

45. $\dfrac{2t^3}{(t-1)(t^2-1)}$ **47.** Sample answer: Time equals distance divided by rate or $\dfrac{d}{r}$. When the distance or the rate is given as a fraction or mixed number, the expression $\dfrac{d}{r}$ becomes a complex fraction. Example: Someone walks $\dfrac{3}{4}$ mile in $10\frac{1}{2}$ minutes; the time in miles per minute is $\dfrac{\frac{3}{4}}{10\frac{1}{2}}$, which simplifies to $\dfrac{1}{14}$ mi/min. **49.** 12 days **51.** B

53. $\dfrac{-d+4}{d-1}$ **55.** $\dfrac{6m^2-1}{15m^3}$ **57.** $\dfrac{b^2+4b+18}{(b+3)(b-2)}$

59. $a+7-\dfrac{1}{a-3}$ **61.** $3y^2+2y-3-\dfrac{1}{y+2}$

63. $3h^2+2h+3-\dfrac{2}{3h-2}$

65. 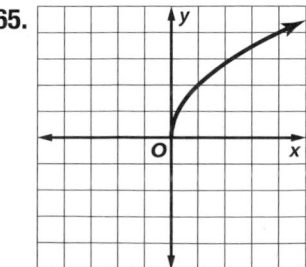 **67.**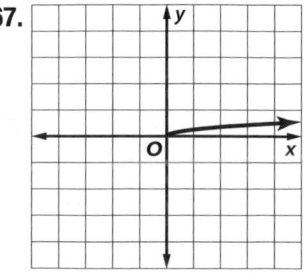

D = {x | x ≥ 0};   D = {x | x ≥ 0};
R = {y | y ≥ 0}   R = {y | y ≥ 0}

69. $(a - 11)(a + 11)$ **71.** $(2y - 5)(2y + 5)$
73. $4(t - 1)(t + 1)(t^2 + 1)$ **75.** 4   **77.** 7   **79.** −11.5

Lesson 11-8

1. −2   **3.** $-\dfrac{4}{5}$ **5.** −3   **7.** $\dfrac{15}{38}$ hour or about 0.4 hour

9 $\dfrac{8}{n} = \dfrac{3}{n - 5}$

$8(n - 5) = 3n$
$8n - 40 = 3n$
$-40 = -5n$
$8 = n$

11. $\dfrac{2}{3}$ **13.** −13   **15.** 0   **17.** −2, 3   **19.** no solution;
extraneous: 1   **21.** $\dfrac{15}{8}$ hours or $1\dfrac{7}{8}$ hours

23. 26.2 hours   **25a.** line   **25b.** $f(x) = \dfrac{(x + 5)(x - 6)}{x - 6}$
$= x + 5$ **25c.** −5   **27a.** parabola   **27b.** $f(x) = x^2 + 6x + 12$ **27c.** no real zeros   **29a.** $s = r - w; s = r + w$ **29b.** $d = t(r - w), d = t(r + w); t = \dfrac{d}{r - w}$,
$t = \dfrac{d}{r + w}$

31   **a.** $h = \dfrac{kp}{c}$

b. $8 = \dfrac{k(100)}{5}$

$8 = 20k$

$\dfrac{2}{5} = k$

$h = \dfrac{\frac{2}{5}p}{c}$

$= \dfrac{\frac{2}{5}(100)}{10}$

$= 4$ hours

33. 0, −4; extraneous: 1   **35.** $\dfrac{50}{11}$
37. The extraneous solution of a rational equation is the excluded value of one of the expressions in the equation.   **39.** Sample answer: $\dfrac{x}{x - 8} = 0$ **41.** D
43. D   **45.** $\dfrac{d}{c}$ **47.** $\dfrac{b - 3}{2b}$ **49.** $4h^2$ **51.** $(x - 4)(x + 2)$
53. linear   **55.** quadratic   **57.** $0.5Bb + 0.5b^2; \dfrac{1}{2}$
59. $\{r \mid r > 49\}$ **61.** 0.3   **63.** 0.75

Chapter 11   Study Guide and Review

1. false; $x^2 - 25$ **3.** true   **5.** true   **7.** false; complex
fraction   **9.** false; inverse variation   **11.** $\dfrac{1}{3}$
13. $-\dfrac{9}{16}$ **15.** 3   **17.** 2

19. 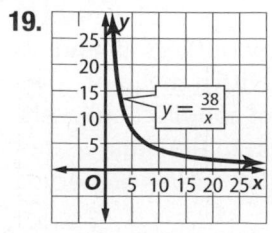 The vertical asymptote is at $x = 0$ and the horizontal asymptote is at $y = 0$.

$y = \dfrac{38}{x}$

21. $\dfrac{1}{x + 8}$ **23.** $\dfrac{y - 5}{y - 2}$ **25.** $\dfrac{1}{2y(y + 2)}$ **27.** $-\dfrac{3}{2}, \dfrac{1}{3}$
29. $\dfrac{3}{x^2 - 3x}$ **31.** $\dfrac{3}{(b + 4)(b - 2)}$ **33.** xy **35.** $x^2 + 4x - 2$ **37.** $4y + 1 + \dfrac{8}{12y - 1}$ **39.** $\dfrac{3a}{b}$ **41.** $\dfrac{-y^2 + 2y - 9}{(y + 1)(y - 3)}$
43. 4 ft   **45.** $\dfrac{x^2 + 8x - 65}{x^2 + 8x + 12}$ **47.** $\dfrac{y^2 + 11y + 10}{y^2 + 6y + 8}$ **49.** $\dfrac{2}{5}$,
extraneous: 2   **51.** 24   **53.** −2, 4

CHAPTER 12
Statistics and Probability

Chapter 12   Get Ready

1. $\dfrac{3}{7}$ **3.** $\dfrac{4}{7}$ **5.** $\dfrac{1}{6}$ **7.** $\dfrac{5}{6}$ **9.** $\dfrac{7}{256}$ **11.** $\dfrac{84}{625}$ **13.** 82.4%
15. 85.6%   **17.** 35%

Lesson 12-1

1. Sample: the guests who receive the stickered plates; population: all of the guests; simple: all of the guests have an equal chance of receiving a stickered plate.   **3.** Unbiased; sample answer: The sample is random, and does not favor one outcome over another.   **5.** Survey; sample answer: The data is obtained from opinions given by people that return the questionnaire.

7   Sample answer: The survey question does not encourage any particular answer, does not cause a strong reaction, does not address more than one issue, and is not confusing. Therefore, the question is unbiased.

9. Biased; sample answer: The survey question encourages the members of the sample to choose Suzanne as class president.   **11.** Biased; sample answer: The normal advertising strategy was used in states that have a cooler climate than the states where the new advertising strategy was used.
13. Sample: the cards that Greg selects; population: all of Greg's baseball cards; stratified: Greg divides the cards by teams before the sample is selected.
15. Sample: the five students who win the raffle; population: all of the students at the meeting; simple: each student has an equal chance of winning.   **17.** Biased; sample answer: The sample is too small and doesn't represent the population.
19. Unbiased; sample answer: The participants are randomly selected. The fact that the sample is selected at a gas station has no bearing on the conclusion because many different kinds of people enter gasoline stations.   **21.** Survey; sample answer: The data is obtained from opinions given by guests that return the paper.

23 This situation describes an experiment. Sample answer: The control group is the team in short sleeves, while the experimental group is wearing long sleeves. The different affects of the clothes worn by each team in the heat are analyzed. It is not an observational study because the members of the teams *are* being affected by the study. It is not a survey since the team members are *not* being asked for responses.

25. unbiased   **27.** Biased; sample answer: The question addresses more than one issue.
29. Unbiased; sample answer: The only difference between the two groups is the recreation room.
31. unbiased   **33.** Experiment; sample answer: The new color will be tested on a sample group. These members will be affected by the study, so it is an experiment.   **35.** Survey; sample answer: Data are collected from responses given by members of a population, so it is a survey.

37 Both; both Amy's and Esteban's methods can result in a more accurate survey. By increasing the number of people in the population to be included, Amy's survey will more closely approximate the entire population. By using more random samples, Esteban's survey will also more closely approximate the entire population.

39. Sample answer: The survey of 150 people at the concert would most likely represent the population. While each survey may have regional bias, more people were surveyed at the concert. The larger the sample size, the more likely it will reflect the population.   **41.** Sample answer: They need accurate surveys to make decisions about how to market and sell products that will earn the company the most profit. They also make decisions about marketing and advertising and how to reach their target audience. Finally, they make decisions about the types of products they will develop or continue to sell.   **43.** C   **45.** B

47. 6   **49.** 2   **51.** $\frac{9}{2}$ **53.** 0   **55a.** $P = \dfrac{2559.7075}{\pi r^3}$
55b. 4.67 in.

57.

Height (cm) vs Time (sec)

1. sample: 1000 college students; population: all college students in the United States; sample statistic: mean of the money spent on books in a year by the sample; population parameter: mean of money spent on books by all college students in the United States

3 Find the mean.
$$\bar{x} = \frac{10 + 12 + 0 + 6 + 9 + 15 + 12 + 10 + 11 + 20}{10}$$
or 10.5
Find the sum of the square of the differences, $(\bar{x} - x_n)^2$.
$$(10 - 10.5)^2 = 0.25$$
$$(12 - 10.5)^2 = 2.25$$
$$(0 - 10.5)^2 = 110.25$$
$$(6 - 10.5)^2 = 20.25$$
$$(9 - 10.5)^2 = 2.25$$
$$(15 - 10.5)^2 = 20.25$$
$$(12 - 10.5)^2 = 2.25$$
$$(10 - 10.5)^2 = 0.25$$
$$(11 - 10.5)^2 = 0.25$$
$$+ (20 - 10.5)^2 = 90.25$$
$$= 248.50$$
Find the variance.
$$\sigma^2 = \frac{(\bar{x} - x_1)^2 + (\bar{x} - x_2)^2 + \ldots + (\bar{x} - x_n)^2}{n}$$
$$= \frac{248.50}{10} \text{ or } 24.85$$
Find the standard deviation.
$$\sigma = \sqrt{\sigma^2}$$
$$= \sqrt{24.85} \text{ or about } 4.98$$
Sample answer: The mean is 10.5 and the standard deviation is about 4.98. The standard deviation is relatively high due to outliers 0 and 20.

5. sample: 1003 voters in Mercy County; population: all voters in Mercy County; sample statistic: the number of people in the sample who would vote for the incumbent candidate; population parameter: the number of people in the county who would vote for the incumbent candidate   **7.** sample: stratified random sample of 2500; population: high school students in the country; sample statistic: how much money the 2500 students spent each month; population parameter: how much money all the students in the country spent each month   **9.** 14; Sample answer: On average, each day's attendance is 14 people from the mean of 94 people. The mean absolute deviation is affected by outliers 45 and 166.   **11.** 2.1; Sample answer: With a mean of 77.875, the standard deviation of about 2.1 suggests that there is very little deviation to the data. Therefore, you can conclude that Carla's archery scores are pretty consistent.

13 a. Movie A:

Find the mean.

$\bar{x} =$
$$\frac{7+8+7+6+8+6+7+8+6+8+8+6+7+7+8+8}{16}$$
or about 7.2

Find the sum of the square of the differences, $(\bar{x} - x_n)^2$.

$$(7 - 7.2)^2 = 0.04$$
$$(8 - 7.2)^2 = 0.64$$
$$(7 - 7.2)^2 = 0.04$$
$$(6 - 7.2)^2 = 1.44$$
$$(8 - 7.2)^2 = 0.64$$
$$(6 - 7.2)^2 = 1.44$$
$$(7 - 7.2)^2 = 0.04$$
$$(8 - 7.2)^2 = 0.64$$
$$(6 - 7.2)^2 = 1.44$$
$$(8 - 7.2)^2 = 0.64$$
$$(8 - 7.2)^2 = 0.64$$
$$(6 - 7.2)^2 = 1.44$$
$$(7 - 7.2)^2 = 0.04$$
$$(7 - 7.2)^2 = 0.04$$
$$(8 - 7.2)^2 = 0.64$$
$$+ (8 - 7.2)^2 = 0.64$$
$$= 10.44$$

Find the variance.

$$\sigma^2 = \frac{(\bar{x} - x_1)^2 + (\bar{x} - x_2)^2 + \dots + (\bar{x} - x_n)^2}{n}$$
$$= \frac{10.44}{16} \text{ or } 0.6525$$

Find the standard deviation.

$$\sigma = \sqrt{\sigma^2}$$
$$= \sqrt{0.6525} \text{ or about } 0.81$$

Movie B:

Find the mean.

$\bar{x} =$
$$\frac{9+3+8+2+5+10+3+8+10+9+9+10+6+4+9+3}{16}$$
or about 6.8

Find the sum of the square of the differences, $(\bar{x} - x_n)^2$.

$$(9 - 6.8)^2 = 4.84$$
$$(3 - 6.8)^2 = 14.44$$
$$(8 - 6.8)^2 = 1.44$$
$$(2 - 6.8)^2 = 23.04$$
$$(5 - 6.8)^2 = 3.24$$
$$(10 - 6.8)^2 = 10.24$$
$$(3 - 6.8)^2 = 14.44$$
$$(8 - 6.8)^2 = 1.44$$
$$(10 - 6.8)^2 = 10.24$$
$$(9 - 6.8)^2 = 4.84$$
$$(9 - 6.8)^2 = 4.84$$
$$(10 - 6.8)^2 = 10.24$$
$$(6 - 6.8)^2 = 0.64$$
$$(4 - 6.8)^2 = 7.84$$
$$(9 - 6.8)^2 = 4.84$$
$$+ (3 - 6.8)^2 = 14.44$$
$$= 131.04$$

Find the variance.
$$\sigma^2 = \frac{(\bar{x} - x_1)^2 + (\bar{x} - x_2)^2 + \dots + (\bar{x} - x_n)^2}{n}$$
$$= \frac{131.04}{16} \text{ or } 8.19$$

Find the standard deviation.
$$\sigma = \sqrt{\sigma^2}$$
$$= \sqrt{8.19} \text{ or about } 2.86$$

Sample answer: Movie A had a mean of about 7.2 with a standard deviation of about 0.81. Movie B had a mean of about 6.8 with a standard deviation of about 2.86. While both movies had a mean rating of close to 7, the ratings for Movie B had more variability. In other words, some people really liked it while others didn't like it at all. The ratings for Movie A were much more consistent.

b. Movie A: Sample answer: The reviews are consistent and all are between 6 and 8. You would prefer this movie if you want to see a movie that, while it isn't anyone's favorite, you can assume that it will be decent.

Movie B: Sample answer: While some people didn't like the movie at all, others loved it. You would prefer this movie if you hope to be one of the people that love it.

15 a. Step 1: To find the mean, add the numbers and then divide by how many numbers are in the data set.

$\bar{x} = (956 + 966 + 971 + 981 + 986 + 994 + 1001 + 1014 + 1020 + 1023 + 1034 + 1066 + 1076 + 1077 + 1083)/15$
$$= \frac{15248}{15}$$
$$= 1016.5 \text{ seconds or } 16.9 \text{ minutes}$$

Step 2: To find the mean absolute deviation, first find the sum of the absolute values of the differences between each value in the set of data and the mean.

$|956 - 1016.5| + |966 - 1016.5| + |971 - 1016.5| + |981 - 1016.5| + |986 - 1016.5| + |994 - 1016.5| + |1001 - 1016.5| + |1014 - 1016.5| + |1020 - 1016.5| + |1023 - 1016.5| + |1034 - 1016.5| + |1066 - 1016.5| + |1076 - 1016.5| + |1077 - 1016.5| + |1083 - 1016.5| = 499.5$

Step 3: Then divide the sum by the number of values in the data set.
$$\frac{499.5}{15} \approx 33.3 \text{ seconds}$$

b. The sample is the top 15 runners. The population is all the people who ran.

c. The data is quantitative. No, since the sample is the top 15 runners in the race, it is not random. So, it would not be accurate to apply the mean and standard deviation of the running times to the population.

17. Sometimes; if the samples are truly random, they would rarely contain identical elements and the mean and standard deviation would differ. If the sample produces identical elements, the mean and standard deviation would be the same.
19. Sample answer: Any set of data with identical terms. For example: 5, 5, 5, 5, 5, 5, 5.   **21.** Both are calculated statistical values that show how each data value deviates from the mean of the data set. The mean absolute deviation is calculated by taking the mean of the absolute values of the differences between each number and the mean of the data set. To find the standard deviation, you square each difference and then take the square root of the mean of the squares.   **23.** 125   **25.** C
27. Biased; because they are at a rock concert, they are more likely to select a rock music station.

29. $\dfrac{x+3}{x-2}$ **31a.** $y = 880(1.06)^t$ **31b.** about 1576 members   **33.** 31; 21; 15; 36   **35.** 11; 21; 18; 25
37. 10; 11; 9.5; 13

Lesson 12-3

1.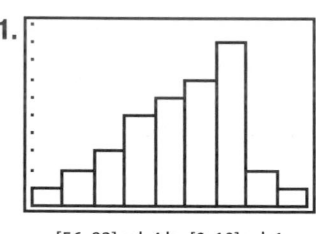

[56, 92] scl: 4 by [0, 10] scl: 1

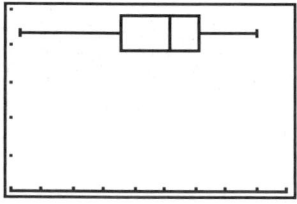

[56, 92] scl: 4 by [0, 5] scl: 1

negatively skewed

3. Sample answer: The distribution is skewed, so use the five-number summary. The range is 92 − 52 or 40. The median is 65, and half of the data are between 59.5 and 74.

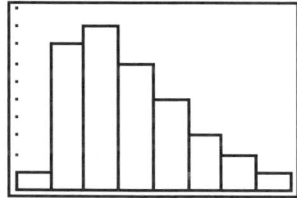

[48, 96] scl: 6 by [0, 10] scl: 1

5.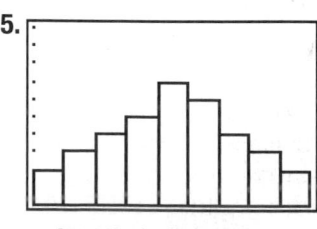

[24, 78] scl: 6 by [0, 10] scl: 1

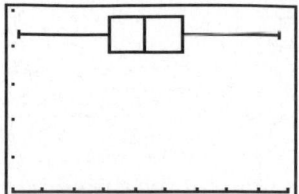

[24, 78] scl: 6 by [0, 5] scl: 1

symmetric

(7) Use a graphing calculator to enter the data into L1 and create a histogram. Adjust the window to the dimensions shown. The graph is high on the right. Therefore, the distribution is negatively skewed.

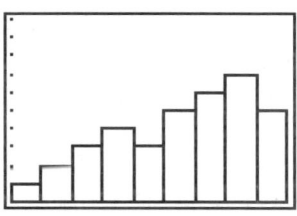

[10, 55] scl: 5 by [0, 10] scl: 1

The distribution is negatively skewed, so use the five-number summary.

```
1-Var Stats
↑n=36
 minX=12
 Q₁=28
 Med=39.5
 Q₃=48
 maxX=53
```

The range is 53 − 12 or 41. The median of the numbers is 39.5, and half the numbers are between 28 and 48.

9. Sample answer: The distribution is symmetric, so use the mean and standard deviation to describe the center and spread. The mean temperature is 52.8° with standard deviation of about 4.22°.

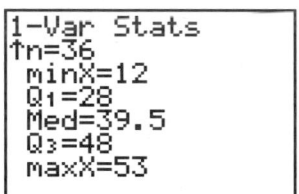

[42, 62] scl: 2 by [0, 5] scl: 1

(11) a. Enter the list of prices as L1 and create a box-and-whisker plot. Adjust the window to the dimensions shown. The right whisker is longer than the left. The median is closer to the shorter whisker. Thus, the distribution is positively skewed.

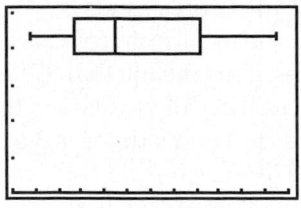

[7, 19] scl: 1 by [0, 5] scl: 1

The distribution is positively skewed, so use the five-number summary.

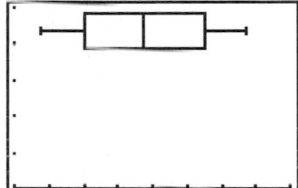

```
1-Var Stats
↑n=20
  minX=7.75
  Q₁=9.625
  Med=11.5
  Q₃=15.125
  maxX=18.5
■
```

The range of prices is $18.50 − $7.75 or $10.75. The median price is $11.50, and half of the prices are between $9.625 and $15.125.

b. Delete all the entries in L1 that are greater than $15 and create another box-and-whisker plot. The lengths of the whiskers are approximately equal, and the median is in the middle of the data. Thus, the distribution is symmetric.

[7, 15] scl: 1 by [0, 5] scl: 1

The distribution is symmetric, so use the mean and standard deviation to describe the center and the spread of the new data.

```
1-Var Stats
  x̄=10.66666667
  Σx=160
  Σx²=1757.625
  Sx=1.907847204
  σx=1.843155507
↓n=15
```

The mean is about $10.67 with standard deviation of about $1.84.

13. i **15.** Sample answer: A bimodal distribution is a distribution of data that is characterized by having data divided into two clusters, thus producing two modes, and having two peaks. The distribution can be described by summarizing the center and spread of each cluster of data.

17. Sample answer: In a symmetrical distribution, the majority of the data is located near the center of the distribution. The mean of the distribution is also located near the center of the distribution. Therefore, the mean and standard deviation should be used to describe the data. In a skewed

distribution, the majority of the data lies either on the right or left side of the distribution. Since the distribution has a tail or may have outliers, the mean is pulled away from the majority of the data. The median is less affected. Therefore, the five-number summary should be used to describe the data. **19.** F **21a.** $3w = 2\ell + 3$; $4\ell = 12 + P$
21b. 21 in., 15 in. **21c.** 315 in^2 **23.** sample: random sample of 100 seniors; population: all seniors at North Boyton High School; sample statistic: the mean amount of money the sample spent on prom; population parameter: the mean amount of money seniors at North Boyton High School spent on prom **25.** unbiased **27.** $\dfrac{11a + 7}{a + 1}$
29. $f^{-1}(x) = -\dfrac{1}{5}x + \dfrac{17}{5}$ **31.** $f^{-1}(x) = -7x - 7$
33. $f^{-1}(x) = -\dfrac{5}{3}x + 20$ **35.** $\dfrac{4}{11}$ **37.** $\dfrac{7}{11}$ **39.** $\dfrac{1}{2}$

Lesson 12-4

1. 3.5, 3.5, 1, 8, 2.4
(3) Find the mean, median, mode, range, and standard deviation of the original set.
Mean: 5.75
Median: 5.5
Mode: 3
Range: 10
Standard Deviation: 3.14
Multiply the mean, median, mode, range, and standard deviation by 3.
Mean: $3 \times 5.75 \approx 17.3$
Median: $3 \times 5.5 = 16.5$
Mode: $3 \times 3 = 9$
Range: $3 \times 10 = 30$
Standard Deviation: $3 \times 3.14 \approx 9.4$

5a.

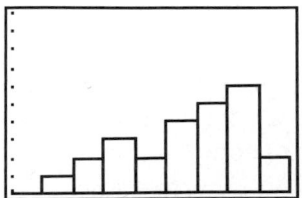

Kyle's Distances

[17, 19.25] scl: 0.25 by [0, 10] scl: 1

Mark's Distances

[17, 19.25] scl: 0.25 by [0, 10] scl: 1

Kyle, positively skewed; Mark, negatively skewed
5b. Sample answer: The distributions are skewed, so use the five-number summaries. Kyle's upper quartile is 17.98, while Mark's lower quartile is 18.065. This means that 75% of Mark's distances

are greater than 75% of Kyle's distances. Therefore, we can conclude that overall, Mark's distances are higher than Kyle's distances. **7.** 60.9, 60, 60, 14, 4.7 **9.** 22.5, 21.5, no mode, 24, 7.4 **11.** 36.8, 38, 12, 56, 20.0 **13.** 26.8, 27.2, 29.6, 10.4, 3.5
15a.

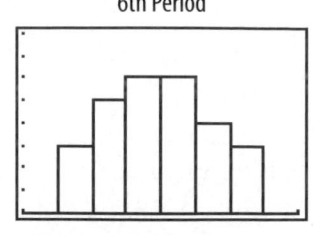

1st period, positively skewed; 6th period, symmetric **15b.** Sample answer: One distribution is symmetric and the other is skewed, so use the five-number summaries. The lower quartile for 1st period is 291 pages, while the minimum for 6th period is 294 pages. This means that the lower 25% of data for 1st period is lower than any data from 6th period. The range for 1st period is $578 - 206$ or 372 pages. The range for 6th period is $506 - 294$ or 212 pages. The median for 1st period is about 351 pages, while the median for 6th period is 392 pages. This means, that while the median for 6th period is greater, 1st period's pages have a greater range and include greater values than 1st period.

17 a. Enter Leon's times as L_1 and Cassie's times as L_2. Then use STAT PLOT to create a histogram for each list.

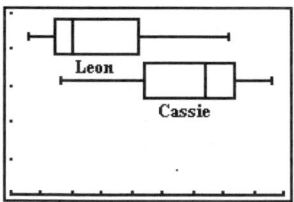

[1.5, 6] scl: 0.5 by [0, 5] scl: 1

For Leon's times, the right whisker is longer than the left, and the median is closer to the left whisker. The distribution is positively skewed. For Cassie's times, the left whisker is longer than the right and the median is closer to the right whisker. The distribution is negatively skewed.

b. One distribution is positively skewed and the other is negatively skewed, so use the five-number summaries to compare the data.

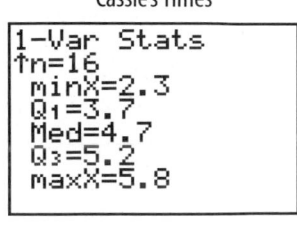

The lower quartile for Leon's times is 2.2 minutes, while the minimum for Cassie's times is 2.3 minutes. This means that 25% of Leon's times are less than all of Cassie's times. The upper quartile for Leon's times is 3.6 minutes, while the lower quartile for Cassie's times is 3.7 minutes. This means that 75% of Leon's times are less than 75% of Cassie's time. Overall, we can conclude that Leon completed the brainteasers faster than Cassie.

19a. 52.96, 53, 53, 19, 6.07 **19b.** 47.96, 48, 48, 19, 6.07

21 Find the total cost of all 20 SUVs and luxury cars.
$20 \times \$39{,}250 = \$785{,}000$
Next subtract the price reduction of the 15 SUVs.
$\$785{,}000 - (15 \times 2000) = \$755{,}000$
Finally, divide this new total by the 20 cars.
$\$755{,}000 \div 20 = \$37{,}750$
The new average price for all the vehicles is $37,750.

23. Sample answer: Histograms show the frequency of values occurring within set intervals. This makes the shape of the distribution easy to recognize. However, no specific values of the data set can be identified from looking at a histogram, and the overall spread of the data can be difficult to determine. The box-and-whisker plots show the data divided into four sections. This aids when comparing the spread of one set of data to another. However, the box-and-whisker plots are limited because they cannot display the data any more specifically than showing it divided into four sections. **25.** Sample answer: When two distributions are symmetric, the first thing to determine is how close the averages are and how spread out each set of data is. The mean and standard deviation are the best values to use for this comparison. When distributions are skewed, we also want to determine the direction and the degree to which it is skewed. The mean and standard deviation cannot provide any information in this regard, but we can get this information by comparing the range, quartiles, and medians found in the five-number summaries. So, if one or both sets of data are skewed, it is best to compare their five-number summaries.
27. $m\angle A = 36°, c \approx 9.9, a \approx 5.8$ **29.** D **31.** 5.58; Sample answer: The standard deviation is relatively high compared to the mean of 6.4 due to the outlier 23. If this outlier were removed, the new mean of the data would be about 5.2 with a standard deviation of about 3.51. **33.** 4 **35.** $(x + 3)(x - 7)$ **37.** $(x - 8)(x - 4)$ **39.** $(x + 10)(x + 2)$ **41.** $4w^3 + 14w^2 + 10w$ **43.** 4 **45.** 0 **47.** 1

1 We are asked to find the probability of selecting a customer who is older than 12 but younger than 46. Therefore, we need to consider selecting someone who is age 13–17, 18–23, 24–31, and 32–45.

P(older than 12 but younger than 46)

$= \dfrac{\text{sum of customers 13–45}}{\text{total number of customers}}$

$= \dfrac{48 + 42 + 27 + 40}{13 + 28 + 48 + 42 + 27 + 40 + 33 + 23}$

$= \dfrac{157}{254}$

The probability of selecting a customer older than 12 but younger than 46 is $\dfrac{157}{254}$. The correct answer is B.

3 a. P(heart) $= \dfrac{\text{frequency of hearts}}{\text{total of frequencies}}$

$= \dfrac{5}{7 + 4 + 5 + 9}$

$= \dfrac{5}{25}$ or $\dfrac{1}{5}$

The experimental probability of drawing a heart is $\dfrac{1}{5}$ or 20%.

b. We are asked to find the probability of drawing a black card. Therefore, we need to consider drawing a club and drawing a spade.

P(black) $= \dfrac{\text{sum of clubs and spades}}{\text{sum of all the frequencies}}$

$= \dfrac{7 + 9}{7 + 4 + 5 + 9}$

$= \dfrac{16}{25}$

The experimental probability of drawing a black card is $\dfrac{16}{25}$ or 64%.

c. Combine the frequencies for each suit for the first and second test.

Suit	clubs	diamonds	hearts	spades
Frequency	12	12	11	15

P(spade) $= \dfrac{\text{number of spades drawn}}{\text{total number of cards drawn}}$

$= \dfrac{15}{12 + 12 + 11 + 15}$

$= \dfrac{15}{50}$ or $\dfrac{3}{10}$

The experimental probability of drawing a spade for both tests is $\dfrac{3}{10}$ or 30%.

5a. Sample answer: The theoretical probability that the player gets a hit is 30%, and the theoretical probability that he does not get a hit is 70%. Use a random number generator to generate integers 1 through 10. The integers 1–3 will represent a hit, and the integers 4–10 will represent the player not getting a hit. The simulation will consist of

50 trials. **5b.** Sample answer: P(hit) = 28%, P(not a hit) = 72%

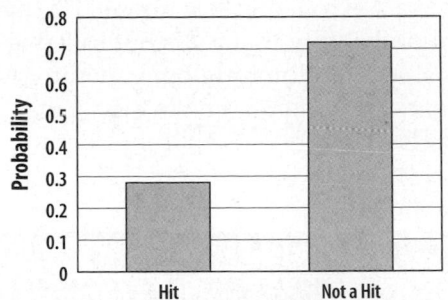

7a. Sample answer: The theoretical probability that a spin results in each prize is 20%. Use a random number generator to generate integers 1 through 5. The integer 1 will represent a hot pretzel, the integer 2 will represent a burger, the integer 3 will represent a large drink, the integer 4 will represent nachos, and the integer 5 will represent a small popcorn. The simulation will consist of 50 trials.
7b. P(hot pretzel) = 20%, P(burger) = 20%, P(large drink) = 28%, P(nachos) = 14%, P(small popcorn) = 18%

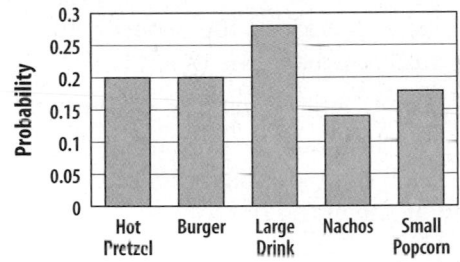

9 a. We are asked to find the experimental probability that a participant was a winner in the second group.

$P = \dfrac{\text{number of winners in second group}}{\text{number of participants in second group}}$

$= \dfrac{15}{50}$ or $\dfrac{3}{10}$

The probability of a participant in the second group being a winner is $\dfrac{3}{10}$ or 30%.

b. We are asked to find the experimental probability that a participant in any of the three groups was a winner.

$P = \dfrac{\text{sum of winners in all three groups}}{\text{sum of participants in all three groups}}$

$= \dfrac{13 + 15 + 19}{50 + 50 + 50}$

$= \dfrac{47}{150}$

The experimental probability of winning for all three groups is $\dfrac{47}{150}$ or about 31.3%.

c. Yes; sample answer: In total, there were 150 participants with a success rate of about 31%. This is lower than the 33% success rate required by the owners.

11a. $\frac{1}{6}$ **11b.** Sample answer: The theoretical probability that a 1 is rolled is $\frac{1}{6}$ or about 17%, and the theoretical probability that a 1 is not rolled is $\frac{5}{6}$ or about 83%. Use a random number generator to generate integers 1 through 6. The integer 1 will represent a 1. The integers 2–6 will represent the other numbers on the die.

11c. Sample answer:

Rolling a 1		
Trials	Frequency	Experimental Probability
20	2	$\frac{1}{10}$ or 10%
50	11	$\frac{11}{50}$ or 22%
100	19	$\frac{19}{100}$ or 19%
200	37	$\frac{37}{200}$ or 18.5%

11d. Sample answer: The experimental probability of rolling a 1 is getting closer to the theoretical probability of $\frac{1}{6}$ or 17%.   **11e.** Sample answer: As the number of trials increases, the experimental probability should approach the theoretical probability of $\frac{1}{6}$ or 17%.   **13.** Sample answer: A simulation can be designed to estimate the probability that a basketball player will make her next free throw if she made 70% of her free throws the previous season. The theoretical probability that she makes her next free throw is 70%, and the theoretical probability that she misses is 30%. Use a random number generator to generate integers 1 through 10. The integers 1–7 will represent a made free throw, and the integers 8–10 will represent a miss. The simulation will consist of 30 trials.

15. Sample answer: Jeremy's concern about the fairness of the coin should be dependent on the number of times he tossed it. If he only completed 4, or even 20 trials, then the sample size is not large enough to warrant concern. However, if he completed 100 or more trials, then he should be concerned since, due to the Law of Large Numbers, the experimental probability should be closer to the theoretical probability of $\frac{1}{2}$ or 50%.

17. A   **19.** G   **21.** 13.9, 13.5, 9, 12, 3.8   **23.** 7.9, 7, 5, 12, 4.0   **25.** 60   **27.** $xy = 12.4$; -20 **29.** 7.1 ft   **31.** 44%   **33.** 75%   **35.** 27%

Lesson 12-6

1. 720   **3.** combination   **5.** 42
7 Use the combination formula with $n = 6$ and $r = 4$.
$$C(n, r) = \frac{n!}{(n - r)!r!}$$

$$C(6, 4) = \frac{6!}{(6 - 4)!4!}$$
$$= \frac{6!}{2!4!}$$
$$= \frac{6 \cdot 5 \cdot \cancel{4} \cdot \cancel{3} \cdot \cancel{2} \cdot \cancel{1}}{2 \cdot 1 \cdot \cancel{4} \cdot \cancel{3} \cdot \cancel{2} \cdot \cancel{1}}$$
$$= \frac{30}{2} \text{ or } 15$$

9. 3003   **11.** 40,320   **13.** 5   **15.** 210   **17.** 10   **19.** 1
21. permutation   **23.** combination   **25.** 1320

27 First find the total number of outcomes. Since order is not important, this is a combination. Find the number of combinations of 18 songs chosen 8 at a time.
$$C(n, r) = \frac{n!}{(n - r)!r!}$$
$$C(18, 8) = \frac{18!}{(18 - 8)!8!}$$
$$= \frac{18!}{10!8!} \text{ or } 43,758$$

There are 43,758 possible outcomes.
Next, find the number of successes. Of the 43,758 combinations of 8 songs, only one can be all country songs since there are only 8 country songs from which to choose.
Last, find the probability.
$$P = \frac{\text{number of successes}}{\text{number of possible outcomes}} \text{ or } \frac{1}{43,758}$$
The probability that the first 8 songs played on Kenneth's playlist are country songs is $\frac{1}{43,758}$.

29. 281,887,200   **31.** 455   **33.** 40,320

35 **a.** Find the total number of pairings. Since order is not important, this is a combination. Find the number of combinations of 32 bowlers chosen 2 at a time.
$$C(32, 2) = \frac{32!}{(32 - 2)!2!}$$
$$= \frac{32!}{30!2!} \text{ or } 496$$

Since order is not important, only one pair could be Chris and Kelly. Therefore, the probability that Chris and Kelly bowl against each other in any lane is $\frac{1}{496}$.

b. There are to be 2 bowlers per lane. This will require 16 lanes since 32 bowlers will divide into 16 pairs. The probability of being assigned to the last lane is $\frac{1}{16}$. The probability of Chris and Kelly being selected to bowl against each other in the last lane is the probability of being paired together times the probability of being assigned to the last lane.
$P(\text{Chris vs Kelly and last lane}) = \frac{1}{496} \times \frac{1}{16}$ or $\frac{1}{7936}$

37a. 1,352,078   **37b.** about 99.97%   **37c.** Sample answer: A random number generator can be used with 1–4 representing freshmen, 5–9 representing sophomores, 10–15 representing juniors, and 16–23 representing seniors. **37d.** Sample answer: Assume the first number generated is a 3 and the first selection is a freshman. The random number generator can be reconfigured to 1–3 representing freshmen, 4–8 representing sophomores, 9–14 representing juniors, and 15–22 representing seniors. An easier method would be to use the original number generator and ignore any repeats of the number 3. Follow this process for each remaining selection. **39.** Ming; since order is not important, combinations should have been used. **41.** Determining class rank in a senior class; this is the only situation in which order matters. **43.** Sample answer: choosing 3 clubs out of 8 to join **45.** G **47.** $\frac{3}{20}$ **49.** 64.8% **51.** 2 bags **53.** $\frac{2}{n}$ **55.** $\frac{n-4}{n+4}$ **57.** $\frac{12ag}{5b}$ **59.** $\frac{6}{23}$ **61.** $\frac{8}{23}$ **63.** $\frac{17}{23}$

Lesson 12-7

1. dependent; $\frac{28}{253}$ or about 11% **3.** independent; $\frac{4}{9}$ or about 44% **5.** mutually exclusive; $\frac{2}{13}$ or about 15% **7.** not mutually exclusive; $\frac{4}{13}$ or about 31% **9.** independent; $\frac{1}{16}$ or about 6%

11. These events are dependent because the chocolate is not being replaced.
First, milk chocolate: $P = \frac{10}{24}$
Second, white chocolate: $P = \frac{6}{23}$
$P(\text{milk, then white}) = \left(\frac{10}{24}\right)\left(\frac{6}{23}\right)$
$= \frac{60}{552}$
$= \frac{5}{46}$ or about 11%

13. independent; $\frac{100}{361}$ or about 28% **15.** mutually exclusive; $\frac{3}{4}$ or 75% **17.** not mutually exclusive; $\frac{739}{750}$ or about 98.5% **19.** $\frac{8}{87}$ or about 9%

21. $\frac{42}{145}$ or about 29%

23. choose white sock: $P = \frac{14}{24}$
choose white sock: $P = \frac{13}{23}$
$P(\text{white, then white}) = \left(\frac{14}{24}\right)\left(\frac{13}{23}\right)$
$= \frac{182}{552}$
$= \frac{91}{276}$ or about 33%

25. $\frac{1}{2}$ or 50% **27.** $\frac{4}{13}$ or about 31% **29.** $\frac{7}{13}$ or about 54% **31a.** 345 **31b.** 159 **31c.** $\frac{227}{345}$ or about 66% **31d.** $\frac{2}{23}$ or about 9%

33. a.
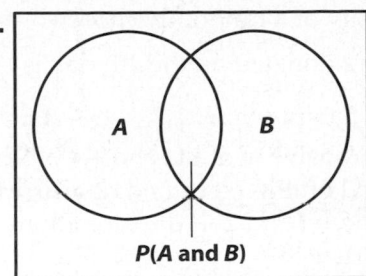
P(A and B)

b. Divide the overlap of the two circles by the $P(A)$ circle

c. $P(\text{red}) = \frac{8}{12} = \frac{2}{3}$
$P(\text{red, green}) = \left(\frac{2}{3}\right)\left(\frac{4}{11}\right)$
$= \frac{8}{33}$

d. $P(\text{green} \mid \text{red}) = \dfrac{\frac{8}{33}}{\frac{2}{3}} = \frac{8}{33} \cdot \frac{3}{2}$ or $\frac{4}{11}$

e. $P(B \mid A) = \dfrac{P(A \text{ and } B)}{P(A)}$

f. $P(B \mid A) = \dfrac{P(A \text{ and } B)}{P(A)} = \dfrac{0.20}{0.80} = 0.25$ or 25%

35. 21 **37.** Sample answer: Choosing a CD to listen to, putting it back, and then choosing another CD to listen to would represent independent events since the CD was placed back before the second CD was chosen. Choosing a pair of jeans to wear would represent dependent events if I did not like the first pair chosen, and I did not put them back. **39.** C **41.** 2 ft **43a.** combination **43b.** 220 **43c.** $\frac{210}{1320}$ or 16% **45.** 8 **47.** $-\frac{3}{2}$ **49.** about 57 pieces **51.** −12 **53.** −16 **55.** −63

Lesson 12-8

1a. 4.7% **1b.** 57.4%

3a.

X	P(X)
$1000	0.0125
$5000	0.005
$25,000	0.001
$100,000	0.0001
$0	0.9814

3b. $72.50 **3c.** Sample answer: The game show producers expect to lose about $72.50 each time a certain game is played.

5. a. We are to find the probability of the chosen product having been donated on the 4th day. There were 45 packages donated on the 4th day.
$P(\text{soup}) = \dfrac{\text{number of donated packages on day 4}}{\text{total number of packages}}$
$= \dfrac{45}{36 + 22 + 12 + 45}$
$= \dfrac{45}{115}$ or $\dfrac{9}{23}$

The probability of a randomly chosen product being donated on the 4th day is $\frac{9}{23}$ or about 39.1%.

b. To find the probability of the chosen product being donated on the 1st or 2nd day, first find the sum of the 1st and 2nd day donations.
P(1st or 2nd day)
$$= \frac{\text{sum of donations on 1st and 2nd days}}{\text{total number of packages}}$$
$$= \frac{36 + 22}{115}$$
$$= \frac{58}{115}$$
The probability of a randomly chosen product being donated on the 1st or 2nd day is $\frac{58}{115}$ or about 50.4%.

7a. All of the values are between 0 and 1; $0.29 + 0.43 + 0.17 + 0.11 + 0 = 1$.   **7b.** 71%

7c.

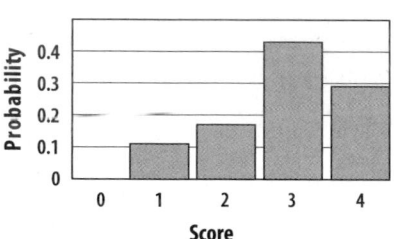

9 **a.** Find the probability associated with each prize. Find the probability of winning $0 by subtracting the winning something from 1.

X	P(X)
$200	0.002
$1000	0.0002
$5000	0.00004
$25,000	0.00001
$0	0.99775

b. Calculate the expected value.
$E(X) = [X_1 \cdot P(X_1)] + [X_2 \cdot P(X_2)] + \ldots + [X_n \cdot P(X_n)]$
$= 200(0.002) + 1000(0.0002) + 5000(0.00004) + 25{,}000(0.00001) + 0(0.99775)$
$= 0.4 + 0.2 + 0.2 + 0.25 + 0$
$= 1.05$
The expected value is 1.05 or $1.05.

c. Sample answer: The makers of the contest expect to lose $1.05 for each ticket valued at $1.00. Therefore, they can expect to lose money by doing this contest.

11 **a.** We are to find the probability of the driver having an accident. This will be the sum of the probabilities for all the variables.
$P(\text{accident}) = P(\text{driving record}) + P(\text{vehicle type}) + P(\text{age}) + P(\text{gender}) + P(\text{residence})$
$= \frac{1}{50} + \frac{1}{500} + \frac{1}{250} + \frac{1}{200} + \frac{1}{1000}$
$= \frac{4}{125}$
The probability of the driver having an accident is $\frac{4}{125}$ or 3.2%.

b. The expected payout can be found by calculating the expected value. Each value of X_n will be $10,000.
$E(X) = [X_1 \cdot P(X_1)] + [X_2 \cdot P(X_2)] + \ldots + [X_n \cdot P(X_n)]$
$= 10{,}000(0.02) + 10{,}000(0.002) + 10{,}000(0.004) + 10{,}000(0.005) + 10{,}000(0.001)$
$= 200 + 20 + 40 + 50 + 10$
$= 320$
The expected payout for this driver would be $320.

c. Sample answer: The company should charge at least $320 to cover their expected costs. However, in order to cover other costs such as paying employees and other business expenses and still earn a profit, the company should charge around $600 for the policy.

13a.

X = Sum	2	3	4	5	6	7	8	9	10	11	12
Frequency	2	3	4	6	7	8	6	5	4	2	1
Probability	$\frac{1}{24}$	$\frac{1}{16}$	$\frac{1}{12}$	$\frac{1}{8}$	$\frac{7}{48}$	$\frac{1}{6}$	$\frac{1}{8}$	$\frac{5}{48}$	$\frac{1}{12}$	$\frac{1}{24}$	$\frac{1}{48}$

13b.

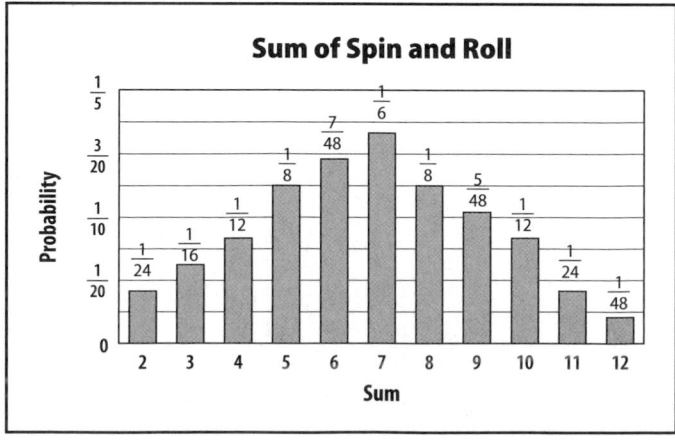

Sum of Spin and Roll

13c. 6.75   **13d.** Sample answer: Use values 1–48 in a random number generator. Set up the following representations. This way, the exact probabilities of each sum are accurately represented.

X = Sum	2	3	4	5	6	7	8	9	10	11	12
Range	1–2	3–5	6–9	10–15	16–22	23–30	31–36	37–41	42–45	46–47	48
Tally	2	3	4	2	7	12	8	7	3	2	0

13e.

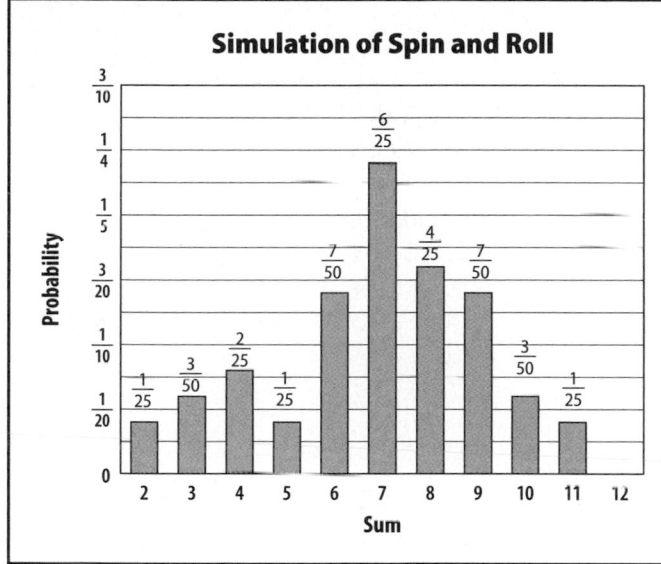

Simulation of Spin and Roll

Sample answer: The graph in part b is the theoretical probability, while the graph in part f is the experimental probability. The graphs will be similar in shape, but the experimental probability graph will vary and more than likely will never completely mirror the theoretical probability graph.

15. The sum 7 is most likely to happen.

X = Sum of Dice	2	3	4	5	6	7	8	9	10	11	12
Probability	$\frac{1}{36}$	$\frac{1}{18}$	$\frac{1}{12}$	$\frac{1}{9}$	$\frac{5}{36}$	$\frac{1}{6}$	$\frac{5}{36}$	$\frac{1}{9}$	$\frac{1}{12}$	$\frac{1}{18}$	$\frac{1}{36}$

Sum of Numbers Showing on the Dice

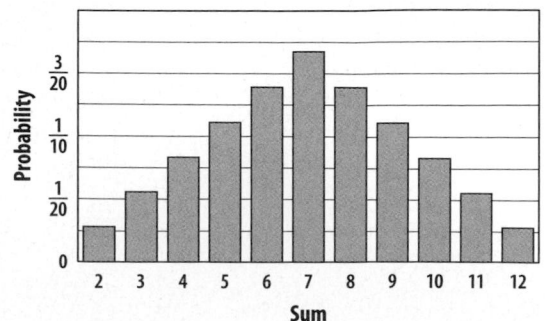

17. True; sample answer: While a discrete random variable is a countable variable, it can still take on an infinite number of values. For example, the number of hits a Web site receives is countable, but can be infinite.   **19.** False; sample answer: The expected value of a random variable is the weighted average of the variable. The expected values of a random variable does not even have to be a possible value. For example, the expected value of rolling a die is 3.5, which is not possible.

21. A   **23.** H   **25.** $\frac{10}{253}$ or about 4%   **27.** $\frac{6a + 4}{a^2}$

29. $\frac{9d + 10}{(d + 6)(d + 5)}$ **31.** $\frac{9h(h - 2)}{(h + 6)(h - 3)}$

33. $\sin A = \frac{4}{5}$; $\cos A = \frac{3}{5}$; $\tan A = \frac{4}{3}$ **35.** $\sin A = \frac{4}{5}$; $\cos A = \frac{3}{5}$; $\tan A = \frac{4}{3}$ **37.** $\frac{t\sqrt{t}}{3\sqrt{2}}$ **39.** $\frac{9 + 3\sqrt{5}}{2}$

41. $2 + \sqrt{3}$

Chapter 12   Study Guide and Review

1. permutation   **3.** population   **5.** random variable
7. Survey; sample answer: The data are obtained from opinions given by the girls that return the questionnaire.   **9.** 0.88; Sample answer: On average, the number of sidewalks that Ben shovels each day is 0.88 away from the mean of 3.4.
11. The day customers had a mean of about 9.6 times per month with a standard deviation of about 3.5. The night customers had a mean of about 11.7 times per month with a standard deviation of about 2.5. The night customers had a higher average and their data values were more consistent.

13.

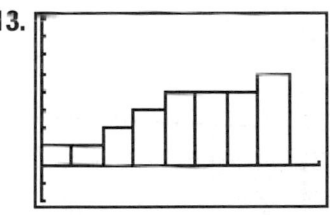

[0, 90] scl: 10 by [−2, 8] scl: 1

negatively skewed
15. 22.5, 21.5, no mode, 24, 7.4
17a. Ms. Miller:

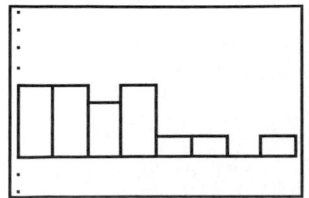

[9, 17] scl: 1 by [−2, 8] scl: 1

Ms. Anderson:

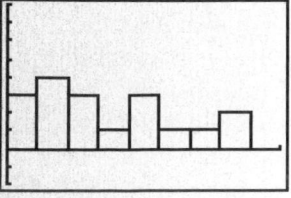

[0, 9] scl: 1 by [−2, 8] scl: 1

both positively skewed

17b. Sample answer: Use the five-number summaries. The range for Ms. Miller is 7. The median is 11. Half of the data are between 10 and 12. The range for Ms. Anderson is 7. The median is 2. Half of the data are between 1 and 4. All of the data in Ms. Anderson's distribution are less than all of the data in Ms. Miller's distribution. Therefore, Ms. Miller will more than likely hand out more disciplinary actions than Ms. Anderson. **19a.** Sample answer: The theoretical probability that the bus is late is 60%, and the theoretical probability that the bus is not late is 40%. Use a random number generator to generate integers 1 through 5. The integers 1–3 will represent the bus being late, and the integers 4–5 will represent the bus not being late. The simulation will consist of 50 trials. **19b.** Sample answer: $P(\text{late}) = 58\%$

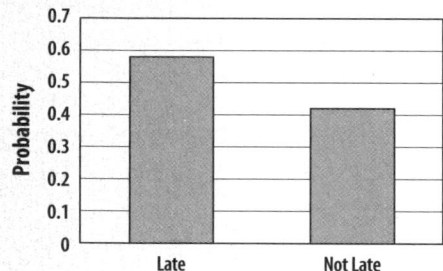

21. permutation   **23.** permutation   **25.** $\frac{2}{35}$
27. $\frac{1}{204}$ **29.** $\frac{22}{325}$ **31.** $\frac{1}{2}$ **33.** For each X, the probability is greater than or equal to 0 and less than or equal to 1, $0.18 + 0.36 + 0.34 + 0.08 + 0.04 = 1$, so the sum of the probabilities is 1.

Glossary/Glosario

 MultilingualeGlossary

Go to connectED.mcgraw-hill.com for a glossary of terms in these additional languages:

Arabic	Chinese	Hmong	Spanish	Vietnamese
Bengali	English	Korean	Tagalog	
Brazilian Portugese	Haitian Creole	Russian	Urdu	

English

Español

A

absolute value (p. P11) The distance a number is from zero on the number line.

valor aboluto Es la distancia que dista de cero en una recta numerica.

absolute value function (p. 599) A function written as $f(x) = |x|$, in which $f(x) \geq 0$ for all values of x.

función del valor absoluto Una función que se escribe $f(x) = |x|$, donde $f(x) \geq 0$, para todos los valores de x.

additive identity (p. 16) For any number a, $a + 0 = 0 + a = a$.

identidad de la adición Para cualquier número a, $a + 0 = 0 + a = a$.

additive inverse (p. P11) Two integers, x and $-x$, are called additive inverses. The sum of any number and its additive inverse is zero.

inverso aditivo Dos enteros x y $-x$ reciben el nobre de inversos aditivos. La suma de cualquier número y su inverso aditivo es cero.

algebraic expression (p. 5) An expression consisting of one or more numbers and variables along with one or more arithmetic operations.

expresión algebraica Una expresión que consiste en uno o más números y variables, junto con una o más operaciones aritméticas.

area (p. P26) The measure of the surface enclosed by a geometric figure.

área La medida de la superficie incluida por una figura geométrica.

arithmetic sequence (p. 189) A numerical pattern that increases or decreases at a constant rate or value. The difference between successive terms of the sequence is constant.

sucesión aritmética Un patrón numérico que aumenta o disminuye a una tasa o valor constante. La diferencia entre términos consecutivos de la sucesión es siempre la misma.

asymptote (p. 685) A line that a graph approaches.

asíntota Una línea a que un gráfico acerca.

augmented matrix (p. 370) A coefficient matrix with an extra column containing the constant terms.

matriz aumentada una matriz del coeficiente con una columna adicional que contiene los términos de la constante.

axis of symmetry (p. 543) The vertical line containing the vertex of a parabola.

eje de simetría La recta vertical que pasa por el vértice de una parábola.

bar graph (p. P41) A graphic form using bars to make comparisons of statistics.

gráfico de barra Forma gráfica usando barras para comparar estadísticas.

base (p. 5) In an expression of the form x^n, the base is x.

base En una expresión de la forma x^n, la base es x.

best-fit line (p. 255) The line that most closely approximates the data in a scatter plot.

recta de ajuste óptimo La recta que mejor aproxima los datos de una gráfica de dispersión.

bias (p. 748) An error that results in a misrepresentation of members of a population.

sesgo Error que resulta en la representación errónea de los miembros de una población.

binomial (p. 465) The sum of two monomials.

binomio La suma de dos monomios.

bivariate data (p. 247) Data with two variables.

datos bivariate Datos con dos variables.

boundary (p. 317) A line or curve that separates the coordinate plane into regions.

frontera Recta o curva que divide el plano de coordenadas en regiones.

box-and-whisker plot (p. P43) A diagram that divides a set of data into four parts using the median and quartiles. A box is drawn around the quartile values and whiskers extend from each quartile to the extreme data points.

diagrama de caja y patillas Diagram que divide un conjunto de datos en cuatro partes usando la mediana y los cuartiles. Se dibuja una caja alrededor de los cuartiles y se extienden patillas de cada uno de ellos a los valores extremos.

center (p. P24) The given point from which all points on the circle are the same distance.

centro Punto dado del cual equidistan todos los puntos de un circulo.

circle (p. P24) The set of all points in a plane that are the same distance from a given point called the center.

círculo Conjunto de todos los puntos del plano que están a la misma distancia de un punto dado del plano llamado centro.

circle graph (p. P42) A type of statistical graph used to compare parts of a whole.

gráfico del círculo Tipo de gráfica estadística que se usa para comparar las partes de un todo.

circumference (p. P24) The distance around a circle.

circunferencia Longitud del contorno de un círculo.

closed (p. 634) A set is closed under an operation if for any numbers in the set, the result of the operation is also in the set.

cerrado Un conjunto es cerrado bajo una operación si para cualquier número en el conjunto, el resultado de la operación es también en el conjunto.

closed half-plane (p. 317) The solution of a linear inequality that includes the boundary line.

mitad-plano cerrado La solución de una desigualdad linear que incluye la línea de límite.

coefficient (p. 28) The numerical factor of a term.

coeficiente Factor numérico de un término.

combination (p. 787) An arrangement or listing in which order is not important.

combinación Arreglo o lista en que el orden no es importante.

common difference (p. 189) The difference between the terms in an arithmetic sequence.

diferencia común Diferencia entre términos consecutivos de una sucesión aritmética.

common ratio (p. 438) The ratio of successive terms of a geometric sequence.

razón común El razón de términos sucesivos de una secuencia geométrica.

Glossary/Glosario

complements (p. P33) One of two parts of a probability making a whole.

completing the square (p. 574) To add a constant term to a binomial of the form $x^2 + bx$ so that the resulting trinomial is a perfect square.

complex fraction (p. 720) A fraction that has one or more fractions in the numerator or denominator.

compound event (p. 793) Two or more simple events.

compound inequality (p. 306) Two or more inequalities that are connected by the words *and* or *or*.

compound interest (p. 433) A special application of exponential growth.

conditional probability (p. 799) The probability of an event under the condition that some preceding event has occurred.

conjugates (p. 630) Binomials of the form $a\sqrt{b} + c\sqrt{d}$ and $a\sqrt{b} - c\sqrt{d}$.

consecutive integers (p. 92) Integers in counting order.

consistent (p. 335) A system of equations that has at least one ordered pair that satisfies both equations.

constant (pp. 155, 391) A monomial that is a real number.

constant function (p. 217) A linear function of the form $y = b$.

constant of variation (p. 182) The number k in equations of the form $y = kx$.

continuous function (p. 48) A function that can be graphed with a line or a smooth curve.

convenience sample (p. 747) A sample that includes members of a population that are easily accessed.

converse (p. 649) The statement formed by exchanging the phrases after *if* and *then* of an if-then statement.

coordinate (p. P8) The number that corresponds to a point on a number line.

coordinate plane (p. 40) The plane containing the x- and y-axes.

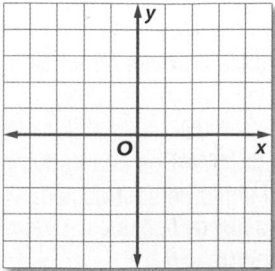

complementos Una de dos partes de una probabilidad que forma un todo.

completar el cuadrado Adición de un término constante a un binomio de la forma $x^2 + bx$, para que el trinomio resultante sea un cuadrado perfecto.

fracción compleja Fracción con una o más fracciones en el numerador o denominador.

evento compuesto Dos o más eventos simples.

desigualdad compuesta Dos o más desigualdades que están unidas por las palabras *y* u *o*.

interés compuesto Aplicación especial de crecimiento exponencial.

probabilidad condicional La probabilidad de un acontecimiento bajo condición que ha ocurrido un cierto acontecimiento precedente.

conjugados Binomios de la forma $a\sqrt{b} + c\sqrt{d}$ y $a\sqrt{b} - c\sqrt{d}$.

enteros consecutivos Enteros en el orden de contar.

consistente Sistema de ecuaciones para el cual existe al menos un par ordenado que satisface ambas ecuaciones.

constante Monomio que es un número real.

función constante Función lineal de la forma $f(x) = b$.

constante de variación El número k en ecuaciones de la forma $y = kx$.

función continua Función cuya gráfica puedes ser una recta o una curva suave.

muestra de conveniencia Muestra que incluye miembros de una población fácilmente accesibles.

recíproco Enunciado que se obtiene al inter cambiar la hipótesis y la conclusión de un enucnciado condicional dado.

coordenada Número que corresponde a un punto en una recta numérica.

plano de coordenadas Plano que contiene los ejes x y y.

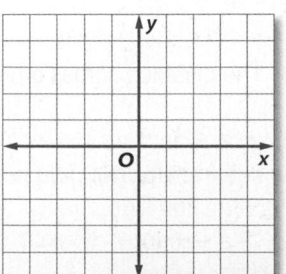

coordinate system (p. 40) The grid formed by the intersection of two number lines, the horizontal axis and the vertical axis.

correlation coefficient (p. 255) A value that shows how close data points are to a line.

cosine (p. 656) For an acute angle of a right triangle, the ratio of the measure of the leg adjacent to the acute angle to the measure of the hypotenuse.

counterexample (p. 201) A specific case in which a statement is false.

cube root (p. 407) If $a^3 = b$, then a is the cube root of b.

sistema de coordenadas Cuadriculado formado por la intersección de dos rectas numéricas: los ejes x y y.

coeficiente de correlación Un valor que demostraciones cómo los puntos de referencias cercanos están a una línea.

coseno Para un ángulo agudo de un triángulo derecho, el razón de la medida de la pierna adyacente al ángulo agudo de la medida de la hipotenusa.

contraejemplo Ejemplo específico de la falsedad de un enunciado.

raíz cúbica Si $a^3 = b$, entonces a es la raíz cúbica de b.

D

---

deductive reasoning (p. 196) The process of using facts, rules, definitions, or properties to reach a valid conclusion.

decreasing (p. 57) The graph of a function goes down on a portion of its domain when viewed from left to right.

defining a variable (p. P5) Choosing a variable to represent one of the unspecified numbers in a problem and using it to write expressions for the other unspecified numbers in the problem.

degree of a monomial (p. 465) The sum of the exponents of all its variables.

degree of a polynomial (p. 465) The greatest degree of any term in the polynomial.

dependent (p. 335) A system of equations that has an infinite number of solutions.

dependent events (p. 794) Two or more events in which the outcome of one event affects the outcome of the other events.

dependent variable (p. 42) The variable in a relation with a value that depends on the value of the independent variable.

diameter (p. P24) The distance across a circle through its center.

difference of two squares (p. 516) Two perfect squares separated by a subtraction sign.
$a^2 - b^2 = (a + b)(a - b)$ or
$a^2 - b^2 = (a - b)(a + b)$.

razonamiento deductivo Proceso de usar hechos, reglas, definiciones o propiedades para sacar conclusiones válidas.

decreciente El gráfico de una función va abajo en una porción de su dominio cuando está visto de izquierda a derecha.

definir una variable Consiste en escoger una variable para representar uno de los números desconocidos en un problema y luego usarla para escribir expresiones para otros números desconocidos en el problema.

grado de un monomio Suma de los exponentes de todas sus variables.

grado de un polinomio El grado mayor de cualquier término del polinomio.

dependiente Sistema de ecuaciones que posee un número infinito de soluciones.

eventos dependientes Dos o más eventos en que el resultado de un evento afecta el resultado de los otros eventos.

variable dependiente La variable de una relación cuyo valor depende del valor de la variable independiente.

diámetro La distancia a través de un círculo a través de su centro.

diferencia de cuadrados Dos cuadrados perfectos separados por el signo de sustracción.
$a^2 - b^2 = (a + b)(a - b)$ or
$a^2 - b^2 = (a - b)(a + b)$.

dilation (p. 566) A transformation that alters the size of a figure but not its shape.

dimension (p. 370) The number of rows, m, and the number of columns, n, of a matrix written as $m \times n$.

dimensional analysis (p. 128) The process of carrying units throughout a computation.

direct variation (p. 182) An equation of the form $y = kx$, where $k \neq 0$.

discrete function (p. 48) A function of points that are not connected.

discrete random variable (p. 803) A variable with a value that is a finite number of possible outcomes.

discriminant (p. 586) In the Quadratic Formula, the expression under the radical sign, $b^2 - 4ac$.

Distance Formula (p. 654) The distance d between any two points with coordinates (x_1, y_1) and (x_2, y_2) is given by the formula
$$d = \sqrt{(x_2 - x_1)^2 + (y_2 - y_1)^2}.$$

distribution (p. 764) A graph or table that shows the theoretical frequency of each possible data value.

domain (p. 40) The set of the first numbers of the ordered pairs in a relation.

double root (p. 556) The roots of a quadratic function that are the same number.

homotecia Transformación que altera el tamaño de una figure, pero no su forma.

dimension El número de filas, de m, y del número de la columna, n, de una matriz escrita como $m \times n$.

análisis dimensional Proceso de tomar en cuenta las unidades de medida al hacer cálculos.

variación directa Una ecuación de la forma $y = kx$, donde $k \neq 0$.

función discreta Función de puntos desconectados.

variable aleatoria discreta Variable cuyo valor es un número finito de posibles resultados.

discriminante En la fórmula cuadrática, la expresión debajo del signo radical, $b^2 - 4ac$.

Fórmula de la distancia La distancia d entre cualquier par de puntos con coordenadas (x_1, y_1) y (x_2, y_2) viene dada por la fórmula
$$d = \sqrt{(x_2 - x_1)^2 + (y_2 - y_1)^2}.$$

distrubución Un gráfico o una tabla que muestra la frecuencia teórica de cada valor de datos posible.

dominio Conjunto de los primeros números de los pares ordenados de una relación.

raíces dobles Las raíces de una función cuadrática que son el mismo número.

E

---

element (p. 370) Each entry in a matrix.

elimination (p. 350) The use of addition or subtraction to eliminate one variable and solve a system of equations.

end behavior (p. 57) Describes how the values of a function behave at each end of the graph.

equally likely (p. P33) The outcomes of an experiment are equally likely if there are n outcomes and the probability of each is $\frac{1}{n}$.

equation (p. 33) A mathematical sentence that contains an equals sign, $=$.

equivalent equations (p. 83) Equations that have the same solution.

equivalent expressions (p. 16) Expressions that denote the same value for all values of the variable(s).

elemento Cada entrada de una matriz.

eliminación El uso de la adición o la sustracción para eliminar una variable y resolver así un sistema de ecuaciones.

comportamiento extremo Describe como los valores de una función se comportan en el cada fin del gráfico.

igualmente probablemente Los resultados de un experimento son igualmente probables si hay resultados de n y la probabilidad de cada uno es $\frac{1}{n}$.

ecuación Enunciado matemático que contiene el signo de igualdad, $=$.

ecuaciones equivalentes Ecuaciones que poseen la misma solución.

expresiones equivalentes Expresiones que denotan el mismo valor para todos los valores de la(s) variable(s).

evaluate (p. 10) To find the value of an expression.

excluded values (p. 684) Any values of a variable that result in a denominator of 0 must be excluded from the domain of that variable.

expected value (p. 805) The weighted average of all outcomes.

experiment (p. 749) Data are recorded from outcomes involving characteristics of a sample.

experimental probability (p. 780) What actually occurs when conducting a probability experiment, or the ratio of relative frequency to the total number of events or trials.

exponent (p. 5) In an expression of the form x^n, the exponent is n. It indicates the number of times x is used as a factor.

exponential decay (p. 424) When an initial amount decreases by the same percent over a given period of time.

exponential equation (p. 409) An equation in which the variables occur as exponents.

exponential function (p. 424) A function that can be described by an equation of the form $y = a^x$, where $a > 0$ and $a \neq 1$.

exponential growth (p. 424) When an initial amount increases by the same percent over a given period of time.

extraneous solutions (pp. 643, 727) Results that are not solutions to the original equation.

extremes (p. 112) In the ratio $\frac{a}{b} = \frac{c}{d}$, a and d are the extremes.

evaluar Calcular el valor de una expresión.

valores excluidos Cualquier valor de una variable cuyo resultado sea un denominador igual a cero, debe excluirse del dominio de dicha variable.

valor previsto El promédio cargado de todos los resultados.

experimento Los datos se registran de los resultados que implican características de una muestra.

probabilidad experimental Lo que realmente sucede cuando se realiza un experimento probabilístico o la razón de la frecuencia relativa al número total de eventos o pruebas.

exponente En una expresión de la forma x^n, el exponente es n. Éste indica cuántas veces se usa x como factor.

desintegración exponencial La cantidad inicial disminuye según el mismo porcentaje a lo largo de un período de tiempo dado.

ecuación exponencial Ecuación en que las variables aparecen en los exponentes.

función exponencial Función que puede describirse mediante una ecuación de la forma $y = a^x$, donde $a > 0$ y $a \neq 1$.

crecimiento exponencial La cantidad inicial aumenta según el mismo porcentaje a lo largo de un período de tiempo dado.

soluciones extrañas Resultados que no son soluciones de la ecuación original.

extremos En la razón $\frac{a}{b} = \frac{c}{d}$, a y d son los extremos.

F

factorial (p. 786) The expression $n!$, read n factorial, where n is greater than zero, is the product of all positive integers beginning with n and counting backward to 1.

factoring (p. 494) To express a polynomial as the product of monomials and polynomials.

factoring by grouping (p. 495) The use of the Distributive Property to factor some polynomials having four or more terms.

factors (p. 5) In an algebraic expression, the quantities being multiplied are called factors.

factorial La expresión $n!$, que se lee n factorial, donde n que es mayor que cero, es el producto de todos los números naturales, comenzando con n y contando hacia atrás hasta llegar al 1.

factorización La escritura de un polinomio como producto de monomios y polinomios.

factorización por agrupamiento Uso de la Propiedad distributiva para factorizar polinomios que poseen cuatro o más términos.

factores En una expresión algebraica, los factores son las cantidades que se multiplican.

Glossary/Glosario

family of graphs (p. 163) Graphs and equations of graphs that have at least one characteristic in common.

FOIL method (p. 481) To multiply two binomials, find the sum of the products of the First terms, the Outer terms, the Inner terms, and the Last terms.

formula (p. 76) An equation that states a rule for the relationship between certain quantities.

four-step problem-solving plan (p. P5)
Step 1 Explore the problem.
Step 2 Plan the solution.
Step 3 Solve the problem.
Step 4 Check the solution.

frequency table (p. P41) A chart that indicates the number of values in each interval.

function (p. 47) A relation in which each element of the domain is paired with exactly one element of the range.

function notation (p. 50) A way to name a function that is defined by an equation In function notation, the equation $y = 3x - 8$ is written as $f(x) = 3x - 8$.

Fundamental Counting Principle (p. P34) If an event M can occur in m ways and is followed by an event N that can occur in n ways, then the event M followed by the event N can occur in $m \times n$ ways.

familia de gráficas Gráficas y ecuaciones de gráficas que tienen al menos una característica común.

método FOIL Para multiplicar dos binomios, busca la suma de los productos de los primeros (First) términos, los términos exteriores (Outer), los términos interiores (Inner) y los últimos términos (Last).

fórmula Ecuación que establece una relación entre ciertas cantidades.

plan de cuatro pasos para resolver problemas
Paso 1 Explora el problema.
Paso 2 Planifica la solución.
Paso 3 Resuelve el problema.
Paso 4 Examina la solución.

Tabla de frecuencias Tabla que indica el número de valores en cada intervalo.

función Una relación en que a cada elemento del dominio le corresponde un único elemento del rango.

notación funcional Una manera de nombrar una función definida por una ecuación. En notación funcional, la ecuación $y = 3x - 8$ se escribe $f(x) = 3x - 8$.

Principio fundamental de contar Si un evento M puede ocurrir de m maneras y lo sigue un evento N que puede ocurrir de n maneras, entonces el evento M seguido del evento N puede ocurrir de $m \times n$ maneras.

G

---

general equation for exponential decay (p. 433) $y = C(1 - r)^t$, where y is the final amount, C is the initial amount, r is the rate of decay expressed as a decimal, and t is time.

general equation for exponential growth (p. 432) $y = C(1 + r)^t$, where y is the final amount, C is the initial amount, r is the rate of change expressed as a decimal, and t is time.

geometric sequence (p. 438) A sequence in which each term after the first is found by multiplying the previous term by a constant r, called the common ratio.

graph (p. P8) To draw, or plot, the points named by certain numbers or ordered pairs on a number line or coordinate plane.

greatest integer function (p. 598) A step function, written as $f(x) = [\![x]\!]$, where $f(x)$ is the greatest integer less than or equal to x.

ecuación general de desintegración exponencial $y = C(1 - r)^t$, donde y es la cantidad final, C es la cantidad inicial, r es la tasa de desintegración escrita como decimal y t es el tiempo.

ecuación general de crecimiento exponencial $y = C(1 + r)^t$, donde y es la cantidad final, C es la cantidad inicial, r es la tasa de cambio del crecimiento escrita como decimal y t es el tiempo.

secuencia geométrica Una secuencia en la cual cada término después de que la primera sea encontrada multiplicando el término anterior por un r constante, llamado el razón común.

graficar Marcar los puntos que denotan ciertos números en una recta numérica o ciertos pares ordenados en un plano de coordenadas.

La función más grande del número entero Una función del paso, escrita como $f(x) = [\![x]\!]$, donde está el número entero $f(x)$ es el número más grande menos que o igual a x.

half-plane (p. 317) The region of the graph of an inequality on one side of a boundary.

histogram (p. P41) A graphical display that uses bars to display numerical data that have been organized into equal intervals.

hypotenuse (p. 648) The side opposite the right angle in a right triangle.

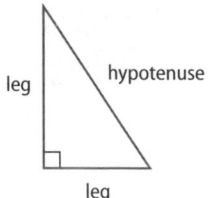

semiplano Región de la gráfica de una desigualdad en un lado de la frontera.

histograma Una exhibición gráfica que utiliza barras para exhibir los datos numéricos que se han organizado en intervalos iguales.

hipotenusa Lado opuesto al ángulo recto en un triángulo rectángulo.

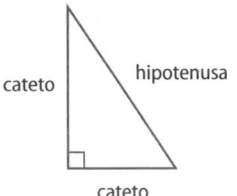

identity (pp. 35, 98) An equation that is true for every value of the variable.

identity function (p. 224) The function $y = x$.

identity matrix (p. 371) A square matrix that, when multiplied by another matrix, equals that same matrix. If A is any $n \times n$ matrix and I is the $n \times n$ identity matrix, then $A \cdot I = A$ and $I \cdot A = A$.

inconsistent (p. 335) A system of equations with no ordered pair that satisfy both equations.

increasing (p. 57) The graph of a function goes up on a portion of its domain when viewed from left to right.

independent (p. 335) A system of equations with exactly one solution.

independent events (p. 793) Two or more events in which the outcome of one event does not affect the outcome of the other events.

independent variable (p. 42) The variable in a function with a value that is subject to choice.

inductive reasoning (p. 196) A conclusion based on a pattern of examples.

inequality (p. 285) An open sentence that contains the symbol $<$, $\leq$, $>$, or $\geq$.

integers (p. P7) The set $\{\ldots, -2, -1, 0, 1, 2, \ldots\}$.

identidad Ecuación que es verdad para cada valor de la variable.

unción identidad La función $y = x$.

matriz de la identidad Una matriz cuadrada que, cuando es multiplicada por otra matriz, iguala que la misma matriz. Si A es alguna de la matriz $n \times n$ e I es la matriz de la identidad de $n \times n$, entonces $A \cdot I = A$ e $I \cdot A = A$.

inconsistente Un sistema de ecuaciones para el cual no existe par ordenado alguno que satisfaga ambas ecuaciones.

crecciente El gráfico de una función va arriba en una porción de su dominio cuando está visto de izquierda a derecha.

independiente Un sistema de ecuaciones que posee una única solución.

eventos independientes El resultado de un evento no afecta el resultado del otro evento.

variable independiente La variable de una función sujeta a elección.

razonamiento inductivo Conclusión basada en un patrón de ejemplos.

desigualdad Enunciado abierto que contiene uno o más de los símbolos $<$, $\leq$, $>$, o $\geq$.

enteros El conjunto $\{\ldots, -2, -1, 0, 1, 2, \ldots\}$.

interquartile range (p. P38) The range of the middle half of a set of data. It is the difference between the upper quartile and the lower quartile.

intersection (p. 306) The graph of a compound inequality containing *and*; the solution is the set of elements common to both inequalities.

inverse cosine (p. 658) If $\angle A$ is an acute angle and the cosine of A is x, then the inverse cosine of x is the measure of $\angle A$.

inverse sine (p. 658) If $\angle A$ is an acute angle and the sine of A is x, then the inverse sine of x is the measure of $\angle A$.

inverse tangent (p. 658) If $\angle A$ is an acute angle and the tangent of A is x, then the inverse tangent of x is the measure of $\angle A$.

inverse variation (p. 676) An equation of the form $xy = k$, where $k \neq 0$.

irrational numbers (p. P7) Numbers that cannot be expressed as terminating or repeating decimals.

amplitud intercuartílica Amplitude de la mitad central de un conjunto de datos. Es la diferenccia entre el cuartil superior y el inferior.

intersección Gráfica de una desigualdad compuesta que contiene la palabra y; la solución es el conjunto de soluciones de ambas desigualdades.

cosino inverso Si el $\angle A$ es un ángulo agudo y el coseno de A es x, entonces el coseno inverso de x es la medida de $\angle A$.

seno inverso Si $\angle A$ es un ángulo agudo y el seno de A es x, entonces el seno inverso de x es la medida de $\angle A$.

tangente inverso Si el $\angle A$ es un ángulo agudo y la tangente de A es x, entonces la tangente inversa de x es la medida de $\angle A$.

variación inversa Ecuación de la forma $xy = k$, donde $k \neq 0$.

números irracionales Números que no pueden escribirse como decimales terminales o periódicos.

L

leading coefficient (p. 466) The coefficient of the term with the highest degree in a polynomial.

loact common denominator (LCD) (p. 714) The least common multiple of the denominators of two or more fractions.

least common multiple (LCM) (p. 713) The least number that is a common multiple of two or more numbers.

legs (p. 648) The sides of a right triangle that form the right angle.

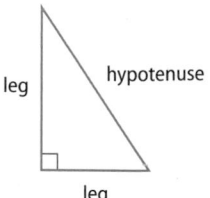

like terms (p. 27) Terms that contain the same variables, with corresponding variables having the same exponent.

linear equation (p. 155) An equation in the form $Ax + By = C$, with a graph that is a straight line.

linear extrapolation (p. 228) The use of a linear equation to predict values that are outside the range of data.

linear function (p. 163) A function with ordered pairs that satisfy a linear equation.

coeficiente inicial El coeficicnte del término con el grado más alto (el primer coeficiente inicial) en un polinomio.

mínimo denominador común (mcd) El mínimo común múltiplo de los denominadores de dos o más fracciones.

mínimo común múltiplo (mcm) El número menor que es múltiplo común de dos o más números.

catetos Lados de un triángulo rectángulo que forman el ángulo recto del mismo.

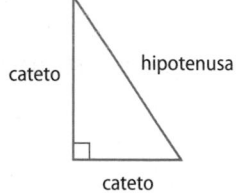

términos semejantes Expresiones que tienen las mismas variables, con las variables correspondientes elevadas a los mismos exponentes.

ecuación lineal Ecuación de la forma $Ax + By = C$, cuya gráfica es una recta.

extrapolación lineal Uso de una ecuación lineal para predecir valores fuera de la amplitud de los datos.

función lineal Función cuyos pares ordenados satisfacen una ecuación lineal.

linear interpolation (p. 249) The use of a linear equation to predict values that are inside of the data range.

linear regression (p. 255) An algorithm to find a precise line of fit for a set of data.

linear transformation (p. 771) One or more operations performed on a set of data that can be written as a linear function.

line of fit (p. 248) A line that describes the trend of the data in a scatter plot.

literal equation (p. 127) A formula or equation with several variables.

lower quartile (p. P38) Divides the lower half of the data into two equal parts.

interpolación lineal Uso de una ecuación lineal para predecir valores dentro de la amplitud de los datos.

regresión lineal Un algoritmo para encontrar una línea exacta del ajuste para un sistema de datos.

transformación lineal Una o más operaciones que se hacen en un conjunto de datos y que se pueden escribir como una función lineal.

recta de ajuste Recta que describe la tendencia de los datos en una gráfica de dispersión.

ecuación literal Un fórmula o ecuación con varias variables.

cuartil inferior Éste divide en dos partes iguales la mitad inferior de un conjunto de datos.

M

mapping (p. 40) Illustrates how each element of the domain is paired with an element in the range.

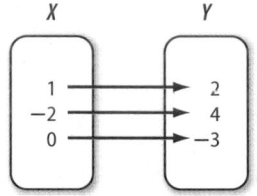

aplicaciones Ilustra la correspondencia entre cada elemento del dominio con un elemento del rango.

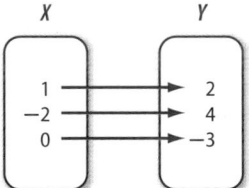

matrix (p. 370) Any rectangular arrangement of numbers in rows and columns.

maximum (p. 58) The highest point on the graph of a curve.

mean (p. P37) The sum of numbers in a set of data divided by the number of items in the data set.

mean absolute deviation (p. 758) The average of the absolute values of differences between the mean and each value in a data set. It is used to predict errors and to judge equality.

means (p. 112) The middle terms of the proportion.

measures of central tendency (p. P37) Numbers or pieces of data that can represent the whole set of data.

measures of position (p. P38) Measures that compare the position of a value relative to other values in a set.

measures of variation (p. P38) Used to describe the distribution of statistical data.

median (p. P37) The middle number in a set of data when the data are arranged in numerical order. If the data set has an even number, the median is the mean of the two middle numbers.

matriz Disposción rectangular de numeros colocados en filas y columnas.

máximo El punto más alto en la gráfica de una curva.

media La suma de los números de un conjunto de datos dividida entre el numero total de artículos.

desviación absoluta media El promedio de los valores absolutos de diferencias entre el medio y cada valor de un conjunto de datos. Ha usado para predecir errores y para juzgar igualdad.

medios Los términos centrales de una proporción.

medidas de tendencia central Números o fragmentos que pueden representar el conjunto de datos total de datos.

medidas de la posición Las medidas que comparar la posición de un valor relativo a otros valores de un conjunto.

medidas de variación Números que se usan para describir la distribución o separación de un conjunto de datos.

mediana El número central de conjunto de datos, una vezque los datos han sido ordenados numéricamente. Si hay un número par de datos, la mediana es el promedio de los datos centrales.

median fit line (p. 258) A type of best-fit line that is calculated using the medians of the coordinates of the data points.

línea apta del punto medio Tipo de mejor-cupo la línea se calcula que usando los puntos medios de los coordenadas de los puntos de referencias.

metric (p. 118) A rule for assigning a number to some characteristic or attribute.

métrico Una regla para asignar un número a alguna característica o atribuye.

midpoint (p. 654) The point halfway between the endpoints of a segment.

punto medio Punto que divide a un segmento separándolo en dos segmentos congruentes.

minimum (p. 58) The lowest point on the graph of a curve.

mínimo El punto más bajo en la gráfica de una curva.

mixed expression (p. 720) An expression that contains the sum of a monomial and a rational expression.

expresión mixta Expresión que contiene la suma de un monomio y una expresión racional.

mixture problems (p. 132) Problems in which two or more parts are combined into a whole.

problemas de mezclas Problemas en que dos o más partes se combinan en un todo.

mode (p. P37) The number(s) that appear most often in a set of data.

moda El número(s) que aparece más frecuencia en un conjunto de datos.

monomial (p. 391) A number, a variable, or a product of a number and one or more variables.

monomio Número, variable o producto de un número por una o más variables.

multiplicative identity (p. 17) For any number a, $a \cdot 1 = 1 \cdot a = a$.

identidad de la multiplicación Para cualquier número $a \cdot 1 = 1 \cdot a = a$.

multiplicative inverses (pp. P18, 17) Two numbers with a product of 1.

inversos multiplicativos Dos números cuyo producto es igual a 1.

multi-step equation (p. 91) Equations with more than one operation.

ecuaciones de varios pasos Ecuaciones con más de una operación.

mutually exclusive events (p. 706) Events that cannot occur at the same time.

mutuamente exclusivos Eventos que no pueden ocurrir simultáneamente.

N

nth root (p. 640) If $a^n = b$ for a positive integer n, then a is an nth root of b.

raíz enésima Si $a^n = b$ para cualquier entero positivo n, entonces a se llama una raíz enésima de b.

natural numbers (p. P7) The set $\{1, 2, 3, \ldots\}$.

números naturales El conjunto $\{1, 2, 3, \ldots\}$.

negative (p. 57) A function is negative on a portion of its domain where its graph lies below the x-axis.

negativo Una función es negativa en una porción de su dominio donde su gráfico está debajo del eje-x.

negative correlation (p. 247) In a scatter plot, as x increases, y decreases.

correlación negativa En una gráfica de dispersión, a medida que x aumenta, y disminuye.

negative exponent (p. 400) For any real number $a \neq 0$ and any integer n, $a^{-n} = \frac{1}{a^n}$ and $\frac{1}{a^{-n}} = a^n$.

exponente negativo Para números reales, si $a \neq 0$, y cualquier número entero n, entonces $a^{-n} = \frac{1}{a^n}$ and $\frac{1}{a^{-n}} = a^n$.

negative number (p. P7) Any value less than zero.

número negativo Cualquier valor menor que cero.

nonlinear function (p. 50) A function with a graph that is not a straight line.

función no lineal Una función con un gráfica que no es una línea recta.

number theory (p. 92) The study of numbers and the relationships between them.

teoría del número El estudio de números y de las relaciones entre ellas.

O

observational study (p. 749) Members of a sample are measured or observed without being affected by the study.

estudio de observación Los miembros de una muestra son medidos o son observados sin ser afectado por el estudio.

odds (p. P35) The ratio of the probability of the success of an event to the probability of its complement.

probabilidades El cociente de la probabilidad del éxito de un acontecimiento a la probabilidad de su complemento.

open half-plane (p. 317) The solution of a linear inequality that does not include the boundary line.

abra el mitad-plano La solución de una desigualdad linear que no incluya la línea de límite.

open sentence (p. 33) A mathematical statement with one or more variables.

enunciado abierto Un enunciado matemático que contiene una o más variables.

opposites (p. P11) Two numbers with the same absolute value but different signs.

opuestos Dos números que tienen el mismo valor absoluto, pero que tienen distintos signos.

ordered pair (p. 40) A set of numbers or coordinates used to locate any point on a coordinate plane, written in the form (x, y).

par ordenado Un par de números que se usa para ubicar cualquier punto de un plano de coordenadas y que se escribe en la forma (x, y).

order of magnitude (p. 401) The order of magnitude of a quantity is the number rounded to the nearest power of 10.

orden de magnitud de una cantidad Un número redondeado a la potencia más cercana de 10.

order of operations (p. 10)
1. Evaluate expressions inside grouping symbols.
2. Evaluate all powers.
3. Do all multiplications and/or divisions from left to right.
4. Do all additions and/or subtractions from left to right.

orden de las operaciones
1. Evalúa las expresiones dentro de los símbolos de agrupamiento.
2. Evalúa todas las potencias.
3. Multiplica o divide de izquierda a derecha.
4. Suma o resta de izquierda a derecha.

origin (p. 40) The point where the two axes intersect at their zero points.

origen Punto donde se intersecan los dos ejes en sus puntos cero.

outliers (p. P39) Data that are more than 1.5 times the interquartile range beyond the quartiles.

valores atípicos Datos que distan de los cuartiles más de 1.5 veces la amplitude intercuartílica.

P

parabola (p. 543) The graph of a quadratic function.

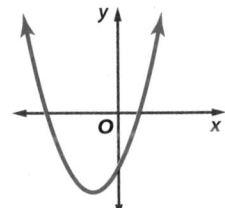

parábola La gráfica de una función cuadrática.

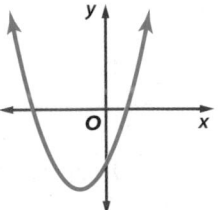

parallel lines (p. 239) Lines in the same plane that do not intersect and either have the same slope or are vertical lines.

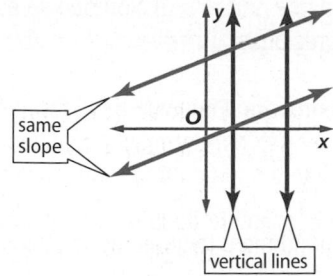

rectas paralelas Rectas en el mismo plano que no se intersecan y que tienen pendientes iguales, o las mismas rectas verticales.

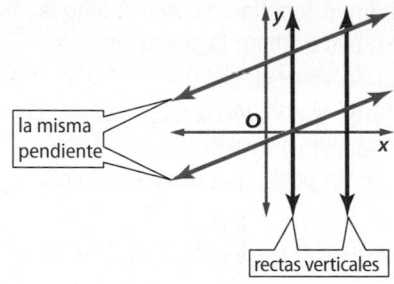

parameter (p. 757) A measure that describes a characteristic of the population as a whole.

parámetro Una medida que se describe característica de la población en su totalidad.

parent function (p. 163) The simplest of functions in a family.

función basíca La función más fundamental de un familia de funciones.

parent graph (p. 163) The simplest of the graphs in a family of graphs.

gráfica madre La gráfica más sencilla en una familia de gráficas.

percent (p. P20) A ratio that compares a number to 100.

porcentaje Razón que compara un numero con 100.

percent of change (p. 119) When an increase or decrease is expressed as a percent.

porcentaje de cambio Cuando un aumento o disminución se escribe como un tanto por ciento.

percent of decrease (p. 119) The ratio of an amount of decrease to the previous amount, expressed as a percent.

porcentaje de disminución Razón de la cantidad de disminución a la cantidad original, escrita como un tanto por ciento.

percent of increase (p. 119) The ratio of an amount of increase to the previous amount, expressed as a percent.

porcentaje de aumento Razón de la cantidad de aumento a la cantidad original, escrita como un tanto por ciento.

percent proportion (p. P20)
$$\frac{\text{part}}{\text{whole}} = \frac{\text{percent}}{100} \text{ or } \frac{a}{b} = \frac{P}{100}$$

proporción porcentual
$$\frac{\text{parte}}{\text{todo}} = \frac{\text{por ciento}}{100} \text{ or } \frac{a}{b} = \frac{P}{100}$$

perimeter (p. P23) The distance around a geometric figure.

perímetro Longitud alrededor una figura geométrica.

perfect square (p. P7) A number with a square root that is a rational number.

cuadrado perfecto Número cuya raíz cuadrada es un número racional.

perfect square trinomial (p. 522) A trinomial that is the square of a binomial.
$(a + b)^2 = (a + b)(a + b) = a^2 + 2ab + b^2$ or
$(a - b)^2 = (a - b)(a - b) = a^2 - 2ab + b^2$

trinomio cuadrado perfecto Un trinomio que es el cuadrado de un binomio.
$(a + b)^2 = (a + b)(a + b) = a^2 + 2ab + b^2$ or
$(a - b)^2 = (a - b)(a - b) = a^2 - 2ab + b^2$

permutation (p. 786) An arrangement or listing in which order is important.

permutación Arreglo o lista en que el orden es importante.

perpendicular lines (p. 240) Lines that intersect to form a right angle.

recta perpendicular Recta que se intersecta formando un ángulo recto

piecewise-linear function (p. 598) A function written using two or more linear expressions.

piecewise-defined function (p. 599) A function that is written using two or more expressions.

point-slope form (p. 233) An equation of the form $y - y_1 = m(x - x_1)$, where m is the slope and (x_1, y_1) is a given point on a nonvertical line.

polynomial (p. 465) A monomial or sum of monomials.

population (p. 747) A large group of data usually represented by a sample.

positive (p. 57) A function is positive on a portion of its domain where its graph lies above the x-axis.

positive correlation (p. 247) In a scatter plot, as x increases, y increases.

positive number (p. P7) Any value that is greater than zero.

power (p. 5) An expression of the form x^n, read x to the nth power.

prime polynomial (p. 512) A polynomial that cannot be written as a product of two polynomials with integral coefficients.

principal square root (p. P7) The nonnegative square root of a number.

probability (p. P33) The ratio of the number of favorable equally likely outcomes to the number of possible equally likely outcomes.

probability distribution (p. 804) The probability of every possible value of the random variable x.

probability graph (p. 804) A way to give the probability distribution for a random variable and obtain other data.

probability model (p. 781) A mathematical model used to represent the outcomes of an experiment.

product (p. 5) In an algebraic expression, the result of quantities being multiplied is called the product.

product rule (p. 677) If (x_1, y_1) and (x_2, y_2) are solutions to an inverse variation, then $y_1 x_1 = y_2 x_2$.

proportion (p. 111) An equation of the form $\frac{a}{b} = \frac{c}{d}$, where $b, d \neq 0$, stating that two ratios are equivalent.

función lineal por partes Función que se escribe usando dos o más expresiones lineal.

función definida por partes Función que se escribe usando dos o más expresiones.

forma punto-pendiente Ecuación de la forma $y - y_1 = m(x - x_1)$, donde m es la pendiente y (x_1, y_1) es un punto dado de una recta no vertical.

polinomio Un monomio o la suma de monomios.

población Grupo grande de datos, representado por lo general por una muestra.

positiva Una función es positiva en una porción de su dominio donde su gráfico está encima del eje-x.

correlación positiva En una gráfica de dispersión, a medida que x aumenta, y aumenta.

número positivos Cualquier valor mayor que cero.

potencia Una expresión de la forma x^n, se lee x a la enésima potencia.

polinomio primo Polinomio que no puede escribirse como producto de dos polinomios con coeficientes enteros.

raíz cuadrada principal La raíz cuadrada no negativa de un número.

probabilidad La razón del número de maneras en que puede ocurrir el evento al numero de resultados posibles.

distribución de probabilidad Probabilidad de cada valor posible de una variable aleatoria x.

gráfico probabilístico Una manera de exhibir la distribución de probabilidad de una variable aleatoria y obtener otros datos.

modelo de probabilidad Modelo matemático que se usa para representar los resultados de un experimento.

producto En una expresión algebraica, se llama producto al resultado de las cantidades que se multiplican.

regla del producto Si (x_1, y_1) y (x_2, y_2) son soluciones de una variación inversa, entonces $y_1 x_1 = y_2 x_2$.

proporción Ecuación de la forma $\frac{a}{b} = \frac{c}{d}$, donde $b, d \neq 0$, que afirma la equivalencia de dos razones.

Pythagorean Theorem (p. 648) If a and b are the measures of the legs of a right triangle and c is the measure of the hypotenuse, then $c^2 = a^2 + b^2$.

Pythagorean triple (p. 649) Counting numbers that satisfy the Pythagorean Theorem.

Teorema de Pitágoras Si a y b son las longitudes de los catetos de un triángulo rectángulo y si c es la longitud de la hipotenusa, entonces $c^2 = a^2 + b^2$.

Triple pitagórico Números de contar que satisfacen el Teorema de Pitágoras.

Q

quadratic equation (pp. 506, 555) An equation of the form $ax^2 + bx + c = 0$, where $a \neq 0$.

quadratic expression (p. 481) An expression in one variable with a degree of 2 written in the form $ax^2 + bx + c$.

Quadratic Formula (p. 583) The solutions of a quadratic equation in the form $ax^2 + bx + c = 0$, where $a \neq 0$, are given by the formula
$$x = \frac{-b \pm \sqrt{b^2 - 4ac}}{2a}.$$

quadratic function (p. 543) An equation of the form $y = ax^2 + bx + c$, where $a \neq 0$.

quartile (p. P38) The values that divide a set of data into four equal parts.

ecuación cuadrática Ecuación de la forma $ax^2 + bx + c = 0$, donde $a \neq 0$.

expression cuadratica Una expresión en una variable con un grado de 2, escritos en la forma $ax^2 + bx + c$.

Fórmula cuadrática Las soluciones de una ecuación cuadrática de la forma $ax^2 + bx + c = 0$, donde $a \neq 0$, vienen dadas por la fórmula
$$x = \frac{-b \pm \sqrt{b^2 - 4ac}}{2a}.$$

función cuadrática Función de la forma $y = ax^2 + bx + c$, donde $a \neq 0$.

cuartile Valores que dividen en conjunto de datos en cuarto partes iguales.

R

radical equations (p. 642) Equations that contain radicals with variables in the radicand.

radical expression (p. 628) An expression that contains a square root.

radical function (p. 621) A function that contains radicals with variables in the radicand.

radicand (p. 621) The expression that is under the radical sign.

radius (p. P24) Distance from the center to any point on the circle.

random variable (p. 803) A variable with a value that is the numerical outcome of a random event.

range 1. (p. 40) The set of second numbers of the ordered pairs in a relation. **2.** (p. P38) The difference between the greatest and least data values.

rate (p. 113) The ratio of two measurements having different units of measure.

rate of change (p. 172) How a quantity is changing with respect to a change in another quantity.

ecuaciones radicales Ecuaciones que contienen radicales con variables en el radicando.

expresión radical Expresión que contiene una raíz cuadrada.

ecuaciones radicales Ecuaciones que contienen radicales con variables en el radicando.

radicando La expresión debajo del signo radical.

radio Distancia del centro cualquier punto de un círculo.

variable aleatoria Una variable cuyos valores son los resultados numéricos de un evento aleatorio.

rango 1. Conjunto de los segundos números de los pares ordenados de una relación. **2.** La diferencia entre los valores de datos más grande o menos.

tasa Razón de dos medidas que tienen distintas unidades de medida.

tasa de cambio Cómo cambia una cantidad con respecto a un cambio en otra cantidad.

rate problems (pp. 134, 729) Problems in which an object moves at a certain speed, or rate.

ratio (p. 111) A comparison of two numbers by division.

rational equations (p. 726) Equations that contain rational expressions.

rational exponent (p. 406) For any positive real number b and any integers m and $n > 1$, $b^{\frac{m}{n}} = \left(\sqrt[n]{b}\right)^m$ or $\sqrt[n]{b^m}$. $\frac{m}{n}$ is a rational exponent.

rational expression (p. 690) An algebraic fraction with a numerator and denominator that are polynomials.

rational function (p. 684) An equation of the form $f(x) = \frac{p(x)}{q(x)}$, where $p(x)$ and $q(x)$ are polynomial functions, and $q(x) \neq 0$.

rationalizing the denominator (p. 630) A method used to eliminate radicals from the denominator of a fraction.

rational numbers (p. P7) The set of numbers expressed in the form of a fraction $\frac{a}{b}$, where a and b are integers and $b \neq 0$.

real numbers (p. P7) The set of rational numbers and the set of irrational numbers together.

reciprocal (pp. P18, 17) The multiplicative inverse of a number.

recursive formula (p. 445) Each term is formulated from one or more previous terms.

reflection (p. 566) A transformation where a figure, line, or curve, is flipped across a line.

relation (p. 40) A set of ordered pairs.

relative frequency (p. 802) The number of times an outcome occurred in a probability experiment.

replacement set (p. 33) A set of numbers from which replacements for a variable may be chosen.

residual (p. 256) The difference between an observed y-value and its predicted y-value on a regression line.

root (p. 163) The solutions of a quadratic equation.

row reduction (p. 371) The process of performing elementary row operations on an augmented matrix to solve a system.

problemas de tasas Problemas en el que un objeto se mueue a una velocidad o tasa determinada.

razón Comparación de dos números mediante división.

ecuaciones racionales Ecuaciones que contienen expresiones racionales.

exponent racional Para cualquier número real no nulo b y cualquier entero m y $n > 1$, $b^{\frac{m}{n}} = \left(\sqrt[n]{b}\right)^m$ or $\sqrt[n]{b^m}$. $\frac{m}{n}$ es un exponent racional.

expresión racional Fracción algebraica cuyo numerador y denominador son polinomios.

función racional Ecuación de la forma $f(x) = \frac{p(x)}{q(x)}$, donde $p(x)$ y $q(x)$ son funciones polinomiales y $q(x) \neq 0$.

racionalizar el denominador Método que se usa para eliminar radicales del denominador de una fracción.

números racionales Conjunto de los números que pueden escribirse en forma de fracción $\frac{a}{b}$, donde a y b son enteros y $b \neq 0$.

números reales El conjunto de los números racionales junto con el conjunto de los números irracionales.

recíproco Inverso multiplicativo de un número.

fórmula recursiva Cada tórmino proviene de uno o más terminos anteriores.

reflexión Transformación en que cadapunto de una figura se aplica a través de una recta de simetría a su imagen correspondiente.

relación Conjunto de pares ordenados.

frecuencia relativa Número de veces que aparece un resultado en un experimento probabilístico.

conjunto de sustitución Conjunto de números del cual se pueden escoger sustituciones para una variable.

residual Diferencia entre el valor observado de y y el valor redicho de y en la recta de regresion.

raíces Las soluciones de una ecuación cuadrática

reducción de la fila El proceso de realizar operaciones elementales de la fila en una matriz aumentada para solucionar un sistema.

sample (p. 747) Some portion of a larger group selected to represent that group.

sample space (p. P33) The list of all possible outcomes.

scale (p. 114) The relationship between the measurements on a drawing or model and the measurements of the real object.

scale model (p. 114) A model used to represent an object that is too large or too small to be built at actual size.

scatter plot (p. 247) A scatter plot shows the relationship between a set of data with two variables, graphed as ordered pairs on a coordinate plane.

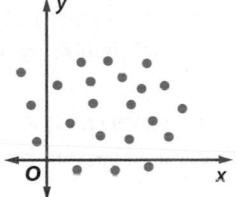

scientific notation (p. 414) A number in scientific notation is expressed as $a \times 10^n$, where $1 \leq a < 10$ and n is an integer.

self-selected sample (p. 747) Members volunteer to be included in the sample.

sequence (p. 189) A set of numbers in a specific order.

set-builder notation (p. 286) A concise way of writing a solution set. For example, $\{t \mid t < 17\}$ represents the set of all numbers t such that t is less than 17.

simple random sample (p. 747) A sample that is as likely to be chosen as any other from the population.

simplest form (p. 27) An expression is in simplest form when it is replaced by an equivalent expression having no like terms or parentheses.

simulation (p. 781) Using an object to act out an event that would be difficult or impractical to perform.

sine (p. 656) For an acute angle of a right triangle, the ratio of the measure of the leg opposite the acute angle to the measure of the hypotenuse.

muestra Porción de un grupo más grande que se escoge para representarlo.

espacio muestral Lista de todos los resultados posibles.

escala Relación entre las medidas de un dibujo o modelo y las medidas de la figura verdadera.

modelo a escala Modelo que se usa para representar un figura que es demasiado grande o pequeña como para ser construida de tamaño natural.

gráfica de dispersión Es un diagrama que muestra la relación entre un conjunto de datos con dos variables, graficados como pares ordenados en un plano coordenadas.

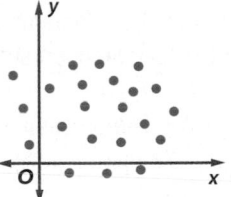

notación científica Un numero en notación científica se escribe con $a \times 10^n$, donde $1 \leq a < 10$ y n es un número entero.

muestra auto-seleccionados Los miembros se ofrecen voluntariamente para ser incluidos en la muestra.

sucesión Conjunto de números en un orden específico.

notación de construcción de conjuntos Manera concisa de escribir un conjunto solución. Por ejemplo, $\{t \mid t < 17\}$ representa el conjunto de todos los números t que son menores o iguales que 17.

muestra aleatoria simple Muestra de una población que tiene la misma probabilidad de escogerse que cualquier otra.

forma reducida Una expresión está reducida cuando se puede sustituir por una expresión equivalente que no tiene ni términos semejantes ni paréntesis.

simulación Uso de un objeto para representar un evento que pudiera ser difícil o poco práctico de ejecutar.

seno La razón entre la medida del cateto opuesto al ángulo agudo y la medida de la hipotenusa de un triángulo rectángulo.

slope (p. 174) The ratio of the change in the *y*-coordinates (risc) to the corresponding change in the *x*-coordinates (run) as you move from one point to another along a line.

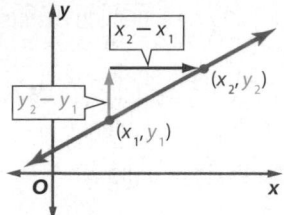

pendiente Razón del cambio en la coordenada *y* (elevación) al cambio correspondiente en la coordenada *x* (desplazamiento) a medida que uno se mueve de un punto a otro en una recta.

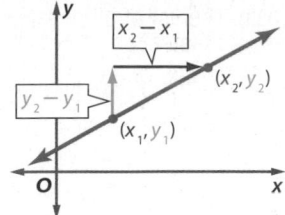

slope-intercept form (p. 216) An equation of the form $y = mx + b$, where *m* is the slope and *b* is the *y*-intercept.

forma pendiente-intersección Ecuación de la forma $y = mx + b$, donde *m* es la pendiente y *b* es la intersección *y*.

solution (p. 33) A replacement value for the variable in an open sentence.

solución Valor de sustitución de la variable en un enunciado abierto.

solution set (p. 33) The set of elements from the replacement set that make an open sentence true.

conjunto solución Conjunto de elementos del conjunto de sustitución que hacen verdadero un enunciado abierto.

solve an equation (p. 83) The process of finding all values of the variable that make the equation a true statement.

resolver una ecuación Proceso en que se hallan todos los valores de la variable que hacen verdadera la ecuación.

solving an open sentence (p. 33) Finding a replacement value for the variable that results in a true sentence or an ordered pair that results in a true statement when substituted into the equation.

resolver un enunciado abierto Hallar un valor de sustitución de la variable que resulte en un enunciado verdadero o un par ordenado que resulte en una proposición verdadera cuando se lo sustituye en la ecuación.

solving a triangle (p. 657) Finding the measures of all of the angles and sides of a triangle.

resolver un triángulo Hallar las medidas de todos los lados y todos los ángulos de un triángulo.

square root (p. P7) One of two equal factors of a number.

raíz cuadrada Uno de dos factores iguales de un número.

square root function (p. 621) Function that contains the square root of a variable.

función radical Función que contiene la raíz cuadrada de una variable.

standard deviation (p. 759) The square root of the variance.

desviación típica Calculada como la raíz cuadrada de la varianza.

standard form (p. 155) The standard form of a linear equation is $Ax + By = C$, where $A \geq 0$, *A* and *B* are not both zero, and *A*, *B*, and *C* are integers with a greatest common factor of 1.

forma estándar La forma estándar de una ecuación lineal es $Ax + By = C$, donde $A \geq 0$, ni *A* ni *B* son ambos cero, y *A*, *B*, y *C* son enteros cuyo máximo común divisor es 1.

standard form of a polynomial (p. 466) A polynomial that is written with the terms in order from greatest degree to least degree.

forma de estándar de un polinomio Un polinomio que se escribe con los términos en orden del grado más grande a menos grado.

statistic (p. 757) A quantity calculated from a sample.

estadística Una cantidad calculaba de una muestra.

statistical inference (p. 757) The statistics of a sample are used to draw conclusions about the population.

inferencia estadística La estadística de una muestra se utiliza para dibujar conclusiones sobre la población.

stratified sample (p. 747) A sample in which the population is first divided into similar, nonoverlapping groups; a simple random sample is then selected from each group.

muestra aleatoria estratificada Muestra en que la población se divide en grupos similares que no se sobreponen; luego se selecciona una muestra aleatoria simple, de cada grupo.

stem-and-leaf plot (p. P42) A system used to condense a set of data where the greatest place value of the data forms the stem and the next greatest place value forms the leaves.

diagrama de tallo y hojas Sistema que se usa para condensar un conjunto de datos, en que el valor de posición máximo de los datos forma el tallo y el segundo valor de posiciós máximo forma las hojas. El valor de posición máximo de los datos forma eld tallo y el segundo valor de posición máximo forma las hojas.

step function (p. 598) A function with a graph that is a series of horizontal line segments.

funcion escalonada Función cuya gráfica es una serie de segmentos de recto.

substitution (p. 344) Use algebraic methods to find an exact solution of a system of equations.

sustitución Usa métodos algebraicos para hallar una solución exacta a un sistema de ecuaciones.

surface area (p. P31) The sum of the areas of all the surfaces of a three-dimensional figure.

área de superficie Suma de las áreas de todas las superficies (caras) de una figura tridimensional.

survey (p. 749) Data are from responses given by a sample of the population.

encuesta Datos son de las respuestas dadas por una muestra de la población.

symmetry (p. 57) A geometric property of figures that can be folded and each half matches the other exactly.

simetría Propiedad geométrica de figuras que pueden plegarse de modo que cada mitad corresponde exactamente a la otra.

system of equations (p. 335) A set of equations with the same variables.

sistema de ecuaciones Conjunto de ecuaciones con las mismas variables.

system of inequalities (p. 372) A set of two or more inequalities with the same variables.

sistema de desigualdades Conjunto de dos o más desigualdades con las mismas variables

systematic sample (p. 747) A sample in which the items in the sample are selected according to a specified time or item interval.

muestra aleatoria sistemática Muestra en que los elementos de la muestra se escogen según un intervalo de tiempo o elemento específico.

T

tangent (p. 656) For an acute angle of a right triangle, the ratio of the measure of the leg opposite the acute angle to the measure of the leg adjacent to the acute angle.

tangente La razón entre la medida del cateto opuesto al ángulo agudo y la medida del cateto adyacente al ángulo agudo de un triángulo rectángulo.

term (p. 5) A number, a variable, or a product or quotient of numbers and variables.

término Número, variable o producto, o cociente de números y variables.

terms of a sequence (p. 189) The numbers in a sequence.

términos Los números de una sucesión.

theoretical probability (p. 780) What should occur in a probability experiment.

probabilidad teórica Lo que debería ocurrir en un experimento probabilístico.

transformation (p. 564) A movement of a geometric figure.

transformación Desplazamiento de una figura geométrica.

translation (p. 564) A transformation where a figure is slid from one position to another without being turned.

translación Transformación en que una figura se desliza sin girar, de una posición a otra.

tree diagram (p. P34) A diagram used to show the total number of possible outcomes.

trigonometric ratio (p. 656) A ratio of the lengths of sides of a right triangle.

trigonometry (p. 656) The study of the properties of triangles and trigonometric functions and their applications.

trinomials (p. 465) The sum of three monomials.

diagrama de árbol Diagrama que se usa para mostrar el número total de resultados posibles.

razón trigonométrica Razón entre las longitudes de dos lados de un triángulo rectángulo.

trigonométria Estudio de las relaciones entre los lados y ángulos de un triángulo rectángulo.

trinomios Suma de tres monomios.

U

uniform motion problems (p. 134) Problems in which an object moves at a certain speed, or rate.

union (p. 307) The graph of a compound inequality containing or; the solution is a solution of either inequality, not necessarily both.

unit analysis (p. 128) The process of including units of measurement when computing.

unit rate (p. 113) A ratio of two quantities, the second of which is one unit.

univariate data (p. P37) Data with one variable.

upper quartile (p. P38) The median of the upper half of a set of data.

problemas de movimiento uniforme Problemas en que el cuerpo se mueve a cierta velocidad o tasa.

unión Gráfica de una desigualdad compuesta que contiene la palabra o; la solución es el conjunto de soluciones de por lo menos una de las desigualdades, no necesariamente ambas.

análisis de la unidad Proceso de incluir unidades de medida al computar.

tasa unitaria Tasa reducida que tiene denominador igual a 1.

datos univariate Datos con una variable.

cuartil superior Mediana de la mitad superior de un conjunto de datos.

V

variable 1. (p. 5) Symbols used to represent unspecified numbers or values. 2. (p. P37) a characteristic of a group of people or objects that can assume different values

variance (p. 759) The mean of the squares of the deviations from the arithmetic mean.

vertex (p. 543) The maximum or minimum point of a parabola.

vertex form (p. 568) A quadratic function in the form $f(x) = a(x - h)^2 + k$.

vertical line test (p. 49) If any vertical line passes through no more than one point of the graph of a relation, then the relation is a function.

volume (p. P29) The measure of space occupied by a solid region.

variable 1. Símbolos que se usan para representar números o valores no especificados. 2. una característica de un grupo de personas u objetos que pueden asumir valores diferentes

varianza Media de los cuadrados de las desviaciones de la media aritmética.

vértice Punto máximo o mínimo de una parábola.

forma de vértice Una función cuadrática de la forma $f(x) = a(x - h)^2 + k$.

prueba de la recta vertical Si cualquier recta vertical pasa por un sólo punto de la gráfica de una relación, entonces la relación es una función.

volumen Medida del espacio que ocupa un solido.

weighted average (p. 132) The sum of the product of the number of units and the value per unit divided by the sum of the number of units, represented by M.

promedio ponderado Suma del producto del número de unidades por el valor unitario dividida entre la suma del número de unidades y la cual se denota por M.

whole numbers (p. P7) The set {0, 1, 2, 3, …}.

números enteros El conjunto {0, 1, 2, 3, …}.

work problems (p. 728) Rational equations are used to solve problems involving work rates.

problemas de trabajo Las ecuaciones racionales se usan para resolver problemas de tasas de trabajo.

x-axis (p. 40) The horizontal number line on a coordinate plane.

eje x Recta numérica horizontal que forma parte de un plano de coordenadas.

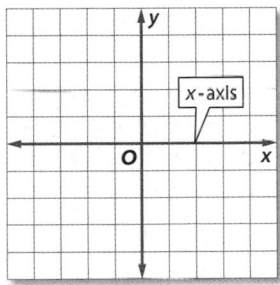

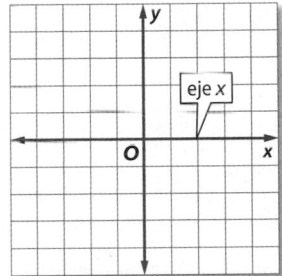

x-coordinate (p. 40) The first number in an ordered pair.

coordenada x El primer número de un par ordenado.

x-intercept (p. 56) The x-coordinate of a point where a graph crosses the x-axis.

intersección x La coordenada x de un punto donde la gráfica corte al eje de x.

y-axis (p. 40) The vertical number line on a coordinate plane.

eje y Recta numérica vertical que forma parte de un plano de coordenadas.

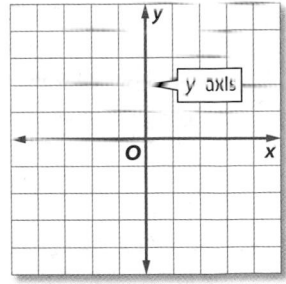

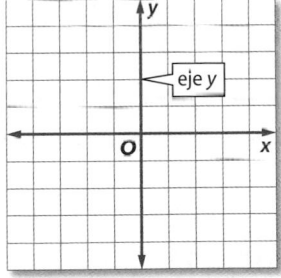

y-coordinate (p. 40) The second number in an ordered pair.

coordenada y El segundo número de un par ordenado.

y-intercept (p. 56) The y-coordinate of a point where a graph crosses the y-axis.

intersección y La coordenada y de un punto donde la grafica corta al eje de y.

zero (p. 163) The x-intercepts of the graph of a function; the values of x for which $f(x) = 0$.

cero Las intersecciones x de la grafica de una función; los puntos x para los que $f(x) = 0$.

zero exponent (p. 399) For any nonzero number a, $a^0 = 1$. Any nonzero number raised to the zero power is equal to 1.

exponente cero Para cualquier número distinto a cero a, $a^0 = 1$. Cualquier número distinto a cero levantado al potente cero es igual a 1.

Index

638, 682, 778
of rhombi, 637
of squares, 232, 527, 647
of trapezoids, 143, 689
of triangles, 13, 76, 333, 500, 577, 637, 778

Arithmetic sequences, 189–194, 203, 438
identifying, 190

Assessment. *See* Chapter Test; Guided Practice; Mid-Chapter Quiz; Prerequisite Skills; Quick Check; Spiral Review; Standardized Test Practice

Associative Property
of addition, 18, 28
of multiplication, 18, 28

Asymptote, 685–688

Augmented matrix, 370

Averages,
weighted, 132–138. *See also* Mean

Axis of symmetry, 543–552

B

Bar graph, P41, 804

Bases, 5, 391

Best-fit lines, 255–257
median-fit, 257–260, 272
regression equations, 596–597

Bias, 748

Biased samples, 748

Binomials, 465,
additive inverse, 467
conjugates, 630
difference of squares, 516–521, 530
dividing by a binomial, 707
factoring, 493
multiplying, 478–479, 480–481

Bivariate data, 247

Boundaries, 317–318

Box-and-whisker plot, P43, 765, 767, 774, 775, 814

C

Calculator. *See* Graphing Technology Labs; TI-Nspire Technology Labs

Careers. *See* Real-World Careers

Causation, 254

Celsius, 207

Center, P24

Central tendency. *See* Measures of central tendency

Challenge. *See* Higher-Order Thinking Problems

Chapter 0
adding and subtracting rational numbers, P13–P16
area, P26–P28
measures of center, variation, and position, P37–P40, 773
multiplying and dividing rational numbers, P17–P19
operations with integers, P11–P12
percent proportion, P20–P22
perimeter, P23–P25
plan for problem solving, P5–P6
real numbers, P7–P10
representing data, P41–P44
simple probability and odds, P33–P36
surface area, P31–P32
volume, P29–P30

Check for Reasonableness, 250, 251, 340, 355, 368, 415, 508, 576, 578, 685, 688, 691, 715, 729

Circle graph, P42–P43

Circles, P24
area of, 393, 529, 553, 625
center of, P24
circumference, P24, 109, 187, 529
diameter, P24
radius, P24

Circumference, P24, 109, 187, 529

Closed, 467, 634

Closed half-plane, 317

Closure Property, 21, 634
irrational numbers, 634
rational numbers, 634
whole numbers, 21

Coefficients, 28, 351, 392
correlation, 255
leading, 466–469, 567, 575
negative, 298

Combinations, 787

Common Core State Standards, Mathematical Content, P41, 5, 10, 16, 23, 25, 33, 40, 47, 55, 56, 75, 81, 83, 90, 91, 97, 103, 111, 118, 119, 126, 132, 153, 155, 163, 169, 171, 172, 182, 189, 196, 197, 216, 224, 226, 233, 239, 247, 254, 255, 263, 271, 285, 291, 292, 298, 306, 312, 317, 323, 335, 342, 344, 350, 357, 364, 370, 372, 377, 391, 398, 406, 414, 422, 424, 430, 432, 437, 438, 444, 445, 463, 465, 472, 478, 480, 486, 493, 494, 501, 503, 510, 516, 522, 543, 554, 555, 562, 564, 572, 574, 580, 583, 590, 596, 598, 606, 619, 621, 627, 628, 634, 635, 640, 642, 684, 726, 733, 757, 764, 771, 801, 810

Common Core State Standards, Mathematical Practice,
Arguments, 16, 17, 21, 125, 137, 179, 196, 226, 231, 252, 261, 269, 312, 315, 335, 340, 412, 414, 419, 445, 449, 508, 570, 594, 638, 661, 684, 688, 731, 769, 784, 806, 807
Critique, 8, 33, 38, 47, 60, 101, 103, 201, 237, 244, 361, 442, 465, 470, 499, 514, 555, 559, 642, 645, 681, 753, 755, 762
Modeling, 5, 6, 36, 43, 44, 78, 118, 121, 122, 132, 136, 163, 166, 193, 229, 247, 285, 288, 320, 338, 339, 360, 364, 367, 374, 411, 418, 427, 432, 448, 475, 510, 513, 558, 574, 576, 587, 598, 601, 642, 720, 723, 726, 760, 780, 781, 783, 789, 797, 803
Perseverance, 14, 25, 26, 30, 53, 130, 194, 225, 243, 306, 310, 348, 351, 396, 428, 476, 516, 520, 578, 648, 652, 703, 718, 724, 791
Precision, 20, 23, 45, 83, 87, 107, 111, 114, 115, 126, 128, 129, 182, 184, 226, 228, 242, 263, 264, 292, 295, 372, 375, 414, 417, 434, 435, 522, 527, 555, 556, 577, 583, 584, 621, 631, 695, 701, 702, 712, 714, 757
Reasoning, 12, 16, 19, 29, 51, 75, 79, 86, 94, 159, 160, 172, 185, 216, 220, 221, 233, 236, 267, 285, 289, 302, 344, 347, 353, 364, 365, 368, 398, 441, 469, 494, 497, 526, 543, 549, 635, 644, 694, 717, 726, 729, 730, 757, 761, 769, 810
Regularity, 25, 81, 91, 93, 95, 119, 121, 155, 161, 173, 189, 190, 216, 222, 306, 314, 335, 391, 393, 404, 423, 444, 484, 486, 487, 503, 504, 528, 564, 569, 628, 706, 710, 777, 786, 799
Sense-Making, P5, 7, 25, 26, 37, 40, 41, 52, 56, 59, 97, 100, 106, 132, 135, 170, 177, 182, 186, 197, 225,

J

K

Mutually exclusive events, 795

N

Natural numbers, P7

Negative correlation, 247, 248, 255

Negative exponents, 400–401

Negative numbers, P7

Nonlinear functions, 50, 543, 601. *See also* Functions

values of, 50

Nonproportional relationships, 199, 201, 203, 206

Normal curve, 810–811

Normal distribution, 810–811

Notation

function, 50, 200–201

permutations, 787

scientific, 414–420, 451, 453

set-builder notation, 286

Note taking. *See* Foldables® Study Organizer

*n*th roots, 407, 640–641

*n*th terms, 190–191

Number line, P15

inequalities, 286–287, 307–310

Number Theory, 93, 100, 101, 302, 309, 360, 514, 558, 577, 608, 609

Numbers

complex, 555

integers, P7, P11–P12, 92–93, 141

irrational, P7, 584, 634

mixed, 720

natural, P7

negative, P7

positive, P7

properties of, 28, 64

Pythagorean triples, 649

rational, P13–P16, P17–P19, 634

real, P7–P10

whole, P7, 373

Number Sense. *See* Higher-Order Thinking Problems

Numerical expressions, 12

evaluating, 10–12, 25

grouping symbols, 11

writing, 12

Numerical fractions, 720–725, 735, 738

O

Observational study, 749

Octagons, 21, 36

Odds, P35–P36

Open Ended. *See* Higher-Order Thinking Problems

Open half-plane, 317

Open sentence, 33

Operations

closed, 634

Opposite reciprocals, 240

Opposites, P11, 692

Order of magnitude, 401–402

Order of operations, 10–15, 34, 63

Ordered pair, 40–41, 335

Ordinate. *See* *y*-coordinate

Origin, 40

Outliers, P39, 765

P

Parabolas, 543–550

axis of symmetry, 543–550

graphing, 543

shading inside and outside, 561

vertices of, 543–550

Parallel lines, 239–245, 272, 274, 336

slope of, 239

Parallelograms, 436, 506

Parameter, 216, 757, 758

Parent function, 163, 627

Part. *See* Percents

Percent proportion, P20–P22

Percentiles, 125

Percents, P20

of change, 119–124, 143

of decrease, 119–124, 143

of increase, 119–124, 143

to fractions, P20–P22

mixture problems, 133

probabilities as, 803

proportion, P20–P22

Perfect square trinomial, 487,

522–529, 574

Perfect squares, P7

Perimeter, P23–P25

of rectangles, 37, 80, 213, 413, 508, 638

of squares, 471, 624

of triangles, 15, 20, 96, 110, 469

Permutations, 786–792

Perpendicular lines, 240–244, 272, 274

Personal Tutor. *See* ConnectED

Piecewise-defined functions, 598, 600–601

Piecewise-linear functions, 600

Pisano, Leonardo, 358

Planes

coordinate, 40

half, 317–318

Plus or minus **symbol,** 525

Point-slope form, 233–238, 239, 272

Polynomials, 463, 465

adding, 463–464, 465–471, 531

binomials, 465, 478–479, 480–481, 493, 516–521, 530, 630, 707

closed, 467

complex fractions involving, 722

degree of, 465–469

descending order, 466

dividing, 699–700, 706–711, 735, 737

by a binomial, 467, 707

by a monomial, 706

expressions, 474

factored form, 481

least common multiples of, 713

modeling, 463

multiplying, 478–479, 480–485, 531

by a monomial, 472–477, 531

operations with, 463–471, 478–485, 531, 699–700, 706–711, 735, 737

predictions using, 468

prime, 512–514

special products, 486–491, 532

standard form, 466–469

subtracting, 463–464, 465–471

trinomials, 465, 501–502, 503–509, 510–511, 530, 533

Population, 747, 758

Positive correlation, 247, 248, 255

Positive numbers, P7

Power of a Power Property, 392, 393

Index

Index

Symbols

$\neq$	is not equal to		AB	measure of $\overline{AB}$
$\approx$	is approximately equal to		$\angle$	angle
$\sim$	is similar to		$\triangle$	triangle
$>, \geq$	is greater than, is greater than or equal to		$^\circ$	degree
$<, \leq$	is less than, is less than or equal to		π	pi
$-a$	opposite or additive inverse of a		$\sin x$	sine of x
$\|a\|$	absolute value of a		$\cos x$	cosine of x
$\sqrt{a}$	principal square root of a		$\tan x$	tangent of x
$a:b$	ratio of a to b		$!$	factorial
(x, y)	ordered pair		$P(a)$	probability of a
$f(x)$	f of x, the value of f at x		$P(n, r)$	permutation of n objects taken r at a time
$\overline{AB}$	line segment AB		$C(n, r)$	combination of n objects taken r at a time

Algebraic Properties and Key Concepts

Identity	For any number a, $a + 0 = 0 + a = a$ and $a \cdot 1 = 1 \cdot a = a$.
Substitution (=)	If $a = b$, then a may be replaced by b.
Reflexive (=)	$a = a$
Symmetric (=)	If $a = b$, then $b = a$.
Transitive (=)	If $a = b$ and $b = c$, then $a = c$.
Commutative	For any numbers a and b, $a + b = b + a$ and $a \cdot b = b \cdot a$.
Associative	For any numbers a, b, and c, $(a + b) + c = a + (b + c)$ and $(a \cdot b) \cdot c = a \cdot (b \cdot c)$.
Distributive	For any numbers a, b, and c, $a(b + c) = ab + ac$ and $a(b - c) = ab - ac$.
Additive Inverse	For any number a, there is exactly one number $-a$ such that $a + (-a) = 0$.
Multiplicative Inverse	For any number $\frac{a}{b}$, where a, $b \neq 0$, there is exactly one number $\frac{b}{a}$ such that $\frac{a}{b} \cdot \frac{b}{a} = 1$.
Multiplicative (0)	For any number a, $a \cdot 0 = 0 \cdot a = 0$.
Addition (=)	For any numbers a, b, and c, if $a = b$, then $a + c = b + c$.
Subtraction (=)	For any numbers a, b, and c, if $a = b$, then $a - c = b - c$.
Multiplication and Division (=)	For any numbers a, b, and c, with $c \neq 0$, if $a = b$, then $ac = bc$ and $\frac{a}{c} = \frac{b}{c}$.
Addition (>)*	For any numbers a, b, and c, if $a > b$, then $a + c > b + c$.
Subtraction (>)*	For any numbers a, b, and c, if $a > b$, then $a - c > b - c$.
Multiplication and Division (>)*	For any numbers a, b, and c, 1. if $a > b$ and $c > 0$, then $ac > bc$ and $\frac{a}{c} > \frac{b}{c}$. 2. if $a > b$ and $c < 0$, then $ac < bc$ and $\frac{a}{c} < \frac{b}{c}$.
Zero Product	For any real numbers a and b, if $ab = 0$, then $a = 0$, $b = 0$, or both a and b equal 0.
Square of a Sum	$(a + b)^2 = (a + b)(a + b) = a^2 + 2ab + b^2$
Square of a Difference	$(a - b)^2 = (a - b)(a - b) = a^2 - 2ab + b^2$
Product of a Sum and a Difference	$(a + b)(a - b) = (a - b)(a + b) = a^2 - b^2$

* These properties are also true for $<$, $\geq$, and $\leq$.

Formulas

Slope	$m = \dfrac{y_2 - y_1}{x_2 - x_1}$
Distance on a coordinate plane	$d = \sqrt{(x_2 - x_1)^2 + (y_2 - y_1)^2}$
Midpoint on a coordinate plane	$M = \left(\dfrac{x_1 + x_2}{2}, \dfrac{y_1 + y_2}{2} \right)$
Pythagorean Theorem	$a^2 + b^2 = c^2$
Quadratic Formula	$x = \dfrac{-b \pm \sqrt{b^2 - 4ac}}{2a}$
Perimeter of a rectangle	$P = 2\ell + 2w$ or $P = 2(\ell + w)$
Circumference of a circle	$C = 2\pi r$ or $C = \pi d$

Area

rectangle	$A = \ell w$	trapezoid	$A = \frac{1}{2}h(b_1 + b_2)$
parallelogram	$A = bh$	circle	$A = \pi r^2$
triangle	$A = \frac{1}{2}bh$		

Surface Area

cube	$S = 6s^2$	regular pyramid	$S = \frac{1}{2}P\ell + B$
prism	$S = Ph + 2B$	cone	$S = \pi r\ell + \pi r^2$
cylinder	$S = 2\pi rh + 2\pi r^2$		

Volume

cube	$V = s^3$	regular pyramid	$V = \frac{1}{3}Bh$
prism	$V = Bh$	cone	$V = \frac{1}{3}\pi r^2 h$
cylinder	$V = \pi r^2 h$		

Measures

Metric	Customary
Length	
1 kilometer (km) = 1000 meters (m)	1 mile (mi) = 1760 yards (yd)
1 meter = 100 centimeters (cm)	1 mile = 5280 feet (ft)
1 centimeter = 10 millimeters (mm)	1 yard = 3 feet
	1 foot = 12 inches (in.)
	1 yard = 36 inches
Volume and Capacity	
1 liter (L) = 1000 milliliters (mL)	1 gallon (gal) = 4 quarts (qt)
1 kiloliter (kL) = 1000 liters	1 gallon = 128 fluid ounces (fl oz)
	1 quart = 2 pints (pt)
	1 pint = 2 cups (c)
	1 cup = 8 fluid ounces
Weight and Mass	
1 kilogram (kg) = 1000 grams (g)	1 ton (T) = 2000 pounds (lb)
1 gram = 1000 milligrams (mg)	1 pound = 16 ounces (oz)
1 metric ton (t) = 1000 kilograms	